Intermediate Algebra

SECOND EDITION

K. Elayn Martin-Gay

University of New Orleans

Prentice Hall

PRENTICE HALL
Upper Saddle River, New Jersey 07458

Library of Congress Cataloging-in-Publication Data

Martin-Gay, K. Elayn
    Intermediate algebra/K. Elayn Martin-Gay.—2nd ed.
      p. cm.
    Includes index
    ISBN 0-13-067667-5—ISBN 0-13-067670-5—ISBN 0-13-008746-7
    1. Algebra.   I. Title.

QA152.3.M36 2003
512.9—dc21   2002025777

Executive Acquisition Editor: Karin E. Wagner
Editor in Chief: Christine Hoag
Project Manager: Mary Beckwith
Vice President/Director of Production and Manufacturing: David W. Riccardi
Executive Managing Editor: Kathleen Schiaparelli
Senior Managing Editor: Linda Mihatov Behrens
Production Management: Elm Street Publishing Services, Inc.
Production Assistant: Nancy Bauer
Manufacturing Buyer: Alan Fischer
Manufacturing Manager: Trudy Pisciotti
Executive Marketing Manager: Eilish Collins Main
Marketing Assistant: Annett Uebel
Development Editor: Kathy Sessa-Federico
Editor in Chief, Development: Carol Trueheart
Media Project Manager, Developmental Math: Audra J. Walsh
Editorial Assistant: Heather Balderson
Art Director: Maureen Eide
Assistant to the Art Director: John Christiana
Interior Designer: Circa 86
Cover Designer: Jack Robol
Art Editor: Thomas Benfatti
Creative Director: Carole Anson
Director of Creative Services: Paul Belfanti
Photo Researcher: Melinda Alexander
Photo Editor: Beth Boyd
Cover Art: Seaform Set detail. Handblown glass, by Dale Chihuly. Photo by Scott M. Leen.
Art Studio: Scientific Illustrators
Compositor: Preparé/Emilcomp

Photo Credits appear on page I8, which constitutes a continuation of the copyright page.

Printed in the United States of America
10 9 8 7 6 5 4 3 2

ISBN: 0-13-067667-5 (paperback) 0-13-008746-7 (case bound)

Pearson Education Ltd.
Pearson Education Australia Pty., Limited
Pearson Education Singapore, Pte. Ltd.
Pearson Education North Asia Ltd.
Pearson Education Canada, Ltd.
Pearson Educacíon de Mexico, S.A. de C.V.
Pearson Education—Japan
Pearson Education Malaysia, Pte. Ltd.

CONTENTS

Preface   vii

CHAPTER 1   Real Numbers and Algebraic Expressions   1

CHAPTER 1 PRETEST   2
1.1   Tips for Success in Mathematics   3
1.2   Sets of Numbers   9
1.3   Properties of Real Numbers   19
1.4   Operations on Real Numbers   29

Integrated Review—Real Numbers   39

1.5   Order of Operations and Algebraic Expressions   41
1.6   Exponents and Scientific Notation   49
1.7   More Work with Exponents and Scientific Notation   59

Chapter 1 Activity: Searching for Patterns   66

CHAPTER 1 VOCABULARY CHECK   67
CHAPTER 1 HIGHLIGHTS   67
CHAPTER 1 REVIEW   71
CHAPTER 1 TEST   77

CHAPTER 2   Equations, Inequalities, and Problem Solving   79

CHAPTER 2 PRETEST   80
2.1   Linear Equations in One Variable   81
2.2   An Introduction to Problem Solving   91
2.3   Formulas and Problem Solving   105
2.4   Linear Inequalities and Problem Solving   115

Integrated Review—Linear Equations and Inequalities   129

2.5   Sets and Compound Inequalities   131
2.6   Absolute Value Equations and Inequalities   139

Chapter 2 Activity: Room Redecorating   150

CHAPTER 2 VOCABULARY CHECK   151
CHAPTER 2 HIGHLIGHTS   151
CHAPTER 2 REVIEW   157
CHAPTER 2 TEST   163
CUMULATIVE REVIEW   165

CHAPTER 3   Graphs and Functions   167

CHAPTER 3 PRETEST   168
3.1   Graphing Linear Equations   169
3.2   The Slope of a Line   185
3.3   The Slope-Intercept Form   197
3.4   More Equations of Lines   205

Integrated Review—Linear Equations in Two Variables   215

3.5   Introduction to Functions   217
3.6   Graphing Linear Inequalities   231

Chapter 3 Activity: Measuring Slope   238

CHAPTER 3 VOCABULARY CHECK   239
CHAPTER 3 HIGHLIGHTS   239
CHAPTER 3 REVIEW   245
CHAPTER 3 TEST   251
CUMULATIVE REVIEW   255

CHAPTER 4   **Systems of Equations and Inequalities   257**

CHAPTER 4 PRETEST   258
4.1   Solving Systems of Linear Equations in Two Variables   259
4.2   Solving Systems of Linear Equations in Three Variables   273
4.3   Systems of Linear Equations and Problem Solving   281

Integrated Review—Systems of Linear Equations   295

4.4   Solving Systems of Equations Using Matrices   297
4.5   Systems of Linear Inequalities   303

Chapter 4 Activity: Locating Lightning Strikes   310

CHAPTER 4 VOCABULARY CHECK   311
CHAPTER 4 HIGHLIGHTS   311
CHAPTER 4 REVIEW   317
CHAPTER 4 TEST   321
CUMULATIVE REVIEW   323

CHAPTER 5   **Polynomials and Polynomial Functions   325**

CHAPTER 5 PRETEST   326
5.1   Adding and Subtracting Polynomials   327
5.2   Multiplying Polynomials   339
5.3   Dividing Polynomials   349
5.4   The Greatest Common Factor and Factoring by Grouping   359
5.5   Factoring Trinomials   367
5.6   Factoring by Special Products   377

Integrated Review—Operations on Polynomials and Factoring Strategies   385

5.7   Solving Equations by Factoring and Solving Problems   387
5.8   An Introduction to Graphing Polynomial Functions   397

Chapter 5 Activity: Finding the Largest Area   406

CHAPTER 5 VOCABULARY CHECK   407
CHAPTER 5 HIGHLIGHTS   407
CHAPTER 5 REVIEW   411
CHAPTER 5 TEST   417
CUMULATIVE REVIEW   419

CHAPTER 6   **Rational Expressions   421**

CHAPTER 6 PRETEST   422
6.1   Multiplying and Dividing Rational Expressions   423
6.2   Adding and Subtracting Rational Expressions   435
6.3   Simplifying Complex Fractions   445
6.4   Solving Equations Containing Rational Expressions   453

Integrated Review—Expressions and Equations Containing Rational Expressions   461

6.5   Rational Equations and Problem Solving   463
6.6   Variation and Problem Solving   475

Chapter 6 Activity: Estimating Population Sizes   485

CHAPTER 6 VOCABULARY CHECK   487
CHAPTER 6 HIGHLIGHTS   487
CHAPTER 6 REVIEW   491
CHAPTER 6 TEST   497
CUMULATIVE REVIEW   499

CHAPTER 7   Rational Exponents, Radicals, and Complex Numbers   501

CHAPTER 7 PRETEST   502
7.1   Radical Expressions   503
7.2   Rational Exponents   513
7.3   Simplifying Radical Expressions   521
7.4   Adding, Subtracting, and Multiplying Radical Expressions   529
7.5   Rationalizing Numerators and Denominators of Radical Expressions   537

Integrated Review—Radicals and Rational Exponents   547

7.6   Radical Equations and Problem Solving   549
7.7   Complex Numbers   561

Chapter 7 Activity: Calculating the Length and Period of a Pendulum   570

CHAPTER 7 VOCABULARY CHECK   571
CHAPTER 7 HIGHLIGHTS   571
CHAPTER 7 REVIEW   575
CHAPTER 7 TEST   581
CUMULATIVE REVIEW   583

CHAPTER 8   Quadratic Equations and Functions   585

CHAPTER 8 PRETEST   586
8.1   Solving Quadratic Equations by Completing the Square   587
8.2   Solving Quadratic Equations by Using the Quadratic Formula   599
8.3   Solving Equations by Using Quadratic Methods   611

Integrated Review—Summary on Solving Quadratic Equations   623

8.4   Nonlinear Inequalities in One Variable   625
8.5   Quadratic Functions and Their Graphs   633
8.6   Further Graphing of Quadratic Functions   645

Chapter 8 Activity: Recognizing Linear and Quadratic Models   656

CHAPTER 8 VOCABULARY CHECK   657
CHAPTER 8 HIGHLIGHTS   657
CHAPTER 8 REVIEW   661
CHAPTER 8 TEST   665
CUMULATIVE REVIEW   667

CHAPTER 9   Conic Sections   669

CHAPTER 9 PRETEST   670
9.1   The Parabola and the Circle   671
9.2   The Ellipse and the Hyperbola   687

Integrated Review—Graphing Conic Sections   697

9.3   Graphing Nonlinear Functions   699
9.4   Solving Nonlinear Systems of Equations   705
9.5   Nonlinear Inequalities and Systems of Inequalities   711

Chapter 9 Activity: Modeling Conic Sections 718

CHAPTER 9 VOCABULARY CHECK   719
CHAPTER 9 HIGHLIGHTS   719
CHAPTER 9 REVIEW   723
CHAPTER 9 TEST   729
CUMULATIVE REVIEW   733

CHAPTER 10 Exponential and Logarithmic Functions   735

CHAPTER 10 PRETEST   736
10.1   The Algebra of Functions   737
10.2   Inverse Functions   743
10.3   Exponential Functions   755
10.4   Logarithmic Functions   765
10.5   Properties of Logarithms   775

Integrated Review—Functions and Properties of Logarithms   783

10.6   Common Logarithms, Natural Logarithms, and Change of Base   785
10.7   Exponential and Logarithmic Equations and Problem Solving   793

Chapter 10 Activity: Modeling Temperature   802

CHAPTER 10 VOCABULARY CHECK   803
CHAPTER 10 HIGHLIGHTS   803
CHAPTER 10 REVIEW   807
CHAPTER 10 TEST   813
CUMULATIVE REVIEW   817

Appendix A   An Introduction to Using a Graphing Utility   819

Appendix B   Solving Systems of Equations Using Determinants   825

Appendix C   Fractions   833

Appendix D   Review of Angles, Lines, and Special Triangles   841

Appendix E   Review of Geometric Figures   849

Answers to Selected Exercises   A1

Solutions to Selected Exercises   A43

Index   I1

Photo Credits   I8

PREFACE

About This Book

Intermediate Algebra, Second Edition was written to provide a **solid foundation** in algebra as well as to help develop problem-solving skills. Specific care was taken to ensure that students have the most **up-to-date relevant** text preparation for their next mathematics course or for non-mathematical courses that require an understanding of algebraic fundamentals. I have tried to achieve this by writing a user-friendly text that is keyed to objectives and contains many worked-out examples. As suggested by the AMATYC Crossroads Document and the NCTM Standards (plus Addenda), real-life and real-data applications, data interpretation, conceptual understanding, problem solving, writing, cooperative learning, appropriate use of technology, mental mathematics, number sense, critical thinking, and geometric concepts are emphasized and integrated throughout the book.

The many factors that contributed to the success of the first edition have been retained. In preparing the Second Edition, I considered comments and suggestions of colleagues, students, and many users of the prior edition throughout the country.

Intermediate Algebra, Second Edition is part of a series of texts that can include *Basic College Mathematics,* Second Edition; *Prealgebra,* Third Edition; *Introductory Algebra,* Second Edition; and a combined text, *Algebra A Combined Approach,* Second Edition. Throughout the series pedagogical features are designed to develop student proficiency in algebra and problem solving, and to prepare students for future courses.

Key Pedagogical Features and Changes in the Second Edition

Readability and Connections   I have tried to make the writing style as clear as possible while still retaining the mathematical integrity of the content. When a new topic is presented, an effort has been made to relate the new ideas to those that students may already know. Constant reinforcement and connections within problem-solving strategies, data interpretation, geometry, patterns, graphs, and situations from everyday life can help students gradually master both new and old information. In addition, each section begins with a list of objectives covered in the section. Clear organization of section material based on objectives further enhances readability.

Problem-Solving Process   This is formally introduced in Chapter 2 with a four-step process that is integrated throughout the text. The four steps are **Understand**, **Translate**, **Solve**, and **Interpret**. The repeated use of these steps in a variety of examples shows their wide applicability. Reinforcing the steps can increase students' comfort level and confidence in tackling problems.

Applications and Connections   Every effort was made to include as many interesting and relevant real-life applications as possible throughout the text in both worked-out examples and exercise sets. In the Second Edition, the applications have been thoroughly revised and updated, and the number of applications has increased. The applications help to motivate students and strengthen their understanding of mathematics in the real world. They show connections to a wide range of fields including agriculture, allied health, anthropology, art, astronomy, biology, business, chemistry, construction, consumer affairs, earth science, education, entertainment, environmental issues, finance, geography, government, history, medicine, music, nutrition, physics, sports, travel, and weather. Many of the applications are based on recent real data. Sources for data include newspapers, magazines, publicly held companies, government agencies, special-interest groups, research organizations, and reference books. Opportunities for obtaining your own real data are also included. See the Applications Index on page xvi.

Practice Problems   Throughout the text, each worked-out example has a parallel Practice Problem placed next to the example in the margin. These invite students to be actively involved in the learning process before beginning the end-of-section exercise set. Practice Problems immediately reinforce a skill after it is developed. Answers appear at the bottom of the page for quick reference.

Concept Checks   These margin exercises are appropriately placed throughout the text. They allow students to gauge their grasp of an idea as it is being explained in the text. Concept Checks stress conceptual understanding at the point of use and help suppress misconceived notions before they start. Answers appear at the bottom of the page.

 Increased Integration of Geometry Concepts   In addition to the traditional topics in intermediate algebra courses, this text contains a strong emphasis on problem solving and geometric concepts, which are integrated throughout. The geometry concepts presented are those most important to a student's understanding of algebra, and I have included many applications and exercises devoted to this topic. These are marked with the the geometry icon. Also, geometric figures, a review of angles, lines, and special triangles, are covered in the appendices.

Helpful Hints   Helpful Hints contain practical advice on applying mathematical concepts. These are found throughout the text and strategically placed where students are most likely to need immediate reinforcement. Helpful Hints are highlighted for quick reference.

Visual Reinforcement of Concepts   The Second Edition contains a wealth of graphics, models, photographs, and illustrations to visually clarify and reinforce concepts. These include new and updated bar graphs, line graphs, calculator screens, application illustrations, and geometric figures.

Calculator and Graphing Calculator Explorations   These optional explorations offer point-of-use intruction, through examples and exercises, on the proper use of scientific and graphing calculators as tools in the mathematical problem-solving process. Placed appropriately throughout the text, Calculator and Graphing Calculator Explorations also reinforce concepts learned in the corresponding section and motivate discovery-based learning.

Additional exercises building on the skill developed in the Explorations may be found in exercise sets throughout the text. Exercises requiring a calculator are marked with the ▦ icon. Exercises requiring a graphing calculator are marked with the ▦ icon. An Introduction to Using a Graphing Utility is included in the appendix.

Study Skills Reminders   New Study Skills Reminder boxes are integrated throughout the text. They are strategically placed to constantly remind and encourage students as they hone their study skills. A new **Section 1.1**, Tips on Success in Mathematics, provides an overview of the Study Skills needed to succeed in math. These are reinforced by the Study Skills Reminder boxes throughout the text.

Focus On   Appropriately placed throughout each chapter, these are divided into Focus on Mathematical Connections, Focus on Business and Career, Focus on the Real World, and Focus on History. They are written to help students develop effective habits for engaging in investigations of other branches of mathematics, understanding the importance of mathematics in various careers and in the world of business, and seeing the relevance of mathematics in both the present and past through critical thinking exercises and group activities.

Chapter Highlights   Found at the end of each chapter, these contain key definitions, concepts, and examples to help students understand and retain what they have learned and help them organize their notes and study for tests.

Chapter Activity   These features occur once per chapter at the end of the chapter, often serving as a chapter wrap-up. For individual or group completion, the Chapter Activity, usually hands-on or data-based, complements and extends to concepts of the chapter, allowing students to make decisions and interpretations and to think and write about algebra.

Integrated Reviews   These "mid-chapter reviews" are appropriately placed once per chapter. Integrated Reviews allow students to review and assimilate the many different skills learned separately over several sections before moving on to related material in the chapter.

Pretests   Each chapter begins with a pretest that is designed to help students identify areas where they need to pay special attention in the upcoming chapter.

Chapter Review and Test   The end of each chapter contains a review of topics introduced in the chapter. The Chapter Review offers exercises that are keyed to sections of the chapter. The Chapter Test is a practice test and is not keyed to sections of the chapter.

Cumulative Review   These features are found at the end of each chapter (except Chapter 1). Each problem contained in the Cumulative Review is an earlier worked example in the text that is referenced in the back of the book along with the answer. Students who need to see a complete worked-out solution, with explanation, can do so by turning to the appropriate example in the text.

Student Resource Icons   At the beginning of each section, videotape and CD, tutorial software, Prentice Hall Tutor Center, and solutions manual icons are displayed. These icons help reinforce that these learning aids are available should students wish to use them to help them review concepts and skills at their own pace. These items have direct correlation to the text and emphasize the text's methods of solution.

Functional Use of Color and New Design   Elements of this text are highlighted with color or design to make it easier for students to read and study. Special care has been taken to use color within solutions to examples or in the art to **help clarify, distinguish, or connect concepts**.

Exercise Sets   Each text section ends with an Exercise Set. Each exercise in the set, except those found in parts labeled Review and Preview or Combining Concepts, is keyed to one of the objectives of the section. Wherever possible, a specific example is also referenced. In addition to the approximately 4300 exercises in end-of-section exercise sets, exercises may also be found in the Pretests, Integrated Reviews, Chapter Reviews, Chapter Tests, and Cumulative Reviews.

Exercises and examples marked with a video icon have been worked out step-by-step by the author in the videos that accompany this text.

Throughout the exercises in the text there is an emphasis on data and graphical interpretation via tables, charts, and graphs. The ability to interpret data and read and create a variety of types of graphs is developed gradually so students become comfortable with it. Similarly, geometric concepts—such as perimeter and area—are integrated throughout the text. Exercises and examples marked with a geometry icon △ have been identified for convenience.

Each exercise set contains one or more of the following features.

Mental Math   Found at the beginning of an exercise set, these mental warmups reinforce concepts found in the accompanying section and increase students' confidence before they tackle an exercise set. By relying on their own mental skills, students increase not only their confidence in themselves but also their number sense and estimation ability.

Review and Preview   These exercises occur in each exercise set (except for those in Chapter 1) after the exercises keyed to the objectives of the section. Review and Preview problems are keyed to previous sections and review concepts learned earlier in the text that are needed in the next section or in the next chapter. These exercises show the links between earlier topics and later material.

Combining Concepts   These exercises are found at the end of each exercise set after the Review and Preview exercises. Combining Concepts exercises require students to combine several concepts from that section or to take the concepts of the section a step further by combining them with concepts learned in previous sections. For instance, sometimes students are required to combine the concepts of the section with the problem-solving process they learned in Chapter 2 to try their hand at solving an application problem.

Writing Exercises   These exercises occur in almost every exercise set and are marked with an icon. They require students to assimilate information and provide a written response to explain concepts or justify their thinking. Guidelines recommended by the American Mathematical Association of Two Year Colleges (AMATYC) and other professional groups recommend incorporating writing in mathematics courses to reinforce concepts.

Vocabulary Checks   Vocabulary Checks, **new to this edition**, provide an opportunity for students to become more familiar with the use of mathematical terms as they strengthen their verbal skills. These appear at the end of the chapter before the Chapter Highlights.

Data and Graphical Interpretation   There is an emphasis on data interpretation in exercises via tables and graphs. The ability to interpret data and read and create a variety of types of graphs is developed gradually so students become comfortable with it.

Internet Excursions   These exercises occur once per chapter. Internet Excursions require students to use the Internet as a data-collection tool to complete the exercises, allowing students first-hand experience with manipulating and working with real data.

Key Content Features in the Second Edition

Overview   This new edition retains many of the factors that have contributed to its success. Even so, **every section of the text was carefully re-examined**. Throughout the new edition you will find numerous new applications, examples, and many real-life applications and exercises. For example, look at the exercise sets of Sections 2.2, 3.1, 4.3, 5.1, 6.5, 7.6, and 9.1. Some sections have internal re-organization to better clarify and enhance the presentation.

Chapter 1 now begins with Tips for Success in Mathematics (Section 1.1). **New applications** and real data enhance the chapter.

New Study Skills Reminder boxes have been inserted throughout the text. These boxes reinforce the tips from Section 1.1. They are placed strategically to encourage students to hone their study skills.

Increased Integration of Geometry Concepts In addition to the traditional topics in intermediate algebra courses, this text contains a strong emphasis on problem solving, and geometric concepts are integrated throughout. The geometry concepts presented are those most important to a student's understanding of algebra, and I have included many **applications and exercises** devoted to this topic. These are marked with a geometry icon. Also, geometric figures and a review of angles, lines, and special triangles are covered in the appendices.

New Examples Detailed step-by-step examples were added, deleted, replaced, or updated as needed. Many of these reflect real life.

Exercise Sets Revised and Updated The exercise sets have been carefully examined and extensively revised. The **real-world and real-data applications** have been thoroughly updated and many new applications are included. In addition, an **increased number of challenging problems** have been included in the new edition. **Writing exercises** are now included in most exercise sets and new **Vocabulary Checks** have been added to the end of the chapter to help students become proficient in the language of mathematics.

Enhanced Supplements Package The Second Edition is supported by a wealth of supplements designed for **added effectiveness and efficiency**. New items include the MathPro 5 on-line tutorial with diagnostic and unique video clip feature, a new computerized testing system (TestGen-EQ with Quiz-Master), Prentice Hall Tutor Center, digitized videos on CD, and Instructor's CD Series. Please see the list of supplements for descriptions.

Options for On-line and Distance Learning

For maximum convenience, Prentice Hall offers on-line interactivity and delivery options for a variety of distance learning needs. Instructors may access or adopt these in conjunction with this text.

The **Companion Web** site includes basic distance learning access to provide links to the text's Internet Excursions and a selection of on-line self quizzes. Email is available.

WebCT WebCT includes distance learning access to content found in the Martin-Gay companion Web site plus more. WebCT provides tools to create, manage, and use on-line course materials. Save time and take advantage of items such as on-line help, communication tools, and access to instructor and student manuals. Your college may already have WebCT's software installed on their server or you may choose to download it. Contact your local Prentice Hall sales representative for details.

BlackBoard Visit http://www.prenhall.com/demo. For distance learning access to content and features from the Martin-Gay companion Web site plus more, Blackboard provides simple templates and tools.

Course Compass™ Powered by BlackBoard. Visit http://www.prenhall.com/demo.

Supplements for the Instructor

Printed Supplements

Annotated Instructor's Edition (0-13-067670-5)

- Answers to all exercises printed on the same text page.
- Teaching Tips throughout the text placed at key points in the margin.

Instructor's Solution Manual (0-13-067668-3)

- Solutions to even-numbered section exercises.
- Solutions to every (even and odd) Mental Math exercise.
- Solutions to every (even and odd) Practice Problem (margin exercise).
- Solutions to every (even and odd) exercise found in the Pretests, Integrated Reviews (mid-chapter reviews), Chapter Reviews, Chapter Tests, and Cumulative Reviews.

Instructor's Resource Manual with Tests (0-13-067669-1)

- Notes to the Instructor that include an introduction to Interactive Learning, Interpreting Graphs and Data, Alternative Assessment, Using Technology, and Helping Students Succeed.
- Two free-response Pretests per chapter.
- Eight Chapter Tests per chapter (3 multiple-choice, 5 free-response).
- Two Cumulative Review Tests (one multiple-choice, one free-response) every two chapters.
- Eight Final Exams (3 multiple-choice, 5 free-response).
- Twenty additional exercises per section for added test exercises if needed.
- Group Activities (an average of two per chapter; providing short group activities in a convenient, ready-to-use format).
- Answers to all items.

Media Supplements

TestGen-EQ with QuizMaster CD-ROM (Windows/Macintosh) (0-13-067671-3)

- Algorithmically driven, text-specific testing program.
- Networkable for administering tests and capturing grades on-line.
- Edit and add your own questions to create a nearly unlimited number of tests and worksheets.
- Use the new "Function Plotter" to create graphs.
- Tests can be easily exported to HTML so they can be posted to the Web for student practice.
- Includes an email function for network users, enabling instructors to send a message to a specific student or an entire group.
- Network-based reports and summaries for a class or student and for cumulative or selected scores are available.

MathPro 5 Instructor Version

- On-line, customizable tutorial, diagnostic, and assessment program for anytime, anywhere tutorial support.
- Text specific at the learning objective level.

- Diagnostic option identifies student skills, provides individual learning plan, and tutorial reinforcement.
- Integration of TestGen-EQ allows for testing to operate within the tutorial environment.
- Course management tracking of tutorial and testing activity.

MathPro Explorer 4.0

- Network Version IBM/Mac 0-13-067660-8.
- Enables instructors to create either customized or algorithmically generated practice quizzes from any section of a chapter.
- Includes email function for network users, enabling instructors to send a message to a specific student or to an entire group.
- Network-based reports and summaries for a class or student and for cumulative or selected scores.

Instructor's CD Series

- Written and presented by Elayn Martin-Gay.
- Contains suggestions for presenting course material, utilizing the integrated resource package, time-saving tips, and much more.

Companion Web Site http://www.prenhall.com/martin-gay_interm

- Create a customized on-line syllabus with Syllabus Manager.
- Links related to the Internet Excursions in each chapter allow students to find and retrieve real data for use in guided problem solving.
- Assign quizzes or monitor student self quizzes by having students email results, such as true/false reading quizzes or vocabulary check quizzes.
- Destination links provide additional opportunities to explore related sites.

Supplements for the Student

Printed Supplements

Student's Solution Manual (0-13-067678-0)

- Solutions to odd-numbered section exercises.
- Solutions to every (even and odd) Mental Math exercise.
- Solutions to every (even and odd) Practice Problem (margin exercise).
- Solutions to every (even and odd) exercise found in the Pretests, Integrated Reviews (mid-chapter reviews), Chapter Reviews, Chapter Tests, and Cumulative Reviews.

Media Supplements

MathPro 5 (Student Version)

- Online, customizable tutorial, diagnostic, and assessment software.
- Text specific to the learning objective level, providing anytime, anywhere tutorial support.
- Algorithmically driven for virtually unlimited practice problems with immediate feedback.
- "Watch" screen videoclips by K. Elayn Martin-Gay.
- Step-by Step solutions.
- Summary of Progress.

MathPro 4.0 Explorer Student Version (0-13-067677-2)

- Available on CD-ROM for stand alone use or can be networked in the school laboratory.
- Text specific tutorial exercises and instructions at the objective level.

- Algorithmically generated Practice Problems.
- "Watch" screen videoclips by K. Elayn Martin-Gay.

Videotape Series (0-13-067676-4)

- Written and presented by Elayn Martin-Gay.
- Keyed to each section of the text.
- Step-by-step solutions to exercises from each section of the text. Exercises that are worked in the videos are marked with a video icon.

New Digitized Lecture Videos on CD-ROM (0-13-067675-6)

- The entire set of *Intermediate Algebra*, Second Edition lecture videotapes in digital form.
- Convenient access anytime to video tutorial support from a computer at home or on campus.
- Available shrink-wrapped with the text or stand-alone.

New Prentice Hall Tutor Center

- Staffed with developmental math instructors and open 5 days a week, 7 hours per day.
- Obtain help for examples and exercises in Martin-Gay, *Intermediate Algebra*, Second Edition via toll-free telephone, fax, or email.
- The Prentice Hall Tutor Center is accessed through a registration number that may be bundled with a new text or purchased separately with a used book. Visit http://www.prenhall.com/tutorcenter to learn more.

Companion Web Site www.prenhall.com/martin-gay_interm

- Links related to the Internet Excursions in each chapter allow you to collect data to solve specific internet exercises.

Acknowledgments

First, as usual, I would like to thank my husband, Clayton, for his constant encouragement. I would also like to thank my children, Eric and Bryan, for their sense of humor and especially for asking Dad to cook the bacon that I always used to burn.

I would also like to thank my extended family for their invaluable help and also their sense of humor. Their contributions are too numerous to list. They are Rod, Karen, and Adam Pasch; Michael, Christopher, Matthew, and Jessica Callac; Stuart, Earline, Melissa, Mandy, Bailey, and Ethan Martin; Mark, Sabrina, and Madison Martin; Leo and Barbara Miller; and Jewett Gay.

I would like to thank the following reviewers for their input and suggestions:

Karen J. Bright, *Southern Alabama Community College*
Betty Dennison, *Roane State Community College*
Matthew Gardner, *North Hennepin Community College*
Joye Elaine Gowan, *Roane State Community College*
Debra R. Hill, *University of North Carolina–Charlotte*
Jeff Sexton, *Roane State Community College*
Lee Ann Spahr, *Durham Technical Community College*

There were many people who helped me develop this text and I will attempt to thank some of them here. Laurie Semarne was invaluable for contributing to the overall accuracy of this text. Emily Keaton and Kathy Sessa-Federico were also invaluable for their many suggestions and contributions during the development and writing of this first edition. Ingrid Mount at Elm Street Publishing Services provided guidance throughout the production process. I thank Jenny Crawford and Richard Semmler for all their work on the solutions, text, and accuracy. Lastly, a special thank you to my project manager Mary Beckwith and executive editor Karin Wagner, for their support and assistance throughout the development and production of this text and to all the staff at Prentice Hall: Linda Behrens, Alan Fischer, Maureen Eide, Grace Hazeldine, Tom Benfatti, Eilish Main, John Tweeddale, Chris Hoag, Paul Corey, and Tim Bozik.

K. Elayn Martin-Gay

About the Author

K. Elayn Martin-Gay has taught mathematics at the University of New Orleans for more than 20 years and has received numerous teaching awards including the local University Alumni Association's Award for Excellence in Teaching.

Over the years, Elayn has developed a videotaped lecture series to help her students understand algebra better. This highly successful video material is the basis for her books: *Basic College Mathematics*, Second Edition; *Prealgebra*, Third Edition; *Introductory Algebra*, Second Edition; *Intermediate Algebra*, Second Edition; *Algebra A Combined Approach*, Second Edition; and her hardback series: *Beginning Algebra*, Third Edition; *Intermediate Algebra*, Third Edition; *Beginning and Intermediate Algebra*, Second Edition; and *Intermediate Algebra: A Graphing Approach*, Second Edition.

**To my mother, Barbara M. Miller,
and her husband, Leo Miller,
and to the memory of my father,
Robert J. Martin**

APPLICATIONS INDEX

A

Architecture
circular pool with fountain, 685
height/width of bridge arch, 731
Astronomy
astronomical units, 113
distance between Jupiter and the sun, 56
Drake Equation, 113
eccentricities of planets' orbits, 696
escape velocity, 511, 512
pulsars, 57
Automobiles
car and light truck sales, 290
driving speeds, 615–16
miles driven, 158
radius of curvature, 582
rate of speed, 283–84
rental charges, 291
Saab production, 268
Saturn vehicle sales, 656
sports utility vehicles, 337

B

Business
airline revenues, 180
average cost of production, 460
beginning salaries, 126
big Mac price in Russia, 172
biotechnology spending, 229
boat manufacturing, 237
break-even point, 104, 158, 287–88, 293, 317, 322
car sales, 164
cellular telephone subscribers/users, 251, 520, 758–59, 763
cigarette consumption, 103, 121, 126
computer support specialists, 203
cost functions, revenue functions, profit functions, 742
cost of production, 482
delivery company overhead, 560
demand equation, 527, 598
digital camera sales, 337
DVD player sales, 213
equilibrium point, 272
fastest-growing occupations, 440
financial ratios, 622
Internet shopping revenues, 762
labor statistics, 214
linear modeling, 244
manufacturing costs, 184, 491, 653
market equilibrium, 710
maximum profits, 653
minimum wages, 183
motel occupancy, 322
net earnings, 610
net profit margin, 357
office supply prices, 290
online shopping, 98
original price of computer, 323
price-sales relationship, 212
profit function, 336
profits, 164, 213, 336, 632
revenue, 357
revenue/cost function, 293
sales prediction, 213
supply and demand, 272
switchboard operators, 100
telephone operators, 79
total cost of production, 336
total revenues, 57, 336, 433
trademark registrations, 365
unit costs, 429

Wal-Mart stores, 656
women/men over 65 in labor force, 291
workforce size, 101

C

Chemistry
alcohol solution, 318
mixing solutions, 284–85, 292
phenobarbital solution, 289
plant food solution, 100
Construction
pipe fitting, 579
pitch of roof, 193
sewer pipe slope, 196
slope measurement, 238
wall framing, 391–92

D

Distance. See Time and distance

E

Education
ACT assessment, 294
average income and level of, 203
computers for instructional use, 365
learning curve, 799
SAT scores, 101
student travel destinations, 290
teacher assistants, 213
test scores, 125
tuition projections, 167, 204
Electricity
cable length, 394
current in circuit, 363
homes heated by, 465
resistance and current, 482
Electronics
angular frequency of oscillations, 520
combined resistance, 474, 496
Engineering
bridge over bayou, 685
Entertainment
drama club ticket sales, 290
movie going, 169
theater costs, 77
ticket price at Disney World, 169, 199
Environment
carbon dioxide concentration, 763
greenhouse effect, 585
methane emissions, 655
nuclear waste, 762
pollutant removal costs, 433
pollution production, 481
radioactive decay rate, 762, 764, 774
radioactive material in milk, 760
recycled newsprint, 103

F

Finance and economics
average annual earnings, 158
average monthly checking account balances, 569
car rental plans, 126
car savings, 160
compound interest, 110, 113, 159, 164, 255, 661, 759, 763, 796, 798, 809, 812, 814
continuous compounding of interest, 787, 790, 806, 811, 818
doubling of investment, 795

drafting supply prices, 291
energy audit, 122
fax machine price, 99
frame purchases, 290
gold price, 46
income with commission calculation, 120–21
interest rates, 318, 592–93, 597, 598
investment amount, 110
investment risk over time, 38
lottery prize division, 434
online spending, 58
original price of computer, 94–95, 166
parking garage charges, 126
personal bankruptcy petitions, 294
room redecorating, 150
savings account amount, 107–8, 110
simple interest, 363
space shuttle costs, 113
stock closing price, 99
stock markets, 38
telephone billing plans, 126
textbook price, 99
travel spending, 164
U.S. federal government income/expenditures, 272
value of imports per person, 65
Food
beef supply, 268
butter consumption, 291
butterfat milk, 289
calorie count of, 112
candy mixture, 290, 318
cheese production, 149
coffee beverages, 58
commercially prepared breakfasts, 98
fruit drinks, 322
ice cream container, 159
milk consumption, 127
nutrition labels, 114
pork and chicken consumption, 137
poultry consumption, 229
Recommended Dietary Allowance (RDA), 610
red meat and poultry consumption, 272, 281–82
sweeteners consumption, 245
wedding cake, 395

G

Geometry
angles
complementary, 101
measures, 101, 258, 286–87, 291, 292, 293, 817
supplementary, 101
with supplement equal to twice the complement, 103
circle
area of, 46, 164, 228, 347, 383, 484
center of, 670, 684
circumference of, 75, 484
fractional part of, 37, 74
radius of, 577, 598, 670, 684
cone
radius of, 546
surface area of, 527
volume of, 483
conic section modeling, 718
cube
volume of, 228, 373, 383

cylinder
area of, 363
radius of, 112
surface area of, 479, 668
volume of, 111, 347, 383, 414, 484
cylindrical solid
dimensions of, 102
eccentricities of conic sections, 695
golden ratio, 608
height of triangle, 80
heptagon
perimeter of, 493
hexagon
lengths of sides of, 357
investigations of, 14
irregular shapes
area of, 348
isosceles right triangle
length of legs/sides of, 319, 598, 607, 662
parallelogram
area of, 65, 434
height of, 357
quadrilateral
angle measures of, 293, 296
lengths of sides of, 292
perimeter of, 441
rectangle
area of, 395, 412, 432, 534
dimensions of, 100, 110, 111, 152, 158, 159, 318, 395, 415, 607, 710
length and width of, 99, 395, 586, 608, 727
maximum area of, 654
perimeter of, 46, 338, 395, 411, 534, 607
width of, 357
rectangular solid
surface area of, 336
volume of, 105
right triangle
length of legs of, 395
length of unknown leg of, 552–53, 556, 557
shaded region
area of, 383, 386, 413, 418
sphere
circumference of, 112
radius of, 546, 557
surface area of, 496
volume of, 111, 112
square
area of, 363, 607
dimensions of, 100, 110, 598
length of side of, 607, 624
perimeter of, 441
trapezoid
area of, 106, 534
perimeter of, 534
triangle
angle measures of, 100
area of, 432
base and height of, 608
height of, 110
Heron's formula for area of, 559
length of unknown side of, 666
lengths of sides of, 95–96, 99, 291
perimeter of, 338, 511, 534
unknown length of, 578, 579, 582
volume
of cube, 65

M

Medicine
  basal metabolic rate (BMR), 520
  body-mass index (BMI), 473, 474
  flu epidemic, 798
  health maintenance organization
      enrollment, 337
  heartworm preventive for dog, 229
  medical assistants, 203
  target heart rate, 138
Miscellaneous
  aircraft carriers, 289
  armed forces active duty personnel
      (U.S.), 245
  black-and-white
      television/microwave oven
      ownership, 302
  board lengths, 357
  bookshelf manufacturing, 46
  cardboard dimensions, 620
  coin types, 318
  consecutive integers, 96, 104, 158,
      395, 495
  convention locations, 290
  conveyor belts, 472
  difference in elevations, 37, 74
  distance across lake, 685
  distance seen from height above
      ground, 559
  dog food consumption, 620
  expected height of boy given
      weight, 799
  federally owned land, 98
  fencing around land, 97
  garden sprinkler radius, 620
  goldfish pond fill time, 471
  Hooke's law, 476
  horsepower, 483
  length of bent wire, 511
  life expectancy at birth for
      females, 17
  mail handling, 77
  median retirement age for men
      (U.S.), 180
  men in Navy, 470
  milk carton fill times, 471
  Mosteller formula for body surface
      area, 511
  perimeter of floor plan, 97
  picture frame, 406
  pond fill time, 619
  prison and jail inmates, 610, 655
  rabbit diet, 292
  speaker wire, 394
  spinner and fair game, 37
  state and federal prison inmates, 459
  study and work schedule, 237
  tank emptying time, 495
  tank fill time, 471, 619
  time management, 309
  tourist destinations, 158
  unknown numbers, 93, 282–83, 290,
      291, 292, 296, 318, 319, 324,
      415, 464, 495, 620, 632, 654,
      710
  water consumed by camel, 470
  wind force against sail, 483

  women in Coast Guard, 470
  working hours, 182, 183
  work problems, 466–67, 470, 471,
      472, 483, 489, 495, 498,
      613–14, 620, 624, 662, 666
  zip codes, 101

P

Physics
  speed of wave traveling over
      string, 520
Politics
  gubernatorial salaries, 99
Population
  Arizona, 800
  Australia, 812
  bison, 762
  Brazil, 798
  decrease in, 101
  estimating size of, 485–86
  finding for given cities, 92
  France, 812
  of given cities, 99
  Hungary, 798
  increase in, 99
  Japan and Oceania, 65
  mallard ducks, 812
  Mexico, 812
  Nevada, 800
  over 65 (U.S.), 48
  over 85 (U.S.), 199
  Paraguay, 798
  prairie dog, 815
  Russia, 798
  Sierra Leone, 812
  United States, 762
  wolf, 798
  wood duck, 815
  world, 57, 815
Psychology
  IQ scores, 558

R

Real estate
  building appreciation, 249
  median price of existing homes, 213
  sales predictions, 208–9

S

Science
  atmospheric pressure, 799
  bulb hours, 102
  Cassini spacecraft, 111
  center of mass, 210
  Clarke belt, 111
  communications satellites, 113
  density of object, 65
  diffusion, 528
  Doppler effect, 421, 451, 473
  earthquake magnitudes, 98
  height of woman given height of
      femur or tibia bone, 229
  length of pendulum, 558, 559, 570
  light intensity, 482, 484
  mass of water in Lake Superior, 65
  Mount Vesuvius, 501

  period of pendulum, 558, 559, 570
  resulting force, 558
  Richter scale and magnitude of
      earthquake, 735, 791, 801
  solar wind, 46
  sound intensity, 782
  surface area of volcano, 527
  volcano height above sea level, 37
  whale weights, 102
Sports
  auto racing speed, 621
  baseball opening-day player
      payroll, 56, 99, 253
  earned run average, 451, 498
  home runs, 103
  NBA top scorer, 292
  Olympic medals won, 103
  scores and silver medal, 160
  speed skating score, 125
  stadium seats, 99
  3–point baskets, 172
  women's basketball top scorer, 292
  X-Games, 289

T

Technology
  CD player reliability, 815
  cell phone subscribers, 102
  computer depreciation, 249
  e-mail checking, 98–99
  Internet access company
      subscribers, 99
  radio stations on Internet, 337
  women using Internet, 290
Temperature
  Boyle's law, 478, 496
  at center of sun, 57
  Charles's law, 481
  daily low, 609
  Fahrenheit/Celsius interconversions,
      42–43, 46, 110, 159, 738
  glass liquidity, 126
  modeling, 802
  stibnite melting point, 126
Time and distance
  average speed, 110, 159
  camel's traveling time, 470
  cyclist's rate of travel, 472
  cyclist's speeds, 290
  distance fallen and velocity of
      object, 558
  distance of rockets, 471
  distance of run, 472
  distance saved, 603, 607
  distance traveled, 661
  drill retrieval rate, 112
  driving speed, 619, 668
  driving time, 110
  feet traveled by freely falling
      object, 597, 598, 608
  flatland and mountain rate of truck
      travel, 472
  height of cliff over gorge, 498
  height of object, 325, 331, 336, 365,
      375, 395, 396, 415, 418, 608,
      632, 662, 664, 668
  jogger's speed, 619

  jogging and biking speed, 619
  landing time, 604
  line of sight, 481
  mamba snake travel time, 471
  maximum height, 650, 653, 666
  rate of speed, 318
  return time of rocket, 390
  rock's rate of travel, 212
  rowing in still water/speed of
      current, 289
  rowing time, 472
  seagull's flight, 196
  skid mark and speed of car, 557
  speed of bicyclist, 471
  speed of boat, 470, 496
  speed of cars, 472, 482, 500
  speed of current, 467–68, 470
  speed of jet, 495
  speed of plane, 289, 290, 470, 471
  speed of trains, 471, 496
  speed of truck, 470
  speed of walker, 496
  speed of wind, 496
  tortoise travel time, 471
  travel time by jet and car, 472
  treadmill speed, 624
Transportation
  aircraft seats, 99
  bus driver's median weekly
      earnings, 294
  commercial airplane deliveries, 213
  miles of highways, roads, and
      streets, 57
  motor vehicle safety statistics, 306
  New York City subway system, 57
  passenger traffic at airport, 610
  road slope, 193
  slope of descent, 193
  slope of jet's climb, 193
  subways, 1
  teenage deaths from motor vehicle
      accidents, 306

W

Weather
  lightning strike locations,
      257, 310
  rainfall data, 294
  tornadoes, 98
  tornado-related deaths, 100
Weight
  of ball, 481
  and beam support, 480
  elevator limit, 126
  equivalent, on Earth and
      Jupiter, 250
  and inverse variation, 482
  of laundry, 159
  plane's takeoff weight, 126
  postal, 126
  supported by circular column,
      480, 482
  supported by rectangular
      beam, 483

HIGHLIGHTS OF INTERMEDIATE ALGEBRA, SECOND EDITION

Intermediate Algebra, Second Edition is the primary learning tool in a fully integrated learning package to help you succeed in this course. Author K. Elayn Martin-Gay focuses on enhancing the traditional emphasis of mastering the basics with innovative pedagogy and a meaningful learning program. There are three goals that drive her authorship:

▲ **Master and apply skills and concepts**

▲ **Build confidence**

▲ **Increase motivation**

Take a few moments now to examine some of the features that have been incorporated into *Intermediate Algebra, Second Edition* to help students excel.

Graphs and Functions

CHAPTER 3

The linear equations we explored in Chapter 2 are statements about a single variable. This chapter examines statements about two variables: linear equations and inequalities in two variables. We focus particularly on graphs of these equations and inequalities, which lead to the notion of relation and to the notion of function, perhaps the single most important and useful concept in all of mathematics.

3.1 Graphing Linear Equations
3.2 The Slope of a Line
3.3 The Slope-Intercept Form
3.4 More Equations of Lines
    Integrated Review—Linear Equations in Two Variables
3.5 Introduction to Functions
3.6 Graphing Linear Inequalities

At the beginning of the 20th century, there were approximately 237,600 students enrolled in the 977 institutions of higher education in the United States. At that time, only 19% of bachelor's degree recipients were women. By the year 2010, the projected 4700 degree-granting colleges and universities in the United States will have an estimated 17,490,000 students. Of these, roughly 59% of bachelor's degree recipients are expected to be women. The phenomenal growth of colleges and universities can also be seen in the average tuition costs at these institutions of higher learning. For instance, the average annual tuition at a private four-year college or university has increased from $1809 in 1970 to $15,380 in 2000, an increase of about 750%! In Exercises 33 and 34 on page 204 we will use linear equations to predict the future cost of annual tuition at both two-year and four-year public colleges and universities.

◀ **REAL WORLD APPLICATIONS**

Chapter-opening real-world applications introduce you to everyday situations that are applicable to the mathematics you will learn in the upcoming chapter, showing the relevance of mathematics in daily life.

Become a Confident Problem Solver

A goal of this text is to help you develop problem-solving abilities.

EXAMPLE 12   Finding the Annual Consumption

In the United States, the annual consumption of cigarettes is declining. The consumption c in billions of cigarettes per year since the year 1985 can be approximated by the formula

$$c = -14.25t + 598.69$$

where t is the number of years after 1985. Use this formula to predict the first year that the consumption of cigarettes will be less than 200 billion per year.

Solution:

1. UNDERSTAND. Read and reread the problem. To become familiar with the given formula, let's find the cigarette consumption after 20 years, which would be the year 1985 + 20, or 2005. To do so, we substitute 20 for t in the given formula.

   $$c = -14.25(20) + 598.69 = 313.69$$

   Thus, in 2005, we predict cigarette consumption to be about 313.69 billion.
   Variables have already been assigned in the given formula. For review, they are

   c = the annual consumption of cigarettes in the United States in billions of cigarettes

   t = the number of years after 1985

2. TRANSLATE. We are looking for the first year that the consumption of cigarettes c is less than 200. Since we are finding years t, we substitute the expression in the formula given for c, or

   $$-14.25t + 598.69 < 200$$

3. SOLVE the inequality.

   $$-14.25t + 598.69 < 200 \qquad \text{Subtract 598.69 from both sides.}$$
   $$-14.25t < -398.69 \qquad \text{Divide both sides by } -14.25 \text{ and round the result.}$$
   $$t > 27.98$$

4. INTERPRET.

   Check: We substitute a number greater than 27.98 and see that c is less than 200.

   State: The annual consumption of cigarettes will be less than 200 billion for more than 27.98 years after 1985, or in approximately $28 + 1985 = 2013$. ●

Page 121

◀ GENERAL STRATEGY FOR PROBLEM-SOLVING

Save time by having a plan. This text's organization can help you. Note the outlined problem-solving steps, *Understand, Translate, Solve,* and *Interpret.*

Problem solving is introduced early, emphasized and integrated throughout the book. The author provides patient explanations and illustrates how to apply the problem-solving procedure to the in-text examples.

GEOMETRY ▶

Geometric concepts are integrated throughout the text. Examples and exercises involving geometric concepts are now identified with a triangle icon. △ The text includes appendices on geometry as well.

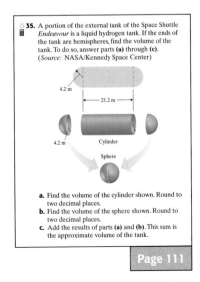

△ **35.** A portion of the external tank of the Space Shuttle *Endeavour* is a liquid hydrogen tank. If the ends of the tank are hemispheres, find the volume of the tank. To do so, answer parts **(a)** through **(c)**. (*Source:* NASA/Kennedy Space Center)

4.2 m

21.2 m

4.2 m   Cylinder

Sphere

a. Find the volume of the cylinder shown. Round to two decimal places.
b. Find the volume of the sphere shown. Round to two decimal places.
c. Add the results of parts **(a)** and **(b)**. This sum is the approximate volume of the tank.

Page 111

Master and Apply Basic Skills and Concepts

K. Elayn Martin-Gay provides thorough explanations of key concepts and enlivens the content by integrating successful and innovative pedagogy.*Intermediate Algebra, Second Edition* integrates skill building throughout the text and provides problem-solving strategies and hints along the way. These features have been included to enhance your understanding of algebraic concepts.

Concept Check

In the definition of slope, we state that $x_2 \neq x_1$. Explain why.

Page 186

◀ CONCEPT CHECKS

Concept Checks are special margin exercises found in most sections. Work these to help gauge your grasp of the concept being developed in the text.

Page 195

Combining Concepts

69. Each line on the graph has negative slope.

  a. Find the slope of each line.

  b. Use the results of part (a) to fill in the blank:
    For lines with negative slopes, the steeper line has the
    _____ (greater/lesser) slope.

COMBINING CONCEPTS ▶

Combining Concepts exercises are found at the end of each exercise set. Solving these exercises will expose you to the way mathematical ideas build upon each other.

PRACTICE PROBLEMS ▶

Practice Problems occur in the margins next to every Example. Work these problems after an example to immediately reinforce your understanding.

Page 184

Practice Problem 7

Southwest Regional is an established office product maintenance company that has enjoyed constant growth in new maintenance contracts since 1985. In 1993, the company obtained 15 new contracts and in 2000, the company obtained 36 new contracts. Use these figures to predict the number of new contracts this company can expect in 2007.

EXAMPLE 7 Predicting Sales

Southern Star Realty is an established real estate company that has enjoyed constant growth in sales since 1993. In 1995 the company sold 200 houses, and in 2000 the company sold 275 houses. Use these figures to predict the number of houses this company will sell in 2005.

Solution:

1. UNDERSTAND. Read and reread the problem. Then let

    x = the number of years after 1993 and

    y = the number of houses sold in the year corresponding to x.

  The information provided then gives the ordered pairs $(2, 200)$ and $(7, 275)$. To better visualize the sales of Southern Star Realty, we graph the line that passes through the points $(2, 200)$ and $(7, 275)$.

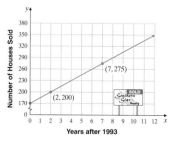

2. TRANSLATE. We write an equation of the line that passes through the points $(2, 200)$ and $(7, 275)$. To do so, we first find the slope of the line.

$$m = \frac{275 - 200}{7 - 2} = \frac{75}{5} = 15$$

Page 208

62. Explain why we generally use three points to graph a line, when only two points are needed.

▲ WRITING EXERCISES

New Writing Exercises, marked by an icon, ＼ are now found in most practice sets.

Test Yourself and Check Your Understanding

Good exercise sets and an abundance of worked-out examples are essential for building student confidence. The exercises you will find in this worktext are intended to help you build skills and understand concepts as well as motivate and challenge you. In addition, features like Chapter Highlights, Chapter Reviews, Chapter Tests, and Cumulative Reviews are found at the end of each chapter to help you study and organize your notes.

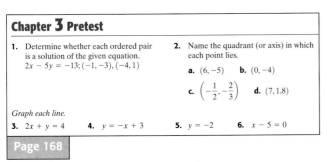

Chapter 3 Pretest

1. Determine whether each ordered pair is a solution of the given equation.
   $2x - 5y = -13$; $(-1, -3)$, $(-4, 1)$

2. Name the quadrant (or axis) in which each point lies.
   a. $(6, -5)$ **b.** $(0, -4)$
   c. $\left(-\frac{1}{2}, -\frac{2}{3}\right)$ **d.** $(7, 1.8)$

Graph each line.
3. $2x + y = 4$ 4. $y = -x + 3$ 5. $y = -2$ 6. $x - 5 = 0$

Page 168

◄ PRETESTS

Pretests open each chapter. Take a Pretest to evaluate where you need the most help before beginning a new chapter.

Page 129

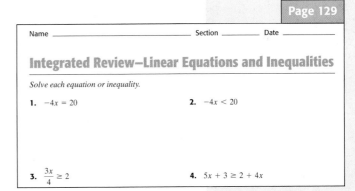

Name _____ Section _____ Date _____

Integrated Review—Linear Equations and Inequalities

Solve each equation or inequality.

1. $-4x = 20$ 2. $-4x < 20$

3. $\frac{3x}{4} \geq 2$ 4. $5x + 3 \geq 2 + 4x$

INTEGRATED REVIEWS ►

Integrated Reviews serve as mid-chapter reviews and help you to assimilate the new skills you have learned separately over several sections.

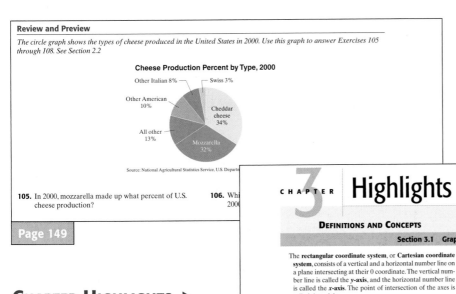

Review and Preview

The circle graph shows the types of cheese produced in the United States in 2000. Use this graph to answer Exercises 105 through 108. See Section 2.2

Cheese Production Percent by Type, 2000

Other Italian 8% — Swiss 3%
Other American 10%
Cheddar cheese 34%
All other 13%
Mozzarella 32%

Source: National Agricultural Statistics Service, U.S. Depart...

105. In 2000, mozzarella made up what percent of U.S. cheese production?

106. Whi... 2000

Page 149

◄ REVIEW AND PREVIEW

Review and Preview exercises review concepts learned earlier in the text that are needed in the next section or chapters.

Page 239

CHAPTER **3** Highlights

DEFINITIONS AND CONCEPTS	EXAMPLES

Section 3.1   Graphing Linear Equations

The **rectangular coordinate system**, or **Cartesian coordinate system**, consists of a vertical and a horizontal number line on a plane intersecting at their 0 coordinate. The vertical number line is called the **y-axis**, and the horizontal number line is called the **x-axis**. The point of intersection of the axes is called the **origin**.

To **plot** or **graph** an ordered pair means to find its corresponding point on a rectangular coordinate system.

An ordered pair is a **solution** of an equation in two variables if replacing the variables by the corresponding coordinates results in a true statement.

$(-2, 5)$
5 Units
2 Units
Origin

Plot or graph the ordered pair $(-2, 5)$.
Start at the origin. Move 2 units to the left along the x-axis, then 5 units upward parallel to the y-axis.
Determine whether $(-2, 3)$ is a solution of $3x + 2y = 0$.

$$3(-2) + 2(3) = 0$$
$$-6 + 6 = 0$$
$$0 = 0 \quad \text{True.}$$

$(-2, 3)$ is a solution.

CHAPTER HIGHLIGHTS ►

Found at the end of every chapter, the Chapter Highlights contain key definitions, concepts, and examples to help students understand and retain what they have learned.

Increase Motivation

Throughout *Intermediate Algebra, Second Edition*, K. Elayn Martin-Gay provides interesting real-world applications to strengthen your understanding of the relevance of math in everyday life. When a new topic is presented, an effort has been made to relate the new ideas to those that students may already know. The Second Edition increases emphasis on visualization to clarify and reinforce key concepts.

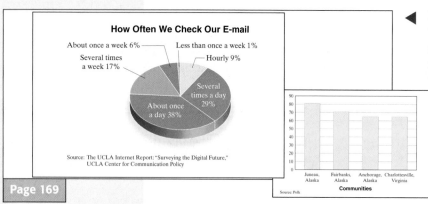

Page 169

Page 172

◄ Real data is integrated throughout the worktext, drawn from current and familiar sources.

GRAPHING CALCULATOR EXPLORATIONS

Many graphing calculators have a TRACE feature. This feature allows you to trace along a graph and see the corresponding x- and y-coordinates appear on the screen. Use this feature for the following exercises.

Graph each equation and then use the TRACE feature to complete each ordered pair solution. (Many times the tracer will not show the exact x- or y-value asked for. In each case, trace as closely as you can to the given x- or y-coordinate and approximate the other, unknown coordinate to one decimal place.)

1. $y = 2.3x + 6.7$; $x = 5.1, y = ?$

2. $y = -4.8x + 2.9$; $x = -1.8, y = ?$

3. $y = -5.9x - 1.6$; $x = ?, y = 7.2$

4. $y = 0.4x - 8.6$; $x = ?, y = -4.4$

5. $y = 5.2x - 3.3$; $x = 2.3, y = ?$ $x = ?, y = 36$

6. $y = -6.2x - 8.3$; $x = 3.2, y = ?$ $x = ?, y = 12$

Page 209

CALCULATOR EXPLORATIONS

▲ Optional Calculator Explorations and exercises appear in appropriate sections.

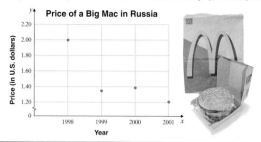

Price of a Big Mac in Russia

Year	Price (in U.S. dollars)
1998	2.00
1999	1.35
2000	1.39
2001	1.21

(Sources: McDonald's, *The Economist)*

Solution: To graph the paired data in the table, we use the first column for the x- (or horizontal) axis and the second column for the y- (or vertical) axis.

▲ Graphics, models, and illustrations provide visual reinforcement.

FOCUS ON BOXES ►

Focus On boxes found throughout each chapter help you see the relevance of math through critical-thinking exercises and group activities. Try these on your own or with a classmate. Focus On covers the areas of: History, Mathematical Connections, Real World, and Business and Career.

FOCUS ON The Real World

ANOTHER MATHEMATICAL MODEL

Sometimes mathematical models other than linear models are appropriate for data. Suppose that an equation of the form $y = ax^2 + bx + c$ is an appropriate model for the ordered pairs (x_1, y_1), (x_2, y_2), and (x_3, y_3). Then it is necessary to find the values of a, b, and c such that the given ordered pairs are solutions of the equation $y = ax^2 + bx + c$. To do so, substitute each ordered pair into the equation. Each time, the result is an equation in three unknowns: a, b, and c. Solving the resulting system of three linear equation in three unknowns will give the required values of a, b, and c.

GROUP ACTIVITY

1. The table gives the total beef supply (in billions of pounds) in the United States in each of the years listed.
   a. Write the data as ordered pairs of the form (x, y), where y is the beef supply (in billions of pounds) in the year x ($x = 0$ represents 1998).
   b. Find the values of a, b, and c such that the equation $y = ax^2 + bx + c$ models this data.
   c. Verify that the model you found in part (b) gives each of the ordered pair solutions from part (a).
   d. According to the model, what was the U.S. beef supply in 2001?

Total U.S. Beef Supply

Year	Beef Supply (billions of pounds)
1998	28.9
2000	30.3
2002	28.8

(Source: Economic Research Service, U.S. Department of Agriculture)

Page 268

Build Confidence

Several features of this text can be helpful in building your confidence and mathematical competence. As you study, also notice the connections the author makes to relate new material to ideas that you may already know.

1.1 Tips for Success in Mathematics

Before reading this section, remember that your instructor is your best source for information. Please see your instructor for any additional help or information.

A Getting Ready for This Course

Now that you have decided to take this course, remember that a **positive attitude** will make all the difference in the world. Your belief that you can succeed is just as important as your commitment to this course. Make sure that you are ready for this course by having the time and positive attitude that it takes to succeed.

Next make sure that you have scheduled your math course at a time that will give you the best chance for success. For example, if you are also working, you may want to check with your employer to make sure that your work hours will not conflict with your course schedule.

Before your first class period, double-check your schedule and allow your-

OBJECTIVES

- A Get ready for this course.
- B Understand some general tips for success.
- C Understand how to use this text.
- D Get help as soon as you need it.
- E Learn how to prepare for and take an exam.
- F Develop good time management

SSM · TUTOR CENTER · SG · CD & VIDEO · MATH PRO · WEB

Page 3

◀ TIPS FOR SUCCESS

New coverage of study skills in Section 1.1 reinforces this important component to success in this course.

Page 144

STUDY SKILLS REMINDERS ▶

New Study Skills Reminders are integrated throughout the book to reinforce section 1.1 and encourage the development of strong study skills.

STUDY SKILLS REMINDER

Are you organized?

Have you ever had trouble finding a completed assignment? When it's time to study for a test, are your notes neat and organized? Have you ever had trouble reading your own mathematics handwriting? (Be honest—I have had trouble reading my own handwriting before.)

When any of these things happen, it's time to get organized. Here are a few suggestions:

Write your notes and complete your homework assignments in a notebook with pockets (spiral or ring binder). Take class notes in this notebook, and then follow the notes with your completed homework assignment. When you receive graded papers or handouts, place them in the notebook pocket so that you will not lose them.

Place a mark (possibly an exclamation point) beside any note(s) that seem especially important to you. Also place a mark (possibly a question mark) beside any note(s) or homework that you are having trouble with. Don't forget to see your instructor, a tutor, or your fellow classmates to help you understand the concepts or exercises you have marked.

Also, if you are having trouble reading your own handwriting, *slow down* and write your mathematics work clearly!

Mental Math

Find the slope and the y-intercept of each line.

1. $y = -4x + 12$
2. $y = \frac{2}{3}x - \frac{7}{2}$
3. $y = 5x$
4. $y = -x$
5. $y = \frac{1}{2}x + 6$
6. $y = -\frac{2}{3}x + 5$

Page 201

◀ MENTAL MATH

Mental Math warm-up exercises reinforce concepts found in the accompanying section and can increase your confidence before beginning an exercise set.

Page 187

HELPFUL HINTS ▶

Found throughout the text, these contain practical advice on applying mathematical concepts. They are strategically placed where you are most likely to need immediate reinforcement.

Helpful Hint

When we are trying to find the slope of a line through two given points, it makes no difference which given point is called (x_1, y_1) and which is called (x_2, y_2). Once an x-coordinate is called x_1, however, make sure its corresponding y-coordinate is called y_1.

Chapter 3 VOCABULARY CHECK

Fill in each blank with one of the words or phrases listed below.

relation	line	function	standard	slope	domain
slope-intercept	x	y	range	parallel	linear function
point-slope	perpendicular	linear inequality			

1. A _____ is a set of ordered pairs.
2. The graph of every linear equation in two variables is a _____.
3. The statement $-x + 2y > 0$ is called a _____ in two variables.
4. _____ form of linear equation in two variables is $Ax + By = C$.
5. The _____ of a relation is the set of all second components of the ordered pairs of the relation.
6. _____ lines have the same slope and different y-intercepts.
7. _____ form of a linear equation in two variables is $y = mx + b$.
8. A _____ is a relation in which each first component in the ordered pairs corresponds to exactly one second component.

◀ VOCABULARY CHECKS

New Vocabulary Checks allow you to write your answers to questions about chapter content and strengthen verbal skills.

Page 239

Enrich Your Learning

Seek out these additional Student Resources to match your personal learning style.

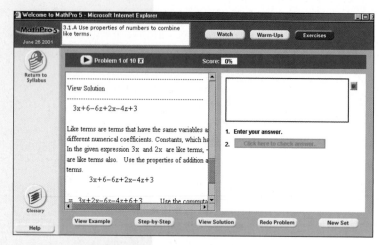

◄ MathPro 5 is the online customizable tutorial, diagnostic and assessment software. It is text-specific to the learning objective level and provides anytime, anywhere tutorial support. It provides:
- Diagnostic review of student skills
- Virtually unlimited practice problems with immediate feedback
- Video clips by K. Elayn Martin-Gay
- Step-by-step solutions
- Summary of progress

MathPro 4 is available on CD-ROM for standalone use or can be networked in the school laboratory.

Text-specific videos, available on CD or VHS, are hosted by the award-winning teacher and author of *Intermediate Algebra*. They cover each objective in every chapter section as a supplementary review. ►

◄ Prentice Hall Tutor Center provides text-specific tutoring via phone, fax, and e-mail. Visit
http://prenhall.com/tutorcenter
for details.

ALSO AVAILABLE:

▲ Student Solutions Manual
▲ How to Study Math

▲ Math on the Internet
▲ *The New York Times/ Themes of the Times*

Ask your instructor or bookstore about these additional study aids.

Real Numbers and Algebraic Expressions

CHAPTER

1

In arithmetic, we add, subtract, multiply, divide, raise to powers, and take roots of numbers. In algebra, we add, subtract, multiply, divide, raise to powers, and take roots of variables. Letters, such as *x*, that represent numbers are called variables. Understanding algebraic expressions made up of combinations of variables and numbers depends on your understanding of arithmetic expressions. This chapter reviews the arithmetic operations on real numbers and the corresponding algebraic expressions. After this review, we will be prepared to explore how widely useful these algebraic expressions are for problem solving in diverse situations.

1.1 Tips for Success in Mathematics

1.2 Sets of Numbers

1.3 Properties of Real Numbers

1.4 Operations on Real Numbers

Integrated Review—Real Numbers

1.5 Order of Operations and Algebraic Expressions

1.6 Exponents and Scientific Notation

1.7 More Work with Exponents and Scientific Notation

The world's first subway system was built in London, England, in 1863. The growing congestion on the streets of London led to relocating the existing steam-operated metropolitan railway underground. Thirty-four years later, America's first subway was built in Boston, Massachusetts. The Boston subway started out with single electrified streetcars running on rails underground. New York City soon followed suit and completed its own subway system in 1904. The New York subway's underground electric trains replaced surface electric train routes, elevated railways, and horse car tracks. Today, New York's subway system is one of the largest and most complex in the world. New York City Transit operates the world's largest fleet of subway cars, a total of 5871. It also has 26 subway lines, 685 miles of subway track, and 490 subway stations. In Exercise 87 on page 57, we will use scientific notation to write the very large number of passengers who ride on New York City subways each year.

Name _____ Section _____ Date _____

1. _____

2. _____

3. _____

4. _____

5. _____

6. _____

7. _____

8. _____

9. _____

10. _____

11. _____

12. _____

13. _____

14. _____

15. _____

16. _____

17. _____

18. _____

19. _____

20. _____

Chapter 1 Pretest

1. Evaluate $2x - 3y^2$ when $x = 5$ and $y = 2$.

2. Translate the following phrase into an algebraic expression. Use x to represent the unknown number. Seven more than twice a number.

Insert $<, >,$ or $=$ between each pair of numbers to form a true statement.

3. $\dfrac{2}{3}$ $\quad$ $\dfrac{10}{17}$

4. -7 $\quad$ -6

Write the opposite (or additive inverse) of each number if one exists.

5. 9.25

6. $-\dfrac{7}{8}$

Simplify.

7. $-|-21|$

8. $-13 - (-22)$

9. $\dfrac{-5.1}{1.7}$

10. -5^2

11. $\sqrt{\dfrac{25}{121}}$

12. $9 - [(2 - 6) + (3 - 17)]$

13. $2(3x - 5) - 4(2 - x)$

Simplify. Write answers using positive exponents only.

14. $(-5b^6)(4b^8)$

15. $\dfrac{-45a^2b^7c}{-9ab^{12}c^2}$

16. $\dfrac{z^{-6}z^{12}}{z^{-9}}$

17. $5(y^4z)^{-3}$

18. $\left(\dfrac{2x^{-2}y^3}{6xz^5}\right)^{-4}$

19. Evaluate: $2x^0 + 3$

20. Write the following number in scientific notation: $65,400,000,000$

1.1   Tips for Success in Mathematics

Before reading this section, remember that your instructor is your best source for information. Please see your instructor for any additional help or information.

A   Getting Ready for This Course

Now that you have decided to take this course, remember that a **positive attitude** will make all the difference in the world. Your belief that you can succeed is just as important as your commitment to this course. Make sure that you are ready for this course by having the time and positive attitude that it takes to succeed.

Next, make sure that you have scheduled your math course at a time that will give you the best chance for success. For example, if you are also working, you may want to check with your employer to make sure that your work hours will not conflict with your course schedule.

Before your first class period, double-check your schedule and allow yourself extra time to arrive in case of traffic or in case you have trouble locating your classroom. Make sure that you bring at least your textbook, paper, and a writing instrument with you. Are you required to have a lab manual, graph paper, calculator, or some other supply besides this text? If so, bring this material with you also.

B   General Tips for Success

Below are some general tips that will increase your chance for success in a mathematics class. Many of these tips will also help you in other courses you may be registered for.

Exchange names and phone numbers with at least one other person in class. This contact person can be a great help in case you miss the class assignment or want to discuss math concepts or exercises that you find difficult.

Choose to attend all class periods. If possible, sit near the front of the classroom. This way, you will see and hear the presentation better. It may also be easier for you to participate in classroom activities.

Do your homework. You've probably heard the phrase "practice makes perfect" in relation to music and sports. It also applies to mathematics. You will find that the more time you spend solving mathematics problems, the easier the process becomes. Be sure to schedule enough time to complete your assignments before the next class period.

Check your work. Review the steps you made while working a problem. Learn to check your answers in the original problems. You may also compare your answers to the answers to selected exercises listed in the back of the book. If you have made a mistake, try to figure out what went wrong. Then correct your mistake. If you can't find your mistake, don't erase your work or throw it away. Bring your work to your instructor, a tutor in a math lab, or a classmate. It is easier for someone to find where you had trouble if they look at your original work.

Learn from your mistakes. Everyone, even your instructor, makes mistakes. (That definitely includes me—Elayn Martin-Gay. You usually don't see my mistakes because many other people double-check my work in this text. If I make a mistake on a videotape, it is edited out so that you are not confused by it.) Use your errors to learn and to become a better math student. The key is finding and understanding your errors. Was your mistake a careless one or did you make it because you can't read your own "math" writing? If so, try to work more slowly or write more neatly and make a conscious effort to carefully check your work. Did you make a mistake because

OBJECTIVES

A   Get ready for this course.

B   Understand some general tips for success.

C   Understand how to use this text.

D   Get help as soon as you need it.

E   Learn how to prepare for and take an exam.

F   Develop good time management.

SSM   TUTOR CENTER   SG   CD & VIDEO   MATH PRO   WEB

you don't understand a concept? Take the time to review the concept or ask questions to better understand the concept.

Know how to get help if you need it. It's OK to ask for help. In fact, it's a good idea to ask for help whenever there is something that you don't understand. Make sure you know when your instructor has office hours and how to find his or her office. Find out if math tutoring services are available on your campus. Check out the hours, location, and requirements of the tutoring service. Know whether videotapes or software are available and how to access these resources.

Organize your class materials, including homework assignments, graded quizzes and tests, and notes from your class or lab. All of these items will make valuable references throughout your course and as you study for upcoming tests and your final exam. Make sure that you can locate any of these materials when you need them.

Read your textbook before class. Reading a mathematics textbook is unlike entertainment reading such as reading a newspaper. Your pace will be much slower. It is helpful to have a pencil and paper with you when you read. Try to work out examples on your own as you encounter them in your text. You may also write down any questions that you want to ask in class. I know that when you read a mathematics textbook, sometimes some of the information in a section will still be unclear. But once you hear a lecture or watch a video on that section, you will understand it much more easily than if you had not read your text.

Don't be afraid to ask questions. From experience, I can tell you that you are not the only person in class with questions. Other students are normally grateful that someone has spoken up.

Hand in assignments on time. This way you can be sure that you will not lose points needlessly for being late. Show every step of a problem and be neat and organized. Also be sure that you understand which problems are assigned for homework. You can always double-check this assignment with another student in your class.

C Using This Text

There are many helpful resources that are available to you in this text. It is important that you become familiar with and use these resources. This should increase your chances for success in this course. For example:

- Each example in the section has a parallel Practice Problem. As you read a section, try each Practice Problem after you've finished the corresponding example. This "learn-by-doing" approach will help you grasp ideas before you move on to other concepts.

- The main section of exercises in an exercise set are referenced by an objective, such as **A** or **B**, and an example(s). Use this referencing if you have trouble completing an assignment from the exercise set.

- If you need extra help in a particular section, check at the beginning of the section to see what videotapes and software are available.

- Make sure that you understand the meaning of the icons that are beside many exercises. The video icon tells you that the corresponding exercise may be viewed on the videotape that corresponds to that section. The pencil icon tells you that this exercise is a writing exercise in which you should answer in complete sentences. The △ icon simply tells you that this exercise involves geometry.

- Integrated Reviews in each chapter offer you a chance to practice—in one place—the many concepts that you have learned separately over several sections.

- There are many opportunities at the end of each chapter to help you understand the concepts of the chapter.

   Chapter Highlights contain chapter summaries with examples.
   Chapter Reviews contain review problems organized by section.
   Chapter Tests are sample tests to help you prepare for an exam.
   Cumulative Reviews are reviews consisting of material from the beginning of the book to the end of that particular chapter.

   See the Preface at the beginning of this text for a more thorough explanation of the features of this text.

Ⓓ Getting Help

If you have trouble completing assignments or understanding the mathematics, get help as soon as you need it! This tip is presented as an objective on its own because it is *so* important. In mathematics, usually the material presented in one section builds on your understanding of the previous section. What does this mean? It means that if you don't understand the concepts covered during a class period, there is a good chance that you will not understand the concepts covered during the next class period. If this happens to you, get help as soon as you can.

   Where can you get help? Many suggestions have been made in this section on where to get help, and now it is up to you to do it. Try your instructor, a tutoring center, or math lab, or you may want to form a study group with fellow classmates. If you do decide to see your instructor or go to a tutoring center, make sure that you have a neat notebook and be ready with your questions.

Ⓔ Preparing for and Taking an Exam

Make sure that you allow yourself plenty of time to prepare for a test. If you think that you are a little "math anxious," it may be that you are not preparing for a test in a way that will ensure success. The way that you prepare for a test in mathematics is important. To prepare for a test,

1. Review your previous homework assignments.

2. Review any notes from class and section-level quizzes you may have taken. (If this is a final exam, also review chapter tests you have taken.)

3. Review concepts and definitions by reading the Highlights at the end of each chapter.

4. Practice working exercises by completing the Chapter Review found at the end of each chapter. (If this is a final exam, work a Cumulative Review. There is one found at the end of each chapter (except Chapter 1). Choose the review found at the end of the latest chapter that you have covered in your course.) **Don't stop here!**

5. It is important that you place yourself in conditions similar to test conditions to see how you will perform. In other words, once you feel that you know the material, get out a few blank sheets of paper and take a sample test. There is a Chapter Test available at the end of each chapter, or you can work selected problems from the Chapter Review, or your instructor may provide you with a review sheet. During this sample test, do not use your notes or your textbook. Then check your sample test. If you are not satisfied with the results, study the areas that you are weak in and try again.

6. On the day of the test, allow yourself plenty of time to arrive to where you will be taking your exam.

When taking your test,

1. Read the directions on the test carefully.

2. Read each problem carefully as you take your test. Make sure that you answer the question asked.

3. Watch your time and pace yourself so that you may attempt each problem on your test.

4. If you have time, check your work and answers.

5. Do not turn your test in early. If you have extra time, spend it double-checking your work.

Ⓕ Managing Your Time

As a college student, you know the demands that classes, homework, work, and family place on your time. Some days you probably wonder how you'll ever get everything done. One key to managing your time is developing a schedule. Here are some hints for making a schedule:

1. Make a list of all of your weekly commitments for the term. Include classes, work, regular meetings, extracurricular activities, etc. You may also find it helpful to list such things as doing laundry, regular workouts, grocery shopping, etc.

2. Next, estimate the time needed for each item on the list. Also make a note of how often you will need to do each item. Don't forget to include time estimates for reading, studying, and homework you do outside of your classes. You may want to ask your instructor for help estimating the time needed for this item.

3. In the exercise set below, you are asked to block out a typical week on the schedule grid given. Start with items with fixed time slots, like classes and work.

4. Next, include the items on your list with flexible time slots. Think carefully about how best to schedule some items such as study time.

5. Don't fill up every time slot on the schedule. Remember that you need to allow time for eating, sleeping, and relaxing! You should also allow a little extra time in case things take longer than planned.

6. If you find that your weekly schedule is too full for you to handle, you may need to make some changes in your workload, class load, or in other areas of your life. You may want to talk to your advisor, manager or supervisor at work, or someone in your college's academic counseling center for help with such decisions.

Note: In this chapter, we begin a feature called Study Skills Reminder. The purpose of this feature is to remind you of some of the information given in this section and to further expand on some topics in this section.

EXERCISE SET 1.1

1. What is your instructor's name?

2. What are your instructor's office location and office hours?

3. What is the best way to contact your instructor?

4. What does the ＼ icon mean?

5. What does the ▦ icon mean?

6. What does the △ icon mean?

7. Do you have the name and contact information of at least one other student in class?

8. Will your instructor allow you to use a calculator in this class?

9. Are videotapes and/or tutorial software available to you? If so, where?

10. Is there a tutoring service available? If so, what are its hours?

11. Have you attempted this course before? If so, write down ways that you may improve your chances of success during this attempt.

12. List some steps that you may take in case you begin having trouble understanding the material or completing an assignment.

13. Read or reread objective **F** and fill out the schedule grid below.

	Monday	Tuesday	Wednesday	Thursday	Friday	Saturday	Sunday
7:00 a.m.							
8:00 a.m.							
9:00 a.m.							
10:00 a.m.							
11:00 a.m.							
12:00 p.m.							
1:00 p.m.							
2:00 p.m.							
3:00 p.m.							
4:00 p.m.							
5:00 p.m.							
6:00 p.m.							
7:00 p.m.							
8:00 p.m.							
9:00 p.m.							

1.2 Sets of Numbers

Recall that letters that represent numbers are called **variables**. An **algebraic expression** is formed by numbers and variables connected by the operations of addition, subtraction, multiplication, division, raising to powers, or taking roots. For example,

$$2x, \qquad \frac{x+5}{6}, \qquad \sqrt{y} - 1.6, \qquad \text{and} \qquad z^3$$

are algebraic expressions or, more simply, expressions. (Recall that the expression $2x$ means $2 \cdot x$.)

Algebraic expressions occur often during problem solving. For example, the B747-400 aircraft costs $6964 per hour to operate. The algebraic expression

$$6964t$$

gives the total cost to operate the aircraft for t hours. (*Source: The World Almanac*, 2002) To find the cost to operate the aircraft for 5.2 hours, for example, we replace the variable t with 5.2 and perform the indicated operation. This process is called **evaluating** an expression, and the result is called the **value** of the expression for the given replacement value.

In our example, when $t = 5.2$ hours,

$$6964\,t = 6964(5.2) = 36{,}212.8$$

Thus, it costs $36,212.80 to operate the B747-400 aircraft for 5.2 hours.

When evaluating an expression to solve a problem, we often need to think about the kind of number that is appropriate for the solution. For example, if we are asked to determine the maximum number of parking spaces for a parking lot to be constructed, an answer of $98\frac{1}{10}$ is not appropriate because $\frac{1}{10}$ of a parking space is not realistic.

Ⓐ Identifying Common Sets of Numbers

Let's review some common sets of numbers and their graphs on a **number line**. To construct a number line, we draw a line and label a point 0 with which we associate the number 0. This point is called the **origin**. If we choose a point to the right of 0 and label it 1, the distance from 0 to 1 is called the **unit distance** and can be used to locate more points. The **positive numbers** lie to the right of the origin, and the **negative numbers** lie to the left of the origin. The number 0 is neither positive nor negative.

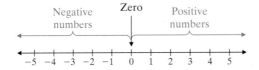

OBJECTIVES

Ⓐ Identify natural numbers, whole numbers, integers, rational, and irrational real numbers.

Ⓑ Write phrases as algebraic expressions.

SSM
TUTOR CENTER   SG   CD & VIDEO   MATH PRO   WEB

Concept Check

Use the definitions of positive numbers, negative numbers, and zero to describe the meaning of *nonnegative numbers*.

Try the Concept Check in the margin.

A number is **graphed** on a number line by shading the point on the number line that corresponds to the number. Some common sets of numbers and their graphs are shown next.

Identifying Numbers

Natural numbers: $\{1, 2, 3, \dots\}$

Whole numbers: $\{0, 1, 2, 3, \dots\}$

Integers: $\{\dots, -3, -2, -1, 0, 1, 2, 3, \dots\}$

Each listing of three dots, $\dots$, is called an **ellipsis** and means to continue in the same pattern.

A **set** is a collection of objects. The objects of a set are called its **elements**. When the elements of a set are listed, such as those displayed in the box, the set is written in **roster** form. A set can also be written in **set builder notation**, which describes the members of the set but does not list them. The following set is written in set builder notation.

$$\{x \mid x \text{ is a natural number less than } 3\}$$

The set of all x such that x is a natural number less than 3

This same set written in roster form is $\{1, 2\}$.

A set that contains *no* elements is called the **empty set** or **null set**, symbolized by $\{\ \}$ or $\varnothing$.

$$\{x \mid x \text{ is a month with 32 days}\} \text{ is } \varnothing \text{ or } \{\ \}$$

because no month has 32 days. The set has no elements.

> **Helpful Hint**
>
> Use $\{\ \}$ or $\varnothing$ alone to write the empty set. $\{\varnothing\}$ is *not* the empty set because it has one element: $\varnothing$.

Practice Problems 1–2

Write each set in roster form.

1. $\{x \mid x$ is a whole number between 0 and 4$\}$

2. $\{x \mid x$ is a natural number greater than 80$\}$

EXAMPLES   Write each set in roster form.

1. $\{x \mid x$ is a whole number between 1 and 6$\}$

   $\{2, 3, 4, 5\}$

2. $\{x \mid x$ is a natural number greater than 100$\}$

   $\{101, 102, 103, \dots\}$

The symbol $\in$ is used to denote that an element is in a particular set. The symbol $\in$ is read as "is an element of." For example, the true statement "3 is an element of $\{1, 2, 3, 4, 5\}$" can be written in symbols as

$$3 \in \{1, 2, 3, 4, 5\}$$

The symbol $\notin$ is read as "is not an element of." In symbols, we write the true statement "p is not an element of $\{a, 5, g, j, q\}$" as

$$p \notin \{a, 5, g, j, q\}$$

Answers

1. $\{1, 2, 3\}$,   **2.** $\{81, 82, 83, \dots\}$

Concept Check: a number that is zero or positive

EXAMPLES   Determine whether each statement is true or false.

3. $3 \in \{x \mid x \text{ is a natural number}\}$ True, since 3 is a natural number and therefore an element of the set.

4. $7 \notin \{1, 2, 3\}$ True, since 7 is not an element of the set $\{1, 2, 3\}$. ●

Practice Problems 3–4

Determine whether each statement is true or false.

3. $0 \in \{x \mid x \text{ is a natural number}\}$

4. $9 \notin \{4, 6, 8, 10\}$

We can use set builder notation to describe three other common sets of numbers.

Identifying Numbers

Real numbers: $\{x \mid x \text{ corresponds to a point on the number line}\}$

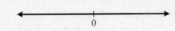

Rational numbers: $\left\{ \dfrac{a}{b} \,\middle|\, a \text{ and } b \text{ are integers and } b \neq 0 \right\}$

Irrational numbers: $\{x \mid x \text{ is a real number and } x \text{ is not a rational number}\}$

Notice that every integer is also a rational number since each integer can be written as the quotient of itself and 1:

$$3 = \frac{3}{1}, \qquad 0 = \frac{0}{1}, \qquad -8 = \frac{-8}{1}$$

Not every rational number, however, is an integer. The rational number $\frac{2}{3}$, for example, is not an integer. Some square roots are rational numbers and some are irrational numbers. For example, $\sqrt{2}$, $\sqrt{3}$, and $\sqrt{7}$ are irrational numbers but $\sqrt{25}$ is a rational number because $\sqrt{25} = 5 = \frac{5}{1}$. The number π is an irrational number. To help you make the distinction between rational and irrational numbers, here are a few examples of each.

Rational Numbers		Irrational Numbers
Number	Equivalent Quotient of Integers, $\dfrac{a}{b}$	
$-\dfrac{2}{3}$	$\dfrac{-2}{3}$ or $\dfrac{2}{-3}$	$\sqrt{5}$
$\sqrt{36}$	$\dfrac{6}{1}$	$\dfrac{\sqrt{6}}{7}$
5	$\dfrac{5}{1}$	$-\sqrt{3}$
0	$\dfrac{0}{1}$	π
1.2	$\dfrac{12}{10}$	$\dfrac{2}{\sqrt{3}}$
$3\dfrac{7}{8}$	$\dfrac{31}{8}$	

Answers

3. false,   **4.** true

Every rational number can be written as a decimal that either repeats or terminates. For example,

$$\frac{1}{2} = 0.5 \qquad\qquad \frac{5}{4} = 1.25$$

$$\frac{2}{3} = 0.6666666\ldots = 0.\overline{6} \qquad\qquad \frac{1}{11} = 0.090909\ldots = 0.\overline{09}$$

An irrational number written as a decimal neither terminates nor repeats. When we perform calculations with irrational numbers, we often use decimal approximations that have been rounded. For example, consider the following irrational numbers along with a four-decimal-place approximation of each:

$$\pi \approx 3.1416 \qquad \sqrt{2} \approx 1.4142$$

Earlier we mentioned that every integer is also a rational number. In other words, all the elements of the set of integers are also elements of the set of rational numbers. When this happens, we say that the set of integers, set I, is a **subset** of the set of rational numbers, set Q. The natural numbers, whole numbers, integers, rational numbers, and irrational numbers are each a subset of the set of real numbers. The relationships among these sets of numbers are shown in the following diagram.

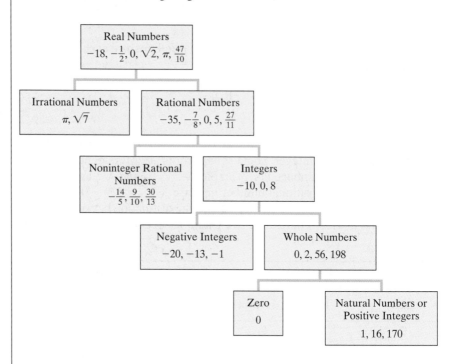

Practice Problems 5–7

Determine whether each statement is true or false.

5.   0 is a real number.
6.   Every integer is a rational number.
7. $\sqrt{3}$ is a rational number.

Answers

5. true,   **6.** true,   **7.** false

EXAMPLES   Determine whether each statement is true or false.

5.   3 is a real number.

True. Every whole number is a real number.

6.   Every rational number is an integer.

False. The number $\frac{2}{3}$, for example, is a rational number, but it is not an integer.

7. $\frac{1}{5}$ is an irrational number.

False. The number $\frac{1}{5}$ is a rational number since it is in the form $\frac{a}{b}$ with a and b integers and $b \neq 0$.

B   **Writing Phrases as Algebraic Expressions**

Often, solving problems involves translating a phrase into an algebraic expression. The following is a list of key words and phrases and their translations.

Addition	Subtraction	Multiplication	Division
sum	difference of	product	quotient
plus	minus	times	divide
added to	subtracted from	multiply	into
more than	less than	twice	ratio
increased by	decreased by	of	
total	less		

EXAMPLES

Write each phrase as an algebraic expression. Use the variable x to represent each unknown number.

8. Eight times a number $8 \cdot x$ or $8x$
9. Three more than eight times a number $8x + 3$
10. The quotient of a number and -7 $x \div -7$ or $\dfrac{x}{-7}$
11. One and six-tenths subtracted from
    twice a number $2x - 1.6$ or $2x - 1\dfrac{6}{10}$
12. Six less than a number $x - 6$
13. Twice the sum of four and a number $2(4 + x)$

Practice Problems 8–13

Write each phrase as an algebraic expression. Use the variable x to represent each unknown number.

8. Twice a number
9. Five more than six times a number
10. The quotient of six and a number
11. One-fourth subtracted from three times a number
12. Eleven less than a number
13. Three times the difference of a number and ten

Answers

8. $2x$,   9. $6x + 5$,   10. $\dfrac{6}{x}$ or $6 \div x$,

11. $3x - \dfrac{1}{4}$,   12. $x - 11$,   13. $3(x - 10)$

FOCUS ON **Mathematical Connections**

GEOMETRY INVESTIGATIONS

Recall that the perimeter of a figure is the distance around the outside of the figure. For a rectangle with length l and width w, the perimeter of the rectangle is given by the expression $2l + 2w$.

Area is a measure of the surface of a region. For example, we measure a plot of land or the floor space of a home by area. For a rectangle with length l and width w, the area of the rectangle is given by the expression lw.

A circular cylinder can be formed by rolling a rectangle into a tube. The surface area of the cylinder (excluding the two ends of the cylinder) is the same as the area of the rectangle used to form the cylinder. Recall that volume is a measure of the space inside a three-dimensional region. The volume of a circular cylinder with height h and radius r is given by the expression $\pi r^2 h$.

GROUP ACTIVITY

1. Work together to discover whether two rectangles with the same perimeter always have the same area. Explain your results. Give examples.

2. Do figures with the same surface area always have the same volume? To see, take two $8\frac{1}{2}$-by-11-inch sheets of paper and construct two cylinders using the following figures as a guide. Verify that both cylinders have the same surface area. Measure the height and radius of each resulting cylinder. Then find the volume of each cylinder to the nearest tenth of a cubic inch. Explain your results.

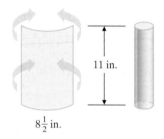

11 in.

$8\frac{1}{2}$ in.

Cylinder 1

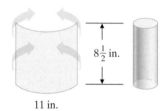

$8\frac{1}{2}$ in.

11 in.

Cylinder 2

EXERCISE SET 1.2

(A) *Write each set in roster form. See Examples 1 and 2.*

1. $\{x | x$ is a natural number less than $6\}$

2. $\{x | x$ is a natural number greater than $6\}$

3. $\{x | x$ is a natural number between 10 and $17\}$

4. $\{x | x$ is an odd natural number$\}$

5. $\{x | x$ is a whole number that is not a natural number$\}$

6. $\{x | x$ is a natural number less than $1\}$

7. $\{x | x$ is an even whole number less than $9\}$

8. $\{x | x$ is an odd whole number less than $9\}$

List the elements of the set $\left\{3, 0, \sqrt{7}, \sqrt{36}, \dfrac{2}{5}, -134\right\}$ *that are also elements of the given set. See Examples 3 and 4.*

9. Whole numbers

10. Integers

11. Natural numbers

12. Rational numbers

13. Irrational numbers

14. Real numbers

Place $\in$ *or* $\notin$ *in the space provided to make each statement true. See Examples 3 through 7.*

15. -11 ___ $\{x | x$ is an integer$\}$

16. -6 ___ $\{2, 4, 6, \dots\}$

17. 0 ___ $\{x | x$ is a positive integer$\}$

18. 12 ___ $\{1, 2, 3, \dots\}$

19. 12 ___ $\{1, 3, 5, \dots\}$

20. $\dfrac{1}{2}$ ___ $\{x | x$ is an irrational number$\}$

Determine whether each statement is true or false. See Examples 5 through 7.

21. Every whole number is a real number.

22. Every irrational number is a real number.

23. Some real numbers are irrational numbers.

24. Some real numbers are whole numbers.

25. Every whole number is a natural number.

26. Every irrational number is a rational number.

(B) *Write each phrase as an algebraic expression. Use the variable x to represent each unknown number. See Examples 8 through 13.*

27. Twice a number

28. Six times a number

29. Ten less than a number

30. A number minus seven

31. The sum of a number and two

32. The difference of twenty-five and a number

33. A number divided by eleven

34. The quotient of a number and thirteen

35. Four subtracted from a number

36. Seventeen subtracted from a number

37. A number plus twenty

38. Fifteen plus a number

39. A number less than ten

40. Twelve less than a number

41. Nine times a number

42. Nine minus a number

43. Nine added to a number

44. Nine divided by a number

45. Five more than twice a number

46. One more than six times a number

47. Twelve minus three times a number

48. Four subtracted from three times a number

49. One plus twice a number

50. Three less than twice a number

51. Ten subtracted from five times a number

52. Four minus three times a number

53. Twice the sum of a number and three

54. The quotient of four and the sum of a number and one

55. The quotient of five and the difference of four and a number

56. Eight times the difference of a number and nine

57. The following bar graph shows the U.S. life expectancy at birth for females born in the years shown. Use the graph to calculate the *increase* in life expectancy over each ten-year period shown.

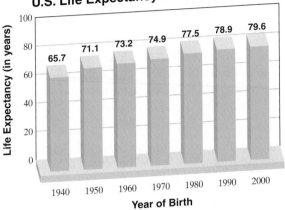

U.S. Life Expectancy at Birth for Females

Source: Social Security Administration

Year	Increase in Life Expectancy (in years) from 10 Years Earlier
1950	
1960	
1970	
1980	
1990	
2000	

58. In your own words, explain why every natural number is also a rational number but not every rational number is a natural number.

59. In your own words, explain why every irrational number is a real number but not every real number is an irrational number.

CRITICAL THINKING
WHAT IS CRITICAL THINKING?

Although exact definitions often vary, critical thinking usually refers to evaluating, analyzing, and interpreting information to make a decision, draw a conclusion, reach a goal, make a prediction, or form an opinion. It often involves problem solving, communication, and reasoning skills. Critical thinking is more than a technique that helps you pass your courses—critical thinking skills are life skills. Developing these skills can help you solve problems in your workplace and in everyday life. For instance, well-developed critical thinking skills would be useful in the following situation:

> Suppose you work as a medical lab technician. Your lab supervisor has decided that some lab equipment should be replaced. She asks you to collect information on several different models from equipment manufacturers. Your assignment is to study the data and then make a recommendation on which model the lab should buy.

HOW CAN CRITICAL THINKING BE DEVELOPED?

Just as physical exercise can help to develop and strengthen certain muscles of the body, mental exercise can help to develop critical thinking skills. Mathematics is ideal for helping to develop such skills because it requires using logic and reasoning, recognizing patterns, making conjectures and educated guesses, and drawing conclusions. You will find many opportunities to build your critical thinking skills throughout Intermediate Algebra:

- In real-life application problems (see Exercise 89 in Section 1.5)

- In writing exercises marked with the ✎ icon (see Exercise 55 in Section 1.3)

- In the Combining Concepts subsection of the exercise sets (see Exercise 117 in Section 1.4)

- In the Chapter Activities (see the Chapter 1 Activity at the end of this chapter)

- In the Critical Thinking and Group Activities questions found in Focus On features like this one throughout the book.

1.3 Properties of Real Numbers

(A) Writing Sentences as Equations

When writing sentences as equations, we use the symbol = to translate the phrase "**is equal to**." All of the following key words and phrases also mean equality.

Equality

equals	is/was	represents	is the same as
gives	yields	amounts to	is equal to

EXAMPLES   Write each sentence as an equation.

1. The sum of x and 5 is 20.

$$x + 5 = 20$$

2. The difference of 8 and x is the same as the product of 2 and x.

$$8 - x = 2 \cdot x$$

3. The quotient of z and 9 amounts to 9 plus z.

$$z \div 9 = 9 + z$$

or $$\frac{z}{9} = 9 + z$$

●

O B J E C T I V E S

(A) Write sentences as equations.

(B) Use inequality symbols.

(C) Find the opposite, or additive inverse, and the reciprocal, or multiplicative inverse, of a number.

(D) Identify and use the commutative, associative, and distributive properties.

SSM
TUTOR CENTER   SG   CD & VIDEO   MATH PRO   WEB

Practice Problems 1–3

Write each sentence as an equation.

1. The difference of x and 7 is 45.

2. The product of 5 and x amounts to the sum of x and 14.

3. The quotient of y and 23 is the same as 20 subtracted from y.

(B) Using Inequality Symbols

If we want to write in symbols that two numbers are not equal, we can use the symbol ≠, which means "**is not equal to**." For example,

$$3 \neq 2$$

Graphing two numbers on a number line gives us a way to compare two numbers. For any two real numbers graphed on a number line, the number to the left is less than the number to the right. This means that the number to the right is greater than the number to the left.

The symbol < means "**is less than**." Since −4 is to the left of −1 on the number line, we write $-4 < -1$. The symbol > means "**is greater than**." Since −1 is to the right of −4 on the number line, we write $-1 > -4$.

$$-4 < -1 \text{ or } -1 > -4$$

Notice that since $-4 < -1$, then we also know that $-1 > -4$. This is true for any two numbers, say, a and b.

If $a < b$, then also $b > a$.

Answers

1. $x - 7 = 45$,   **2.** $5x = x + 14$,

3. $\dfrac{y}{23} = y - 20$

Practice Problems 4–11

Insert $<$, $>$, or $=$ between each pair of numbers to form a true statement.

4. 7 -7
5. -1 11
6. -10 -12
7. -3.25 -3.025
8. 7.206   7.2060
9. 18.6 -14.2
10. $\dfrac{4}{7}$ $\dfrac{5}{7}$
11. $\dfrac{3}{8}$ $\dfrac{1}{3}$

EXAMPLES

Insert $<$, $>$, or $=$ between each pair of numbers to form a true statement.

4. -5 5 $\qquad$ -5 is to the left of 5 on a number line, so $-5 < 5$.
5. 3 -7 $\qquad$ 3 is to the right of -7, so $3 > -7$.
6. -16 -6 $\qquad$ -16 is to the left of -6, so $-16 < -6$.
7. -2.5 -2.1 $\qquad$ -2.5 is to the left of -2.1, so $-2.5 < -2.1$.
8. 6.36   6.360 $\qquad$ The true statement is $6.36 = 6.360$.
9. 4.3 -5.2 $\qquad$ 4.3 is to the right of -5.2, so $4.3 > -5.2$.
10. $\dfrac{5}{8}$ $\dfrac{3}{8}$ $\qquad$ The denominators are the same, so $\dfrac{5}{8} > \dfrac{3}{8}$ since $5 > 3$.
11. $\dfrac{2}{3}$ $\dfrac{3}{4}$ $\qquad$ By dividing, we see that $\dfrac{3}{4} = 0.75$ and $\dfrac{2}{3} = 0.666\ldots$.

Thus $\dfrac{2}{3} < \dfrac{3}{4}$ since $0.666\ldots < 0.75$. $\qquad$ ●

> **Helpful Hint**
>
> When inserting the $>$ or $<$ symbol, think of the symbols as arrowheads that "point" toward the smaller number when the statement is true.

In addition to $<$ and $>$, there are the inequality symbols $\leq$ and $\geq$. The symbol $\leq$ means "**is less than or equal to**," and the symbol $\geq$ means "**is greater than or equal to**."

Practice Problems 12–15

Determine whether each statement is true or false.

12. $-11 \leq 16$
13. $-7 \leq -7$
14. $-7 \geq -7$
15. $-25 \geq -30$

EXAMPLES

Determine whether each statement is true or false.

12. $-9 \leq 7$ $\qquad$ True, since $-9 < 7$ is true.
13. $-5 \leq -5$ $\qquad$ True, since $-5 = -5$ is true.
14. $-5 \geq -5$ $\qquad$ True, since $-5 = -5$ is true.
15. $-24 \geq -20$ $\qquad$ False, since neither $-24 > -20$ nor $-24 = -20$ is true. $\qquad$ ●

Ⓒ Finding Opposites and Reciprocals

Of all the real numbers, two of them stand out as extraordinary: 0 and 1. Zero is the only real number that can be added to *any* real number and result in the same real number. Also, 1 is the only real number that can be multiplied by *any* real number and result in the same real number. This is why 0 is called the **additive identity** and 1 is called the **multiplicative identity**.

> **Identity Properties**
>
> For any real number a,
>
> $\qquad$ Identity Property of 0: $a + 0 = 0 + a = a$
>
> Also,
>
> $\qquad$ Identity Property of 1: $a \cdot 1 = 1 \cdot a = a$

Answers

4. $>$, **5.** $<$, **6.** $>$, **7.** $<$, **8.** $=$, **9.** $>$, **10.** $<$, **11.** $>$, **12.** true, **13.** true, **14.** true, **15.** true

We use the identity property of 1 when we say that x, for example, means $1 \cdot x$ or $1x$. We also use this property when we write equivalent expressions. For example,

$$\underbrace{\frac{2}{3} = \frac{2}{3} \cdot 1}_{\text{identity property of 1}} = \frac{2}{3} \cdot \frac{5}{5} = \frac{10}{15} \qquad \frac{5}{5} \text{ is another name for 1.}$$

Two numbers whose sum is the additive identity 0 are called **opposites** or **additive inverses** of each other. Each real number has a unique opposite.

Opposites or Additive Inverses

If a is a real number, then the unique **opposite**, or **additive inverse**, of a is written as $-a$ and the following is true:

$$a + (-a) = 0$$

On the number line, we picture a real number and its opposite as being the same distance from 0 but on opposite sides of 0.

The opposite of 6 is -6.

The opposite of $\frac{2}{3}$ is $-\frac{2}{3}$.

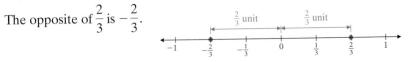

The opposite of -4 is 4.

We stated that the opposite or additive inverse of a number a is $-a$. This means that the opposite of -4 is $-(-4)$. But we stated above that the opposite of -4 is 4. This means that $-(-4) = 4$, and in general, we have the following property.

Double Negative Property

For every real number a, we have $-(-a) = a$.

EXAMPLES   Find the opposite, or additive inverse, or each number.

16. 8 The opposite of 8 is -8.

17. $-\dfrac{1}{5}$ The opposite of $-\dfrac{1}{5}$ is $-\left(-\dfrac{1}{5}\right)$ or $\dfrac{1}{5}$.

18. 0 The opposite of 0 is -0, or 0.

19. -3.5 The opposite of -3.5 is $-(-3.5)$ or 3.5.

Two numbers whose product is 1 are called **reciprocals** or **multiplicative inverses** of each other. Just as each real number has a unique opposite, each nonzero real number also has a unique reciprocal.

Practice Problems 16–19

Find the opposite, or additive inverse, of each number.

16. 7 17. $\dfrac{2}{3}$

18. $-\dfrac{5}{7}$ 19. -4.7

Answers

16. -7,   **17.** $-\dfrac{2}{3}$,   **18.** $\dfrac{5}{7}$,   **19.** 4.7

Reciprocals or Multiplicative Inverses

If a is a nonzero real number, then its **reciprocal**, or **multiplicative inverse**, is $\dfrac{1}{a}$ and the following is true:

$$a \cdot \frac{1}{a} = 1$$

Practice Problems 20–22

Find the reciprocal, or multiplicative inverse, of each number.

20. 13
21. −5
22. $\dfrac{2}{3}$

EXAMPLES

Find the reciprocal, or multiplicative inverse, of each number.

20. 11   The reciprocal of 11 is $\dfrac{1}{11}$.

21. −9   The reciprocal of −9 is $-\dfrac{1}{9}$.

22. $\dfrac{7}{4}$ The reciprocal of $\dfrac{7}{4}$ is $\dfrac{4}{7}$ $\left(\text{since } \dfrac{7}{4} \cdot \dfrac{4}{7} = 1\right)$.

 Helpful Hint

The number 0 has no reciprocal. Why? There is no number that when multiplied by 0 gives a product of 1.

Try the Concept Check in the margin.

Concept Check

Can a number's additive inverse and multiplicative inverse ever be the same? Explain.

D Using the Commutative, Associative, and Distributive Properties

In addition to these special real numbers, all real numbers have certain properties that allow us to write equivalent expressions—that is, expressions that have the same value. These properties will be especially useful in Chapter 2 when we solve equations.

The **commutative properties** state that the order in which two real numbers are added or multiplied does not affect their sum or product.

Commutative Properties

For any real numbers a and b,

     Addition: $a + b = b + a$

     Multiplication: $a \cdot b = b \cdot a$

For example,

$$7 + 11 = 18 \quad \text{and} \quad 11 + 7 = 18 \qquad \text{Addition}$$

$$7 \cdot 11 = 77 \quad \text{and} \quad 11 \cdot 7 = 77 \qquad \text{Multiplication}$$

The **associative properties** state that regrouping numbers that are added or multiplied does not affect their sum or product.

Answers

20. $\dfrac{1}{13}$,   **21.** $-\dfrac{1}{5}$,   **22.** $\dfrac{3}{2}$

Concept Check: no; answers may vary

Associative Properties

For real numbers a, b, and c,

Addition: $(a + b) + c = a + (b + c)$

Multiplication: $(a \cdot b) \cdot c = a \cdot (b \cdot c)$

For example,

$$2 + (3 + 7) = 2 + 10 = 12 \qquad \text{Addition}$$
$$(2 + 3) + 7 = 5 + 7 = 12$$
$$2 \cdot (3 \cdot 7) = 2 \cdot 21 = 42 \qquad \text{Multiplication}$$
$$(2 \cdot 3) \cdot 7 = 6 \cdot 7 = 42$$

EXAMPLE 23

Use the commutative property of addition to write an expression equivalent to $7x + 5$.

Solution: $7x + 5 = 5 + 7x$

EXAMPLE 24

Use the associative property of multiplication to write an expression equivalent to $4 \cdot (9y)$. Then simplify this equivalent expression.

Solution: $4 \cdot (9y) = (4 \cdot 9)y = 36y$

The **distributive property** states that multiplication distributes over addition.

Distributive Properties

For real numbers a, b, and c,

$$a(b + c) = ab + ac$$

For example,

$$3(6 + 2) = 3(8) = 24$$
$$3(6 + 2) = 3(6) + 3(2) = 18 + 6 = 24$$

EXAMPLES   Use the distributive property to multiply.

25. $3(2x - y) = 3 \cdot 2x - 3 \cdot y = 6x - 3y$

26. $-4(y + 5) = -4 \cdot y + (-4) \cdot 5 = -4y - 20$

27. $0.7x(y - 2) = 0.7x \cdot y - 0.7x \cdot 2 = 0.7xy - 1.4x$

Try the Concept Check in the margin.

Practice Problem 23

Use the commutative property of addition to write an expression equivalent to $9 + 4x$.

Practice Problem 24

Use the associative property of multiplication to write an expression equivalent to $5 \cdot (6x)$. Then simplify this equivalent expression.

Practice Problems 25–27

Use the distributive property to multiply.

25. $7(4x - y)$
26. $-8(3 + x)$
27. $5x(y - 4)$

Concept Check

Is the statement below true? Why or why not?

$$6(2a)(3b) = 6(2a) \cdot 6(3b)$$

Answers

23. $4x + 9$, **24.** $(5 \cdot 6)x = 30x$,
25. $28x - 7y$, **26.** $-24 - 8x$,
27. $5xy - 20x$

Concept Check: no; $6(2a)(3b) = 6(6ab) = 36ab$

STUDY SKILLS REMINDER

Are you preparing for a test on Chapter 1?

Below I have listed some common trouble areas for students in Chapter 1. After studying for your test—but before taking your test—read these.

■ Do you remember the meaning of a negative exponent?

$$7^{-2} = \frac{1}{7^2} = \frac{1}{49}$$

■ Don't forget the order of operations and the distributive property.

$$7 - \overset{\frown}{3(2x} - 6y) + 5 = 7 - 6x + 18y + 5 \qquad \text{Use the distributive property.}$$
$$\underset{\text{Notice the sign.}}{\uparrow}$$
$$= -6x + 18y + 12 \qquad \text{Combine like terms.}$$

■ Don't forget the difference between $(-3)^{-2}$ and -3^{-2}.

$$(-3)^{-2} = \frac{1}{(-3)^2} = \frac{1}{9}$$

$$-3^{-2} = -1 \cdot 3^{-2} = \frac{-1}{3^2} = \frac{-1}{9} \text{ or } -\frac{1}{9}$$

■ Remember that

$$\frac{0}{8} = 0 \text{ while } \frac{8}{0} \text{ is undefined.}$$

■ Don't forget the difference between reciprocal and opposite.

The opposite of $-\frac{3}{5}$ is $\frac{3}{5}$.

The reciprocal of $-\frac{3}{5}$ is $-\frac{5}{3}$.

Remember: This is simply a checklist of common trouble areas. For a review of Chapter 1, see the Highlights and Chapter Review at the end of Chapter 1.

EXERCISE SET 1.3

A *Write each sentence as an equation. See Examples 1 through 3.*

1. The product of 4 and c is 7.

2. The sum of 10 and x is -12.

3. Twice x plus 5 is the same as -14.

4. The difference of y and 3 amounts to 12.

5. The quotient of n and 5 is 4 times n.

6. The quotient of 8 and y is 3 more than y.

7. The difference of z and 2 is the same as the product of z and 2.

8. Five added to twice q is the same as 4 more than q.

B *Insert $<$, $>$, or $=$ between each pair of numbers to form a true statement. See Examples 4 through 11.*

9. 0 -2

10. -5 0

11. $\dfrac{12}{3}$ $\dfrac{8}{2}$

12. $\dfrac{20}{5}$ $\dfrac{20}{4}$

13. -7.9 -7.09

14. -13.07 -13.7

15. 7.4   7.40

16. $\dfrac{12}{4}$ $\dfrac{15}{5}$

17. 8.6 -3.5

18. -4.7 3.8

19. $\dfrac{7}{11}$ $\dfrac{9}{11}$

20. $\dfrac{9}{20}$ $\dfrac{3}{20}$

21. $\dfrac{1}{2}$ $\dfrac{5}{8}$

22. $\dfrac{3}{4}$ $\dfrac{7}{8}$

23. -16 -17

24. -14 -24

Determine whether each statement is true or false. See Examples 12 through 15.

25. $-6 \leq 0$

26. $0 \leq -4$

27. $-3 \geq -3$

28. $-8 \leq -8$

29. $-14 \geq -1$

30. $-14 \leq -1$

31. $-3 \leq -3$

32. $-8 \geq -8$

C *Write the opposite (or additive inverse) of each number. See Examples 16 through 19.*

33. -6.2 **34.** -7.8 **35.** $\dfrac{4}{7}$ **36.** $\dfrac{9}{5}$

37. $-\dfrac{2}{3}$ **38.** $-\dfrac{14}{3}$ **39.** 0 **40.** 10.3

Write the reciprocal (or multiplicative inverse) of each number if one exists. See Examples 20 through 22.

41. 5 **42.** 9 **43.** -8 **44.** -4 **45.** $-\dfrac{1}{4}$

46. $\dfrac{1}{9}$ **47.** 0 **48.** $\dfrac{0}{6}$ **49.** $\dfrac{7}{8}$ **50.** $-\dfrac{23}{5}$

Fill in the chart. See Examples 16 through 22.

	Number	Opposite	Reciprocal
51.	5		
52.	-3		
53.	$\dfrac{2}{3}$		
54.	$-\dfrac{7}{11}$		

55. Name the only real number that has no reciprocal, and explain why this is so.

56. Name the only real number that is its own opposite, and explain why this is so.

D *Use a commutative property to write an equivalent expression. See Example 23.*

57. $7x + y$ **58.** $3a + 2b$ **59.** $z \cdot w$

60. $r \cdot s$

61. $\dfrac{1}{3} \cdot \dfrac{x}{5}$

62. $\dfrac{x}{2} \cdot \dfrac{9}{10}$

Use an associative property to write an equivalent expression. See Example 24.

 63. $5 \cdot (7x)$

64. $3 \cdot (10z)$

65. $(x + 1.2) + y$

66. $5q + (2r + s)$

67. $(14z) \cdot y$

68. $(9.2x) \cdot y$

Use the distributive property to multiply. See Examples 25 through 27.

69. $3(x + 5)$ **70.** $7(y + 2)$ **71.** $8(2a + b)$ **72.** $9(c + 7d)$ **73.** $-2(6x + 5y + 2z)$

74. $-5(3a + b + 9c)$ **75.** $4(z - 6)$ **76.** $2(7 - y)$ **77.** $-6x(y - 4)$ **78.** $-11y(z - 2)$

79. $0.4(2x + 5y)$ **80.** $0.5(3a - 4b)$ **81.** $\dfrac{1}{2}(4x - 9y)$ **82.** $\dfrac{1}{3}(4x - 9y)$

Complete each statement to illustrate the given property.

83. $3x + 6 =$ _____ Commutative property of addition   **84.** $8 + 0 =$ _____ Additive identity property

85. $\dfrac{2}{3} + \left(-\dfrac{2}{3}\right) =$ _____ Additive inverse property   **86.** $4(x + 3) =$ _____ Distributive property

87. $7 \cdot 1 =$ _____ Multiplicative identity property

88. $0 \cdot (-5.4) =$ _____ Multiplication property of zero

89. $10(2y) =$ _____ Associative property

90. $9y + (x + 3z) =$ _____ Associative property

 Combining Concepts

In each statement, a property of real numbers has been incorrectly applied. Correct the right-hand side of each statement.

91. $3(x + 4) = 3x + 4$

92. $5(7y) = (5 \cdot 7)(5 \cdot y)$

93. $4 + 8y = 4y + 8$

94. Is subtraction commutative? Explain why or why not.

95. Is division commutative? Explain why or why not.

96. Evaluate $12 - (5 - 3)$ and $(12 - 5) - 3$. Use these two expressions and discuss whether subtraction is associative.

97. Evaluate $24 \div (6 \div 3)$ and $(24 \div 6) \div 3$. Use these two expressions and discuss whether division is associative.

△ **98.** To demonstrate the distributive property geometrically, represent the area of the larger rectangle in two ways: First as length a times width $b + c$, and second as the sum of the areas of the smaller rectangles.

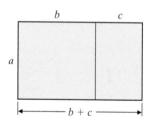

1.4 Operations on Real Numbers

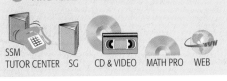

OBJECTIVES

Ⓐ Find the absolute value of a number.

Ⓑ Add and subtract real numbers.

Ⓒ Multiply and divide real numbers.

Ⓓ Simplify expressions containing exponents.

Ⓔ Find roots of numbers.

SSM
TUTOR CENTER   SG   CD & VIDEO   MATH PRO   WEB

Ⓐ Finding the Absolute Value of a Number

In Section 1.2, we used the number line to compare two real numbers. The number line can also be used to visualize distance, which leads to the concept of absolute value. The **absolute value** of a number is the distance between the number and 0 on the number line. The symbol for absolute value is $|\ \ |$. For example, since -4 and 4 are both 4 units from 0 on the number line, each has an absolute value of 4.

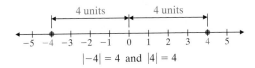

$$4 \text{ units} \qquad 4 \text{ units}$$

$$-5 \;-4\; -3\; -2\; -1\quad 0\quad 1\quad 2\quad 3\quad 4\quad 5$$

$$|-4| = 4 \text{ and } |4| = 4$$

An equivalent definition of the absolute value of a real number a is given next.

Absolute Value

The absolute value of a, written as $|a|$, is

$$|a| = \begin{cases} a \text{ if } a \text{ is 0 or a positive number} \\ -a \text{ if } a \text{ is a negative number} \end{cases}$$

↑
the opposite of

EXAMPLES   Find each absolute value.

 1. $|3| = 3$

 2. $|0| = 0$

3. $|-4| = -(-4) = 4$
  ↑
  the opposite of

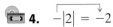 **4.** $-|2| = -2$

 **5.** $-|-8| = -8$ Since $|-8|$ is 8, we have $-|-8| = -8$.

> **Helpful Hint**
>
> Since distance is always positive or zero, the absolute value of a number is always positive or zero.

Try the Concept Check in the margin.

Practice Problems 1–5

Find each absolute value.

1. $|7|$ 2. $|-1|$
3. $|-9|$ 4. $-|5|$
5. $-|-3|$

Concept Check

Explain how you know that $|14| = -14$ is a false statement.

Answers

1. 7, **2.** 1, **3.** 9, **4.** −5, **5.** −3

Concept Check: $|14| = 14$ since the absolute value of a number is the distance between the number and 0 and distance cannot be negative.

Ⓑ Adding and Subtracting Real Numbers

When solving problems, we often need to add real numbers. For example, if the New Orleans Saints lose 5 yards in one play, then lose another 7 yards in the next play, their total loss may be described by $-5 + (-7)$.

The addition of two real numbers may be summarized by the following.

Adding Real Numbers

1. To add two numbers with the *same* sign, add their absolute values and attach their common sign.
2. To add two numbers with *different* signs, subtract the smaller absolute value from the larger absolute value and attach the sign of the number with the larger absolute value.

For example, to add $-5 + (-7)$, we first add their absolute values.

$$|-5| = 5, \quad |-7| = 7, \quad \text{and} \quad 5 + 7 = 12$$

Next, we attach their common negative sign.

$$-5 + (-7) = -12$$

(This represents a total loss of 12 yards for the New Orleans Saints in the example above.)

To find $(-4) + 3$, we first subtract their absolute values.

$$|-4| = 4, \quad |3| = 3, \quad \text{and} \quad 4 - 3 = 1$$

Next, we attach the sign of the number with the larger absolute value.

$$(-4) + 3 = -1$$

Practice Problems 6–11

Add.

6. $-7 + (-10)$
7. $8 + (-12)$
8. $-14 + 20$
9. $-4.6 + (-1.9)$
10. $-\dfrac{2}{3} + \dfrac{1}{6}$
11. $-\dfrac{1}{7} + \dfrac{1}{2}$

EXAMPLES Add.

6. $-3 + (-11) = -(3 + 11) = -14$ Add their absolute values, or $3 + 11 = 14$. Then attach the common negative sign.

7. $3 + (-7) = -4$ Subtract their absolute values, or $7 - 3 = 4$. Since -7 has the larger absolute value, the answer is -4.

8. $-10 + 15 = 5$

9. $-8.3 + (-1.9) = -10.2$

10. $-\dfrac{1}{4} + \dfrac{1}{2} = -\dfrac{1}{4} + \dfrac{2}{4} = \dfrac{1}{4}$

11. $-\dfrac{2}{3} + \dfrac{3}{7} = -\dfrac{14}{21} + \dfrac{9}{21} = -\dfrac{5}{21}$

Subtraction of two real numbers may be defined in terms of addition.

Subtracting Real Numbers

If a and b are real numbers, then the difference of a and b, written $a - b$, is defined by

$$a - b = a + (-b)$$

In other words, to subtract a second real number from a first, we add the first number and the opposite of the second number.

EXAMPLES   Subtract.

Add the opposite.

12. $2 - 8 = 2 + (-8) = -6$

Add the opposite.

13. $-8 - (-1) = -8 + (1) = -7$

14. $10.7 - (-9.8) = 10.7 + 9.8 = 20.5$

15. $-\dfrac{2}{3} - \dfrac{1}{4} = -\dfrac{2}{3} + \left(-\dfrac{1}{4}\right)$

$= -\dfrac{2 \cdot 4}{3 \cdot 4} + \left(-\dfrac{1 \cdot 3}{4 \cdot 3}\right) = -\dfrac{8}{12} + \left(-\dfrac{3}{12}\right) = -\dfrac{11}{12}$

⬤

To add or subtract three or more real numbers, we add or subtract from left to right.

EXAMPLES   Simplify each expression.

16. $11 + 2 - 7 = 13 - 7 = 13 + (-7) = 6$

17. $-5 - 4 + 2 = -5 + (-4) + 2 = -9 + 2 = -7$

⬤

ⓒ Multiplying and Dividing Real Numbers

To discover sign patterns when you multiply real numbers, recall that multiplication by a positive integer is the same as repeated addition. For example,

$3(2) = 2 + 2 + 2 = 6$
$3(-2) = (-2) + (-2) + (-2) = -6$

Notice here that $3(-2) = -6$. This illustrates that the product of two numbers with different signs is negative. We summarize sign patterns for multiplying any two real numbers as follows.

Multiplying Two Real Numbers

1. The product of two numbers with the *same* sign is positive.
2. The product of two numbers with *different* signs is negative.

Also recall that the product of zero and any real number a is zero.

$0 \cdot a = 0$

EXAMPLES   Multiply.

18. $-8(-1) = 8$

19. $2\left(-\dfrac{1}{6}\right) = \dfrac{2}{1} \cdot \left(-\dfrac{1}{6}\right) = -\dfrac{2}{6} \text{ or } -\dfrac{1}{3}$

20. $-1.2(0.3) = -0.36$

21. $7(-6) = -42$

22. $-\dfrac{1}{3}\left(-\dfrac{1}{2}\right) = \dfrac{1}{6}$

23. $(-4.6)(-2.5) = 11.5$

24. $0(-6) = 0$

Practice Problems 12–15

Subtract.

12. $7 - 14$
13. $-10 - (-2)$
14. $13.3 - (-8.9)$
15. $-\dfrac{1}{3} - \dfrac{1}{2}$

Practice Problems 16–17

Simplify each expression.

16. $18 + 3 - 4$
17. $-3 - 11 + 7$

Practice Problems 18–24

Multiply.

18. $-4(-2)$
19. $5\left(-\dfrac{1}{10}\right)$
20. $-3.2(0.1)$
21. $8(-6)$
22. $-\dfrac{2}{5}\left(-\dfrac{1}{3}\right)$
23. $(-1.3)(-1.5)$
24. $0(-10)$

Answers

12. -7, **13.** -8, **14.** 22.2, **15.** $-\dfrac{5}{6}$, **16.** 17,

17. -7, **18.** 8, **19.** $-\dfrac{1}{2}$, **20.** -0.32, **21.** -48,

22. $\dfrac{2}{15}$, **23.** 1.95, **24.** 0

⬤

Recall that $\dfrac{8}{4} = 2$ because $2 \cdot 4 = 8$. Likewise, $\dfrac{8}{-4} = -2$ because $(-2)(-4) = 8$. Also, $\dfrac{-8}{4} = -2$ because $(-2)4 = -8$, and $\dfrac{-8}{-4} = 2$ because $2(-4) = -8$. From these examples, we can see that the sign patterns for division are the same as for multiplication.

Dividing Two Real Numbers

1. The quotient of two numbers with the *same* sign is positive.
2. The quotient of two numbers with *different* signs is negative.

Notice from the previous reasoning that we cannot divide by 0. Why? If $\dfrac{5}{0}$ did exist, it would equal a number such that the number times 0 would equal 5. There is no such number, so we cannot define division by 0. We say, for example, that $\dfrac{5}{0}$ is **undefined**.

Practice Problems 25–30

Divide.

25. $\dfrac{45}{-9}$

26. $\dfrac{-16}{-4}$

27. $\dfrac{25}{-5}$

28. $\dfrac{-3}{0}$

29. $\dfrac{0}{-3}$

30. $\dfrac{-1}{-4}$

EXAMPLES   Divide.

25. $\dfrac{20}{-4} = -5$

26. $\dfrac{-9}{-3} = 3$

27. $\dfrac{-40}{10} = -4$

28. $\dfrac{-8}{0}$ is undefined.

29. $\dfrac{0}{-8} = 0$

30. $\dfrac{-10}{-80} = 0.125$

With sign rules for division, we can understand why the positioning of the negative sign in a fraction does not change the value of the fraction. For example,

$$\frac{-12}{3} = -4, \qquad \frac{12}{-3} = -4, \qquad \text{and} \qquad -\frac{12}{3} = -4$$

Since all these fractions equal -4, we can say that

$$\frac{-12}{3} = \frac{12}{-3} = -\frac{12}{3}$$

In general, the following holds true:

If a and b are real numbers and $b \neq 0$, then

$$\frac{a}{-b} = \frac{-a}{b} = -\frac{a}{b}$$

Answers

25. -5, **26.** 4, **27.** -5, **28.** undefined,
29. 0, **30.** 0.25

Also recall that division by a nonzero real number b is the same as multiplication by its reciprocal $\frac{1}{b}$. In other words,

$$a \div b = a \cdot \frac{1}{b}$$

EXAMPLES   Divide.

31. $-\frac{1}{10} \div \left(-\frac{2}{5}\right) = -\frac{1}{10} \cdot \left(-\frac{5}{2}\right) = \frac{5}{20} \text{ or } \frac{1}{4}$

32. $-\frac{1}{4} \div \frac{3}{7} = -\frac{1}{4} \cdot \frac{7}{3} = -\frac{7}{12}$ ●

D Simplifying Expressions Containing Exponents

Recall that when two numbers are multiplied, they are called **factors**. For example, in $3 \cdot 5 = 15$, the 3 and 5 are factors.

A natural number *exponent* is a shorthand notation for repeated multiplication of the same factor. This repeated factor is called the **base**, and the number of times it is used as a factor is indicated by the **exponent**. For example,

$$\overset{\text{exponent}}{4^3} = \underbrace{4 \cdot 4 \cdot 4}_{4 \text{ is a factor 3 times.}} = 64$$
base

Exponents

If a is a real number and n is a natural number, then the **nth power of a**, or **a raised to the nth power**, written as a^n, is the product of n factors, each of which is a.

$$\overset{\text{exponent}}{a^n} = \underbrace{a \cdot a \cdot a \cdot a \cdots a}_{a \text{ is a factor } n \text{ times.}}$$
base

It is not necessary to write an exponent of 1. For example, 3 is assumed to be 3^1.

EXAMPLES   Find the value of each expression.

33. $3^2 = 3 \cdot 3 = 9$

34. $-5^2 = -(5 \cdot 5) = -25$

35. $-5^3 = -(5 \cdot 5 \cdot 5) = -125$

36. $\left(\frac{1}{2}\right)^4 = \left(\frac{1}{2}\right)\left(\frac{1}{2}\right)\left(\frac{1}{2}\right)\left(\frac{1}{2}\right) = \frac{1}{16}$

37. $(-5)^2 = (-5)(-5) = 25$

38. $(-5)^3 = (-5)(-5)(-5) = -125$ ●

Practice Problems 31–32

Divide.

31. $-\frac{3}{4} \div \left(-\frac{3}{8}\right)$ **32.** $-\frac{1}{11} \div \frac{2}{7}$

Practice Problems 33–38

Simplify each expression.

33. 4^2 **34.** -2^2

35. -2^3 **36.** $\left(\frac{1}{3}\right)^4$

37. $(-2)^2$ **38.** $(-2)^3$

Answers

31. 2, **32.** $-\frac{7}{22}$, **33.** 16, **34.** -4, **35.** -8, **36.** $\frac{1}{81}$, **37.** 4, **38.** -8

Concept Check

When $(-8.2)^7$ is evaluated, will the value be positive or negative? How can you tell without making any calculations?

Try the Concept Check in the margin.

> **Helpful Hint**
>
> Be very careful when finding the value of expressions such as -5^2 and $(-5)^2$.
>
> $$-5^2 = -(5 \cdot 5) = -25 \quad \text{and} \quad (-5)^2 = (-5)(-5) = 25$$
>
> Without parentheses, the base to square is 5, not -5.

E Finding the Root of a Number

The opposite of squaring a number is taking the **square root** of a number. For example, since the square of 4, or 4^2, is 16, we say that a square root of 16 is 4. The notation $\sqrt{a}$ is used to denote the **positive**, or **principal square root** of a nonnegative number a. We then have in symbols that

$$\sqrt{16} = 4$$

Practice Problems 39–41

Find each root.
39. $\sqrt{36}$
40. $\sqrt{4}$
41. $\sqrt{\dfrac{1}{9}}$

EXAMPLES   Find each root.

39. $\sqrt{9} = 3$, since 3 is positive and $3^2 = 9$.

40. $\sqrt{25} = 5$, since $5^2 = 25$.

41. $\sqrt{\dfrac{1}{4}} = \dfrac{1}{2}$, since $\left(\dfrac{1}{2}\right)^2 = \dfrac{1}{4}$.

We can find roots other than square roots. Since 2 cubed, written as 2^3, is 8, we say that the **cube root** of 8 is 2. This is written as

$$\sqrt[3]{8} = 2$$

Also, since $3^4 = 81$ and 3 is positive,

$$\sqrt[4]{81} = 3$$

Practice Problems 42–44

Find each root.
42. $\sqrt[3]{8}$
43. $\sqrt[4]{1}$
44. $\sqrt[5]{32}$

EXAMPLES   Find each root.

42. $\sqrt[3]{27} = 3$, since $3^3 = 27$.

43. $\sqrt[5]{1} = 1$, since $1^5 = 1$.

44. $\sqrt[4]{16} = 2$, since 2 is positive and $2^4 = 16$.

Of course, as mentioned in Section 1.2, not all roots simplify to rational numbers. We study radicals further in Chapter 7.

Answers

39. 6, **40.** 2, **41.** $\dfrac{1}{3}$, **42.** 2, **43.** 1, **44.** 2

Concept Check: negative; the exponent is an odd number

EXERCISE SET 1.4

A *Find each absolute value. See Examples 1 through 5.*

1. $|2|$

2. $|8|$

3. $|-4|$

4. $|-6|$

5. $|0|$

6. $|-1|$

7. $-|3|$

8. $-|11|$

9. $-|-2|$

10. $-|-14|$

B *Add or subtract as indicated. See Examples 6 through 17.*

11. $-3 + 8$

12. $-5 + (-9)$

 13. $-14 + (-10)$

14. $12 + (-7)$

15. $-4.3 - 6.7$

16. $-8.2 - (-6.6)$

17. $13 - 17$

18. $15 - (-1)$

19. $\frac{11}{15} - \left(-\frac{3}{5}\right)$

20. $\frac{7}{10} - \frac{4}{5}$

21. $19 - 10 - 11$

22. $-13 - 4 + 9$

23. $-14 - 7$

24. $-6 - 31$

25. $-\frac{4}{5} - \left(-\frac{3}{10}\right)$

26. $-\frac{5}{2} - \left(-\frac{2}{3}\right)$

27. Subtract 14 from 8.

28. Subtract 9 from -3.

29. $-4 + 7$

30. $-9 + 15$

31. $-9 + (-3)$

32. $-17 + (-2)$

33. $-4 - (-19)$

34. $-5 - (-17)$

35. $6.3 - 18.5$

36. $15.9 - 21.7$

37. $16 - 8 - 9$

38. $-14 - 3 + 6$

39. $-5 + (-7) - 10$

40. $-8 + (-10) - 6$

C *Multiply or divide as indicated. See Examples 18 through 32.*

41. $-5 \cdot 12$

42. $-3 \cdot 8$

43. $6(-3)$

44. $5(-4)$

45. $-8(-10)$

46. $-4(-11)$

47. $-9 \cdot 8$

48. $-6 \cdot 7$

49. $-17 \cdot 0$

50. $-5 \cdot 0$

51. $0(-1)$

52. $0(-34)$

53. $\frac{-9}{3}$

54. $\frac{-20}{5}$

55. $\frac{16}{-2}$

56. $\frac{35}{-7}$

57. $\frac{-12}{-4}$

58. $\frac{-36}{-6}$

59. $-18 \div 6$

60. $-42 \div 6$

61. $\dfrac{0}{-5}$

62. $\dfrac{0}{-11}$

63. $\dfrac{-18}{0}$

64. $\dfrac{-22}{0}$

65. $-4(-2)(-1)$

66. $-5(-3)(-2)$

67. $-7(-1)(5)$

68. $-6(2)(-3)$

69. $\dfrac{-6}{7} \div 2$

70. $\dfrac{-9}{13} \div (-3)$

71. $-\dfrac{2}{7} \cdot \left(-\dfrac{1}{6}\right)$

72. $\dfrac{5}{9} \cdot \left(-\dfrac{3}{5}\right)$

73. $-\dfrac{1}{6} \div \dfrac{9}{10}$

74. $\dfrac{4}{7} \div \left(-\dfrac{1}{8}\right)$

75. $-\dfrac{2}{3} \cdot \left(\dfrac{6}{4}\right)$

76. $\dfrac{5}{6} \cdot \left(\dfrac{-12}{15}\right)$

77. $-2(-3.6)$

78. $-5(-4.2)$

79. $-6(-5)(0)$

80. $4(-3)(0)$

81. $\dfrac{-5.2}{-1.3}$

82. $\dfrac{-7}{-1.4}$

83. $-25 \div (-5)$

84. $-88 \div (-11)$

85. $\dfrac{3}{5} \div \left(-\dfrac{2}{5}\right)$

86. $\dfrac{2}{7} \div \left(-\dfrac{1}{14}\right)$

87. $9.1 \div (-1.3)$

88. $22.5 \div (-2.5)$

D *Find the value of each expression. See Examples 33 through 38.*

89. -7^2

90. $(-7)^2$

91. $(-6)^2$

92. -6^2

93. $(-2)^3$

94. -2^3

95. $\left(-\dfrac{1}{3}\right)^3$

96. $\left(-\dfrac{1}{2}\right)^4$

97. Explain why -3^2 and $(-3)^2$ simplify to different numbers.

98. Explain why -3^3 and $(-3)^3$ simplify to the same number.

E *Find each root. See Examples 39 through 44.*

99. $\sqrt{49}$

100. $\sqrt{81}$

101. $\sqrt{64}$

102. $\sqrt{100}$

103. $\sqrt{\dfrac{1}{9}}$

104. $\sqrt{\dfrac{1}{25}}$

105. $\sqrt{\dfrac{1}{16}}$

106. $\sqrt{\dfrac{1}{49}}$

107. $\sqrt[3]{64}$

108. $\sqrt[5]{32}$

109. $\sqrt[4]{81}$

110. $\sqrt[3]{1}$

111. $\sqrt[3]{8}$

112. $\sqrt[3]{125}$

36

Each circle below represents a whole, or 1. Determine the unknown fractional part of each circle.

113.

114.

115. Most of Mauna Kea, a volcano on Hawaii, lies below sea level. If this volcano begins at 5998 meters below sea level and then rises 10,203 meters, find the height of the volcano above sea level.

116. The highest point on land on Earth is the top of Mt. Everest, in the Himalayas, at an elevation of 29,028 feet above sea level. The lowest point on land is the Dead Sea, between Israel and Jordan, at 1312 feet below sea level. Find the difference in elevations. (*Source:* National Geographic Society)

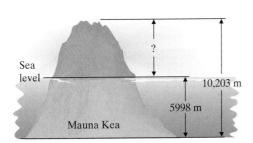

A fair game is one in which each team or player has the same chance of winning. Suppose that a game consists of three players taking turns spinning a spinner. If the spinner lands on yellow, player 1 gets a point. If the spinner lands on red, player 2 gets a point, and if the spinner lands on blue, player 3 gets a point. After 12 spins, the player with the most points wins. Use this information to answer Exercises 117 through 121.

A **B** **C** **D**

117. Which spinner would lead to a fair game?

118. If you are player 2 and want to win the game, which spinner would you choose?

119. If you are player 1 and want to lose the game, which spinner would you choose?

120. Is it possible for the game to end in a three-way tie? If so, list the possible ending scores.

121. Is it possible for the game to end in a two-way tie? If so, list the possible ending scores.

Use a calculator to approximate each square root. Round to four decimal places.

122. $\sqrt{10}$ **123.** $\sqrt{273}$ **124.** $\sqrt{7.9}$ **125.** $\sqrt{19.6}$

Investment firms often advertise their gains and losses in the form of bar graphs such as the one that follows. This graph shows investment risk over time for the S&P 500 Index by showing average annual compound returns for 1 year, 5 years, 15 years, and 25 years. For example, after 1 year, the annual compound return in percent for an investor is anywhere from a gain of 181.5% to a loss of 64%. Use this graph to answer Exercises 126 through 130.

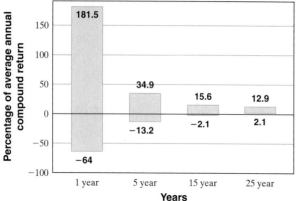

Investment Risk Over Time

Source: Investment Digest, VALIC, Fall, 1995

126. A person investing in the S&P 500 Index may expect at most an average annual gain of what percent after 15 years?

127. A person investing in the S&P 500 Index may expect to lose at most an average per year of what percent after 5 years?

128. Find the difference in percent of the highest average annual return and the lowest average annual return after 15 years.

129. Find the difference in percent of the highest average annual return and the lowest average annual return after 25 years.

130. Do you think that the type of investment shown in the figure is recommended for short-term investments or long-term investments? Explain your answer.

Internet Excursions

Go To: http://www.prenhall.com/martin-gay_interm   What's Related

Publicly held corporations sell shares of their company's stock on a stock exchange such as the New York Stock Exchange (NYSE). Many sites on the World Wide Web allow you to track stock prices. By going to the World Wide Web site listed above, you will gain access to the CNN Financial Network Web site or a related site. You will be able to find current information about the activity on the various stock markets in the United States.

131. Record the date and time of your visit to this Web site. Submit a request for a listing of the NYSE gainers. Make a list of the five stocks with the largest gains. Which stock made the largest gain? The gain listed is from the previous day's closing price. If you had purchased 275 shares of the stock with the largest gain at the previous day's close, how much money would you have gained?

132. Record the date and time of your visit to this Web site. Submit a request for a listing of the NYSE losers. Make a list of the five stocks with the biggest losses. Which stock had the largest loss? The loss listed is from the previous day's closing price. If you had purchased 425 shares of the stock with the largest loss at the previous day's close, how much money would you have lost?

Integrated Review—Real Numbers

Write each set by listing its elements.

1. $\{x \mid x$ is a natural number less than 4$\}$

2. $\{x \mid x$ is an odd whole number less than 6$\}$

3. $\{x \mid x$ is an even natural number greater than 7$\}$

4. $\{x \mid x$ is a whole number between 10 and 15$\}$

Write each phrase as an algebraic expression. Let x represent the unknown number.

5. Twice the difference of a number and three

6. The quotient of six and the sum of a number and ten

Insert $<, >,$ *or* $=$ *between each pair of numbers to form a true statement.*

7. $-4 \qquad -6$

8. $8.6 \qquad 8.600$

9. $\dfrac{9}{10} \qquad \dfrac{11}{10}$

10. $-6.1 \qquad -6.01$

Write each sentence as an equation.

11. The product of 5 and x is the same as 20.

12. The sum of a and 12 amounts to 14.

13. The quotient of y and 10 is the same as the product of y and 10.

14. The sum of x and 1 equals the difference of x and 1.

Perform each indicated operation.

15. $-4 + 7$

16. $-11 + 20$

17. $-4(7)$

18. $-11(20)$

19. $-8 - (-13)$

20. $-12 - 16$

21. $\dfrac{-20}{-4}$

22. $\dfrac{-18}{6}$

23. -5^2

24. $(-5)^2$

25. $-6 - 1 + 20$

26. $18 - 4 - 19$

27. $\dfrac{0}{-3}$

28. $\dfrac{5}{0}$

29. $-4(3)(2)$

30. $-5(-1)(6)$

1. _____
2. _____
3. _____
4. _____
5. _____
6. _____
7. _____
8. _____
9. _____
10. _____
11. _____
12. _____
13. _____
14. _____
15. _____
16. _____
17. _____
18. _____
19. _____
20. _____
21. _____
22. _____
23. _____
24. _____
25. _____
26. _____
27. _____
28. _____
29. _____
30. _____

31. _____

32. _____

33. _____

34. _____

35. _____

36. _____

37. _____

38. _____

43. _____

44. _____

45. _____

46. _____

47. _____

48. _____

49. _____

50. _____

31. $-\dfrac{1}{2} \cdot \dfrac{6}{7}$ **32.** $\dfrac{4}{5} \cdot \left(-\dfrac{1}{8}\right)$ **33.** $\dfrac{3}{10} - \dfrac{4}{5}$ **34.** $-\dfrac{2}{3} - \dfrac{1}{4}$

35. $\dfrac{1.6}{-0.2}$ **36.** $\dfrac{-4.8}{16}$ **37.** $6.7 - (-1.3)$ **38.** $-4.6 + 9$

Fill in the chart.

	Number	Opposite	Reciprocal
39.	-6		
40.	4		
41.			$\dfrac{7}{5}$
42.		$\dfrac{2}{3}$	

Use the distribution property to multiply.

43. $9(m + 5)$ **44.** $11(7 + r)$ **45.** $-3(2y - 3x)$

46. $-8(4m - 7n)$ **47.** $-0.2(3a - 7)$ **48.** $-0.6(2n - 5)$

49. $\dfrac{1}{5}(10x - 19y + 20)$ **50.** $\dfrac{1}{2}(10x - 19y + 20)$

1.5 Order of Operations and Algebraic Expressions

Ⓐ Using the Order of Operations

The expression $3 + 2 \cdot 30$ represents the total number of compact disks (CDs) shown.

CD-R
30 pack

CD-R
30 pack

Expressions containing more than one operation are written to follow a particular agreed-upon **order of operations**. For example, when we write $3 + 2 \cdot 10$, we mean to multiply first, and then add.

Order of Operations

Simplify expressions using the following order. If grouping symbols such as parentheses are present, simplify expressions within those first, starting with the innermost set. If fraction bars are present, simplify the numerator and the denominator separately.

1. Evaluate exponential expressions.

2. Perform multiplications or divisions in order from left to right.

3. Perform additions or subtractions in order from left to right.

EXAMPLE 1   Simplify: $3 + 2 \cdot 30$

Solution: First we multiply; then we add.

$$3 + 2 \cdot 30 = 3 + 60 = 63$$

EXAMPLE 2   Simplify: $2(1 - 4)^2$

Solution:

$$
\begin{aligned}
2(1 - 4)^2 &= 2(-3)^2 && \text{Simplify inside parentheses first.} \\
&= 2(9) && \text{Write } (-3)^2 \text{ as 9.} \\
&= 18 && \text{Multiply.}
\end{aligned}
$$

EXAMPLE 3   Simplify: $\dfrac{|-2|^3 + 1}{-7 - \sqrt{4}}$

Solution: We simplify the numerator and the denominator separately. Then we divide.

$$
\begin{aligned}
\frac{|-2|^3 + 1}{-7 - \sqrt{4}} &= \frac{2^3 + 1}{-7 - 2} && \text{Write } |-2| \text{ as 2 and } \sqrt{4} \text{ as 2.} \\[2mm]
&= \frac{8 + 1}{-9} && \text{Write } 2^3 \text{ as 8.} \\[2mm]
&= \frac{9}{-9} = -1 && \text{Simplify the numerator; then divide.}
\end{aligned}
$$

Practice Problem 1

Simplify: $15 - 2 \cdot 5$

Practice Problem 2

Simplify: $5(2 - 6)^2$

Practice Problem 3

Simplify: $\dfrac{|-3|^2 + 5}{\sqrt{9} - 10}$

Answers

1. 5, **2.** 80, **3.** -2

Practice Problem 4

Simplify: $\dfrac{(8-3)-(-6)}{4-(-1)}$

Practice Problem 5

Simplify:

$7 - [2(1-3) + 5(10-12)]$

Concept Check

True or false? If two different people use the order of operations to simplify a numerical expression and neither makes a calculation error, it is not possible that they each obtain a different result. Explain.

Practice Problem 6

Use the algebraic expression given in Example 6 to complete the following table.

Degrees Fahrenheit	x	-13	0	41
Degrees Celsius	$\dfrac{5(x-32)}{9}$			

Answers

4. $\dfrac{11}{5}$, **5.** 21, **6.** $-25; -\dfrac{160}{9}$ or $-17\dfrac{7}{9}; 5$

Concept Check: true; answers may vary

EXAMPLE 4 Simplify: $\dfrac{(6+2)-(-4)}{2-(-3)}$

Solution:

$$\dfrac{(6+2)-(-4)}{2-(-3)} = \dfrac{8-(-4)}{2-(-3)} \quad \text{Simplify inside parentheses first.}$$

$$= \dfrac{8+4}{2+3} \quad \text{Write subtraction as equivalent addition.}$$

$$= \dfrac{12}{5} \quad \text{Add in both the numerator and the denominator.}$$

Besides parentheses, other symbols used for grouping expressions are brackets [ ] and braces { }. These other grouping symbols are commonly used when we group expressions that already contain parentheses.

EXAMPLE 5 Simplify: $3 - [(4-6) + 2(5-9)]$

Solution:

$$3 - [(4-6) + 2(5-9)] = 3 - [-2 + 2(-4)] \quad \text{Simplify within the innermost sets of parentheses.}$$

$$= 3 - [-2 + (-8)]$$

$$= 3 - [-10]$$

$$= 13$$

Helpful Hint

When grouping symbols occur within grouping symbols, remember to perform operations on the innermost set first.

Try the Concept Check in the margin.

B **Evaluating Algebraic Expressions**

Recall from Section 1.2 that an algebraic expression is formed by numbers and variables connected by the operations of addition, subtraction, multiplication, division, raising to powers, or taking roots. Also, if numbers are substituted for the variables in an algebraic expression and the operations performed, the result is called the value of the expression for the given replacement values. This entire process is called **evaluating an expression**.

EXAMPLE 6 **Converting Degrees Fahrenheit to Degrees Celsius**

The algebraic expression $\dfrac{5(x-32)}{9}$ represents the equivalent temperature in degrees Celsius when x is degrees Fahrenheit. Complete the following table by evaluating this expression at the given values of x.

Degrees Fahrenheit	x	-4	10	32
Degrees Celsius	$\dfrac{5(x-32)}{9}$			

Solution: To complete the table, we evaluate $\dfrac{5(x-32)}{9}$ at each given replacement value.

When $x = -4$,

$$\frac{5(x-32)}{9} = \frac{5(-4-32)}{9} = \frac{5(-36)}{9} = -20$$

When $x = 10$,

$$\frac{5(x-32)}{9} = \frac{5(10-32)}{9} = \frac{5(-22)}{9} = \frac{-110}{9} \text{ or } -12\frac{2}{9}$$

When $x = 32$,

$$\frac{5(x-32)}{9} = \frac{5(32-32)}{9} = \frac{5 \cdot 0}{9} = 0$$

The completed table is:

Degrees Fahrenheit	x	-4	10	32
Degrees Celsius	$\dfrac{5(x-32)}{9}$	-20	$\dfrac{-110}{9}$ or $-12\dfrac{2}{9}$	0

Thus, $-4°F$ is equivalent to $-20°C$, $10°F$ is equivalent to $-\dfrac{110}{9}°C$ or $-12\dfrac{2}{9}°C$, and $32°F$ is equivalent to $0°C$.   ●

(C) Simplifying Algebraic Expressions by Combining Like Terms

Often, an expression may be **simplified** by removing grouping symbols and combining any like terms. The **terms** of an expression are the addends of the expression. For example, in the expression $3x^2 + 4x$, the terms are $3x^2$ and $4x$.

Expression	Terms
$-2x + y$	$-2x,\quad y$
$3x^2 - \dfrac{y}{5} + 7$	$3x^2,\quad -\dfrac{y}{5},\quad 7$

Terms with the same variable(s) raised to the same power(s) are called **like terms**. We can add or subtract like terms by using the distributive property. This process is called **combining like terms**.

Practice Problems 7–9

Simplify by combining like terms.

7. $9x - 15x + 7$
8. $8y + y$
9. $4x + 12x - 9 - 10$

EXAMPLES   Simplify by combining like terms.

7. $3x - 5x + 4 = (3 - 5)x + 4$ Use the distributive property.

$$= -2x + 4$$

8. $y + 3y = 1y + 3y = (1 + 3)y$

$$= 4y$$

9. $7x + 9x + 6 - 10 = (7 + 9)x + (6 - 10)$

$$= 16x - 4$$ ●

The associative and commutative properties may sometimes be needed to rearrange and group like terms when we simplify expressions.

Practice Problems 10–11

Simplify.

10. $-4x + 7 - 5x - 8$
11. $5y - 6y + 2 - 11 + y$

EXAMPLES   Simplify.

10. $-7x + 5 + 3x - 2 = -7x + 3x + 5 - 2$ Use the commutative property.

$$= (-7 + 3)x + (5 - 2)$$ Use the distributive property.

$$= -4x + 3$$ Simplify.

11. $3y - 2y + 5 - 7 + y = 3y - 2y + y + 5 - 7$ Use the commutative property.

$$= (3 - 2 + 1)y + (5 - 7)$$ Use the distributive property.

$$= 2y - 2$$ Simplify. ●

Practice Problems 12–14

Simplify by using the distributive property to multiply and then combining like terms.

12. $-3(y + 1)$
13. $8x + 2 - 4(x - 9)$
14. $(3.2x - 4.1) - (-x + 7.6)$
15. $\dfrac{1}{5}(15m - 40n) -$

$$\dfrac{1}{4}(8m - 4n + 1) + \dfrac{1}{5}$$

EXAMPLES

Simplify by using the distributive property to multiply and then combining like terms.

12. $-2(x + 3) = -2(x) + (-2)(3) = -2x - 6$

13. $7x + 3 - 5(x - 4) = 7x + 3 - 5x + 20$ Use the distributive property.

$$= 2x + 23$$ Combine like terms.

14. $(2.1x - 5.6) - (-x - 5.3) = (2.1x - 5.6) - 1(-x - 5.3)$

$$= 2.1x - 5.6 + 1x + 5.3$$ Use the distributive property.

$$= 3.1x - 0.3$$ Combine like terms.

15. $\dfrac{1}{2}(4a - 6b) - \dfrac{1}{3}(9a + 12b - 1) + \dfrac{1}{4}$

$$= 2a - 3b - 3a - 4b + \dfrac{1}{3} + \dfrac{1}{4}$$ Use the distributive property.

$$= -a - 7b + \dfrac{7}{12}$$ Combine like terms. ●

Try the Concept Check in the margin.

Concept Check

Find and correct the error in the following:
$x - 4(x - 5) = x - 4x - 20$
$$= -3x - 20$$

Answers

7. $-6x + 7$, **8.** $9y$, **9.** $16x - 19$, **10.** $-9x - 1$,
11. -9, **12.** $-3y - 3$, **13.** $4x + 38$,

14. $4.2x - 11.7$, **15.** $m - 7n - \dfrac{1}{20}$

Concept Check:
$x - 4(x - 5) = x - 4x + 20$
$$= -3x + 20$$

EXERCISE SET 1.5

A *Simplify each expression. See Examples 1 through 5.*

1. $3(5 - 7)^4$

2. $7(3 - 8)^2$

3. $-3^2 + 2^3$

4. $-5^2 - 2^4$

5. $\dfrac{3 - (-12)}{-5}$

6. $\dfrac{-4 - (-8)}{-4}$

7. $|3.6 - 7.2| + |3.6 + 7.2|$

8. $|8.6 - 1.9| - |2.1 + 5.3|$

9. $(-3)^2 + 2^3$

10. $(-15)^2 - 2^4$

11. $-3[6 - (-2)]$

12. $-5[8 - (-3)]$

13. $-9 \cdot 8 + 5(-6)$

14. $-6 \cdot 6 + 9(-5)$

15. $4[8 - (2 - 4)]$

16. $3[11 - (1 - 3)]$

17. $-8\left(-\dfrac{3}{4}\right) - 8$

18. $-10\left(-\dfrac{2}{5}\right) - 10$

19. $2 - [(7 - 6) + (9 - 19)]$

20. $8 - [(4 - 7) + (8 - 1)]$

21. $5^2 - 3^4$

22. $6^2 - 5^3$

23. $2 \cdot 7 - 4 \cdot 5$

24. $(2 \cdot 7) - (4 \cdot 5)$

25. $2 \cdot (7 - 4 \cdot 5)$

26. $(2 \cdot 7 - 4) \cdot 5$

27. $18 - 3(-4) + 7$

28. $25 - 2(-3) + 10$

29. $\dfrac{(-9 + 6)(-1^2)}{-2 - 2}$

30. $\dfrac{(-1 - 2)(-3^2)}{-6 - 3}$

31. $(\sqrt[3]{8})(-4) - (\sqrt{9})(-5)$

32. $(\sqrt[3]{27})(-5) - (\sqrt{25})(-3)$

33. $12 + \{6 - [5 - 2(-5)]\}$

34. $18 + \{9 - [1 - 6(-3)]\}$

35. $25 - [(3 - 5) + (14 - 18)]^2$

36. $10 - [(4 - 5)^2 + (12 - 14)]^4$

37. $\dfrac{(3 - \sqrt{9}) - (-5 - 1.3)}{-3}$

38. $\dfrac{-\sqrt{16} - (6 - 2.4)}{-2}$

39. $\dfrac{|3 - 9| - |-5|}{-3}$

40. $\dfrac{|-14| - |2 - 7|}{-15}$

41. $\dfrac{3(-2 + 1)}{5} - \dfrac{-7(2 - 4)}{1 - (-2)}$

42. $\dfrac{-1 - 2}{2(-3) + 10} - \dfrac{2(-5)}{-1(8) + 1}$

43. $\dfrac{\dfrac{1}{3} \cdot 9 - 7}{3 + \dfrac{1}{2} \cdot 4}$

44. $\dfrac{\dfrac{1}{5} \cdot 20 - 6}{10 + \dfrac{1}{4} \cdot 12}$

45. $3\{-2 + 5[1 - 2(-2 + 5)]\}$

46. $2\{-1 + 3[7 - 4(-10 + 12)]\}$

47. $-150(3.25 - 1.68)$ **48.** $-290(9.61 - 6.27)$ **49.** $\left(\dfrac{5.6 - 8.4}{1.9 - 2.7}\right)^2$ **50.** $\left(\dfrac{9.4 - 10.8}{8.7 - 7.9}\right)^2$

B *Complete each table. See Example 6.*

△ **51.** The algebraic expression $8 + 2y$ represents the perimeter of a rectangle with width 4 and length y.

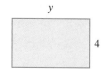

y

4

a. Complete the table by evaluating this expression at the given values of y.

Length	y	5	7	10	100
Perimeter	$8 + 2y$				

b. Use the results of the table in (a) to answer the following question. As the width of a rectangle remains the same and the length increases, does the perimeter increase or decrease? Explain how you arrived at your answer.

△ **52.** The algebraic expression πr^2 represents the area of a circle with radius r.

r

a. Complete the table by evaluating this expression at the given values of r. (Use 3.14 for π.)

Radius	r	2	3	7	10
Area	πr^2				

b. As the radius of a circle increases, does its area increase or decrease? Explain your answer.

53. The algebraic expression $\dfrac{100x + 5000}{x}$ represents the cost per bookshelf (in dollars) of producing x bookshelves.

a. Complete the table.

Number of Bookshelves	x	10	100	1000
Cost per Bookshelf	$\dfrac{100x + 5000}{x}$			

b. As the number of bookshelves manufactured increases, does the cost per bookshelf increase or decrease? Why do you think that this is so?

54. If C is degrees Celsius, the algebraic expression $1.8C + 32$ represents the equivalent temperature in degrees Fahrenheit.

a. Complete the table.

Degrees Celsius	C	-10	0	50
Degrees Fahrenheit	$1.8C + 32$			

b. As degrees Celsius increase, do degrees Fahrenheit increase or decrease? Explain your answer.

55. The average price for an ounce of gold in the U.S. during September 2001 was $283.47. The algebraic expression $283.47z$ gives the average cost of z ounces of gold during this period. Find the average cost if 8.4 ounces of gold had been purchased during this time. (*Source:* www.kitco.com)

56. On September 27, 2001, the velocity of the solar wind in Earth's upper atmosphere was 558.1 kilometers per second. At this speed, the algebraic expression $558.1t$ gives the total distance covered in t seconds. Find the distance covered by a proton traveling in the solar wind in 5 seconds. (*Source:* NOAA Space Environment Center)

Simplify. See Examples 7 through 11.

57. $6x + 2x$ **58.** $9y - 11y$ **59.** $7x + x$ **60.** $2y + y$ **61.** $19y - y$

62. $14x - x$ **63.** $6x - 4x + 10x$ **64.** $13y - 2y - 23y$ **65.** $9x - 8 - 10x$ **66.** $14x - 1 - 20x$

67. $10a + 7 + 4a + 8$ **68.** $9a + 6 + 2a + 9$ **69.** $-9 + 4x + 18 - 10x$ **70.** $5y - 14 + 7y - 20y$

71. $3a - 4b + a - 9b$ **72.** $11x - y + 11x - 6y$ **73.** $x - y + x - y$ **74.** $a - b + 3a - 3b$

75. $1.5x + 2.3 - 0.7x - 5.9$ **76.** $6.3y - 9.7 + 2.2y - 11.1$ **77.** $\frac{3}{4}b - \frac{1}{2} + \frac{1}{6}b - \frac{2}{3}$ **78.** $\frac{7}{8}a - \frac{11}{12} - \frac{1}{2}a + \frac{5}{6}$

Simplify. See Examples 12 through 15.

79. $2(3x + 7)$ **80.** $4(5y + 12)$ **81.** $-5(x - 1)$ **82.** $-6(b - 6)$

83. $3(2a - 3b + 4)$ **84.** $5(3x - 4y + 1)$ **85.** $5k - (3k - 10)$ **86.** $-11c - (4 - 2c)$

87. $(3x + 4) - (6x - 1)$ **88.** $(8 - 5y) - (4 + 3y)$ **89.** $3(x - 2) + x + 15$ **90.** $-4(y + 3) - 7y + 1$

91. $-(n + 5) + (5n - 3)$ **92.** $-(8 - t) + (2t - 6)$ **93.** $4(6n - 3) - 3(8n + 4)$ **94.** $5(2z - 6) + 10(3 - z)$

95. $\frac{1}{4}(8x - 4) - \frac{1}{5}(20x - 6y)$ **96.** $\frac{1}{2}(10x - 2) - \frac{1}{6}(60x - 5y)$

97. $3x - 2(x - 5) + x$ **98.** $7n + 3(2n - 6) - 2$

99. $\frac{1}{6}(24a - 18b) - \frac{1}{7}(7a - 21b - 2) - \frac{1}{5}$

100. $\frac{1}{3}(6x - 33y) - \frac{1}{8}(24x - 40y + 1) - \frac{1}{3}$

 101. $-1.2(5.7x - 3.6) + 8.75x$

102. $5.8(-9.6 - 31.2y) - 18.65$

 103. $8.1z + 7.3(z + 5.2) - 6.85$

104. $6.5y - 4.4(1.8y - 3.3) + 10.95$

Combining Concepts

Insert parentheses so that when simplified each expression is equal to the given number.

105. $2 + 7 \cdot 1 + 3; \quad 36$

106. $6 - 5 \cdot 2 + 2; \quad -6$

The following graph is called a broken-line graph, or simply a line graph. This particular graph shows the past, present, and future predicted U.S. population over 65. Just as with a bar graph, to find the population over 65 for a particular year, read the height of the corresponding point. To read the height, follow the point horizontally to the left until you reach the vertical axis. Use this graph to answer Exercises 107 through 112.

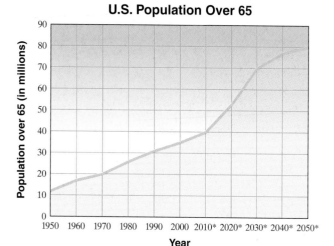

U.S. Population Over 65

Source: U.S. Census Bureau   * Projected

107. Estimate the population over 65 in the year 1970.

108. Estimate the predicted population over 65 in the year 2050.

109. Estimate the predicted population over 65 in the year 2030.

110. Estimate the population over 65 in the year 2000.

111. Is the population over 65 increasing as time passes or decreasing? Explain how you arrived at your answer.

112. The percent of Americans over 65 in 1950 was 8.1%. The percent of Americans over 65 in 2050 is expected to be 2.5 times the percent over 65 in 1950. Estimate the percent of Americans expected to be over age 65 in 2050.

Simplify. Round each result to the nearest ten thousandth.

 113. $\dfrac{-1.682 - 17.895}{(-7.102)(-4.691)}$

 114. $\dfrac{(-5.161)(3.222)}{(7.955 - 19.676)}$

1.6   Exponents and Scientific Notation

OBJECTIVES

Ⓐ   Use the product rule for exponents.

Ⓑ   Simplify expressions raised to the zero power.

Ⓒ   Use the quotient rule for exponents.

Ⓓ   Simplify expressions raised to negative powers.

Ⓔ   Convert between scientific notation and standard notation.

SSM
TUTOR CENTER   SG   CD & VIDEO   MATH PRO   WEB

Ⓐ   Using the Product Rule

Recall that exponents may be used to write repeated factors in a more compact form. As we have seen in the previous sections, exponents can be used when the repeated factor is a number or a variable. For example,

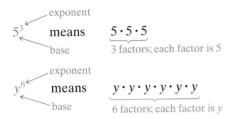

Expressions that contain exponents such as 5^3 and y^6 are called **exponential expressions**.

Exponential expressions can be multiplied, divided, added, subtracted, and themselves raised to powers. In this section, we review operations on exponential expressions.

We review multiplication first. To multiply x^2 by x^3, we use the definition of a^n:

$$x^2 \cdot x^3 = \underbrace{(x \cdot x)(x \cdot x \cdot x)}_{x \text{ is a factor 5 times.}}$$

$$= x^5$$

Notice that the result is exactly the same if we add the exponents.

$$x^2 \cdot x^3 = x^{2+3} = x^5$$

This suggests the following rule.

Product Rule for Exponents

If m and n are integers and a is a real number, then

$$a^m \cdot a^n = a^{m+n}$$

In other words, the *product* of exponential expressions with a common base is the common base raised to a power equal to the *sum* of the exponents of the factors.

EXAMPLES   Use the product rule to simplify.

1. $2^2 \cdot 2^5 = 2^{2+5} = 2^7$

2. $x^7 \cdot x^3 = x^{7+3} = x^{10}$

3. $y \cdot y^2 \cdot y^4 = (y^1 \cdot y^2) \cdot y^4$
$$= y^3 \cdot y^4$$
$$= y^7$$

EXAMPLES   Use the product rule to simplify.

4. $(3x^6)(5x) = 3(5)x^6x^1 = 15x^7$ Use properties of multiplication to group like bases.

5. $(-2x^3p^2)(4xp^{10}) = -2(4)x^3x^1p^2p^{10} = -8x^4p^{12}$

Practice Problems 1–3

Use the product rule to simplify.

1. $5^2 \cdot 5^6$ 2. $x^5 \cdot x^9$
3. $y \cdot y^4 \cdot y^3$

Practice Problems 4–5

Use the product rule to simplify.

4. $(7y^5)(6y)$ 5. $(-3x^2y^7)(5xy^6)$

Answers

1. 5^8, **2.** x^{14}, **3.** y^8, **4.** $42y^6$, **5.** $-15x^3y^{13}$

B Simplifying Expressions Raised to the Zero Power

The definition of a^n does not include the possibility that n might be 0. But if it did, then, by the product rule,

$$\underbrace{a^0 \cdot a^n = a^{0+n} = a^n = \underbrace{1 \cdot a^n}}$$

From this, we reasonably define that $a^0 = 1$, as long as a does not equal 0.

Zero Exponent

If a does not equal 0, then $a^0 = 1$.

Practice Problems 6–9

Evaluate each expression.

6. 8^0
7. -8^0
8. $(y - 3)^0$
9. $5x^0$

EXAMPLES Evaluate each expression.

6. $7^0 = 1$
7. $-7^0 = -(7^0) = -(1) = -1$ Without parentheses, only 7 is raised to the 0 power.
8. $(2x + 5)^0 = 1$
9. $2x^0 = 2(1) = 2$ ●

C Using the Quotient Rule

To find quotients of exponential expressions, we again begin with the definition of a^n to simplify $\dfrac{x^9}{x^2}$. For example,

$$\frac{x^9}{x^2} = \frac{x \cdot x \cdot x \cdot x \cdot x \cdot x \cdot x \cdot x \cdot x}{x \cdot x} = x^7$$

(Assume for the next two sections that denominators containing variables are not 0.) Notice that the result is exactly the same if we subtract the exponents.

$$\frac{x^9}{x^2} = x^{9-2} = x^7$$

This suggests the following rule.

Quotient Rule for Exponents

If a is a nonzero real number and m and n are integers, then

$$\frac{a^m}{a^n} = a^{m-n}$$

In other words, the *quotient* of exponential expressions with a common base is the common base raised to a power equal to the *difference* of the exponents.

Practice Problems 10–13

Use the quotient rule to simplify.

10. $\dfrac{y^6}{y^2}$ 11. $\dfrac{6^{10}}{6^2}$

12. $\dfrac{36x^5}{9x}$ 13. $\dfrac{10a^7b^9}{15a^5b^9}$

EXAMPLES Use the quotient rule to simplify.

10. $\dfrac{x^7}{x^4} = x^{7-4} = x^3$

11. $\dfrac{5^8}{5^2} = 5^{8-2} = 5^6$

12. $\dfrac{20x^6}{4x^5} = 5x^{6-5} = 5x^1$, or $5x$

Answers

6. 1, **7.** −1, **8.** 1, **9.** 5, **10.** y^4,

11. 6^8, **12.** $4x^4$

13. $\dfrac{12y^{10}z^7}{14y^8z^7} = \dfrac{6}{7}y^{10-8} \cdot z^{7-7} = \dfrac{6}{7}y^2z^0 = \dfrac{6}{7}y^2$, or $\dfrac{6y^2}{7}$

Ⓓ Simplifying Expressions Raised to Negative Powers

When the exponent of the denominator is larger than the exponent of the numerator, applying the quotient rule gives a negative exponent. For example,

$$\frac{x^3}{x^5} = x^{3-5} = x^{-2}$$

However, using the definition of a^n gives us

$$\frac{x^3}{x^5} = \frac{x \cdot x \cdot x}{x \cdot x \cdot x \cdot x \cdot x} = \frac{1}{x^2}$$

From this, we reasonably define $x^{-2} = \dfrac{1}{x^2}$ or, in general, $a^{-n} = \dfrac{1}{a^n}$.

Negative Exponents

If a is a real number other than 0 and n is a positive integer, then

$$a^{-n} = \frac{1}{a^n}$$

EXAMPLES   Simplify and write with positive exponents only.

14. $5^{-2} = \dfrac{1}{5^2} = \dfrac{1}{25}$

15. $2x^{-3} = 2 \cdot \dfrac{1}{x^3} = \dfrac{2}{x^3}$ *Without parentheses, only x is raised to the -3 power.*

16. $(3x)^{-1} = \dfrac{1}{(3x)^1} = \dfrac{1}{3x}$ *With parentheses, both 3 and x are raised to the -1 power.*

17. $\dfrac{m^5}{m^{15}} = m^{5-15} = m^{-10} = \dfrac{1}{m^{10}}$

18. $\dfrac{3^3}{3^6} = 3^{3-6} = 3^{-3} = \dfrac{1}{3^3} = \dfrac{1}{27}$

19. $2^{-1} + 3^{-2} = \dfrac{1}{2^1} + \dfrac{1}{3^2} = \dfrac{1}{2} + \dfrac{1}{9} = \dfrac{9}{18} + \dfrac{2}{18} = \dfrac{11}{18}$

20. $\dfrac{1}{t^{-5}} = \dfrac{1}{\dfrac{1}{t^5}} = 1 \div \dfrac{1}{t^5} = 1 \cdot \dfrac{t^5}{1} = t^5$

Helpful Hint

Notice that when a factor containing an exponent is moved from the numerator to the denominator or from the denominator to the numerator, the sign of its exponent changes.

$$x^{-3} = \frac{1}{x^3} \qquad 5^{-2} = \frac{1}{5^2} = \frac{1}{25}$$

$$\frac{1}{y^{-4}} = y^4 \qquad \frac{1}{2^{-3}} = 2^3 = 8$$

Practice Problems 14–20

Simplify and write with positive exponents only.

14. 7^{-2}

15. $5x^{-4}$

16. $(2x)^{-1}$

17. $\dfrac{x^3}{x^{10}}$

18. $\dfrac{4^2}{4^5}$

19. $3^{-1} + 2^{-2}$

20. $\dfrac{1}{y^{-4}}$

Answers

13. $\dfrac{2}{3}a^2$,   **14.** $\dfrac{1}{49}$,   **15.** $\dfrac{5}{x^4}$,   **16.** $\dfrac{1}{2x}$,   **17.** $\dfrac{1}{x^7}$,

18. $\dfrac{1}{64}$,   **19.** $\dfrac{7}{12}$,   **20.** y^4

Practice Problems 21–25

Simplify and write using positive exponents only.

21. $\dfrac{y^{-10}}{y^3}$ **22.** $\dfrac{q^5}{q^{-4}}$

23. $\dfrac{5^{-4}}{5^{-2}}$ **24.** $\dfrac{10x^{-8}y^5}{20xy^{-5}}$

25. $\dfrac{(4x^{-1})(x^5)}{x^7}$

Concept Check

Find and correct the error in the following:

$$\frac{y^{-6}}{y^{-2}} = y^{-6-2} = y^{-8} = \frac{1}{y^8}$$

Practice Problems 26–27

Simplify. Assume that n and m are nonzero integers and that x is not 0.

26. $x^{3m} \cdot x^n$

27. $\dfrac{x^{2m-2}}{x^{m-6}}$

EXAMPLES   Simplify and write using positive exponents only.

21. $\dfrac{x^{-9}}{x^2} = x^{-9-2} = x^{-11} = \dfrac{1}{x^{11}}$

22. $\dfrac{p^4}{p^{-3}} = p^{4-(-3)} = p^7$

23. $\dfrac{2^{-3}}{2^{-1}} = 2^{-3-(-1)} = 2^{-2} = \dfrac{1}{2^2} = \dfrac{1}{4}$

24. $\dfrac{2x^{-7}y^2}{10xy^{-5}} = \dfrac{x^{-7-1} \cdot y^{2-(-5)}}{5} = \dfrac{x^{-8}y^7}{5} = \dfrac{y^7}{5x^8}$

25. $\dfrac{(3x^{-3})(x^2)}{x^6} = \dfrac{3x^{-3+2}}{x^6} = \dfrac{3x^{-1}}{x^6} = 3x^{-1-6} = 3x^{-7} = \dfrac{3}{x^7}$

Try the Concept Check in the margin.

EXAMPLES

Simplify. Assume that a and t are nonzero integers and that x is not 0.

26. $x^{2a} \cdot x^3 = x^{2a+3}$ Use the product rule.

27. $\dfrac{x^{2t-1}}{x^{t-5}} = x^{(2t-1)-(t-5)}$ Use the quotient rule.

$\phantom{\dfrac{x^{2t-1}}{x^{t-5}}} = x^{2t-1-t+5} = x^{t+4}$

E   Converting Between Scientific Notation and Standard Notation

Very large and very small numbers occur frequently in nature. For example, the distance between Earth and the sun is approximately 150,000,000 kilometers. A helium atom has a diameter of 0.000000022 centimeters. It can be tedious to write these very large and very small numbers in standard notation like this. **Scientific notation** is a convenient shorthand notation for writing very large and very small numbers.

Helium atom
0.000000022 cm

150,000,000 km

Scientific Notation

A positive number is written in **scientific notation** if it is written as the product of a number a, where $1 \le a < 10$, and an integer power n of 10: $a \times 10^n$.

For example,

2.03×10^2 7.362×10^7 8.1×10^{-5}

Writing a Number in Scientific Notation

Step 1. Move the decimal point in the original number until the new number has a value between 1 and 10.

Step 2. Count the number of decimal places the decimal point was moved in Step 1. If the original number is 10 or greater, the count is positive. If the original number is less than 1, the count is negative.

Step 3. Write the product of the new number in Step 1 by 10 raised to an exponent equal to the count found in Step 2.

EXAMPLE 28   Write 730,000 in scientific notation.

Solution:

Step 1. Move the decimal point until the number is between 1 and 10.

730,000.

Step 2. The decimal point is moved 5 places and the original number is 10 or greater, so the count is positive 5.

Step 3. $730{,}000 = 7.3 \times 10^5$

EXAMPLE 29   Write 0.00000104 in scientific notation.

Solution:

Step 1. Move the decimal point until the number is between 1 and 10.

0.000001 04

Step 2. The decimal point is moved 6 places and the original number is less than 1, so the count is −6.

Step 3. $0.00000104 = 1.04 \times 10^{-6}$

To write a scientific notation number in standard form, we reverse the preceding steps.

Writing a Scientific Notation Number in Standard Notation

Move the decimal point in the number the same number of places as the exponent on 10. If the exponent is positive, move the decimal point to the right. If the exponent is negative, move the decimal point to the left.

EXAMPLES   Write each number in standard notation.

30. $7.7 \times 10^8 = 770{,}000{,}000$ Since the exponent is positive, move the decimal point 8 places to the right. Add zeros as needed.

31. $1.025 \times 10^{-3} = 0.001025$ Since the exponent is negative, move the decimal point 3 places to the left. Add zeros as needed.

Try the Concept Check in the margin.

Practice Problem 28

Write 1,760,000 in scientific notation.

Practice Problem 29

Write 0.00028 in scientific notation.

Practice Problems 30–31

Write each number in standard notation.

30. 8.6×10^7

31. 3.022×10^{-4}

Concept Check

Which of the following numbers have values that are less than 1?

a. 3.5×10^{-5}

b. 3.5×10^5

c. -3.5×10^5

d. -3.5×10^{-5}

Answers

28. 1.76×10^6,   **29.** 2.8×10^{-4},
30. 86,000,000,   **31.** 0.0003022
Concept Check: a, c, d

CALCULATOR EXPLORATIONS

Multiply 5,000,000 by 700,000 on your calculator. The display should read $\boxed{3.5 \quad 12}$ or $\boxed{3.5\,\text{E}\,12}$, which is the product written in scientific notation. Both these notations mean 3.5×10^{12}.

To enter a number written in scientific notation on a calculator, find the key marked $\boxed{\text{EE}}$. (On some calculators, this key may be marked $\boxed{\text{EXP}}$.)

To enter 7.26×10^{13}, press the keys

$$\boxed{7.26} \;\; \boxed{\text{EE}} \;\; \boxed{13}$$

The display will read $\boxed{7.26 \quad 13}$ or $\boxed{7.26\,\text{E}\,13}$.

Use your calculator to perform each indicated operation.

1. Multiply 3×10^{11} and 2×10^{32}.
2. Divide 6×10^{14} by 3×10^{9}.
3. Multiply 5.2×10^{23} and 7.3×10^{4}.
4. Divide 4.38×10^{41} by 3×10^{17}.

Name _____ Section _____ Date _____

Mental Math

Use positive exponents to rewrite each expression.

1. $5x^{-1}y^{-2}$ **2.** $7xy^{-4}$ **3.** $a^2b^{-1}c^{-5}$ **4.** $a^{-4}b^2c^{-6}$ **5.** $\dfrac{y^{-2}}{x^{-4}}$ **6.** $\dfrac{x^{-7}}{z^{-3}}$

EXERCISE SET 1.6

A *Use the product rule to simplify each expression. See Examples 1 through 5.*

1. $4^2 \cdot 4^3$ **2.** $3^3 \cdot 3^5$ **3.** $x^5 \cdot x^3$ **4.** $a^2 \cdot a^9$ **5.** $-7x^3 \cdot 20x^9$

6. $-3y \cdot -9y^4$ **7.** $(4xy)(-5x)$ **8.** $(7xy)(7aby)$ **9.** $(-4x^3p^2)(4y^3x^3)$ **10.** $(-6a^2b^3)(-3ab^3)$

11. $x^7 \cdot x^8$ **12.** $y^6 \cdot y$ **13.** $2x^3 \cdot 5x^7$ **14.** $-3z^4 \cdot 10z^7$

B *Evaluate each expression. See Examples 6 through 9.*

15. -8^0 **16.** $(-9)^0$ **17.** $(4x + 5)^0$ **18.** $8x^0 + 1$ **19.** $(5x)^0 + 5x^0$

20. $4y^0 - (4y)^0$ **21.** $4x^0 + 5$ **22.** $-5x^0$ **23.** $3^0 - 3t^0$ **24.** $4^0 + 4x^0$

25. Explain why $(-5)^0$ simplifies to 1 but -5^0 simplifies to -1.

26. Explain why both $4x^0 - 3y^0$ and $(4x - 3y)^0$ simplify to 1.

C *Use the quotient rule to simplify. See Examples 10 through 13.*

27. $\dfrac{a^5}{a^2}$ **28.** $\dfrac{x^9}{x^4}$ **29.** $\dfrac{x^9y^6}{x^8y^6}$ **30.** $\dfrac{a^{12}b^2}{a^9b}$ **31.** $-\dfrac{26z^{11}}{2z^7}$

32. $\dfrac{16x^5}{8x}$ **33.** $\dfrac{-36a^5b^7c^{10}}{6ab^3c^4}$ **34.** $\dfrac{49a^3bc^{14}}{-7abc^8}$ **35.** $\dfrac{z^{12}}{z^{15}}$ **36.** $\dfrac{x^{11}}{x^{20}}$

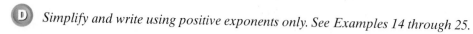

D *Simplify and write using positive exponents only. See Examples 14 through 25.*

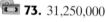 **37.** 4^{-2} **38.** 2^{-3} **39.** $\dfrac{x^7}{x^{15}}$ **40.** $\dfrac{z}{z^3}$ **41.** $5a^{-4}$ **42.** $10b^{-1}$

43. $\dfrac{x^{-2}}{x^5}$ **44.** $\dfrac{y^{-6}}{y^{-9}}$ **45.** $\dfrac{8r^4}{2r^{-4}}$ **46.** $\dfrac{3s^3}{15s^{-3}}$ **47.** $\dfrac{x^{-9}x^4}{x^{-5}}$

48. $\dfrac{y^{-7}y}{y^8}$ **49.** $4^{-1} + 3^{-2}$ **50.** $1^{-3} - 4^{-2}$ **51.** $\dfrac{y^{-3}}{y^{-7}}$ **52.** $\dfrac{z^{-12}}{z^{10}}$

53. $3x^{-1}$ **54.** $(4x)^{-1}$ 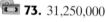 **55.** $\dfrac{r^4}{r^{-4}}$ **56.** $\dfrac{x^{-5}}{x^3}$ **57.** $\dfrac{x^{-7}y^{-2}}{x^2y^2}$

58. $\dfrac{a^{-5}b^7}{a^{-2}b^{-3}}$ **59.** $\dfrac{2a^{-6}b^2}{18ab^{-5}}$ **60.** $\dfrac{18ab^{-6}}{3a^{-3}b^6}$ **61.** $\dfrac{(24x^8)(x)}{20x^{-7}}$ **62.** $\dfrac{(30z^2)(z^5)}{55z^{-4}}$

Simplify. Assume that variables in the exponents represent nonzero integers and that x, y, and z are not 0. See Examples 26 and 27.

63. $x^5 \cdot x^{7a}$ **64.** $y^{2p} \cdot y^{9p}$ **65.** $\dfrac{x^{3t-1}}{x^t}$ **66.** $\dfrac{y^{4p-2}}{y^{3p}}$ **67.** $x^{4a} \cdot x^7$

68. $x^{9y} \cdot x^{-7y}$ **69.** $\dfrac{z^{6x}}{z^7}$ **70.** $\dfrac{y^6}{y^{4z}}$ **71.** $\dfrac{x^{3t} \cdot x^{4t-1}}{x^t}$ **72.** $\dfrac{z^{5x} \cdot z^{x-7}}{z^x}$

E *Write each number in scientific notation. See Examples 28 and 29.*

 73. 31,250,000   **74.** 678,000   **75.** 0.016   **76.** 0.007613   **77.** 67,413

78. 36,800,000   **79.** 0.0125   **80.** 0.00084   **81.** 0.000053   **82.** 98,700,000,000

Write each number in scientific notation.

83. The approximate distance between Jupiter and the sun is 778,300,000 kilometers. (*Source:* National Space Data Center)

84. For the 2001 Major League Baseball season, the Cleveland Indians' opening day payroll was $91,975,000. (*Source:* Associated Press)

85. The estimated world population in 2001 was 6,137,000,000. (*Source:* Population Reference Bureau)

86. Total revenues for Microsoft in fiscal year 2001 were $25,296,000,000. (*Source:* Microsoft Corporation)

87. The New York City subway system carries a total of 1,300,000,000 passengers annually. (*Source:* New York City Transit, Metropolitan Transportation Authority)

88. The temperature of the center of the sun is about 27,000,000°F.

89. A pulsar is a rotating neutron star that gives off sharp, regular pulses of radio waves. For one particular pulsar, the rate of pulses is every 0.001 second.

△ **90.** To convert from cubic inches to cubic meters, multiply by 0.0000164.

Write each number in standard notation. See Examples 30 and 31.

 91. 3.6×10^{-9} **92.** 2.7×10^{-5} 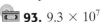 **93.** 9.3×10^{7} **94.** 6.378×10^{8} **95.** 1.278×10^{6}

96. 7.6×10^{4} **97.** 7.35×10^{12} **98.** 1.66×10^{-5} **99.** 4.03×10^{-7} **100.** 8.007×10^{8}

Write each number in standard notation.

101. The estimated world population in 1 A.D. was 2.0×10^{8}. (*Source: 2002 World Almanac and Book of Facts*)

102. There are 3.949×10^{6} miles of highways, roads, and streets in the United States. (*Source:* Bureau of Transportation Statistics)

103. In 2005, teenagers and children are expected to spend 4.9×10^9 dollars on purchases and transactions made online. (*Source:* Jupiter Research)

104. Each day, an estimated 2.0×10^7 adults in America drink gourmet coffee beverages. (*Source:* National Coffee Association)

 Combining Concepts

105. Explain how to convert a number from standard notation to scientific notation.

106. Explain how to convert a number from scientific notation to standard notation.

1.7 More Work with Exponents and Scientific Notation

OBJECTIVES

 Use the power rules for exponents.

 Use exponent rules and definitions to simplify exponential expressions.

 Use scientific notation to compute.

SSM
TUTOR CENTER   SG   CD & VIDEO   MATH PRO   WEB

A Using the Power Rules

The volume of the cube shown whose side measures x^2 units is $(x^2)^3$ cubic units. To simplify an expression such as $(x^2)^3$, we use the definition of a^n:

$$(x^2)^3 = \underbrace{(x^2)(x^2)(x^2)}_{x^2 \text{ is a factor 3 times.}} = x^{2+2+2} = x^6$$

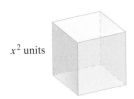

x^2 units

Notice that the result is exactly the same if the exponents are multiplied.

$$(x^2)^3 = x^{2\cdot3} = x^6$$

This suggests that an expression raised to a power that is then all raised to another power is equal to the original expression raised to the product of the powers. Two additional rules for exponents are given in the following box.

Power Rule and Power of a Product or Quotient Rules for Exponents

If a and b are real numbers and m and n are integers, then

$$(a^m)^n = a^{m\cdot n} \qquad \text{Power rule}$$

$$(ab)^m = a^m b^m \qquad \text{Power of a product}$$

$$\left(\frac{a}{b}\right)^n = \frac{a^n}{b^n} \qquad (b \neq 0) \qquad \text{Power of a quotient}$$

EXAMPLES

Use the power rule to simplify each expression. Write each answer using only positive exponents.

1. $(x^5)^7 = x^{5\cdot7} = x^{35}$

2. $(2^2)^3 = 2^{2\cdot3} = 2^6 = 64$

3. $(5^{-1})^2 = 5^{-1\cdot2} = 5^{-2} = \frac{1}{5^2} = \frac{1}{25}$

4. $(y^{-3})^{-4} = y^{-3(-4)} = y^{12}$

Practice Problems 1–4

Use the power rule to simplify each expression. Write each answer using positive exponents only.

1. $(y^2)^8$ 2. $(3^3)^2$
3. $(6^2)^{-1}$ 4. $(x^{-5})^{-7}$

Answers

1. y^{16}, **2.** 729, **3.** $\frac{1}{36}$, **4.** x^{35}

Practice Problems 5–9

Use the power rules to simplify each expression. Write each answer using positive exponents only.

5. $(3x^4)^3$ 6. $\left(\dfrac{4}{5}\right)^2$

7. $\left(\dfrac{4m^5}{n^3}\right)^3$ 8. $\left(\dfrac{2^{-1}}{y}\right)^{-3}$

9. $(a^{-4}b^3c^{-2})^6$

EXAMPLES

Use the power rules to simplify each expression. Write each answer using positive exponents only.

5. $(5x^2)^3 = 5^3 \cdot (x^2)^3 = 5^3 \cdot x^{2\cdot3} = 125x^6$

6. $\left(\dfrac{2}{3}\right)^3 = \dfrac{2^3}{3^3} = \dfrac{8}{27}$

7. $\left(\dfrac{3p^4}{q^5}\right)^2 = \dfrac{(3p^4)^2}{(q^5)^2} = \dfrac{3^2 \cdot (p^4)^2}{(q^5)^2} = \dfrac{9p^8}{q^{10}}$

8. $\left(\dfrac{2^{-3}}{y}\right)^{-2} = \dfrac{(2^{-3})^{-2}}{y^{-2}}$

$= \dfrac{2^6}{y^{-2}} = 64y^2$ Use the negative exponent rule.

9. $(x^{-5}y^2z^{-1})^7 = (x^{-5})^7 \cdot (y^2)^7 \cdot (z^{-1})^7$

$= x^{-35}y^{14}z^{-7} = \dfrac{y^{14}}{x^{35}z^7}$

B Using Exponent Rules to Simplify Expressions

In the next few examples, we practice the use of several of the rules and definitions for exponents. The following is a summary of these rules and definitions.

Summary of Rules for Exponents

If a and b are real numbers and m and n are integers, then

Product rule	$a^m \cdot a^n = a^{m+n}$	
Zero exponent	$a^0 = 1$	$(a \neq 0)$
Negative exponent	$a^{-n} = \dfrac{1}{a^n}$	$(a \neq 0)$
Quotient rule	$\dfrac{a^m}{a^n} = a^{m-n}$	$(a \neq 0)$
Power rule	$(a^m)^n = a^{m\cdot n}$	
Power of a product	$(ab)^m = a^m \cdot b^m$	
Power of a quotient	$\left(\dfrac{a}{b}\right)^m = \dfrac{a^m}{b^m}$	$(b \neq 0)$

Answers

5. $27x^{12}$, **6.** $\dfrac{16}{25}$, **7.** $\dfrac{64m^{15}}{n^9}$, **8.** $8y^3$, **9.** $\dfrac{b^{18}}{a^{24}c^{12}}$

EXAMPLES

Simplify each expression. Write each answer using positive exponents only.

10. $(2x^0y^{-3})^{-2} = 2^{-2}(x^0)^{-2}(y^{-3})^{-2}$

$\qquad = 2^{-2}x^0y^6$

$\qquad = \dfrac{1(y^6)}{2^2}$ Write x^0 as 1.

$\qquad = \dfrac{y^6}{4}$

11. $\left(\dfrac{x^{-5}}{x^{-2}}\right)^{-3} = \dfrac{(x^{-5})^{-3}}{(x^{-2})^{-3}} = \dfrac{x^{15}}{x^6} = x^{15-6} = x^9$

12. $\left(\dfrac{2}{7}\right)^{-2} = \dfrac{2^{-2}}{7^{-2}} = \dfrac{7^2}{2^2} = \dfrac{49}{4}$

13. $\dfrac{5^{-2}x^{-3}y^{11}}{x^2y^{-5}} = (5^{-2})\left(\dfrac{x^{-3}}{x^2}\right)\left(\dfrac{y^{11}}{y^{-5}}\right) = 5^{-2}x^{-3-2}y^{11-(-5)} = 5^{-2}x^{-5}y^{16}$

$\qquad = \dfrac{y^{16}}{5^2x^5} = \dfrac{y^{16}}{25x^5}$

EXAMPLES

Simplify each expression. Write each answer using positive exponents only.

14. $\left(\dfrac{3x^2y}{y^{-9}z}\right)^{-2} = \left(\dfrac{3x^2y^{10}}{z}\right)^{-2} = \dfrac{3^{-2}x^{-4}y^{-20}}{z^{-2}} = \dfrac{z^2}{3^2x^4y^{20}} = \dfrac{z^2}{9x^4y^{20}}$

15. $\left(\dfrac{3a^2}{2x^{-1}}\right)^3\left(\dfrac{x^{-3}}{4a^{-2}}\right)^{-1} = \dfrac{27a^6}{8x^{-3}}\cdot\dfrac{x^3}{4^{-1}a^2}$

$\qquad = \dfrac{27\cdot 4\cdot a^6x^3x^3}{8\cdot a^2} = \dfrac{27a^4x^6}{2}$

EXAMPLES

Simplify. Assume that a and b are integers and that x and y are not 0.

16. $x^{-b}(2x^b)^2 = x^{-b}2^2x^{2b} = 4x^{-b+2b} = 4x^b$

17. $\dfrac{(y^{3a})^2}{y^{a-6}} = \dfrac{y^{2(3a)}}{y^{a-6}} = \dfrac{y^{6a}}{y^{a-6}} = y^{6a-(a-6)} = y^{6a-a+6} = y^{5a+6}$

C   Using Scientific Notation to Compute

To perform operations on numbers written in scientific notation, we use the properties of exponents.

Practice Problems 10–13

Simplify each expression. Write each answer using positive exponents only.

10. $(7xy^{-2})^{-2}$

11. $\left(\dfrac{y^{-7}}{y^{-10}}\right)^{-3}$

12. $\left(\dfrac{3}{5}\right)^{-2}$

13. $\dfrac{6^{-2}x^{-4}y^{10}}{x^2y^{-6}}$

Practice Problems 14–15

Simplify each expression. Write each answer using positive exponents only.

14. $\left(\dfrac{4a^3b^2}{b^{-6}c}\right)^{-2}$

15. $\left(\dfrac{4x^3}{3y^{-1}}\right)^3\left(\dfrac{y^{-2}}{3x^{-1}}\right)^{-1}$

Practice Problems 16–17

Simplify. Assume that m and n are integers and that x and y are not 0.

16. $x^{-n}(3x^n)^2$

17. $\dfrac{(y^{2m})^2}{y^{m-3}}$

Answers

10. $\dfrac{y^4}{49x^2}$, **11.** $\dfrac{1}{y^9}$, **12.** $\dfrac{25}{9}$, **13.** $\dfrac{y^{16}}{36x^6}$,

14. $\dfrac{c^2}{16a^6b^{16}}$, **15.** $\dfrac{64x^8y^5}{9}$, **16.** $9x^n$, **17.** y^{3m+3}

Practice Problems 18–19

Perform each indicated operation. Write each answer in scientific notation.

18. $(9.6 \times 10^6)(4 \times 10^{-8})$

19. $\dfrac{4.2 \times 10^7}{7 \times 10^{-3}}$

EXAMPLES

Perform each indicated operation. Write each answer in scientific notation.

18. $(8.1 \times 10^5)(5 \times 10^{-7}) = 8.1 \times 5 \times 10^5 \times 10^{-7}$

$$= 40.5 \times 10^{-2}$$
$$= (4.05 \times 10^1) \times 10^{-2}$$
$$= 4.05 \times 10^{-1}$$

19. $\dfrac{1.2 \times 10^4}{3 \times 10^{-2}} = \left(\dfrac{1.2}{3}\right)\left(\dfrac{10^4}{10^{-2}}\right) = 0.4 \times 10^{4-(-2)}$

$$= 0.4 \times 10^6 = (4 \times 10^{-1}) \times 10^6 = 4 \times 10^5$$

Practice Problem 20

Use scientific notation to simplify:

$$\frac{3000 \times 0.000012}{400}$$

EXAMPLE 20   Use scientific notation to simplify: $\dfrac{2000 \times 0.000021}{700}$

Solution:

$$\frac{2000 \times 0.000021}{700} = \frac{(2 \times 10^3)(2.1 \times 10^{-5})}{7 \times 10^2} = \frac{2(2.1)}{7} \cdot \frac{10^3 \cdot 10^{-5}}{10^2}$$
$$= 0.6 \times 10^{-4}$$
$$= (6 \times 10^{-1}) \times 10^{-4}$$
$$= 6 \times 10^{-5}$$

Answers

18. 3.84×10^{-1},   **19.** 6×10^9,   **20.** 9×10^{-5}

Mental Math

Simplify. See Examples 1 through 4.

1. $(x^4)^5$ **2.** $(5^6)^2$ **3.** $x^4 \cdot x^5$ **4.** $x^7 \cdot x^8$ **5.** $(y^6)^7$

6. $(x^3)^4$ **7.** $(z^4)^9$ **8.** $(z^3)^7$ **9.** $(z^{-6})^{-3}$ **10.** $(y^{-4})^{-2}$

EXERCISE SET 1.7

(A) *Simplify. Write each answer using positive exponents only. See Examples 1 through 9.*

1. $(3^{-1})^2$ **2.** $(2^{-2})^2$ **3.** $(x^4)^{-9}$ **4.** $(y^7)^{-3}$ **5.** $(y)^{-5}$

6. $(z^{-1})^{10}$ **7.** $(3x^2y^3)^2$ **8.** $(4x^3yz)^2$ **9.** $\left(\dfrac{2x^5}{y^{-3}}\right)^4$ **10.** $\left(\dfrac{3a^{-4}}{b^7}\right)^3$

11. $(a^2bc^{-3})^{-6}$ **12.** $(6x^{-6}y^7z^0)^{-2}$ **13.** $\left(\dfrac{x^7y^{-3}}{z^{-4}}\right)^{-5}$ **14.** $\left(\dfrac{a^{-2}b^{-5}}{c^{-11}}\right)^{-6}$ **15.** $(5^{-1})^3$

16. $(8^2)^{-1}$ **17.** $(x^7)^{-9}$ **18.** $(y^{-4})^5$ **19.** $\left(\dfrac{7}{8}\right)^3$ **20.** $\left(\dfrac{4}{3}\right)^2$

21. $(4x^2)^2$ **22.** $(-8x^3)^2$ **23.** $(-2^{-2}y)^3$ **24.** $(-4^{-6}y^{-6})^{-4}$ **25.** $\left(\dfrac{4^{-4}}{y^3x}\right)^{-2}$

26. $\left(\dfrac{7^{-3}}{ab^2}\right)^{-2}$ **27.** $\left(\dfrac{2x^{-3}}{y^{-1}}\right)^{-3}$ **28.** $\left(\dfrac{n^5}{2m^{-2}}\right)^{-4}$

(B) *Simplify. Write each answer using positive exponents only. See Examples 10 through 15.*

29. $\left(\dfrac{a^{-4}}{a^{-5}}\right)^{-2}$ **30.** $\left(\dfrac{x^{-9}}{x^{-4}}\right)^{-3}$ **31.** $\left(\dfrac{2a^{-2}b^5}{4a^2b^7}\right)^{-2}$ **32.** $\left(\dfrac{5x^7y^4}{10x^3y^{-2}}\right)^{-3}$

33. $\dfrac{4^{-1}x^2yz}{x^{-2}yz^3}$ **34.** $\dfrac{8^{-2}x^{-3}y^{11}}{x^2y^{-5}}$ **35.** $\left(\dfrac{6p^6}{p^{12}}\right)^2$ **36.** $\left(\dfrac{4p^6}{p^9}\right)^3$

37. $(-8y^3xa^{-2})^{-3}$ **38.** $(-xy^0x^2a^3)^{-3}$ **39.** $\left(\dfrac{x^{-2}y^{-2}}{a^{-3}}\right)^{-7}$ **40.** $\left(\dfrac{x^{-1}y^{-2}}{5^{-3}}\right)^{-5}$

41. $\left(\dfrac{3x^5}{6x^4}\right)^4$

42. $\left(\dfrac{8^{-3}}{y^2}\right)^{-2}$

43. $\left(\dfrac{1}{4}\right)^{-3}$

44. $\left(\dfrac{1}{8}\right)^{-2}$

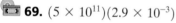 **45.** $\dfrac{(y^3)^{-4}}{y^3}$

46. $\dfrac{2(y^3)^{-3}}{y^{-3}}$

47. $\dfrac{3^{-2}a^{-5}b^6}{4^{-2}a^{-7}b^{-3}}$

48. $\dfrac{2^{-3}m^{-4}n^{-5}}{5^{-2}m^{-5}n}$

49. $(4x^6y^5)^{-2}(6x^4y^3)$

50. $(5xy)^3(z^{-2})^{-3}$

51. $x^6(x^6bc)^{-6}$

52. $2(y^2b)^{-4}$

53. $\dfrac{2^{-3}x^2y^{-5}}{5^{-2}x^7y^{-1}}$

54. $\dfrac{7^{-1}a^{-3}b^5}{a^2b^{-2}}$

55. $\left(\dfrac{2x^2}{y^4}\right)^3\left(\dfrac{2x^5}{y}\right)^{-2}$

56. $\left(\dfrac{3z^{-2}}{y}\right)^2\left(\dfrac{9y^{-4}}{z^{-3}}\right)^{-1}$

Simplify. Assume that variables in the exponents represent nonzero integers and that all other variables are not 0. See Examples 16 and 17.

57. $(x^{3a+6})^3$

58. $(x^{2b+7})^2$

59. $\dfrac{x^{4a}(x^{4a})^3}{x^{4a-2}}$

60. $\dfrac{x^{-5y+2}x^{2y}}{x}$

61. $(b^{5x-2})^{2x}$

62. $(c^{2a+3})^3$

63. $\dfrac{(y^{2a})^8}{y^{a-3}}$

64. $\dfrac{(y^{4a})^7}{y^{2a-1}}$

65. $\left(\dfrac{2x^{3t}}{x^{2t-1}}\right)^4$

66. $\left(\dfrac{3y^{5a}}{y^{-a+1}}\right)^2$

67. $\dfrac{25x^{2a+b}y^{2a-b}}{5x^{a-b}y^{a+b}}$

68. $\dfrac{16x^{3a-b}y^{4a+b}}{2x^{a-2b}y^{a+3b}}$

C *Perform each indicated operation. Write each answer in scientific notation. See Examples 18 through 20.*

69. $(5 \times 10^{11})(2.9 \times 10^{-3})$

70. $(3.6 \times 10^{-12})(6 \times 10^9)$

71. $(2 \times 10^5)^3$

72. $(3 \times 10^{-7})^3$

73. $\dfrac{3.6 \times 10^{-4}}{9 \times 10^2}$

74. $\dfrac{1.2 \times 10^9}{2 \times 10^{-5}}$

75. $\dfrac{0.0069}{0.023}$

76. $\dfrac{0.00048}{0.0016}$

77. $\dfrac{18,200 \times 100}{91,000}$

78. $\dfrac{0.0003 \times 0.0024}{0.0006 \times 20}$

79. $\dfrac{6000 \times 0.006}{0.009 \times 400}$

80. $\dfrac{0.00016 \times 300}{0.064 \times 100}$

81. $\dfrac{0.00064 \times 2000}{16,000}$

82. $\dfrac{0.00072 \times 0.003}{0.00024}$

83. $\dfrac{66,000 \times 0.001}{0.002 \times 0.003}$

84. $\dfrac{0.0007 \times 11,000}{0.001 \times 0.0001}$

85. $\dfrac{1.25 \times 10^{15}}{(2.2 \times 10^{-2})(6.4 \times 10^{-5})}$

86. $\dfrac{(2.6 \times 10^{-3})(4.8 \times 10^{-4})}{1.3 \times 10^{-12}}$

Solve.

87. A computer can add two numbers in about 10^{-8} second. Express in scientific notation how long it would take this computer to do this task 200,000 times.

88. To convert from square inches to square meters, multiply by 6.452×10^{-4}. The area of the following square is 4×10^{-2} square inches. Convert this area to square meters.

4×10^{-2} sq in.

89. To convert from cubic inches to cubic meters, multiply by 1.64×10^{-5}. A grain of salt is in the shape of a cube. If an average size of a grain of salt is 3.8×10^{-6} cubic inches, convert this volume to cubic meters.

 Combining Concepts

90. Each side of the cube shown is $\dfrac{2x^{-2}}{y}$ meters. Find its volume.

$\dfrac{2x^{-2}}{y}$ m

91. The lot shown is in the shape of a parallelogram with base $\dfrac{3x^{-1}}{y^{-3}}$ feet and height $5x^{-7}$ feet. Find its area.

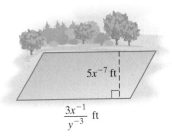

$5x^{-7}$ ft

$\dfrac{3x^{-1}}{y^{-3}}$ ft

92. The density D of an object is equivalent to the quotient of its mass M and volume V. Thus $D = \dfrac{M}{V}$.

Express in scientific notation the density of an object whose mass is 500,000 pounds and whose volume is 250 cubic feet.

93. The density of ordinary water is 3.12×10^{-2} tons per cubic foot. The volume of water in the largest of the Great Lakes, Lake Superior, is 4.269×10^{14} cubic feet. Use the formula $D = \dfrac{M}{V}$ (see Exercise 92) to find the mass (in tons) of the water in Lake Superior. Express your answer in scientific notation. (*Source:* National Ocean Service)

94. Is there a number a such that $a^{-1} = a^1$? If so, give the value of a.

95. Is there a number a such that a^{-2} is a negative number? If so, give the value of a.

96. Explain whether 0.4×10^{-5} is written in scientific notation.

97. The estimated population of the United States in 2001 was 2.854×10^8 people. The land area of the United States is 3.536×10^6 square miles. Find the population density (number of people per square mile) for the United States in 2001. Round to the nearest whole number. (*Source:* U.S. Census Bureau)

98. In August 2001, the value of goods and services imported into the United States was $\$1.116 \times 10^{11}$. The estimated population of the United States in 2001 was 2.854×10^8 people. Find the average value of imports per person in the United States for August 2001. Round to the nearest dollar. (*Sources:* U.S. Census Bureau, Bureau of Economic Analysis)

99. In 2001, the population of Japan was 1.271×10^8 people. At the same time, the population of Oceania (including the countries of Australia, Fiji, New Zealand, etc.) was 3.1×10^7 people. How many times greater was the population of Japan than the population of Oceania? (*Source:* Population Reference Bureau)

This activity may be completed by working in groups or individually.

Professor Jakow Trachtenberg was the founder of the Mathematical Institute in Zurich, Switzerland, in 1950. During World War II, he spent seven years in various Nazi concentration camps. To keep himself occupied and mentally sharp while a prisoner, he developed his own speed system of basic mathematics. Trachtenberg worked out his system of simplified mental arithmetic entirely without the use of pencil and paper. After escaping from a labor camp in Trieste in 1945, Trachtenberg fled to Switzerland and completed his speed system of mathematics. He began teaching his system of addition, subtraction, multiplication, and division to children who had trouble learning conventionally taught mathematics. His success lead him to found the Mathematical Institute in Zurich.

Trachtenberg's speed system is based on certain patterns that occur when basic operations are performed. For example, he developed a different rule for multiplying any number by each of the factors 2 through 12. The following steps describe how to quickly multiply any number by 11.

TRACHTENBERG'S RULE FOR MULTIPLICATION BY ELEVEN

a. The last digit of the product is the last digit of the number being multiplied by 11.
b. To find the middle digits in the product, add each digit of the number being multiplied by 11 to its immediate "neighbor" to the right. (It may be necessary to add a carried 1 to the result at any stage.)
c. The first digit of the product is the first digit of the number being multiplied by 11. (Adding a carried 1 to this digit may be necessary.)

Example $35{,}924 \times 11$

```
      35,924
  ×       11
           4   (last digit of 35,924)
           6   (sum of 2 and its neighbor 4 in 35,924)
           1   (sum of 9 and its neighbor 2; carry 1)
           5   (sum of 5 and its neighbor 9 + 1; carry 1)
           9   (sum of 3 and its neighbor 5 + 1)
           3   (first digit of 35,924)
     395,164   Product
```

1. Use the Trachtenberg rule to find each product. Then check each result with a calculator.
   a. 7234×11
   b. $362{,}713 \times 11$
   c. $4{,}386{,}275 \times 11$
   d. 5845×11

2. Develop a Trachtenberg-like rule for finding the product of any number and 2. Begin by looking for patterns when numbers like 14, 19, 28, 374, 621, and 314,672 are multiplied by 2. What relationship do you observe between the digits of the product and the digits of the number being multiplied by 2?

3. Develop a Trachtenberg-like rule for finding the product of any number and 12. Begin by looking for patterns when numbers like 18, 28, 71, 89, 123, and 456 are multiplied by 12. Investigate additional products as necessary. What relationship do you observe between the digits of the product and the digits of the number being multiplied by 12? (*Hint:* To make your rule work out, you may find it helpful to add a place-holding 0 in front of each number being multiplied by 12. For instance, you may want to consider the numbers 018, 028, 071, 089, 0123, and 0456.)

Chapter 1 VOCABULARY CHECK

Fill in each blank with one of the words or phrases listed below.

distributive	absolute value	inequality	algebraic expression
real	opposite	commutative	exponent
reciprocals	associative	whole	variable

1. A(n) _____ is formed by numbers and variables connected by the operations of addition, subtraction, multiplication, division, raising to powers, and/or taking roots.
2. The _____ of a number a is $-a$.
3. $3(x - 6) = 3x - 18$ by the _____ property.
4. The _____ of a number is the distance between that number and 0 on the number line.
5. A(n) _____ is a shorthand notation for repeated multiplication of the same factor.
6. A letter that represents a number is called a _____.
7. The symbols $<$ and $>$ are called _____ symbols.
8. If a is not 0, then a and $\frac{1}{a}$ are called _____.
9. $A + B = B + A$ by the _____ property.
10. $(A + B) + C = A + (B + C)$ by the _____ property.
11. The numbers $0, 1, 2, 3, \ldots$ are called _____ numbers.
12. If a number corresponds to a point on the number line, we know that number is a _____ number.

CHAPTER 1 Highlights

DEFINITIONS AND CONCEPTS

EXAMPLES

Section 1.2 Sets of Numbers

Letters that represent numbers are called **variables**.

An **algebraic expression** is formed by numbers and variables connected by the operations of addition, subtraction, multiplication, division, raising to powers, or taking roots.

Natural numbers: $\{1, 2, 3, \ldots\}$

Whole numbers: $\{0, 1, 2, 3, \ldots\}$

Integers: $\{\ldots, -3, -2, -1, 0, 1, 2, 3, \ldots\}$
Each listing of three dots is called an **ellipsis**. The members of a set are called its **elements**.

Set builder notation describes the elements of a set but does not list them.

Real numbers: $\{x \mid x$ corresponds to a point on the number line$\}$

Rational numbers: $\left\{ \frac{a}{b} \mid a \text{ and } b \text{ are integers and } b \neq 0 \right\}$

Irrational numbers: $\{x \mid x$ is a real number and x is not a rational number$\}$

If all the elements of set A are also in set B, we say that set A is a **subset** of set B.

$x, \quad a, \quad m, \quad y$

$$7y, \quad -3, \quad \frac{x^2 - 9}{-2} + 14x, \quad \sqrt{3} + \sqrt{m}$$

Given the set $\left\{ -9.6, -5, -\sqrt{2}, 0, \frac{2}{5}, 101 \right\}$, list the elements that belong to the set of

Natural numbers	101
Whole numbers	$0, 101$
Integers	$-5, 0, 101$
Real numbers	$-9.6, -5, -\sqrt{2}, 0, \frac{2}{5}, 101$
Rational numbers	$-9.6, -5, 0, \frac{2}{5}, 101$
Irrational numbers	$-\sqrt{2}$

Write the set $\{x \mid x$ is an integer between -2 and $5\}$ in roster form.

$$\{-1, 0, 1, 2, 3, 4\}$$

The set of integers is a subset of the set of rational numbers.

Section 1.3   Properties of Real Numbers

SYMBOLS

$=$	is equal to
$\neq$	is not equal to
$>$	is greater than
$<$	is less than
$\geq$	is greater than or equal to
$\leq$	is less than or equal to

$$-5 = -5$$
$$-5 \neq -3$$
$$1.7 > 1.2$$
$$-1.7 < -1.2$$
$$\frac{5}{3} \geq \frac{5}{3}$$
$$-\frac{1}{2} \leq \frac{1}{2}$$

IDENTITY

$$a + 0 = a \qquad 0 + a = a$$
$$a \cdot 1 = a \qquad 1 \cdot a = a$$

$$3 + 0 = 3 \qquad 0 + 3 = 3$$
$$-1.8 \cdot 1 = -1.8 \qquad 1 \cdot -1.8 = -1.8$$

INVERSE

$$a + (-a) = 0 \qquad -a + a = 0$$
$$a \cdot \frac{1}{a} = 1 \qquad \frac{1}{a} \cdot a = 1$$

$$7 + (-7) = 0 \qquad -7 + 7 = 0$$
$$5 \cdot \frac{1}{5} = 1 \qquad \frac{1}{5} \cdot 5 = 1$$

COMMUTATIVE

$$a + b = b + a$$
$$a \cdot b = b \cdot a$$

$$x + 7 = 7 + x$$
$$9 \cdot y = y \cdot 9$$

ASSOCIATIVE

$$(a + b) + c = a + (b + c)$$
$$(a \cdot b) \cdot c = a \cdot (b \cdot c)$$

$$(3 + 1) + 10 = 3 + (1 + 10)$$
$$(3 \cdot 1) \cdot 10 = 3 \cdot (1 \cdot 10)$$

DISTRIBUTIVE

$$a(b + c) = ab + ac$$

$$6(x + 5) = 6 \cdot x + 6 \cdot 5$$
$$= 6x + 30$$

Section 1.4   Operations on Real Numbers

ABSOLUTE VALUE

$$|a| = \begin{cases} a \text{ if } a \text{ is 0 or a positive number} \\ -a \text{ if } a \text{ is a negative number} \end{cases}$$

$$|3| = 3, \quad |0| = 0, \quad |-7.2| = 7.2$$

ADDING REAL NUMBERS

1. To add two numbers with the same sign, add their absolute values and attach their common sign.
2. To add two numbers with different signs, subtract the smaller absolute value from the larger absolute value and attach the sign of the number with the larger absolute value.

$$\frac{2}{7} + \frac{1}{7} = \frac{3}{7}$$
$$-5 + (-2.6) = -7.6$$
$$-18 + 6 = -12$$
$$20.8 + (-10.2) = 10.6$$

| **DEFINITIONS AND CONCEPTS** | **EXAMPLES** |

Section 1.4   Operations on Real Numbers *(continued)*

SUBTRACTING REAL NUMBERS

$$a - b = a + (-b)$$

$$18 - 21 = 18 + (-21) = -3$$

MULTIPLYING AND DIVIDING REAL NUMBERS

The product or quotient of two numbers with the same sign is positive.

$$(-8)(-4) = 32 \qquad \frac{-8}{-4} = 2$$

$$8 \cdot 4 = 32 \qquad \frac{8}{4} = 2$$

The product or quotient of two numbers with different signs is negative.

$$-17 \cdot 2 = -34 \qquad \frac{-14}{2} = -7$$

$$4(-1.6) = -6.4 \qquad \frac{22}{-2} = -11$$

A natural number **exponent** is a shorthand notation for repeated multiplication of the same factor.

$$3^4 = 3 \cdot 3 \cdot 3 \cdot 3 = 81$$

The notation $\sqrt{a}$ is used to denote the **positive**, or **principal square root** of a nonnegative number a.

$$\sqrt{a} = b \text{ if } b^2 = a \text{ and } b \text{ is positive}$$

$$\sqrt{49} = 7$$

Also,

$$\sqrt[3]{a} = b \text{ if } b^3 = a$$
$$\sqrt[4]{a} = b \text{ if } b^4 = a \text{ and } b \text{ is positive}$$

$$\sqrt[3]{64} = 4$$

$$\sqrt[4]{16} = 2$$

Section 1.5   Order of Operations and Algebraic Expressions

ORDER OF OPERATIONS

Simplify expressions using the order that follows. If grouping symbols such as parentheses are present, simplify expressions within those first, starting with the innermost set. If fraction bars are present, simplify the numerator and denominator separately.

Simplify: $\dfrac{42 - 2(3^2 - \sqrt{16})}{-8}$

1. Evaluate exponential expressions.
2. Multiply or divide in order from left to right.
3. Add or subtract in order from left to right.

$$\frac{42 - 2(3^2 - \sqrt{16})}{-8} = \frac{42 - 2(9 - 4)}{-8}$$

$$= \frac{42 - 2(5)}{-8}$$

$$= \frac{42 - 10}{-8}$$

$$= \frac{32}{-8} = -4$$

To **evaluate** an algebraic expression containing variables, substitute the given numbers for the variables and simplify. The result is called the **value** of the expression.

Evaluate: $2.7x$ when $x = 3$

$$2.7x = 2.7(3)$$

$$= 8.1$$

Section 1.6   Exponents and Scientific Notation

PRODUCT RULE $a^m \cdot a^n = a^{m+n}$

ZERO EXPONENT $a^0 = 1$ $(a \neq 0)$

QUOTIENT RULE $\dfrac{a^m}{a^n} = a^{m-n}$ $(a \neq 0)$

NEGATIVE EXPONENT $a^{-n} = \dfrac{1}{a^n}$ $(a \neq 0)$

A positive number is written in **scientific notation** if it is written as the product of a number a, where $1 \leq a < 10$, and an integer power of 10: $a \times 10^n$.

$x^2 \cdot x^3 = x^5$

$7^0 = 1, (-10)^0 = 1$

$\dfrac{y^{10}}{y^4} = y^{10-4} = y^6$

$3^{-2} = \dfrac{1}{3^2} = \dfrac{1}{9}, \dfrac{x^{-5}}{x^{-7}} = x^{-5-(-7)} = x^2$

Numbers written in scientific notation:

$$568{,}000 = 5.68 \times 10^5$$

$$0.0002117 = 2.117 \times 10^{-4}$$

Section 1.7   More Work with Exponents and Scientific Notation

POWER RULES

$(a^m)^n = a^{m \cdot n}$

$(ab)^m = a^m b^m$

$\left(\dfrac{a}{b}\right)^n = \dfrac{a^n}{b^n}$

$(7^8)^2 = 7^{16}$

$(2y)^3 = 2^3 y^3 = 8y^3$

$\left(\dfrac{5x^{-3}}{x^2}\right)^{-2} = \dfrac{5^{-2} x^6}{x^{-4}}$

$= 5^{-2} \cdot x^{6-(-4)}$

$= \dfrac{x^{10}}{5^2}, \text{or } \dfrac{x^{10}}{25}$

Chapter 1 Review

(1.2) Write each as an expression. Use x to represent the number.

1. The quotient of a number and seven

2. The product of a number and seven

3. Four times the sum of a number and ten

4. The difference of three times a number, and nine

Write each set in roster form.

5. $\{x | x$ is an odd integer between -2 and $4\}$

6. $\{x | x$ is an even integer between -3 and $7\}$

7. $\{x | x$ is a negative whole number$\}$

8. $\{x | x$ is a natural number that is not a rational number$\}$

9. $\{x | x$ is a whole number greater than $5\}$

10. $\{x | x$ is an integer less than $3\}$

Determine whether each statement is true or false if $A = \{6, 10, 12\}, B = \{5, 9, 11\}, C = \{\ldots, -3, -2, -1, 0, 1, 2, 3, \ldots\}$, $D = \{2, 4, 6, \ldots, 16\}, E = \{x | x$ is a rational number$\}, F = \{\ \}, G = \{x | x$ is an irrational number$\}$, and $H = \{x | x$ is a real number$\}$.

11. $10 \in D$

12. $59 \in B$

13. $\sqrt{169} \notin A$

14. $0 \notin F$

15. $\pi \in E$

16. $\pi \in H$

17. $\sqrt{4} \in G$

18. $-9 \in C$

List the elements of the set $\left\{5, -\dfrac{2}{3}, \dfrac{8}{2}, \sqrt{9}, 0.3, \sqrt{7}, 1\dfrac{5}{8}, -1, \pi\right\}$ that are also elements of each given set.

19. Whole numbers

20. Natural numbers

21. Rational numbers

22. Irrational numbers

23. Real numbers

24. Integers

25. Twelve is the product of *x* and negative 4.

26. The sum of *n* and twice *n* is negative fifteen.

27. Four times the sum of *y* and three is −1.

28. The difference of *t* and five, multiplied by six is four.

29. Seven subtracted from *z* is six.

30. Ten less than the product of *x* and nine is five.

31. The difference of *x* and 5 is the same as 12.

32. The opposite of four is equal to the product of *y* and seven.

33. Two-thirds is equal to twice the sum of *n* and one-fourth.

34. The sum of *t* and six amounts to negative twelve.

Find the opposite, or additive inverse, of each number.

35. $-\dfrac{3}{4}$ **36.** 0.6   **37.** 0   **38.** 1

Find the reciprocal, or multiplicative inverse, of each number.

39. $-\dfrac{3}{4}$ **40.** 0.6   **41.** 0   **42.** 1

Name each property illustrated.

43. $(M + 5) + P = M + (5 + P)$

44. $5(3x - 4) = 15x - 20$

45. $(-4) + 4 = 0$

46. $(3 + x) + 7 = 7 + (3 + x)$

47. $(XY)Z = (YZ)X$

48. $\left(-\dfrac{3}{5}\right) \cdot \left(-\dfrac{5}{3}\right) = 1$

49. $T \cdot 0 = 0$

50. $(ab)c = a(bc)$

51. $A + 0 = A$

52. $8 \cdot 1 = 8$

Complete each equation using the given property.

53. $5(x - 3z) =$ _____ Distributive property

54. $(7 + y) + (3 + x) =$ _____ Commutative property

55. $0 =$ _____ Additive inverse property

56. $1 =$ _____ Multiplicative inverse property

57. $[(3.4)(0.7)]5 =$ _____ Associative property

58. $7 =$ _____ Additive identity property

Insert $<, >,$ or $=$ to make each statement true.

59. -9 _____ -12

60. 0 _____ -6

61. -3 _____ -1

62. 7 _____ $|-7|$

63. -5 _____ $-(-5)$

64. $-(-2)$ _____ -2

Use the distributive property to multiply.

65. $-3(2x - 7y)$

66. $-9(10a + 4b)$

67. $\dfrac{1}{2}(18m - 8n + 1)$

68. $\dfrac{1}{3}(12x - 33y + 2)$

(1.4) *Simplify.*

69. $-7 + 3$

70. $-10 + (-25)$

71. $5(-0.4)$

72. $(-3.1)(-0.1)$

73. $-7 - (-15)$

74. $9 - (-4.3)$

75. $\sqrt{16} - 2^3$

76. $\sqrt[3]{27} - 5^2$

77. $(-24) \div 0$

78. $0 \div (-45)$

79. $(-36) \div (-9)$

80. $(60) \div (-12)$

81. $\left(-\dfrac{4}{5}\right) - \left(-\dfrac{2}{3}\right)$

82. $\left(\dfrac{5}{4}\right) - \left(-2\dfrac{3}{4}\right)$

△ **83.** Determine the unknown fractional part.

84. The Bertha Rogers gas well in Washita County, Oklahoma, is the deepest well in the United States. From the surface, this now-capped well extends 31,441 feet into the earth. The elevation of the nearby Cordell Municipal Airport is 1589 feet above sea level. Assuming that the surface elevation of the well is the same as at the Cordell Municipal Airport, find the elevation relative to sea level of the *bottom* of the Bertha Rogers gas well. (*Sources:* U.S. Geological Survey, Oklahoma Department of Transportation)

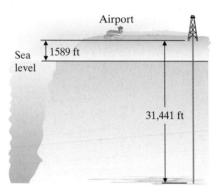

(1.5) *Simplify.*

85. $-5 + 7 - 3 - (-10)$

86. $8 - (-3) + (-4) + 6$

87. $3(4 - 5)^4$

88. $6(7 - 10)^2$

89. $\left(-\dfrac{8}{15}\right) \cdot \left(-\dfrac{2}{3}\right)^2$

90. $\left(-\dfrac{3}{4}\right)^2 \cdot \left(-\dfrac{10}{21}\right)$

91. $-\dfrac{6}{15} \div \dfrac{8}{25}$

92. $\dfrac{4}{9} \div -\dfrac{8}{45}$

93. $-\dfrac{3}{8} + 3(2) \div 6$

94. $5(-2) - (-3) - \dfrac{1}{6} + \dfrac{2}{3}$

95. $|2^3 - 3^2| - |5 - 7|$

96. $|5^2 - 2^2| + |9 \div (-3)|$

97. $(2^3 - 3^2) - (5 - 7)$

98. $(5^2 - 2^2) + [9 \div (-3)]$

99. $\dfrac{(8 - 10)^3 - (-4)^2}{2 + 8(2) \div 4}$

100. $\dfrac{(2+4)^2 + (-1)^5}{12 \div 2 \cdot 3 - 3}$

101. $\dfrac{(4-9)+4-9}{10-12 \div 4 \cdot 8}$

102. $\dfrac{3-7-(7-3)}{15+30 \div 6 \cdot 2}$

103. $\dfrac{\sqrt{25}}{4+3 \cdot 7}$

104. $\dfrac{\sqrt{64}}{24-8 \cdot 2}$

△ *The algebraic expression $2\pi r$ represents the circumference of (distance around) a circle of radius r.*

△ **105.** Complete the table by evaluating the expression at the given values of *r*. (Use 3.14 for π.)

Radius	r	1	10	100
Circumference	$2\pi r$			

106. As the radius of a circle increases, does the circumference of the circle increase or decrease?

△

Simplify.

107. $14x - 3 - 11x - 10$

108. $81y + 19 - y - 20$

109. $7a - 3(2a - y) + 4y - 6$

110. $9b - 4(8b - x) + 9x - 10$

111. $\dfrac{1}{5}(15m - 5n) - (3m - 5) + 2n$

112. $-(2x - 9) + \dfrac{1}{4}(8x + 4y) - 13$

(1.6) *Evaluate.*

113. $(-2)^2$

114. $(-3)^4$

115. -2^2

116. -3^4

117. 8^0

118. -9^0

119. -4^{-2}

120. $(-4)^{-2}$

Simplify each expression. Write each answer with positive exponents only.

121. $-xy^2 \cdot y^3 \cdot xy^2z$

122. $(-4xy)(-3xy^2b)$

123. $a^{-14} \cdot a^5$

124. $\dfrac{a^{16}}{a^{17}}$

125. $\dfrac{x^{-7}}{x^4}$

126. $\dfrac{9a(a^{-3})}{18a^{15}}$

127. $\dfrac{y^{6p-3}}{y^{6p+2}}$

128. $(3x^{2a+b}y^{-3b})^2$

Write each number in scientific notation.

129. 36,890,000

130. -0.000362

Write each number without exponents.

131. 1.678×10^{-6}

132. 4.1×10^5

(1.7) *Simplify. Write each answer with positive exponents only.*

133. $(8^5)^3$

134. $\left(\dfrac{a}{4}\right)^2$

135. $(3x)^3$

136. $(-4x)^{-2}$

137. $\left(\dfrac{6x}{5}\right)^2$

138. $(8^6)^{-3}$

139. $\left(\dfrac{4}{3}\right)^{-2}$

140. $(-2x^3)^{-3}$

141. $\left(\dfrac{8p^6}{4p^4}\right)^{-2}$

142. $(-3x^{-2}y^2)^3$

143. $\left(\dfrac{x^{-5}y^{-3}}{z^3}\right)^{-5}$

144. $\dfrac{4^{-1}x^3yz}{x^{-2}yx^4}$

145. $(5xyz)^{-4}(x^{-2})^{-3}$

146. $\dfrac{2(3yz)^{-3}}{y^{-3}}$

Simplify each expression.

147. $x^{4a}(3x^{5a})^3$

148. $\dfrac{4y^{3x-3}}{2y^{2x+4}}$

Chapter 1 Test

Determine whether each statement is true or false.

1. $-2.3 > 2.33$

2. $-6^2 = (-6)^2$

3. $-5 - 8 = -(5 - 8)$

4. $(-2)(-3)(0) = \dfrac{(-4)}{0}$

5. All natural numbers are integers.

6. All rational numbers are integers.

Simplify.

7. $5 - 12 \div 3(2)$

8. $|4 - 6|^3 - (1 - 6^2)$

9. $(4 - 9)^3 - |-4 - 6|^2$

10. $[3|4 - 5|^5 - (-9)] \div (-6)$

11. $\dfrac{6(7 - 9)^3 + (-2)}{(-2)(-5)(-5)}$

Evaluate each expression when $q = 4$, $r = -2$, and $t = 1$.

12. $q^2 - r^2$

13. $\dfrac{5t - 3q}{3r - 1}$

14. The algebraic expression $5.75x$ represents the total cost for x adults to attend the theater.
   a. Complete the table that follows.

   b. As the number of adults increases does the total cost increase or decrease?

Adults	x	1	3	10	20
Total Cost	$5.75x$				

Write each statement as an equation.

15. Three times, the quotient of n and five is the opposite of n.

16. Twenty is equal to six subtracted from twice x.

17. Negative two is equal to x divided by the sum of x and five.

Answers

1. _____

2. _____

3. _____

4. _____

5. _____

6. _____

7. _____

8. _____

9. _____

10. _____

11. _____

12. _____

13. _____

14. **a.** see table

   b. _____

15. _____

16. _____

17. _____

18. _____

19. _____

20. _____

21. _____

22. _____

23. _____

24. _____

25. _____

26. _____

27. _____

28. _____

29. _____

30. _____

31. _____

32. _____

33. _____

34. _____

Name each property illustrated.

18. $6(x - 4) = 6x - 24$

19. $(4 + x) + z = 4 + (x + z)$

20. $(-7) + 7 = 0$

21. $(-18)(0) = 0$

Simplify.

22. $9x - 3(x - 4y + 2) + 20$

23. $\dfrac{1}{3}(15x - 27y + 1) - (x - 2)$

Simplify. Write answers using positive exponents only.

24. $(-9x)^{-2}$

25. $\dfrac{6^{-1}a^2b^{-3}}{3^{-2}a^{-5}b^2}$

26. $\left(\dfrac{-xy^{-5}z}{xy^3}\right)^{-5}$

27. $\dfrac{27x^{-5}y^5}{18x^{-6}y^2} \cdot \dfrac{x^4y^{-2}}{x^{-2}y^3}$

28. $\dfrac{(x^w)^2}{(x^{w-4})^{-2}}$

Write each number in scientific notation.

29. 630,000,000

30. 0.01200

31. Write 5.0×10^{-6} without exponents.

Use scientific notation to find the quotient. Express the quotient in scientific notation.

32. $\dfrac{(0.00012)(144,000)}{0.0003}$

33. $\dfrac{(0.0024)(0.00012)}{0.00032}$

34. During the first half of fiscal year 2001, the United States Postal Service handled approximately 578,000,000 pieces of Priority Mail. Write this number in scientific notation. (*Source:* United States Postal Service)

Equations, Inequalities, and Problem Solving

Mathematics is a tool for solving problems in such diverse fields as transportation, engineering, economics, medicine, business, and biology. We solve problems using mathematics by modeling real-world phenomena with mathematical equations or inequalities. Our ability to solve problems using mathematics, then, depends in part on our ability to solve equations and inequalities. In this chapter, we solve linear equations and inequalities in one variable and graph their solutions on number lines.

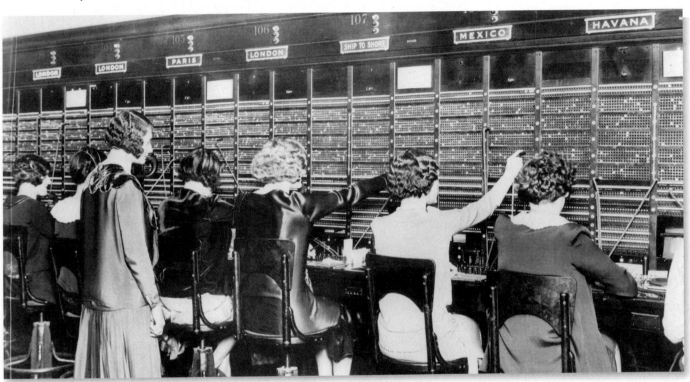

2.1 Linear Equations in One Variable

2.2 An Introduction to Problem Solving

2.3 Formulas and Problem Solving

2.4 Linear Inequalities and Problem Solving

Integrated Review—Linear Equations and Inequalities

2.5 Sets and Compound Inequalities

2.6 Absolute Value Equations and Inequalities

Alexander Graham Bell patented his telephone invention in 1876. He introduced it to the world later that year at the 1876 Centennial Exposition in Philadelphia. The following year he formed the Bell Telephone Company, and the modern telecommunications industry was born. At first, telephone service was available only on a subscription basis. Citizens paid to be connected to their neighbors within a local exchange. As exchanges were connected between cities, the telephone system grew. By 1885, there were 140,000 telephone subscribers in the United States. Well into the 20th century, telephone calls were connected manually at banks of switchboards by legions of telephone operators. The invention of automatic switching equipment allowed telephone calls to be connected first mechanically and later electronically. This development freed telephone operators to concentrate on providing services such as collect calls, third-party billing, and directory assistance to telephone customers. In Exercise 35 on page 100, we will analyze the change in employment of modern telephone operators.

Name _____ Section _____ Date _____

1. _____

2. _____

3. _____

4. _____

5. _____

6. _____

7. _____

8. _____

9. _____

10. _____

11. _____

12. _____

13. _____

14. _____

15. _____

16. _____

17. _____

18. _____

19. _____

20. _____

Chapter 2 Pretest

Solve each equation.

1. $2x - 17 = 21$

2. $3x - 2 + 14 = 8x + 4 - x$

3. $\dfrac{3y}{5} + 1 = \dfrac{4y}{3} - 2$

4. $7(t - 1) + 6 = 7t + 10$

5. $|8 - 3d| = 5$

6. $|2x - 1| = |-x + 4|$

Solve each equation for the specified variable.

7. $7y + 5x = 6$ for y

△ **8.** $S = 2LW + 2LH + 2WH$ for L

Solve each inequality.

9. $x + 12 \le -8$

10. $\dfrac{5}{7}y > 20$

11. $2(x - 9) \le 4x + 6$

12. $\dfrac{5x - 3}{6} - \dfrac{x + 4}{3} < -2$

13. $x \ge -1$ and $x \ge 2$

14. $x + 2 \ge -3$ and $x + 1 \le 5$

15. $3x - 2 < 1$ or $2x < 10$

16. $|x + 9| \le 6$

17. $|3 - x| \ge 1$

Solve.

18. Find two numbers such that the second number is 4 less than twice the first number and the sum of the two numbers is 50.

19. Find 18% of 900.

△ **20.** If the area of a triangular sign is 4 square feet and its base is 2 feet, find the height of the sign.

2.1   Linear Equations in One Variable

Ⓐ Deciding Whether a Number Is a Solution of an Equation

An **equation** is a statement that two expressions are equal. To solve problems, we need to be able to solve equations. In this section, we will solve a special type of equation called a *linear equation in one variable*.

> **Linear Equation in One Variable**
>
> A **linear equation in one variable** is an equation that can be written in the form
>
> $$ax + b = c$$
>
> where a, b, and c are real numbers and $a \neq 0$. For example,
>
> $$3x = -15 \qquad 7 - y = 3y \qquad 4n - 9n + 6 = 0 \qquad z = -2$$

When a variable in an equation is replaced by a number and the resulting equation is true, then that number is called a **solution** of the equation. For example, 1 is a solution of the equation $3x + 4 = 7$ since $3(1) + 4 = 7$ is a true statement. But 2 is not a solution of this equation since $3(2) + 4 = 7$ is *not* a true statement. The **solution set** of an equation is the set of solutions of the equation. For example, the solution set of $3x + 4 = 7$ is $\{1\}$.

EXAMPLE 1   Determine whether -15 is a solution of $x - 9 = -24$.

Solution: We replace x with -15 and see whether a true statement results.

$$x - 9 = -24$$
$$-15 - 9 \stackrel{?}{=} -24 \qquad \text{Replace } x \text{ with } -15.$$
$$-24 = -24 \qquad \text{True.}$$

Since a true statement results, -15 is a solution.   ●

EXAMPLE 2   Determine whether 5 is a solution of $2x - 3 = x + 3$.

Solution:
$$2x - 3 = x + 3$$
$$2 \cdot 5 - 3 \stackrel{?}{=} 5 + 3 \qquad \text{Replace } x \text{ with } 5.$$
$$7 = 8 \qquad \text{False.}$$

Since a false statement results, 5 is not a solution.   ●

Ⓑ Using the Properties of Equality

To **solve an equation** is to find the solution set of an equation. Equations with the same solution set are called **equivalent equations**. For example,

$$3x + 4 = 7 \qquad 3x = 3 \qquad x = 1$$

are equivalent equations because they all have the same solution set, namely, $\{1\}$. To solve an equation in x, we start with the given equation and write a series of simpler equivalent equations until we obtain an equation of the form

$$x = \textbf{number}$$

Practice Problem 1

Determine whether -7 is a solution of $14 - x = 21$.

Practice Problem 2

Determine whether 8 is a solution of $x - 10 = 2x - 14$.

Answers

1. -7 is a solution,   **2.** 8 is not a solution

To write equivalent equations, we use two important properties.

Addition Property of Equality

If a, b, and c, are real numbers, then

$$a = b \quad \text{and} \quad a + c = b + c$$

are equivalent equations.

Multiplication Property of Equality

If $c \neq 0$, then

$$a = b \quad \text{and} \quad ac = bc$$

are equivalent equations.

The **addition property of equality** guarantees that the same number may be added to both sides of an equation and the result is an equivalent equation. Recall that we define subtraction in terms of addition.

$$7 - 10 = 7 + (-10) = -3$$

This means that the addition property also says we can *subtract* the same number from both sides and the result is an equivalent equation.

The **multiplication property of equality** guarantees that both sides of an equation may be multiplied by the same nonzero number and the result is an equivalent equation. Recall that we define division in terms of multiplication. This means that the multiplication property also says we can *divide* the same nonzero number from both sides and the result is an equivalent equation.

For example, to solve $2x + 5 = 9$, we use the addition and multiplication properties of equality to get x alone—that is, to write an equivalent equation of the form

$$x = \text{number}$$

We will do this in the next example.

Practice Problem 3

Solve: $3x + 6 = 21$

EXAMPLE 3   Solve: $2x + 5 = 9$

Solution: First we use the addition property of equality and subtract 5 from both sides.

$$\begin{aligned}
2x + 5 &= 9 \\
2x + 5 - 5 &= 9 - 5 \qquad \text{Subtract 5 from both sides.} \\
2x &= 4 \qquad \text{Simplify.}
\end{aligned}$$

Now we use the multiplication property of equality and divide both sides by 2.

$$\frac{2x}{2} = \frac{4}{2} \qquad \text{Divide both sides by 2.}$$

$$x = 2 \qquad \text{Simplify.}$$

Check: To check, we replace x in the original equation with 2.

$$\begin{aligned}
2x + 5 &= 9 \qquad \text{Original equation} \\
2(2) + 5 &\stackrel{?}{=} 9 \qquad \text{Replace } x \text{ with 2.} \\
4 + 5 &\stackrel{?}{=} 9 \\
9 &= 9 \qquad \text{True.}
\end{aligned}$$

Answer

3. $\{5\}$

The solution set is $\{2\}$.

EXAMPLE 4   Solve: $0.6 = 2 - 3.5c$

Solution: We use both the addition property and the multiplication property of equality.

$$0.6 = 2 - 3.5c$$
$$0.6 - 2 = 2 - 3.5c - 2 \qquad \text{Subtract 2 from both sides.}$$
$$-1.4 = -3.5c \qquad \text{Simplify.}$$
$$\frac{-1.4}{-3.5} = \frac{-3.5c}{-3.5} \qquad \text{Divide both sides by } -3.5.$$
$$0.4 = c \qquad \text{Simplify } \frac{-1.4}{-3.5}.$$

Check:

$$0.6 = 2 - 3.5c$$
$$0.6 \stackrel{?}{=} 2 - 3.5(0.4) \qquad \text{Replace } c \text{ with } 0.4.$$
$$0.6 \stackrel{?}{=} 2 - 1.4 \qquad \text{Multiply.}$$
$$0.6 = 0.6 \qquad \text{True.}$$

The solution set is $\{0.4\}$.

Ⓒ Solving Linear Equations by Combining Like Terms

Often, an equation can be simplified by removing any grouping symbols and combining any like terms.

EXAMPLE 5   Solve: $-6x - 1 + 5x = 3$

Solution: First we simplify the left side of this equation by combining the like terms $-6x$ and $5x$. Then we use the addition property of equality and add 1 to both sides of the equation.

$$-6x - 1 + 5x = 3$$
$$-x - 1 = 3 \qquad \text{Combine like terms.}$$
$$-x - 1 + 1 = 3 + 1 \qquad \text{Add 1 to both sides.}$$
$$-x = 4 \qquad \text{Simplify.}$$

Notice that this equation is not solved for x since we have $-x$, or $-1x$, not x. To get x alone, we divide both sides by -1.

$$\frac{-x}{-1} = \frac{4}{-1} \qquad \text{Divide both sides by } -1.$$
$$x = -4 \qquad \text{Simplify.}$$

Check to see that the solution set is $\{-4\}$.

If an equation contains parentheses, we use the distributive property to remove them.

EXAMPLE 6   Solve: $2(x - 3) = 5x - 9$

Solution: First we use the distributive property.

$$2(x - 3) = 5x - 9$$
$$2x - 6 = 5x - 9 \qquad \text{Use the distributive property.}$$

Practice Problem 4

Solve: $4.5 = 3 + 2.5x$

> **Helpful Hint**
>
> Don't forget that
> $$0.4 = c \text{ and } c = 0.4$$
> are equivalent equations. We may solve an equation so that the variable is alone on either side of the equation.

Practice Problem 5

Solve: $-2x + 2 - 4x = 20$

Practice Problem 6

Solve: $4(x - 2) = 6x - 10$

Answers

4. $\{0.6\}$,   **5.** $\{-3\}$,   **6.** $\{1\}$

Next we get variable terms on the same side of the equation by using the addition property of equality.

$$2x - 6 - 5x = 5x - 9 - 5x \qquad \text{Subtract } 5x \text{ from both sides.}$$
$$-3x - 6 = -9 \qquad \text{Simplify.}$$
$$-3x - 6 + 6 = -9 + 6 \qquad \text{Add 6 to both sides.}$$
$$-3x = -3 \qquad \text{Simplify.}$$
$$\frac{-3x}{-3} = \frac{-3}{-3} \qquad \text{Divide both sides by } -3.$$
$$x = 1$$

Check to see that $\{1\}$ is the solution set. ●

> **Helpful Hint**
>
> When we multiply both sides of an equation by a number, the distributive property tells us that each term of the equation is multiplied by that number.

D Solving Linear Equations Containing Fractions or Decimals

If an equation contains fractions, we first clear the equation of fractions by multiplying both sides of the equation by the *least common denominator* (LCD) of all fractions in the equation.

EXAMPLE 7 Solve: $\dfrac{y}{3} - \dfrac{y}{4} = \dfrac{1}{6}$

Solution: First we clear the equation of fractions by multiplying both sides of the equation by 12, the LCD of the denominators 3, 4, and 6.

$$\frac{y}{3} - \frac{y}{4} = \frac{1}{6}$$
$$12\left(\frac{y}{3} - \frac{y}{4}\right) = 12\left(\frac{1}{6}\right) \qquad \text{Multiply both sides by the LCD, 12.}$$
$$12\left(\frac{y}{3}\right) - 12\left(\frac{y}{4}\right) = 2 \qquad \text{Use the distributive property.}$$
$$4y - 3y = 2 \qquad \text{Simplify.}$$
$$y = 2 \qquad \text{Simplify.}$$

Check: To check, we replace y with 2 in the original equation.

$$\frac{y}{3} - \frac{y}{4} = \frac{1}{6} \qquad \text{Original equation}$$
$$\frac{2}{3} - \frac{2}{4} \overset{?}{=} \frac{1}{6} \qquad \text{Replace } y \text{ with 2.}$$
$$\frac{8}{12} - \frac{6}{12} \overset{?}{=} \frac{1}{6} \qquad \text{Write fractions with the LCD.}$$
$$\frac{2}{12} \overset{?}{=} \frac{1}{6} \qquad \text{Subtract.}$$
$$\frac{1}{6} = \frac{1}{6} \qquad \text{Simplify.}$$

Since a true statement results, the solution set is $\{2\}$. ●

As a general guideline, the following steps may be used to solve a linear equation in one variable.

Practice Problem 7

Solve: $\dfrac{x}{6} - \dfrac{x}{8} = \dfrac{1}{8}$

Answer

7. $\{3\}$

Solving a Linear Equation in One Variable

Step 1. Clear the equation of fractions or decimals by multiplying both sides of the equation by an appropriate nonzero number.

Step 2. Use the distributive property to remove grouping symbols such as parentheses.

Step 3. Combine like terms on each side of the equation.

Step 4. Use the addition property of equality to rewrite the equation as an equivalent equation, with variable terms on one side and numbers on the other side.

Step 5. Use the multiplication property of equality to get the variable alone.

Step 6. Check the proposed solution in the original equation.

EXAMPLE 8   Solve: $\dfrac{x+5}{2} + \dfrac{1}{2} = \dfrac{1}{8}(15x+3)$

Solution: To begin, we multiply both sides of the equation by 8, the LCD of 2 and 8. This will clear the equation of fractions.

$$8\left(\frac{x+5}{2} + \frac{1}{2}\right) = 8\left[\frac{1}{8}(15x+3)\right] \quad \text{Multiply both sides by 8.}$$

$$8\left(\frac{x+5}{2}\right) + 8\left(\frac{1}{2}\right) = 8 \cdot \frac{1}{8}(15x+3) \quad \text{Use the distributive property.}$$

$$4(x+5) + 4 = 15x + 3 \quad \text{Multiply.}$$

$$4x + 20 + 4 = 15x + 3 \quad \text{Use the distributive property to remove parentheses.}$$

$$4x + 24 = 15x + 3 \quad \text{Combine like terms.}$$

$$4x - 15x = 3 - 24 \quad \text{Subtract 15x and 24 from both sides.}$$

$$-11x = -21 \quad \text{Simplify.}$$

$$\frac{-11x}{-11} = \frac{-21}{-11} \quad \text{Divide both sides by -11.}$$

$$x = \frac{21}{11} \quad \text{Simplify.}$$

To check, verify that replacing x with $\dfrac{21}{11}$ makes the original equation true. The solution set is $\left\{\dfrac{21}{11}\right\}$.

If an equation contains decimals, you may want to first clear the equation of decimals by multiplying by an appropriate power of 10.

EXAMPLE 9   Solve: $0.3x + 0.1 = 0.27x - 0.02$

Solution: To clear this equation of decimals, we multiply both sides of the equation by 100. Recall that multiplying a number by 100 moves its decimal point two places to the right.

$$100(0.3x + 0.1) = 100(0.27x - 0.02)$$

$$100(0.3x) + 100(0.1) = 100(0.27x) - 100(0.02) \quad \text{Use the distributive property.}$$

$$30x + 10 = 27x - 2 \quad \text{Multiply.}$$

$$30x - 27x = -2 - 10 \quad \text{Subtract 27x and 10 from both sides.}$$

$$3x = -12 \quad \text{Simplify.}$$

Practice Problem 8

Solve: $\dfrac{x-1}{3} + \dfrac{2}{3} = x - \dfrac{2x+3}{9}$

Practice Problem 9

Solve: $0.2x + 0.1 = 0.12x - 0.06$

Answers

8. $\left\{\dfrac{3}{2}\right\}$,   **9.** $\{-2\}$

$$\frac{3x}{3} = \frac{-12}{3}$$ Divide both sides by 3.

$$x = -4$$ Simplify.

Check to see that the solution set is $\{-4\}$.

Try the Concept Check in the margin.

E Recognizing Identities and Equations with No Solution

So far, each linear equation that we have solved has had a single solution. We will now look at two other types of equations: *contradictions* and *identities*.

An equation in one variable that has no solution is called a **contradiction**, and an equation in one variable that has every number (for which the equation is defined) as a solution is called an **identity**. The next examples show how to recognize contradictions and identities.

EXAMPLE 10 Solve: $3x + 5 = 3(x + 2)$

Solution: First we use the distributive property to remove parentheses.

$$3x + 5 = 3(x + 2)$$
$$3x + 5 = 3x + 6$$ Use the distributive property.
$$3x + 5 - 3x = 3x + 6 - 3x$$ Subtract $3x$ from both sides.
$$5 = 6$$

The equation $5 = 6$ is a false statement no matter what value the variable x might have. Thus the original equation has no solution. Its solution set is written either as $\{\ \}$ or $\varnothing$. This equation is a contradiction.

EXAMPLE 11 Solve: $6x - 4 = 2 + 6(x - 1)$

Solution: First we use the distributive property to remove parentheses.

$$6x - 4 = 2 + 6(x - 1)$$
$$6x - 4 = 2 + 6x - 6$$ Use the distributive property.
$$6x - 4 = 6x - 4$$ Combine like terms.

At this point we might notice that both sides of the equation are the same, so replacing x by any real number gives a true statement. Thus the solution set of this equation is the set of real numbers, and the equation is an identity. Continuing to "solve" $6x - 4 = 6x - 4$, we eventually arrive at the same conclusion.

$$6x - 4 + 4 = 6x - 4 + 4$$ Add 4 to both sides.
$$6x = 6x$$ Simplify.
$$6x - 6x = 6x - 6x$$ Subtract $6x$ from both sides.
$$0 = 0$$ Simplify.

Since $0 = 0$ is a true statement for every value of x, the solution set is the set of all real numbers, which can be written as $\{x \mid x \text{ is a real number}\}$. The equation is called an identity.

Concept Check

Explain what is wrong with the following:
$$3x - 5 = 16$$
$$3x = 11$$
$$\frac{3x}{3} = \frac{11}{3}$$
$$x = \frac{11}{3}$$

Practice Problem 10

Solve: $5x - 1 = 5(x + 3)$

Practice Problem 11

Solve: $-4(x - 1) = -4x - 9 + 13$

Answers

10. $\varnothing$, **11.** $\{x \mid x \text{ is a real number}\}$

Concept Check:
$$3x - 5 = 16$$
$$3x = 21$$
$$x = 7$$
Therefore the correct solution set is $\{7\}$.

Helpful Hint

For linear equations, *any* false statement such as $5 = 6, 0 = 1$, or $-2 = 2$ informs us that the original equation has no solution. Also, *any* true statement such as $0 = 0, 2 = 2$, or $-5 = -5$ informs us that the original equation is an identity.

Mental Math

Solve each equation.

1. $3x = 18$

2. $2x = 60$

3. $x - 7 = 10$

4. $x - 2 = 15$

5. $\dfrac{x}{2} = 4$

6. $\dfrac{x}{3} = 5$

7. $x + 1 = 11$

8. $x + 4 = 20$

EXERCISE SET 2.1

A *Determine whether each number is a solution of the given equation. See Examples 1 and 2.*

1. $-24; \dfrac{x}{-6} = 4$

2. $15; \dfrac{x}{-3} = -5$

3. $-3; x - 17 = 20$

4. $-8; x - 10 = -2$

5. $-2; 5 + 3x = -1$

6. $-1; 6 - 2x = 4$

7. $5; x - 7 = x + 2$

8. $5; x - 1 = x - 1$

9. $5; 4(x - 3) = 12$

10. $12; 5(x - 6) = 30$

11. $-8; 4x - 2 = 5x + 6$

12. $2; 7x + 1 = 6x - 1$

B *Solve each equation and check. See Examples 3 and 4.*

13. $-5x = -30$

14. $-2x = 18$

15. $10 = x + 12$

16. $25 = y + 30$

17. $x + 2.8 = 1.9$

18. $y - 8.6 = -6.3$

19. $5x - 4 = 26$

20. $2y - 3 = 11$

21. $-4.1 - 7z = 3.6$

22. $10.3 - 6x = -2.3$

23. $5y + 12 = 2y - 3$

24. $4x + 14 = 6x + 8$

© Solve each equation and check. See Examples 5 and 6.

25. $8x - 5x + 3 = x - 7 + 10$

26. $6 + 3x + x = -x + 2 - 26$

27. $5x + 12 = 2(2x + 7)$

28. $2(x + 3) = x + 5$

29. $3(x - 6) = 5x$

30. $6x = 4(5 + x)$

31. $3x - 4 - 5x = x + 4 + x$

32. $13x - 15x + 8 = 4x + 2 - 24$

33. $-2(5y - 1) - y = -4(y - 3)$

34. $-3(2w - 7) - 10 = 9 - 2(5w + 4)$

35. $y + 0.2 = 0.6(y + 3)$

36. $-(w + 0.2) = 0.3(4 - w)$

37. a. Simplify the expression $4(x + 1) + 1$.
   b. Solve the equation $4(x + 1) + 1 = -7$.
   c. Explain the difference between solving an equation for a variable and simplifying an expression.

38. Explain why the multiplication property of equality does not include multiplying both sides of an equation by 0. (*Hint:* Write down a false statement and then multiply both sides by 0. Is the result true or false? What does this mean?)

Ⓓ Solve each equation and check. See Examples 7 through 9.

39. $\dfrac{x}{2} + \dfrac{2}{3} = \dfrac{3}{4}$

40. $\dfrac{x}{2} + \dfrac{x}{3} = \dfrac{5}{2}$

41. $\dfrac{3t}{4} - \dfrac{t}{2} = 1$

42. $\dfrac{4r}{5} - 7 = \dfrac{r}{10}$

43. $\dfrac{n - 3}{4} + \dfrac{n + 5}{7} = \dfrac{5}{14}$

44. $\dfrac{2 + h}{9} + \dfrac{h - 1}{3} = \dfrac{1}{3}$

45. $0.6x - 10 = 1.4x - 14$

46. $0.3x + 2.4 = 0.1x + 4$

47. $\dfrac{3x - 1}{9} + x = \dfrac{3x + 1}{3} + 4$

48. $\dfrac{2z + 7}{8} - 2 = z + \dfrac{z - 1}{2}$

88

49. $1.5(4 - x) = 1.3(2 - x)$

50. $2.4(2x + 3) = -0.1(2x + 3)$

E *Solve each equation and check. Some of these equations are identities and some are contradictions. See Examples 3 through 11.*

51. $4(n + 3) = 2(6 + 2n)$

52. $6(4n + 4) = 8(3 + 3n)$

53. $3(x + 1) + 5 = 3x + 2$

54. $5x - (x + 4) = 4 + 4(x + 2)$

55. $2y + 5(y - 4) = 4y - 2(y - 10)$

56. $9c - 3(6 - 5c) = c - 2(3c + 9)$

57. $2(x - 8) + x = 3(x - 6) + 2$

58. $4(x + 5) = 3(x - 4) + x$

59. $\dfrac{3}{8} + \dfrac{b}{3} = \dfrac{5}{12}$

60. $\dfrac{a}{2} + \dfrac{7}{4} = 5$

61. $x - 10 = -6x + 4$

62. $4x - 7 = 2x - 7$

63. $5(x - 2) + 2x = 7(x + 4)$

64. $3x + 2(x + 4) = 5(x + 1) + 3$

65. $\dfrac{1}{4}(a + 2) = \dfrac{1}{6}(5 - a)$

66. $\dfrac{1}{3}(8 + 2c) = \dfrac{1}{5}(3c - 5)$

67. $6x - 2(x - 3) = 4(x + 1) + 4$

68. $10x - 2(x + 4) = 8(x - 2) + 6$

69. $\dfrac{m - 4}{3} - \dfrac{3m - 1}{5} = 1$

70. $\dfrac{n + 1}{8} - \dfrac{2 - n}{3} = \dfrac{5}{6}$

71. $-(3x - 5) - (2x - 6) + 1 = -5(x - 1) - (3x + 2) + 3$

72. $-4(2x - 3) - (10x + 7) - 2 = -(12x - 5) - (4x + 9) - 1$

73. $\frac{1}{3}(y + 4) + 6 = \frac{1}{4}(3y - 1) - 2$

74. $\frac{1}{5}(2y - 1) - 2 = \frac{1}{2}(3y - 5) + 3$

75. In your own words, explain why the equation $x + 7 = x + 6$ has no solution while the solution set of the equation $x + 7 = x + 7$ contains all real numbers.

76. In your own words, explain why the equation $x = -x$ has one solution—namely, 0—while the solution set of the equation $x = x$ is all real numbers.

Review and Preview

Translate each phrase into an expression. Use the variable x to represent each unknown number. See Section 1.2.

77. The quotient of 8 and a number

78. The sum of 8 and a number

79. The product of 8 and a number

80. The difference of 8 and a number

81. Two more than three times a number

82. Five subtracted from twice a number

 Combining Concepts

Solve and check.

83. $-9.112y = -47.537304$

84. $2.86z - 8.1258 = -3.75$

85. $x(x - 6) + 7 = x(x + 1)$

86. $7x^2 + 2x - 3 = 6x(x + 4) + x^2$

Find the value of K such that the equations are equivalent.

87. $3.2x + 4 = 5.4x - 7$

$3.2x = 5.4x + K$

88. $\frac{x}{6} + 4 = \frac{x}{3}$

$x + K = 2x$

2.2 An Introduction to Problem Solving

OBJECTIVES

 Write algebraic expressions that can be simplified.

 Apply the steps for problem solving.

SSM
TUTOR CENTER   SG   CD & VIDEO   MATH PRO   WEB

A Writing Algebraic Expressions

In order to prepare for problem solving, we practice writing algebraic expressions that can be simplified.

Our first example involves consecutive integers and perimeter. Recall that *consecutive integers* are integers that follow one another in order. Study the examples of consecutive, even, and odd integers and their representations.

Consecutive Integers:

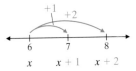

$x \quad x+1 \quad x+2$

Consecutive Even Integers:

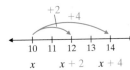

$x \quad x+2 \quad x+4$

Consecutive Odd Integers:

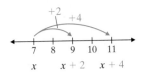

$x \quad x+2 \quad x+4$

EXAMPLE 1   Write the following as algebraic expressions. Then simplify.

a. The sum of two consecutive integers, if x is the first consecutive integer.

△ **b.** The perimeter of the triangle with sides of length x, $5x$, and $6x - 3$.

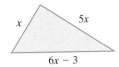

Solution:

a. Recall that if x is the first integer, then the next consecutive integer is 1 more, or $x + 1$.

In words: first integer   plus   next consecutive integer

Translate: x + $(x + 1)$

Then $x + (x + 1) = x + x + 1$

$\qquad = 2x + 1$ Simplify by combining like terms.

b. The perimeter of a triangle is the sum of the lengths of the sides.

In words: side   +   side   +   side

Translate: x + $5x$ + $(6x - 3)$

Then $x + 5x + (6x - 3) = x + 5x + 6x - 3$

$\qquad = 12x - 3$ Simplify.

Practice Problem 2

In 2000, the three internet access companies with the most subscribers were Earthlink, MSN, and AOL. Earthlink had 700,000 more subscribers than MSN and AOL had 3,700,000 more than five times the subscribers that MSN had. Write the sum of the subscribers of these three companies as an algebraic expression. Let x be the number of MSN subscribers. (In Exercise 28, we will find the actual number of subscribers for each company.) (*Source:* Jupiter Media Metrix)

EXAMPLE 2

The most populous city in the United States is New York, although it is only the fifth most populous city in the world. Tokyo is the most populous city in the world followed by Mexico City. Mexico City's population is 13.2 million more than New York's and Tokyo's population is twice New York's, increased by 0.7 million. Write the sum of the populations of these three cities as an algebraic expression. Let x be the population of New York (in millions). (*Source:* Planet101.com)

Solution:

If x = the population of New York (in millions) then

$x + 13.2$ = the population of Mexico City (in millions) and

$2x + 0.7$ = the population of Tokyo (in millions)

In words: population of New York   +   population of Mexico City   +   population of Tokyo

Translate: x + $(x + 13.2)$ + $(2x + 0.7)$

Then $x + (x + 13.2) + (2x + 0.7) = x + x + 2x + 13.2 + 0.7$

$$= 4x + 13.9$$

In Exercise 27, we will find the actual populations of these cities.

B Solving Problems

Our main purpose for studying algebra is to solve problems. The following problem-solving strategy will be used throughout this text and may also be used to solve real-life problems that occur outside the mathematics classroom.

> **Helpful Hint**
>
> You may want to begin this section by studying key words and phrases and their translations in Sections 1.2 Objective B and 1.3 Objective A.

General Strategy for Problem Solving

1. UNDERSTAND the problem. During this step, become comfortable with the problem. Some ways of doing this are:
   Read and reread the problem.
   Choose a variable to represent the unknown.
   Construct a drawing.
   Propose a solution and check. Pay careful attention to how you check your proposed solution. This will help when writing an equation to model the problem.
2. TRANSLATE the problem into an equation.
3. SOLVE the equation.
4. INTERPRET the results: If possible, check to see whether your answer is reasonable. Then *check* the proposed solution in the stated problem and *state* your conclusion.

Answer

2. $7x + 4,400,000$

Let's review this strategy by solving a problem involving unknown numbers.

EXAMPLE 3   Finding Unknown Numbers

Find two numbers such that the second number is 3 more than twice the first number and the sum of the two numbers is 72.

Practice Problem 3

One number is three times another number. If their sum is 148, find the two numbers.

Solution:

1. UNDERSTAND the problem. First let's read and reread the problem and then propose a solution. For example, if the first number is 25, then the second number is 3 more than twice 25, or 53. The sum of 25 and 53 is 78, not the required sum, but we have gained some valuable information about the problem. First, we know that the first number is less than 25 since our guess led to a sum greater than the required sum. Also, we have gained some information as to how to model the problem.

 Helpful Hint

The purpose of guessing a solution is not to guess correctly but to gain confidence and to help understand the problem and how to model it.

Next let's assign a variable and use this variable to represent any other unknown quantities. If we let

the first number $= x$, then
the second number $= 2x + 3$

> 3 more than
> twice the second number

2. TRANSLATE the problem into an equation. To do so, we use the fact that the sum of the numbers is 72. First let's write this relationship in words and then translate to an equation.

In words:	first number	added to	second number	is	72
	↓	↓	↓	↓	↓
Translate:	x	$+$	$(2x + 3)$	$=$	72

3. SOLVE the equation.

$$x + (2x + 3) = 72$$
$$x + 2x + 3 = 72 \quad \text{Remove parentheses.}$$
$$3x + 3 = 72 \quad \text{Combine like terms.}$$
$$3x = 69 \quad \text{Subtract 3 from both sides.}$$
$$x = 23 \quad \text{Divide both sides by 3.}$$

4. INTERPRET. Here, we *check* our work and *state* the solution. Recall that if the first number $x = 23$, then the second number $2x + 3 = 2 \cdot 23 + 3 = 49$.

Check: Is the second number 3 more than twice the first number? Yes, since 3 more than twice 23 is $46 + 3$, or 49. Also, their sum, $23 + 49 = 72$, is the required sum.

State: The two numbers are 23 and 49.

Many of today's rates and statistics are given as percents. Interest rates, tax rates, nutrition labeling, and percent of households in a given category are just a few examples. Before we practice solving problems containing percents, let's take a moment to review the meaning of percent and how to find a percent of a number.

Answer

3. 37 and 111

The word *percent* means *per hundred*, and the symbol % is used to denote percent. This means that 23% is 23 per hundred, or $\frac{23}{100}$. Also,

$$41\% = \frac{41}{100} = 0.41$$

To find a percent of a number, we multiply.

$$16\% \text{ of } 25 = 16\% \cdot 25 = 0.16 \cdot 25 = 4$$

Thus, 16% of 25 is 4.

Next, we solve a problem containing a percent.

Try the Concept Check in the margin.

Concept Check

Suppose you are finding 112% of a number x. Which of the following is a correct description of the result? Explain.

a. The result is less than x.
b. The result is equal to x.
c. The result is greater than x.

Practice Problem 4

The price of a home was just decreased by 6%. If the decreased price is $83,660, find the original price of the home.

EXAMPLE 4 Finding the Original Price of a Computer

Suppose that a computer store just announced an 8% decrease in the price of a particular computer model. If this computer sells for $2162 after the decrease, find its original price.

Solution:

1. UNDERSTAND. Read and reread the problem. Recall that a percent decrease means a percent of the original price. Let's guess that the original price of the computer is $2500. The amount of decrease is then 8% of $2500, or $(0.08)(\$2500) = \$200$. This means that the new price of the computer is the original price minus the decrease, or $\$2500 - \$200 = \$2300$. Our guess is incorrect, but we now have an idea of how to model this problem. In our model, we will let $x =$ the original price of the computer.

2. TRANSLATE.

In words:	original price of computer	minus	8% of original price	is	new price
	↓	↓	↓	↓	↓
Translate:	x	$-$	$0.08x$	$=$	2162

3. SOLVE the equation.

$$x - 0.08x = 2162$$
$$0.92x = 2162 \qquad \text{Combine like terms.}$$
$$x = \frac{2162}{0.92} = 2350 \qquad \text{Divide both sides by 0.92.}$$

Answers

4. $89,000

Concept Check: c; the result is greater than x

4.   INTERPRET.

Check: The amount $2350 is a reasonable price for a computer. If the original price of the computer was $2350, the new price is

$$\$2350 - (0.08)(\$2350) = \$2350 - \$188$$
$$= \$2162 \qquad \text{The given new price}$$

State: The original price of the computer was $2350.

△ **EXAMPLE 5   Finding the Lengths of a Triangle's Sides**

A pennant in the shape of an isosceles triangle is to be constructed for the Slidell High School Athletic Club and sold at a fund-raiser. The company manufacturing the pennant charges according to perimeter, and the athletic club has determined that a perimeter of 149 centimeters should make a nice profit. If each equal side of the triangle is twice the length of the third side, increased by 12 centimeters, find the lengths of the sides of the triangular pennant.

52 cm   20 cm   52 cm   2x + 12   x   2x + 12

Practice Problem 5 △

A rectangle has a perimeter of 106 meters. Its length is 5 meters more than twice its width. Find the length and the width of the rectangle.

Solution:

1.   UNDERSTAND. Read and reread the problem. Recall that the perimeter of a triangle is the distance around. Let's guess that the third side of the triangular pennant is 20 centimeters. This means that each equal side is twice 20 centimeters, increased by 12 centimeters, or $2(20) + 12 = 52$ centimeters.

This gives a perimeter of $20 + 52 + 52 = 124$ centimeters. Our guess is incorrect, but we now have a better understanding of how to model this problem. Now we let

the third side of the triangle $= x$

the first side	=	twice	the third side	increased by 12
	=	2	x	+ 12,

or $2x + 12$

the second side $=$ $2x + 12$

2.   TRANSLATE.

In words:	first side	+	second side	+	third side	=	149
Translate:	$(2x + 12)$	+	$(2x + 12)$	+	x	=	149

3.   SOLVE the equation.

$$(2x + 12) + (2x + 12) + x = 149$$
$$2x + 12 + 2x + 12 + x = 149 \qquad \text{Remove parentheses.}$$
$$5x + 24 = 149 \qquad \text{Combine like terms.}$$
$$5x = 125 \qquad \text{Subtract 24 from both sides.}$$
$$x = 25 \qquad \text{Divide both sides by 5.}$$

Answer

5. length: 16 m, width: 37 m

4. INTERPRET. If the third side is 25 centimeters, then the first side is $2(25) + 12 = 62$ centimeters and the second side is 62 centimeters also.

Check: The first and second sides are each twice 25 centimeters increased by 12 centimeters or 62 centimeters. Also, the perimeter is $25 + 62 + 62 = 149$ centimeters, the required perimeter.

State: The dimensions of the triangle are 25 centimeters, 62 centimeters, and 62 centimeters.

Practice Problem 6

Find two consecutive integers whose sum is 251.

EXAMPLE 6   Finding Consecutive Integers

Kelsey Ohleger was helping her friend Benji Burnstine study for an exam. Kelsey told Benji that her two latest art history quiz scores are two consecutive even integers whose sum is 174. Help Benji find the scores.

Solution:

1. UNDERSTAND. Read and reread the problem. Since we are looking for consecutive even integers, let

x = the first integer. Then
$x + 2$ = the next consecutive even integer.

2. TRANSLATE.

In words: first integer + next even integer = 174

Translate: x + $(x + 2)$ = 174

3. SOLVE.

$$x + (x + 2) = 174$$
$$2x + 2 = 174 \quad \text{Combine like terms.}$$
$$2x = 172 \quad \text{Subtract 2 from both sides.}$$
$$x = 86 \quad \text{Divide both sides by 2.}$$

4. INTERPRET. If $x = 86$, then $x + 2 = 86 + 2$ or 88.

Check: The numbers 86 and 88 are two consecutive even integers. Their sum is 174, the required sum.

State: Kelsey's art history quiz scores are 86 and 88.

Answer

6. 125 and 126

Name _____ Section _____ Date _____

EXERCISE SET 2.2

A *Write the following as algebraic expressions. Then simplify. See Examples 1 and 2.*

△ **1.** The perimeter of the square with side length y.

△ **2.** The perimeter of the rectangle with length x and width $x - 5$.

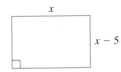

3. The sum of three consecutive integers if the first is z.

4. The sum of three consecutive odd integers if the first integer is x.

5. The total amount of money (in cents) in x nickels and $(x + 3)$ dimes. (*Hint:* the value of a nickel is 5 cents and the value of a dime is 10 cents.)

6. The total amount of money (in cents) in y quarters and $(2y - 1)$ nickels. (Use the hint for Exercise 5.)

△ **7.** A piece of land along Bayou Liberty is to be fenced and subdivided as shown so that each rectangle has the same dimensions. Express the total amount of fencing needed as an algebraic expression in x.

△ **8.** Write the perimeter of the floor plan shown as an algebraic expression in x.

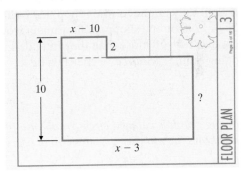

B *Solve. See Example 3.*

9. Four times the difference of a number and 2 is the same as 2 increased by six times the number. Find the number.

10. Twice the sum of a number and 3 is the same as 1 subtracted from the number. Find the number.

11. One number is five times another number. If the sum of the two numbers is 270, find the numbers.

12. One number is 6 less than another number. If the sum of the two numbers is 150, find the numbers.

Solve. See Examples 4 through 6.

13. The United States consists of 2271 million acres of land. Approximately 29% of this land is federally owned. Find the number of acres that are federally owned. (*Source:* U.S. General Services Administration)

14. The state of Nevada contains the most federally owned acres of land in the United States. If 90% of the state's 70 million acres of land is federally owned, find the number of federally owned acres. (*Source:* U.S. General Services Administration)

15. In 2000, a total of 2342 earthquakes occurred in the United States. Of these, 85% were minor tremors with magnitudes of 3.9 or less on the Richter scale. How many minor earthquakes occurred in the United States in 2000? Round to the nearest whole. (*Source:* U.S. Geological Survey National Earthquake Information Center)

16. Of the 1071 tornadoes that occurred in the United States during 2000, 22.5% occurred during the month of May. How many tornadoes occurred during May 2000? Round to the nearest whole. (*Source:* Storm Prediction Center)

17. According to a recent survey, $33\frac{1}{3}$% of online shoppers in the United States say that they spend more than they intended when shopping online. In a group of 1290 online shoppers, how many would you expect don't spend more than they intended when shopping online? (*Source:* Cyber Dialogue, 04/11/2001)

18. Only 6% of Americans eat a commercially prepared breakfast five times or more per week. As of 2000, Cincinnati, Ohio, had a population of 331,285 people. How many of these people would you expect do not eat a commercially prepared breakfast fives or more times per week? Round to the nearest whole. (*Sources*: National Restaurant Association, U.S. Census Bureau)

The following graph is called a circle graph or a pie chart. The circle represents a whole, or in this case, 100%. This particular graph shows the number of times e-mail users check their e-mail. Use this graph to answer Exercises 19 through 22.

How Often We Check Our E-mail

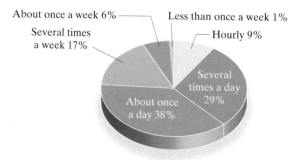

Source: The UCLA Internet Report: "Surveying the Digital Future," UCLA Center for Communication Policy

19. What percent of e-mail users check their e-mail several times per week?

20. Among e-mail users, what is the most popular frequency for checking e-mail?

21. If it is estimated that Fort Wayne, Indiana, has 112,500 e-mail users, how many of these would you expect check their e-mail about once a week?

22. If it is estimated that the city of New Orleans has 265,000 e-mail users, how many of these would you expect check their e-mail either several times a day or hourly?

23. INVESCO Field at Mile High, home to the Denver Broncos, has 11,675 more seats than Heinz Field, home to the Pittsburgh Steelers. Together, these two stadiums can seat a total of 140,575 NFL fans. How many seats does each stadium have? (*Sources:* Denver Broncos, Pittsburgh Steelers)

24. For the 2001 Major League Baseball season, the opening day payroll for the Minnesota Twins was $46,718,000 less than the opening day payroll for the Colorado Rockies. The total of the opening day payrolls for these two teams was $95,418,000. What was the opening day payroll for each team? (*Source:* Associated Press)

△ **25.** The perimeter of the triangle in Example 1b in this section is 483 feet. Find the length of each side.

△ **26.** The perimeter of the rectangle in Practice Problem 1b in this section is 127 meters. Find its length and width.

27. The sum of the population of New York, Tokyo, and Mexico City is 72.3 million. Use this information and Example 2 in this section to find the population of each city. (*Source:* Planet101.com)

28. The total subscribers for the internet access companies Earthlink, AOL, and MSN is 32,400,000. Use this information and Practice Problem 2 in this section to find the number of subscribers for each company. (*Source:* Jupiter Media Metrix)

29. The B767-300ER aircraft has 104 more seats than the B737-200 aircraft. If their total number of seats is 328, find the number of seats for each aircraft. (*Source:* Air Transport Association of America)

30. The governor of Connecticut makes $29,000 less per year than the governor of Delaware. If the total of these salaries is $185,000, find the salary of each governor. (*Source: 2001 World Almanac*)

31. A new fax machine was recently purchased for an office in Hopedale for $464.40 including tax. If the tax rate in Hopedale is 8%, find the price of the fax machine before taxes.

32. A premedical student at a local university was complaining that she had just paid $86.11 for her human anatomy book, including tax. Find the price of the book before taxes if the tax rate at this university is 9%.

33. In 2001, the population of Morocco was 29.2 million. This represented an increase in population of 2% from a year earlier. What was the population of Morocco in 2000? (*Source:* Population Reference Bureau)

34. Palm, Inc. is a worldwide supplier of handheld computing devices. On October 25, 2001, shares of Palm stock closed at $2.44 per share. This represents a 97.4% decrease in stock price from the closing price on March 2, 2000, the day when Palm stock first traded on the NASDAQ exchange. Find the closing price on March 2, 2000. (*Source:* Financial Insight Systems, Inc.)

35. According to government statistics, the number of switchboard operators in the United States is expected to decrease to 185,000 by the year 2008. This represents a decrease of 13.9% from the number of switchboard operators in 1998. (*Source:* U.S. Bureau of Labor Statistics)

   a. Find the number of switchboard operators in 1998. Round to the nearest whole number.
   b. In your own words, explain why you think that the need for switchboard operators is decreasing.

36. The number of deaths caused by tornadoes descreased 59.2% from the 1950s to the 1999s. There were 579 deaths from tornadoes in the 1990s. (*Source:* National Weather Service)

   a. Find the number of deaths caused by tornadoes in the 1950s. Round to the nearest whole number.
   b. In your own words, explain why you think that the number of tornado-related deaths has decreased so much since the 1950s.

37. The sum of three consecutive integers is 228. Find the integers.

38. The sum of three consecutive odd integers is 327. Find the integers.

Recall that the sum of the angle measures of a triangle is 180°.

△ **39** Find the measures of the angles of a triangle if the measure of one angle is twice the measure of a second angle and the third angle measures 3 times the second angle decreased by 12.

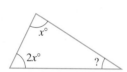

△ **40.** Find the angles of a triangle whose two base angles are equal and whose third angle is 10° less than three times a base angle.

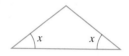

△ **41.** The official manual for traffic signs is the *Manual on Uniform Traffic Control Devices* published by the Government Printing Office. The rectangular sign below has a length 12 inches more than twice its height. If the perimeter of the sign is 312 inches, find its dimensions.

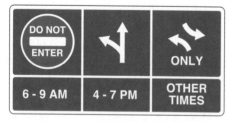

42. Two frames are needed with the same outside perimeter: one frame in the shape of a square and one in the shape of an equilateral triangle. Each side of the triangle is 6 centimeters longer than each side of the square. Find the dimensions of each frame. (An equilateral triangle has sides that are the same length.)

43. In a blueprint of a rectangular room, the length is to be 2 centimeters greater than twice its width. Find the dimensions if the perimeter is to be 40 centimeters.

44. A plant food solution contains 5 cups of water for every 1 cup of concentrate. If the solution contains 78 cups of these two ingredients, find the number of cups of concentrate in the solution.

45. In 2001, the population of South Africa was 43,600,000 people. From 2001 to 2050, South Africa's population is expected to decrease by 25%. Find the expected population of South Africa in 2050. (*Source:* Population Reference Bureau)

46. Dana, an auto parts supplier headquartered in Toledo, Ohio, recently announced it would be cutting 11,000 jobs worldwide. This is equivalent to 15% of Dana's workforce. Find the size of Dana's workforce prior to this round of job layoffs. Round to the nearest whole. (*Source:* Dana Corporation)

47. The zip codes of three Nevada locations—Fallon, Fernley, and Gardnerville Ranchos—are three consecutive even integers. If twice the first integer added to the third is 268,222, find each zip code.

48. During a recent year, the average SAT scores in math for the states of Alabama, Louisiana, and Michigan were 3 consecutive integers. If the sum of the first integer, second integer, and three times the third integer is 2637, find each score.

Recall that two angles are complements of each other if their sum is 90°. Two angles are supplements of each other if their sum is 180°. Find the measure of each angle.

△ **49.**

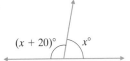

△ **50.**

△ **51.**

△ **52.**

△ **53.** One angle is three times its supplement increased by 20°. Find the measures of the two supplementary angles.

△ **54.** One angle is twice its complement increased by 30°. Find the measure of the two complementary angles.

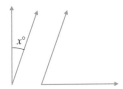

55. The external tank of a NASA Space Shuttle contains the propellants used for the first 8.5 minutes after launch. Its height is 5 times the sum of its width and 1. If the sum of the height and width is 55.4 meters, find the dimensions of this tank. (*Source:* NASA/Kennedy Space Center)

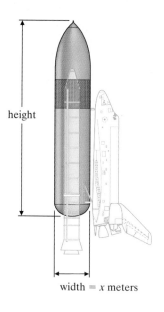

height

width = *x* meters

56. The blue whale is the largest of whales. Its average weight is 3 times the difference of the average weight of a humpback whale and 5 tons. If the total of the average weights of a blue whale and a humpback whale is 117 tons, find the average weight of each type of whale.

57. Incandescent, fluorescent, and halogen bulbs are lasting longer today than ever before. On average, the number of bulb hours for a fluorescent bulb is 25 times the number of bulb hours for a halogen bulb. The number of bulb hours for an incandescent bulb is 2,500 less than the halogen bulb. If the total number of bulb hours for the three types of bulbs is 105,500, find the number of bulb hours for each type. (*Source:* Popular Science Magazine)

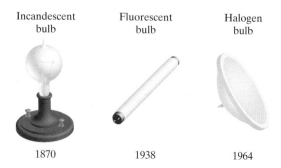

Incandescent bulb	Fluorescent bulb	Halogen bulb
1870	1938	1964

58. Finland, Sweden, and Austria have the greatest percent of their population as cell phone subscribers in the world. In Sweden, 7% more of its population are cell phone subscribers than in Austria. In Finland, 13% more of its population are cell phone subscribers than in Austria. If the sum of their percents is 179%, find the percent of population that are cell phone subscribers for each country. (Currently, 29% of the U.S. population are cell phone subscribers). (*Source:* EMC World Cellular Database; Cellular Telecommunications Industry Asso.)

59. During the 2001 Major League Baseball season, the numbers of home runs hit by Jim Thome of the Cleveland Indians, Rafael Palmeiro of the Texas Rangers, and Richie Sexson of the Milwaukee Brewers were three consecutive odd integers. Of these three players, Thome hit the most home runs, and Sexson hit the fewest home runs. The total number of home runs hit by these three players over the course of the season was 141. How many home runs did each player hit during the 2001 season? (*Source:* Major League Baseball)

60. During the 2000 Olympic Games in Sydney, Australia, the total number of medals won by Germany, Australia, and the People's Republic of China were three consecutive integers. Of these three countries, Germany won the fewest medals and China won the most. If the sum of the first integer, twice the second integer, and four times the third integer is 409, find the number of medals won by each country. (*Source:* International Olympic Committee)

Review and Preview

Find the value of each expression for the given values. See Section 1.5.

61. $2a + b - c$; $a = 5, b = -1$, and $c = 3$

62. $-3a + 2c - b$; $a = -2, b = 6$, and $c = -7$

63. $4ab - 3bc$; $a = -5, b = -8$, and $c = 2$

64. $ab + 6bc$; $a = 0, b = -1$, and $c = 9$

65. $n^2 - m^2$; $n = -3$ and $m = -8$

66. $2n^2 + 3m^2$; $n = -2$ and $m = 7$

 67. $P + PRT$; $P = 3000, R = 0.0325$, and $T = 2$

68. $\frac{1}{3}lwh$; $l = 37.8, w = 5.6$, and $h = 7.9$

Combining Concepts

69. Newsprint is either discarded or recycled. Americans recycle about 27% of all newsprint, but an amount of newsprint equivalent to 30 million trees is discarded every year. About how many trees' worth of newsprint is *recycled* in the United States each year? (*Source:* The Earth Works Group)

70. Find an angle such that its supplement is equal to twice its complement increased by 50°.

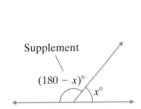 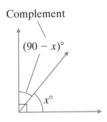

Supplement
$(180 - x)°$
$x°$

Complement
$(90 - x)°$
$x°$

71. The average annual number of cigarettes smoked by an American adult continues to decline. For the years 1991–2000, the equation $y = -64.45x + 2795.5$ approximates this data. Here, x is the number of years after 1990 and y is the average annual number of cigarettes smoked.

a. If this trend continues, find the year in which the average annual number of cigarettes smoked is 0. To do this, let $y = 0$ and solve for x.

b. Predict the average annual number of cigarettes smoked by an American adult in 2005. To do so, let $x = 15$ (Since $2005 - 1990 = 15$) and find y.

c. Use the result of part b to predict the average *daily* number of cigarettes smoked by an American adult in 2005. Round to the nearest whole. Do you think this number represents the average daily number of cigarettes smoked by an adult smoked? Why or why not?

72. Determine whether there are three consecutive integers such that their sum is three times the second integer.

73. Determine whether there are two consecutive odd integers such that 7 times the first exceeds 5 times the second by 54.

To break even in a manufacturing business, income or revenue R must equal the cost of production C. Use this information to answer Exercises 74 through 77.

74. The cost C to produce x number of skateboards is $C = 100 + 20x$. The skateboards are sold wholesale for $24 each, so revenue R is given by $R = 24x$. Find how many skateboards the manufacturer needs to produce and sell to break even. (*Hint:* Set the cost expression equal to the revenue expression and solve for x.)

75. The revenue R from selling x number of computer boards is given by $R = 60x$, and the cost C of producing them is given by $C = 50x + 5000$. Find how many boards must be sold to break even. Find how much money is needed to produce the break-even number of boards.

76. In your own words, explain what happens if a company makes and sells fewer products than the break-even number.

77. In your own words, explain what happens if more products than the break-even number are made and sold.

2.3 Formulas and Problem Solving

OBJECTIVES

Ⓐ Solve a formula for a specified variable.

Ⓑ Use formulas to solve problems.

SSM   TUTOR CENTER   SG   CD & VIDEO   MATH PRO   WEB

Ⓐ Solving Formulas for Specified Variables

Solving problems that we encounter in the real world sometimes requires us to express relationships among measured quantities. A **formula** is an equation that describes a known relationship among measured phenomena, such as time, area, and gravity. Some examples of formulas follow.

Formula	Meaning
$I = Prt$	Interest = principal · rate · time
$A = lw$	Area of a rectangle = length · width
$d = rt$	Distance = rate · time
$C = 2\pi r$	Circumference of a circle = 2 · π · radius
$V = lwh$	Volume of a rectangular solid = length · width · height

Other formulas are listed on the inside front cover of this text. Notice that the formula for the volume of a rectangular solid, $V = lwh$, is solved for V since V is by itself on one side of the equation with no Vs on the other side of the equation. Suppose that the volume of a rectangular solid is known as well as its width and its length, and we wish to find its height. One way to find its height is to begin by solving the formula $V = lwh$ for h.

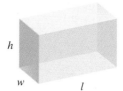

EXAMPLE 1   Solve $V = lwh$ for h.

Solution: To solve $V = lwh$ for h, we want to get h alone on one side of the equation. To do so, we divide both sides of the equation by lw.

$$V = lwh$$

$$\frac{V}{lw} = \frac{lw\,h}{lw} \qquad \text{Divide both sides by } lw.$$

$$\frac{V}{lw} = h \qquad \text{Simplify.}$$

Thus we see that to find the height of a rectangular solid, we divide its volume by the product of its length and its width.   ●

The following steps may be used to solve formulas and equations for a specified variable.

Practice Problem 1

Solve $d = rt$ for r.

Answer

1. $r = \dfrac{d}{t}$

Solving an Equation for a Specified Variable

Step 1. Clear the equation of fractions or decimals by multiplying each side of the equation by an appropriate nonzero number.

Step 2. Use the distributive property to remove grouping symbols such as parentheses.

Step 3. Combine like terms on each side of the equation.

Step 4. Use the addition property of equality to rewrite the equation as an equivalent equation with terms containing the specified variable on one side and all other terms on the other side.

Step 5. Use the distributive property and the multiplication property of equality to get the specified variable alone.

Practice Problem 2

Solve $2y + 5x = 10$ for y.

EXAMPLE 2 Solve $3y - 2x = 7$ for y.

Solution: This is a linear equation in two variables. Often an equation such as this is solved for y to reveal some properties about the graph of this equation, which we will learn more about in Chapter 3. Since there are no fractions or grouping symbols, we begin with Step 4 and get the term containing the specified variable y on one side by adding $2x$ to both sides of the equation.

$$3y - 2x = 7$$
$$3y - 2x + 2x = 7 + 2x \qquad \text{Add } 2x \text{ to both sides.}$$
$$3y = 7 + 2x$$

To solve for y, we divide both sides by 3.

$$\frac{3y}{3} = \frac{7 + 2x}{3} \qquad \text{Divide both sides by 3.}$$

$$y = \frac{7 + 2x}{3} \quad \text{or} \quad y = \frac{7}{3} + \frac{2x}{3}$$

Practice Problem 3

Solve $A = \frac{1}{2}(B + b)h$ for B.

EXAMPLE 3 Solve $A = \frac{1}{2}(B + b)h$ for b.

Solution: Since this formula for finding the area of a trapezoid contains fractions, we begin by multiplying both sides of the equation by the LCD, 2.

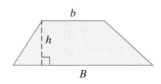

$$A = \frac{1}{2}(B + b)h$$

$$2 \cdot A = 2 \cdot \frac{1}{2}(B + b)h \qquad \text{Multiply both sides by 2.}$$

$$2A = (B + b)h \qquad \text{Simplify.}$$

$$2A = Bh + bh \qquad \text{Use the distributive property.}$$

$$2A - Bh = bh \qquad \text{Get the term containing } b \text{ alone by subtracting } Bh \text{ from both sides.}$$

Answers

2. $y = \dfrac{10 - 5x}{2}$, **3.** $B = \dfrac{2A - bh}{h}$

$$\frac{2A - Bh}{h} = \frac{bh}{h}$$ Divide both sides by h.

$$\frac{2A - Bh}{h} = b \quad \text{or} \quad b = \frac{2A - Bh}{h}$$

●

Helpful Hint

Remember that we may get the specified variable alone on either side of the equation.

B   Using Formulas to Solve Problems

In this section, we also solve problems that can be modeled by known formulas. We use the same problem-solving steps that were introduced in the previous section.

Formulas are very useful in problem solving. For example, the compound interest formula

$$A = P\left(1 + \frac{r}{n}\right)^{nt}$$

is used by banks to compute the amount A in an account that pays compound interest. The variable P represents the principal or amount invested in the account, r is the annual rate of interest, t is the time in years, and n is the number of times compounded per year.

EXAMPLE 4   Finding the Amount in a Savings Account

Marial Callier just received an inheritance of $10,000 and plans to place all the money in a savings account that pays 5% compounded quarterly to help her son go to college in 3 years. How much money will be in the account in 3 years?

Solution:

1. UNDERSTAND. Read and reread the problem. The appropriate formula needed to solve this problem is the compound interest formula

$$A = P\left(1 + \frac{r}{n}\right)^{nt}$$

Make sure that you understand the meaning of all the variables in this formula:

A = amount in the account after t years

P = principal or amount invested

t = time in years

r = annual rate of interest

n = number of times compounded per year

2. TRANSLATE. Use the compound interest formula and let $P = \$10,000, r = 5\% = 0.05, t = 3$ years, and $n = 4$ since the account is compounded quarterly, or 4 times a year.

Formula: $A = P\left(1 + \dfrac{r}{n}\right)^{nt}$

Substitute: $A = 10,000\left(1 + \dfrac{0.05}{4}\right)^{4 \cdot 3}$

Practice Problem 4

If $5000 is invested in an account paying 4% compounded monthly, determine how much money will be in the account in 2 years. Use the formula from Example 4.

Answer

4. $5415.71

3. SOLVE. We simplify the right side of the equation.

$$A = 10,000\left(1 + \frac{0.05}{4}\right)^{4\cdot3}$$

$$A = 10,000(1.0125)^{12} \qquad \text{Simplify } \frac{1 + 0.05}{4} \text{ and write } 4 \cdot 3 \text{ as } 12.$$

$$A \approx 10,000(1.160754518) \qquad \text{Approximate } (1.0125)^{12}.$$

$$A \approx 11,607.55 \qquad \text{Multiply and round to two decimal places.}$$

4. INTERPRET.

Check: Repeat your calculations to make sure that you made no error. Notice that $11,607.55 is a reasonable amount to have in the account after 3 years.

State: In 3 years, the account will contain $11,607.55. ●

GRAPHING CALCULATOR EXPLORATIONS

To solve Example 4, we approximated the expression

$$10,000\left(1 + \frac{0.05}{4}\right)^{4\cdot3}.$$

Use the following keystrokes to evaluate this expression using a graphing calculator. Notice the use of parentheses.

```
10000(1+(.05/4))
^(4*3)
            11607.54518
```

Mental Math

Solve each equation for the specified variable. See Examples 1 through 3.

1. $2x + y = 5$ for y

2. $7x - y = 3$ for y

3. $a - 5b = 8$ for a

4. $7r + s = 10$ for s

5. $5j + k - h = 6$ for k

6. $w - 4y + z = 0$ for z

EXERCISE SET 2.3

 *Solve each equation for the specified variable. See Examples 1 through 3.*

1. $d = rt$ for t

2. $W = gh$ for g

3. $I = Prt$ for r

△ **4.** $C = 2\pi r$ for r

△ **5.** $P = a + b + c$ for c

△ **6.** $a^2 + b^2 = c^2$ for b^2

7. $9x - 4y = 16$ for y

8. $2x + 3y = 17$ for y

△ **9.** $P = 2l + 2w$ for l

△ **10.** $P = 2l + 2w$ for w

11. $E = I(r + R)$ for r

12. $A = P(1 + rt)$ for t

13. $5x + 4y = 20$ for y

14. $-9x - 5y = 18$ for y

△ **15.** $S = 2LW + 2LH + 2WH$ for H

△ **16.** $S = 2\pi r^2 + 2\pi rh$ for h

△ **17.** $C = 2\pi r$ for r

△ **18.** $A = \pi r^2$ for π

19. $C = \dfrac{5}{9}(F - 32)$ for F

20. $F = \dfrac{9}{5}C + 32$ for C

Solve. Round all dollar amounts to two decimal places. See Example 4.

21. Complete the table and find the balance *A* if $3500 is invested at an annual rate of 3% for 10 years and compounded *n* times a year.

n	1	2	4	12	365
A					

22. Complete the table and find the balance *A* if $5000 is invested at an annual rate of 6% for 15 years and compounded *n* times a year.

n	1	2	4	12	365
A					

23. A principal of $6000 is invested in an account paying an annual rate of 4%. Find the amount in the account after 5 years if the account is compounded
a. semiannually.
b. quarterly.
c. monthly.

24. A principal of $25,000 is invested in an account paying an annual rate of 5%. Find the amount in the account after 2 years if the account is compounded
a. semiannually.
b. quarterly.
c. monthly.

25. One day's high temperature in Phoenix, Arizona, was recorded as 104°F. Write 104°F as degrees Celsius. [Use the formula $C = \frac{5}{9}(F - 32)$.]

26. One year's low temperature in Nome, Alaska, was recorded as −15°C. Write −15°C as degrees Fahrenheit. (Use the formula $F = \frac{9}{5}C + 32$.)

27. Omaha, Nebraska, is about 90 miles from Lincoln, Nebraska. Irania Schmidt must go to the law library in Lincoln to get a document for the law firm she works for. Find how long it takes her to drive *round-trip* if she averages 50 mph.

28. It took the Selby family $5\frac{1}{2}$ hours round-trip to drive from their house to their beach house 154 miles away. Find their average speed.

△**29.** A package of floor tiles contains 24 one-foot-square tiles. Find how many packages should be bought to cover a square ballroom floor whose side measures 64 feet.

64 ft
64 ft

△**30.** One-foot-square ceiling tiles are sold in packages of 50. Find how many packages must be bought for a rectangular ceiling 18 feet by 12 feet.

31. If the area of a triangular kite is 18 square feet and its base is 4 feet, find the height of the kite.

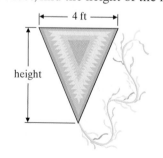

4 ft

height

32. Bryan, Eric, Mandy, and Melissa would like to go to Disneyland in 3 years. The total cost should be $4500. If each invests $1000 in a savings account paying 5.5% interest, compounded semiannually, will they have enough in 3 years?

110

△ **33.** A gallon of latex paint can cover 500 square feet. Find how many gallon containers of paint should be bought to paint two coats on the walls of a rectangular room whose dimensions are 14 feet by 16 feet. (Assume 8-foot ceilings and disregard any openings such as windows or doors.)

△ **34.** A gallon of enamel paint can cover 300 square feet. Find how many gallon containers of paint should be bought to paint three coats on a wall measuring 21 feet by 8 feet.

△ **35.** A portion of the external tank of the Space Shuttle *Endeavour* is a liquid hydrogen tank. If the ends of the tank are hemispheres, find the volume of the tank. To do so, answer parts **(a)** through **(c)**. (*Source:* NASA/Kennedy Space Center)

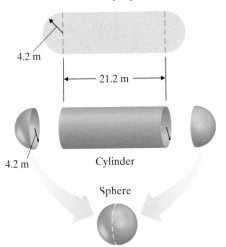

a. Find the volume of the cylinder shown. Round to two decimal places.
b. Find the volume of the sphere shown. Round to two decimal places.
c. Add the results of parts **(a)** and **(b)**. This sum is the approximate volume of the tank.

36. The Cassini spacecraft mission to Saturn was launched October 15, 1997. It will take more than six and a half years to reach Saturn, arriving in July 2004. During its mission, Cassini will travel a total distance of 2 billion miles in 80.5 months. Find the average speed of the spacecraft in miles per hour. (*Hint:* Convert 80.5 months to hours using 1 month = 30 days and then use the formula $d = rt$.) (*Source:* NASA Jet Propulsion Laboratory)

△ **37.** In 1945, Arthur C. Clarke, a scientist and science-fiction writer, predicted that an artificial satellite placed at a height of 22,248 miles directly above the equator would orbit the globe at the same speed with which Earth was rotating. This belt along the equator is known as the Clarke belt. Use the formula for the circumference of a circle and approximate the "length" of the Clarke belt. (*Hint:* Recall that the radius of Earth is approximately 4000 miles. Round to the nearest whole mile.)

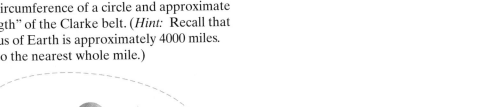

△ **38.** The *Endeavour* Space Shuttle has a cargo bay that is in the shape of a cylinder whose length is 18.3 meters and whose diameter is 4.6 meters. Find its volume.

△ **39.** The deepest hole in the ocean floor is beneath the Pacific Ocean and is called Hole 504B. It is located off the coast of Ecuador. Scientists are drilling it to learn more about Earth's history. Currently, the hole is in the shape of a cylinder whose volume is approximately 3800 cubic feet and whose length is 1.3 miles. Find the radius of the hole to the nearest hundredth of a foot. (*Hint:* Make sure the same units of measurement are used.)

40. The deepest man-made hole is called the Kola Superdeep Borehole. It is approximately 8 miles deep and is located near a small Russian town in the Arctic Circle. If it takes 7.5 hours to remove the drill from the bottom of the hole, find the rate that the drill can be retrieved in feet per second. Round to the nearest tenth. (*Hint:* Write 8 miles as feet, 7.5 hours as seconds, and then use the formula $d = rt$.)

△ **41.** Eartha is the world's largest globe. It is located at the headquarters of DeLorme, a mapmaking company in Yarmouth, Maine. Eartha is 41.125 feet in diameter. Find its exact circumference (distance around) and then approximate its circumference using 3.14 for π. (*Source:* DeLorme)

△ **42.** Eartha is in the shape of a sphere. Its radius is about 20.6 feet. Approximate Eartha's volume to the nearest cubic foot. Using the approximation 3.14 for π. (*Source:* DeLorme)

The calorie count of a serving of food can be computed based on its composition of carbohydrate, fat, and protein. The calorie count C for a serving of food can be computed using the formula $C = 4h + 9f + 4p$, where h is the number of grams of carbohydrate contained in the serving, f is the number of grams of fat contained in the serving, and p is the number of grams of protein contained in the serving.

43. Solve this formula for f, the number of grams of fat contained in a serving of food.

44. Solve this formula for h, the number of grams of carbohydrate contained in a serving of food.

45. A serving of cashews contains 14 grams of fat, 7 grams of carbohydrate, and 6 grams of protein. How many calories are in this serving of cashews?

46. A serving of chocolate candies contains 9 grams of fat, 30 grams of carbohydrate, and 2 grams of protein. How many calories are in this serving of chocolate candies?

47. A serving of raisins contains 130 calories and 31 grams of carbohydrate. If raisins are a fat-free food, how much protein is provided by this serving of raisins?

48. A serving of yogurt contains 120 calories, 21 grams of carbohydrate, and 5 grams of protein. How much fat is provided by this serving of yogurt? Round to the nearest tenth of a gram.

Review and Preview

Determine which numbers in the set $\{-3, -2, -1, 0, 1, 2, 3\}$ are solutions of each inequality. See Sections 1.3 and 2.1.

49. $x < 0$

50. $x > 1$

51. $x + 5 \leq 6$

52. $x - 3 \geq -7$

53. In your own words, explain what real numbers are solutions of $x < 0$.

54. In your own words, explain what real numbers are solutions of $x > 1$.

Combining Concepts

55. Solar System distances are so great that units other than miles or kilometers are often used. For example, the astronomical unit (AU) is the average distance between Earth and the sun, or 92,900,000 miles. Use this information to convert each planet's distance in miles from the sun to astronomical units. Round to three decimal places. (*Source:* National Space Science Data Center)

Planet	Miles from the Sun	AU from the Sun	Planet	Miles from the Sun	AU from the Sun
Mercury	36 million		Saturn	886.1 million	
Venus	67.2 million		Uranus	1783 million	
Earth	92.9 million		Neptune	2793 million	
Mars	141.5 million		Pluto	3670 million	
Jupiter	483.3 million				

56. An orbit such as Clarke's belt in Exercise 37 is called a geostationary orbit. In your own words, why do you think that communications satellites are placed in geostationary orbits?

57. How much do you think it costs each American to build a Space Shuttle? Write down your estimate. The Space Shuttle *Endeavour* was completed in 1992 and cost approximately $1.7 billion. If the population of the United States in 1992 was 250 million, find the cost per person to build the *Endeavour*. How close was your estimate?

58. Find *how much interest* $10,000 earns in 2 years in a certificate of deposit account paying 8.5% interest compounded quarterly.

59. If you are investing money in a savings account paying a rate of r, which account should you choose—an account compounded 4 times a year or 12 times a year? Explain your choice.

60. To borrow money at a rate of r, which loan plan should you choose—one compounding 4 times a year or 12 times a year? Explain your choice.

61. The Drake Equation is a formula used to estimate the number of technological civilizations that might exist in our own Milky Way Galaxy. The Drake Equation is given as $N = R^* \times f_p \times n_e \times f_l \times f_i \times f_c \times L$. Solve the Drake Equation for the variable n_e. (*Note:* Descriptions of the meaning of each variable in this equation, as well as Drake Equation calculators, exist online. For more information, try doing a Web search on "Drake Equation.")

NUTRITION LABELS

Since 1994, the Food and Drug Administration (FDA) of the Department of Health and Human Services and the Food Safety and Inspection Service of the U.S. Department of Agriculture (USDA) have required nutrition labels like the one below on most food packaging. The labels were designed to help consumers make healthful food choices by giving standardized nutrition information.

Nutrition Facts		
Serving Size	(36g)	
Servings Per Container	6	
Amount Per Serving		
Calories 150	Calories from Fat 50	
		% Daily Value*
Total Fat 6g		**9%**
Saturated Fat 2g		9%
Cholesterol 0mg		**0%**
Sodium 80mg		**3%**
Potassium 40mg		**1%**
Total Carbohydrate 24g		**8%**
Dietary Fiber 0g		0%
Sugars 11g		
Protein 2g		

Vitamin A	25%	Vitamin C	25%
Calcium	50%	Iron	25%
Vitamin D	25%	Vitamin E	25%
Thiamin	25%	Riboflavin	25%
Niacin	25%	Vitamin B$_6$	25%
Folate	25%	Vitamin B$_{12}$	25%
Biotin	25%	Pant. Acid	25%
Phosphorus	30%	Iodine	25%
Magnesium	25%	Zinc	25%
Copper	25%		

* Percent Daily Values are based on a 2,000 calorie diet. Your daily values may be higher or lower depending on your calorie needs:

One key feature of this labeling is the column of % Daily Value figures. Most of these values are based on a 2000-calorie diet. Critics complain that this diet applies to only a small segment of the population and should be more versatile. However, FDA and USDA officials responded that these are based on 2000 calories as a guideline only to help consumers gauge the relative amount of a nutrient contained by a food product. For instance, a food with 10 grams of saturated fat per serving could be mistaken for a food low in saturated fat. However, in a 2000-calorie diet, 10 grams of saturated fat represents 50% of the allowable daily intake. This percentage signals to consumers that this product is relatively high in saturated fat. Similarly, a food with 125 mg of sodium per serving could be mistaken for a high-sodium food. However, because a person should ingest no more than 2400 mg of sodium per day, a corresponding % Daily Value figure of about 5% conveys to a consumer that a serving of food with 125 mg of sodium is relatively low in sodium.

The % Daily Value figures that depend on the number of calories consumed per day include total fat, saturated fat, carbohydrate, protein, and dietary fiber. The % Daily Value figures for nutrients such as cholesterol, sodium, and potassium do not depend on calories.

For diets that include more or less than 2000 calories, the daily allowable amount of a nutrient that depends on calories can be figured with the following guidelines:

- The daily allowable amount of total fat is based on 30% of calories.

- The daily allowable amount of saturated fat is based on 10% of calories.

- The daily allowable amount of carbohydrate is based on 60% of calories.

- The daily allowable amount of protein is based on 10% of calories.

- The daily allowable amount of dietary fiber is based on 11.5 grams of fiber per 1000 calories.

Additionally, each gram of protein and carbohydrate contains 4 calories. A gram of fat contains 9 calories.

The above information can be used to calculate % Daily Value figures for diets with other calorie levels. For instance, the daily allowable amount of total fat in a 2200-calorie diet is $0.30(2200) = 660$ calories from fat. For the nutrition label shown, one serving contains 6 grams of fat or $6(9) = 54$ calories from fat. The % Daily Value for total fat that one serving of this food provides in a 2200-calorie diet is $54 \div 660 \approx 0.08$ or 8%.

GROUP ACTIVITY

1. Calculate the daily allowable amounts of total fat, saturated fat, carbohydrate, protein, and dietary fiber in 1500-calorie, 1800-calorie, 2500-calorie, and 2800-calorie diets. Summarize your results in a table.

2. Choose five different food products having Nutrition Facts labels. For each product, use the information given on the label to calculate the % Daily Value figures for total fat, saturated fat, carbohydrate, protein, and dietary fiber for (a) a 1500-calorie diet, (b) an 1800-calorie diet, (c) a 2500-calorie diet, and (d) a 2800-calorie diet. Create a chart showing your results for each food product.

3. Use the data given in the food product nutrition labels used in Question 2 to estimate the daily allowable amounts of cholesterol, sodium, and potassium. Recall that the allowable amounts for these nutrients do not depend on calorie intake.

2.4 Linear Inequalities and Problem Solving

Relationships among measurable quantities are not always described by equations. For example, suppose that a salesperson earns a base of $600 per month plus a commission of 20% of sales. Suppose we want to find the minimum amount of sales needed to receive a total income of *at least* $1500 per month. Here, the phrase "at least" implies that an income of $1500 *or more* is acceptable. In symbols, we can write

income ≥ 1500

This is an example of an inequality, which we will solve in Example 11.

A *linear inequality* is similar to a linear equation except that the equality symbol is replaced with an inequality symbol, such as $<, >, \leq,$ or $\geq$.

Linear Inequality in One Variable

A **linear inequality in one variable** is an inequality that can be written in the form

$ax + b < c$

where a, b, and c are real numbers and $a \neq 0$.
For example,

$3x + 5 \geq 4$ $2y < 0$ $4n \geq n - 3$

$3(x - 4) < 5x$ $\dfrac{x}{3} \leq 5$

In this section, when we make definitions, state properties, or list steps about an inequality containing the symbol $<$, we mean that the definition, property, or steps apply to an inequality containing the symbols $>, \leq,$ and $\geq$ also.

Ⓐ Using Interval Notation

A **solution** of an inequality is a value of the variable that makes the inequality a true statement. The **solution set** of an inequality is the set of all solutions. Notice that the solution set of the inequality $x > 2$, for example, contains all numbers greater than 2. Its graph is an interval on the number line since an infinite number of values satisfy the variable. If we use open/closed-circle notation, the graph of $\{x | x > 2\}$ looks like:

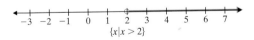

$\{x | x > 2\}$

In this text, a different graphing notation will be used to help us understand **interval notation**. Instead of an open circle, we use a parenthesis; instead of a closed circle, we use a bracket. With this new notation, the graph of $\{x | x > 2\}$ now looks like:

and can be represented in interval notation as $(2, \infty)$. The symbol ∞ is read "infinity" and indicates that the interval includes *all* numbers greater than 2. The left parenthesis indicates that 2 *is not* included in the interval. Using a left bracket, [, would indicate that 2 *is* included in the interval. The following table shows three equivalent ways to describe an interval: in set notation, as a graph, and in interval notation.

Set Notation	Graph	Interval Notation
$\{x \mid x < a\}$		$(-\infty, a)$
$\{x \mid x > a\}$		(a, ∞)
$\{x \mid x \leq a\}$		$(-\infty, a]$
$\{x \mid x \geq a\}$		$[a, \infty)$
$\{x \mid a < x < b\}$		(a, b)
$\{x \mid a \leq x \leq b\}$		$[a, b]$
$\{x \mid a < x \leq b\}$		$(a, b]$
$\{x \mid a \leq x < b\}$		$[a, b)$

Practice Problems 1-3

Graph each set on a number line and then write it in interval notation.

1. $\{x \mid x > -3\}$

2. $\{x \mid x \leq 0\}$

3. $\{x \mid -0.5 \leq x < 2\}$

Helpful Hint

Notice that a parenthesis is always used to enclose ∞ and $-\infty$.

EXAMPLES

Graph each set on a number line and then write it in interval notation.

1. $\{x \mid x \geq 2\}$ 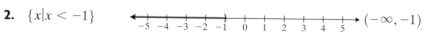 $[2, \infty)$

2. $\{x \mid x < -1\}$ 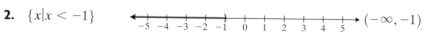 $(-\infty, -1)$

3. $\{x \mid 0.5 < x \leq 3\}$ 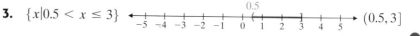 $(0.5, 3]$

Try the Concept Check in the margin.

Concept Check

Explain what is wrong with writing the interval $(5, \infty]$.

Answers

1. $(-3, \infty)$,

2. $(-\infty, 0]$,

3. $[-0.5, 2)$,

Concept Check: should be $(5, \infty)$ since a parenthesis is always used to enclose ∞

B Using the Addition Property of Inequality

Interval notation can be used to write solutions of linear inequalities. To solve a linear inequality, we use a process similar to the one used to solve a linear equation. We use properties of inequalities to write equivalent inequalities until the variable is alone on one side of the inequality.

Addition Property of Inequality

If a, b, and c are real numbers, then

$$a < b \quad \text{and} \quad a + c < b + c$$

are equivalent inequalities.

In other words, we may add the same real number to both sides of an inequality, and the resulting inequality will have the same solution set. This property also allows us to subtract the same real number from both sides.

EXAMPLE 4   Solve: $x - 2 < 5$. Graph the solution set.

Solution:

$$x - 2 < 5$$
$$x - 2 + 2 < 5 + 2 \qquad \text{Add 2 to both sides.}$$
$$x < 7 \qquad \text{Simplify.}$$

The solution set is $\{x \mid x < 7\}$, which in interval notation is $(-\infty, 7)$. The graph of the solution set is

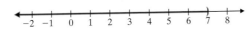

Helpful Hint

In Example 4, the solution set is $\{x \mid x < 7\}$. This means that *all* numbers less than 7 are solutions. For example, 6.9, 0, $-\pi$, 1, and -56.7 are solutions, just to name a few. To see this, replace x in $x - 2 < 5$ with each of these numbers and see that the result is a true inequality.

EXAMPLE 5   Solve: $4x - 2 < 5x$. Graph the solution set.

Solution: To get x alone on one side of the inequality, we subtract $4x$ from both sides.

$$4x - 2 < 5x$$
$$4x - 2 - 4x < 5x - 4x \qquad \text{Subtract } 4x \text{ from both sides.}$$
$$-2 < x \quad \text{or} \quad x > -2 \qquad \text{Simplify.}$$

Helpful Hint

Don't forget that $-2 < x$ means the same as $x > -2$.

The solution set is $\{x \mid x > -2\}$, which in interval notation is $(-2, \infty)$. The graph is

Practice Problem 4

Solve: $x + 3 < 1$. Graph the solution set.

Practice Problem 5

Solve: $3x - 4 < 4x$. Graph the solution set.

Answers

4. $\{x \mid x < -2\}, (-\infty, -2),$

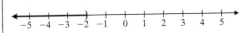

5. $\{x \mid x > -4\}, (-4, \infty),$

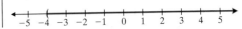

Practice Problem 6

Solve: $5x - 1 \geq 4x + 4$. Graph the solution set.

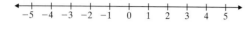

EXAMPLE 6   Solve: $3x + 4 \geq 2x - 6$. Graph the solution set.

Solution:

$$3x + 4 \geq 2x - 6$$
$$3x + 4 - 2x \geq 2x - 6 - 2x \qquad \text{Subtract } 2x \text{ from both sides.}$$
$$x + 4 \geq -6 \qquad \text{Combine like terms.}$$
$$x + 4 - 4 \geq -6 - 4 \qquad \text{Subtract 4 from both sides.}$$
$$x \geq -10 \qquad \text{Simplify.}$$

The solution set is $\{x | x \geq -10\}$, which in interval notation is $[-10, \infty)$. The graph of the solution set is

(number line graphed from −11 to −1, bracket at −10 shading right)

c   Using the Multiplication Property of Inequality

Next, we introduce and use the multiplication property of inequality to solve linear inequalities. To understand this property, let's start with the true statement $-3 < 7$ and multiply both sides by 2.

$$-3 < 7$$
$$-3(2) < 7(2) \qquad \text{Multiply both sides by 2.}$$
$$-6 < 14 \qquad \text{True.}$$

The statement remains true.

Notice what happens if both sides of $-3 < 7$ are multiplied by -2.

$$-3 < 7$$
$$-3(-2) < 7(-2)$$
$$6 < -14 \qquad \text{False.}$$

The inequality $6 < -14$ is a false statement. However, *if the direction of the inequality sign is reversed*, the result is

$$6 > -14 \qquad \text{True.}$$

These examples suggest the following property.

Multiplication Property of Inequality

If a, b, and c are real numbers and c is **positive**, then $a < b$ and $ac < bc$ are equivalent inequalities.

If a, b, and c are real numbers and c is **negative**, then $a < b$ and $ac > bc$ are equivalent inequalities.

In other words, we may multiply both sides of an inequality by the same positive real number, and the result is an equivalent inequality. We may also multiply both sides of an inequality by the same *negative number* and *reverse the direction of the inequality symbol*, and the result is an equivalent inequality. The multiplication property holds for division also since division is defined in terms of multiplication.

Helpful Hint

Whenever both sides of an inequality are multiplied or divided by a negative number, the direction of the inequality symbol *must be* reversed to form an equivalent inequality.

Answer

6. $\{x | x \geq 5\}, [5, \infty),$

EXAMPLE 7   Solve: $\frac{1}{4}x \le \frac{3}{2}$. Graph the solution set.

Solution:

> **Helpful Hint**
> The inequality symbol is the same since we are multiplying by a *positive* number.

The solution set is $\{x \mid x \le 6\}$, which in interval notation is $(-\infty, 6]$. The graph of this solution set is

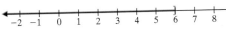

EXAMPLE 8   Solve: $-2.3x < 6.9$. Graph the solution set.

Solution:

> **Helpful Hint**
> The inequality symbol is *reversed* since we divided by a *negative* number.

The solution set is $\{x \mid x > -3\}$, which is $(-3, \infty)$ in interval notation. The graph of the solution set is

Try the Concept Check in the margin.

To solve linear inequalities in general, we follow steps similar to those for solving linear equations.

Solving a Linear Inequality in One Variable

Step 1.   Clear the equation of fractions or decimals by multiplying both sides of the inequality by an appropriate number.

Step 2.   Use the distributive property to remove grouping symbols such as parentheses.

Step 3.   Combine like terms on each side of the inequality.

Step 4.   Use the addition property of inequality to write the inequality as an equivalent inequality with variable terms on one side and numbers on the other side.

Step 5.   Use the multiplication property of inequality to get the variable alone on one side of the inequality.

D   **Using Both Properties of Inequality**

Many problems require us to use both properties of inequality.

Practice Problem 7

Solve: $\frac{1}{6}x \le \frac{2}{3}$. Graph the solution set.

Practice Problem 8

Solve: $-1.1x < 5.5$. Graph the solution set.

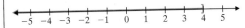

Concept Check

In which of the following inequalities must the inequality symbol be reversed during the solution process?

a. $-2x > 7$

b. $2x - 3 > 10$

c. $-x + 4 + 3x < 5$

d. $-x + 4 < 5$

Answers

7. $\{x \mid x \le 4\}, (-\infty, 4,]$

8. $\{x \mid x > -5\}, (-5, \infty)$,

Concept Check: a, d

Practice Problem 9

Solve: $6 - 2x \leq 8x - 14$. Write the solution set in interval notation.

Practice Problem 10

Solve: $\frac{3}{4}(x + 2) \geq x - 6$. Write the solution set in interval notation.

EXAMPLE 9

Solve: $5 - x \leq 4x - 15$. Write the solution set in interval notation.

Solution:

$$5 - x \leq 4x - 15$$
$$5 - x + x \leq 4x - 15 + x \quad \text{Add } x \text{ to both sides.}$$
$$5 \leq 5x - 15 \quad \text{Combine like terms.}$$
$$5 + 15 \leq 5x - 15 + 15 \quad \text{Add 15 to both sides.}$$
$$20 \leq 5x \quad \text{Combine like terms.}$$
$$\frac{20}{5} \leq \frac{5x}{5} \quad \text{Divide both sides by 5.}$$
$$4 \leq x \quad \text{or} \quad x \geq 4 \quad \text{Simplify.}$$

The solution set is $[4, \infty)$.

EXAMPLE 10

Solve: $\frac{2}{5}(x - 6) \geq x - 1$. Write the solution set in interval notation.

Solution:

$$\frac{2}{5}(x - 6) \geq x - 1$$
$$5\left[\frac{2}{5}(x - 6)\right] \geq 5(x - 1) \quad \text{Multiply both sides by 5 to eliminate fractions.}$$
$$2x - 12 \geq 5x - 5 \quad \text{Use the distributive property.}$$
$$-3x - 12 \geq -5 \quad \text{Subtract } 5x \text{ from both sides.}$$
$$-3x \geq 7 \quad \text{Add 12 to both sides.}$$
$$\frac{-3x}{-3} \leq \frac{7}{-3} \quad \text{Divide both sides by } -3 \text{ and reverse the inequality symbol.}$$
$$x \leq -\frac{7}{3} \quad \text{Simplify.}$$

The solution set is $\left(-\infty, -\frac{7}{3}\right]$.

E Linear Inequalities and Problem Solving

Problems containing words such as "at least," "at most," "between," "no more than," and "no less than" usually indicate that an inequality is to be solved instead of an equation. In solving applications involving linear inequalities, we use the same four-step strategy as when we solved applications involving linear equations.

Practice Problem 11

A salesperson earns $1000 a month plus a commission of 15% of sales. Find the minimum amount of sales needed to receive a total income of at least $4000 per month.

EXAMPLE 11 Calculating Income with Commission

A salesperson earns $600 per month plus a commission of 20% of sales. Find the minimum amount of sales needed to receive a total income of at least $1500 per month.

Solution:

1. UNDERSTAND. Read and reread the problem. Let

    x = amount of sales

2. TRANSLATE. As stated in the beginning of this section, we want the income to be greater than or equal to $1500. To write an inequality, notice that the salesperson's income consists of $600 plus a commission (20% of sales).

Answers

9. $[2, \infty)$, **10.** $(-\infty, 30]$, **11.** $20,000

In words: 600   +   commission (20% of sales $\geq$ 1500

$\downarrow$ $\downarrow$ $\downarrow$

Translate: 600   + $0.20x$ $\geq$ 1500

3. SOLVE the inequality for x.

$$600 + 0.20x \geq 1500$$
$$600 + 0.20x - 600 \geq 1500 - 600$$
$$0.20x \geq 900$$
$$x \geq 4500$$

4. INTERPRET.

Check: The income for sales of $4500 is

$600 + 0.20(4500)$, or 1500

Thus, if sales are greater than or equal to $4500, income is greater than or equal to $1500.

State: The minimum amount of sales needed for the salesperson to earn at least $1500 per month is $4500. ●

EXAMPLE 12   Finding the Annual Consumption

In the United States, the annual consumption of cigarettes is declining. The consumption c in billions of cigarettes per year since the year 1985 can be approximated by the formula

$$c = -14.25t + 598.69$$

where t is the number of years after 1985. Use this formula to predict the first year that the consumption of cigarettes will be less than 200 billion per year.

Solution:

1. UNDERSTAND. Read and reread the problem. To become familiar with the given formula, let's find the cigarette consumption after 20 years, which would be the year $1985 + 20$, or 2005. To do so, we substitute 20 for t in the given formula.

$$c = -14.25(20) + 598.69 = 313.69$$

Thus, in 2005, we predict cigarette consumption to be about 313.69 billion. Variables have already been assigned in the given formula. For review, they are

c = the annual consumption of cigarettes in the United States in billions of cigarettes

t = the number of years after 1985

2. TRANSLATE. We are looking for the first year that the consumption of cigarettes c is less than 200. Since we are finding years t, we substitute the expression in the formula given for c, or

$$-14.25t + 598.69 < 200$$

3. SOLVE the inequality.

$$-14.25t + 598.69 < 200 \qquad \text{Subtract 598.69 from both sides.}$$
$$-14.25t < -398.69 \qquad \text{Divide both sides by } -14.25 \text{ and round the result.}$$
$$t > 27.98$$

4. INTERPRET.

Check: We substitute a number greater than 27.98 and see that c is less than 200.

State: The annual consumption of cigarettes will be less than 200 billion for more than 27.98 years after 1985, or in approximately $28 + 1985 = 2013$.

Practice Problem 12

Use the formula given in Example 12 to predict when the consumption of cigarettes will be less than 100 billion per year.

Answer

12. after the year 2020

FOCUS ON **The Real World**

ENERGY AUDIT

Have you ever been surprised by high electric bills? Has it made you wonder where all of your electricity expenditure is going or how to lower your bill? If so, one approach to learning more about your electricity consumption is performing an energy audit of your home or apartment. Once you understand your patterns of electricity usage, you can make informed decisions on where to cut back or whether or not to replace an older appliance with a newer, more energy efficient one.

To perform your own energy audit, fill out the table below. First, make a list of all the electrical appliances in your home. Be sure to include components of your heating, hot water, and/or air-conditioning systems, major appliances, indoor and outdoor lights, computer and audio-visual components, and small kitchen or personal care appliances. Don't forget to include easily overlooked items such as room space heaters, ceiling fans, and water bed heaters.

Next, estimate how many hours each item is run per 30-day month. For items used nearly every day, estimate daily usage in hours and multiply by 30. For items used less often, estimate how many hours they are used per week. Then divide by 7 and multiply by 30 to get an estimate for a 30-day month.

For each item on the list, record its wattage. This information can usually be found on its serial number plate. Wattage is abbreviated W, so look for a number like 13W or 200W. If wattage is not listed on the plate, look for information on voltage (abbreviated V for volts) and amperage (abbreviated A for amps). Wattage can be estimated by multiplying volts times amps. (*Note:* Sometimes a range of values is listed for voltage. If the range includes 120 V, use 120 in the wattage calculation. Otherwise, use the maximum value of the voltage range for the wattage calculation.)

Fill in the fourth column of the table by multiplying the number of hours each appliance is run during a month by its wattage to find watt-hours. Then fill in the fifth column of the table by dividing watt-hours by 1000 to find kilowatt-hours. For the last column, consult your electricity bill to find the price charged by your electric company per kilowatt-hour (often abbreviated KWH). Alternatively, contact the local electric company to ask its standard charge per kilowatt-hour for residential customers. Fill in the last column of the table by figuring the cost to run an appliance for one month: Multiply the number of kilowatt-hours per month by what the electric company charges per kilowatt-hour.

Appliance	Hours Run per 30-Day Month	Wattage (or use volts × amps)	Watt-Hours (hours × wattage)	Kilowatt-Hours (watt-hours ÷ 1000)	Cost to Run Appliance for One Month (kilowatt-hours × cost per KWH)

CRITICAL THINKING

Do an energy audit of your home, apartment, or dormitory room.

1. Which item is the most expensive to operate over the course of the month? Does this surprise you? Why or why not?

2. Which item is the least expensive to operate? Does this surprise you? Why or why not?

3. Are there any items on the list whose usage could be cut back to save energy costs? Which ones would be the most viable choices? How much could usage be cut and how much money would that save? Explain.

4. Are there any appliances that could be replaced with more energy-efficient models? Conduct research to find a more recent model that would be a better choice. If the usage of the new model is the same as the old model, how much money could be saved each month by switching? How long will it take for the monthly energy cost savings to "pay off" the price of buying the new appliance? Explain.

5. In what other ways could you lower your electric bill?

Name _____ Section _____ Date _____

Mental Math

Solve each inequality.

1. $x - 2 < 4$

2. $x - 1 > 6$

3. $x + 5 \geq 15$

4. $x + 1 \leq 8$

5. $3x > 12$

6. $5x < 20$

7. $\dfrac{x}{2} \leq 1$

8. $\dfrac{x}{4} \geq 2$

EXERCISE SET 2.4

Ⓐ *Graph the solution set of each inequality on a number line and then write it in interval notation. See Examples 1 through 3.*

1. $\{x \mid x < -3\}$

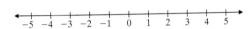

2. $\{x \mid x \geq -7\}$

3. $\{x \mid x \geq 0.3\}$

4. $\{x \mid x < -0.2\}$

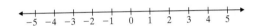

5. $\{x \mid 5 < x\}$

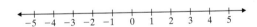

6. $\{x \mid -7 \geq x\}$

7. $\{x \mid -2 < x < 5\}$

8. $\{x \mid -5 \leq x \leq -1\}$

9. $\{x \mid 5 > x > -1\}$

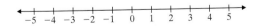

10. $\{x \mid -3 \geq x \geq -7\}$

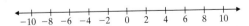

11. When graphing the solution set of an inequality, explain how you know whether to use a parenthesis or a bracket.

12. Explain what is wrong with the interval notation $(-6, -\infty)$.

123

Solve. Graph the solution set and write it in interval notation. See Examples 4 through 6.

13. $x - 7 \geq -9$

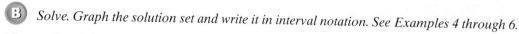

14. $x + 2 \leq -1$

15. $7x < 6x + 1$

16. $11x < 10x + 5$

17. $8x - 7 \leq 7x - 5$

18. $7x - 1 \geq 6x - 1$

C *Solve. Graph the solution set and then write it in interval notation. See Examples 7 and 8.*

19. $\dfrac{3}{4}x \geq 2$

20. $\dfrac{5}{6}x \geq -8$

21. $5x < -23.5$

22. $4x > -11.2$

23. $-3x \geq 9$

24. $-4x \geq 15$

25. $-x < -4$

26. $-x > -2$

D *Solve. Write the solution set using interval notation. See Examples 9 and 10.*

27. $-2x + 7 \geq 9$

28. $8 - 5x \leq 23$

29. $15 + 2x \geq 4x - 7$

30. $20 + x < 6x$

31. $3(x - 5) < 2(2x - 1)$

32. $5(x + 4) \leq 4(2x + 3)$

33. $\dfrac{1}{2} + \dfrac{2}{3} \geq \dfrac{x}{6}$

34. $\dfrac{3}{4} - \dfrac{2}{3} > \dfrac{x}{6}$

35. $-5x + 4 \leq -4(x - 1)$

36. $-6x + 2 < -3(x + 4)$

37. $\dfrac{1}{4}(x - 7) \geq x + 2$

38. $\dfrac{3}{5}(x + 1) \leq x + 1$

39. $0.8x + 0.6x \geq 4.2$

40. $0.7x - x > 0.45$

41. $4(2x + 1) > 4$

42. $6(2 - x) \geq 12$

43. $\dfrac{5x + 1}{7} - \dfrac{2x - 6}{4} \geq -4$

44. $\dfrac{1 - 2x}{3} + \dfrac{3x + 7}{7} > 1$

45. $4(x - 6) + 2x - 4 \geq 3(x - 7) + 10x$

46. $7(2x + 3) + 4x \leq 7 + 5(3x - 4)$

47. $-3(2x - 1) < -4[2 + 3(x + 2)]$

48. $-2(4x + 2) > -5[1 + 2(x - 1)]$

49. $14 - (5x - 6) \geq -6(x + 1) - 5$

50. $13y - (9y + 2) \leq 5(y - 6) + 10$

51. $\dfrac{1}{2}(3x - 4) \leq \dfrac{3}{4}(x - 6) + 1$

52. $\dfrac{2}{3}(x + 3) < \dfrac{1}{6}(2x - 8) + 2$

E *Solve. See Examples 11 and 12.*

53. Shureka Washburn has scores of 72, 67, 82, and 79 on her algebra tests. Use an inequality to find the minimum score she must make on the final exam to pass the course with an average of 60 or higher, given that the final exam counts as two tests.

54. In a Winter Olympics 5000-meter speed-skating event, Hans Holden scored times of 6.85, 7.04, and 6.92 minutes on his first three trials. Use an inequality to find the maximum time he can score on his last trial so that his average time is under 7.0 minutes.

55. A small plane's maximum takeoff weight is 2000 pounds. Six passengers weigh an average of 160 pounds each. Use an inequality to find the maximum weight of luggage and cargo the plane can carry.

56. A clerk must use the elevator to move boxes of paper. The elevator's weight limit is 1500 pounds. If each box of paper weighs 66 pounds and the clerk weighs 147 pounds, use an inequality to find the maximum number of boxes she can move on the elevator at one time.

57. To mail an envelope first class, the U.S. Post Office charges 34 cents for the first ounce and 23 cents per ounce for each additional ounce. Use an inequality to find the maximum weight that can be mailed for $3.10.

58. A shopping mall parking garage charges $1 for the first half-hour and 60 cents for each additional half-hour or a portion of a half-hour. Use an inequality to find how long you can park if you have only $4.00 in cash.

59. Northeast Telephone Company offers two billing plans for local calls. Plan 1 charges $25 per month for unlimited calls, and plan 2 charges $13 per month plus 6 cents per call. Use an inequality to find the number of monthly calls for which plan 1 is more economical than plan 2.

60. A car rental company offers two subcompact rental plans. Plan A charges $32 per day for unlimited mileage, and plan B charges $24 per day plus 15 cents per mile. Use an inequality to find the number of daily miles for which plan A is more economical than plan B.

61. At room temperature, glass used in windows actually has some properties of a liquid. It has a very slow, viscous flow. (Viscosity is the property of a fluid that resists internal flow. For example, lemonade flows more easily than fudge syrup. Fudge syrup has a higher viscosity than lemonade.) Glass does not become a true liquid until temperatures are greater than or equal to 500°C. Find the Fahrenheit temperatures for which glass is a liquid. (Use the formula $F = \frac{9}{5}C + 32$.)

62. Stibnite is a silvery white mineral with a metallic luster. It is one of the few minerals that melts easily in a match flame or at temperatures of approximately 977°F or greater. Find the Celsius temperatures for which stibnite melts. [Use the formula $C = \frac{5}{9}(F - 32)$.]

63. Although beginning salaries vary greatly according to your field of study, the equation

$$s = 651.2t + 27{,}821$$

can be used to approximate and to predict average beginning salaries for candidates with bachelor's degrees. The variable s is the starting salary and t is the number of years after 1989.

a. Approximate when beginning salaries for candidates will be greater than $35,000.

b. Determine the year you plan to graduate from college. Use this year to find the corresponding value of t and approximate your beginning salary.

64. Use the formula in Example 12 to estimate the years that the consumption of cigarettes will be less than 50 billion per year.

The average consumption per person per year of whole milk w can be approximated by the equation

$$w = -1.43t + 79.37$$

where t is the number of years after 1993 and w is measured in pounds. The average consumption of skim milk s per person per year can be approximated by the equation

$$s = 1.18t + 28.31$$

where t is the number of years after 1993 and s is measured in pounds. The consumption of whole milk is shown on the graph in blue and the consumption of skim milk is shown on the graph in red. Use this information to answer Exercises 65 through 73.

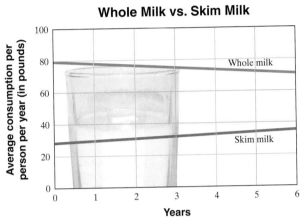

Whole Milk vs. Skim Milk

Source: Based on data from Economic Research Service, U.S. Department of Agriculture, *Agricultural Outlook*, October 2001

65. Is the consumption of whole milk increasing or decreasing over time? Explain how you arrived at your answer.

66. Is the consumption of skim milk increasing or decreasing over time? Explain how you arrived at your answer.

67. Predict the consumption of whole milk in 2007. (*Hint:* Find the value of *t* that corresponds to 2007.)

68. Predict the consumption of skim milk in 2007. (*Hint:* Find the value of *t* that corresponds to 2007.)

69. Determine when the consumption of whole milk will be less than 55 pounds per person per year.

70. Determine when the consumption of skim milk will be greater than 45 pounds per person per year.

71. For 1993 through 1999, the consumption of whole milk was greater than the consumption of skim milk. Explain how this can be determined from the graph.

72. How will the two lines in the graph appear when the consumption of whole milk is the same as the consumption of skim milk?

73. The consumption of whole milk will be the same as the consumption of skim milk when *w* = *s*. Find when this will occur by substituting the given equivalent expression for *w* and the given

equivalent expression for *s* and solving for *t*. Round the value of *t* to the nearest whole and estimate the year when this will occur.

Review and Preview

List or describe the integers that make both inequalities true.

74. $x < 5$ and $x > 1$ **75.** $x \geq 0$ and $x \leq 7$ **76.** $x \geq -2$ and $x \geq 2$ **77.** $x < 6$ and $x < -5$

Graph each set on a number line and write it in interval notation. See Section 2.4.

78. $\{x \mid 0 \leq x \leq 5\}$

79. $\{x \mid -7 < x \leq 1\}$

80. $\left\{x \mid -\dfrac{1}{2} < x < \dfrac{3}{2}\right\}$

81. $\{x \mid -2.5 \leq x < 5.3\}$

Combining Concepts

Solve each inequality.

82. $4(x - 1) \geq 4x - 8$

83. $3x + 1 < 3(x - 2)$

84. $7x < 7(x - 2)$

85. $8(x + 3) \leq 7(x + 5) + x$

86. Explain how solving a linear inequality is similar to solving a linear equation.

87. Explain how solving a linear inequality is different from solving a linear equation.

Integrated Review—Linear Equations and Inequalities

Solve each equation or inequality.

1. $-4x = 20$

2. $-4x < 20$

3. $\dfrac{3x}{4} \geq 2$

4. $5x + 3 \geq 2 + 4x$

5. $6(y - 4) = 3(y - 8)$

6. $-4x \leq \dfrac{2}{5}$

7. $-3x \geq \dfrac{1}{2}$

8. $5(y + 4) = 4(y + 5)$

9. $7x < 7(x - 2)$

10. $\dfrac{-5x + 11}{2} \leq 7$

11. $-5x + 1.5 = -19.5$

12. $-5x + 4 = -26$

13. $5 + 2x - x = -x + 3 - 14$

14. $12x + 14 < 11x - 2$

Answers

1. _____

2. _____

3. _____

4. _____

5. _____

6. _____

7. _____

8. _____

9. _____

10. _____

11. _____

12. _____

13. _____

14. _____

129

15. _____

16. _____

17. _____

18. _____

19. _____

20. _____

21. _____

22. _____

23. _____

24. _____

25. _____

26. _____

27. _____

28. _____

15. $\dfrac{x}{5} - \dfrac{x}{4} = \dfrac{x-2}{2}$

16. $12x - 12 = 8(x - 1)$

17. $2(x - 3) > 70$

18. $-3x - 4.7 = 11.8$

19. $-2(b - 4) - (3b - 1) = 5b + 3$

20. $8(x + 3) < 7(x + 5) + x$

21. $\dfrac{3t + 1}{8} = \dfrac{5 + 2t}{7} + 2$

22. $4(x - 6) - x = 8(x - 3) - 5x$

23. $\dfrac{x + 3}{12} + \dfrac{x - 5}{15} < \dfrac{2}{3}$

24. $\dfrac{y}{3} + \dfrac{y}{5} = \dfrac{y + 3}{10}$

25. $5(x - 6) + 2x > 3(2x - 1) - 4$

26. $14(x - 1) - 7x \le 2(3x - 6) + 4$

27. $\dfrac{1}{4}(3x + 2) - x \ge \dfrac{3}{8}(x - 5) + 2$

28. $\dfrac{1}{3}(x - 10) - 4x > \dfrac{5}{6}(2x + 1) - 1$

2.5 Sets and Compound Inequalities

Two inequalities joined by the words **and** or **or** are called **compound inequalities**.

Compound Inequalities

$$x + 3 < 8 \text{ and } x > 2$$
$$\frac{2x}{3} \geq 5 \text{ or } -x + 10 < 7$$

OBJECTIVES

Ⓐ Find the intersection of two sets.

Ⓑ Solve compound inequalities containing "**and**."

Ⓒ Find the union of two sets.

Ⓓ Solve compound inequalities containing "**or**."

SSM TUTOR CENTER   SG   CD & VIDEO   MATH PRO   WEB

Ⓐ Finding the Intersection of Two Sets

The solution set of a compound inequality formed by the word **and** is the **intersection** of the solution sets of the two inequalities.

> **Intersection of Two Sets**
>
> The intersection of two sets, A and B, is the set of all elements common to both sets. A intersect B is denoted by
>
> $$A \cap B$$

EXAMPLE 1 Find the intersection: $\{2, 4, 6, 8\} \cap \{3, 4, 5, 6\}$

Solution: The numbers 4 and 6 are in both sets. The intersection is $\{4, 6.\}$ ●

Ⓑ Solving Compound Inequalities Containing "and"

A value of x is a solution of a compound inequality formed by the word **and** if it is a solution of *both* inequalities. For example, the solution set of the compound inequality $x \leq 5$ and $x \geq 3$ contains all values of x that make the inequality $x \leq 5$ a true statement **and** the inequality $x \geq 3$ a true statement. The first graph shown here is the graph of $x \leq 5$, the second graph is the graph of $x \geq 3$, and the third graph shows the intersection of the two graphs. The third graph is the graph of $x \leq 5$ **and** $x \geq 3$.

$\{x | x \leq 5\}$ $(-\infty, 5]$

$\{x | x \geq 3\}$ $[3, \infty)$

$\{x | x \leq 5 \text{ and } x \geq 3\}$ $[3, 5]$

In interval notation, the set $\{x | x \leq 5 \text{ and } x \geq 3\}$ is written as $[3, 5]$.

EXAMPLE 2 Solve: $x - 7 < 2$ and $2x + 1 < 9$

Solution: First we solve each inequality separately.

$$x - 7 < 2 \text{ and } 2x + 1 < 9$$
$$x < 9 \text{ and } \quad 2x < 8$$
$$x < 9 \text{ and } \quad\quad x < 4$$

Now we can graph the two intervals on two number lines and find their intersection.

Practice Problem 1

Find the intersection:

$$\{1, 2, 3, 4, 5\} \cap \{3, 4, 5, 6\}$$

Practice Problem 2

Solve: $x + 5 < 9$ and $3x - 1 < 2$

Answers

1. $\{3, 4, 5\}$, **2.** $(-\infty, 1)$

$\{x|x < 9\}$ $(-\infty, 9)$

$\{x|x < 4\}$ $(-\infty, 4)$

$\{x|x < 9 \text{ and } x < 4\}$ $(-\infty, 4)$
$= \{x|x < 4\}$

The solution set is $(-\infty, 4)$.

Practice Problem 3

Solve: $4x \geq 0$ and $2x + 4 \geq 2$

EXAMPLE 3 Solve: $2x \geq 0$ and $4x - 1 \leq -9$

Solution: First we solve each inequality separately.

$$2x \geq 0 \text{ and } 4x - 1 \leq -9$$
$$x \geq 0 \text{ and } \quad 4x \leq -8$$
$$x \geq 0 \text{ and } \quad x \leq -2$$

Now we can graph the two intervals and find their intersection.

$\{x|x \geq 0\}$ $[0, \infty)$

$\{x|x \leq -2\}$ $(-\infty, -2]$

$\{x|x \geq 0 \text{ and } x \leq -2\}$
$= \emptyset$

There is no number that is greater than or equal to 0 **and** less than or equal to -2. The solution set is $\emptyset$.

Some compound inequalities containing the word **and** can be written in a more compact form. The compound inequality $2 \leq x$ and $x \leq 6$ can be written as

$$2 \leq x \leq 6$$

Recall from Section 2.4 that the graph of $2 \leq x \leq 6$ is all numbers between 2 and 6, including 2 and 6.

The set $\{x|2 \leq x \leq 6\}$ written in interval notation is $[2, 6]$.

To solve a compound inequality like $2 < 4 - x < 7$, we get x alone in the middle. Since a compound inequality is really two inequalities in one statement, we must perform the same operation to all three parts of the inequality.

Practice Problem 4

Solve: $5 < 1 - x < 9$

EXAMPLE 4 Solve: $2 < 4 - x < 7$

Solution: To get x alone, we first subtract 4 from all three parts.

Helpful Hint

Don't forget to reverse both inequality symbols.

$$2 < 4 - x < 7$$
$$2 - 4 < 4 - x - 4 < 7 - 4 \quad \text{Subtract 4 from all three parts.}$$
$$-2 < -x < 3 \quad \text{Simplify.}$$
$$\frac{-2}{-1} > \frac{-x}{-1} > \frac{3}{-1} \quad \text{Divide all three parts by } -1 \text{ and reverse the inequality symbols.}$$
$$2 > x > -3$$

This is equivalent to $-3 < x < 2$, and its graph is shown.

The solution set in interval notation is $(-3, 2)$.

Answers

3. $[0, \infty)$, **4.** $(-8, -4)$

EXAMPLE 5   Solve: $-1 \le \dfrac{2x}{3} + 5 \le 2$

Solution: First we clear the inequality of fractions by multiplying all three parts by the LCD, 3.

$$-1 \le \frac{2x}{3} + 5 \le 2$$

$$3(-1) \le 3\left(\frac{2x}{3} + 5\right) \le 3(2) \qquad \text{Multiply all three parts by the LCD, 3.}$$

$$-3 \le 2x + 15 \le 6 \qquad \text{Use the distributive property and multiply.}$$

$$-3 - 15 \le 2x + 15 - 15 \le 6 - 15 \qquad \text{Subtract 15 from all three parts.}$$

$$-18 \le 2x \le -9 \qquad \text{Simplify.}$$

$$\frac{-18}{2} \le \frac{2x}{2} \le \frac{-9}{2} \qquad \text{Divide all three parts by 2.}$$

$$-9 \le x \le -\frac{9}{2} \qquad \text{Simplify.}$$

The graph of the solution is shown.

The solution set in interval notation is $\left[-9, -\dfrac{9}{2}\right]$.

Ⓒ Finding the Union of Two Sets

The solution set of a compound inequality formed by the word **or** is the **union** of the solution sets of the two inequalities.

> **Union of Two Sets**
>
> The union of two sets, A and B, is the set of elements that belong to *either* of the sets. A union B is denoted by
>
> $$A \cup B$$

EXAMPLE 6   Find the union: $\{2, 4, 6, 8\} \cup \{3, 4, 5, 6\}$

Solution: The numbers that are in either set or both sets are $\{2, 3, 4, 5, 6, 8\}$. This set is the union.

Ⓓ Solving Compound Inequalities Containing "or"

A value of x is a solution of a compound inequality formed by the word **or** if it is a solution of **either** inequality. For example, the solution set of the compound inequality $x \le 1$ **or** $x \ge 3$ contains all numbers that make the inequality $x \le 1$ a true statement **or** the inequality $x \ge 3$ a true statement.

Practice Problem 5

Solve: $-3 \le \dfrac{x}{2} + 1 \le 5$

Practice Problem 6

Find the union:

$$\{1, 2, 3, 4, 5\} \cup \{3, 4, 5, 6\}$$

Answers

5. $[-8, 8]$,   **6.** $\{1, 2, 3, 4, 5, 6\}$

$$(-\infty, 1]$$

$$[3, \infty)$$

$$(-\infty, 1] \cup [3, \infty)$$

In interval notation, the set $\{x | x \leq 1 \text{ or } x \geq 3\}$ is written as $(-\infty, 1] \cup [3, \infty)$.

Practice Problem 7

Solve: $3x - 2 \geq 10 \text{ or } x - 6 \leq -4$

EXAMPLE 7   Solve: $5x - 3 \leq 10 \text{ or } x + 1 \geq 5$

Solution: First we solve each inequality separately.

$$5x - 3 \leq 10 \text{ or } x + 1 \geq 5$$
$$5x \leq 13 \text{ or } \quad x \geq 4$$
$$x \leq \frac{13}{5} \text{ or } \quad x \geq 4$$

Now we can graph each interval and find their union.

$$\left\{ x \middle| x \leq \frac{13}{5} \right\} \qquad \left(-\infty, \frac{13}{5} \right]$$

$$\{x | x \geq 4\} \qquad [4, \infty)$$

$$\left\{ x \middle| x \leq \frac{13}{5} \text{ or } x \geq 4 \right\}$$

$$\left(-\infty, \frac{13}{5} \right] \cup [4, \infty)$$

The solution set is $\left(-\infty, \frac{13}{5} \right] \cup [4, \infty)$.

Practice Problem 8

Solve: $x - 7 \leq -1 \text{ or } 2x - 6 \geq 2$

EXAMPLE 8   Solve: $-2x - 5 < -3 \text{ or } 6x < 0$

Solution: First we solve each inequality separately.

$$-2x - 5 < -3 \quad \text{or} \quad 6x < 0$$
$$-2x < 2 \quad \text{or} \quad x < 0$$
$$x > -1 \quad \text{or} \quad x < 0$$

Now we can graph each interval and find their union.

$$\{x | x > -1\} \qquad (-1, \infty)$$

$$\{x | x < 0\} \qquad (-\infty, 0)$$

$$\{x | x > -1 \text{ or } x < 0\} \qquad (-\infty, \infty)$$
$$= \text{all real numbers}$$

The solution set is $(-\infty, \infty)$.

Concept Check

Which of the following is *not* a correct way to represent the set of all numbers between -3 and 5?

a. $\{x | -3 < x < 5\}$
b. $-3 < x \text{ or } x < 5$
c. $(-3, 5)$
d. $x > -3 \text{ and } x < 5$

Try the Concept Check in the margin.

Answers

7. $(-\infty, 2] \cup [4, \infty)$,   **8.** $(-\infty, \infty)$

Concept Check: b is not correct

EXERCISE SET 2.5

A If $A = \{x | x \text{ is an even integer}\}$, $B = \{x | x \text{ is an odd integer}\}$, $C = \{2, 3, 4, 5\}$, and $D = \{4, 5, 6, 7\}$, list the elements of each set. See Example 1.

1. $A \cap C$ **2.** $B \cap D$ **3.** $A \cap B$ **4.** $C \cap D$ **5.** $B \cap C$ **6.** $A \cap D$

B Solve each compound inequality. Graph the two inequalities on the first two number lines and the solution set on the third number line. See Examples 2 and 3.

7. $x < 1$ and $x > -3$

8. $x \leq 0$ and $x \geq -2$

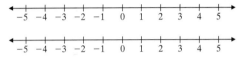

9. $x \leq -3$ and $x \geq -2$

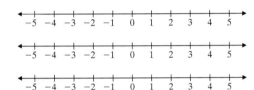

10. $x < 2$ and $x > 4$

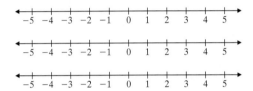

11. $x < -1$ and $x < 1$

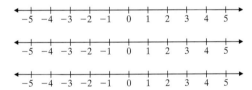

12. $x \geq -4$ and $x > 1$

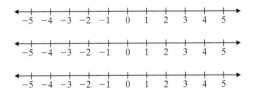

Solve each compound inequality. See Examples 2 and 3.

13. $x < 5$ and $x > -2$

14. $x \leq 7$ and $x \leq 1$

 15. $x + 1 \geq 7$ and $3x - 1 \geq 5$

16. $-2x < -8$ and $x - 5 < 5$

17. $4x + 2 \leq -10$ and $2x \leq 0$

18. $x + 4 > 0$ and $4x > 0$

19. $x + 3 \geq 3$ and $x + 3 \leq 2$

20. $2x - 1 \geq 3$ and $-x > 2$

Solve each compound inequality. See Examples 4 and 5.

21. $5 < x - 6 < 11$

22. $-2 \leq x + 3 \leq 0$

23. $-2 \leq 3x - 5 \leq 7$

24. $1 < 4 + 2x < 7$

 **25.** $1 \leq \dfrac{2}{3}x + 3 \leq 4$

26. $-2 < \dfrac{1}{2}x - 5 < 1$

27. $-5 \leq \dfrac{x + 1}{4} \leq -2$

28. $-4 \leq \dfrac{2x + 5}{3} \leq 1$

29. $0 \leq 2x - 3 \leq 9$

30. $3 < 5x + 1 < 11$

31. $-6 < 3(x - 2) \leq 8$

32. $-5 < 2(x + 4) < 8$

C *If $A = \{1, 2, 3, 4, 5, 6, 7, 8\}$, $B = \{1, 5\}$, $C = \{2, 4, 6, 8\}$, and $D = \{6\}$, list the elements of each set. See Example 6.*

33. $A \cup B$ **34.** $A \cup C$ **35.** $B \cup D$ **36.** $B \cup C$ **37.** $C \cup D$ **38.** $D \cup B$

D *Solve each compound inequality. Graph the two given inequalities on the first two number lines and the solution set on the third number line. See Examples 7 and 8.*

39. $x < 4$ or $x < 5$

40. $x \geq -2$ or $x \leq 2$

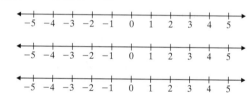

41. $x \leq -4$ or $x \geq 1$

42. $x < 0$ or $x < 1$

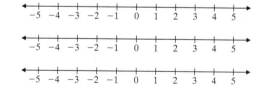

43. $x > 0$ or $x < 3$

44. $x \geq -3$ or $x \leq -4$

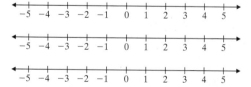

Solve each compound inequality. See Examples 7 and 8.

45. $x < -1$ or $x > 0$

46. $x \leq 1$ or $x \leq -3$

47. $-2x \leq -4$ or $5x - 20 \geq 5$

48. $x + 4 < 0$ or $6x > -12$

49. $3(x - 1) < 12$ or $x + 7 > 10$

50. $5(x - 1) \geq -5$ or $5 - x \leq 11$

 51. $3x + 2 \leq 5$ or $7x > 29$

52. $-x < 7$ or $3x + 1 < -20$

53. $3x \geq 5$ or $-x - 6 < 1$

54. $\dfrac{3}{8}x + 1 \leq 0$ or $-2x < -4$

55. $6x - 4 > 2x$ or $4x - 1 < x + 5$

56. $6x - 2 > 5x + 3$ or $4x - 3 < x$

Review and Preview

Evaluate. See Section 1.5.

57. $|-7| - |19|$

58. $|-7 - 19|$

59. $-(-6) - |-10|$

60. $|-4| - (-4) + |-20|$

Find by inspection all values for x that make each equation true.

61. $|x| = 7$

62. $|x| = 5$

63. $|x| = 0$

64. $|x| = -2$

Combining Concepts

Use the graph to answer Exercises 65 and 66.

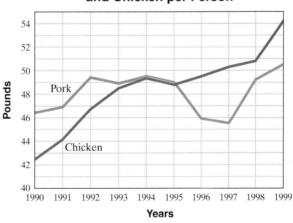

U.S. Consumption of Pork and Chicken per Person

Source: Based on data from Economic Research Service, U.S. Department of Agriculture, *Agricultural Outlook*, October 2001

65. For what years was the consumption of pork greater than 48 pounds per person *and* the consumption of chicken greater than 48 pounds per person?

66. For what years was the consumption of pork less than 47 pounds per person *or* the consumption of chicken greater than 50 pounds per person?

67. In your own words, describe how to find the union of two sets.

68. In your own words, describe how to find the intersection of two sets.

To solve a compound inequality such as $x - 6 < 3x < 2x + 5$, we solve

$$x - 6 < 3x \quad \text{and} \quad 3x < 2x + 5$$

Use this information to solve the inequalities in Exercises 69 through 72.

69. $x - 6 < 3x < 2x + 5$

70. $2x - 3 < 3x + 1 < 4x - 5$

71. $x + 3 < 2x + 1 < 4x + 6$

72. $-3(x - 2) \le 3 - 2x \le 10 - 3x$

Internet Excursions

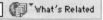 Go To: http://www.prenhall.com/martin-gay_interm   What's Related

A target heart rate is the number of heartbeats per minute a person should maintain during aerobic exercise to get maximum cardiovascular benefits. Many health experts recommend keeping your heart rate while exercising within a certain interval: between a lower and upper limit on target heart rate, called the target heart rate zone. Each person has a different target heart rate zone, depending on his or her age. By visiting the World Wide Web address listed above, you will gain access to a Web site where you can learn how to calculate your target heart rate zone.

73. Using the description of calculating the target heart rate zone given on this Web site, write an algebraic expression that represents the lower limit on a person's target heart rate. Similarly, write an algebraic expression that represents the upper limit on a person's target heart rate. Then use these expressions to write a compound inequality that describes the range in which a person's heart rate should fall while exercising, that is, his or her target heart rate zone. Be sure to define all variables used.

74. Use your compound inequality from Exercise 73 to find your own target heart rate zone. Then use the link to the handy target heart rate calculator shown on this Web site to check your work.

2.6 Absolute Value Equations and Inequalities

In Chapter 1, we defined the absolute value of a number as its distance from 0 on a number line.

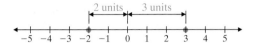

$$|-2| = 2 \quad \text{and} \quad |3| = 3$$

In this section, we concentrate on solving equations and inequalities containing the absolute value of a variable or a variable expression. Examples of absolute value equations and inequalities are

$$|x| = 3 \quad -5 \geq |2y + 7| \quad |z - 6.7| = |3z + 1.2| \quad |x - 3| > 7$$

Absolute value equations and inequalities are extremely useful in data analysis, especially for calculating acceptable measurement error and errors that result from the way numbers are sometimes represented in computers.

A Solving Absolute Value Equations

To begin, let's solve a few absolute value equations by inspection.

EXAMPLE 1 Solve: $|x| = 3$

Solution: The solution set of this equation will contain all numbers whose distance from 0 is 3 units. Two numbers are 3 units away from 0 on the number line: 3 and -3.

Check: To check, let $x = 3$ and $x = -3$ in the original equation.

$$|x| = 3$$
$$|3| \overset{?}{=} 3 \quad \text{Let } x = 3.$$
$$3 = 3 \quad \text{True.}$$

$$|x| = 3$$
$$|-3| \overset{?}{=} 3 \quad \text{Let } x = -3.$$
$$3 = 3 \quad \text{True.}$$

Both solutions check. Thus the solution set of the equation $|x| = 3$ is $\{3, -3\}$. ●

Practice Problem 1

Solve: $|y| = 5$

EXAMPLE 2 Solve: $|x| = -2$

Solution: The absolute value of a number is never negative, so this equation has no solution. The solution set is $\{\ \}$ or $\emptyset$. ●

Practice Problem 2

Solve: $|p| = -4$

EXAMPLE 3 Solve: $|y| = 0$

Solution: We are looking for all numbers whose distance from 0 is zero units. The only number is 0. The solution set is $\{0\}$. ●

From the above examples, we have the following.

Practice Problem 3

Solve: $|x| = 0$

Answers

1. $\{-5, 5\}$, **2.** $\emptyset$, **3.** $\{0\}$

Absolute Value Property

Solve $|X| = a$ as follows.
If a is positive, then solve $X = a$ or $X = -a$.
If a is 0, then $X = 0$.
If a is negative, the equation $|X| = a$ has no solution.

Helpful Hint

For the equation $|X| = a$ in the box above, X can be a single variable or a variable expression.

When we are solving absolute value equations, if $|X|$ is not alone on one side of the equation we first use properties of equality to get $|X|$ alone.

Practice Problem 4

Solve: $3|y| - 4 = 17$

EXAMPLE 4   Solve: $2|x| + 25 = 37$

Solution: First we get $|x|$ alone.

$$2|x| + 25 = 37$$
$$2|x| = 12 \quad \text{Subtract 25 from both sides.}$$
$$|x| = 6 \quad \text{Divide both sides by 2.}$$
$$x = 6 \quad \text{or} \quad x = -6 \quad \text{Use the absolute value property.}$$

The solution set is $\{-6, 6\}$.

If the expression inside the absolute value bars is more complicated than a single variable x, we can still use the absolute value property.

Practice Problem 5

Solve: $|x - 4| = 11$

EXAMPLE 5   Solve: $|w + 3| = 7$

Solution: If we think of the expression $w + 3$ as X in the absolute value property, we have that

$$|w + 3| = 7$$
$$w + 3 = 7 \quad \text{or} \quad w + 3 = -7 \quad \text{Use the absolute value property.}$$
$$w = 4 \quad \text{or} \quad w = -10$$

The solution set is $\{4, -10\}$.

Don't forget that to use the absolute value property you must first make sure that the absolute value expression is alone on one side of the equation.

Helpful Hint

If the equation has a single absolute value expression containing variables, get the absolute value expression alone. Then use the absolute value property.

Practice Problem 6

Solve: $|4x + 2| + 1 = 7$

Answers

4. $\{-7, 7\}$,   **5.** $\{15, -7\}$,   **6.** $\{1, -2\}$

EXAMPLE 6   Solve: $|2x - 1| + 5 = 6$

Solution: We want the absolute value expression alone on one side of the equation, so we begin by subtracting 5 from both sides. Then we use the absolute value property.

$$|2x - 1| + 5 = 6$$
$$|2x - 1| = 1$$

Subtract 5 from both sides.

$$2x - 1 = 1 \quad \text{or} \quad 2x - 1 = -1$$

Use the absolute value property.

$$2x = 2 \quad \text{or} \quad 2x = 0$$
$$x = 1 \quad \text{or} \quad x = 0$$

Solve.

The solution set is $\{0, 1\}$.

Given two absolute value expressions, we might ask, when are the absolute values of two expressions equal? To see the answer, notice that

$$|2| = |2| \quad |-2| = |-2| \quad |-2| = |2| \quad |2| = |-2|$$

same   same   opposites   opposites

Two absolute value expressions are equal when the expressions inside the absolute value bars are equal to or are opposites of each other.

EXAMPLE 7   Solve: $|3x + 2| = |5x - 8|$

Solution: This equation is true if the expressions inside the absolute value bars are equal to or are opposites of each other.

$$3x + 2 = 5x - 8 \quad \text{or} \quad 3x + 2 = -(5x - 8)$$

Next we solve each equation.

$$3x + 2 = 5x - 8 \quad \text{or} \quad 3x + 2 = -5x + 8$$
$$-2x + 2 = -8 \quad \text{or} \quad 8x + 2 = 8$$
$$-2x = -10 \quad \text{or} \quad 8x = 6$$
$$x = 5 \quad \text{or} \quad x = \frac{3}{4}$$

Check to see that replacing x with 5 or with $\frac{3}{4}$ results in a true statement.

The solution set is $\left\{ \frac{3}{4}, 5 \right\}$.

EXAMPLE 8   Solve: $|x - 3| = |5 - x|$

Solution:

$$x - 3 = 5 - x \quad \text{or} \quad x - 3 = -(5 - x)$$
$$2x - 3 = 5 \quad \text{or} \quad x - 3 = -5 + x$$
$$2x = 8 \quad \text{or} \quad x - 3 - x = -5 + x - x$$
$$x = 4 \quad \text{or} \quad -3 = -5 \quad \text{False.}$$

Recall from Section 2.1 that when an equation simplifies to a false statement, the equation has no solution. Thus the only solution for the original absolute value equation is 4, and the solution set is $\{4\}$.

Try the Concept Check in the margin.

B   **Solving Absolute Value Inequalities**

To begin, let's solve a few absolute value inequalities by inspection.

Practice Problem 7

Solve: $|4x - 5| = |3x + 5|$

Practice Problem 8

Solve: $|x + 2| = |4 - x|$

Concept Check

True or false? Absolute value equations always have two solutions. Explain your answer.

Answers

7. $\{0, 10\}$,   **8.** $\{1\}$

Concept Check: false; answers may vary

Practice Problem 9

Solve $|x| < 4$ using a number line.

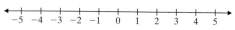

Practice Problem 10

Solve $|x| \geq 5$ using a number line.

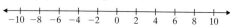

Practice Problem 11

Solve: $|x + 2| > 4$. Graph the solution set.

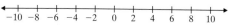

Answers

9.

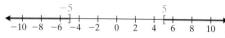

10.

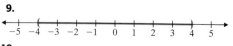

11. $(-\infty, -6) \cup (2, \infty)$

EXAMPLE 9   Solve $|x| < 2$ using a number line.

Solution: The solution set contains all numbers whose distance from 0 is less than 2 units on the number line.

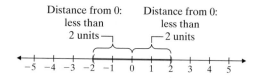

The solution set is $\{x | -2 < x < 2\}$, or $(-2, 2)$ in interval notation. ●

EXAMPLE 10   Solve $|x| \geq 3$ using a number line.

Solution: The solution set contains all numbers whose distance from 0 is 3 or more units. Thus the graph of the solution set contains 3 and all points to the right of 3 on the number line or -3 and all points to the left of -3 on the number line.

This solution set is $\{x | x \leq -3 \text{ or } x \geq 3\}$. In interval notation, the solution set is $(-\infty, -3] \cup [3, \infty)$, since **or** means union. ●

The following box summarizes solving absolute value equations and inequalities.

Solving Absolute Value Equations and Inequalities

If a is a positive number,

To solve $|X| = a$, solve $X = a$ or $X = -a$.

To solve $|X| < a$, solve $-a < X < a$.

To solve $|X| > a$, solve $X < -a$ or $X > a$.

EXAMPLE 11   Solve: $|x - 3| > 7$

Solution: Since 7 is positive, to solve $|x - 3| > 7$, we solve the compound inequality $x - 3 < -7$ or $x - 3 > 7$.

$$x - 3 < -7 \quad \text{or} \quad x - 3 > 7$$
$$x < -4 \quad \text{or} \quad x > 10 \quad \text{Add 3 to both sides.}$$

The solution set is $\{x | x < -4 \text{ or } x > 10\}$ or $(-\infty, -4) \cup (10, \infty)$ in interval notation. Its graph is shown.

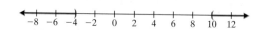

Let's review the differences in solving absolute value equations and inequalities by solving an absolute value equation.

EXAMPLE 12   Solve: $|x + 1| = 6$

Solution: This is an equation, so we solve

$$x + 1 = 6 \quad \text{or} \quad x + 1 = -6$$
$$x = 5 \quad \text{or} \quad x = -7$$

The solution set is $\{-7, 5\}$. Its graph is shown.

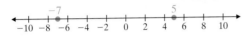

EXAMPLE 13   Solve: $|x - 6| \le 2$

Solution: To solve $|x - 6| \le 2$, we solve

$$-2 \le x - 6 \le 2$$
$$-2 + 6 \le x - 6 + 6 \le 2 + 6 \qquad \text{Add 6 to all three parts.}$$
$$4 \le x \le 8 \qquad \text{Simplify.}$$

The solution set is $\{x | 4 \le x \le 8\}$, or $[4, 8]$ in interval notation. Its graph is shown.

Helpful Hint

Before using an absolute value inequality property, get an absolute value expression alone on one side of the inequality.

EXAMPLE 14   Solve: $|5x + 1| + 1 \le 10$

Solution: First we get the absolute value expression alone by subtracting 1 from both sides.

$$|5x + 1| + 1 \le 10$$
$$|5x + 1| \le 10 - 1 \qquad \text{Subtract 1 from both sides.}$$
$$|5x + 1| \le 9 \qquad \text{Simplify.}$$

Since 9 is positive, to solve $|5x + 1| \le 9$, we solve

$$-9 \le 5x + 1 \le 9$$
$$-9 - 1 \le 5x + 1 - 1 \le 9 - 1 \qquad \text{Subtract 1 from all three parts.}$$
$$-10 \le 5x \le 8 \qquad \text{Simplify.}$$
$$-2 \le x \le \frac{8}{5} \qquad \text{Divide all three parts by 5.}$$

The solution set is $\left[-2, \frac{8}{5} \right]$.

The next few examples are special cases of absolute value inequalities.

Practice Problem 12

Solve: $|x - 3| = 5$. Graph the solution set.

Practice Problem 13

Solve: $|x - 2| \le 1$. Graph the solution set.

Practice Problem 14

Solve: $|2x - 5| + 2 \le 9$

Answers

12. $\{-2, 8\}$

13.

14. $[-1, 6]$

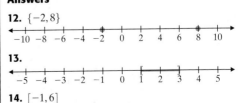

Practice Problem 15

Solve: $|x| < -1$

Practice Problem 16

Solve: $|x + 1| \geq -3$

Concept Check

Without taking any solution steps, how do you know that the absolute value inequality $|3x - 2| > -9$ has a solution? What is its solution?

Answers

15. $\emptyset$, **16.** $(-\infty, \infty)$

Concept Check: $(-\infty, \infty)$ since the absolute value is always nonnegative

EXAMPLE 15   Solve: $|x| \leq -3$

Solution: The absolute value of a number is never negative. Thus it will then never be less than or equal to -3. The solution set is $\{\ \}$ or $\emptyset$. ●

EXAMPLE 16   Solve: $|x - 1| > -2$

Solution: The absolute value of a number is always nonnegative. Thus it will always be greater than -2. The solution set contains all real numbers, or $(-\infty, \infty)$. ●

Try the Concept Check in the margin.

STUDY SKILLS REMINDER

Are you organized?

Have you ever had trouble finding a completed assignment? When it's time to study for a test, are your notes neat and organized? Have you ever had trouble reading your own mathematics handwriting? (Be honest—I have had trouble reading my own handwriting before.)

When any of these things happen, it's time to get organized. Here are a few suggestions:

Write your notes and complete your homework assignments in a notebook with pockets (spiral or ring binder). Take class notes in this notebook, and then follow the notes with your completed homework assignment. When you receive graded papers or handouts, place them in the notebook pocket so that you will not lose them.

Place a mark (possibly an exclamation point) beside any note(s) that seem especially important to you. Also place a mark (possibly a question mark) beside any note(s) or homework that you are having trouble with. Don't forget to see your instructor, a tutor, or your fellow classmates to help you understand the concepts or exercises you have marked.

Also, if you are having trouble reading your own handwriting, *slow down* and write your mathematics work clearly!

Name _____ Section _____ Date _____

Mental Math

Simplify each expression.

1. $|-7|$ **2.** $|-8|$ **3.** $-|5|$ **4.** $-|10|$ **5.** $-|-6|$

6. $-|-3|$ **7.** $|-3| + |-2| + |-7|$ **8.** $|-1| + |-6| + |-8|$

EXERCISE SET 2.6

 **A**   *Solve. See Examples 1 through 6.*

1. $|x| = 7$ **2.** $|y| = 15$ **3.** $|x| = -4$ **4.** $|x| = -20$ **5.** $|3x| = 12.6$

6. $|6n| = 12.6$ **7.** $3|x| - 5 = 7$ **8.** $5|x| - 12 = 8$ **9.** $-6|x| + 44 = -10$ **10.** $-4|x| + 18 = -22$

11. $|x - 9| = 14$ **12.** $|x + 2| = 8$ **13.** $|2x - 5| = 9$ **14.** $|6 + 2n| = 4$ **15.** $\left|\dfrac{x}{2} - 3\right| = 1$

16. $\left|\dfrac{n}{3} + 2\right| = 4$ **17.** $|z| + 4 = 9$ **18.** $|x| + 1 = 3$ **19.** $|3x| + 5 = 14$ **20.** $|2x| - 6 = 4$

21. $\left|\dfrac{4x - 6}{3}\right| = 6$ **22.** $\left|\dfrac{2x + 1}{5}\right| = 7$ **23.** $|2x| = 0$ **24.** $|7z| = 0$ **25.** $|4n + 1| + 10 = 4$

26. $|3z - 2| + 8 = 1$ **27.** $3|x - 1| + 19 = 23$ **28.** $5|x + 1| - 1 = 3$

Solve. See Examples 7 and 8.

29. $|5x - 7| = |3x + 11|$ **30.** $|9y + 1| = |6y + 4|$ **31.** $|z + 8| = |z - 3|$

32. $|2x - 5| = |2x + 5|$ **33.** $|2y - 3| = |9 - 4y|$ **34.** $|5z - 1| = |7 - z|$

35. $\left|\dfrac{3}{4}x - 2\right| = \left|\dfrac{1}{4}x + 6\right|$

36. $\left|\dfrac{2}{3}x - 5\right| = \left|\dfrac{1}{3}x + 4\right|$

37. $|2x - 6| = |10 - 2x|$

38. $|4n + 5| = |4n + 3|$

39. $|x + 4| = |7 - x|$

40. $|8 - y| = |y + 2|$

41. $\left|\dfrac{2x + 1}{5}\right| = \left|\dfrac{3x - 7}{3}\right|$

42. $\left|\dfrac{5x - 1}{2}\right| = \left|\dfrac{4x + 5}{6}\right|$

43. $|5x + 1| = |4x - 7|$

44. $|3 + 6n| = |4n + 11|$

B *Solve. Graph the solution set. See Examples 9 through 16.*

45. $|x| \leq 4$

46. $|x| < 6$

47. $|x| > 3$

48. $|y| \geq 4$

49. $|x + 3| < 2$

50. $|x + 4| < 6$

51. $|y - 6| \geq 7$

52. $|x - 3| \geq 10$

53. $\left|\dfrac{x + 2}{3}\right| < 1$

54. $\left|\dfrac{x - 6}{4}\right| < 1$

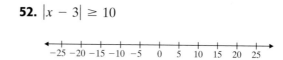

55. $|x| + 7 \leq 12$

56. $|x| + 6 \leq 7$

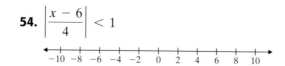

146

57. $|x| + 2 > 6$

58. $|x| - 1 > 3$

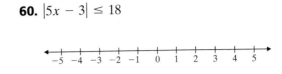

59. $|2x + 7| \leq 13$

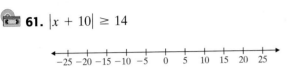

60. $|5x - 3| \leq 18$

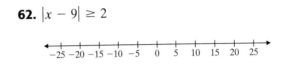

61. $|x + 10| \geq 14$

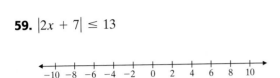

62. $|x - 9| \geq 2$

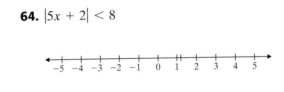

63. $|2x - 7| \leq 11$

64. $|5x + 2| < 8$

65. $|x| > -4$

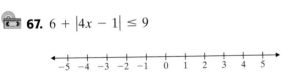

66. $|x| \leq -7$

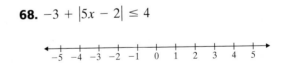

67. $6 + |4x - 1| \leq 9$

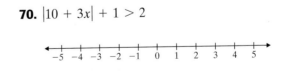

68. $-3 + |5x - 2| \leq 4$

69. $|6x - 8| + 3 > 7$

70. $|10 + 3x| + 1 > 2$

71. $|5x + 3| < -6$

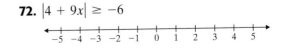

72. $|4 + 9x| \geq -6$

73. $\left|\dfrac{x+6}{3}\right| > 2$

74. $\left|\dfrac{7+x}{2}\right| \geq 4$

Solve each equation or inequality for x. See Examples 1 through 16.

75. $|x| = 13$

76. $|x| < 13$

77. $|x| > 13$

78. $|3x| = 12$

79. $|x| + 12 = 9$

80. $|x| - 4 = -9$

81. $2|x| - 9 \leq 11$

82. $4|x| - 2 \geq 6$

83. $|2x - 3| < 7$

84. $|2x - 3| > 7$

85. $|2x - 3| = 7$

86. $|5 - 6x| = 29$

87. $|x - 5| \geq 12$

88. $|x + 4| \geq 20$

89. $|9 + 4x| = 0$

90. $|9 + 4x| \geq 0$

91. $|2x + 1| + 4 < 7$

92. $8 + |5x - 3| \geq 11$

93. $\left|\dfrac{1}{3}x + 1\right| > 5$

94. $\left|\dfrac{1}{4}x - 2\right| < 1$

95. $|3x - 5| + 4 = 5$

96. $|x - 1| + 7 = 11$

97. $|x + 11| = -1$

98. $|4x - 4| = -3$

99. $\left|\dfrac{2x - 1}{3}\right| = 6$

100. $\left|\dfrac{6 - x}{4}\right| = 5$

101. $\left|\dfrac{3x - 5}{6}\right| > 5$

102. $\left|\dfrac{4x - 7}{5}\right| < 2$

103. $|6x - 3| = |4x + 5|$

104. $|3x + 1| = |4x + 10|$

Review and Preview

The circle graph shows the types of cheese produced in the United States in 2000. Use this graph to answer Exercises 105 through 108. See Section 2.2.

Cheese Production Percent by Type, 2000

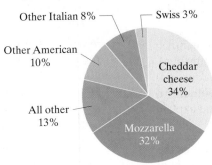

Source: National Agricultural Statistics Service, U.S. Department of Agriculture

105. In 2000, mozzarella made up what percent of U.S. cheese production?

106. Which cheese had the highest U.S. production in 2000?

107. A circle contains 360°. Find the number of degrees found in the 8% sector for other Italian cheese.

108. In 2000, the total production of cheese in the United States was 8,255,000,000 pounds. Find the amount of Swiss cheese produced during that year.

Consider the equation $3x - 4y = 12$. For each value of x or y given, find the corresponding value of the other variable that makes the statement true. See Section 2.3.

109. If $x = 2$, find y.   **110.** If $y = -1$, find x.   **111.** If $y = -3$, find x.   **112.** If $x = 4$, find y.

 Combining Concepts

113. Write an absolute value equation representing all numbers x whose distance from 0 is 5 units.

114. Write an absolute value equation representing all numbers x whose distance from 0 is 2 units.

115. Write an absolute value inequality representing all numbers x whose distance from 0 is less than 7 units.

116. Write an absolute value inequality representing all numbers x whose distance from 0 is greater than 4 units.

117. Write $-5 \le x \le 5$ as an equivalent inequality containing an absolute value. Explain your answer.

118. Write $x > 1$ or $x < -1$ as an equivalent inequality containing an absolute value. Explain your answer.

The expression $|x_T - x|$ is defined to be the absolute error in x, where x_T is the true value of a quantity and x is the measured value or value as stored in a computer.

119. If the true value of a quantity is 3.5 and the absolute error must be less than 0.05, find the acceptable measured values.

120. If the true value of a quantity is 0.2 and the approximate value stored in a computer is $\frac{51}{256}$, find the absolute error.

This activity may be completed by working in groups or individually.

Have you ever stopped to think about how much math and geometry is involved in redecorating a room? In this project, you will plan for a redecorating project by estimating the necessary amount of materials and their costs. Be sure to show and explain all of your work.

1. Choose an actual room that you can use as the model for your redecorating project. You could use your math classroom, your living room, your dormitory room, or any room that can be easily measured. Measure each dimension of the room with a tape measure. Make a sketch of the room and label each dimension with its measurement. Be sure to include the sizes and locations of items such as doors and windows.

2. **Paint** The first task in your redecorating project will be painting the walls of the room.

    a. To estimate the amount of paint you will need, first find the total wall area to be painted. Be sure to subtract the areas of any regions, such as doors or windows, that will not be painted. (*Hint:* Use the geometric formulas on the inside front cover of this text to help you find the necessary areas.)

    b. Using newspaper flyers or a trip to a local paint store, choose a paint that would be suitable for your redecorating project. List its price per gallon and its normal surface coverage. Using the information on coverage, estimate how many gallons of paint will be needed for the project. (*Note:* If surface coverage data, normally found on the paint can label, is unavailable, use the guideline that a gallon of paint generally covers 400 square feet of wall area.)

    c. How much will all of the paint that is needed cost?

3. **Wallpaper Border** The second task in your redecorating project will be installing a wallpaper border all the way around the room just below the ceiling.

    a. Calculate the length of the border needed.

    b. Using newspaper flyers or a trip to a local paint/decorating store, choose a border that would be suitable for your redecorating project. List its price per roll and the length of border on each roll. How many rolls of wallpaper border will be needed for the project?

    c. Find the cost of the border necessary for the project.

4. **Wall-to-Wall Carpeting** The third task in your redecorating project will be installing wall-to-wall carpeting.

    a. How much carpeting will be needed?

    b. Using newspaper flyers or a trip to a local carpeting store, choose a wall-to-wall carpet that would be suitable for your redecorating project. List its price per square unit.

    c. How much will carpeting the room cost? Be sure to include any fixed fees for installation or delivery, and so on.

5. Redecorating projects often involve buying additional utensils and supplies such as paint brushes, wallpaper adhesive, and so on. A reasonable cost estimate for these extras is 20% of the cost of the basic project supplies (paint, border, and carpeting in this case). Taking these extras into consideration, what is your estimate of the total cost of the redecorating project?

Chapter 2 VOCABULARY CHECK

Fill in each blank with one of the words or phrases listed below.

contradiction formula consecutive integers linear equation in one variable identity solution

absolute value linear inequality in one variable compound inequality intersection union

1. The statement "$x < 5$ or $x > 7$" is called a(n) _____.
2. An equation in one variable that has no solution is called a(n) _____.
3. The _____ of two sets is the set of all elements common to both sets.
4. The _____ of two sets is the set of all elements that belong to either of the sets.
5. An equation in one variable that has every number (for which the equation is defined) as a solution is called a(n) _____.
6. The equation $d = rt$ is also called a(n) _____.
7. A number's distance from 0 is called its _____.
8. When a variable in an equation is replaced by a number and the resulting equation is true, then that number is called a(n) _____ of the equation.
9. The integers 17, 18, 19 are examples of _____.
10. The statement $5x - 0.2 < 7$ is an example of a(n) _____.
11. The statement $5x - 0.2 = 7$ is an example of a(n) _____.

C H A P T E R

Highlights

DEFINITIONS AND CONCEPTS	**EXAMPLES**
Section 2.1 Linear Equations in One Variable	

An **equation** is a statement that two expressions are equal.

A **linear equation in one variable** is an equation that can be written in the form $ax + b = c$, where a, b, and c are real numbers and a is not 0.

A **solution** of an equation is a value for the variable that makes the equation a true statement.

$5 = 5$ $\quad 7x + 2 = -14$ $\quad 3(x - 1)^2 = 9x^2 - 6$

$7x + 2 = -14$ $\quad x = -3$

$5(2y - 7) = -2(8y - 1)$

Determine whether -1 is a solution of
$\quad 3(x - 1) = 4x - 2$.

$$3(-1 - 1) \stackrel{?}{=} 4(-1) - 2$$

$$3(-2) \stackrel{?}{=} -4 - 2$$

$$-6 = -6 \qquad \text{True.}$$

Thus, -1 is a solution.

Equivalent equations have the same solution.

$x - 12 = 14$ and $x = 26$ are equivalent equations.

The **addition property of equality** guarantees that the same number may be added to (or subtracted from) both sides of an equation and the result is an equivalent equation.

The **multiplication property of equality** guarantees that both sides of an equation may be multiplied by (or divided by) the same nonzero number and the result is an equivalent equation.

Solve: $-3x - 2 = 10$

$$-3x - 2 + 2 = 10 + 2 \qquad \text{Add 2 to both sides.}$$

$$-3x = 12$$

$$\frac{-3x}{-3} = \frac{12}{-3} \qquad \text{Divide both sides by } -3.$$

$$x = -4$$

| **DEFINITIONS AND CONCEPTS** | **EXAMPLES** |

Section 2.1   Linear Equations in One Variable *(continued)*

SOLVING A LINEAR EQUATION IN ONE VARIABLE

Solve: $x - \dfrac{x-2}{6} = \dfrac{x-7}{3} + \dfrac{2}{3}$

Step 1. Clear the equation of fractions and decimals.

1. $6\left(x - \dfrac{x-2}{6}\right) = 6\left(\dfrac{x-7}{3} + \dfrac{2}{3}\right)$ Multiply both sides by 6.

$6x - (x-2) = 2(x-7) + 2(2)$ Use the distributive property.

Step 2. Remove grouping symbols such as parentheses.

2. $6x - x + 2 = 2x - 14 + 4$

Step 3. Simplify by combining like terms.

3. $5x + 2 = 2x - 10$

Step 4. Write variable terms on one side and numbers on the other side by using the addition property of equality.

4. $5x + 2 - 2 = 2x - 10 - 2$ Subtract 2 from both sides.
$5x = 2x - 12$
$5x - 2x = 2x - 12 - 2x$ Subtract $2x$ from both sides.
$3x = -12$

Step 5. Get the variable alone by using the multiplication property of equality.

5. $\dfrac{3x}{3} = \dfrac{-12}{3}$ Divide both sides by 3.
$x = -4$

Step 6. Check the proposed solution in the original equation.

6. $-4 - \dfrac{-4-2}{6} \overset{?}{=} \dfrac{-4-7}{3} + \dfrac{2}{3}$ Replace x with -4 in the original equation.

$-4 - \dfrac{-6}{6} \overset{?}{=} \dfrac{-11}{3} + \dfrac{2}{3}$

$-4 - (-1) \overset{?}{=} \dfrac{-9}{3}$

$-3 = -3$ True.

Section 2.2   An Introduction to Problem Solving

PROBLEM-SOLVING STRATEGY

Colorado is shaped like a rectangle whose length is about 1.3 times its width. If the perimeter of Colorado is 2070 kilometers, find its dimensions.

1. UNDERSTAND the problem.

1. Read and reread the problem. Guess a solution and check your guess.

Let x = width of Colorado in kilometers. Then $1.3x$ = length of Colorado in kilometers.

$1.3x$

2. TRANSLATE the problem.

2. In words:

| twice the length | + | twice the width | = | perimeter |

Translate: $2(1.3x)$ + $2x$ = 2070

3. SOLVE the equation.

3. $2.6x + 2x = 2070$
$4.6x = 2070$
$x = 450$

4. INTERPRET the results.

4. If x = 450 kilometers, then $1.3x = 1.3(450) = 585$ kilometers. *Check*: The perimeter of a rectangle whose width is 450 kilometers and length is 585 kilometers is $2(450) + 2(585) = 2070$ kilometers, the required perimeter. *State*: The dimensions of Colorado are 450 kilometers by 585 kilometers.

DEFINITIONS AND CONCEPTS	**EXAMPLES**

Section 2.3   Formulas and Problem Solving

An equation that describes a known relationship among quantities is called a **formula**.

$A = \pi r^2$ (area of a circle)

$I = Prt$ (interest = principal · rate · time)

To solve a formula for a specified variable, use the steps for solving an equation. Treat the specified variable as the only variable of the equation.

Solve $A = 2HW + 2LW + 2LH$ for H.

$A - 2LW = 2HW + 2LH$ Subtract $2LW$ from both sides.

$A - 2LW = H(2W + 2L)$ Use the distributive property.

$\dfrac{A - 2LW}{2W + 2L} = \dfrac{H(2W + 2L)}{2W + 2L}$ Divide both sides by $2W + 2L$.

$\dfrac{A - 2LW}{2W + 2L} = H$ Simplify.

Section 2.4   Linear Inequalities and Problem Solving

A **linear inequality in one variable** is an inequality that can be written in the form $ax + b < c$, where a, b, and c are real numbers and $a \neq 0$. (The inequality symbols $\leq$, $>$, and $\geq$ also apply here.)

$5x - 2 \leq -7 \qquad 3y > 1 \qquad \dfrac{z}{7} < -9(z - 3)$

The **addition property of inequality** guarantees that the same number may be added to (or subtracted from) both sides of an inequality, and the resulting inequality will have the same solution set.

$x - 9 \leq -16$

$x - 9 + 9 \leq -16 + 9$ Add 9 to both sides.

$x \leq -7$

The **multiplication property of inequality** guarantees that both sides of an inequality may be multiplied by (or divided by) the same **positive** number, and the resulting inequality will have the same solution set.

$6x < -66$ Divide both sides by 6. Do not reverse the direction of the inequality symbol.

$\dfrac{6x}{6} < \dfrac{-66}{6}$

$x < -11$

We may also multiply (or divide) both sides of an inequality by the same **negative** number and **reverse the direction of the inequality symbol**, and the result will be an inequality with the same solution set.

$-6x < -66$ Divide both sides by -6. Reverse the direction of the inequality symbol.

$\dfrac{-6x}{-6} > \dfrac{-66}{-6}$

$x > 11$

SOLVING A LINEAR INEQUALITY IN ONE VARIABLE

Solve: $\dfrac{3}{7}(x - 4) \geq x + 2$

Step 1. Clear the equation of fractions and decimals.

1. $7\left[\dfrac{3}{7}(x - 4)\right] \geq 7(x + 2)$ Multiply both sides by 7.

$3(x - 4) \geq 7(x + 2)$

Step 2. Remove grouping symbols such as parentheses.

Step 3. Simplify by combining like terms.

2. $3x - 12 \geq 7x + 14$ Use the distributive property.

Step 4. Write variable terms on one side and numbers on the other side using the addition property of inequality.

4. $-4x - 12 \geq 14$ Subtract $7x$ from both sides.

$-4x \geq 26$ Add 12 to both sides.

Step 5. Get the variable alone using the multiplication property of inequality.

$\dfrac{-4x}{-4} \leq \dfrac{26}{-4}$ Divide both sides by -4. Reverse the direction of the inequality symbol.

$x \leq -\dfrac{13}{2}$

| DEFINITIONS AND CONCEPTS | EXAMPLES |

Section 2.5   Sets and Compound Inequalities

Two inequalities joined by the words **and** or **or** are called **compound inequalities**.

$$x - 7 \le 4 \text{ and } x \ge -21$$

$$2x + 7 > x - 3 \text{ or } 5x + 2 > -3$$

The solution set of a compound inequality formed by the word **and** is the **intersection** ∩ of the solution sets of the two inequalities.

Solve:

$$x < 5 \text{ and } x < 3$$

$\{x|x < 5\}$ $(-\infty, 5)$

$\{x|x < 3\}$ $(-\infty, 3)$

The solution set of a compound inequality formed by the word **or** is the **union** ∪ of the solution sets of the two inequalities.

$\{x|x < 3$ and $x < 5\}$ $(-\infty, 3)$

Solve:

$$x - 2 \ge -3 \quad \text{or} \quad 2x \le -4$$

$$x \ge -1 \quad \text{or} \quad x \le -2$$

$\{x|x \ge -1\}$ $[-1, \infty)$

$\{x|x \le -2\}$ $(-\infty, -2]$

$\{x|x \le -2$ or $x \ge -1\}$ $(-\infty, -2] \cup [-1, \infty)$

Section 2.6   Absolute Value Equations and Inequalities

If a is a positive number, then $|X| = a$ is equivalent to $X = a$ or $X = -a$.

Solve: $|5y - 1| - 7 = 4$

$$|5y - 1| = 11 \qquad \text{Add 7 to both sides.}$$

$$5y - 1 = 11 \quad \text{or} \quad 5y - 1 = -11$$

$$5y = 12 \quad \text{or} \quad 5y = -10 \qquad \text{Add 1 to both sides.}$$

$$y = \frac{12}{5} \quad \text{or} \quad 5y = -2 \qquad \text{Divide both sides by 5.}$$

The solution set is $\left\{-2, \dfrac{12}{5}\right\}$.

If a is negative, then $|X| = a$ has no solution.

Solve: $\left|\dfrac{x}{2} - 7\right| = -1$

The solution set is { }, or ∅.

Section 2.6   Absolute Value Equations and Inequalities *(continued)*

If an absolute value equation is of the form $|X| = |Y|$, solve $X = Y$ or $X = -Y$.

Solve: $|x - 7| = |2x + 1|$

$$x - 7 = 2x + 1 \qquad \text{or} \qquad x - 7 = -(2x + 1)$$
$$x = 2x + 8 \qquad \text{or} \qquad x - 7 = -2x - 1$$
$$-x = 8 \qquad \text{or} \qquad x = -2x + 6$$
$$x = -8 \qquad \text{or} \qquad 3x = 6$$
$$x = 2$$

The solution set is $\{-8, 2\}$.

If a is a positive number, then $|X| < a$ is equivalent to $-a < X < a$.

Solve: $|y - 5| \leq 3$

$$-3 \leq y - 5 \leq 3$$
$$-3 + 5 \leq y - 5 + 5 \leq 3 + 5 \qquad \text{Add 5 to all three parts.}$$
$$2 \leq y \leq 8$$

The solution set is $[2, 8]$.

If a is a positive number, then $|X| > a$ is equivalent to $X < -a$ or $X > a$.

Solve: $\left| \dfrac{x}{2} - 3 \right| > 7$

$$\dfrac{x}{2} - 3 < -7 \qquad \text{or} \qquad \dfrac{x}{2} - 3 > 7 \qquad \text{Multiply both sides by 2.}$$
$$x - 6 < -14 \qquad \text{or} \qquad x - 6 > 14$$
$$x < -8 \qquad \text{or} \qquad x > 20 \qquad \text{Add 6 to both sides.}$$

The solution set is $(-\infty, -8) \cup (20, \infty)$.

155

Are you preparing for a test on Chapter 2?

Below I have listed some common trouble areas for students in Chapter 2. After studying for your test—but before taking your test—read these.

- Remember to reverse the direction of the inequality symbol when multiplying or dividing both sides of an inequality by a negative number.

$$-11x < 33 \qquad \text{Direction of arrow is reversed.}$$
$$\frac{-11x}{-11} > \frac{33}{-11}$$
$$x > -3$$

- Remember the differences when solving absolute value equations and inequalities.

$$|x + 1| = 3$$
$$x + 1 = 3 \text{ or } x + 1 = -3$$
$$x = 2 \text{ or } x = -4$$
$$\{2, -4\}$$

$$|x + 1| < 3$$
$$-3 < x + 1 < 3$$
$$-3 - 1 < x < 3 - 1$$
$$-4 < x < 2$$
$$(-4, 2)$$

$$|x + 1| > 3$$
$$x + 1 < -3 \text{ or } x + 1 > 3$$
$$x < -4 \text{ or } x > 2$$
$$(-\infty, -4) \cup (2, \infty)$$

- Remember that an equation is not solved for a specified variable unless the variable is alone on one side of an equation *and* the other side contains *no* specified variables.

$$y = 10x + 6 - y \qquad \text{Equation is } not \text{ solved for } y.$$
$$2y = 10x + 6 \qquad \text{Add } y \text{ to both sides.}$$
$$y = 5x + 3 \qquad \text{Divide both sides by 2.}$$

Remember: This is simply a checklist of common trouble areas. For a review of Chapter 2, see the Highlights and Chapter Review at the end of this chapter.

Solve.

48. A principal of $3000 is invested in an account paying an annual percentage rate of 3%. Find the amount (to the nearest cent) in the account after 7 years if the amount is compounded

   a. semiannually.

   b. weekly.

49. The high temperature in Slidell, Louisiana, one day was 90° Fahrenheit. Convert this temperature to degrees Celsius.

△**50.** Angie Applegate has a photograph for which the length is 2 inches longer than the width. If she increases each dimension by 4 inches, the area is increased by 88 square inches. Find the original dimensions.

△**51.** One-square-foot floor tiles come 24 to a package. Find how many packages are needed to cover a rectangular floor 18 feet by 21 feet.

△**52.** Determine which container holds more ice cream, an 8 inch × 5 inch × 3 inch box or a cylinder with radius of 3 inches and height of 6 inches.

53. Erasmos Gonzalez left Los Angeles at 11 A.M. and drove nonstop to San Diego, 130 miles away. If he arrived at 1:15 P.M., find his average speed, rounded to the nearest mile per hour.

(2.4) *Solve each linear inequality. Write your answers in interval notation.*

54. $3(x - 5) > -(x + 3)$

55. $-2(x + 7) \geq 3(x + 2)$

56. $4x - (5 + 2x) < 3x - 1$

57. $3(x - 8) < 7x + 2(5 - x)$

58. $24 \geq 6x - 2(3x - 5) + 2x$

59. $48 + x \geq 5(2x + 4) - 2x$

60. $\dfrac{x}{3} + \dfrac{1}{2} > \dfrac{2}{3}$

61. $x + \dfrac{3}{4} < \dfrac{-x}{2} + \dfrac{9}{4}$

62. $\dfrac{x - 5}{2} \leq \dfrac{3}{8}(2x + 6)$

63. $\dfrac{3(x - 2)}{5} > \dfrac{-5(x - 2)}{3}$

Solve.

64. George Boros can pay his housekeeper $15 per week to do his laundry, or he can have the laundromat do it at a cost of 50 cents per pound for the first 10 pounds and 40 cents for each additional pound. Use an inequality to find the weight at which it is more economical to use the housekeeper than the laundromat.

65. Ceramic firing temperatures usually range from 500° to 1000° Fahrenheit. Use a compound inequality to convert this range to the Celsius scale. Round to the nearest degree.

66. In the Olympic gymnastics competition, Nana must average a score of 9.65 to win the silver medal. Seven of the eight judges have reported scores of 9.5, 9.7, 9.9, 9.7, 9.7, 9.6, and 9.5. Use an inequality to find the minimum score that Nana must receive from the last judge to win the silver medal.

67. Carol would like to pay cash for a car when she graduates from college and estimates that she can afford a car that costs between $4000 and $8000. She has saved $500 so far and plans to earn the rest of the money by working the next two summers. If Carol plans to save the same amount each summer, use a compound inequality to find the range of money she must save each summer to buy the car.

(2.5) *Solve each inequality. Write your answers in interval notation.*

68. $1 \leq 4x - 7 \leq 3$

69. $-2 \leq 8 + 5x < -1$

70. $-3 < 4(2x - 1) < 12$

71. $-6 < x - (3 - 4x) < -3$

72. $\dfrac{1}{6} < \dfrac{4x - 3}{3} \leq \dfrac{4}{5}$

73. $0 \leq \dfrac{2(3x + 4)}{5} \leq 3$

74. $x \leq 2$ and $x > -5$

75. $x \leq 2$ or $x > -5$

76. $3x - 5 > 6$ or $-x < -5$

77. $-2x \leq 6$ and $-2x + 3 < -7$

(2.6) *Solve each absolute value equation.*

78. $|x - 7| = 9$

79. $|8 - x| = 3$

80. $|2x + 9| = 9$

81. $|-3x + 4| = 7$

82. $|3x - 2| + 6 = 10$

83. $5 + |6x + 1| = 5$

84. $-5 = |4x - 3|$

85. $|5 - 6x| + 8 = 3$

86. $|7x| - 26 = -5$

87. $-8 = |x - 3| - 10$

88. $\left| \dfrac{3x - 7}{4} \right| = 2$

89. $\left| \dfrac{9 - 2x}{5} \right| = -3$

90. $|6x + 1| = |15 + 4x|$

91. $|x - 3| = |7 + 2x|$

Solve each absolute value inequality. Graph the solution set and write it in interval notation.

92. $|5x - 1| < 9$

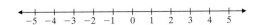

93. $|6 + 4x| \geq 10$

94. $|3x| - 8 > 1$

95. $9 + |5x| < 24$

96. $|6x - 5| \leq -1$

97. $|6x - 5| \geq -1$

98. $\left|3x + \dfrac{2}{5}\right| \geq 4$

99. $\left|\dfrac{4x - 3}{5}\right| < 1$

100. $\left|\dfrac{x}{3} + 6\right| - 8 > -5$

101. $\left|\dfrac{4(x - 1)}{7}\right| + 10 < 2$

Have you decided to successfully complete this course?

Ask yourself if one of your current goals is to successfully complete this course.

If it is not a goal of yours, ask yourself why? One common reason is fear of failure. Amazingly enough, fear of failure alone can be strong enough to keep many of us from doing our best in any endeavor. Another common reason is that you simply haven't taken the time to make successfully completing this course one of your goals.

If you are taking this mathematics course, then successfully completing this course probably should be one of your goals. To make it a goal, start by writing this goal in your mathematics notebook. Then read or reread Section 1.1 and make a commitment to try the suggestions in this section.

If successfully completing this course is already a goal of yours, also read or reread Section 1.1 and try some suggestions in this section so that you are actively working toward your goal.

Good luck and don't forget that a positive attitude will make a big difference!

Chapter 2 Test

Solve each equation.

1. $8x + 14 = 5x + 44$

2. $3(x + 2) = 11 - 2(2 - x)$

3. $3(y - 4) + y = 2(6 + 2y)$

4. $7n - 6 + n = 2(4n - 3)$

5. $\dfrac{z}{2} + \dfrac{z}{3} = 10$

6. $\dfrac{7w}{4} + 5 = \dfrac{3w}{10} + 1$

7. $|6x - 5| = 1$

8. $|8 - 2t| = -6$

9. $|2x - 3| = |4x + 5|$

Solve each equation for the specified variable.

10. $3x - 4y = 8$ for y

11. $4(2n - 3m) - 3(5n - 7m) = 0$ for n

12. $S = gt^2 + gvt$ for g

13. $F = \dfrac{9}{5}C + 32$ for C

Solve each inequality. Write your answers in interval notation.

14. $3(2x - 7) - 4x > -(x + 6)$

15. $8 - \dfrac{x}{2} \le 7$

16. $-3 < 2(x - 3) \le 4$

17. $|3x + 1| > 5$

18. $|x - 6| + 4 \le 9$

19. $x \ge 5$ and $x \ge 4$

1. _____

2. _____

3. _____

4. _____

5. _____

6. _____

7. _____

8. _____

9. _____

10. _____

11. _____

12. _____

13. _____

14. _____

15. _____

16. _____

17. _____

18. _____

19. _____

20. $x \geq 5$ or $x \geq 4$

21. $-x > 1$ and $3x + 3 \geq x - 3$

22. $6x + 1 > 5x + 4$ or $1 - x > -4$

23. Find 12% of 80.

Solve.

24. In 2000, Ford sold 7,424,000 new vehicles worldwide. This represents an 8.8% increase over the number of new vehicles sold by Ford in 1998. Use this information to find the number of new Ford vehicles sold in 1998. Round to the nearest thousand. (_Source:_ Ford Motor Company)

△ **25.** A circular dog pen has a circumference of 78.5 feet. Approximate π by 3.14 and estimate how many hunting dogs could be safely kept in the pen if each dog needs at least 60 square feet of room.

26. The company that makes Photoray sunglasses figures that the cost C to make x number of sunglasses weekly is given by $C = 3910 + 2.8x$, and the weekly revenue R is given by $R = 7.4x$. Use an inequality to find the number of sunglasses that must be made and sold to make a profit. (Recall that revenue must exceed cost in order to make a profit.)

27. Find the amount of money in an account after 10 years if a principal of $2500 is invested at 3.5% interest compounded quarterly. (Round to the nearest cent.)

28. The top three states where international travelers spend the most money are Florida, California, and New York. International travelers spent $4 billion more money in California than New York, and in Florida they spent $1 billion less than twice the amount spent in New York. If total international spending in these three states is $39 billion, find the amount spent in each state. (_Source:_ Travel Industry Asso. of America)

Cumulative Review

1. Write the set $\{x|x$ is a natural number greater than 100$\}$ in roster form.

Write each sentence as an equation.

2. The sum of x and 5 is 20.

3. The quotient of z and 9 amounts to 9 plus z.

Find the opposite or additive inverse of each number.

4. 8

5. $-\dfrac{1}{5}$

Add.

6. $-3 + (-11)$

7. $-10 + 15$

8. $-\dfrac{2}{3} + \dfrac{3}{7}$

Divide.

9. $\dfrac{20}{-4}$

10. $\dfrac{0}{-8}$

11. $\dfrac{-10}{-80}$

12. Simplify: $3 + 2 \cdot 30$

Simplify by combining like terms.

13. $3x - 5x + 4$

14. $y + 3y$

Use the quotient rule to simplify.

15. $\dfrac{x^7}{x^4}$

16. $\dfrac{20x^6}{4x^5}$

1. _____

2. _____

3. _____

4. _____

5. _____

6. _____

7. _____

8. _____

9. _____

10. _____

11. _____

12. _____

13. _____

14. _____

15. _____

16. _____

Simplify each expression. Write each answer using positive exponents only.

17. _____

17. $\left(\dfrac{3x^2y}{y^{-9}z}\right)^{-2}$

18. $\left(\dfrac{3a^2}{2x^{-1}}\right)^3\left(\dfrac{x^{-3}}{4a^{-2}}\right)^{-1}$

18. _____

19. Solve: $2x + 5 = 9$

19. _____

20. Solve: $6x - 4 = 2 + 6(x - 1)$

20. _____

21. Suppose that Best Buy just announced an 8% decrease in the price of a particular computer model. If this computer sells for $2162 after the decrease, find its original price.

22. Solve $V = LWH$ for H.

21. _____

22. _____

23. Solve $\dfrac{1}{4}x \le \dfrac{3}{2}$. Graph the solution set.

23. _____

24. Solve: $-2x - 5 < -3$ or $6x < 0$

24. _____

25. Solve: $|x| = 3$

25. _____

Graphs and Functions

The linear equations we explored in Chapter 2 are statements about a single variable. This chapter examines statements about two variables: linear equations and inequalities in two variables. We focus particularly on graphs of these equations and inequalities, which lead to the notion of relation and to the notion of function, perhaps the single most important and useful concept in all of mathematics.

CHAPTER 3

3.1 Graphing Linear Equations

3.2 The Slope of a Line

3.3 The Slope-Intercept Form

3.4 More Equations of Lines

Integrated Review—Linear Equations in Two Variables

3.5 Introduction to Functions

3.6 Graphing Linear Inequalities

At the beginning of the 20th century, there were approximately 237,600 students enrolled in the 977 institutions of higher education in the United States. At that time, only 19% of bachelor's degree recipients were women. By the year 2010, the projected 4700 degree-granting colleges and universities in the United States will have an estimated 17,490,000 students. Of these, roughly 59% of bachelor's degree recipients are expected to be women. The phenomenal growth of colleges and universities can also be seen in the average tuition costs at these institutions of higher learning. For instance, the average annual tuition at a private four-year college or university has increased from $1809 in 1970 to $15,380 in 2000, an increase of about 750%! In Exercises 33 and 34 on page 204 we will use linear equations to predict the future cost of annual tuition at both two-year and four-year public colleges and universities.

Name _____ Section _____ Date _____

Chapter 3 Pretest

1. _____

2. **a.** _____

b. _____

c. _____

d. _____

3. see graph

4. see graph

5. see graph

6. see graph

7. _____

8. _____

9. _____

10. _____

11. _____

12. _____

13. _____

14. _____

15. _____

16. _____

17. _____

18. _____

19. see graph

20. see graph

1. Determine whether each ordered pair is a solution of the given equation. $2x - 5y = -13; (-1, -3), (-4, 1)$

2. Name the quadrant (or axis) in which each point lies.

a. $(6, -5)$ **b.** $(0, -4)$

c. $\left(-\dfrac{1}{2}, -\dfrac{2}{3}\right)$ **d.** $(7, 1.8)$

Graph each line.

3. $2x + y = 4$ **4.** $y = -x + 3$ **5.** $y = -2$ **6.** $x - 5 = 0$

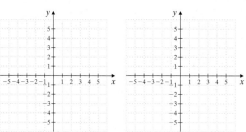 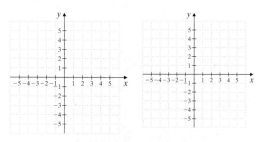

7. Find the slope of the line that passes through $(5, 4)$ and $(-6, 2)$.

8. Find the slope and the y-intercept of the line $4x - 5y = 2$.

9. Find the slope of the line $y = 5$.

Use the slope-intercept form of a linear equation to write the equation of each line with the given slope and y-intercept.

10. Slope $\dfrac{1}{3}$; y-intercept 6

11. Slope -7; y-intercept 0

Find an equation of each line satisfying the conditions given. Write the equations in standard form.

12. Slope 2; through $(-3, -7)$

13. Through $(5, 4)$ and $(-1, 6)$

14. Horizontal; through $(9, 10)$

15. Perpendicular to $3y - x = 6$; through $(8, 0)$

16. Find the domain and the range of the given relation. Also determine whether the relation is a function. $\{(-2, 5), (3, -7), (2, 5)\}$

If $f(x) = 7x - 1$ and $g(x) = 2x^2 + x - 5$, find the following.

17. $f(-3)$

18. $g(0)$

Graph each inequality.

19. $x - y < 4$

20. The intersection of $x \leq 3$ and $y > -1$

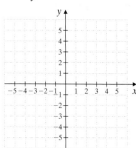 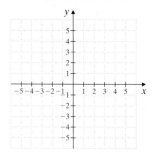

3.1 Graphing Linear Equations

Graphs are widely used today in newspapers, magazines, and all forms of newsletters. A few examples of graphs are shown here.

OBJECTIVES

(A) Plot ordered pairs on a rectangular coordinate system.

(B) Determine whether an ordered pair of numbers is a solution of an equation in two variables.

(C) Graph linear equations.

(D) Graph vertical and horizontal lines.

SSM
TUTOR CENTER   SG   CD & VIDEO   MATH PRO   WEB

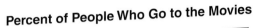

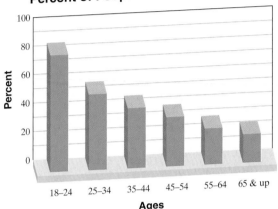

Percent of People Who Go to the Movies

Source: TELENATION/Market Facts, Inc.

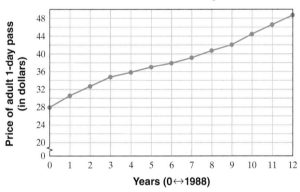

Ticket Price at Disney World

Source: The Walt Disney Company

Note: The notation 0↔1988 means that year 0 corresponds to the year 1988, 1 corresponds to 1989, and so on.

To help us understand how to read these graphs, we will review their origin—the rectangular coordinate system.

(A) Plotting Ordered Pairs on a Rectangular Coordinate System

One way to locate points on a plane is by using a **rectangular coordinate system**, which is also called a **Cartesian coordinate system** after its inventor, René Descartes (1596–1650). A rectangular coordinate system consists of two number lines that intersect at right angles at their 0 coordinates. We position these axes on paper such that one number line is horizontal and the other number line is then vertical. The horizontal number line is called the **x-axis** (or the axis of the **abscissa**), and the vertical number line is called the **y-axis** (or the axis of the **ordinate**). The point of intersection of these axes is named the **origin**.

Notice that the axes divide the plane into four regions. These regions are called **quadrants**. The top-right region is quadrant I. Quadrants II, III, and IV are numbered counterclockwise from the first quadrant as shown. The *x*-axis and the *y*-axis are not in any quadrant.

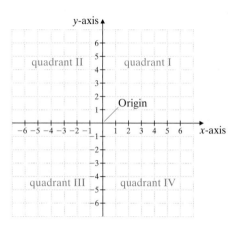

Each point in the plane can be located, or **plotted**, by describing its position in terms of distances along each axis from the origin. An **ordered pair**, represented by the notation (x, y), records these distances. For example, the location of point A in the figure below is described as 2 units to the left of the origin along the *x*-axis and 5 units upward parallel to the *y*-axis. Thus, we identify point A with the ordered pair $(-2, 5)$. Notice that the *order* of these numbers is *critical*. The *x*-value -2 is called the **x-coordinate** and is associated with the *x*-axis. The *y*-value 5 is called the **y-coordinate** and is associated with the *y*-axis.

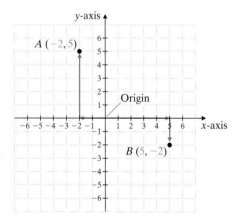

Compare the location of point A with the location of point B, which corresponds to the ordered pair $(5, -2)$. The *x*-coordinate 5 indicates that we move 5 units to the right of the origin along the *x*-axis. The *y*-coordinate -2 indicates that we move 2 units down parallel to the *y*-axis. Point A is in a different position than point B. Two ordered pairs are considered equal and correspond to the same point if and only if their *x*-coordinates are equal and their *y*-coordinates are equal.

Keep in mind that *each ordered pair corresponds to exactly one point in the real plane and that each point in the plane corresponds to exactly one ordered pair*. Thus, we may refer to the ordered pair (x, y) as the **point** (x, y).

EXAMPLE 1

Plot each ordered pair on a rectangular coordinate system and name the quadrant in which the point is located.

a. $(2, -1)$ **b.** $(0, 5)$ **c.** $(-3, 5)$

d. $(-2, 0)$ **e.** $\left(-\dfrac{1}{2}, -4\right)$ **f.** $(1.5, 1.5)$

Solution: The six points are graphed as shown in the figure.

a. $(2, -1)$ lies in quadrant IV.   **b.** $(0, 5)$ is not in any quadrant.
c. $(-3, 5)$ lies in quadrant II.   **d.** $(-2, 0)$ is not in any quadrant.
e. $\left(-\dfrac{1}{2}, -4\right)$ is in quadrant III.   **f.** $(1.5, 1.5)$ is in quadrant I.

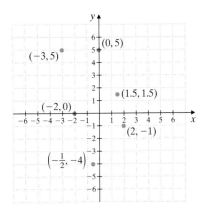

Notice that the y-coordinate of any point on the x-axis is 0. For example, the point with coordinates $(-2, 0)$ lies on the x-axis. Also, the x-coordinate of any point on the y-axis is 0. For example, the point with coordinates $(0, 5)$ lies on the y-axis. A point on an axis is called a **quadrantel** point.

Try the Concept Check in the margin.

Many types of real-world data occur in pairs. For example, the data pairs below were used for the Disney World ticket graph at the beginning of this section. The graph of paired data, such as the one below, is called a **scatter diagram**. Such diagrams are used to look for patterns and relationships in paired data.

Paired Data	
Year, x	**Price (in dollars), y**
0	28
1	30.65
2	32.75
3	34.85
4	35.9
5	37.1
6	38
7	39.22
8	40.81
9	42.14
10	44.52
11	46.64
12	48.76

Note: The notation $0 \leftrightarrow 1988$ means that year 0 corresponds to the year 1988, 1 corresponds to 1989, and so on.

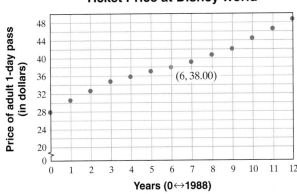

Ticket Price at Disney World

Source: The Walt Disney Company

Helpful Hint

Notice, for example, the paired data $(6, 38.00)$ and its corresponding plotted point, both in blue.

Practice Problem 1

Plot each ordered pair on a rectangular coordinate system and name the quadrant in which the point is located.

a. $(3, -2)$ b. $(0, 3)$
c. $(-4, 1)$ d. $(-1, 0)$
e. $\left(-2\dfrac{1}{2}, -3\right)$ f. $(3.5, 4.5)$

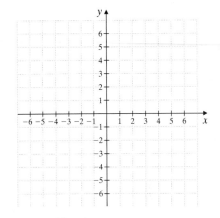

Concept Check

Which of the following correctly describes the location of the point $(3, -6)$ in a rectangular coordinate system?

a. 3 units to the left of the y-axis and 6 units above the x-axis

b. 3 units above the x-axis and 6 units to the left of the y-axis

c. 3 units to the right of the y-axis and 6 units below the x-axis

d. 3 units below the x-axis and 6 units to the right of the y-axis

Answers

1.

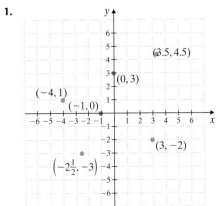

a. quadrant IV,   b. not in any quadrant,
c. quadrant II,   d. not in any quadrant,
e. quadrant III,   f. quadrant I

Concept Check: c

Practice Problem 2

Create a scatter diagram for the given paired data.

3-Point Baskets Made by Lisa Leslie of the Los Angeles Sparks	
WNBA Season	3-Point Baskets Made
1997	12
1998	9
1999	22
2000	7
2001	22

(*Source:* WNBA)

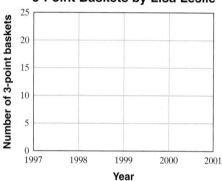

Practice Problem 3

Determine whether $(0, -6)$, $(1, 4)$, and $(-1, -4)$ are solutions of the equation $2x + y = -6$.

Answers

2.

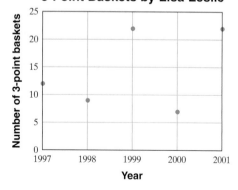

3. yes; no; yes

EXAMPLE 2   Create a scatter diagram for the given paired data.

Price of a Big Mac in Russia

Year	Price (in U.S. dollars)
1998	2.00
1999	1.35
2000	1.39
2001	1.21

(*Sources:* McDonald's, *The Economist*)

Solution: To graph the paired data in the table, we use the first column for the *x*- (or horizontal) axis and the second column for the *y*- (or vertical) axis.

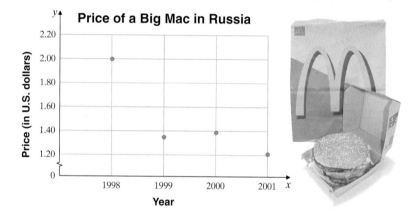

B   **Determining Whether an Ordered Pair of Numbers Is a Solution of an Equation**

A **solution** of an equation in two variables consists of two numbers that can be written as an ordered pair of numbers. Unless we are told otherwise, we will assume that variable values are written as ordered pairs in alphabetical order (that is, *x* first and then *y*).

EXAMPLE 3

Determine whether $(0, -12)$, $(1, 9)$, and $(2, -6)$ are solutions of the equation $3x - y = 12$.

Solution: To check each ordered pair, we replace *x* with the *x*-coordinate and *y* with the *y*-coordinate and see whether a true statement results.

Let $x = 0$ and $y = -12$.

$$3x - y = 12$$
$$3(0) - (-12) \stackrel{?}{=} 12$$
$$0 + 12 \stackrel{?}{=} 12$$
$$12 = 12 \quad \text{True.}$$

Let $x = 1$ and $y = 9$.

$$3x - y = 12$$
$$3(1) - 9 \stackrel{?}{=} 12$$
$$3 - 9 \stackrel{?}{=} 12$$
$$-6 = 12 \quad \text{False.}$$

Let $x = 2$ and $y = -6$.

$$3x - y = 12$$
$$3(2) - (-6) \stackrel{?}{=} 12$$
$$6 + 6 \stackrel{?}{=} 12$$
$$12 = 12 \quad \text{True.}$$

We see that $(1, 9)$ is not a solution but both $(2, -6)$ and $(0, -12)$ are solutions.   ●

C   Graphing Linear Equations

As we saw in Example 3, some linear equations have more than one ordered pair solution. In fact, the equation $3x - y = 12$ has an infinite number of ordered pair solutions. Since it is impossible to list all solutions, we visualize them by graphing them.

A few more ordered pairs that satisfy $3x - y = 12$ are $(4, 0)$, $(3, -3)$, $(5, 3)$, and $(1, -9)$. These ordered pair solutions, along with the ordered pair solutions from Example 3, are plotted on the following graph. The graph of $3x - y = 12$ is the single line containing these points. Every ordered pair solution of the equation corresponds to a point on this line, and every point on this line corresponds to an ordered pair solution.

x	y
4	0
1	−9
0	−12
2	−6
5	3
3	−3

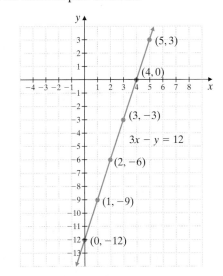

The equation $3x - y = 12$ is called a *linear equation in two variables*, and *the graph of every linear equation in two variables is a line.*

Linear Equation in Two Variables

A **linear equation in two variables** is an equation that can be written in the form

$$Ax + By = C$$

where A, B, and C are real numbers, and A and B are not both 0. The graph of a linear equation in two variables is a line.

A linear equation written in the form $Ax + By = C$ is said to be written in **standard form**. Some examples are

$$3x - y = 12$$
$$-2.1x + 5.6y = 0$$

Helpful Hint

Remember that in a linear equation in standard form, all of the variable terms are on one side of the equation and the constant is on the other side.

Recall from geometry that a line is determined by two points. This means that to graph a linear equation in two variables, just two solutions are needed. We will find a third solution, just to check our work. To find ordered pair solutions of linear equations in two variables, we can choose an x-value and find its corresponding y-value, or we can choose a y-value and find its corresponding x-value. The number 0 is often a convenient value to choose for x and also for y.

Practice Problem 4

Graph: $3x - 2y = 6$

EXAMPLE 4   Graph: $5x - 2y = 10$

Solution: First we find three ordered pair solutions, and then we plot the ordered pairs. The line through the plotted points is the graph. Let's let x be 0, let y be 0, and then let x be 1 to find our three ordered pair solutions.

Let $x = 0$.
$$5x - 2y = 10$$
$$5 \cdot 0 - 2y = 10$$
$$-2y = 10 \quad \text{Simplify.}$$
$$y = -5 \quad \text{Divide by} -2.$$

Let $y = 0$.
$$5x - 2y = 10$$
$$5x - 2 \cdot 0 = 10$$
$$5x = 10 \quad \text{Simplify.}$$
$$x = 2 \quad \text{Divide by 5.}$$

Let $x = 1$.
$$5x - 2y = 10$$
$$5 \cdot 1 - 2y = 10$$
$$5 - 2y = 10 \quad \text{Multiply.}$$
$$-2y = 5 \quad \text{Subtract 5.}$$
$$y = -\frac{5}{2}, \text{or} -2\frac{1}{2}$$

The three ordered pair solutions—$(0, -5)$, $(2, 0)$, and $\left(1, -2\frac{1}{2}\right)$—are listed in the table, and the graph of $5x - 2y = 10$ is shown.

x	y
0	-5
2	0
1	$-2\frac{1}{2}$

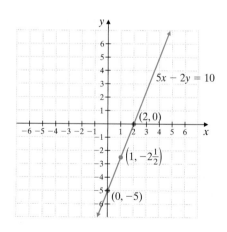

Answer

4.

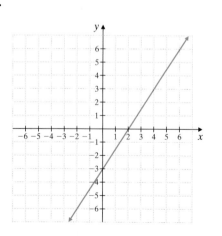

Notice that the graph in Example 4 crosses the x-axis at the point $(2, 0)$. This point is called the **x-intercept**. This graph also crosses the y-axis at the point $(0, -5)$. This point is called the **y-intercept**.

In general, to find the y-intercept of the graph of an equation, let $x = 0$ since any point on the y-axis has an x-coordinate of 0. To find the x-intercept of a line, let $y = 0$ since any point on the x-axis has a y-coordinate of 0.

Finding x- and y-Intercepts

To find an x-intercept, let $y = 0$ and solve for x.
To find a y-intercept, let $x = 0$ and solve for y.

EXAMPLE 5   Find the intercepts and graph: $x + 4y = -4$

Solution: To find the y-intercept, we let $x = 0$ and solve for y. To find the x-intercept, we let $y = 0$ and solve for x. Let's let $x = 0$, $y = 0$, and then let $x = 2$ to find our third check point.

Let $x = 0$.	Let $y = 0$.	Let $x = 2$.
$x + 4y = -4$	$x + 4y = -4$	$x + 4y = -4$
$0 + 4y = -4$	$x + 4 \cdot 0 = -4$	$2 + 4y = -4$
$4y = -4$	$x = -4$	$4y = -6$
$y = -1$		$y = -\dfrac{6}{4} = -1\dfrac{1}{2}$
$(0, -1)$	$(-4, 0)$	$\left(2, -1\dfrac{1}{2}\right)$

The ordered pairs are $(0, -1)$, $(-4, 0)$, and $\left(2, -1\dfrac{1}{2}\right)$. We plot these points to obtain the graph shown.

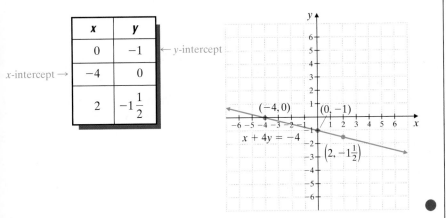

x	y	
0	-1	← y-intercept
-4	0	← x-intercept
2	$-1\dfrac{1}{2}$	

EXAMPLE 6   Find the intercepts and graph: $x = -2y$

Solution: We let $y = 0$ to find the x-intercept and $x = 0$ to find the y-intercept.

Let $y = 0$.	Let $x = 0$.
$x = -2y$	$x = -2y$
$x = -2(0)$	$0 = -2y$
$x = 0$	$0 = y$

Both the x-intercept and y-intercept are $(0, 0)$. In other words, when $x = 0$, then $y = 0$, which gives the ordered pair $(0, 0)$. Also, when $y = 0$, then $x = 0$, which gives the same ordered pair $(0, 0)$. This happens when the graph passes through the origin. Since two points are needed to determine a line, we must find at least one more ordered pair that satisfies $x = -2y$. Let's let $y = -1$ to find a second ordered pair solution and let $y = 1$ to find a third check point.

Practice Problem 5

Find the intercepts and graph:
$x + 3y = -6$

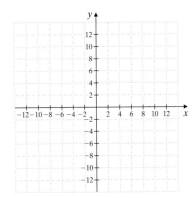

Practice Problem 6

Find the intercepts and graph: $y = 2x$

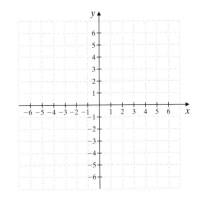

Answers

5.

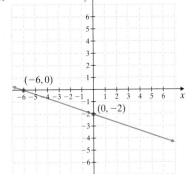

6.

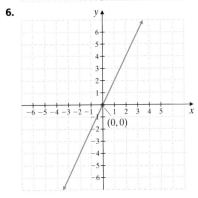

Let $y = -1$.

$x = -2(-1)$

$x = 2$

Let $y = 1$.

$x = -2(1)$

$x = -2$

Helpful Hint

Since the equation $x = -2y$ is solved for x, we choose y-values in order to find second and third points. This way, we simply need to evaluate an expression to find the x-value, as shown in Example 6.

The ordered pairs are $(0, 0)$, $(2, -1)$, and $(-2, 1)$. We plot these points to obtain the graph shown.

	x	y	
x-intercept →	0	0	← y-intercept
	2	-1	
	-2	1	

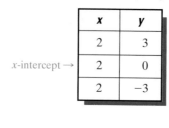

Practice Problem 7

Graph: $x = -1$

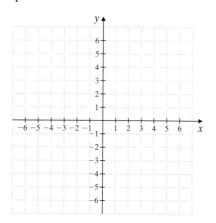

Answer

7.

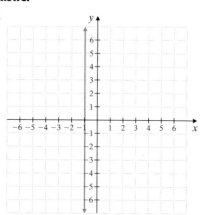

D Graphing Vertical and Horizontal Lines

The equation $x = c$, where c is a real number constant, is a linear equation in two variables because it can be written in the form $x + 0y = c$. The graph of this equation is a vertical line, as shown in the next example.

EXAMPLE 7 Graph: $x = 2$

Solution: The equation $x = 2$ can be written as $x + 0y = 2$. Notice that for any y-value chosen, x is 2. No other value for x satisfies $x + 0y = 2$. Any ordered pair whose x-coordinate is 2 is a solution to $x + 0y = 2$ because 2 added to 0 times any value of y is $2 + 0$, or 2. We will use the ordered pairs $(2, 3)$, $(2, 0)$ and $(2, -3)$ to graph $x = 2$.

	x	y
	2	3
x-intercept →	2	0
	2	-3

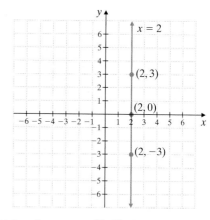

The graph is a vertical line with x-intercept $(2, 0)$. It has no y-intercept because x is never 0.

EXAMPLE 8   Graph: $y = -3$

Solution: The equation $y = -3$ can be written as $0x + y = -3$. For any x-value chosen, y is -3. If we choose $4, 0$, and -2 as x-values, the ordered pair solutions are $(4, -3)$, $(0, -3)$, and $(-2, -3)$. We will use these ordered pairs to graph $y = -3$.

x	y
4	-3
0	-3	← y-intercept
-2	-3

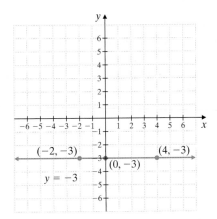

The graph is a horizontal line with y-intercept -3 and no x-intercept.   ●

From Examples 7 and 8, we have the following generalization.

Graphing Vertical and Horizontal Lines

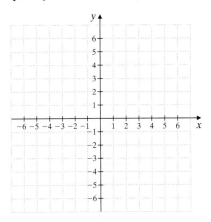

The graph of $x = c$, where c is a real number, is a vertical line with x-intercept $(c, 0)$.
The graph of $y = c$, where c is a real number, is a horizontal line with y-intercept $(0, c)$.

Practice Problem 8

Graph: $y = 2$

Answer

8.

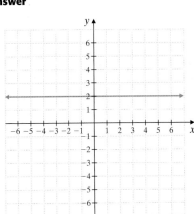

GRAPHING CALCULATOR EXPLORATIONS

In this section, we begin a study of graphing calculators and graphing software packages for computers.

These graphers use the same point-plotting technique that we introduced in this section. The advantage of this graphing technology is, of course, that graphing calculators and computers can find and plot ordered pair solutions much faster than we can. Note, however, that the features described in these boxes may not be available on all graphing calculators.

The rectangular screen where a portion of the rectangular coordinate system is displayed is called a **window**. We call it a **standard window** for graphing when both the x- and y-axes display coordinates between -10 and 10. This information is often displayed in the window menu on a graphing calculator as

Xmin = -10
Xmax = 10
Xscl = 1 The scale on the x-axis is one unit per tick mark.
Ymin = -10
Ymax = 10
Yscl = 1 The scale on the y-axis is one unit per tick mark.

To use a graphing calculator to graph the equation $y = -5x + 4$, press the $\boxed{\text{Y=}}$ key and enter the keystrokes

(Check your owner's manual to make sure the "negative" key is pressed here and not the "subtraction" key.)

The top row should now read $Y_1 = -5x + 4$. Next press the $\boxed{\text{GRAPH}}$ key, and the display should look like this:

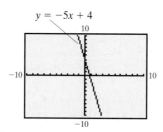

Use a standard window and graph each linear equation. (Unless otherwise stated, we will use a standard window when graphing.)

1. $y = 6x - 1$

2. $y = 3x - 2$

3. $y = -3.2x + 7.9$

4. $y = -x + 5.85$

5. $y = \dfrac{1}{4}x - \dfrac{2}{3}$ (Parentheses may need to be inserted around $\frac{1}{4}$.)

6. $y = \dfrac{2}{3}x - \dfrac{1}{5}$ (Parentheses may need to be inserted around $\frac{2}{3}$.)

Name _____ Section _____ Date _____

Mental Math

Determine the coordinates of each point on the graph.

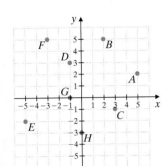

1. Point *A*

2. Point *B*

3. Point *C*

4. Point *D*

5. Point *E*

6. Point *F*

7. Point *G*

8. Point *H*

EXERCISE SET 3.1

A Plot each ordered pair on a rectangular coordinate system and name the quadrant (or axis) in which the point lies.
See Example 1.

1. $(3, 2)$

$(-5, 3)$

$\left(5\frac{1}{2}, -4\right)$

$(0, 3.5)$

$(-2, -4)$

2. $(2, -1)$

$(-3, -1)$

$\left(-2, 6\frac{1}{3}\right)$

$(-2, 4)$

$(-4.2, 0)$

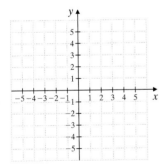

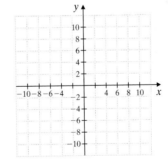

Given that x is a positive number and y is a positive number, determine the quadrant (or axis) in which each point lies.
See Example 1.

3. $(x, -y)$

4. $(-x, y)$

5. $(x, 0)$

6. $(0, -y)$

7. $(-x, -y)$

8. $(0, 0)$

Create a scatter diagram for the given paired data. See Example 2.

9. Airline Revenues from Passengers in the United States

Year	Revenue (in billions of dollars)
1996	75
1997	79
1998	81
1999	84
2000	94

(*Source:* Air Transport Association of America)

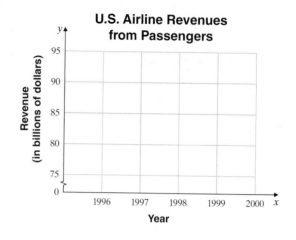

10. Median Age at Retirement for U.S. Men

Year	Age (in years)
1950	67
1960	65
1970	63
1980	63
1990	63
2000	62

(*Source:* U.S. Bureau of Labor Statistics)

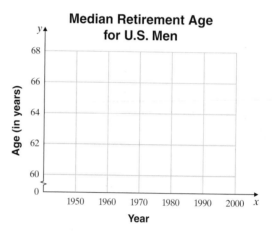

B *Determine whether each ordered pair is a solution of the given equation. See Example 3.*

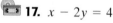 **11.** $y = 3x - 5; (0, 5), (-1, -8)$

12. $y = -2x + 7; (1, 5), (-2, 3)$

13. $-6x + 5y = -6; (1, 0), \left(2, \dfrac{6}{5}\right)$

14. $5x - 3y = 9; (0, 3), \left(\dfrac{12}{5}, -1\right)$

15. $y = -3; (1, -3), (-3, 6)$

16. $y = 2; (2, 5), (0, 2)$

C **D** *Graph each linear equation. See Examples 4 through 8.*

 17. $x - 2y = 4$

18. $y - 2x = 4$

19. $3x + 2y = 6$

20. $2x + 4y = 8$

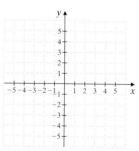

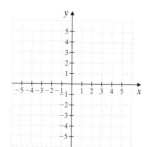

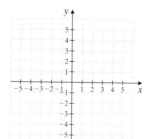

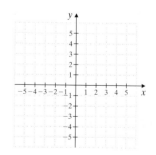

21. $x = 4$ 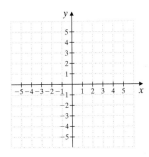 **22.** $y = 5$ **23.** $x - 3y = 6$ **24.** $x - 2y = 4$

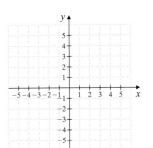   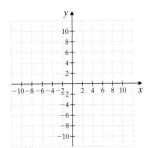   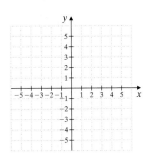

25. $y = 3x$ **26.** $y = -4x$ **27.** $y = -2$ **28.** $x = -3$

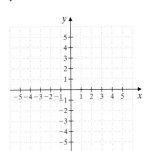   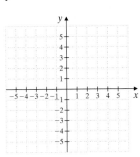   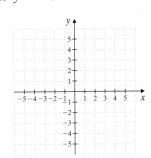   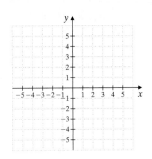

29. $4x + 5y = 15$ **30.** $2x + 3y = 9$ **31.** $5y = x - 10$ **32.** $3y = x - 3$

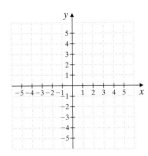   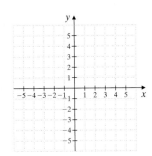   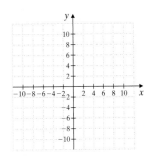   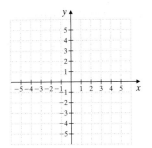

33. $x = \dfrac{1}{2}$ **34.** $y = -\dfrac{5}{2}$ **35.** $y = \dfrac{1}{2}x$ **36.** $x = \dfrac{1}{2}y$

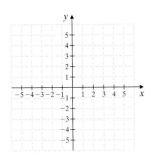   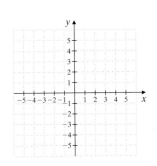   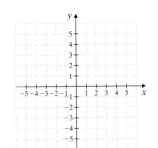   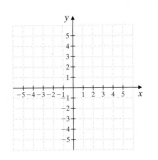

37. $y = -4x + 1$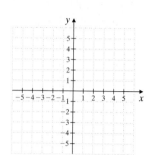

38. $y = -3x + 1$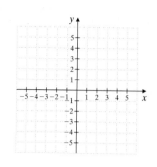

39. $2y - 6 = 0$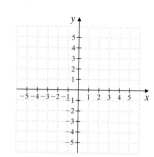

40. $3x + 6 = 0$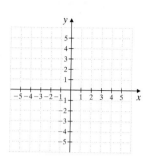

Review and Preview

Match each equation with its graph.

A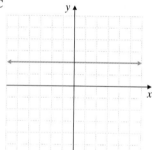

B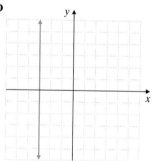

41. $y = 2$

42. $x = -3$

43. $x - 2 = 0$

44. $y + 1 = 0$

C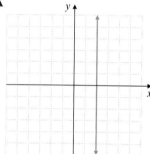

D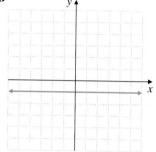

Simplify. See Section 1.5.

45. $\dfrac{-6 - 3}{2 - 8}$

46. $\dfrac{4 - 5}{-1 - 0}$

47. $\dfrac{-8 - (-2)}{-3 - (-2)}$

48. $\dfrac{12 - 3}{10 - 9}$

49. $\dfrac{0 - 6}{5 - 0}$

50. $\dfrac{2 - 2}{3 - 5}$

 Combining Concepts

For exercises 51 through 54, match each description with the graph that best illustrates it.

51. Moe worked 40 hours per week until the fall semester started. He quit and didn't work again until he worked 60 hours a week during the Christmas break.

52. Kawana worked 40 hours a week for her father during the summer. She slowly cut back her hours to not working at all during the fall semester. During the Christmas break, she started working again and increased her hours to 60 hours per week.

53. Wendy worked from July through February, never quitting. She worked between 10 and 30 hours per week.

54. Bartholomew worked from July through February, never quitting. He worked between 10 and 30 hours per week except during Christmas. At that time, he worked 40 hours per week.

A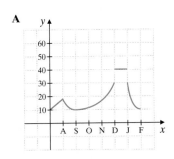

B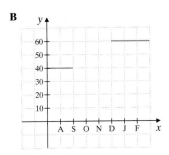

C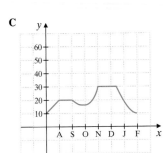

D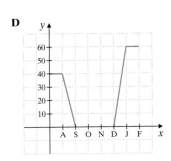

This broken line graph shows hourly minimum wages and the years it increased. Use this graph for Exercises 55 through 58.

The Federal Hourly Minimum Wage Since Its Inception

* Passed by Congress

55. What was the first year that the minimum hourly wage rose above $4.00?

56. What was the first year that the minimum hourly wage rose above $5.00?

57. Why do you think that this graph is shaped the way it is?

58. The federal hourly minimum wage started in 1938 at $0.25. How much will it have increased by in 2002?

183

59. Broyhill Furniture found that it takes 2 hours to manufacture each table for one of its special dining room sets. Each chair takes 3 hours to manufacture. A total of 1500 hours is available to produce tables and chairs of this style. The linear equation that models this situation is $2x + 3y = 1500$, where x represents the number of tables produced and y the number of chairs produced.
  a. Complete the ordered pair solution $(0, \quad)$ of this equation. Describe the manufacturing situation to which this solution corresponds.
  b. Complete the ordered pair solution $(\quad, 0)$ for this equation. Describe the manufacturing situation to which this solution corresponds.
  c. If 50 tables are produced, find the greatest number of chairs the company can make.

60. While manufacturing two different camera models, Kodak found that the basic model costs \$55 to produce, whereas the deluxe model costs \$75. The weekly budget for those two models is limited to \$33,000 in production costs. The linear equation that models this situation is $55x + 75y = 33,000$, where x represents the number of basic models and y the number of deluxe models.
  a. Complete the ordered pair solution $(0, \quad)$ of this equation. Describe the manufacturing situation to which this solution corresponds.
  b. Complete the ordered pair solution $(\quad, 0)$ of this equation. Describe the manufacturing situation to which this solution corresponds.
  c. If 350 deluxe models are produced, find the greatest number of basic models that can be made in one week.

61. On the same set of axes, graph $y = 2x$, $y = 2x - 5$, and $y = 2x + 5$. What patterns do you see in these graphs?

62. Explain why we generally use three points to graph a line, when only two points are needed.

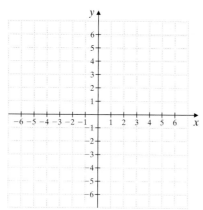

Use a grapher to verify the graph of each exercise.

63. Exercise 25

64. Exercise 26

65. Exercise 37

66. Exercise 38

67. Discuss whether a vertical line ever has a y-intercept.

68. Discuss whether a horizontal line ever has an x-intercept.

3.2 The Slope of a Line

You may have noticed by now that different lines often tilt differently. It is very important in many fields to be able to measure and compare the tilt, or **slope**, of lines. For example, a wheelchair ramp with a slope of $\frac{1}{12}$ means that the ramp rises 1 foot for every 12 horizontal feet. A road with a slope or grade of 8% $\left(\text{or } \dfrac{8}{100}\right)$ means that the road rises 8 feet for every 100 horizontal feet.

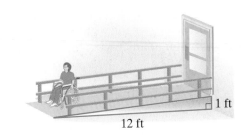
1 ft
12 ft

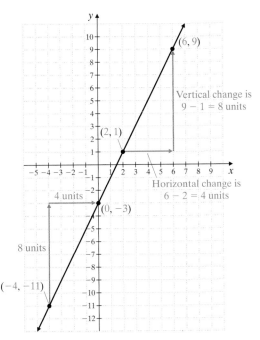
8 ft
100 ft

We measure the slope of a line as a ratio of **vertical change** to **horizontal change**. Slope is usually designated by the letter m.

Ⓐ Finding Slope from Two Points

Suppose that we want to measure the slope of the following line.

The vertical change between both pairs of points on the line is 8 units per horizontal change of 4 units. Thus

$$\text{slope } m = \frac{\text{change in } y \text{ (vertical change)}}{\text{change in } x \text{ (horizontal change)}} = \frac{8}{4} = 2$$

Notice that slope is a **rate of change** between points. A slope of 2 or $\frac{2}{1}$ means that between pairs of points on the line, the rate of change is a vertical change of 2 units per horizontal change of 1 unit.

In general, consider the line that passes through the points (x_1, y_1) and (x_2, y_2). (The notation x_1 is read "x-sub-one.") The vertical change, or **rise**, between these points is the difference in the y-coordinates: $y_2 - y_1$. The horizontal change, or **run**, between the points is the difference of the x-coordinates: $x_2 - x_1$.

Concept Check

In the definition of slope, we state that $x_2 \neq x_1$. Explain why.

Practice Problem 1

Find the slope of the line containing the points $(-1, -2)$ and $(2, 5)$. Graph the line.

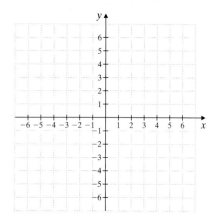

Slope of a Line

Given a line passing through points (x_1, y_1) and (x_2, y_2) the **slope** m of the line is

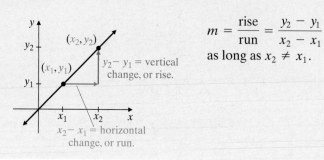

$$m = \frac{\text{rise}}{\text{run}} = \frac{y_2 - y_1}{x_2 - x_1}$$

as long as $x_2 \neq x_1$.

Try the Concept Check in the margin.

EXAMPLE 1

Find the slope of the line containing the points $(0, 3)$ and $(2, 5)$. Graph the line.

Solution: We use the slope formula. It does not matter which point we call (x_1, y_1) and which point we call (x_2, y_2). We'll let $(x_1, y_1) = (0, 3)$ and $(x_2, y_2) = (2, 5)$.

$$m = \frac{y_2 - y_1}{x_2 - x_1}$$

$$= \frac{5 - 3}{2 - 0} = \frac{2}{2} = 1$$

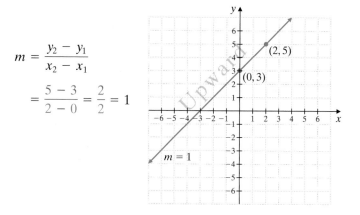

Answers

1. $m = \dfrac{7}{3}$,

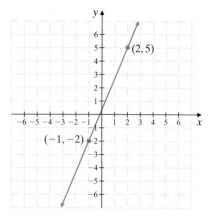

Notice in this example that the slope is positive and that the graph of the line containing $(0, 3)$ and $(2, 5)$ moves upward—that is, the y-values increase—as we go from left to right. ●

Concept Check: so that the denominator is never 0

When we are trying to find the slope of a line through two given points, it makes no difference which given point is called (x_1, y_1) and which is called (x_2, y_2). Once an x-coordinate is called x_1, however, make sure its corresponding y-coordinate is called y_1.

EXAMPLE 2

Find the slope of the line containing the points $(5, -4)$ and $(-3, 3)$. Graph the line.

Solution: We use the slope formula, and let $(x_1, y_1) = (5, -4)$ and $(x_2, y_2) = (-3, 3)$.

$$m = \frac{y_2 - y_1}{x_2 - x_1}$$
$$= \frac{3 - (-4)}{-3 - 5} = \frac{7}{-8} = -\frac{7}{8}$$

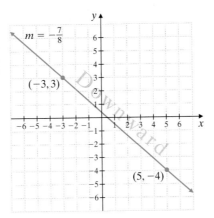

Notice in this example that the slope is negative and that the graph of the line through $(5, -4)$ and $(-3, 3)$ moves downward—that is, the y-values decrease—as we go from left to right.

Try the Concept Check in the margin.

B Finding Slope from an Equation

As we have seen, the slope of a line is defined by two points on the line. Thus, if we know the equation of a line, we can find its slope.

EXAMPLE 3 Find the slope of the line $y = 3x + 2$.

Solution: We must find two points on the line defined by $y = 3x + 2$ to find its slope. We will let $x = 0$ and then $x = 1$ to find the required points.

Let $x = 0$.
$$y = 3x + 2$$
$$y = 3 \cdot 0 + 2$$
$$y = 2$$

Let $x = 1$.
$$y = 3x + 2$$
$$y = 3 \cdot 1 + 2$$
$$y = 5$$

Now we use the points $(0, 2)$ and $(1, 5)$ to find the slope. We'll let (x_1, y_1) be $(0, 2)$ and (x_2, y_2) be $(1, 5)$. Then

$$m = \frac{y_2 - y_1}{x_2 - x_1} = \frac{5 - 2}{1 - 0} = \frac{3}{1} = 3$$

Practice Problem 2

Find the slope of the line containing the points $(1, -1)$ and $(-2, 4)$. Graph the line.

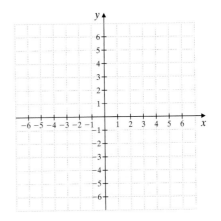

Concept Check

Find and correct the error in the following calculation of slope of the line containing the points $(12, 2)$ and $(4, 7)$.

$$m = \frac{12 - 4}{2 - 7} = \frac{8}{-5} = -\frac{8}{5}$$

Practice Problem 3

Find the slope of the line $y = 2x + 4$.

Answers

2. $m = -\dfrac{5}{3}$,

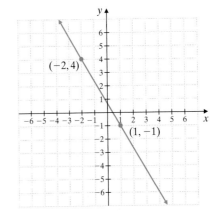

3. $m = 2$

Concept Check: $m = \dfrac{2 - 7}{12 - 4} = \dfrac{-5}{8} = -\dfrac{5}{8}$

Analyzing the results of Example 3, you may notice a striking pattern:

The slope of $y = 3x + 2$ is 3, the same as the coefficient of x.
The y-intercept is 2, the same as the constant term.

When a linear equation is written in the form $y = mx + b$, m is the slope of the line and $(0, b)$ is its y-intercept. The form $y = mx + b$ is appropriately called the *slope-intercept form*.

> **Slope-Intercept Form**
>
> When a linear equation in two variables is written in **slope-intercept form**,
>
> slope $\quad$ y-intercept $(0, b)$
> $\quad\quad\downarrow\quad\quad\quad\downarrow$
> $$y = mx + b$$
>
> then m is the slope of the line and $(0, b)$ is the y-intercept of the line.

Practice Problem 4

Find the slope and the y-intercept of the line $2x - 4y = 8$.

EXAMPLE 4 Find the slope and the y-intercept of the line $3x - 4y = 4$.

Solution: We write the equation in slope-intercept form by solving for y.

$$3x - 4y = 4$$
$$-4y = -3x + 4 \qquad \text{Subtract } 3x \text{ from both sides.}$$
$$\frac{-4y}{-4} = \frac{-3x}{-4} + \frac{4}{-4} \qquad \text{Divide both sides by } -4.$$
$$y = \frac{3}{4}x - 1 \qquad \text{Simplify.}$$

The coefficient of x, $\frac{3}{4}$, is the slope, and $(0, -1)$, is the y-intercept. ●

The graphs of $y = \frac{1}{2}x + 1$ and $y = 5x + 1$ are shown below. Recall that the graph of $y = \frac{1}{2}x + 1$ has a slope of $\frac{1}{2}$ and that the graph of $y = 5x + 1$ has a slope of 5.

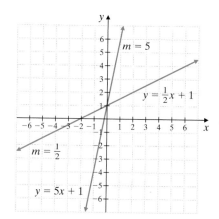

Notice that the line with the slope of 5 is steeper than the line with the slope of $\frac{1}{2}$. This is true in general for positive slopes.

For a line with positive slope m, as m increases the line becomes steeper.

Answer

4. slope: $\frac{1}{2}$; y-intercept: $(0, -2)$

(c) Finding Slopes of Horizontal and Vertical Lines

Next we find the slopes of two special types of lines: vertical lines and horizontal lines.

EXAMPLE 5   Find the slope of the line $x = -5$.

Solution: Recall that the graph of $x = -5$ is a vertical line with x-intercept $(-5, 0)$. To find the slope, we find two ordered pair solutions of $x = -5$. Of course, solutions of $x = -5$ must have an x-value of -5. We will let $(x_1, y_1) = (-5, 0)$ and $(x_2, y_2) = (-5, 4)$. Then

$$m = \frac{y_2 - y_1}{x_2 - x_1} = \frac{4 - 0}{-5 - (-5)} = \frac{4}{0}$$

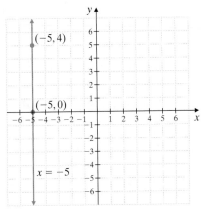

Since $\frac{4}{0}$ is undefined, we say that the slope of the vertical line $x = -5$ is undefined. ●

EXAMPLE 6   Find the slope of the line $y = 2$.

Solution: Recall that the graph of $y = 2$ is a horizontal line with y-intercept $(0, 2)$. To find the slope, we find two points on the line, such as $(0, 2)$ and $(1, 2)$, and use these points to find the slope.

$$m = \frac{2 - 2}{1 - 0} = \frac{0}{1} = 0$$

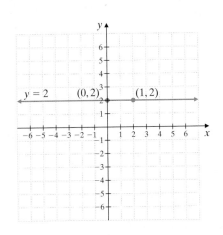

The slope of the horizontal line $y = 2$ is 0. ●

From the above examples, we have the following generalization.

> The slope of any vertical line is undefined.
> The slope of any horizontal line is 0.

Practice Problem 5

Find the slope of the line $x = 3$.

Practice Problem 6

Find the slope of the line $y = -3$.

> **Helpful Hint**
>
> Slope of 0 and undefined slope are not the same. Vertical lines have undefined slope, whereas horizontal lines have slope of 0.

The following four graphs summarize the overall appearance of lines with positive, negative, zero, and undefined slopes.

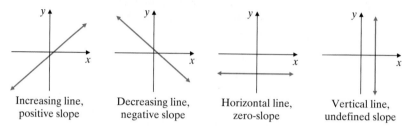

| Increasing line, positive slope | Decreasing line, negative slope | Horizontal line, zero-slope | Vertical line, undefined slope |

D ## Comparing Slopes of Parallel and Perpendicular Lines

Slopes of lines can help us determine whether lines are parallel. Parallel lines are distinct lines with the same steepness, so it follows that they have the same slope.

Parallel Lines

Two nonvertical lines are parallel if they have the same slope and different y-intercepts.

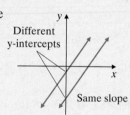

How do the slopes of perpendicular lines compare? (Two lines intersecting at right angles are called **perpendicular lines**.) Suppose that a line has a slope of $\frac{a}{b}$. If the line is rotated $90°$, the rise and run are now switched, except that the run is now negative. This means that the new slope is $-\frac{b}{a}$. Notice that

$$\left(\frac{a}{b}\right) \cdot \left(-\frac{b}{a}\right) = -1$$

This is how we tell whether two lines are perpendicular.

Perpendicular Lines

Two nonvertical lines are perpendicular if the product of their slopes is -1.

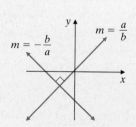

In other words, two nonvertical lines are perpendicular if the slope of one is the negative reciprocal of the slope of the other.

EXAMPLE 7   Determine whether the two lines are parallel, perpendicular, or neither.

$$3x + 7y = 4$$
$$6x + 14y = 7$$

Solution: We find the slope of each line by solving each equation for y.

$$3x + 7y = 4 \qquad\qquad 6x + 14y = 7$$
$$7y = -3x + 4 \qquad\qquad 14y = -6x + 7$$
$$\frac{7y}{7} = \frac{-3x}{7} + \frac{4}{7} \qquad\qquad \frac{14y}{14} = \frac{-6x}{14} + \frac{7}{14}$$
$$y = -\frac{3}{7}x + \frac{4}{7} \qquad\qquad y = -\frac{3}{7}x + \frac{1}{2}$$

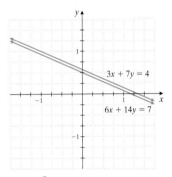

$\qquad\quad$ slope $\quad$ y-intercept $\left(0, \frac{4}{7}\right)$ $\qquad$ slope $\quad$ y-intercept $\left(0, \frac{1}{2}\right)$

The slopes of both lines are $-\frac{3}{7}$. The y-intercepts are different. Therefore, the lines are parallel. ●

EXAMPLE 8

Determine whether the two lines are parallel, perpendicular, or neither.

$$-x + 3y = 2$$
$$2x + 6y = 5$$

Solution: When we solve each equation for y, we have

$$-x + 3y = 2 \qquad\qquad 2x + 6y = 5$$
$$3y = x + 2 \qquad\qquad 6y = -2x + 5$$
$$\frac{3y}{3} = \frac{x}{3} + \frac{2}{3} \qquad\qquad \frac{6y}{6} = \frac{-2x}{6} + \frac{5}{6}$$
$$y = \frac{1}{3}x + \frac{2}{3} \qquad\qquad y = -\frac{1}{3}x + \frac{5}{6}$$

$\quad$ slope $\quad$ y-intercept $\left(0, \frac{2}{3}\right)$ $\quad$ slope $\quad$ y-intercept $\left(0, \frac{5}{6}\right)$

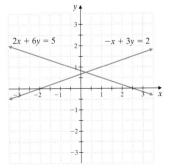

Practice Problem 7

Determine whether the two lines are parallel, perpendicular, or neither.

$$2x + 5y = 1$$
$$4x + 10y = 3$$

Practice Problem 8

Determine whether the two lines are parallel, perpendicular, or neither.

$$x - 4y = 3$$
$$3x + 12y = 7$$

Answers

7. parallel,   **8.** neither parallel nor perpendicular

Concept Check

What is *different* about the equations of two parallel lines?

Answer

Concept Check: *y*-intercepts are different

The slopes are not the same and their product is not −1 $\left[\left(\dfrac{1}{3}\right)\cdot\left(-\dfrac{1}{3}\right) = -\dfrac{1}{9}\right]$. Therefore, the lines are neither parallel nor perpendicular. ●

Try the Concept Check in the margin.

CALCULATOR EXPLORATIONS

It is possible to use a grapher to sketch the graph of more than one equation on the same set of axes. For example, let's graph the equations $y = 2x - 3$ and $y = 2x + 5$ on the same set of axes.

To graph on the same set of axes, press the $\boxed{Y=}$ key and enter the equations on the first two lines.

$Y_1 = 2x - 3$

$Y_2 = 2x + 5$

Then press the $\boxed{\text{GRAPH}}$ key as usual. The screen should look like this:

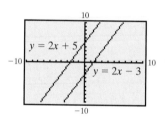

Notice the slopes and *y*-intercepts of the graphs. Since their slopes are the same and they have different *y*-intercepts, we have parallel lines, as shown.

Graph each pair of equations on the same set of axes. Describe the similarities and differences in their graphs.

1. $y = 3x, y = 3x + 4$

2. $y = 5x, y = 5x - 2$

3. $y = -\dfrac{2}{3}x + 1, y = -\dfrac{2}{3} + 6$

4. $y = -\dfrac{1}{4}x - 3, y = -\dfrac{1}{4}x + 6$

5. $y = 4.61x - 1.86, y = 4.61x + 2.11$

6. $y = 3.78x + 1.92, y = 3.78x + 8.08$

Mental Math

Determine whether a line with the given slope slants upward, downward, horizontally, or vertically from left to right.

1. $m = \dfrac{7}{6}$ **2.** $m = -3$ **3.** $m = 0$ **4.** m is undefined.

EXERCISE SET 3.2

Ⓐ *Find the slope of the line containing each pair of points. See Examples 1 and 2.*

1. $(3, 2), (8, 11)$ **2.** $(1, 6), (7, 11)$ **3.** $(3, 1), (1, 8)$ **4.** $(2, 9), (6, 4)$

5. $(-2, 8), (4, 3)$ **6.** $(3, 7), (-2, 11)$ **7.** $(-2, -6), (4, -4)$ **8.** $(-3, -4), (-1, 6)$

9. $(-3, -1), (-12, 11)$ **10.** $(3, -1), (-6, 5)$ **11.** $(-2, 5), (3, 5)$ **12.** $(4, 2), (4, 0)$

Find each slope. See Examples 1 and 2.

13. Find the pitch, or slope, of the roof shown.

14. Upon takeoff, a Delta Airlines jet climbs to 3 miles as it passes over 25 miles of land below it. Find the slope of its climb.

15. Find the grade, or slope, of the road shown.

16. Driving down Bald Mountain in Wyoming, Bob Dean finds that he descends 1600 feet in elevation by the time he is 2.5 miles (horizontally) away from the high point on the mountain road. Find the slope of his descent. (*Hint:* 1 mile = 5280 feet.)

Ⓑ *Find the slope and the y-intercept of each line. See Examples 3 and 4.*

17. $y = 5x - 2$ **18.** $y = -2x + 6$ **19.** $2x + y = 7$ **20.** $-5x + y = 10$ **21.** $2x - 3y = 10$

22. $-3x - 4y = 6$ **23.** $y = \dfrac{1}{2}x$ **24.** $y = -\dfrac{1}{4}x$ **25.** $3x + 9 = y$ **26.** $2y - 7 = x$

Match each graph with its equation.

A
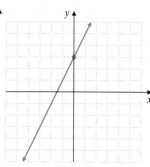

B
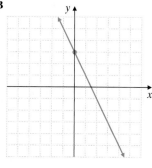

27. $y = 2x + 3$ **28.** $y = 2x - 3$

29. $y = -2x + 3$ **30.** $y = -2x - 3$

C
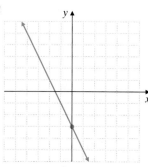

D
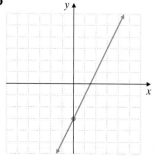

B **C** *Find the slope of each line. See Examples 3 through 6.*

31. $x = 1$ **32.** $y = -2$ **33.** $y = -x + 5$ **34.** $y = x + 2$

35. $-6x + 5y = 30$ **36.** $4x - 7y = 28$ **37.** $x = 4$ **38.** $y = -3$

39. $y = 7x$ **40.** $y = \dfrac{1}{7}x$ **41.** $x + 2 = 0$ **42.** $y - 7 = 0$

Two lines are graphed on each set of axes. For each graph, determine whether l_1 or l_2 has the greater slope.

43.
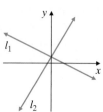

44.
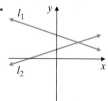

45.
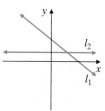

46.
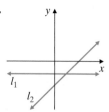

47.
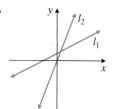

48.
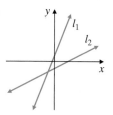

Determine whether each pair of lines is parallel, perpendicular, or neither. See Examples 7 and 8.

49. $y = -3x + 6$ **50.** $y = 5x - 6$
$y = 3x + 5$ $y = 5x + 2$

51. $-4x + 2y = 5$
$2x - y = 7$

52. $2x - y = -10$
$2x + 4y = 2$

53. $-2x + 3y = 1$
$3x + 2y = 12$

54. $x + 4y = 7$
$2x - 5y = 0$

55. $y = -9x + 3$
$y = \dfrac{3}{2}x - 7$

56. $y = 2x - 12$
$y = \dfrac{1}{2}x - 6$

57. $y = 12x + 6$
$y = 12x - 2$

58. $y = -5x + 8$
$y = -5x - 8$

59. Find the slope of a line parallel to the line
$y = -\dfrac{7}{2}x - 6.$

60. Find the slope of a line perpendicular to the line
$y = -\dfrac{7}{2}x - 6.$

61. Find the slope of a line parallel to the line
$5x - 2y = 6.$

62. Find the slope of a line parallel to the line
$-3x + 4y = 10.$

Review and Preview

Solve. See Section 2.6.
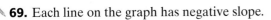

63. $|x - 3| = 6$

64. $|x + 2| < 4$

65. $|2x + 5| > 3$

66. $|5x| = 10$

67. $|3x - 4| \le 2$

68. $|7x - 2| \ge 5$

Combining Concepts

69. Each line on the graph has negative slope.

  a. Find the slope of each line.

  b. Use the results of part (a) to fill in the blank:
    For lines with negative slopes, the steeper line has the
    _____ (greater/lesser) slope.

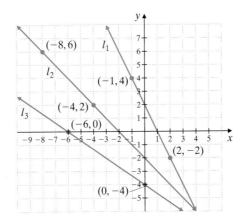

The following graph shows the altitude of a seagull in flight over a time period of 30 seconds. Use this graph to answer Exercises 70 through 73.

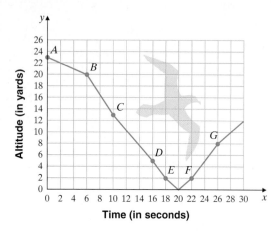

70. Find the coordinates of point *B*.

71. Find the coordinates of point *C*.

72. Find the rate of change of altitude between points *B* and *C*. (Recall that the rate of change between points is the slope between points. This rate of change will be in yards per second.)

73. Find the rate of change of altitude (in yards per second) between points *F* and *G*.

74. Professional plumbers suggest that a sewer pipe should be sloped 0.25 inch for every foot. Find the recommended slope for a sewer pipe. (*Source: Rules of Thumb* by Tom Parker, Houghton Mifflin Company)

75. Explain whether two lines, both with positive slopes, can be perpendicular.

76. Explain how merely looking at a line can tell us whether its slope is negative, positive, undefined, or zero.

77. a. On a single screen of a graphing calculator, graph $y = \frac{1}{2}x + 1$, $y = x + 1$, and $y = 2x + 1$. Notice the change in slope for each graph.

b. On a single screen of a graphing calculator, graph $y = -\frac{1}{2}x + 1$, $y = -x + 1$, and $y = -2x + 1$. Notice the change in slope for each graph.

c. Determine whether the following statement is true or false for slope *m* of a given line: As $|m|$ becomes greater, the line becomes steeper.

3.3 The Slope-Intercept Form

Ⓐ Graphing a Line Using Its Slope and *y*-Intercept

In the last section, we learned that the slope-intercept form of a linear equation is $y = mx + b$. When an equation is written in this form, the slope of the line is the same as the coefficient m of x. Also, the y-intercept of the line is $(0, b)$. For example, the slope of the line defined by $y = 2x + 3$ is 2 and its y-intercept is $(0, 3)$.

We may also use the slope-intercept form to graph a linear equation.

EXAMPLE 1   Graph: $y = \dfrac{1}{4}x - 3$

Solution: Recall that the slope of the graph of $y = \dfrac{1}{4}x - 3$ is $\dfrac{1}{4}$ and the y-intercept is $(0, -3)$. To graph the line, we first plot the y-intercept $(0, -3)$. To find another point on the line, we recall that slope is $\dfrac{\text{rise}}{\text{run}} = \dfrac{1}{4}$. We may then plot another point by starting at $(0, -3)$, rising 1 unit up, and then running 4 units to the right. We are now at the point $(4, -2)$. The graph of $y = \dfrac{1}{4}x - 3$ is the line through points $(0, -3)$ and $(4, -2)$, as shown.

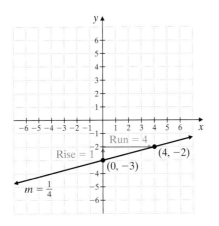

EXAMPLE 2   Graph: $2x + y = 3$

Solution: First, we solve the equation for y to write it in slope-intercept form. In slope-intercept form, the equation is $y = -2x + 3$. Next we plot the y-intercept $(0, 3)$. To find another point on the line, we use the slope -2, which can be written as $\dfrac{\text{rise}}{\text{run}} = \dfrac{-2}{1}$. We start at $(0, 3)$ and move vertically 2 units down, since the numerator of the slope is -2; then we move horizontally 1 unit to the right since the denominator of the slope is 1. We arrive at the point $(1, 1)$. The line through $(1, 1)$ and $(0, 3)$ will have the required slope of -2.

OBJECTIVES

Ⓐ Graph a line using its slope and
    y-intercept.

Ⓑ Use the slope-intercept form to write
    an equation of the line.

Ⓒ Interpret the slope-intercept form in
    an application.

SSM
TUTOR CENTER   SG   CD & VIDEO   MATH PRO   WEB

Practice Problem 1

Graph: $y = \dfrac{2}{3}x + 1$

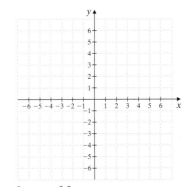

Practice Problem 2

Graph: $3x + y = -2$

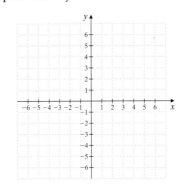

Answers

1.

2. (answer on next page)

The slope -2 can also be written as $\dfrac{2}{-1}$, so to find another point for Example 2 we could start at $(0, 3)$ and move 2 units up and then 1 unit left. We would stop at the point $(-1, 5)$. The line through $(-1, 5)$ and $(0, 3)$ will have the required slope and will be the same line as shown previously through $(1, 1)$ and $(0, 3)$.

B Using the Slope-Intercept Form to Write an Equation

Given the slope and y-intercept of a line, we may write its equation as well as graph the line.

EXAMPLE 3

Write an equation of the line with y-intercept $(0, -3)$ and slope of $\dfrac{1}{4}$.

Solution: We are given the slope and the y-intercept. We let $m = \dfrac{1}{4}$ and $b = -3$, and write the equation in slope-intercept form, $y = mx + b$.

$$y = mx + b$$

$$y = \frac{1}{4}x + (-3) \quad \text{Let } m = \frac{1}{4} \text{ and } b = -3.$$

$$y = \frac{1}{4}x - 3 \quad \text{Simplify.}$$

Notice that the graph of this equation has slope $\dfrac{1}{4}$ and y-intercept $(0, -3)$ as desired.

Try the Concept Check in the margin.

C Interpreting the Slope-Intercept Form

Recall from Section 3.1 the graph of an adult one-day pass price for Disney World. Notice that the graph resembles the graph of a line. Often, businesses depend on equations that "closely fit" lines like this one to model the data and predict future trends. For example, by a method called least squares regression, the linear equation $y = 1.568x + 29.00$ approximates the data shown, where x is the number of years since 1988 and y is the ticket price for that year.

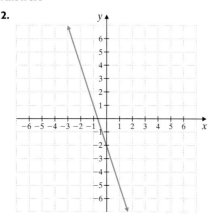

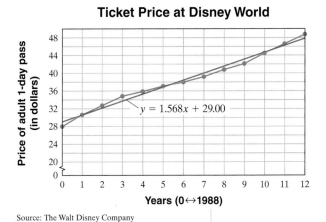

Ticket Price at Disney World

$y = 1.568x + 29.00$

Years (0↔1988)

Source: The Walt Disney Company

Helpful Hint

The notation $0 \leftrightarrow 1988$ means that the number 0 corresponds to the year 1988, 1 corresponds to the year 1989, and so on.

EXAMPLE 4   Predicting Future Prices

The adult one-day pass price y for Disney World is given by

$$y = 1.568x + 29.00$$

where x is the number of years since 1988.

a. Use this equation to predict the ticket price for 2004.
b. What does the slope of this equation mean?
c. What does the y-intercept of this equation mean?

Solution:

a. To predict the price of a pass in 2004, we need to find y when x is 16. (Since year 1988 corresponds to $x = 0$, year 2004 corresponds to $x = 2004 - 1988 = 16$.)

$$y = 1.568x + 29.00$$
$$= 1.568(16) + 29.00 \quad \text{Let } x = 16.$$
$$= 54.088$$

We predict that in 2004 the price of an adult one-day pass to Disney World will be about $54.09.

b. The slope of $y = 1.568x + 29.00$ is 1.568. We can think of this number as $\dfrac{\text{rise}}{\text{run}}$ or $\dfrac{1.568}{1}$. This means that the ticket price increases on the average by $1.568 every 1 year.

c. The y-intercept of $y = 1.568x + 29.00$ is 29.00. Notice that it corresponds to the point of the graph $(0, 29.00)$

      ↑ ↖
    year price

This means that at year $x = 0$, or 1988, the ticket price was about $29.00. ●

Practice Problem 4

For the period 1980 through 2020, the number of people y age 85 or older living in the United States is given by the equation $y = 110,520x + 2,127,400$, where x is the number of years since 1980. (*Source:* Based on data and estimates from the U.S. Bureau of the Census)

a. Estimate the number of people age 85 or older living in the United States in 2010.
b. What does the slope of this equation mean?
c. What does the y-intercept of this equation mean?

Answers

a. 5,443,000   **b.** The number of people age 85 or older in the United States increases at a rate of 110,520 per year.   **c.** At year $x = 0$, or 1980, there were 2,127,400 people age 85 or older in the United States.

GRAPHING CALCULATOR EXPLORATIONS

You may have noticed by now that to use the $\boxed{Y=}$ key on a grapher to graph an equation, the equation must be solved for y.

Graph each equation by first solving the equation for y.

1. $x = 3.5y$

2. $-2.7y = x$

3. $5.78x + 2.31y = 10.98$

4. $-7.22x + 3.89y = 12.57$

5. $y - x = 3.78$

6. $3y - 5x = 6x - 4$

7. $y - 5.6x = 7.7x + 1.5$

8. $y + 2.6x = -3.2$

Name _____ Section _____ Date _____

Mental Math

Find the slope and the y-intercept of each line.

1. $y = -4x + 12$

2. $y = \dfrac{2}{3}x - \dfrac{7}{2}$

3. $y = 5x$

4. $y = -x$

5. $y = \dfrac{1}{2}x + 6$

6. $y = -\dfrac{2}{3}x + 5$

EXERCISE SET 3.3

Ⓐ *Graph each line passing through the given point with the given slope. See Examples 1 and 2.*

1. Through $(1, 3)$ with slope $\dfrac{3}{2}$

2. Through $(-2, -4)$ with slope $\dfrac{2}{5}$

3. Through $(0, 0)$ with slope 5

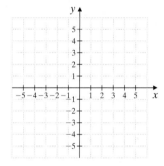
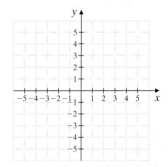
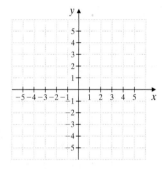

4. Through $(-5, 2)$ with slope 2

5. Through $(0, 7)$ with slope -1

6. Through $(3, 0)$ with slope -3

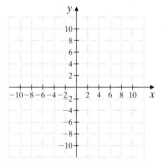
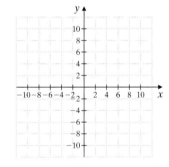
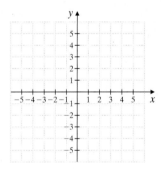

Graph each linear equation using the slope and y-intercept. See Examples 1 and 2.

7. $y = -2x$

8. $y = 2x$

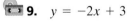 **9.** $y = -2x + 3$

10. $y = 2x + 6$

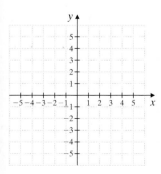
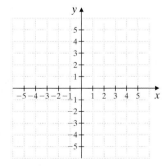
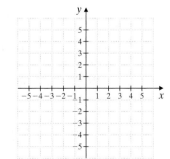
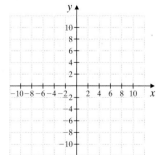

11. $y = \dfrac{1}{2}x$

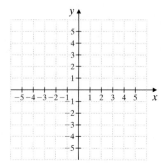

12. $y = \dfrac{1}{3}x$

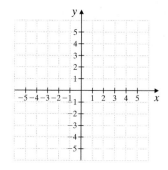

13. $y = \dfrac{1}{2}x - 4$

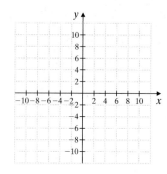

14. $y = \dfrac{1}{3}x - 2$

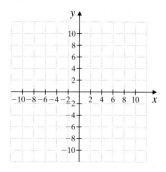

15. $x - y = 3$

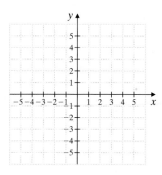

16. $x - y = -4$

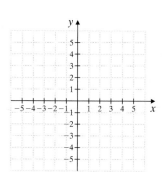

17. $x + 2y = 8$

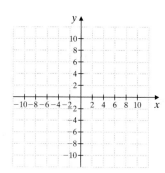

18. $x - 3y = 3$

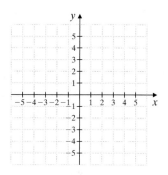

The graph of $y = 5x$ follows. Use this graph to match each linear equation with its graph. See Examples 1 and 2.

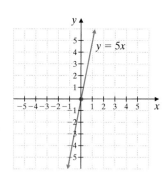

19. $y = 5x - 3$

20. $y = 5x - 2$

21. $y = 5x + 1$

22. $y = 5x + 3$

A

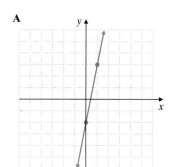

B

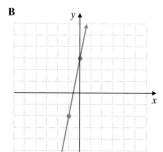

C

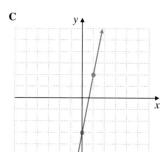

D

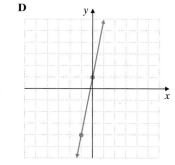

Use the slope-intercept form of a linear equation to write the equation of each line with the given slope and y-intercept. See Example 3.

23. Slope -1; y-intercept $(0, 1)$

24. Slope $\dfrac{1}{2}$; y-intercept $(0, -6)$

25. Slope 2; y-intercept $\left(0, \dfrac{3}{4}\right)$

26. Slope -3; y-intercept $\left(0, -\dfrac{1}{5}\right)$

27. Slope $\dfrac{2}{7}$; y-intercept $(0, 0)$

28. Slope $-\dfrac{4}{5}$; y-intercept $(0, 0)$

C *Solve. See Example 4.*

29. The annual average income y of an American man with an associate's degree is given by the linear equation $y = 1765.1x + 35{,}815.0$, where x is the number of years after 1995. (*Source*: Based on data from the U.S. Bureau of the Census, 1995–2000)
 a. Find the average income of an American man with an associate's degree in 2000.
 b. Find and interpret the slope of the equation.
 c. Find and interpret the y-intercept of the equation.

30. The annual average income y of an American woman with a bachelor's degree is given by the linear equation $y = 1733.8x + 26{,}914.7$, where x is the number of years after 1995. (*Source*: Based on data from the U.S. Bureau of the Census, 1995–2000)
 a. Find the average income of an American woman with a bachelor's degree in 2000.
 b. Find and interpret the slope of the equation.
 c. Find and interpret the y-intercept of the equation.

31. One of the top ten occupations in terms of job growth in the next few years is expected to be computer support specialist. The number of people y, in thousands, employed as computer support specialists in the United States can be estimated by the linear equation $220x - 5y = -2145$, where x is the number of years after 1998. (*Source*: Based on projections from the U.S. Bureau of Labor Statistics, 1998–2008)
 a. Find the slope and y-intercept of the linear equation.
 b. What does the slope mean in this context?
 c. What does the y-intercept mean in this context?

32. One of the faster growing occupations over the next few years is expected to be medical assistant. The number of people y, in thousands, employed as medical assistants in the United States can be estimated by the linear equation $-146x + 10y = 2520$, where x is the number of years after 1998. (*Source*: Based on projections from the U.S. Bureau of Labor Statistics, 1998–2008)
 a. Find the slope and y-intercept of the linear equation.
 b. What does the slope mean in this context?
 c. What does the y-intercept mean in this context?

33. The yearly cost of undergraduate tuition and required fees for attending a public four-year college full-time can be estimated by the linear equation $y = 136.2x + 2827.6$, where x is the number of years after 1996 and y is the total cost in dollars. (*Source:* Based on data from The College Board, 1996–2000)

   a. Use this equation to approximate the yearly cost of attending a four-year public college in 2010.

   b. Use this equation to predict in what year the yearly cost of tuition and required fees will exceed $5000. (*Hint:* Let $y = 5000$ and solve for x.)

   c. Use this equation to approximate the yearly cost of attending a four-year college in the present year. If you attend a four-year college, is this amount greater or less than the amount currently charged by the college you attend?

34. The yearly cost of tuition and required fees for attending a public two-year college full-time can be estimated by the linear equation $y = 68.3x + 1372$, where x is the number of years after 1996 and y is the total cost in dollars. (*Source:* Based on data from The College Board, 1996–2000)

   a. Use this equation to approximate the yearly cost of attending a two-year public college in 2010.

   b. Use this equation to predict in what year the yearly cost of tuition and required fees will exceed $3000. (*Hint:* Let $y = 3000$ and solve for x.)

   c. Use this equation to approximate the yearly cost of attending a two-year college in the present year. If you attend a two-year college, is this amount greater or less than the amount currently charged by the college you attend?

Review and Preview

Simplify and solve for y. See Section 2.3.

35. $y - 2 = 5(x + 6)$ **36.** $y - 0 = -3[x - (-10)]$ **37.** $y - (-1) = 2(x - 0)$ **38.** $y - 9 = -8[x - (-4)]$

 Combining Concepts

39. In your own words, explain how to graph an equation using its slope and y-intercept.

40. Suppose that the revenue of a company has increased at a steady rate of $42,000 per year since 1995. Also the company's revenue in 1995 was $2,900,000. Write an equation that describes the company's revenue since 1995.

41. Suppose that a bird dives off a 500-foot cliff and descends at a rate of 7 feet per second. Write an equation that describes the bird's height at any time x.

3.4 More Equations of Lines

Ⓐ Using the Point-Slope Form to Write an Equation

When the slope of a line and a point on the line are known, the equation of the line can also be found. To do this, we use the slope formula to write the slope of a line that passes through points (x, y), and (x_1, y_1). We have

$$m = \frac{y - y_1}{x - x_1}$$

We multiply both sides of this equation by $x - x_1$ to obtain

$$y - y_1 = m(x - x_1)$$

This form is called the *point-slope form* of the equation of a line.

Point-Slope Form of the Equation of a Line

The **point-slope form** of the equation of a line is

$$\overset{\text{slope}}{\underset{\text{point}}{y - y_1 = m(x - x_1)}}$$

where m is the slope of the line and (x_1, y_1) is a point on the line.

EXAMPLE 1

Write an equation of the line with slope -3 and containing the point $(1, -5)$.

Solution: Because we know the slope and a point on the line, we use the point-slope form with $m = -3$ and $(x_1, y_1) = (1, -5)$.

$$\begin{aligned} y - y_1 &= m(x - x_1) &&\text{Point-slope form} \\ y - (-5) &= -3(x - 1) &&\text{Let } m = -3 \text{ and } (x_1, y_1) = (1, -5). \\ y + 5 &= -3x + 3 &&\text{Use the distributive property.} \\ y &= -3x - 2 \end{aligned}$$

The equation is $y = -3x - 2$.   ●

EXAMPLE 2

Write an equation of the line through points $(4, 0)$ and $(-4, -5)$.

Solution: First we find the slope of the line.

$$m = \frac{-5 - 0}{-4 - 4} = \frac{-5}{-8} = \frac{5}{8}$$

Next we make use of the point-slope form. We replace (x_1, y_1) by either $(4, 0)$ or $(-4, -5)$ in the point-slope equation. We will choose the point $(4, 0)$. The line through $(4, 0)$ with slope $\frac{5}{8}$ is as follows.

Practice Problem 1

Write an equation of the line with slope -2 and containing the point $(2, -4)$. Write the equation in slope-intercept form, $y = mx + b$.

Practice Problem 2

Write an equation of the line through points $(3, 0)$ and $(-2, 4)$. Write the equation in slope-intercept form, $y = mx + b$.

Answers

1. $y = -2x$,   **2.** $y = -\frac{4}{5}x + \frac{12}{5}$

$$y - y_1 = m(x - x_1) \quad \text{Point-slope form}$$

$$y - 0 = \frac{5}{8}(x - 4) \quad \text{Let } m = \frac{5}{8} \text{ and } (x_1, y_1) = (4, 0).$$

$$y = \frac{5}{8}x - \frac{5}{8} \cdot 4 \quad \text{Use the distributive property.}$$

$$y = \frac{5}{8}x - \frac{5}{2} \quad \text{Simplify.}$$

The equation is $y = \frac{5}{8}x - \frac{5}{2}$. If we had chosen to use the point $(-4, -5)$, we would have obtained $y - (-5) = \frac{5}{8}[x - (-4)]$, which also simplifies to $y = \frac{5}{8}x - \frac{5}{2}$.

Helpful Hint

If two points of a line are given, either one may be used with the point-slope form to write an equation of the line.

B Writing Equations of Vertical and Horizontal Lines

A few special types of linear equations are those whose graphs are vertical and horizontal lines.

Practice Problem 3

Write an equation of the horizontal line containing the point $(-1, 6)$.

EXAMPLE 3

Write an equation of the horizontal line containing the point $(2, 3)$.

Solution: Recall, from Section 3.1, that a horizontal line has an equation of the form $y = b$. Since the line contains the point $(2, 3)$, the equation is $y = 3$.

Practice Problem 4

Write an equation of the line containing the point $(4, 7)$ with undefined slope.

EXAMPLE 4

Write an equation of the line containing the point $(2, 3)$ with undefined slope.

Solution: Since the line has undefined slope, the line must be vertical. A vertical line has an equation of the form $x = c$, and since the line contains the point $(2, 3)$, the equation is $x = 2$.

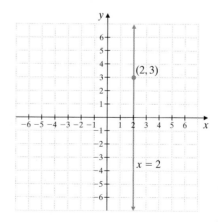

C Writing Equations of Parallel and Perpendicular Lines

Next, we write equations of parallel and perpendicular lines.

EXAMPLE 5

Write an equation of the line containing the point $(4, 4)$ and parallel to the line $2x + y = -6$.

Solution: Because the line we want to find is *parallel* to the line $2x + y = -6$, the two lines must have equal slopes. So we first find the slope of $2x + y = -6$ by solving it for y to write it in the form $y = mx + b$. Here $y = -2x - 6$, so the slope is -2.

Now we use the point-slope form to write the equation of a line through $(4, 4)$ with slope -2.

$$y - y_1 = m(x - x_1)$$
$$y - 4 = -2(x - 4) \qquad \text{Let } m = -2, x_1 = 4, \text{ and } y_1 = 4.$$
$$y - 4 = -2x + 8 \qquad \text{Use the distributive property.}$$
$$y = -2x + 12$$

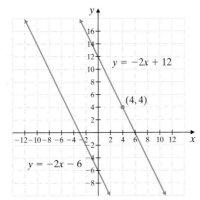

The equation, $y = -2x - 6$, and the new equation, $y = -2x + 12$, have the same slope but different y-intercepts so their graphs are parallel. Also, the graph of $y = -2x + 12$ contains the point $(4, 4)$, as desired. ●

EXAMPLE 6

Write an equation of the line containing the point $(-2, 1)$ and perpendicular to the line $3x + 5y = 4$.

Solution: First we find the slope of $3x + 5y = 4$ by solving it for y.

$$5y = -3x + 4$$
$$y = -\frac{3}{5}x + \frac{4}{5}$$

The slope of the given line is $-\frac{3}{5}$. A line perpendicular to this line will have a slope that is the negative reciprocal of $-\frac{3}{5}$, or $\frac{5}{3}$. We use the point-slope form to write an equation of a new line through $(-2, 1)$ with slope $\frac{5}{3}$.

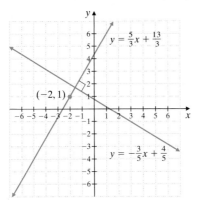

Practice Problem 5

Write an equation of the line containing the point $(-1, 2)$ and parallel to the line $3x + y = 5$. Write the equation in the form $y = mx + b$.

Practice Problem 6

Write an equation of the line containing the point $(3, 4)$ and perpendicular to the line $2x + 4y = 5$. Write the equation in standard form.

Answers

5. $y = -3x - 1$,   **6.** $2x - y = 2$

$$y - 1 = \frac{5}{3}[x - (-2)]$$

$$y - 1 = \frac{5}{3}(x + 2) \qquad \text{Simplify.}$$

$$y - 1 = \frac{5}{3}x + \frac{10}{3} \qquad \text{Use the distributive property.}$$

$$y = \frac{5}{3}x + \frac{13}{3} \qquad \text{Add 1 to both sides.}$$

The equation $y = -\frac{3}{5}x + \frac{4}{5}$ and the new equation $y = \frac{5}{3}x + \frac{13}{3}$ have negative reciprocal slopes, so their graphs are perpendicular. Also, the graph of

$y = \frac{5}{3}x + \frac{13}{3}$ contains the point $(-2, 1)$, as desired.   ●

D   Using the Point-Slope Form in Applications

The point-slope form of an equation is very useful for solving real-world problems.

EXAMPLE 7   Predicting Sales

Southern Star Realty is an established real estate company that has enjoyed constant growth in sales since 1993. In 1995 the company sold 200 houses, and in 2000 the company sold 275 houses. Use these figures to predict the number of houses this company will sell in 2005.

Solution:

1. UNDERSTAND. Read and reread the problem. Then let

    $x =$ the number of years after 1993 and
    $y =$ the number of houses sold in the year corresponding to x.

    The information provided then gives the ordered pairs $(2, 200)$ and $(7, 275)$. To better visualize the sales of Southern Star Realty, we graph the line that passes through the points $(2, 200)$ and $(7, 275)$.

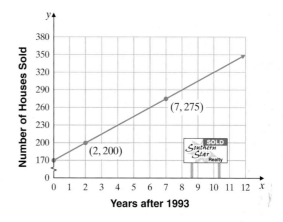

2. TRANSLATE. We write an equation of the line that passes through the points $(2, 200)$ and $(7, 275)$. To do so, we first find the slope of the line.

    $$m = \frac{275 - 200}{7 - 2} = \frac{75}{5} = 15$$

Practice Problem 7

Southwest Regional is an established office product maintenance company that has enjoyed constant growth in new maintenance contracts since 1985. In 1993, the company obtained 15 new contracts and in 2000, the company obtained 36 new contracts. Use these figures to predict the number of new contracts this company can expect in 2007.

Answer

7. 57 new contracts

Then, using the point-slope form to write the equation, we have

$$y - y_1 = m(x - x_1)$$
$$y - 200 = 15(x - 2) \qquad \text{Let } m = 15 \text{ and } (x_1, y_1) = (2, 200).$$
$$y - 200 = 15x - 30 \qquad \text{Multiply.}$$
$$y = 15x + 170 \qquad \text{Add 200 to both sides.}$$

3. SOLVE. To predict the number of houses sold in 2005, we use $y = 15x + 170$ and complete the ordered pair $(12, \quad)$ since $2005 - 1993 = 12$.

$$y = 15(12) + 170 \qquad \text{Let } x = 12.$$
$$y = 350$$

4. INTERPRET.

Check: Verify that the point $(12, 350)$ is a point on the line graphed in Step 1.

State: Southern Star Realty should expect to sell 350 houses in 2005.   ●

GRAPHING CALCULATOR EXPLORATIONS

Many graphing calculators have a TRACE feature. This feature allows you to trace along a graph and see the corresponding x- and y-coordinates appear on the screen. Use this feature for the following exercises.

Graph each equation and then use the TRACE feature to complete each ordered pair solution. (Many times the tracer will not show the exact x- or y-value asked for. In each case, trace as closely as you can to the given x- or y-coordinate and approximate the other, unknown coordinate to one decimal place.)

1. $y = 2.3x + 6.7$;
   $x = 5.1, y = ?$

2. $y = -4.8x + 2.9$;
   $x = -1.8, y = ?$

3. $y = -5.9x - 1.6$;
   $x = ?, y = 7.2$

4. $y = 0.4x - 8.6$;
   $x = ?, y = -4.4$

5. $y = 5.2x - 3.3$;
   $x = 2.3, y = ?$
   $x = ?, y = 36$

6. $y = -6.2x - 8.3$;
   $x = 3.2, y = ?$
   $x = ?, y = 12$

FOCUS ON **The Real World**

CENTERS OF MASS

The **center of mass**, also known as **center of gravity**, of an object is the point at which the mass of the object may be considered to be concentrated. For a two-dimensional object or surface, such as a flat board, the center of mass can be described as the point on which the surface would balance.

The idea of center of mass is an important one in many disciplines, especially physics and its applications. The following list describes situations in which an object's center of mass is important.

- Geographers are sometimes concerned with pinpointing the *geographic center* of a county, state, country, or continent. A geographic center is actually the center of mass of a geographic region if it is considered as a two-dimensional surface. The geographic center of the 48 contiguous United States is near Lebanon, Kansas. The geographic center of the North American continent is 6 miles west of Balta, North Dakota.

- A top-loading washing machine is designed so its center of mass is located within its agitator post. During the spin cycle, the washer tub spins around its center of mass. If clothes aren't carefully distributed within the tub, they can bunch up and throw off the center of mass. This causes the machine to vibrate and sometimes jump or shake wildly. Some washing machine models will stop operating when the loads become unbalanced in this way.

- Single-hulled boats are normally designed so that their centers of mass are below the water line. This provides a boat with stability. Otherwise, with a center of mass above the water line, the boat would have a tendency to tip over in the water.

- Most small airplanes must be carefully loaded so as not to affect the location of the airplane's center of mass. The center of mass of an airplane must be near the center of the wings. Pilots of small aircraft usually try to balance the weight of their cargo around the airplane's center of mass.

GROUP ACTIVITY

Attach a piece of graph paper to a piece of cardboard. Cut out a triangle and label the vertices with their coordinates. Lay the triangle on a horizontal table top with the graph paper face down. Slide the triangle toward the edge of the table until it is balanced on the edge, just about to tip over the side of the table. Firmly hold the triangle in place while another group member uses the straight edge of the table to draw a line on the graph paper side of the triangle marking the position of the table edge. Rotate the triangle a quarter turn and rebalance the triangle on the edge of the table. Draw a second line on the graph paper side of the triangle marking the position of the table edge.

1. The point where the two lines drawn on the triangle intersect is the center of mass of the triangle. Find the coordinates of this point.

2. Verify that the point you have located is roughly the center of mass of the triangle by balancing it at this point on the tip of a pencil or pen.

3. List the coordinates of the vertices of your triangle. What is the relationship between the coordinates of the center of mass and the coordinates of the triangle's vertices? (*Hint:* You may find it helpful to examine the sum of the *x*-coordinates and the sum of the *y*-coordinates of the vertices of the triangle.)

4. Test your observation in Question 3. Cut out another triangle. Label its vertices and, using your observation from Question 3, predict the location of the center of mass of the triangle. Use the balancing procedure to find the center of mass. How close was your prediction?

Mental Math

Find the slope of and a point on the line described by each equation.

1. $y - 4 = -2(x - 1)$

2. $y - 6 = -3(x - 4)$

3. $y - 0 = \dfrac{1}{4}(x - 2)$

4. $y - 1 = -\dfrac{2}{3}(x - 0)$

5. $y + 2 = 5(x - 3)$

6. $y - 7 = 4(x + 6)$

EXERCISE SET 3.4

Ⓐ *Write an equation of each line with the given slope and containing the given point. Write the equation in the form $y = mx + b$. See Example 1.*

1. Slope 3; through $(1, 2)$

2. Slope 4; through $(5, 1)$

3. Slope -2; through $(1, -3)$

4. Slope -4; through $(2, -4)$

5. Slope $\dfrac{1}{2}$; through $(-6, 2)$

6. Slope $\dfrac{2}{3}$; through $(-9, 4)$

7. Slope $-\dfrac{9}{10}$; through $(-3, 0)$

8. Slope $-\dfrac{1}{5}$; through $(4, -6)$

9. Slope 2; through $(-2, 3)$

10. Slope 3; through $(-4, 2)$

11. Slope $-\dfrac{4}{3}$; through $(-5, 0)$

12. Slope $-\dfrac{3}{5}$; through $(4, -1)$

Write an equation of the line passing through the given points. Write the equation in the form $y = mx + b$. See Example 2.

13. $(2, 0)$ and $(4, 6)$

14. $(3, 0)$ and $(7, 8)$

15. $(-2, 5)$ and $(-6, 13)$

16. $(7, -4)$ and $(2, 6)$

17. $(-2, -4)$ and $(-4, -3)$

18. $(-9, -2)$ and $(-3, 10)$

19. $(-3, -8)$ and $(-6, -9)$

20. $(8, -3)$ and $(4, -8)$

21. $(-7, -4)$ and $(0, -6)$

22. $(2, -8)$ and $(-4, -3)$

23. $\left(\dfrac{3}{5}, \dfrac{4}{10}\right)$ and $\left(-\dfrac{1}{5}, \dfrac{7}{10}\right)$

24. $\left(\dfrac{1}{2}, -\dfrac{1}{4}\right)$ and $\left(\dfrac{3}{2}, \dfrac{3}{4}\right)$

Write an equation of each line. See Examples 3 and 4.

25. Vertical; through $(2, 6)$

26. Slope 0; through $(-2, -4)$

27. Horizontal; through $(-3, 1)$

28. Vertical; through $(4, 7)$

29. Undefined slope; through $(0, 5)$

30. Horizontal; through $(0, 5)$

C *Write an equation of each line. Write the equation in the form $x = a$, $y = b$, or $y = mx + b$. See Examples 5 and 6.*

31. Through $(3, 8)$; parallel to $y = 4x - 2$

32. Through $(1, 5)$; parallel to $y = 3x - 4$

33. Through $(2, -5)$; perpendicular to $y = -2x - 6$

34. Through $(-4, 8)$; perpendicular to $y = -4x - 1$

35. Through $(1, 4)$; parallel to $y = 7$

36. Through $(-2, 6)$; perpendicular to $y = 7$

37. Through $(-2, -3)$; parallel to $3x + 2y = 5$

38. Through $(-2, -3)$; perpendicular to $3x + 2y = 5$

39. Through $(3, 5)$; perpendicular to $2x - y = 8$

40. Through $(6, 1)$; parallel to $8x - y = 9$

41. Through $(-1, -5)$; perpendicular to $x = 3$

42. Through $(4, -6)$; parallel to $x = -2$

43. Through $(6, -2)$; parallel to $2x + 4y = 9$

44. Through $(8, -3)$; parallel to $6x + 2y = 5$

45. Through $(-1, 5)$; perpendicular to $x - 4y = 4$

46. Through $(2, -3)$; perpendicular to $x - 5y = 10$

D *Solve. See Example 7.*

47. A rock is dropped from the top of a 400-foot building. After 1 second, the rock is traveling 32 feet per second. After 3 seconds, the rock is traveling 96 feet per second. Let y be the rate of descent and x be the number of seconds since the rock was dropped.
a. Write a linear equation that relates time x to rate y. [*Hint:* Use the ordered pairs $(1, 32)$ and $(3, 96)$.]
b. Use this equation to determine the rate of travel of the rock 4 seconds after it was dropped.

48. The Whammo Company has learned that by pricing a newly released Frisbee at $6, sales will reach 2000 per day. Raising the price to $8 will cause the sales to fall to 1500 per day. Assume that the ratio of change in price to change in daily sales is constant, and let x be the price of the Frisbee and y be number of sales.
a. Find the linear equation that models the price–sales relationship for this Frisbee. [*Hint:* The line must pass through $(6, 2000)$ and $(8, 1500)$.]
b. Use this equation to predict the daily sales of Frisbees if the price is set at $7.50.

49. A fruit company recently released a new applesauce. By the end of its first year, profits on this product amounted to $30,000. The anticipated profit for the end of the fourth year is $66,000. The ratio of change in time to change in profit is constant. Let x be years and y be profit.

a. Write a linear equation that relates profit and time.

b. Use this equation to predict the company's profit at the end of the seventh year.

c. Predict when the profit should reach $126,000.

50. The Pool Fun Company has learned that, by pricing a newly released Fun Noodle at $3, sales will reach 10,000 Fun Noodles per day during the summer. Raising the price to $5 will cause the sales to fall to 8000 Fun Noodles per day. Let x be price and y be the number sold.

a. Assume that the relationship between sales price and number of Fun Noodles sold is linear and write an equation describing this relationship.

b. Use this equation to predict the daily sales of Fun Noodles if the price is $3.50.

51. In 1996, the median price of an existing home in the United States was $115,800. In 2000, the median price of an existing home was $142,200. Let y be the median price of an existing home in the year x, where $x = 0$ represents 1996. (*Source:* National Association of REALTORS®)

a. Write a linear equation that models the median existing home price in terms of the year x. [*Hint:* The line must pass through $(0, 115{,}800)$ and $(4, 142{,}200)$.]

b. Use this equation to predict the median existing home price for 2005.

52. The number of commercial airplanes delivered to customers by Boeing in 1997 was 374. In 2000, Boeing delivered a total of 489 commercial airplanes to customers. Let y be the number of Boeing commercial aircraft delivered to customers in the year x, where $x = 0$ represents 1997. (*Source:* The Boeing Company)

a. Write a linear equation that models the number of Boeing commercial aircraft delivered to customers in terms of the year x. [*Hint:* The line must pass through $(0, 374)$ and $(3, 489)$.]

b. Use this equation to predict the number of commercial aircraft Boeing delivers to customers in 2006.

53. The number of DVD players sold in the United States in 1998 was 1,089,261. In 2000, there were 8,498,545 DVD players sold in the United States. Let y be the number of DVD players sold in the year x, where $x = 0$ represents 1998. (*Source:* Consumer Electronics Association)

a. Write a linear equation that models the number of DVD players sold in the United States in the year x. [*Hint:* The line must pass through $(0, 1{,}089{,}261)$ and $(2, 8{,}498{,}545)$.]

b. Use this equation to estimate the number of DVD players that will be sold in 2005. [*Hint:* The line must pass through $(0, 1{,}089{,}261)$ and $(2, 8{,}498{,}545)$.]

54. The number of people employed in the United States as teacher assistants was 1192 thousand in 1998. By 2008, this number is expected to rise to 1567 thousand. Let y be the number of teacher assistants (in thousands) employed in the United States in the year x, where $x = 0$ represents 1998. (*Source:* U.S. Bureau of Labor Statistics)

a. Write a linear equation that models the number of people (in thousands) employed as teacher assistants in the year x. [*Hint:* The line must pass through $(0, 1192)$ and $(10, 1567)$.]

b. Use this equation to estimate the number of people who will be employed as teacher assistants in 2007.

Review and Preview

Complete each ordered pair for the given equation. See Section 3.1.

55. $y = 7x + 3;\ (4,\quad)$

56. $y = 2x - 6;\ (2,\quad)$

57. $y = 4.2x;\ (-2,\quad)$

58. $y = -1.3x;\ (6,\quad)$

59. $y = x^2 + 2x + 1;\ (1,\quad)$

60. $y = x^2 - 6x + 4;\ (0,\quad)$

Combining Concepts

Answer true or false.

61. A vertical line is always perpendicular to a horizontal line.

62. A vertical line is always parallel to a vertical line.

Write an equation of each line.

63. Through $(5, -6)$; perpendicular to $y = 9$

64. Through $(-3, -5)$; parallel to $y = 9$

Use a grapher with a TRACE feature to see the results of each exercise.

65. Exercise 31: Graph the equation and verify that it passes through $(3, 8)$ and is parallel to $y = 4x - 2$.

66. Exercise 32: Graph the equation and verify that it passes through $(1, 5)$ and is parallel to $y = 3x - 4$.

Internet Excursions

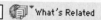

Go To: http://www.prenhall.com/martin-gay_interm   What's Related

The U.S. Bureau of Labor Statistics (BLS) is the principal fact-finding agency for the federal government in the broad field of labor economics and statistics. The BLS regularly makes data such as unemployment figures, average earnings, and job growth numbers available to the American public, government, and businesses. The World Wide Web address listed above will provide you with access to the BLS page of Labor Force Statistics from the Current Population Survey, or a related site. You will be able to research information needed to answer the following questions.

67. From the listing of links, choose a table or data set that interests you. Write down two ordered pairs from that set of data. Describe what each ordered pair represents. Then use the ordered pairs to find an equation for the line between these two points.

68. Now choose a different table or data set of interest. Write down two ordered pairs and describe what each ordered pair represents. Use the ordered pairs to find an equation for the line between these two points. Then make a prediction using your equation and explain its significance.

Integrated Review—Linear Equations in Two Variables

Below is a review of equations of lines.

Forms of Linear Equations

$Ax + By = C$ **Standard form** of a linear equation
A and B are not both 0.

$y = mx + b$ **Slope-intercept form** of a linear equation
The slope is m, and the y-intercept is $(0, b)$.

$y - y_1 = m(x - x_1)$ **Point-slope form** of a linear equation
The slope is m, and (x_1, y_1) is a point on the line.

$y = c$ **Horizontal line**
The slope is 0, and the y-intercept is $(0, c)$.

$x = c$ **Vertical line**
The slope is undefined and the x-intercept is $(c, 0)$.

Parallel and Perpendicular Lines

Nonvertical parallel lines have the same slope.
The product of the slopes of two nonvertical
perpendicular lines is -1.

Graph each linear equation.

1. $y = -2x$

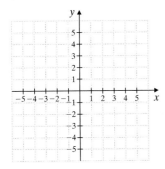

2. $3x - 2y = 6$

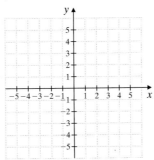

3. $x = -3$

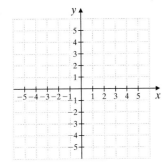

4. $y = 1.5$

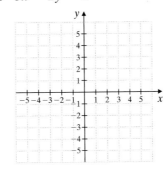

Find the slope of the line containing each pair of points.

5. $(-2, -5), (3, -5)$

6. $(5, 2), (0, 5)$

7. _____

8. _____

9. _____

10. _____

11. _____

12. _____

13. _____

14. _____

15. _____

16. _____

17. _____

18. _____

19. _____

20. _____

21. _____

22. _____

23. _____

24. _____

Find the slope and y-intercept of each line.

7. $y = 3x - 5$

8. $5x - 2y = 7$

Determine whether each pair of lines is parallel, perpendicular, or neither.

9. $y = 8x - 6$

$y = 8x + 6$

10. $y = \dfrac{2}{3}x + 1$

$2y + 3x = 1$

Find the equation of each line. Write the equation in the form $x = a$, $y = b$, or $y = mx + b$.

11. Through $(1, 6)$ and $(5, 2)$

12. Vertical line; through $(-2, -10)$

13. Horizontal line; through $(1, 0)$

14. Through $(2, -8)$ and $(-6, -5)$

15. Through $(-2, 4)$ with slope -5

16. Slope -4; y-intercept $\left(0, \dfrac{1}{3}\right)$

17. Slope $\dfrac{1}{2}$; y-intercept $(0, -1)$

18. Through $\left(\dfrac{1}{2}, 0\right)$ with slope 3

19. Through $(-1, -5)$; parallel to
$3x - y = 5$

20. Through $(0, 4)$; perpendicular to
$4x - 5y = 10$

21. Through $(2, -3)$; perpendicular to
$4x + y = \dfrac{2}{3}$

22. Through $(-1, 0)$; parallel to
$5x + 2y = 2$

23. Undefined slope; through $(-1, 3)$

24. $m = 0$; through $(-1, 3)$

3.5 Introduction to Functions

A Defining Relation, Domain, and Range

Equations in two variables, such as $y = 2x + 1$, describe **relations** between x-values and y-values. For example, if $x = 1$, then this equation describes how to find the y-value related to $x = 1$. In words, the equation $y = 2x + 1$ says that twice the x-value increased by 1 gives the corresponding y-value. The x-value of 1 corresponds to the y-value of $2(1) + 1 = 3$ for this equation, and we have the ordered pair $(1, 3)$.

There are other ways of describing relations or correspondences between two numbers or, in general, a first set (sometimes called the set of *inputs*) and a second set (sometimes called the set of *outputs*). For example,

First Set: Input	Correspondence	Second Set: Output
People in a certain city	Each person's age	The set of nonnegative integers

A few examples of ordered pairs from this relation might be (Ana, 4); (Bob, 36); (Trey, 21); and so on.

Below are just a few other ways of describing relations between two sets and the ordered pairs that they generate.

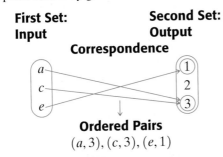

Ordered Pairs
$(a, 3), (c, 3), (e, 1)$

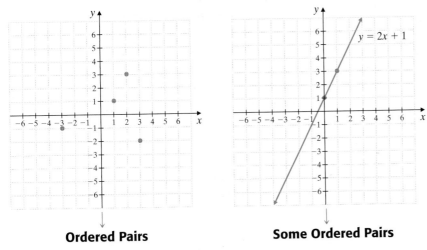

Ordered Pairs

$(-3, -1), (1, 1), (2, 3), (3, -2)$

Some Ordered Pairs

$(1, 3), (0, 1),$ and so on

Relation, Domain, and Range

A **relation** is a set of ordered pairs.
The **domain** of the relation is the set of all first components of the ordered pairs.
The **range** of the relation is the set of all second components of the ordered pairs.

For example, the domain for our middle relation on the previous page is $\{a, c, e\}$ and the range is $\{1, 3\}$. Notice that the range does not include the element 2 of the second set. This is because no element of the first set is assigned to this element. If a relation is defined in terms of x- and y-values, we will agree that the domain corresponds to x-values and that the range corresponds to y-values.

Practice Problems 1-3

Determine the domain and range of each relation.

1. $\{(1, 6), (2, 8), (0, 3), (0, -2)\}$

2.

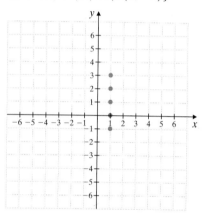

3.

Input: States	Output: Number of Representatives
Arkansas	4
Texas	30
Oklahoma	10 8
South Carolina	6

EXAMPLES Determine the domain and range of each relation.

1. $\{(2, 3), (2, 4), (0, -1), (3, -1)\}$

The domain is the set of all first coordinates of the ordered pairs, $\{2, 0, 3\}$. The range is the set of all second coordinates, $\{3, 4, -1\}$.

2.

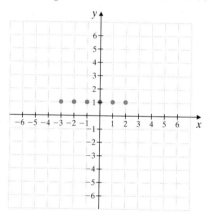

The relation is $\{(-3, 1), (-2, 1), (-1, 1), (0, 1), (1, 1), (2, 1)\}$.
The domain is $\{-3, -2, -1, 0, 1, 2\}$.
The range is $\{1\}$.

3.

Input: Cities	Output: Population (in thousands)
Erie	109 200
Miami	359 52
Escondido	117
Waco	182 104
Gary	

The domain is the first set, {Erie, Escondido, Gary, Miami, Waco}.
The range is the numbers in the second set that correspond to elements in the first set, $\{104, 109, 117, 359\}$. ●

B Identifying Functions

Now we consider a special kind of relation called a *function*.

Function

A **function** is a relation in which each first component in the ordered pairs corresponds to *exactly one* second component.

Helpful Hint

A function is a special type of relation, so all functions are relations. But not all relations are functions.

EXAMPLES   Determine whether each relation is also a function.

4. $\{(-2,5), (2,7), (-3,5), (9,9)\}$

Although the ordered pairs $(-2,5)$ and $(-3,5)$ have the same y-value, each x-value is assigned to only one y-value, so this set of ordered pairs is a function.

5.

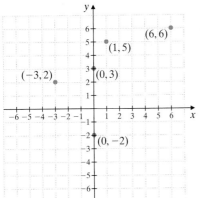

The x-value 0 is assigned to two y-values, -2 and 3, in this graph, so this relation is not a function.

6.

Input	Correspondence	Output
People in a certain city	Each person's age	The set of nonnegative integers

This relation is a function because although two different people may have the same age, each person has only one age. This means that each element in the first set is assigned to only one element in the second set. ●

Try the Concept Check in the margin.

We will call an equation such as $y = 2x + 1$ a relation since this equation defines a set of ordered pair solutions.

EXAMPLE 7

Determine whether the relation $y = 2x + 1$ is also a function.

Solution: The relation $y = 2x + 1$ is a function if each x-value corresponds to just one y-value. For each x-value substituted in the equation $y = 2x + 1$, the multiplication and addition performed gives a single result, so only one y-value will be associated with each x-value. Thus, $y = 2x + 1$ is a function. ●

EXAMPLE 8   Determine whether the relation $x = y^2$ is also a function.

Solution: In $x = y^2$, if $y = 3$, then $x = 9$. Also, if $y = -3$, then $x = 9$. In other words, we have the ordered pairs $(9,3)$ and $(9,-3)$. Since the x-value 9 corresponds to two y-values, 3 and -3, $x = y^2$ is not a function. ●

Practice Problems 4–6

Determine whether each relation is also a function.

4. $\{(-3,7), (1,7), (2,2)\}$

5.

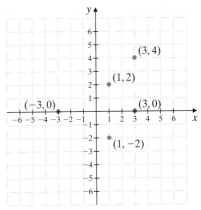

6.

Input	Correspondence	Output
People in a certain state	County/ parish that a person lives in	Counties of that state

Concept Check

Explain why a function can contain both the ordered pairs $(1,3)$ and $(2,3)$ but not both $(3,1)$ and $(3,2)$.

Practice Problem 7

Determine whether the relation $y = 3x + 2$ is also a function.

Practice Problem 8

Determine whether the relation $x = y^2 + 1$ is also a function.

Answers

4. function,   **5.** not a function,   **6.** function,
7. yes,   **8.** no

Concept Check: Two different ordered pairs can have the same y-value, but not the same x-value in a function.

Practice Problems 9–13

Use the vertical line test to determine which are graphs of functions.

9.

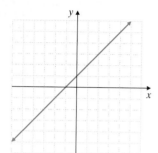

10.

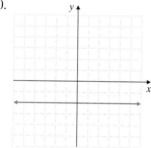

11.

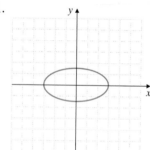

12.

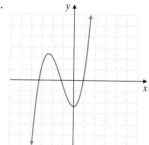

13.

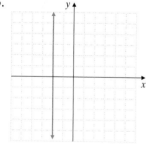

Answers

9. function, **10.** function, **11.** not a function, **12.** function, **13.** not a function

---

Ⓒ Using the Vertical Line Test

As we have seen, not all relations are functions. Consider the graphs of $y = 2x + 1$ and $x = y^2$ shown next. On the graph of $y = 2x + 1$, notice that each x-value corresponds to only one y-value. Recall from Example 7 that $y = 2x + 1$ is a function.

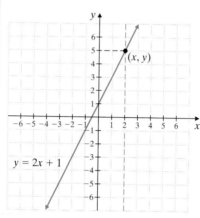

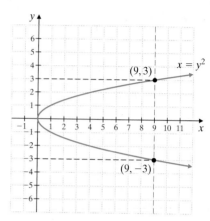

On the graph of $x = y^2$, the x-value 9, for example, corresponds to two y-values, 3 and -3, as shown by the vertical line. Recall from Example 8 that $x = y^2$ is not a function.

Graphs can be used to help determine whether a relation is also a function by the following **vertical line test**.

Vertical Line Test

If no vertical line can be drawn so that it intersects a graph more than once, the graph is the graph of a function. If such a line can be drawn, the graph is not that of a function.

EXAMPLES

Use the vertical line test to determine which are graphs of functions.

9.

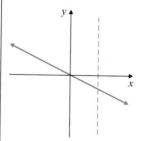

This is the graph of a function since no vertical line will intersect this graph more than once.

10.

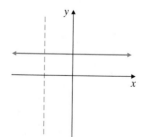

This is the graph of a function.

11.

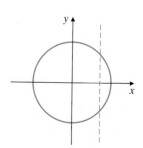

This is not the graph of a function. Note that vertical lines can be drawn that intersect the graph in two points.

12.

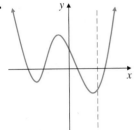

This is the graph
of a function.

13.

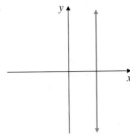

This is not the graph
of a function. A vertical line
can be drawn that intersects
this line at every point.

Try the Concept Check in the margin.

D **Finding Domain and Range from a Graph**

Next we practice finding the domain and range of a relation from its graph.

EXAMPLES   Find the domain and range of each relation.

14.

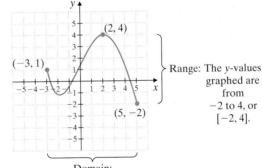

Range: The y-values
graphed are
from
-2 to 4, or
$[-2, 4]$.

Domain:
The x-values graphed are
from -3 to 5, or $[-3, 5]$.

15.

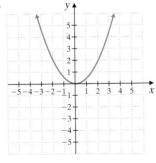

Range:
$[0, \infty)$

Domain: $(-\infty, \infty)$

Determine which equations represent
functions. Explain your answer.
a. $y = 14$
b. $x = -5$
c. $x + y = 6$

Practice Problems 14–17

Find the domain and range of each
relation.

14.

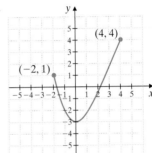

15.

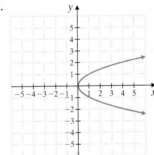

16.

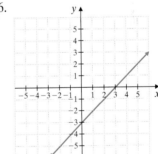

17.

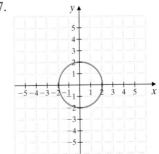

Answers

14. domain: $[-2, 4]$; range: $[-3, 4]$,   **15.** domain:
$[0, \infty)$; range: $(-\infty, \infty)$,   **16.** domain: $(-\infty, \infty)$;
range: $(-\infty, \infty)$,   **17.** domain: $[-2, 2]$; range: $[-2, 2]$

Concept Check: a, c

16.

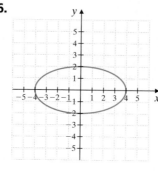 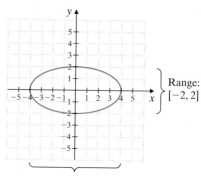

Domain: $[-4, 4]$

Range: $[-2, 2]$

17.

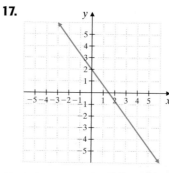 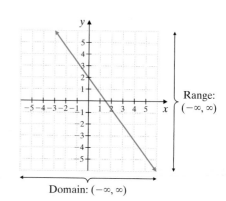

Domain: $(-\infty, \infty)$

Range: $(-\infty, \infty)$

E Using Function Notation

Many times letters such as f, g, and h are used to name functions. To denote that y is a function of x, we can write

$$y = f(x)$$

This means that **y is a function of x** or that y *depends on x*. For this reason, y is called the **dependent variable** and x the **independent variable**. The notation $f(x)$ is read "f of x" and is called **function notation**.

For example, to use function notation with the function $y = 4x + 3$, we write $f(x) = 4x + 3$. The notation $f(1)$ means to replace x with 1 and find the resulting y- or function value. Since

$$f(x) = 4x + 3$$

then

$$f(1) = 4(1) + 3 = 7$$

This means that when $x = 1$, y or $f(x) = 7$. The corresponding ordered pair is $(1, 7)$. Here, the input is 1 and the output is $f(1)$ or 7. Now let's find $f(2)$, $f(0)$, and $f(-1)$.

$$
\begin{array}{ccc}
f(x) = 4x + 3 & f(x) = 4x + 3 & f(x) = 4x + 3 \\
f(2) = 4(2) + 3 & f(0) = 4(0) + 3 & f(-1) = 4(-1) + 3 \\
= 8 + 3 & = 0 + 3 & = -4 + 3 \\
= 11 & = 3 & = -1
\end{array}
$$

Ordered Pairs:

$$(2, 11) \qquad\qquad (0, 3) \qquad\qquad (-1, -1)$$

Helpful Hint

Note that $f(x)$ is a special symbol in mathematics used to denote a function. The symbol $f(x)$ is read "f of x." It does *not* mean $f \cdot x$ (f times x).

EXAMPLES   Find each function value.

18. If $g(x) = 3x - 2$, find $g(1)$.

$$g(1) = 3(1) - 2 = 1$$

19. If $g(x) = 3x - 2$, find $g(0)$.

$$g(0) = 3(0) - 2 = -2$$

20. If $f(x) = 7x^2 - 3x + 1$, find $f(1)$.

$$f(1) = 7(1)^2 - 3(1) + 1 = 5$$

21. If $f(x) = 7x^2 - 3x + 1$, find $f(-2)$.

$$f(-2) = 7(-2)^2 - 3(-2) + 1 = 35$$

Try the Concept Check in the margin.

F   Graphing Linear Functions

Recall that the graph of a linear equation in two variables is a line, and a line that is not vertical will always pass the vertical line test. Thus, *all linear equations are functions except those whose graphs are vertical lines*. We call such functions *linear functions*.

Linear Function

A **linear function** is a function that can be written in the form

$$f(x) = mx + b$$

EXAMPLE 22   Graph the function $f(x) = 2x + 1$.

Solution: Since $y = f(x)$, we can replace $f(x)$ with y and graph as usual. The graph of $y = 2x + 1$ has slope 2 and y-intercept $(0, 1)$ Its graph is shown.

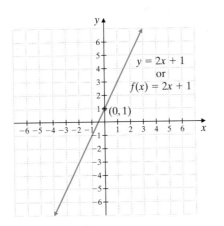

$y = 2x + 1$
or
$f(x) = 2x + 1$

$(0, 1)$

Practice Problems 18–21

Find each function value.

18. If $g(x) = 4x + 5$, find $g(0)$.
19. If $g(x) = 4x + 5$, find $g(-5)$.
20. If $f(x) = 3x^2 - x + 2$, find $f(2)$.
21. If $f(x) = 3x^2 - x + 2$, find $f(-1)$.

Concept Check

Suppose $y = f(x)$ and we are told that $f(3) = 9$. Which is not true?

a. When $x = 3$, $y = 9$.
b. A possible function is $f(x) = x^2$.
c. A point on the graph of the function is $(3, 9)$.
d. A possible function is $f(x) = 2x + 4$.

Practice Problem 22

Graph the function $f(x) = 3x - 2$.

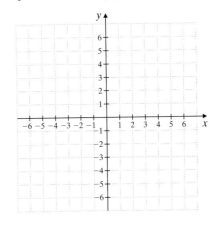

Answers

18. $g(0) = 5$,   **19.** $g(-5) = -15$,
20. $f(2) = 12$,   **21.** $f(-1) = 6$,
22.

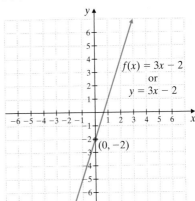

$f(x) = 3x - 2$
or
$y = 3x - 2$

$(0, -2)$

Concept Check: d

FOCUS ON Mathematical Connections

PERPENDICULAR BISECTORS

A **perpendicular bisector** is a line that is perpendicular to a given line segment and divides the segment into two equal lengths. A perpendicular bisector crosses the line segment at the point that is located exactly halfway between the two endpoints of the line segment. That point is called the **midpoint** of the line segment. If a line segment has the endpoints (x_1, y_1) and (x_2, y_2), then the midpoint of this line segment is the point with coordinates $\left(\dfrac{x_1 + x_2}{2}, \dfrac{y_1 + y_2}{2} \right)$.

An example of a line segment and its perpendicular bisector is shown in the figure.

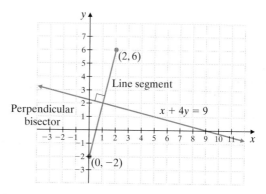

To find the equation of a line segment's perpendicular bisector, follow these steps:

Step 1. Find the midpoint of the line segment. (See the formula in Section 9.1.)

Step 2. Find the slope of the line segment.

Step 3. Find the slope of a line that is perpendicular to the line segment.

Step 4. Use the midpoint and the slope of the perpendicular line to find the equation of the perpendicular bisector.

CRITICAL THINKING

Use the steps given above and what you have learned in this chapter to find the equation of the perpendicular bisector of each line segment whose endpoints are given.

1. $(3, -1), (-5, 1)$
2. $(-6, -3), (-8, -1)$
3. $(-2, 6), (-22, -4)$
4. $(5, 8), (7, 2)$
5. $(2, 3), (-4, 7)$
6. $(-6, 8), (-4, -2)$

Name _____ Section _____ Date _____

EXERCISE SET 3.5

A **B** *Find the domain and the range of each relation. Also determine whether the relation is a function. See Examples 1 through 6.*

1. $\{(-1, 7), (0, 6), (-2, 2), (5, 6)\}$

2. $\{(4, 9), (-4, 9), (2, 3), (10, -5)\}$

3. $\{(-2, 4), (6, 4), (-2, -3), (-7, -8)\}$

4. $\{(6, 6), (5, 6), (5, -2), (7, 6)\}$

5. $\{(1, 1), (1, 2), (1, 3), (1, 4)\}$

6. $\{(1, 1), (2, 1), (3, 1), (4, 1)\}$

7. $\left\{\left(\frac{3}{2}, \frac{1}{2}\right), \left(1\frac{1}{2}, -7\right), \left(0, \frac{4}{5}\right)\right\}$

8. $\{(\pi, 0), (0, \pi), (-2, 4), (4, -2)\}$

9. $\{(-3, -3), (0, 0), (3, 3)\}$

10. $\left\{\left(\frac{1}{2}, \frac{1}{4}\right), \left(0, \frac{7}{8}\right), (0.5, \pi)\right\}$

11.

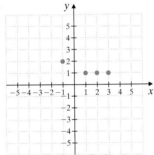

12.

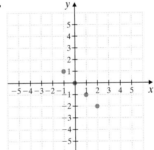

13.

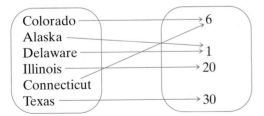

14.

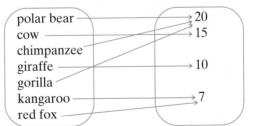

225

15. Input: Degrees Fahrenheit / Output: Degrees Celsius

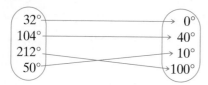

16. Input: Words / Output: Number of Letters

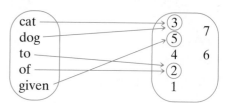

17. Input: / Output:

18. Input: / Output:

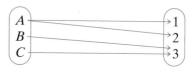

Determine whether each relation is a function. See Examples 4 through 6.

19.

First Set: Input	Correspondence	Second Set: Output
Class of algebra students	Grade average	Set of nonnegative numbers

20.

First Set: Input	Correspondence	Second Set: Output
People in New Orleans (population 500,000)	Birthdate	Days of the year

Determine whether each relation is also a function. See Examples 7 and 8.

 21. $y = x + 1$

22. $y = x - 1$

23. $x = 2y^2$

24. $y = x^2$

25. $y - x = 7$

26. $2x - 3y = 9$

 Use the vertical line test to determine whether each graph is the graph of a function. See Examples 9 through 13.

27.

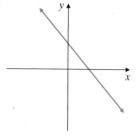

28.

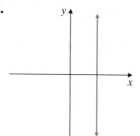

29.

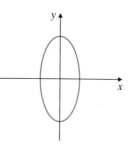

30.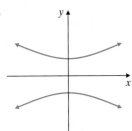

31.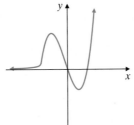

32.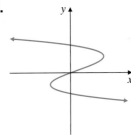

D *Find the domain and the range of each relation. Use the vertical line test to determine whether each graph is the graph of a function. See Examples 14 through 17.*

33.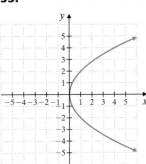

34.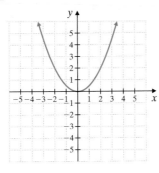

35.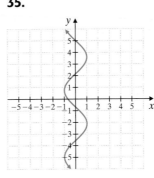

36.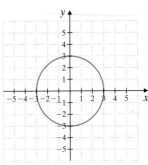

37.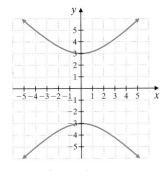

38.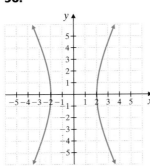

39.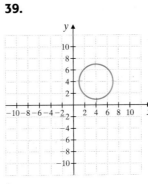

40.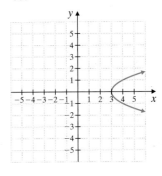

41.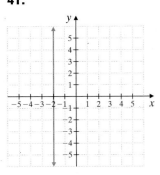

42.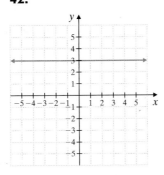

43.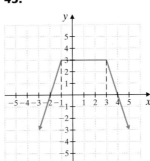

44.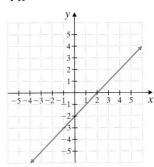

Ⓔ If $f(x) = 3x + 3$, $g(x) = 4x^2 - 6x + 3$, and $h(x) = 5x^2 - 7$, find each function value. See Examples 18 through 21.

45. $f(4)$ **46.** $f(-1)$ 💳 **47.** $h(-3)$ **48.** $h(0)$

49. $g(2)$ **50.** $g(1)$ 💳 **51.** $g(0)$ **52.** $h(-2)$

For each function, find the indicated values. See Examples 18 through 21.

53. $f(x) = \dfrac{1}{2}x$;

   a. $f(0)$

   b. $f(2)$

   c. $f(-2)$

54. $g(x) = -\dfrac{1}{3}x$;

   a. $g(0)$

   b. $g(-1)$

   c. $g(3)$

55. $f(x) = -5$;

   a. $f(2)$

   b. $f(0)$

   c. $f(606)$

56. $h(x) = 7$;

   a. $h(7)$

   b. $h(542)$

   c. $h\left(-\dfrac{3}{4}\right)$

The function $A(r) = \pi r^2$ may be used to find the area of a circle if we are given its radius. Use this function to answer Exercises 57 and 58.

△ **57.** Find the area of a circle whose radius is 5 centimeters. (Do not approximate π.)

△ **58.** Find the area of a circular garden whose radius is 8 feet. (Do not approximate π.)

The function $V(x) = x^3$ may be used to find the volume of a cube if we are given the length x of a side. Use this function to answer Exercises 59 and 60.

▦ **59.** Find the volume of a cube whose side
△   is 14 inches.

▦ **60.** Find the volume of a die whose side is
△   1.7 centimeters.

Forensic scientists use the following functions to find the height of a woman if they are given the height of her femur bone (f) or her tibia bone (t) in centimeters.

$$H(f) = 2.59f + 47.24$$
$$H(t) = 2.72t + 61.28$$

Use these functions to answer Exercises 61 and 62.

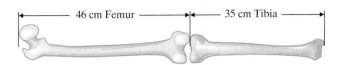

46 cm Femur — 35 cm Tibia

61. Find the height of a woman whose femur measures 46 centimeters.

62. Find the height of a woman whose tibia measures 35 centimeters.

The dosage in milligrams D of Ivermectin, a heartworm preventive, for a dog who weighs x pounds is given by

$$D(x) = \frac{136}{25}x$$

Use this function to answer Exercises 63 and 64.

63. Find the proper dosage for a dog that weighs 30 pounds.

64. Find the proper dosage for a dog that weighs 50 pounds.

65. The per capita consumption (in pounds) of all poultry in the United States is given by the function $C(x) = 1.69x + 87.54$, where x is the number of years since 1995. (*Source:* Based on actual and estimated data from the Economic Research Service, U.S. Department of Agriculture, 1995–2002)
a. Find and interpret $C(5)$.
b. Estimate the per capita consumption of all poultry in the United States in 2002.

66. The amount of money (in billions of dollars) spent by U.S. biotechnology companies on research and development annually is given by the function $R(x) = 0.8x + 5.0$, where x is the number of years since 1993. (*Source:* Based on data from Ernst & Young LLP, Annual Biotechnology Industry Reports, 1993–2000)
a. Find and interpret $R(7)$.
b. Estimate the amount of money spent on biotechnology research and development in 1998.

F *Graph each linear function. See Example 22.*

67. $f(x) = 2x + 3$

68. $f(x) = 5x - 1$

69. $f(x) = -3x$

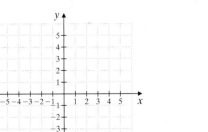

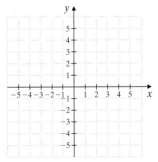

70. $f(x) = -4x$

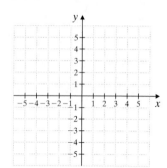

71. $f(x) = -x + 2$

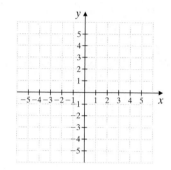

72. $f(x) = -x + 1$

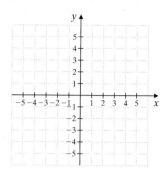

Review and Preview

Solve. See Section 2.4.

73. $2x - 7 \le 21$

74. $-3x + 1 > 0$

75. $5(x - 2) \ge 3(x - 1)$

76. $-2(x + 1) \le -x + 10$

77. $\dfrac{x}{2} + \dfrac{1}{4} < \dfrac{1}{8}$

78. $\dfrac{x}{5} - \dfrac{3}{10} \ge \dfrac{x}{2} - 1$

Combining Concepts

79. If $f(x) = 1.3x^2 - 2.6x + 5.1$, the following.
  a. $f(2)$ **b.** $f(-2)$ **c.** $f(3.1)$

For each function, find the indicated values.

80. $f(x) = 2x + 7$;
  a. $f(2)$ **b.** $f(a)$

81. $f(x) = x^2 - 12$;
  a. $f(12)$ **b.** $f(a)$

82. Describe a function whose domain is the set of people in your home town.

83. Describe a function whose domain is the set of people in your algebra class.

84. Since $y = x + 7$ describes a function, rewrite the equation using function notation.

85. In your own words, explain how to find the domain of a function given its graph.

86. Explain the vertical line test and how it is used.

3.6 Graphing Linear Inequalities

Ⓐ Graphing Linear Inequalities

Recall that the graph of a linear equation in two variables is the graph of all ordered pairs that satisfy the equation and that the graph is a line. Here we graph **linear inequalities** in two variables; that is, we graph all the ordered pairs that satisfy the inequality.

If the equal sign in a linear equation in two variables is replaced with an inequality symbol, the result is a linear inequality in two variables. Some examples are

$$3x + 5y \geq 6 \qquad\qquad 2x - 4y < -3$$
$$4x > 2 \qquad\qquad y \leq 5$$

To graph the linear inequality $x + y < 3$, we first graph the related equation $x + y = 3$. The resulting **boundary line** contains all ordered pairs whose coordinates add up to 3. The line separates the plane into two regions called **half-planes**. All points above the boundary line $x + y = 3$ have coordinates that satisfy the inequality $x + y > 3$, and all points below the line have coordinates that satisfy the inequality $x + y < 3$.

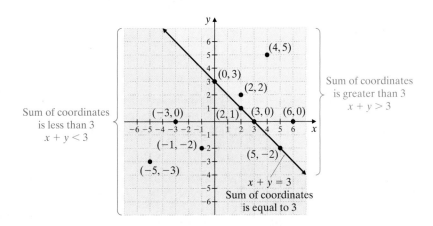

The graph, or **solution region**, for $x + y < 3$, then, is the half-plane below the boundary line and is shown shaded in the figure below. The boundary line is shown dashed since it is not a part of the solution region. The ordered pairs on this line satisfy $x + y = 3$, but not $x + y < 3$.

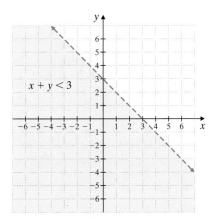

The following steps may be used to graph linear inequalities in two variables.

Practice Problem 1

Graph: $x + 3y > 4$

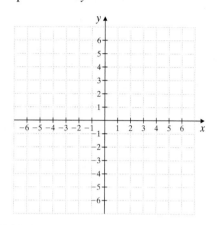

Practice Problem 2

Graph: $x \leq 2y$

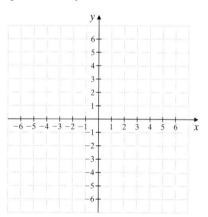

Answer

1.

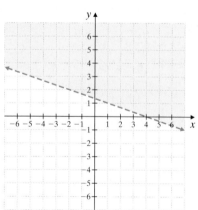

2. (answer on next page)

Graphing a Linear Inequality in Two Variables

Step 1. Graph the boundary line found by replacing the inequality sign with an equal sign. If the inequality sign is $<$ or $>$, graph a dashed line indicating that points on the line are not solutions of the inequality. If the inequality sign is $\leq$ or $\geq$, graph a solid line indicating that points on the line are solutions of the inequality.

Step 2. Choose a **test point** *not on the boundary line* and substitute the coordinates of this test point into the *original inequality*.

Step 3. If a true statement is obtained in *Step 2*, shade the half-plane that contains the test point. If a false statement is obtained, shade the half-plane that does not contain the test point.

EXAMPLE 1   Graph: $2x - y < 6$

Solution: The boundary line for this inequality is the graph of $2x - y = 6$. We graph a dashed boundary line because the inequality symbol is $<$. Next we choose a test point on either side of the boundary line. The point $(0, 0)$ is not on the boundary line, so we use this point. Replacing x with 0 and y with 0 in the *original inequality* $2x - y < 6$ leads to the following:

$$2x - y < 6$$
$$2(0) - 0 < 6 \quad \text{Let } x = 0 \text{ and } y = 0.$$
$$0 < 6 \quad \text{True.}$$

Because $(0, 0)$ satisfies the inequality, so does every point on the same side of the boundary line as $(0, 0)$. We shade the half-plane that contains $(0, 0)$, as shown. Every point in the shaded half-plane satisfies the original inequality.

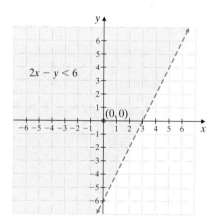

EXAMPLE 2   Graph: $3x \geq y$

Solution: The boundary line is the graph of $3x = y$. We graph a solid boundary line because the inequality symbol is $\geq$. We test a point not on the boundary line to determine which half-plane contains points that satisfy the inequality. Let's choose $(0, 1)$ as our test point.

$$3x \geq y$$
$$3(0) \geq 1 \quad \text{Let } x = 0 \text{ and } y = 1.$$
$$0 \geq 1 \quad \text{False.}$$

This point does not satisfy the inequality, so the correct half-plane is on the opposite side of the boundary line from $(0, 1)$. The graph of $3x \geq y$ is the boundary line together with the shaded region, as shown.

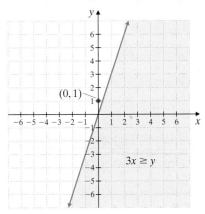

Try the Concept Check in the margin.

B **Graphing Intersections and Unions**

The intersections and the unions of linear inequalities can also be graphed, as shown in the next two examples.

EXAMPLE 3   Graph the intersection of $x \geq 1$ and $y \geq 2x - 1$.

Solution: First we graph each inequality. The intersection of the two graphs is all points common to both regions, as shown by the *heaviest* shading in the third graph.

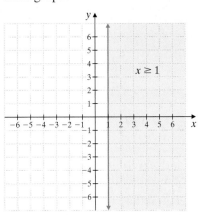

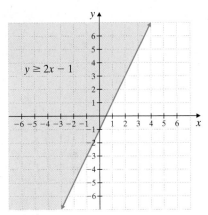

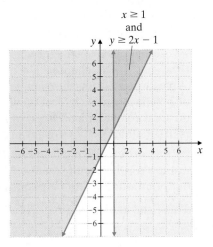

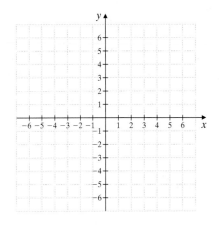
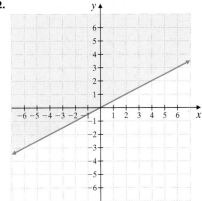
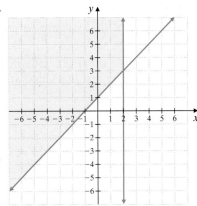

Practice Problem 4

Graph the union of $x + 2y \le 4$ or $y \ge -1$.

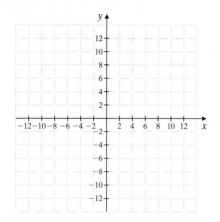

Answer

4.

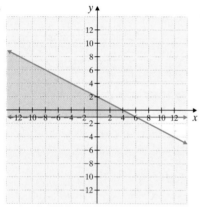

EXAMPLE 4 Graph the union of $2x + y \ge -8$ or $y \le -2$.

Solution: First we graph each inequality. The union of the two inequalities is both shaded regions, including the solid boundary lines, as shown in the third graph.

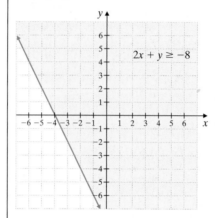

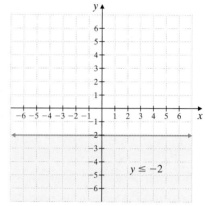

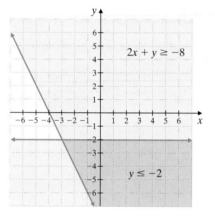

STUDY SKILLS REMINDER

Tips for studying for an exam

To prepare for an exam, try the following study techniques.

- Start the study process days before your exam.
- Make sure that you are current and up-to-date on your assignments.
- If there is a topic that you are unsure of, use one of the many resources that are available to you. For example,

> See your instructor.
> Visit a learning resource center on campus where math tutors are available.
> Read the textbook material and examples on the topic.
> View a videotape on the topic.

- Reread your notes and carefully review the Chapter Highlights at the end of the chapter.
- Work the review exercises at the end of the chapter and check your answers. Make sure that you correct any missed exercises. If you have trouble on a topic, use a resource listed at the left.
- Find a quiet place to take the Chapter Test found at the end of the chapter. Do not use any resources when taking this sample test. This way you will have a clear indication of how prepared you are for your exam. Check your answers and make sure that you correct any missed exercises.
- Get lots of rest the night before the exam. It's hard to show how well you know the material if your brain is foggy from lack of sleep.

Good luck and keep a positive attitude.

Name _____   Section _____   Date _____

EXERCISE SET 3.6

A *Graph each inequality. See Examples 1 and 2.*

1. $x < 2$

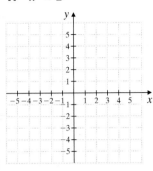

2. $x > -3$

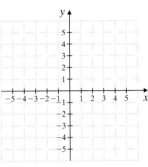

3. $x - y \geq 7$

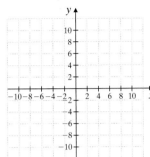

4. $3x + y \leq 1$

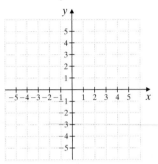

5. $3x + y > 6$

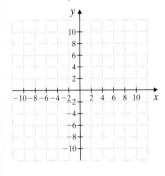

6. $2x + y > 2$

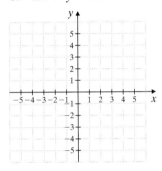

7. $y \leq -2x$

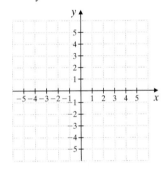

8. $y \leq 3x$

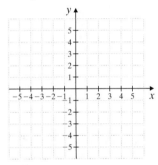

9. $2x + 4y \geq 8$

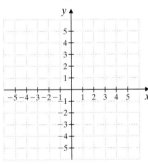

10. $2x + 6y \leq 12$

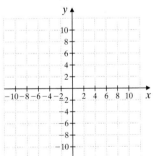

11. $5x + 3y > -15$

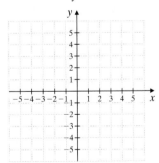

12. $2x + 5y < -20$

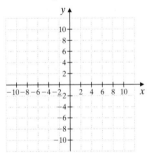

13. Explain when a dashed boundary line should be used in the graph of an inequality.

14. Explain why, after the boundary line is sketched, we test a point on either side of this boundary in the original inequality.

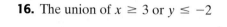

B *Graph each union or intersection. See Examples 3 and 4.*

15. The intersection of $x \geq 3$ and $y \leq -2$

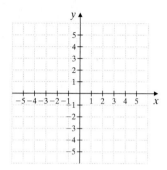

16. The union of $x \geq 3$ or $y \leq -2$

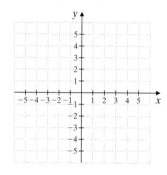

17. The union of $x \leq -2$ or $y \geq 4$

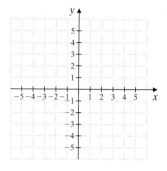

18. The intersection of $x \leq -2$ and $y \geq 4$

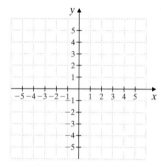

19. The intersection of $x - y < 3$ and $x > 4$

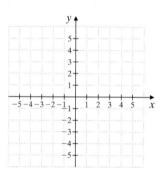

20. The intersection of $2x > y$ and $y > x + 2$

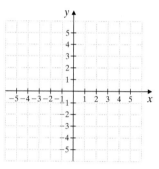

21. The union of $x + y \leq 3$ or $x - y \geq 5$

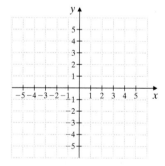

22. The union of $x - y \leq 3$ or $x + y > -1$

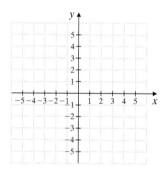

23. The union of $x - y \geq 2$ or $y < 5$

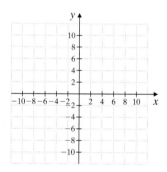

24. The union of $x - y < 3$ or $x > 4$

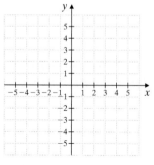

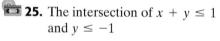 **25.** The intersection of $x + y \leq 1$ and $y \leq -1$

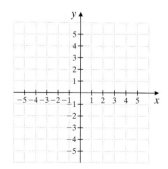

26. The intersection of $y \geq x$ and $2x - 4y \geq 6$

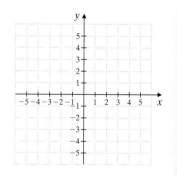

Match each inequality with its graph.

A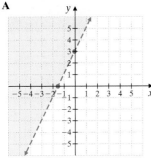

B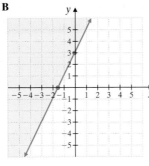

C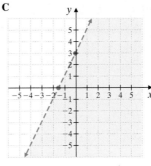

D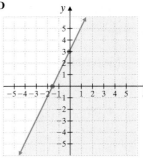

27. $y \leq 2x + 3$

28. $y < 2x + 3$

29. $y > 2x + 3$

30. $y \geq 2x + 3$

Review and Preview

Determine whether the given ordered pair is a solution of both equations. See Section 3.1.

31. $(3, -1); \quad x - y = 4$
$\qquad\qquad x + 2y = 1$

32. $(0, 2); \quad x + 3y = 6$
$\qquad\qquad 4x - y = -2$

33. $(-4, 0); 3x + 2y = -12$
$\qquad\qquad\quad x = 4y$

34. $(-5, 2); \quad x + y = -3$
$\qquad\qquad\quad 2x - y = -8$

 Combining Concepts

Solve.

35. Chris-Craft manufactures boats out of Fiberglas and wood. Fiberglas hulls require 2 hours of work, whereas wood hulls require 4 hours of work. Employees work at most 40 hours a week. The following inequalities model these restrictions, where x represents the number of Fiberglas hulls produced and y represents the number of wood hulls produced.

$$x \geq 0$$
$$y \geq 0$$
$$2x + 4y \leq 40$$

Graph the intersection of these inequalities.

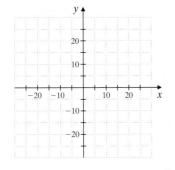

36. Rheem Abo-Zahrah decides that she will study at most 20 hours every week and that she must work at least 10 hours every week. Let x represent the hours studying and y represent the hours working. Write two inequalities that model this situation and graph their intersection.

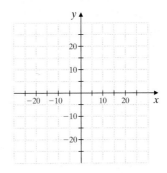

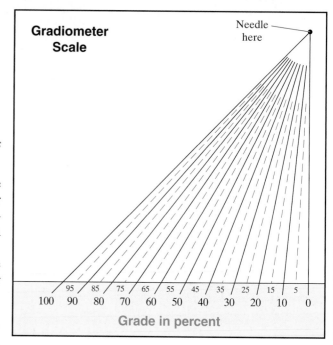

Gradiometer Scale

Needle here

95 85 75 65 55 45 35 25 15 5
100 90 80 70 60 50 40 30 20 10 0

Grade in percent

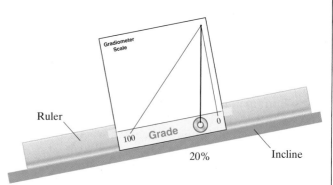

CHAPTER 3 ACTIVITY — **Measuring Slope**

Materials:

- cardboard
- tape or glue
- ruler
- metal washer
- scissors
- string
- rug needle

This activity may be completed by working in groups or individually.

The grade of an incline is the same as its slope given as a percent. A 6% grade means that for every horizontal run of 100 units, there is a vertical rise of 6 units. This can also be written as a slope of 0.06.

A gradiometer is a device that measures the grade of an incline. You can build your own gradiometer by following these steps

- Cut out the gradiometer scale given at the right.

- Attach the scale to a piece of rigid cardboard. Trim the cardboard so it is even with the bottom edge of the scale.

- Thread the rug needle with string. Poke the needle through the gradiometer scale and cardboard at the large dot in the upper-right corner.

- Tie a large knot in the portion of the string hanging out the back. Pull the string from the front until the knot blocks the hole.

- Tie the washer to the portion of the string hanging across the front of the scale so that, when the gradiometer is held upright, the washer hangs in the portion marked "Grade (percent)" at the bottom.

- Attach the gradiometer to a ruler (roughly in the middle) so the bottom edge of the gradiometer is aligned with the bottom edge of the ruler.

- To use your gradiometer to measure the grade of an incline, place the bottom edge of the ruler along the incline. The point at which the string attached to the washer crosses the scale at the bottom of the gradiometer corresponds to the grade of the incline. (See figure.)

1. Refer to the graph on page 197. Express the slope of the line as a percent (grade). Use your gradiometer to measure the grade of the line. How close is your gradiometer reading to the actual grade? (*Hint:* You will need to hold the textbook upright on a flat surface to measure the grade of the line.)

2. Use your gradiometer to measure the grade of three inclines in your classroom. Interpret each measurement. (*Hint:* Consider measuring the inclines of desks, lecterns, ramps, or steps).

3. Notice that your gradiometer directly measures only positive slopes—inclines that rise from left to right. Explain how you could use your gradiometer to measure a negative slope (an incline that falls from left to right).

4. (Optional) According to the Americans with Disabilities Act (1990), handicapped-accessible ramps should have a grade of no more than 8.3%. Ramps with vertical rises greater than 6 inches and grades greater than 5% must provide handrails. Use your gradiometer to measure the grades of several wheelchair ramps on campus. Do they comply with the Americans with Disabilities Act guidelines?

Chapter 3 VOCABULARY CHECK

Fill in each blank with one of the words or phrases listed below.

relation	line	function	standard	slope	domain
slope-intercept	x	y	range	parallel	linear function
point-slope	perpendicular	linear inequality			

1. A _____ is a set of ordered pairs.
2. The graph of every linear equation in two variables is a _____ .
3. The statement $-x + 2y > 0$ is called a _____ in two variables.
4. _____ form of linear equation in two variables is $Ax + By = C$.
5. The _____ of a relation is the set of all second components of the ordered pairs of the relation.
6. _____ lines have the same slope and different y-intercepts.
7. _____ form of a linear equation in two variables is $y = mx + b$.
8. A _____ is a relation in which each first component in the ordered pairs corresponds to exactly one second component.
9. In the equation $y = 4x - 2$, the coefficient of x is the _____ of its corresponding graph.
10. Two lines are _____ if the product of their slopes is -1.
11. To find the x-intercept of a linear equation, let _____ $= 0$ and solve for the other variable.
12. The _____ of a relation is the set of all first components of the ordered pairs of the relation.
13. A _____ is a function that can be written in the form $f(x) = mx + b$.
14. To find the y-intercept of a linear equation, let _____ $= 0$ and solve for the other variable.
15. The equation $y - 8 = -5(x + 1)$ is written in _____ form.

CHAPTER 3

Highlights

DEFINITIONS AND CONCEPTS	**EXAMPLES**

Section 3.1   Graphing Linear Equations

The **rectangular coordinate system**, or **Cartesian coordinate system**, consists of a vertical and a horizontal number line on a plane intersecting at their 0 coordinate. The vertical number line is called the **y-axis**, and the horizontal number line is called the **x-axis**. The point of intersection of the axes is called the **origin**.

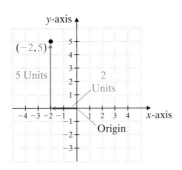

To **plot** or **graph** an ordered pair means to find its corresponding point on a rectangular coordinate system.

Plot or graph the ordered pair $(-2, 5)$.
Start at the origin. Move 2 units to the left along the x-axis, then 5 units upward parallel to the y-axis.

An ordered pair is a **solution** of an equation in two variables if replacing the variables by the corresponding coordinates results in a true statement.

Determine whether $(-2, 3)$ is a solution of $3x + 2y = 0$.

$$3(-2) + 2(3) = 0$$
$$-6 + 6 = 0$$
$$0 = 0 \quad \text{True.}$$

$(-2, 3)$ is a solution.

| **DEFINITIONS AND CONCEPTS** | **EXAMPLES** |

Section 3.1   Graphing Linear Equations *(continued)*

A **linear equation in two variables** is an equation that can be written in the form $Ax + By = C$, where A, B, and C are real numbers and A and B are not both 0. The form $Ax + By = C$ is called **standard form**.

$y = -2x + 5, \quad x = 7,$
$y - 3 = 0, \quad 6x - 4y = 10$
$6x - 4y = 10$ is in standard form.

The **graph of a linear equation** in two variables is a line. To graph a linear equation in two variables, find three ordered pair solutions. Plot the solution points, and draw the line connecting the points.

Graph: $\quad 3x + y = -6$

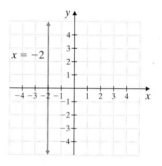

x	y
0	-6
-2	0
-3	3

To find an x-intercept, let $y = 0$ and solve for x.
To find a y-intercept, let $x = 0$ and solve for y.

The graph of $x = c$ is a vertical line with x-intercept c.
The graph of $y = c$ is a horizontal line with y-intercept c.

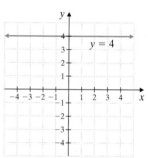

Section 3.2   The Slope of a Line

The **slope** m of the line through (x_1, y_1) and (x_2, y_2) is given by

$$m = \frac{y_2 - y_1}{x_2 - x_1}$$

as long as $x_2 \neq x_1$

Find the slope of the line through $(-1, 7)$ and $(-2, -3)$.

$$m = \frac{y_2 - y_1}{x_2 - x_1} = \frac{-3 - 7}{-2 - (-1)} = \frac{-10}{-1} = 10$$

The **slope-intercept form** of a linear equation is

$$y = mx + b$$

where m is the slope of the line and $(0, b)$ is the y-intercept.

Find the slope and y-intercept of $-3x + 2y = -8$.

$$2y = 3x - 8$$

$$\frac{2y}{2} = \frac{3x}{2} - \frac{8}{2}$$

$$y = \frac{3}{2}x - 4$$

The slope of the line is $\frac{3}{2}$, and the y-intercept is $(0, -4)$.

DEFINITIONS AND CONCEPTS	**EXAMPLES**

Section 3.2   The Slope of a Line *(continued)*

The slope of a horizontal line is 0.	The slope of $y = -2$ is 0.
The slope of a vertical line is undefined.	The slope of $x = 5$ is undefined.
Nonvertical parallel lines have the same slope.	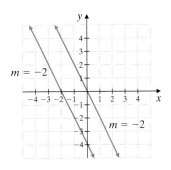
If the product of the slopes of two lines is -1, then the lines are perpendicular.	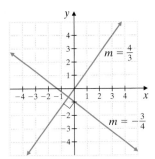

Section 3.3   The Slope-Intercept Form

We can use the slope-intercept form to write an equation of a line given its slope and y-intercept.	Write an equation of the line with y-intercept $(0, -1)$ and slope $\dfrac{2}{3}$. $$y = mx + b$$ $$y = \frac{2}{3}x - 1$$

Section 3.4   More Equations of Lines

The **point-slope form** of the equation of a line is $$y - y_1 = m(x - x_1)$$ where m is the slope of the line and (x_1, y_1) is a point on the line.	Find an equation of the line with slope 2 containing the point $(1, -4)$. Write the equation in standard form: $Ax + By = C$. $$y - y_1 = m(x - x_1)$$ $$y - (-4) = 2(x - 1)$$ $$y + 4 = 2x - 2$$ $$-2x + y = -6 \qquad \text{Standard form}$$

Section 3.5   Introduction to Functions

A **relation** is a set of ordered pairs. The **domain** of the relation is the set of all first components of the ordered pairs. The **range** of the relation is the set of all second components of the ordered pairs.

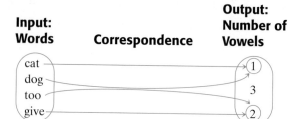

Domain: {cat, dog, too, give}
Range: {1, 2}

A **function** is a relation in which each element of the first set corresponds to exactly one element of the second set.

The previous relation is a function. Each word contains one exact number of vowels.

VERTICAL LINE TEST

If no vertical line can be drawn so that it intersects a graph more than once, the graph is the graph of a function. If such a line can be drawn, the graph is not that of a function.

Find the domain and the range of the relation. Also determine whether the relation is a function.

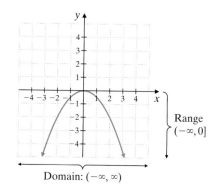

Range $(-\infty, 0]$

Domain: $(-\infty, \infty)$

By the vertical line test, this is the graph of a function.

The symbol $f(x)$ means **function of x** and is called **function notation**.

If $f(x) = 2x^2 - 5$, find $f(-3)$.

$$f(-3) = 2(-3)^2 - 5 = 2(9) - 5 = 13$$

A **linear function** is a function that can be written in the form

$$f(x) = mx + b$$

To graph a linear function, use the slope and y-intercept.

$$f(x) = -3, g(x) = 5x, h(x) = -\frac{1}{3}x - 7$$

Graph: $f(x) = -2x$
(or $y = -2x + 0$)

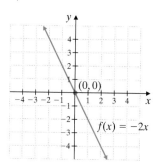

The slope is $\dfrac{2}{-1}$.

The y-intercept is $(0, 0)$.

DEFINITIONS AND CONCEPTS	**EXAMPLES**

Section 3.6 Graphing Linear Inequalities

If the equal sign in a linear equation in two variables is replaced with an inequality symbol, the result is a **linear inequality in two variables**.

$x \leq -5y, \quad y \geq 2,$
$3x - 2y > 7, \quad x < -5$

GRAPHING A LINEAR INEQUALITY

Step 1. Graph the **boundary line** by graphing the related equation. Draw a solid line if the inequality symbol is $\leq$ or $\geq$. Draw a dashed line if the inequality symbol is $<$ or $>$.

Step 2. Choose a **test point** not on the line. Substitute its coordinates into the original inequality.

Step 3. If the resulting inequality is true, shade the **half-plane** that contains the test point. If the inequality is not true, shade the half-plane that does not contain the test point.

Graph: $2x - 4y > 4$

1. Graph $2x - 4y = 4$. Draw a dashed line because the inequality symbol is $>$.

2. Check the test point $(0, 0)$ in the inequality $2x - 4y > 4$.

$\qquad 2 \cdot 0 - 4 \cdot 0 > 4 \qquad$ Let $x = 0$ and $y = 0$.

$\qquad \qquad \qquad 0 > 4 \qquad$ False.

3. The inequality is false, so shade the half-plane that does not contain $(0, 0)$.

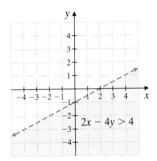

STUDY SKILLS REMINDER

Are you preparing for a test on Chapter 3?

Below I have listed some common trouble areas for students in Chapter 3. After studying for your test—but before taking your test—read these.

- Don't forget that the graph of an ordered pair is a *single* point in the rectangular coordinate plane.
- Remember that the slope of a horizontal line is 0 while a vertical line has undefined slope or no slope.
- For a linear equation such as $2y = 3x - 6$, the slope is not the coefficient of x unless the equation is solved for y. Solving this equation for y, we have $y = \frac{3}{2}x - 3$. The slope is $\frac{3}{2}$ and the y-intercept is $(0, -3)$.

- Parallel lines have the same slope while perpendicular lines have negative reciprocal slopes.

Slope	Parallel line	Perpendicular line
$m = 6$	$m = 6$	$m = -\dfrac{1}{6}$
$m = -\dfrac{2}{3}$	$m = -\dfrac{2}{3}$	$m = \dfrac{3}{2}$

- Don't forget that the statement $f(2) = 3$ corresponds to the ordered pair $(2, 3)$.

Remember: this is simply a checklist of common trouble areas. For a review of Chapter 3, see the Highlights and Chapter Review at the end of this chapter.

LINEAR MODELING

As we saw in Section 3.3, businesses often depend on equations that "closely fit" data. To *model* the data means to find an equation that describes the relationship between the paired data of two variables, such as time in years and profit. A model that accurately summarizes the relationship between two variables can be used to replace a potentially lengthy listing of the raw data. An accurate model might also be used to predict future trends by answering questions such as "If the trend seen in our company's performance in the last several years continues, what level of profit can we reasonably expect in 3 years?"

There are several ways to find a linear equation that models a set of data. If only two ordered pair data points are involved, an exact equation that contains both points can be found using the methods of Section 3.4. When more than two ordered pair data points are involved, it may be impossible to find a linear equation that contains all of the data points. In this case, the graph of the **best fit equation** should have a majority of the plotted ordered pair data points on the graph or close to it. In statistics, a technique called least squares regression is used to determine an equation that best fits a set of data. Various graphing utilities have built-in capabilities for finding an equation (called a regression equation) that best fits a set of ordered pair data points. Regression capabilities are often found with a graphing utility's statistics features.* A best fit equation can also be estimated using an algebraic method, which is outlined in the Group Activity below. In either case, a useful first step when finding a linear equation that models a set of data is creating a scatter diagram of the ordered pair data points to verify that a linear equation is an appropriate model.

GROUP ACTIVITY

Phoenix Sky Harbor International Airport, located in Phoenix, Arizona, is the fifth busiest airport in the United States in terms of aircraft operations. In terms of passenger traffic, Phoenix Sky Harbor is the ninth busiest airport in the United States. The table shows the total passenger traffic (in millions) for the years 1994 through 2000. Use the table along with your answers to the questions below to find a linear equation $y = mx + b$ that represents total passenger traffic y (in millions) at Phoenix Sky Harbor International Airport, where x represents the number of years after 1994.

Year	1994	1995	1996	1997	1998	1999	2000
Total Passenger	26	28	30	31	32	34	36
Traffic (in millions)							

(*Source:* The City of Phoenix Aviation Department)

1. Create a scatter diagram of the paired data given in the table. Does a linear model seem appropriate for the data?

2. Use a straightedge to draw on your graph what appears to be the line that "best fits" the data you plotted.

3. Estimate the coordinates of two points that fall on your best fit line. Use these points to find the equation of the line that passes through both points.

4. Use this equation to find the value of y for x = 11. Interpret the meaning of this pair of data.

5. How could this equation be useful to those who operate Phoenix Sky Harbor International Airport?

6. Compare your group's linear equation with other groups' equations. Are they the same or different? Explain why.

7. (Optional) Enter the data from the table into a graphing utility and use the linear regression feature to find a linear equation that models the data. Compare this equation with the one you found in Question 3. How are they alike or different?

8. (Optional) Using corporation annual reports or articles from magazines or newspapers, search for a set of business-related data that could be modeled with a linear equation. Explain how modeling this data could be useful to a business. Then find the best fit equation for the data.

*To find out more about using a graphing utility to find a regression equation, consult the user's manual for your graphing utility.

Chapter 3 Review

(3.1) *Plot the points and name the quadrant in which each point lies.*

1. $A(2, -1), B(-2, 1), C(0, 3), D(-3, -5)$

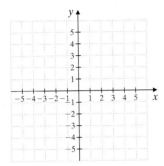

2. $A(-3, 4), B(4, -3), C(-2, 0), D(-4, 1)$

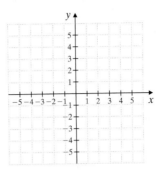

Create a scatter diagram for the given paired data.

3.

Per Person Consumption of Sweeteners in the United States	
Year	**Per Person Consumption (pounds per person)**
1995	144
1996	145
1997	150
1998	151
1999	153
2000	150

(*Source:* Economic Research Service, U.S. Department of Agriculture)

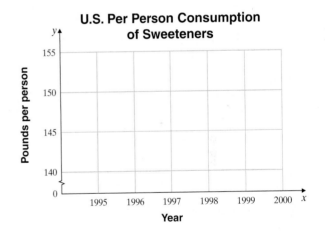

4.

U.S. Armed Forces Active Duty Personnel	
Year	**Armed Forces (in millions)**
1950	1.5
1960	2.5
1970	3
1980	2
1990	2
2000	1

(*Source:* 1998 World Almanac)

Determine whether each ordered pair is a solution to the given equation.

5. $7x - 8y = 56; (0, 56), (8, 0)$

6. $-2x + 5y = 10; (-5, 0), (1, 1)$

7. $x = 13; (13, 5), (13, 13)$

8. $y = 2; (7, 2), (2, 7)$

Graph each linear equation.

9. $3x - y = 3$

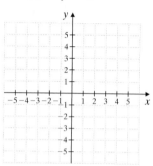

10. $2x - y = 4$

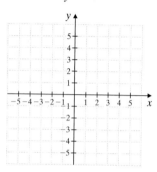

11. $4x + 5y = 20$

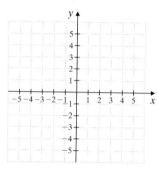

12. $3x - 2y = -9$

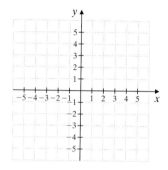

13. $y = 5$

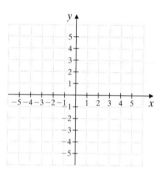

14. $x = -2$

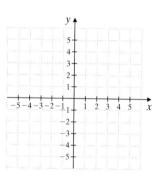

15. $y = \dfrac{1}{3}x$

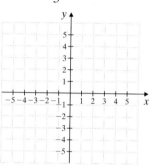

16. $x = -4y$

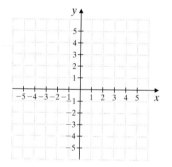

(3.2) *Find the slope of the line through each pair of points.*

17. $(2, 8)$ and $(6, -4)$

18. $(-3, 9)$ and $(5, 13)$

19. $(-7, -4)$ and $(-3, 6)$

20. $(7, -2)$ and $(-5, 7)$

Determine the slope of each line.

21.

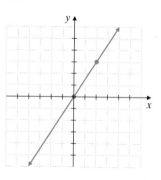

22.

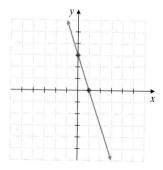

23.

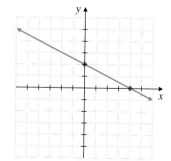

24.

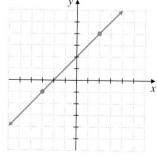

Two lines are graphed on each set of axes. Determine whether l_1 or l_2 has the greater slope.

25.

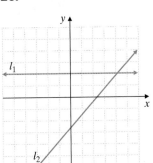

26.

27.

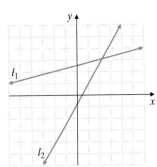

28.

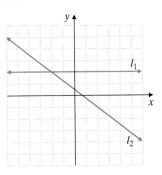

Find the slope and y-intercept of each line.

29. $y = -3x + \dfrac{1}{2}$

30. $y = 2x + 4$

31. $6x - 15y = 20$

32. $4x + 14y = 21$

Find the slope of each line.

33. $y - 3 = 0$

34. $x = -5$

Determine whether each pair of lines is parallel, perpendicular, or neither.

35. $y = -2x + 6$
     $y = 2x - 1$

36. $-x + 3y = 2$
     $6x - 18y = 3$

37. $y = \dfrac{3}{4}x + 1$
     $y = -\dfrac{4}{3}x + 1$

38. $x - 2y = 6$
     $4x + y = 8$

(3.3) *Graph each line passing through the given point with the given slope.*

39. Through $(2, -3)$ with slope $\dfrac{2}{3}$

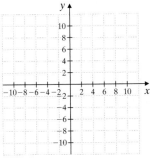

40. Through $(1, -4)$ with slope $\dfrac{1}{2}$

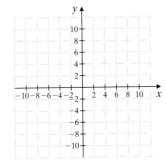

41. Through $(0, 1)$ with slope 2

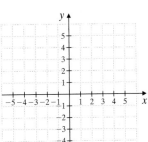

42. Through $(-2, 0)$ with slope -3

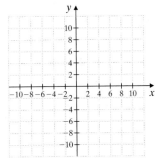

Graph each linear equation using the slope and y-intercept.

43. $y = -x + 1$

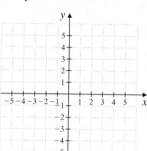

44. $y = 4x - 3$

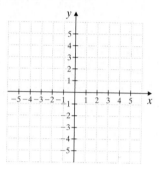

45. $3x - y = 6$

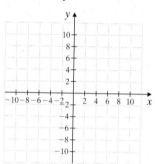

46. $y = -5x$

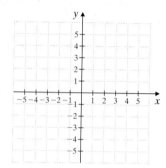

47. The cost C, in dollars, of renting a minivan for a day is given by the linear equation $C = 0.3x + 42$, where x is number of miles driven.
  a. Find the cost of renting the minivan for a day and driving it 150 miles.

  b. Find and interpret the slope of this equation.
  c. Find and interpret the y-intercept of this equation.

(3.4) *Write an equation of the line satisfying each set of conditions.*

48. Horizontal; through $(3, -1)$

49. Vertical; through $(-2, -4)$

50. Slope undefined; through $(-4, -3)$

51. Slope 0; through $(2, 5)$

Write the equation of the line satisfying each set of conditions. Write the equation in the form $y = mx + b$.

52. Through $(-3, 5)$; slope 3

53. Slope 2; through $(5, -2)$

54. Through $(-6, -1)$ and $(-4, -2)$

55. Through $(-5, 3)$ and $(-4, -8)$

56. Through $(2, -6)$; parallel to $y = -2x + 3$

57. Through $(-4, -2)$; parallel to $y = -\dfrac{3}{2}x + 1$

58. Through $(-6, -1)$; perpendicular to $4x + 3y = 5$

59. Through $(-4, 5)$; perpendicular to $2x - 3y = 6$

60. The value of a computer bought in 1996 continues to depreciate, or decrease, as time passes. Two years after the computer was bought, it was worth $2600; four years after it was bought, it was worth $1000.
   a. Assuming that this relationship between the number of years past 1996 and value of computer is linear, write an equation describing this relationship. [*Hint*: Use ordered pairs of the form (years past 1996, value of computer).]
   b. Use this equation to estimate the value of the computer in 2001.

61. The value of a building bought in 1980 continues to appreciate, or increase, as time passes. Seven years after the building was bought, it was worth $165,000; 12 years after it was bought, it was worth $180,000.
   a. Assuming that this relationship between the number of years past 1980 and value of the building is linear, write an equation describing this relationship. [*Hint:* Use ordered pairs of the form (years past 1980, value of building).]
   b. Use this equation to estimate the value of the building in 2005.

(3.5) *Find the domain and range for each relation. Then determine whether the relation is also a function.*

62. $\left\{ \left(-\frac{1}{2}, \frac{3}{4} \right), (6, 0.65), (0, -12), (25, 25) \right\}$

63. $\left\{ \left(\frac{3}{4}, -\frac{1}{2} \right), (0.65, 6), (-12, 0), (25, 25) \right\}$

64.

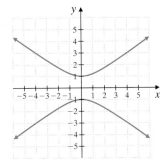

65.

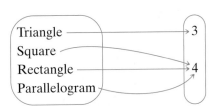

66.

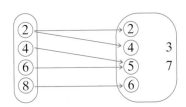

67.

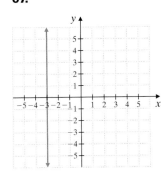

68.

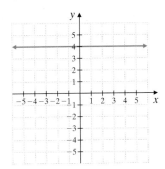

69.

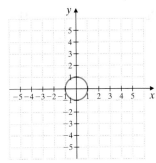

If $f(x) = x - 5$, $g(x) = -3x$, and $h(x) = 2x^2 - 6x + 1$, find each function value.

70. $f(2)$

71. $g(0)$

72. $g(-6)$

73. $h(-1)$ **74.** $h(1)$ **75.** $f(5)$

The function $J(x) = 2.54x$ may be used to calculate the weight of an object on Jupiter (J) given its weight on Earth (x).

76. If a person weighs 150 pounds on Earth, find the equivalent weight on Jupiter.

77. A 2000-pound probe on Earth weighs how many pounds on Jupiter?

Graph each linear function.

78. $f(x) = x + 2$

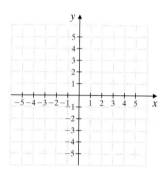

79. $f(x) = -\dfrac{1}{2}x + 3$

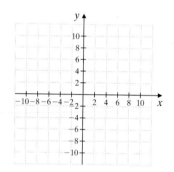

(3.6) *Graph each linear inequality.*

80. $3x + y > 4$

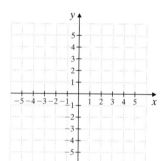

81. $\dfrac{1}{2}x - y < 2$

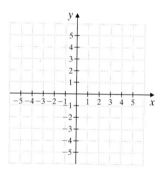

82. $5x - 2y \leq 9$

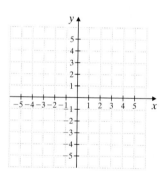

83. $3y \geq x$

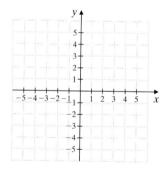

84. $y < 1$

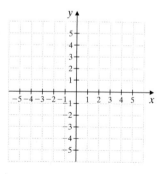

85. $x > -2$

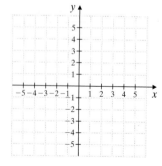

86. Graph the union of $y > 2x + 3$ or $x \leq -3$.

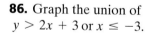
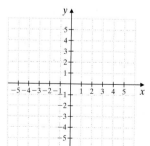

87. Graph the intersection of $2x < 3y + 8$ and $y \geq -2$.

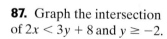
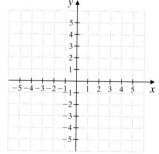

Name _____ Section _____ Date _____

Chapter 3 Test

1. Plot the points, and name the quadrant in which each is located:
$A(6, -2), B(4, 0), C(-1, 6)$.

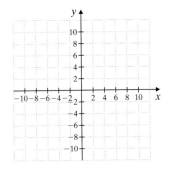

2. Create a scatter diagram for the paired data.

U.S. Cellular Phone Subscribers

Year	Number (in millions)
1994	24
1995	34
1996	44
1997	55
1998	69
1999	86
2000	109

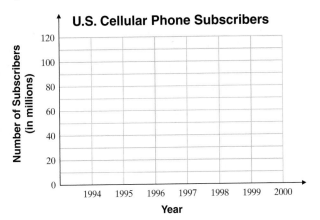

Graph each linear equation.

3. $-3x + y = -3$

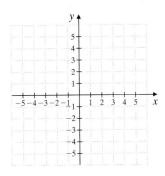

4. $2x - 3y = -6$

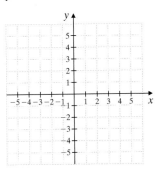

5. $4x + 6y = 8$

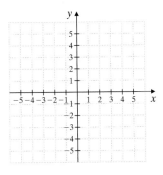

6. $y = -3$

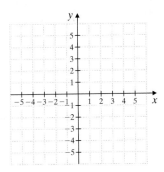

7. _____

8. _____

9. _____

10. _____

11. _____

12. _____

13. _____

14. _____

15. _____

16. _____

17. _____

18. _____

19. _____

20. _____

7. Find the slope of the line that passes through $(5, -8)$ and $(-7, 10)$.

8. Find the slope and the y-intercept of the line $3x + 12y = 8$.

Match each equation with its graph.

9. $f(x) = 3x + 1$

10. $f(x) = 3x - 2$

11. $f(x) = 3x + 2$

12. $f(x) = 3x - 5$

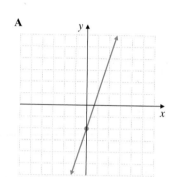

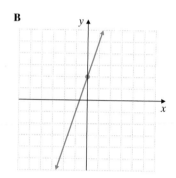

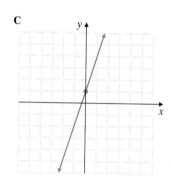

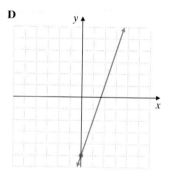

Find an equation of the line satisfying each set of conditions. Write the equations in the form $x = a$, $y = b$, or $y = mx + b$.

13. Horizontal; through $(2, -8)$

14. Vertical; through $(-4, -3)$

15. Perpendicular to $x = 5$; through $(3, -2)$

16. Through $(4, -1)$; slope -3

17. Through $(0, -2)$; slope 5

18. Through $(4, -2)$ and $(6, -3)$

19. Through $(-1, 2)$; perpendicular to $3x - y = 4$

20. Parallel to $2y + x = 3$; through $(3, -2)$

7. _____

8. _____

9. _____

10. _____

11. _____

12. _____

13. _____

14. _____

15. _____

16. _____

17. _____

18. _____

19. _____

20. _____

7. Find the slope of the line that passes through $(5, -8)$ and $(-7, 10)$.

8. Find the slope and the y-intercept of the line $3x + 12y = 8$.

Match each equation with its graph.

9. $f(x) = 3x + 1$

10. $f(x) = 3x - 2$

11. $f(x) = 3x + 2$

12. $f(x) = 3x - 5$

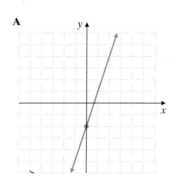

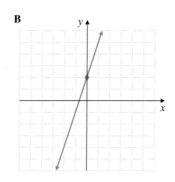

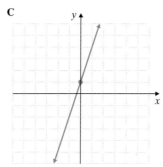

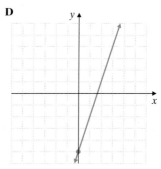

Find an equation of the line satisfying each set of conditions. Write the equations in the form $x = a, y = b,$ *or* $y = mx + b.$

13. Horizontal; through $(2, -8)$

14. Vertical; through $(-4, -3)$

15. Perpendicular to $x = 5$; through $(3, -2)$

16. Through $(4, -1)$; slope -3

17. Through $(0, -2)$; slope 5

18. Through $(4, -2)$ and $(6, -3)$

19. Through $(-1, 2)$; perpendicular to $3x - y = 4$

20. Parallel to $2y + x = 3$; through $(3, -2)$

Name _____ Section _____ Date _____

Chapter 3 Test

1. Plot the points, and name the quadrant in which each is located:
$A(6, -2), B(4, 0), C(-1, 6)$.

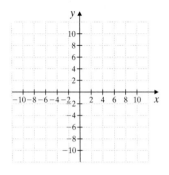

2. Create a scatter diagram for the paired data.

U.S. Cellular Phone Subscribers

Year	Number (in millions)
1994	24
1995	34
1996	44
1997	55
1998	69
1999	86
2000	109

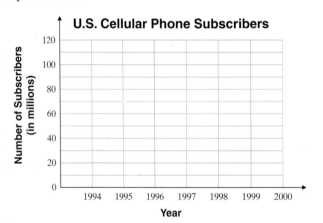

Graph each linear equation.

3. $-3x + y = -3$

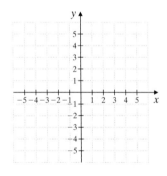

4. $2x - 3y = -6$

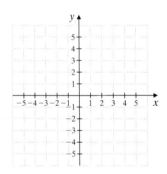

5. $4x + 6y = 8$

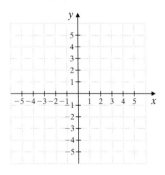

6. $y = -3$

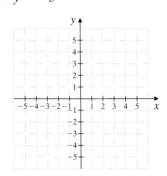

21. Line L_1 has the equation $2x - 5y = 8$. Line L_2 passes through the points $(1, 4)$ and $(-1, -1)$. Determine whether these lines are parallel, perpendicular, or neither.

22. For the 2000 Major League Baseball season, the following linear equation describes the relationship between a team's opening-day player payroll x (in millions of dollars) and the number of games y that team won during the regular season: $y = 0.154x + 72.379$. Round answers to the nearest wholes. (*Sources*: Based on data from The Associated Press and Major League Baseball)

a. According to this equation, how many games would have been won during the 2000 season by a team with an opening-day payroll of $60 million?

b. The Arizona Diamondbacks had an opening-day payroll of $81 million in 2000. According to this equation, how many games would they have won during the season?

c. According to this equation, what opening-day payroll would have been necessary in 2000 to have won 90 games during the season?

d. Find and interpret the slope of the equation.

Graph each inequality.

23. $x \leq -4$

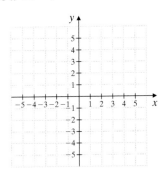

24. $2x - y > 5$

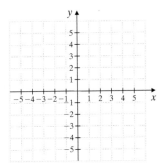

25. The intersection of $2x + 4y < 6$ and $y \leq -4$

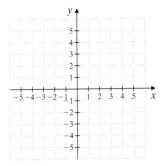

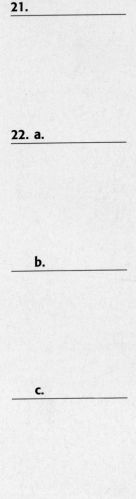

21. _____

22. a. _____

b. _____

c. _____

d. _____

23. see graph

24. see graph

25. see graph

253

26. _____

27. _____

28. _____

29. _____

Find the domain and range of each relation. Also determine whether the relation is a function.

26.

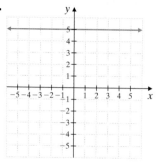

27.

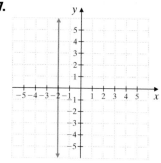

28.

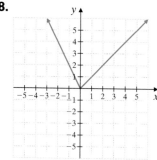

29.

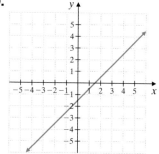

STUDY SKILLS REMINDER

How are your homework assignments going?

It is so important in mathematics to keep up with homework. Why? Many concepts build on each other. Often, your understanding of a day's lecture in mathematics depends on an understanding of the previous day's material.

Remember that completing your homework assignment involves a lot more than attempting a few of the problems assigned.

To complete a homework assignment, remember these four things:

1. Attempt all of it.

2. Check it.

3. Correct it.

4. If needed, ask questions about it.

Cumulative Review

Write each set in roster form.

1. $\{x|x$ is a whole number between 1 and 6$\}$

2. $\{x|x$ is a natural number greater than 100$\}$

Find the reciprocal, or multiplicative inverse, of each number.

3. -9

4. $\dfrac{7}{4}$

Find each absolute value.

5. $|3|$

6. $-|2|$

7. Simplify: $3 - [(4 - 6) + 2(5 - 9)]$

Use the product rule to simplify.

8. $2^2 \cdot 2^5$

9. $y \cdot y^2 \cdot y^4$

Use the power rules to simplify. Write each result using positive exponents.

10. $(5x^2)^3$

11. $\left(\dfrac{2^{-3}}{y}\right)^{-2}$

12. Solve: $0.6 = 2 - 3.5c$

13. Find two numbers such that the second number is 3 more than twice the first number and the sum of the two numbers is 72.

14. Marial Callier just received an inheritance of $10,000 and plans to place all the money in a savings account that pays 5% compounded quarterly to help her son go to college in 3 years. How much money will be in the account in 3 years?

Answers

1. _____

2. _____

3. _____

4. _____

5. _____

6. _____

7. _____

8. _____

9. _____

10. _____

11. _____

12. _____

13. _____

14. _____

15. _____

16. _____

17. _____

18. _____

19. _____

20. see graph _____

21. _____

22. _____

23. _____

24. _____

25. _____

Graph each set on a number line and then write it in interval notation.

15. $\{x|x \geq 2\}$

16. $\{x|0.5 < x \leq 3\}$

17. Solve: $x - 7 < 2$ and $2x + 1 < 9$

18. Solve: $|x - 3| = |5 - x|$

19. Solve: $|x - 3| > 7$

20. Find the intercepts and graph:
$x + 4y = -4$

21. Find the slope of the line $y = 3x + 2$.

22. Write an equation of the line with y-intercept $(0, -3)$ and slope of $\dfrac{1}{4}$.

23. Find an equation of the horizontal line containing the point $(2, 3)$.

Find the following.

24. If $f(x) = 7x^2 - 3x + 1$, find $f(1)$.

25. If $g(x) = 3x - 2$, find $g(0)$.

Systems of Equations and Inequalities

In this chapter, two or more equations in two or more variables are solved simultaneously. Such a collection of equations is called a **system of equations**. Systems of equations are good mathematical models for many real-world problems because these problems may involve several related patterns. We will study various methods for solving systems of equations and will conclude with a look at systems of inequalities.

4.1 Solving Systems of Linear Equations in Two Variables

4.2 Solving Systems of Linear Equations in Three Variables

4.3 Systems of Linear Equations and Problem Solving

Integrated Review—Systems of Linear Equations

4.4 Solving Systems of Equations Using Matrices

4.5 Systems of Linear Inequalities

Lightning, most often produced during thunderstorms, is a rapid discharge of high-current electricity into the atmosphere. At any given moment around the world, there are about 2000 thunderstorms in progress producing approximately 100 lightning flashes per second. In the United States, lightning causes an average of 75 fatalities per year. An estimated 5% of all residential insurance claims in the United States are due to lightning damage, totaling more than $1 billion per year. In addition, roughly 30% of all power outages in the United States are lightning related. Because of lightning's potentially destructive nature, meteorologists track lightning activity by recording and plotting the positions of lightning strikes. In the Chapter Activity on page 310, we will see how systems of equations can be used to pinpoint the location of a lightning strike.

Name _____ Section _____ Date _____

1. _____

2. _____

3. _____

4. _____

5. _____

6. _____

7. _____

8. _____

9. _____

10. _____

11. _____

12. _____

13. see graph _____

14. see graph _____

15. _____

16. _____

Chapter 4 Pretest

Solve each system of equations by graphing.

1. $\begin{cases} 2x + y = -1 \\ x + y = -3 \end{cases}$

2. $\begin{cases} x - y = 2 \\ 3x - y = 0 \end{cases}$

Solve each system of equations by the substitution method.

3. $\begin{cases} 2x - 3y = -8 \\ y = x + 1 \end{cases}$

4. $\begin{cases} \dfrac{x}{6} + \dfrac{y}{2} = -2 \\ x - \dfrac{y}{2} = 2 \end{cases}$

Solve each system of equations by the elimination method.

5. $\begin{cases} x - 4y = -29 \\ 5x + y = 44 \end{cases}$

6. $\begin{cases} 2x + 3y = -34 \\ 3x - 5y = 25 \end{cases}$

Solve each system.

7. $\begin{cases} x - y = 3 \\ 7y = -7 \\ 2x + y + 3z = -6 \end{cases}$

8. $\begin{cases} 3x - y + 2z = -1 \\ 2x + 5y - z = 4 \\ 4x + 6y + z = 2 \end{cases}$

Use matrices to solve each system.

9. $\begin{cases} 3x + 4y = -10 \\ x - y = 20 \end{cases}$

10. $\begin{cases} -2x + y = 6 \\ 4x - 2y = 12 \end{cases}$

11. $\begin{cases} 3x + 2z = -5 \\ x + y + z = -2 \\ -x + 2y - 3z = -5 \end{cases}$

12. $\begin{cases} x - 6y + z = -48 \\ 5x + y + 3z = 8 \\ 2x - y - z = -8 \end{cases}$

Graph the solution of each system of linear inequalities.

13. $\begin{cases} y \le x + 1 \\ y > 3x - 2 \end{cases}$

14. $\begin{cases} -6x + 3y \ge 0 \\ y \le 3 \end{cases}$

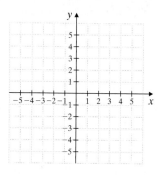

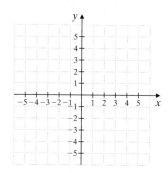

Solve

15. Six times one number minus a second is 12, and the sum of the numbers is 16. Find the numbers.

16. The measure of the largest angle of a triangle is five times the measure of the smallest angle, and the measure of the remaining angle is 40° more than the measure of the smallest angle. Find the measure of each angle.

4.1 Solving Systems of Linear Equations in Two Variables

Recall from Chapter 3 that the graph of a linear equation in two variables is a line. Two or more linear equations form a **system of linear equations**. Some examples of systems of linear equations in two variables follow.

$$\begin{cases} x - 2y = -7 \\ 3x + y = 0 \end{cases} \quad \begin{cases} x = 5 \\ x + \dfrac{y}{2} = 9 \end{cases} \quad \begin{cases} x - 3 = 2y + 6 \\ y = 1 \end{cases}$$

OBJECTIVES

- **A** Determine whether an ordered pair is a solution of a system of two linear equations.
- **B** Solve a system of two equations by graphing.
- **C** Solve a system using substitution.
- **D** Solve a system using elimination.

SSM   TUTOR CENTER   SG   CD & VIDEO   MATH PRO   WEB

A Determining Whether an Ordered Pair Is a Solution

Recall that a solution of an equation in two variables is an ordered pair (x, y) that makes the equation true. A **solution of a system** of two equations in two variables is an ordered pair (x, y) that makes both equations true.

EXAMPLE 1

Determine whether the ordered pair $(-1, 1)$ is a solution of the system.

$$\begin{cases} -x + y = 2 \\ 2x - y = -3 \end{cases}$$

Solution: We replace x with -1 and y with 1 in each equation.

$$\begin{array}{ll} -x + y = 2 & \text{First equation} \\ -(-1) + (1) = 2 & \text{Let } x = -1 \text{ and } y = -1. \\ 1 + 1 = 2 & \\ 2 = 2 & \text{True.} \end{array}$$

$$\begin{array}{ll} 2x - y = -3 & \text{Second equation} \\ 2(-1) - (1) = -3 & \text{Let } x = -1 \text{ and } y = -1. \\ -2 - 1 = -3 & \\ -3 = -3 & \text{True.} \end{array}$$

Since $(-1, 1)$ makes both equations true, it is a solution. ●

EXAMPLE 2

Determine whether the ordered pair $(-2, 3)$ is a solution of the system.

$$\begin{cases} 5x + 3y = -1 \\ x - y = 1 \end{cases}$$

Solution: We replace x with -2 and y with 3 in each equation.

$$\begin{array}{ll} 5x + 3y = -1 & \text{First equation} \\ 5(-2) + 3(3) = -1 & \text{Let } x = -2 \text{ and } y = -3. \\ -10 + 9 = -1 & \\ -1 = -1 & \text{True.} \end{array}$$

$$\begin{array}{ll} x - y = 1 & \text{Second equation} \\ (-2) - (3) = 1) & \text{Let } x = -2 \text{ and } y = 3. \\ -5 = 1 & \text{False.} \end{array}$$

Since the ordered pair $(-2, 3)$ does not make both equations true, it is not a solution of the system.

Practice Problem 1

Determine whether the ordered pair $(4, 1)$ is a solution of the system.

$$\begin{cases} x - y = 3 \\ 2x - 3y = 5 \end{cases}$$

Practice Problem 2

Determine whether the ordered pair $(-3, 3)$ is a solution of the system.

$$\begin{cases} 3x - y = -12 \\ x - y = 0 \end{cases}$$

Answers

1. Yes, a solution,   **2.** No, not a solution

Practice Problems 3–4

Solve each system by graphing. If the system has just one solution, estimate the solution.

3. $\begin{cases} x - y = 2 \\ x + 3y = 6 \end{cases}$

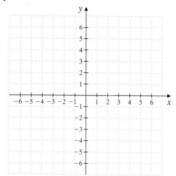

4. $\begin{cases} y = -3x \\ 6x + 2y = 4 \end{cases}$

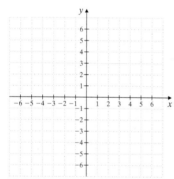

Answers

3. $(3, 1)$,

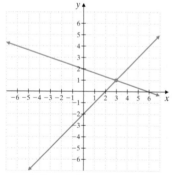

4. no solution, or $\varnothing$

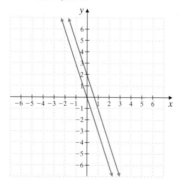

B Solving a System by Graphing

The graph of each linear equation in a system is a line. Each point on each line corresponds to an ordered pair solution of its equation. If the lines intersect, the point of intersection lies on both lines and corresponds to an ordered pair solution of both equations. In other words, the point of intersection corresponds to an ordered pair solution of the system. Therefore, we can estimate the solutions of a system by graphing the equations on the same rectangular coordinate system and estimating the coordinates of any points of intersection.

EXAMPLE 3

Solve the system by graphing. If the system has just one solution, estimate the solution.

$$\begin{cases} x + y = 2 \\ 3x - y = -2 \end{cases}$$

Solution: First we graph the linear equations on the same rectangular coordinate system.

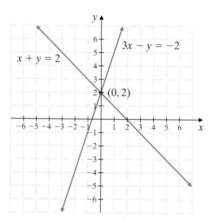

These lines intersect at one point as shown. The coordinates of the point of intersection appear to be $(0, 2)$. We check this estimated solution by replacing x with 0 and y with 2 in *both* equations.

$x + y = 2$	First equation	$3x - y = -2$	Second equation
$0 + 2 = 2$	Let $x = 0$ and $y = 2$.	$3(0) - 2 = -2$	Let $x = 0$ and $y = 2$.
$2 = 2$	True.	$-2 = -2$	True.

The ordered pair $(0, 2)$ is the solution of the system. A system that has at least one solution, such as this one, is said to be **consistent**. ●

EXAMPLE 4

Solve the system by graphing. If the system has just one solution, estimate the solution.

$$\begin{cases} x - 2y = 4 \\ x \quad\;\; = 2y \end{cases}$$

Solution: We graph each linear equation.

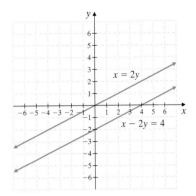

The lines appear to be parallel. To be sure, let's write each equation in slope-intercept form, $y = mx + b$. To do so, we solve for y.

$x - 2y = 4$ First equation

$-2y = -x + 4$ Substract x from both sides.

$y = \dfrac{1}{2}x - 2$ Divide both sides by -2.

$x = 2y$ Second equation

$\dfrac{1}{2}x = y$ Divide both sides by 2.

$y = \dfrac{1}{2}x$

The graphs of these equations have the same slope, $\dfrac{1}{2}$, but different y-intercepts, so these lines are parallel. Therefore, the system has no solution since the equations have no common solution (there are no intersection points). A system that has no solution is said to be **inconsistent**. ●

> **Helpful Hint**
>
> - If a system of equations has *at least one solution*, the system is *consistent*.
> - If a system of equations has *no solution*, the system is *inconsistent*.

The pairs of equations in Example 3 and 4 are called independent because their graphs differ. In Example 5, we see an example of dependent equations.

Try the Concept Check in the margin.

EXAMPLE 5

Solve the system by graphing. If the system has just one solution, estimate the solution.

$$\begin{cases} 2x + 4y = 10 \\ x + 2y = 5 \end{cases}$$

Solution: We graph each linear equation.

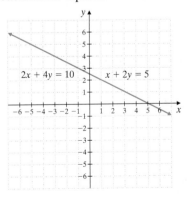

Concept Check

How can you tell just by looking at the following system that it has no solution?

$$\begin{cases} y = 3x + 5 \\ y = 3x - 7 \end{cases}$$

Practice Problem 5

Solve the system by graphing.

$$\begin{cases} -2x + y = 1 \\ 4x - 2y = -2 \end{cases}$$

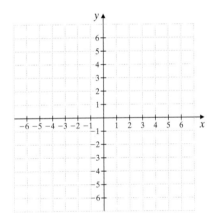

Answers

5.

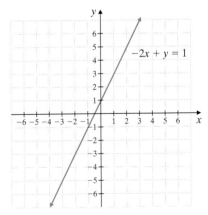

Concept Check: answers may vary

We see that the graphs of the equations are the same line. To confirm this, notice that if both sides of the second equation are multiplied by 2, the result is the first equation. This means that the equations have identical solutions. Any ordered pair solution of one equation satisfies the other equation also. These equations are said to be **dependent equations**. The solution set of the system is $\{(x, y)|x + 2y = 5\}$ or, equivalently, $\{(x, y)|2x + 4y = 10\}$ since the lines describe identical ordered pairs. Written the second way, the solution set is read "the set of all ordered pairs (x, y), such that $2x + 4y = 10$." There are an infinite number of solutions to this system.   ●

Helpful Hint

- If the graphs of two equations *differ*, they are *independent* equations.
- If the graphs of two equations are the *same*, they are *dependent* equations.

Concept Check

How can you tell just by looking at the following system that it has infinitely many solutions?

$$\begin{cases} x + y = 5 \\ 2x + 2y = 10 \end{cases}$$

Try the Concept Check in the margin.

We can summarize the information discovered in Examples 3 through 5 as follows.

Possible Solutions to Systems of Two Linear Equations

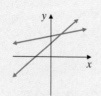

One solution:
Independent equations
Consistent system

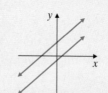

No solution:
Independent equations
Inconsistent system

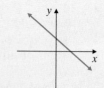

Infinite number of solutions:
Dependent equations
Consistent system

C   Solving a System Using Substitution

Graphing the equations of a system by hand is often a good method for finding approximate solutions of a system, but it is not a reliable method for finding exact solutions. To find an exact solution, we need to use *algebra*. One such *algebraic* method is called the **substitution method**.

Solving a System of Two Equations Using the Substitution Method

Step 1. Solve one of the equations for one of its variables.

Step 2. Substitute the expression for the variable found in Step 1 into the other equation.

Step 3. Find the value of one variable by solving the equation from Step 2.

Step 4. Find the value of the other variable by substituting the value found in Step 3 into the equation from Step 1.

Step 5. Check the ordered pair solution in *both* original equations.

Answer

Concept Check: answers may vary

EXAMPLE 6   Use the substitution method to solve the system:

$$\begin{cases} 2x + 4y = -6 & \text{First equation} \\ x = 2y - 5 & \text{Second equation} \end{cases}$$

Solution: In the second equation, we are told that x is equal to $2y - 5$. Since they are equal, we can *substitute* $2y - 5$ for x in the first equation. This will give us an equation in one variable, which we can solve for y.

$$2x + 4y = -6 \qquad \text{First equation}$$

$$2(2y - 5) + 4y = -6 \qquad \text{Substitute } 2y - 5 \text{ for } x.$$

$$4y - 10 + 4y = -6$$

$$8y = 4$$

$$y = \frac{4}{8} = \frac{1}{2} \qquad \text{Solve for } y.$$

The y-coordinate of the solution is $\frac{1}{2}$. To find the x-coordinate, we replace y with $\frac{1}{2}$ in the second equation,

$$x = 2y - 5$$

$$x = 2y - 5$$

$$x = 2\left(\frac{1}{2}\right) - 5 = 1 - 5 = -4$$

The ordered pair solution is $\left(-4, \frac{1}{2}\right)$. Check to see that $\left(-4, \frac{1}{2}\right)$ satisfies both equations of the system.   ●

EXAMPLE 7   Use the substitution method to solve the system:

$$\begin{cases} -\dfrac{x}{6} + \dfrac{y}{2} = \dfrac{1}{2} \\ \dfrac{x}{3} - \dfrac{y}{6} = -\dfrac{3}{4} \end{cases}$$

Solution: First we multiply each equation by its least common denominator to clear the system of fractions. We multiply the first equation by 6 and the second equation by 12.

$$\begin{cases} 6\left(-\dfrac{x}{6} + \dfrac{y}{2}\right) = 6\left(\dfrac{1}{2}\right) \\ 12\left(\dfrac{x}{3} - \dfrac{y}{6}\right) = 12\left(-\dfrac{3}{4}\right) \end{cases} \text{ simplifies to } \begin{cases} -x + 3y = 3 & \text{First equation} \\ 4x - 2y = -9 & \text{Second equation} \end{cases}$$

We now solve the first equation for x.

$$-x + 3y = 3 \qquad \text{First equation}$$

$$3y - 3 = x \qquad \text{Solve for } x.$$

Practice Problem 6

Use the substitution method to solve the system:

$$\begin{cases} 6x - 4y = 10 \\ y = 3x - 3 \end{cases}$$

Practice Problem 7

Use the substitution method to solve the

system: $\begin{cases} -\dfrac{x}{2} + \dfrac{y}{4} = \dfrac{1}{2} \\ \dfrac{x}{2} + \dfrac{y}{2} = -\dfrac{1}{8} \end{cases}$

Answers

6. $\left(\dfrac{1}{3}, -2\right)$,   **7.** $\left(-\dfrac{3}{4}, \dfrac{1}{2}\right)$

Next we replace x with $3y - 3$ in the second equation.

$$4x - 2y = -9 \qquad \text{Second equation}$$

$$4(3y - 3) - 2y = -9$$

$$12y - 12 - 2y = -9$$

$$10y = 3$$

$$y = \frac{3}{10} \qquad \text{Solve for } y.$$

The y-coordinate is $\dfrac{3}{10}$. To find the x-coordinate, we replace y with $\dfrac{3}{10}$ in the equation $x = 3y - 3$. Then

$$x = 3\left(\frac{3}{10}\right) - 3 = \frac{9}{10} - 3 = \frac{9}{10} - \frac{30}{10} = -\frac{21}{10}$$

The ordered pair solution is $\left(-\dfrac{21}{10}, \dfrac{3}{10}\right)$. Check to see that this solution satisfies both original equations. ●

> **Helpful Hint**
>
> If a system of equations contains equations with fractions, the first step is to clear the equations of fractions.

Ⓓ Solving a System Using Elimination

The **elimination method**, or **addition method**, is a second algebraic technique for solving systems of equations. For this method, we rely on a version of the addition property of equality, which states that "equals added to equals are equal."

If $A = B$ and $C = D$ then $A + C = B + D$

Solving a System of Two Linear Equations Using the Elimination Method

Step 1. Rewrite each equation in standard form, $Ax + By = C$.

Step 2. If necessary, multiply one or both equations by some nonzero number so that the coefficient of one variable in one equation is the opposite of the coefficient of that variable in the other equation.

Step 3. Add the equations. Your chosen variable should be eliminated.

Step 4. Find the value of the remaining variable by solving the equation from Step 3.

Step 5. Find the value of the other variable by substituting the value found in Step 4 into either original equation.

Step 6. Check the proposed ordered pair solution in *both* original equations.

EXAMPLE 8   Use the elimination method to solve the system:

$$\begin{cases} x - 5y = -12 & \text{First equation} \\ -x + y = 4 & \text{Second equation} \end{cases}$$

Solution: Since the left side of each equation is equal to the right side, we add equal quantities by adding the left sides of the equations and the right sides of the equations. This sum gives us an equation in one variable, y, which we can solve for y.

$$\begin{aligned} x - 5y &= -12 & \text{First equation} \\ \underline{-x + y = 4} & & \text{Second equation} \\ -4y &= -8 & \text{Add.} \\ y &= 2 & \text{Solve for } y. \end{aligned}$$

The y-coordinate of the solution is 2. To find the corresponding x-coordinate, we replace y with 2 in either original equation of the system. Let's use the second equation.

$$\begin{aligned} -x + y &= 4 & \text{Second equation} \\ -x + 2 &= 4 & \text{Let } y = 2. \\ -x &= 2 \\ x &= -2 \end{aligned}$$

The ordered pair solution is $(-2, 2)$. Check to see that $(-2, 2)$ satisfies both equations of the system. ●

EXAMPLE 9   Use the elimination method to solve the system:

$$\begin{cases} 3x + \dfrac{y}{2} = 2 \\ 6x + y = 5 \end{cases}$$

Solution: If we add the two equations, the sum will still be an equation in two variables. Notice, however, that if we multiply both sides of the first equation by -2, the coefficients of x in the two equations will be opposites. Then

$$\begin{cases} -2\left(3x + \dfrac{y}{2}\right) = -2(2) \\ 6x + y = 5 \end{cases} \quad \text{simplifies to} \quad \begin{cases} -6x - y = -4 \\ 6x + y = 5 \end{cases}$$

Now we can add the left sides and add the right sides.

$$\begin{aligned} -6x - y &= -4 \\ \underline{6x + y = 5} \\ 0 &= 1 \qquad \text{False.} \end{aligned}$$

The resulting equation, $0 = 1$, is false for all values of y or x. Thus, the system has no solution. The solution set is $\{\ \}$ or $\varnothing$. This system is inconsistent, and the graphs of the equations are parallel lines.

Practice Problem 8

Use the elimination method to solve the system: $\begin{cases} 3x - y = 1 \\ 4x + y = 6 \end{cases}$

Practice Problem 9

Use the elimination method to solve the system: $\begin{cases} \dfrac{x}{3} + 2y = -1 \\ x + 6y = 2 \end{cases}$

Answers

● **8.** $(1, 2)$,   **9.** no solution or $\varnothing$

Practice Problem 10

Use the elimination method to solve the system: $\begin{cases} 2x - 5y = 6 \\ 3x - 4y = 9 \end{cases}$

EXAMPLE 10 Use the elimination method to solve the system:

$$\begin{cases} 3x - 2y = 10 \\ 4x - 3y = 15 \end{cases}$$

Solution: To eliminate y, our first step is to multiply both sides of the first equation by 3 and both sides of the second equation by -2. Then

$$\begin{cases} 3(3x - 2y) = 3(10) \\ -2(4x - 3y) = -2(15) \end{cases} \text{ simplifies to } \begin{cases} 9x - 6y = 30 \\ -8x + 6y = -30 \end{cases}$$

Next we add the left sides and add the right sides.

$$\begin{array}{r} 9x - 6y = 30 \\ -8x + 6y = -30 \\ \hline x \quad\quad = 0 \end{array}$$

To find y, we let $x = 0$ in either equation of the system

$$\begin{array}{ll} 3x - 2y = 10 & \text{First equation} \\ 3(0) - 2y = 10 & \text{Let } x = 0. \\ -2y = 10 & \\ y = -5 & \end{array}$$

The ordered pair solution is $(0, -5)$. Check to see that $(0, -5)$ satisfies both equations. ●

Practice Problem 11

Use the elimination method to solve the system: $\begin{cases} 4x - 7y = 10 \\ -8x + 14y = -20 \end{cases}$

EXAMPLE 11 Use the elimination method to solve the system:

$$\begin{cases} -5x - 3y = 9 \\ 10x + 6y = -18 \end{cases}$$

Solution: To eliminate x, our first step is to multiply both sides of the first equation by 2. Then

$$\begin{cases} 2(-5x - 3y) = 2(9) \\ 10x + 6y = -18 \end{cases} \text{ simplifies to } \begin{cases} -10x - 6y = 18 \\ 10x + 6y = -18 \end{cases}$$

Next we add the equations.

$$\begin{array}{r} -10x - 6y = 18 \\ 10x + 6y = -18 \\ \hline 0 = 0 \end{array}$$

The resulting equation, $0 = 0$, is true for all possible values of y or x. Notice in the original system that if both sides of the first equation are multiplied by -2, the result is the second equation. This means that the two equations are equivalent. They have the same solution set and there are an infinite number of solutions. Thus, the equations of this system are dependent, and the solution set of the system is

$$\{(x, y) \mid -5x - 3y = 9\}$$

or, equivalently,

$$\{(x, y) \mid 10x + 6y = -18\}$$

Answers

10. $(3, 0)$, **11.** $\{(x, y) \mid 4x - 7y = 10\}$

GRAPHING CALCULATOR EXPLORATIONS

We may use a grapher to approximate solutions of systems of equations by graphing both equations on the same set of axes and approximating any points of intersection. For example, let's approximate the solution of the system

$$\begin{cases} y = -2.6x + 5.6 \\ y = 4.3x - 4.9 \end{cases}$$

We use a standard window and graph the equations on a single screen.

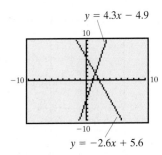

The two lines intersect. To approximate the point of intersection, we trace to the point of intersection and use an INTERSECT feature of the grapher, a ZOOM IN feature of the grapher, or redefine the window to $[0, 3]$ by $[0, 3]$. If we redefine the window to $[0, 3]$ by $[0, 3]$, the screen should look like the following:

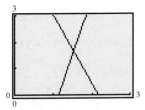

By tracing along the curves, we can see that the point of intersection has an x-value between 1.5 and 1.532. We can continue to zoom and trace or redefine the window until the coordinates of the point of intersection can be determined to the nearest hundredth. The approximate point of intersection is $(1.52, 1.64)$.

Solve each system of equations. Approximate each solution to two decimal places.

1. $y = -1.65x + 3.65$
$y = 4.56x - 9.44$

2. $y = 7.61x + 3.48$
$y = -1.26x - 6.43$

3. $2.33x - 4.72y = 10.61$
$5.86x - 6.22y = -8.89$

4. $-7.89x - 5.68y = 3.26$
$-3.65x + 4.98y = 11.77$

FOCUS ON **The Real World**

ANOTHER MATHEMATICAL MODEL

Sometimes mathematical models other than linear models are appropriate for data. Suppose that an equation of the form $y = ax^2 + bx + c$ is an appropriate model for the ordered pairs $(x_1, y_1), (x_2, y_2),$ and (x_3, y_3). Then it is necessary to find the values of $a, b,$ and c such that the given ordered pairs are solutions of the equation $y = ax^2 + bx + c$. To do so, substitute each ordered pair into the equation. Each time, the result is an equation in three unknowns: $a, b,$ and c. Solving the resulting system of three linear equation in three unknowns will give the required values of $a, b,$ and c.

GROUP ACTIVITY

1. The table gives the total beef supply (in billions of pounds) in the United States in each of the years listed.

   a. Write the data as ordered pairs of the form (x, y), where y is the beef supply (in billions of pounds) in the year x ($x = 0$ represents 1998).

   b. Find the values of a, b, and c such that the equation $y = ax^2 + bx + c$ models this data.

   c. Verify that the model you found in part (b) gives each of the ordered pair solutions from part (a).

   d. According to the model, what was the U.S. beef supply in 2001?

Total U.S. Beef Supply

Year	Beef Supply (billions of pounds)
1998	28.9
2000	30.3
2002	28.8

(*Source:* Economic Research Service, U.S. Department of Agriculture)

2. The table gives Saab production figures for each of the years listed.

   a. Write the data as ordered pairs of the form (x, y), where y is Saab production in the year x ($x = 0$ represents 1996).

   b. Find the values of a, b, and c such that the equation $y = ax^2 + bx + c$ models this data.

   c. According to the model, what was the total Saab production in 2001?

Total Saab Production

Year	Number of Saab Vehicles Produced
1996	28,439
1998	30,756
2000	39,479

(*Source:* Automotive Intelligence, www.autointell.com)

3. a. Make up an equation of the form $y = ax^2 + bx + c$.

   b. Find three ordered pair solutions of the equation.

   c. Without revealing your equation from part (a), exchange lists of ordered pair solutions with another group.

   d. Use the method described above to find the values of $a, b,$ and c such that the equation $y = ax^2 + bx + c$ has the ordered pair solutions you received from the other group.

   e. Check with the other group to see if your equation from part (d) is the correct one.

Name _____ Section _____ Date _____

Mental Math

Match each graph with the solution of the corresponding system.

A 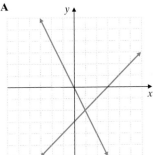 **B** 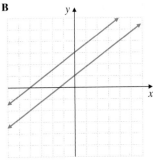 **C** 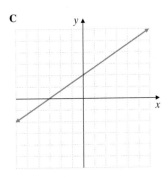 **D**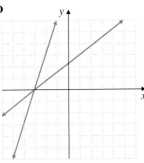

1. No solution

2. Infinite number of solutions

3. $(1, -2)$

4. $(-3, 0)$

EXERCISE SET 4.1

Ⓐ *Determine whether the given ordered pair is a solution of the system. See Examples 1 and 2.*

1. $\begin{cases} x - y = 3 \\ 2x - 4y = 8 \end{cases}$ $(2, -1)$

2. $\begin{cases} x - y = -4 \\ 2x + 10y = 4 \end{cases}$ $(-3, 1)$

3. $\begin{cases} 2x - 3y = -9 \\ 4x + 2y = -2 \end{cases}$ $(3, 5)$

4. $\begin{cases} 2x - 5y = -2 \\ 3x + 4y = 4 \end{cases}$ $(4, 2)$

5. $\begin{cases} y = -5x \\ x = -2 \end{cases}$ $(-2, 10)$

6. $\begin{cases} y = 6 \\ x = -2y \end{cases}$ $(-12, 6)$

Ⓑ *Solve each system by graphing. See Examples 3 through 5.*

7. $\begin{cases} x + y = 1 \\ x - 2y = 4 \end{cases}$

8. $\begin{cases} 2x - y = 8 \\ x + 3y = 11 \end{cases}$

9. $\begin{cases} 2y - 4 = 0 \\ x + 2y = 5 \end{cases}$

10. $\begin{cases} 4x - y = 6 \\ x - y = 0 \end{cases}$

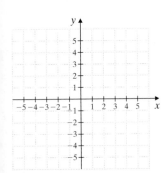 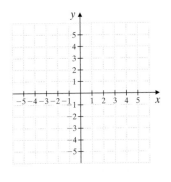 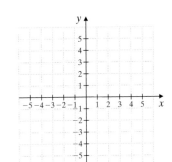 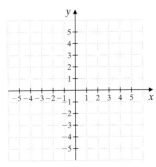

269

11. $\begin{cases} 3x - y = 4 \\ 6x - 2y = 4 \end{cases}$

12. $\begin{cases} -x + 3y = 6 \\ 3x - 9y = 9 \end{cases}$

13. $\begin{cases} y = -3x \\ 2x - y = -5 \end{cases}$

14. $\begin{cases} y = -2x \\ -3x + y = 10 \end{cases}$

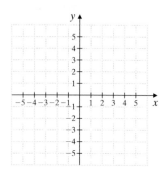
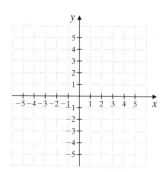
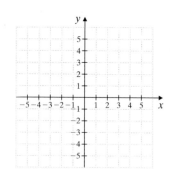
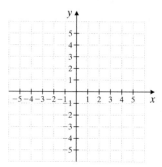

C *Use the substitution method to solve each system of equations. See Examples 6 and 7.*

15. $\begin{cases} x + y = 10 \\ y = 4x \end{cases}$

16. $\begin{cases} 5x + 2y = -17 \\ x = 3y \end{cases}$

17. $\begin{cases} 4x - y = 9 \\ 2x + 3y = -27 \end{cases}$

18. $\begin{cases} 3x - y = 6 \\ -4x + 2y = -8 \end{cases}$

19. $\begin{cases} \dfrac{1}{2}x + \dfrac{3}{4}y = -\dfrac{1}{4} \\ \dfrac{3}{4}x - \dfrac{1}{4}y = 1 \end{cases}$

20. $\begin{cases} \dfrac{2}{5}x + \dfrac{1}{5}y = -1 \\ x + \dfrac{2}{5}y = -\dfrac{8}{5} \end{cases}$

21. $\begin{cases} \dfrac{x}{3} + y = \dfrac{4}{3} \\ -x + 2y = 11 \end{cases}$

22. $\begin{cases} \dfrac{x}{8} - \dfrac{y}{2} = 1 \\ \dfrac{x}{3} - y = 2 \end{cases}$

23. $\begin{cases} 2x - y = -1 \\ y = -2x \end{cases}$

24. $\begin{cases} x = \dfrac{1}{5}y \\ x - y = -4 \end{cases}$

25. $\begin{cases} 2x = 6 \\ y = 5 - x \end{cases}$

26. $\begin{cases} x = 3y + 4 \\ -y = 5 \end{cases}$

D *Use the elimination method to solve each system of equations. See Examples 8 through 11.*

27. $\begin{cases} 2x - 4y = 0 \\ x + 2y = 5 \end{cases}$

28. $\begin{cases} 2x - 3y = 0 \\ 2x + 6y = 3 \end{cases}$

29. $\begin{cases} 5x + 2y = 1 \\ x - 3y = 7 \end{cases}$

30. $\begin{cases} 6x - y = -5 \\ 4x - 2y = 6 \end{cases}$

31. $\begin{cases} 5x - 2y = 27 \\ -3x + 5y = 18 \end{cases}$

32. $\begin{cases} 3x + 4y = 2 \\ 2x + 5y = -1 \end{cases}$

33. $\begin{cases} 3x - 5y = 11 \\ 2x - 6y = 2 \end{cases}$

34. $\begin{cases} 6x - 3y = -3 \\ 4x + 5y = -9 \end{cases}$

35. $\begin{cases} x - 2y = 4 \\ 2x - 4y = 4 \end{cases}$

36. $\begin{cases} -x + 3y = 6 \\ 3x - 9y = 9 \end{cases}$

37. $\begin{cases} 3x + y = 1 \\ 2y = 2 - 6x \end{cases}$

38. $\begin{cases} y = 2x - 5 \\ 8x - 4y = 20 \end{cases}$

39. $\begin{cases} x = 3y - 2 \\ 5x - 15y = 0 \end{cases}$

40. $\begin{cases} x = 3y - 1 \\ 2x - 6y = -2 \end{cases}$

41. $\begin{cases} 4x + 2y = 5 \\ 2x + y = -1 \end{cases}$

42. $\begin{cases} 3x + 6y = 15 \\ 2x + 4y = 3 \end{cases}$

43. $\begin{cases} \dfrac{3}{4}x + \dfrac{5}{2}y = 11 \\ \dfrac{1}{16}x - \dfrac{3}{4}y = -1 \end{cases}$

44. $\begin{cases} \dfrac{2}{3}x + \dfrac{1}{4}y = -\dfrac{3}{2} \\ \dfrac{1}{2}x - \dfrac{1}{4}y = -2 \end{cases}$

45. $\begin{cases} \dfrac{2}{3}x - \dfrac{3}{4}y = -1 \\ -\dfrac{1}{6}x + \dfrac{3}{8}y = 1 \end{cases}$

46. $\begin{cases} \dfrac{1}{2}x - \dfrac{1}{3}y = -3 \\ \dfrac{1}{8}x + \dfrac{1}{6}y = 0 \end{cases}$

47. $\begin{cases} 0.7x - 0.2y = -1.6 \\ 0.2x - y = -1.4 \end{cases}$

48. $\begin{cases} -0.7x + 0.6y = 1.3 \\ 0.5x - 0.3y = -0.8 \end{cases}$

49. $\begin{cases} 10y - 2x = 1 \\ 5y = 4 - 6x \end{cases}$

50. $\begin{cases} 3x + 4y = 0 \\ 7x = 3y \end{cases}$

51. $\begin{cases} x = 3y + 2 \\ 5x - 15y = 10 \end{cases}$

52. $\begin{cases} y = \dfrac{1}{7}x + 3 \\ x - 7y = -21 \end{cases}$

Review and Preview

Determine whether the given replacement values make each equation true or false. See Section 2.1.

53. $3x - 4y + 2y = 5$;
$x = 1$, $y = 2$, and $z = 5$

54. $x + 2y - z = 7$;
$x = 2$, $y = -3$, and $z = 3$

55. $-x - 5y + 3z = 15$;
$x = 0$, $y = -1$, and $z = 5$

56. $-4x + y - 8z = 4$;
$x = 1$, $y = 0$, and $z = -1$

Add the equations. See this section.

57. $\begin{cases} 3x + 2y - 5z = 10 \\ -3x + 4y + z = 15 \end{cases}$

58. $\begin{cases} x + 4y - 5z = 20 \\ 2x - 4y - 2z = -17 \end{cases}$

59. $\begin{cases} 10x + 5y + 6z = 14 \\ -9x + 5y - 6z = -12 \end{cases}$

60. $\begin{cases} -9x - 8y - z = 31 \\ 9x + 4y - z = 12 \end{cases}$

Combining Concepts

61. Can a system consisting of two linear equations have exactly two solutions? Explain why or why not.

62. Suppose the graph of the equations in a system of two equations in two variables consists of a circle and a line. Discuss the possible number of solutions for this system.

The concept of supply and demand is used often in business. In general, as the unit price of a commodity increases, the demand for that commodity decreases. Also, as a commodity's unit price increases, the manufacturer normally increases the supply. The point where supply is equal to demand is called the equilibrium point. The following shows the graph of a demand equation and the graph of a supply equation for previously rented DVDs. The x-axis represents the number of DVDs in thousands, and the y-axis represents the cost of a DVD. Use this graph to answer Exercises 63 through 66.

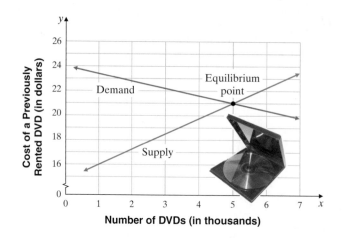

Number of DVDs (in thousands)

63. Find the number of DVDs and the price per DVD when supply equals demand.

64. When x is between 3 and 4, is supply greater than demand or is demand greater than supply?

65. When x is greater than 7, is supply greater than demand or is demand greater than supply?

66. For what x-values are the y-values corresponding to the supply equation greater than the y-values corresponding to the demand equation?

67. The amount y of red meat consumed per person in the United States (in pounds) in the year x can be modeled by the linear equation $y = -x + 124.6$. The amount y of all poultry consumed per person in the United States (in pounds) in the year x can be modeled by the linear equation $y = 0.9x + 93$. In both models, $x = 0$ represents the year 1998. (*Source:* Based on data and forecasts from the Economic Research Service, U.S. Department of Agriculture, 1998–2002)
   a. What does the slope of each equation tell you about the patterns of red meat and poultry consumption in the United States?
   b. Solve this system of equations. (Round your final results to the nearest whole numbers.)
   c. Explain the meaning of your answer to part (b).

68. The amount of U.S. federal government income y (in billions of dollars) for the fiscal year x, from 1995 through 2000 ($x = 0$ represents 1995), can be modeled by the linear equation $y = 132.4x + 1328.3$. The amount of U.S. federal government expenditures y (in billions of dollars) for the same period can be modeled by the linear equation $y = 52.9x + 1504.6$. Did expenses ever equal income during this period? If so, in what year? (*Source:* Based on data from Financial Management Service, U.S. Department of the Treasury, 1995–2000)

4.2 Solving Systems of Linear Equations in Three Variables

OBJECTIVE

Ⓐ Solve a system of three linear equations in three variables.

SSM   TUTOR CENTER   SG   CD & VIDEO   MATH PRO   WEB

In this section, we solve systems of linear equations in three variables. We call the equation $3x - y + z = -15$, for example, a **linear equation in three variables** since there are three variables and each variable is raised only to the power 1. A solution of this equation is an **ordered triple (x, y, z)** that makes the equation a true statement.

For example, the ordered triple $(2, 0, -21)$ is a solution of $3x - y + z = -15$ since replacing x with 2, y with 0, and z with -21 yields the true statement

$$3(2) - 0 + (-21) = -15$$

The graph of this equation is a plane in three-dimensional space, just as the graph of a linear equation in two variables is a line in two-dimensional space.

Although we will not discuss the techniques for graphing equations in three variables, visualizing the possible patterns of intersecting planes gives us insight into the possible patterns of solutions of a system of three three-variable linear equations. There are four possible patterns.

1. Three planes have a single point in common. This point represents the single solution of the system. This system is **consistent**.

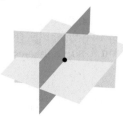

2. Three planes intersect at no point common to all three. This system has no solution. A few ways that this can occur are shown. This system is **inconsistent**.

3. Three planes intersect at all the points of a single line. The system has infinitely many solutions. This system is **consistent**.

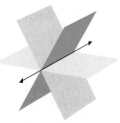

4. Three planes coincide at all points on the plane. The system is consistent, and the equations are **dependent**.

Solving a System of Three Linear Equations by the Elimination Method

Step 1. Write each equation in standard form, $Ax + By + Cz = D$.

Step 2. Choose a pair of equations and use them to eliminate a variable.

Step 3. Choose any other pair of equations and eliminate the *same variable* as in Step 2.

Step 4. Two equations in two variables should be obtained from Step 2 and Step 3. Use methods from Section 4.1 to solve this system for both variables.

Step 5. To solve for the third variable, substitute the values of the variables found in Step 4 into any of the original equations containing the third variable.

Step 6. Check the ordered triple solution in *all three* original equations.

(A) Solving a System of Three Linear Equations in Three Variables

Just as with systems of two equations in two variables, we can use the elimination or substitution method to solve a system of three equations in three variables. To use the elimination method, we eliminate a variable and obtain a system of two equations in two variables. Then we use the methods we learned in the previous section to solve the system of two equations.

EXAMPLE 1 Solve the system:

$$\begin{cases} 3x - y + z = -15 & \text{Equation (1)} \\ x + 2y - z = 1 & \text{Equation (2)} \\ 2x + 3y - 2z = 0 & \text{Equation (3)} \end{cases}$$

Solution: We add equations (1) and (2) to eliminate z.

$$\begin{array}{r} 3x - y + z = -15 \\ x + 2y - z = 1 \\ \hline 4x + y \phantom{{}- z} = -14 \end{array} \quad \text{Equation (4)}$$

Next we add two *other* equations and *eliminate z again*. To do so, we multiply both sides of equation (1) by 2 and add this resulting equation to equation (3). Then

$$\begin{cases} 2(3x - y + z) = 2(-15) \\ 2x + 3y - 2z = 0 \end{cases} \begin{array}{c} \text{simplifies} \\ \text{to} \end{array} \begin{cases} 6x - 2y + 2z = -30 \\ 2x + 3y - 2z = 0 \\ \hline 8x + y \phantom{{}- 2z} = -30 \end{cases}$$

Equation (5)

Now we solve equations (4) and (5) for x and y. To solve by elimination, we multiply both sides of equation (4) by -1 and add this resulting equation to equation (5). Then

$$\begin{cases} -1(4x + y) = -1(-14) \\ 8x + y = -30 \end{cases} \begin{array}{c} \text{simplifies} \\ \text{to} \end{array} \begin{cases} -4x - y = 14 \\ 8x + y = -30 \\ \hline 4x \phantom{{}- y} = -16 \\ x = -4 \end{cases}$$

Add the equations.
Solve for x.

Practice Problem 1

Solve the system:

$$\begin{cases} 2x - y + 3z = 13 \\ x + y - z = -2 \\ 3x + 2y + 2z = 13 \end{cases}$$

Answer

1. $(1, 1, 4)$

We now replace x with -4 in equation (4) or (5).

$$4x + y = -14 \qquad \text{Equation (4)}$$
$$4(-4) + y = -14 \qquad \text{Let } x = -4.$$
$$y = 2 \qquad \text{Solve for } y.$$

Finally, we replace x with -4 and y with 2 in equation (1), (2), or (3).

$$x + 2y - z = 1 \qquad \text{Equation (2)}$$
$$-4 + 2(2) - z = 1 \qquad \text{Let } x = -4 \text{ and } y = 2.$$
$$-4 + 4 - z = 1$$
$$-z = 1$$
$$z = -1$$

The ordered triple solution is $(-4, 2, -1)$. To check, we let $x = -4$, $y = 2$, and $z = -1$ in *all three* original equations of the system.

Equation (1)

$$3x - y + z = -15$$
$$3(-4) - 2 + (-1) = -15$$
$$-12 - 2 - 1 = -15$$
$$-15 = -15$$
True.

Equation (2)

$$x + 2y - z = 1$$
$$-4 + 2(2) - (-1) = 1$$
$$-4 + 4 + 1 = 1$$
$$1 = 1$$
True.

Equation (3)

$$2x + 3y - 2z = 0$$
$$2(-4) + 3(2) - 2(-1) = 0$$
$$-8 + 6 + 2 = 0$$
$$0 = 0$$
True.

All three statements are true, so the ordered triple solution is $(-4, 2, -1)$. ●

EXAMPLE 2   Solve the system:

$$\begin{cases} 2x - 4y + 8z = 2 & (1) \\ -x - 3y + z = 11 & (2) \\ x - 2y + 4z = 0 & (3) \end{cases}$$

Solution: When we add equations (2) and (3) to eliminate x, the new equation is

$$-5y + 5z = 11 \qquad (4)$$

To eliminate x again, we multiply both sides of equation (2) by 2 and add the resulting equation to equation (1). Then

$$\begin{cases} 2x - 4y + 8z = 2 \\ 2(-x - 3y + z) = 2(11) \end{cases} \begin{array}{c} \text{simplifies} \\ \text{to} \end{array} \begin{cases} 2x - 4y + 8z = 2 \\ \underline{-2x - 6y + 2z = 22} \\ -10y + 10z = 24 \quad (5) \end{cases}$$

Next we solve for y and z using equations (4) and (5). To do so, we multiply both sides of equation (4) by -2 and add the resulting equation to equation (5).

$$\begin{cases} -2(-5y + 5z) = -2(11) \\ -10y + 10z = 24 \end{cases} \begin{array}{c} \text{simplifies} \\ \text{to} \end{array} \begin{cases} 10y - 10z = -22 \\ \underline{-10y + 10z = 24} \\ 0 = 2 \quad \text{False.} \end{cases}$$

Since the statement is false, this system is inconsistent and has no solution. The solution set is the empty set $\{\ \}$ or $\varnothing$. ●

Practice Problem 2

Solve the system:

$$\begin{cases} 2x + 4y - 2z = 3 \\ -x + y - z = 6 \\ x + 2y - z = 1 \end{cases}$$

Answer

2. $\varnothing$

Concept Check

In the system

$$\begin{cases} x + y + z = 6 & \text{Equation (1)} \\ 2x - y + z = 3 & \text{Equation (2)} \\ x + 2y + 3z = 14 & \text{Equation (3)} \end{cases}$$

equations (1) and (2) are used to eliminate y. Which action could be used to finish solving? Why?
(a) Use (1) and (2) to eliminate z.
(b) Use (2) and (3) to eliminate y.
(c) Use (1) and (3) to eliminate x.

Practice Problem 3

Solve the system:

$$\begin{cases} 3x + 2y = -1 \\ 6x - 2z = 4 \\ y - 3z = 2 \end{cases}$$

Try the Concept Check in the margin.

EXAMPLE 3 Solve the system:

$$\begin{cases} 2x + 4y = 1 & (1) \\ 4x - 4z = -1 & (2) \\ y - 4z = -3 & (3) \end{cases}$$

Solution: Notice that equation (2) has no term containing the variable y. Let's eliminate y using equations (1) and (3). We multiply both sides of equation (3) by -4 and add the resulting equation to equation (1). Then

$$\begin{cases} 2x + 4y = 1 \\ -4(y - 4z) = -4(-3) \end{cases} \quad \begin{matrix} \text{simplifies} \\ \text{to} \end{matrix} \quad \begin{cases} 2x + 4y = 1 \\ \underline{-4y + 16z = 12} \\ 2x + 16z = 13 \quad (4) \end{cases}$$

Next we solve for z using equations (4) and (2). We multiply both sides of equation (4) by -2 and add the resulting equation to equation (2).

$$\begin{cases} -2(2x + 16z) = -2(13) \\ 4x - 4z = -1 \end{cases} \quad \begin{matrix} \text{simplifies} \\ \text{to} \end{matrix} \quad \begin{cases} -4x - 32z = -26 \\ \underline{4x - 4z = -1} \\ -36z = -27 \\ z = \dfrac{3}{4} \end{cases}$$

Now we replace z with $\dfrac{3}{4}$ in equation (3) and solve for y.

$$y - 4\left(\dfrac{3}{4}\right) = -3 \qquad \text{Let } z = \dfrac{3}{4} \text{ in equation (3).}$$

$$y - 3 = -3$$

$$y = 0$$

Finally, we replace y with 0 in equation (1) and solve for x.

$$2x + 4(0) = 1 \qquad \text{Let } y = 0 \text{ in equation (1).}$$

$$2x = 1$$

$$x = \dfrac{1}{2}$$

The ordered triple solution is $\left(\dfrac{1}{2}, 0, \dfrac{3}{4}\right)$. Check to see that this solution satisfies *all three* equations of the system.   ●

Answers

3. $\left(\dfrac{1}{3}, -1, -1\right)$

Concept Check: b

EXAMPLE 4   Solve the system:

$$\begin{cases} x - 5y - 2z = 6 & (1) \\ -2x + 10y + 4z = -12 & (2) \\ \dfrac{1}{2}x - \dfrac{5}{2}y - z = 3 & (3) \end{cases}$$

Solution: We multiply both sides of equation (3) by 2 to eliminate fractions, and we multiply both sides of equation (2) by $-\dfrac{1}{2}$ so that the coefficient of x is 1. The resulting system is then

$$\begin{cases} x - 5y - 2z = 6 & (1) \\ x - 5y - 2z = 6 & \text{Multiply (2) by } -\dfrac{1}{2}. \\ x - 5y - 2z = 6 & \text{Multiply (3) by 2.} \end{cases}$$

All three resulting equations are identical, and therefore equations (1), (2), and (3) are all equivalent. There are infinitely many solutions of this system. The equations are dependent. The solution set can be written as $\{(x, y, z) | x - 5y - 2z = 6\}$.   ●

As mentioned earlier, we can also use the substitution method to solve a system of linear equations in three variables.

EXAMPLE 5   Solve the system:

$$\begin{cases} x - 4y - 5z = 35 & (1) \\ x - 3y = 0 & (2) \\ -y + z = -25 & (3) \end{cases}$$

Solution: Notice in equations (2) and (3) that a variable is missing. Also notice that both equations contain the variable y. Let's use the substitution method by solving equation (2) for x and equation (3) for z and substituting the results in equation (1).

$$\begin{aligned} x - 3y &= 0 & (2) \\ x &= 3y & \text{Solve equation (2) for } x. \\ -y + z &= -55 & (3) \\ z &= y - 55 & \text{Solve equation (3) for } z. \end{aligned}$$

Now substitute $3y$ for x and $y - 55$ for z in equation (1).

$$\begin{aligned} x - 4y - 5z &= 35 & (1) \\ 3y - 4y - 5(y - 55) &= 35 & \text{Let } x = 3y \text{ and } z = y - 55. \\ 3y - 4y - 5y + 275 &= 35 & \text{Use the distributive law and multiply.} \\ -6y + 275 &= 35 & \text{Combine like terms.} \\ -6y &= -240 & \text{Subtract 275 from both sides.} \\ y &= 40 & \text{Solve.} \end{aligned}$$

To find x, recall that $x = 3y$ and substitute 40 for y. Then $x = 3y$ becomes $x = 3 \cdot 40 = 120$. To find z, recall that $z = y - 55$ and substitute 40 for y, also. Then $z = y - 55$ becomes $z = 40 \cdot 55 = -15$. The solution is $(120, 40, -15)$.   ●

Practice Problem 4

Solve the system:

$$\begin{cases} x - 3y + 4z = 2 \\ -2x + 6y - 8z = -4 \\ \dfrac{1}{2}x - \dfrac{3}{2}y + 2z = 1 \end{cases}$$

Practice Problem 5

Solve the system:

$$\begin{cases} 2x + 5y - 3z = 30 & (1) \\ x + y = -3 & (2) \\ 2x - z = 0 & (3) \end{cases}$$

(Hint: Equations (2) and (3) each contain the variable x and have a variable missing.)

> **Helpful Hint**
>
> Do not forget to distribute.

Answers

4. $\{(x, y, z) | x - 3y + 4z = 2\}$,   **5.** $(-5, 2, -10)$

FOCUS ON **Mathematical Connections**

SOLVING NONLINEAR SYSTEMS

Recall that a linear equation in two variables is an equation that can be written in the form $Ax + By = C$. By this definition, we can see that an equation of the form $\dfrac{A}{x} + \dfrac{B}{y} = C$ is clearly not linear. However, with a slight adjustment, we can solve a nonlinear system such as

$$\begin{cases} \dfrac{A}{x} + \dfrac{B}{y} = C \\ \dfrac{D}{x} + \dfrac{E}{y} = F \end{cases}$$

using the methods we already know for solving linear systems.

To solve such a system, first make the following substitutions. Let $w = \dfrac{1}{x}$ and $z = \dfrac{1}{y}$ in both equations. Then

$$\begin{cases} \dfrac{A}{x} + \dfrac{B}{y} = C \\ \dfrac{D}{x} + \dfrac{E}{y} = F \end{cases} \quad \text{becomes} \quad \begin{cases} Aw + Bz = C \\ Dw + Ez = F \end{cases}$$

This new system of equations is linear and can be solved with any of the techniques we already know. Once the values of w and z have been found, simply substitute them into the equations $w = \dfrac{1}{x}$ and $z = \dfrac{1}{y}$. Then solve each equation to find the value of x and the value of y.

CRITICAL THINKING

Apply the method described above to solve each nonlinear system.

1. $\begin{cases} \dfrac{2}{x} + \dfrac{3}{y} = 5 \\ \dfrac{5}{x} - \dfrac{3}{y} = 2 \end{cases}$ **2.** $\begin{cases} x + \dfrac{2}{y} = 7 \\ 3x + \dfrac{3}{y} = 6 \end{cases}$

3. $\begin{cases} \dfrac{3}{x} - \dfrac{2}{y} = -18 \\ \dfrac{2}{x} + \dfrac{3}{y} = 1 \end{cases}$ **4.** $\begin{cases} \dfrac{2}{x} - \dfrac{4}{y} = 5 \\ \dfrac{1}{x} - \dfrac{2}{y} = \dfrac{3}{2} \end{cases}$

EXERCISE SET 4.2

A *Solve each system. See Examples 1 through 4.*

1. $\begin{cases} x + y = 3 \\ 2y = 10 \\ 3x + 2y - 3z = 1 \end{cases}$

2. $\begin{cases} 5x = 5 \\ 2x + y = 4 \\ 3x + y - 4z = -15 \end{cases}$

3. $\begin{cases} 2x + 2y + z = 1 \\ -x + y + 2z = 3 \\ x + 2y + 4z = 0 \end{cases}$

4. $\begin{cases} 2x - 3y + z = 5 \\ x + y + z = 0 \\ 4x + 2y + 4z = 4 \end{cases}$

5. $\begin{cases} x - 2y + z = -5 \\ -3x + 6y - 3z = 15 \\ 2x - 4y + 2z = -10 \end{cases}$

6. $\begin{cases} 3x + y - 2z = 2 \\ -6x - 2y + 4z = -2 \\ 9x + 3y - 6z = 6 \end{cases}$

7. $\begin{cases} 4x - y + 2z = 5 \\ 2y + z = 4 \\ 4x + y + 3z = 10 \end{cases}$

8. $\begin{cases} 5y - 7z = 14 \\ 2x + y + 4z = 10 \\ 2x + 6y - 3z = 30 \end{cases}$

9. $\begin{cases} x + 5z = 0 \\ 5x + y + 10z = -24 \\ y - 3z = 0 \end{cases}$

10. $\begin{cases} 4x - y = 0 \\ 3x - y - z = 6 \\ -x + 5z = 0 \end{cases}$

11. $\begin{cases} 6x - 5z = 17 \\ 5x - y + 3z = -1 \\ 2x + y = -41 \end{cases}$

12. $\begin{cases} x + 2y = 6 \\ 7x + 3y + z = -33 \\ x - z = 16 \end{cases}$

13. $\begin{cases} x + y + z = 8 \\ 2x - y - z = 10 \\ x - 2y - 3z = 22 \end{cases}$

14. $\begin{cases} 5x + y + 3z = 1 \\ x - y + 3z = -7 \\ -x + y = 1 \end{cases}$

15. $\begin{cases} x + 2y - z = 5 \\ 6x + y + z = 7 \\ 2x + 4y - 2z = 5 \end{cases}$

16. $\begin{cases} 4x - y + 3z = 10 \\ x + y - z = 5 \\ 8x - 2y + 6z = 10 \end{cases}$

17. $\begin{cases} 2x - 3y + z = 2 \\ x - 5y + 5z = 3 \\ 3x + y - 3z = 5 \end{cases}$

18. $\begin{cases} 4x + y - z = 8 \\ x - y + 2z = 3 \\ 3x - y + z = 6 \end{cases}$

19. $\begin{cases} -2x - 4y + 6z = -8 \\ x + 2y - 3z = 4 \\ 4x + 8y - 12z = 16 \end{cases}$

20. $\begin{cases} -6x + 12y + 3z = -6 \\ 2x - 4y - z = 2 \\ -x + 2y + \dfrac{z}{2} = -1 \end{cases}$

21. $\begin{cases} 2x + 2y - 3z = 1 \\ y + 2z = -14 \\ 3x - 2y = -1 \end{cases}$

22. $\begin{cases} 7x + 4y = 10 \\ x - 4y + 2z = 6 \\ y - 2z = -1 \end{cases}$

23. $\begin{cases} \dfrac{3}{4}x - \dfrac{1}{3}y + \dfrac{1}{2}z = 9 \\ \dfrac{1}{6}x + \dfrac{1}{3}y - \dfrac{1}{2}z = 2 \\ \dfrac{1}{2}x - y + \dfrac{1}{2}z = 2 \end{cases}$

24. $\begin{cases} \dfrac{1}{3}x - \dfrac{1}{4}y + z = -9 \\ \dfrac{1}{2}x - \dfrac{1}{3}y - \dfrac{1}{4}z = -6 \\ x - \dfrac{1}{2}y - z = -8 \end{cases}$

25. The fraction $\dfrac{1}{24}$ can be written as the following sum:

$$\frac{1}{24} = \frac{x}{8} + \frac{y}{4} + \frac{z}{3}$$

where the numbers x, y, and z are solutions of

$$\begin{cases} x + y + z = 1 \\ 2x - y + z = 0 \\ -x + 2y + 2z = -1 \end{cases}$$

Solve the system and see that the sum of the fractions is $\dfrac{1}{24}$.

26. The fraction $\dfrac{1}{18}$ can be written as the following sum:

$$\frac{1}{18} = \frac{x}{2} + \frac{y}{3} + \frac{z}{9}$$

where the numbers x, y, and z are solutions of

$$\begin{cases} x + 3y + z = -3 \\ -x + y + 2z = -14 \\ 3x + 2y - z = 12 \end{cases}$$

Solve the system and see that the sum of the fractions is $\dfrac{1}{18}$.

Review and Preview

Solve. See Section 2.2.

27. The sum of two numbers is 45 and one number is twice the other. Find the numbers.

28. The difference between two numbers is 5. Twice the smaller number added to five times the larger number is 53. Find the numbers.

Solve. See Section 2.1.

29. $2(x - 1) - 3x = x - 12$

30. $7(2x - 1) + 4 = 11(3x - 2)$

31. $-y - 5(y + 5) = 3y - 10$

32. $z - 3(z + 7) = 6(2z + 1)$

Combining Concepts

33. Write a linear equation in three variables that has $(-1, 2, -4)$ as a solution. (There are many possibilities.) Explain the process you used to write an equation.

34. Write a system of three linear equations in three variables that has $(2, 1, 5)$ as a solution. (There are many possibilities.) Explain the process you used to write an equation.

35. Write a system of linear equations in three variables that has the solution $(-1, 2, -4)$. Explain the process you used to write your system.

36. When solving a system of three equation in three unknowns, explain how to determine that a system has no solution.

Solving systems involving more than three variables can be accomplished with methods similar to those encountered in this section. Apply what you already know to solve each system of equations in four variables.

37. $\begin{cases} x + y - w = 0 \\ y + 2z + w = 3 \\ x - z = 1 \\ 2x - y - w = -1 \end{cases}$

38. $\begin{cases} 5x + 4y = 29 \\ y + z - w = -2 \\ 5x + z = 23 \\ y - z + w = 4 \end{cases}$

39. $\begin{cases} x + y + z + w = 5 \\ 2x + y + z + w = 6 \\ x + y + z = 2 \\ x + y = 0 \end{cases}$

40. $\begin{cases} 2x - z = -1 \\ y + z + w = 9 \\ y - 2w = -6 \\ x + y = 3 \end{cases}$

4.3 Systems of Linear Equations and Problem Solving

Ⓐ Solving Problems Modeled by Systems of Two Equations

Thus far, we have solved problems by writing one-variable equations and solving for the variable. Some of these problems can be solved, perhaps more easily, by writing a system of equations, as illustrated in this section.

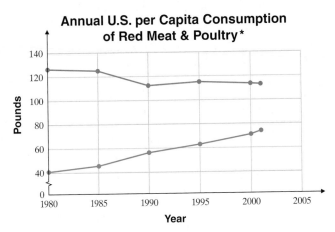

Annual U.S. per Capita Consumption of Red Meat & Poultry*

Source: USDA Economic Research Service

* Excludes shipments to Puerto Rico and other U.S. possessions

EXAMPLE 1   Predicting Equal Consumption of Red Meat and Poultry

America's consumption of red meat has decreased most years since 1980 while consumption of poultry has increased. The function $y = -0.71x + 125.6$ approximates the annual pounds of red meat consumed per capita, where x is the number of years since 1980. The function $y = 1.56x + 39.7$ approximates the annual pounds of poultry consumed per capita, where x is also the number of years since 1980. If this trend continues, determine the year when the annual consumption of red meat and poultry are equal.

Solution:

1. UNDERSTAND. Read and reread the problem and guess a year. Let's guess the year 2010. This year is 30 years since 1980, so $x = 30$. Now let $x = 30$ in each given function.

   Red meat: $y = -0.71x + 125.6 = -0.71(30) + 125.6 = 104.3$ pounds

   Poultry: $y = 1.56x + 39.7 = 1.56(30) + 39.7 = 86.5$ pounds

   Since the projected pounds in 2010 for red meat and poultry are not the same, we guessed incorrectly, but we do have a better understanding of the problem, and we know that the year will be later than 2010.

2. TRANSLATE. We are already given the system of equations.

3. SOLVE. We want to know the year x in which pounds y are the same, so we solve the system:

$$\begin{cases} y = -0.71x + 125.6 \\ y = 1.56x + 39.7 \end{cases}$$

Practice Problem 1

Read Example 1. If we use the years 1999, 2000, and 2001 only to write functions approximating the consumption of red meat and poultry, we have the following.

Red meat: $y = -2x + 125.3$

Poultry: $y = 1.5x + 95.5$

where x is years since 1999 and y is pounds per year consumed.

a. Assuming this trend continues, predict the year when consumption of red meat and poultry will be the same.

b. Does your answer differ from the answer to Example 1? Why or why not.

Answers

1. a 2007, **b.** yes; answers may vary

Since both equations are solved for y, one way to solve is to use the substitution method.

$$y = -0.71x + 125.6 \quad \text{First equation}$$

$$1.56x + 39.7 = -0.71x + 125.6 \quad \text{Let } y = 1.56x + 39.7.$$

$$2.27x = 85.9$$

$$x = \frac{85.9}{2.27} \approx 37.84$$

4. INTERPRET. Since we are only asked to find the year, we need only solve for x.

Check: To check, see whether $x \approx 37.84$ gives approximately the same number of pounds of red meat and poultry.

Red meat: $y = -0.71x + 125.6 = -0.71(37.84) + 125.6 \approx 98.73$ pounds

Poultry: $y = 1.56x + 39.7 = 1.56(37.84) + 39.7 \approx 98.73$ pounds

Since we rounded the number of years, the number of pounds do differ slightly. They differ only by 0.0032, so we can assume that we solved correctly.

State: The consumption of red meat and poultry will be the same about 37.84 years after 1980, or 2017.84. Thus, in the year 2017, we predict the consumption will be the same. ●

Practice Problem 2

A first number is 7 greater than a second number. Twice the first number is 4 more than three times the second. Find the numbers.

EXAMPLE 2 Finding Unknown Numbers

A first number is 4 less than a second number. Four times the first number is 6 more than twice the second. Find the numbers.

Solution:

1. UNDERSTAND. Read and reread the problem and guess a solution. If one number is 10 and this is 4 less than a second number, the second number is 14. Four times the first number is $4(10)$, or 40. This is not equal to 6 more than twice the second number, which is $2(14) + 6$ or 34. Although we guessed incorrectly, we now have a better understanding of the problem.

   Since we are looking for two numbers, we will let

   x = first number

   y = second number

2. TRANSLATE. Since we have assigned two variables to this problem, we will translate the given facts into two equations. For the first statement we have

In words:	the first number	is	4 less than second number
	↓	↓	↓
Translate:	x	$=$	$y - 4$

   Next we translate the second statement into an equation.

In words:	four times the first number	is	6 more than twice the second number
	↓	↓	↓
Translate:	$4x$	$=$	$2y + 6$

3. SOLVE. Now we solve the system

$$\begin{cases} x = y - 4 \\ 4x = 2y + 6 \end{cases}$$

Since the first equation expresses x in terms of y, we will use substitution. We substitute $y - 4$ for x in the second equation and solve for y.

$$4x = 2y + 6 \quad \text{Second equation}$$

$$4(y - 4) = 2y + 6 \quad \text{Let } x = y - 4.$$
$$4y - 16 = 2y + 6$$
$$2y = 22$$
$$y = 11$$

Now we replace y with 11 in the equation $x = y - 4$ and solve for x. Then $x = y - 4$ becomes $x = 11 - 4 = 7$. The ordered pair solution of the system is $(7, 11)$.

4. INTERPRET. Since the solution of the system is $(7, 11)$, the first number we are looking for is 7 and the second number is 11.

Check: Notice that 7 _is_ 4 less than 11, and 4 times 7 _is_ 6 more than twice 11. The proposed numbers, 7 and 11, are correct.

State: The numbers are 7 and 11.

EXAMPLE 3   Finding the Rate of Speed

Two cars leave Indianapolis, one traveling east and the other west. After 3 hours they are 297 miles apart. If one car is traveling 5 mph faster than the other, what is the speed of each?

Solution:

1. UNDERSTAND. Read and reread the problem. Let's guess a solution and use the formula $d = r \cdot t$ to check. Suppose that one car is traveling at a rate of 55 mph. This means that the other car is traveling at a rate of 50 mph since we are told that one car is traveling 5 mph faster than the other. To find the distance apart after 3 hours, we will first find the distance traveled by each car. One car's distance is rate $\cdot$ time $= 55(3) = 165$ miles. The other car's distance is rate $\cdot$ time $= 50(3) = 150$ miles. Since one car is traveling east and the other west, their distance apart is the sum of their distances, or 165 miles $+$ 150 miles $= 315$ miles. Although this distance apart is not the required distance of 297 miles, we now have a better understanding of the problem.

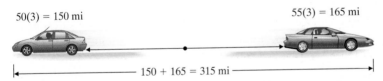

50(3) = 150 mi   55(3) = 165 mi

|← —————————— 150 + 165 = 315 mi —————————— →|

Let's model the problem with a system of equations. We will let

$x =$ speed of one car

$y =$ speed of the other car

We summarize the information on the following chart. Both cars have traveled 3 hours. Since distance $=$ rate $\cdot$ time, their distances are $3x$ and $3y$ miles, respectively.

	Rate $\cdot$	Time $=$	Distance
One Car	x	3	$3x$
Other Car	y	3	$3y$

2. **TRANSLATE.** We can now translate the stated conditions into two equations.

In words:	one car's distance	added to	the other car's distance	is	297
	↓	↓	↓	↓	↓
Translate:	$3x$	$+$	$3y$	$=$	297

In words:	one car's speed	is	5 mph faster than the other
	↓	↓	↓
Translate:	x	$=$	$y + 5$

3. **SOLVE.** Now we solve the system.

$$\begin{cases} 3x + 3y = 297 \\ x = y + 5 \end{cases}$$

Again, the substitution method is appropriate. We replace x with $y + 5$ in the first equation and solve for y.

$$3x + 3y = 297 \quad \text{First equation}$$

$$3(y + 5) + 3y = 297 \quad \text{Let } x = y + 5.$$
$$3y + 15 + 3y = 297$$
$$6y = 282$$
$$y = 47$$

To find x, we replace y with 47 in the equation $x = y + 5$. Then $x = 47 + 5 = 52$. The ordered pair solution of the system is $(52, 47)$.

4. **INTERPRET.** The solution $(52, 47)$ means that the cars are traveling at 52 mph and 47 mph, respectively.

Check: Notice that one car is traveling 5 mph faster than the other. Also, if one car travels 52 mph for 3 hours, the distance is $3(52) = 156$ miles. The other car traveling for 3 hours at 47 mph travels a distance of $3(47) = 141$ miles. The sum of the distances $156 + 141$ is 297 miles, the required distance.

State: The cars are traveling at 52 mph and 47 mph. ●

Practice Problem 4

One solution contains 20% acid and a second solution contains 50% acid. How many ounces of each solution should be mixed in order to have 60 ounces of a 30% acid solution?

EXAMPLE 4 Mixing Solutions

Lynn Pike, a pharmacist, needs 70 liters of a 50% alcohol solution. She has available a 30% alcohol solution and an 80% alcohol solution. How many liters of each solution should she mix to obtain 70 liters of a 50% alcohol solution?

Solution:

1. **UNDERSTAND.** Read and reread the problem. Next, guess the solution. Suppose that we need 20 liters of the 30% solution. Then we need $70 - 20 = 50$ liters of the 80% solution. To see if this gives us 70 liters of a 50% alcohol solution, let's find the amount of pure alcohol in each solution.

number of liters	$\times$	alcohol strength	$=$	amount of pure alcohol
↓		↓		↓
20 liters	$\times$	0.30	$=$	6 liters
50 liters	$\times$	0.80	$=$	40 liters
70 liters	$\times$	0.50	$=$	35 liters

Answer

4. 40 oz of 20% solution; 20 oz of 50% solution

Since 6 liters + 40 liters = 46 liters and not 35 liters, our guess is incorrect, but we have gained some insight as to how to model and check this problem.

We will let

x = amount of 30% solution, in liters

y = amount of 80% solution, in liters

and use a table to organize the given data.

	Number of Liters	Alcohol Strength	Amount of Pure Alcohol
30% Solution	x	30%	$0.30x$
80% Solution	y	80%	$0.80y$
50% Solution Needed	70	50%	$(0.50)(70)$

2. TRANSLATE. We translate the stated conditions into two equations.

In words: amount of 30% solution   +   amount of 80% solution   =   70

Translate: x + y =   70

In words: amount of pure alcohol in 30% solution   +   amount of pure alcohol in 80% solution   =   amount of pure alcohol in 50% solution

Translate: $0.30x$ + $0.80y$ = $(0.50)(70)$

3. SOLVE. Now we solve the system

$$\begin{cases} x + y = 70 \\ 0.30x + 0.80y = (0.50)(70) \end{cases}$$

To solve this system, we use the elimination method. We multiply both sides of the first equation by -3 and both sides of the second equation by 10. Then

$$\begin{cases} -3(x + y) = -3(70) \\ 10(0.30x + 0.80y) = 10(0.50)(70) \end{cases}$$ simplifies to $$\begin{cases} -3x - 3y = -210 \\ \underline{3x + 8y = 350} \\ 5y = 140 \\ y = 28 \end{cases}$$

Now we replace y with 28 in the equation $x + y = 70$ and find that $x + 28 = 70$, or $x = 42$. The ordered pair solution of the system is $(42, 28)$.

4. INTERPRET.

Check: We check the solution in the same way that we checked our guess.

State: The pharmacist needs to mix 42 liters of 30% solution and 28 liters of 80% solution to obtain 70 liters of 50% solution.

●

Try the Concept Check in the margin.

B **Solving Problems Modeled by Systems of Three Equations**

To introduce problem solving with systems of three linear equations in three variables, we solve a problem about triangles.

Concept Check

Suppose you mix an amount of 25% acid solution with an amount of 60% acid solution. You then calculate the acid strength of the resulting acid mixture. For which of the following results should you suspect an error in your calculation? Why?

a. 14%

b. 32%

c. 55%

Answer

Concept Check: a; answers may vary

Practice Problem 5

The measure of the largest angle of a tri-angle is 90° more than the measure of the smallest angle, and the measure of the re-maining angle is 30° more than the mea-sure of the smallest angle. Find the measure of each angle.

EXAMPLE 5   Finding Angle Measures

The measure of the largest angle of a triangle is 80° more than the mea-sure of the smallest angle, and the measure of the remaining angle is 10° more than the measure of the smallest angle. Find the measure of each angle.

Solution:

1. UNDERSTAND. Read and reread the problem. Recall that the sum of the measures of the angles of a triangle is 180°. Then guess a solution. If the smallest angle measures 20°, the measure of the largest angle is 80° more, or $20° + 80° = 100°$. The measure of the remaining angle is 10° more than the measure of the smallest angle, or $20° + 10° = 30°$. The sum of these three angles is $20° + 100° + 30° = 150°$, not the re-quired 180°. We now know that the measure of the smallest angle is greater than 20°.

    To model this problem we will let

    x = degree measure of the smallest angle

    y = degree measure of the largest angle

    z = degree measure of the remaining angle

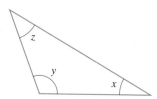

2. TRANSLATE. We translate the given information into three equations.

    In words: the sum of the measures   =   180

    Translate: $x + y + z$ =   180

    In words: the largest angle   is   80 more than the smallest angle

    Translate: y = $x + 80$

    In words: the remaining angle   is   10 more than the smallest angle

    Translate: z = $x + 10$

3. SOLVE. We solve the system

    $$\begin{cases} x + y + z = 180 \\ y = x + 80 \\ z = x + 10 \end{cases}$$

    Since y and z are both expressed in terms of x, we will solve using the substitution method. We substitute $y = x + 80$ and $z = x + 10$ in the first equation.

Answer

5. 20°; 50°; 110°

Then

$$x + y + z = 180 \qquad \text{First equation}$$

$$x + (x + 80) + (x + 10) = 180 \qquad \text{Let } y = x + 80 \text{ and } z = x + 10.$$

$$3x + 90 = 180$$

$$3x = 90$$

$$x = 30$$

Then $y = x + 80 = 30 + 80 = 110$, and $z = x + 10 = 30 + 10 = 40$.
The ordered triple solution is $(30, 110, 40)$.

4. INTERPRET.

Check: Notice that $30° + 40° + 110° = 180°$. Also, the measure of the largest angle, $110°$, is $80°$ more than the measure of the smallest angle, $30°$. The measure of the remaining angle, $40°$, is $10°$ more than the measure of the smallest angle, $30°$.

State: The angles measure $30°, 110°,$ and $40°$.

Ⓒ Solving Problems with Cost and Revenue Functions

Recall that businesses are often computing cost and revenue functions or equations to predict sales, to determine whether prices need to be adjusted, and to see whether the company is making or losing money. Recall also that the value at which revenue equals cost is called the break-even point. When revenue is less than cost, the company is losing money; when revenue is greater than cost, the company is making money.

EXAMPLE 6   Finding a Break-Even Point

A manufacturing company recently purchased $3000 worth of new equipment to create new personalized stationery to sell to its customers. The cost of producing a package of personalized stationery is $3.00, and it is sold for $5.50. Find the number of packages that must be sold for the company to break even.

Solution:

1. UNDERSTAND. Read and reread the problem.

Notice that the cost to the company will include a one-time cost of $3000 for the equipment and then $3.00 per package produced. The revenue will be $5.50 per package sold.

To model this problem, we will let

x = number of packages of personalized stationery

$C(x)$ = total cost for producing x packages of stationery

$R(x)$ = total revenue for selling x packages of stationery

Practice Problem 6

A company that manufactures boxes recently purchased $2,000 worth of new equipment to make gift boxes to sell to its customers. The cost of producing a package of gift boxes is $1.50 and it is sold for $4.00. Find the number of packages that must be sold for the company to break even.

Answer

6. 800 packages

2. TRANSLATE. The revenue equation is

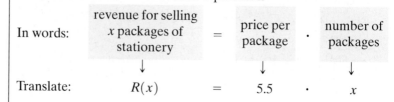

In words: revenue for selling
              x packages of   =   price per   ·   number of
              stationery   package   packages
                  ↓   ↓   ↓
Translate: $R(x)$ =   5.5   · x

The cost equation is

In words: cost for producing   cost   number   cost for
           x packages of   =   per   ·   of   +   equip-
           stationery   package   packages   ment
                ↓   ↓   ↓   ↓
Translate: $C(x)$ =   3   · x +   3000

Since the break-even point is when $R(x) = C(x)$, we solve the equation

$$5.5x = 3x + 3000$$

3. SOLVE.

$$5.5x = 3x + 3000$$
$$2.5x = 3000 \qquad \text{Subtract } 3x \text{ from both sides.}$$
$$x = 1200 \qquad \text{Divide both sides by 2.5.}$$

4. INTERPRET.

Check: To see whether the break-even point occurs when 1200 packages are produced and sold, we check to see if revenue equals cost when $x = 1200$. When $x = 1200$,

$$R(x) = 5.5x = 5.5(1200) = 6600$$
$$C(x) = 3x + 3000 = 3(1200) + 3000 = 6600$$

Since $R(1200) = C(1200) = 6600$, the break-even point is 1200.

State: The company must sell 1200 packages of stationery to break even. The graph of this system is shown.

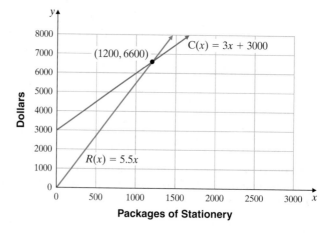

EXERCISE SET 4.3

A *Solve. See Examples 2 through 4.*

1. One number is two more than a second number. Twice the first is 4 less than 3 times the second. Find the numbers.

2. Three times one number minus a second is 8, and the sum of the numbers is 12. Find the numbers.

3. The U.S.A. has the world's only "large deck" aircraft carriers which can hold up to 72 aircraft. The Enterprise class carrier is longest in length while the Nimitz class carrier is the second longest. The total length of these two carriers is 2193 feet while the difference of their lengths is only 9 feet. (*Source: U.S.A. Today, May, 2001*)

a. Find the length of each class carrier.

b. If a football field has a length of 100 yards, determine the length of the Enterprise class carrier in terms of number of football fields.

4. The rate of growth of participation (age 6 and older) in sports featured in the X-Games is surpassing that for older sports such as football and baseball. The most popular X-Game sport is in-line roller skating, followed by skateboarding. In 2000, the total number of participants in both sports was 40.6 million. If the number of participants in roller skating was 5.8 million more than twice the number of participants in skateboarding, find the number of participants in each sport. (*Source:* January 2001 Sporting Goods Manufacturers Association Sports Participation Topline Report)

5. A Delta 727 traveled 560 mph with the wind and 480 mph against the wind. Find the speed of the plane in still air and the speed of the wind.

6. Terry Watkins can row about 10.6 kilometers in 1 hour downstream and 6.8 kilometers upstream in 1 hour. Find how fast he can row in still water, and find the speed of the current.

7. Find how many quarts of 4% butterfat milk and 1% butterfat milk should be mixed to yield 60 quarts of 2% butterfat milk.

8. A pharmacist needs 500 milliliters of a 20% phenobarbital solution but has only 5% and 25% phenobarbital solutions available. Find how many milliliters of each she should mix to get the desired solution.

9. In a recent year, the United Kingdom was the most popular host country for U.S. students traveling abroad to study. Spain was the second most popular destination. A total of 40,012 students visited one of the two countries. If 15,428 more U.S. students studied in the United Kingdom than in Spain, find how many students studied abroad in each country. (*Source:* Institute of International Education, Open Doors 2000)

10. In 2000, Washington D.C. had the most conventions while Orlando was the second most popular convention location. A total of 353 conventions were held in one of the two destinations. If Washington D.C. had 27 more conventions than Orlando, find the number of conventions held in each location. (*Source: Successful Meetings* magazine)

11. Karen Karlin bought some large frames for $15 each and some small frames for $8 each at a closeout sale. If she bought 22 frames for $239, find how many of each type she bought.

12. Hilton University Drama Club sold 311 tickets for a play. Student tickets cost 50 cents each; nonstudent tickets cost $1.50. If total receipts were $385.50, find how many tickets of each type were sold.

13. One number is two less than a second number. Twice the first is 4 more than 3 times the second. Find the numbers.

14. Twice one number plus a second number is 42, and the first number minus the second number is -6. Find the numbers.

15. In the United States, the percent of women using the Internet is increasing faster than the percent of men. For the years 1996–2001, the function $y = 7x + 18.7$ can be used to estimate the percent of females using the Internet while the function $y = 6x + 27.7$ can be used to estimate the percent of males. For both functions, x is the number of years since 1996. If this trend continues, predict the year in which the percent of females using the Internet is equal to the percent of males. (*Source:* Pew Internet & American Life Project)

16. The percent of car-vehicle sales is decreasing while the percent of light-truck (pickups, sport-utility vans and minivans) vehicle sales is increasing. For the years 1997–2000, the function $y = -x + 54.5$ can be used to estimate the percent of vehicle sales being cars while the function $y = x + 45.5$ can be used to estimate the percent of vehicle sales being light trucks. For both functions, x is the number of years since 1997.
   a. If this trend continues, predict the year in which the percent of car sales equals the percent of light-truck sales.
   b. Before the actual 2001 vehicle sales data was published, USA today predicted that light-truck sales would likely be greater than car sales in the year 2001. Does your prediction from part a agree with this statement? (*Source: USA Today* and Autodata)

17. An office supply store in San Diego sells 7 writing tablets and 4 pens for $6.40. Also, 2 tablets and 19 pens cost $5.40. Find the price of each.

18. A Candy Barrel shop manager mixes M&M's worth $2.00 per pound with trail mix worth $1.50 per pound. Find how many pounds of each she should use to get 50 pounds of a party mix worth $1.80 per pound.

19. An airplane takes 3 hours to travel a distance of 2160 miles with the wind. The return trip takes 4 hours against the wind. Find the speed of the plane in still air and the speed of the wind.

20. Two cyclists start at the same point and travel in opposite directions. One travels 4 mph faster than the other. In 4 hours they are 112 miles apart. Find how fast each is traveling.

21. The percent of men 65 years of age or older in the labor force has decreased most years since 1900 while the percent of women 65 years of age or older has slightly increased. For the years 1900–2000, the function $y = -0.52x + 64.5$ approximates the percent of men 65 years of age or older in the labor force while the function $y = 0.02x + 7.2$ approximates the percent of women 65 years of age or older in the labor force. For both functions, x is the number of years after 1900. (*Source: The World Almanac*, 2002)

  a. Explain how the decrease of men described can be verified by their given function while the slight increase of women described can be verified by their given function.

  b. If this trend continues, determine the year when the percent of men and woman 65 years of age or older in the labor force are the same.

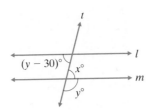

22. The annual U.S. per capita consumption of butter has remained about the same since 1980 while consumption of margarine has decreased. For the years 1995–1999, the function $y = 0.05x + 4.3$ approximates the annual U.S. per capita consumption of butter in pounds and the function $y = -0.31x + 9.3$ approximates the annual U.S. per capita consumption of margarine in pounds. For both functions, x is the number of years after 1995. If this trend continues, determine the year when the pounds of butter consumed equals the pounds of margarine consumed.

△ **23.** The perimeter of a triangle is 93 centimeters. If two sides are equally long and the third side is 9 centimeters longer than the others, find the lengths of the three sides.

24. Jack Reinholt, a car salesman, has a choice of two pay arrangements: a weekly salary of $200 plus 5% commission on sales, or a straight 15% commission. Find the amount of weekly sales for which Jack's earnings are the same regardless of the pay arrangement.

25. Hertz car rental agency charges $25 daily plus 10 cents per mile. Budget charges $20 daily plus 25 cents per mile. Find the daily mileage for which the Budget charge for the day is twice that of the Hertz charge for the day.

26. Carroll Blakemore, a drafting student, bought three templates and a pencil one day for $6.45. Another day he bought two pads of paper and four pencils for $7.50. If the price of a pad of paper is three times the price of a pencil, find the price of each type of item.

△ **27.** In the figure, line l and line m are parallel lines cut by transversal t. Find the values of x and y.

△ **28.** Find the values of x and y in the following isosceles triangle.

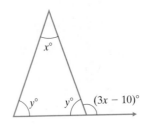

Solve. See Example 5.

29. Rabbits in a lab are to be kept on a strict daily diet that includes 30 grams of protein, 16 grams of fat, and 24 grams of carbohydrates. The scientist has only three food mixes available with the following grams of nutrients per unit.

	Protein	Fat	Carbohydrate
Mix A	4	6	3
Mix B	6	1	2
Mix C	4	1	12

Find how many units of each mix are needed daily to meet each rabbit's dietary need.

30. Gerry Gundersen mixes different solutions with concentrations of 25%, 40%, and 50% to get 200 liters of a 32% solution. If he uses twice as much of the 25% solution as of the 40% solution, find how many liters of each kind he uses.

△ **31.** The perimeter of a quadrilateral (four-sided polygon) is 29 inches. The longest side is twice as long as the shortest side. The other two sides are equally long and are 2 inches longer than the shortest side. Find the length of all four sides.

△ **32.** The measure of the largest angle of a triangle is 90° more than the measure of the smallest angle, and the measure of the remaining angle is 30° more than the measure of the smallest angle. Find the measure of each angle.

33. The sum of three numbers is 40. One number is five more than a second number. It is also twice the third. Find the numbers.

34. The sum of the digits of a three-digit number is 15. The tens-place digit is twice the hundreds-place digit, and the ones-place digit is 1 less than the hundreds-place digit. Find the three-digit number.

35. In 2001, the WNBA's top scorer was Katie Smith of the Minnesota Lynx. She scored a total of 739 points during the regular season. The number of two-point field goals Smith made was 51 less than twice the number of three-point field goals she made. The number of free throws (each worth one point) she made was 8 more than twice the number of two-point field goals she made. Find how many free throws, two-point field goals, and three-point field goals Katie Smith made during the 2001 regular season. (*Source:* Women's National Basketball Association)

36. During the 2000–2001 regular NBA season, the top-scoring player was Jerry Stackhouse of the Detroit Pistons. Stackhouse scored a total of 2380 points during the regular season. The number of free throws (each worth one point) he made was 2 more than four times the number of three-point field goals he made. Stackhouse also made 58 more free throws than two-point field goals. How many free throws, two-point field goals, and three-point field goals did Jerry Stackhouse make during the 2000–2001 season? (*Source:* National Basketball Association)

37. Find the values of $x, y,$ and z in the following triangle.

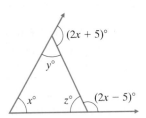

△ 38. The sum of the measures of the angles of a quadrilateral is 360°. Find the value of $x, y,$ and z in the following quadrilateral.

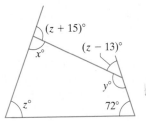

C Given the cost function $C(x)$ and the revenue function $R(x)$, find the number of units x that must be sold to break even. See Example 6.

39. $C(x) = 30x + 10{,}000$
$R(x) = 46x$

40. $C(x) = 12x + 15{,}000$
$R(x) = 32x$

41. $C(x) = 1.2x + 1500$
$R(x) = 1.7x$

42. $C(x) = 0.8x + 900$
$R(x) = 2x$

43. $C(x) = 75x + 160{,}000$
$R(x) = 200x$

44. $C(x) = 105x + 70{,}000$
$R(x) = 245x$

45. The planning department of Abstract Office Supplies has been asked to determine whether the company should introduce a new computer desk next year. The department estimates that $6000 of new manufacturing equipment will need to be purchased and that the cost of constructing each desk will be $200. The department also estimates that the revenue from each desk will be $450.
a. Determine the revenue function $R(x)$ from the sale of x desks.
b. Determine the cost function $C(x)$ for manufacturing x desks.
c. Find the break-even point.

46. Baskets, Inc., is planning to introduce a new woven basket. The company estimates that $500 worth of new equipment will be needed to manufacture this new type of basket and that it will cost $15 per basket to manufacture. The company also estimates that the revenue from each basket will be $31.
a. Determine the revenue function $R(x)$ from the sale of x baskets.
b. Determine the cost function $C(x)$ for manufacturing x baskets.
c. Find the break-even point.

Review and Preview

Multiply both sides of equation (1) by 2, and add the resulting equation to equation (2). See Section 4.2.

47. $3x - y + z = 2$ (1)
$-x + 2y + 3z = 6$ (2)

48. $2x + y + 3z = 7$ (1)
$-4x + y + 2z = 4$ (2)

Multiply both sides of equation (1) by −3, and add the resulting equation to equation (2). See Section 4.2.

49. $x + 2y - z = 0$ (1)
$3x + y - z = 2$ (2)

50. $2x - 3y + 2z = 5$ (1)
$x - 9y + z = -1$ (2)

51. The number of personal bankruptcy petitions filed in the United States has been on the rise since the early 1980s. In 2001, the number of petitions filed was 200,000 more than four times the number of petitions filed in 1980. This is equivalent to an increase of 1,100,000 petitions filed from 1980 to 2001. Find how many personal bankruptcy petitions were filed in each year. (*Source:* Based on data from the Administrative Office of the United States Courts)

52. In 2000, the median weekly earnings for male bus drivers in the United States was $100 more than the median weekly earnings for female bus drivers. The median weekly earnings for female bus drivers was 0.75 times that of their male counterparts. Also in 2000, the median weekly earnings for female financial managers in the United States was $400 less than the median weekly earnings for male financial managers. The median weekly earnings for male financial managers was 1.5 times that of their female counterparts. (*Source:* Based on data from the U.S. Bureau of Labor Statistics)

  a. Find the median weekly earnings for female bus drivers in the United States in 2000.

  b. Find the median weekly earnings for female financial managers in the United States in 2000.

  c. Of the four groups of workers described in the problem, which group makes the greatest weekly earnings? Which group makes the least weekly earnings?

53. Find the values of a, b, and c such that the equation $y = ax^2 + bx + c$ has ordered pair solutions $(1, 6)$, $(-1, -2)$, and $(0, -1)$. To do so, substitute each ordered pair solution into the equation. Each time, the result is an equation in three unknowns: a, b, and c. Then solve the resulting system of three linear equations in three unknowns, a, b, and c.

54. Find the values of a, b, and c such that the equation $y = ax^2 + bx + c$ has ordered pair solutions $(1, 2)$, $(2, 3)$, and $(-1, 6)$. (*Hint:* See Exercise 53.)

55. Data (x, y) for the total number y (in thousands) of college-bound students who took the ACT assessment in the year x are $(5, 945)$, $(6, 925)$, and $(11, 1070)$, where $x = 5$ represents 1995 and $x = 9$ represents 2001. Find the values of a, b, and c such that the equation $y = ax^2 + bx + c$ models this data. According to your model, how many students will take the ACT in 2007? (*Source:* ACT, Inc.)

56. Monthly normal rainfall data (x, y) for Portland, Oregon, are $(4, 2.47)$, $(7, 0.6)$, $(8, 1.1)$, where x represents time in months (with $x = 1$ representing January) and y represents rainfall in inches. Find the values of a, b, and c rounded to 2 decimal places such that the equation $y = ax^2 + bx + c$ models this data. According to your model, how much rain should Portland expect during September? (*Source:* National Climatic Data Center)

Integrated Review—Systems of Linear Equations

The graphs of various systems of equations are shown. Match each graph with the solution of its corresponding system.

A

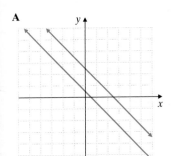

B

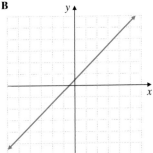

C

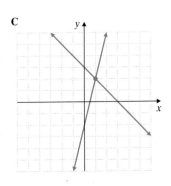

D

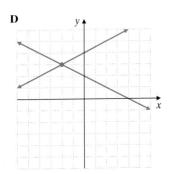

1. Solution: $(1, 2)$

2. Solution: $(-2, 3)$

3. No solution

4. Infinite number of solutions

Solve each system by elimination or substitution.

5. $\begin{cases} x + y = 4 \\ \quad\;\; y = 3x \end{cases}$

6. $\begin{cases} x - y = -4 \\ \quad\;\; y = 4x \end{cases}$

7. $\begin{cases} x + y = 1 \\ x - 2y = 4 \end{cases}$

8. $\begin{cases} 2x - y = 8 \\ x + 3y = 11 \end{cases}$

9. $\begin{cases} 2x + 5y = 8 \\ 6x + y = 10 \end{cases}$

10. $\begin{cases} x - 4y = -5 \\ -3x - 8y = 0 \end{cases}$

11. _____

12. _____

13. _____

14. _____

15. _____

16. _____

17. _____

18. _____

19. _____

20. _____

11. $\begin{cases} 4x - 7y = 7 \\ 12x - 21y = 24 \end{cases}$

12. $\begin{cases} 2x - 5y = 3 \\ -4x + 10y = -6 \end{cases}$

13. $\begin{cases} x + y = 2 \\ -3y + z = -7 \\ 2x + y - z = -1 \end{cases}$

14. $\begin{cases} y + 2z = -3 \\ x - 2y = 7 \\ 2x - y + z = 5 \end{cases}$

15. $\begin{cases} 2x + 4y - 6z = 3 \\ -x + y - z = 6 \\ x + 2y - 3z = 1 \end{cases}$

16. $\begin{cases} x - y + 3z = 2 \\ -2x + 2y - 6z = -4 \\ 3x - 3y + 9z = 6 \end{cases}$

17. $\begin{cases} x + y - 4z = 5 \\ x - y + 2z = -2 \\ 3x + 2y + 4z = 18 \end{cases}$

18. $\begin{cases} 2x - y + 3z = 2 \\ x + y - 6z = 0 \\ 3x + 4y - 3z = 6 \end{cases}$

19. A first number is 8 less than a second number. Twice the first number is 11 more than the second number. Find the numbers.

△ **20.** The sum of the measures of the angles of a quadrilateral is 360°. The two smallest angles of the quadrilateral have the same measure. The third angle measures 30° more than the measure of one of the smallest angles and the fourth angle measures 50° more than the measure of one of the smallest angles. Find the measure of each angle.

4.4 Solving Systems of Equations Using Matrices

By now, you may have noticed that the solution of a system of equations depends on the coefficients of the equations in the system and not on the variables. In this section, we introduce how to solve a system of equations using a **matrix**.

OBJECTIVES

Ⓐ Use matrices to solve a system of two equations.

Ⓑ Use matrices to solve a system of three equations.

SSM   TUTOR CENTER   SG   CD & VIDEO   MATH PRO   WEB

Ⓐ Using Matrices to Solve a System of Two Equations

A **matrix** (plural: **matrices**) is a rectangular array of numbers. The following are examples of matrices.

$$\begin{bmatrix} 1 & 0 \\ 0 & 1 \end{bmatrix} \qquad \begin{bmatrix} 2 & 1 & 3 & -1 \\ 0 & -1 & 4 & 5 \\ -6 & 2 & 1 & 0 \end{bmatrix} \qquad \begin{bmatrix} a & b & c \\ d & e & f \end{bmatrix}$$

The numbers aligned horizontally in a matrix are in the same **row**. The numbers aligned vertically are in the same **column**.

Row 1 →
Row 2 →
$$\begin{bmatrix} 2 & 1 & 0 \\ -1 & 6 & 2 \end{bmatrix}$$

This matrix has 2 rows and 3 columns. It is called a 2 × 3 (read "two by three") matrix.

Column 1
Column 2
Column 3

To see the relationship between systems of equations and matrices, study the example below.

System of Equations

$$\begin{cases} 2x - 3y = 6 & \text{Equation 1} \\ x + y = 0 & \text{Equation 2} \end{cases}$$

Corresponding Matrix

$$\begin{bmatrix} 2 & -3 & \vdots & 6 \\ 1 & 1 & \vdots & 0 \end{bmatrix} \begin{matrix} \text{Row 1} \\ \text{Row 2} \end{matrix}$$

Notice that the rows of the matrix correspond to the equations in the system. The coefficients of the variables are placed to the left of a vertical dashed line. The constants are placed to the right. Each of these numbers in the matrix is called an **element**.

> **Helpful Hint**
>
> Before writing the corresponding matrix associated with a system of equations, make sure that the equations are written in standard form.

The method of solving systems by matrices is to write this matrix as an equivalent matrix from which we can easily identify the solution. Two matrices are equivalent if they represent systems that have the same solution set. The following **row operations** can be performed on matrices, and the result is an equivalent matrix.

Elementary Row Operations

1. Any two rows in a matrix may be interchanged.
2. The elements of any row may be multiplied (or divided) by the same nonzero number.
3. The elements of any row may be multiplied (or divided) by a nonzero number and added to their corresponding elements in any other row.

Helpful Hint

Notice that these *row* operations are the same operations that we can perform on *equations* in a system.

Practice Problem 1

Use matrices to solve the system:

$$\begin{cases} x + 2y = -4 \\ 2x - 3y = 13 \end{cases}$$

EXAMPLE 1 Use matrices to solve the system:

$$\begin{cases} x + 3y = 5 \\ 2x - y = -4 \end{cases}$$

Solution: The corresponding matrix is $\left[\begin{array}{cc|c} 1 & 3 & 5 \\ 2 & -1 & -4 \end{array}\right]$. We use elementary row operations to write an equivalent matrix that looks like $\left[\begin{array}{cc|c} 1 & a & b \\ 0 & 1 & c \end{array}\right]$.

For the matrix given, the element in the first row, first column is already 1, as desired. Next we write an equivalent matrix with a 0 below the 1. To do this, we multiply row 1 by -2 and add to row 2. *We will change only row 2.*

$$\left[\begin{array}{cc|c} 1 & 3 & 5 \\ -2(1) + 2 & -2(3) + (-1) & -2(5) + (-4) \end{array}\right]$$

Row 1 Row 2   Row 1 Row 2   Row 1 Row 2
element element element element   element element

simplifies to

$$\left[\begin{array}{cc|c} 1 & 3 & 5 \\ 0 & -7 & -14 \end{array}\right]$$

Now we change the -7 to a 1 by use of an elementary row operation. We divide row 2 by -7, then

$$\left[\begin{array}{cc|c} 1 & 3 & 5 \\ \frac{0}{-7} & \frac{-7}{-7} & \frac{-14}{-7} \end{array}\right] \text{ simplifies to } \left[\begin{array}{cc|c} 1 & 3 & 5 \\ 0 & 1 & 2 \end{array}\right]$$

This last matrix corresponds to the system

$$\begin{cases} x + 3y = 5 \\ y = 2 \end{cases}$$

Thus we know that y is 2. To find x, we let $y = 2$ in the first equation, $x + 3y = 5$.

$$x + 3y = 5 \qquad \text{First equation}$$

$$x + 3(2) = 5 \qquad \text{Let } y = 2.$$

$$x = -1$$

Answer

1. $(2, -3)$

The ordered pair solution is $(-1, 2)$. Check to see that this ordered pair satisfies both original equations. ●

EXAMPLE 2   Use matrices to solve the system:

$$\begin{cases} 2x - y = 3 \\ 4x - 2y = 5 \end{cases}$$

Solution: The corresponding matrix is $\begin{bmatrix} 2 & -1 & | & 3 \\ 4 & -2 & | & 5 \end{bmatrix}$. To get 1 in the row 1, column 1 position, we divide the elements of row 1 by 2.

$$\begin{bmatrix} \dfrac{2}{2} & -\dfrac{1}{2} & | & \dfrac{3}{2} \\ 4 & -2 & | & 5 \end{bmatrix} \quad \text{simplifies to} \quad \begin{bmatrix} 1 & -\dfrac{1}{2} & | & \dfrac{3}{2} \\ 4 & -2 & | & 5 \end{bmatrix}$$

To get 0 under the 1, we multiply the elements of row 1 by -4 and add the new elements to the elements of row 2.

$$\begin{bmatrix} 1 & -\dfrac{1}{2} & | & \dfrac{3}{2} \\ -4(1) + 4 & -4\left(-\dfrac{1}{2}\right) - 2 & | & -4\left(\dfrac{3}{2}\right) + 5 \end{bmatrix} \quad \text{simplifies to}$$

$$\begin{bmatrix} 1 & -\dfrac{1}{2} & | & \dfrac{3}{2} \\ 0 & 0 & | & -1 \end{bmatrix}$$

The corresponding system is $\begin{cases} x - \dfrac{1}{2}y = \dfrac{3}{2} \\ 0 = -1 \end{cases}$. The equation $0 = -1$ is false for all y or x values; hence the system is inconsistent and has no solution. ●

Try the Concept Check in the margin.

B Using Matrices to Solve a System of Three Equations

To solve a system of three equations in three variables using matrices, we will write the corresponding matrix in the form

$$\begin{bmatrix} 1 & a & b & | & d \\ 0 & 1 & c & | & e \\ 0 & 0 & 1 & | & f \end{bmatrix}$$

EXAMPLE 3   Use matrices to solve the system:

$$\begin{cases} x + 2y + z = 2 \\ -2x - y + 2z = 5 \\ x + 3y - 2z = -8 \end{cases}$$

Solution: The corresponding matrix is $\begin{bmatrix} 1 & 2 & 1 & | & 2 \\ -2 & -1 & 2 & | & 5 \\ 1 & 3 & -2 & | & -8 \end{bmatrix}$.

Our goal is to write an equivalent matrix with 1s along the diagonal (see the numbers in red) and 0s below the 1s. The element in row 1, column 1 is already 1. Next we get 0s for each element in the rest of column 1. To do this, first we multiply the elements of row 1 by 2 and add the new elements to row 2. Also,

Practice Problem 2

Use matrices to solve the system:

$$\begin{cases} -3x + y = 0 \\ -6x + 2y = 2 \end{cases}$$

Concept Check

Consider the system

$$\begin{cases} 2x - 3y = 8 \\ x + 5y = -3 \end{cases}$$

What is wrong with its corresponding matrix shown below?

$$\begin{bmatrix} 2 & -3 & | & 8 \\ 0 & 5 & | & -3 \end{bmatrix}$$

Practice Problem 3

Use matrices to solve the system:

$$\begin{cases} x + 3y + z = 5 \\ -3x + y - 3z = 5 \\ x + 2y - 2z = 9 \end{cases}$$

Answers

2. no solution,   **3.** $(1, 2, -2)$

Concept Check: matrix should be $\begin{bmatrix} 2 & -3 & | & 8 \\ 1 & 5 & | & -3 \end{bmatrix}$

we multiply the elements of row 1 by -1 and add the new elements to the elements of row 3. *We do not change row 1.* Then

$$\begin{bmatrix} 1 & 2 & 1 & \vdots & 2 \\ 2(1)-2 & 2(2)-1 & 2(1)+2 & \vdots & 2(2)+5 \\ -1(1)+1 & -1(2)+3 & -1(1)-2 & \vdots & -1(2)-8 \end{bmatrix} \text{ simplifies to}$$

$$\begin{bmatrix} 1 & 2 & 1 & \vdots & 2 \\ 0 & 3 & 4 & \vdots & 9 \\ 0 & 1 & -3 & \vdots & -10 \end{bmatrix}$$

We continue down the diagonal and use elementary row operations to get 1 where the element 3 is now. To do this, we interchange rows 2 and 3.

$$\begin{bmatrix} 1 & 2 & 1 & \vdots & 2 \\ 0 & 3 & 4 & \vdots & 9 \\ 0 & 1 & -3 & \vdots & -10 \end{bmatrix} \text{ is equivalent to } \begin{bmatrix} 1 & 2 & 1 & \vdots & 2 \\ 0 & 1 & -3 & \vdots & -10 \\ 0 & 3 & 4 & \vdots & 9 \end{bmatrix}$$

Next we want the new row 3, column 2 element to be 0. We multiply the elements of row 2 by -3 and add the result to the elements of row 3.

$$\begin{bmatrix} 1 & 2 & 1 & \vdots & 2 \\ 0 & 1 & -3 & \vdots & -10 \\ -3(0)+0 & -3(1)+3 & -3(-3)+4 & \vdots & -3(-10)+9 \end{bmatrix} \text{ simplifies to}$$

$$\begin{bmatrix} 1 & 2 & 1 & \vdots & 2 \\ 0 & 1 & -3 & \vdots & -10 \\ 0 & 0 & 13 & \vdots & 39 \end{bmatrix}$$

Finally, we divide the elements of row 3 by 13 so that the final diagonal element is 1.

$$\begin{bmatrix} 1 & 2 & 1 & \vdots & 2 \\ 0 & 1 & -3 & \vdots & -10 \\ \dfrac{0}{13} & \dfrac{0}{13} & \dfrac{13}{13} & \vdots & \dfrac{39}{13} \end{bmatrix} \text{ simplifies to } \begin{bmatrix} 1 & 2 & 1 & \vdots & 2 \\ 0 & 1 & -3 & \vdots & -10 \\ 0 & 0 & 1 & \vdots & 3 \end{bmatrix}$$

This matrix corresponds to the system

$$\begin{cases} x + 2y + z = 2 \\ y - 3z = -10 \\ z = 3 \end{cases}$$

We identify the z-coordinate of the solution as 3. Next we replace z with 3 in the second equation and solve for y.

$$y - 3z = -10 \qquad \text{Second equation}$$
$$y - 3(3) = -10 \qquad \text{Let } z = 3.$$
$$y = -1$$

To find x, we let $z = 3$ and $y = -1$ in the first equation.

$$x + 2y + z = 2 \qquad \text{First equation}$$
$$x + 2(-1) + 3 = 2 \qquad \text{Let } z = 3 \text{ and } y = -1.$$
$$x = 1$$

The ordered triple solution is $(1, -1, 3)$. Check to see that it satisfies all three equations in the original system. ●

EXERCISE SET 4.4

A *Use matrices to solve each system of linear equations. See Examples 1 and 2.*

1. $\begin{cases} x + y = 1 \\ x - 2y = 4 \end{cases}$

2. $\begin{cases} 2x - y = 8 \\ x + 3y = 11 \end{cases}$

3. $\begin{cases} x + 3y = 2 \\ x + 2y = 0 \end{cases}$

4. $\begin{cases} 4x - y = 5 \\ 3x - 3 = 0 \end{cases}$

5. $\begin{cases} x - 2y = 4 \\ 2x - 4y = 4 \end{cases}$

6. $\begin{cases} -x + 3y = 6 \\ 3x - 9y = 9 \end{cases}$

7. $\begin{cases} 3x - 3y = 9 \\ 2x - 2y = 6 \end{cases}$

8. $\begin{cases} 9x - 3y = 6 \\ -18x + 6y = -12 \end{cases}$

9. $\begin{cases} x - 4 = 0 \\ x + y = 1 \end{cases}$

10. $\begin{cases} 3y = 6 \\ x + y = 7 \end{cases}$

B *Use matrices to solve each system of linear equations. See Example 3.*

11. $\begin{cases} x + y = 3 \\ 2y = 10 \\ 3x + 2y - 4z = 12 \end{cases}$

12. $\begin{cases} 5x = 5 \\ 2x + y = 4 \\ 3x + y - 5z = -15 \end{cases}$

13. $\begin{cases} 2y - z = -7 \\ x + 4y + z = -4 \\ 5x - y + 2z = 13 \end{cases}$

14. $\begin{cases} 4y + 3z = -2 \\ 5x - 4y = 1 \\ -5x + 4y + z = -3 \end{cases}$

15. $\begin{cases} x + y + z = 2 \\ 2x - z = 5 \\ 3y + z = 2 \end{cases}$

16. $\begin{cases} x + 2y + z = 5 \\ x - y - z = 3 \\ y + z = 2 \end{cases}$

17. $\begin{cases} 4x + y + z = 3 \\ -x + y - 2z = -11 \\ x + 2y + 2z = -1 \end{cases}$

18. $\begin{cases} x + y + z = 9 \\ 3x - y + z = -1 \\ -2x + 2y - 3z = -2 \end{cases}$

Review and Preview

Evaluate. See Section 1.4.

19. $(-1)(-5) - (6)(3)$

20. $(2)(-8) - (-4)(1)$

21. $(4)(-10) - (2)(-2)$

22. $(-7)(3) - (-2)(-6)$

23. $(-3)(-3) - (-1)(-9)$

24. $(5)(6) - (10)(10)$

 Combining Concepts

25. The percent y of U.S. households that owned a black-and-white television set between the years 1980 and 1993 can be modeled by the linear equation $2.3x + y = 52$, where x represents the number of years after 1980. Similarly, the percent y of U.S. households that owned a microwave oven during this same period can be modeled by the linear equation $-5.4x + y = 14$. (*Source:* Based on data from the Energy Information Administration, U.S. Department of Energy)

a. Determine the year in which the percent of households owning black-and-white television sets was the same as the percent of households owning microwave ovens. Use matrix methods to estimate the year in which this occurred.

b. Did more households own black-and-white television sets or microwave ovens in 1980? In 1993? What trends do these models show? Does this seem to make sense? Why or why not?

c. According to the models, when will the percent of households owning black-and-white television sets reach 0%?

4.5 Systems of Linear Inequalities

(A) Graphing Systems of Linear Inequalities

In Section 3.6 we solved linear inequalities in two variables. Just as two linear equations make a system of linear equations, two linear inequalities make a **system of linear inequalities**. Systems of inequalities are very important in a process called linear programming. Many businesses use linear programming to find the most profitable way to use limited resources such as employees, machines, or buildings.

A **solution of a system of linear inequalities** is an ordered pair that satisfies each inequality in the system. The set of all such ordered pairs is the solution set of the system. Graphing this set gives us a picture of the solution set. We can graph a system of inequalities by graphing each inequality in the system and identifying the region of overlap.

Graphing the Solutions of a System of Linear Inequalities

Step 1.   Graph each inequality in the system on the same set of axes.

Step 2.   The solutions of the system are the points common to the graphs of all the inequalities in the system.

EXAMPLE 1   Graph the solutions of the system: $\begin{cases} 3x \geq y \\ x + 2y \leq 8 \end{cases}$

Solution: We begin by graphing each inequality on the *same* set of axes. The graph of the solutions of the system is the region contained in the graphs of both inequalities. In other words, it is their intersection.

First let's graph $3x \geq y$. The boundary line is the graph of $3x = y$. We sketch a solid boundary line since the inequality $3x \geq y$ means $3x > y$ or $3x = y$. The test point $(1, 0)$ satisfies the inequality, so we shade the half-plane that includes $(1, 0)$.

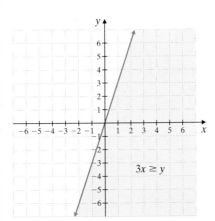

$3x \geq y$

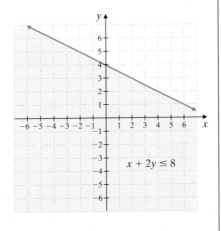

$x + 2y \leq 8$

Next we sketch a solid boundary line $x + 2y = 8$ on the same set of axes. The test point $(0, 0)$ satisfies the inequality $x + 2y \leq 8$, so we shade the half-plane that includes $(0, 0)$. (For clarity, the graph of $x + 2y \leq 8$ is shown here on a separate set of axes.)

An ordered pair solution of the system must satisfy both inequalities. These solutions are points that lie in both shaded regions. The solution of the

OBJECTIVE

(A) Graph a system of linear inequalities

SSM
TUTOR CENTER   SG   CD & VIDEO   MATH PRO   WEB

Practice Problem 1

Graph the solutions of the system:

$$\begin{cases} 2x \leq y \\ x + 4y \geq 4 \end{cases}$$

Answer

1.

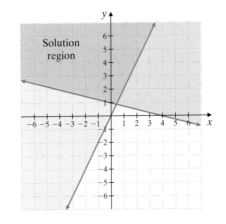

system is the darkest shaded region. This solution includes parts of both boundary lines.

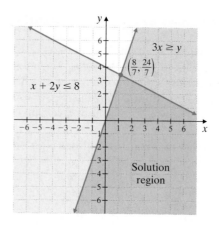

Practice Problem 2

Graph the solutions of the system:

$$\begin{cases} -x + y < 3 \\ y < 1 \\ 2x + y > -2 \end{cases}$$

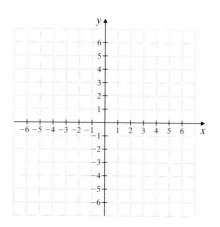

In linear programming, it is sometimes necessary to find the coordinates of the **corner point**: the point at which the two boundary lines intersect. To find the corner point for the system of Example 1, we solve the related linear system

$$\begin{cases} 3x = y \\ x + 2y = 8 \end{cases}$$

using either the substitution or the elimination method. The lines intersect at $\left(\dfrac{8}{7}, \dfrac{24}{7} \right)$, the corner point of the graph.

EXAMPLE 2   Graph the solutions of the system: $\begin{cases} x - y < 2 \\ x + 2y > -1 \\ y < 2 \end{cases}$

Solution: First we graph all three inequalities on the same set of axes. All boundary lines are dashed lines since the inequality symbols are $<$ and $>$. The solution of the system is the region shown by the darkest shading. In this example, the boundary lines are *not* a part of the solution.

Answer

2.

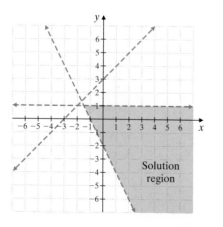

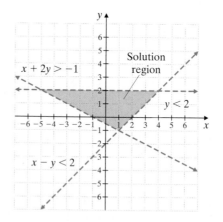

Try the Concept Check in the margin.

EXAMPLE 3   Graph the solutions of the system: $\begin{cases} -3x + 4y \le 12 \\ \quad\quad x \le 3 \\ \quad\quad x \ge 0 \\ \quad\quad y \ge 0 \end{cases}$

Solution: We graph the inequalities on the same set of axes. The intersection of the inequalities is the solution region. It is the only region shaded in this graph and includes the portions of all four boundary lines that border the shaded region.

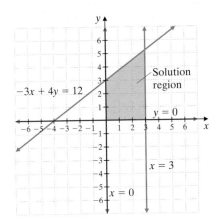

Concept Check

Describe the solution of the system of inequalities:

$\begin{cases} x \le 2 \\ x \ge 2 \end{cases}$

Practice Problem 3

Graph the solutions of the system:

$\begin{cases} 2x - 3y \le 6 \\ \quad\quad y \ge 0 \\ \quad\quad y \le 4 \\ \quad\quad x \ge 0 \end{cases}$

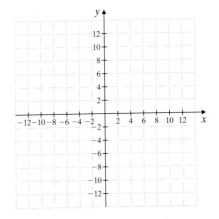

Answers

3.

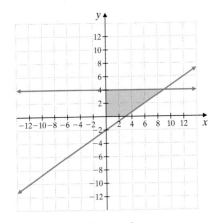

Concept Check: the line $x = 2$

FOCUS ON **The Real World**

LINEAR MODELING

In Chapter 3, we learned several ways to find a linear model when given either two ordered pairs or an ordered pair and slope. Another way to find a linear model of the form $y = mx + b$ for two ordered pairs (x_1, y_1) and (x_2, y_2) is to solve the following system of linear equations for m and b:

$$\begin{cases} y_1 = mx_1 + b \\ y_2 = mx_2 + b \end{cases}$$

For example, suppose a researcher wishes to find a linear model for the number of traffic accidents involving teenagers. The researcher locates statistics stating that there were 9920 teenage deaths from motor vehicle accidents in the United States in 1979. By 1999, this number had decreased to 5749 teenage deaths in motor vehicle accidents. (*Source:* Insurance Institute for Highway Safety)

This data gives two ordered pairs: (1979, 9920) and (1999, 5749). Alternatively, the ordered pairs could be written as (9, 9920) and (29, 5749), where the x-coordinate represents the number of years after 1970. (Adjusting data given as years in this way often simplifies calculations.) By substituting the coordinates of the second set of ordered pairs into the general linear system, we obtain the system

$$\begin{cases} 9920 = 9m + b \\ 5749 = 29m + b \end{cases}$$

The solution of this system is $m = -208.55$ and $b = 11,796.95$. We can use these values to write the model the researcher wished to find: $y = -208.55x + 11,796.95$, where y is the number of teenage deaths in motor vehicle accidents x years after 1990.

Internet Excursions

Go To: http://www.prenhall.com/martin-gay_interm   What's Related

The Insurance Institute for Highway Safety (IIHS) is a nonprofit research organization supported by automobile insurers that collects, studies, and distributes motor vehicle safety statistics. This page offers access to a wide variety of traffic fatality statistics organized by categories such as alcohol-related fatalities, fatalities of children, and fatalities involving bicycles. (Alternatively, you can visit the IIHS homepage at http://www.hwysafety.org and look for the Fatality Facts option.)

1. Browse the list to find a set of data that interests you. Make a list of the ordered pairs that make up the set of data.

2. Create a scatter diagram of the data. Does the data appear approximately linear? If not, is there a portion of the data that appears approximately linear? If so, indicate which portion is approximately linear. If not, start over with Question 1.

3. Pick two ordered pairs from the linear portion of the data. Use these ordered pairs to form a system of linear equations.

4. Solve the system from Question 3. Find the linear equation that models your data.

5. Add the graph of your linear model to the scatter diagram from Question 2. How well does your model "fit" the data?

6. What trend does your model describe over the linear portion of your data?

EXERCISE SET 4.5

A *Graph the solutions of each system of linear inequalities. See Examples 1 through 3.*

1. $\begin{cases} y \geq x + 1 \\ y \geq 3 - x \end{cases}$

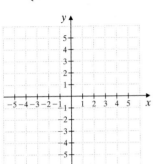

2. $\begin{cases} y \geq x - 3 \\ y \geq -1 - x \end{cases}$

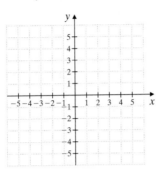

3. $\begin{cases} y < 3x - 4 \\ y \leq x + 2 \end{cases}$

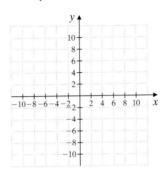

4. $\begin{cases} y \leq 2x + 1 \\ y > x + 2 \end{cases}$

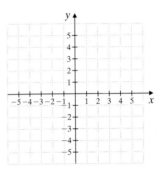

5. $\begin{cases} y \leq -2x - 2 \\ y \geq x + 4 \end{cases}$

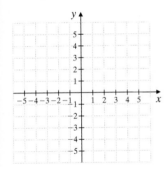

6. $\begin{cases} y \leq 2x + 4 \\ y \geq -x - 5 \end{cases}$

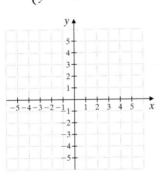

7. $\begin{cases} y \geq -x + 2 \\ y \leq 2x + 5 \end{cases}$

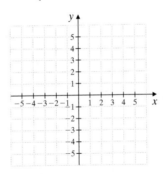

8. $\begin{cases} y \geq x - 5 \\ y \leq -3x + 3 \end{cases}$

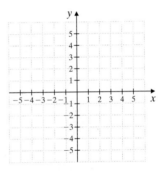

9. $\begin{cases} x \geq 3y \\ x + 3y \leq 6 \end{cases}$

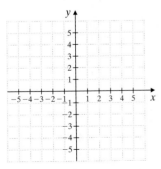

10. $\begin{cases} -2x < y \\ x + 2y < 3 \end{cases}$

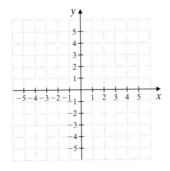

11. $\begin{cases} x \leq 2 \\ y \geq -3 \end{cases}$

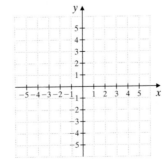

12. $\begin{cases} x \geq -3 \\ y \geq -2 \end{cases}$

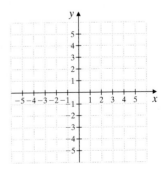

13. $\begin{cases} y \geq 1 \\ x < -3 \end{cases}$

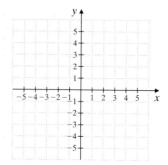

14. $\begin{cases} y > 2 \\ x \geq -1 \end{cases}$

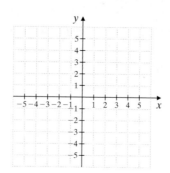

15. $\begin{cases} y + 2x \geq 0 \\ 5x - 3y \leq 12 \\ y \leq 2 \end{cases}$

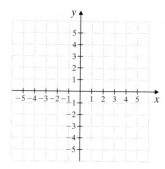

16. $\begin{cases} y + 2x \leq 0 \\ 5x + 3y \geq -2 \\ y \leq 4 \end{cases}$

17. $\begin{cases} 3x - 4y \geq -6 \\ 2x + y \leq 7 \\ y \geq -3 \end{cases}$

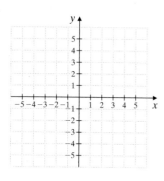

18. $\begin{cases} 4x - y \geq -2 \\ 2x + 3y \leq -8 \\ y \geq -5 \end{cases}$

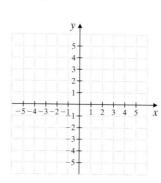

19. $\begin{cases} 2x + y \leq 5 \\ x \leq 3 \\ x \geq 0 \\ y \geq 0 \end{cases}$

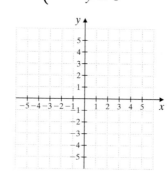

20. $\begin{cases} 3x + y \leq 4 \\ x \leq 4 \\ x \geq 0 \\ y \geq 0 \end{cases}$

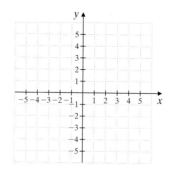

Match each system of inequalities to the corresponding graph.

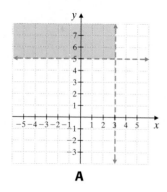

A

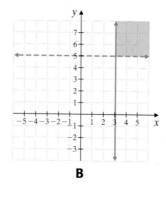

B

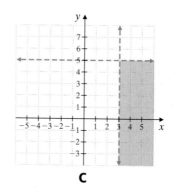

C

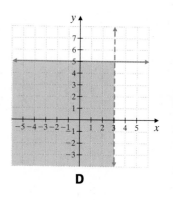

D

21. $\begin{cases} y < 5 \\ x > 3 \end{cases}$

22. $\begin{cases} y > 5 \\ x < 3 \end{cases}$

23. $\begin{cases} y \leq 5 \\ x < 3 \end{cases}$

24. $\begin{cases} y > 5 \\ x \geq 3 \end{cases}$

Review and Preview

Evaluate each expression. See Section 1.7.

25. $(-3)^2$

26. $(-5)^3$

27. $\left(\dfrac{2}{3}\right)^2$

28. $\left(\dfrac{3}{4}\right)^3$

Perform each indicated operation. See Section 1.4.

29. $(-2)^2 - (-3) + 2(-1)$

30. $5^2 - 11 + 3(-5)$

31. $8^2 + (-13) - 4(-2)$

32. $(-12)^2 + (-1)(2) - 6$

Combining Concepts

33. Tony Noellert budgets his time at work today. Part of the day he can write bills; the rest of the day he can use to write purchase orders. The total time available is at most 8 hours. Less than 3 hours is to be spent writing bills.

   a. Write a system of inequalities to describe the situation. (Let x = hours available for writing bills and y = hours available for writing purchase orders.)

   b. Graph the solutions of the system.

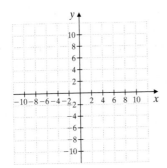

34. Explain how to decide which region to shade to show the solution region of the following system.

$$\begin{cases} x \geq 3 \\ y \geq -2 \end{cases}$$

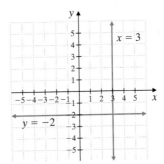

MATERIALS:

- calculator
- graphing utility (optional)

This activity may be completed by working in groups or individually.

Weather-recording stations use a directional antenna to detect and measure the electromagnetic field emitted by a lightning bolt. The antenna can determine the angle between a fixed point and the position of the lightning strike but cannot determine the distance to the lightning strike. However, the angle measured by the antenna can be used to find the slope of the line connecting the positions of the weather station and the lightning strike. From there, the equation of the line connecting the positions of the weather station and the lightning strike may be found. If two such lines may be found—that is, if another weather station's antenna detects the same lightning flash—the coordinates of the lightning strike's position may be pinpointed.

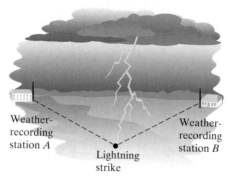

Weather-recording station *A*

Weather-recording station *B*

Lightning strike

1. A weather-recording station *A* is located at the coordinates $(35, 28)$. A second weather-recording station *B* is located at the coordinates $(52, 12)$. Plot the position of the two weather-recording stations.

2. A lightning strike is detected by both stations. Station *A* uses a measured angle to find the slope of the line from the station to the lightning strike as $m = -1.732$. Station *B* computes a slope of $m = 0.577$ from the angle it measured. Use this information to find the equations of the lines connecting each station to the position of the lightning strike.

3. Solve the resulting system of equations in each of the following ways (or work with other students in your class so that each student solves the system in one of the following ways):

   a. Using a graph. Graph the two equations on your plot of the positions of the two weather-recording stations. Estimate the coordinates of their point of intersection.

   b. Using either the method of substitution or of elimination (whichever you prefer)

   c. Using matrices

   d. (Optional) Using a graphing utility to graph the lines and use the utility's intersect feature to estimate the coordinates of their point of intersection.

4. Compare the results from each method. What are the coordinates of the lightning strike? Which method do you prefer? Why?

Chapter 4   VOCABULARY CHECK

Fill in each blank with one of the words or phrases listed below.

matrix   consistent   system of equations

solution   inconsistent   square

1. Two or more linear equations in two variables form a _____ .

2. A _____ of a system of two equations in two variables is an ordered pair that makes both equations true.

3. A(n) _____ system of equations has at least one solution.

4. If a matrix has the same number of rows and columns, it is called a _____ matrix.

5. A(n) _____ system of equations has no solution.

6. A _____ is a rectangular array of numbers.

C H A P T E R

Highlights

DEFINITIONS AND CONCEPTS	**EXAMPLES**

Section 4.1   Solving Systems of Linear Equations in Two Variables

A **system of linear equations** consists of two or more linear equations.

$$\begin{cases} x - 3y = 6 \\ y = \dfrac{1}{2}x \end{cases} \qquad \begin{cases} x + 2y - z = 1 \\ 3x - y + 4z = 0 \\ 5y + z = 6 \end{cases}$$

A **solution** of a system of two equations in two variables is an ordered pair (x, y) that makes both equations true.

Determine whether $(2, -5)$ is a solution of the system.

$$\begin{cases} x + y = -3 \\ 2x - 3y = 19 \end{cases}$$

Replace x with 2 and y with -5 in both equations.

$$x + y = -3 \qquad\qquad\quad 2x - 3y = 19$$
$$2 + (-5) = -3 \qquad\qquad 2(2) - 3(-5) = 19$$
$$-3 = -3 \quad \text{True.} \qquad\qquad 4 + 15 = 19$$
$$\qquad\qquad\qquad\qquad\qquad\qquad 19 = 19 \quad \text{True.}$$

$(2, -5)$ is a solution of the system.

Geometrically, a solution of a system in two variables is a point common to the graphs of the equations.

Solve by graphing: $\begin{cases} y = 2x - 1 \\ x + 2y = 13 \end{cases}$

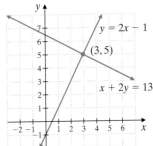

Section 4.1   Solving Systems of Linear Equations in Two Variables (continued)

A system of equations with at least one solution is a **consistent system**. A system that has no solution is an **inconsistent system**.

If the graphs of two linear equations are identical, the equations are **dependent**.

If their graphs are different, the equations are **independent**.

One solution:
Independent equations
Consistent system

No solution:
Independent equations
Inconsistent system

Infinite number of solutions:
Dependent equations
Consistent system

SOLVING A SYSTEM OF LINEAR EQUATIONS BY THE SUBSTITUTION METHOD

Step 1. Solve one equation for a variable.

Step 2. Substitute the expression for the variable into the other equation.

Step 3. Solve the equation from Step 2 to find the value of one variable.

Step 4. Substitute the value from Step 3 in either original equation to find the value of the other variable.

Step 5. Check the solution in both equations.

Solve by substitution:

$$\begin{cases} y = x + 2 \\ 3x - 2y = -5 \end{cases}$$

Substitute $x + 2$ for y in the second equation.

$$3x - 2y = -5 \qquad \text{Second equation}$$

$$3x - 2(x + 2) = -5 \qquad \text{Let } y = x + 2.$$

$$3x - 2x - 4 = -5$$

$$x - 4 = -5 \qquad \text{Simplify.}$$

$$x = -1 \qquad \text{Add 4.}$$

To find y, let $x = -1$ in $y = x + 2$, so $y = -1 + 2 = 1$. The solution $(-1, 1)$ checks.

SOLVING A SYSTEM OF LINEAR EQUATIONS BY THE ELIMINATION METHOD

Step 1. Rewrite each equation in standard form, $Ax + By = C$.

Step 2. Multiply one or both equations by a nonzero number so that the coefficients of a variable are opposites.

Step 3. Add the equations.

Step 4. Find the value of the remaining variable by solving the resulting equation.

Step 5. Substitute the value from Step 4 into either original equation to find the value of the other variable.

Step 6. Check the solution in both equations.

Solve by elimination:

$$\begin{cases} x - 3y = -3 \\ -2x + y = 6 \end{cases}$$

Multiply both sides of the first equation by 2.

$$\begin{array}{r} 2x - 6y = -6 \\ -2x + y = 6 \\ \hline -5y = 0 \qquad \text{Add.} \\ y = 0 \qquad \text{Divide by } -5. \end{array}$$

To find x, let $y = 0$ in an original equation.

$$x - 3y = -3$$

$$x - 3 \cdot 0 = -3$$

$$x = -3$$

The solution $(-3, 0)$ checks.

DEFINITIONS AND CONCEPTS

EXAMPLES

Section 4.2   Solving Systems of Linear Equations in Three Variables

A **solution** of an equation in three variables x, y, and z is an **ordered triple** (x, y, z) that makes the equation a true statement.

Verify that $(-2, 1, 3)$ is a solution of $2x + 3y - 2z = -7$. Replace x with -2, y with 1, and z with 3.

$$2(-2) + 3(1) - 2(3) = -7$$
$$-4 + 3 - 6 = -7$$
$$-7 = -7 \quad \text{True.}$$

$(-2, 1, 3)$ is a solution.

SOLVING A SYSTEM OF THREE LINEAR EQUATIONS BY THE ELIMINATION METHOD

Step 1. Write each equation in standard form, $Ax + By + Cz = D$.

Step 2. Choose a pair of equations and use them to eliminate a variable.

Step 3. Choose any other pair of equations and eliminate the same variable.

Step 4. Solve the system of two equations in two variables from Steps 2 and 3.

Step 5. Solve for the third variable by substituting the values of the variables from Step 4 into any of the original equations.

Step 6. Check the solution in all three original equations.

Solve:

$$\begin{cases} 2x + y - z = 0 & (1) \\ x - y - 2z = -6 & (2) \\ -3x - 2y + 3z = -22 & (3) \end{cases}$$

1. Each equation is written in standard form.

2. $\begin{aligned} 2x + y - z &= 0 \quad (1) \\ \underline{x - y - 2z} &= \underline{-6} \quad (2) \\ 3x \qquad - 3z &= -6 \quad (4) \quad \text{Add.} \end{aligned}$

3. Eliminate y from equations (1) and (3) also.

$$\begin{aligned} 4x + 2y - 2z &= 0 & &\text{Multiply equation} \\ \underline{-3x - 2y + 3z} &= \underline{-22} \quad (3) & &(1) \text{ by } 2. \\ x \qquad + z &= -22 \quad (5) & &\text{Add.} \end{aligned}$$

4. Solve.

$$\begin{cases} 3x - 3z = -6 & (4) \\ x + z = -22 & (5) \end{cases}$$

$$\begin{aligned} x - z &= -2 & &\text{Divide equation (4) by 3.} \\ \underline{x + z} &= \underline{-22} & &(5) \\ 2x \quad &= -24 \\ x \quad &= -12 \end{aligned}$$

To find z, use equation (5).

$$x + z = -22$$
$$-12 + z = -22$$
$$z = -10$$

5. To find y, use equation (1).

$$2x + y - z = 0$$
$$2(-12) + y - (-10) = 0$$
$$-24 + y + 10 = 0$$
$$y = 14$$

6. The solution $(-12, 14, -10)$ checks.

DEFINITIONS AND CONCEPTS	**EXAMPLES**

Section 4.3   Systems of Linear Equations and Problem Solving

	Two numbers have a sum of 11. Twice one number is 3 less than 3 times the other. Find the numbers.
1. UNDERSTAND the problem.	**1.** Read and reread.
	x = one number
	y = other number
2. TRANSLATE.	**2.** In words:

	sum of numbers	is	11
	$\downarrow$	$\downarrow$	$\downarrow$
Translate:	$x + y$	$=$	11

In words:

	twice one number	is	3 less than 3 times the other number
	$\downarrow$	$\downarrow$	$\downarrow$
Translate:	$2x$	$=$	$3y - 3$

3. SOLVE.

3. Solve the system: $\begin{cases} x + y = 11 \\ 2x = 3y - 3 \end{cases}$

In the first equation, $x = 11 - y$. Substitute into the other equation.

$$2x = 3y - 3$$
$$2(11 - y) = 3y - 3$$
$$22 - 2y = 3y - 3$$
$$-5y = -25$$
$$y = 5$$

Replace y with 5 in the equation $x = 11 - y$. Then $x = 11 - 5 = 6$. The solution is $(6, 5)$.

4. INTERPRET.

4. *Check:*   See that $6 + 5 = 11$ is the required sum and that twice 6 is 3 times 5 less 3. *State:*   The numbers are 6 and 5.

Section 4.4   Solving Systems of Equations Using Matrices

A **matrix** is a rectangular array of numbers.

$$\begin{bmatrix} -7 & 0 & 3 \\ 1 & 2 & 4 \end{bmatrix} \quad \begin{bmatrix} a & b & c \\ d & e & f \\ g & h & i \end{bmatrix}$$

The **matrix** corresponding to a system is composed of the coefficients of the variables and the constants of the system.

The matrix corresponding to the system

$$\begin{cases} x - y = 1 \\ 2x + y = 11 \end{cases} \text{ is } \begin{bmatrix} 1 & -1 & \vdots & 1 \\ 2 & 1 & \vdots & 11 \end{bmatrix}$$

| **DEFINITIONS AND CONCEPTS** | **EXAMPLES** |

Section 4.4 Solving Systems of Equations Using Matrices *(continued)*

The following **row operations** can be performed on matrices, and the result is an equivalent matrix.

Elementary row operations:

1. Interchange any two rows.
2. Multiply (or divide) the elements of one row by the same nonzero number.
3. Multiply (or divide) the elements of one row by the same nonzero number and add them to their corresponding elements in any other row.

Use matrices to solve: $\begin{cases} x - y = 1 \\ 2x + y = 11 \end{cases}$

The corresponding matrix is

$$\left[\begin{array}{cc|c} 1 & -1 & 1 \\ 2 & 1 & 11 \end{array} \right]$$

Use row operations to write an equivalent matrix with 1s along the diagonal and 0s below each 1 in the diagonal. Multiply row 1 by -2 and add to row 2. Change row 2 only.

$$\left[\begin{array}{cc|c} 1 & -1 & 1 \\ -2(1) + 2 & -2(-1) + 1 & -2(1) + 11 \end{array} \right]$$

simplifies to $\left[\begin{array}{cc|c} 1 & -1 & 1 \\ 0 & 3 & 9 \end{array} \right]$

Divide row 2 by 3.

$$\left[\begin{array}{cc|c} 1 & -1 & 1 \\ \frac{0}{3} & \frac{3}{3} & \frac{9}{3} \end{array} \right] \text{ simplifies to } \left[\begin{array}{cc|c} 1 & -1 & 1 \\ 0 & 1 & 3 \end{array} \right]$$

This matrix corresponds to the system

$$\begin{cases} x - y = 1 \\ y = 3 \end{cases}$$

Let $y = 3$ in the first equation.

$$x - 3 = 1$$
$$x = 4$$

The ordered pair solution is $(4, 3)$.

Section 4.5 Systems of Linear Inequalities

A **system of linear inequalities** consists of two or more linear inequalities.

To graph a system of inequalities, graph each inequality in the system. The overlapping region is the solution of the system.

$$\begin{cases} x - y \geq 3 \\ y \leq -2x \end{cases}$$

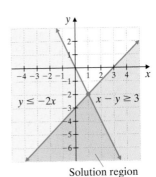

Solution region

STUDY SKILLS REMINDER

How well do you know your textbook?

See if you can answer the questions below.

1. What does the 🔓 icon mean?

2. What does the ╲ icon mean?

3. What does the △ icon mean?

4. Where can you find a review for each chapter? What answers to this review can be found in the back of your text?

5. Each chapter contains an overview of the chapter along with examples. What is this feature called?

6. Does this text contain any solutions to exercises? If so, where?

STUDY SKILLS REMINDER

What should you do the day of an exam?

On the day of an exam, try the following:

- Allow yourself plenty of time to arrive.

- Read the directions on the test carefully.

- Read each problem carefully as you take your test. Make sure that you answer the question asked.

- Watch your time and pace yourself so that you may attempt each problem on your test.

- If you have time, check your work and answers.

- Do not turn your test in early. If you have extra time, spend it double-checking your work.

Good luck!

Chapter 4 Review

(4.1) *Solve each system of equations in two variables by each method: (a) graphing, (b) substitution, and (c) elimination.*

1. $\begin{cases} 3x + 10y = 1 \\ x + 2y = -1 \end{cases}$

2. $\begin{cases} y = \dfrac{1}{2}x + \dfrac{2}{3} \\ 4x + 6y = 4 \end{cases}$

3. $\begin{cases} 2x - 4y = 22 \\ 5x - 10y = 15 \end{cases}$

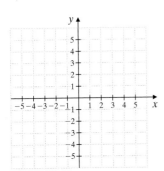
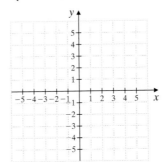
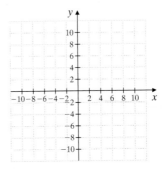

4. $\begin{cases} 3x - 6y = 12 \\ 2y = x - 4 \end{cases}$

5. $\begin{cases} \dfrac{1}{2}x - \dfrac{3}{4}y = -\dfrac{1}{2} \\ \dfrac{1}{8}x + \dfrac{3}{4}y = \dfrac{19}{8} \end{cases}$

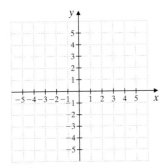
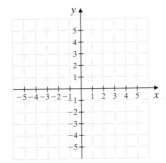

6. The revenue equation for a certain style of backpack is $y = 32x$, where x is the number of backpacks sold and y is the income in dollars for selling x backpacks. The cost equation for these units is $y = 15x + 25,500$, where x is the number of backpacks manufactured and y is the cost in dollars for manufacturing x backpacks. Find the number of units to be sold for the company to break even. (*Hint:* Solve the system of equations formed by the two given equations.)

(4.2) *Solve each system of equations in three variables.*

7. $\begin{cases} x \qquad + z = 4 \\ 2x - y \qquad = 4 \\ x + y - z = 0 \end{cases}$

8. $\begin{cases} 2x + 5y \qquad = 4 \\ x - 5y + z = -1 \\ 4x \qquad - z = 11 \end{cases}$

9. $\begin{cases} 4y + 2z = 5 \\ 2x + 8y \qquad = 5 \\ 6x + \qquad 4z = 1 \end{cases}$

10. $\begin{cases} 5x + 7y = 9 \\ 14y - z = 28 \\ 4x + 2z = -4 \end{cases}$

11. $\begin{cases} 3x - 2y + 2z = 5 \\ -x + 6y + z = 4 \\ 3x + 14y + 7z = 20 \end{cases}$

12. $\begin{cases} x + 2y + 3z = 11 \\ y + 2z = 3 \\ 2x + 2z = 10 \end{cases}$

13. $\begin{cases} 7x - 3y + 2z = 0 \\ 4x - 4y - z = 2 \\ 5x + 2y + 3z = 1 \end{cases}$

14. $\begin{cases} x - 3y - 5z = -5 \\ 4x - 2y + 3z = 13 \\ 5x + 3y + 4z = 22 \end{cases}$

(4.3) *Use systems of equations to solve.*

15. The sum of three numbers is 98. The sum of the first and second is two more than the third number, and the second is four times the first. Find the numbers.

16. One number is three times a second number, and twice the sum of the numbers is 168. Find the numbers.

17. Two cars leave Chicago, one traveling east and the other west. After 4 hours they are 492 miles apart. If one car is traveling 7 mph faster than the other, find the speed of each.

△ **18.** The foundation for a rectangular Hardware Warehouse has a length three times the width and is 296 feet around. Find the dimensions of the building.

19. James Callahan has available a 10% alcohol solution and a 60% alcohol solution. Find how many liters of each solution he should mix to make 50 liters of a 40% alcohol solution.

20. An employee at See's Candy Store needs a special mixture of candy. She has creme-filled chocolates that sell for $3.00 per pound, chocolate-covered nuts that sell for $2.70 per pound, and chocolate-covered raisins that sell for $2.25 per pound. She wants to have twice as many raisins as nuts in the mixture. Find how many pounds of each she should use to make 45 pounds worth $2.80 per pound.

21. Chris Kringler has $2.77 in her coin jar—all in pennies, nickels, and dimes. If she has 53 coins in all and four more nickels than dimes, find how many of each type of coin she has.

22. If $10,000 and $4000 are invested such that $1250 in interest is earned in one year, and if the rate of interest on the larger investment is 2% more than that on the smaller investment, find the rates of interest.

△23. The perimeter of an isosceles (two sides equal) triangle is 73 centimeters. If the unequal side is 7 centimeters longer than the two equal sides, find the lengths of the three sides.

24. The sum of three numbers is 295. One number is five more than a second and twice the third. Find the numbers.

(4.4) *Use matrices to solve each system.*

25. $\begin{cases} 3x + 10y = 1 \\ x + 2y = -1 \end{cases}$

26. $\begin{cases} 3x - 6y = 12 \\ 2y = x - 4 \end{cases}$

27. $\begin{cases} 3x - 2y = -8 \\ 6x + 5y = 11 \end{cases}$

28. $\begin{cases} 6x - 6y = -5 \\ 10x - 2y = 1 \end{cases}$

29. $\begin{cases} 3x - 6y = 0 \\ 2x + 4y = 5 \end{cases}$

30. $\begin{cases} 5x - 3y = 10 \\ -2x + y = -1 \end{cases}$

31. $\begin{cases} 0.2x - 0.3y = -0.7 \\ 0.5x + 0.3y = 1.4 \end{cases}$

32. $\begin{cases} 3x + 2y = 8 \\ 3x - y = 5 \end{cases}$

33. $\begin{cases} x + z = 4 \\ 2x - y = 0 \\ x + y - z = 0 \end{cases}$

34. $\begin{cases} 2x + 5y = 4 \\ x - 5y + z = -1 \\ 4x - z = 11 \end{cases}$

35. $\begin{cases} 3x - y = 11 \\ x + 2z = 13 \\ y - z = -7 \end{cases}$

36. $\begin{cases} 5x + 7y + 3z = 9 \\ 14y - z = 28 \\ 4x + 2z = -4 \end{cases}$

37. $\begin{cases} 7x - 3y + 2z = 0 \\ 4x - 4y - z = 2 \\ 5x + 2y + 3z = 1 \end{cases}$

38. $\begin{cases} x + 2y + 3z = 14 \\ y + 2z = 3 \\ 2x - 2z = 10 \end{cases}$

(4.5) *Graph the solution of each system of linear inequalities.*

39. $\begin{cases} y \geq 2x - 3 \\ y \leq -2x + 1 \end{cases}$

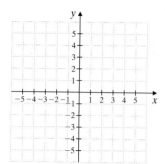

40. $\begin{cases} y \leq -3x - 3 \\ y \leq 2x + 7 \end{cases}$

41. $\begin{cases} x + 2y > 0 \\ x - y \leq 6 \end{cases}$

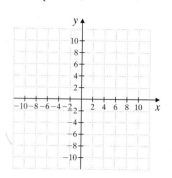

42. $\begin{cases} x - 2y \geq 7 \\ x + y \leq -5 \end{cases}$

43. $\begin{cases} 3x - 2y \leq 4 \\ 2x + y \geq 5 \\ y \leq 4 \end{cases}$

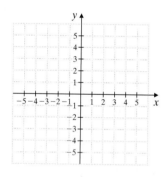

44. $\begin{cases} 4x - y \leq 0 \\ 3x - 2y \geq -5 \\ y \geq -4 \end{cases}$

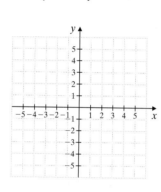

45. $\begin{cases} x + 2y \leq 5 \\ x \leq 2 \\ x \geq 0 \\ y \geq 0 \end{cases}$

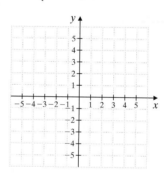

46. $\begin{cases} x + 3y \leq 7 \\ y \leq 5 \\ x \geq 0 \\ y \geq 0 \end{cases}$

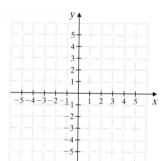

Chapter 4 Test

Solve each system of equations graphically and then solve by the elimination method or the substitution method.

1. $\begin{cases} 2x - y = -1 \\ 5x + 4y = 17 \end{cases}$

2. $\begin{cases} 7x - 14y = 5 \\ x \quad\quad = 2y \end{cases}$

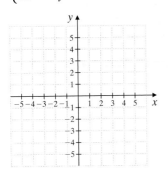

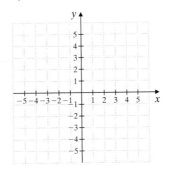

Solve each system.

3. $\begin{cases} 4x - 7y = 29 \\ 2x + 5y = -11 \end{cases}$

4. $\begin{cases} 15x + 6y = 15 \\ 10x + 4y = 10 \end{cases}$

5. $\begin{cases} 2x - 3y \quad\quad = 4 \\ 3y + 2z = 2 \\ x \quad\quad - z = -5 \end{cases}$

6. $\begin{cases} 3x - 2y - z = -1 \\ 2x - 2y \quad\quad = 4 \\ 2x \quad\quad - 2z = -12 \end{cases}$

7. $\begin{cases} \dfrac{x}{2} + \dfrac{y}{4} = -\dfrac{3}{4} \\ x + \dfrac{3}{4}y = -4 \end{cases}$

8. $\begin{cases} 3x - y = 7 \\ 2x + 5y = -1 \end{cases}$

9. $\begin{cases} 4x - 3y = -6 \\ -2x + y = 0 \end{cases}$

10. $\begin{cases} x + y + z = 4 \\ 2x + 5y \quad\quad = 1 \\ x - y - 2z = 0 \end{cases}$

11. $\begin{cases} 3x + 2y + 3z = 3 \\ x \quad\quad - z = 9 \\ 4y + z = -4 \end{cases}$

Use matrices to solve each system.

12. $\begin{cases} x - y = -2 \\ 3x - 3y = -6 \end{cases}$

13. $\begin{cases} x + 2y = -1 \\ 2x + 5y = -5 \end{cases}$

14.
$$\begin{cases} x - y - z = 0 \\ 3x - y - 5z = -2 \\ 2x + 3y = -5 \end{cases}$$

15.
$$\begin{cases} 2x - y + 3z = 4 \\ 3x - 3z = -2 \\ -5x + y = 0 \end{cases}$$

16. A motel in New Orleans charges $90 per day for double occupancy and $80 per day for single occupancy. If 80 rooms are occupied for a total of $6930, how many rooms of each kind are occupied?

17. The research department of a company that manufactures children's fruit drinks is experimenting with a new flavor. A 17.5% fructose solution is needed, but only 10% and 20% solutions are available. How many gallons of a 10% fructose solution should be mixed with a 20% fructose solution to obtain 20 gallons of a 17.5% fructose solution?

18. Frame Masters, Inc., recently purchased $5500 worth of new equipment to produce a new style of eyeglass frame. The marketing department of Frame Masters estimates that the cost of producing this new frame is $18 and that the frame will be sold to stores for $38. Find the number of frames that must be sold in order to break even.

Graph the solutions of each system of linear inequalities.

19.
$$\begin{cases} 2y - x \geq 1 \\ x + y \geq -4 \\ y \leq 2 \end{cases}$$

20.
$$\begin{cases} y + 2x \leq 4 \\ y \leq 2 \\ y \geq 0 \\ x \geq 0 \end{cases}$$

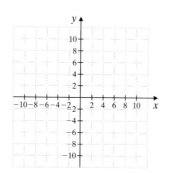

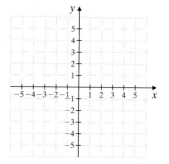

Cumulative Review

Determine whether each statement is true or false.

1. $7 \notin \{1, 2, 3\}$

2. $\dfrac{1}{5}$ is an irrational number.

Insert $<, >$, *or* $=$ *between each pair of numbers to form a true statement.*

3. $-5 \qquad 5$

4. $-2.5 \qquad -2.1$

5. $\dfrac{2}{3} \qquad \dfrac{3}{4}$

Find each absolute value.

6. $|-4|$

7. $-|-8|$

8. Simplify: $\dfrac{|-2|^3 + 1}{-7 - \sqrt{4}}$

Simplify and write with positive exponents only.

9. $(3x)^{-1}$

10. $2^{-1} + 3^{-2}$

11. Determine whether -15 is a solution of $x - 9 = -24$.

12. Suppose that Service Merchandise just announced an 8% decrease in the price of their Compaq Presario computers. If one particular computer model sells for $2162 after the decrease, find the original price of this computer.

△ **13.** Solve $A = \dfrac{1}{2}(B + b)h$ for b.

14. Solve: $4x - 2 < 5x$

15. Find the intersection: $\{2, 4, 6, 8\} \cap \{3, 4, 5, 6\}$

16. Solve: $|y| = 0$

17. see graph _____

18. _____

19. _____

20. _____

21. see graph _____

22. _____

23. _____

24. _____

17. Graph: $y = -3$

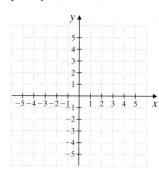

18. Find the slope and the y-intercept of the line $3x - 4y = 4$.

19. Write an equation of the line with slope -3 containing the point $(1, -5)$.

20. Determine whether the relation $y = 2x + 1$ is also a function.

21. Graph the intersection of $x \geq 1$ and $y \geq 2x - 1$.

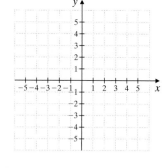

22. Use the elimination method to solve the system: $\begin{cases} 3x - 2y = 10 \\ 4x - 3y = 15 \end{cases}$

23. Solve the system:
$$\begin{cases} 2x - 4y + 8z = 2 \\ -x - 3y + z = 11 \\ x - 2y + 4z = 0 \end{cases}$$

24. A first number is 4 less than a second number. Four times the first number is 6 more than twice the second. Find the numbers.

Polynomials and Polynomial Functions

Linear equations are important for solving problems. They are not sufficient, however, to solve all problems. Many real-world phenomena are modeled by polynomials. In the first portion of this chapter we will study operations on polynomials. We then look at how polynomials can be used in problem solving. We conclude with a study of graphs of polynomial functions.

5.1 Adding and Subtracting Polynomials

5.2 Multiplying Polynomials

5.3 Dividing Polynomials

5.4 The Greatest Common Factor and Factoring by Grouping

5.5 Factoring Trinomials

5.6 Factoring by Special Products

Integrated Review—Operations on Polynomials and Factoring Strategies

5.7 Solving Equations by Factoring and Solving Problems

5.8 An Introduction to Graphing Polynomial Functions

The Eiffel Tower in Paris, France, is recognized throughout the world. It was built in 1889 for that year's World's Fair (held in Paris) as well as to commemorate the 100th anniversary of the French Revolution. Alexandre Gustave Eiffel built the 984-foot-tall tower with 7000 tons of wrought iron. Workers on the mammoth structure assembled over 18,000 individual parts with more than 2.5 million rivets. At the time of its construction, the Eiffel Tower was the tallest man-made structure in the world and remained so until the Empire State Building was completed in 1931 in New York City. In Exercise 76 on page 336, we will use a polynomial function to find the height of an object thrown from the top of the Eiffel Tower.

Name _____ Section _____ Date _____

1. _____

2. _____

3. _____

4. _____

5. _____

6. _____

7. _____

8. _____

9. _____

10. _____

11. _____

12. _____

13. _____

14. _____

15. _____

16. _____

17. _____

18. _____

19. _____

20. see graph

Chapter 5 Pretest

1. Find the degree of the polynomial $2x^4 - 3xy^5 + y^3$.

2. If $P(x) = -x^2 + 2x + 6$, find $P(-1)$.

Perform each indicated operation.

3. $(-2x + 7) + (3x^2 + 6x - 6)$

4. $(-8y^2 - 3y + 5) - (4y^2 - 6y - 1)$

5. $(2x - 1)(3x + 5)$

6. $(8y + 3)^2$

7. $(2m + 5)(2m - 5)$

8. $\dfrac{6t^3 - 4t^2 + 5t}{2t}$

9. $(2x^3 - 5x^2 - 10x - 4) \div (2x + 3)$

Factor each polynomial completely.

10. $6x^4 - 12x^3 + 10x^2$

11. $3ac - 4ad + 6bc - 8bd$

12. $x^2 + 2x - 63$

13. $6x^2 + 19x + 3$

14. $(x - 4)^2 - 7(x - 4) + 10$

15. $8t^3 - 1$

16. $12x^3 - 27xy^2$

Solve each equation.

17. $2n^2 + 13n = -15$

18. $x^3 + x^2 = 25x + 25$

19. One number exceeds another by seven, and their product is 120. Find the numbers.

20. Graph the function $f(x) = x^2 - 6x + 8$. Find and label the vertex and intercepts.

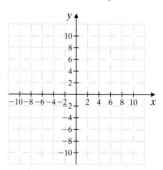

5.1   Adding and Subtracting Polynomials

A   Defining a Polynomial and Related Terms

A **term** is a number or the product of a number and one or more variables raised to powers. The **numerical coefficient**, or simply the **coefficient**, is the numerical factor of a term.

Term	Numerical Coefficient
$-12x^5$	-12
x^3y	1
$-z$	-1
2	2

If a term contains only a number, it is called a **constant term**, or simply a **constant**.

A **polynomial** is a finite sum of terms in which all variables are raised to nonnegative integer powers and no variables appear in any denominator.

Polynomials	Not Polynomials	
$4x^5y + 7xz$	$5x^{-3} + 2x$	Negative integer exponent
$-5x^3 + 2x + \dfrac{2}{3}$	$\dfrac{6}{x^2} - 5x + 1$	Variable in denominator

A polynomial that contains only one variable is called a **polynomial in one variable**. For example, $3x^2 - 2x + 7$ is a **polynomial in x**. This polynomial in x is written in **descending order** since the terms are listed in descending order of the variable's exponents. (The term 7 can be thought of as $7x^0$.) The following examples are polynomials in one variable written in descending order:

$$4x^3 - 7x^2 + 5 \qquad y^2 - 4 \qquad 8a^4 - 7a^2 + 4a$$

A **monomial** is a polynomial consisting of one term. A **binomial** is a polynomial consisting of two terms. A **trinomial** is a polynomial consisting of three terms.

Monomials	Binomials	Trinomials
ax^2	$x + y$	$x^2 + 4xy + y^2$
$-3x$	$6y^2 - 2$	$-x^4 + 3x^3 + 1$
4	$\dfrac{5}{7}z^3 - 2z$	$8y^2 - 2y - 10$

By definition, all monomials, binomials, and trinomials are also polynomials.

Each term of a polynomial has a **degree**.

Degree of a Term

The **degree of a term** is the sum of the exponents on the *variables* contained in the term.

Practice Problems 1–5

Find the degree of each term.

1. $2x^3$
2. 7^2x^4
3. x
4. $15xy^2z^4$
5. 9

Practice Problems 6–8

Find the degree of each polynomial and indicate whether the polynomial is a monomial, binomial, or trinomial.

6. $4x^5 + 7x^3 - 1$
7. $-2xy^2z$
8. $y^3 + 6y$

Practice Problem 9

Find the degree of the polynomial $7x^2y - 6x^2yz + 2 - 4y^3$.

EXAMPLES   Find the degree of each term.

1. $3x^2$ — The exponent on x is 2, so the degree of the term is 2.
2. -2^3x^5 — The exponent on x is 5, so the degree of the term is 5. (Recall that the degree is the sum of the exponents on *only* the *variables*.)
3. y — The degree of y, or y^1, is 1.
4. $12x^2yz^3$ — The degree is the sum of the exponents on the variables, or $2 + 1 + 3 = 6$.
5. 5 — The degree of 5, which can be written as $5x^0$, is 0. ●

From the preceding examples, we can say that the degree of a constant is 0. Also, the term 0 has no degree.

Each polynomial also has a degree.

> ### Degree of a Polynomial
>
> The **degree of a polynomial** is the largest degree of any of its terms.

EXAMPLES

Find the degree of each polynomial and also indicate whether the polynomial is a monomial, binomial, or trinomial.

	Polynomial	Degree	Classification
6.	$7x^3 - 3x + 2$	3	Trinomial
7.	$-xyz$	$1 + 1 + 1 = 3$	Monomial
8.	$x^4 - 16$	4	Binomial

EXAMPLE 9   Find the degree of the polynomial

$$3xy + x^2y^2 - 5x^2 - 6.$$

Solution: The degree of each term is

$$3xy + x^2y^2 - 5x^2 - 6$$

Degree: 2   4   2   0

The largest degree of any term is 4, so the degree of this polynomial is 4. ●

B   Combining Like Terms

Before we add polynomials, recall from Section 1.5 that terms are considered to be **like terms** if they contain exactly the same variables raised to exactly the same powers.

Like Terms	**Unlike Terms**
$-5x^2, -x^2$	$4x^2, 3x$
$7xy^3z, -2xzy^3$	$12x^2y^3, -2xy^3$

To simplify a polynomial, we **combine like terms** by using the distributive property. For example, by the distributive property,

$$5x + 7x = (5 + 7)x = 12x$$

EXAMPLES   Simplify each polynomial by combining like terms.

10. $-12x^2 + 7x^2 - 6x = (-12 + 7)x^2 - 6x = -5x^2 - 6x$

11. $3xy - 2x + 5xy - x = 3xy + 5xy - 2x - x$

$$= (3 + 5)xy + (-2 - 1)x$$
$$= 8xy - 3x$$

Practice Problems 10–11

Simplify each polynomial by combining like terms.

10. $10x^3 - 12x^3 - 3x$

11. $-6ab + 2a + 12ab - a$

(C) Adding Polynomials

Now we have reviewed the skills we need to add polynomials.

Adding Polynomials

To add polynomials, combine all like terms.

EXAMPLE 12   Add $11x^3 - 12x^2 + x - 3$ and $x^3 - 10x + 5$.

Solution:

$$(11x^3 - 12x^2 + x - 3) + (x^3 - 10x + 5)$$
$$= 11x^3 + x^3 - 12x^2 + x - 10x - 3 + 5 \quad \text{Group like terms.}$$
$$= 12x^3 - 12x^2 - 9x + 2 \quad \text{Combine like terms.}$$

Practice Problem 12

Add $14x^4 - 6x^3 + x^2 - 6$ and $x^3 - 5x^2 + 1$.

   Sometimes it is more convenient to add polynomials vertically. To do this, we line up like terms beneath one another and then add like terms.

EXAMPLE 13   Add $11x^3 - 12x^2 + x - 3$ and $x^3 - 10x + 5$ vertically.

Solution:

$$
\begin{array}{r}
11x^3 - 12x^2 + x - 3 \\
\underline{x^3 \qquad\quad - 10x + 5} \\
12x^3 - 12x^2 - 9x + 2
\end{array}
$$

Line up like terms.
Combine like terms.

This example is the same as Example 12, only here we added vertically.

Practice Problem 13

Add $10y^3 - y^2 + 4y - 11$ and $y^3 - 4y^2 + 3y$ vertically.

EXAMPLE 14   Add: $(7x^3y - xy^3 + 11) + (6x^3y - 4)$

Solution: To add these polynomials, we remove the parentheses and group like terms.

$$(7x^3y - xy^3 + 11) + (6x^3y - 4)$$
$$= 7x^3y - xy^3 + 11 + 6x^3y - 4 \quad \text{Remove parentheses.}$$
$$= 7x^3y + 6x^3y - xy^3 + 11 - 4 \quad \text{Group like terms.}$$
$$= 13x^3y - xy^3 + 7 \quad \text{Combine like terms.}$$

Practice Problem 14

Add: $(4x^2y - xy^2 + 5) + (-6x^2y - 1)$

(D) Subtracting Polynomials

The definition of subtraction of real numbers can be extended to apply to polynomials. To subtract a number, we add its opposite:

$$a - b = a + (-b)$$

Likewise, to subtract a polynomial we add its opposite. In other words, if P and Q are polynomials, then

$$P - Q = P + (-Q)$$

Answers

10. $-2x^3 - 3x$,   **11.** $6ab + a$,
12. $14x^4 - 5x^3 - 4x^2 - 5$,
13. $11y^3 - 5y^2 + 7y - 11$,   **14.** $-2x^2y - xy^2 + 4$

The polynomial $-Q$ is the **opposite**, or **additive inverse**, of the polynomial Q. We can find $-Q$ by changing the sign of each term of Q.

Try the Concept Check in the margin.

Concept Check

Which polynomial is the opposite of, $16x^3 - 5x + 7$?

a. $-16x^3 - 5x + 7$
b. $-16x^3 + 5x - 7$
c. $16x^3 + 5x + 7$
d. $-16x^3 + 5x + 7$

Subtracting Polynomials

To subtract polynomials, change the signs of the terms of the polynomial being subtracted and then add.

Review the example below.

To subtract, change the signs; then add.

$$(3x^2 + 4x - 7) - (3x^2 - 2x - 5) = (3x^2 + 4x - 7) + (-3x^2 + 2x + 5)$$
$$= 3x^2 + 4x - 7 - 3x^2 + 2x + 5$$
$$= 6x - 2 \qquad \text{Combine like terms.}$$

Practice Problem 15

Subtract:
$(7x^4 - 8x^2 + x) - (9x^4 + x^2 - 18)$

EXAMPLE 15   Subtract: $(12z^5 - 12z^3 + z) - (-3z^4 + z^3 + 12z)$

Solution: First we change the sign of each term of the second polynomial, and then we add the result to the first polynomial.

$$(12z^5 - 12z^3 + z) - (-3z^4 + z^3 + 12z)$$
$$= 12z^5 - 12z^3 + z + 3z^4 - z^3 - 12z \qquad \text{Change signs and add.}$$
$$= 12z^5 + 3z^4 - 12z^3 - z^3 + z - 12z \qquad \text{Group like terms.}$$
$$= 12z^5 + 3z^4 - 13z^3 - 11z \qquad \text{Combine like terms.} \quad \bullet$$

Practice Problem 16

Subtract $(2y^4 + 4y) - (6y^4 + 7y^3 - 3y)$ vertically.

EXAMPLE 16   Subtract $(10x^3 - 7x^2) - (4x^3 - 3x^2 + 2)$ vertically.

Solution: To subtract these polynomials, we add the opposite of the second polynomial to the first one.

$$
\begin{array}{ll}
10x^3 - 7x^2 & \text{is equivalent to} \\
-(4x^3 - 3x^2 + 2) &
\end{array}
\qquad
\begin{array}{l}
10x^3 - 7x^2 \\
\underline{-4x^3 + 3x^2 - 2} \\
6x^3 - 4x^2 - 2 \quad \text{Add.} \quad \bullet
\end{array}
$$

Concept Check

Why is the following subtraction incorrect?
$(7z - 5) - (3z - 4)$
$= 7z - 5 - 3z - 4$
$= 4z - 9$

Try the Concept Check in the margin.

Practice Problem 17

Subtract $3a^2b^3 - 4ab^2 + 6a$ from $7a^2b^3 - ab^2$.

EXAMPLE 17   Subtract $4x^3y^2 - 3x^2y^2 + 2y^2$ from $10x^3y^2 - 7x^2y^2$.

Solution: Notice the order of the numbers, and then write "Subtract $4x^3y^2 - 3x^2y^2 + 2y^2$ from $10x^3y^2 - 7x^2y^2$" as a mathematical expression. (For example, if we subtract 2 from 8, we would write $8 - 2 = 6$.)

$$(10x^3y^2 - 7x^2y^2) - (4x^3y^2 - 3x^2y^2 + 2y^2)$$
$$= 10x^3y^2 - 7x^2y^2 - 4x^3y^2 + 3x^2y^2 - 2y^2 \qquad \text{Remove parentheses.}$$
$$= 6x^3y^2 - 4x^2y^2 - 2y^2 \qquad \text{Combine like terms.} \quad \bullet$$

Ⓔ   **Evaluating Polynomial Functions**

Recall function notation first introduced in Section 3.5. At times it is convenient to use function notation to represent polynomials. For example, we may write $P(x)$ to represent the polynomial $3x^2 - 2x - 5$. In symbols, we would write

$$P(x) = 3x^2 - 2x - 5$$

Answers

15. $-2x^4 - 9x^2 + x + 18$,
16. $-4y^4 - 7y^3 + 7y$,   **17.** $4a^2b^3 + 3ab^2 - 6a$

Concept Check: b

Concept Check: With parentheses removed, the expression should be
$7z - 5 - 3z + 4 = 4z - 1$

This function is called a **polynomial function** because the expression $3x^2 - 2x - 5$ is a polynomial.

> **Helpful Hint**
>
> Recall that the symbol $P(x)$ *does not mean* P times x. It is a special symbol used to denote a function.

EXAMPLES   If $P(x) = 3x^2 - 2x - 5$, find each function value.

18. $P(1) = 3(1)^2 - 2(1) - 5 = -4$ Let $x = 1$ in the function $P(x)$.

19. $P(-2) = 3(-2)^2 - 2(-2) - 5 = 11$ Let $x = -2$ in the function $P(x)$. ●

Many real-world phenomena are modeled by polynomial functions. If the polynomial function model is given, we can often find the solution of a problem by evaluating the function at a certain value.

EXAMPLE 20   Finding the Height of an Object

The world's highest bridge, Royal Gorge suspension bridge in Colorado, is 1053 feet above the Arkansas River. An object is dropped from the top of this bridge. Neglecting air resistance, the height of the object at time t seconds is given by the polynomial function $P(t) = -16t^2 + 1053$. Find the height of the object when $t = 1$ second and when $t = 8$ seconds.

Solution: To find the height of the object at 1 second, we find $P(1)$.

$$P(t) = -16t^2 + 1053$$
$$P(1) = -16(1)^2 + 1053$$
$$P(1) = 1037$$

When $t = 1$ second, the height of the object is 1037 feet.

To find the height of the object at 8 seconds, we find $P(8)$.

$$P(t) = -16t^2 + 1037$$
$$P(8) = -16(8)^2 + 1037$$
$$P(8) = -1024 + 1037$$
$$P(8) = 13$$

When $t = 8$ seconds, the height of the object is 13 feet. Notice that as time t increases, the height of the object decreases. ●

Practice Problems 18–19

If $P(x) = 5x^2 - 3x + 7$, find each function value.

18. $P(2)$

19. $P(-1)$

Practice Problem 20

Use the polynomial function in Example 20 to find the height of the object when $t = 3$ seconds and $t = 7$ seconds.

Answers

18. $P(2) = 21$,   **19.** $P(-1) = 15$,   **20.** at 3 sec, height is 909 ft; at 7 sec, height is 269 ft

GRAPHING CALCULATOR EXPLORATIONS

A graphing calculator may be used to check addition and subtraction of polynomials in one variable. For example, to check the polynomial subtraction statement

$$(3x^2 - 6x + 9) - (x^2 - 5x + 6) = 2x^2 - x + 3$$

graph both

$$Y_1 = (3x^2 - 6x + 9) - (x^2 - 5x + 6) \qquad \text{Left side of equation}$$

and

$$Y_2 = 2x^2 - x + 3 \qquad \text{Right side of equation}$$

on the same screen and see that their graphs coincide. (*Note:* If the graphs do not coincide, we can be sure that a mistake has been made either in combining polynomials or in calculator keystrokes. However, if the graphs appear to coincide, we cannot be sure that our work is correct. This is because it is possible for the graphs to differ so slightly that we do not notice it.)

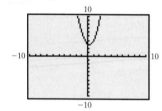

The graphs of Y_1 and Y_2 are shown. The graphs appear to coincide, so the subtraction statement

$$(3x^2 - 6x + 9) - (x^2 - 5x + 6) = 2x^2 - x + 3$$

appears to be correct.

Perform each indicated operation. Then use the procedure described above to check your work.

1. $(2x^2 + 7x + 6) + (x^3 - 6x^2 - 14)$ **2.** $(-14x^3 - x + 2) + (-x^3 + 3x^2 + 4x)$

3. $(1.8x^2 - 6.8x - 1.7) - (3.9x^2 - 3.6x)$ **4.** $(-4.8x^2 + 12.5x - 7.8) - (3.1x^2 - 7.8x)$

5. $(1.29x - 5.68) + (7.69x^2 - 2.55x + 10.98)$ **6.** $(-0.98x^2 - 1.56x + 5.57) + (4.36x - 3.71)$

Mental Math

Add or subtract as indicated.

1. $7x + 3x$ **2.** $8x - 2x$ **3.** $14y - 9y$ **4.** $14y + 9y$ **5.** $3z - 12z$ **6.** $2z - 6z$

EXERCISE SET 5.1

(A) *Find the degree of each term. See Examples 1 through 5.*

1. 4 **2.** 7 **3.** $5x^2$ **4.** $-z^3$ **5.** $-3xy^2$ **6.** $12x^3z$

Find the degree of each polynomial and indicate whether the polynomial is a monomial, binomial, trinomial, or none of these. See Examples 6 through 9.

7. $6x + 3$ **8.** $7x - 8$ **9.** $3x^2 - 2x + 5$ **10.** $5x^2 - 3x^2y - 2x^3$

11. $-xyz$ **12.** -9 **13.** $x^2y - 4xy^2 + 5x + y$ **14.** $-2x^2y - 3y^2 + 4x + y^5$

15. In your own words, describe how to find the degree of a term.

16. In your own words, describe how to find the degree of a polynomial.

(B) *Simplify each polynomial by combining like terms. See Examples 10 and 11.*

17. $5y + y$ **18.** $-x + 3x$ **19.** $4x + 7x - 3$

20. $-8y + 9y + 4y^2$ **21.** $4xy + 2x - 3xy - 1$ **22.** $-8xy^2 + 4x - x + 2xy^2$

(C) *Add. See Examples 12 through 14.*

23. $(9y^2 - 8) + (9y^2 - 9)$ **24.** $(x^2 + 4x - 7) + (8x^2 + 9x - 7)$

25. $(x^2 + xy - y^2)$ and $(2x^2 - 4xy + 7y^2)$ **26.** $(4x^3 - 6x^2 + 5x + 7)$ and $(2x^2 + 6x - 3)$

27. $\begin{array}{r} x^2 - 6x + 3 \\ + \quad (2x + 5) \\ \hline \end{array}$ **28.** $\begin{array}{r} -2x^2 + 3x - 9 \\ + \quad (2x - 3) \\ \hline \end{array}$

29. $3x^2 + 15x + 8$
$+ (2x^2 + 7x + 8)$
$\overline{}$

30. $9x^2 + 9x - 4$
$+ (7x^2 - 3x - 4)$
$\overline{}$

31. $(-3x + 8) + (-3x^2 + 3x - 5)$

32. $(5y^2 - 2y + 4) + (3y + 7)$

33. $(5y^4 - 7y^2 + x^2 - 3) + (-3y^4 + 2y^2 + 4)$

34. $(8x^4 - 14x^2 + 6) + (-12x^6 - 21x^4 - 9x^2)$

35. $(5x - 11) + (-x - 2)$

36. $(3x^2 - 2x) + (5x^2 - 9x)$

37. $(3x^3 - b + 2a - 6) + (-4x^3 + b + 6a - 6)$

38. $(5x^2 - 6) + (2x^2 - 4x + 8)$

39. $(-3 + 4x^2 + 7xy^2) + (2x^3 - x^2 + xy^2)$

40. $(-3x^2y + 4) + (-7x^2y - 8y)$

41. $(7x^3y - 4xy + 8) + (5x^3y + 4xy + 8x)$

42. $(9xyz + 4x - y) + (-9xyz - 3x + y + 2)$

43. $(0.6x^3 + 1.2x^2 - 4.5x + 9.1) + (3.9x^3 - x^2 + 0.7x)$

44. $(9.3y^2 - y + 12.8) + (2.6y^2 + 4.4y - 8.9)$

D *Subtract. See Examples 15 through 17.*

45. $(9y^2 - 7y + 5) - (8y^2 - 7y + 2)$

46. $(2x^2 + 3x + 12) - (5x - 7)$

47. Subtract $(6x^2 - 3x)$ from $(4x^2 + 2x)$.

48. Subtract $(y^2 + x - y)$ from $(y^2 + x - 3)$.

49.
$$\begin{array}{r} 3x^2 - 4x + 8 \\ - \quad (5x^2 - 7) \\ \hline \end{array}$$

50.
$$\begin{array}{r} -3x^2 - 4x + 8 \\ - \quad (5x + 12) \\ \hline \end{array}$$

51.
$$\begin{array}{r} 6y^2 - 6y + 4 \\ -(-y^2 - 6y + 7) \\ \hline \end{array}$$

52.
$$\begin{array}{r} -4x^3 + 4x^2 - 4x \\ -(2x^3 - 2x^2 + 3x) \\ \hline \end{array}$$

53. $(4x^2 - 6x + 2) - (-x^2 + 3x + 5)$

54. $(5x^2 + x + 9) - (2x^2 - 9)$

55. $(7x^2 + x + 1) - (6x^2 + x - 1)$

56. $(4x - 4) - (-x - 4)$

57. $(9x^3 - 2x^2 + 4x - 7) - (2x^3 - 6x^2 - 4x + 3)$

58. $(3x^2 + 6xy + 3y^2) - (8x^2 - 6xy - y^2)$

59. Subtract $(y^2 + 4yx + 7)$ from $(-19y^2 + 7yx + 7)$.

60. Subtract $(x^2y - 4)$ from $(3x^2 - 4x^2y + 5)$.

61. Subtract $(3x + 7)$ from the sum of $(7x^2 + 4x + 9)$ and $(8x^2 + 7x - 8)$.

62. Subtract $(9x + 8)$ from the sum of $(3x^2 - 2x - x^3 + 2)$ and $(5x^2 - 8x - x^3 + 4)$.

63. $(14ab - 10a^2b + 6b^2) - (18a^2 - 20a^2b - 6b^2)$

64. $(13x^2 - 26x^2y^2 + 4) - (19x^2 + x^2y^2 - 11)$

65. $\left(\dfrac{2}{3}x^2 - \dfrac{1}{6}x + \dfrac{5}{6}\right) - \left(\dfrac{1}{3}x^2 + \dfrac{5}{6}x - \dfrac{1}{6}\right)$

66. $\left(\dfrac{3}{16}x^2 + \dfrac{5}{8}x - \dfrac{1}{4}\right) - \left(\dfrac{5}{16}x^2 - \dfrac{3}{8}x + \dfrac{3}{4}\right)$

E *If $P(x) = x^2 + x + 1$ and $Q(x) = 5x^2 - 1$, find each function value. See Examples 18 and 19.*

67. $P(7)$ **68.** $Q(4)$ **69.** $Q(-10)$ **70.** $P(-4)$ **71.** $P(0)$ **72.** $Q(0)$

Solve. See Example 20.

The surface area of a rectangular box is given by the polynomial

$$2HL + 2LW + 2HW$$

and is measured in square units. In business, surface area is often calculated to help determine cost of materials.

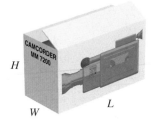

△ **73.** A rectangular box is to be constructed to hold a new camcorder. The box is to have dimensions 5 inches by 4 inches by 9 inches. Find the surface area of the box.

△ **74.** Suppose it has been determined that a box of dimensions 4 inches by 4 inches by 8.5 inches can be used to contain the camcorder in Exercise 73. Find the surface area of this box and calculate the square inches of material saved by using this box instead of the box in Exercise 73.

75. A projectile is fired upward from the ground with an initial velocity of 300 feet per second. Neglecting air resistance, the height of the projectile at any time t can be described by the polynomial function $P(t) = -16t^2 + 300t$. Find the height of the projectile at each given time.
  a. $t = 1$ second
  b. $t = 2$ seconds
  c. $t = 3$ seconds
  d. $t = 4$ seconds
  e. Explain why the height increases and then decreases as time passes.
  f. Approximate (to the nearest second) how long before the object hits the ground.

76. An object is thrown upward with an initial velocity of 25 feet per second from the top of the 984-foot-high Eiffel Tower in Paris, France. The height of the object at any time t can be described by the polynomial function $P(t) = -16t^2 + 25t + 984$. Find the height of the projectile at each given time. (*Source:* Council on Tall Buildings and Urban Habitat, Lehigh University)
  a. $t = 1$ second
  b. $t = 3$ seconds
  c. $t = 5$ seconds
  d. Approximate (to the nearest second) how long before the object hits the ground.

77. The polynomial function $P(x) = 45x - 100,000$ models the relationship between the number of computer briefcases x that a company sells and the profit the company makes, $P(x)$. Find $P(4000)$, the profit from selling 4000 computer briefcases.

78. The total cost (in dollars) for MCD, Inc., Manufacturing Company to produce x blank audiocassette tapes per week is given by the polynomial function $C(x) = 0.8x + 10,000$. Find the total cost of producing 20,000 tapes per week.

79. The total revenues (in dollars) for MCD, Inc., Manufacturing Company to sell x blank audiocasette tapes per week is given by the polynomial function $R(x) = 2x$. Find the total revenue from selling 20,000 tapes per week.

80. In business, profit equals revenue minus cost, or $P(x) = R(x) - C(x)$. Find the profit function for MCD, Inc. by subtracting the given functions in Exercises 78 and 79.

Review and Preview

Multiply. See Section 1.3.

81. $5(3x - 2)$

82. $-7(2z - 6y)$

83. $-2(x^2 - 5x + 6)$

84. $5(-3y^2 - 2y + 7)$

 Combining Concepts

85. The function $f(x) = 136.7x^2 + 327.6x + 21.6$ can be used to approximate the increasing number of radio stations on the Internet during the years 1996–2001 where x is the number of years after 1996 and $f(x)$ is the number of stations. Round answers to the nearest whole. (*Source:* BRS Media, Inc.)

   a. Approximate the number of radio stations on the Internet in 1996.

   b. Approximate the number of radio stations on the Internet in 2000.

   c. Use the function to predict the number of radio stations on the Internet in 2004.

   d. From parts (a), (b), and (c), determine whether the number of radio stations on the Internet is increasing at a steady rate. Explain why or why not.

Radio Stations on the Internet

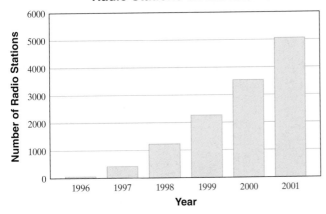

86. The function $f(x) = -x^2 + 11.4x + 49.8$ can be used to approximate the number of Americans enrolled in health maintenance organizations (HMOs) during the period 1995–2000, where x is the number of years after 1995 and $f(x)$ is the number of millions of Americans. Round answers to the nearest tenth of a million. (*Source:* Based on data from *Health, United States, 2001*, National Center for Health Statistics)

   a. Approximate the number of Americans enrolled in HMOs in 1995.

   b. Approximate the number of Americans enrolled in HMOs in 2000.

   c. Use the function to predict the number of Americans enrolled in HMOs in 2005.

   d. From parts (a), (b), and (c), determine whether the number of Americans enrolled in HMOs is changing at a steady rate. Explain why or why not.

87. Sport utility vehicle (SUV) sales in the U.S. have increased since 1992. The function $f(x) = 0.014x^2 + 0.12x + 0.85$ can be used to approximate the number of SUV sales during the years 1990–2000 where x is the number of years after 1990 and $f(x)$ is the SUV sales (in millions). Round answers to the nearest tenth of a million. (*Source:* Wards Communications)

a. Approximate the number of SUV's sold in 1999.

b. Use the function to predict the number of SUV's sold in 2005.

88. Digital camera sales have increased since 1996. The function $f(x) = 34.7x^2 + 68.2x + 377.3$ can be used to approximate the revenue from selling digital cameras for the years 1996–2001 where x is the number of years after 1996 and $f(x)$ is the sales revenue (in million of dollars). Round answers to the nearest whole million. (*Source:* International Data Corporation)

a. Approximate the revenue from digital camera sales in 2000.

b. Use the function to predict the revenue from digital camera sales in 2004.

If $P(x) = 3x + 3, Q(x) = 4x^2 - 6x + 3, and R(x) = 5x^2 - 7,$ find each function.

89. $P(x) + Q(x)$

90. $Q(x) - R(x)$

91. If $P(x) = 2x - 3$, find $P(a)$, $P(-x)$, and $P(x + h)$.

Perform each indicated operation.

92. $(8x^{2y} - 7x^y + 3) + (-4x^{2y} + 9x^y - 14)$

93. $(14z^{5x} + 3z^{2x} + z) - (2z^{5x} - 10z^{2x} + 3z)$

Find each perimeter.

△ **94.**

$(x + y)$ units

$(3x^2 - x + 2y)$ units

△ **95.**

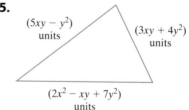

$(5xy - y^2)$ units

$(3xy + 4y^2)$ units

$(2x^2 - xy + 7y^2)$ units

5.2 Multiplying Polynomials

Ⓐ Multiplying Any Two Polynomials

OBJECTIVES

Ⓐ Multiply any two polynomials.

Ⓑ Multiply binomials.

Ⓒ Square binomials.

Ⓓ Multiply the sum and difference of two terms.

Ⓔ Multiply three or more polynomials.

Properties of real numbers and exponents are used continually in the process of multiplying polynomials. To multiply monomials, for example, we apply the commutative and associative properties of real numbers and the product rule for exponents.

EXAMPLES Multiply.

Group like bases and apply the product rule for exponents.

1. $(2x^3)(5x^6) = 2(5)(x^3)(x^6) = 10x^{3+6} = 10x^9$
2. $(7y^4z^4)(-xy^{11}z^5) = 7(-1)x(y^4y^{11})(z^4z^5) = -7xy^{4+11}z^{4+5} = -7xy^{15}z^9$

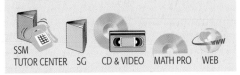

SSM TUTOR CENTER · SG · CD & VIDEO · MATH PRO · WEB

> **Helpful Hint**
>
> See Sections 1.6 and 1.7 to review exponential expressions further.

To multiply a monomial by a polynomial other than a monomial, we use an expanded form of the distributive property.

$$a(b + c + d + \cdots + z) = ab + ac + ad + \cdots + az$$

Notice that the monomial a is multiplied by each term of the polynomial.

EXAMPLES Multiply.

3. $2x(5x - 4) = 2x(5x) + 2x(-4)$ Use the distribute property.
   $= 10x^2 - 8x$ Multiply.

4. $-3x^2(4x^2 - 6x + 1) = -3x^2(4x^2) + (-3x^2)(-6x) + (-3x^2)(1)$
   $= -12x^4 + 18x^3 - 3x^2$

5. $-xy(7x^2y + 3xy - 11) = -xy(7x^2y) + (-xy)(3xy) + (-xy)(-11)$
   $= -7x^3y^2 - 3x^2y^2 + 11xy$

Try the Concept Check in the margin.

To multiply any two polynomials, we can use the following.

Multiplying Any Two Polynomials

To multiply any two polynomials, use the distributive property and multiply each term of one polynomial by each term of the other polynomial. Then combine any like terms.

EXAMPLE 6 Multiply: $(x + 3)(2x + 5)$

Solution: We multiply each term of $(x + 3)$ by $(2x + 5)$.

$(x + 3)(2x + 5) = x(2x + 5) + 3(2x + 5)$ Use the distributive property.
$= 2x^2 + 5x + 6x + 15$ Use the distributive property again.
$= 2x^2 + 11x + 15$ Combine like terms.

Practice Problems 1–2

Multiply.

1. $(7y^2)(4y^5)$
2. $(-a^2b^3c)(10ab^2c^{12})$

Practice Problems 3–5

Multiply.

3. $4x(3x - 2)$
4. $-2y^3(5y^2 - 2y + 6)$
5. $-a^2b(4a^3 - 2ab + b^2)$

Concept Check

Find the error:
$4x(x - 5) + 2x$
$= 4x(x) + 4x(-5) + 4x(2x)$
$= 4x^2 - 20x + 8x^2$
$= 12x^2 - 20x$

Practice Problem 6

Multiply: $(x + 2)(3x + 1)$

Answers

1. $28y^7$, 2. $-10a^3b^5c^{13}$, 3. $12x^2 - 8x$,
4. $-10y^5 + 4y^4 - 12y^3$, 5. $-4a^5b + 2a^3b^2 - a^2b^3$,
6. $3x^2 + 7x + 2$

Concept Check: $4x(x - 5) + 2x$
$= 4x(x) + 4x(-5) + 2x$
$= 4x^2 - 20x + 2x$
$= 4x^2 - 18x$

Practice Problem 7

Multiply: $(5x - 1)(2x^2 - x + 4)$

EXAMPLE 7 Multiply: $(2x - 3)(5x^2 - 6x + 7)$

Solution: We multiply each term of $(2x - 3)$ by each term of $(5x^2 - 6x + 7)$.

$$(2x - 3)(5x^2 - 6x + 7) = 2x(5x^2 - 6x + 7) + (-3)(5x^2 - 6x + 7)$$
$$= 10x^3 - 12x^2 + 14x - 15x^2 + 18x - 21$$
$$= 10x^3 - 27x^2 + 32x - 21 \text{ Combine like terms.} \bullet$$

Sometimes polynomials are easier to multiply vertically, in the same way we multiply real numbers. When multiplying vertically, we line up like terms in the **partial products** vertically. This makes combining like terms easier.

Practice Problem 8

Multiply vertically:
$$(3x^2 + 5)(x^2 - 6x + 1)$$

EXAMPLE 8 Multiply vertically: $(4x^2 + 7)(x^2 + 2x + 8)$

Solution:

$$
\begin{array}{r}
x^2 + 2x + 8 \\
4x^2 + 7 \\
\hline
7x^2 + 14x + 56 \quad 7(x^2 + 2x + 8) \\
4x^4 + 8x^3 + 32x^2 \quad\quad 4x^2(x^2 + 2x + 8) \\
\hline
4x^4 + 8x^3 + 39x^2 + 14x + 56 \quad \text{Combine like terms.} \bullet
\end{array}
$$

B **Multiplying Binomials**

When multiplying a binomial by a binomial, we can follow a special order for multiplying terms, called the **FOIL** order. The letters of FOIL stand for "First–**O**uter–**I**nner–**L**ast." To illustrate this method, let's multiply $(2x - 3)$ by $(3x + 1)$.

Multiply the **F**irst terms of each binomial. $(2x - 3)(3x + 1)$ **F** $2x(3x) = 6x^2$

Multiply the **O**uter terms of each binomial. $(2x - 3)(3x + 1)$ **O** $2x(1) = 2x$

Multiply the **I**nner terms of each binomial. $(2x - 3)(3x + 1)$ **I** $-3(3x) = -9x$

Multiply the **L**ast terms of each binomial. $(2x - 3)(3x + 1)$ **L** $-3(1) = -3$

Combine like terms.

$$6x^2 + 2x - 9x - 3 = 6x^2 - 7x - 3$$

Practice Problem 9

Use the FOIL order to multiply $(x - 7)(x + 5)$.

Answers

7. $10x^3 - 7x^2 + 21x - 4$,
8. $3x^4 - 18x^3 + 8x^2 - 30x + 5$,
9. $x^2 - 2x - 35$

EXAMPLE 9 Use the FOIL order to multiply $(x - 1)(x + 2)$.

Solution:

$$
\begin{array}{cccc}
\text{First} & \text{Outer} & \text{Inner} & \text{Last}
\end{array}
$$

$$(x - 1)(x + 2) = x \cdot x + 2 \cdot x + (-1)x + (-1)(2)$$
$$= x^2 + 2x - x - 2$$
$$= x^2 + x - 2 \quad \text{Combine like terms.} \bullet$$

EXAMPLES   Use the FOIL order to multiply.

Practice Problems 10–11

Use the FOIL order to multiply.

$$\begin{array}{cccc} \text{First} & \text{Outer} & \text{Inner} & \text{Last} \\ \downarrow & \downarrow & \downarrow & \downarrow \end{array}$$

10. $(2x - 7)(3x - 4) = 2x(3x) + 2x(-4) + (-7)(3x) + (-7)(-4)$
$$= 6x^2 - 8x - 21x + 28$$
$$= 6x^2 - 29x + 28$$

$$\begin{array}{cccc} \text{F} & \text{O} & \text{I} & \text{L} \\ \downarrow & \downarrow & \downarrow & \downarrow \end{array}$$

11. $(3x + y)(5x - 2y) = 15x^2 - 6xy + 5xy - 2y^2$
$$= 15x^2 - xy - 2y^2$$ ●

10. $(4x - 3)(x - 6)$

11. $(6x + 5y)(2x - y)$

Ⓒ Squaring Binomials

The **square of a binomial** is a special case of the product of two binomials. By the FOIL order for multiplying two binomials, we have

$$(a + b)^2 = (a + b)(a + b)$$

$$\begin{array}{cccc} \text{F} & \text{O} & \text{I} & \text{L} \\ \downarrow & \downarrow & \downarrow & \downarrow \end{array}$$
$$= a^2 + ab + ba + b^2$$
$$= a^2 + 2ab + b^2$$

We can visualize this product geometrically by analyzing areas.

Area of square in the margin: $(a + b)^2$
Sum of areas of smaller rectangles: $a^2 + 2ab + b^2$
Thus, $(a + b)^2 = a^2 + 2ab + b^2$

The same pattern occurs for the square of a difference. In general, we have the following.

Square of a Binomial

$$(a + b)^2 = a^2 + 2ab + b^2 \qquad (a - b)^2 = a^2 - 2ab + b^2$$

In other words, a binomial squared is the sum of the first term squared, twice the product of both terms, and the second term squared.

EXAMPLES   Multiply.

Practice Problems 12–15

Multiply.

$$\begin{array}{cccccccc} (a & + & b)^2 & = & a^2 & + & 2 & \cdot & a & \cdot & b & + & b^2 \\ \downarrow & & \downarrow & & \downarrow & & \downarrow & & \downarrow & & \downarrow & & \downarrow \end{array}$$

12. $(x + 5)^2 = x^2 + 2 \cdot x \cdot 5 + 5^2 = x^2 + 10x + 25$
13. $(x - 9)^2 = x^2 - 2 \cdot x \cdot 9 + 9^2 = x^2 - 18x + 81$
14. $(3x + 2z)^2 = (3x)^2 + 2(3x)(2z) + (2z)^2 = 9x^2 + 12xz + 4z^2$
15. $(4m^2 - 3n)^2 = (4m^2)^2 - 2(4m^2)(3n) + (3n)^2 = 16m^4 - 24m^2n + 9n^2$ ●

12. $(x + 3)^2$
13. $(y - 6)^2$
14. $(2x + 5y)^2$
15. $(6a^2 - 2b)^2$

Helpful Hint

Note that $(a + b)^2 = a^2 + 2ab + b^2$, not $a^2 + b^2$. Also,
$(a - b)^2 = a^2 - 2ab + b^2$, not $a^2 - b^2$.

Answers

10. $4x^2 - 27x + 18$,   **11.** $12x^2 + 4xy - 5y^2$,
12. $x^2 + 6x + 9$,   **13.** $y^2 - 12y + 36$,
14. $4x^2 + 20xy + 25y^2$,   **15.** $36a^4 - 24a^2b + 4b^2$

D Multiplying the Sum and Difference of Two Terms

Another special product applies to the sum and difference of the same two terms. Multiply $(a + b)(a - b)$ to see a pattern.

$$(a + b)(a - b) = a^2 - ab + ba - b^2$$
$$= a^2 - b^2$$

> **Product of the Sum and Difference of Two Terms**
>
> $$(a + b)(a - b) = a^2 - b^2$$

In other words, the product of the sum and difference of the same two terms is the difference of the first term squared and the second term squared.

Practice Problems 16–19

Multiply.

16. $(x + 4)(x - 4)$
17. $(3m - 6)(3m + 6)$
18. $(a^2 + 5y)(a^2 - 5y)$
19. $\left(4y^2 - \dfrac{1}{3}\right)\left(4y^2 + \dfrac{1}{3}\right)$

EXAMPLES   Multiply.

$$\underset{\downarrow\ \ \ \downarrow}{(a + b)}\ \underset{\downarrow\ \ \ \downarrow}{(a - b)}\ =\ \underset{\downarrow\ \ \ \downarrow}{a^2 - b^2}$$

16. $(x + 3)(x - 3) = x^2 - 3^2 = x^2 - 9$
17. $(4y - 1)(4y + 1) = (4y)^2 - 1^2 = 16y^2 - 1$
18. $(x^2 + 2y)(x^2 - 2y) = (x^2)^2 - (2y)^2 = x^4 - 4y^2$
19. $\left(3m^2 - \dfrac{1}{2}\right)\left(3m^2 + \dfrac{1}{2}\right) = (3m^2)^2 - \left(\dfrac{1}{2}\right)^2 = 9m^4 - \dfrac{1}{4}$ ●

Practice Problem 20

Multiply:
$[(2x + 3y) - 2][(2x + 3y) + 2]$

EXAMPLE 20   Multiply: $[(5x - 2y) - 1][(5x - 2y) - 1]$

Solution: We can think of $(5x - 2y)$ as the first term and 1 as the second term. Then we can apply the method for the product of the sum and difference of two terms.

$$\underset{a}{[\overbrace{(5x - 2y)}} \underset{-\ b}{-\ 1][}\underset{a}{\overbrace{(5x - 2y)}} \underset{+\ b}{+\ 1]} = \underset{a^2}{\overbrace{(5x - 2y)^2}} \underset{-\ b^2}{-\ 1^2}$$
$$= (5x)^2 - 2(5x)(2y) + (2y)^2 - 1 \qquad \text{Square } (5x - 2y).$$
$$= 25x^2 - 20xy + 4y^2 - 1 \qquad ●$$

E Multiplying Three or More Polynomials

To multiply three or more polynomials, more than one method may be needed.

Practice Problem 21

Multiply: $(y - 2)(y + 2)(y^2 - 4)$

EXAMPLE 21   Multiply: $(x - 3)(x + 3)(x^2 - 9)$

Solution: We multiply the first two binomials, the sum and difference of two terms. Then we multiply the resulting two binomials, the square of a binomial.

$$(x - 3)(x + 3)(x^2 - 9) = (x^2 - 9)(x^2 - 9) \qquad \text{Multiply } (x - 3)(x + 3)$$
$$= (x^2 - 9)^2$$
$$= x^4 - 18x^2 + 81 \qquad \text{Square } (x^2 - 9). \quad ●$$

Answers

16. $x^2 - 16$,   17. $9m^2 - 36$,   18. $a^4 - 25y^2$,
19. $16y^4 - \dfrac{1}{9}$,   20. $4x^2 + 12xy + 9y^2 - 4$,
21. $y^4 - 8y^2 + 16$

GRAPHING CALCULATOR EXPLORATIONS

In the previous section, we used a graphing calculator to check addition and subtraction of polynomials in one variable. In this section, we use the same method to check multiplication of polynomials in one variable. For example, to see that

$$(x - 2)(x + 1) = x^2 - x - 2$$

graph both $Y_1 = (x - 2)(x + 1)$ and $Y_2 = x^2 - x - 2$ on the same screen and see whether their graphs coincide.

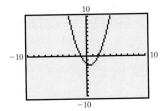

By tracing along both graphs, we see that the graphs of Y_1 and Y_2 appear to coincide, and thus $(x - 2)(x + 1) = x^2 - x - 2$ appears to be correct.

Multiply. Then use a graphing calculator to check the results.

1. $(x + 4)(x - 4)$

2. $(x + 3)(x + 3)$

3. $(3x - 7)^2$

4. $(5x - 2)^2$

5. $(5x + 1)(x^2 - 3x - 2)$

6. $(7x + 4)(2x^2 + 3x - 5)$

FOCUS ON **Mathematical Connections**

FINITE DIFFERENCES

When polynomial functions are evaluated at successive integer values, a list of values called a **sequence** is generated. The differences between successive pairs of numbers in such a sequence have special properties. Let's investigate these properties, beginning with a first-degree polynomial function, the linear function.

Notice in the table below on the left that *first differences* are the differences between the successive pairs of numbers in the original sequence. Find the first differences for any other linear function and fill in the table on the right. What do you notice? (*Note*: You may wish to try several different linear functions.)

x	Original Sequence $f(x) = 3x + 4$	First Differences		x	Original Sequence $f(x) =$	First Differences
8	28					
7	25	3				
6	22	3				
5	19	3				
4	16	3				
3	13	3				
2	10	3				
1	7	3				

Now let's look at differences for a second-degree polynomial. Notice in the table below on the left that *second differences* are the differences between successive pairs of first differences. Find first and second differences for any other second-degree polynomial function and fill in the table on the right. What do you notice? (*Note*: You may wish to try several different second-degree polynomial functions.)

x	Original Sequence $f(x) = 2x^2 - 3x + 4$	First Differences	Second Differences		x	Original Sequence $f(x) =$	First Differences	Second Differences
8	108							
7	81	27	4					
6	58	23	4					
5	39	19	4					
4	24	15	4					
3	13	11	4					
2	6	7	4					
1	3	3						

CRITICAL THINKING

1. As you might guess, third differences are the differences between successive pairs of second differences. Find the first, second, and third differences for any two third-degree polynomial functions. What do you notice?

2. What would you expect to be true about the differences for a fourth-degree polynomial function?

3. What would you expect to be true about the differences for an *n*th-degree polynomial function?

EXERCISE SET 5.2

 A *Multiply. See Examples 1 through 8.*

1. $(-4x^3)(3x^2)$

2. $(-6a)(4a)$

3. $3x(4x + 7)$

4. $5x(6x - 4)$

5. $-6xy(4x + y)$

6. $-8y(6xy + 4x)$

7. $-4ab(xa^2 + ya^2 - 3)$

8. $-6b^2z(z^2a + baz - 3b)$

9. $(x - 3)(2x + 4)$

10. $(y + 5)(3y - 2)$

11. $(2x + 3)(x^3 - x + 2)$

12. $(a + 2)(3a^2 - a + 5)$

13. $\begin{array}{r} 3x - 2 \\ \underline{5x + 1} \end{array}$

14. $\begin{array}{r} 2z - 4 \\ \underline{6z - 2} \end{array}$

15. $\begin{array}{r} 3m^2 + 2m - 1 \\ \underline{5m + 2} \end{array}$

16. $\begin{array}{r} 2x^2 - 3x - 4 \\ \underline{x + 5} \end{array}$

17. $\begin{array}{r} 3x^2 + 4x - 4 \\ \underline{3x + 6} \end{array}$

18. $\begin{array}{r} 6x^2 + 2x - 1 \\ \underline{3x - 6} \end{array}$

19. $-6a^2b^2(5a^2b^2 - 6a - 6b)$

20. $7x^2y^3(-3ax - 4xy + z)$

21. $(2x^3 + 5)(5x^2 + 4x + 1)$

22. $(3y^3 - 1)(3y^3 - 6y + 1)$

23. $(3x^2 + 2x - 1)^2$

24. $(4x^2 + 4x - 4)^2$

25. $(3x + 1)(4x^2 - 2x + 5)$

26. $(2x - 1)(5x^2 - x - 2)$

27. Explain how to multiply a polynomial by a polynomial.

28. Explain why $(3x + 2)^2$ does not equal $9x^2 + 4$.

Use the FOIL order to multiply. See Examples 9 through 11.

29. $(x - 3)(x + 4)$ **30.** $(c - 3)(c + 1)$ **31.** $(5x + 8y)(2x - y)$ **32.** $(2n - 9m)(n - 7m)$

33. $(3x - 1)(x + 3)$ **34.** $(5d - 3)(d + 6)$ **35.** $(a - 4)(2a - 4)$ **36.** $(2x - 3)(x + 1)$

37. $(y - 4)(y - 3)$ **38.** $(c - 8)(c + 2)$ **39.** $(3x + 1)(3x + 5)$ **40.** $(4x - 5)(5x + 6)$

41. $\left(4x + \dfrac{1}{3}\right)\left(4x - \dfrac{1}{2}\right)$ **42.** $\left(4y - \dfrac{1}{3}\right)\left(3y - \dfrac{1}{8}\right)$ **43.** $(5x^2 - 2y^2)(x^2 - 3y^2)$ **44.** $(4x^2 - 5y^2)(x^2 - 2y^2)$

C   D   *Use special products to multiply. See Examples 12 through 20.*

45. $(x + 4)^2$ **46.** $(x - 5)^2$ **47.** $(6y - 1)(6y + 1)$ **48.** $(x - 9)(x + 9)$

49. $(3x - y)^2$ **50.** $(4x - z)^2$ **51.** $(7ab + 3c)(7ab - 3c)$

52. $(3xy - 2b)(3xy + 2b)$ **53.** $(m - 4)^2$ **54.** $(x + 2)^2$

55. $(3x + 1)^2$ **56.** $(4x + 6)^2$ **57.** $(3b - 6y)(3b + 6y)$

58. $(2x - 4y)(2x + 4y)$

59. $(7x - 3)(7x + 3)$

60. $(4x + 1)(4x - 1)$

61. $\left(3x + \dfrac{1}{2}\right)\left(3x - \dfrac{1}{2}\right)$

62. $\left(2x - \dfrac{1}{3}\right)\left(2x + \dfrac{1}{3}\right)$

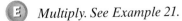 **63.** $(6x + 1)^2$

64. $(4x + 7)^2$

65. $(x^2 + 2y)(x^2 - 2y)$

66. $(3x + 2y)(3x - 2y)$

67. $[3 + (4b + 1)]^2$

68. $[5 - (3b - 3)]^2$

69. $[(2s - 3) - 1][(2s - 3) + 1]$

70. $[(2y + 5) + 6][(2y + 5) - 6]$

71. $[(xy + 4) - 6]^2$

72. $[(2a^2 + 4a) + 1]^2$

E *Multiply. See Example 21.*

73. $(x + y)(2x - 1)(x + 1)$

74. $(z + 2)(z - 3)(2z + 1)$

75. $(x - 2)^4$

76. $(x - 1)^4$

77. $(x - 5)(x + 5)(x^2 + 25)$

78. $(x + 3)(x - 3)(x^2 + 9)$

Review and Preview

Simplify. See Section 1.6.

79. $\dfrac{6x^3}{3x}$

80. $\dfrac{4x^7}{x^2}$

81. $\dfrac{20a^3b^5}{18ab^2}$

82. $\dfrac{15x^7y^2}{6xy^2}$

83. $\dfrac{8m^4n}{12mn}$

84. $\dfrac{6n^6p}{8np}$

◆ Combining Concepts

△ **85.** Find the area of the circle. Do not approximate π.

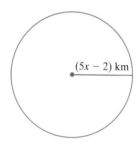
(5x − 2) km

△ **86.** Find the volume of the cylinder. Do not approximate π.

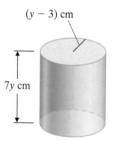
(y − 3) cm

7y cm

Find the area of each shaded region.

△ **87.**

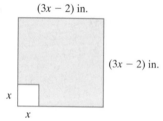
(3x − 2) in.

(3x − 2) in.

x

x

△ **88.**

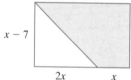
x − 7

2x   x

Multiply. Assume that variables represent positive integers.

89. $5x^2y^n(6y^{n+1} - 2)$ **90.** $-3yz^n(2y^3z^{2n} - 1)$ **91.** $(x^a + 5)(x^{2a} - 3)$ **92.** $(x^a + y^{2b})(x^a - y^{2b})$

93. Perform each indicated operation. Explain the difference between the two problems.
   a. $(3x + 5) + (3x + 7)$
   b. $(3x + 5)(3x + 7)$

94. Explain when the FOIL method can be used to multiply polynomials.

If $R(x) = x + 5$, $Q(x) = x^2 - 2$, and $P(x) = 5x$, find each function.

95. $P(x) \cdot R(x)$ **96.** $P(x) \cdot Q(x)$

If $f(x) = x^2 - 3x$, find each function value.

97. $f(a)$ **98.** $f(a + h)$

5.3 Dividing Polynomials

Now that we have added, subtracted, and multiplied polynomials, we will learn how to divide them.

(A) Dividing a Polynomial by a Monomial

Recall the following addition fact for fractions with a common denominator:

$$\frac{a}{c} + \frac{b}{c} = \frac{a + b}{c}$$

If $a, b,$ and c are monomials, we can read this equation from right to left and gain insight into how to divide a polynomial by a monomial.

> **Dividing a Polynomial by a Monomial**
>
> To divide a polynomial by a monomial, divide each term in the polynomial by the monomial.
>
> $$\frac{a + b}{c} = \frac{a}{c} + \frac{b}{c}, \quad c \neq 0$$

EXAMPLE 1   Divide $10x^3 - 5x^2 + 20x$ by $5x$.

Solution: We divide each term of $10x^3 - 5x^2 + 20x$ by $5x$ and simplify.

$$\frac{10x^3 - 5x^2 + 20x}{5x} = \frac{10x^3}{5x} - \frac{5x^2}{5x} + \frac{20x}{5x} = 2x^2 - x + 4$$

To check, see that $(\text{quotient})(\text{divisor}) = \text{dividend}$, or

$$(2x^2 - x + 4)(5x) = 10x^3 - 5x^2 + 20x$$

EXAMPLE 2   Divide: $\dfrac{3x^5y^2 - 15x^3y - x^2y - 6x}{x^2y}$

Solution: We divide each term in the numerator by x^2y.

$$\frac{3x^5y^2 - 15x^3y - x^2y - 6x}{x^2y} = \frac{3x^5y^2}{x^2y} - \frac{15x^3y}{x^2y} - \frac{x^2y}{x^2y} - \frac{6x}{x^2y}$$

$$= 3x^3y - 15x - 1 - \frac{6}{xy}$$

(B) Dividing by a Polynomial

To divide a polynomial by a polynomial other than a monomial, we use **long division**. Polynomial long division is similar to long division of real numbers. We review long division of real numbers by dividing 7 into 296.

$$
\begin{array}{r}
42 \\
\text{Divisor: } 7\overline{)296} \\
\underline{-28} \qquad 4(7) = 28 \\
16 \qquad \text{Subtract and bring down the next digit in the dividend.} \\
\underline{-14} \qquad 2(7) = 14 \\
2 \qquad \text{Subtract. The remainder is 2.}
\end{array}
$$

Practice Problem 1

Divide $16y^3 - 8y^2 + 6y$ by $2y$.

Practice Problem 2

Divide: $\dfrac{9a^3b^3 - 6a^2b^2 + a^2b - 4a}{a^2b}$

Answers

1. $8y^2 - 4y + 3$,   **2.** $9ab^2 - 6b + 1 - \dfrac{4}{ab}$

The quotient is

$$42\frac{2}{7} \quad \begin{array}{l}\text{remainder}\\ \text{divisor}\end{array}$$

To check, notice that $42(7) + 2 = 296$, which is the dividend.
This same division process can be applied to polynomials, as shown next.

Practice Problem 3

Divide $6x^2 + 11x - 2$ by $x + 2$.

EXAMPLE 3 Divide $2x^2 - x - 10$ by $x + 2$.

Solution: $2x^2 - x - 10$ is the dividend, and $x + 2$ is the divisor.

Step 1. Divide $2x^2$ by x.

$$x + 2 \overline{)\begin{array}{c}2x\\ 2x^2 - x - 10\end{array}} \qquad \frac{2x^2}{x} = 2x, \text{ so } 2x \text{ is the first term of the quotient.}$$

Step 2. Multiply $2x(x + 2)$.

$$\begin{array}{r}2x\phantom{{}-x-10}\\ x + 2 \overline{)\begin{array}{c}2x^2 - \phantom{{}}x - 10\end{array}}\\ 2x^2 + 4x\phantom{{}-10}\end{array} \qquad \begin{array}{l}2x(x + 2)\\ \text{Like terms are lined up vertically.}\end{array}$$

Step 3. Subtract $(2x^2 + 4x)$ from $(2x^2 - x - 10)$ by changing the signs of $(2x^2 + 4x)$ and adding.

$$\begin{array}{r}2x\phantom{{}-x-10}\\ x + 2 \overline{)\begin{array}{c}2x^2 - \phantom{{}}x - 10\end{array}}\\ \underline{-2x^2 - 4x\phantom{{}-10}}\\ -5x\phantom{{}-10}\end{array}$$

Step 4. Bring down the next term, -10, and start the process over.

$$\begin{array}{r}2x\phantom{{}-x-10}\\ x + 2 \overline{)\begin{array}{c}2x^2 - \phantom{{}}x - 10\end{array}}\\ \underline{-2x^2 - 4x\phantom{{}-10}}\\ -5x - 10\end{array}$$

Step 5. Divide $-5x$ by x.

$$\begin{array}{r}2x-5\phantom{{}-10}\\ x + 2 \overline{)\begin{array}{c}2x^2 - \phantom{{}}x - 10\end{array}}\\ \underline{-2x^2 - 4x\phantom{{}-10}}\\ -5x - 10\end{array} \qquad \frac{-5x}{x} = -5 \text{ so } -5 \text{ is the second term of the quotient.}$$

Step 6. Multiply $-5(x + 2)$.

$$\begin{array}{r}2x - 5\\ x + 2 \overline{)\begin{array}{c}2x^2 - \phantom{{}}x - 10\end{array}}\\ \underline{-2x^2 - 4x\phantom{{}-10}}\\ -5x - 10\\ \underline{-5x - 10}\end{array} \qquad \begin{array}{l}-5(x + 2)\\ \text{Like terms are lined up vertically.}\end{array}$$

Step 7. Subtract $(-5x - 10)$ from $(-5x - 10)$.

$$\begin{array}{r}2x - 5\\ x + 2 \overline{)\begin{array}{c}2x^2 - \phantom{{}}x - 10\end{array}}\\ \underline{-2x^2 - 4x\phantom{{}-10}}\\ -5x - 10\\ \underline{+5x + 10}\\ 0\end{array}$$

Answer

3. $6x - 1$

Then $\dfrac{2x^2 - x - 10}{x + 2} = 2x - 5$. There is no remainder.

Check this result by multiplying $2x - 5$ by $x + 2$. Their product is

$$(2x - 5)(x + 2) = 2x^2 - x - 10 \quad \text{The dividend}$$

EXAMPLE 4   Divide: $(6x^2 - 19x + 12) \div (3x - 5)$

Solution:

$$\begin{array}{r} 2x \\ 3x - 5\overline{)6x^2 - 19x + 12} \\ \underline{6x^2 - 10x} \\ -9x + 12 \end{array}$$

Divide: $\dfrac{6x^2}{3x} = 2x$

Multiply: $2x(3x - 5)$

Subtract: $6x^2 - 19x - (6x^2 - 10x) = -9x$

Bring down the next term, $+12$.

$$\begin{array}{r} 2x - 3 \\ 3x - 5\overline{)6x^2 - 19x + 12} \\ \underline{6x^2 - 10x} \\ -9x + 12 \\ \underline{-9x + 15} \\ -3 \end{array}$$

Divide: $\dfrac{-9x}{3x} = -3$

Multiply: $-3(3x - 5)$

Subtract: $-9x + 12 - (-9x + 15) = -3$

Check:

$$\boxed{\text{divisor}} \cdot \boxed{\text{quotient}} + \boxed{\text{remainder}}$$

$$(3x - 5)(2x - 3) + (-3) = 6x^2 - 19x + 15 - 3$$
$$= 6x^2 - 19x + 12 \quad \text{The dividend}$$

The division checks, so

$$\dfrac{6x^2 - 19x + 12}{3x - 5} = 2x - 3 - \dfrac{3}{3x - 5}$$

Helpful Hint

This fraction is the remainder over the divisor.

EXAMPLE 5   Divide $3x^4 + 2x^3 - 8x + 6$ by $x^2 - 1$.

Solution: Before dividing, we represent any "missing powers" by the product of 0 and the variable raised to the missing power. There is no x^2-term in the dividend, so we include $0x^2$ to represent the missing term. Also, there is no x term in the divisor, so we include $0x$ in the divisor.

$$\begin{array}{r} 3x^2 + 2x + 3 \\ x^2 + 0x - 1\overline{)3x^4 + 2x^3 + 0x^2 - 8x + 6} \\ \underline{3x^4 + 0x^3 - 3x^2} \\ 2x^3 + 3x^2 - 8x \\ \underline{2x^3 + 0x^2 - 2x} \\ 3x^2 - 6x + 6 \\ \underline{3x^2 + 0x - 3} \\ -6x + 9 \end{array}$$

$\dfrac{3x^4}{x^2} = 3x^2$

$3x^2(x^2 + 0x - 1)$

Subtract. Bring down $-8x$.

$2x^3/x^2 = 2x$, a term of the quotient

$2x(x^2 + 0x - 1)$

Subtract. Bring down 6.

$3x^2/x^2 = 3$, a term of the quotient

$3(x^2 + 0x - 1)$

Subtract.

Practice Problem 4

Divide: $(10x^2 - 17x + 5) \div (5x - 1)$

Practice Problem 5

Divide $3x^4 + 4x^2 - 6x + 1$ by $x^2 + 1$.

Answers

4. $2x - 3 + \dfrac{2}{5x - 1}$,   **5.** $3x^2 + 1 - \dfrac{6x}{x^2 + 1}$

The division process is finished when the degree of the remainder polynomial is less than the degree of the divisor. Thus,

$$\frac{3x^4 + 2x^3 - 8x + 6}{x^2 - 1} = 3x^2 + 2x + 3 + \frac{-6x + 9}{x^2 - 1}$$

Practice Problem 6

Divide $64x^3 - 27$ by $4x - 3$.

EXAMPLE 6   Divide $27x^3 + 8$ by $3x + 2$.

Solution: We replace the missing terms in the dividend with $0x^2$ and $0x$.

$$
\begin{array}{r}
9x^2 - 6x + 4 \\
3x + 2\overline{)27x^3 + 0x^2 + 0x + 8} \\
\underline{27x^3 + 18x^2} \\
-18x^2 + 0x \\
\underline{-18x^2 - 12x} \\
12x + 8 \\
\underline{12x + 8} \\
\end{array}
$$

$9x^2(3x + 2)$

Subtract. Bring down $0x$.

$-6x(3x + 2)$

Subtract. Bring down 8.

$4(3x + 2)$

Thus, $\dfrac{27x^3 + 8}{3x + 2} = 9x^2 - 6x + 4.$

Try the Concept Check in the margin.

Concept Check

In a division problem, the divisor is $4x^3 - 5$. The division process can be stopped when which of these possible remainder polynomials is reached?

a. $2x^4 + x^2 - 3$

b. $x^3 - 5^2$

c. $4x^2 + 25$

(c) Using Synthetic Division

When a polynomial is to be divided by a binomial of the form $x - c$, a shortcut process called **synthetic division** may be used. On the left is an example of long division, and on the right is the same example showing the coefficients of the variables only.

$$
\begin{array}{r}
2x^2 + 5x + 2 \\
x - 3\overline{)2x^3 - x^2 - 13x + 1} \\
\underline{2x^3 - 6x^2} \\
5x^2 - 13x \\
\underline{5x^2 - 15x} \\
2x + 1 \\
\underline{2x - 6} \\
7
\end{array}
\qquad
\begin{array}{r}
2\quad 5\quad 2 \\
1 - 3\overline{)2 - 1 - 13 + 1} \\
\underline{2 - 6} \\
5 - 13 \\
\underline{5 - 15} \\
2 + 1 \\
\underline{2 - 6} \\
7
\end{array}
$$

Notice that as long as we keep coefficients of powers of x in the same column, we can perform division of polynomials by performing algebraic operations on the coefficients only. This shorter process of dividing with coefficients only in a special format is called synthetic division. To find $(2x^3 - x^2 - 13x + 1) \div (x - 3)$ by synthetic division, follow the next example.

Practice Problem 7

Use synthetic division to divide $3x^3 - 2x^2 + 5x + 4$ by $x - 2$.

EXAMPLE 7

Use synthetic division to divide $2x^3 - x^2 - 13x + 1$ by $x - 3$.

Solution: To use synthetic division, the divisor must be in the form $x - c$. Since we are dividing by $x - 3$, c is 3. We write down 3 and the coefficients of the dividend.

Answers

6. $16x^2 + 12x + 9$,   **7.** $3x^2 + 4x + 13 + \dfrac{30}{x - 2}$

Concept Check: c

```
 c
  ↘
  3| 2  -1  -13  1
     ↓
  ─────────────────
     2
```
Next, draw a line and bring down the first coefficient of the dividend.

```
  3| 2  -1  -13  1
        6
  ─────────────────
    2
```
Multiply 3·2 and write down the product, 6.

```
  3| 2  -1  -13  1
        6
  ─────────────────
     2   5
```
Add −1 + 6. Write down the sum, 5.

```
  3| 2  -1  -13  1
        6   15
  ─────────────────
     2   5   2
```
3·5 = 15
−13 + 15 = 2

```
  3| 2  -1  -13  1
        6   15   6
  ─────────────────
     2   5   2   7
```
3·2 = 6
1 + 6 = 7

The quotient is found in the bottom row. The numbers 2, 5, and 2 are the coefficients of the quotient polynomial, and the number 7 is the remainder. The degree of the quotient polynomial is one less than the degree of the dividend. In our example, the degree of the dividend is 3, so the degree of the quotient polynomial is 2. As we found when we performed the long division, the quotient is

$$2x^2 + 5x + 2, \quad \text{remainder } 7$$

or

$$2x^2 + 5x + 2 + \frac{7}{x - 3}$$

When using synthetic division, if there are missing powers of the variable, insert 0s as coefficients.

EXAMPLE 8

Use synthetic division to divide $x^4 - 2x^3 - 11x^2 + 34$ by $x + 2$.

Solution: The divisior is $x + 2$, which in the form $x - c$ is $x - (-2)$. Thus, c is -2. There is no x-term in the dividend, so we insert a coefficient of 0. The dividend coefficients are 1, −2, −11, 0, and 34.

```
 c
  ↘
 -2| 1  -2  -11   0   34
        -2    8   6  -12
  ──────────────────────
    1  -4   -3   6   22
```

The dividend is a fourth-degree polynomial, so the quotient polynomial is a third-degree polynomial. The quotient is $x^3 - 4x^2 - 3x + 6$ with a remainder of 22. Thus,

$$\frac{x^4 - 2x^3 - 11x^2 + 34}{x + 2} = x^3 - 4x^2 - 3x + 6 + \frac{22}{x + 2}$$

Practice Problem 8

Use synthetic division to divide $x^4 + 3x^3 - 5x + 4$ by $x + 1$.

Answer

8. $x^3 + 2x^2 - 2x - 3 + \dfrac{7}{x + 1}$

Before dividing by synthetic division, write the dividend in descending order of variable exponents. Any "missing powers" of the variable must be represented by 0 times the variable raised to the missing power.

Try the Concept Check in the margin.

Concept Check

Which division problems are candidates for the synthetic division process?

a. $(3x^2 + 5) \div (x + 4)$
b. $(x^3 - x^2 + 2) \div (3x^3 - 2)$
c. $(y^4 + y - 3) \div (x^2 + 1)$
d. $x^5 \div (x - 5)$

Answer

Concept Check: a and d

STUDY SKILLS REMINDER

Are you getting all the mathematics help that you need?

Remember that, in addition to your instructor, there are many places to get help with your mathematics course. For example, see which of the list below are available.

- This text has an accompanying video lesson for every section in this text.
- The back of this book contains answers to odd-numbered exercises and selected solutions.
- MathPro is available with this text. It is a tutorial software program with lessons corresponding to each section in the text.
- A student solutions manual is available that contains worked-out solutions to odd-numbered exercises as well as solutions to every exercise in the Chapter Pretests, Integrated Reviews, Chapter Reviews, Chapter Tests, and Cumulative Reviews.
- Don't forget to check with your instructor for other local resources available to you, such as a tutor center.

EXERCISE SET 5.3

(A) *Divide. See Examples 1 and 2.*

1. $4a^2 + 8a$ by $2a$

2. $6x^4 - 3x^3$ by $3x^2$

3. $\dfrac{12a^5b^2 + 16a^4b}{4a^4b}$

4. $\dfrac{4x^3y + 12x^2y^2 - 4xy^3}{4xy}$

5. $\dfrac{4x^2y^2 + 6xy^2 - 4y^2}{2x^2y}$

6. $\dfrac{6x^5 + 74x^4 + 24x^3}{2x^3}$

(B) *Divide. See Examples 3 through 6.*

7. $(x^2 + 3x + 2) \div (x + 2)$

8. $(y^2 + 7y + 10) \div (y + 5)$

9. $(2x^2 - 6x - 8) \div (x + 1)$

10. $(3x^2 + 19x + 20) \div (x + 5)$

11. $2x^2 + 3x - 2$ by $2x + 4$

12. $6x^2 - 17x - 3$ by $3x - 9$

13. $(4x^3 + 7x^2 + 8x + 20) \div (2x + 4)$

14. $(18x^3 + x^2 - 90x - 5) \div (9x^2 - 45)$

15. $(6x^3 + 2x^2 - 18x - 6) \div (3x + 1)$

16. $(10x^3 - 15x^2 + 4x - 6) \div (2x - 3)$

17. $(2x^3 - 6x^2 - 4) \div (x - 4)$

18. $(3x^3 + 4x - 10) \div (x + 2)$

19. $(10x^3 - 5x^2 - 12x + 1) \div (2x - 1)$

20. $(20x^3 - 8x^2 + 5x - 5) \div (5x - 2)$

21. $(3x^5 - x^3 + 4x^2 - 12x - 8) \div (x^2 - 2)$

22. $(2x^5 - 6x^4 + x^3 - 4x + 3) \div (x^2 - 3)$

23. $\left(2x^4 + \dfrac{1}{2}x^3 + x^2 + x\right) \div (x - 2)$

24. $\left(x^4 - \dfrac{2}{3}x^3 + x\right) \div (x - 3)$

C Use synthetic division to divide. See Examples 7 and 8.

25. $\dfrac{x^2 + 3x - 40}{x - 5}$

26. $\dfrac{x^2 - 14x + 24}{x - 2}$

27. $\dfrac{x^2 + 5x - 6}{x + 6}$

28. $\dfrac{x^2 + 12x + 32}{x + 4}$

29. $\dfrac{x^3 - 7x^2 - 13x + 5}{x - 2}$

30. $\dfrac{x^3 + 6x^2 + 4x - 7}{x + 5}$

31. $\dfrac{4x^2 - 9}{x - 2}$

32. $\dfrac{3x^2 - 4}{x - 1}$

33. $\dfrac{2x^4 - 13x^3 + 16x^2 - 9x + 20}{x - 5}$

34. $\dfrac{3x^4 + 5x^3 - x^2 + x - 2}{x + 2}$

35. $\dfrac{7x^2 - 4x + 12 + 3x^3}{x + 1}$

36. $\dfrac{x^4 + 4x^3 - x^2 - 16x - 4}{x - 2}$

37. $\dfrac{3x^3 + 2x^2 - 4x + 1}{x - \dfrac{1}{3}}$

38. $\dfrac{9y^3 + 9y^2 - y + 2}{y + \dfrac{2}{3}}$

39. $\dfrac{x^3 - 1}{x - 1}$

40. $\dfrac{y^3 - 8}{y - 2}$

Review and Preview

Multiply. See Section 5.2.

41. $6x(x + 3) + 5(x + 3)$

42. $7y(y - 1) + 2(y - 1)$

Solve each inequality. See Section 2.7.

43. $|x + 5| < 4$

44. $|x - 1| \leq 8$

45. $|2x + 7| \geq 9$

46. $|4x + 2| > 10$

47. A board of length $(3x^4 + 6x^2 - 18)$ meters is to be cut into three pieces of the same length. Find the length of each piece.

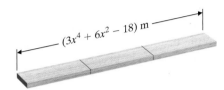

48. The perimeter of a regular hexagon is given to be $(12x^5 - 48x^3 + 3)$ miles. Find the length of each side.

49. If the area of the rectangle is $(15x^2 - 29x - 14)$ square inches, and its length is $(5x + 2)$ inches, find its width.

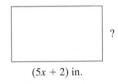

$(5x + 2)$ in.

50. If the area of a parallelogram is $(2x^2 - 17x + 35)$ square centimeters and its base is $(2x - 7)$ centimeters, find its height.

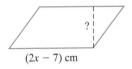

$(2x - 7)$ cm

51. Find $P(1)$ for the polynomial function $P(x) = 3x^3 + 2x^2 - 4x + 3$. Next, divide $3x^3 + 2x^2 - 4x + 3$ by $x - 1$. Compare the remainder in the division with $P(1)$. In your own words, explain your findings.

52. Find $P(-2)$ for the polynomial function $P(x) = x^3 - 4x^2 - 3x + 5$. Next, divide $x^3 - 4x^2 - 3x + 5$ by $x + 2$. Compare the remainder in the division with $P(-2)$. In your own words, explain your findings.

53. If a polynomial is divided by $x + 3$, the quotient is $x^2 - x + 10$ and the remainder is -2. Find the original polynomial.

54. Explain how to divide a polynomial in x by $(x - c)$ using synthetic division.

55. In your own words, explain how to check whether the division below is correct.

$$\frac{12x^2 - 22x - 14}{3x - 7} = 4x + 2$$

Now use your procedure and check the division.

56. Explain an advantage of using synthetic division instead of long division.

57. Gateway, Inc., is a direct marketer of personal computers. Gateway's annual net profit can be modeled by the polynomial function $P(x) = -54x^3 + 305x^2 - 363x + 245$, where $P(x)$ is net profit in millions of dollars in the year x. Gateway's annual revenue can be modeled by the function $R(x) = 1140x + 5295$, where $R(x)$ is revenue in millions of dollars and x is the number of years after 1996. (*Source:* Gateway, Inc., 1996–2000)

a. Suppose that a market analyst has found the model $P(x)$, and another analyst at the same firm has found the model $R(x)$. The analysts have been asked by their manager to work together to find a model for Gateway's net profit margin. The analysts know that a company's net profit margin is the ratio of its net profit to its revenue. Describe how these two analysts could collaborate to find a function $m(x)$ that models Gateway's net profit margin based on the work they have done independently.

b. Without actually finding $m(x)$, give a general description of what you would expect the form of the result to be.

MUHAMMAD AL-KHWARIZMI

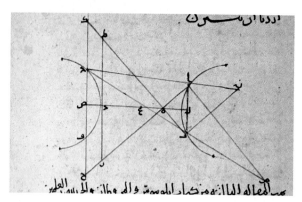

Muhammad ibn Musa al-Khwarizmi was an Arabic mathematician who lived from around 800 A.D. to about 847 A.D. He was originally from what is today Uzbekistan. He later became a scholar at the House of Wisdom in Baghdad (in what is today Iraq). In 830 A.D., al-Khwarizmi wrote the first known Arabic text on the algebra of polynomials. This text, *al-Kitab al-mukhtasar fi hisab al-jabr wa'l-muqabala* (translated in English as *The compendious book on calculation by completion and balancing* or *A summary of calculating through the reunion of broken parts*), discusses first-degree and second-degree polynomial equations and their application to practical matters. Al-Khwarizmi's *Al-jabr wa'l-muqabala* is believed to be a summary of Arab mathematics and the mathematical techniques and algebraic theories known to the Arabs through Hindu, Greek, Hebrew, and Babylonian influences.

Al-Khwarizmi and his *Al-jabr wa'l-muqabala* are important to the history of mathematics for several reasons. First, it is through this textbook that many diverse lines of mathematical thought were merged into what was to become a more unified theory of algebra. Second, a later translation of *Al-jabr wa'l-muqabala* into Latin was responsible for spreading algebraic concepts to Europe during the Middle Ages. Third, al-Khwarizmi's textbook is actually responsible for the naming of algebra. The word *algebra* is a corruption of the Arabic word *al-jabr* in the text's title. Finally, the Latin form of al-Khwarizmi's name is the source of the word *algorithm* in the English language.

Al-Khwarizmi was the first to encourage writing out mathematical calculations rather than relying solely on calculating with an abacus. He is also thought to be responsible for spreading the use of the modern decimal system for writing numbers and may have introduced the use of zero as a placeholder in the decimal system.

Muhammad ibn Musa al-Khwarizmi's many contributions to mathematics in general and algebra in particular were rewarded when a crater of the moon, Crater Al-Khwarizmi, was named in his honor by the International Astronomical Union in 1973.

5.4 The Greatest Common Factor and Factoring by Grouping

Factoring is the reverse process of multiplying. It is the process of writing a polynomial as a product

$$6x^2 + 13x - 5 = (3x - 1)(2x + 5)$$

(Factoring / Multiplying)

In the next few sections, we review techniques for factoring polynomials.

OBJECTIVES

A Factor out the greatest common factor of a polynomials terms.

B Factor polynomials by grouping.

SSM   TUTOR CENTER   SG   CD & VIDEO   MATH PRO   WEB

A Factoring Out the Greatest Common Factor

To factor a polynomial, we first **factor out** the greatest common factor of its terms, using the distributive property. The **greatest common factor** (GCF) of the terms of a polynomial is the product of the GCF of the numerical coefficients and each GCF of the powers of a common variable.

Let's find the GCF of $20x^3y$, $10x^2y^2$, and $35x^3$.

The GCF of the numerical coefficients 20, 10, and 35 is 5, the largest integer that is a factor of each integer. The GCF of the variable factors x^3, x^2, and x^3 is x^2 because x^2 is the largest factor common to all three powers of x. The variable y is not a common factor because it does not appear in all three monomials. The GCF is thus

$$5 \cdot x^2, \quad \text{or} \quad 5x^2$$

EXAMPLE 1 Factor: $8x + 4$

Solution: The greatest common factor of the terms $8x$ and 4 is 4.

$$8x + 4 = 4 \cdot 2x + 4 \cdot 1 \qquad \text{Factor out 4 from each term.}$$
$$= 4(2x + 1) \qquad \text{Use the distributive property.}$$

The factored form of $8x + 4$ is $4(2x + 1)$. To check, multiply $4(2x + 1)$ to see that the product is $8x + 4$. ●

EXAMPLES Factor.

2. $6x^2 + 3x^3 = 3x^2 \cdot 2 + 3x^2 \cdot x$ The GCF of 6 and 3 is 3 and the GCF of x^2 and x^3 is x^2. Thus, the GCF of the terms is $3x^2$.

$$= 3x^2(2 + x) \qquad \text{Use the distributive property.}$$

3. $3y + 1$ There is no common factor other than 1.

4. $17x^3y^2 - 34x^4y^2 = 17x^3y^2 \cdot 1 - 17x^3y^2 \cdot 2x$ Factor out the greatest common factor, $17x^3y^2$.

$$= 17x^3y^2(1 - 2x) \qquad \text{Use the distributive property.}$$

Helpful Hint

If the greatest common factor happens to be one of the terms in the polynomial, a factor of 1 will remain for this term when the greatest common factor is factored out. For example, in the polynomial $21x^2 + 7x$, the greatest common factor of $21x^2$ and $7x$ is $7x$, so

$$21x^2 + 7x = 7x(3x) + 7x(1) = 7x(3x + 1)$$

Practice Problem 1

Factor: $9x + 3$

Practice Problems 2–4

Factor each polynomial.

2. $20y^2 - 4y^3$

3. $6a - 7$

4. $6a^4b^2 - 3a^2b^2$

Answers

1. $3(3x + 1)$, **2.** $4y^2(5 - y)$, **3.** $6a - 7$,
4. $3a^2b^2(2a^2 - 1)$

Concept Check

Which factorization of $12x^2 + 9x - 3$ is correct?

a. $3(4x^2 + 3x + 1)$
b. $3(4x^2 + 3x - 1)$
c. $3(4x^2 + 3x - 3)$
d. $3(4x^2 + 3x)$

Practice Problem 5

Factor: $-2x^2y - 4xy + 10y$

Practice Problem 6

Factor: $3(x + 7) + 5y(x + 7)$

Practice Problem 7

Factor: $6a(2a + 3b) - (2a + 3b)$

Practice Problem 8

Factor: $xy - 5y + 3x - 15$

Answers

5. $-2y(x^2 + 2x - 5)$, **6.** $(x + 7)(3 + 5y)$,
7. $(2a + 3b)(6a - 1)$, **8.** $(x - 5)(y + 3)$

Concept Check: b

Try the Concept Check in the margin.

> **Helpful Hint**
>
> To check that the greatest common factor has been factored out correctly, multiply the factors together and see that their product is the original polynomial.

EXAMPLE 5   Factor: $-3x^3y + 2x^2y - 5xy$

Solution: Two possibilities are shown for factoring this polynomial.

First, the common factor xy is factored out.
$$-3x^3y + 2x^2y - 5xy = xy(-3x^2 + 2x - 5)$$
Also, the common factor $-xy$ can be factored out as shown.
$$-3x^3y + 2x^2y - 5xy = -xy(3x^2) + (-xy)(-2x) + (-xy)(5)$$
$$= -xy(3x^2 - 2x + 5)$$

Both of these are correct.

EXAMPLE 6   Factor: $2(x - 5) + 3a(x - 5)$

Solution: The greatest common factor is the binomial factor $(x - 5)$.
$$2(x - 5) + 3a(x - 5) = (x - 5)(2 + 3a)$$

EXAMPLE 7   Factor: $7x(x^2 + 5y) - (x^2 + 5y)$

Solution:

> **Helpful Hint**
>
> Notice that we wrote $-(x^2 + 5y)$ as $-1(x^2 + 5y)$ to aid in factoring.

$$7x(x^2 + 5y) - (x^2 + 5y) = 7x(x^2 + 5y) - 1(x^2 + 5y)$$
$$= (x^2 + 5y)(7x - 1)$$

B Factoring by Grouping

Sometimes it is possible to factor a polynomial by grouping the terms of the polynomial and looking for common factors in each group. This method of factoring is called **factoring by grouping**.

EXAMPLE 8   Factor: $ab - 6a + 2b - 12$

Solution:

$$ab - 6a + 2b - 12 = (ab - 6a) + (2b - 12)$$ Group pairs of terms.
$$= a(b - 6) + 2(b - 6)$$ Factor each binomial.
$$= (b - 6)(a + 2)$$ Factor out the greatest common factor, $(b - 6)$.

To check, multiply $(b - 6)$ and $(a + 2)$ to see that the product is $ab - 6a + 2b - 12$.

Helpful Hint

Notice that the polynomial $a(b - 6) + 2(b - 6)$ is *not* in factored form. It is a *sum*, not a *product*. The factored form is $(b - 6)(a + 2)$.

EXAMPLE 9   Factor: $x^3 + 5x^2 + 3x + 15$

Solution:

$$\begin{aligned}
x^3 + 5x^2 + 3x + 15 &= (x^3 + 5x^2) + (3x + 15) &&\text{Group pairs of terms.}\\
&= x^2(x + 5) + 3(x + 5) &&\text{Factor each binomial.}\\
&= (x + 5)(x^2 + 3) &&\text{Factor out the common}\\
&&&\text{factor, } (x + 5).
\end{aligned}$$

Practice Problem 9

Factor: $y^3 + 6y^2 + 4y + 24$

EXAMPLE 10   Factor: $m^2n^2 + m^2 - 2n^2 - 2$

Solution:

$$\begin{aligned}
m^2n^2 + m^2 - 2n^2 - 2 &= (m^2n^2 + m^2) + (-2n^2 - 2) &&\text{Group pairs of terms.}\\
&= m^2(n^2 + 1) - 2(n^2 + 1) &&\text{Factor each binomial.}\\
&= (n^2 + 1)(m^2 - 2) &&\text{Factor out the com-}\\
&&&\text{mon factor, } (n^2 + 1).
\end{aligned}$$

Practice Problem 10

Factor: $a^2b^2 + a^2 - 3b^2 - 3$

EXAMPLE 11   Factor: $xy + 2x - y - 2$

Solution:

$$\begin{aligned}
xy + 2x - y - 2 &= (xy + 2x) + (-y - 2) &&\text{Group pairs of terms.}\\
&= x(y + 2) - 1(y + 2) &&\text{Factor each binomial.}\\
&= (y + 2)(x - 1) &&\text{Factor out the common}\\
&&&\text{factor, } y + 2.
\end{aligned}$$

Practice Problem 11

Factor: $ab + 5a - b - 5$

Answers

9. $(y + 6)(y^2 + 4)$,   **10.** $(b^2 + 1)(a^2 - 3)$,
11. $(b + 5)(a - 1)$

FOCUS ON **Business and Career**

BUSINESS TERMS

For most businesses, a financial goal is to "make money." But what does that mean from a mathematical point of view? To find out, we must first discuss some common business terms.

- **Revenue** is the amount of money a business takes in. A company's annual revenue is the amount of money it collects during its fiscal, or business, year. For most companies, the largest source of revenue is from the sales of their products or services. For instance, a computer manufacturer's annual revenue is the amount of money it collects during the year from selling computers to customers. Large companies may also have revenues from investment interest or leases. When revenue can be expressed as a function of another variable, it is often denoted as $R(x)$.

- **Expenses** are the costs of doing business. For instance, a large part of a computer manufacturer's expenses include the cost of the computer components it buys from wholesalers to use in the manufacturing or assembling process. Other expenses include salaries, mortgage payments, equipment, taxes, advertising, and so on. Some

businesses refer to their expenses simply as cost. When cost can be expressed as a function of another variable, it is often denoted as $C(x)$.

- **Net income/loss** is the difference between a company's annual revenues and expenses. This difference may also be referred to as net earnings. Positive net earnings—that is, a positive difference—result in a net income or net profit. Posting a net income can be interpreted as "making money." Negative net earnings—that is, a negative difference—result in a net loss. Posting a net loss can be interpreted as "losing money." A profit function can be expressed as $P(x) = R(x) - C(x)$. In this case, a negative profit is interpreted as a net loss.

GROUP ACTIVITY

Locate several corporate annual reports. Using the data in the reports, verify that the net income or net earnings given in a report was calculated as the difference between revenue and expenses. If this was not the case, can you tell what caused the variation? If so, explain.

Mental Math

Find the greatest common factor of each list of monomials.

1. $6, 12$

2. $9, 27$

3. $15x, 10$

4. $9x, 12$

5. $13x, 2x$

6. $4y, 5y$

7. $7x, 14x$

8. $8z, 4z$

EXERCISE SET 5.4

Ⓐ *Factor out the greatest common factor. See Examples 1 through 7.*

1. $18x - 12$

2. $21x + 14$

3. $4y^2 - 16xy^3$

4. $3z - 21xz^4$

 **5.** $6x^5 - 8x^4 + 2x^3$

6. $9x + 3x^2 - 6x^3$

7. $8a^3b^3 - 4a^2b^2 + 4ab + 16ab^2$

8. $12a^3b - 6ab + 18ab^2 - 18a^2b$

9. $6(x + 3) + 5a(x + 3)$

10. $2(x - 4) + 3y(x - 4)$

 **11.** $2x(z + 7) + (z + 7)$

12. $x(y - 2) + (y - 2)$

13. $3x(x^2 + 5) - 2(x^2 + 5)$

14. $4x(2y + 3) - 5(2y + 3)$

△ **15.** The area of the material needed to manufacture a tin can is given by the polynomial $2\pi r^2 + 2\pi rh$,

where the radius is r and height is h. Factor this expression.

16. The amount E of current in an electrical circuit is given by the formula $IR_1 + IR_2 = E$. Write an equivalent equation by factoring the expression $IR_1 + IR_2$.

17. At the end of T years, the amount of money A in a savings account earning simple interest from an initial investment of P dollars at rate R is given by the formula $A = P + Prt$. Write an equivalent equation by factoring the expression $P + Prt$.

△ **18.** An open-topped box has a square base and a height of 10 inches. If each of the bottom edges of the box has length x inches, find the amount of material needed to construct the box. Write the answer in factored form.

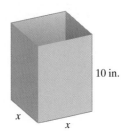

10 in.

x

x

19. When $3x^2 - 9x + 3$ is factored, the result is $3(x^2 - 3x + 1)$. Explain why it is necessary to include the term 1 in this factored form.

20. Construct a trinomial whose greatest common factor is $5x^2y^3$.

B *Factor each polynomial by grouping. See Examples 8 through 11.*

21. $ab + 3a + 2b + 6$

22. $ab + 2a + 5b + 10$

23. $ac + 4a - 2c - 8$

24. $bc + 8b - 3c - 24$

25. $2xy - 3x - 4y + 6$

26. $12xy - 18x - 10y + 15$

27. $12xy - 8x - 3y + 2$

28. $20xy - 15x - 4y + 3$

29. $x^3 + 3x^2 + 4x + 12$

30. $x^3 + 4x^2 + 3x + 12$

31. $x^3 - x^2 - 2x + 2$

32. $x^3 - 2x^2 - 3x + 6$

33. $2x^2 + 3xy + 4x + 6y$

34. $3x^2 + 12x + 4xy + 16y$

35. $5x^2 + 5xy - 3x - 3y$

36. $4x^2 + 2xy - 10x - 5y$

37. $6xy + 10x + 9y + 15$

38. $15xy + 20x + 6y + 8$

39. $xy + 3y - 5x - 15$

40. $xy + 4y - 3x - 12$

41. $9abc^2 + 6a^2bc - 6ab + 3bc$

42. $4a^2b^2c - 6ab^2c - 4ac + 8a$

Review and Preview

Find each product by using the FOIL order of multiplying binomials. See Section 5.2.

43. $(x + 2)(x - 5)$

44. $(x - 7)(x - 1)$

45. $(x + 3)(x + 2)$

46. $(x - 4)(x + 2)$

47. $(y - 3)(y - 1)$

48. $(s + 8)(s + 10)$

49. A factored polynomial can be in many forms. For example, a factored form of $xy - 3x - 2y + 6$ is $(x - 2)(y - 3)$. Which of the following (if any) is not a factored form of $xy - 3x - 2y + 6$?
 a. $(2 - x)(3 - y)$
 b. $(-2 + x)(-3 + y)$
 c. $(y - 3)(x - 2)$
 d. $(-x + 2)(-y + 3)$

50. The polynomial function $f(x) = 2900x^2 - 3500x + 120{,}000$ models the number of applications for trademark registrations for the years 1990–2000 where x represents the number of years after 1990 and $f(x)$ is the number of trademark registrations. Write an equivalent expression for $f(x)$ by factoring the greatest common factor from the terms of $2900x^2 - 3500x + 120{,}000$. (*Source:* International Trademark Association)

51. The number of computers for instructional use in public schools is increasing. The polynomial function $f(x) = 98x^2 + 514x + 4746$ models the number of these computers for the years 1994–2000, where x represents the number of years after 1994 and $f(x)$ is the number of computers used for instruction. Write an equivalent expression for $f(x)$ by factoring the greatest common factor from the terms of $98x^2 + 514x + 4746$.

Factor out the greatest common factor. Assume that variables used as exponents represent positive integers.

52. $x^{3n} - 2x^{2n} + 5x^n$ **53.** $3y^n + 3y^{2n} + 5y^{8n}$ **54.** $6x^{8a} - 2x^{5a} - 4x^{3a}$ **55.** $3x^{5a} - 6x^{3a} + 9x^{2a}$

56. An object is thrown upward from the ground with an initial velocity of 64 feet per second. The height $h(t)$ in feet of the object after t seconds is given by the polynomial function

$$h(t) = -16t^2 + 64t$$

 a. Write an equivalent factored expression for the function $h(t)$ by factoring $-16t^2 + 64t$.
 b. Find $h(1)$ by using

$$h(t) = -16t^2 + 64t$$

 and then by using the factored form of $h(t)$.
 c. Explain why the values found in part (b) are the same.

57. An object is dropped from the gondola of a hot-air balloon at a height of 224 feet. The height $h(t)$ in feet of the object after t seconds is given by the polynomial function

$$h(t) = -16t^2 + 224$$

224 ft

 a. Write an equivalent factored expression for the function $h(t)$ by factoring $-16t^2 + 224$.
 b. Find $h(2)$ by using

$$h(t) = -16t^2 + 224$$

 and then by using the factored form of the function.
 c. Explain why the values found in part (b) are the same.

Are you satisfied with your performance on a particular quiz or exam?

If not, analyze your quiz or exam like you would a good mystery novel. Look for common themes in your errors.

Were most of your errors a result of

- *Carelessness*? If your errors were careless, did you turn in your work before the allotted time expired? If so, resolve next time to use the entire time allotted. Any extra time can be spent checking your work.

- *Running out of time*? If so, make a point to better manage your time on your next exam. A few suggestions are to work any questions that you are unsure of last and to check your work after all of the questions have been answered.

- *Not understanding a concept*? If so, review that concept and correct your work. Remember next time to make sure that all concepts on a quiz or exam are understood before the exam.

5.5 Factoring Trinomials

Ⓐ Factoring Trinomials of the Form $x^2 + bx + c$

In the previous section, we used factoring by grouping to factor four-term polynomials. In this section, we present techniques for factoring trinomials. Since $(x - 2)(x + 5) = x^2 + 3x - 10$, we say that $(x - 2)(x + 5)$ is a factored form of $x^2 + 3x - 10$. Taking a close look at how $(x - 2)$ and $(x + 5)$ are multiplied suggests a pattern for factoring trinomials of the form $x^2 + bx + c$.

$$(x - 2)(x + 5) = x^2 + 3x - 10$$

The pattern for factoring is summarized next.

> **Factoring a Trinomial of the Form $x^2 + bx + c$**
>
> Find two numbers whose product is c and whose sum is b. The factored form of $x^2 + bx + c$ is
>
> $$(x + \text{one number})(x + \text{other number})$$

EXAMPLE 1   Factor: $x^2 + 10x + 16$

Solution: We look for two integers whose product is 16 and whose sum is 10. Since our integers must have a positive product and a positive sum, we look at only positive factors of 16.

Positive Factors of 16	Sum of Factors	
1, 16	$1 + 16 = 17$	
4, 4	$4 + 4 = 8$	
2, 8	$2 + 8 = 10$	Correct pair

The correct pair of numbers is 2 and 8 because their product is 16 and their sum is 10. Thus,

$$x^2 + 10x + 16 = (x + 2)(x + 8)$$

To check, see that $(x + 2)(x + 8) = x^2 + 10x + 16$.   ●

EXAMPLE 2   Factor: $x^2 - 12x + 35$

Solution: We need to find two integers whose product is 35 and whose sum is -12. Since our integers must have a positive product and a negative sum, we consider only negative factors of 35.

Negative Factors of 35	Sum of Factors	
$-1, -35$	$-1 + (-35) = -36$	
$-5, -7$	$-5 + (-7) = -12$	Correct pair

The numbers are -5 and -7.

$$x^2 - 12x + 35 = [x + (-5)][x + (-7)]$$
$$= (x - 5)(x - 7)$$

To check, see that $(x - 5)(x - 7) = x^2 - 12x + 35$.   ●

Practice Problem 1

Factor: $x^2 + 8x + 15$

Practice Problem 2

Factor: $x^2 - 10x + 24$

Answers

1. $(x + 5)(x + 3)$,   **2.** $(x - 4)(x - 6)$

Practice Problem 3

Factor: $6x^3 + 24x^2 - 30x$

EXAMPLE 3   Factor: $5x^3 - 30x^2 - 35x$

Solution: First we factor out the greatest common factor, $5x$.

$$5x^3 - 30x^2 - 35x = 5x(x^2 - 6x - 7)$$

Next we factor $x^2 - 6x - 7$ by finding two numbers whose product is -7 and whose sum is -6. The numbers are 1 and -7.

$$5x^3 - 30x^2 - 35x = 5x(x^2 - 6x - 7)$$
$$= 5x(x + 1)(x - 7)$$ ●

> **Helpful Hint**
>
> If the polynomial to be factored contains a common factor that is factored out, don't forget to include that common factor in the final factored form of the original polynomial.

Practice Problem 4

Factor: $3y^2 + 6y + 6$

EXAMPLE 4   Factor: $2n^2 - 38n + 80$

Solution: The terms of this polynomial have a greatest common factor of 2, which we factor out first.

$$2n^2 - 38n + 80 = 2(n^2 - 19n + 40)$$

Next we factor $n^2 - 19n + 40$ by finding two numbers whose product is 40 and whose sum is -19. Both numbers must be negative since their product is positive and their sum is negative. Possibilities are

$$-1 \text{ and } -40, \quad -2 \text{ and } -20, \quad -4 \text{ and } -10, \quad -5 \text{ and } -8$$

None of the pairs has a sum of -19, so no further factoring with integers is possible. The factored form of $2n^2 - 38n + 80$ is

$$2n^2 - 38n + 80 = 2(n^2 - 19n + 40)$$ ●

We call a polynomial such as $n^2 - 19n + 40$, which cannot be factored further, a **prime polynomial.**

B   Factoring Trinomials of the Form $ax^2 + bx + c$

Next, we factor trinomials of the form $ax^2 + bx + c$, where the coefficient a of x^2 is not 1. Don't forget that the first step in factoring any polynomial is to factor out the greatest common factor of its terms.

Factoring $ax^2 + bx + c$ by Trial and Check

Practice Problem 5

Factor: $3x^2 + 13x + 4$

Answers

3. $6x(x - 1)(x + 5)$,   **4.** $3(y^2 + 2y + 2)$,
5. $(3x + 1)(x + 4)$

EXAMPLE 5   Factor: $2x^2 + 11x + 15$

Solution: Factors of $2x^2$ are $2x$ and x. Let's try these factors as first terms of the binomials.

$$2x^2 + 11x + 15 = (2x + \quad)(x + \quad)$$

Next we try combinations of factors of 15 until the correct middle term, $11x$, is obtained. We will try only positive factors of 15 since the coefficient of the middle term, 11, is positive. Positive factors of 15 are 1 and 15 and 3 and 5.

$(2x + 1)(x + 15)$
$\underbrace{\quad}_{1x}$
$\underbrace{\quad}$
$\dfrac{30x}{31x}$ Incorrect middle term

$(2x + 15)(x + 1)$
$\underbrace{\quad}_{15x}$
$\underbrace{\quad}$
$\dfrac{2x}{17x}$ Incorrect middle term

$(2x + 3)(x + 5)$
$\underbrace{\quad}_{3x}$
$\underbrace{\quad}$
$\dfrac{10x}{13x}$ Incorrect middle term

$(2x + 5)(x + 3)$
$\underbrace{\quad}_{5x}$
$\underbrace{\quad}$
$\dfrac{6x}{11x}$ Correct middle term

Thus, the factored form of $2x^2 + 11x + 15$ is $(2x + 5)(x + 3)$. ●

Factoring a Trinomial of the Form $ax^2 + bx + c$

Step 1. Write all pairs of factors of ax^2.

Step 2. Write all pairs of factors of c, the constant term.

Step 3. Try various combinations of these factors until the correct middle term bx is found.

Step 4. If no combination exists, the polynomial is **prime**.

EXAMPLE 6 Factor: $3x^2 - x - 4$

Solution: Factors of $3x^2$: $3x \cdot x$

Factors of -4: $-1 \cdot 4$, $1 \cdot -4$, $-2 \cdot 2$, $2 \cdot -2$

Let's try possible combinations of these factors.

$(3x - 1)(x + 4)$
$\underbrace{\quad}_{-1x}$
$\underbrace{\quad}$
$\dfrac{12x}{11x}$ Incorrect middle term

$(3x + 4)(x - 1)$
$\underbrace{\quad}_{4x}$
$\underbrace{\quad}$
$\dfrac{-3x}{1x}$ Incorrect middle term

$(3x - 4)(x + 1)$
$\underbrace{\quad}_{-4x}$
$\underbrace{\quad}$
$\dfrac{3x}{-1x}$ Correct middle term

Thus, $3x^2 - x - 4 = (3x - 4)(x + 1)$.

Practice Problem 6

Factor: $5x^2 + 13x - 6$

Answer

● **6.** $(x + 3)(5x - 2)$

Helpful Hint

A positive constant in a trinomial tells us to look for two numbers with the same sign. The sign of the coefficient of the middle term tells us whether the signs are both positive or both negative.

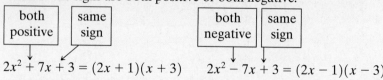

A negative constant in a trinomial tells us to look for two numbers with opposite signs.

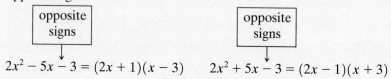

Practice Problem 7

Factor: $24x^2y^2 - 42xy^2 + 9y^2$

EXAMPLE 7   Factor: $12x^3y - 22x^2y + 8xy$

Solution: First we factor out the greatest common factor of the terms of this trinomial, $2xy$.

$$12x^3y - 22x^2y + 8xy = 2xy(6x^2 - 11x + 4)$$

Now we try to factor the trinomial $6x^2 - 11x + 4$.

Factors of $6x^2$: $2x \cdot 3x$, $6x \cdot x$

Let's try $2x$ and $3x$.

$$2xy(6x^2 - 11x + 4) = 2xy(2x + \quad)(3x + \quad)$$

The constant term, 4, is positive and the coefficient of the middle term, -11, is negative, so we factor 4 into negative factors only.

Negative factors of 4: $-4(-1)$, $-2(-2)$

Let's try -4 and -1.

$$2xy(2x \underbrace{- 4)(3x}_{-12x} - 1)$$
$$\frac{-2x}{-14x} \qquad \text{Incorrect middle term}$$

This combination cannot be correct because one of the factors, $(2x - 4)$, has a common factor of 2. This cannot happen if the polynomial $6x^2 - 11x + 4$ has no common factors.

Now let's try -1 and -4.

$$2xy(2x \underbrace{- 1)(3x}_{-3x} - 4)$$
$$\frac{-8x}{-11x} \qquad \text{Correct middle term}$$

Thus,

$$12x^3y - 22x^2y + 8xy = 2xy(2x - 1)(3x - 4)$$

Answer

7. $3y^2(2x - 3)(4x - 1)$

If this combination had not worked, we would try -2 and -2 as factors of 4 and then $6x$ and x as factors of $6x^2$.

Helpful Hint

If a trinomial has no common factor (other than 1), then none of its binomial factors will contain a common factor (other than 1).

EXAMPLE 8   Factor: $16x^2 + 24xy + 9y^2$

Solution: No greatest common factor can be factored out of this trinomial.

Factors of $16x^2$: $16x \cdot x$, $8x \cdot 2x$, $4x \cdot 4x$
Factors of $9y^2$: $y \cdot 9y$, $3y \cdot 3y$

We try possible combinations until the correct factorization is found.

$$16x^2 + 24xy + 9y^2 = (4x + 3y)(4x + 3y) \quad \text{or} \quad (4x + 3y)^2$$

The trinomial $16x^2 + 24xy + 9y^2$ in Example 8 is an example of a **perfect square trinomial** since its factors are two identical binomials. In the next section, we examine a special method for factoring perfect square trinomials.

Factoring $ax^2 + bx + c$ by Grouping

There is another method we can use when factoring trinomials of the form $ax^2 + bx + c$: Write the trinomial as a four-term polynomial, and then factor by grouping.

Factoring a Trinomial of the Form $ax^2 + bx + c$ by Grouping

Step 1.   Find two numbers whose product is $a \cdot c$ and whose sum is b.

Step 2.   Write the term bx as a sum by using the factors found in Step 1.

Step 3.   Factor by grouping.

EXAMPLE 9   Factor: $6x^2 + 13x + 6$

Solution: In this trinomial, $a = 6, b = 13$, and $c = 6$.

Step 1.   Find two numbers whose product is $a \cdot c$, or $6 \cdot 6 = 36$, and whose sum is b, 13. The two numbers are 4 and 9.

Step 2.   Write the middle term $13x$ as the sum $4x + 9x$.

$$6x^2 + 13x + 6 = 6x^2 + 4x + 9x + 6$$

Step 3.   Factor $6x^2 + 4x + 9x + 6$ by grouping.

$$(6x^2 + 4x) + (9x + 6) = 2x(3x + 2) + 3(3x + 2)$$
$$= (3x + 2)(2x + 3)$$

Try the Concept Check in the margin.

c   Factoring by Substitution

A complicated-looking polynomial may be a simpler trinomial "in disguise." Revealing the simpler trinomial is possible by substitution.

Practice Problem 8

Factor: $4x^2 + 28xy + 49y^2$

Practice Problem 9

Factor: $12x^2 + 11x + 2$

Concept Check

Name one way that a factorization can be checked.

Answers

8. $(2x + 7y)^2$,   **9.** $(3x + 2)(4x + 1)$

Concept Check: Answers may vary. A sample is: By multiplying the factors to see that the product is the original polynomial.

Practice Problem 10

Factor: $3(z + 2)^2 - 19(z + 2) + 6$

EXAMPLE 10   Factor: $2(a + 3)^2 - 5(a + 3) - 7$

Solution: The quantity $(a + 3)$ is in two of the terms of this polynomial. If we *substitute* x for $(a + 3)$, the result is the following simpler trinomial.

$$2(a + 3)^2 \quad - \quad 5(a + 3) \quad - 7 \qquad \text{Original trinomial}$$
$$= \quad 2(x)^2 \quad - \quad 5(x) \quad - 7 \qquad \text{Substitute } x \text{ for } (a + 3).$$

Now we can factor $2x^2 - 5x - 7$.

$$2x^2 - 5x - 7 = (2x - 7)(x + 1)$$

But the quantity in the original polynomial was $(a + 3)$, not x. Thus we need to reverse the substitution and replace x with $(a + 3)$.

$$(2x - 7)(x + 1) \qquad \text{Factored expression}$$
$$= [2(a + 3) - 7][(a + 3) + 1] \qquad \text{Substitute } (a + 3) \text{ for } x.$$
$$= (2a + 6 - 7)(a + 3 + 1) \qquad \text{Remove inside parentheses.}$$
$$= (2a - 1)(a + 4) \qquad \text{Simplify.}$$

Thus, $2(a + 3)^2 - 5(a + 3) - 7 = (2a - 1)(a + 4)$. ●

Practice Problem 11

Factor: $14x^4 + 23x^2 + 3$

EXAMPLE 11   Factor: $5x^4 + 29x^2 - 42$

Solution: Again, substitution may help us factor this polynomial more easily. We will let $y = x^2$, so $y^2 = (x^2)^2$, or x^4. Then

$$5x^4 + 29x^2 - 42$$
becomes
$$5y^2 + 29y - 42$$

which factors as

$$5y^2 + 29y - 42 = (5y - 6)(y + 7)$$

Now we replace y with x^2 to get

$$(5x^2 - 6)(x^2 + 7) \qquad ●$$

Answers

10. $(3z + 5)(z - 4)$,
11. $(2x^2 + 3)(7x^2 + 1)$

Name _____ Section _____ Date _____

Mental Math

1. Find two numbers whose product is 10 and whose sum is 7.

2. Find two numbers whose product is 12 and whose sum is 8.

3. Find two numbers whose product is 24 and whose sum is 11.

4. Find two numbers whose product is 30 and whose sum is 13.

EXERCISE SET 5.5

Ⓐ *Factor each trinomial. See Examples 1 through 4.*

1. $x^2 + 9x + 18$

2. $x^2 + 9x + 20$

3. $x^2 - 12x + 32$

4. $x^2 - 12x + 27$

5. $x^2 + 10x - 24$

6. $x^2 + 3x - 54$

7. $x^2 - 2x - 24$

8. $x^2 - 9x - 36$

9. $3x^2 - 18x + 24$

10. $x^2y^2 + 4xy^2 + 3y^2$

11. $4x^2z + 28xz + 40z$

12. $5x^2 - 45x + 70$

13. $2x^2 + 30x - 108$

14. $3x^2 + 12x - 96$

15. $x^2 - 24x - 81$

16. $x^2 - 48x - 100$

17. $x^2 - 15x - 54$

18. $x^2 - 15x + 54$

19. $3x^2 - 6x + 3$

20. $2x^2 + 4x + 2$

21. $2x^2 + 2x - 12$

22. $3x^2 + 6x - 45$

23. $x^2 + 6xy + 5y^2$

24. $x^2 + 6xy + 8y^2$

△**25.** The volume $V(x)$ of a box in terms of its height x is given by the function $V(x) = x^3 + 2x^2 - 8x$. Factor this expression for $V(x)$.

△**26.** Based on your results from Exercise 25, find the length and width of the box if the height is 5 inches and the dimensions of the box are whole numbers.

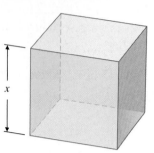

27. Find all positive and negative integers b such that $x^2 + bx + 6$ is factorable.

28. Find all positive and negative integers b such that $x^2 + bx - 10$ is factorable.

B *Factor each trinomial. See Examples 5 through 9.*

29. $5x^2 + 16x + 3$

30. $3x^2 + 8x + 4$

31. $2x^2 - 11x + 12$

32. $3x^2 - 19x + 20$

33. $2x^2 + 25x - 20$

34. $6x^2 - 13x - 8$

35. $4x^2 - 12x + 9$

36. $25x^2 - 30x + 9$

37. $12x^2 + 10x - 50$

38. $12y^2 - 48y + 45$

39. $3y^4 - y^3 - 10y^2$

40. $2x^2z + 5xz - 12z$

41. $6x^3 + 8x^2 + 24x$

42. $18y^3 + 12y^2 + 2y$

43. $x^2 + 8xz + 7z^2$

44. $a^2 - 2ab - 15b^2$

45. $2x^2 - 5xy - 3y^2$

46. $6x^2 + 11xy + 4y^2$

47. $x^2 - x - 12$

48. $x^2 + 4x - 5$

49. $28y^2 + 22y + 4$

50. $24y^3 - 2y^2 - y$

51. $2x^2 + 15x - 27$

52. $3x^2 + 14x + 15$

53. $3x^2 - 5x - 2$ **54.** $5x^2 - 14x - 3$ **55.** $8x^2 - 26x + 15$ **56.** $12x^2 - 17x + 6$

57. $18x^4 + 21x^3 + 6x^2$ **58.** $20x^5 + 54x^4 + 10x^3$ **59.** $3a^2 + 12ab + 12b^2$ **60.** $2x^2 + 16xy + 32y^2$

61. $6x^3 - x^2 - x$ **62.** $12x^3 + x^2 - x$ **63.** $12a^2 - 29ab + 15b^2$ **64.** $16y^2 + 6yx - 27x^2$

65. $9x^2 + 30x + 25$ **66.** $4x^2 + 6x + 9$ **67.** $3x^2y - 11xy + 8y$ **68.** $5xy^2 - 9xy + 4x$

C *Use substitution to factor each polynomial completely. See Examples 10 and 11.*

69. $x^4 + x^2 - 6$ **70.** $x^4 - x^2 - 20$ **71.** $(5x + 1)^2 + 8(5x + 1) + 7$

72. $(3x - 1)^2 + 5(3x - 1) + 6$ **73.** $x^6 - 7x^3 + 12$ **74.** $x^6 - 4x^3 - 12$

75. $(a + 5)^2 - 5(a + 5) - 24$ **76.** $(3c + 6)^2 + 12(3c + 6) - 28$ **77.** $(x - 4)^2 + 3(x - 4) - 18$

78. $(x - 3)^2 - 2(x - 3) - 8$ **79.** $2x^6 + 3x^3 - 9$ **80.** $3x^6 - 14x^3 + 8$

81. $2(x + 4)^2 + 3(x + 4) - 5$ **82.** $3(x + 3)^2 + 2(x + 3) - 5$ **83.** $x^4 - 5x^2 - 6$

84. $x^4 - 5x^2 + 6$

Review and Preview

Multiply. See Section 5.2.

85. $(x - 3)(x + 3)$

86. $(x - 4)(x + 4)$

87. $(2x + 1)^2$

88. $(3x + 5)^2$

89. $(x - 2)(x^2 + 2x + 4)$

90. $(y + 1)(y^2 - y + 1)$

Combining Concepts

91. Suppose that a movie is being filmed in New York City. An action shot requires an object to be thrown upward with an initial velocity of 80 feet per second off the top of 1 Madison Square Plaza, a height of 576 feet. The height $h(t)$ in feet of the object after t seconds is given by the function $h(t) = -16t^2 + 80t + 576$. (*Source:* The World Almanac, 2001)
  a. Find the height of the object at $t = 0$ seconds, $t = 2$ seconds, $t = 4$ seconds, and $t = 6$ seconds.
  b. Explain why the height of the object increases and then decreases as time passes.
  c. Factor the polynomial $-16t^2 + 80t + 576$.

92. Suppose that an object is thrown upward with an initial velocity of 64 feet per second off the edge of a 960-foot-cliff. The height $h(t)$ in feet of the object after t seconds is given by the function

$$h(t) = -16t^2 + 64t + 960$$

  a. Find the height of the object at $t = 0$ seconds, $t = 3$ seconds, $t = 6$ seconds, and $t = 9$ seconds.
  b. Explain why the height of the object increases and then decreases as time passes.
  c. Factor the polynomial $-16t^2 + 64t + 960$.

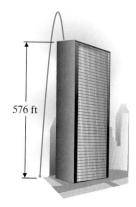

576 ft

Factor. Assume that variables used as exponents represent positive integers.

93. $x^{2n} + 10x^n + 16$

94. $x^{2n} - 7x^n + 12$

95. $x^{2n} - 3x^n - 18$

96. $x^{2n} + 7x^n - 18$

97. $2x^{2n} + 11x^n + 5$

98. $3x^{2n} - 8x^n + 4$

99. $4x^{2n} - 12x^n + 9$

100. $9x^{2n} + 24x^n + 16$

Recall that a graphing calculator may be used to check addition, subtraction, and multiplication of polynomials. In the same manner, a graphing calculator may be used to check factoring of polynomials in one variable. For example, to see that

$$2x^3 - 9x^2 - 5x = x(2x + 1)(x - 5)$$

graph $Y_1 = 2x^3 - 9x^2 - 5x$ and $Y_2 = x(2x + 1)(x - 5)$. Then trace along both graphs to see that they coincide. Factor the following and use this method to check your results.

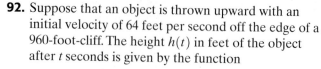

101. $x^4 + 6x^3 + 5x^2$

102. $x^3 + 6x^2 + 8x$

103. $30x^3 + 9x^2 - 3x$

104. $-6x^4 + 10x^3 - 4x^2$

5.6 Factoring by Special Products

OBJECTIVES

Ⓐ Factor a perfect square trinomial.

Ⓑ Factor the difference of two squares.

Ⓒ Factor the sum or difference of two cubes.

SSM
TUTOR CENTER   SG   CD & VIDEO   MATH PRO   WEB

Ⓐ Factoring Perfect Square Trinomials

In the previous section, we considered a variety of ways to factor trinomials of the form $ax^2 + bx + c$. In Example 8, we factored $16x^2 + 24xy + 9y^2$ as

$$16x^2 + 24xy + 9y^2 = (4x + 3y)^2$$

Recall that we called $16x^2 + 24xy + 9y^2$ a perfect square trinomial because its factors are two identical binomials. A trinomial is a perfect square trinomial if it can be written so that its first term is the square of some quantity a, its last term is the square of some quantity b, and its middle term is twice the product of the quantities a and b.

The following special formulas can be used to factor perfect square trinomials.

Perfect Square Trinomials

$$a^2 + 2ab + b^2 = (a + b)^2$$
$$a^2 - 2ab + b^2 = (a - b)^2$$

Notice that these equations are the same special products from Section 5.2 for the square of a binomial.

From

$$a^2 + 2ab + b^2 = (a + b)^2$$

we see that

$$16x^2 + 24xy + 9y^2 = (4x)^2 + 2(4x)(3y) + (3y)^2 = (4x + 3y)^2$$

EXAMPLE 1   Factor: $m^2 + 10m + 25$

Solution: Notice that the first term is a square: $m^2 = (m)^2$, the last term is a square: $25 = 5^2$, and $10m = 2 \cdot 5 \cdot m$.

This is a perfect square trinomial. Thus,

$$m^2 + 10m + 25 = m^2 + 2(m)(5) + 5^2 = (m + 5)^2$$

●

EXAMPLES   Factor each trinomial.

2. $4x^2 + 4x + 1 = (2x)^2 + 2 \cdot 2x \cdot 1 + 1^2$ *See whether it is a perfect square trinomial.*

$\qquad = (2x + 1)^2$ *Factor.*

3. $9x^2 - 12x + 4 = (3x)^2 - 2(3x)(2) + 2^2$ *See whether it is a perfect square trinomial.*

$\qquad = (3x - 2)^2$ *Factor.*

●

EXAMPLE 4   Factor: $3a^2x - 12abx + 12b^2x$

Solution: The terms of this trinomial have a greatest common factor of $3x$, which we factor out first.

$$3a^2x - 12abx + 12b^2x = 3x(a^2 - 4ab + 4b^2)$$

The polynomial $a^2 - 4ab + 4b^2$ is a perfect square trinomial. Notice that the first term is a square: $a^2 = (a)^2$, the last term is a square: $4b^2 = (2b)^2$, and $4ab = 2(a)(2b)$. The factoring can now be completed as

$$3x(a^2 - 4ab + 4b^2) = 3x(a - 2b)^2$$

●

Practice Problem 1

Factor: $x^2 + 8x + 16$

Practice Problems 2–3

Factor.

2. $9x^2 + 6x + 1$

3. $25x^2 - 20x + 4$

Practice Problem 4

Factor: $4x^3 - 32x^2y + 64xy^2$

Answers

1. $(x + 4)^2$,   **2.** $(3x + 1)^2$,   **3.** $(5x - 2)^2$,
4. $4x(x - 4y)^2$

> **Helpful Hint**
>
> If you recognize a trinomial as a perfect square trinomial, use the special formulas to factor. However, methods for factoring trinomials in general from Section 5.5 will also result in the correct factored form.

B Factoring the Difference of Two Squares

We now factor special types of binomials, beginning with the **difference of two squares**. The special product pattern presented in Section 5.2 for the product of a sum and a difference of two terms is used again here. However, the emphasis is now on factoring rather than on multiplying.

> **Difference of Two Squares**
>
> $$a^2 - b^2 = (a + b)(a - b)$$

Notice that a binomial is a difference of two squares when it is the difference of the square of some quantity a and the square of some quantity b.

Practice Problems 5–8

Factor.

5. $x^2 - 49$

6. $4y^2 - 81$

7. $12 - 3a^2$

8. $y^2 - \dfrac{1}{25}$

EXAMPLES   Factor.

5. $\begin{aligned} x^2 - 9 &= x^2 - 3^2 \\ &= (x + 3)(x - 3) \end{aligned}$

6. $\begin{aligned} 16y^2 - 9 &= (4y)^2 - 3^2 \\ &= (4y + 3)(4y - 3) \end{aligned}$

7. $\begin{aligned} 50 - 8y^2 &= 2(25 - 4y^2) && \text{Factor out the common factor of 2.} \\ &= 2[5^2 - (2y)^2] \\ &= 2(5 + 2y)(5 - 2y) \end{aligned}$

8. $\begin{aligned} x^2 - \dfrac{1}{4} &= x^2 - \left(\dfrac{1}{2}\right)^2 \\ &= \left(x + \dfrac{1}{2}\right)\left(x - \dfrac{1}{2}\right) \end{aligned}$

The binomial $x^2 + 9$ is a **sum of two squares** and cannot be factored by using real numbers. *In general, except for factoring out a greatest common factor, the sum of two squares usually cannot be factored by using real numbers.*

> **Helpful Hint**
>
> The sum of two squares whose greatest common factor is 1 usually cannot be factored by using real numbers.

Practice Problem 9

Factor: $a^4 - 81$

EXAMPLE 9   Factor: $p^4 - 16$

Solution: $\begin{aligned} p^4 - 16 &= (p^2)^2 - 4^2 \\ &= (p^2 + 4)(p^2 - 4) \end{aligned}$

The binomial factor $p^2 + 4$ cannot be factored by using real numbers, but the binomial factor $p^2 - 4$ is a difference of squares.

$$(p^2 + 4)(p^2 - 4) = (p^2 + 4)(p + 2)(p - 2)$$

Answers

5. $(x + 7)(x - 7)$,   6. $(2y + 9)(2y - 9)$,

7. $3(2 + a)(2 - a)$,   8. $\left(y + \dfrac{1}{5}\right)\left(y - \dfrac{1}{5}\right)$,

9. $(a^2 + 9)(a + 3)(a - 3)$

Try the Concept Check in the margin.

Concept Check

Is $(x - 4)(y^2 - 9)$ completely factored? Why or why not?

Practice Problem 10

Factor: $(x + 1)^2 - 9$

EXAMPLE 10   Factor: $(x + 3)^2 - 36$

Solution:

$$(x + 3)^2 - 36 = (x + 3)^2 - 6^2$$
$$= [(x + 3) + 6][(x + 3) - 6] \quad \text{Factor as the difference of two squares.}$$
$$= [x + 3 + 6][x + 3 - 6] \quad \text{Remove parentheses.}$$
$$= (x + 9)(x - 3) \quad \text{Simplify.} \quad \bullet$$

EXAMPLE 11   Factor: $x^2 + 4x + 4 - y^2$

Solution: Factoring by grouping comes to mind since the sum of the first three terms of this polynomial is a perfect square trinomial.

$$x^2 + 4x + 4 - y^2 = (x^2 + 4x + 4) - y^2 \quad \text{Group the first three terms.}$$
$$= (x + 2)^2 - y^2 \quad \text{Factor the perfect square trinomial.}$$

This is not completely factored yet since we have a *difference*, not a *product*. Since $(x + 2)^2 - y^2$ is a difference of squares, we have

$$(x + 2)^2 - y^2 = [(x + 2) + y][(x + 2) - y]$$
$$= (x + 2 + y)(x + 2 - y) \quad \bullet$$

Practice Problem 11

Factor: $a^2 + 2a + 1 - b^2$

(c)   Factoring the Sum or Difference of Two Cubes

Although the sum of two squares usually cannot be factored, the sum of two cubes, as well as the difference of two cubes, can be factored as follows.

> **Sum and Difference of Two Cubes**
>
> $$a^3 + b^3 = (a + b)(a^2 - ab + b^2)$$
> $$a^3 - b^3 = (a - b)(a^2 + ab + b^2)$$

To check the first pattern, let's find the product of $(a + b)$ and $(a^2 - ab + b^2)$.

$$(a + b)(a^2 - ab + b^2) = a(a^2 - ab + b^2) + b(a^2 - ab + b^2)$$
$$= a^3 - a^2b + ab^2 + a^2b - ab^2 + b^3$$
$$= a^3 + b^3$$

Answers

10. $(x - 2)(x + 4)$,
11. $(a + 1 + b)(a + 1 - b)$

Concept Check: no; $(y^2 - 9)$ can be factored

Practice Problem 12

Factor: $x^3 + 27$

Practice Problem 13

Factor: $x^3 + 64y^3$

Practice Problem 14

Factor: $y^3 - 8$

Practice Problem 15

Factor: $27a^2 - b^3a^2$

EXAMPLE 12   Factor: $x^3 + 8$

Solution: First we write the binomial in the form $a^3 + b^3$. Then we use the formula

$$a^3 + b^3 = (a + b)(a^2 - a \cdot b + b^2), \quad \text{where } a \text{ is } x \text{ and } b \text{ is 2.}$$

$$x^3 + 8 = x^3 + 2^3 = (x + 2)(x^2 - x \cdot 2 + 2^2)$$

Thus, $x^3 + 8 = (x + 2)(x^2 - 2x + 4)$

EXAMPLE 13   Factor: $p^3 + 27q^3$

Solution: $p^3 + 27q^3 = p^3 + (3q)^3$

$$= (p + 3q)[p^2 - (p)(3q) + (3q)^2]$$

$$= (p + 3q)(p^2 - 3pq + 9q^2)$$

EXAMPLE 14   Factor: $y^3 - 64$

Solution: This is a difference of cubes since $y^3 - 64 = y^3 - 4^3$.

From $\quad a^3 - b^3 = (a - b)(a^2 + a \cdot b + b^2) \quad$ we have that

$$y^3 - 4^3 = (y - 4)(y^2 + y \cdot 4 + 4^2)$$

$$= (y - 4)(y^2 + 4y + 16)$$

Helpful Hint

When factoring sums or differences of cubes, be sure to notice the sign patterns.

$$\overset{\text{Same sign}}{x^3 + y^3} = (x + y)(x^2 - xy + y^2)$$

Opposite sign   Always positive

$$\overset{\text{Same sign}}{x^3 - y^3} = (x - y)(x^2 + xy + y^2)$$

Opposite sign   Always positive

EXAMPLE 15   Factor: $125q^2 - n^3q^2$

Solution: First we factor out a common factor of q^2.

$$125q^2 - n^3q^2 = q^2(125 - n^3)$$

$$= q^2(5^3 - n^3)$$

Opposite sign   Positive

$$= q^2(5 - n)[5^2 + (5)(n) + (n^2)]$$

$$= q^2(5 - n)(25 + 5n + n^2)$$

Thus, $125q^2 - n^3q^2 = q^2(5 - n)(25 + 5n + n^2)$. The trinomial $25 + 5n + n^2$ cannot be factored further.

EXERCISE SET 5.6

 Factor. See Examples 1 through 4.

1. $x^2 + 6x + 9$

2. $x^2 - 10x + 25$

3. $4x^2 - 12x + 9$

4. $25x^2 + 10x + 1$

 5. $3x^2 - 24x + 48$

6. $x^3 + 14x^2 + 49x$

7. $9y^2x^2 + 12yx^2 + 4x^2$

8. $32x^2 - 16xy + 2y^2$

9. $4a^2 + 12a + 9$

10. $9a^2 - 30a + 25$

B *Factor. See Examples 5 through 11.*

11. $x^2 - 25$

12. $y^2 - 100$

13. $9 - 4z^2$

14. $16x^2 - y^2$

15. $(y + 2)^2 - 49$

16. $(x - 1)^2 - z^2$

17. $64x^2 - 100$

18. $4x^2 - 36$

19. $18x^2y - 2y$

20. $12xy^2 - 108x$

21. $9x^2 - 49$

22. $25x^2 - 4$

23. $x^4 - 81$

24. $x^4 - 256$

25. $(x + 2y)^2 - 9$

26. $(3x + y)^2 - 25$

27. $x^2 + 16x + 64 - x^4$

28. $x^2 + 20x + 100 - x^4$

29. $x^2 - 10x + 25 - y^2$

30. $x^2 - 18x + 81 - y^2$

31. $4x^2 + 4x + 1 - z^2$

32. $9y^2 + 12y + 4 - x^2$

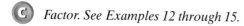

Factor. See Examples 12 through 15.

33. $x^3 + 27$

34. $y^3 + 1$

35. $z^3 - 1$

36. $x^3 - 8$

37. $m^3 + n^3$

38. $r^3 + 125$

39. $x^3y^2 - 27y^2$

40. $64 - p^3$

41. $a^3b + 8b^4$

42. $8ab^3 + 27a^4$

43. $125y^3 - 8x^3$

44. $54y^3 - 128$

45. $x^6 - y^3$

46. $x^3 - y^6$

47. $8x^3 + 27y^3$

48. $125x^3 + 8y^3$

49. $x^3 - 1$

50. $x^3 - 8$

51. $x^3 + 125$

52. $x^3 + 216$

53. $3x^6y^2 + 81y^2$

54. $x^2y^9 + x^2y^3$

Review and Preview

Solve each equation. See Section 2.1.

55. $x - 5 = 0$

56. $x + 7 = 0$

57. $3x + 1 = 0$

58. $5x - 15 = 0$

59. $-2x = 0$

60. $3x = 0$

61. $-5x + 25 = 0$

62. $-4x - 16 = 0$

△ **63.** A manufacturer of metal washers needs to determine the cross-sectional area of each washer. If the outer radius of the washer is R and the radius of the hole is r, express the area of the washer as a polynomial. Factor this polynomial completely.

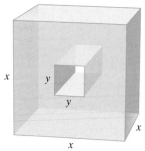

△ **64.** Express the area of the shaded region as a polynomial. Factor the polynomial completely.

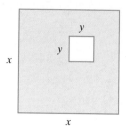

Express the volume of each solid as a polynomial. To do so, subtract the volume of the "hole" from the volume of the larger solid. Then factor the resulting polynomial.

△ **65.**

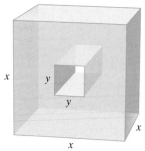

△ **66.**

Find the value of c that makes each trinomial a perfect square trinomial.

67. $x^2 + 6x + c$
68. $y^2 + 10y + c$
69. $m^2 - 14m + c$
70. $n^2 - 2n + c$

✎ **71.** Factor $x^6 - 1$ completely, using the following methods from this chapter.
  a. Factor the expression by treating it as the difference of two squares, $(x^3)^2 - 1^2$.
  b. Factor the expression treating it as the difference of two cubes, $(x^2)^3 - 1^3$.

✎ **c.** Are the answers to parts (a) and (b) the same? Why or why not?

Factor. Assume that variables used as exponents represent positive integers.

72. $x^{2n} - 25$
73. $x^{2n} - 36$
74. $36x^{2n} - 49$

75. $25x^{2n} - 81$
76. $x^{4n} - 16$
77. $x^{4n} - 625$

Integrated Review—Operations on Polynomials and Factoring Strategies

Operations on Polynomials

Perform each indicated operation.

1. $(-y^2 + 6y - 1) + (3y^2 - 4y - 10)$ **2.** $(5z^4 - 6z^2 + z + 1) - (7z^4 - 2z + 1)$

3. Subtract $(x - 5)$ from $(x^2 - 6x + 2)$.   **4.** $(2x^2 + 6x - 5) + (5x^2 - 10x)$

5. $(5x - 3)^2$ **6.** $(5x^2 - 14x - 3) \div (5x + 1)$

7. $(2x^4 - 3x^2 + 5x - 2) \div (x + 2)$ **8.** $(4x - 1)(x^2 - 3x - 2)$

Factoring Strategies

The key to proficiency in factoring polynomials is to practice until you are comfortable with each technique. A strategy for factoring polynomials completely is given next.

Factoring a Polynomial

Step 1. Are there any common factors? If so, factor out the greatest common factor.

Step 2. How many terms are in the polynomial?

    a. If there are *two* terms, decide if one of the following formulas may be applied:

        i. Difference of two squares: $a^2 - b^2 = (a - b)(a + b)$

        ii. Difference of two cubes: $a^3 - b^3 = (a - b)(a^2 + ab + b^2)$

        iii. Sum of two cubes: $a^3 + b^3 = (a + b)(a^2 - ab + b^2)$

    b. If there are *three* terms, try one of the following:

        i. Perfect square trinomial: $a^2 + 2ab + b^2 = (a + b)^2$
$$a^2 - 2ab + b^2 = (a - b)^2$$

        ii. If not a perfect square trinomial, factor by using the methods presented in Section 5.5.

    c. If there are *four* or more terms, try factoring by grouping.

Step 3. See whether any factors in the factored polynomial can be factored further.

12. _____

13. _____

14. _____

15. _____

16. _____

17. _____

18. _____

Factor completely.

9. $x^2 - 8x + 16 - y^2$ **10.** $12x^2 - 22x - 20$

11. $x^4 - x$ **12.** $(2x + 1)^2 - 3(2x + 1) + 2$

13. $14x^2y - 2xy$ **14.** $24ab^2 - 6ab$

15. $4x^2 - 16$ **16.** $9x^2 - 81$ **17.** $3x^2 - 8x - 11$

18. $5x^2 - 2x - 3$ **19.** $4x^2 + 8x - 12$ **20.** $6x^2 - 6x - 12$

21. $4x^2 + 36x + 81$ **22.** $25x^2 + 40x + 16$ **23.** $8x^3 + 125y^3$

24. $27x^3 - 64y^3$ **25.** $64x^2y^3 - 8x^2$ **26.** $27x^5y^4 - 216x^2y$

27. $(x + 5)^3 + y^3$ **28.** $(y - 1)^3 + 27x^3$

29. $(5a - 3)^2 - 6(5a - 3) + 9$ **30.** $(4r + 1)^2 + 8(4r + 1) + 16$

31. $7x^2 - 63x$ **32.** $20x^2 + 23x + 6$ **33.** $ab - 6a + 7b - 42$

34. $20x^2 - 220x + 600$ **35.** $x^4 - 1$ **36.** $15x^2 - 20x$

37. $10x^2 - 7x - 33$ **38.** $45m^3n^3 - 27m^2n^2$ **39.** $5a^3b^3 - 50a^3b$

40. $x^4 + x$ **41.** $16x^2 + 25$ **42.** $20x^3 + 20y^3$

43. $10x^3 - 210x^2 + 1100x$ **44.** $9y^2 - 42y + 49$ **45.** $64a^3b^4 - 27a^3b$

46. $y^4 - 16$ **47.** $2x^3 - 54$ **48.** $2sr + 10s - r - 5$

49. $3y^5 - 5y^4 + 6y - 10$ **50.** $64a^2 + b^2$ **51.** $100z^3 + 100$

52. $250x^4 - 16x$ **53.** $4b^2 - 36b + 81$ **54.** $2a^5 - a^4 + 6a - 3$

55. $(y - 6)^2 + 3(y - 6) + 2$ **56.** $(c + 2)^2 - 6(c + 2) + 5$

△ **57.** Express the area of the shaded region
as a polynomial. Factor the polynomial
completely.

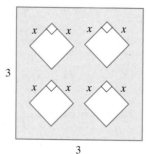

24. _____

25. _____

26. _____

27. _____

28. _____
29. _____
30. _____
31. _____
32. _____
33. _____
34. _____

35. _____
36. _____
37. _____
38. _____
39. _____

40. _____
41. _____

42. _____

43. _____
44. _____

45. _____

46. _____

47. _____
48. _____
49. _____
50. _____

51. _____

52. _____
53. _____
54. _____
55. _____
56. _____

57. _____

5.7 Solving Equations by Factoring and Solving Problems

OBJECTIVES

Ⓐ Solve polynomial equations by factoring.

Ⓑ Solve problems that can be modeled by polynomial equations.

SSM   TUTOR CENTER   SG   CD & VIDEO   MATH PRO   WEB

Ⓐ Solving Polynomial Equations by Factoring

In this section, your efforts to learn factoring will start to pay off. We use factoring to solve polynomial equations.

A **polynomial equation** is the result of setting two polynomials equal to each other. Examples are shown below.

$$3x^3 - 2x^2 = x^2 + 2x \qquad 2.6x + 7 = -1.3$$
$$-5x^2 - 5 = -9x^2 - 2x + 1$$

A polynomial equation is in **standard form** if one side of the equation is 0, as in the following examples.

$$3x^3 - 3x^2 - 2x + 1 = 0 \qquad 2.6x + 8.3 = 0$$
$$4x^2 + 2x - 6 = 0$$

The degree of a simplified polynomial equation in standard form is the same as the highest degree of any of its terms. A polynomial equation of degree 2 is also called a **quadratic equation**.

A solution of a polynomial equation in one variable is a value of the variable that makes the equation true. The method presented in this section for solving polynomial equations is called the **factoring method**. This method is based on the **zero-factor property**.

Zero-Factor Property

If a and b are real numbers and $a \cdot b = 0$, then $a = 0$ or $b = 0$. This property is true for three or more factors also.

In other words, if the product of two or more real numbers is zero, then at least one of the numbers must be zero.

EXAMPLE 1 Solve: $(x + 2)(x - 6) = 0$

Solution: By the zero-factor property, $(x + 2)(x - 6) = 0$ only if
$$x + 2 = 0 \qquad \text{or} \quad x - 6 = 0.$$

$$x + 2 = 0 \qquad \text{or} \quad x - 6 = 0 \qquad \text{Use the zero-factor property.}$$
$$x = -2 \qquad\qquad x = 6 \qquad \text{Solve each linear equation.}$$

To check, let $x = -2$ and then let $x = 6$ in the original equation.

$$\text{Let } x = -2. \qquad\qquad \text{Let } x = 6.$$
$$(x + 2)(x - 6) = 0 \qquad (x + 2)(x - 6) = 0$$
$$(-2 + 2)(-2 - 6) = 0 \qquad (6 + 2)(6 - 6) = 0$$
$$(0)(-8) = 0 \qquad\qquad (8)(0) = 0$$
$$0 = 0 \quad \text{True.} \qquad\qquad 0 = 0 \quad \text{True.}$$

Both -2 and 6 check, so the solution set is $\{-2, 6\}$.

Practice Problem 1

Solve: $(x - 3)(x + 5) = 0$

Answer

1. $\{-5, 3\}$

Practice Problem 2

Solve: $3x^2 + 5x - 2 = 0$

EXAMPLE 2 Solve: $2x^2 + 9x - 5 = 0$

Solution: To use the zero-factor property, one side of the equation must be 0, and the other side must be in factored form.

$$2x^2 + 9x - 5 = 0$$
$$(2x - 1)(x + 5) = 0 \qquad \text{Factor.}$$
$$2x - 1 = 0 \quad \text{or} \quad x + 5 = 0 \qquad \text{Set each factor equal to 0.}$$
$$2x = 1 \qquad\qquad x = -5 \qquad \text{Solve each linear equation.}$$
$$x = \frac{1}{2}$$

To check, let $x = \dfrac{1}{2}$ in the original equation; then let $x = -5$ in the original equation. The solution set is $\left\{-5, \dfrac{1}{2}\right\}$.

●

Solving a Polynomial Equation by Factoring

Step 1. Write the equation in standard form so that one side of the equation is 0.

Step 2. Factor the polynomial completely.

Step 3. Set each factor containing a variable equal to 0.

Step 4. Solve the resulting equations.

Step 5. Check each solution in the original equation.

Since it is not always possible to factor a polynomial, not all polynomial equations can be solved by factoring. Other methods of solving polynomial equations are presented in Chapter 8.

Practice Problem 3

Solve: $x(5x - 7) = -2$

EXAMPLE 3 Solve: $x(2x - 7) = 4$

Solution: We first write the equation in standard form; then we factor.

$$x(2x - 7) = 4$$
$$2x^2 - 7x = 4 \qquad \text{Multiply. Write in standard form.}$$
$$2x^2 - 7x - 4 = 0$$
$$(2x + 1)(x - 4) = 0 \qquad \text{Factor.}$$
$$2x + 1 = 0 \quad \text{or} \quad x - 4 = 0 \qquad \text{Set each factor equal to 0.}$$
$$2x = -1 \qquad\qquad x = 4 \qquad \text{Solve.}$$
$$x = -\frac{1}{2}$$

Check both solutions in the original equation. The solution set is $\left\{-\dfrac{1}{2}, 4\right\}$.

●

Helpful Hint

To apply the zero-factor property, one side of the equation must be 0, and the other side of the equation must be factored. To solve the equation $x(2x - 7) = 4$, for example, you may *not* set each factor equal to 4.

Answers

2. $\left\{-2, \dfrac{1}{3}\right\}$, **3.** $\left\{\dfrac{2}{5}, 1\right\}$

EXAMPLE 4   Solve: $3(x^2 + 4) + 5 = -6(x^2 + 2x) + 13$

Solution: We rewrite the equation so that one side is 0.

$$3(x^2 + 4) + 5 = -6(x^2 + 2x) + 13$$

$$3x^2 + 12 + 5 = -6x^2 - 12x + 13 \qquad \text{Use the distributive property.}$$

$$9x^2 + 12x + 4 = 0 \qquad \text{Rewrite the equation so that one side is 0.}$$

$$(3x + 2)(3x + 2) = 0 \qquad \text{Factor.}$$

$$\begin{array}{llll} 3x + 2 = 0 & \text{or} & 3x + 2 = 0 & \text{Set each factor equal to 0.} \\ 3x = -2 & & 3x = -2 & \text{Solve each equation.} \\ x = -\dfrac{2}{3} & & x = -\dfrac{2}{3} & \end{array}$$

Check by substituting $-\dfrac{2}{3}$ into the original equation. The solution set is $\left\{-\dfrac{2}{3}\right\}$. ●

If the equation contains fractions, we clear the equation of fractions as a first step.

EXAMPLE 5   Solve: $2x^2 = \dfrac{17}{3}x + 1$

Solution:

$$2x^2 = \dfrac{17}{3}x + 1$$

$$3(2x^2) = 3\left(\dfrac{17}{3}x + 1\right) \qquad \text{Clear the equation of fractions.}$$

$$6x^2 = 17x + 3 \qquad \text{Use the distributive property.}$$

$$6x^2 - 17x - 3 = 0 \qquad \text{Rewrite the equation in standard form.}$$

$$(6x + 1)(x - 3) = 0 \qquad \text{Factor.}$$

$$\begin{array}{llll} 6x + 1 = 0 & \text{or} & x - 3 = 0 & \text{Set each factor equal to 0.} \\ 6x = -1 & & x = 3 & \text{Solve each equation.} \\ x = -\dfrac{1}{6} & & & \end{array}$$

Check by substituting into the original equation. The solution set is $\left\{-\dfrac{1}{6}, 3\right\}$. ●

EXAMPLE 6   Solve: $x^3 = 4x$

Solution:

$$x^3 = 4x$$

$$x^3 - 4x = 0 \qquad \text{Rewrite the equation so that one side is 0.}$$

$$x(x^2 - 4) = 0 \qquad \text{Factor out the greatest common factor.}$$

$$x(x + 2)(x - 2) = 0 \qquad \text{Factor the difference of squares.}$$

$$\begin{array}{lllll} x = 0 & \text{or} & x + 2 = 0 & \text{or} & x - 2 = 0 & \text{Set each factor equal to 0.} \\ & & x = -2 & & x = 2 & \text{Solve each equation.} \end{array}$$

Check by substituting into the original equation. The solution set is $\{-2, 0, 2\}$. ●

Notice that the *third*-degree equation of Example 6 yielded *three* solutions.

Practice Problem 4

Solve:
$2(x^2 + 5) + 10 = -2(x^2 + 10x) - 5$

Practice Problem 5

Solve: $2x^2 + \dfrac{5}{2}x = 3$

Practice Problem 6

Solve: $x^3 = x^2 + 6x$

Answers

4. $\left\{-\dfrac{5}{2}\right\}$,   **5.** $\left\{-2, \dfrac{3}{4}\right\}$,   **6.** $\{-2, 0, 3\}$

Concept Check

Which solution strategies are incorrect? Why?

a. Solve $(y - 2)(y + 2) = 4$ by setting each factor equal to 4.

b. Solve $(x + 1)(x + 3) = 0$ by setting each factor equal to 0.

c. Solve $z^2 + 5z + 6 = 0$ by factoring $z^2 + 5z + 6$ and setting each factor equal to 0.

d. Solve $x^2 + 6x + 8 = 10$ by factoring $x^2 + 6x + 8$ and setting each factor equal to 0.

Practice Problem 7

A model rocket is launched from the ground. Its height h in feet at time t seconds is approximated by the equation

$$h = -16t^2 + 112t$$

Find how long it takes the rocket to return to the ground.

Try the Concept Check in the margin.

B Solving Problems Modeled by Polynomial Equations

Some problems may be modeled by polynomial equations. To solve these problems, we use the same problem-solving steps that were introduced in Section 2.2. When solving these problems, keep in mind that a solution of an equation that models a problem is not always a solution to the problem. For example, a person's weight or the length of a side of a geometric figure is always a positive number. Discard solutions that do not make sense as solutions of the problem.

EXAMPLE 7 Finding the Return Time of a Rocket

An Alpha III model rocket is launched from the ground with an A8-3 engine. Without a parachute the height h in feet of the rocket at time t seconds is approximated by the equation

$$h = -16t^2 + 144t$$

Find how long it takes the rocket to return to the ground.

Solution:

1. UNDERSTAND. Read and reread the problem. The equation $h = -16t^2 + 144t$ models the height of the rocket. Familiarize yourself with this equation by finding a few values.

   When $t = 1$ second, the height of the rocket is

   $$h = -16(1)^2 + 144(1) = 128 \text{ feet}$$

   When $t = 2$ seconds, the height of the rocket is

   $$h = -16(2)^2 + 144(2) = 224 \text{ feet}$$

2. TRANSLATE. To find how long it takes the rocket to return to the ground, we want to know what value of t makes the height h equal to 0. That is, we want to solve for t when $h = 0$.

   $$-16t^2 + 144t = 0$$

3. SOLVE the quadratic equation by factoring.

   $$-16t^2 + 144t = 0$$
   $$-16t(t - 9) = 0$$
   $$-16t = 0 \quad \text{or} \quad t - 9 = 0$$
   $$t = 0 \qquad\qquad t = 9$$

4. INTERPRET. The height h is 0 feet at time 0 seconds (when the rocket is launched) and at time 9 seconds.

Check: See that the height of the rocket at 9 seconds equals 0.

$$h = -16(9)^2 + 144(9) = -1296 + 1296 = 0$$

State: The rocket returns to the ground 9 seconds after it is launched. ●

Answers

7. 7 sec

Concept Check: a and d; the zero-factor property works only if one side of the equation is 0

Some of the exercises at the end of this section make use of the **Pythagorean theorem**. Before we review this theorem, recall that a **right triangle** is a triangle that contains a 90° angle, or right angle. The **hypotenuse** of a right triangle is the side opposite the right angle and is the longest side of the triangle. The **legs** of a right triangle are the other sides of the triangle.

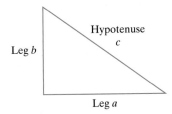

Pythagorean Theorem

In a right triangle, the sum of the squares of the lengths of the two legs is equal to the square of the length of the hypotenuse.

$$(\text{leg})^2 + (\text{leg})^2 = (\text{hypotenuse})^2 \qquad \text{or} \qquad a^2 + b^2 = c^2$$

◣ EXAMPLE 8   Using the Pythagorean Theorem

While framing an addition to an existing home, Kim Menzies, a carpenter, used the Pythagorean theorem to determine whether a wall was "square"—that is, whether the wall formed a right angle with the floor. He used a triangle whose sides are three consecutive integers. Find a right triangle whose sides are three consecutive integers.

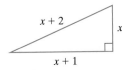

Solution:

1. UNDERSTAND. Read and reread the problem. Let x, $x + 1$, and $x + 2$ be three consecutive integers. Since these integers represent lengths of the sides of a right triangle, we have

    $x =$ one leg,
    $x + 1 =$ other leg, and
    $x + 2 =$ hypotenuse (longest side)

2. TRANSLATE. By the Pythagorean theorem, we have

In words: $(\text{leg})^2 + (\text{leg})^2 = (\text{hypotenuse})^2$

Translate: $(x)^2 + (x + 1)^2 = (x + 2)^2$

Practice Problem 8   △

Find a right triangle whose sides are three consecutive even integers.

Answer

8. 6, 8, and 10 units

3. SOLVE the equation.

$$x^2 + (x + 1)^2 = (x + 2)^2$$
$$x^2 + x^2 + 2x + 1 = x^2 + 4x + 4 \qquad \text{Multiply.}$$
$$2x^2 + 2x + 1 = x^2 + 4x + 4$$
$$x^2 - 2x - 3 = 0 \qquad \text{Write in standard form.}$$
$$(x - 3)(x + 1) = 0$$
$$x - 3 = 0 \quad \text{or} \quad x + 1 = 0$$
$$x = 3 \qquad\qquad x = -1$$

4. INTERPRET. Discard $x = -1$ since length cannot be negative. If $x = 3$, then $x + 1 = 4$ and $x + 2 = 5$.

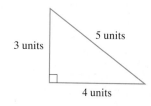

Check: To check, see that $(\text{leg})^2 + (\text{leg})^2 = (\text{hypotenuse})^2$.

$$3^2 + 4^2 = 5^2$$
$$9 + 16 = 25 \qquad \text{True.}$$

State: The lengths of the sides of the right triangle are 3, 4, and 5 units. Kim used this information by marking off lengths of 3 and 4 feet on the floor and framing, respectively. If the diagonal length between these marks was 5 feet, the wall was square. If not, adjustments were made.

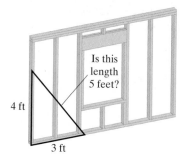

Mental Math

Solve each equation for the variable.

1. $(x - 3)(x + 5) = 0$

2. $(y + 5)(y + 3) = 0$

3. $(z - 3)(z + 7) = 0$

4. $(c - 2)(c - 4) = 0$

5. $x(x - 9) = 0$

6. $w(w + 7) = 0$

EXERCISE SET 5.7

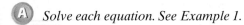 Solve each equation. See Example 1.

1. $(x + 3)(3x - 4) = 0$

2. $(5x + 1)(x - 2) = 0$

3. $3(2x - 5)(4x + 3) = 0$

4. $8(3x - 4)(2x - 7) = 0$

Solve each equation. See Examples 2 through 5.

5. $x^2 + 11x + 24 = 0$

6. $y^2 - 10y + 24 = 0$

7. $12x^2 + 5x - 2 = 0$

8. $3y^2 - y - 14 = 0$

9. $z^2 + 9 = 10z$

10. $n^2 + n = 72$

11. $x(5x + 2) = 3$

12. $n(2n - 3) = 2$

13. $x^2 - 6x = x(8 + x)$

14. $n(3 + n) = n^2 + 4n$

15. $\dfrac{z^2}{6} - \dfrac{z}{2} - 3 = 0$

16. $\dfrac{c^2}{20} - \dfrac{c}{4} + \dfrac{1}{5} = 0$

17. $\dfrac{x^2}{2} + \dfrac{x}{20} = \dfrac{1}{10}$

18. $\dfrac{y^2}{30} = \dfrac{y}{15} + \dfrac{1}{2}$

19. $\dfrac{4t^2}{5} = \dfrac{t}{5} + \dfrac{3}{10}$

20. $\dfrac{5x^2}{6} - \dfrac{7x}{2} + \dfrac{2}{3} = 0$

Solve each equation. See Example 6.

21. $(x + 2)(x - 7)(3x - 8) = 0$

22. $(4x + 9)(x - 4)(x + 1) = 0$

23. $y^3 = 9y$

24. $n^3 = 16n$

25. $x^3 - x = 2x^2 - 2$

26. $m^3 = m^2 + 12m$

27. Explain how solving $2(x - 3)(x - 1) = 0$ differs from solving $2x(x - 3)(x - 1) = 0$.

28. Explain why the zero-factor property works for more than two numbers whose product is 0.

Solve each equation. See Examples 1 through 6.

29. $(2x + 7)(x - 10) = 0$

30. $(x + 4)(5x - 1) = 0$

31. $3x(x - 5) = 0$

32. $4x(2x + 3) = 0$

33. $x^2 - 2x - 15 = 0$

34. $x^2 + 6x - 7 = 0$

35. $12x^2 + 2x - 2 = 0$

36. $8x^2 + 13x + 5 = 0$

37. $w^2 - 5w = 36$

38. $x^2 + 32 = 12x$

39. $25x^2 - 40x + 16 = 0$

40. $9n^2 + 30n + 25 = 0$

41. $2r^3 + 6r^2 = 20r$

42. $-2t^3 = 108t - 30t^2$

43. $z(5z - 4)(z + 3) = 0$

44. $2r(r - 3)(5r + 4) = 0$

45. $2z(z + 6) = 2z^2 + 12z - 8$

46. $3c^2 - 8c + 2 = c(3c - 8)$

47. $-3(x - 4) + x = 5(3 - x)$

48. $-4(a + 1) - 3a = -7(2a - 3)$

Solve. See Examples 7 and 8.

49. One number exceeds another by five, and their product is 66. Find the numbers.

50. If the sum of two numbers is 4 and their product is $\frac{15}{4}$, find the numbers.

51. An electrician needs to run a cable from the top of a 60-foot tower to a transmitter box located 45 feet away from the base of the tower. Find how long he should cut the cable.

52. A stereo-system installer needs to run speaker wire above the ceiling along the two diagonals of a rectangular room whose dimensions are 40 feet by 75 feet. Find how much speaker wire she needs.

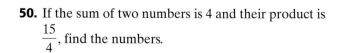

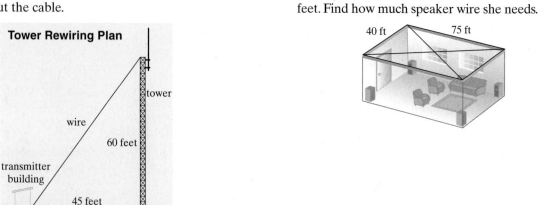

53. The shorter leg of a right triangle is two feet less than the other leg. Find the length of the two legs if the hypotenuse is 10 feet.

54. The shorter leg of a right triangle is 3 centimeters less than the other leg. Find the length of the two legs if the hypotenuse is 15 centimeters.

55. The sum of the squares of two consecutive even integers is 340. Find the integers.

56. The sum of the squares of two consecutive odd integers is 202. Find the integers.

57. While hovering near the top of Ribbon Falls in Yosemite National Park at 1600 feet, a helicopter pilot accidentally drops his sunglasses. The height h in feet of the sunglasses after t seconds is given by the polynomial equation

$$h = -16t^2 + 1600$$

When will the sunglasses hit the ground? (*Hint:* Replace h with 0 and solve for t.)

58. After t seconds, the height h in feet of a model rocket launched from the ground into the air is given by the equation

$$h = -16t^2 + 80t$$

Find how long it takes the rocket to reach a height of 96 feet. (*Hint:* Replace h with 96 and solve for t.)

59. The floor of a shed has an area of 91 square feet. The floor is in the shape of a rectangle whose length is 6 feet more than the width. Find the length and the width of the floor of the shed.

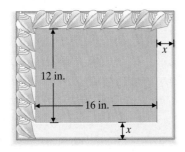

60. A vegetable garden with an area of 143 square feet is to be fertilized. If the width of the garden is 2 feet less than the length, find the dimensions of the garden.

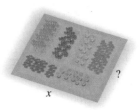

61. Marie Mulroney has a rectangular board 12 inches by 16 inches around which she wants to put a uniform border of shells. If she has enough shells for a border whose area is 128 square inches, determine the width of the border.

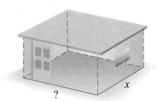

12 in.

16 in.

x

62. A gardener has a rose garden that measures 30 feet by 20 feet. He wants to put a uniform border of pine bark around the outside of the garden. Find how wide the border should be if he has enough pine bark to cover 336 square feet.

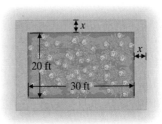

x

20 ft

30 ft

x

63. The function $W(x) = 0.5x^2$ gives the number of servings of wedding cake that can be obtained from a two-layer x-inch square wedding cake tier. What size square wedding cake tier is needed to serve 50 people? (*Source:* Based on data from the *Wilton 2001 Yearbook of Cake Decorating*)

64. Use the function in Exercise 63 to determine what size square wedding cake tier is needed to serve 200 people.

65. Suppose that a movie is being filmed in New York City. An action shot requires an object to be thrown upward with an initial velocity of 80 feet per second off the top of 1 Madison Square Plaza, a height of 576 feet. The height $h(t)$ in feet of the object after t seconds is given by the function

$$h(t) = -16t^2 + 80t + 576.$$

Determine how long before the object strikes the ground. (See Exercise 91, Section 5.5) (*Source: The World Almanac, 2001*)

576 ft

66. Suppose that an object is thrown upward with an initial velocity of 64 feet per second off the edge of a 960-foot-cliff. The height $h(t)$ in feet of the object after t seconds is given by the function

$$h(t) = -16t^2 + 64t + 960$$

Determine how long before the object strikes the ground. (See Exercise 92, Section 5.5)

Review and Preview

Write the x- and y-intercepts for each graph. See Section 3.1.

67.

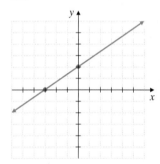

68.

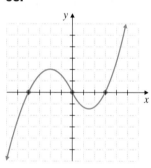

69.

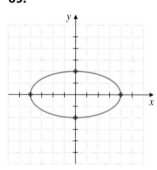

70.

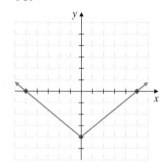

Combining Concepts

Solve.

71. $(x^2 + x - 6)(3x^2 - 14x - 5) = 0$

72. $(x^2 - 9)(x^2 + 8x + 16) = 0$

73. Is the following step correct? Why or why not?
$$x(x - 3) = 5$$
$$x = 5 \text{ or } x - 3 = 5$$

Write a quadratic equation that has the given numbers as solutions.

74. $5, 3$

75. $6, 7$

5.8 An Introduction to Graphing Polynomial Functions

Ⓐ Analyzing Graphs of Polynomial Functions

In Section 5.1, we introduced polynomial functions. Some polynomial functions are given special names according to their degree. For example,

$f(x) = 2x - 6$ is called a **linear function**; its **degree is one**.

$f(x) = 5x^2 - x + 3$ is called a **quadratic function**; its **degree is two**.

$f(x) = 7x^3 + 3x^2 - 1$ is called a **cubic function**; its **degree is three**.

$f(x) = -8x^4 - 3x^3 + 2x^2 + 20$ is called a **quartic function**; its **degree is four**.

All the above functions are also polynomial functions.

EXAMPLE 1 Given the graph of the function g below to the left:

  a. Find the domain and the range of the function.
  b. List the x- and y-intercepts.
  c. Find the coordinates of the point with the greatest y-value.
  d. Find the coordinates of the point with the smallest y-value.

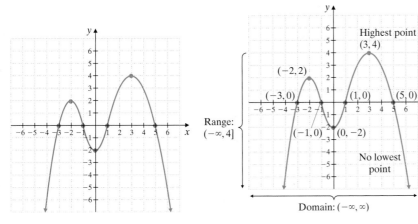

Solution:

a. The domain is the set of all real numbers, or in interval notation, $(-\infty, \infty)$. The range is $(-\infty, 4]$.

b. The x-intercepts are $(-3, 0), (-1, 0), (1, 0)$, and $(5, 0)$. The y-intercept is $(0, -2)$.

c. The point with the greatest y-value corresponds to the "highest" point. This is the point with coordinates $(3, 4)$. (This means that for all real number values for x, the greatest y-value, or value of $g(x)$, is 4.)

d. The point with the smallest y-value corresponds to the "lowest" point. This graph contains no "lowest" point, so there is no point with a smallest y-value. ●

The graph of any polynomial function (linear, quadratic, cubic, and so on) can be sketched by plotting a sufficient number of ordered pairs that satisfy the function and connecting them to form a smooth curve. The graphs of all polynomial functions will pass the vertical line test since they are graphs of functions.

OBJECTIVES

Ⓐ Analyze the graph of a polynomial function.

Ⓑ Graph quadratic functions.

Ⓒ Find the vertex of a parabola by using the vertex formula.

SSM   TUTOR CENTER   SG   CD & VIDEO   MATH PRO   WEB

Practice Problem 1

Given the graph of the function f:

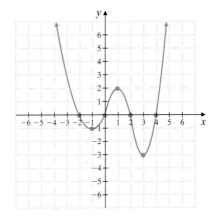

a. Find the domain and the range of the function.

b. List the x- and y-intercepts.

c. Find the coordinates of the point with the greatest y-value.

d. Find the coordinates of the point with the smallest y-value.

Answers

1. a. domain: $(-\infty, \infty)$; range: $[-3, \infty)$,
b. $(-2, 0), (0, 0), (2, 0), (4, 0)$, **c.** no point with a greatest y-value, **d.** $(3, -3)$

B Graphing Quadratic Functions

Since we know how to graph linear functions (see Section 3.1), we will now graph quadratic functions and discuss special characteristics of their graphs.

Quadratic Function

A **quadratic function** is a function that can be written in the form

$$f(x) = ax^2 + bx + c$$

where $a, b,$ and c are real numbers and $a \neq 0$.

We know that an equation of the form $f(x) = ax^2 + bx + c$ may be written as $y = ax^2 + bx + c$. Thus, both $f(x) = ax^2 + bx + c$ and $y = ax^2 + bx + c$ define quadratic functions as long as a is not 0.

Practice Problem 2

Graph the function $f(x) = -x^2$.

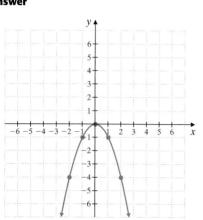

EXAMPLE 2   Graph the function $f(x) = x^2$ by plotting points.

Solution: This function is not linear, and its graph is not a line. We begin by finding ordered pair solutions. Then we plot the points and draw a smooth curve through them.

If $x = -3$, then $f(-3) = (-3)^2$, or 9.

If $x = -2$, then $f(-2) = (-2)^2$, or 4.

If $x = -1$, then $f(-1) = (-1)^2$, or 1.

If $x = 0$, then $f(0) = 0^2$, or 0.

If $x = 1$, then $f(1) = 1^2$, or 1.

If $x = 2$, then $f(2) = 2^2$, or 4.

If $x = 3$, then $f(3) = 3^2$, or 9.

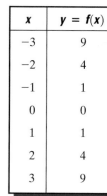

x	$y = f(x)$
-3	9
-2	4
-1	1
0	0
1	1
2	4
3	9

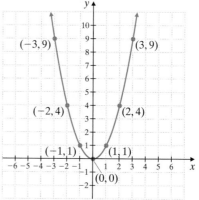

Answer

2.

Notice that the graph of Example 2 passes the vertical line test, as it should since it is a function. This curve is called a **parabola**. The highest point on a parabola that opens downward or the lowest point on a parabola that opens upward is called the **vertex** of the parabola. The vertex of this parabola is $(0, 0)$, the lowest point on the graph. If we fold the graph along the y-axis, we can see that the two sides of the graph coincide. This means that this curve is symmetric about the y-axis, and the y-axis, or the line $x = 0$, is called the **axis of symmetry**. The graph of every quadratic function is a parabola and has an axis of symmetry: the vertical line that passes through the vertex of the parabola.

EXAMPLE 3

Graph the quadratic function $f(x) = -x^2 + 2x - 3$ by plotting points.

Solution: To graph, we choose values for x and find corresponding values of $f(x)$ or y. Then we plot the points and draw a smooth curve through them.

x	$y = f(x)$
-2	-11
-1	-6
0	-3
1	-2
2	-3
3	-6

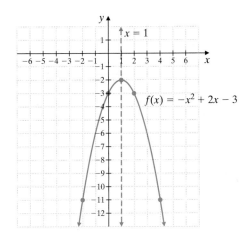

The vertex of this parabola is $(1, -2)$, the highest point on the graph. The vertical line $x = 1$ is the axis of symmetry. Recall that to find the x-intercepts of a graph, we let $y = 0$. Using function notation, this is the same as letting $f(x) = 0$. Since this graph has no x-intercepts, it means that $0 = -x^2 + 2x - 3$ has no real number solutions.   ●

Notice that the parabola $f(x) = -x^2 + 2x - 3$ opens downward, whereas $f(x) = x^2$ opens upward. When the equation of a quadratic function is written in the form $f(x) = ax^2 + bx + c$, the coefficient of the squared variable, a, determines whether the parabola opens downward or upward. If $a > 0$, the parabola opens upward, and if $a < 0$, the parabola opens downward.

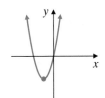

$f(x) = ax^2 + bx + c$,
$a > 0$, opens upward

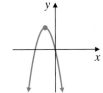

$f(x) = ax^2 + bx + c$,
$a < 0$, opens downward

Ⓒ **Finding the Vertex of a Parabola**

In both $f(x) = x^2$ and $f(x) = -x^2 + 2x - 3$, the vertex happens to be one of the points we chose to plot. Since this is not always the case, and since plotting the vertex allows us to draw the graph quickly, we need a consistent method for finding the vertex. One method is to use the following formula, which we shall derive in Chapter 8.

Practice Problem 3

Graph the quadratic function
$f(x) = -x^2 - 2x - 3$.

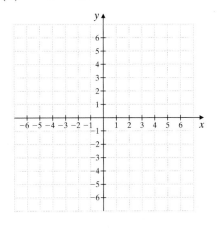

Answer

3.

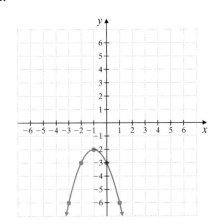

Vertex Formula

The graph of $f(x) = ax^2 + bx + c, a \neq 0$, is a parabola with vertex

$$\left(\frac{-b}{2a}, f\left(\frac{-b}{2a}\right) \right)$$

We can also find the x- and y-intercepts of a parabola to aid in graphing. Recall that x-intercepts of the graph of any equation may be found by letting $y = 0$ or $f(x) = 0$ in the equation and solving for x. Also, y-intercepts may be found by letting $x = 0$ in the equation and solving for y or $f(x)$.

Practice Problem 4

Graph $f(x) = x^2 - 2x - 3$. Find the vertex and any intercepts.

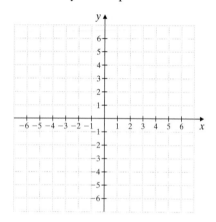

EXAMPLE 4

Graph $f(x) = x^2 + 2x - 3$. Find the vertex and any intercepts.

Solution: To find the vertex, we use the vertex formula. For the function $f(x) = x^2 + 2x - 3, a = 1$ and $b = 2$. Thus,

$$x = \frac{-b}{2a} = \frac{-2}{2(1)} = -1 \quad \text{and} \quad \begin{aligned} f(-1) &= (-1)^2 + 2(-1) - 3 \qquad \text{Find } f(-1). \\ &= 1 - 2 - 3 \\ &= -4 \end{aligned}$$

The vertex is $(-1, -4)$, and since $a = 1$ is greater than 0, this parabola opens upward. Graph the vertex and notice that this parabola will have two x-intercepts because its vertex lies below the x-axis and it opens upward. To find the x-intercepts, we let y or $f(x) = 0$ and solve for x.

To find the y-intercept, let $x = 0$.

$$\begin{aligned} f(x) &= x^2 + 2x - 3 \\ 0 &= x^2 + 2x - 3 \\ 0 &= (x + 3)(x - 1) \\ x + 3 = 0 \quad &\text{or} \quad x - 1 = 0 \\ x = -3 \qquad & \qquad x = 1 \end{aligned} \qquad\qquad \begin{aligned} f(x) &= x^2 + 2x - 3 \\ f(0) &= 0^2 + 2(0) - 3 \\ f(0) &= -3 \end{aligned}$$

The x-intercepts are $(-3, 0)$ and $(1, 0)$. The y-intercept is $(0, -3)$.

Now we plot these points and connect them with a smooth curve.

Answer

4.

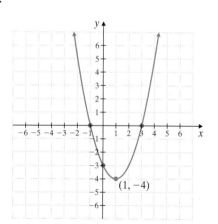

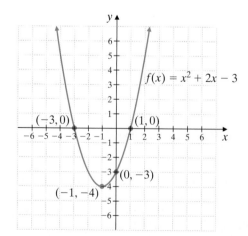

Helpful Hint

Not all graphs of parabolas have x-intercepts. To see this, first plot the vertex of the parabola and decide whether the parabola opens upward or downward. Then use this information to decide whether the graph of the parabola has x-intercepts.

EXAMPLE 5

Graph $f(x) = 3x^2 - 12x + 13$. Find the vertex and any intercepts.

Solution: To find the vertex, we use the vertex formula. For the function $y = 3x^2 - 12x + 13$, $a = 3$ and $b = -12$. Thus,

$$x = \frac{-b}{2a} = \frac{-(-12)}{2(3)} = \frac{12}{6} = 2$$

and

$$f(2) = 3(2)^2 - 12(2) + 13 \qquad \text{Find } f(2).$$

$$= 3(4) - 24 + 13$$

$$= 1$$

The vertex is $(2, 1)$. Also, this parabola opens upward since $a = 3$ is greater than 0. Graph the vertex and notice that this parabola has no x-intercepts: Its vertex lies above the x-axis, and it opens upward.
   To find the y-intercept, let $x = 0$.

$$f(0) = 3(0)^2 - 12(0) + 13$$

$$= 0 - 0 + 13$$

$$= 13$$

The y-intercept is $(0, 13)$. Use this information along with symmetry of a parabola to sketch the graph of $f(x) = 3x^2 - 12x + 13$.

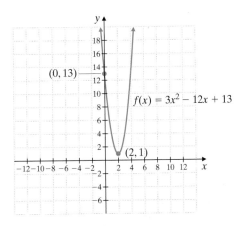

In Section 8.5, we study the graphing of quadratic functions further.

Practice Problem 5

Graph $f(x) = 2x^2 + 4x + 4$. Find the vertex and any intercepts.

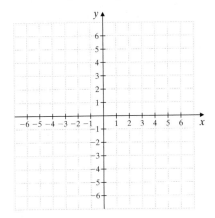

Answer

5.

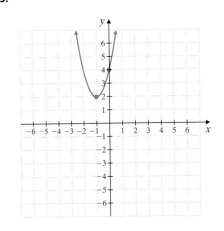

GRAPHING CALCULATOR EXPLORATIONS

We can use a graphing calculator to approximate real number solutions of any quadratic equation in standard form, whether the associated polynomial is factorable or not. For example, let's solve the quadratic equation $x^2 - 2x - 4 = 0$. The solutions of this equation will be the x-intercepts of the graph of the function $f(x) = x^2 - 2x - 4$. (Recall that to find x-intercepts, we let $f(x) = 0$, or $y = 0$.) When we use a standard window, the graph of this function looks like this:

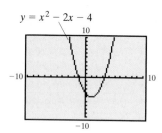

$$y = x^2 - 2x - 4$$

The graph appears to have one x-intercept between -2 and -1 and one between 3 and 4. To find the x-intercept between 3 and 4 to the nearest hundredth, we can use a ROOT feature, a ZOOM feature (which magnifies a portion of the graph around the cursor), or we can redefine our window. If we redefine our window according to the following values,

Xmin = 2	Ymin = -1
Xmax = 5	Ymax = 1
Xscl = 1	Yscl = 1

the resulting screen is as follows.

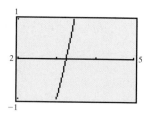

By using the TRACE feature, we can now see that one of the intercepts is between 3.21 and 3.25. To approximate to the nearest hundredth, we zoom again or redefine the window according to these values:

Xmin = 3.2	Ymin = -0.1
Xmax = 3.3	Ymax = 0.1
Xscl = 1	Yscl = 1

If we use the TRACE feature again, we see that, to the nearest thousandth, the x-intercept is $(3.236, 0)$. By repeating this process, we can approximate the other x-intercept to be $(-1.236, 0)$.

To check, we find $f(3.236)$ and $f(-1.236)$. Both of these values should be close to 0. (They will not be exactly 0 since we approximated these solutions.)

$$f(3.236) = -0.000304 \quad \text{and}$$
$$f(-1.236) = -0.000304$$

Solve each quadratic equation by graphing a related function and approximating the x-intercepts to the nearest thousandth.

1. $x^2 + 3x - 2 = 0$

2. $5x^2 - 7x + 1 = 0$

3. $2.3x^2 - 4.4x - 5.6 = 0$

4. $0.2x^2 + 6.2x + 2.1 = 0$

5. $0.09x^2 - 0.13x - 0.08 = 0$

6. $x^2 + 0.08x - 0.01 = 0$

Name _____ Section _____ Date _____

Mental Math

State whether the graph of each quadratic function, a parabola, opens upward or downward.

1. $f(x) = 2x^2 + 7x + 10$ **2.** $f(x) = -3x^2 - 5x$ **3.** $f(x) = -x^2 + 5$ **4.** $f(x) = x^2 + 3x + 7$

EXERCISE SET 5.8

 For the graph of each function f, answer the following. See Example 1.

     a. Find the domain and the range of the function.
     b. List the x- and y-intercepts.
     c. Find the coordinates of the point with the greatest y-value.
     d. Find the coordinates of the point with the smallest y-value.

1.

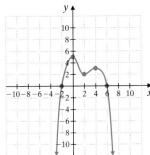

2.

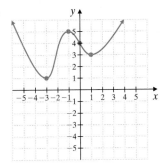

3. The graph in Example 4 of this section

4. The graph in Example 5 of this section

5. The graph in Example 6 of this section

6. The graph in Example 7 of this section

 Graph each quadratic function by plotting points. See Examples 2 and 3.

7. $f(x) = 2x^2$

8. $f(x) = -3x^2$

9. $f(x) = x^2 + 1$

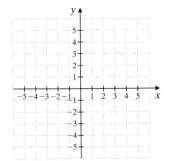   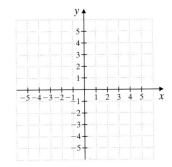   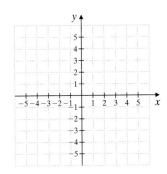

10. $f(x) = x^2 - 2$

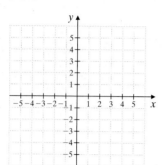

11. $f(x) = -x^2$

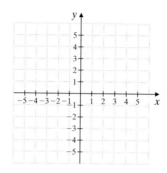

12. $f(x) = \dfrac{1}{2}x^2$

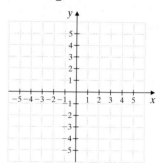

C *Graph each quadratic function. Find and label the vertex and intercepts. See Examples 4 and 5.*

13. $f(x) = x^2 + 8x + 7$

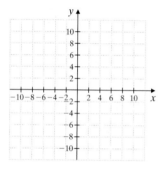

14. $f(x) = x^2 + 6x + 5$

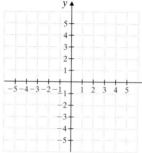

15. $f(x) = x^2 - 2x - 24$

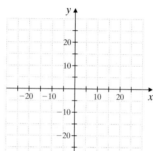

16. $f(x) = x^2 - 12x + 35$

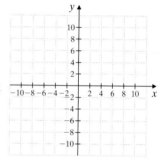

17. $f(x) = 2x^2 - 6x$

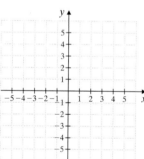

18. $f(x) = -3x^2 + 6x$

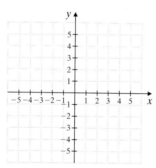

19. $f(x) = x^2 + 1$

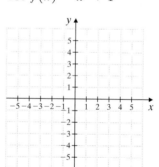

20. $f(x) = x^2 + 4$

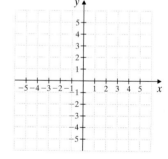

21. If the vertex of a parabola lies below the x-axis and the parabola opens upward, how many x-intercepts will the graph have?

22. If the vertex of a parabola lies below the x-axis and the parabola opens downward, how many x-intercepts will the graph have?

23. If the vertex of a parabola lies above the x-axis and the parabola opens upward, how many x-intercepts will the graph have?

24. If the vertex of a parabola lies above the x-axis and the parabola opens downward, how many x-intercepts will the graph have?

25. Can the graph of a function ever have more than one y-intercept point? Why or why not?

26. In general, is there a limit to the number of x-intercepts for the graph of a function?

Review and Preview

Simplify each fraction. See Sections 1.6 and 1.7.

27. $-\dfrac{8}{10}$

28. $-\dfrac{45}{100}$

29. $\dfrac{x^7 y^{10}}{x^3 y^{15}}$

30. $\dfrac{a^{14} b^2}{ab^4}$

31. $\dfrac{7n^{-9} m^{-2}}{14nm^{-5}}$

32. $\dfrac{20x^{-3} y^5}{25 y^{-2} x}$

Match each polynomial function (A–D) with its graph.

A. $f(x) = (x - 2)(x + 5)$ **B.** $f(x) = (x + 1)(x - 6)$ **C.** $f(x) = 2x^2 + 9x + 4$ **D.** $f(x) = 2x^2 - 7x - 4$

33.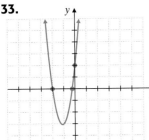

34.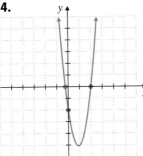

35.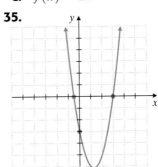

36.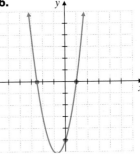

Combining Concepts

Use a graphing calculator to verify the graph in each exercise.

 37. Exercise 13

38. Exercise 14

Internet Excursions

Go To: http://www.prenhall.com/martin-gay_interm   What's Related

This World Wide Web address will give you access to a Web site where you can graph a second-degree polynomial or quadratic function of the form $f(x) = (a_2)x^2 + (a_1)x + (a_0)$. Specify the coefficients $a_2, a_1,$ and a_0 in order to answer the questions below.

39. Leaving the values of a_1 and a_0 set equal to 0, investigate the effect of the value of a_2, the coefficient of the x^2-term of the quadratic function, on the graph of the function. Describe the shape of the graph for larger versus smaller positive values of a_2. Similarly describe the shape of the graph for larger versus smaller negative values of a_2. What patterns do you notice?

40. Leaving the value of a_1 set equal to 0 and the value of a_2 set equal to any value except 0, investigate the effect of the value of a_0, the constant term of the quadratic function, on the graph of the function. As you change the value of a_0, describe how the graph changes. What relationship do you notice between the value of a_0 and the position of the graph?

This activity may be completed by working in groups or individually.

A picture framer has a piece of wood that measures 1 inch wide by 50 inches long. She would like to make a picture frame with the largest possible interior area. Complete the following activity to help her determine the dimensions of the frame that she should use to achieve her goal.

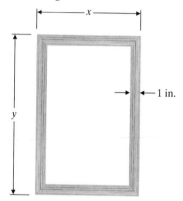

1. Use the situation shown in the figure to write an equation in x and y for the *outer* perimeter of the frame. (Remember that the outer perimeter will equal 50 inches.)
2. Use your equation from Question 1 to help you find the value of y for each value of x given in the table. Complete the y column of the table. (*Note*: The first two columns of the table give possible combinations for the outer dimensions of the frame.)
3. How is the interior width of the frame related to the exterior width of the frame? How is the interior height of the frame related to the exterior height of the frame? Use these relationships to complete the two columns of the table labeled "Interior Width" and "Interior Height."
4. Complete the last column of the table labeled "Interior Area" by using the columns of dimensions for the interior width and height.
5. From the table, what appears to be the largest interior area of the frame? Which exterior dimensions of the frame provide this area?
6. Use the patterns in the table to write an algebraic expression in terms of x for the interior width of the frame.
7. Use the patterns in the table to write an algebraic expression in terms of y for the interior height of the frame.
8. Use the perimeter equation from Question 1 to rewrite the algebraic expression for the interior height of the frame in terms of x.

9. Find a function A that gives the interior area of the frame in terms of its exterior width x. (*Hint*: Study the patterns in the table. How could the expressions from Questions 6 and 8 be used to write this function?)
10. Graph the function A. Locate and label the point from the table that represents the maximum interior area. Describe the location of the point in relation to the rest of the graph.

Frame's Interior Dimensions				
x	**y**	**Interior Width**	**Interior Height**	**Interior Area**
2.0				
2.5				
3.0				
3.5				
4.0				
4.5				
5.0				
5.5				
6.0				
6.5				
7.0				
7.5				
8.0				
8.5				
9.0				
9.5				
10.0				
10.5				
11.0				
11.5				
12.0				
12.5				
13.0				
13.5				
14.0				
14.5				
15.0				

Chapter 5   VOCABULARY CHECK

Fill in each blank with one of the words or phrases listed below.

| quadratic equation | synthetic division | polynomial | FOIL | 0 | monomial |
| binomial | trinomial | degree of a polynomial | degree of a term | | factoring |

1. A _____ is a finite sum of terms in which all variables are raised to nonnegative integer powers and no variables appear in any denominator.

2. _____ is the process of writing a polynomial as a product.

3. The _____ is the sum of the exponents on the variables contained in the term.

4. A _____ is a polynomial with one term.

5. A _____ is a polynomial with three terms.

6. A polynomial equation of degree 2 is also called a _____.

7. The _____ is the largest degree of all of its terms.

8. A _____ is a polynomial with two terms.

9. If a and b are real numbers and $a \cdot b =$ _____, then $a = 0$ or $b = 0$.

10. The _____ order may be used when multiplying two binomials.

11. A shortcut method called _____ may be used to divide a polynomial by a binomial of the form $x - c$.

C H A P T E R

Highlights

DEFINITIONS AND CONCEPTS	EXAMPLES

Section 5.1   Adding and Subtracting Polynomials

A **polynomial** is a finite sum of terms in which all variables have exponents raised to nonnegative integer powers and no variables appear in any denominator.	$1.3x^2$ Monomial $-\dfrac{1}{3}y + 5$ Binomial $6z^2 - 5z + 7$ Trinomial
To add polynomials, combine all like terms.	Add: $(3y^2x - 2yx + 11) + (-5y^2x - 7)$ $= -2y^2x - 2yx + 4$
To subtract polynomials, change the signs of the terms of the polynomial being subtracted; then add.	Subtract: $(-2z^3 - z + 1) - (3z^3 + z - 6)$ $= -2z^3 - z + 1 - 3z^3 - z + 6$ $= -5z^3 - 2z + 7$
A function P is a **polynomial function** if $P(x)$ is a polynomial.	For the polynomial function $P(x) = -x^2 + 6x - 12$ find $P(-2)$. $P(-2) = -(-2)^2 + 6(-2) - 12 = -28$

DEFINITIONS AND CONCEPTS	**EXAMPLES**

Section 5.2   Multiplying Polynomials

MULTIPLYING TWO POLYNOMIALS

Use the distributive property and multiply each term of one polynomial by each term of the other polynomial; then combine like terms.

SPECIAL PRODUCTS

$(a + b)^2 = a^2 + 2ab + b^2$

$(a - b)^2 = a^2 - 2ab + b^2$

$(a + b)(a - b) = a^2 - b^2$

The **FOIL order** may be used when multiplying two binomials.

Multiply.

$$(x^2 - 2x)(3x^2 - 5x + 1)$$
$$= 3x^4 - 5x^3 + x^2 - 6x^3 + 10x^2 - 2x$$
$$= 3x^4 - 11x^3 + 11x^2 - 2x$$

$$(3m + 2n)^2 = 9m^2 + 12mn + 4n^2$$
$$(z^2 - 5)^2 = z^4 - 10z^2 + 25$$
$$(7y + 1)(7y - 1) = 49y^2 - 1$$

Multiply.

$$(x^2 + 5)(2x^2 - 9)$$

$$\quad\quad\text{F}\quad\quad\text{O}\quad\quad\text{I}\quad\quad\text{L}$$
$$\quad\quad\downarrow\quad\quad\downarrow\quad\quad\downarrow\quad\quad\downarrow$$
$$= x^2(2x^2) + x^2(-9) + 5(2x^2) + 5(-9)$$
$$= 2x^4 - 9x^2 + 10x^2 - 45$$
$$= 2x^4 + x^2 - 45$$

Section 5.3   Dividing Polynomials

DIVIDING A POLYNOMIAL BY A MONOMIAL

Divide each term in the polynomial by the monomial.

$$\frac{12a^5b^3 - 6a^2b^2 + ab}{6a^2b^2}$$

$$= \frac{12a^5b^3}{6a^2b^2} - \frac{6a^2b^2}{6a^2b^2} + \frac{ab}{6a^2b^2}$$

$$= 2a^3b - 1 + \frac{1}{6ab}$$

DIVIDING A POLYNOMIAL BY A POLYNOMIAL OTHER THAN A MONOMIAL

Use **long division.**

Divide $2x^3 - x^2 - 8x - 1$ by $x - 2$.

$$
\begin{array}{r}
2x^2 + 3x - 2 \\
x - 2 \overline{) 2x^3 - x^2 - 8x - 1} \\
\underline{2x^3 - 4x^2} \\
3x^2 - 8x - 1 \\
\underline{3x^2 - 6x} \\
-2x - 1 \\
\underline{-2x + 4} \\
-5
\end{array}
$$

A shortcut method called **synthetic division** may be used to divide a polynomial by a binomial of the form $x - c$.

The quotient is $2x^2 + 3x - 2 - \dfrac{5}{x - 2}$.

Use synthetic division to divide $2x^3 - x^2 - 8x - 1$ by $x - 2$.

$$
\begin{array}{c|cccc}
2 & 2 & -1 & -8 & -1 \\
  &   & 4 & 6 & -4 \\
\hline
  & 2 & 3 & -2 & -5
\end{array}
$$

The quotient is $2x^2 + 3x - 2 - \dfrac{5}{x - 2}$.

DEFINITIONS AND CONCEPTS	**EXAMPLES**

Section 5.4   The Greatest Common Factor and Factoring by Grouping

The greatest common factor of the terms of a polynomial is the product of the greatest common factor of the numerical coefficients and the greatest common factor of the variable factors.	Factor: $14xy^3 - 2xy^2 = 2 \cdot 7 \cdot x \cdot y^3 - 2 \cdot x \cdot y^2$ The greatest common factor is $2 \cdot x \cdot y^2$, or $2xy^2$. $14xy^3 - 2xy^2 = 2xy^2(7y - 1)$
FACTORING A POLYNOMIAL BY GROUPING Group the terms so that each group has a common factor. Factor out these common factors. Then see if the new groups have a common factor.	Factor: $x^4y - 5x^3 + 2xy - 10$ $\quad = x^3(xy - 5) + 2(xy - 5)$ $\quad = (xy - 5)(x^3 + 2)$

Section 5.5   Factoring Trinomials

FACTORING $ax^2 + bx + c$ **Step 1.** Write all pairs of factors of ax^2. **Step 2.** Write all pairs of factors of c. **Step 3.** Try combinations of these factors until the middle term bx is found.	Factor: $28x^2 - 27x - 10$ Factors of $28x^2$: $28x$ and x, $2x$ and $14x$, $4x$ and $7x$. Factors of -10: -2 and 5, 2 and -5, -10 and 1, 10 and -1. $28x^2 - 27x - 10 = (7x + 2)(4x - 5)$

Section 5.6   Factoring by Special Products

PERFECT SQUARE TRINOMIAL $a^2 + 2ab + b^2 = (a + b)^2$ $a^2 - 2ab + b^2 = (a - b)^2$	Factor. $25x^2 + 30x + 9 = (5x + 3)^2$ $49z^2 - 28z + 4 = (7z - 2)^2$
DIFFERENCE OF TWO SQUARES $a^2 - b^2 = (a + b)(a - b)$	Factor. $36x^2 - y^2 = (6x + y)(6x - y)$
SUM AND DIFFERENCE OF TWO CUBES $a^3 + b^3 = (a + b)(a^2 - ab + b^2)$ $a^3 - b^3 = (a - b)(a^2 + ab + b^2)$	Factor. $8y^3 + 1 = (2y + 1)(4y^2 - 2y + 1)$ $27p^3 - 64q^3 = (3p - 4q)(9p^2 + 12pq + 16q^2)$

Section 5.7   Solving Equations by Factoring and Solving Problems

SOLVING A POLYNOMIAL EQUATION BY FACTORING **Step 1.** Write the equation so that one side is 0. **Step 2.** Factor the polynomial completely. **Step 3.** Set each factor equal to 0. **Step 4.** Solve the resulting equations. **Step 5.** Check each solution.	Solve: $2x^3 - 5x^2 = 3x$ $2x^3 - 5x^2 - 3x = 0$ $x(2x + 1)(x - 3) = 0$ $x = 0$ or $2x + 1 = 0$ or $x - 3 = 0$ $x = 0 \qquad\qquad x = -\dfrac{1}{2} \qquad\qquad x = 3$

Section 5.8   An Introduction to Graphing Polynomial Functions

A **quadratic function** is a function that can be written in the form $f(x) = ax^2 + bx + c, \quad a \neq 0$ The graph of this quadratic function is a **parabola** with **vertex** $\left(\dfrac{-b}{2a}, f\left(\dfrac{-b}{2a}\right)\right)$.	Find the vertex of the graph of the quadratic function $f(x) = 2x^2 - 8x + 1$ Here $a = 2$ and $b = -8$. $\dfrac{-b}{2a} = \dfrac{-(-8)}{2 \cdot 2} = 2$ $f(2) = 2 \cdot 2^2 - 8 \cdot (2) + 1 = -7$ The vertex has coordinates $(2, -7)$.

STUDY SKILLS REMINDER

Are you preparing for a test on Chapter 5?

Below I have listed some *common trouble areas* for students in Chapter 5. After studying for your test—but before taking your test—read these.

■ Don't forget to watch your signs when subtracting polynomials.

$$(7x^3 - 6x^2 + 2x) - (9x^3 + 7x^2 - 20x - 20)$$
$$= 7x^3 - 6x^2 + 2x - 9x^3 - 7x^2 + 20x + 20 = -2x^3 - 13x^2 + 22x + 20$$

■ Can you evaluate $P(-1)$ if $P(x) = -16x^2 + 2x$?

$$P(-1) = -16(-1)^2 + 2(-1) = -16 \cdot 1 + (-2) = -18$$

■ Don't forget how to square a binomial.

$$(3x + 5y)^2 = (3x)^2 + 2(3x)(5y) + (5y)^2 = 9x^2 + 30xy + 25y^2$$

■ Remember that the first step to factoring a polynomial is to factor out any common factors. Also, always check to see if a factor can be factored further.

$$\text{Factor:} \quad 4x^4 - 64 = 4(x^4 - 16)$$

$$= 4(x^2 + 4)(x^2 - 4)$$
$$= 4(x^2 + 4)(x + 2)(x - 2)$$

■ When factoring the sum or difference of two cubes, it may be helpful to first write each term as a quantity cubed.

$$\text{Factor:} \quad 27x^3 - 8y^3 = (3x)^3 - (2y)^3 = (3x - 2y)[(3x)^2 + (3x)(2y) + (2y)^2]$$

$$= (3x - 2y)(9x^2 + 6xy + 4y^2)$$

■ Remember that to use the zero-factor property to solve a quadratic equation, one side of the equation must be 0 and the other side must be a factored polynomial.

$$\text{Solve } x(5x + 3) = 2 \quad \text{Cannot use zero-factor property.}$$
$$5x^2 + 3x - 2 = 0 \quad \text{Multiply and subtract 2 from both sides.}$$
$$(5x - 2)(x + 1) = 0 \quad \text{Now you can use zero-factor property.}$$
$$5x - 2 = 0 \quad \text{or} \quad x + 1 = 0 \quad \text{Set each factor equal to 0.}$$
$$x = \frac{2}{5} \quad \text{or} \quad x = -1 \quad \text{Solve.}$$

Remember: This is simply a checklist of common trouble areas. For a review of Chapter 5, see the Highlights and Chapter Review at the end of this chapter.

Chapter 5 Review

(5.1) *Find the degree of each polynomial.*

1. $x^2y - 3xy^3z + 5x + 7y$

2. $3x + 2$

Simplify by combining like terms.

3. $4x + 8x - 6x^2 - 6x^2y$

4. $-8xy^3 + 4xy^3 - 3x^3y$

Add or subtract as indicated.

5. $(3x + 7y) + (4x^2 - 3x + 7) + (y - 1)$

6. $(4x^2 - 6xy + 9y^2) - (8x^2 - 6xy - y^2)$

7. $(3x^2 - 4b + 28) + (9x^2 - 30) - (4x^2 - 6b + 20)$

8. Add $(9xy + 4x^2 + 18)$ and $(7xy - 4x^3 - 9x)$.

9. Subtract $(x - 7)$ from the sum of $(3x^2y - 7xy - 4)$ and $(9x^2y + x)$.

10. $\begin{aligned} x^2 - 5x + 7 \\ - \quad\quad (x + 4) \\ \hline \end{aligned}$

11. $\begin{aligned} x^3 + 2xy^2 - y \\ + (x - 4xy^2 \quad - 7) \\ \hline \end{aligned}$

If $P(x) = 9x^2 - 7x + 8$, find each function value.

12. $P(6)$

13. $P(-2)$

14. $P(-3)$

If $P(x) = 2x - 1$ and $Q(x) = x^2 + 2x - 5$, find each function value.

15. $P(x) + Q(x)$

16. $2P(x) - Q(x)$

△ **17.** Find the perimeter of the rectangle.

$x^2y + 5$
cm

$2x^2y - 6x + 1$
cm

(5.2) *Multiply.*

18. $-6x(4x^2 - 6x + 1)$

19. $-4ab^2(3ab^3 + 7ab + 1)$

20. $(x - 4)(2x + 9)$

21. $(-3xa + 4b)^2$

22. $(9x^2 + 4x + 1)(4x - 3)$

23. $(5x - 9y)(3x + 9y)$

24. $\left(x - \dfrac{1}{3}\right)\left(x + \dfrac{2}{3}\right)$

25. $(x^2 + 9x + 1)^2$

26. $(2x - 1)(x^2 + 2x - 5)$

Use special products to multiply.

27. $(3x - y)^2$

28. $(4x + 9)^2$

29. $(x + 3y)(x - 3y)$

30. $[4 + (3a - b)][4 - (3a - b)]$ △**31.** Find the area of the rectangle.

$3y - 7z$ units

$3y + 7z$ units

(5.3) *Divide.*

32. $(4xy + 2x^2 - 9) \div (4xy)$

33. $12xb^2 + 16xb^4$ by $4xb^3$

34. $(3x^4 - 25x^2 - 20) \div (x - 3)$

35. $(-x^2 + 2x^4 + 5x - 12) \div (x - 3)$

36. $(2x^4 - x^3 + 2x^2 - 3x + 1) \div \left(x - \dfrac{1}{2}\right)$

37. $(x^3 + 3x^2 - 2x + 2) \div \left(x - \dfrac{1}{2}\right)$

38. $(3x^4 + 5x^3 + 7x^2 + 3x - 2) \div (x^2 + x + 2)$

39. $(9x^4 - 6x^3 + 3x^2 - 12x - 30) \div (3x^2 - 2x - 5)$

Use synthetic division to find each quotient.

40. $(3x^3 + 12x - 4) \div (x - 2)$

41. $(4x^3 + 2x^2 - 4x - 2) \div \left(x + \dfrac{3}{2}\right)$

42. $(x^5 - 1) \div (x + 1)$

43. $(x^3 - 81) \div (x - 3)$

44. $(x^3 - x^2 + 3x^4 - 2) \div (x - 4)$

45. $(3x^4 - 2x^2 + 10) \div (x + 2)$

(5.4) *Factor out the greatest common factor.*

46. $16x^3 - 24x^2$

47. $36y - 24y^2$

48. $6ab^2 + 8ab - 4a^2b^2$

49. $14a^2b^2 - 21ab^2 + 7ab$

50. $6a(a + 3b) - 5(a + 3b)$

51. $4x(x - 2y) - 5(x - 2y)$

Factor by grouping.

52. $xy - 6y + 3x - 18$

53. $ab - 8b + 4a - 32$

54. $pq - 3p - 5q + 15$

55. $x^3 - x^2 - 2x + 2$

△ **56.** A smaller square is cut from a larger rectangle. Write the area of the shaded region as a factored polynomial.

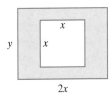

(5.5) *Factor each polynomial completely.*

57. $x^2 - 14x - 72$

58. $x^2 + 16x - 80$

59. $2x^2 - 18x + 28$

60. $3x^2 + 33x + 54$

61. $2x^3 - 7x^2 - 9x$

62. $3x^2 + 2x - 16$

63. $6x^2 + 17x + 10$

64. $15x^2 - 91x + 6$

65. $4x^2 + 2x - 12$

66. $9x^2 - 12x - 12$

67. $y^2(x + 6)^2 - 2y(x + 6)^2 - 3(x + 6)^2$

68. $(x + 5)^2 + 6(x + 5) + 8$

69. $x^4 - 6x^2 - 16$

70. $x^4 + 8x^2 - 20$

(5.6) *Factor each polynomial completely.*

71. $x^2 - 100$

72. $x^2 - 81$

73. $2x^2 - 32$

74. $6x^2 - 54$

75. $81 - x^4$

76. $16 - y^4$

77. $(y + 2)^2 - 25$

78. $(x - 3)^2 - 16$

79. $x^3 + 216$

80. $y^3 + 512$

81. $8 - 27y^3$

82. $1 - 64y^3$

83. $6x^4y + 48xy$

84. $2x^5 + 16x^2y^3$

85. $x^2 - 2x + 1 - y^2$

86. $x^2 - 6x + 9 - 4y^2$

87. $4x^2 + 12x + 9$

88. $16a^2 - 40ab + 25b^2$

△ **89.** The volume of the cylindrical shell is $\pi R^2 h - \pi r^2 h$ cubic units. Write this volume as a factored expression.

(5.7) *Solve each polynomial equation for the variable.*

90. $(3x - 1)(x + 7) = 0$

91. $3(x + 5)(8x - 3) = 0$

92. $5x(x - 4)(2x - 9) = 0$

93. $6(x + 3)(x - 4)(5x + 1) = 0$

94. $2x^2 = 12x$

95. $4x^3 - 36x = 0$

96. $(1 - x)(3x + 2) = -4x$

97. $2x(x - 12) = -40$

98. $3x^2 + 2x = 12 - 7x$

99. $2x^2 + 3x = 35$

100. $x^3 - 18x = 3x^2$

101. $19x^2 - 42x = -x^3$

102. $12x = 6x^3 + 6x^2$

103. $8x^3 + 10x^2 = 3x$

104. The sum of a number and twice its square is 105. Find the number.

△ **105.** The length of a rectangular piece of carpet is 2 meters less than 5 times its width. Find the dimensions of the carpet if its area is 16 square meters.

106. A scene from an adventure film calls for a stunt dummy to be dropped from above the second-story platform of the Eiffel Tower, a distance of 400 feet. Its height h in feet at the time t seconds is given by $h = -16t^2 + 400$. Determine how long before the stunt dummy reaches the ground.

400 ft

For Exercises 107 through 110, refer to the following graph.

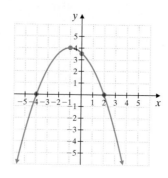

107. Find the domain and the range of the function.

108. List the *x*- and *y*-intercepts.

109. Find the coordinates of the point with the greatest *y*-value.

110. List the *x*-values for which the *y*-values are greater than 0.

Graph each polynomial function defined by the equation. Find all intercepts. If the function is a quadratic function, find the vertex.

111. $f(x) = x^2 + 6x + 9$　**112.** $f(x) = x^2 - 5x + 4$　**113.** $f(x) = 2x^2 - 4x + 5$　**114.** $f(x) = x^2 - 2x + 3$

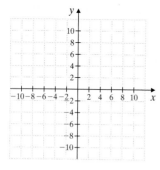　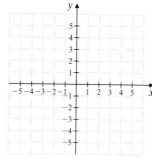　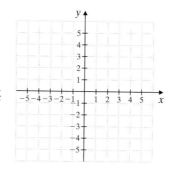　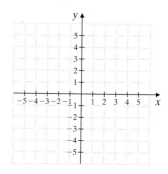

Chapter 5 Test

Perform each indicated operation.

1. $(4x^3 - 3x - 4) - (9x^3 + 8x + 5)$ **2.** $-3xy(4x + y)$

3. $(3x + 4)(4x - 7)$ **4.** $(5a - 2b)(5a + 2b)$ **5.** $(6m + n)^2$

6. $(2x - 1)(x^2 - 6x + 4)$ **7.** $(4x^2y + 9x + z) \div (3xz)$

8. $(4x^5 - 2x^4 + 4x^2 - 6x + 3) \div (2x - 1)$ **9.** Use synthetic division to divide $4x^4 - 3x^3 + 2x^2 - x - 1$ by $x + 3$.

Factor each polynomial completely.

10. $16x^3y - 12x^2y^4$ **11.** $x^2 - 13x - 30$ **12.** $4y^2 + 20y + 25$

13. $6x^2 - 15x - 9$ **14.** $4x^2 - 25$ **15.** $x^3 + 64$

16. $3x^2y - 27y^3$ **17.** $6x^2 + 24$ **18.** $x^2y - 9y - 3x^2 + 27$

Solve each equation for the variable.

19. $3(n - 4)(7n + 8) = 0$ **20.** $(x + 2)(x - 2) = 5(x + 4)$

Answers
1.
2.
3.
4.
5.
6.
7.
8.
9.
10.
11.
12.
13.
14.
15.
16.
17.
18.
19.
20.

21. $2x^3 + 5x^2 - 8x - 20 = 0$

△ **22.** Write the area of the shaded region as a factored polynomial.

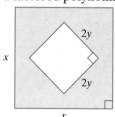

▦ **23.** A pebble is hurled upward from the top of the 880-foot-tall Canada Trust Tower with an initial velocity of 96 feet per second. Neglecting air resistance, the height $h(t)$ in feet of the pebble after t seconds is given by the polynomial function

$$h(t) = -16t^2 + 96t + 880$$

▦ **a.** Find the height of the pebble when $t = 1$.

b. Find the height of the pebble when $t = 5.1$.

c. Factor the polynomial $-16t^2 + 96t + 880$.

d. When will the pebble hit the ground?

Graph. Find and label the vertex and x- and y-intercepts.

24. $f(x) = x^2 - 4x - 5$

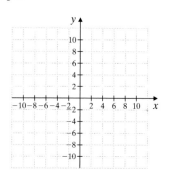

Name _____ Section _____ Date _____

Cumulative Review

Simplify.

1. $-2(x + 3)$

2. $7x + 3 - 5(x - 4)$

3. Use scientific notation to simplify: $\dfrac{2000 \times 0.000021}{700}$

4. Solve: $2(x - 3) = 5x - 9$

5. Solve $3y - 2x = 7$ for y.

6. Solve: $5 - x \le 4x - 15$. Write the solution set in interval notation.

7. Solve $-1 \le \dfrac{2x}{3} + 5 \le 2$. Write the solution in interval notation.

8. Solve: $|3x + 2| = |5x - 8|$

9. Find the slope of the line $x = -5$.

10. Graph: $y = \dfrac{1}{4}x - 3$

11. Write an equation of the line containing the point $(-2, 1)$ and perpendicular to the line $3x + 5y = 4$.

12. Determine the domain and range of the relation: $\{(2, 3), (2, 4), (0, -1), (3, -1)\}$

Find each function value.

13. If $g(x) = 3x - 2$, find $g(0)$.

14. If $f(x) = 7x^2 - 3x + 1$, find $f(1)$.

Answers

1. _____

2. _____

3. _____

4. _____

5. _____

6. _____

7. _____

8. _____

9. _____

10. see graph

11. _____

12. _____

13. _____

14. _____

15. see graph

15. Graph: $2x - y < 6$

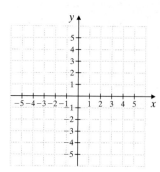

16. Use the substitution method to solve
the system: $\begin{cases} 2x + 4y = -6 \\ x = 2y - 5 \end{cases}$

16. _____

17. _____

17. Solve the system:

$$\begin{cases} 3x - y + z = -15 \\ x + 2y - z = 1 \\ 2x + 3y - 2z = 0 \end{cases}$$

18. Use matrices to solve the system:

$$\begin{cases} x + 3y = 5 \\ 2x - y = -4 \end{cases}$$

18. _____

19. see graph

19. Graph the solutions of the system

$$\begin{cases} x - y < 2 \\ x + 2y > -1 \\ y < 2 \end{cases}$$

20. Find the degree of the polynomial:

$$3xy + x^2y^2 - 5x^2 - 6$$

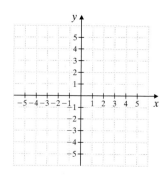

20. _____

21. _____

Multiply.

21. $2x(5x - 4)$

22. $-xy(7x^2y + 3xy - 11)$

22. _____

23. _____

23. Divide $10x^3 - 5x^2 + 20x$ by $5x$.

Rational Expressions

Dividing one polynomial by another in Section 5.3, we found quotients of polynomials. When the remainder part of the quotient was not 0, the remainder was a fraction, such as $\dfrac{x}{x^2 + 1}$. This fraction is not a polynomial, since it cannot be written as the sum of whole number powers. Instead, it is called a *rational expression*. In this chapter, we study these algebraic forms, the operations that can be performed on them, and the *rational functions* they generate.

6.1 Multiplying and Dividing Rational Expressions

6.2 Adding and Subtracting Rational Expressions

6.3 Simplifying Complex Fractions

6.4 Solving Equations Containing Rational Expressions

Integrated Review—Expressions and Equations Containing Rational Expressions

6.5 Rational Equations and Problem Solving

6.6 Variation and Problem Solving

Have you ever watched an auto race, either in person or on television, and noticed that the pitch of the racecar's engine sounds higher as it approaches on the track and drops lower once it passes and moves away? If so, then you have experienced the Doppler effect. This apparent change in pitch of a moving sound can also be heard when a vehicle sounding its horn or siren, or playing its stereo passes you on the street. This effect was discovered by Austrian physicist Christian Doppler in 1842. The Doppler effect has applications in many diverse fields: from meteorology to medicine to astronomy. In Exercise 66 on page 473, we will analyze situations involving the Doppler effect with a rational equation.

For more on the Doppler effect, including an audio demonstration, visit http://www.haystack.mit.edu/midas/doppler.html on the Internet.

Name _____ Section _____ Date _____

Chapter **6** Pretest

1. Find all numbers for which the rational expression $\dfrac{x-1}{x^2-5x+6}$ is undefined.

Simplify each rational expression.

2. $\dfrac{2x-4x^2}{2x}$

3. $\dfrac{3x^2+7x+2}{3x^2-11x-4}$

Perform each indicated operation.

4. $\dfrac{3x-6}{8}\cdot\dfrac{4}{2-x}$

5. $\dfrac{x^2+3x+2}{x^2+4x+3}\cdot\dfrac{2x^2-x-21}{2x^2-x-10}$

6. $\dfrac{8m^2n^3}{m^2-25}\div\dfrac{4mn}{m+5}$

7. $\dfrac{2y-y^2}{y^3-8}\div\dfrac{y}{y^2+2y+4}$

8. $\dfrac{7}{x-7}+\dfrac{x}{x-7}$

9. $\dfrac{4}{3a^2}-\dfrac{2}{5a}$

10. $\dfrac{y+4}{y-3}-\dfrac{y+1}{y+3}$

11. $\dfrac{5}{x^2+7x+6}+\dfrac{x}{2x^2+13x+6}$

Simplify each complex fraction.

12. $\dfrac{\dfrac{9}{2x}}{\dfrac{3}{8x}}$

13. $\dfrac{\dfrac{4}{y}+\dfrac{5}{y^2}}{\dfrac{16}{y^2}-\dfrac{25}{y}}$

14. $\dfrac{x^{-2}}{x^{-1}+y^{-1}}$

Solve each equation for x.

15. $\dfrac{x}{6}-\dfrac{x}{5}=-2$

16. $\dfrac{x+3}{x^2+9x+18}=\dfrac{1}{2x+6}-\dfrac{1}{x+6}$

17. Solve for L: $S=\dfrac{n(a+L)}{2}$

Solve.

18. The sum of a number and 6 times its reciprocal is 5. Find the number(s).

19. Suppose y varies directly as x. If $y=6$ when $x=24$, find y when $x=30$.

20. Suppose that W is inversely proportional to V. If $W=100$ when $V=8$, find W when $V=25$.

6.1 Multiplying and Dividing Rational Expressions

Recall that a *rational number*, or *fraction*, is a number that can be written as the quotient $\frac{p}{q}$ of two integers p and q as long as q is not 0. A **rational expression** is an expression that can be written as the quotient $\frac{P}{Q}$ of two polynomials P and Q as long as Q is not 0. Examples are

$$\frac{3x+7}{2} \qquad \frac{5x^2-3}{x-1} \qquad \frac{7x-2}{2x^2+7x+6}$$

Ⓐ Finding Values for Which a Rational Expression Is Undefined

As with numerical fractions, a rational expression is **undefined** if the denominator is 0. If a variable in a rational expression is replaced with a number that makes the denominator 0, we say that the rational expression is **undefined** for this value of the variable. For example, the rational expression

$$\frac{x^2+2}{x-3} \quad \text{is undefined when } x \text{ is 3}$$

because replacing x with 3 results in a denominator of 0.

$$\frac{x^2+2}{x-3} = \frac{3^2+2}{3-3} = \frac{11}{0} \qquad \text{This rational expression is undefined when } x = 3.$$

EXAMPLES

Find all numbers for which each rational expression is undefined.

1. $\dfrac{5x^2-3}{x-1}$ is undefined when the denominator

$x-1$ is 0.

$x-1=0$ or $x=1$ Set the denominator equal to 0 and solve.

If x is replaced with 1, the rational expression is undefined.

2. $\dfrac{7x-2}{x^2-2x-15}$ is undefined when the denominator is 0.

$x^2-2x-15=0$ Set the denominator equal to 0 and solve.

$(x-5)(x+3)=0$

$x-5=0$ or $x+3=0$

$x=5$ or $x=-3$

If x is replaced with 5 or with -3, the rational expression is undefined.

3. $\dfrac{5x-1}{3}$ is undefined when the denominator is 0. No matter what value x is replaced with, the denominator—3—is never 0. No real number makes this rational expression undefined. That is, this expression is defined for all real numbers. ●

Try the Concept Check in the margin.

Practice Problems 1–3

Find all numbers for which each rational expression is undefined.

1. $\dfrac{x^2+1}{x-6}$

2. $\dfrac{5x+4}{x^2-3x-10}$

3. $\dfrac{x^2-9}{4}$

Concept Check

For which of these values (if any) is the rational expression $\dfrac{x-3}{x^2+2}$ undefined? Explain.

a. 2

b. 3

c. -2

d. 0

e. None of these

Answers

1. 6, **2.** 5, -2, **3.** no real number

Concept Check: e

Ⓑ Simplifying Rational Expressions

Recall that a fraction is in lowest terms or simplest form if the numerator and denominator have no common factors other than 1 (or -1). For example, $\dfrac{3}{13}$ is in lowest terms since 3 and 13 have no common factors other than 1 (or -1).

To **simplify** a rational expression, or to write it in lowest terms, we use the fundamental principle of rational expressions.

Fundamental Principle of Rational Expressions

For any rational expression $\dfrac{P}{Q}$ and any polynomial R, $R \neq 0$,

$$\frac{PR}{QR} = \frac{P}{Q}$$

Thus, the fundamental principle says that multiplying or dividing the numerator and denominator of a rational expression by the same nonzero polynomial yields an equivalent rational expression.

To simplify a rational expression such as $\dfrac{(x+2)^2}{x^2-4}$, factor the numerator and the denominator and then use the fundamental principle of rational expressions to divide out common factors.

$$\frac{(x+2)^2}{x^2-4} = \frac{(x+2)(x+2)}{(x+2)(x-2)}$$

$$= \frac{x+2}{x-2}$$

This means that the rational expression $\dfrac{(x+2)^2}{x^2-4}$ has the same value as the rational expression $\dfrac{x+2}{x-2}$ for all values of x except 2 and -2. (Remember that when x is 2, the denominators of both rational expressions are 0 and that when x is -2, the original rational expression has a denominator of 0.)

As we simplify rational expressions, we will assume that the simplified rational expression is equivalent to the original rational expression for all real numbers except those for which either denominator is 0.

In general, the following steps may be used to simplify rational expressions or to write a rational expression in lowest terms.

Simplifying or Writing a Rational Expression in Lowest Terms

Step 1.   Completely factor the numerator and denominator of the rational expression.

Step 2.   Apply the fundamental principle of rational expressions to divide out factors common to both the numerator and denominator.

For now, we assume that variables in a rational expression do not represent values that make the denominator 0.

EXAMPLES   Simplify each rational expression.

4. $\dfrac{24x^6y^5}{8x^7y} = \dfrac{(8x^6y)\,3y^4}{(8x^6y)\,x}$ Factor the numerator and denominator.

$\qquad\quad = \dfrac{3y^4}{x}$ Apply the fundamental principle and divide out common factors.

5. $\dfrac{2x^2}{10x^3 - 2x^2} = \dfrac{2x^2 \cdot 1}{2x^2(5x - 1)}$ Factor the numerator and denominator.

$\qquad\qquad\quad = \dfrac{1}{5x - 1}$ Apply the fundamental principle and divide out common factors. ●

EXAMPLES   Simplify each rational expression.

6. $\dfrac{2 + x}{x + 2} = \dfrac{x + 2}{x + 2} = 1$ By the commutative property of addition, $2 + x = x + 2$.

7. $\dfrac{2 - x}{x - 2}$

The terms in the numerator of $\dfrac{2 - x}{x - 2}$ differ by sign from the terms of the denominator, so the polynomials are opposites of each other and the expression simplifies to -1. To see this, we factor out -1 from the numerator or the denominator. If -1 is factored from the numerator, then

$$\dfrac{2 - x}{x - 2} = \dfrac{-1(-2 + x)}{x - 2} = \dfrac{-1(x - 2)}{x - 2} = \dfrac{-1}{1} = -1$$

If -1 is factored from the denominator, the result is the same.

$$\dfrac{2 - x}{x - 2} = \dfrac{2 - x}{-1(-x + 2)} = \dfrac{2 - x}{-1(2 - x)} = \dfrac{1}{-1} = -1$$

Helpful Hint

When the numerator and the denominator of a rational expression are opposites of each other, the expression simplifies to -1.

8. $\dfrac{18 - 2x^2}{x^2 - 2x - 3} = \dfrac{2(9 - x^2)}{(x + 1)(x - 3)}$ Factor.

$\qquad\qquad\qquad = \dfrac{2(3 + x)(3 - x)}{(x + 1)(x - 3)}$ Factor completely.

Notice the opposites $3 - x$ and $x - 3$. We write $3 - x$ as $-1(x - 3)$ and simplify.

$$\dfrac{2(3 + x)(3 - x)}{(x + 1)(x - 3)} = \dfrac{2(3 + x) \cdot -1(x - 3)}{(x + 1)(x - 3)} = -\dfrac{2(3 + x)}{x + 1}$$ ●

Practice Problems 4–5

Simplify each rational expression.

4. $\dfrac{20a^7b^4}{5a^3b^5}$

5. $\dfrac{3y^3}{6y^4 - 3y^3}$

Practice Problems 6–8

Simplify each rational expression.

6. $\dfrac{5 + x}{x + 5}$

7. $\dfrac{5 - x}{x - 5}$

8. $\dfrac{3 - 3x^2}{x^2 + x - 2}$

Answers

4. $\dfrac{4a^4}{b}$,   **5.** $\dfrac{1}{2y - 1}$,   **6.** 1,   **7.** -1,

8. $-\dfrac{3(x + 1)}{x + 2}$

Concept Check

Which of the following expressions are equivalent to $\dfrac{x}{8-x}$?

a. $\dfrac{-x}{x-8}$ b. $\dfrac{-x}{8-x}$

c. $\dfrac{x}{x-8}$ d. $\dfrac{-x}{-8+x}$

Practice Problems 9–10

Simplify each rational expression.

9. $\dfrac{x^3+27}{x+3}$

10. $\dfrac{3x^2+6}{x^3-3x^2+2x-6}$

Concept Check

Does $\dfrac{n}{n+2}$ simplify to $\dfrac{1}{2}$? Why or why not?

Answers

9. x^2-3x+9, 10. $\dfrac{3}{x-3}$

Concept Check: a and d

Concept Check: no; answers may vary

Helpful Hint

Recall that for a fraction,

$$\frac{a}{-b}=\frac{-a}{b}=-\frac{a}{b}$$

For example,

$$\frac{-(x+1)}{(x+2)}=\frac{(x+1)}{-(x+2)}=-\frac{x+1}{x+2}$$

Try the Concept Check in the margin.

EXAMPLES   Simplify each rational expression.

9. $\dfrac{x^3+8}{x+2}=\dfrac{(x+2)(x^2-2x+4)}{x+2}$ Factor the sum of the two cubes.

$=x^2-2x+4$ Divide out common factors.

10. $\dfrac{2y^2+2}{y^3-5y^2+y-5}=\dfrac{2(y^2+1)}{(y^3-5y^2)+(y-5)}$ Factor the numerator.

$=\dfrac{2(y^2+1)}{y^2(y-5)+1(y-5)}$ Factor the denominator by grouping.

$=\dfrac{2(y^2+1)}{(y-5)(y^2+1)}$

$=\dfrac{2}{y-5}$ Divide out common factors.

Try the Concept Check in the margin.

Multiplying Rational Expressions

Arithmetic operations on rational expressions are performed in the same way as they are on rational numbers. To multiply rational expressions, we multiply numerators and multiply denominators.

When multiplying rational expressions, we will factor each numerator and denominator first. This will help when we apply the fundamental principle to simplify the product.

The following steps may be used to multiply rational expressions.

Multiplying Rational Expressions

Step 1. Completely factor each numerator and denominator.

Step 2. Multiply the numerators and multiply the denominators.

Step 3. Simplify the product by applying the fundamental principle and dividing the numerator and denominator by their common factors.

EXAMPLES   Multiply.

11. $\dfrac{3n+1}{2n} \cdot \dfrac{2n-4}{3n^2-2n-1} = \dfrac{3n+1}{2n} \cdot \dfrac{2(n-2)}{(3n+1)(n-1)}$ Factor.

$\qquad = \dfrac{(3n+1)\cdot 2\,(n-2)}{2\,n(3n+1)(n-1)}$ Multiply.

$\qquad = \dfrac{n-2}{n(n-1)}$ Divide out common factors.

12. $\dfrac{x^3-1}{-3x+3} \cdot \dfrac{15x^2}{x^2+x+1} = \dfrac{(x-1)(x^2+x+1)}{-3(x-1)} \cdot \dfrac{15x^2}{x^2+x+1}$ Factor.

$\qquad = \dfrac{(x-1)(x^2+x+1)\cdot 3\cdot 5x^2}{-1\cdot 3(x-1)(x^2+x+1)}$ Factor.

$\qquad = \dfrac{5x^2}{-1}$ Divide out common factors.

$\qquad = -5x^2$ ●

(D) Dividing Rational Expressions

Recall that two numbers are reciprocals of each other if their product is 1. Similarly, if $\dfrac{P}{Q}$ is a rational expression and $P \neq 0$, then $\dfrac{Q}{P}$ is its **reciprocal**, since

$$\frac{P}{Q} \cdot \frac{Q}{P} = \frac{P\cdot Q}{Q\cdot P} = 1$$

The following are examples of expressions and their reciprocals.

Expression	**Reciprocal**
$\dfrac{3}{x}$	$\dfrac{x}{3}$
$\dfrac{2+x^2}{4x-3}$	$\dfrac{4x-3}{2+x^2}$
x^3	$\dfrac{1}{x^3}$
0	no reciprocal

Dividing Rational Expressions

To divide by a rational expression, multiply by its reciprocal. Then simplify if possible.

EXAMPLES   Divide.

13. $\dfrac{3x}{5y} \div \dfrac{9y}{x^5} = \dfrac{3x}{5y} \cdot \dfrac{x^5}{9y}$ Multiply by the reciprocal of the divisor.

$\qquad = \dfrac{x^6}{15y^2}$ Simplify.

Practice Problems 11–12

Multiply.

11. $\dfrac{2x-3}{5x} \cdot \dfrac{5x+5}{2x^2-x-3}$

12. $\dfrac{x^3+27}{-2x-6} \cdot \dfrac{4x^3}{x^2-3x+9}$

Practice Problems 13–14

Divide.

13. $\dfrac{7x}{2y^2} \div \dfrac{8y}{3x^4}$

14. $\dfrac{12y^3}{5y^2-5} \div \dfrac{6}{1-y}$

Answers

11. $\dfrac{1}{x}$,   **12.** $-2x^3$,   **13.** $\dfrac{21x^5}{16y^3}$,   **14.** $-\dfrac{2y^3}{5(y+1)}$

14. $\dfrac{8m^2}{3m^2 - 12} \div \dfrac{40}{2 - m} = \dfrac{8m^2}{3m^2 - 12} \cdot \dfrac{2 - m}{40}$ Multiply by the reciprocal of the divisor.

$$= \dfrac{8m^2(2 - m)}{3(m + 2)(m - 2) \cdot 40}$$ Factor and multiply.

$$= \dfrac{8\ m^2 \cdot -1(m - 2)}{3(m + 2)(m - 2) \cdot 8 \cdot 5}$$ Write $(2 - m)$ as $-1(m - 2)$.

$$= -\dfrac{m^2}{15(m + 2)}$$ Simplify.

Helpful Hint

When dividing rational expressions, do not divide out common factors until the division problem is rewritten as a multiplication problem.

Practice Problem 15

Perform each indicated operation.

$$\dfrac{(x + 3)^2}{x^2 - 9} \cdot \dfrac{2x - 6}{5x} \div \dfrac{x^2 + 7x + 12}{x}$$

EXAMPLE 15   Perform each indicated operation.

$$\dfrac{x^2 - 25}{(x + 5)^2} \cdot \dfrac{3x + 15}{4x} \div \dfrac{x^2 - 3x - 10}{x}$$

Solution: $\dfrac{x^2 - 25}{(x + 5)^2} \cdot \dfrac{3x + 15}{4x} \div \dfrac{x^2 - 3x - 10}{x}$

$$= \dfrac{x^2 - 25}{(x + 5)^2} \cdot \dfrac{3x + 15}{4x} \cdot \dfrac{x}{x^2 - 3x - 10}$$ To divide, multiply by the reciprocal.

$$= \dfrac{(x + 5)(x - 5)}{(x + 5)(x + 5)} \cdot \dfrac{3(x + 5)}{4\ x} \cdot \dfrac{x}{(x - 5)(x + 2)}$$

$$= \dfrac{3}{4(x + 2)}$$

E   Applications with Rational Functions

Rational expressions are sometimes used to describe functions. For example, we call the function $f(x) = \dfrac{x^2 + 2}{x - 3}$ a **rational function** since $\dfrac{x^2 + 2}{x - 3}$ is a rational expression in one variable.

The domain of a rational function such as $f(x) = \dfrac{x^2 + 2}{x - 3}$ is the set of all possible replacement values for x. In other words, since the rational expression $\dfrac{x^2 + 2}{x - 3}$ is not defined when $x = 3$, we say that the domain of $f(x) = \dfrac{x^2 + 2}{x - 3}$ is all real numbers except 3. We can write the domain as

$$\{x \mid x \text{ is a real number and } x \neq 3\}$$

See the Graphing Calculator Explorations on page 430 for further domain exercises.

Answer

15. $\dfrac{2}{5(x + 4)}$

EXAMPLE 16   Finding Unit Cost

For the ICL Production Company, the rational function $C(x) = \dfrac{2.6x + 10,000}{x}$ describes the company's cost per disc of pressing x compact discs. Find the cost per disc for pressing:

a. 100 compact discs

b. 1000 compact discs

Solution:

a. $C(100) = \dfrac{2.6(100) + 10,000}{100} = \dfrac{10,260}{100} = 102.6$

The cost per disc for pressing 100 compact discs is $102.60.

b. $C(1000) = \dfrac{2.6(1000) + 10,000}{1000} = \dfrac{12,600}{1000} = 12.6$

The cost per disc for pressing 1000 compact discs is $12.60. Notice that as more compact discs are produced, the cost per disc decreases.

Practice Problem 16

A company's cost per book for printing x particular books is given by the rational function $C(x) = \dfrac{0.8x + 5000}{x}$.

Find the cost per book for printing:

a.   100 books

b.   1000 books

Answers

16. a. $50.80,   **b.** $5.80

STUDY SKILLS REMINDER

Are you satisfied with your performance in this course thus far?

If not, ask yourself the following questions:

- Am I attending all class periods and arriving on time?
- Am I working and checking my homework assignments?
- Am I getting help when I need it?
- In addition to my instructor, am I using the supplements to this text that could help me? For example, the tutorial video lessons? Math-Pro, the tutorial software?
- Am I satisfied with my performance on quizzes and tests?

If you answered no to *any* of these questions, read or reread Section 1.1 for suggestions in these areas. Also, you may want to contact your instructor for additional feedback.

Recall that since the rational expression $\dfrac{7x - 2}{(x - 2)(x + 5)}$ is not defined when $x = 2$ or when $x = -5$, we say that the domain of the rational function $f(x) = \dfrac{7x - 2}{(x - 2)(x + 5)}$ is all real numbers except 2 and -5. This domain can be written as $\{x \mid x \text{ is a real number and } x \neq 2, x \neq -5\}$. This means that the graph of f should not cross the vertical lines $x = 2$ and $x = -5$. The graph of f in *connected* mode is shown below. In connected mode the grapher tries to connect all dots of the graph so that the result is a smooth curve. This is what has happened in the graph. Notice that the graph appears to contain vertical lines at $x = 2$ and at $x = -5$. We know that this cannot happen because the function is not defined at $x = 2$ and at $x = -5$. We also know that this cannot happen because the graph of this function would not pass the vertical line test.

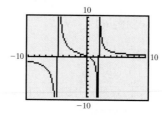

If we graph f in *dot* mode, the graph appears as below on the left. In dot mode the grapher will not connect dots with a smooth curve. Notice that the vertical lines have disappeared, and we have a better picture of the graph. It actually appears more like the hand-drawn graph to its right. By using a TABLE feature, a CALCULATE VALUE feature, or by tracing, we can see that the function is not defined at $x = 2$ and at $x = -5$.

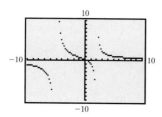

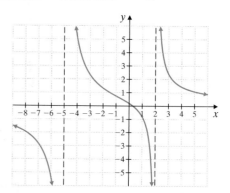

Find the domain of each rational function. Then graph each rational function and use the graph to confirm the domain.

1. $f(x) = \dfrac{5x}{x - 6}$

2. $f(x) = \dfrac{x}{x + 4}$

3. $f(x) = \dfrac{x + 1}{x^2 - 4}$

4. $g(x) = \dfrac{5x}{x^2 - 9}$

5. $h(x) = \dfrac{x^2}{2x^2 + 7x - 4}$

6. $f(x) = \dfrac{3x + 2}{4x^2 - 19x - 5}$

7. $g(x) = \dfrac{x^2 + x + 1}{5}$

8. $h(x) = \dfrac{x^2 + 25}{2}$

Mental Math

Multiply.

1. $\dfrac{x}{5} \cdot \dfrac{y}{2}$ 2. $\dfrac{y}{6} \cdot \dfrac{z}{5}$ 3. $\dfrac{2}{x} \cdot \dfrac{y}{3}$ 4. $\dfrac{a}{5} \cdot \dfrac{7}{b}$ 5. $\dfrac{m}{6} \cdot \dfrac{m}{6}$ 6. $\dfrac{9}{x} \cdot \dfrac{8}{x}$

EXERCISE SET 6.1

A *Find all numbers for which each rational expression is undefined. See Examples 1 through 3.*

1. $\dfrac{x+3}{x-2}$ 2. $\dfrac{x+5}{x-1}$ 3. $\dfrac{2x}{5x+1}$ 4. $\dfrac{5x}{7x+2}$

5. $\dfrac{x^2+1}{3x}$ 6. $\dfrac{x^2+7}{5x}$ 7. $\dfrac{x-7}{4}$ 8. $\dfrac{4-3x}{2}$

9. $\dfrac{3+2x}{x^3+x^2-2x}$ 10. $\dfrac{5-3x}{2x^3-14x^2+20x}$ 11. $\dfrac{x+3}{x^2-4}$ 12. $\dfrac{5}{x^2-7x}$

B *Simplify each rational expression. See Examples 4 through 8.*

13. $\dfrac{10x^3}{18x}$ 14. $-\dfrac{48a^7}{16a^{10}}$ 15. $\dfrac{9x^6y^3}{18x^2y^5}$ 16. $\dfrac{10ab^5}{15a^3b^5}$ 17. $\dfrac{8x-16x^2}{8x}$

18. $\dfrac{3x-6x^2}{3x}$ 19. $\dfrac{x^2-9}{x-3}$ 20. $\dfrac{x^2-25}{x+5}$ 21. $\dfrac{9y-18}{7y-14}$ 22. $\dfrac{6y-18}{2y-6}$

23. $\dfrac{x^2+6x-40}{x+10}$ 24. $\dfrac{x^2-8x+16}{x-4}$ 25. $\dfrac{x-9}{9-x}$ 26. $\dfrac{x-4}{4-x}$

27. $\dfrac{x^2-49}{7-x}$ 28. $\dfrac{x^2-y^2}{y-x}$ 29. $\dfrac{2x^2-7x-4}{x^2-5x+4}$ 30. $\dfrac{3x^2-11x+10}{x^2-7x+10}$

Simplify each rational expression. See Examples 9 and 10.

31. $\dfrac{x^3 - 125}{2x - 10}$

32. $\dfrac{4x + 4}{x^3 + 1}$

33. $\dfrac{3x^2 - 5x - 2}{6x^3 + 2x^2 + 3x + 1}$

34. $\dfrac{2x^2 - x - 3}{2x^3 - 3x^2 + 2x - 3}$

35. $\dfrac{9x^2 - 15x + 25}{27x^3 + 125}$

36. $\dfrac{8x^3 - 27}{4x^2 + 6x + 9}$

C *Multiply and simplify. See Examples 11 and 12.*

37. $\dfrac{4}{x} \cdot \dfrac{x^2}{8}$

38. $\dfrac{x}{3} \cdot \dfrac{9}{x^3}$

39. $\dfrac{2x - 4}{15} \cdot \dfrac{6}{2 - x}$

40. $\dfrac{10 - 2x}{7} \cdot \dfrac{14}{5x - 25}$

41. $\dfrac{18a - 12a^2}{4a^2 + 4a + 1} \cdot \dfrac{4a^2 + 8a + 3}{4a^2 - 9}$

42. $\dfrac{a - 5b}{a^2 + ab} \cdot \dfrac{b^2 - a^2}{10b - 2a}$

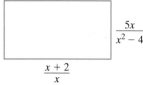 **43.** $\dfrac{9x + 9}{4x + 8} \cdot \dfrac{2x + 4}{3x^2 - 3}$

44. $\dfrac{2x^2 - 2}{10x + 30} \cdot \dfrac{12x + 36}{3x - 3}$

45. $\dfrac{2x^3 - 16}{6x^2 + 6x - 36} \cdot \dfrac{9x + 18}{3x^2 + 6x + 12}$

46. $\dfrac{x^2 - 3x + 9}{5x^2 - 20x - 105} \cdot \dfrac{x^2 - 49}{x^3 + 27}$

47. $\dfrac{a^3 + a^2b + a + b}{5a^3 + 5a} \cdot \dfrac{6a^2}{2a^2 - 2b^2}$

48. $\dfrac{4a^2 - 8a}{ab - 2b + 3a - 6} \cdot \dfrac{8b + 24}{3a + 6}$

49. $\dfrac{x^2 - 6x - 16}{2x^2 - 128} \cdot \dfrac{x^2 + 16x + 64}{3x^2 + 30x + 48}$

50. $\dfrac{2x^2 + 12x - 32}{x^2 + 16x + 64} \cdot \dfrac{x^2 + 10x + 16}{x^2 - 3x - 10}$

△ **51.** Find the area of the rectangle.

$\dfrac{5x}{x^2 - 4}$

$\dfrac{x + 2}{x}$

△ **52.** Find the area of the triangle.

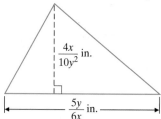

$\dfrac{4x}{10y^2}$ in.

$\dfrac{5y}{6x}$ in.

D *Divide and simplify. See Examples 13 and 14.*

53. $\dfrac{4}{x} \div \dfrac{8}{x^2}$

54. $\dfrac{x}{3} \div \dfrac{x^3}{9}$

55. $\dfrac{2x}{5} \div \dfrac{6x + 12}{5x + 10}$

56. $\dfrac{7}{3x} \div \dfrac{14 - 7x}{18 - 9x}$

57. $\dfrac{a+b}{ab} \div \dfrac{a^2-b^2}{4a^3b}$ **58.** $\dfrac{6a^2b^2}{a^2-4} \div \dfrac{3ab^2}{a-2}$ **59.** $\dfrac{x^2-6x+9}{x^2-x-6} \div \dfrac{x^2-9}{4}$ **60.** $\dfrac{x^2-4}{3x+6} \div \dfrac{2x^2-8x+8}{x^2+4x+4}$

61. $\dfrac{x^2-6x-16}{2x^2-128} \div \dfrac{x^2+10x+16}{x^2+16x+64}$ **62.** $\dfrac{a^2-a-6}{a^2-81} \div \dfrac{a^2-7a-18}{4a+36}$ **63.** $\dfrac{3x-x^2}{x^3-27} \div \dfrac{x}{x^2+3x+9}$

64. $\dfrac{x^2-3x}{x^3-27} \div \dfrac{2x}{2x^2+6x+18}$ **65.** $\dfrac{8b+24}{3a+6} \div \dfrac{ab-2b+3a-6}{a^2-4a+4}$ **66.** $\dfrac{2a^2-2b^2}{a^3+a^2b+a+b} \div \dfrac{6a^2}{a^3+a}$

Perform each indicated operation. See Example 15.

67. $\dfrac{4}{x} \div \dfrac{3xy}{x^2} \cdot \dfrac{6x^2}{x^4}$

68. $\dfrac{4}{x} \cdot \dfrac{3xy}{x^2} \div \dfrac{6x^2}{x^4}$

69. $\dfrac{3x^2-5x-2}{y^2+y-2} \cdot \dfrac{y^2+4y-5}{12x^2+7x+1} \div \dfrac{5x^2-9x-2}{8x^2-2x-1}$

70. $\dfrac{x^2+x-2}{3y^2-5y-2} \cdot \dfrac{12y^2+y-1}{x^2+4x-5} \div \dfrac{8y^2-6y+1}{5y^2-9y-2}$

E *Find each function value. See Example 16.*

71. If $f(x) = \dfrac{x+8}{2x-1}$, find $f(2)$, $f(0)$, and $f(-1)$.

72. If $f(x) = \dfrac{x-2}{-5+x}$, find $f(-5)$, $f(0)$, and $f(10)$.

73. The total revenue from the sale of a popular book is approximated by the rational function
$R(x) = \dfrac{1000x^2}{x^2+4}$, where x is the number of years
since publication and $R(x)$ is the total revenue in millions of dollars.

  a. Find the total revenue at the end of the first year.
  b. Find the total revenue at the end of the second year.
  c. Find the revenue during the second year only.
  d. Find the domain of function R.

74. The function $f(x) = \dfrac{100{,}000x}{100-x}$ models the cost in
dollars for removing x percent of the pollutants from a bayou in which a nearby company dumped creosol.

  a. Find the cost of removing 20% of the pollutants from the bayou. [*Hint:* Find $f(20)$.]
  b. Find the cost of removing 60% of the pollutants and then 80% of the pollutants.
  c. Find $f(90)$, then $f(95)$, and then $f(99)$. What happens to the cost as x approaches 100%?
  d. Find the domain of function f.

Review and Preview

Perform each indicated operation. See Section 1.4.

75. $\dfrac{4}{5} + \dfrac{3}{5}$

76. $\dfrac{4}{10} - \dfrac{7}{10}$

77. $\dfrac{5}{28} - \dfrac{2}{21}$

78. $\dfrac{5}{13} + \dfrac{2}{7}$

79. $\dfrac{3}{8} + \dfrac{1}{2} - \dfrac{3}{16}$

80. $\dfrac{2}{9} - \dfrac{1}{6} + \dfrac{2}{3}$

 ## Combining Concepts

△ **81.** A parallelogram has an area of $\dfrac{x^2 + x - 2}{x^3}$ square feet and a height of $\dfrac{x^2}{x - 1}$ feet. Express the length of its base as a rational expression in x. (*Hint:* Since $A = b \cdot h$, then $b = \dfrac{A}{h}$ or $b = A \div h$.)

82. A lottery prize of $\dfrac{15x^3}{y^2}$ dollars is to be divided among $5x$ people. Express the amount of money each person is to receive as a rational expression in x and y.

83. In your own words explain how to simplify a rational expression.

84. In your own words, explain the difference between multiplying rational expressions and dividing rational expressions.

85. Decide whether each rational expression equals 1, −1, or neither.

a. $\dfrac{x + 5}{5 + x}$

b. $\dfrac{x - 5}{5 - x}$

c. $\dfrac{x + 5}{x - 5}$

d. $\dfrac{-x - 5}{x + 5}$

e. $\dfrac{x - 5}{-x + 5}$

f. $\dfrac{-5 + x}{x - 5}$

86. In our definition of division for

$$\dfrac{P}{Q} \div \dfrac{R}{S}$$

we stated that $Q \neq 0$, $S \neq 0$, and $R \neq 0$. Explain why R cannot equal 0.

87. Find the polynomial in the second numerator such that the following statement is true.

$$\dfrac{x^2 - 4}{x^2 - 7x + 10} \cdot \dfrac{?}{2x^2 + 11x + 14} = 1$$

Simplify. Assume that no denominator is 0.

88. $\dfrac{p^x - 4}{4 - p^x}$

89. $\dfrac{3 + q^n}{q^n + 3}$

90. $\dfrac{x^n + 4}{x^{2n} - 16}$

91. $\dfrac{x^{2k} - 9}{3 + x^k}$

6.2 Adding and Subtracting Rational Expressions

(A) Adding or Subtracting Rational Expressions with the Same Denominator

We add or subtract rational expressions just as we add or subtract fractions.

> **Adding or Subtracting Rational Expressions with the Same Denominator**
>
> To add or subtract rational expressions with the same denominator, add or subtract the numerators. Write the result over the common denominator.

EXAMPLES   Add.

1. $\dfrac{5}{7} + \dfrac{x}{7} = \dfrac{5+x}{7}$ Add the numerators and write the result over the common denominator.

2. $\dfrac{x}{4} + \dfrac{5x}{4} = \dfrac{x+5x}{4} = \dfrac{6x}{4} = \dfrac{3x}{2}$

EXAMPLES   Subtract.

3. $\dfrac{x^2}{x+7} - \dfrac{49}{x+7} = \dfrac{x^2-49}{x+7}$ Subtract the numerators and write the result over the common denominator.

$ = \dfrac{(x+7)(x-7)}{x+7}$ Factor the numerator.

$ = x - 7$ Simplify.

Helpful Hint

Be sure to insert parentheses here so that the entire numerator is subtracted.

4. $\dfrac{x}{3y^2} - \dfrac{x+1}{3y^2} = \dfrac{x-(x+1)}{3y^2}$ Subtract the numerators.

$ = \dfrac{x-x-1}{3y^2}$ Use the distributive property.

$ = -\dfrac{1}{3y^2}$ Simplify.

Try the Concept Check in the margin.

(B) Finding the LCD of Rational Expressions

To add or subtract rational expressions with unlike, or different, denominators, we first write the rational expressions as equivalent rational expressions with common denominators.

The **least common denominator (LCD)** is usually the easiest common denominator to work with.

Practice Problems 1–2

Add.

1. $\dfrac{9}{11} + \dfrac{y}{11}$ 2. $\dfrac{x}{6} + \dfrac{7x}{6}$

Practice Problems 3–4

Subtract.

3. $\dfrac{x^2}{x+3} - \dfrac{9}{x+3}$ 4. $\dfrac{a}{5b^3} - \dfrac{a+2}{5b^3}$

Concept Check

Find and correct the error.
$$\dfrac{3+2y}{y^2-1} - \dfrac{y+3}{y^2-1}$$
$$= \dfrac{3+2y-y+3}{y^2-1}$$
$$= \dfrac{y+6}{y^2-1}$$

Answers

1. $\dfrac{9+y}{11}$,   **2.** $\dfrac{4x}{3}$,   **3.** $x-3$,   **4.** $-\dfrac{2}{5b^3}$

Concept Check:
$$\dfrac{3+2y}{y^2-1} - \dfrac{y+3}{y^2-1} = \dfrac{3+2y-y-3}{y^2-1} = \dfrac{y}{y^2-1}$$

The following steps can be used to find the LCD.

Finding the Least Common Denominator (LCD)

Step 1. Factor each denominator completely.

Step 2. The LCD is the product of all unique factors each raised to the greatest power that appears in any factored denominator.

Practice Problem 5

Find the LCD of the rational expressions in each list.

a. $\dfrac{7}{20a^2b^3}, \dfrac{9}{15ab^4}$

b. $\dfrac{6x}{x-2}, \dfrac{5}{x+2}$

c. $\dfrac{x+4}{x^2-36}, \dfrac{x}{x^2+12x+36}, \dfrac{x^3}{3x^2+19x+6}$

d. $\dfrac{6}{x^2-1}, \dfrac{7}{2-2x}$

EXAMPLE 5   Find the LCD of the rational expressions in each list.

a. $\dfrac{2}{3x^5y^2}, \dfrac{3z}{5xy^3}$

b. $\dfrac{7}{z+1}, \dfrac{z}{z-1}$

c. $\dfrac{m-1}{m^2-25}, \dfrac{2m}{2m^2-9m-5}, \dfrac{7}{m^2-10m+25}$

d. $\dfrac{x}{x^2-4}, \dfrac{11}{6-3x}$

Solution:

a. First we factor each denominator.

$$3x^5y^2 = 3 \cdot x^5 \cdot y^2$$
$$5xy^3 = 5 \cdot x \cdot y^3$$
$$\text{LCD} = 3 \cdot 5 \cdot x^5 \cdot y^3 = 15x^5y^3$$

> **Helpful Hint**
>
> The greatest power of x is 5, so we have a factor of x^5. The greatest power of y is 3, so we have a factor of y^3.

b. The denominators $z + 1$ and $z - 1$ do not factor further.

$$(z + 1) = (z + 1)$$
$$(z - 1) = (z - 1)$$
$$\text{LCD} = (z + 1)(z - 1)$$

c. We first factor each denominator.

$$m^2 - 25 = (m + 5)(m - 5)$$
$$2m^2 - 9m - 5 = (2m + 1)(m - 5)$$
$$m^2 - 10m + 25 = (m - 5)(m - 5)$$
$$\text{LCD} = (m + 5)(2m + 1)(m - 5)^2$$

d. We factor each denominator.

$$x^2 - 4 = (x + 2)(x - 2)$$
$$6 - 3x = 3(2 - x) = 3(-1)(x - 2)$$
$$\text{LCD} = 3(-1)(x + 2)(x - 2)$$
$$= -3(x + 2)(x - 2)$$

> **Helpful Hint**
>
> $(x - 2)$ and $(2 - x)$ are opposite factors. Notice that a -1 was factored from $(2 - x)$ so that the factors are identical.

Answers

5. a. $60a^2b^4$,   **b.** $(x - 2)(x + 2)$,
c. $(x - 6)(3x + 1)(x + 6)^2$,
d. $-2(x + 1)(x - 1)$

Helpful Hint

If opposite factors occur, do not use both in the LCD. Instead, factor -1 from one of the opposite factors so that the factors are then identical.

Ⓒ Adding or Subtracting Rational Expressions with Different Denominators

To add or subtract rational expressions with different denominators, we write each rational expression as an equivalent rational expression with the LCD as the denominator. To do this, we use the multiplication property of 1 and multiply the numerator and the denominator by the same factor so that the denominator becomes the LCD.

Adding or Subtracting Rational Expressions with Different Denominators

Step 1. Find the LCD of the rational expressions.

Step 2. Write each rational expression as an equivalent rational expression whose denominator is the LCD found in Step 1.

Step 3. Add or subtract numerators, and write the result over the common denominator.

Step 4. Simplify the resulting rational expression.

EXAMPLE 6   Add: $\dfrac{2}{x^2} + \dfrac{5}{3x^3}$

Solution: The LCD is $3x^3$, so we write each rational expression as an equivalent rational expression with denominator $3x^3$.

$$\frac{2}{x^2} + \frac{5}{3x^3} = \frac{2 \cdot 3x}{x^2 \cdot 3x} + \frac{5}{3x^3} \quad \text{The second expression already has a denominator of } 3x^3.$$

$$= \frac{6x}{3x^3} + \frac{5}{3x^3}$$

$$= \frac{6x + 5}{3x^3} \quad \text{Add the numerators.}$$

EXAMPLE 7   Add: $\dfrac{3}{x + 2} + \dfrac{2x}{x - 2}$

Solution: The LCD is the product of the two denominators: $(x + 2)(x - 2)$.

$$\frac{3}{x + 2} + \frac{2x}{x - 2} = \frac{3 \cdot (x - 2)}{(x + 2) \cdot (x - 2)} + \frac{2x \cdot (x + 2)}{(x - 2) \cdot (x + 2)} \quad \begin{array}{l}\text{Write equiva-}\\\text{lent rational}\\\text{expressions.}\end{array}$$

$$= \frac{3(x - 2) + 2x(x + 2)}{(x + 2)(x - 2)} \quad \text{Add the numerators.}$$

$$= \frac{3x - 6 + 2x^2 + 4x}{(x + 2)(x - 2)} \quad \text{Multiply in the numerator.}$$

$$= \frac{2x^2 + 7x - 6}{(x + 2)(x - 2)} \quad \text{Simplify.}$$

Practice Problem 6

Add: $\dfrac{7}{a^3} + \dfrac{9}{2a^4}$

Practice Problem 7

Add: $\dfrac{1}{x + 5} + \dfrac{6x}{x - 5}$

Answers

6. $\dfrac{14a + 9}{2a^4}$,   **7.** $\dfrac{6x^2 + 31x - 5}{(x + 5)(x - 5)}$

Practice Problem 8

Subtract: $\dfrac{m}{m-6} - \dfrac{8}{6-m}$

EXAMPLE 8 Subtract: $\dfrac{2x-6}{x-1} - \dfrac{4}{1-x}$

Solution: The LCD is either $x-1$ or $1-x$. To get a common denominator of $x-1$, we factor -1 from the denominator of the second rational expression.

$$\frac{2x-6}{x-1} - \frac{4}{1-x} = \frac{2x-6}{x-1} - \frac{4}{-1(x-1)} \qquad \text{Write } 1-x \text{ as } -1(x-1).$$

$$= \frac{2x-6}{x-1} - \frac{-1\cdot 4}{x-1} \qquad \text{Write } \frac{4}{-1(x-1)} \text{ as } \frac{-1\cdot 4}{x-1}.$$

$$= \frac{2x-6-(-4)}{x-1}$$

$$= \frac{2x-6+4}{x-1} \qquad \text{Simplify.}$$

$$= \frac{2x-2}{x-1}$$

$$= \frac{2(x-1)}{x-1}$$

$$= 2$$

Practice Problem 9

Subtract: $\dfrac{2x}{x^2-9} - \dfrac{3}{x^2-4x+3}$

EXAMPLE 9 Subtract: $\dfrac{5k}{k^2-4} - \dfrac{2}{k^2+k-2}$

Solution: Factor each denominator to find the LCD.

$$\frac{5k}{k^2-4} - \frac{2}{k^2+k-2} = \frac{5k}{(k+2)(k-2)} - \frac{2}{(k+2)(k-1)}$$

The LCD is $(k+2)(k-2)(k-1)$. We write equivalent rational expressions with the LCD as the denominators.

$$\frac{5k}{(k+2)(k-2)} - \frac{2}{(k+2)(k-1)}$$

$$= \frac{5k\cdot(k-1)}{(k+2)(k-2)\cdot(k-1)} - \frac{2\cdot(k-2)}{(k+2)(k-1)\cdot(k-2)}$$

$$= \frac{5k(k-1) - 2(k-2)}{(k+2)(k-2)(k-1)} \qquad \text{Subtract the numerators.}$$

$$= \frac{5k^2-5k-2k+4}{(k+2)(k-2)(k-1)} \qquad \text{Multiply in the numerator.}$$

$$= \frac{5k^2-7k+4}{(k+2)(k-2)(k-1)} \qquad \text{Simplify.}$$

Answers

8. $\dfrac{m+8}{m-6}$, **9.** $\dfrac{2x^2-5x-9}{(x+3)(x-3)(x-1)}$

EXAMPLE 10   Add: $\dfrac{2x-1}{2x^2-9x-5}+\dfrac{x+3}{6x^2-x-2}$ Factor the denominators.

Solution:

$$\dfrac{2x-1}{2x^2-9x-5}+\dfrac{x+3}{6x^2-x-2}=\dfrac{2x-1}{(2x+1)(x-5)}+\dfrac{x+3}{(2x+1)(3x-2)}$$

The LCD is $(2x+1)(x-5)(3x-2)$.

$$=\dfrac{(2x-1)\cdot(3x-2)}{(2x+1)(x-5)\cdot(3x-2)}+\dfrac{(x+3)\cdot(x-5)}{(2x+1)(3x-2)\cdot(x-5)}$$

$$=\dfrac{(2x-1)(3x-2)+(x+3)(x-5)}{(2x+1)(x-5)(3x-2)}\quad\text{Add the numerators.}$$

$$=\dfrac{6x^2-7x+2+x^2-2x-15}{(2x+1)(x-5)(3x-2)}\quad\text{Multiply in the numerator.}$$

$$=\dfrac{7x^2-9x-13}{(2x+1)(x-5)(3x-2)}\quad\text{Simplify.}$$

EXAMPLE 11   Perform each indicated operation:

$$\dfrac{7}{x-1}+\dfrac{10x}{x^2-1}-\dfrac{5}{x+1}$$

Solution:

$$\dfrac{7}{x-1}+\dfrac{10x}{x^2-1}-\dfrac{5}{x+1}=\dfrac{7}{x-1}+\dfrac{10x}{(x-1)(x+1)}-\dfrac{5}{x+1}$$

The LCD is $(x-1)(x+1)$.

$$=\dfrac{7\cdot(x+1)}{(x-1)\cdot(x+1)}+\dfrac{10x}{(x-1)\cdot(x+1)}-\dfrac{5\cdot(x-1)}{(x+1)\cdot(x-1)}$$

$$=\dfrac{7(x+1)+10x-5(x-1)}{(x-1)(x+1)}\quad\text{Add and subtract the numerators.}$$

$$=\dfrac{7x+7+10x-5x+5}{(x-1)(x+1)}\quad\text{Multiply in the numerator.}$$

$$=\dfrac{12x+12}{(x-1)(x+1)}\quad\text{Simplify.}$$

$$=\dfrac{12(x+1)}{(x-1)(x+1)}\quad\text{Factor the numerator.}$$

$$=\dfrac{12}{x-1}\quad\text{Divide out common factors.}$$

Practice Problem 10

Add: $\dfrac{x+1}{x^2+x-12}+\dfrac{2x-1}{x^2+6x+8}$

Practice Problem 11

Perform each indicated operation.

$$\dfrac{6}{x-5}+\dfrac{x-35}{x^2-5x}-\dfrac{2}{x}$$

GRAPHING CALCULATOR EXPLORATIONS

A grapher can be used to support the results of operations on rational expressions. For example, to verify the result of Example 7, graph

$$Y_1=\dfrac{3}{x+2}+\dfrac{2x}{x-2}\quad\text{and}\quad Y_2=\dfrac{2x^2+7x-6}{(x+2)(x-2)}$$

on the same set of axes. The graphs should be the same. Use a TABLE feature or a TRACE feature to see that this is true.

Answers

10. $\dfrac{3x^2-4x+5}{(x+2)(x-3)(x+4)}$,   **11.** $\dfrac{5}{x}$

FOCUS ON **Business and Career**

FASTEST-GROWING OCCUPATIONS

According to U.S. Bureau of Labor Statistics projections, the careers listed below will be the top ten fastest-growing jobs into the next century, according to the percent increase in the number of jobs.

Employment (in thousands)

Occupation	1998	2008	% Change
Computer engineers	299	622	108%
Computer support specialists	429	869	
Database administrators	87	155	
Desktop publishing specialists	26	44	
Medical assistants	252	398	
Paralegals and legal assistants	136	220	
Personal care and home health aides	746	1179	
Physician assistants	66	98	
Social and human service assistants	268	410	
Systems analysts	617	1194	

(*Source:* U.S. Bureau of Labor Statistics)

What do all of these fast-growing occupations have in common? They all require a knowledge of math! For some careers, such as desktop publishing specialists, medical assistants, and computer engineers, the ways math is used on the job may be obvious. For other occupations, the use of math may not be quite as apparent. However, tasks common to many jobs like filling in a time sheet, writing up an expense or mileage report, planning a budget, figuring a bill, ordering supplies, and even making a work schedule all require math.

GROUP ACTIVITY

1. Find the percent change in the number of jobs available from 1998 to 2008 for each occupation in the list.
2. Rank these top-ten occupations according to percent growth, from greatest to least.
3. Which occupation will be the fastest growing during this period?
4. How many occupations will have more than double the number of positions in 2008 than in 1998?
5. Which of the listed occupations will be the slowest growing during this period?

Name _____ Section _____ Date _____

EXERCISE SET 6.2

A *Add or subtract as indicated. Simplify each answer. See Examples 1 through 4.*

1. $\dfrac{2}{x} - \dfrac{5}{x}$

2. $\dfrac{4}{x^2} + \dfrac{2}{x^2}$

 3. $\dfrac{2}{x-2} + \dfrac{x}{x-2}$

4. $\dfrac{x}{5-x} + \dfrac{7}{5-x}$

5. $\dfrac{x^2}{x+2} - \dfrac{4}{x+2}$

6. $\dfrac{x^2}{x+6} - \dfrac{36}{x+6}$

7. $\dfrac{2x-6}{x^2+x-6} + \dfrac{3-3x}{x^2+x-6}$

8. $\dfrac{5x+2}{x^2+2x-8} + \dfrac{2-4x}{x^2+2x-8}$

9. $\dfrac{x-5}{2x} - \dfrac{x+5}{2x}$

10. $\dfrac{x+4}{4x} - \dfrac{x-4}{4x}$

△ **11.** Find the perimeter of the square.

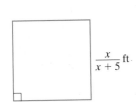

$\dfrac{x}{x+5}$ ft.

△ **12.** Find the perimeter of the quadrilateral.

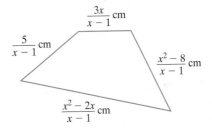

$\dfrac{3x}{x-1}$ cm

$\dfrac{5}{x-1}$ cm

$\dfrac{x^2-8}{x-1}$ cm

$\dfrac{x^2-2x}{x-1}$ cm

B *Find the LCD of the rational expressions in each list. See Example 5.*

13. $\dfrac{2}{7}, \dfrac{3}{5x}$

14. $\dfrac{4}{5y}, \dfrac{3}{4y^2}$

15. $\dfrac{3}{x}, \dfrac{2}{x+1}$

16. $\dfrac{5}{2x}, \dfrac{7}{2+x}$

17. $\dfrac{12}{x+7}, \dfrac{8}{x-7}$

18. $\dfrac{1}{2x-1}, \dfrac{x}{2x+1}$

19. $\dfrac{5}{3x+6}, \dfrac{2x}{2x-4}$

20. $\dfrac{2}{3a+9}, \dfrac{5}{5a-15}$

21. $\dfrac{2a}{a^2-b^2}, \dfrac{1}{a^2-2ab+b^2}$

22. $\dfrac{2a}{a^2 + 8a + 16}, \dfrac{7a}{a^2 + a - 12}$

23. $\dfrac{x}{x^2 - 9}, \dfrac{5}{x}, \dfrac{7}{12 - 4x}$

24. $\dfrac{9}{x^2 - 25}, \dfrac{1}{50 - 10x}, \dfrac{6}{x}$

25. When is the LCD of two rational expressions equal to the product of their denominators? (*Hint:* What is the LCD of $\dfrac{1}{x}$ and $\dfrac{7}{x + 5}$?)

26. When is the LCD of two rational expressions with different denominators equal to one of the denominators? (*Hint:* What is the LCD of $\dfrac{3x}{x + 2}$ and $\dfrac{7x + 1}{(x + 2)^3}$?)

C *Add or subtract as indicated. Simplify each answer. See Examples 6 through 10.*

27. $\dfrac{4}{3x} + \dfrac{3}{2x}$

28. $\dfrac{10}{7x} - \dfrac{5}{2x}$

29. $\dfrac{5}{2y^2} - \dfrac{2}{7y}$

30. $\dfrac{4}{11x^4} - \dfrac{1}{4x^2}$

31. $\dfrac{x - 3}{x + 4} - \dfrac{x + 2}{x - 4}$

32. $\dfrac{x - 1}{x - 5} - \dfrac{x + 2}{x + 5}$

33. $\dfrac{1}{x - 5} + \dfrac{2x - 19}{x^2 - x - 20}$

34. $\dfrac{4x - 2}{x^2 - x - 20} - \dfrac{2}{x + 4}$

35. $\dfrac{3}{2x + 10} + \dfrac{8}{3x + 15}$

36. $\dfrac{10}{3x - 3} + \dfrac{1}{7x - 7}$

37. $\dfrac{-2}{x^2 - 3x} - \dfrac{1}{x^3 - 3x^2}$

38. $\dfrac{-3}{2a + 8} - \dfrac{8}{a^2 + 4a}$

39. $\dfrac{1}{a - b} + \dfrac{1}{b - a}$

40. $\dfrac{1}{a - 3} - \dfrac{1}{3 - a}$

41. $\dfrac{5}{x - 2} + \dfrac{x + 4}{2 - x}$

42. $\dfrac{3}{5 - x} + \dfrac{x + 2}{x - 5}$

43. $\dfrac{y + 1}{y^2 - 6y + 8} - \dfrac{3}{y^2 - 16}$

44. $\dfrac{x + 2}{x^2 - 36} - \dfrac{x}{x^2 + 9x + 18}$

45. $\dfrac{7}{x^2 - x - 2} + \dfrac{x}{x^2 + 4x + 3}$

46. $\dfrac{a}{a^2 + 10a + 25} + \dfrac{4}{a^2 + 6a + 5}$

47. $\dfrac{x + 4}{3x^2 + 11x + 6} + \dfrac{x}{2x^2 + x - 15}$

48. $\dfrac{x + 3}{5x^2 + 12x + 4} + \dfrac{6}{x^2 - x - 6}$

49. $\dfrac{2}{a^2 + 2a + 1} + \dfrac{3}{a^2 - 1}$

50. $\dfrac{9x + 2}{3x^2 - 2x - 8} + \dfrac{7}{3x^2 + x - 4}$

51. $\dfrac{ab}{a^2 - b^2} + \dfrac{b}{2a + 2b}$

52. $\dfrac{2}{3x - 15} + \dfrac{x}{25 - x^2}$

53. $\dfrac{5}{x^2 - 4} - \dfrac{4}{x^2 + 4x + 4}$

54. $\dfrac{3z}{z^2 - 9} - \dfrac{2}{3 - z}$

Perform each indicated operation. Simplify each answer. See Example 11.

55. $\dfrac{2}{x + 1} - \dfrac{3x}{3x + 3} + \dfrac{1}{2x + 2}$

56. $\dfrac{5}{3x - 6} - \dfrac{x}{x - 2} + \dfrac{3 + 2x}{5x - 10}$

57. $\dfrac{3}{x + 3} + \dfrac{5}{x^2 + 6x + 9} - \dfrac{x}{x^2 - 9}$

58. $\dfrac{x + 2}{x^2 - 2x - 3} + \dfrac{x}{x - 3} - \dfrac{4}{x + 1}$

59. $\dfrac{x}{x^2 - 9} + \dfrac{3}{x^2 - 6x + 9} - \dfrac{1}{x + 3}$

60. $\dfrac{3}{x^2 - 9} - \dfrac{x}{x^2 - 6x + 9} + \dfrac{1}{x + 3}$

61. $\left(\dfrac{1}{x} + \dfrac{2}{3} \right) - \left(\dfrac{1}{x} - \dfrac{2}{3} \right)$

62. $\left(\dfrac{1}{2} + \dfrac{2}{x} \right) - \left(\dfrac{1}{2} - \dfrac{1}{x} \right)$

Review and Preview

Use the distributive property to multiply each expression. See Section 1.3.

63. $12\left(\dfrac{2}{3} + \dfrac{1}{6}\right)$

64. $14\left(\dfrac{1}{7} + \dfrac{3}{14}\right)$

65. $x^2\left(\dfrac{4}{x^2} + 1\right)$

66. $5y^2\left(\dfrac{1}{y^2} - \dfrac{1}{5}\right)$

Combining Concepts

67. In your own words, explain how to add rational expressions with different denominators.

68. In your own words, explain how to multiply rational expressions.

69. In your own words, explain how to divide rational expressions.

70. In your own words, explain how to subtract rational expressions with different denominators.

Perform the indicated operation. If possible, simplify your answer.

71. $\left(\dfrac{2}{3} - \dfrac{1}{x}\right) \cdot \left(\dfrac{3}{x} + \dfrac{1}{2}\right)$

72. $\left(\dfrac{2}{3} - \dfrac{1}{x}\right) \div \left(\dfrac{3}{x} + \dfrac{1}{2}\right)$

73. $\left(\dfrac{2a}{3}\right)^2 \div \left(\dfrac{a^2}{a+1} - \dfrac{1}{a+1}\right)$

74. $\left(\dfrac{x+2}{2x} - \dfrac{x-2}{2x}\right) \cdot \left(\dfrac{5x}{4}\right)^2$

75. $\left(\dfrac{2x}{3}\right)^2 \div \left(\dfrac{x}{3}\right)^2$

76. $\left(\dfrac{2x}{3}\right)^2 \cdot \left(\dfrac{3}{x}\right)^2$

77. $\left(\dfrac{x}{x+1} - \dfrac{x}{x-1}\right) \div \dfrac{x}{2x+2}$

78. $\dfrac{x}{2x+2} \div \left(\dfrac{x}{x+1} + \dfrac{x}{x-1}\right)$

79. $\dfrac{4}{x} \cdot \left(\dfrac{2}{x+2} - \dfrac{2}{x-2}\right)$

80. $\dfrac{1}{x+1} \cdot \left(\dfrac{5}{x} + \dfrac{2}{x-3}\right)$

Perform each indicated operation. (Hint: First write each expression with positive exponents.)

81. $x^{-1} + (2x)^{-1}$

82. $y^{-1} + (4y)^{-1}$

83. $4x^{-2} - 3x^{-1}$

84. $(4x)^{-2} - (3x)^{-1}$

Use a graphing calculator to support the results of each exercise.

85. Exercise 3

86. Exercise 4

444

6.3 Simplifying Complex Fractions

A rational expression whose numerator, denominator, or both contain one or more rational expressions is called a **complex rational expression** or a **complex fraction**. Examples are

$$\frac{\dfrac{1}{a}}{\dfrac{b}{2}} \qquad \frac{\dfrac{x}{2y^2}}{\dfrac{6x-2}{9y}} \qquad \frac{x+\dfrac{1}{y}}{y+1}$$

The parts of a complex fraction are

$$\left.\begin{array}{c}\dfrac{x}{y+2}\\[2mm]\end{array}\right\} \leftarrow \text{Numerator of complex fraction}$$

$\leftarrow$ Main fraction bar

$$\left.7+\dfrac{1}{y}\right\} \leftarrow \text{Denominator of complex fraction}$$

Our goal in this section is to simplify complex fractions. A complex fraction is simplified when it is in the form $\dfrac{P}{Q}$, where P and Q are polynomials that have no common factors. Two methods of simplifying complex fractions are introduced.

A **Method 1: Simplifying a Complex Fraction by Simplifying the Numerator and Denominator and Then Dividing**

In the first method we study, we simplify complex fractions by simplifying and dividing.

Simplifying a Complex Fraction: Method 1

Step 1. Simplify the numerator and the denominator of the complex fraction so that each is a single fraction.

Step 2. Perform the indicated division by multiplying the numerator of the complex fraction by the reciprocal of the denominator of the complex fraction.

Step 3. Simplify if possible.

EXAMPLE 1   Simplify: $\dfrac{\dfrac{2x}{27y^2}}{\dfrac{6x^2}{9}}$

Solution: The numerator of the complex fraction is already a single fraction, and so is the denominator. Thus we perform the indicated division by multiplying the numerator, $\dfrac{2x}{27y^2}$, by the reciprocal of the denominator, $\dfrac{6x^2}{9}$. Then we simplify.

OBJECTIVES

A Simplify complex fractions by simplifying the numerator and denominator and then dividing.

B Simplify complex fractions by multiplying by the least common denominator (LCD).

C Simplify expressions with negative exponents.

SSM   SG   CD & VIDEO   MATH PRO   WEB
TUTOR CENTER

Practice Problem 1

Use Method 1 to simplify: $\dfrac{\dfrac{3a}{25b^2}}{\dfrac{9a^2}{5b}}$

Answer

1. $\dfrac{1}{15ab}$

$$\frac{\dfrac{2x}{27y^2}}{\dfrac{6x^2}{9}} = \frac{2x}{27y^2} \div \frac{6x^2}{9}$$

$$= \frac{2x}{27y^2} \cdot \frac{9}{6x^2} \qquad \text{Multiply by the reciprocal of } \frac{6x^2}{9}.$$

$$= \frac{2x \cdot 9}{27y^2 \cdot 6x^2} = \frac{1}{9xy^2} \quad \text{Simplify.}$$

Practice Problem 2

Use Method 1 to simplify: $\dfrac{\dfrac{6x}{x-5}}{\dfrac{12}{x+5}}$

Concept Check

Which of the following are equivalent $\dfrac{\dfrac{1}{x}}{\dfrac{3}{y}}$?

a. $\dfrac{1}{x} \div \dfrac{3}{y}$

b. $\dfrac{1}{x} \cdot \dfrac{y}{3}$

c. $\dfrac{1}{x} \div \dfrac{y}{3}$

Practice Problem 3

Use Method 1 to simplify: $\dfrac{\dfrac{x}{y^2} - \dfrac{1}{y}}{\dfrac{y}{x^2} - \dfrac{1}{x}}$

EXAMPLE 2 Simplify: $\dfrac{\dfrac{5x}{x+2}}{\dfrac{10}{x-2}}$

Solution: $\dfrac{\dfrac{5x}{x+2}}{\dfrac{10}{x-2}} = \dfrac{5x}{x+2} \cdot \dfrac{x-2}{10} \qquad \text{Multiply by the reciprocal of } \dfrac{10}{x-2}.$

$$= \frac{5\,x(x-2)}{2 \cdot 5\,(x+2)}$$

$$= \frac{x(x-2)}{2(x+2)} \qquad \text{Simplify.}$$

Try the Concept Check in the margin.

EXAMPLE 3 Simplify: $\dfrac{\dfrac{x}{y^2} + \dfrac{1}{y}}{\dfrac{y}{x^2} + \dfrac{1}{x}}$

Solution: First we simplify the numerator and the denominator of the complex fraction separately so that each is a single fraction.

$$\frac{\dfrac{x}{y^2} + \dfrac{1}{y}}{\dfrac{y}{x^2} + \dfrac{1}{x}} = \frac{\dfrac{x}{y^2} + \dfrac{1 \cdot y}{y \cdot y}}{\dfrac{y}{x^2} + \dfrac{1 \cdot x}{x \cdot x}} \qquad \begin{array}{l}\text{The LCD is } y^2.\\[1em]\text{The LCD is } x^2.\end{array}$$

$$= \frac{\dfrac{x+y}{y^2}}{\dfrac{y+x}{x^2}} \qquad \begin{array}{l}\text{Add.}\\[1em]\text{Add.}\end{array}$$

$$= \frac{x+y}{y^2} \cdot \frac{x^2}{y+x} \qquad \text{Multiply by the reciprocal of } \frac{y+x}{x^2}.$$

$$= \frac{x^2(x+y)}{y^2(y+x)}$$

$$= \frac{x^2}{y^2} \qquad \text{Simplify.}$$

B ## Method 2: Simplifying a Complex Fraction by Multiplying the Numerator and Denominator by the LCD

With this method, we multiply the numerator and the denominator of the complex fraction by the least common denominator (LCD) of all fractions in the complex fraction.

Answers

2. $\dfrac{x(x+5)}{2(x-5)}$, 3. $-\dfrac{x^2}{y^2}$

Concept Check: a and b

Simplifying a Complex Fraction: Method 2

Step 1. Multiply the numerator and the denominator of the complex fraction by the LCD of the fractions in both the numerator and the denominator.

Step 2. Simplify.

EXAMPLE 4 Simplify: $\dfrac{\dfrac{5x}{x+2}}{\dfrac{10}{x-2}}$

Solution: The least common denominator of $\dfrac{5x}{x+2}$ and $\dfrac{10}{x-2}$ is $(x+2)(x-2)$. We multiply both the numerator, $\dfrac{5x}{x+2}$, and the denominator, $\dfrac{10}{x-2}$, by this LCD.

$$\dfrac{\dfrac{5x}{x+2}}{\dfrac{10}{x-2}} = \dfrac{\left(\dfrac{5x}{x+2}\right)\cdot(x+2)(x-2)}{\left(\dfrac{10}{x-2}\right)\cdot(x+2)(x-2)}$$ Multiply the numerator and denominator by the LCD.

$$= \dfrac{5x\cdot(x-2)}{2\cdot5\cdot(x+2)}$$ Simplify.

$$= \dfrac{x(x-2)}{2(x+2)}$$ Simplify.

EXAMPLE 5 Simplify: $\dfrac{\dfrac{x}{y^2}+\dfrac{1}{y}}{\dfrac{y}{x^2}+\dfrac{1}{x}}$

Solution: The least common denominator of $\dfrac{x}{y^2},\dfrac{1}{y},\dfrac{y}{x^2}$, and $\dfrac{1}{x}$ is x^2y^2.

$$\dfrac{\dfrac{x}{y^2}+\dfrac{1}{y}}{\dfrac{y}{x^2}+\dfrac{1}{x}} = \dfrac{\left(\dfrac{x}{y^2}+\dfrac{1}{y}\right)\cdot x^2y^2}{\left(\dfrac{y}{x^2}+\dfrac{1}{x}\right)\cdot x^2y^2}$$ Multiply the numerator and denominator by the LCD.

$$= \dfrac{\dfrac{x}{y^2}\cdot x^2y^2+\dfrac{1}{y}\cdot x^2y^2}{\dfrac{y}{x^2}\cdot x^2y^2+\dfrac{1}{x}\cdot x^2y^2}$$ Use the distributive property.

$$= \dfrac{x^3+x^2y}{y^3+xy^2}$$ Simplify.

$$= \dfrac{x^2(x+y)}{y^2(y+x)}$$ Factor.

$$= \dfrac{x^2}{y^2}$$ Simplify.

Practice Problem 4

Use Method 2 to simplify: $\dfrac{\dfrac{6x}{x-5}}{\dfrac{12}{x+5}}$

Practice Problem 5

Use Method 2 to simplify: $\dfrac{\dfrac{x}{y^2}-\dfrac{1}{y}}{\dfrac{y}{x^2}-\dfrac{1}{x}}$

Answers

4. $\dfrac{x(x+5)}{2(x-5)}$, **5.** $-\dfrac{x^2}{y^2}$

C Simplifying Expressions with Negative Exponents

Some expressions containing negative exponents can be written as complex fractions. To simplify these expressions, we first write them as equivalent expressions with positive exponents.

Practice Problem 6

Simplify: $\dfrac{2x^{-1} + 3y^{-1}}{x^{-1} - 2y^{-1}}$

EXAMPLE 6   Simplify: $\dfrac{x^{-1} + 2xy^{-1}}{x^{-2} - x^{-2}y^{-1}}$

Solution: This fraction does not appear to be a complex fraction. However, if we write it by using only positive exponents we see that it is a complex fraction.

$$\frac{x^{-1} + 2xy^{-1}}{x^{-2} - x^{-2}y^{-1}} = \frac{\dfrac{1}{x} + \dfrac{2x}{y}}{\dfrac{1}{x^2} - \dfrac{1}{x^2 y}}$$

The LCD of $\dfrac{1}{x}, \dfrac{2x}{y}, \dfrac{1}{x^2}$, and $\dfrac{1}{x^2 y}$ is $x^2 y$. We multiply both the numerator and denominator by $x^2 y$.

$$\frac{\dfrac{1}{x} + \dfrac{2y}{y}}{\dfrac{1}{x^2} - \dfrac{1}{x^2 y}} = \frac{\left(\dfrac{1}{x} + \dfrac{2x}{y}\right) \cdot x^2 y}{\left(\dfrac{1}{x^2} - \dfrac{1}{x^2 y}\right) \cdot x^2 y}$$

$$= \frac{\dfrac{1}{x} \cdot x^2 y + \dfrac{2x}{y} \cdot x^2 y}{\dfrac{1}{x^2} \cdot x^2 y - \dfrac{1}{x^2 y} \cdot x^2 y} \qquad \text{Use the distributive property.}$$

$$= \frac{xy + 2x^3}{y - 1} \qquad \text{Simplify.}$$

$$\text{or } \frac{x(y + 2x^2)}{y - 1}$$

Practice Problem 7

Simplify: $\dfrac{5 - 3x^{-1}}{2 + (3x)^{-1}}$

EXAMPLE 7   Simplify: $\dfrac{(2x)^{-1} + 1}{2x^{-1} - 1}$

Solution: $\dfrac{(2x)^{-1} + 1}{2x^{-1} - 1} = \dfrac{\dfrac{1}{2x} + 1}{\dfrac{2}{x} - 1}$ Write using positive exponents.

> **Helpful Hint**
> Don't forget that $(2x)^{-1} = \dfrac{1}{2x}$, but
> $2x^{-1} = 2 \cdot \dfrac{1}{x} = \dfrac{2}{x}$.

$$= \frac{\left(\dfrac{1}{2x} + 1\right) \cdot 2x}{\left(\dfrac{2}{x} - 1\right) \cdot 2x} \qquad \text{The LDC of } \dfrac{1}{2x} \text{ and } \dfrac{2}{x} \text{ is } 2x.$$

$$= \frac{\dfrac{1}{2x} \cdot 2x + 1 \cdot 2x}{\dfrac{2}{x} \cdot 2x - 1 \cdot 2x} \qquad \text{Use distributive property.}$$

$$= \frac{1 + 2x}{4 - 2x} \qquad \text{Simplify.}$$

Answers

6. $\dfrac{2y + 3x}{y - 2x}$,   7. $\dfrac{15x - 9}{6x + 1}$

EXERCISE SET 6.3

 B *Simplify each complex fraction. See Examples 1 through 5.*

1. $\dfrac{\dfrac{10}{3x}}{\dfrac{5}{6x}}$

2. $\dfrac{\dfrac{15}{2x}}{\dfrac{5}{6x}}$

3. $\dfrac{1 + \dfrac{2}{5}}{2 + \dfrac{3}{5}}$

4. $\dfrac{2 + \dfrac{1}{7}}{3 - \dfrac{4}{7}}$

5. $\dfrac{\dfrac{4}{x - 1}}{\dfrac{x}{x - 1}}$

6. $\dfrac{\dfrac{x}{x + 2}}{\dfrac{2}{x + 2}}$

7. $\dfrac{1 - \dfrac{2}{x}}{x + \dfrac{4}{9x}}$

8. $\dfrac{5 - \dfrac{3}{x}}{x + \dfrac{2}{3x}}$

9. $\dfrac{\dfrac{4x^2 - y^2}{xy}}{\dfrac{2}{y} - \dfrac{1}{x}}$

10. $\dfrac{\dfrac{x^2 - 9y^2}{xy}}{\dfrac{1}{y} - \dfrac{3}{x}}$

11. $\dfrac{\dfrac{x + 1}{3}}{\dfrac{2x - 1}{6}}$

12. $\dfrac{\dfrac{x + 3}{12}}{\dfrac{4x - 5}{15}}$

13. $\dfrac{\dfrac{2}{x} + \dfrac{3}{x^2}}{\dfrac{4}{x^2} - \dfrac{9}{x}}$

14. $\dfrac{\dfrac{2}{x^2} + \dfrac{1}{x}}{\dfrac{4}{x^2} - \dfrac{1}{x}}$

15. $\dfrac{\dfrac{1}{x} + \dfrac{2}{x^2}}{x + \dfrac{8}{x^2}}$

16. $\dfrac{\dfrac{1}{y} + \dfrac{3}{y^2}}{y + \dfrac{27}{y^2}}$

17. $\dfrac{\dfrac{4}{5 - x} + \dfrac{5}{x - 5}}{\dfrac{2}{x} + \dfrac{3}{x - 5}}$

18. $\dfrac{\dfrac{3}{x - 4} - \dfrac{2}{4 - x}}{\dfrac{2}{x - 4} - \dfrac{2}{x}}$

19. $\dfrac{\dfrac{x + 2}{x} - \dfrac{2}{x - 1}}{\dfrac{x + 1}{x} + \dfrac{x + 1}{x - 1}}$

20. $\dfrac{\dfrac{5}{a + 2} - \dfrac{1}{a - 2}}{\dfrac{3}{2 + a} + \dfrac{6}{2 - a}}$

21. $\dfrac{\dfrac{2}{x} + 3}{\dfrac{4}{x^2} - 9}$

22. $\dfrac{2 + \dfrac{1}{x}}{4x - \dfrac{1}{x}}$

23. $\dfrac{1 - \dfrac{x}{y}}{\dfrac{x^2}{y^2} - 1}$

24. $\dfrac{1 - \dfrac{2}{x}}{x - \dfrac{4}{x}}$

25. $\dfrac{\dfrac{-2x}{x - y}}{\dfrac{y}{x^2}}$

26. $\dfrac{\dfrac{7y}{x^2 + xy}}{\dfrac{y^2}{x^2}}$

27. $\dfrac{\dfrac{2}{x} + \dfrac{1}{x^2}}{\dfrac{y}{x^2}}$

28. $\dfrac{\dfrac{5}{x^2} - \dfrac{2}{x}}{\dfrac{1}{x} + 2}$

29. $\dfrac{\dfrac{x}{9} - \dfrac{1}{x}}{1 + \dfrac{3}{x}}$

30. $\dfrac{\dfrac{x}{4} - \dfrac{4}{x}}{1 - \dfrac{4}{x}}$

31. $\dfrac{\dfrac{x - 1}{x^2 - 4}}{1 + \dfrac{1}{x - 2}}$

32. $\dfrac{\dfrac{2}{x + 5} + \dfrac{4}{x + 3}}{\dfrac{3x + 13}{x^2 + 8x + 15}}$

C *Simplify. See Examples 6 and 7.*

33. $\dfrac{x^{-1}}{x^{-2} + y^{-2}}$

34. $\dfrac{a^{-3} + b^{-1}}{a^{-2}}$

35. $\dfrac{2a^{-1} + 3b^{-2}}{a^{-1} - b^{-1}}$

36. $\dfrac{x^{-1} + y^{-1}}{3x^{-2} + 5y^{-2}}$

37. $\dfrac{1}{x - x^{-1}}$

38. $\dfrac{x^{-2}}{x + 3x^{-1}}$

39. $\dfrac{a^{-1} + 1}{a^{-1} - 1}$

40. $\dfrac{a^{-1} - 4}{4 + a^{-1}}$

41. $\dfrac{3x^{-1} + (2y)^{-1}}{x^{-2}}$

42. $\dfrac{5x^{-2} - 3y^{-1}}{x^{-1} + y^{-1}}$

43. $\dfrac{2a^{-1} + (2a)^{-1}}{a^{-1} + 2a^{-2}}$

44. $\dfrac{a^{-1} + 2a^{-2}}{2a^{-1} + (2a)^{-1}}$

45. $\dfrac{5x^{-1} + 2y^{-1}}{x^{-2}y^{-2}}$

46. $\dfrac{x^{-2}y^{-2}}{5x^{-1} + 2y^{-1}}$

47. $\dfrac{5x^{-1} - 2y^{-1}}{25x^{-2} - 4y^{-2}}$

48. $\dfrac{3x^{-1} + 3y^{-1}}{4x^{-2} - 9y^{-2}}$

Review and Preview

Solve each equation for x. See Sections 2.1 and 5.7.

49. $7x + 2 = x - 3$

50. $4 - 2x = 17 - 5x$

51. $x^2 = 4x - 4$

52. $5x^2 + 10x = 15$

53. $\dfrac{x}{3} - 5 = 13$

54. $\dfrac{2x}{9} + 1 = \dfrac{7}{9}$

55. When the source of a sound is traveling toward a listener, the pitch that the listener hears due to the Doppler effect is given by the complex rational compression $\dfrac{a}{1 - \dfrac{s}{770}}$, where a is the actual pitch of the sound and s is the speed of the sound source. Simplify this expression.

56. In baseball, the earned run average (ERA) statistic gives the average number of earned runs scored on a pitcher per game. It is computed with the following expression: $\dfrac{E}{\dfrac{I}{9}}$, where E is the number of earned runs scored on a pitcher and I is the total number of innings pitched by the pitcher. Simplify this expression.

57. Which of the following are equivalent to $\dfrac{\dfrac{1}{x}}{\dfrac{3}{y}}$?

a. $\dfrac{1}{x} \div \dfrac{3}{y}$

b. $\dfrac{1}{x} \cdot \dfrac{y}{3}$

c. $\dfrac{1}{x} \div \dfrac{y}{3}$

58. In your own words, explain one method for simplifying a complex fraction.

Simplify.

59. $\dfrac{1}{1 + (1 + x)^{-1}}$

60. $\dfrac{(x + 2)^{-1} + (x - 2)^{-1}}{(x^2 - 4)^{-1}}$

61. $\dfrac{x}{1 - \dfrac{1}{1 + \dfrac{1}{x}}}$

62. $\dfrac{x}{1 - \dfrac{1}{1 - \dfrac{1}{x}}}$

63. $\dfrac{\dfrac{2}{y^2} - \dfrac{5}{xy} - \dfrac{3}{x^2}}{\dfrac{2}{y^2} + \dfrac{7}{xy} + \dfrac{3}{x^2}}$

64. $\dfrac{\dfrac{2}{x^2} - \dfrac{1}{xy} - \dfrac{1}{y^2}}{\dfrac{1}{x^2} - \dfrac{3}{xy} + \dfrac{2}{y^2}}$

PROBLEM SOLVING

When you are faced with solving a problem in real life that involves math, it may not always be immediately obvious how to approach the problem or what type of math is needed. Here are some problem-solving strategies that may be helpful:

■ Break a complicated problem up into simpler problems or steps. For instance, this strategy might be useful if you need to find the area or volume of a complicated geometric shape.

■ In problems that involve considering many different scenarios, try listing the possibilities. This strategy would help in a situation such as finding the number of four-person committees that could be formed from a pool of 12 club members.

■ Look at several specific examples of a general problem. Then see if a pattern emerges from these results that can be used to solve the problem.

■ Make a table of values related to the problem. If done methodically, this may reveal patterns that can be used to solve the problem.

■ Work backward from the desired results until suitable initial conditions for the problem are reached.

■ When not all of the necessary initial conditions are given in a problem, make some assumptions. Try to use reasonable estimates or educated guesses for the missing information so that your solution is meaningful for the problem.

■ Try a simpler or similar problem. Starting with smaller numbers or "round" numbers may help you see how to handle the problem.

■ Make a guess and check to see if it satisfies the problem. If not, revise your guess and check again. Continue guessing and checking until a guess checks or until this process gives enough insight into the problem that another method can be used.

■ Draw a diagram. This strategy can be helpful in many different problem-solving situations. This can help you visualize a physical relationship that might make the problem easier or clearer.

CRITICAL THINKING

Try solving the following problems. Explain how you found each solution. You may find some of the strategies described on the left useful.

1. Jamie Webb can slice through a loaf of cinnamon bread in 15 seconds. How long will it take her to cut a 12-inch loaf into half-inch slices?

2. How many times does the digit "5" appear in the numbers from 500 (inclusive) through 600 (inclusive)?

3. Two cars are racing on an oval racetrack. One car can complete a lap in 60 seconds. The other car can complete a lap in 65 seconds. If both cars cross the start/finish line at the same time; how long will it be before the faster car passes the slower car on the track?

4. At a carnival, there are three gambling games. Game A costs $10 to play; Game B costs $50 to play; and Game C costs $100 to play. For each game, the player must pay the game operator the required fee. If the player loses the game, he or she receives nothing. If the player wins the game, he or she receives twice what was paid to play the game. A gambler plays Game A four times, Game B three times, and Game C twice. She wins each type of game once and ends up with $180. How much money did she start out with?

5. A couple has three children. If each of their children has three children, and each of their children has three children, and so on, how many great-great-great-grandchildren will the original couple have?

6. Numbering the pages in a book requires a total of 1350 digits. If the book numbering starts on page 1, how many pages are in the book?

6.4 Solving Equations Containing Rational Expressions

A Solving Equations Containing Rational Expressions

OBJECTIVE

Ⓐ Solve equations containing rational expressions.

SSM
TUTOR CENTER   SG   CD & VIDEO   MATH PRO   WEB

In this section, we solve rational equations. A *rational equation* is an equation containing at least one rational expression. Before beginning this section, make sure that you understand the difference between an *equation* and an *expression*. An **equation** contains an equal sign and an **expression** does not.

Equation	Expression
$\dfrac{x}{2} + \dfrac{x}{6} = \dfrac{2}{3}$	$\dfrac{x}{2} + \dfrac{x}{6}$

Solving an Equation Containing Rational Expressions

To solve an *equation* containing rational expressions, first clear the equation of fractions by multiplying both sides of the equation by the LCD of all rational expressions. Then solve as usual.

Helpful Hint

The method described above is for equations only. It may *not* be used for performing operations on expressions.

Try the Concept Check in the margin.

EXAMPLE 1 Solve: $\dfrac{4x}{5} + \dfrac{3}{2} = \dfrac{3x}{10}$

Solution: The LCD of $\dfrac{4x}{5}, \dfrac{3}{2}$, and $\dfrac{3x}{10}$ is 10. We multiply both sides of the equation by 10.

$$\frac{4x}{5} + \frac{3}{2} = \frac{3x}{10}$$

$$10\left(\frac{4x}{5} + \frac{3}{2}\right) = 10\left(\frac{3x}{10}\right) \qquad \text{Multiply both sides by the LCD.}$$

$$10 \cdot \frac{4x}{5} + 10 \cdot \frac{3}{2} = 10 \cdot \frac{3x}{10} \qquad \text{Use the distributive property.}$$

$$8x + 15 = 3x \qquad \text{Simplify.}$$

$$15 = -5x \qquad \text{Subtract } 8x \text{ from both sides.}$$

$$-3 = x \qquad \text{Solve.}$$

We verify this solution by replacing x with -3 in the original equation.

Check:
$$\frac{4x}{5} + \frac{3}{2} = \frac{3x}{10}$$

$$\frac{4(-3)}{5} + \frac{3}{2} \stackrel{?}{=} \frac{3(-3)}{10}$$

$$\frac{-12}{5} + \frac{3}{2} \stackrel{?}{=} \frac{-9}{10}$$

$$-\frac{24}{10} + \frac{15}{10} \stackrel{?}{=} -\frac{9}{10}$$

$$-\frac{9}{10} = -\frac{9}{10} \qquad \text{True.}$$

The solution set is $\{-3\}$.

Concept Check

True or false? Clearing fractions is valid when solving an equation and when simplifying rational expressions. Explain.

Practice Problem 1

Solve: $\dfrac{5x}{6} + \dfrac{1}{2} = \dfrac{x}{3}$

Answers

1. $\{-1\}$

Concept Check: false; answers may vary

EXAMPLE 2 Solve: $\dfrac{3}{x} - \dfrac{x + 21}{3x} = \dfrac{5}{3}$

Solution: The LCD of the denominators x, $3x$, and 3 is $3x$. We multiply both sides by $3x$.

$$\frac{3}{x} - \frac{x + 21}{3x} = \frac{5}{3}$$

$$3x\left(\frac{3}{x} - \frac{x + 21}{3x}\right) = 3x\left(\frac{5}{3}\right) \qquad \text{Multiply both sides by the LCD.}$$

$$3x \cdot \frac{3}{x} - 3x \cdot \frac{x + 21}{3x} = 3x \cdot \frac{5}{3} \qquad \text{Use the distributive property.}$$

$$9 - (x + 21) = 5x \qquad \text{Simplify.}$$

$$9 - x - 21 = 5x$$

$$-12 = 6x$$

$$-2 = x \qquad \text{Solve.}$$

The proposed solution is -2.

Check: We check the proposed solution in the original equation.

$$\frac{3}{x} - \frac{x + 21}{3x} = \frac{5}{3}$$

$$\frac{3}{-2} - \frac{-2 + 21}{3(-2)} \overset{?}{=} \frac{5}{3}$$

$$-\frac{9}{6} + \frac{19}{6} \overset{?}{=} \frac{5}{3}$$

$$\frac{10}{6} = \frac{5}{3} \qquad \text{True.}$$

The solution set is $\{-2\}$.

The important difference about the equations in this section is that the denominator of a rational expression may contain a variable. Recall that a rational expression is undefined for values of the variable that make the denominator 0. If a proposed solution makes the denominator 0, then it must be rejected as a solution of the original equation. Such proposed solutions are called **extraneous solutions.**

EXAMPLE 3 Solve: $\dfrac{x + 6}{x - 2} = \dfrac{2(x + 2)}{x - 2}$

Solution: First multiply both sides of the equation by the LCD, $x - 2$.

$$\frac{x + 6}{x - 2} = \frac{2(x + 2)}{x - 2}$$

$$(x - 2) \cdot \frac{x + 6}{x - 2} = (x - 2) \cdot \frac{2(x + 2)}{x - 2} \qquad \text{Multiply both sides by } x - 2.$$

$$x + 6 = 2(x + 2) \qquad \text{Simplify.}$$

$$x + 6 = 2x + 4 \qquad \text{Use the distributive property.}$$

$$2 = x \qquad \text{Solve.}$$

Check: The proposed solution is 2. Notice that 2 makes the denominator 0 in the original equation. This can also be seen in a check. Check the proposed solution 2 in the original equation.

$$\frac{x + 6}{x - 2} = \frac{2(x + 2)}{x - 2}$$

$$\frac{2 + 6}{6 - 2} = \frac{2(2 + 2)}{2 - 2}$$

$$\frac{8}{0} = \frac{2(4)}{0}$$

The denominators are 0, so 2 is not a solution of the original equation. The solution set is $\varnothing$ or $\{\ \}$. ●

EXAMPLE 4   Solve: $\dfrac{2x}{2x - 1} + \dfrac{1}{x} = \dfrac{1}{2x - 1}$

Solution: The LCD is $x(2x - 1)$. Multiply both sides by $x(2x - 1)$. By the distributive property, this is the same as multiplying each term by $x(2x - 1)$.

$$x(2x - 1) \cdot \frac{2x}{2x - 1} + x(2x - 1) \cdot \frac{1}{x} = x(2x - 1) \cdot \frac{1}{2x - 1}$$

$$x(2x) + (2x - 1) = x$$

$$2x^2 + 2x - 1 - x = 0$$

$$2x^2 + x - 1 = 0$$

$$(x + 1)(2x - 1) = 0$$

$$x + 1 = 0 \quad \text{or} \quad 2x - 1 = 0$$

$$x = -1 \qquad\qquad x = \frac{1}{2}$$

The number $\dfrac{1}{2}$ makes the denominator $2x - 1$ equal 0, so it is not a solution. The solution set is $\{-1\}$. ●

EXAMPLE 5   Solve: $\dfrac{2x}{x - 3} + \dfrac{6 - 2x}{x^2 - 9} = \dfrac{x}{x + 3}$

Solution: We factor the second denominator to find that the LCD is $(x + 3)(x - 3)$. We multiply both sides of the equation by $(x + 3)(x - 3)$. By the distributive property, this is the same as multiplying each term by $(x + 3)(x - 3)$.

$$\frac{2x}{x - 3} + \frac{6 - 2x}{x^2 - 9} = \frac{x}{x + 3}$$

$$(x + 3)(x - 3) \cdot \frac{2x}{x - 3} + (x + 3)(x - 3) \cdot \frac{6 - 2x}{(x + 3)(x - 3)}$$

$$= (x + 3)(x - 3)\left(\frac{x}{x + 3}\right)$$

$$2x(x + 3) + (6 - 2x) = x(x - 3) \qquad \text{Simplify.}$$

$$2x^2 + 6x + 6 - 2x = x^2 - 3x \qquad \text{Use the distributive property.}$$

Practice Problem 4

Solve: $\dfrac{3x}{3x - 1} + \dfrac{1}{x} = \dfrac{1}{3x - 1}$

Practice Problem 5

Solve: $\dfrac{2x}{x - 4} + \dfrac{10 - 5x}{x^2 - 16} = \dfrac{x}{x + 4}$

Answers

4. $\{-1\}$, **5.** $\{-5, -2\}$

Next we solve this quadratic equation by the factoring method. To do so, we first write the equation so that one side is 0.

$$x^2 + 7x + 6 = 0$$
$$(x + 6)(x + 1) = 0 \qquad \text{Factor.}$$
$$x = -6 \quad \text{or} \quad x = -1 \qquad \text{Set each factor equal to 0 and solve.}$$

Neither -6 nor -1 makes any denominator 0. The solution set is $\{-6, -1\}$.

STUDY SKILLS REMINDER

Are you preparing for a test on Chapter 6?

Below I have listed some common trouble areas for students in Chapter 6. After studying for your test—but before taking your test—read these.

■ Make sure you know the difference in the following:

Simplify: $\dfrac{\dfrac{3}{x}}{\dfrac{1}{x} - \dfrac{5}{y}}$

Solve: $\dfrac{5x}{6} - \dfrac{1}{2} = \dfrac{5x}{12}$

Subtract: $\dfrac{1}{2x} - \dfrac{7}{x - 3}$

Multiply numerator and denominator by the LCD.

Multiply both sides by the LCD.

Write each expression as an equivalent expression with the LCD.

$$\dfrac{\dfrac{3}{x} \cdot xy}{\dfrac{1}{x} \cdot xy - \dfrac{5}{y} \cdot xy}$$
$$= \dfrac{3y}{y - 5x}$$

$$12 \cdot \dfrac{5x}{6} - 12 \cdot \dfrac{1}{2} = 12 \cdot \dfrac{5x}{12}$$
$$2 \cdot 5x - 6 = 5x$$
$$10x - 6 = 5x$$
$$5x = 6$$
$$x = \dfrac{6}{5}$$

$$\dfrac{1 \cdot (x - 3)}{2x \cdot (x - 3)} - \dfrac{7 \cdot 2x}{(x - 3) \cdot 2x}$$
$$= \dfrac{x - 3}{2x(x - 3)} - \dfrac{14x}{2x(x - 3)}$$
$$= \dfrac{-13x - 3}{2x(x - 3)}$$

Remember: This is simply a checklist of common trouble areas. For a review of Chapter 6, see the Highlights and Chapter Review at the end of this chapter.

Mental Math

Determine whether each is an equation or an expression. Do not solve or simplify.

1. $\dfrac{x}{2} = \dfrac{3x}{5} + \dfrac{x}{6}$

2. $\dfrac{3x}{5} + \dfrac{x}{6}$

3. $\dfrac{x}{x-1} + \dfrac{2x}{x+1}$

4. $\dfrac{x}{x-1} + \dfrac{2x}{x+1} = 5$

5. $\dfrac{y+7}{2} = \dfrac{y+1}{6} + \dfrac{1}{y}$

6. $\dfrac{y+1}{6} + \dfrac{1}{y}$

EXERCISE SET 6.4

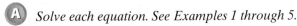 **A** *Solve each equation. See Examples 1 through 5.*

1. $\dfrac{x}{2} - \dfrac{x}{3} = 12$

2. $x = \dfrac{x}{2} - 4$

3. $\dfrac{5}{x} = \dfrac{20}{12}$

4. $\dfrac{2}{x} = \dfrac{10}{5}$

5. $1 - \dfrac{4}{a} = 5$

6. $7 + \dfrac{6}{a} = 5$

7. $\dfrac{x}{3} = \dfrac{1}{6} + \dfrac{x}{4}$

8. $\dfrac{x}{2} = \dfrac{21}{10} - \dfrac{x}{5}$

9. $\dfrac{2}{x} + \dfrac{1}{2} = \dfrac{5}{x}$

10. $\dfrac{5}{3x} + 1 = \dfrac{7}{6}$

11. $\dfrac{x+3}{x} = \dfrac{5}{x}$

12. $\dfrac{4-3x}{2x} = -\dfrac{8}{2x}$

13. $\dfrac{5}{x-2} - \dfrac{2}{x+4} = -\dfrac{4}{x^2+2x-8}$

14. $\dfrac{1}{x-1} + \dfrac{1}{x+1} = \dfrac{2}{x^2-1}$

15. $\dfrac{1}{x-1} = \dfrac{2}{x+1}$

16. $\dfrac{6}{x+3} = \dfrac{4}{x-3}$

17. $\dfrac{1}{x-4} - \dfrac{3x}{x^2-16} = \dfrac{2}{x+4}$

18. $\dfrac{3}{2x+3} - \dfrac{1}{2x-3} = \dfrac{4}{4x^2-9}$

19. $\dfrac{1}{x-4} = \dfrac{8}{x^2-16}$

20. $\dfrac{2}{x^2-4} = \dfrac{1}{2x-4}$

21. $\dfrac{1}{x-2} - \dfrac{2}{x^2-2x} = 1$

22. $\dfrac{12}{3x^2+12x} = 1 - \dfrac{1}{x+4}$

457

23. $\dfrac{1}{2x} - \dfrac{1}{x+1} = \dfrac{1}{3x^2 + 3x}$

24. $\dfrac{2}{x-5} + \dfrac{1}{2x} = \dfrac{5}{3x^2 - 15x}$

25. $\dfrac{1}{x} - \dfrac{x}{25} = 0$

26. $\dfrac{x}{4} + \dfrac{5}{x} = 3$

27. $5 - \dfrac{2}{2y-5} = \dfrac{3}{2y-5}$

28. $1 - \dfrac{5}{y+7} = \dfrac{4}{y+7}$

29. $\dfrac{x+3}{x+2} = \dfrac{1}{x+2}$

30. $\dfrac{2x+1}{4-x} = \dfrac{9}{4-x}$

31. $\dfrac{1}{a-3} + \dfrac{2}{a+3} = \dfrac{1}{a^2-9}$

32. $\dfrac{12}{9-a^2} + \dfrac{3}{3+a} = \dfrac{2}{3-a}$

33. $\dfrac{64}{x^2-16} + 1 = \dfrac{2x}{x-4}$

34. $2 + \dfrac{3}{x} = \dfrac{2x}{x+3}$

35. $\dfrac{-15}{4y+1} + 4 = y$

36. $\dfrac{36}{x^2-9} + 1 = \dfrac{2x}{x+3}$

37. $\dfrac{28}{x^2-9} + \dfrac{2x}{x-3} + \dfrac{6}{x+3} = 0$

38. $\dfrac{x^2-20}{x^2-7x+12} = \dfrac{3}{x-3} + \dfrac{5}{x-4}$

39. $\dfrac{x+2}{x^2+7x+10} = \dfrac{1}{3x+6} - \dfrac{1}{x+5}$

40. $\dfrac{3}{2x-5} + \dfrac{2}{2x+3} = 0$

Review and Preview

Write each sentence as an equation and solve. See Section 2.2.

41. Four more than 3 times a number is 19. Find the number.

42. The sum of two consecutive integers is 147. Find the integers.

△ **43.** The length of a rectangle is 5 inches more than the width. Its perimeter is 50 inches. Find the length and width.

44. The sum of a number and its reciprocal is $\frac{5}{2}$. Find the number and its reciprocal.

The following graph is from a survey of state and federal prisons. Use this histogram to answer Exercises 45 through 49.

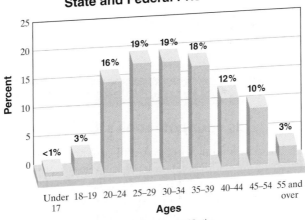

State and Federal Prison Inmates

Source: Bureau of Justice Statistics, U.S. Department of Justice

45. What percent of state and federal prison inmates are age 45 to 54?

46. What percent of state and federal prison inmates are 55 years old or older?

47. What age category shows the highest percent of prison inmates?

48. What percent of state and federal prison inmates are 20 to 34 years old?

49. At the end of 2000, there were 35,047 inmates under the jurisdiction of state and federal correctional authorities in the state of Louisiana. Approximately how many 25- to 29-year-old inmates would you expect to have been held in Louisiana at the end of 2000? Round to nearest whole. (*Source:* U.S. Bureau of Justice Statistics)

50. Use the data from Exercise 49 to answer the following.
  a. Approximate the number of 35- to 39- year-old inmates you might expect to have been held in Louisiana at the end of 2000. Round to the nearest whole.
  b. Is your answer to part a greater than or less than your answer to Exercise 49? Is this reasonable? Why or why not.

Combining Concepts

51. In your own words, explain the differences between equations and expressions.

52. In your own words, explain why it is necessary to check solutions to equations containing rational expressions.

Solve each equation. Begin by writing each equation with positive exponents only.

53. $x^{-2} - 19x^{-1} + 48 = 0$

54. $x^{-2} - 5x^{-1} - 36 = 0$

Solve each equation. Round solutions to two decimal places.

55. $\dfrac{1.4}{x-2.6} = \dfrac{-3.5}{x+7.1}$

56. $\dfrac{10.6}{y} - 14.7 = \dfrac{9.92}{3.2} + 7.6$

57. The average cost of producing x game disks for a computer is given by the function $f(x) = 3.3 + \dfrac{5400}{x}$. Find the number of game disks that must be produced for the average cost to be \$5.10.

58. The average cost of producing x electric pencil sharpeners is given by the function $f(x) = 20 + \dfrac{4000}{x}$. Find the number of electric pencil sharpeners that must be produced for the average cost to be \$25.

Use a grapher to verify the solution of each given exercise.

59. Exercise 21

60. Exercise 20

STUDY SKILLS REMINDER

Is your notebook still organized?

Is your notebook still organized? If it's not, it's not too late to start organizing it. Start writing your notes and completing your homework assignment in a notebook with pockets (spiral or ring binder). Take class notes in this notebook, and then follow the notes with your completed homework assignment. When you receive graded papers or handouts, place them in the notebook pocket so that you will not lose them.

Remember to mark (possibly with an exclamation point) any note(s) that seem extra important to you. Also remember to mark (possibly with a question mark) any notes or homework that you are having trouble with. Don't forget to see your instructor or a math tutor to help you with the concepts or exercises that you are having trouble understanding.

Also don't forget to write neatly and keep a positive attitude.

Integrated Review–Expressions and Equations Containing Rational Expressions

It is very important that you understand the difference between an expression and an equation containing rational expressions. An equation contains an equal sign; an expression does not.

Expression to be Simplified	**Equation to be Solved**

$$\frac{x}{2} + \frac{x}{6}$$

Write both rational expressions with the LCD, 6, as the denominator.

$$\frac{x}{2} + \frac{x}{6} = \frac{x \cdot 3}{2 \cdot 3} + \frac{x}{6}$$
$$= \frac{3x}{6} + \frac{x}{6}$$
$$= \frac{4x}{6} = \frac{2x}{3}$$

$$\frac{x}{2} + \frac{x}{6} = \frac{2}{3}$$

Multiply both sides by the LCD, 6.

$$6\left(\frac{1}{2} + \frac{x}{6}\right) = 6\left(\frac{2}{3}\right)$$

$$3 + x = 4$$

$$x = 1$$

Check to see that the solution is 1.

Helpful Hint

Remember: Equations can be cleared of fractions, expressions cannot.

Perform each indicated operation and either simplify the expression, or solve the equation for the variable.

1. $\dfrac{x}{2} = \dfrac{1}{8} + \dfrac{x}{4}$

2. $\dfrac{x}{4} = \dfrac{3}{2} + \dfrac{x}{10}$

3. $\dfrac{1}{8} + \dfrac{x}{4}$

4. $\dfrac{3}{2} + \dfrac{x}{10}$

5. $\dfrac{4}{x + 2} - \dfrac{2}{x - 1}$

6. $\dfrac{5}{x - 2} - \dfrac{10}{x + 4}$

7. $\dfrac{4}{x + 2} = \dfrac{2}{x - 1}$

8. $\dfrac{5}{x - 2} = \dfrac{10}{x + 4}$

9. $\dfrac{2}{x^2 - 4} = \dfrac{1}{x + 2} - \dfrac{3}{x - 2}$

10. $\dfrac{3}{x^2 - 25} = \dfrac{1}{x + 5} + \dfrac{2}{x - 5}$

11. $\dfrac{5}{x^2 - 3x} + \dfrac{4}{2x - 6}$

12. $\dfrac{5}{x^2 - 3x} \div \dfrac{4}{2x - 6}$

13. $\dfrac{x - 1}{x + 1} + \dfrac{x + 7}{x - 1} = \dfrac{4}{x^2 - 1}$

14. $\left(1 - \dfrac{y}{x}\right) \div \left(1 - \dfrac{x}{y}\right)$

15. _____

16. _____

17. _____

18. _____

19. _____

20. _____

21. _____

22. _____

23. _____

24. _____

25. _____

26. _____

27. a. _____

   b. _____

   c. _____

28. a. _____

   b. _____

   c. _____

15. $\dfrac{a^2 - 9}{a - 6} \cdot \dfrac{a^2 - 5a - 6}{a^2 - a - 6}$

16. $\dfrac{2}{a - 6} + \dfrac{3a}{a^2 - 5a - 6} - \dfrac{a}{5a + 5}$

17. $\dfrac{2x + 3}{3x - 2} = \dfrac{4x + 1}{6x + 1}$

18. $\dfrac{5x - 3}{2x} = \dfrac{10x + 3}{4x + 1}$

19. $\dfrac{a}{9a^2 - 1} + \dfrac{2}{6a - 2}$

20. $\dfrac{3}{4a - 8} - \dfrac{a + 2}{a^2 - 2a}$

21. $-\dfrac{3}{x^2} - \dfrac{1}{x} + 2 = 0$

22. $\dfrac{x}{2x + 6} + \dfrac{5}{x^2 - 9}$

23. $\dfrac{x - 8}{x^2 - x - 2} + \dfrac{2}{x - 2}$

24. $\dfrac{x - 8}{x^2 - x - 2} + \dfrac{2}{x - 2} = \dfrac{3}{x + 1}$

25. $\dfrac{3}{a} - 5 = \dfrac{7}{a} - 1$

26. $\dfrac{7}{3z - 9} + \dfrac{5}{z}$

Use $\dfrac{x}{5} - \dfrac{x}{4} = \dfrac{1}{10}$ *and* $\dfrac{x}{5} - \dfrac{x}{4} + \dfrac{1}{10}$ *for Exercises 27 and 28.*

27. a. Which one above is an expression?

   b. Describe the first step to simplify this expression.

   c. Simplify the expression.

28. a. Which one above is an equation?

   b. Describe the first step to solve this equation.

   c. Solve the equation.

6.5 Rational Equations and Problem Solving

Ⓐ Solving Rational Equations for a Specified Variable

In Section 2.3, we solved equations for a specified variable. In this section, we continue practicing this skill by solving equations containing rational expressions for a specified variable. The steps given in Section 2.3 for solving equations for a specified variable are repeated here.

Solving an Equation for a Specified Variable

Step 1. Clear the equation of fractions or rational expressions by multiplying each side of the equation by the least common denominator (LCD) of all denominators in the equation.

Step 2. Use the distributive property to remove grouping symbols such as parentheses.

Step 3. Combine like terms on each side of the equation.

Step 4. Use the addition property of equality to rewrite the equation as an equivalent equation with terms containing the specified variable on one side and all other terms on the other side.

Step 5. Use the distributive property and the multiplication property of equality to get the specified variable alone.

EXAMPLE 1   Solve $\dfrac{1}{x} + \dfrac{1}{y} = \dfrac{1}{z}$ for x.

Solution: To clear this equation of fractions, we multiply both sides of the equation by xyz, the LCD of $\dfrac{1}{x}, \dfrac{1}{y}$, and $\dfrac{1}{z}$.

$$\frac{1}{x} + \frac{1}{y} = \frac{1}{z}$$

$$xyz\left(\frac{1}{x} + \frac{1}{y}\right) = xyz\left(\frac{1}{z}\right) \qquad \text{Multiply both sides by } xyz.$$

$$xyz\left(\frac{1}{x}\right) + xyz\left(\frac{1}{y}\right) = xyz\left(\frac{1}{z}\right) \qquad \text{Use the distributive property.}$$

$$yz + xz = xy \qquad \text{Simplify.}$$

Notice the two terms that contain the specified variable, x.

Next, we subtract xz from both sides so that all terms containing the specified variable x are on one side of the equation and all other terms are on the other side.

$$yz = xy - xz$$

Now we use the distributive property to factor x from $xy - xz$ and then the multiplication property of equality to solve for x.

$$yz = x(y - z)$$

$$\frac{yz}{y - z} = x \quad \text{or} \quad x = \frac{yz}{y - z} \qquad \text{Divide both sides by } y - z.$$
●

Ⓑ Solving Number Problems Modeled by Rational Equations

Problem solving sometimes involves modeling a described situation with an equation containing rational expressions. In Examples 2 through 5, we practice solving such problems and use the problem-solving steps first introduced in Section 2.2.

OBJECTIVES

Ⓐ Solve an equation containing rational expressions for a specified variable.

Ⓑ Solve number problems by writing equations containing rational expressions.

Ⓒ Solve problems modeled by proportions.

Ⓓ Solve problems about work.

Ⓔ Solve problems about distance, rate, and time.

SSM
TUTOR CENTER   SG   CD & VIDEO   MATH PRO   WEB

Practice Problem 1

Solve $\dfrac{1}{x} + \dfrac{1}{y} = \dfrac{1}{z}$ for y.

Answer

1. $y = \dfrac{xz}{x - z}$

Practice Problem 2

Find the number that, when added to the numerator and subtracted from the denominator of $\frac{1}{20}$, results in a fraction equivalent to $\frac{2}{5}$.

EXAMPLE 2 Finding an Unknown Number

If a certain number is subtracted from the numerator and added to the denominator of $\frac{9}{19}$, the new fraction is equivalent to $\frac{1}{3}$. Find the number.

Solution:

1. UNDERSTAND the problem. Read and reread the problem and try guessing the solution. For example, If the unknown number is 3, we have

$$\frac{9-3}{19+3} = \frac{6}{22} = \frac{3}{11} \neq \frac{1}{3}$$

Thus, $\frac{9-3}{19+3} \neq \frac{1}{3}$ and 3 is not the correct number. Remember that the purpose of this step is not to guess the correct solution but to gain an understanding of the problem posed.

We will let n = the number to be subtracted from the numerator and added to the denominator.

2. TRANSLATE the problem.

In words:	when the number is subtracted from the numerator and added to the denominator of the fraction $\frac{9}{19}$	this is equiavalent to	$\frac{1}{3}$
	↓	↓	↓
Translate:	$\frac{9-n}{19+n}$	=	$\frac{1}{3}$

3. SOLVE the equation for n.

$$\frac{9-n}{19+n} = \frac{1}{3}$$

To solve for n, we begin by multiplying both sides by the LCD, $3(19+n)$.

$$3(19+n)\cdot\frac{9-n}{19+n} = 3(19+n)\cdot\frac{1}{3} \quad \text{Multiply both sides by the LCD.}$$
$$3(9-n) = 19+n \quad \text{Simplify.}$$
$$27-3n = 19+n$$
$$8 = 4n$$
$$2 = n \quad \text{Solve.}$$

4. INTERPRET the results.

Check: If we subtract 2 from the numerator and add 2 to the denominator of $\frac{9}{19}$, we have $\frac{9-2}{19+2} = \frac{7}{21} = \frac{1}{3}$, and the problem checks.

State: The unknown number is 2. ●

Solving Problems Modeled by Proportions

A **ratio** is the quotient of two number or two quantities. Since rational expressions are quotients of quantities, rational expressions are ratios, also. A **proportion** is a mathematical statement that two ratios are equal.

Let's review two methods for solving a proportion such as $\frac{x-3}{10} = \frac{7}{15}$.

We can multiply both sides of the equation by the LCD, 30.

$$30 \cdot \frac{x-3}{10} = 30 \cdot \frac{7}{15}$$

$$3(x-3) = 2 \cdot 7 \qquad \text{Multiply.}$$

$$3x - 9 = 14 \qquad \text{Use the distributive property.}$$

$$3x = 23$$

$$x = \frac{23}{3}$$

We can also solve a proportion by setting cross products equal. Here, we are using the fact that if $\frac{a}{b} = \frac{c}{d}$, then $ad = bc$.

$$\frac{x-3}{10} \diagdown\!\!\!\diagup \frac{7}{15}$$

$$15(x-3) = 10 \cdot 7 \qquad \text{Set cross products equal.}$$

$$15x - 45 = 70 \qquad \text{Use the distributive property.}$$

$$15x = 115$$

$$x = \frac{115}{15} \quad \text{or} \quad \frac{23}{3}$$

A ratio of two different quantities is called a **rate**. For example $\dfrac{3 \text{ miles}}{2 \text{ hours}}$ or 1.5 miles/hour is a rate. The proportions we write to solve problems will sometimes include rates. When this happens, make sure that the rates contain units written in the same order.

EXAMPLE 3

In the U.S., 7 out of every 25 homes are heated by electricity. At this rate, how many homes in a community of 36,000 homes would you predict are heated by electricity? (*Source:* 2000 Census Survey)

Solution:

1. UNDERSTAND. Read and reread the problem. Try to estimate a reasonable solution. For example, since 7 is less than $\frac{1}{3}$ of 25, we might reason that the solution would be less than $\frac{1}{3}$ of 36,000 or 12,000.

   Let's let x = number of homes in the community heated by electricity

2. TRANSLATE.

   homes heated by electricity $\longrightarrow$ $\dfrac{7}{25} = \dfrac{x}{36{,}000}$ $\longleftarrow$ homes heated by electricity
   total homes $\longrightarrow$ $\phantom{\dfrac{7}{25} = \dfrac{x}{36{,}000}}$ $\longleftarrow$ total homes

3. SOLVE. To solve this proportion we will set cross products equal.

$$\frac{7}{25} \diagdown\!\!\!\diagup \frac{x}{36{,}000}$$

$$7 \cdot 36{,}000 = 25x$$

$$\frac{252{,}000}{25} = x$$

$$10{,}080 = x$$

Practice Problem 3

In the U.S., 1 out of 50 homes is heated by wood. At this rate, how many homes in a community of 36,000 homes are heated by wood? (*Source:* 2000 Census Survey)

Answer

3. 720 homes

4. INTERPRET.

Check: To check, replace x with 10,080 in the proportion and see that a true statement results. Notice that our answer is reasonable since it is less than 12,000 as we stated above.

State: We predict that 10,080 homes are heated by electricity. ●

D **Solving Problems About Work**

The following work example leads to an equation containing rational expressions.

EXAMPLE 4 Calculating Work Hours

Melissa Scarlatti can clean the house in 4 hours, whereas her husband, Zack, can do the same job in 5 hours. They have agreed to clean together so that they can finish in time to watch a movie on TV that starts in 2 hours. How long will it take them to clean the house together? Can they finish before the movie starts?

Solution:

1. Read and reread the problem. The key idea here is the relationship between the *time* (in hours) it takes to complete the job and the *part of the job* completed in 1 unit of time (1 hour). For example, if the *time* it takes Melissa to complete the job is 4 hours, the *part of the job* she can complete in 1 hour is $\frac{1}{4}$. Similarly, Zack can complete $\frac{1}{5}$ of the job in 1 hour.

We will let t = the *time* in hours it takes Melissa and Zack to clean the house together. Then $\frac{1}{t}$ represents the *part of the job* they complete in 1 hour. We summarize the given information on a chart.

	Hours to Complete the Job	Part of Job Completed in 1 Hour
Melissa Alone	4	$\frac{1}{4}$
Zack Alone	5	$\frac{1}{5}$
Together	t	$\frac{1}{t}$

2. TRANSLATE.

In words:	part of job Melissa can complete in 1 hour	added to	part of job Zack can complete in 1 hour	is equal to	part of job they can complete together in 1 hour
	↓	↓	↓	↓	↓
Translate:	$\frac{1}{4}$	$+$	$\frac{1}{5}$	$=$	$\frac{1}{t}$

Practice Problem 4

Greg Guillot can paint a room alone in 3 hours. His brother Phillip can do the same job alone in 5 hours. How long would it take them to paint the room if they work together?

Answer

4. $1\frac{7}{8}$ hr

3. SOLVE.

$$\frac{1}{4} + \frac{1}{5} = \frac{1}{t}$$

$$20t\left(\frac{1}{4} + \frac{1}{5}\right) = 20t\left(\frac{1}{t}\right) \quad \text{Multiply both sides by the LCD, } 20t.$$

$$5t + 4t = 20$$

$$9t = 20$$

$$t = \frac{20}{9} \quad \text{or} \quad 2\frac{2}{9} \quad \text{Solve.}$$

4. INTERPRET.

Check: The proposed solution is $2\frac{2}{9}$. That is, Melissa and Zack would take $2\frac{2}{9}$ hours to clean the house together. This proposed solution is reasonable since $2\frac{2}{9}$ hours is more than half of Melissa's time and less than half of Zack's time. Check this solution in the originally stated problem.

State: Melissa and Zack can clean the house together in $2\frac{2}{9}$ hours. They cannot complete the job before the movie starts. ●

E **Solving Problems About Distance, Rate, and Time**

EXAMPLE 5   Finding the Speed of a Current

Steve Deitmer takes $1\frac{1}{2}$ times as long to go 72 miles upstream in his boat as he does to return. If the boat cruises at 30 mph in still water, what is the speed of the current?

Practice Problem 5

A fisherman traveling on the Pearl River takes $\frac{2}{3}$ times as long to travel 60 miles downstream in his boat than to return. If the boat's speed is 25 mph in still water, find the speed of the current.

Solution:

1. UNDERSTAND. Read and reread the problem. Guess a solution. Suppose that the current is 4 mph. The speed of the boat upstream is slowed down by the current: $30 - 4$, or 26 mph, and the speed of the boat downstream is speeded up by the current: $30 + 4$, or 34 mph. Next let's find out how long it takes to travel 72 miles upstream and 72 miles downstream. To do so, we use the formula $d = r \cdot t$, or $\frac{d}{r} = t$.

Upstream

$$\frac{d}{r} = t$$

$$\frac{72}{26} = t$$

$$2\frac{10}{13} = t$$

Downstream

$$\frac{d}{r} = t$$

$$\frac{72}{34} = t$$

$$2\frac{2}{17} = t$$

Answer

5. 5 mph

Since the time upstream $\left(2\frac{10}{13} \text{ hours}\right)$ is not $1\frac{1}{2}$ times the time downstream $\left(2\frac{2}{17} \text{ hours}\right)$, our guess is not correct. We do, however, have a better understanding of the problem.

We will let

$$x = \text{the speed of the current}$$
$$30 + x = \text{the speed of the boat downstream}$$
$$30 - x = \text{the speed of the boat upstream}$$

This information is summarized in the following chart, where we use the formula $\frac{d}{t} = t$.

	Distance	Rate	Time $\left(\dfrac{d}{r}\right)$
Upstream	72	$30 - x$	$\dfrac{72}{30 - x}$
Downstream	72	$30 + x$	$\dfrac{72}{30 + x}$

2. TRANSLATE. Since the time spent traveling upstream is $1\frac{1}{2}$ times the time spent traveling downstream, we have

In words:

time upstream	is	$1\frac{1}{2}$	times	time downstream
↓	↓	↓	↓	↓

Translate: $\quad \dfrac{72}{30 - x} \quad = \quad \dfrac{3}{2} \quad \cdot \quad \dfrac{72}{30 + x}$

3. SOLVE. $\dfrac{72}{30 - x} = \dfrac{3}{2} \cdot \dfrac{72}{30 + x}$

First we multiply both sides by the LCD, $2(30 + x)(30 - x)$.

$$2(30 + x)(30 - x) \cdot \frac{72}{30 - x} = 2(30 + x)(30 - x)\left(\frac{3}{2} \cdot \frac{72}{30 + x}\right)$$

$$72 \cdot 2(30 + x) = 3 \cdot 72 \cdot (30 - x) \qquad \text{Simplify.}$$
$$2(30 + x) = 3(30 - x) \qquad \text{Divide both sides by 72.}$$
$$60 + 2x = 90 - 3x \qquad \text{Use the distributive property.}$$
$$5x = 30$$
$$x = 6 \qquad \text{Solve.}$$

4. INTERPRET.

Check: Check the proposed solution of 6 mph in the originally-stated problem.
State: The current's speed is 6 mph.

EXERCISE SET 6.5

(A) Solve each equation for the specified variable. See Example 1.

1. $F = \dfrac{9}{5}C + 32$ for C (Meteorology) △ **2.** $V = \dfrac{1}{3}\pi r^2 h$ for h (Volume) **3.** $Q = \dfrac{A - I}{L}$ for I (Finance)

4. $P = 1 - \dfrac{C}{S}$ for S (Finance) 🔒 **5.** $\dfrac{1}{R} = \dfrac{1}{R_1} + \dfrac{1}{R_2}$ for R (Electronics) **6.** $\dfrac{1}{R} = \dfrac{1}{R_1} + \dfrac{1}{R_2}$ for R_1 (Electronics)

7. $S = \dfrac{n(a + L)}{2}$ for n (Sequences) **8.** $S = \dfrac{n(a + L)}{2}$ for a (Sequences) △ **9.** $A = \dfrac{h(a + b)}{2}$ for b (Area)

△ **10.** $A = \dfrac{h(a + b)}{2}$ for h (Area) **11.** $\dfrac{P_1 V_1}{T_1} = \dfrac{P_2 V_2}{T_2}$ for T_2 (Chemistry) **12.** $H = \dfrac{kA(T_1 - T_2)}{L}$ for T_2 (Physics)

13. $f = \dfrac{f_1 f_2}{f_1 + f_2}$ for f_2 **14.** $I = \dfrac{E}{R + r}$ for r (Electronics) **15.** $\lambda = \dfrac{2L}{n}$ for L

16. $S = \dfrac{a_1 - a_n r}{1 - r}$ for a_1 (Sequences) **17.** $\dfrac{\theta}{\omega} = \dfrac{2L}{c}$ for c **18.** $F = \dfrac{-GMm}{r^2}$ for M (Physics)

(B) Solve. See Example 2.

19. The sum of a number, and 5 times its reciprocal, is 6. Find the number(s).

20. The quotient of a number, and 9 times its reciprocal, is 1. Find the number(s).

21. If a number is added to the numerator of $\dfrac{12}{41}$ and twice the number is added to the denominator of $\dfrac{12}{41}$, the resulting fraction is equivalent to $\dfrac{1}{3}$. Find the number.

22. If a number is subtracted from the numerator of $\dfrac{13}{8}$ and added to the denominator of $\dfrac{13}{8}$, the resulting fraction is equivalent to $\dfrac{2}{5}$. Find the number.

Solve. See Example 3.

23. An Arabian camel can drink 15 gallons of water in 10 minutes. At this rate, how much water can the camel drink in 3 minutes? (*Source:* Grolier, Inc.)

24. An Arabian camel can travel 20 miles in 8 hours, carrying a 300-pound load on its back. At this rate, how far can the camel travel in 10 hours? (*Source:* Grolier, Inc.)

25. In 2000, 10.2 out of every 100 Coast Guard personnel were women. If there are 35,712 total Coast Guard personnel on active duty, estimate the number of women. Round to the nearest whole. (*Source: The World Almanac*, 2002)

26. In 2001, 43 out of every 50 Navy personnel were men. If there are 375,618 total Navy personnel on active duty, estimate the number of men. Round to the nearest whole. (*Source: The World Almanac*, 2002)

D *Solve. See Example 4.*

27. An experienced roofer can roof a house in 26 hours. A beginning roofer needs 39 hours to complete the same job. Find how long it takes for the two to do the job together.

28. Alan Cantrell can word process a research paper in 6 hours. With Steve Isaac's help, the paper can be processed in 4 hours. Find how long it takes Steve to word process the paper alone.

29. Three postal workers can sort a stack of mail in 20 minutes, 30 minutes, and 60 minutes, respectively. Find how long it takes them to sort the mail if all three work together.

30. A new printing press can print newspapers twice as fast as the old one can. The old one can print the afternoon edition in 4 hours. Find how long it takes to print the afternoon edition if both printers are operating.

E *Solve. See Example 5.*

31. Mattie Evans drove 150 miles in the same amount of time that it took a turbopropeller plane to travel 600 miles. The speed of the plane was 150 mph faster than the speed of the car. Find the speed of the plane.

32. An F-100 plane and a Toyota truck leave the same town at sunrise and head for a town 450 miles away. The speed of the plane is three times the speed of the truck, and the plane arrives 6 hours ahead of the truck. Find the speed of the truck.

33. The speed of Lazy River's current is 5 mph. If a boat travels 20 miles downstream in the same time that it takes to travel 10 miles upstream, find the speed of the boat in still water.

34. The speed of a boat in still water is 24 mph. If the boat travels 54 miles upstream in the same time that it takes to travel 90 miles downstream, find the speed of the current.

35. The sum of the reciprocals of two consecutive integers is $-\dfrac{15}{56}$. Find the two integers.

36. The sum of the reciprocals of two consecutive odd integers is $\dfrac{20}{99}$. Find the two integers.

37. One hose can fill a goldfish pond in 45 minutes, and two hoses can fill the same pond in 20 minutes. Find how long it takes the second hose alone to fill the pond.

38. If Sarah Clark can do a job in 5 hours and Dick Belli and Sarah working together can do the same job in 2 hours, find how long it takes Dick to do the job alone.

39. Two trains going in opposite directions leave at the same time. One train travels 15 mph faster than the other. In 6 hours the trains are 630 miles apart. Find the speed of each.

40. The speed of a bicyclist is 10 mph faster than the speed of a walker. If the bicyclist travels 26 miles in the same amount of time that the walker travels 6 miles, find the speed of the bicyclist.

41. A giant tortoise can travel 0.17 miles in 1 hour. At this rate, how long would it take the tortoise to travel 1 mile? Round to the nearest tenth of an hour. (*Source: The World Almanac, 2002*)

42. A black mamba snake can travel 88 feet in 3 seconds. At this rate, how long does it take to travel 300 feet (the length of a football field)? Round to the nearest tenth of a second. (*Source: The World Almanac, 2002*)

43. Moo Dairy has three machines to fill half-gallon milk cartons. The machines can fill the daily quota in 5 hours, 6 hours, and 7.5 hours, respectively. Find how long it takes to fill the daily quota if all three machines are running.

44. The inlet pipe of an oil tank can fill the tank in 1 hour, 30 minutes. The outlet pipe can empty the tank in 1 hour. Find how long it takes to empty a full tank if both pipes are open.

45. A plane flies 465 miles with the wind and 345 miles against the wind in the same length of time. If the speed of the wind is 20 mph, find the speed of the plane in still air.

46. Two rockets are launched. The first travels at 9000 mph. Fifteen minutes later the second is launched at 10,000 mph. Find the distance at which both rockets are an equal distance from Earth.

47. Two joggers, one averaging 8 mph and one averaging 6 mph, start from a designated initial point. The slower jogger arrives at the end of the run a half-hour after the other jogger. Find the distance of the run.

48. A semi truck travels 300 miles through the flatland in the same amount of time that it travels 180 miles through the Great Smoky Mountains. The rate of the truck is 20 miles per hour slower in the mountains than in the flatland. Find both the flatland rate and mountain rate.

49. Smith Engineering is in the process of reviewing the salaries of their surveyors. During this review, the company found that an experienced surveyor can survey a roadbed in 4 hours. An apprentice surveyor needs 5 hours to survey the same stretch of road. If the two work together, find how long it takes them to complete the job.

50. The numerator of a fraction is 4 less than the denominator. If both the numerator and the denominator are increased by 2, the resulting fraction is equivalent to $\frac{2}{3}$. Find the fraction.

51. The denominator of a fraction is 1 more than the numerator. If both the numerator and the denominator are decreased by 3, the resulting fraction is equivalent to $\frac{4}{5}$. Find the fraction.

52. An amateur cyclist training for a road race rode the first 20-mile portion of his workout at a constant rate. For the 16-mile cooldown portion of his workout, he reduced his speed by 2 miles per hour. Each portion of the workout took equal time. Find the cyclist's rate during the first portion and his rate during the cooldown portion.

53. In 2 minutes, a conveyor belt can move 300 pounds of recyclable aluminum from the delivery truck to a storage area. A smaller belt can move the same quantity of cans the same distance in 6 minutes. If both belts are used, find how long it takes to move the cans to the storage area.

54. Gary Marcus and Tony Alva work at Lombardo's Pipe and Concrete. Mr. Lombardo is preparing an estimate for a customer. He knows that Gary can lay a slab of concrete in 6 hours. Tony can lay the same size slab in 4 hours. If both work on the job and the cost of labor is $45.00 per hour, determine what the labor estimate should be.

55. While road testing a new make of car, the editor of a consumer magazine finds that she can go 10 miles into a 3-mile-per-hour wind in the same amount of time that she can go 11 miles with a 3-mile-per-hour wind behind her. Find the speed of the car in still air.

56. Mr. Dodson can paint his house by himself in four days. His son will need an additional day to complete the job if he works by himself. If they work together, find how long it takes to paint the house.

57. The world record for the largest white bass caught is held by Ronald Sprouse of Virginia. The bass weighed 6 pounds 13 ounces. If Ronald rows to his favorite fishing spot 9 miles downstream in the same amount of time that he rows 3 miles upstream and if the current is 6 mph, find how long it takes him to cover the 12 miles.

58. A marketing manager travels 1080 miles in a corporate jet and then an additional 240 miles by car. If the car ride takes 1 hour longer, and if the rate of the jet is 6 times the rate of the car, find the time the manager travels by jet and find the time she travels by car.

59. An experienced bricklayer can construct a small wall in 3 hours. An apprentice can complete the job in 6 hours. Find how long it takes if they work together.

Review and Preview

Solve each equation for x. See Section 2.1.

60. $\dfrac{x}{5} = \dfrac{x+2}{3}$

61. $\dfrac{x}{4} = \dfrac{x+3}{6}$

62. $\dfrac{x-3}{2} = \dfrac{x-5}{6}$

63. $\dfrac{x-6}{4} = \dfrac{x-2}{5}$

Combining Concepts

Calculating body-mass index (BMI) is a way to gauge whether a person should lose weight. Doctors recommend that body-mass index values fall between 19 and 25. The formula for body-mass index B is $B = \dfrac{705w}{h^2}$, where w is weight in pounds and h is height in inches. Use this formula to answer Exercises 64 and 65.

64. A patient is 5 ft 8 in. tall. What should his or her weight be to have a body-mass index of 25? Round to the nearest whole pound.

65. A doctor recorded a body-mass index of 47 on a patient's chart. Later, a nurse notices that the doctor recorded the patient's weight as 240 pounds but neglected to record the patient's height. Explain how the nurse can use the information from the chart to find the patient's height. Then find the height.

In physics, when the source of a sound is traveling toward an observer, the relationship between the actual pitch a of the sound and the pitch h that the observer hears due to the Doppler effect is described by the formula $h = \dfrac{a}{1 - \dfrac{s}{770}}$, where s is the speed of the sound source in miles per hour. Use this formula to answer Exercise 66.

66. An emergency vehicle has a single-tone siren with the pitch of the musical note E. As it approaches an observer standing by the road, the vehicle is traveling 50 mph. Is the pitch that the observer hears due to the Doppler effect lower or higher than the actual pitch? To which musical note is the pitch that the observer hears closest?

Pitch of an Octave of Musical Notes in Hertz (Hz)	
Note	**Pitch**
Middle C	261.63
D	293.66
E	329.63
F	349.23
G	392.00
A	440.00
B	493.88

Note: Greater numbers indicate higher pitches (acoustically).
(*Source:* American Standards Association)

In electronics, the relationship among the resistances R_1 and R_2 of two resistors wired in a parallel circuit and their combined resistance R is described by the formula $\dfrac{1}{R} = \dfrac{1}{R_1} + \dfrac{1}{R_2}$. Use this formula to solve Exercises 67 through 69.

67. If the combined resistance is 2 ohms and one of the two resistances is 3 ohms, find the other resistance.

68. Find the combined resistance of two resistors of 12 ohms each when they are wired in a parallel circuit.

69. The relationship among resistance of two resistors wired in a parallel circuit and their combined resistance may be extended to three resistors of resistances R_1, R_2, and R_3. Write an equation you think may describe the relationship, and use it to find the combined resistance if R_1 is 5, R_2 is 6, and R_3 is 2.

Internet Excursions

 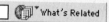

Body-mass index (BMI) determines a person's risk for weight-related health problems. One way to calculate BMI is to use the formula given with Exercises 64 and 65. Another way is to use one of the many sites on the World Wide Web that offer an interactive BMI calculator. The World Wide Web address listed above will direct you to a Web site with an interactive BMI calculator. You can calculate BMI by entering height in feet and inches, and weight in pounds.

70. Use the interactive BMI calculator to find the BMI for a person who is 5 ft 5 in. and weighs 145 pounds. Then verify your result using the formula given with Exercises 64 and 65.

71. Use the interactive BMI calculator to find your own BMI.

6.6 Variation and Problem Solving

A Solving Problems Involving Direct Variation

A very familiar example of **direct variation** is the relationship of the circumference C of a circle to its radius r. The formula $C = 2\pi r$ expresses that the circumference is always 2π times the radius. In other words, C is always a constant multiple (2π) of r. Because it is, we say that C *varies directly as r*, that C *varies directly with r*, or that C *is directly proportional to r*.

> **Direct Variation**
>
> **y varies directly as x**, or **y is directly proportional to x**, if there is a nonzero constant k such that
>
> $$y = kx$$
>
> The number k is called the **constant of variation** or the **constant of proportionality**.

In the above definition, the relationship described between x and y is a linear one. In other words, the graph of $y = kx$ is a line. The slope of the line is k, and the line passes through the origin.

For example, the graph of the direct variation equation $C = 2\pi r$ is shown. The horizontal axis represents the radius r, and the vertical axis is the circumference C. From the graph we can read that when the radius is 6 units, the circumference is approximately 38 units. Also, when the circumference is 45 units, the radius is between 7 and 8 units. Notice that as the radius increases, the circumference increases.

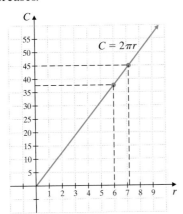

EXAMPLE 1

Suppose that y varies directly as x. If y is 5 when x is 30, find the constant of variation and the direct variation equation.

Solution: Since y varies directly as x, we write $y = kx$. If $y = 5$ when $x = 30$, we have that

$$y = kx$$
$$5 = k(30) \qquad \text{Replace } y \text{ with 5 and } x \text{ with 30.}$$
$$\frac{1}{6} = k \qquad \text{Solve for } k.$$

The constant of variation is $\frac{1}{6}$. After finding the constant of variation k, the direct variation equation can be written as $y = \frac{1}{6}x$.

Practice Problem 1

Suppose that y varies directly as x. If y is 24 when x is 8, find the constant of variation and the direct variation equation.

Answer

1. $k = 3$; $y = 3x$

Practice Problem 2

Use Hooke's law as stated in Example 2. If a 56-pound weight attached to a spring stretches the spring 8 inches, find the distance that an 85-pound weight attached to the spring stretches the spring.

EXAMPLE 2 Using Direct Variation and Hooke's Law

Hooke's law states that the distance a spring stretches is directly proportional to the weight attached to the spring. If a 40-pound weight attached to the spring stretches the spring 5 inches, find the distance that a 65-pound weight attached to the spring stretches the spring.

Solution:

1. UNDERSTAND. Read and reread the problem. Notice that we are given that the distance a spring stretches is *directly proportional* to the weight attached. We let

   d = the distance stretched

   w = the weight attached

   The constant of variation is represented by k.

2. TRANSLATE. Because d is directly proportional to w, we write
   $$d = kw$$

3. SOLVE. When a weight of 40 pounds is attached, the spring stretches 5 inches. That is, when $w = 40$, $d = 5$.

   $$5 = k(40) \quad \text{Replace } d \text{ with 5 and } w \text{ with 40.}$$

   $$\frac{1}{8} = k \quad \text{Solve for } k.$$

   Now when we replace k with $\frac{1}{8}$ in the equation $d = kw$, we have

   $$d = \frac{1}{8}w$$

   To find the stretch when a weight of 65 pounds is attached, we replace w with 65 to find d.

   $$d = \frac{1}{8}(65)$$

   $$= \frac{65}{8} = 8\frac{1}{8} \text{ or } 8.125$$

4. INTERPRET.

Check: Check the proposed solution of 8.125 inches in the original problem.

State: The spring stetches 8.125 inches when a 65-pound weight is attached.

Answer

2. $12\frac{1}{7}$in.

B   Solving Problems Involving Inverse Variation

When y is proportional to the *reciprocal* of another variable x, we say that *y varies inversely as x*, or that *y is inversely proportional to x*. An example of the **inverse variation** relationship is the relationship between the pressure that a gas exerts and the volume of its container. As the volume of a container decreases, the pressure of the gas it contains increases.

Inverse Variation

y **varies inversely as** *x*, or *y* **is inversely proportional to** *x*, if there is a nonzero constant k such that

$$y = \frac{k}{x}$$

The number k is called the **constant of variation** or the **constant of proportionality**.

Notice that $y = \dfrac{k}{x}$ is an equation containing a rational expression. Its graph for $k > 0$ and $x > 0$ is shown. From the graph, we can see that as x increases, y decreases.

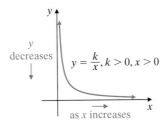

EXAMPLE 3

Suppose that u varies inversely as w. If u is 3 when w is 5, find the constant of variation and the inverse variation equation.

Solution: Since u varies inversely as w, we have $u = \dfrac{k}{w}$. We let $u = 3$ and $w = 5$, and we solve for k.

$$u = \frac{k}{w}$$

$$3 = \frac{k}{5} \qquad \text{Let } u = 3 \text{ and } w = 5.$$

$$15 = k \qquad \text{Multiply both sides by 5.}$$

The constant of variation k is 15. This gives the inverse variation equation

$$u = \frac{15}{w}$$

Practice Problem 3

Suppose that y varies inversely as x. If y is 6 when x is 3, find the constant of variation and the inverse variation equation.

Answer

3. $k = 18$; $y = \dfrac{18}{x}$

Practice Problem 4

The speed r at which one needs to drive in order to travel a constant distance is inversely proportional to the time t. A fixed distance can be driven in 5 hours at a rate of 24 mph. Find the rate needed to drive the same distance in 4 hours.

EXAMPLE 4 Using Inverse Variation and Boyle's Law

Boyle's law says that if the temperature stays the same, the pressure P of a gas is inversely proportional to the volume V. If a cylinder in a steam engine has a pressure of 960 kilopascals when the volume is 1.4 cubic meters, find the pressure when the volume increases to 2.5 cubic meters.

Solution:

1. UNDERSTAND. Read and reread the problem. Notice that we are given that the pressure of a gas is *inversely proportional* to the volume. We will let P = the pressure and V = the volume. The constant of variation is represented by k.

2. TRANSLATE. Because P is inversely proportional to V, we write

$$P = \frac{k}{V}$$

When P = 960 kilopascals, the volume V = 1.4 cubic meters. We use this information to find k.

$$960 = \frac{k}{1.4} \qquad \text{Let } P = 960 \text{ and } V = 1.4.$$

$$1344 = k \qquad \text{Multiply both sides by 1.4.}$$

Thus, the value of k is 1344. Replacing k with 1344 in the variation equation, we have

$$P = \frac{1344}{V}$$

Next we find P when V is 2.5 cubic meters.

3. SOLVE.

$$P = \frac{1344}{2.5} \qquad \text{Let } V = 2.5$$

$$= 537.6$$

4. INTERPRET. *Check* the proposed solution in the original problem.

State: When the volume is 2.5 cubic meters, the pressure is 537.6 kilopascals. ●

ⓒ Solving Problems Involving Joint Variation

Sometimes the ratio of a variable to the product of many other variables is constant. For example, the ratio of distance traveled to the product of speed and time traveled is constantly 1:

$$\frac{d}{rt} = 1 \qquad \text{or} \qquad d = rt$$

Such a relationship is called **joint variation**.

Answer

4. 30 mph

Joint Variation

If the ratio of a variable y to the product of two or more variables is constant, then **y varies jointly as**, or **is jointly proportional to**, the other variables. If

$$y = kxz$$

then the number k is the **constant of variation** or the **constant of proportionality**.

Try the Concept Check in the margin.

Concept Check

Which type of variation is represented by the equation $xy = 8$? Explain.

a. Direct variation

b. Inverse variation

c. Joint variation

⟋ EXAMPLE 5

The lateral surface area of a cylinder varies jointly as its radius and height. Express surface area S in terms of radius r and height h.

Solution: Because the surface area varies jointly as the radius r and the height h, we equate S to a constant multiple of r and h:

$$S = krh$$

Note: From actual values of $S, r,$ and h, it can be determined that the constant k is 2π, and we then have the formula $S = 2\pi rh$. ●

(D)   Solving Problems Involving Combined Variation

Some examples of variation involve combinations of direct, inverse, and joint variation. We will call these variations **combined variation**.

EXAMPLE 6

Suppose that y varies directly as the square of x. If y is 24 when x is 2, find the constant of variation and the variation equation.

Solution: Since y varies directly as the square of x, we have

$$y = kx^2$$

Now let $y = 24$ and $x = 2$ and solve for k.

$$y = kx^2$$
$$24 = k \cdot 2^2$$
$$24 = 4k$$
$$6 = k$$

The constant of variation is 6, so the variation equation is

$$y = 6x^2$$

Practice Problem 5   △

The area of a triangle varies jointly as its base and height. Express the area in terms of base b and height h.

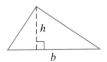

Practice Problem 6

Suppose that y varies inversely as the square of x. If y is 24 when x is 2, find the constant of variation and the variation equation.

Answers

5. $A = kbh$,   **6.** $k = 96$; $y = \dfrac{96}{x^2}$

Concept Check: b; answers may vary

Practice Problem 7   △

The maximum weight that a rectangular beam can support varies jointly as its width and the square of its height and inversely as its length. If a beam $\frac{1}{3}$ foot wide, 1 foot high, and 10 feet long can support 3 tons, find how much weight a similar beam can support if it is 1 foot wide, $\frac{1}{3}$ foot high, and 9 feet long.

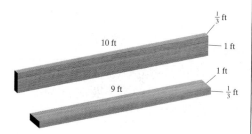

EXAMPLE 7   Using Combined Variation

The maximum weight that a circular column can support is directly proportional to the fourth power of its diameter and inversely proportional to the square of its height. A 2-meter-wide column that is 8 meters in height can support 1 ton. Find the weight that a 1-meter-wide column that is 4 meters in height can support.

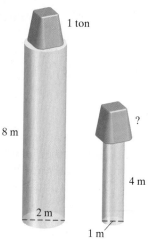

Solution:

1. UNDERSTAND. Read and reread the problem. Let w = weight, d = diameter, h = height, and k = the constant of variation.

2. TRANSLATE. Since w is directly proportional to d^4 and inversely proportional to h^2, we have

    $$w = \frac{kd^4}{h^2}$$

3. SOLVE. To find k, we are given that a 2-meter-wide column that is 8 meters in height can support 1 ton. That is, $w = 1$ when $d = 2$ and $h = 8$, or

    $$1 = \frac{k \cdot 2^4}{8^2} \qquad \text{Let } w = 1, d = 2, \text{ and } h = 8.$$

    $$1 = \frac{k \cdot 16}{64}$$

    $$4 = k \qquad \text{Solve for } k.$$

    Now we replace k with 4 in the equation $w = \frac{kd^4}{h^2}$:

    $$w = \frac{4d^4}{h^2}$$

    To find weight, w, for a 1-meter-wide column that is 4 meters in height, we let $d = 1$ and $h = 4$.

    $$w = \frac{4 \cdot 1^4}{4^2}$$

    $$w = \frac{4}{16} = \frac{1}{4}$$

4. INTERPRET. *Check* the proposed solution in the original problem.

State: The 1-meter-wide column that is 4 meters in height can hold $\frac{1}{4}$ ton of weight. ●

Answer

7. $1\frac{1}{9}$ tons

EXERCISE SET 6.6

A *If y varies directly as x, find the constant of variation and the direct variation equation for each situation. See Example 1.*

1. $y = 4$ when $x = 20$

2. $y = 5$ when $x = 30$

3. $y = 6$ when $x = 4$

4. $y = 12$ when $x = 8$

5. $y = 7$ when $x = \dfrac{1}{2}$

6. $y = 11$ when $x = \dfrac{1}{3}$

7. $y = 0.2$ when $x = 0.8$

8. $y = 0.4$ when $x = 2.5$

Solve. See Example 2.

9. The weight of a synthetic ball varies directly with the cube of its radius. A ball with a radius of 2 inches weighs 1.20 pounds. Find the weight of a ball of the same material with a 3-inch radius.

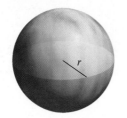

10. At sea, the distance to the horizon is directly proportional to the square root of the elevation of the observer. If a person who is 36 feet above the water can see 7.4 miles, find how far a person 64 feet above the water can see. Round to the nearest tenth of a mile.

11. The amount P of pollution varies directly with the population N of people. Kansas City has a population of 442,000 and produces 260,000 tons of pollutants. Find how many tons of pollution we should expect St. Louis to produce, if we know that its population is 348,000. Round to the nearest whole ton. (*Population Source: The World Almanac, 2002*)

12. Charles's law states that if the pressure P stays the same, the volume V of a gas is directly proportional to its temperature T. If a balloon is filled with 20 cubic meters of a gas at a temperature of 300 K, find the new volume if the temperature rises to 360 K while the pressure stays the same.

B *If y varies inversely as x, find the constant of variation and the inverse variation equation for each situation. See Example 3.*

13. $y = 6$ when $x = 5$

14. $y = 20$ when $x = 9$

15. $y = 100$ when $x = 7$

16. $y = 63$ when $x = 3$

17. $y = \dfrac{1}{8}$ when $x = 16$

18. $y = \dfrac{1}{10}$ when $x = 40$

19. $y = 0.2$ when $x = 0.7$

20. $y = 0.6$ when $x = 0.3$

Solve. See Example 4.

21. Pairs of markings a set distance apart are made on highways so that police can detect drivers exceeding the speed limit. Over a fixed distance, the speed R varies inversely with the time T. In one particular pair of markings, R is 45 mph when T is 6 seconds. Find the speed of a car that travels the given distance in 5 seconds.

22. The weight of an object on or above the surface of Earth varies inversely as the square of the distance between the object and Earth's center. If a person weighs 160 pounds on Earth's surface, find the individual's weight if he moves 200 miles above Earth. Round to the nearest whole pound. (Assume that Earth's radius is 4000 miles.)

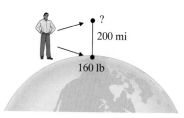

23. If the voltage V in an electric circuit is held constant, the current I is inversely proportional to the resistance R. If the current is 40 amperes when the resistance is 270 ohms, find the current when the resistance is 150 ohms.

24. Because it is more efficient to produce larger numbers of items, the cost of producing Dysan computer disks is inversely proportional to the number produced. If 4000 can be produced at a cost of $1.20 each, find the cost per disk when 6000 are produced.

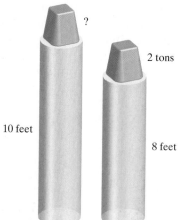

25. The intensity I of light varies inversely as the square of the distance d from the light source. If the distance from the light source is doubled (see the figure), determine what happens to the intensity of light at the new location.

△ **26.** The maximum weight that a circular column can hold is inversely proportional to the square of its height. If an 8-foot column can hold 2 tons, find how much weight a 10-foot column can hold.

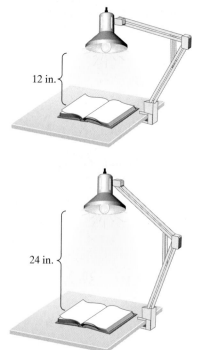

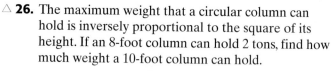

27. x varies jointly as y and z.

28. P varies jointly as R and the square of S.

29. r varies jointly as s and the cube of t.

30. a varies jointly as b and c.

For each statement, find the constant of variation and the variation equation. See Examples 5 and 6.

31. y varies directly as the cube of x; $y = 9$ when $x = 3$

32. y varies directly as the cube of x; $y = 32$ when $x = 4$

33. y varies directly as the square root of x; $y = 0.4$ when $x = 4$

34. y varies directly as the square root of x; $y = 2.1$ when $x = 9$

35. y varies inversely as the square of x; $y = 0.052$ when $x = 5$

36. y varies inversely as the square of x; $y = 0.011$ when $x = 10$

37. y varies jointly as x and the cube of z; $y = 120$ when $x = 5$ and $z = 2$

38. y varies jointly as x and the square of z; $y = 360$ when $x = 4$ and $z = 3$

Solve. See Example 6.

39. The maximum weight that a rectangular beam can support varies jointly as its width and the square of its height and inversely as its length. If a beam $\frac{1}{2}$ foot wide, $\frac{1}{3}$ foot high, and 10 feet long can support 12 tons find how much a similar beam can support if the beam is $\frac{2}{3}$ foot wide, $\frac{1}{2}$ foot high, and 16 feet long.

40. The number of cars manufactured on an assembly line at a General Motors plant varies jointly as the number of workers and the time they work. If 200 workers can produce 60 cars in 2 hours, find how many cars 240 workers should be able to make in 3 hours.

△ **41.** The volume of a cone varies jointly as its height and the square of its radius. If the volume of a cone is 32π cubic inches when the radius is 4 inches and the height is 6 inches, find the volume of a cone when the radius is 3 inches and the height is 5 inches.

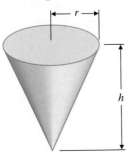

△ **42.** When a wind blows perpendicularly against a flat surface, its force is jointly proportional to the surface area and the speed of the wind. A sail whose surface area is 12 square feet experiences a 20-pound force when the wind speed is 10 miles per hour. Find the force on an 8-square-foot sail if the wind speed is 12 miles per hour.

43. The intensity of light (in foot-candles) varies inversely as the square of x, the distance in feet from the light source. The intensity of light 2 feet from the source is 80 foot-candles. How far away is the source if the intensity of light is 5 foot-candles?

44. The horsepower that can be safely transmitted to a shaft varies jointly as the shaft's angular speed of rotation (in revolutions per minute) and the cube of its diameter. A 2-inch shaft making 120 revolutions per minute safely transmits 40 horsepower. Find how much horsepower can be safely transmitted by a 3-inch shaft making 80 revolutions per minute.

Review and Preview

Find the exact circumference and area of each circle. See the inside cover for a list of geometric formulas.

△ **45.**

6 cm

△ **46.**

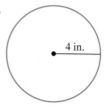

4 in.

△ **47.**

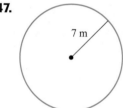

7 m

△ **48.**

9 cm

Find the slope of the line containing each pair of points. See Section 3.2.

49. $(3, 6), (-2, 6)$ **50.** $(-5, -2), (0, 7)$ **51.** $(4, -1), (5, -2)$ **52.** $(2, 1), (2, -3)$

Combining Concepts

△ **53.** The volume of a cylinder varies jointly as the height and the square of the radius. If the height is halved and the radius is doubled, determine what happens to the volume.

54. The horsepower to drive a boat varies directly as the cube of the speed of the boat. If the speed of the boat is to double, determine the corresponding increase in horsepower required.

55. Suppose that y varies directly as x^2. If x is doubled, what is the effect on y?

56. Suppose that y varies directly as x. If x is doubled, what is the effect on y?

MATERIALS:

- dried beans
- felt-tip marker
- bowl or bag
- measuring cup or small paper cup

This activity may be completed by working in groups or individually.

In wildlife management, conservationists sometimes need to know the size of a certain wildlife population—the number of a certain species of fish in a lake or the number of certain birds in a geographic region. In manufacturing, it might be necessary to estimate the number of parts in a large storage bin. In either case, actually counting the number of animals or parts would be either too difficult or too time consuming.

When it is necessary to know the approximate size of a population, it is sometimes useful to use sampling techniques to estimate the population size. There are several different ways to take samples, but no matter which way is used, a rational equation is solved to estimate the size of the population. One sampling method is the **capture–recapture method**. In this method, an initial sample is taken (or captured) from the population and then tagged or marked. The tagged sample is counted and then returned to the population. After the marked portion of the population has been allowed to thoroughly mix with the unmarked portion, a second, or recapture, sample is taken. If the population has been well mixed and the second sample is taken randomly, the fraction of marked units in the recapture sample should approximate the fraction of marked units in the entire population. This information leads to the following equation that can be solved for the size of the population x:

$$\frac{\text{number of marked}}{x} = \frac{\text{number of marked}}{\text{total size of}}$$

Another sampling method is the **addition method**. In this method, rather than capturing, marking, and returning an initial sample, a known number of nearly identical marked units are *added* to the original population. It is important to understand that these marked units were not part of the original population. Once the marked units have been thoroughly mixed with the original unmarked population, a sample is taken. This method leads to the following equation that can be solved for the size of the *original* population x:

$$\frac{\substack{\text{number of marked units} \\ \text{added to original population}}}{\substack{x + \text{number of marked units} \\ \text{added to original population}}} = \frac{\substack{\text{number of marked} \\ \text{units in sample}}}{\substack{\text{total size} \\ \text{of sample}}}$$

The addition method is very useful in situations where it would be (1) unwise to mark or damage existing members of the population, (2) difficult to take an initial sample from the population to be reintroduced after tagging, and/or (3) feasible to introduce nearly identical but marked units into the original population. In some situations, such as in wildlife management, where it would be impossible or unwise to introduce nearly identical units into an existing population, the capture–recapture method is the better choice.

Method 1

1. Fill a bowl or bag with an unknown number of dried beans. Your goal is to estimate the size of this population of beans.

2. Using a measuring cup or small paper cup, take a sample of beans from the bowl. Count and record the number of beans in the sample. Mark each bean with several Xs using a felt-tip marker. Then return the marked beans to the bowl.

3. Thoroughly mix the beans. Take a sample from the bowl. Count and record the total number of beans in the sample and the number of marked beans in the sample. Then return the bean sample to the bowl. Repeat this sampling process a total of five times to complete the table for Method 1.

4. Which sampling method does this procedure represent?

5. For each sample in the table, calculate an estimate of the number of beans.

6. If you had to give a single estimate of the number of beans, what would it be? Explain.

(continued)

7. Count the total bean population. How close are your individual estimates? Single estimate?

Method 1			
Sample	Total Number of Beans in Sample	Number of Marked Beans in Sample	Population Estimate
1			
2			
3			
4			
5			

Method 2

8. Gather or make a small supply of marked beans. Count and record this number of marked beans.

9. Start with a new and different population of dried beans in a bowl or bag. (*Note:* If reusing beans from Method 1, make sure that any and all marked beans have been removed.) Your goal is to estimate the size of this population of beans.

10. Add the marked beans you gathered to the bowl.

11. Thoroughly mix the beans. Take a sample from the bowl. Count and record the total number of beans in the sample and the number of marked beans in the sample. Then return the bean sample to the bowl. Repeat this sampling process a total of five times to complete the table for Method 2.

12. Which sampling method does this procedure represent?

13. For each sample in the table, calculate an estimate of the number of beans.

14. If you had to give a single estimate of the number of beans, what would it be? Explain.

15. Count the original bean population. How close are your individual estimates? Single estimate?

Method 2			
Sample	Total Number of Beans in Sample	Number of Marked Beans in Sample	Population Estimate
1			
2			
3			
4			
5			

Fill in each blank with one of the words or phrases listed below.

rational expression	equation	complex fraction	opposites	directly
least common denominator	expression	inversely	jointly	

1. A rational expression whose numerator, denominator, or both contain one or more rational expressions is called a

_____.

2. In the equation $y = kx$, y varies _____ as x.

3. In the equation $y = \dfrac{k}{x}$, y varies _____ as x.

4. The _____ of a list of rational expressions is a polynomial of least degree whose factors include the denominator factors in the list.

5. In the equation $y = kxz$, y varies _____ as x and z.

6. The expressions $(x - 5)$ and $(5 - x)$ are called _____.

7. A _____ is an expression that can be written as the quotient of $\dfrac{P}{Q}$ of two polynomials P and Q as long as Q is not 0.

8. Which is an expression and which is an equation? An example of an _____ is $\dfrac{2}{x} + \dfrac{2}{x^2} = 7$ and an example of an _____ is $\dfrac{2}{x} + \dfrac{5}{x^2}$.

CHAPTER 6

Highlights

DEFINITIONS AND CONCEPTS	EXAMPLES

Section 6.1   Multiplying and Dividing Rational Expressions

A **rational expression** is the quotient $\dfrac{P}{Q}$ of two polynomials P and Q, as long as Q is not 0.	$\dfrac{2x - 6}{7}, \dfrac{t^2 - 3t + 5}{t - 1}$

SIMPLIFYING A RATIONAL EXPRESSION

Step 1. Completely factor the numerator and the denominator.

Step 2. Apply the fundamental principle of rational expressions.

Simplify.

$$\frac{2x^2 + 9x - 5}{x^2 - 25} = \frac{(2x - 1)(x + 5)}{(x - 5)(x + 5)}$$
$$= \frac{2x - 1}{x - 5}$$

MULTIPLYING RATIONAL EXPRESSIONS

Step 1. Completely factor numerators and denominators.

Step 2. Multiply the numerators and multiply the denominators.

Step 3. Apply the fundamental principle of rational expressions.

Multiply: $\dfrac{x^3 + 8}{12x - 18} \cdot \dfrac{14x^2 - 21x}{x^2 + 2x}$

$$= \frac{(x + 2)(x^2 - 2x + 4)}{6(2x - 3)} \cdot \frac{7x(2x - 3)}{x(x + 2)}$$
$$= \frac{7(x^2 - 2x + 4)}{6}$$

DIVIDING RATIONAL EXPRESSIONS

Multiply the first rational expression by the reciprocal of the second rational expression.

Divide: $\dfrac{x^2 + 6x + 9}{5xy - 5y} \div \dfrac{x + 3}{10y}$

$$= \frac{(x + 3)(x + 3)}{5y(x - 1)} \cdot \frac{2 \cdot 5y}{x + 3}$$
$$= \frac{2(x + 3)}{x - 1}$$

A **rational function** is a function described by a rational expression.

$$f(x) = \frac{2x - 6}{7}, \quad h(t) = \frac{t^2 - 3t + 5}{t - 1}$$

DEFINITIONS AND CONCEPTS	EXAMPLES

Section 6.2   Adding and Subtacting Rational Expressions

ADDING OR SUBTRACTING RATIONAL EXPRESSIONS

Step 1. Find the LCD.

Step 2. Write each rational expression as an equivalent rational expression whose denominator is the LCD.

Step 3. Add or subtract numerators and write the sum or difference over the common denominator.

Step 4. Simplify the result.

Subtract: $\dfrac{3}{x+2} - \dfrac{x+1}{x-3}$

$$= \dfrac{3\cdot(x-3)}{(x+2)\cdot(x-3)} - \dfrac{(x+1)\cdot(x+2)}{(x-3)\cdot(x+2)}$$

$$= \dfrac{3(x-3) - (x+1)(x+2)}{(x+2)(x-3)}$$

$$= \dfrac{3x-9 - (x^2+3x+2)}{(x+2)(x-3)}$$

$$= \dfrac{3x-9 - x^2 - 3x - 2}{(x+2)(x-3)}$$

$$= \dfrac{-x^2 - 11}{(x+2)(x-3)}$$

Section 6.3   Simplifying Complex Fractions

Method 1: Simplify the numerator and the denominator so that each is a single fraction. Then perform the indicated division and simplify if possible.

Method 2: Multiply the numerator and the denominator of the complex fraction by the LCD of the fractions in both the numerator and the denominator. Then simplify if possible.

Simplify: $\dfrac{\dfrac{x+2}{x}}{x - \dfrac{4}{x}}$

Method 1: $\dfrac{\dfrac{x+2}{x}}{\dfrac{x\cdot x}{1\cdot x} - \dfrac{4}{x}} = \dfrac{\dfrac{x+2}{x}}{\dfrac{x^2-4}{x}}$

$$= \dfrac{x+2}{x} \cdot \dfrac{x}{(x+2)(x-2)} = \dfrac{1}{x-2}$$

Method 2: $\dfrac{\left(\dfrac{x+2}{x}\right)\cdot x}{\left(x - \dfrac{4}{x}\right)\cdot x} = \dfrac{x+2}{x\cdot x - \dfrac{4}{x}\cdot x}$

$$= \dfrac{x+2}{x^2-4} = \dfrac{x+2}{(x+2)(x-2)} = \dfrac{1}{x-2}$$

Section 6.4   Solving Equations Containing Rational Expressions

SOLVING AN EQUATION CONTAINING RATIONAL EXPRESSIONS

Multiply both sides of the equation by the LCD of all rational expressions. Then use the distributive property and simplify. Solve the resulting equation and then check each proposed solution to see whether it makes any denominator 0. Discard any solutions that do.

Solve: $x - \dfrac{3}{x} = \dfrac{1}{2}$

$$2x\left(x - \dfrac{3}{x}\right) = 2x\left(\dfrac{1}{2}\right) \qquad \text{The LCD is } 2x.$$

$$2x\cdot x - 2x\left(\dfrac{3}{x}\right) = 2x\left(\dfrac{1}{2}\right) \qquad \text{Distribute.}$$

$$2x^2 - 6 = x$$

$$2x^2 - x - 6 = 0 \qquad \text{Subtract } x \text{ from both sides.}$$

$$(2x+3)(x-2) = 0 \qquad \text{Factor.}$$

$$x = -\dfrac{3}{2} \quad \text{or} \quad x = 2$$

Both $-\dfrac{3}{2}$ and 2 check. The solution set is $\left\{2, -\dfrac{3}{2}\right\}$.

DEFINITIONS AND CONCEPTS	**EXAMPLES**

Section 6.5   Rational Equations and Problem Solving

SOLVING AN EQUATION FOR A SPECIFIED VARIABLE

Treat the specified variable as the only variable of the equation and solve as usual.

Solve for x.

$$A = \frac{2x + 3y}{5}$$

$$5A = 2x + 3y \qquad \text{Multiply both sides by 5.}$$

$$5A - 3y = 2x \qquad \text{Subtract } 3y \text{ from both sides.}$$

$$\frac{5A - 3y}{2} = x \qquad \text{Divide both sides by 2.}$$

Jeanee and David Dillon volunteer every year to clean a strip of Lake Ponchartrain beach. Jeanee can clean all the trash in this area of beach in 6 hours; David takes 5 hours. Find how long it will take them to clean the area of beach together.

1. UNDERSTAND.

1. Read and reread the problem. Let x = time in hours that it takes Jeanee and David to clean the beach together.

	Hours to Complete	**Part Completed in 1 Hour**
Jeanee Alone	6	$\frac{1}{6}$
David Alone	5	$\frac{1}{5}$
Together	x	$\frac{1}{x}$

2. TRANSLATE.

2. In words:

part Jeanee can complete in 1 hour	+	part David can complete in 1 hour	=	part they can complete together in 1 hour
↓		↓		↓

Translate: $\quad \dfrac{1}{6} \quad + \quad \dfrac{1}{5} \quad = \quad \dfrac{1}{x}$

3. SOLVE.

3.

$$\frac{1}{6} + \frac{1}{5} = \frac{1}{x}$$

$$5x + 6x = 30 \qquad \text{Multiply both sides by } 30x.$$

$$11x = 30$$

$$x = \frac{30}{11} \quad \text{or} \quad 2\frac{8}{11}$$

4. INTERPRET.

4. *Check* and then *state*. Together, they can clean the beach in $2\frac{8}{11}$ hours.

DEFINITIONS AND CONCEPTS	EXAMPLES

Section 6.6 Variation and Problem Solving

y **varies directly** as *x*, or *y* is **directly proportional** to *x*, if there is a nonzero constant *k* such that

$$y = kx$$

The circumference of a circle *C* varies directly as its radius *r*.

$$C = \underbrace{2\pi}_{k} r$$

y **varies inversely** as *x*, or *y* is **inversely proportional** to *x*, if there is a nonzero constant *k* such that

$$y = \frac{k}{x}$$

Pressure *P* varies inversely with volume *V*.

$$P = \frac{k}{V}$$

y **varies jointly** as *x* and *z*, or *y* is **jointly proportional** to *x* and *z*, if there is a nonzero constant *k* such that

$$y = kxz$$

The lateral surface area *S* of a cylinder varies jointly as its radius *r* and height *h*.

$$S = \underbrace{2\pi}_{k} rh$$

Chapter 6 Review

(6.1) *Find all numbers for which each rational expression is undefined.*

1. $\dfrac{3 - 5x}{7}$

2. $\dfrac{2x + 4}{11}$

3. $\dfrac{-3x^2}{x - 5}$

4. $\dfrac{4x}{3x - 12}$

5. $\dfrac{x^3 + 2}{x^2 + 8x}$

6. $\dfrac{20}{3x^2 - 48}$

Write each rational expression in lowest terms.

7. $\dfrac{15x^4}{45x^2}$

8. $\dfrac{x + 2}{2 + x}$

9. $\dfrac{18m^6 p^2}{10m^4 p}$

10. $\dfrac{x - 12}{12 - x}$

11. $\dfrac{5x - 15}{25x - 75}$

12. $\dfrac{22x + 8}{11x + 4}$

13. $\dfrac{2x}{2x^2 - 2x}$

14. $\dfrac{x + 7}{x^2 - 49}$

15. $\dfrac{2x^2 + 4x - 30}{x^2 + x - 20}$

16. $\dfrac{xy - 3x + 2y - 6}{x^2 + 4x + 4}$

17. The average cost of manufacturing x bookcases is given by the rational function.

$$C(x) = \frac{35x + 4200}{x}$$

   a. Find the average cost per bookcase of manufacturing 50 bookcases.

   b. Find the average cost per bookcase of manufacturing 100 bookcases.

   c. As the number of bookcases increases, does the average cost per bookcase increase or decrease? (See parts (a) and (b).)

Perform each indicated operation. Write your answers in lowest terms.

18. $\dfrac{5}{x^3} \cdot \dfrac{x^2}{15}$

19. $\dfrac{3x^4 y z^3}{15x^2 y^2} \cdot \dfrac{10xy}{z^6}$

20. $\dfrac{4 - x}{5} \cdot \dfrac{15}{2x - 8}$

21. $\dfrac{x^2 - 6x + 9}{2x^2 - 18} \cdot \dfrac{4x + 12}{5x - 15}$

22. $\dfrac{a - 4b}{a^2 + ab} \cdot \dfrac{b^2 - a^2}{8b - 2a}$

23. $\dfrac{x^2 - x - 12}{2x^2 - 32} \cdot \dfrac{x^2 + 8x + 16}{3x^2 + 21x + 36}$

24. $\dfrac{2x^3 + 54}{5x^2 + 5x - 30} \cdot \dfrac{6x + 12}{3x^2 - 9x + 27}$

25. $\dfrac{3}{4x} \div \dfrac{8}{2x^2}$

26. $\dfrac{4x + 8y}{3} \div \dfrac{5x + 10y}{9}$

27. $\dfrac{5ab}{14c^3} \div \dfrac{10a^4b^2}{6ac^5}$

28. $\dfrac{2}{5x} \div \dfrac{4 - 18x}{6 - 27x}$

29. $\dfrac{x^2 - 25}{3} \div \dfrac{x^2 - 10x + 25}{x^2 - x - 20}$

30. $\dfrac{a - 4b}{a^2 + ab} \div \dfrac{20b - 5a}{b^2 - a^2}$

31. $\dfrac{7x + 28}{2x + 4} \div \dfrac{x^2 + 2x - 8}{x^2 - 2x - 8}$

32. $\dfrac{3x + 3}{x - 1} \div \dfrac{x^2 - 6x - 7}{x^2 - 1}$

33. $\dfrac{2x - x^2}{x^3 - 8} \div \dfrac{x^2}{x^2 + 2x + 4}$

34. $\dfrac{5a^2 - 20}{a^3 + 2a^2 + a + 2} \div \dfrac{7a}{a^3 + a}$

35. $\dfrac{2a}{21} \div \dfrac{3a^2}{7} \cdot \dfrac{4}{a}$

36. $\dfrac{5x - 15}{3 - x} \cdot \dfrac{x + 2}{10x + 20} \cdot \dfrac{x^2 - 9}{x^2 - x - 6}$

37. $\dfrac{4a + 8}{5a^2 - 20} \cdot \dfrac{3a^2 - 6a}{a + 3} \div \dfrac{2a^2}{5a + 15}$

(6.2)   *Find the LCD of the rational expressions in each list.*

38. $\dfrac{4}{9}, \dfrac{5}{2}$

39. $\dfrac{5}{4x^2y^5}, \dfrac{3}{10x^2y^4}, \dfrac{x}{6y^4}$

40. $\dfrac{5}{2x}, \dfrac{7}{x - 2}$

41. $\dfrac{3}{5x}, \dfrac{2}{x - 5}$

42. $\dfrac{1}{5x^3}, \dfrac{4}{x^2 + 3x - 28}, \dfrac{11}{10x^2 - 30x}$

Perform each indicated operation. Write your answers in lowest terms.

43. $\dfrac{2}{15} + \dfrac{4}{15}$

44. $\dfrac{4}{x - 4} + \dfrac{x}{x - 4}$

45. $\dfrac{4}{3x^2} + \dfrac{2}{3x^2}$

46. $\dfrac{1}{x-2} - \dfrac{1}{4-2x}$

47. $\dfrac{2x+1}{x^2+x-6} + \dfrac{2-x}{x^2+x-6}$

48. $\dfrac{7}{2x} + \dfrac{5}{6x}$

49. $\dfrac{1}{3x^2y^3} - \dfrac{1}{5x^4y}$

50. $\dfrac{1}{10-x} + \dfrac{x-1}{x-10}$

51. $\dfrac{x-2}{x+1} - \dfrac{x-3}{x-1}$

52. $\dfrac{x}{9-x^2} - \dfrac{2}{5x-15}$

53. $2x+1 - \dfrac{1}{x-3}$

54. $\dfrac{2}{a^2-2a+1} + \dfrac{3}{a^2-1}$

55. $\dfrac{x}{9x^2+12x+16} - \dfrac{3x+4}{27x^3-64}$

Perform each indicated operation. Write your answers in lowest terms.

56. $\dfrac{2}{x-1} - \dfrac{3x}{3x-3} + \dfrac{1}{2x-2}$

57. $\dfrac{2}{x^2-16} - \dfrac{3x}{x^2+8x+16} + \dfrac{3}{x+4}$

△ **58.** Find the perimeter of the heptagon (a polygon with seven sides.)

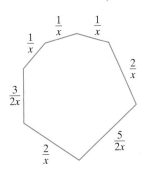

(6.3) *Simplify each complex fraction.*

59. $\dfrac{\dfrac{2x}{5}}{\dfrac{3x}{5}}$

60. $\dfrac{1 - \dfrac{3x}{4}}{2 + \dfrac{x}{4}}$

61. $\dfrac{\dfrac{1}{x} - \dfrac{2}{3x}}{\dfrac{5}{2x} - \dfrac{1}{3}}$

62. $\dfrac{\dfrac{x^2}{15}}{\dfrac{x+1}{5x}}$

63. $\dfrac{\dfrac{3}{y^2}}{\dfrac{6}{y^3}}$

64. $\dfrac{\dfrac{x+2}{3}}{\dfrac{5}{x-2}}$

65. $\dfrac{2-\dfrac{3}{2x}}{x-\dfrac{2}{5x}}$

66. $\dfrac{1+\dfrac{x}{y}}{\dfrac{x^2}{y^2}-1}$

67. $\dfrac{\dfrac{5}{x}+\dfrac{1}{xy}}{\dfrac{3}{x^2}}$

68. $\dfrac{\dfrac{x}{3}-\dfrac{3}{x}}{1+\dfrac{3}{x}}$

69. $\dfrac{\dfrac{1}{x-1}+1}{\dfrac{1}{x+1}-1}$

70. $\dfrac{2}{1-\dfrac{2}{x}}$

71. $\dfrac{1}{1+\dfrac{2}{1-\dfrac{1}{x}}}$

72. $\dfrac{\dfrac{x^2+5x-6}{4x+3}}{\dfrac{(x+6)^2}{8x+6}}$

73. $\dfrac{\dfrac{x-3}{x+3}+\dfrac{x+3}{x-3}}{\dfrac{x-3}{x+3}-\dfrac{x+3}{x-3}}$

74. $\dfrac{\dfrac{3}{x-1}-\dfrac{2}{1-x}}{\dfrac{2}{x-1}-\dfrac{2}{x}}$

(6.4) *Solve each equation.*

75. $\dfrac{2}{5}=\dfrac{x}{15}$

76. $\dfrac{3}{x}+\dfrac{1}{3}=\dfrac{5}{x}$

77. $4+\dfrac{8}{x}=8$

78. $\dfrac{2x+3}{5x-9}=\dfrac{3}{2}$

79. $\dfrac{1}{x-2}-\dfrac{3x}{x^2-4}=\dfrac{2}{x+2}$

80. $\dfrac{7}{x}-\dfrac{x}{7}=0$

81. $\dfrac{x-2}{x^2-7x+10}=\dfrac{1}{5x-10}-\dfrac{1}{x-5}$

Solve each equation or perform each indicated operation. Simplify.

82. $\dfrac{5}{x^2-7x}+\dfrac{4}{2x-14}$

83. $3-\dfrac{5}{x}-\dfrac{2}{x^2}=0$

84. $\dfrac{4}{3-x}-\dfrac{7}{2x-6}+\dfrac{5}{x}$

494

(6.5) *Solve each equation for the specified variable.*

△ **85.** $A = \dfrac{h(a + b)}{2}$ for a

86. $\dfrac{1}{R} = \dfrac{1}{R_1} + \dfrac{1}{R_2}$ for R_2

87. $I = \dfrac{E}{R + r}$ for R

88. $A = P + Prt$ for r

89. $H = \dfrac{kA(T_1 - T_2)}{L}$ for A

Solve.

90. The sum of a number and twice its reciprocal is 3. Find the number(s).

91. If a number is added to the numerator of $\dfrac{3}{7}$, and twice that number is added to the denominator of $\dfrac{3}{7}$, the result is equivalent to $\dfrac{10}{21}$. Find the number.

92. The denominator of a fraction is 2 more than the numerator. If the numerator is decreased by 3 and the denominator is increased by 5, the resulting fraction is equivalent to $\dfrac{2}{3}$. Find the fraction.

93. The sum of the reciprocals of two consecutive even integers is $-\dfrac{9}{40}$. Find the two integers.

94. Three boys can paint a fence in 4 hours, 5 hours, and 6 hours, respectively. Find how long it will take all three boys to paint the fence.

95. If Sue Katz can type a certain number of mailing labels in 6 hours and Tom Neilson and Sue working together can type the same number of mailing labels in 4 hours, find how long it takes Tom alone to type the mailing labels.

96. The inlet pipe of a water tank can fill the tank in 2 hours and 30 minutes. The outlet pipe can empty the tank in 2 hours. Find how long it takes to empty a full tank if both pipes are open.

97. Timmy Garnica drove 210 miles in the same amount of time that it took a DC-10 jet to travel 1715 miles. The speed of the jet was 430 mph faster than the speed of the car. Find the speed of the jet.

98. The combined resistance R of two resistors in parallel with resistances R_1 and R_2 is given by the formula $\frac{1}{R} = \frac{1}{R_1} + \frac{1}{R_2}$. If the combined resistance is $\frac{30}{11}$ ohms and the resistance of one of the two resistors is 5 ohms, find the resistance of the other resistor.

99. The speed of a Ranger boat in still water is 32 mph. If the boat travels 72 miles upstream in the same time that it takes to travel 120 miles downstream, find the current of the stream.

100. A B737 jet flies 445 miles with the wind and 355 miles against the wind in the same length of time. If the speed of the jet in still air is 400 mph, find the speed of the wind.

101. The speed of a jogger is 3 mph faster than the speed of a walker. If the jogger travels 14 miles in the same amount of time that the walker travels 8 miles, find the speed of the walker.

102. Two Amtrak trains traveling on parallel tracks leave Tucson at the same time. In 6 hours the faster train is 382 miles from Tucson and the trains are 112 miles apart. Find how fast each train is traveling.

(6.6) *Solve each variation problem.*

103. A is directly proportional to B. If $A = 6$ when $B = 14$, find A when $B = 21$.

104. C is inversely proportional to D. If $C = 12$ when $D = 8$, find C when $D = 24$.

105. According to Boyle's law, the pressure exerted by a gas is inversely proportional to the volume, as long as the temperature stays the same. If a gas exerts a pressure of 1250 pounds per square inch when the volume is 2 cubic feet, find the volume when the pressure is 800 pounds per square inch.

△ **106.** The surface area of a sphere varies directly as the square of its radius. If the surface area is 36π square inches when the radius is 3 inches, find the surface area when the radius is 4 inches.

Chapter 6 Test

Find all numbers for which each rational expression is undefined.

1. $\dfrac{5x^2}{1-x}$

2. $\dfrac{9x^2-9}{x^2+4x+3}$

Write each rational expression in lowest terms.

3. $\dfrac{5x^7}{3x^4}$

4. $\dfrac{7x-21}{24-8x}$

5. $\dfrac{x^2-4x}{x^2+5x-36}$

Perform each indicated operation. Write your answers in lowest terms.

6. $\dfrac{x}{x-2} \cdot \dfrac{x^2-4}{5x}$

7. $\dfrac{2x^3+16}{6x^2+12x} \cdot \dfrac{5}{x^2-2x+4}$

8. $\dfrac{26ab}{7c} \div \dfrac{13a^2c^5}{14a^4b^3}$

9. $\dfrac{3x^2-12}{x^2+2x-8} \div \dfrac{6x+18}{x+4}$

10. $\dfrac{4x-12}{2x-9} \div \dfrac{3-x}{4x^2-81} \cdot \dfrac{x+3}{5x+15}$

11. $\dfrac{5}{4x^3} + \dfrac{7}{4x^3}$

12. $\dfrac{3+2x}{10-x} + \dfrac{13+x}{x-10}$

13. $\dfrac{3}{x^2-x-6} + \dfrac{2}{x^2-5x+6}$

14. $\dfrac{5}{x-7} - \dfrac{2x}{3x-21} + \dfrac{x}{2x-14}$

15. $\dfrac{3x}{5} \cdot \left(\dfrac{5}{x} - \dfrac{5}{2x} \right)$

Simplify each complex fraction.

16. $\dfrac{\dfrac{4x}{13}}{\dfrac{20x}{13}}$

17. $\dfrac{\dfrac{5}{x} - \dfrac{7}{3x}}{\dfrac{9}{8x} - \dfrac{1}{x}}$

18. $\dfrac{\dfrac{x^2-5x+6}{x+3}}{\dfrac{x^2-4x+4}{x^2-9}}$

Solve each equation.

19. $\dfrac{5x+3}{3x-7} = \dfrac{19}{7}$

20. $\dfrac{5}{x-5} + \dfrac{x}{x+5} = -\dfrac{29}{21}$

21. $\dfrac{x}{x-4} = 3 - \dfrac{4}{x-4}$

22. Solve $\dfrac{x+b}{a} = \dfrac{4x-7a}{b}$ for x.

Answers

1. _____

2. _____

3. _____

4. _____

5. _____

6. _____

7. _____

8. _____

9. _____

10. _____

11. _____

12. _____

13. _____

14. _____

15. _____

16. _____

17. _____

18. _____

19. _____

20. _____

21. _____

22. _____

23. _____

24. _____

25. _____

26. _____

27. _____

28. _____

23. The product of one more than a number and twice the reciprocal of the number is $\dfrac{12}{5}$. Find the number.

24. If Jan Ewing can weed the garden in 2 hours and her husband can weed it in 1 hour and 30 minutes, find how long it takes them to weed the garden together.

25. Suppose that W is inversely proportional to V. If $W = 20$ when $V = 12$, find W when $V = 15$.

26. Suppose that Q is jointly proportional to R and the square of S. If $Q = 24$ when $R = 3$ and $S = 4$, find Q when $R = 2$ and $S = 3$.

27. When an anvil is dropped into a gorge, the speed at which it strikes the ground is directly proportional to the square root of the distance it falls. An anvil that falls 400 feet hits the ground at a speed of 160 feet per second. Find the height of a cliff over the gorge if a dropped anvil hits the ground at a speed of 128 feet per second.

28. In baseball, a pitcher's earned run average statistic A is computed as $A = \dfrac{E}{\frac{I}{9}}$, where E is the number of earned runs scored on the pitcher and I is the total number of innings he or she pitched. During the 2000 season, pitcher Pedro Martinez of the Boston Red Sox led the Major Leagues with an earned run average of 1.74. Martinez pitched a total of 217 innings during the season. How many earned runs were scored on Pedro Martinez? Round to the nearest whole. (*Source:* Major League Baseball)

Cumulative Review

1. Evaluate: $\dfrac{r}{s}$ when $r = 48$ and $s = 6$

Evaluate each expression.

2. -7^0

3. $(2x + 5)^0$

4. Solve: $2x \geq 0$ and $4x - 1 \leq -9$

5. Solve: $|w + 3| = 7$

6. Determine whether $(0, -12), (1, 9)$, and $(2, -6)$ are solutions of the equation $3x - y = 12$.

7. Determine whether the two lines are parallel, perpendicular, or neither.

$$3x + 7y = 4$$
$$6x + 14y = 7$$

8. Write an equation of the line through points $(4, 0)$ and $(-4, -5)$.

Find the domain and range of each relation.

9.

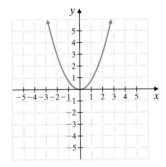

10.

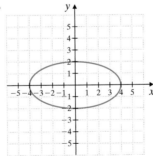

11. Use the substitution method to solve the system:

$$\begin{cases} -\dfrac{x}{6} + \dfrac{y}{2} = \dfrac{1}{2} \\ \dfrac{x}{3} - \dfrac{y}{6} = -\dfrac{3}{4} \end{cases}$$

12. Solve the system:

$$\begin{cases} 2x + 4y \quad\;\; = 1 \\ 4x \quad\;\; - 4z = -1 \\ \quad\;\; y - 4z = -3 \end{cases}$$

Answers

1. _____

2. _____

3. _____

4. _____

5. _____

6. _____

7. _____

8. _____

9. _____

10. _____

11. _____

12. _____

13. _____

14. _____

15. _____

16. see graph _____

17. _____

18. _____

19. _____

20. _____

21. _____

22. _____

23. _____

24. _____

25. _____

13. Two cars leave Indianapolis, one traveling east and the other west. After 3 hours they are 297 miles apart. If one car is traveling 5 mph faster than the other, what is the speed of each?

14. Use matrices to solve the system:

$$\begin{cases} x + 2y + z = 2 \\ -2x - y + 2z = 5 \\ x + 3y - 2z = -8 \end{cases}$$

15. Use Cramer's rule to solve the system:

$$\begin{cases} 5x + y = 5 \\ -7x - 2y = -7 \end{cases}$$

16. Graph the solutions of the system:

$$\begin{cases} 3x \geq y \\ x + 2y \leq 8 \end{cases}$$

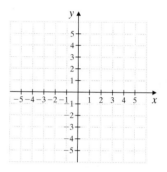

Simplify by combining like terms.

17. $-12x^2 + 7x^2 - 6x$

18. $3xy - 2x + 5xy - x$

19. Multiply: $(4x^2 + 7)(x^2 + 2x + 8)$

20. Divide $2x^2 - x - 10$ by $x + 2$.

21. Factor: $-3x^3y + 2x^2y - 5xy$

22. Factor: $x^2 + 10x + 16$

23. Factor: $p^4 - 16$

24. Solve: $(x + 2)(x - 6) = 0$

25. Simplify: $\dfrac{\dfrac{2x}{27y^2}}{\dfrac{6x^2}{9}}$

Rational Exponents, Radicals, and Complex Numbers

In this chapter, radical notation is reviewed, and then rational exponents are introduced. As the name implies, rational exponents are exponents that are rational numbers. We present an interpretation of rational exponents that is consistent with the meaning and rules already established for integer exponents, and we present two forms of notation for roots: radical and exponent. We conclude this chapter with complex numbers, a natural extension of the real number system.

7.1 Radical Expressions

7.2 Rational Exponents

7.3 Simplifying Radical Expressions

7.4 Adding, Subtracting, and Multiplying Radical Expressions

7.5 Rationalizing Numerators and Denominators of Radical Expressions

Integrated Review—Radicals and Rational Exponents

7.6 Radical Equations and Problem Solving

7.7 Complex Numbers

Mount Vesuvius is the only active volcano on the European continent. Although the Romans thought the volcano was extinct, Vesuvius erupted violently in 79 A.D. The eruption buried the cities of Pompeii, Herculaneum, and Stabiae under up to 60 feet of ash and mud, killing approximately 16,000 people. Although Pompeii was completely engulfed, the city was far from destroyed. The blanket of mud and ash perfectly preserved much of the city in a snapshot of daily life of the ancient Romans. Pompeii lay undisturbed for over 1500 years until the first excavations were made and its archaeological significance was proven. The 79 A.D. eruption of Vesuvius seemed to be the volcano's renewal. It has erupted with varying degrees of violence more than 30 times since Pompeii's burial, most recently in 1944. In Exercise 84 on page 527, we will use a radical expression to find the surface area of Mount Vesuvius.

Name _____ Section _____ Date _____

1. _____

2. _____

3. _____

4. _____

5. _____

6. _____

7. _____

8. _____

9. _____

10. _____

11. _____

12. _____

13. _____

14. _____

15. _____

16. _____

17. _____

18. _____

19. _____

20. _____

Chapter 7 Pretest

Find each root. Assume that all variables represent positive real numbers.

1. $\sqrt{81x^2}$

2. $\sqrt[3]{-27y^9}$

3. $\sqrt[5]{x^{30}}$

4. If $f(x) = \sqrt{7x + 2}$, find $f(2)$.

Simplify. Write with positive exponents.

5. $144^{1/2}$

6. $81^{3/4}$

7. $\dfrac{3}{2m^{-5/6}}$

8. $(9x^4y^{-6})^{3/2}$

Perform each indicated operation. Assume that all variables represent positive real numbers.

9. $\sqrt{13} \cdot \sqrt{3x}$

10. $\sqrt{\dfrac{6}{49}}$

11. $3\sqrt{18} - 6\sqrt{50}$

12. $(4\sqrt{x} + 1)(2\sqrt{x} - 3)$

13. $(\sqrt{5} - y)^2$

Simplify.

14. $\sqrt[3]{135a^2b^{12}}$

Rationalize the denominator. Assume that all variables represent positive real numbers.

15. $\dfrac{2}{\sqrt{12x}}$

16. $\dfrac{14}{2 + \sqrt{11}}$

17. Solve: $\sqrt{2x + 1} - 5 = 0$

Perform each indicated operation and simplify. Write the result in the form $a + bi$.

18. $(10 - 3i) - (6 + 5i)$

19. $(2 - 5i)^2$

20. $\dfrac{3 + i}{5 + i}$

7.1 Radical Expressions

(A) Finding Square Roots

Recall from Section 1.4 that to find a *square root* of a number a, we find a number that was squared to get a.

> ### Square Root
> The number b is a **square root** of a if $b^2 = a$.

EXAMPLES Find the real square roots of each number.

1. 25   Since $5^2 = 25$ and $(-5)^2 = 25$, the square roots of 25 are 5 and -5.

2. 49   Since $7^2 = 49$ and $(-7)^2 = 49$, the square roots of 49 are 7 and -7.

3. -4 There is no real number whose square is -4. The number -4 has no real number square root. ●

Recall that we denote the *nonnegative*, or *principal, square root* with the **radical sign**:

$$\sqrt{25} = 5$$

We denote the *negative square root* with the **negative radical sign**:

$$-\sqrt{25} = -5$$

An expression containing a radical sign is called a **radical expression**. An expression within, or "under," a radical sign is called a **radicand**.

radical expression: $\sqrt{a}$

with *radical sign* pointing to the radical and *radicand* pointing to a.

> ### Principal and Negative Square Roots
> The **principal square root** of a nonnegative number a is its nonnegative square root. The principal square root is written as $\sqrt{a}$. The **negative square root** of a is written as $-\sqrt{a}$.

EXAMPLES

Find each square root. Assume that all variables represent nonnegative real numbers.

4. $\sqrt{36} = 6$ because $6^2 = 36$.

5. $\sqrt{0} = 0$ because $0^2 = 0$.

6. $\sqrt{\dfrac{4}{49}} = \dfrac{2}{7}$ because $\left(\dfrac{2}{7}\right)^2 = \dfrac{4}{49}$.

7. $\sqrt{0.25} = 0.5$ because $(0.5)^2 = 0.25$.

8. $\sqrt{x^6} = x^3$ because $\left(x^3\right)^2 = x^6$.

9. $\sqrt{9x^{10}} = 3x^5$ because $(3x^5)^2 = 9x^{10}$.

10. $-\sqrt{81} = -9$. The negative in front of the radical indicates the negative square root of 81.

11. $\sqrt{-81}$ is not a real number. ●

OBJECTIVES

(A) Find square roots.

(B) Approximate roots using a calculator.

(C) Find cube roots.

(D) Find *n*th roots.

(E) Find $\sqrt[n]{a^n}$ when a is any real number.

(F) Find function values of radical functions.

SSM TUTOR CENTER   SG   CD & VIDEO   MATH PRO   WEB

Practice Problems 1–3

Find the square roots of each number.

1. 36

2. 81

3. -16

Practice Problems 4–11

Find each square root. Assume that all variables represent nonnegative real numbers.

4. $\sqrt{25}$ **5.** $\sqrt{0}$

6. $\sqrt{\dfrac{9}{25}}$ **7.** $\sqrt{0.36}$

8. $\sqrt{x^{10}}$ **9.** $\sqrt{36x^6}$

10. $-\sqrt{25}$ **11.** $\sqrt{-25}$

Answers

1. $6, -6$, **2.** $9, -9$, **3.** no real number square root, **4.** 5, **5.** 0, **6.** $\dfrac{3}{5}$, **7.** 0.6, **8.** x^5, **9.** $6x^3$, **10.** -5, **11.** not a real number

- Remember: $\sqrt{0} = 0$.
- Don't forget that the square root of a negative number is not a real number. For example,

$$\sqrt{-9} \quad \text{is not a real number}$$

because there is no real number that when multiplied by itself would give a product of -9. In Section 7.7, we will see what kind of a number $\sqrt{-9}$ is.

B Approximating Roots

Recall that numbers such as 1, 4, 9, and 25 are called **perfect squares**, since $1 = 1^2, 4 = 2^2, 9 = 3^2$, and $25 = 5^2$. Square roots of perfect square radicands simplify to rational numbers. What happens when we try to simplify a root such as $\sqrt{3}$? Since 3 is not a perfect square, $\sqrt{3}$ is not a rational number. It is called an **irrational number**, and we can find a decimal **approximation** of it. To find decimal approximations, we can use the table in the appendix or a calculator. For example, an approximation for $\sqrt{3}$ is

$$\sqrt{3} \approx 1.732$$

$\uparrow$

approximation symbol

To see if the approximation is reasonable, notice that since

$$1 < 3 < 4, \quad \text{then}$$
$$\sqrt{1} < \sqrt{3} < \sqrt{4}, \quad \text{or}$$
$$1 < \sqrt{3} < 2.$$

We found $\sqrt{3} \approx 1.732$, a number between 1 and 2, so our result is reasonable.

Practice Problem 12

Use a calculator or the appendix to approximate $\sqrt{30}$. Round the approximation to three decimal places and check to see that your approximation is reasonable.

EXAMPLE 12

Use a calculator or the appendix to approximate $\sqrt{20}$. Round the approximation to three decimal places and check to see that your approximation is reasonable.

Solution:

$$\sqrt{20} \approx 4.472$$

Is this reasonable? Since $16 < 20 < 25$, then $\sqrt{16} < \sqrt{20} < \sqrt{25}$, or $4 < \sqrt{20} < 5$. The approximation is between 4 and 5 and is thus reasonable.

C Finding Cube Roots

Finding roots can be extended to other roots such as cube roots. For example, since $2^3 = 8$, we call 2 the *cube root* of 8. In symbols, we write

$$\sqrt[3]{8} = 2$$

Answer

12. 5.477

Cube Root

The **cube root** of a real number a is written as $\sqrt[3]{a}$, and

$\sqrt[3]{a} = b$ only if $b^3 = a$

From this definition, we have

$\sqrt[3]{64} = 4$ since $4^3 = 64$

$\sqrt[3]{-27} = -3$ since $(-3)^3 = -27$

$\sqrt[3]{x^3} = x$ since $x^3 = x^3$

Notice that, unlike with square roots, *it is possible to have a negative radicand when finding a cube root.* This is so because the *cube* of a negative number is a negative number. Therefore, the *cube root* of a negative number is a negative number.

EXAMPLES   Find each cube root.

13. $\sqrt[3]{1} = 1$ because $1^3 = 1$.

14. $\sqrt[3]{-64} = -4$ because $(-4)^3 = -64$.

15. $\sqrt[3]{\dfrac{8}{125}} = \dfrac{2}{5}$ because $\left(\dfrac{2}{5}\right)^3 = \dfrac{8}{125}$.

16. $\sqrt[3]{x^6} = x^2$ because $(x^2)^3 = x^6$.

17. $\sqrt[3]{-8x^9} = -2x^3$ because $(-2x^3)^3 = -8x^9$.

●

Practice Problems 13–17

Find each cube root.

13. $\sqrt[3]{0}$

14. $\sqrt[3]{-8}$

15. $\sqrt[3]{\dfrac{1}{64}}$

16. $\sqrt[3]{x^9}$

17. $\sqrt[3]{-64x^6}$

D   ## Finding nth Roots

Just as we can raise a real number to powers other than 2 or 3, we can find roots other than square roots and cube roots. In fact, we can find the **nth root** of a number, where n is any natural number. In symbols, the nth root of a is written as $\sqrt[n]{a}$, where n is called the **index**. The index 2 is usually omitted for square roots.

Helpful Hint

If the index is even, such as in $\sqrt{}$, $\sqrt[4]{}$, $\sqrt[6]{}$, and so on, the radicand must be nonnegative for the root to be a real number. For example,

$\sqrt[4]{16} = 2$, but $\sqrt[4]{-16}$ is not a real number,

$\sqrt[6]{64} = 2$, but $\sqrt[6]{-64}$ is not a real number.

If the index is odd, such as in $\sqrt[3]{}$, $\sqrt[5]{}$, and so on, the radicand may be any real number. For example,

$\sqrt[3]{64} = 4$ and $\sqrt[3]{-64} = -4$,

$\sqrt[5]{32} = 2$ and $\sqrt[5]{-32} = -2$.

Try the Concept Check in the margin.

Concept Check

Which one is not a real number?

a. $\sqrt[3]{-15}$

b. $\sqrt[4]{-15}$

c. $\sqrt[5]{-15}$

d. $\sqrt{(-15)^2}$

Answers

13. 0,   **14.** -2,   **15.** $\dfrac{1}{4}$,   **16.** x^3,   **17.** $-4x^2$

Concept Check: b

Practice Problems 18–22

Find each root.

18. $\sqrt[4]{16}$ 19. $\sqrt[5]{-32}$

20. $-\sqrt{36}$ 21. $\sqrt[4]{-16}$

22. $\sqrt[3]{8x^6}$

EXAMPLES   Find each root.

18. $\sqrt[4]{81} = 3$ because $3^4 = 81$ and 3 is positive.

19. $\sqrt[5]{-243} = -3$ because $(-3)^5 = -243$.

20. $-\sqrt{25} = -5$ because -5 is the opposite of $\sqrt{25}$.

21. $\sqrt[4]{-81}$ is not a real number. There is no real number that, when raised to the fourth power, is -81.

22. $\sqrt[3]{64x^3} = 4x$ because $(4x)^3 = 64x^3$.   ●

E   ## Finding $\sqrt[n]{a^n}$ When a Is Any Real Number

Recall that the notation $\sqrt{a^2}$ indicates the positive square root of a^2 only. For example,

$$\sqrt{(-5)^2} = \sqrt{25} = 5$$

When variables are present in the radicand and it is *unclear whether the variable represents a positive number or a negative number*, absolute value bars are sometimes needed to ensure that the result is a positive number. For example,

$$\sqrt{x^2} = |x|$$

This ensures that the result is positive. This same situation may occur when the index is any *even* positive integer. When the index is any *odd* positive integer, absolute value bars are not necessary.

> **Finding $\sqrt[n]{a^n}$**
>
> If n is an *even* positive integer, then $\sqrt[n]{a^n} = |a|$.
>
> If n is an *odd* positive integer, then $\sqrt[n]{a^n} = a$.

Practice Problems 23–29

Simplify. Assume that the variables represent any real number.

23. $\sqrt{(-5)^2}$

24. $\sqrt{x^6}$

25. $\sqrt[4]{(x + 6)^4}$

26. $\sqrt[3]{(-3)^3}$

27. $\sqrt[5]{(7x - 1)^5}$

28. $\sqrt{36x^2}$

29. $\sqrt{x^2 + 6x + 9}$

EXAMPLES

Simplify. Assume that the variables represent any real number.

23. $\sqrt{(-3)^2} = |-3| = 3$ When the index is even, the absolute value bars ensure that the result is not negative.

24. $\sqrt{x^2} = |x|$

25. $\sqrt[4]{(x - 2)^4} = |x - 2|$

26. $\sqrt[3]{(-5)^3} = -5$ Absolute value bars are not needed when the index is odd.

27. $\sqrt[5]{(2x - 7)^5} = 2x - 7$

28. $\sqrt{25x^2} = 5|x|$

29. $\sqrt{x^2 + 2x + 1} = \sqrt{(x + 1)^2} = |x + 1|$ ●

Answers

18. 2,   **19.** -2,   **20.** -6,   **21.** not a real number, **22.** $2x^2$,   **23.** 5,   **24.** $|x^3|$,   **25.** $|x + 6|$,   **26.** -3, **27.** $7x - 1$,   **28.** $6|x|$,   **29.** $|x + 3|$

Finding Function Values

Functions of the form

$$f(x) = \sqrt[n]{x}$$

are called **radical functions**. Recall that the domain of a function in x is the set of all possible replacement values of x. This means that if n is even, the domain is the set of all nonnegative numbers, or $\{x \mid x \geq 0\}$. If n is odd, the domain is the set of all real numbers. Keep this in mind as we find function values. In Chapter 9, we will graph these functions and discuss their domains further.

EXAMPLES

If $f(x) = \sqrt{x - 4}$ and $g(x) = \sqrt[3]{x + 2}$, find each function value.

30. $f(8) = \sqrt{8 - 4} = \sqrt{4} = 2$

31. $f(6) = \sqrt{6 - 4} = \sqrt{2}$

32. $g(-1) = \sqrt[3]{-1 + 2} = \sqrt[3]{1} = 1$

33. $g(1) = \sqrt[3]{1 + 2} = \sqrt[3]{3}$

Practice Problems 30–33

If $f(x) = \sqrt{x + 2}$ and $g(x) = \sqrt[3]{x - 1}$, find each function value.

30. $f(7)$

31. $g(9)$

32. $f(0)$

33. $g(10)$

●

Helpful Hint

Notice that for the function $f(x) = \sqrt{x - 4}$, the domain includes all real numbers that make the radicand ≥ 0. To see what numbers these are, solve $x - 4 \geq 0$ and find that $x \geq 4$. The domain is $\{x \mid x \geq 4\}$.

The domain of the cube root function $g(x) = \sqrt[3]{x + 2}$ is the set of real numbers.

See Chapter 9 for further discussions of domains.

Answers

30. 3, **31.** 2, **32.** $\sqrt{2}$, **33.** $\sqrt[3]{9}$

FOCUS ON **History**

HERON OF ALEXANDRIA

Heron (also Hero) was a Greek mathematician and engineer. He lived and worked in Alexandria, Egypt, around 75 A.D. During his prolific work life, Heron developed a rotary steam engine called an aeolipile, a surveying tool called a dioptra, as well as a wind organ and a fire engine. As an engineer, he must have had the need to approximate square roots because he described an iterative method for doing so in his work *Metrica*. Heron's method for approximating a square root can be summarized as follows:

Suppose that x is not a perfect square and a^2 is the nearest perfect square to x. For a rough estimate of the value of $\sqrt{x}$, find the value of

$$y_1 = \frac{1}{2}\left(a + \frac{x}{a}\right).$$ This estimate can be improved

by calculating a second estimate using the first

estimate y_1 in place of a: $y_2 = \frac{1}{2}\left(y_1 + \frac{x}{y_1}\right).$

Repeating this process several times will give more and more accurate estimates of $\sqrt{x}$.

CRITICAL THINKING

1. **a.** Which perfect square is closest to 80?
   b. Use Heron's method for approximating square roots to calculate the first estimate of the square root of 80.
   c. Use the first estimate of the square root of 80 to find a more refined second estimate.
   d. Use a calculator to find the actual value of the square root of 80. List all digits shown on your calculator's display.
   e. Compare the actual value from part (d) to the values of the first and second estimates. What do you notice?
   f. How many iterations of this process are necessary to get an estimate that differs no more than one digit from the actual value recorded in part (d)?

2. Repeat Question 1 for finding an estimate of the square root of 30.

3. Repeat Question 1 for finding an estimate of the square root of 4572.

4. Why would this iterative method have been important to people of Heron's era? Would you say that this method is as important today? Why or why not?

EXERCISE SET 7.1

A *Find the real square roots of each number. See Examples 1 through 3.*

1. 4 **2.** 9 **3.** −25 **4.** −49 **5.** 100 **6.** 64

Find each square root. Assume that all variables represent nonnegative real numbers. See Examples 4 through 11.

7. $\sqrt{100}$ **8.** $\sqrt{400}$ **9.** $\sqrt{\dfrac{1}{4}}$ **10.** $\sqrt{\dfrac{9}{25}}$ **11.** $\sqrt{0.0001}$ **12.** $\sqrt{0.04}$

13. $-\sqrt{36}$ **14.** $-\sqrt{9}$ **15.** $\sqrt{x^{10}}$ **16.** $\sqrt{x^{16}}$ **17.** $\sqrt{16y^6}$ **18.** $\sqrt{64y^{20}}$

B *Use a calculator or the appendix to approximate each square root to three decimal places. Check to see that each approximation is reasonable. See Example 12.*

19. $\sqrt{7}$ **20.** $\sqrt{11}$ **21.** $\sqrt{38}$ **22.** $\sqrt{56}$ **23.** $\sqrt{200}$ **24.** $\sqrt{300}$

C *Find each cube root. See Examples 13 through 17.*

25. $\sqrt[3]{64}$ **26.** $\sqrt[3]{27}$ **27.** $\sqrt[3]{\dfrac{1}{8}}$ **28.** $\sqrt[3]{\dfrac{27}{64}}$ **29.** $\sqrt[3]{-1}$

30. $\sqrt[3]{-125}$ **31.** $\sqrt[3]{x^{12}}$ **32.** $\sqrt[3]{x^{15}}$ **33.** $\sqrt[3]{-27x^9}$ **34.** $\sqrt[3]{-64x^6}$

D *Find each root. Assume that all variables represent nonnegative real numbers. See Examples 18 through 22.*

35. $-\sqrt[4]{16}$ **36.** $\sqrt[3]{-243}$ **37.** $\sqrt[4]{-16}$ **38.** $\sqrt{-16}$ **39.** $\sqrt[5]{-32}$

40. $\sqrt[5]{-1}$

41. $\sqrt[5]{x^{20}}$

42. $\sqrt[4]{x^{20}}$

43. $\sqrt[6]{64x^{12}}$

44. $\sqrt[5]{-32x^{15}}$

45. $\sqrt{81x^4}$

46. $\sqrt[4]{81x^4}$

47. $\sqrt[4]{256x^8}$

48. $\sqrt{256x^8}$

E *Simplify. Assume that the variables represent any real number. See Examples 23 through 29.*

49. $\sqrt{(-8)^2}$

50. $\sqrt{(-7)^2}$

51. $\sqrt[3]{(-8)^3}$

52. $\sqrt[5]{(-7)^5}$

53. $\sqrt{4x^2}$

54. $\sqrt[4]{16x^4}$

55. $\sqrt[3]{x^3}$

56. $\sqrt[5]{x^5}$

57. $\sqrt[4]{(x-2)^4}$

58. $\sqrt[6]{(2x-1)^6}$

59. $\sqrt{x^2 + 4x + 4}$
(*Hint:* Factor the polynomial first.)

60. $\sqrt{x^2 - 8x + 16}$
(*Hint:* Factor the polynomial first.)

F *If $f(x) = \sqrt{2x+3}$ and $g(x) = \sqrt[3]{x-8}$, find each function value. See Examples 30 through 33.*

61. $f(0)$

62. $g(0)$

63. $g(7)$

64. $f(-1)$

65. $g(-19)$

66. $f(3)$

67. $f(2)$

68. $g(1)$

Review and Preview

Simplify each exponential expression. See Sections 1.6 and 1.7.

69. $(-2x^3y^2)^5$

70. $(4y^6z^7)^3$

71. $(-3x^2y^3z^5)(20x^5y^7)$

72. $(-14a^5bc^2)(2abc^4)$

73. $\dfrac{7x^{-1}y}{14(x^5y^2)^{-2}}$

74. $\dfrac{(2a^{-1}b^2)^3}{(8a^2b)^{-2}}$

75. Explain why $\sqrt{-64}$ is not a real number.

76. Explain why $\sqrt[3]{-64}$ is a real number.

For Exercises 77 through 80, do not use a calculator.

77. $\sqrt{160}$ is closest to
    a. 10 **b.** 13 **c.** 20 **d.** 40

78. $\sqrt{1000}$ is closest to
    a. 10 **b.** 30 **c.** 100 **d.** 500

△ **79.** The perimeter of the triangle is closest to
    a. 12 **b.** 18 **c.** 66 **d.** 132

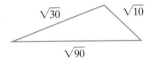

80. The length of the bent wire is closest to
    a. 5 **b.** $\sqrt{28}$ **c.** 7 **d.** 14

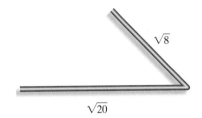

The Mosteller formula for calculating adult body surface area is $B = \sqrt{\dfrac{hw}{3131}}$, *where B is an individual's body surface area in square meters, h is the individual's height in inches, and w is the individual's weight in pounds. Use this information to answer Exercises 81 and 82. Round answers to 2 decimal places.*

△ **81.** Find the body surface area of an individual who is 66 inches tall and who weighs 135 pounds.

△ **82.** Find the body surface area of an individual who is 74 inches tall and who weighs 225 pounds.

83. Suppose that a friend tells you that $\sqrt{13} \approx 5.7$. Without a calculator, how can you convince your friend that he or she must have made an error?

84. Escape velocity is the minimum speed that an object must reach to escape a planet's pull of gravity. Escape velocity v is given by the equation

$v = \sqrt{\dfrac{2Gm}{r}}$, where m is the mass of the planet, r is its radius, and G is the universal gravitational constant, which has a value of $G = 6.67 \times 10^{-11}$ m³/kg·s². The mass of Earth is 5.97×10^{24} kg and its radius is 6.37×10^6 m. Use this information to find the escape velocity for Earth. Round to the nearest whole number. (*Source:* National Space Science Data Center)

Internet Excursions

 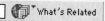

The National Space Science Data Center (NSSDC) is a division of NASA that provides access to information collected from NASA experiments and space flight missions. One of the many areas in which the NSSDC maintains a library of information is the planetary sciences. The given World Wide Web site gives a listing of planetary fact sheets from which the user may choose. (Alternatively, you can visit the NSSDC homepage at http://nssdc.gsfc.nasa.gov and navigate to Planetary Sciences. Then look for the Planetary Fact Sheets option.)

85. Choose any one of the fact sheets for a planet, asteroid, or the sun. Use the information given in the fact sheet and the escape velocity formula given in Exercise 84 to compute the escape velocity for that body. Then compare your calculation to the escape velocity given in the fact sheet. How close is your calculation? (*Note:* You should use the "volumetric mean radius" for the planet's radius in your calculation.)

86. Repeat Exercise 85 for one of the other listed bodies of the solar system.

7.2 Rational Exponents

OBJECTIVES

Ⓐ Understand the meaning of $a^{1/n}$.

Ⓑ Understand the meaning of $a^{m/n}$.

Ⓒ Understand the meaning of $a^{-m/n}$.

Ⓓ Use rules for exponents to simplify expressions that contain rational exponents.

Ⓔ Use rational exponents to simplify radical expressions.

SSM
TUTOR CENTER   SG   CD & VIDEO   MATH PRO   WEB

Ⓐ Understanding $a^{1/n}$

So far in this text, we have not defined expressions with rational exponents such as $3^{1/2}$, $x^{2/3}$, and $-9^{-1/4}$. We will define these expressions so that the rules for exponents shall apply to these rational exponents as well.

Suppose that $x = 5^{1/3}$. Then

$$x^3 = \left(5^{1/3}\right)^3 = 5^{1/3 \cdot 3} = 5^1 \text{ or } 5$$

using rules ↑
for exponents

Since $x^3 = 5$, then x is the number whose cube is 5, or $x = \sqrt[3]{5}$. Notice that we also know that $x = 5^{1/3}$. This means that

$$5^{1/3} = \sqrt[3]{5}$$

Definition of $a^{1/n}$

If n is a positive integer greater than 1 and $\sqrt[n]{a}$ is a real number, then

$$a^{1/n} = \sqrt[n]{a}$$

Notice that the denominator of the rational exponent corresponds to the index of the radical.

EXAMPLES

Use radical notation to rewrite each expression. Simplify if possible.

1. $4^{1/2} = \sqrt{4} = 2$

2. $64^{1/3} = \sqrt[3]{64} = 4$

3. $x^{1/4} = \sqrt[4]{x}$

4. $-9^{1/2} = -\sqrt{9} = -3$

5. $\left(81x^8\right)^{1/4} = \sqrt[4]{81x^8} = 3x^2$

6. $5y^{1/3} = 5\sqrt[3]{y}$

Practice Problems 1–6

Use radical notation to rewrite each expression. Simplify if possible.

1. $25^{1/2}$

2. $27^{1/3}$

3. $x^{1/5}$

4. $-25^{1/2}$

5. $\left(-27y^6\right)^{1/3}$

6. $7x^{1/5}$

Ⓑ Understanding $a^{m/n}$

As we expand our use of exponents to include $\dfrac{m}{n}$, we define their meaning so that rules for exponents still hold true. For example, by properties of exponents,

$$8^{2/3} = \left(8^{1/3}\right)^2 = \left(\sqrt[3]{8}\right)^2 \quad \text{or} \quad 8^{2/3} = \left(8^2\right)^{1/3} = \sqrt[3]{8^2}$$

Definition of $a^{m/n}$

If m and n are positive integers greater than 1 with $\dfrac{m}{n}$ in simplest form, then

$$a^{m/n} = \sqrt[n]{a^m} = \left(\sqrt[n]{a}\right)^m$$

as long as $\sqrt[n]{a}$ is a real number.

Answers

1. 5, **2.** 3, **3.** $\sqrt[5]{x}$, **4.** -5, **5.** $-3y^2$, **6.** $7\sqrt[5]{x}$

Notice that the denominator n of the rational exponent corresponds to the index of the radical. The numerator m of the rational exponent indicates that the base is to be raised to the mth power. This means that

$$8^{2/3} = \sqrt[3]{8^2} = \sqrt[3]{64} = 4 \quad \text{or} \quad 8^{2/3} = (\sqrt[3]{8})^2 = 2^2 = 4$$

Helpful Hint

Most of the time, $(\sqrt[n]{a})^m$ will be easier to calculate than $\sqrt[n]{a^m}$.

Practice Problems 7–11

Use radical notation to rewrite each expression. Simplify if possible.

7. $9^{3/2}$

8. $-256^{3/4}$

9. $(-32)^{2/5}$

10. $\left(\dfrac{1}{4}\right)^{3/2}$

11. $(2x + 1)^{2/7}$

EXAMPLES

Use radical notation to rewrite each expression. Simplify if possible.

7. $4^{3/2} = (\sqrt{4})^3 = 2^3 = 8$

8. $-16^{3/4} = -(\sqrt[4]{16})^3 = -(2)^3 = -8$

9. $(-27)^{2/3} = (\sqrt[3]{-27})^2 = (-3)^2 = 9$

10. $\left(\dfrac{1}{9}\right)^{3/2} = \left(\sqrt{\dfrac{1}{9}}\right)^3 = \left(\dfrac{1}{3}\right)^3 = \dfrac{1}{27}$

11. $(4x - 1)^{3/5} = \sqrt[5]{(4x - 1)^3}$

Helpful Hint

The *denominator* of a rational exponent is the index of the corresponding radical. For example, $x^{1/5} = \sqrt[5]{x}$, and $z^{2/3} = \sqrt[3]{z^2}$ or $z^{2/3} = (\sqrt[3]{z})^2$.

(C) Understanding $a^{-m/n}$

The rational exponents we have given meaning to exclude negative rational numbers. To complete the set of definitions, we define $a^{-m/n}$.

Definition of $a^{-m/n}$

$$a^{-m/n} = \frac{1}{a^{m/n}}$$

as long as $a^{m/n}$ is a nonzero real number.

Practice Problems 12–13

Write each expression with a positive exponent. Then simplify.

12. $27^{-2/3}$

13. $-256^{-3/4}$

EXAMPLES

Write each expression with a positive exponent. Then simplify.

12. $16^{-3/4} = \dfrac{1}{16^{3/4}} = \dfrac{1}{(\sqrt[4]{16})^3} = \dfrac{1}{2^3} = \dfrac{1}{8}$

13. $(-27)^{-2/3} = \dfrac{1}{(-27)^{2/3}} = \dfrac{1}{(\sqrt[3]{-27})^2} = \dfrac{1}{(-3)^2} = \dfrac{1}{9}$

Answers

7. 27, **8.** -64, **9.** 4, **10.** $\dfrac{1}{8}$, **11.** $\sqrt[7]{(2x + 1)^2}$,

12. $\dfrac{1}{9}$, **13.** $-\dfrac{1}{64}$

If an expression contains a negative rational exponent, you may want to first write the expression with a positive exponent, then interpret the rational exponent. Notice that the sign of the base is not affected by the sign of its exponent. For example,

$$9^{-3/2} = \frac{1}{9^{3/2}} = \frac{1}{(\sqrt{9})^3} = \frac{1}{27}$$

Also,

$$(-27)^{-1/3} = \frac{1}{(-27)^{1/3}} = -\frac{1}{3}$$

Try the Concept Check in the margin.

(D) Using Rules for Exponents

It can be shown that the properties of integer exponents hold for rational exponents. By using these properties and definitions, we can now simplify expressions that contain rational exponents. These rules are repeated here for review.

Summary of Exponent Rules

If m and n are rational numbers, and $a, b,$ and c are numbers for which the expressions below exist, then

Product rule for exponents: $\qquad a^m \cdot a^n = a^{m+n}$

Power rule for exponents: $\qquad (a^m)^n = a^{m \cdot n}$

Power rules for products and quotients: $\quad (ab)^n = a^n b^n$ and

$$\left(\frac{a}{c}\right)^n = \frac{a^n}{c^n}, \quad c \neq 0$$

Quotient rule for exponents: $\qquad \dfrac{a^m}{a^n} = a^{m-n}, \quad a \neq 0$

Zero exponent: $\qquad a^0 = 1, \quad a \neq 0$

Negative exponent: $\qquad a^{-n} = \dfrac{1}{a^n}, \quad a \neq 0$

EXAMPLES Use the properties of exponents to simplify.

14. $x^{1/2}x^{1/3} = x^{1/2+1/3} = x^{3/6+2/6} = x^{5/6}$ $\qquad$ Use the product rule.

15. $\dfrac{7^{1/3}}{7^{4/3}} = 7^{1/3-4/3} = 7^{-3/3} = 7^{-1} = \dfrac{1}{7}$ $\qquad$ Use the quotient rule.

16. $y^{-4/7} \cdot y^{6/7} = y^{-4/7+6/7} = y^{2/7}$ $\qquad$ Use the product rule.

17. $(5^{3/8})^4 = 5^{3/8 \cdot 4} = 5^{12/8} = 5^{3/2}$ $\qquad$ Use the power rule.

18. $\dfrac{(2x^{2/5})^5}{x^2} = \dfrac{2^5(x^{2/5})^5}{x^2}$ $\qquad$ Use the power rule.

$\qquad = \dfrac{32x^2}{x^2}$ $\qquad$ Simplify.

$\qquad = 32x^{2-2}$ $\qquad$ Use the quotient rule.

$\qquad = 32x^0$ $\qquad$ Simplify.

$\qquad = 32 \cdot 1 \qquad$ or $\quad 32$ $\qquad$ Substitute 1 for x^0.

Concept Check

Which one is correct?

a. $-8^{2/3} = \dfrac{1}{4}$

b. $8^{-2/3} = -\dfrac{1}{4}$

c. $8^{-2/3} = -4$

d. $-8^{-2/3} = -\dfrac{1}{4}$

Practice Problems 14–18

Use the properties of exponents to simplify.

14. $x^{1/3}x^{1/4}$ $\qquad$ **15.** $\dfrac{9^{2/5}}{9^{12/5}}$

16. $y^{-3/10} \cdot y^{6/10}$ $\qquad$ **17.** $(11^{2/9})^3$

18. $\dfrac{(3x^{2/3})^3}{x^2}$

Answers

14. $x^{7/12}$, **15.** $\dfrac{1}{81}$, **16.** $y^{3/10}$, **17.** $11^{2/3}$, **18.** 27

Concept Check: d

E Using Rational Exponents to Simplify Radical Expressions

We can simplify some radical expressions by first writing the expression with rational exponents. Use the properties of exponents to simplify, and then convert back to radical notation.

Practice Problems 19–21

Use rational exponents to simplify. Assume that all variables represent positive real numbers.

19. $\sqrt[10]{y^5}$

20. $\sqrt[4]{9}$

21. $\sqrt[9]{a^6b^3}$

EXAMPLES

Use rational exponents to simplify. Assume that all variables represent positive real numbers.

19. $\sqrt[8]{x^4} = x^{4/8}$ Write with rational exponents.

        $= x^{1/2}$ Simplify the exponent.

        $= \sqrt{x}$ Write with radical notation.

20. $\sqrt[6]{25} = 25^{1/6}$ Write with rational exponents.

        $= \left(5^2\right)^{1/6}$ Write 25 as 5^2.

        $= 5^{2/6}$ Use the power rule.

        $= 5^{1/3}$ Simplify the exponent.

        $= \sqrt[3]{5}$ Write with radical notation.

21. $\sqrt[6]{r^2s^4} = \left(r^2s^4\right)^{1/6}$ Write with rational exponents.

        $= r^{2/6}s^{4/6}$ Use the power rule.

        $= r^{1/3}s^{2/3}$ Simplify the exponents.

        $= \left(rs^2\right)^{1/3}$ Use $a^nb^n = (ab)^n$.

        $= \sqrt[3]{rs^2}$ Write with radical notation. ●

Practice Problems 22–24

Use rational exponents to write as a single radical.

22. $\sqrt{y} \cdot \sqrt[3]{y}$

23. $\dfrac{\sqrt[3]{x}}{\sqrt[4]{x}}$

24. $\sqrt{5} \cdot \sqrt[3]{2}$

EXAMPLES   Use rational exponents to write as a single radical.

22. $\sqrt{x} \cdot \sqrt[4]{x} = x^{1/2} \cdot x^{1/4} = x^{1/2+1/4}$

             $= x^{3/4} = \sqrt[4]{x^3}$

23. $\dfrac{\sqrt{x}}{\sqrt[3]{x}} = \dfrac{x^{1/2}}{x^{1/3}} = x^{1/2-1/3} = x^{3/6-2/6}$

             $= x^{1/6} = \sqrt[6]{x}$

24. $\sqrt[3]{3} \cdot \sqrt{2} = 3^{1/3} \cdot 2^{1/2}$ Write with rational exponents.

        $= 3^{2/6} \cdot 2^{3/6}$ Write the exponents so that they have the same denominator.

        $= \left(3^2 \cdot 2^3\right)^{1/6}$ Use $a^nb^n = (ab)^n$.

        $= \sqrt[6]{3^2 \cdot 2^3}$ Write with radical notation.

        $= \sqrt[6]{72}$ Multiply $3^2 \cdot 2^3$. ●

Answers

19. $\sqrt{y}$, **20.** $\sqrt[3]{3}$, **21.** $\sqrt[3]{a^2b}$, **22.** $\sqrt[6]{y^5}$,

23. $\sqrt[12]{x}$, **24.** $\sqrt[6]{500}$

Name _____ Section _____ Date _____

EXERCISE SET 7.2

A *Use radical notation to rewrite each expression. Simplify if possible. See Examples 1 through 6.*

1. $49^{1/2}$ **2.** $64^{1/3}$ **3.** $27^{1/3}$ **4.** $8^{1/3}$ **5.** $\left(\dfrac{1}{16}\right)^{1/4}$ **6.** $\left(\dfrac{1}{64}\right)^{1/2}$

7. $169^{1/2}$ **8.** $81^{1/4}$ **9.** $2m^{1/3}$ **10.** $(2m)^{1/3}$ **11.** $(9x^4)^{1/2}$

12. $(16x^8)^{1/2}$ **13.** $(-27)^{1/3}$ **14.** $-64^{1/2}$ **15.** $-16^{1/4}$ **16.** $(-32)^{1/5}$

B *Use radical notation to rewrite each expression. Simplify if possible. See Examples 7 through 11.*

17. $16^{3/4}$ **18.** $4^{5/2}$ **19.** $(-64)^{2/3}$ **20.** $(-8)^{4/3}$ **21.** $(-16)^{3/4}$ **22.** $(-9)^{3/2}$

23. $(2x)^{3/5}$ **24.** $2x^{3/5}$ **25.** $(7x+2)^{2/3}$ **26.** $(x-4)^{3/4}$ **27.** $\left(\dfrac{16}{9}\right)^{3/2}$ **28.** $\left(\dfrac{49}{25}\right)^{3/2}$

C *Write with positive exponents. Simplify if possible. See Examples 12 and 13.*

29. $8^{-4/3}$ **30.** $64^{-2/3}$ **31.** $(-64)^{-2/3}$ **32.** $(-8)^{-4/3}$ **33.** $(-4)^{-3/2}$ **34.** $(-16)^{-5/4}$

35. $x^{-1/4}$ **36.** $y^{-1/6}$ **37.** $\dfrac{1}{a^{-2/3}}$ **38.** $\dfrac{1}{n^{-8/9}}$ **39.** $\dfrac{5}{7x^{-3/4}}$ **40.** $\dfrac{2}{3y^{-5/7}}$

41. Explain how writing x^{-7} with positive exponents is similar to writing $x^{-1/4}$ with positive exponents.

42. Explain how writing $2x^{-5}$ with positive exponents is similar to writing $2x^{-3/4}$ with positive exponents.

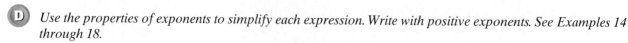

D *Use the properties of exponents to simplify each expression. Write with positive exponents. See Examples 14 through 18.*

43. $a^{2/3}a^{5/3}$

44. $b^{9/5}b^{8/5}$

45. $x^{-2/5} \cdot x^{7/5}$

46. $y^{4/3} \cdot y^{-1/3}$

47. $3^{1/4} \cdot 3^{3/8}$

48. $5^{1/2} \cdot 5^{1/6}$

49. $\dfrac{y^{1/3}}{y^{1/6}}$

50. $\dfrac{x^{3/4}}{x^{1/8}}$

51. $\left(4u^2\right)^{3/2}$

52. $\left(32^{1/5}x^{2/3}\right)^3$

53. $\dfrac{b^{1/2}b^{3/4}}{-b^{1/4}}$

54. $\dfrac{a^{1/4}a^{-1/2}}{a^{2/3}}$

55. $\dfrac{\left(3x^{1/4}\right)^3}{x^{1/12}}$

56. $\dfrac{\left(2x^{1/5}\right)^4}{x^{3/10}}$

57. $\dfrac{\left(y^3z\right)^{1/6}}{y^{-1/2}z^{1/3}}$

58. $\dfrac{\left(m^2n\right)^{1/4}}{m^{-1/2}n^{5/8}}$

59. $\dfrac{\left(x^3y^2\right)^{1/4}}{\left(x^{-5}y^{-1}\right)^{-1/2}}$

60. $\dfrac{\left(a^{-2}b^3\right)^{1/8}}{\left(a^{-3}b\right)^{-1/4}}$

E *Use rational exponents to simplify each radical. Assume that all variables represent positive real numbers. See Examples 19 through 21.*

61. $\sqrt[6]{x^3}$

62. $\sqrt[9]{a^3}$

63. $\sqrt[6]{4}$

64. $\sqrt[4]{36}$

65. $\sqrt[4]{16x^2}$

66. $\sqrt[8]{4y^2}$

67. $\sqrt[4]{(x+3)^2}$

68. $\sqrt[8]{(y+1)^4}$

69. $\sqrt[8]{x^4y^4}$

70. $\sqrt[9]{y^6z^3}$

71. $\sqrt[12]{a^8b^4}$

72. $\sqrt[10]{a^5b^5}$

Use rational expressions to write as a single radical expression. See Examples 22 through 24.

73. $\sqrt[3]{y} \cdot \sqrt[5]{y^2}$

74. $\sqrt[3]{y^2} \cdot \sqrt[6]{y}$

75. $\dfrac{\sqrt[3]{b^2}}{\sqrt[4]{b}}$

76. $\dfrac{\sqrt[4]{a}}{\sqrt[5]{a}}$

77. $\sqrt[3]{x} \cdot \sqrt[4]{x} \cdot \sqrt[8]{x^3}$

78. $\sqrt[6]{y} \cdot \sqrt[3]{y} \cdot \sqrt[5]{y^2}$

79. $\dfrac{\sqrt[3]{a^2}}{\sqrt[6]{a}}$

80. $\dfrac{\sqrt[5]{b^2}}{\sqrt[10]{b^3}}$

81. $\sqrt{3} \cdot \sqrt[3]{4}$

82. $\sqrt[3]{5} \cdot \sqrt{2}$

83. $\sqrt[5]{7} \cdot \sqrt[3]{y}$

84. $\sqrt[4]{5} \cdot \sqrt[3]{x}$

85. $\sqrt{5r} \cdot \sqrt[3]{s}$

86. $\sqrt[3]{b} \cdot \sqrt[5]{4a}$

Review and Preview

Write each integer as a product of two integers such that one of the factors is a perfect square. For example, write 18 as 9·2, because 9 is a perfect square.

87. 75

88. 20

89. 48

90. 45

Write each integer as a product of two integers such that one of the factors is a perfect cube. For example, write 24 as 8·3, because 8 is a perfect cube.

91. 16

92. 56

93. 54

94. 80

Combining Concepts

Basal metabolic rate (BMR) is the number of calories per day a person needs to maintain life. A person's basal metabolic rate $B(w)$ in calories per day can be estimated with the function $B(w) = 70w^{3/4}$, where w is the person's weight in kilograms. Use this information to answer Exercises 95 and 96.

95. Estimate the BMR for a person who weighs 60 kilograms. Round to the nearest calorie.
(*Note:* 60 kilograms is approximately 132 pounds.)

96. Estimate the BMR for a person who weighs 90 kilograms. Round to the nearest calorie.
(*Note:* 90 kilograms is approximately 198 pounds.)

The number of cellular telephone subscriptions in the United States from 1994 through 2000 can be modeled by the function $f(x) = 2.5x^{8/5}$, where y is the number of cellular telephone subscriptions in millions, x years after 1990. (Source: Based on data from the Cellular Telecommunications & Internet Association, 1994–2000) Use this information to answer Exercises 97 and 98.

97. Use this model to estimate the number of cellular telephone subscriptions in the United States in 2000. Round to the nearest tenth of a million.

98. Predict the number of cellular telephone subscriptions in the United States in 2007. Round to the nearest tenth of a million.

Fill in each box with the correct expression.

99. $\square \cdot a^{2/3} = a^{3/3}$, or a

100. $\square \cdot x^{1/8} = x^{4/8}$, or $x^{1/2}$

101. $\dfrac{\square}{x^{-2/5}} = x^{3/5}$

102. $\dfrac{\square}{y^{-3/4}} = y^{4/4}$, or y

Use a calculator to write a four-decimal-place approximation of each number.

103. $8^{1/4}$

104. $18^{3/5}$

105. In physics, the speed of a wave traveling over a stretched string with tension t and density u is given by the expression $\dfrac{\sqrt{t}}{\sqrt{u}}$. Write this expression with rational exponents.

106. In electronics, the angular frequency of oscillations in a certain type of circuit is given by the expression $(LC)^{-1/2}$. Use radical notation to write this expression.

7.3 Simplifying Radical Expressions

Ⓐ Using the Product Rule

It is possible to simplify some radicals that do not evaluate to rational numbers. To do so, we use a product rule and a quotient rule for radicals. To discover the product rule, notice the following pattern:

$$\sqrt{9}\cdot\sqrt{4} = 3\cdot 2 = 6$$
$$\sqrt{9\cdot 4} = \sqrt{36} = 6$$

Since both expressions simplify to 6, it is true that

$$\sqrt{9}\cdot\sqrt{4} = \sqrt{9\cdot 4}$$

This pattern suggests the following product rule for radicals.

Product Rule for Radicals

If $\sqrt[n]{a}$ and $\sqrt[n]{b}$ are real numbers, then

$$\sqrt[n]{a}\cdot\sqrt[n]{b} = \sqrt[n]{ab}$$

Notice that the product rule is the relationship $a^{1/n}\cdot b^{1/n} = (ab)^{1/n}$ stated in radical notation.

EXAMPLES   Use the product rule to multiply.

1. $\sqrt{3}\cdot\sqrt{5} = \sqrt{3\cdot 5} = \sqrt{15}$
2. $\sqrt{21}\cdot\sqrt{x} = \sqrt{21x}$
3. $\sqrt[3]{4}\cdot\sqrt[3]{2} = \sqrt[3]{4\cdot 2} = \sqrt[3]{8} = 2$
4. $\sqrt[4]{5}\cdot\sqrt[4]{2x^3} = \sqrt[4]{5\cdot 2x^3} = \sqrt[4]{10x^3}$
5. $\sqrt{\dfrac{2}{a}}\cdot\sqrt{\dfrac{b}{3}} = \sqrt{\dfrac{2}{a}\cdot\dfrac{b}{3}} = \sqrt{\dfrac{2b}{3a}}$

Ⓑ Using the Quotient Rule

To discover the quotient rule for radicals, notice the following pattern:

$$\sqrt{\dfrac{4}{9}} = \dfrac{2}{3}$$

$$\dfrac{\sqrt{4}}{\sqrt{9}} = \dfrac{2}{3}$$

Since both expressions simplify to $\dfrac{2}{3}$, it is true that

$$\sqrt{\dfrac{4}{9}} = \dfrac{\sqrt{4}}{\sqrt{9}}$$

This pattern suggests the following quotient rule for radicals.

Quotient Rule for Radicals

If $\sqrt[n]{a}$ and $\sqrt[n]{b}$ are real numbers and $\sqrt[n]{b}$ is not zero, then

$$\sqrt[n]{\dfrac{a}{b}} = \dfrac{\sqrt[n]{a}}{\sqrt[n]{b}}$$

Practice Problems 1–5

Use the product rule to multiply.

1. $\sqrt{2}\cdot\sqrt{7}$
2. $\sqrt{17}\cdot\sqrt{y}$
3. $\sqrt[3]{2}\cdot\sqrt[3]{32}$
4. $\sqrt[4]{6}\cdot\sqrt[4]{3x^2}$
5. $\sqrt{\dfrac{3}{x}}\cdot\sqrt{\dfrac{y}{2}}$

Answers

1. $\sqrt{14}$, 2. $\sqrt{17y}$, 3. 4, 4. $\sqrt[4]{18x^2}$, 5. $\sqrt{\dfrac{3y}{2x}}$

Notice that the quotient rule is the relationship $\left(\dfrac{a}{b}\right)^{1/n} = \dfrac{a^{1/n}}{b^{1/n}}$ stated in radical notation. We can use the quotient rule to simplify radical expressions by reading the rule from left to right or to divide radicals by reading the rule from right to left.

For example,

$$\sqrt{\dfrac{x}{16}} = \dfrac{\sqrt{x}}{\sqrt{16}} = \dfrac{\sqrt{x}}{4} \qquad \text{Using } \sqrt[n]{\dfrac{a}{b}} = \dfrac{\sqrt[n]{a}}{\sqrt[n]{b}}$$

$$\dfrac{\sqrt{75}}{\sqrt{3}} = \sqrt{\dfrac{75}{3}} = \sqrt{25} = 5 \quad \text{Using } \dfrac{\sqrt[n]{a}}{\sqrt[n]{b}} = \sqrt[n]{\dfrac{a}{b}}$$

Note: *For the remainder of this chapter, we will assume that variables represent positive real numbers. If this is so, we need not insert absolute value bars when we simplify even roots.*

EXAMPLES Use the quotient rule to simplify.

6. $\sqrt{\dfrac{25}{49}} = \dfrac{\sqrt{25}}{\sqrt{49}} = \dfrac{5}{7}$

7. $\sqrt{\dfrac{x}{9}} = \dfrac{\sqrt{x}}{\sqrt{9}} = \dfrac{\sqrt{x}}{3}$

8. $\sqrt[3]{\dfrac{8}{27}} = \dfrac{\sqrt[3]{8}}{\sqrt[3]{27}} = \dfrac{2}{3}$

9. $\sqrt[4]{\dfrac{3}{16y^4}} = \dfrac{\sqrt[4]{3}}{\sqrt[4]{16y^4}} = \dfrac{\sqrt[4]{3}}{2y}$

(C) Simplifying Radicals

Both the product and quotient rules can be used to simplify a radical. If the product rule is read from right to left, we have that $\sqrt[n]{ab} = \sqrt[n]{a} \cdot \sqrt[n]{b}$. We use this to simplify the following radicals.

EXAMPLE 10 Simplify: $\sqrt{50}$

Solution: We factor 50 such that one factor is the largest perfect square that divides 50. The largest perfect square factor of 50 is 25, so we write 50 as $25 \cdot 2$ and use the product rule for radicals to simplify.

$$\sqrt{50} = \sqrt{25 \cdot 2} = \sqrt{25} \cdot \sqrt{2} = 5\sqrt{2}$$

‎└ the largest perfect square factor of 50

> **Helpful Hint**
> Don't forget that, for example, $5\sqrt{2}$ means $5 \cdot \sqrt{2}$,

EXAMPLES Simplify.

11. $\sqrt[3]{24} = \sqrt[3]{8 \cdot 3} = \sqrt[3]{8} \cdot \sqrt[3]{3} = 2\sqrt[3]{3}$

‎└ the largest perfect cube factor of 24

12. $\sqrt{26}$ The largest perfect square factor of 26 is 1, so $\sqrt{26}$ cannot be simplified further.

13. $\sqrt[4]{32} = \sqrt[4]{16 \cdot 2} = \sqrt[4]{16} \cdot \sqrt[4]{2} = 2\sqrt[4]{2}$

‎└ the largest 4th power factor of 32

Practice Problems 6–9

Use the quotient rule to simplify. Assume that all variables represent positive real numbers.

6. $\sqrt{\dfrac{9}{25}}$

7. $\sqrt{\dfrac{y}{36}}$

8. $\sqrt[3]{\dfrac{27}{64}}$

9. $\sqrt[5]{\dfrac{7}{32x^5}}$

Practice Problem 10

Simplify: $\sqrt{18}$

Practice Problems 11–13

Simplify.

11. $\sqrt[3]{40}$
12. $\sqrt{14}$
13. $\sqrt[4]{162}$

Answers

6. $\dfrac{3}{5}$, **7.** $\dfrac{\sqrt{y}}{6}$, **8.** $\dfrac{3}{4}$, **9.** $\dfrac{\sqrt[5]{7}}{2x}$, **10.** $3\sqrt{2}$,

11. $2\sqrt[3]{5}$, **12.** $\sqrt{14}$, **13.** $3\sqrt[4]{2}$

After simplifying a radical such as a square root, always check the radicand to see that it contains no other perfect square factors. It may, if the largest perfect square factor of the radicand was not originally recognized. For example,

$$\sqrt{200} = \sqrt{4 \cdot 50} = \sqrt{4} \cdot \sqrt{50} = 2\sqrt{50}$$

Notice that the radicand 50 still contains the perfect square factor 25. This is because 4 is not the largest perfect square factor of 200. We continue as follows:

$$2\sqrt{50} = 2\sqrt{25 \cdot 2} = 2 \cdot \sqrt{25} \cdot \sqrt{2} = 2 \cdot 5 \cdot \sqrt{2} = 10\sqrt{2}$$

The radical is now simplified since 2 contains no perfect square factors (other than 1).

> **Helpful Hint**
>
> To recognize the largest perfect power factors of a radicand, it will help if you are familiar with some perfect powers. A few are listed below.
>
> Perfect Squares $\quad$ $\underset{1^2 \ \ 2^2 \ 3^2 \ 4^2 \ \ 5^2 \ \ 6^2 \ \ 7^2 \ \ 8^2 \ \ 9^2 \ \ 10^2 \ \ 11^2 \ \ 12^2}{1, 4, 9, 16, 25, 36, 49, 64, 81, 100, 121, 144}$
>
> Perfect Cubes $\quad$ $\underset{1^3 \ \ 2^3 \ 3^3 \ \ 4^3 \ \ 5^3}{1, 8, 27, 64, 125}$
>
> Perfect 4th powers $\quad$ $\underset{1^4 \ \ 2^4 \ \ 3^4 \ \ 4^4}{1, 16, 81, 256}$

> **Helpful Hint**
>
> We say that a radical of the form $\sqrt[n]{a}$ is simplified when the radicand a contains no factors that are perfect nth powers (other than 1 or -1).

EXAMPLES

Simplify. Assume that all variables represent positive real numbers.

14. $\sqrt{25x^3} = \sqrt{25 \cdot x^2 \cdot x}$ $\qquad$ Find the largest perfect square factor.

$\qquad = \sqrt{25 \cdot x^2} \cdot \sqrt{x}$ $\qquad$ Use the product rule.

$\qquad = 5x\sqrt{x}$ $\qquad$ Simplify.

15. $\sqrt[3]{54x^6y^8} = \sqrt[3]{27 \cdot 2 \cdot x^6 \cdot y^6 \cdot y^2}$ $\qquad$ Factor the radicand and identify perfect cube factors.

$\qquad = \sqrt[3]{27 \cdot x^6 \cdot y^6 \cdot 2y^2}$

$\qquad = \sqrt[3]{27 \cdot x^6 \cdot y^6} \cdot \sqrt[3]{2y^2}$ $\qquad$ Use the product rule.

$\qquad = 3x^2y^2\sqrt[3]{2y^2}$ $\qquad$ Simplify.

16. $\sqrt[4]{81z^{11}} = \sqrt[4]{81 \cdot z^8 \cdot z^3}$ $\qquad$ Factor the radicand and identify perfect 4th power factors.

$\qquad = \sqrt[4]{81 \cdot z^8} \cdot \sqrt[4]{z^3}$ $\qquad$ Use the product rule.

$\qquad = 3z^2\sqrt[4]{z^3}$ $\qquad$ Simplify.

Practice Problems 14–16

Simplify. Assume that all variables represent positive real numbers.

14. $\sqrt{49a^5}$

15. $\sqrt[3]{24x^9y^7}$

16. $\sqrt[4]{16z^9}$

Answers

14. $7a^2\sqrt{a}$, **15.** $2x^3y^2\sqrt[3]{3y}$, **16.** $2z^2\sqrt[4]{z}$

Practice Problems 17–20

Use the quotient rule to divide. Then simplify if possible. Assume that all variables represent positive real numbers.

17. $\dfrac{\sqrt{75}}{\sqrt{3}}$

18. $\dfrac{\sqrt{80y}}{3\sqrt{5}}$

19. $\dfrac{5\sqrt[3]{162x^8}}{\sqrt[3]{3x^2}}$

20. $\dfrac{3\sqrt[4]{243x^9y^6}}{\sqrt[4]{x^{-3}y}}$

EXAMPLES

Use the quotient rule to divide. Then simplify if possible. Assume that all variables represent positive real numbers.

17. $\dfrac{\sqrt{20}}{\sqrt{5}} = \sqrt{\dfrac{20}{5}}$ Use the quotient rule.

     $= \sqrt{4}$ Simplify.

     $= 2$ Simplify.

18. $\dfrac{\sqrt{50x}}{2\sqrt{2}} = \dfrac{1}{2} \cdot \sqrt{\dfrac{50x}{2}}$ Use the quotient rule.

     $= \dfrac{1}{2} \cdot \sqrt{25x}$ Simplify.

     $= \dfrac{1}{2} \cdot \sqrt{25} \cdot \sqrt{x}$ Factor $25x$.

     $= \dfrac{1}{2} \cdot 5 \cdot \sqrt{x}$ Simplify.

     $= \dfrac{5}{2}\sqrt{x}$

19. $\dfrac{7\sqrt[3]{48y^4}}{\sqrt[3]{2y}} = 7\sqrt[3]{\dfrac{48y^4}{2y}} = 7\sqrt[3]{24y^3} = 7\sqrt[3]{8 \cdot y^3 \cdot 3}$

     $= 7\sqrt[3]{8 \cdot y^3} \cdot \sqrt[3]{3} = 7 \cdot 2y\sqrt[3]{3} = 14y\sqrt[3]{3}$

20. $\dfrac{2\sqrt[4]{32a^8b^6}}{\sqrt[4]{a^{-1}b^2}} = 2\sqrt[4]{\dfrac{32a^8b^6}{a^{-1}b^2}} = 2\sqrt[4]{32a^9b^4} = 2\sqrt[4]{16 \cdot a^8 \cdot b^4 \cdot 2 \cdot a}$

     $= 2\sqrt[4]{16 \cdot a^8 \cdot b^4} \cdot \sqrt[4]{2 \cdot a} = 2 \cdot 2a^2b \cdot \sqrt[4]{2a} = 4a^2b\sqrt[4]{2a}$ ●

Concept Check

Find and correct the error:

$$\dfrac{\sqrt[3]{27}}{\sqrt{9}} = \sqrt[3]{\dfrac{27}{9}} = \sqrt[3]{3}$$

Try the Concept Check in the margin.

Answers

17. 5,   18. $\dfrac{4}{3}\sqrt{y}$,   19. $15x^2\sqrt[3]{2}$,   20. $9x^3y\sqrt[4]{3y}$

Concept Check: $\dfrac{\sqrt[3]{27}}{\sqrt{9}} = \dfrac{3}{3} = 1$

Name _____ Section _____ Date _____

EXERCISE SET 7.3

A *Use the product rule to multiply. Assume that all variables represent positive real numbers. See Examples 1 through 5.*

1. $\sqrt{7} \cdot \sqrt{2}$

2. $\sqrt{11} \cdot \sqrt{10}$

3. $\sqrt[4]{8} \cdot \sqrt[4]{2}$

4. $\sqrt[4]{27} \cdot \sqrt[4]{3}$

5. $\sqrt[3]{4} \cdot \sqrt[3]{9}$

6. $\sqrt[3]{10} \cdot \sqrt[3]{5}$

7. $\sqrt{2} \cdot \sqrt{3x}$

8. $\sqrt{3y} \cdot \sqrt{5x}$

9. $\sqrt{\dfrac{7}{x}} \cdot \sqrt{\dfrac{2}{y}}$

10. $\sqrt{\dfrac{6}{m}} \cdot \sqrt{\dfrac{n}{5}}$

11. $\sqrt[4]{4x^3} \cdot \sqrt[4]{5}$

12. $\sqrt[4]{ab^2} \cdot \sqrt[4]{27ab}$

B *Use the quotient rule to simplify. Assume that all variables represent positive real numbers. See Examples 6 through 9.*

13. $\sqrt{\dfrac{6}{49}}$

14. $\sqrt{\dfrac{8}{81}}$

15. $\sqrt{\dfrac{2}{49}}$

16. $\sqrt{\dfrac{5}{121}}$

17. $\sqrt[4]{\dfrac{x^3}{16}}$

18. $\sqrt[4]{\dfrac{y}{81x^4}}$

19. $\sqrt[3]{\dfrac{4}{27}}$ **20.** $\sqrt[3]{\dfrac{3}{64}}$ **21.** $\sqrt[4]{\dfrac{8}{x^8}}$ **22.** $\sqrt[4]{\dfrac{a^3}{81}}$ **23.** $\sqrt[3]{\dfrac{2x}{81y^{12}}}$ **24.** $\sqrt[3]{\dfrac{3}{8x^6}}$

25. $\sqrt{\dfrac{x^2 y}{100}}$ **26.** $\sqrt{\dfrac{y^2 z}{400}}$ 🔒 **27.** $\sqrt{\dfrac{5x^2}{169y^2}}$ **28.** $\sqrt{\dfrac{y^{10}}{225x^6}}$ **29.** $-\sqrt[3]{\dfrac{z^7}{125x^3}}$ **30.** $-\sqrt[3]{\dfrac{1000a}{b^9}}$

C *Simplify. Assume that all variables represent positive real numbers. See Examples 10 through 16.*

🔒 **31.** $\sqrt{32}$ **32.** $\sqrt{27}$ **33.** $\sqrt[3]{192}$ **34.** $\sqrt[3]{108}$ **35.** $5\sqrt{75}$ **36.** $3\sqrt{8}$

37. $\sqrt{24}$ **38.** $\sqrt{20}$ **39.** $\sqrt{100x^5}$ **40.** $\sqrt{64y^9}$ **41.** $\sqrt[3]{16y^7}$ **42.** $\sqrt[3]{64y^9}$

43. $\sqrt[4]{a^8 b^7}$ **44.** $\sqrt[5]{32z^{12}}$ **45.** $\sqrt{y^5}$ **46.** $\sqrt[3]{y^5}$ 🔒 **47.** $\sqrt{25a^2 b^3}$ **48.** $\sqrt{9x^5 y^7}$

49. $\sqrt[5]{-32x^{10}y}$ **50.** $\sqrt[5]{-243z^9}$ **51.** $\sqrt[3]{50x^{14}}$ **52.** $\sqrt[3]{40y^{10}}$ **53.** $-\sqrt{32a^8b^7}$

54. $-\sqrt{20ab^6}$ **55.** $\sqrt{9x^7y^9}$ **56.** $\sqrt{12r^9s^{12}}$ **57.** $\sqrt[3]{125r^9s^{12}}$ **58.** $\sqrt[3]{8a^6b^9}$

Use the quotient rule to divide. Then simplify if possible. Assume that all variables represent positive real numbers. See Examples 17 through 20.

59. $\dfrac{\sqrt{14}}{\sqrt{7}}$ **60.** $\dfrac{\sqrt{45}}{\sqrt{9}}$ **61.** $\dfrac{\sqrt[3]{24}}{\sqrt[3]{3}}$ **62.** $\dfrac{\sqrt[3]{10}}{\sqrt[3]{2}}$ **63.** $\dfrac{5\sqrt[4]{48}}{\sqrt[4]{3}}$

64. $\dfrac{7\sqrt[4]{162}}{\sqrt[4]{2}}$ **65.** $\dfrac{\sqrt{x^5y^3}}{\sqrt{xy}}$ **66.** $\dfrac{\sqrt{a^7b^6}}{\sqrt{a^3b^2}}$ **67.** $\dfrac{8\sqrt[3]{54m^7}}{\sqrt[3]{2m}}$ **68.** $\dfrac{\sqrt[3]{128x^3}}{-3\sqrt[3]{2x}}$

69. $\dfrac{3\sqrt{100x^2}}{2\sqrt{2x^{-1}}}$ **70.** $\dfrac{\sqrt{270y^2}}{5\sqrt{3y^{-4}}}$ **71.** $\dfrac{\sqrt[4]{96a^{10}b^3}}{\sqrt[4]{3a^2b^3}}$ **72.** $\dfrac{\sqrt[5]{64x^{10}y^3}}{\sqrt[5]{2x^3y^{-7}}}$

Review and Preview

Perform each indicated operation. See Sections 1.5 and 5.2.

73. $6x + 8x$

74. $(6x)(8x)$

75. $(2x + 3)(x - 5)$

76. $(2x + 3) + (x - 5)$

77. $9y^2 - 8y^2$

78. $(9y^2)(-8y^2)$

79. $-3(x + 5)$

80. $-3 + x + 5$

81. $(x - 4)^2$

82. $(2x + 1)^2$

 Combining Concepts

83. The formula for the surface area A of a cone with height h and radius r is given by
$$A = \pi r \sqrt{r^2 + h^2}$$

   a. Find the surface area of a cone whose height is 3 centimeters and whose radius is 4 centimeters.

   b. Approximate to two decimal places the surface area of a cone whose height is 7.2 feet and whose radius is 6.8 feet.

84. Before Mount Vesuvius, a volcano in Italy, erupted violently in 79 A.D., its height was 4190 feet. Vesuvius was roughly cone-shaped, and its base had a radius of approximately 25,200 feet. Use the formula for the surface area of a cone, given in Exercise 83, to approximate the surface area this volcano had before it erupted. (*Source:* Global Volcanism Network)

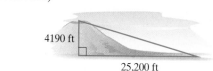

4190 ft

25,200 ft

85. The owner of Knightime Video has determined that the demand equation for renting older released tapes is $F(x) = 0.6\sqrt{49 - x^2}$, where x is the price in dollars per two-day rental and $F(x)$ is the number of times the video is demanded per week.

   a. Approximate to one decimal place the demand per week of an older released video if the rental price is $3 per two-day rental.

   b. Approximate to one decimal place the demand per week of an older released video if the rental price is $5 per two-day rental.

   c. Explain how the owner of the video store can use this equation to predict the number of copies of each tape that should be in stock.

7.4 Adding, Subtracting, and Multiplying Radical Expressions

Ⓐ Adding or Subtracting Radical Expressions

We have learned that the sum or difference of like terms can be simplified. To simplify these sums or differences, we use the distributive property. For example,

$$2x + 3x = (2 + 3)x = 5x$$

The distributive property can also be used to add *like radicals*.

Like Radicals

Radicals with the same index and the same radicand are **like radicals**. The example below shows how to use the distributive property to simplify an expression containing like radicals.

$$2\sqrt{7} + 3\sqrt{7} = (2 + 3)\sqrt{7} = 5\sqrt{7}$$

Like radicals

Helpful Hint

The expression

$$5\sqrt{7} - 3\sqrt{6}$$

does not contain like radicals and cannot be simplified further.

EXAMPLES   Add or subtract as indicated.

1. $4\sqrt{11} + 8\sqrt{11} = (4 + 8)\sqrt{11} = 12\sqrt{11}$

2. $5\sqrt[3]{3x} - 7\sqrt[3]{3x} = (5 - 7)\sqrt[3]{3x} = -2\sqrt[3]{3x}$

3. $2\sqrt{7} + 2\sqrt[3]{7}$ This expression cannot be simplified since $2\sqrt{7}$ and $2\sqrt[3]{7}$ do not contain like radicals. ●

Try the Concept Check in the margin.

When adding or subtracting radicals, always check first to see whether any radicals can be simplified.

EXAMPLES

Add or subtract as indicated. Assume that all variables represent positive real numbers.

4. $\sqrt{20} + 2\sqrt{45} = \sqrt{4 \cdot 5} + 2\sqrt{9 \cdot 5}$ Factor 20 and 45.
$\qquad = \sqrt{4} \cdot \sqrt{5} + 2 \cdot \sqrt{9} \cdot \sqrt{5}$ Use the product rule.
$\qquad = 2 \cdot \sqrt{5} + 2 \cdot 3 \cdot \sqrt{5}$ Simplify $\sqrt{4}$ and $\sqrt{9}$.
$\qquad = 2\sqrt{5} + 6\sqrt{5}$ Add like radicals.
$\qquad = 8\sqrt{5}$

5. $\sqrt[3]{54} - 5\sqrt[3]{16} + \sqrt[3]{2}$
$\qquad = \sqrt[3]{27} \cdot \sqrt[3]{2} - 5 \cdot \sqrt[3]{8} \cdot \sqrt[3]{2} + \sqrt[3]{2}$ Factor and use the product rule.
$\qquad = 3 \cdot \sqrt[3]{2} - 5 \cdot 2 \cdot \sqrt[3]{2} + \sqrt[3]{2}$ Simplify $\sqrt[3]{27}$ and $\sqrt[3]{8}$.
$\qquad = 3\sqrt[3]{2} - 10\sqrt[3]{2} + \sqrt[3]{2}$ Write $5 \cdot 2$ as 10.
$\qquad = -6\sqrt[3]{2}$ Combine like radicals.

6. $\sqrt{27x} - 2\sqrt{9x} + \sqrt{72x}$

$\qquad = \sqrt{9} \cdot \sqrt{3x} - 2 \cdot \sqrt{9} \cdot \sqrt{x} + \sqrt{36} \cdot \sqrt{2x}$ Factor and use the product rule.

$\qquad = 3 \cdot \sqrt{3x} - 2 \cdot 3 \cdot \sqrt{x} + 6 \cdot \sqrt{2x}$ Simplify $\sqrt{9}$ and $\sqrt{36}$.

$\qquad = 3\sqrt{3x} - 6\sqrt{x} + 6\sqrt{2x}$ Write $2 \cdot 3$ as 6.

> **Helpful Hint**
>
> None of these terms contain like radicals. We can simplify no further.

7. $\sqrt[3]{98} + \sqrt{98} = \sqrt[3]{98} + \sqrt{49} \cdot \sqrt{2}$ Factor and use the product rule.

$\qquad\qquad\qquad = \sqrt[3]{98} + 7\sqrt{2}$ No further simplification is possible.

8. $\sqrt[3]{48y^4} + \sqrt[3]{6y^4} = \sqrt[3]{8y^3} \cdot \sqrt[3]{6y} + \sqrt[3]{y^3} \cdot \sqrt[3]{6y}$ Factor and use the product rule.

$\qquad\qquad\qquad\quad = 2y\sqrt[3]{6y} + y\sqrt[3]{6y}$ Simplify $\sqrt[3]{8y^3}$ and $\sqrt[3]{y^3}$.

$\qquad\qquad\qquad\quad = 3y\sqrt[3]{6y}$ Combine like radicals. ●

Practice Problems 9–10

Add or subtract as indicated. Assume that all variables represent positive real numbers.

9. $\dfrac{\sqrt{75}}{9} - \dfrac{\sqrt{3}}{2}$

10. $\sqrt[3]{\dfrac{5x}{27}} + 4\sqrt[3]{5x}$

EXAMPLES

Add or subtract as indicated. Assume that all variables represent positive real numbers.

9. $\dfrac{\sqrt{45}}{4} - \dfrac{\sqrt{5}}{3} = \dfrac{3\sqrt{5}}{4} - \dfrac{\sqrt{5}}{3}$ To subtract, notice that the LCD is 12.

$\qquad\qquad\qquad = \dfrac{3\sqrt{5} \cdot 3}{4 \cdot 3} - \dfrac{\sqrt{5} \cdot 4}{3 \cdot 4}$ Write each expression as an equivalent expression with a denominator of 12.

$\qquad\qquad\qquad = \dfrac{9\sqrt{5}}{12} - \dfrac{4\sqrt{5}}{12}$ Multiply factors in the numerators and the denominators.

$\qquad\qquad\qquad = \dfrac{5\sqrt{5}}{12}$ Subtract.

10. $\sqrt[3]{\dfrac{7x}{8}} + 2\sqrt[3]{7x} = \dfrac{\sqrt[3]{7x}}{\sqrt[3]{8}} + 2\sqrt[3]{7x}$ Use the quotient rule for radicals.

$\qquad\qquad\qquad\quad = \dfrac{\sqrt[3]{7x}}{2} + 2\sqrt[3]{7x}$ Simplify.

$\qquad\qquad\qquad\quad = \dfrac{\sqrt[3]{7x}}{2} + \dfrac{2\sqrt[3]{7x} \cdot 2}{2}$ Write each expression as an equivalent expression with a denominator of 2.

$\qquad\qquad\qquad\quad = \dfrac{\sqrt[3]{7x}}{2} + \dfrac{4\sqrt[3]{7x}}{2}$

$\qquad\qquad\qquad\quad = \dfrac{5\sqrt[3]{7x}}{2}$ Add. ●

B   Multiplying Radical Expressions

We can multiply radical expressions by using many of the same properties used to multiply polynomial expressions. For instance, to multiply $\sqrt{2}(\sqrt{6} - 3\sqrt{2})$, we use the distributive property and multiply $\sqrt{2}$ by each term inside the parentheses.

$\sqrt{2}(\sqrt{6} - 3\sqrt{2}) = \sqrt{2}(\sqrt{6}) - \sqrt{2}(3\sqrt{2})$ Use the distributive property.

$\qquad\qquad\qquad = \sqrt{2 \cdot 6} - 3\sqrt{2 \cdot 2}$

$\qquad\qquad\qquad = \sqrt{2 \cdot 2 \cdot 3} - 3 \cdot 2$ Use the product rule for radicals.

$\qquad\qquad\qquad = 2\sqrt{3} - 6$

Answers

9. $\dfrac{\sqrt{3}}{18}$,   10. $\dfrac{13\sqrt[3]{5x}}{3}$

EXAMPLE 11   Multiply: $\sqrt{3}(5 + \sqrt{30})$

Solution: $\sqrt{3}(5 + \sqrt{30}) = \sqrt{3}(5) + \sqrt{3}(\sqrt{30})$

$\qquad\qquad\qquad = 5\sqrt{3} + \sqrt{3 \cdot 30}$

$\qquad\qquad\qquad = 5\sqrt{3} + \sqrt{3 \cdot 3 \cdot 10}$

$\qquad\qquad\qquad = 5\sqrt{3} + 3\sqrt{10}$

Practice Problem 11

Multiply: $\sqrt{2}(6 + \sqrt{10})$

●

EXAMPLES

Multiply. Assume that all variables represent positive real numbers.

$\qquad\qquad\qquad$ First $\qquad$ Outer $\qquad$ Inner $\qquad$ Last

12. $(\sqrt{5} - \sqrt{6})(\sqrt{7} + 1) = \sqrt{5} \cdot \sqrt{7} + \sqrt{5} \cdot 1 - \sqrt{6} \cdot \sqrt{7} - \sqrt{6} \cdot 1$

$\qquad\qquad\qquad\qquad\qquad\qquad\qquad$ Using the FOIL order.

$\qquad\qquad\qquad\qquad = \sqrt{35} + \sqrt{5} - \sqrt{42} - \sqrt{6}$ $\qquad$ Simplify.

13. $(\sqrt{2x} + 5)(\sqrt{2x} - 5) = (\sqrt{2x})^2 - 5^2$ $\qquad$ Multiply the sum and

$\qquad\qquad\qquad\qquad\qquad\qquad\qquad\qquad$ difference of two terms:

$\qquad\qquad\qquad\qquad\quad = 2x - 25$ $\qquad\qquad\quad$ $(a + b)(a - b) = a^2 - b^2$

14. $(\sqrt{3} - 1)^2 = (\sqrt{3})^2 - 2 \cdot \sqrt{3} \cdot 1 + 1^2$ $\qquad$ Square the binomial:

$\qquad\qquad\quad = 3 - 2\sqrt{3} + 1$ $\qquad\qquad\qquad$ $(a - b)^2 = a^2 - 2ab + b^2$

$\qquad\qquad\quad = 4 - 2\sqrt{3}$ $\qquad\qquad\qquad\qquad$ Square the binomial:

$\qquad\qquad\qquad\qquad\qquad\qquad\qquad\qquad$ $(a + b)^2 = a^2 + 2ab + b^2$

15. $(\underbrace{\sqrt{x - 3}} + 5)^2 = (\underbrace{\sqrt{x - 3}})^2 + 2 \cdot \sqrt{x - 3} \cdot 5 + 5^2$

$\qquad\quad\uparrow\qquad\uparrow\qquad\qquad\uparrow\qquad\uparrow\uparrow\quad\uparrow\qquad\uparrow\quad\uparrow$

$\qquad\quad a\qquad\ b\qquad\qquad a^2\quad + 2 \cdot\ \ a\quad \cdot b\ + b^2$

$\qquad\qquad\qquad = x - 3 + 10\sqrt{x - 3} + 25$ $\qquad$ Simplify.

$\qquad\qquad\qquad = x + 22 + 10\sqrt{x - 3}$ $\qquad\quad$ Combine like terms.

Practice Problems 12–15

Multiply. Assume that all variables represent positive real numbers.

12. $(\sqrt{3} - \sqrt{5})(\sqrt{2} + 7)$
13. $(\sqrt{5y} + 2)(\sqrt{5y} - 2)$
14. $(\sqrt{3} - 7)^2$
15. $(\sqrt{x + 1} + 2)^2$

FOCUS ON **History**

DEVELOPMENT OF THE RADICAL SYMBOL

The first mathematician to use the symbol we use today to denote a square root was Christoff Rudolff (1499–1545). In 1525, Rudolff wrote and published the first German algebra text, *Die Coss*. In it, he used $\sqrt{}$ to represent a square root, the symbol $\sqrt[3]{}$ to represent a cube root, and the symbol $\sqrt[4]{}$ to represent a fourth root. It was another 100 years before the square root symbol was extended with an overbar called a *vinculum*, $\sqrt{}$, to indicate the inclusion of several terms under the radical symbol. This innovation was introduced by René Descartes (1596–1650) in 1637 in his text *La Géométrie*. The modern use of a numeral as part of a radical sign to indicate the index of the radical for higher roots did not appear until 1690, when this notation was used by French mathematician Michel Rolle (1652–1719) in his text *Traité d'Algébre*.

Mental Math

Simplify. Assume that all variables represent positive real numbers.

1. $2\sqrt{3} + 4\sqrt{3}$

2. $5\sqrt{7} + 3\sqrt{7}$

3. $8\sqrt{x} - 5\sqrt{x}$

4. $3\sqrt{y} + 10\sqrt{y}$

5. $7\sqrt[3]{x} + 5\sqrt[3]{x}$

6. $8\sqrt[3]{z} - 2\sqrt[3]{z}$

7. $(\sqrt{3})^2$

8. $(\sqrt{4x+1})^2$

EXERCISE SET 7.4

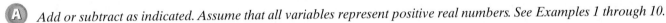

A Add or subtract as indicated. Assume that all variables represent positive real numbers. See Examples 1 through 10.

1. $\sqrt{8} - \sqrt{32}$

2. $\sqrt{27} - \sqrt{75}$

3. $2\sqrt{2x^3} + 4x\sqrt{8x}$

4. $3\sqrt{45x^3} + x\sqrt{5x}$

5. $2\sqrt{50} - 3\sqrt{125} + \sqrt{98}$

6. $4\sqrt{32} - \sqrt{18} + 2\sqrt{128}$

7. $\sqrt[3]{16x} - \sqrt[3]{54x}$

8. $2\sqrt[3]{3a^4} - 3a\sqrt[3]{81a}$

9. $\sqrt{9b^3} - \sqrt{25b^3} + \sqrt{49b^3}$

10. $\sqrt{4x^7} + 9x^2\sqrt{x^3} - 5x\sqrt{x^5}$

11. $\dfrac{5\sqrt{2}}{3} + \dfrac{2\sqrt{2}}{5}$

12. $\dfrac{\sqrt{3}}{2} + \dfrac{4\sqrt{3}}{3}$

13. $\sqrt[3]{\dfrac{11}{8}} - \dfrac{\sqrt[3]{11}}{6}$

14. $\dfrac{2\sqrt[3]{4}}{7} - \dfrac{\sqrt[3]{4}}{14}$

15. $\dfrac{\sqrt{20x}}{9} + \sqrt{\dfrac{5x}{9}}$

16. $\dfrac{3x\sqrt{7}}{5} + \sqrt{\dfrac{7x^2}{100}}$

17. $7\sqrt{9} - 7 + \sqrt{3}$

18. $\sqrt{16} - 5\sqrt{10} + 7$

19. $2 + 3\sqrt{y^2} - 6\sqrt{y^2} + 5$

20. $3\sqrt{7} - \sqrt[3]{x} + 4\sqrt{7} - 3\sqrt[3]{x}$

21. $3\sqrt{108} - 2\sqrt{18} - 3\sqrt{48}$

22. $-\sqrt{75} + \sqrt{12} - 3\sqrt{3}$

23. $-5\sqrt[3]{625} + \sqrt[3]{40}$

24. $-2\sqrt[3]{108} - \sqrt[3]{32}$

25. $\sqrt{9b^3} - \sqrt{25b^3} + \sqrt{16b^3}$

26. $\sqrt{4x^7y^5} + 9x^2\sqrt{x^3y^5} - 5xy\sqrt{x^5y^3}$ **27.** $5y\sqrt{8y} + 2\sqrt{50y^3}$

28. $3\sqrt{8x^2y^3} - 2x\sqrt{32y^3}$

29. $\sqrt[3]{54xy^3} - 5\sqrt[3]{2xy^3} + y\sqrt[3]{128x}$ **30.** $2\sqrt[3]{24x^3y^4} + 4x\sqrt[3]{81y^4}$

31. $6\sqrt[3]{11} + 8\sqrt{11} - 12\sqrt{11}$

32. $3\sqrt[3]{5} + 4\sqrt{5}$

33. $-2\sqrt[4]{x^7} + 3\sqrt[4]{16x^7}$

34. $6\sqrt[3]{24x^3} - 2\sqrt[3]{81x^3} - x\sqrt[3]{3}$

35. $\dfrac{4\sqrt{3}}{3} - \dfrac{\sqrt{12}}{3}$

36. $\dfrac{\sqrt{45}}{10} + \dfrac{7\sqrt{5}}{10}$

37. $\dfrac{\sqrt[3]{8x^4}}{7} + \dfrac{3x\sqrt[3]{x}}{7}$

38. $\dfrac{\sqrt[4]{48}}{5x} - \dfrac{2\sqrt[4]{3}}{10x}$

39. $\sqrt{\dfrac{28}{x^2}} + \sqrt{\dfrac{7}{4x^2}}$

40. $\dfrac{\sqrt{99}}{5x} - \sqrt{\dfrac{44}{x^2}}$

41. $\sqrt[3]{\dfrac{16}{27}} - \dfrac{\sqrt[3]{54}}{6}$

42. $\dfrac{\sqrt[3]{3}}{10} + \sqrt[3]{\dfrac{24}{125}}$

43. $-\dfrac{\sqrt[3]{2x^4}}{9} + \sqrt[3]{\dfrac{250x^4}{27}}$

44. $\dfrac{\sqrt[3]{y^5}}{8} + \dfrac{5y\sqrt[3]{y^2}}{4}$

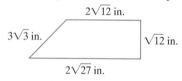 **45.** Find the perimeter of the trapezoid.

$2\sqrt{12}$ in.

$3\sqrt{3}$ in.

$\sqrt{12}$ in.

$2\sqrt{27}$ in.

△ **46.** Find the perimeter of the triangle.

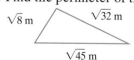

$\sqrt{8}$ m

$\sqrt{32}$ m

$\sqrt{45}$ m

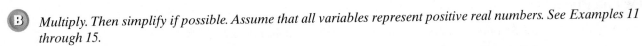

B   *Multiply. Then simplify if possible. Assume that all variables represent positive real numbers. See Examples 11 through 15.*

47. $\sqrt{7}(\sqrt{5} + \sqrt{3})$

48. $\sqrt{5}(\sqrt{15} - \sqrt{35})$

49. $(\sqrt{5} - \sqrt{2})^2$

50. $(3x - \sqrt{2})(3x - \sqrt{2})$

51. $\sqrt{3x}(\sqrt{3} - \sqrt{x})$

52. $\sqrt{5y}(\sqrt{y} + \sqrt{5})$

53. $(2\sqrt{x} - 5)(3\sqrt{x} + 1)$

54. $(8\sqrt{y} + z)(4\sqrt{y} - 1)$

55. $(\sqrt[3]{a} - 4)(\sqrt[3]{a} + 5)$

56. $(\sqrt[3]{a} + 2)(\sqrt[3]{a} + 7)$

57. $6(\sqrt{2} - 2)$

58. $\sqrt{5}(6 - \sqrt{5})$

59. $\sqrt{2}(\sqrt{2} + x\sqrt{6})$

60. $\sqrt{3}(\sqrt{3} - 2\sqrt{5x})$

61. $(2\sqrt{7} + 3\sqrt{5})(\sqrt{7} - 2\sqrt{5})$

62. $(\sqrt{6} - 4\sqrt{2})(3\sqrt{6} + 1)$

63. $(\sqrt{x} - y)(\sqrt{x} + y)$

64. $(3\sqrt{x} + 2)(\sqrt{3x} - 2)$

65. $(\sqrt{3} + x)^2$

66. $(\sqrt{y} - 3x)^2$

67. $(\sqrt{5x} - 3\sqrt{2})(\sqrt{5x} - 3\sqrt{3})$

68. $(5\sqrt{3x} - \sqrt{y})(4\sqrt{x} + 1)$

69. $(\sqrt[3]{4} + 2)(\sqrt[3]{2} - 1)$

70. $(\sqrt[3]{3} + \sqrt[3]{2})(\sqrt[3]{9} - \sqrt[3]{4})$

71. $(\sqrt[3]{x} + 1)(\sqrt[3]{x} - 4\sqrt{x} + 7)$

72. $(\sqrt[3]{3x} + 3)(\sqrt[3]{2x} - 3x - 1)$

73. $(\sqrt{x - 1} + 5)^2$

74. $(\sqrt{3x + 1} + 2)^2$

75. $(\sqrt{2x + 5} - 1)^2$

76. $(\sqrt{x - 6} - 7)^2$

Review and Preview

Factor each numerator and denominator. Then simplify if possible. See Section 6.1.

77. $\dfrac{2x - 14}{2}$

78. $\dfrac{8x - 24y}{4}$

79. $\dfrac{7x - 7y}{x^2 - y^2}$

80. $\dfrac{x^3 - 8}{4x - 8}$

81. $\dfrac{6a^2b - 9ab}{3ab}$

82. $\dfrac{14r - 28r^2s^2}{7rs}$

83. $\dfrac{-4 + 2\sqrt{3}}{6}$

84. $\dfrac{-5 + 10\sqrt{7}}{5}$

 Combining Concepts

△ **85.** Find the perimeter and area of the rectangle.

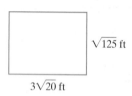

$\sqrt{125}$ ft

$3\sqrt{20}$ ft

△ **86.** Find the area and perimeter of the trapezoid. (*Hint:* The area of a trapezoid is the product of half the height $6\sqrt{3}$ meters and the sum of the bases $2\sqrt{63}$ and $7\sqrt{7}$ meters.)

$2\sqrt{63}$ m

$2\sqrt{27}$ m $6\sqrt{3}$ m

$7\sqrt{7}$ m

87. a. Add: $\sqrt{3} + \sqrt{3}$
  b. Multiply: $\sqrt{3} \cdot \sqrt{3}$
  c. Describe the differences in parts a and b.

88. Multiply: $(\sqrt{2} + \sqrt{3} - 1)^2$

7.5 Rationalizing Numerators and Denominators of Radical Expressions

OBJECTIVES

Ⓐ Rationalize denominators.

Ⓑ Rationalize denominators having two terms.

Ⓒ Rationalize numerators.

SSM
TUTOR CENTER   SG   CD & VIDEO   MATH PRO   WEB

Ⓐ Rationalizing Denominators

Often in mathematics it is helpful to write a radical expression such as $\dfrac{\sqrt{3}}{\sqrt{2}}$ either without a radical in the denominator or without a radical in the numerator. The process of writing this expression as an equivalent expression but without a radical in the denominator is called **rationalizing the denominator**. To rationalize the denominator of $\dfrac{\sqrt{3}}{\sqrt{2}}$, we use the fundamental principle of fractions and multiply the numerator and the denominator by $\sqrt{2}$. Recall that this is the same as multiplying by $\dfrac{\sqrt{2}}{\sqrt{2}}$, which simplifies to 1.

$$\frac{\sqrt{3}}{\sqrt{2}} = \frac{\sqrt{3} \cdot \sqrt{2}}{\sqrt{2} \cdot \sqrt{2}} = \frac{\sqrt{6}}{\sqrt{4}} = \frac{\sqrt{6}}{2}$$

EXAMPLE 1   Rationalize the denominator of $\dfrac{2}{\sqrt{5}}$.

Solution: To rationalize the denominator, we multiply the numerator and denominator by a factor that makes the radicand in the denominator a perfect square.

$$\frac{2}{\sqrt{5}} = \frac{2 \cdot \sqrt{5}}{\sqrt{5} \cdot \sqrt{5}} = \frac{2\sqrt{5}}{5}$$ The denominator is now rationalized.

EXAMPLE 2   Rationalize the denominator of $\dfrac{2\sqrt{16}}{\sqrt{9x}}$.

Solution: First we simplify the radicals; then we rationalize the denominator.

$$\frac{2\sqrt{16}}{\sqrt{9x}} = \frac{2(4)}{\sqrt{9} \cdot \sqrt{x}} = \frac{8}{3\sqrt{x}}$$

To rationalize the denominator, we multiply the numerator and the denominator by $\sqrt{x}$.

$$\frac{8}{3\sqrt{x}} = \frac{8 \cdot \sqrt{x}}{3\sqrt{x} \cdot \sqrt{x}} = \frac{8\sqrt{x}}{3x}$$

EXAMPLE 3   Rationalize the denominator of $\sqrt[3]{\dfrac{1}{2}}$.

Solution: $\sqrt[3]{\dfrac{1}{2}} = \dfrac{\sqrt[3]{1}}{\sqrt[3]{2}} = \dfrac{1}{\sqrt[3]{2}}$

Now we rationalize the denominator. Since $\sqrt[3]{2}$ is a cube root, we want to multiply by a value that will make the radicand 2 a perfect cube. If we multiply by $\sqrt[3]{2^2}$, we get $\sqrt[3]{2^3} = 2$. Thus,

$$\frac{1 \cdot \sqrt[3]{2^2}}{\sqrt[3]{2} \cdot \sqrt[3]{2^2}} = \frac{\sqrt[3]{4}}{\sqrt[3]{2^3}} = \frac{\sqrt[3]{4}}{2}$$ Multiply numerator and denominator by $\sqrt[3]{2^2}$ and then simplify.

Try the Concept Check in the margin.

Practice Problem 1

Rationalize the denominator of $\dfrac{7}{\sqrt{2}}$.

Practice Problem 2

Rationalize the denominator of $\dfrac{2\sqrt{9}}{\sqrt{16y}}$.

Practice Problem 3

Rationalize the denominator of $\sqrt[3]{\dfrac{2}{25}}$.

Concept Check

Determine by which number both the numerator and denominator should be multiplied to rationalize the denominator of the radical expression.

a. $\dfrac{1}{\sqrt[3]{7}}$ b. $\dfrac{1}{\sqrt[4]{8}}$

Answers

1. $\dfrac{7\sqrt{2}}{2}$, **2.** $\dfrac{3\sqrt{y}}{2y}$, **3.** $\dfrac{\sqrt[3]{10}}{5}$

Concept Check: **a.** $\sqrt[3]{7^2}$ or $\sqrt[3]{49}$, **b.** $\sqrt[4]{2}$

Practice Problem 4

Rationalize the denominator of $\sqrt{\dfrac{5m}{11n}}$.
Assume that all variables represent positive real numbers.

Practice Problem 5

Rationalize the denominator of $\dfrac{\sqrt[5]{a^2}}{\sqrt[5]{32b^{12}}}$.
Assume that all variables represent positive real numbers.

EXAMPLE 4

Rationalize the denominator of $\sqrt{\dfrac{7x}{3y}}$. Assume that all variables represent positive real numbers.

Solution:

$$\sqrt{\dfrac{7x}{3y}} = \dfrac{\sqrt{7x}}{\sqrt{3y}}$$
Use the quotient rule. No radical may be simplified further.

$$= \dfrac{\sqrt{7x}\cdot\sqrt{3y}}{\sqrt{3y}\cdot\sqrt{3y}}$$
Multiply numerator and denominator by $\sqrt{3y}$ so that the radicand in the denominator is a perfect square.

$$= \dfrac{\sqrt{21xy}}{3y}$$
Use the product rule in the numerator and denominator. Remember that $\sqrt{3y}\cdot\sqrt{3y} = 3y$. ●

EXAMPLE 5

Rationalize the denominator of $\dfrac{\sqrt[4]{x}}{\sqrt[4]{81y^5}}$. Assume that all variables represent positive real numbers.

Solution: First we simplify each radical if possible.

$$\dfrac{\sqrt[4]{x}}{\sqrt[4]{81y^5}} = \dfrac{\sqrt[4]{x}}{\sqrt[4]{81y^4}\cdot\sqrt[4]{y}}$$
Use the product rule in the denominator.

$$= \dfrac{\sqrt[4]{x}}{3y\sqrt[4]{y}}$$
Write $\sqrt[4]{81y^4}$ as $3y$.

$$= \dfrac{\sqrt[4]{x}\cdot\sqrt[4]{y^3}}{3y\sqrt[4]{y}\cdot\sqrt[4]{y^3}}$$
Multiply numerator and denominator by $\sqrt[4]{y^3}$ so that the radicand in the denominator is a perfect 4th power.

$$= \dfrac{\sqrt[4]{xy^3}}{3y\sqrt[4]{y^4}}$$
Use the product rule in the numerator and denominator.

$$= \dfrac{\sqrt[4]{xy^3}}{3y^2}$$
In the denominator, $\sqrt[4]{y^4} = y$ and $3y\cdot y = 3y^2$. ●

B Rationalizing Denominators Having Two Terms

Remember the product of the sum and difference of two terms?

$$(a + b)(a - b) = a^2 - b^2$$

These two expressions are called **conjugates** of each other.

To rationalize a denominator that is a sum or difference of two terms, we use conjugates. To see how and why this works, let's rationalize the denominator of the expression $\dfrac{5}{\sqrt{3} - 2}$. To do so, we multiply both the numerator and the denominator by $\sqrt{3} + 2$, the *conjugate*, of the denominator $\sqrt{3} - 2$ and see what happens.

$$\dfrac{5}{\sqrt{3} - 2} = \dfrac{5(\sqrt{3} + 2)}{(\sqrt{3} - 2)(\sqrt{3} + 2)}$$

$$= \dfrac{5(\sqrt{3} + 2)}{(\sqrt{3})^2 - 2^2}$$
Multiply the sum and difference of two terms: $(a + b)(a - b) = a^2 - b^2$.

Answers

4. $\dfrac{\sqrt{55mn}}{11n}$, **5.** $\dfrac{\sqrt[5]{a^2b^3}}{2b^3}$

$$= \frac{5(\sqrt{3} + 2)}{3 - 4}$$

$$= \frac{5(\sqrt{3} + 2)}{-1}$$

$$= -5(\sqrt{3} + 2) \quad \text{or} \quad -5\sqrt{3} - 10$$

Notice in the denominator that the product of $(\sqrt{3} - 2)$ and its conjugate, $(\sqrt{3} + 2)$, is -1. In general, the product of an expression and its conjugate will contain no radical terms. This is why, when rationalizing a denominator or a numerator containing two terms, we multiply by its conjugate. Examples of conjugates are

$$\sqrt{a} - \sqrt{b} \quad \text{and} \quad \sqrt{a} + \sqrt{b}$$

$$x + \sqrt{y} \quad \text{and} \quad x - \sqrt{y}$$

EXAMPLE 6   Rationalize the denominator of $\dfrac{2}{3\sqrt{2} + 4}$.

Solution: We multiply the numerator and the denominator by the conjugate of $3\sqrt{2} + 4$.

$$\frac{2}{3\sqrt{2} + 4} = \frac{2(3\sqrt{2} - 4)}{(3\sqrt{2} + 4)(3\sqrt{2} - 4)}$$

$$= \frac{2(3\sqrt{2} - 4)}{(3\sqrt{2})^2 - 4^2} \qquad \begin{array}{l}\text{Multiply the sum and difference of}\\ \text{two terms:}\\ (a + b)(a - b) = a^2 - b^2.\end{array}$$

$$= \frac{2(3\sqrt{2} - 4)}{18 - 16} \qquad \begin{array}{l}\text{Write } (3\sqrt{2})^2 \text{ as } 9 \cdot 2 \text{ or } 18 \text{ and } 4^2\\ \text{as } 16.\end{array}$$

$$= \frac{2(3\sqrt{2} - 4)}{2} \quad \text{or} \quad 3\sqrt{2} - 4 \qquad \bullet$$

As we saw in Example 6, it is often helpful to leave a numerator in factored form to help determine whether the expression can be simplified.

EXAMPLE 7   Rationalize the denominator of $\dfrac{\sqrt{6} + 2}{\sqrt{5} - \sqrt{3}}$.

Solution: We multiply the numerator and the denominator by the conjugate of $\sqrt{5} - \sqrt{3}$.

$$\frac{\sqrt{6} + 2}{\sqrt{5} - \sqrt{3}} = \frac{(\sqrt{6} + 2)(\sqrt{5} + \sqrt{3})}{(\sqrt{5} - \sqrt{3})(\sqrt{5} + \sqrt{3})}$$

$$= \frac{\sqrt{6}\sqrt{5} + \sqrt{6}\sqrt{3} + 2\sqrt{5} + 2\sqrt{3}}{(\sqrt{5})^2 - (\sqrt{3})^2}$$

$$= \frac{\sqrt{30} + \sqrt{18} + 2\sqrt{5} + 2\sqrt{3}}{5 - 3}$$

$$= \frac{\sqrt{30} + 3\sqrt{2} + 2\sqrt{5} + 2\sqrt{3}}{2}$$

Practice Problem 6

Rationalize the denominator of

$$\frac{3}{2\sqrt{5} + 1}.$$

Practice Problem 7

Rationalize the denominator of

$$\frac{\sqrt{5} + 3}{\sqrt{3} - \sqrt{2}}.$$

Answers

6. $\dfrac{3(2\sqrt{5} - 1)}{19}$,   **7.** $\sqrt{15} + \sqrt{10} + 3\sqrt{3} + 3\sqrt{2}$

Practice Problem 8

Rationalize the denominator of $\dfrac{3}{2 - \sqrt{x}}$.
Assume that all variables represent positive real numbers.

EXAMPLE 8

Rationalize the denominator of $\dfrac{2\sqrt{m}}{3\sqrt{x} + \sqrt{m}}$. Assume that all variables represent positive real numbers.

Solution: We multiply by the conjugate of $3\sqrt{x} + \sqrt{m}$ to eliminate the radicals from the denominator.

$$\frac{2\sqrt{m}}{3\sqrt{x} + \sqrt{m}} = \frac{2\sqrt{m}(3\sqrt{x} - \sqrt{m})}{(3\sqrt{x} + \sqrt{m})(3\sqrt{x} - \sqrt{m})} = \frac{6\sqrt{mx} - 2m}{(3\sqrt{x})^2 - (\sqrt{m})^2}$$

$$= \frac{6\sqrt{mx} - 2m}{9x - m}$$

●

Ⓒ Rationalizing Numerators

As mentioned earlier, it is also often helpful to write an expression such as $\dfrac{\sqrt{3}}{\sqrt{2}}$ as an equivalent expression without a radical in the numerator. This process is called **rationalizing the numerator**. To rationalize the numerator of $\dfrac{\sqrt{3}}{\sqrt{2}}$, we multiply the numerator and the denominator by $\sqrt{3}$.

$$\frac{\sqrt{3}}{\sqrt{2}} = \frac{\sqrt{3} \cdot \sqrt{3}}{\sqrt{2} \cdot \sqrt{3}} = \frac{\sqrt{9}}{\sqrt{6}} = \frac{3}{\sqrt{6}}$$

Practice Problem 9

Rationalize the numerator of $\dfrac{\sqrt{18}}{\sqrt{75}}$.

EXAMPLE 9   Rationalize the numerator of $\dfrac{\sqrt{7}}{\sqrt{45}}$.

Solution: First we simplify $\sqrt{45}$.

$$\frac{\sqrt{7}}{\sqrt{45}} = \frac{\sqrt{7}}{\sqrt{9 \cdot 5}} = \frac{\sqrt{7}}{3\sqrt{5}}$$

Next we rationalize the numerator by multiplying the numerator and the denominator by $\sqrt{7}$.

$$\frac{\sqrt{7}}{3\sqrt{5}} = \frac{\sqrt{7} \cdot \sqrt{7}}{3\sqrt{5} \cdot \sqrt{7}} = \frac{7}{3\sqrt{5 \cdot 7}} = \frac{7}{3\sqrt{35}}$$

●

Practice Problem 10

Rationalize the numerator of $\dfrac{\sqrt[3]{3a}}{\sqrt[3]{7b}}$.

EXAMPLE 10   Rationalize the numerator of $\dfrac{\sqrt[3]{2x^2}}{\sqrt[3]{5y}}$.

Solution:

$$\frac{\sqrt[3]{2x^2}}{\sqrt[3]{5y}} = \frac{\sqrt[3]{2x^2} \cdot \sqrt[3]{2^2x}}{\sqrt[3]{5y} \cdot \sqrt[3]{2^2x}} \quad \text{Multiply the numerator and denominator by } \sqrt[3]{2^2x} \text{ so that the radicand in the numerator is a perfect cube.}$$

$$= \frac{\sqrt[3]{2^3x^3}}{\sqrt[3]{5y \cdot 2^2x}} \quad \text{Use the product rule in the numerator and denominator.}$$

$$= \frac{2x}{\sqrt[3]{20xy}} \quad \text{Simplify.}$$

●

Just as for denominators, to rationalize a numerator that is a sum or difference of two terms, we use conjugates.

Answers

8. $\dfrac{6 + 3\sqrt{x}}{4 - x}$, **9.** $\dfrac{6}{5\sqrt{6}}$, **10.** $\dfrac{3a}{\sqrt[3]{63a^2b}}$

EXAMPLE 11

Rationalize the numerator of $\dfrac{\sqrt{x} + 2}{5}$. Assume that all variables represent positive real numbers.

Solution: We multiply the numerator and the denominator by the conjugate of $\sqrt{x} + 2$, the numerator.

$$\frac{\sqrt{x} + 2}{5} = \frac{(\sqrt{x} + 2)(\sqrt{x} - 2)}{5(\sqrt{x} - 2)} \qquad \text{Multiply by } \sqrt{x} - 2, \text{ the conjugate of } \sqrt{x} + 2.$$

$$= \frac{(\sqrt{x})^2 - 2^2}{5(\sqrt{x} - 2)} \qquad (a + b)(a - b) = a^2 - b^2.$$

$$= \frac{x - 4}{5(\sqrt{x} - 2)}$$

●

Practice Problem 11

Rationalize the numerator of $\dfrac{\sqrt{x} + 5}{3}$.

Assume that all variables represent positive real numbers.

Answer

11. $\dfrac{x - 25}{3(\sqrt{x} - 5)}$

STUDY SKILLS REMINDER

How are your homework assignments going?

By now, you should have good homework habits. If not, it's never too late to begin. Why is it so important in mathematics to keep up with homework? You probably now know the answer to that question. You have probably realized by now that many concepts in mathematics build on each other. Your understanding of one chapter in mathematics usually depends on your understanding of the previous chapter's material.

Don't forget that completing your homework assignment involves a lot more than attempting a few of the problems assigned.

To complete a homework assignment, remember these four things:

1. Attempt all of it.
2. Check it.
3. Correct it.
4. If needed, ask questions about it.

Mental Math

Find the conjugate of each expression.

1. $\sqrt{2} + x$

2. $\sqrt{3} + y$

3. $5 - \sqrt{a}$

4. $6 - \sqrt{b}$

5. $7\sqrt{4} + 8\sqrt{x}$

6. $9\sqrt{2} - 6\sqrt{y}$

EXERCISE SET 7.5

A *Rationalize each denominator. Assume that all variables represent positive real numbers. See Examples 1 through 5.*

1. $\dfrac{\sqrt{2}}{\sqrt{7}}$

2. $\dfrac{\sqrt{3}}{\sqrt{2}}$

3. $\sqrt{\dfrac{1}{5}}$

4. $\sqrt{\dfrac{1}{2}}$

5. $\dfrac{4}{\sqrt[3]{3}}$

6. $\dfrac{6}{\sqrt[3]{9}}$

7. $\dfrac{3}{\sqrt{8x}}$

8. $\dfrac{5}{\sqrt{27a}}$

9. $\dfrac{3}{\sqrt[3]{4x^2}}$

10. $\dfrac{5}{\sqrt[3]{3y}}$

11. $\dfrac{9}{\sqrt{3a}}$

12. $\dfrac{x}{\sqrt{5}}$

13. $\dfrac{3}{\sqrt[3]{2}}$

14. $\dfrac{5}{\sqrt[3]{9}}$

15. $\dfrac{2\sqrt{3}}{\sqrt{7}}$

16. $\dfrac{-5\sqrt{2}}{\sqrt{11}}$

17. $\sqrt{\dfrac{2x}{5y}}$

18. $\sqrt{\dfrac{13a}{2b}}$

19. $\sqrt[3]{\dfrac{3}{5}}$

20. $\sqrt[3]{\dfrac{7}{10}}$

21. $\sqrt{\dfrac{3x}{50}}$

22. $\sqrt{\dfrac{11y}{45}}$

23. $\dfrac{1}{\sqrt{12z}}$

24. $\dfrac{1}{\sqrt{32x}}$

25. $\dfrac{\sqrt[3]{2y^2}}{\sqrt[3]{9x^2}}$ **26.** $\dfrac{\sqrt[3]{3x}}{\sqrt[3]{4y^4}}$ **27.** $\sqrt[4]{\dfrac{16}{9x^7}}$ **28.** $\sqrt[5]{\dfrac{32}{m^6n^{13}}}$ **29.** $\dfrac{5a}{\sqrt[5]{8a^9b^{11}}}$ **30.** $\dfrac{9y}{\sqrt[4]{4y^9}}$

B *Rationalize each denominator. Assume that all variables represent positive real numbers. See Examples 6 through 8.*

31. $\dfrac{6}{2-\sqrt{7}}$

32. $\dfrac{3}{\sqrt{7}-4}$

33. $\dfrac{-7}{\sqrt{x}-3}$

34. $\dfrac{-8}{\sqrt{y}+4}$

35. $\dfrac{\sqrt{2}-\sqrt{3}}{\sqrt{2}+\sqrt{3}}$

36. $\dfrac{\sqrt{3}+\sqrt{4}}{\sqrt{2}+\sqrt{3}}$

37. $\dfrac{\sqrt{a}+1}{2\sqrt{a}-\sqrt{b}}$

38. $\dfrac{2\sqrt{a}-3}{2\sqrt{a}-\sqrt{b}}$

39. $\dfrac{8}{1+\sqrt{10}}$

40. $\dfrac{-3}{\sqrt{6}-2}$

41. $\dfrac{\sqrt{x}}{\sqrt{x}+\sqrt{y}}$

42. $\dfrac{2\sqrt{a}}{2\sqrt{x}-\sqrt{y}}$

43. $\dfrac{2\sqrt{3}+\sqrt{6}}{4\sqrt{3}-\sqrt{6}}$

44. $\dfrac{4\sqrt{5}+\sqrt{2}}{2\sqrt{5}-\sqrt{2}}$

C *Rationalize each numerator. Assume that all variables represent positive real numbers. See Examples 9 and 10.*

45. $\sqrt{\dfrac{5}{3}}$ **46.** $\sqrt{\dfrac{3}{2}}$ **47.** $\sqrt{\dfrac{18}{5}}$ **48.** $\sqrt{\dfrac{12}{7}}$ **49.** $\dfrac{\sqrt{4x}}{7}$ **50.** $\dfrac{\sqrt{3x^5}}{6}$

51. $\dfrac{\sqrt[3]{5y^2}}{\sqrt[3]{4x}}$

52. $\dfrac{\sqrt[3]{4x}}{\sqrt[3]{z^4}}$

53. $\sqrt{\dfrac{2}{5}}$

54. $\sqrt{\dfrac{3}{7}}$

55. $\dfrac{\sqrt{2x}}{11}$

56. $\dfrac{\sqrt{y}}{7}$

57. $\sqrt[3]{\dfrac{7}{8}}$

58. $\sqrt[3]{\dfrac{25}{2}}$

59. $\dfrac{\sqrt[3]{3x^5}}{10}$

60. $\sqrt[3]{\dfrac{9y}{7}}$

61. $\sqrt{\dfrac{18x^4y^6}{3z}}$

62. $\sqrt{\dfrac{8x^5y}{2z}}$

63. When rationalizing the denominator of $\dfrac{\sqrt{5}}{\sqrt{7}}$, explain why both the numerator and the denominator must be multiplied by $\sqrt{7}$.

64. When rationalizing the numerator of $\dfrac{\sqrt{5}}{\sqrt{7}}$, explain why both the numerator and the denominator must be multiplied by $\sqrt{5}$.

Rationalize each numerator. Assume that all variables represent positive real numbers. See Example 11.

65. $\dfrac{2 - \sqrt{11}}{6}$

66. $\dfrac{\sqrt{15} + 1}{2}$

67. $\dfrac{2 - \sqrt{7}}{-5}$

68. $\dfrac{\sqrt{5} + 2}{\sqrt{2}}$

69. $\dfrac{\sqrt{x} + 3}{\sqrt{x}}$

70. $\dfrac{5 + \sqrt{2}}{\sqrt{2x}}$

71. $\dfrac{\sqrt{x} + 1}{\sqrt{x} - 1}$

72. $\dfrac{\sqrt{x} + \sqrt{y}}{\sqrt{x} - \sqrt{y}}$

Review and Preview

Solve each equation. See Sections 2.1 and 5.7.

73. $2x - 7 = 3(x - 4)$

74. $9x - 4 = 7(x - 2)$

75. $(x - 6)(2x + 1) = 0$

76. $(y + 2)(5y + 4) = 0$

77. $x^2 - 8x = -12$

78. $x^3 = x$

 ## Combining Concepts

△ **79.** The formula of the radius r of a sphere with surface area A is

$$r = \sqrt{\frac{A}{4\pi}}$$

Rationalize the denominator of the radical expression in this formula.

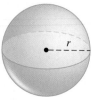

△ **80.** The formula for the radius r of a cone with height 7 centimeters and volume V is

$$r = \sqrt{\frac{3V}{7\pi}}$$

Rationalize the numerator of the radical expression in this formula.

7 cm

r

81. Explain why rationalizing the denominator does not change the value of the original expression.

82. Explain why rationalizing the numerator does not change the value of the original expression.

Name _____ Section _____ Date _____

Integrated Review—Radicals and Rational Exponents

Throughout this review, assume that all variables represent positive real numbers. Find each root.

1. $\sqrt{81}$

2. $\sqrt[3]{-8}$

3. $\sqrt[4]{\dfrac{1}{16}}$

4. $\sqrt{x^6}$

5. $\sqrt[3]{y^9}$

6. $\sqrt{4y^{10}}$

7. $\sqrt[5]{-32y^5}$

8. $\sqrt[4]{81b^{12}}$

Use radical notation to rewrite each expression. Simplify if possible.

9. $36^{1/2}$

10. $(3y)^{1/4}$

11. $64^{-2/3}$

12. $(x+1)^{3/5}$

Use the properties of exponents to simplify each expression. Write with positive exponents.

13. $y^{-1/6} \cdot y^{7/6}$

14. $\dfrac{(2x^{1/3})^4}{x^{5/6}}$

15. $\dfrac{x^{1/4}x^{3/4}}{x^{-1/4}}$

16. $4^{1/3} \cdot 4^{2/5}$

Use rational exponents to simplify each radical.

17. $\sqrt[3]{8x^6}$

18. $\sqrt[12]{a^9b^6}$

Use rational exponents to write each as a single radical expression.

19. $\sqrt[4]{x} \cdot \sqrt{x}$

20. $\sqrt{5} \cdot \sqrt[3]{2}$

Simplify.

21. $\sqrt{40}$

22. $\sqrt[4]{16x^7y^{10}}$

23. $\sqrt[3]{54x^4}$

24. $\sqrt[5]{-64b^{10}}$

1. _____
2. _____
3. _____
4. _____
5. _____
6. _____
7. _____
8. _____
9. _____
10. _____
11. _____
12. _____
13. _____
14. _____
15. _____
16. _____
17. _____
18. _____
19. _____
20. _____
21. _____
22. _____
23. _____
24. _____

25.

26.

27.

28.

29.

30.

31.

32.

33.

34.

35.

36.

37.

38.

39.

40.

Multiply or divide. Then simplify if possible.

25. $\sqrt{5} \cdot \sqrt{x}$

26. $\sqrt[3]{8x} \cdot \sqrt[3]{8x^2}$

27. $\dfrac{\sqrt{98y^6}}{\sqrt{2y}}$

28. $\dfrac{\sqrt[4]{48a^9b^3}}{\sqrt[4]{ab^3}}$

Perform each indicated operation.

29. $\sqrt{20} - \sqrt{75} + 5\sqrt{7}$

30. $\sqrt[3]{54y^4} - y\sqrt[3]{16y}$

31. $\sqrt{3}\,(\sqrt{5} - \sqrt{2})$

32. $(\sqrt{7} + \sqrt{3})^2$

33. $(2x - \sqrt{5})(2x + \sqrt{5})$

34. $(\sqrt{x+1} - 1)^2$

Rationalize each denominator.

35. $\sqrt{\dfrac{7}{3}}$

36. $\dfrac{5}{\sqrt[3]{2x^2}}$

37. $\dfrac{\sqrt{3} - \sqrt{7}}{2\sqrt{3} + \sqrt{7}}$

Rationalize each numerator.

38. $\sqrt{\dfrac{7}{3}}$

39. $\sqrt[3]{\dfrac{9y}{11}}$

40. $\dfrac{\sqrt{x} - 2}{\sqrt{x}}$

7.6 Radical Equations and Problem Solving

Ⓐ Solving Equations That Contain Radical Expressions

In this section, we present techniques to solve equations containing radical expressions such as

$$\sqrt{2x - 3} = 9$$

We use the power rule to help us solve these radical equations.

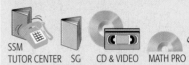

> **Power Rule**
>
> If both sides of an equation are raised to the same power, *all* solutions of the original equation are *among* the solutions of the new equation.

This property *does not* say that raising both sides of an equation to a power yields an equivalent equation. A solution of the new equation *may* or *may* not be a solution of the original equation. Thus, *each solution of the new equation must be checked* to make sure it is a solution of the original equation. Recall that a proposed solution that is not a solution of the original equation is called an extraneous solution.

EXAMPLE 1   Solve: $\sqrt{2x - 3} = 9$

Solution: We use the power rule to square both sides of the equation to eliminate the radical.

$$\sqrt{2x - 3} = 9$$
$$(\sqrt{2x - 3})^2 = 9$$
$$2x - 3 = 81$$
$$2x = 84$$
$$x = 42$$

Now we check the solution in the original equation.

Check:
$$\sqrt{2x - 3} = 9$$
$$\sqrt{2(42) - 3} \overset{?}{=} 9 \qquad \text{Let } x = 42.$$

$$\sqrt{84 - 3} \overset{?}{=} 9$$
$$\sqrt{81} \overset{?}{=} 9$$
$$9 = 9 \qquad \text{True.}$$

The solution checks, so we conclude that the solution set is $\{42\}$. ●

To solve a radical equation, first isolate a radical on one side of the equation.

EXAMPLE 2   Solve: $\sqrt{-10x - 1} + 3x = 0$

Solution: First we isolate the radical on one side of the equation. To do this, we subtract $3x$ from both sides.

$$\sqrt{-10x - 1} + 3x = 0$$
$$\sqrt{-10x - 1} + 3x - 3x = 0 - 3x$$
$$\sqrt{-10x - 1} = -3x$$

Next we use the power rule to eliminate the radical.

$$(\sqrt{-10x - 1})^2 = (-3x)^2$$
$$-10x - 1 = 9x^2$$

Practice Problem 1

Solve: $\sqrt{3x - 2} = 5$

Practice Problem 2

Solve: $\sqrt{9x - 2} - 2x = 0$

Answers

1. $\{9\}$,   2. $\left\{\dfrac{1}{4}, 2\right\}$

Since this is a quadratic equation, we can set the equation equal to 0 and try to solve by factoring.

$$9x^2 + 10x + 1 = 0$$
$$(9x + 1)(x + 1) = 0 \qquad \text{Factor.}$$
$$9x + 1 = 0 \quad \text{or} \quad x + 1 = 0 \qquad \text{Set each factor equal to 0.}$$
$$x = -\frac{1}{9} \qquad\qquad x = -1$$

Check: Let $x = -\frac{1}{9}$. Let $x = -1$.

$$\sqrt{-10x - 1} + 3x = 0 \qquad\qquad \sqrt{-10x - 1} + 3x = 0$$
$$\sqrt{-10\left(-\frac{1}{9}\right) - 1} + 3\left(-\frac{1}{9}\right) \overset{?}{=} 0 \qquad \sqrt{-10(-1) - 1} + 3(-1) \overset{?}{=} 0$$
$$\sqrt{\frac{10}{9} - \frac{9}{9}} - \frac{3}{9} \overset{?}{=} 0 \qquad\qquad \sqrt{10 - 1} - 3 \overset{?}{=} 0$$
$$\sqrt{\frac{1}{9}} - \frac{1}{3} \overset{?}{=} 0 \qquad\qquad \sqrt{9} - 3 \overset{?}{=} 0$$
$$\frac{1}{3} - \frac{1}{3} = 0 \;\; \text{True.} \qquad\qquad 3 - 3 = 0$$

$$\text{True.}$$

Both solutions check. The solution set is $\left\{ -\frac{1}{9}, -1 \right\}$. ●

The following steps may be used to solve a radical equation.

Solving a Radical Equation

Step 1. Isolate one radical on one side of the equation.

Step 2. Raise each side of the equation to a power equal to the index of the radical and simplify.

Step 3. If the equation still contains a radical term, repeat Steps 1 and 2. If not, solve the equation.

Step 4. Check all proposed solutions in the original equation.

Practice Problem 3

Solve: $\sqrt[3]{x - 5} + 2 = 1$

EXAMPLE 3 Solve: $\sqrt[3]{x + 1} + 5 = 3$

Solution: First we isolate the radical by subtracting 5 from both sides of the equation.

$$\sqrt[3]{x + 1} + 5 = 3$$
$$\sqrt[3]{x + 1} = -2$$

Next we raise both sides of the equation to the third power to eliminate the radical.

$$(\sqrt[3]{x + 1})^3 = (-2)^3$$
$$x + 1 = -8$$
$$x = -9$$

The solution checks in the original equation, so the solution set is $\{-9\}$. ●

Answer

3. $\{4\}$

EXAMPLE 4   Solve: $\sqrt{4 - x} = x - 2$

Solution:
$$\sqrt{4 - x} = x - 2$$
$$(\sqrt{4 - x})^2 = (x - 2)^2$$
$$4 - x = x^2 - 4x + 4$$
$$x^2 - 3x = 0 \qquad \text{Write the quadratic equation in standard form.}$$
$$x(x - 3) = 0 \qquad \text{Factor.}$$
$$x = 0 \text{ or } x - 3 = 0 \qquad \text{Set each factor equal to 0.}$$
$$x = 3$$

Check:

$$\sqrt{4 - x} = x - 2$$
$$\sqrt{4 - 0} \overset{?}{=} 0 - 2 \qquad \text{Let } x = 0.$$
$$2 = -2 \qquad \text{False.}$$

$$\sqrt{4 - x} = x - 2$$
$$\sqrt{4 - 3} \overset{?}{=} 3 - 2 \qquad \text{Let } x = 3.$$
$$1 = 1 \qquad \text{True.}$$

The proposed solution 3 checks, but 0 does not. Since 0 is an extraneous solution, the solution set is $\{3\}$.

Helpful Hint

In Example 4, notice that $(x - 2)^2 = x^2 - 4x + 4$. Make sure binomials are squared correctly.

Try the Concept Check in the margin.

EXAMPLE 5   Solve: $\sqrt{2x + 5} + \sqrt{2x} = 3$

Solution: We get one radical alone by subtracting $\sqrt{2x}$ from both sides.

$$\sqrt{2x + 5} + \sqrt{2x} = 3$$
$$\sqrt{2x + 5} = 3 - \sqrt{2x}$$

Now we use the power rule to begin eliminating the radicals. First we square both sides.

$$(\sqrt{2x + 5})^2 = (3 - \sqrt{2x})^2$$
$$2x + 5 = 9 - 6\sqrt{2x} + 2x \qquad \text{Multiply: } (3 - \sqrt{2x})(3 - \sqrt{2x})$$

There is still a radical in the equation, so we get the radical alone again. Then we square both sides.

$$2x + 5 = 9 - 6\sqrt{2x} + 2x$$
$$6\sqrt{2x} = 4 \qquad \text{Get the radical alone.}$$
$$(6\sqrt{2x})^2 = 4^2 \qquad \text{Square both sides of the equation to eliminate the radical.}$$
$$36(2x) = 16$$
$$72x = 16 \qquad \text{Multiply.}$$
$$x = \frac{16}{72} \qquad \text{Solve.}$$
$$x = \frac{2}{9} \qquad \text{Simplify.}$$

Practice Problem 4

Solve: $\sqrt{9 + x} = x + 3$

Concept Check

How can you immediately tell that the equation $\sqrt{2y + 3} = -4$ has no real solution?

Practice Problem 5

Solve: $\sqrt{3x + 1} + \sqrt{3x} = 2$

Answers

4. $\{0\}$,   **5.** $\left\{\dfrac{3}{16}\right\}$

Concept Check: answers may vary

The proposed solution $\frac{2}{9}$ checks in the original equation. The solution set is $\left\{\frac{2}{9}\right\}$.

> **Helpful Hint**
>
> Make sure expressions are squared correctly. In Example 5, we squared $(3 - \sqrt{2x})$ as
> $$(3 - \sqrt{2x})^2 = (3 - \sqrt{2x})(3 - \sqrt{2x})$$
> $$= 3 \cdot 3 - 3\sqrt{2x} - 3\sqrt{2x} + \sqrt{2x} \cdot \sqrt{2x}$$
> $$= 9 - 6\sqrt{2x} + 2x$$

Try the Concept Check in the margin.

Concept Check

What is wrong the following solution?
$$\sqrt{2x + 5} + \sqrt{4 - x} = 8$$
$$(\sqrt{2x + 5} + \sqrt{4 - x})^2 = 8^2$$
$$(2x + 5) + (4 - x) = 64$$
$$x + 9 = 64$$
$$x = 55$$

B ## Using the Pythagorean Theorem

Recall that the Pythagorean theorem states that in a right triangle, the length of the hypotenuse squared equals the sum of the lengths of each of the legs squared.

> **Pythagorean Theorem**
>
> If a and b are the lengths of the legs of a right triangle and c is the length of the hypotenuse, then $a^2 + b^2 = c^2$.
>
>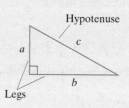

Practice Problem 6

Find the length of the unknown leg of the right triangle.

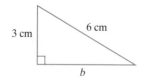

EXAMPLE 6   Find the length of the unknown leg of the right triangle.

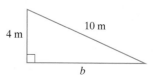

Solution: In the formula $a^2 + b^2 = c^2$, c is the hypotenuse. Here, $c = 10$, the length of the hypotenuse, and $a = 4$. We solve for b. Then $a^2 + b^2 = c^2$ becomes

$$4^2 + b^2 = 10^2$$
$$16 + b^2 = 100$$
$$b^2 = 84 \qquad \text{Subtract 16 from both sides.}$$

Recall from Section 7.1 our definition of square root that if $b^2 = a$, then b is a square root of a. Since b is a length and thus is positive, we have that

$$b = \sqrt{84} = \sqrt{4 \cdot 21} = 2\sqrt{21}$$

The unknown leg of the triangle is exactly $2\sqrt{21}$ meters long. Using a calculator, this is approximately 9.2 meters.

Answers

6. $3\sqrt{3}$ cm

Concept Check: answers may vary

EXAMPLE 7 Calculating Placement of a Wire

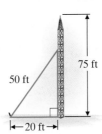

A 50-foot supporting wire is to be attached to a 75-foot antenna. Because of surrounding buildings, sidewalks, and roadways, the wire must be anchored exactly 20 feet from the base of the antenna.

a. How high from the base of the antenna must the wire be attached?

b. Local regulations require that a supporting wire be attached at a height no less than $\frac{3}{5}$ of the total height the antenna. From part (a), have local regulations been met?

Solution:

1. UNDERSTAND. Read and reread the problem. From the diagram we notice that a right triangle is formed with hypotenuse 50 feet and one leg 20 feet. We let x = the height from the base of the antenna to the attached wire.

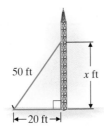

2. TRANSLATE. We'll use the Pythagorean theorem.

$$(a)^2 + (b)^2 = (c)^2$$
$$(20)^2 + x^2 = (50)^2 \qquad a = 20, c = 50$$

3. SOLVE. $(20)^2 + x^2 = (50)^2$

$$400 + x^2 = 2500$$
$$x^2 = 2100 \qquad \text{Subtract 400 from both sides.}$$
$$x = \sqrt{2100}$$
$$= 10\sqrt{21}$$

4. INTERPRET. *Check* the work and *state* the solution.

a. The wire is attached exactly $10\sqrt{21}$ feet from the base of the pole, or approximately 45.8 feet.

b. The supporting wire must be attached at a height no less than $\frac{3}{5}$ of the total height of the antenna. This height is $\frac{3}{5}$ (75 feet), or 45 feet.

Since we know from part (a) that the wire is to be attached at a height of approximately 45.8 feet, local regulations have been met.

●

Practice Problem 7 △

A furniture upholsterer wishes to cut a strip from a piece of fabric that is 45 inches by 45 inches. The strip must be cut on the bias of the fabric. What is the longest strip that can be cut? Give an exact answer and a two-decimal-place approximation.

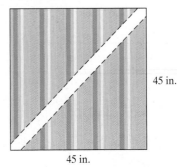

Answer

7. $45\sqrt{2}$ in. ≈ 63.64 in.

GRAPHING CALCULATOR EXPLORATIONS

We can use a graphing calculator to solve radical equations. For example, to use a graphing calculator to approximate the solutions of the equation solved in Example 4, we graph the following:

$$Y_1 = \sqrt{4 - x} \qquad \text{and} \qquad Y_2 = x - 2$$

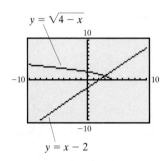

The x-value of the point of intersection is the solution. Use the INTERSECT feature or the ZOOM and TRACE features of your graphing calculator to see that the solution is 3.

Use a graphing calculator to solve each radical equation. Round all solutions to the nearest hundredth.

1. $\sqrt{x + 7} = x$

2. $\sqrt{3x + 5} = 2x$

3. $\sqrt{2x + 1} = \sqrt{2x + 2}$

4. $\sqrt{10x - 1} = \sqrt{-10x + 10} - 1$

5. $1.2x = \sqrt{3.1x + 5}$

6. $\sqrt{1.9x^2 - 2.2} = -0.8x + 3$

Name _____ Section _____ Date _____

 Solve. See Examples 1 and 2.

1. $\sqrt{2x} = 4$ **2.** $\sqrt{3x} = 3$ **3.** $\sqrt{x - 3} = 2$ **4.** $\sqrt{x + 1} = 5$

5. $\sqrt{2x} = -4$ **6.** $\sqrt{5x} = -5$ **7.** $\sqrt{4x - 3} - 5 = 0$ **8.** $\sqrt{x - 3} - 1 = 0$

9. $\sqrt{2x - 3} - 2 = 1$ **10.** $\sqrt{3x + 3} - 4 = 8$

Solve. See Example 3.

11. $\sqrt[3]{6x} = -3$ **12.** $\sqrt[3]{4x} = -2$ **13.** $\sqrt[3]{x - 2} - 3 = 0$ **14.** $\sqrt[3]{2x - 6} - 4 = 0$

Solve. See Examples 4 and 5.

15. $\sqrt{13 - x} = x - 1$ **16.** $\sqrt{2x - 3} = 3 - x$ **17.** $x - \sqrt{4 - 3x} = -8$

18. $2x + \sqrt{x + 1} = 8$ **19.** $\sqrt{y + 5} = 2 - \sqrt{y - 4}$ **20.** $\sqrt{x + 3} + \sqrt{x - 5} = 3$

21. $\sqrt{x - 3} + \sqrt{x + 2} = 5$ **22.** $\sqrt{2x - 4} - \sqrt{3x + 4} = -2$

Solve. See Examples 1 through 5.

23. $\sqrt{3x - 2} = 5$ **24.** $\sqrt{5x - 4} = 9$ **25.** $-\sqrt{2x} + 4 = -6$ **26.** $-\sqrt{3x + 9} = -12$

27. $\sqrt{3x + 1} + 2 = 0$ **28.** $\sqrt{3x + 1} - 2 = 0$ **29.** $\sqrt[4]{4x + 1} - 2 = 0$ **30.** $\sqrt[4]{2x - 9} - 3 = 0$

31. $\sqrt{4x - 3} = 5$

32. $\sqrt{3x + 9} = 12$

33. $\sqrt[3]{6x - 3} - 3 = 0$

34. $\sqrt[3]{3x} + 4 = 7$

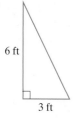 **35.** $\sqrt[3]{2x - 3} - 2 = -5$

36. $\sqrt[3]{x - 4} - 5 = -7$

37. $\sqrt{x + 4} = \sqrt{2x - 5}$

38. $\sqrt{3y + 6} = \sqrt{7y - 6}$

39. $x - \sqrt{1 - x} = -5$

40. $x - \sqrt{x - 2} = 4$

41. $\sqrt[3]{-6x - 1} = \sqrt[3]{-2x - 5}$

42. $x + \sqrt{x + 5} = 7$

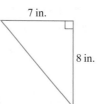 **43.** $\sqrt{5x - 1} - \sqrt{x + 2} = 3$

44. $\sqrt{2x - 1} - 4 = -\sqrt{x - 4}$

45. $\sqrt{2x - 1} = \sqrt{1 - 2x}$

46. $\sqrt{7x - 4} = \sqrt{4 - 7x}$

47. $\sqrt{3x + 4} - 1 = \sqrt{2x + 1}$

48. $\sqrt{x - 2} + 3 = \sqrt{4x + 1}$

49. $\sqrt{y + 3} - \sqrt{y - 3} = 1$

50. $\sqrt{x + 1} - \sqrt{x - 1} = 2$

B *Find the length of the unknown side of each triangle. See Example 6.*

△**51.**

6 ft

3 ft

△**52**

7 in.

8 in.

△**53.**

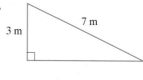

7 m

3 m

△**54.**

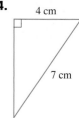

4 cm

7 cm

Find the length of the unknown side of each triangle. Give the exact length and a one-decimal-place approximation. See Example 6.

 55.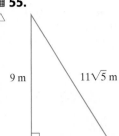

9 m $11\sqrt{5}$ m

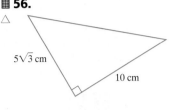 **56.**

$5\sqrt{3}$ cm   10 cm

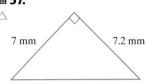 **57.**

7 mm   7.2 mm

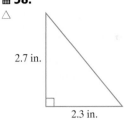 **58.**

2.7 in.   2.3 in.

Solve. Give exact answers and two-decimal-place approximations where appropriate. See Example 7.

59. A wire is needed to support a vertical pole 15 feet high. The cable will be anchored to a stake 8 feet from the base of the pole. How much cable is needed?

15 ft

8 ft

60. The tallest structure in the United States is a TV tower in Blanchard, North Dakota. Its height is 2063 feet. A 2382-foot length of wire is to be used as a guy wire attached to the top of the tower. Approximate to the nearest foot how far from the base of the tower the guy wire must be anchored. (*Source:* U.S. Geological Survey)

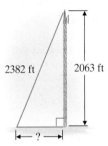

2382 ft   2063 ft

?

△ **61.** A spotlight is mounted on the eaves of a house 12 feet above the ground. A flower bed runs between the house and the sidewalk, so the closest the ladder can be placed to the house is 5 feet. How long a ladder is needed so that an electrician can reach the place where the light is mounted?

12 ft

5 ft

62. A wire is to be attached to support a telephone pole. Because of surrounding buildings, sidewalks, and roadway, the wire must be anchored exactly 15 feet from the base of the pole. Telephone company workers have only 30 feet of cable, and 2 feet of that must be used to attach the cable to the pole and to the stake on the ground. How high from the base of the pole can the wire be attached?

15 ft

557

△ **63.** The radius of the moon is 1080 miles. Use the formula for the radius r of a sphere given its surface area A,

$$r = \sqrt{\frac{A}{4\pi}}$$

to find the surface area of the moon. Round to the nearest square mile. (*Source*: National Space Science Data Center)

64. Police departments find it very useful to be able to approximate driving speeds in skidding accidents. If the road surface is wet concrete, the function $S(x) = \sqrt{10.5x}$ is used, where $S(x)$ is the speed of the car in miles per hour and x is the distance skidded in feet. Find how fast a car was moving if it skidded 280 feet on wet concrete.

65. The formula $v = \sqrt{2gh}$ relates the velocity v, in feet per second, of an object after it falls h feet accelerated by gravity g, in feet per second squared. If g is approximately 32 feet per second squared, find how far an object has fallen if its velocity is 80 feet per second.

66. Two tractors are pulling a tree stump from a field. If two forces A and B pull at right angles (90°) to each other, the size of the resulting force R is given by the formula $R = \sqrt{A^2 + B^2}$. If tractor A is exerting 600 pounds of force and the resulting force is 850 pounds, find how much force tractor B is exerting.

600 lb

In psychology, it has been suggested that the number S of nonsense syllables that a person can repeat consecutively depends on his or her IQ score I according to the equation $S = 2\sqrt{I} - 9$.

67. Use this relationship to estimate the IQ of a person who can repeat 11 nonsense syllables consecutively.

68. Use this relationship to estimate the IQ of a person who can repeat 15 nonsense syllables consecutively.

*The **period** of a pendulum is the time it takes for the pendulum to make one full back-and-forth swing. The period of a pendulum depends on the length of the pendulum. The formula for the period P, in seconds, is $P = 2\pi\sqrt{\dfrac{l}{32}}$ where l is the length of the pendulum in feet. Use this formula for Exercises 69 through 74.*

69. Find the period of a pendulum whose length is 2 feet. Give an exact answer and a two-decimal-place approximation.

70. Klockit sells a 43-inch lyre pendulum. Find the period of this pendulum. Round your answer to 2 decimal places. (Hint: First convert inches to feet.)

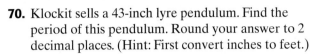

71. Find the length of a pendulum whose period is 4 seconds. Round your answer to 2 decimal places.

72. Find the length of a pendulum whose period is 3 seconds. Round your answer to 3 decimal places.

73. Study the relationship between period and pendulum length in Exercises 69 through 72 and make a conjecture about this relationship.

74. Galileo experimented with pendulums. He supposedly made conjectures about pendulums of equal length with different bob weights. Try this experiment. Make two pendulums 3 feet long. Attach a heavy weight (lead) to one and a light weight (a cork) to the other. Pull both pendulums back the same angle measure and release. Make a conjecture from your observations. (There is more about pendulums in the Chapter 7 Activity.)

If the three lengths of the sides of a triangle are known, Heron's formula can be used to find its area. If a, b, and c are the three lengths of the sides, Heron's formula for area is:

$$A = \sqrt{s(s - a)(s - b)(s - c)}$$

where s is half the perimeter of the triangle, or $s = \dfrac{1}{2}(a + b + c)$. Use this formula to find the area of each triangle. Give an exact answer and then a 2-decimal place approximation.

△ **75.**

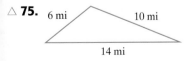

6 mi   10 mi

14 mi

△ **76.**

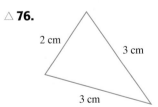

2 cm   3 cm

3 cm

77. Describe when Heron's formula might be useful.

78. In your own words, explain why you think S in *Heron's formula* is called the *semiperimeter*.

The maximum distance $D(h)$ in kilometers that a person can see from a height h kilometers above the ground is given by the function $D(h) = 111.7\sqrt{h}$. Use this function for Exercises 79 and 80. Round your answers to two decimal places.

79. Find the height that would allow a person to see 80 kilometers.

80. Find the height that would allow a person to see 40 kilometers.

Review and Preview

Simplify. See Section 6.3.

81. $\dfrac{\dfrac{x}{6}}{\dfrac{2x}{3} + \dfrac{1}{2}}$

82. $\dfrac{\dfrac{1}{y} + \dfrac{4}{5}}{\dfrac{-3}{20}}$

83. $\dfrac{\dfrac{z}{5} + \dfrac{1}{10}}{\dfrac{z}{20} - \dfrac{z}{5}}$

84. $\dfrac{\dfrac{1}{y} + \dfrac{1}{x}}{\dfrac{1}{y} - \dfrac{1}{x}}$

 Combining Concepts

85. Solve: $\sqrt{\sqrt{x+3} + \sqrt{x}} = \sqrt{3}$

86. Explain why proposed solutions of radical equations must be checked.

87. The cost $C(x)$ in dollars per day to operate a small delivery service is given by $C(x) = 80\sqrt[3]{x} + 500$, where x is the number of deliveries per day. In July, the manager decides that it is necessary to keep delivery costs below \$1620.00. Find the greatest number of deliveries this company can make per day and still keep overhead below \$1620.00.

88. Consider the equations $\sqrt{2x} = 4$ and $\sqrt[3]{2x} = 4$.
a. Explain the difference in solving these equations.
b. Explain the similarity in solving these equations.

7.7 Complex Numbers

A Writing Numbers in the Form *bi*

Our work with radical expressions has excluded expressions such as $\sqrt{-16}$ because $\sqrt{-16}$ is not a real number; there is no real number whose square is -16. In this section, we discuss a number system that includes roots of negative numbers. This number system is the **complex number system**, and it includes the set of real numbers as a subset. The complex number system allows us to solve equations such as $x^2 + 1 = 0$ that have no real number solutions. The set of complex numbers includes the *imaginary unit*.

OBJECTIVES

Ⓐ Write square roots of negative numbers in the form *bi*.

Ⓑ Add or subtract complex numbers.

Ⓒ Multiply complex numbers.

Ⓓ Divide complex numbers.

Ⓔ Raise *i* to powers.

SSM TUTOR CENTER   SG   CD & VIDEO   MATH PRO   WEB

Imaginary Unit

The **imaginary unit**, written i, is the number whose square is -1. That is,

$$i^2 = -1 \quad \text{and} \quad i = \sqrt{-1}$$

To write the square root of a negative number in terms of i, we use the property that if a is a positive number, then

$$\sqrt{-a} = \sqrt{-1} \cdot \sqrt{a}$$
$$= i \cdot \sqrt{a}$$

Using i, we can write $\sqrt{-16}$ as

$$\sqrt{-16} = \sqrt{-1 \cdot 16} = \sqrt{-1} \cdot \sqrt{16} = i \cdot 4 \text{ or } 4i$$

EXAMPLES   Write using i notation.

1. $\sqrt{-36} = \sqrt{-1 \cdot 36} = \sqrt{-1} \cdot \sqrt{36} = i \cdot 6 \text{ or } 6i$

2. $\sqrt{-5} = \sqrt{-1(5)} = \sqrt{-1} \cdot \sqrt{5} = i\sqrt{5}$.

> **Helpful Hint**
> Since $\sqrt{5}i$ can easily be confused with $\sqrt{5i}$, we write $\sqrt{5}i$ as $i\sqrt{5}$.

3. $-\sqrt{-20} = -\sqrt{-1 \cdot 20} = -\sqrt{-1} \cdot \sqrt{4 \cdot 5} = -i \cdot 2\sqrt{5} = -2i\sqrt{5}$ ●

The product rule for radicals does not necessarily hold true for imaginary numbers. *To multiply square roots of negative numbers, first we write each number in terms of the imaginary unit i.* For example, to multiply $\sqrt{-4}$ and $\sqrt{-9}$, we first write each number in the form bi:

$$\sqrt{-4} \cdot \sqrt{-9} = 2i(3i) = 6i^2 = 6(-1) = -6 \quad \text{Correct.}$$

Make sure you notice that the product rule does not work for this example. In other words, $\sqrt{-4} \cdot \sqrt{-9} = \sqrt{(-4)(-9)} = \sqrt{36} = 6$ is Incorrect!

EXAMPLES   Multiply or divide as indicated.

4. $\sqrt{-3} \cdot \sqrt{-5} = i\sqrt{3}(i\sqrt{5}) = i^2\sqrt{15} = -1\sqrt{15} = -\sqrt{15}$

5. $\sqrt{-36} \cdot \sqrt{-1} = 6i(i) = 6i^2 = 6(-1) = -6$

6. $\sqrt{8} \cdot \sqrt{-2} = 2\sqrt{2}(i\sqrt{2}) = 2i(\sqrt{2}\ \sqrt{2}) = 2i(2) = 4i$

7. $\dfrac{\sqrt{-125}}{\sqrt{5}} = \dfrac{i\sqrt{125}}{\sqrt{5}} = i\sqrt{25} = 5i$ ●

Practice Problems 1–3

Write using i notation.

1. $\sqrt{-25}$
2. $\sqrt{-3}$
3. $-\sqrt{-50}$

Practice Problems 4–7

Multiply or divide as indicated.

4. $\sqrt{-2} \cdot \sqrt{-7}$
5. $\sqrt{-25} \cdot \sqrt{-1}$
6. $\sqrt{27} \cdot \sqrt{-3}$
7. $\dfrac{\sqrt{-8}}{\sqrt{2}}$

Answers

1. $5i$, **2.** $i\sqrt{3}$, **3.** $-5i\sqrt{2}$, **4.** $-\sqrt{14}$, **5.** -5, **6.** $9i$, **7.** $2i$

Now that we have practiced working with the imaginary unit, we define *complex numbers*.

Complex Numbers

A **complex number** is a number that can be written in the form $a + bi$, where a and b are real numbers.

Notice that the set of real numbers is a subset of the complex numbers since any real number can be written in the form of a complex number. For example,

$$16 = 16 + 0i$$

In general, a complex number $a + bi$ is a real number if $b = 0$. Also, a complex number is called an **imaginary number** if $a = 0$. For example,

$$3i = 0 + 3i \qquad \text{and} \qquad i\sqrt{7} = 0 + i\sqrt{7}$$

are imaginary numbers.

The following diagram shows the relationship between complex numbers and their subsets.

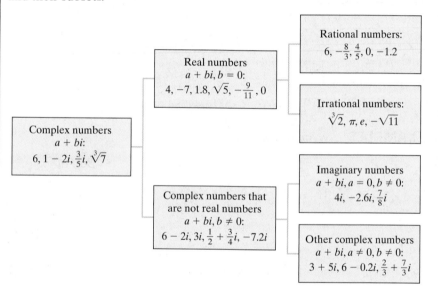

Try the Concept Check in the margin.

Concept Check

True or false? Every complex number is also a real number.

B Adding or Subtracting Complex Numbers

Two complex numbers $a + bi$ and $c + di$ are equal if and only if $a = c$ and $b = d$. Complex numbers can be added or subtracted by adding or subtracting their real parts and then adding or subtracting their imaginary parts.

Sum or Difference of Complex Numbers

If $a + bi$ and $c + di$ are complex numbers, then their sum is

$$(a + bi) + (c + di) = (a + c) + (b + d)i$$

Their difference is

$$(a + bi) - (c + di) = a + bi - c - di = (a - c) + (b - d)i$$

Answer

Concept Check: false

EXAMPLES   Add or subtract as indicated.

8. $(2 + 3i) + (-3 + 2i) = (2 - 3) + (3 + 2)i = -1 + 5i$
9. $5i - (1 - i) = 5i - 1 + i$
$$= -1 + (5 + 1)i$$
$$= -1 + 6i$$
10. $(-3 - 7i) - (-6) = -3 - 7i + 6$
$$= (-3 + 6) - 7i$$
$$= 3 - 7i$$

Practice Problems 8–10

Add or subtract as indicated.

8. $(5 + 2i) + (4 - 3i)$
9. $6i - (2 - i)$
10. $(-2 - 4i) - (-3)$

C Multiplying Complex Numbers

To multiply two complex numbers of the form $a + bi$, we multiply as though they were binomials. Then we use the relationship $i^2 = -1$ to simplify.

EXAMPLES   Multiply.

11. $-7i \cdot 3i = -21i^2$
$$= -21(-1) \qquad \text{Replace } i^2 \text{ with } -1.$$
$$= 21$$
12. $3i(2 - i) = 3i \cdot 2 - 3i \cdot i \qquad \text{Use the distributive property.}$
$$= 6i - 3i^2 \qquad \text{Multiply.}$$
$$= 6i - 3(-1) \qquad \text{Replace } i^2 \text{ with } -1.$$
$$= 6i + 3$$
$$= 3 + 6i \qquad \text{Use the FOIL order. (First, Outer, Inner, Last)}$$
13. $(2 - 5i)(4 + i) = 2(4) + 2(i) - 5i(4) - 5i(i)$
$$\qquad\qquad\qquad \text{F} \qquad \text{O} \qquad \text{I} \qquad \text{L}$$
$$= 8 + 2i - 20i - 5i^2$$
$$= 8 - 18i - 5(-1) \qquad i^2 = -1$$
$$= 8 - 18i + 5$$
$$= 13 - 18i$$
14. $(2 - i)^2 = (2 - i)(2 - i)$
$$= 2(2) - 2(i) - 2(i) + i^2$$
$$= 4 - 4i + (-1) \qquad i^2 = -1$$
$$= 3 - 4i$$
15. $(7 + 3i)(7 - 3i) = 7(7) - 7(3i) + 3i(7) - 3i(3i)$
$$= 49 - 21i + 21i - 9i^2$$
$$= 49 - 9(-1) \qquad i^2 = -1$$
$$= 49 + 9$$
$$= 58$$

Practice Problems 11–15

Multiply.

11. $-5i \cdot 3i$
12. $-2i(6 - 2i)$
13. $(3 - 4i)(6 + i)$
14. $(1 - 2i)^2$
15. $(6 + 5i)(6 - 5i)$

Notice that if you add, subtract, or multiply two complex numbers, the result is a complex number.

D Dividing Complex Numbers

From Example 15, notice that the product of $7 + 3i$ and $7 - 3i$ is a real number. These two complex numbers are called *complex conjugates* of one another. In general, we have the following definition.

Answers
8. $9 - i$, **9.** $-2 + 7i$, **10.** $1 - 4i$, **11.** 15,
12. $-4 - 12i$, **13.** $22 - 21i$, **14.** $-3 - 4i$,
15. 61

Complex Conjugates

The complex numbers $(a + bi)$ and $(a - bi)$ are called **complex conjugates** of each other, and

$$(a + bi)(a - bi) = a^2 + b^2$$

To see that the product of a complex number $a + bi$ and its conjugate $a - bi$ is the real number $a^2 + b^2$, we multiply:

$$(a + bi)(a - bi) = a^2 - abi + abi - b^2i^2$$
$$= a^2 - b^2(-1)$$
$$= a^2 + b^2$$

We will use complex conjugates to divide by a complex number.

Practice Problem 16

Divide and write in the form $a + bi$:
$\dfrac{3 + i}{2 - 3i}$

EXAMPLE 16 Divide and write in the form $a + bi$: $\dfrac{2 + i}{1 - i}$

Solution: We multiply the numerator and the denominator by the complex conjugate of $1 - i$ to eliminate the imaginary number in the denominator.

$$\frac{2 + i}{1 - i} = \frac{(2 + i)(1 + i)}{(1 - i)(1 + i)}$$
$$= \frac{2(1) + 2(i) + 1(i) + i^2}{1^2 - i^2}$$
$$= \frac{2 + 3i - 1}{1 + 1}$$
$$= \frac{1 + 3i}{2} = \frac{1}{2} + \frac{3}{2}i$$

Practice Problem 17

Divide and write in the form $a + bi$: $\dfrac{6}{5i}$

EXAMPLE 17 Divide and write in the form $a + bi$: $\dfrac{7}{3i}$

Solution: We multiply the numerator and the denominator by the conjugate of $3i$. Note that $3i = 0 + 3i$, so its conjugate is $0 - 3i$ or $-3i$.

$$\frac{7}{3i} = \frac{7(-3i)}{(3i)(-3i)} = \frac{-21i}{-9i^2} = \frac{-21i}{-9(-1)} = \frac{-21i}{9} = \frac{-7i}{3} = -\frac{7}{3}i$$

E Finding Powers of i

We can use the fact that $i^2 = -1$ to simplify i^3 and i^4.

$$i^3 = i^2 \cdot i = (-1)i = -i$$
$$i^4 = i^2 \cdot i^2 = (-1) \cdot (-1) = 1$$

We continue this process and use the fact that $i^4 = 1$ and $i^2 = -1$ to simplify i^5 and i^6.

$$i^5 = i^4 \cdot i = 1 \cdot i = i$$
$$i^6 = i^4 \cdot i^2 = 1 \cdot (-1) = -1$$

Answers

16. $\dfrac{3}{13} + \dfrac{11}{13}i$, **17.** $-\dfrac{6}{5}i$

If we continue finding powers of i, we generate the following pattern. Notice that the values $i, -1, -i,$ and 1 repeat as i is raised to higher and higher powers.

$i^1 = i$ $i^5 = i$ $i^9 = i$

$i^2 = -1$ $i^6 = -1$ $i^{10} = -1$

$i^3 = -i$ $i^7 = -i$ $i^{11} = -i$

$i^4 = 1$ $i^8 = 1$ $i^{12} = 1$

This pattern allows us to find other powers of i. To do so, we will use the fact that $i^4 = 1$ and rewrite a power of i in terms of i^4. For example,

$$i^{22} = i^{20} \cdot i^2 = (i^4)^5 \cdot i^2 = 1^5 \cdot (-1) = 1 \cdot (-1) = -1$$

EXAMPLES   Find each power of i.

18. $i^7 = i^4 \cdot i^3 = 1(-i) = -i$

19. $i^{20} = (i^4)^5 = 1^5 = 1$

20. $i^{46} = i^{44} \cdot i^2 = (i^4)^{11} \cdot i^2 = 1^{11}(-1) = -1$

21. $i^{-12} = \dfrac{1}{i^{12}} = \dfrac{1}{(i^4)^3} = \dfrac{1}{(1)^3} = \dfrac{1}{1} = 1$

●

Practice Problems 18–21

Find the powers of i.

18. i^{11}

19. i^{40}

20. i^{50}

21. i^{-10}

Answers

18. $-i$, **19.** 1, **20.** -1, **21.** -1

STUDY SKILLS REMINDER

Are you preparing for a test on Chapter 7?

Below I have listed some common trouble areas for students in Chapter 7. After studying for your test, but before taking your test, read these.

- Remember how to convert an expression with rational expressions to one with radicals and one with radicals to one with rational expressions.

$$7^{2/3} = \sqrt[3]{7^2} \text{ or } (\sqrt[3]{7})^2$$

$$\sqrt[5]{4^3} = 4^{3/5}$$

- Remember the difference between $\sqrt{x} + \sqrt{x}$ and $\sqrt{x} - \sqrt{x}$, $x > 0$.

$$\sqrt{x} + \sqrt{x} = 2\sqrt{x}$$

$$\sqrt{x} \cdot \sqrt{x} = x$$

- Don't forget the difference between rationalizing the denominator of $\sqrt{\dfrac{2}{x}}$ and rationalizing the denominator of $\dfrac{\sqrt{2}}{\sqrt{x} + 1}, x > 0$.

$$\sqrt{\frac{2}{x}} = \frac{\sqrt{2}}{\sqrt{x}} = \frac{\sqrt{2} \cdot \sqrt{x}}{\sqrt{x} \cdot \sqrt{x}} = \frac{\sqrt{2x}}{x}$$

$$\frac{\sqrt{2}}{\sqrt{x} + 1} = \frac{\sqrt{2}(\sqrt{x} - 1)}{(\sqrt{x} + 1)(\sqrt{x} - 1)} = \frac{\sqrt{2}(\sqrt{x} - 1)}{x - 1}$$

Remember: This is simply a checklist of common trouble areas. For a review of Chapter 7, see the Highlights and Chapter Review at the end of this chapter.

Mental Math

Simplify. See Example 1.

1. $\sqrt{-81}$

2. $\sqrt{-49}$

3. $\sqrt{-7}$

4. $\sqrt{-3}$

5. $-\sqrt{16}$

6. $-\sqrt{4}$

7. $\sqrt{-64}$

8. $\sqrt{-100}$

EXERCISE SET 7.7

A *Write using i notation. See Examples 1 through 3.*

1. $\sqrt{-24}$

2. $\sqrt{-32}$

3. $-\sqrt{-36}$

4. $-\sqrt{-121}$

5. $8\sqrt{-63}$

6. $4\sqrt{-20}$

7. $-\sqrt{54}$

8. $\sqrt{-63}$

Multiply or divide as indicated. See Examples 4 through 7.

9. $\sqrt{-2} \cdot \sqrt{-7}$

10. $\sqrt{-11} \cdot \sqrt{-3}$

11. $\sqrt{-5} \cdot \sqrt{-10}$

12. $\sqrt{-2} \cdot \sqrt{-6}$

13. $\sqrt{16} \cdot \sqrt{-1}$

14. $\sqrt{3} \cdot \sqrt{-27}$

15. $\dfrac{\sqrt{-9}}{\sqrt{3}}$

16. $\dfrac{\sqrt{49}}{\sqrt{-10}}$

17. $\dfrac{\sqrt{-80}}{\sqrt{-10}}$

18. $\dfrac{\sqrt{-40}}{\sqrt{-8}}$

B *Add or subtract as indicated. Write your answers in the form a + bi. See Examples 8 through 10.*

19. $(4 - 7i) + (2 + 3i)$

20. $(2 - 4i) - (2 - i)$

21. $(6 + 5i) - (8 - i)$

22. $(8 - 3i) + (-8 + 3i)$

23. $6 - (8 + 4i)$

24. $(9 - 4i) - 9$

25. $(6 - 3i) - (4 - 2i)$

26. $(-2 - 4i) - (6 - 8i)$

27. $(5 - 6i) - 4i$

28. $(6 - 2i) + 7i$

29. $(2 + 4i) + (6 - 5i)$

30. $(5 - 3i) + (7 - 8i)$

Multiply. Write your answers in the form a + bi. See Examples 11 through 15.

31. $6i \cdot 2i$ **32.** $5i \cdot 7i$ **33.** $-9i \cdot 7i$ **34.** $-6i \cdot 4i$ **35.** $-10i \cdot -4i$

36. $-2i \cdot -11i$ **37.** $6i(2 - 3i)$ **38.** $5i(4 - 7i)$ **39.** $-3i(-1 + 9i)$ **40.** $-5i(-2 + i)$

41. $(4 + i)(5 + 2i)$ **42.** $(3 + i)(2 + 4i)$ **43.** $(\sqrt{3} + 2i)(\sqrt{3} - 2i)$ **44.** $(\sqrt{5} - 5i)(\sqrt{5} + 5i)$

45. $(4 - 2i)^2$ **46.** $(6 - 3i)^2$ **47.** $(6 - 2i)(3 + i)$ **48.** $(2 - 4i)(2 - i)$ **49.** $(1 - i)(1 + i)$

50. $(6 + 2i)(6 - 2i)$ **51.** $(9 + 8i)^2$ **52.** $(4 + 7i)^2$ **53.** $(1 - i)^2$ **54.** $(2 - 2i)^2$

D *Divide. Write your answers in the form a + bi. See Examples 16 and 17.*

55. $\dfrac{4}{i}$ **56.** $\dfrac{5}{6i}$ **57.** $\dfrac{7}{4 + 3i}$ **58.** $\dfrac{9}{1 - 2i}$ **59.** $\dfrac{6i}{1 - 2i}$ **60.** $\dfrac{3i}{5 + i}$

61. $\dfrac{3 + 5i}{1 + i}$ **62.** $\dfrac{6 + 2i}{4 - 3i}$ **63.** $\dfrac{4 - 5i}{2i}$ **64.** $\dfrac{6 + 8i}{3i}$ **65.** $\dfrac{16 + 15i}{-3i}$ **66.** $\dfrac{2 - 3i}{-7i}$

67. $\dfrac{2}{3 + i}$ **68.** $\dfrac{5}{3 - 2i}$ **69.** $\dfrac{2 - 3i}{2 + i}$ **70.** $\dfrac{6 + 5i}{6 - 5i}$

E *Find each power of i. See Examples 18 through 21.*

71. i^8 **72.** i^{10} **73.** i^{21} **74.** i^{15} **75.** i^{11} **76.** i^{40}

77. i^{-6} **78.** i^{-9} **79.** $(2i)^6$ **80.** $(5i)^4$ **81.** $(-3i)^5$ **82.** $(-2i)^7$

Thirty people were recently polled about the average monthly balance in their checking account. The results of this poll are shown in the bar graph. Use this graph to answer Exercises 83 through 88. See Section 1.2.

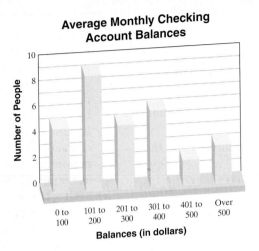

83. How many people polled reported an average checking balance of $201 to $300?

84. How many people polled reported an average checking balance of $0 to $100?

85. How many people polled reported an average checking balance of $200 or less?

86. How many people polled reported an average checking balance of $301 or more?

87. What percent of people polled reported an average checking balance of $201 to $300?

88. What percent of people polled reported an average checking balance of 0 to $100?

Combining Concepts

Write each expression in the form $a + bi$.

89. $i^3 + i^4$

90. $i^8 - i^7$

91. $i^6 + i^8$

92. $i^4 + i^{12}$

93. $2 + \sqrt{-9}$

94. $5 - \sqrt{-16}$

95. $\dfrac{6 + \sqrt{-18}}{3}$

96. $\dfrac{4 - \sqrt{-8}}{2}$

97. $\dfrac{5 - \sqrt{-75}}{10}$

98. Describe how to find the conjugate of a complex number.

99. Explain why the product of a complex number and its complex conjugate is a real number.

Simplify.

100. $(8 - \sqrt{-3}) - (2 + \sqrt{-12})$

101. $(8 - \sqrt{-4}) - (2 + \sqrt{-16})$

102. Determine whether $2i$ is a solution of $x^2 + 4 = 0$.

103. Determine whether $-1 + i$ is a solution of $x^2 + 2x = -2$.

Calculating the Length and Period of a Pendulum

MATERIALS:

- string (at least 1 meter long)
- weight
- meter stick
- stopwatch
- calculator

This activity may be completed by working in groups or individually.

Make a simple pendulum by securely tying the string to a weight.

The formula relating a pendulum's period T (in seconds) to its length l (in centimeters) is

$$T = 2\pi\sqrt{\frac{l}{980}}$$

The **period** of a pendulum is defined as the time it takes the pendulum to complete one full back-and-forth swing. In this activity, you will be measuring your simple pendulum's period with a stopwatch. Because the periods will be only a few seconds long, it will be more accurate for you to time a total of five complete swings and then find the average time of one complete swing.

1. For each of the pendulum (string) lengths given in Table 1, measure the time required for 5 complete swings and record it in the appropriate column. Next, divide this value by 5 to find the measured period of the pendulum for the given length and record it in the Measured Period T_m column in the table. Use the given formula to calculate the theoretical period T for the same pendulum length and record it in the appropriate column. (Round to two decimal places.) Find and record in the last column the difference between the measured period and the theoretical period.

2. For each of the periods T given in Table 2, use the given formula and calculate the theoretical pendulum length l required to yield the given period. Record l in the appropriate column; round to one decimal place. Next, use this length l and measure and record the time for 5 complete swings. Divide this value by 5 to find the measured period T_m, and record it. Then find and record in the last column the difference between the theoretical period and the measured period.

3. Use the general trends you find in the tables to describe the relationship between a pendulum's period and its length.

4. Discuss the differences you found between the values of the theoretical period and the measured period. What factors contributed to these differences?

		Table 1		
Length *l* (centimeters)	Time for 5 Swings (seconds)	Measured Priod T_m (seconds)	Theoretical Period T (seconds)	Difference $\lvert T - T_m \rvert$
30				
55				
70				

		Table 2		
Period *T* (seconds)	Theoretical Length *l* (centimeters)	Time for 5 Swings (seconds)	Measured Period T_m (seconds)	Difference $\lvert T - T_m \rvert$
1				
1.25				
2				

Chapter 7 VOCABULARY CHECK

Fill in each blank with one of the words or phrases listed below.

index	rationalizing	conjugate	principal square root	cube root
complex number	like radicals	radicand	imaginary unit	

1. The _____ of $\sqrt{3} + 2$ is $\sqrt{3} - 2$.
2. The _____ of a nonnegative number a is written as $\sqrt{a}$.
3. The process of writing a radical expression as an equivalent expression but without a radical in the denominator is called _____ the denominator.
4. The _____, written i, is the number whose square is -1.
5. The _____ of a number is written as $\sqrt[3]{a}$.
6. In the notation $\sqrt[n]{a}$, n is called the _____ and a is called the _____.
7. Radicals with the same index and the same radicand are called _____.
8. A _____ is a number that can be written in the form $a + bi$, where a and b are real numbers.

CHAPTER 7 Highlights

DEFINITIONS AND CONCEPTS	EXAMPLES

Section 7.1 Radical Expressions

The **positive**, or **principal**, **square root** of a nonnegative number a is written as $\sqrt{a}$.

$$\sqrt{a} = b \text{ only if } b^2 = a \text{ and } b \geq 0$$

The **negative square root** of a is written as $-\sqrt{a}$.

The **cube root** of a real number a is written as $\sqrt[3]{a}$.

$$\sqrt[3]{a} = b \text{ only if } b^3 = a$$

If n is an even positive integer, then $\sqrt[n]{a^n} = |a|$.
If n is an odd positive integer, then $\sqrt[n]{a^n} = a$.

A **radical function** in x is a function defined by an expression containing a root of x.

$\sqrt{36} = 6 \qquad \sqrt{\dfrac{9}{100}} = \dfrac{3}{10}$
$-\sqrt{36} = -6 \qquad -\sqrt{0.04} = -0.2$
$\sqrt[3]{27} = 3 \qquad \sqrt[3]{-\dfrac{1}{8}} = -\dfrac{1}{2}$
$\sqrt[3]{y^6} = y^2 \qquad \sqrt[3]{64x^9} = 4x^3$
$\sqrt{(-3)^2} = |-3| = 3$
$\sqrt[3]{(-7)^3} = -7$
If $f(x) = \sqrt{x} + 2$,
$\quad f(1) = \sqrt{1} + 2 = 1 + 2 = 3$
$\quad f(3) = \sqrt{3} + 2 \approx 3.73$

Section 7.2 Rational Exponents

$a^{1/n} = \sqrt[n]{a}$ if $\sqrt[n]{a}$ is a real number.

If m and n are positive integers greater than 1 with $\dfrac{m}{n}$ in lowest terms and $\sqrt[n]{a}$ is a real number, then

$$a^{m/n} = (a^{1/n})^m = (\sqrt[n]{a})^m$$

$a^{-m/n} = \dfrac{1}{a^{m/n}}$ as long as $a^{m/n}$ is a nonzero number.

Exponent rules are true for rational exponents.

$81^{1/2} = \sqrt{81} = 9$
$(-8x^3)^{1/3} = \sqrt[3]{-8x^3} = -2x$
$4^{5/2} = (\sqrt{4})^5 = 2^5 = 32$
$27^{2/3} = (\sqrt[3]{27})^2 = 3^2 = 9$
$16^{-3/4} = \dfrac{1}{16^{3/4}} = \dfrac{1}{(\sqrt[4]{16})^3} = \dfrac{1}{2^3} = \dfrac{1}{8}$
$x^{2/3} \cdot x^{-5/6} = x^{2/3-5/6} = x^{-1/6} = \dfrac{1}{x^{1/6}}$
$(8^{14})^{1/7} = 8^2 = 64$
$\dfrac{a^{4/5}}{a^{-2/5}} = a^{4/5-(-2/5)} = a^{6/5}$

DEFINITIONS AND CONCEPTS	**EXAMPLES**

Section 7.3   Simplifying Radical Expressions

PRODUCT AND QUOTIENT RULES

If $\sqrt[n]{a}$ and $\sqrt[n]{b}$ are real numbers,

$$\sqrt[n]{a} \cdot \sqrt[n]{b} = \sqrt[n]{a \cdot b}$$

$$\frac{\sqrt[n]{a}}{\sqrt[n]{b}} = \sqrt[n]{\frac{a}{b}}, \text{provided } \sqrt[n]{b} \neq 0$$

A radical of the form $\sqrt[n]{a}$ is **simplified** when a contains no factors that are perfect nth powers.

Multiply or divide as indicated:

$$\sqrt{11} \cdot \sqrt{3} = \sqrt{33}$$

$$\frac{\sqrt[3]{40x}}{\sqrt[3]{5x}} = \sqrt[3]{8} = 2$$

$$\sqrt{40} = \sqrt{4 \cdot 10} = 2\sqrt{10}$$

$$\sqrt{36x^5} = \sqrt{36x^4 \cdot x} = 6x^2\sqrt{x}$$

$$\sqrt[3]{24x^7y^3} = \sqrt[3]{8x^6y^3 \cdot 3x} = 2x^2y\sqrt[3]{3x}$$

Section 7.4   Adding, Subtracting, and Multiplying Radical Expressions

Radicals with the same index and the same radicand are **like radicals**.

The distributive property can be used to add like radicals.

Radical expressions are multiplied by using many of the same properties used to multiply polynomials.

$$5\sqrt{6} + 2\sqrt{6} = (5 + 2)\sqrt{6} = 7\sqrt{6}$$

$$-\sqrt[3]{3x} - 10\sqrt[3]{3x} + 3\sqrt[3]{10x}$$

$$= (-1 - 10)\sqrt[3]{3x} + 3\sqrt[3]{10x}$$

$$= -11\sqrt[3]{3x} + 3\sqrt[3]{10x}$$

Multiply:

$$(\sqrt{5} - \sqrt{2x})(\sqrt{2} + \sqrt{2x})$$

$$= \sqrt{10} + \sqrt{10x} - \sqrt{4x} - 2x$$

$$= \sqrt{10} + \sqrt{10x} - 2\sqrt{x} - 2x$$

$$(2\sqrt{3} - \sqrt{8x})(2\sqrt{3} + \sqrt{8x})$$

$$= 4(3) - 8x = 12 - 8x$$

Section 7.5   Rationalizing Numerators and Denominators of Radical Expressions

The **conjugate** of $a + b$ is $a - b$.

The process of writing the denominator of a radical expression without a radical is called **rationalizing the denominator**.

The conjugate of $\sqrt{7} + \sqrt{3}$ is $\sqrt{7} - \sqrt{3}$.

Rationalize each denominator:

$$\frac{\sqrt{5}}{\sqrt{3}} = \frac{\sqrt{5} \cdot \sqrt{3}}{\sqrt{3} \cdot \sqrt{3}} = \frac{\sqrt{15}}{3}$$

$$\frac{6}{\sqrt{7} + \sqrt{3}} = \frac{6(\sqrt{7} - \sqrt{3})}{(\sqrt{7} + \sqrt{3})(\sqrt{7} - \sqrt{3})}$$

$$= \frac{6(\sqrt{7} - \sqrt{3})}{7 - 3}$$

$$= \frac{6(\sqrt{7} - \sqrt{3})}{4} = \frac{3(\sqrt{7} - \sqrt{3})}{2}$$

The process of writing the numerator of a radical expression without a radical is called **rationalizing the numerator**.

Rationalize each numerator:

$$\frac{\sqrt[3]{9}}{\sqrt[3]{5}} = \frac{\sqrt[3]{9} \cdot \sqrt[3]{3}}{\sqrt[3]{5} \cdot \sqrt[3]{3}} = \frac{\sqrt[3]{27}}{\sqrt[3]{15}} = \frac{3}{\sqrt[3]{15}}$$

$$\frac{\sqrt{9} + \sqrt{3x}}{12} = \frac{(\sqrt{9} + \sqrt{3x})(\sqrt{9} - \sqrt{3x})}{12(\sqrt{9} - \sqrt{3x})}$$

$$= \frac{9 - 3x}{12(\sqrt{9} - \sqrt{3x})}$$

$$= \frac{3(3 - x)}{3 \cdot 4(3 - \sqrt{3x})} = \frac{3 - x}{4(3 - \sqrt{3x})}$$

DEFINITIONS AND CONCEPTS	**EXAMPLES**

Section 7.6   Radical Equations and Problem Solving

SOLVING A RADICAL EQUATION

Step 1. Write the equation so that one radical is by itself on one side of the equation.

Step 2. Raise each side of the equation to a power equal to the index of the radical and simplify.

Step 3. If the equation still contains a radical, repeat Steps 1 and 2. If not, solve the equation.

Step 4. Check all proposed solutions in the original equation.

Solve: $x = \sqrt{4x + 9} + 3$

1. $x - 3 = \sqrt{4x + 9}$

2. $\quad (x - 3)^2 = (\sqrt{4x + 9})^2$
$x^2 - 6x + 9 = 4x + 9$

3. $x^2 - 10x = 0$
$x(x - 10) = 0$
$x = 0 \quad \text{or} \quad x = 10$

4. The proposed solution 10 checks, but 0 does not. The solution is $\{10\}$.

Section 7.7   Complex Numbers

A **complex number** is a number that can be written in the form $a + bi$, where a and b are real numbers.

$$i^2 = -1 \quad \text{and} \quad i = \sqrt{-1}$$

Simplify: $\sqrt{-9}$

$$\sqrt{-9} = \sqrt{-1 \cdot 9} = \sqrt{-1} \cdot \sqrt{9} = i \cdot 3, \text{or } 3i$$

Complex Numbers	**Written in Form $a + bi$**
12	$12 + 0i$
$-5i$	$0 + (-5)i$
$-2 - 3i$	$-2 + (-3)i$

Multiply.

$$\sqrt{-3} \cdot \sqrt{-7} = i\sqrt{3} \cdot i\sqrt{7}$$
$$= i^2\sqrt{21}$$
$$= -\sqrt{21}$$

To add or subtract complex numbers, add or subtract their real parts and then add or subtract their imaginary parts.

Perform each indicated operation.

$$(-3 + 2i) - (7 - 4i) = -3 + 2i - 7 + 4i$$
$$= -10 + 6i$$

To multiply complex numbers, multiply as though they were binomials.

$$(-7 - 2i)(6 + i) = -42 - 7i - 12i - 2i^2$$
$$= -42 - 19i - 2(-1)$$
$$= -42 - 19i + 2$$
$$= -40 - 19i$$

The complex numbers $(a + bi)$ and $(a - bi)$ are called **complex conjugates**.

The complex conjugate of
$$(3 + 6i) \text{ is } (3 - 6i).$$

Their product is a real number:
$$(3 - 6i)(3 + 6i) = 9 - 36i^2$$
$$= 9 - 36(-1) = 9 + 36 = 45$$

To divide complex numbers, multiply the numerator and the denominator by the conjugate of the denominator.

Divide: $\dfrac{4}{2 - i} = \dfrac{4(2 + i)}{(2 - i)(2 + i)}$

$$= \frac{4(2 + i)}{4 - i^2}$$
$$= \frac{4(2 + i)}{5}$$
$$= \frac{8 + 4i}{5} = \frac{8}{5} + \frac{4}{5}i$$

DIFFUSION

Diffusion is the spontaneous movement of the molecules of a substance from a region of higher concentration to a region of lower concentration until a uniform concentration throughout the region is reached. For example, if a drop of food coloring is added to a glass of water, the molecules of the coloring are diffused so that the entire glass of water is colored evenly without any kind of stirring. Diffusion is also mostly responsible for the spread of the smell of baking brownies throughout a house.

Diffusion is used or seen in important aspects of many disciplines. The following list describes situations in which diffusion plays a role.

- In the commercial production of sugar, sugar can be extracted from sugar cane through a diffusion process.

- Solid-state diffusion plays a role in the manufacturing process of silicon computer chips.

- In biology, the diffusion phenomenon allows water molecules, nutrient molecules, and dissolved gas molecules (such as oxygen and carbon dioxide) to pass through the semipermeable membranes of cell walls.

- During a human pregnancy, the fetus is nourished from the mother's blood supply via diffusion through the placenta. Waste materials from the fetus are also diffused through the placenta to be carried away by the mother's circulatory system.

- The medical treatment known as kidney dialysis, in which waste materials are removed from the blood of a patient without kidney function, is made possible by diffusion.

- A diffusion process is widely used to separate the uranium isotope U-235, which can be used as a fuel in nuclear power plants, from the uranium isotope U-238, which cannot be used to create nuclear energy.

In chemistry, Graham's law states that the diffusion rate of a substance in its gaseous state is inversely proportional to the square root of its molecular weight. Another useful property of diffusion is that the distance a material diffuses over time is directly proportional to the square root of the time.

CRITICAL THINKING

1. Write an equation for the relationship described by Graham's law. Be sure to define the variables and constants that you use.

2. According to Graham's law, which molecule will diffuse more rapidly: a molecule with a molecular weight of 58.4 or a molecule with a molecular weight of 180.2? Explain your reasoning.

3. Write an equation for the relationship between the distance that a material diffuses and time. Again, be sure to define the variables and constants that you use.

4. Suppose it takes sugar 1 week to diffuse a distance of 1 cm from its starting point in a particular liquid. How long will it take the sugar to diffuse a total of 3 cm from its starting point in the liquid?

Chapter 7 Review

(7.1) *Find each root. Assume that all variables represent positive real numbers.*

1. $\sqrt{81}$ **2.** $\sqrt[4]{81}$ **3.** $\sqrt[3]{-8}$ **4.** $\sqrt[4]{-16}$ **5.** $-\sqrt{\dfrac{1}{49}}$

6. $\sqrt{x^{64}}$ **7.** $-\sqrt{36}$ **8.** $\sqrt[3]{64}$ **9.** $\sqrt[3]{-a^6 b^9}$ **10.** $\sqrt{16a^4 b^{12}}$

11. $\sqrt[5]{32a^5 b^{10}}$ **12.** $\sqrt[5]{-32x^{15} y^{20}}$ **13.** $\sqrt{\dfrac{x^{12}}{36y^2}}$ **14.** $\sqrt[3]{\dfrac{27y^3}{z^{12}}}$

Simplify. Use absolute value bars when necessary.

15. $\sqrt{x^2}$ **16.** $\sqrt[4]{(x^2 - 4)^4}$ **17.** $\sqrt[3]{(-27)^3}$ **18.** $\sqrt[5]{(-5)^5}$ **19.** $-\sqrt[5]{x^5}$

20. $\sqrt[4]{16(2y + z)^4}$ **21.** $\sqrt{25(x - y)^2}$ **22.** $\sqrt[5]{y^5}$ **23.** $\sqrt[6]{x^6}$

24. If $f(x) = \sqrt{x} + 3$, find $f(0)$ and $f(9)$.   **25.** If $g(x) = \sqrt[3]{x} - 3$, find $g(11)$ and $g(20)$.

(7.2) *Evaluate.*

26. $\left(\dfrac{1}{81}\right)^{1/4}$

27. $\left(-\dfrac{1}{27}\right)^{1/3}$

28. $(-27)^{-1/3}$

29. $(-64)^{-1/3}$

30. $-9^{3/2}$

31. $64^{-1/3}$

32. $(-25)^{5/2}$

33. $\left(\dfrac{25}{49}\right)^{-3/2}$

34. $\left(\dfrac{8}{27}\right)^{-2/3}$

35. $\left(-\dfrac{1}{36}\right)^{-1/4}$

Write with rational exponents.

36. $\sqrt[3]{x^2}$

37. $\sqrt[5]{5x^2y^3}$

Write using radical notation.

38. $y^{4/5}$

39. $5\left(xy^2z^5\right)^{1/3}$

40. $(x+2y)^{-1/2}$

Simplify each expression. Assume that all variables represent positive real numbers. Write with only positive exponents.

41. $a^{1/3}a^{4/3}a^{1/2}$

42. $\dfrac{b^{1/3}}{b^{4/3}}$

43. $\left(a^{1/2}a^{-2}\right)^3$

44. $\left(x^{-3}y^6\right)^{1/3}$

45. $\left(\dfrac{b^{3/4}}{a^{-1/2}}\right)^8$

46. $\dfrac{x^{1/4}x^{-1/2}}{x^{2/3}}$

47. $\left(\dfrac{49c^{5/3}}{a^{-1/4}b^{5/6}}\right)^{-1}$

48. $a^{-1/4}\left(a^{5/4}-a^{9/4}\right)$

Use a calculator and write a three-decimal-place approximation of each number.

49. $\sqrt{20}$ **50.** $\sqrt[3]{-39}$ **51.** $\sqrt[4]{726}$ **52.** $56^{1/3}$ **53.** $-78^{3/4}$ **54.** $105^{-2/3}$

Use rational exponents to write each as a single radical.

55. $\sqrt[3]{2} \cdot \sqrt{7}$

56. $\sqrt[3]{3} \cdot \sqrt[4]{x}$

(7.3) *Perform each indicated operation and then simplify if possible. Assume that all variables represent positive real numbers.*

57. $\sqrt{3} \cdot \sqrt{8}$ **58.** $\sqrt[3]{7y} \cdot \sqrt[3]{x^2z}$ **59.** $\dfrac{\sqrt{44x^3}}{\sqrt{11x}}$ **60.** $\dfrac{\sqrt[4]{a^6b^{13}}}{\sqrt[4]{a^2b}}$

Simplify.

61. $\sqrt{60}$ **62.** $-\sqrt{75}$ **63.** $\sqrt[3]{162}$ **64.** $\sqrt[3]{-32}$ **65.** $\sqrt{36x^7}$

66. $\sqrt[3]{24a^5b^7}$ **67.** $\sqrt{\dfrac{p^{17}}{121}}$ **68.** $\sqrt[3]{\dfrac{y^5}{27x^6}}$ **69.** $\sqrt[4]{\dfrac{xy^6}{81}}$ **70.** $\sqrt{\dfrac{2x^3}{49y^4}}$

△ **71.** The formula for the radius r of a circle of area A is

$$r = \sqrt{\dfrac{A}{\pi}}$$

a. Find the exact radius of a circle whose area is 25 square meters.

b. Approximate to two decimal places the radius of a circle whose area is 104 square inches.

577

(7.4) *Perform each indicated operation. Assume that all variables represent positive real numbers.*

72. $\sqrt{20} + \sqrt{45} - 7\sqrt{5}$

73. $x\sqrt{75x} - \sqrt{27x^3}$

74. $\sqrt[3]{128} + \sqrt[3]{250}$

75. $3\sqrt[4]{32a^5} - a\sqrt[4]{162a}$

76. $\dfrac{5}{\sqrt{4}} + \dfrac{\sqrt{3}}{3}$

77. $\sqrt{\dfrac{8}{x^2}} - \sqrt{\dfrac{50}{16x^2}}$

78. $2\sqrt{50} - 3\sqrt{125} + \sqrt{98}$

79. $2a\sqrt[4]{32b^5} - 3b\sqrt[4]{162a^4b} + \sqrt[4]{2a^4b^5}$

Multiply and then simplify if possible. Assume that all variables represent positive real numbers.

80. $\sqrt{3}(\sqrt{27} - \sqrt{3})$

81. $(\sqrt{x} - 3)^2$

82. $(\sqrt{5} - 5)(2\sqrt{5} + 2)$

83. $(2\sqrt{x} - 3\sqrt{y})(2\sqrt{x} + 3\sqrt{y})$

84. $(\sqrt{a} + 3)(\sqrt{a} - 3)$

85. $(\sqrt[3]{a} + 2)^2$

86. $(\sqrt[3]{5x} + 9)(\sqrt[3]{5x} - 9)$

87. $(\sqrt[3]{a} + 4)(\sqrt[3]{a^2} - 4\sqrt[3]{a} + 16)$

(7.5) *Rationalize each numerator. Assume that all variables represent positive real numbers.*

88. $\dfrac{3}{\sqrt{7}}$

89. $\sqrt{\dfrac{x}{12}}$

90. $\dfrac{5}{\sqrt[3]{4}}$

91. $\sqrt{\dfrac{24x^5}{3y^2}}$

92. $\sqrt[3]{\dfrac{15x^6y^7}{z^2}}$

93. $\sqrt[4]{\dfrac{81}{8x^{10}}}$

94. $\dfrac{3}{\sqrt{y} - 2}$ **95.** $\dfrac{\sqrt{2} - \sqrt{3}}{\sqrt{2} + \sqrt{3}}$

Rationalize each numerator. Assume that all variables represent positive real numbers.

96. $\dfrac{\sqrt{11}}{3}$ **97.** $\sqrt{\dfrac{18}{y}}$ **98.** $\dfrac{\sqrt[3]{9}}{7}$ **99.** $\sqrt{\dfrac{24x^5}{3y^2}}$ **100.** $\sqrt[3]{\dfrac{xy^2}{10z}}$ **101.** $\dfrac{\sqrt{x} + 5}{-3}$

(7.6) *Solve each equation.*

102. $\sqrt{y - 7} = 5$

103. $\sqrt{2x} + 10 = 4$

104. $\sqrt[3]{2x - 6} = 4$

105. $\sqrt{x + 6} = \sqrt{x + 2}$

106. $2x - 5\sqrt{x} = 3$

107. $\sqrt{x + 9} = 2 + \sqrt{x - 7}$

Find each unknown length.

△ **108.**

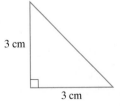

△ **109.**

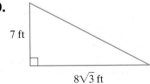

△ **110.** Craig and Daniel Cantwell want to determine the distance x across a pond on their property. They are able to measure the distances shown on the following diagram. Find how wide the lake is at the crossing point indicated by the triangle to the nearest tenth of a foot.

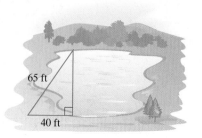

65 ft

40 ft

111. Andrea Roberts, a pipefitter, needs to connect two underground pipelines that are offset by 3 feet, as pictured in the diagram. Neglecting the joints needed to join the pipes, find the length of the shortest possible connecting pipe rounded to the nearest hundredth of a foot.

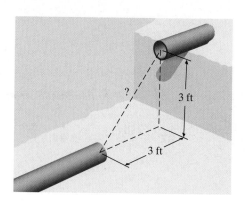

?

3 ft

3 ft

(7.7) *Perform each indicated operation and simplify. Write the results in the form $a + bi$.*

112. $\sqrt{-8}$

113. $-\sqrt{-6}$

114. $\sqrt{-4} + \sqrt{-16}$

115. $\sqrt{-2} \cdot \sqrt{-5}$

116. $(12 - 6i) + (3 + 2i)$

117. $(-8 - 7i) - (5 - 4i)$

118. $(2i)^6$

119. $2i(2 - 5i)$

120. $-3i(6 - 4i)$

121. $(3 + 2i)(1 + i)$

122. $(2 - 3i)^2$

123. $(\sqrt{6} - 9i)(\sqrt{6} + 9i)$

124. $\dfrac{2 + 3i}{2i}$

125. $\dfrac{1 + i}{-3i}$

Chapter 7 Test

Raise to the power or find the root. Assume that all variables represent positive real numbers. Write with only positive exponents.

1. $\sqrt{216}$

2. $-\sqrt[4]{x^{64}}$

3. $\left(\dfrac{1}{125}\right)^{1/3}$

4. $\left(\dfrac{1}{125}\right)^{-1/3}$

5. $\left(\dfrac{8x^3}{27}\right)^{2/3}$

6. $\sqrt[3]{-a^{18}b^9}$

7. $\left(\dfrac{64c^{4/3}}{a^{-2/3}b^{5/6}}\right)^{1/2}$

8. $a^{-2/3}(a^{5/4} - a^3)$

Find each root. Use absolute value bars when necessary.

9. $\sqrt[4]{(4xy)^4}$

10. $\sqrt[3]{(-27)^3}$

Rationalize each denominator. Assume that all variables represent positive real numbers.

11. $\sqrt{\dfrac{9}{y}}$

12. $\dfrac{4 - \sqrt{x}}{4 + 2\sqrt{x}}$

13. $\sqrt[3]{\dfrac{8}{9x}}$

14. Rationalize the numerator of
$\dfrac{\sqrt{6} + x}{8}$ and simplify.

Perform each indicated operation. Assume that all variables represent positive real numbers.

15. $\sqrt{125x^3} - 3\sqrt{20x^3}$

16. $\sqrt{3}(\sqrt{16} - \sqrt{2})$

17. $(\sqrt{x} + 1)^2$

18. $(\sqrt{2} - 4)(\sqrt{3} + 1)$

19. $(\sqrt{5} + 5)(\sqrt{5} - 5)$

Answers
1.
2.
3.
4.
5.
6.
7.
8.
9.
10.
11.
12.
13.
14.
15.
16.
17.
18.
19.

■ *Use a calculator to approximate each number to three decimal places.*

20. $\sqrt{561}$

21. $386^{-2/3}$

Solve.

22. $x = \sqrt{x - 2} + 2$

23. $\sqrt{x^2 - 7} + 3 = 0$

24. $\sqrt{x + 5} = \sqrt{2x - 1}$

Perform each indicated operation and simplify. Write the results in the form a + bi.

25. $\sqrt{-2}$

26. $-\sqrt{-8}$

27. $(12 - 6i) - (12 - 3i)$

28. $(6 - 2i)(6 + 2i)$

29. $(4 + 3i)^2$

30. $\dfrac{1 + 4i}{1 - i}$

△ **31.** Find x.

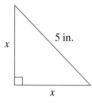

5 in.

x

x

32. If $g(x) = \sqrt{x + 2}$, find $g(0)$ and $g(23)$.

Solve.

33. The function $V = \sqrt{2.5r}$ can be used to estimate the maximum safe velocity, V, in miles per hour, at which a car can travel if it is driven along a curved road with a *radius of curvature, r,* in feet. To the nearest whole number, find the maximum safe speed if a cloverleaf exit on an expressway has a radius of curvature of 300 feet.

34. Use the formula from Exercise 33 to find the radius of curvature if the safe velocity is 30 miles per hour.

Name _____ Section _____ Date _____

Cumulative Review

Simplify. Assume that a and b are integers and that x and y are not 0.

1. $x^{-b}(2x^b)^2$

2. $\dfrac{(y^{3a})^2}{y^{a-6}}$

3. Solve: $|x + 1| = 6$

4. Write an equation of the line containing the point $(4, 4)$ and parallel to the line $2x + y = -6$.

5. Use the elimination method to solve the system: $\begin{cases} x - 5y = -12 \\ -x + \ y = 4 \end{cases}$

6. Solve the system:
$$\begin{cases} x - \ 5y - 2z = 6 \\ -2x + 10y + 4z = -12 \\ \dfrac{1}{2}x - \dfrac{5}{2}y - \ z = 3 \end{cases}$$

7. Use matrices to solve the system:
$$\begin{cases} 2x - \ y = 3 \\ 4x - 2y = 5 \end{cases}$$

8. Add:
$(7x^3y - xy^3 + 11) + (6x^3y - 4)$

If $P(x) = 3x^2 - 2x - 5$, find the following.

9. $P(1)$

10. $P(-2)$

Multiply.

11. $(x + 5)^2$

12. $(4m^2 - 3n)^2$

13. Use synthetic division to divide
$2x^3 - x^2 - 13x + 1$ by $x - 3$.

Factor.

14. $ab - 6a + 2b - 12$

15. $2n^2 - 38n + 80$

16. $16x^2 + 24xy + 9y^2$

17. $50 - 8y^2$

18. Solve: $x(2x - 7) = 4$

19. Graph $f(x) = x^2 + 2x - 3$. Find the vertex and any intercepts.

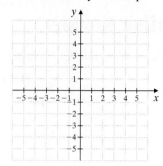

Simplify each rational expression.

20. $\dfrac{24x^6y^5}{8x^7y}$

21. $\dfrac{2x^2}{10x^3 - 2x^2}$

22. Add: $\dfrac{5}{7} + \dfrac{x}{7}$

23. Solve: $\dfrac{3}{x} - \dfrac{x + 21}{3x} = \dfrac{5}{3}$

Write each expression with a positive exponent, and then simplify.

24. $16^{-3/4}$

25. $(-27)^{-2/3}$

18. _____

19. see graph _____

20. _____

21. _____

22. _____

23. _____

24. _____

25. _____

Quadratic Equations and Functions

An important part of algebra is learning to model and solve problems. Often, the model of a problem is a quadratic equation or a function containing a second-degree polynomial. In this chapter, we continue the work from Chapter 5, solving polynomial equations in one variable by factoring. Two other methods of solving quadratic equations are analyzed in this chapter, with methods of solving nonlinear inequalities in one variable and the graphs of quadratic functions.

8.1 Solving Quadratic Equations by Completing the Square

8.2 Solving Quadratic Equations by Using the Quadratic Formula

8.3 Solving Equations by Using Quadratic Methods

Integrated Review—Summary on Solving Quadratic Equations

8.4 Nonlinear Inequalities in One Variable

8.5 Quadratic Functions and Their Graphs

8.6 Further Graphing of Quadratic Functions

The surface of Earth is heated by the sun and then that heat is slowly radiated into outer space. Sometimes, certain gases in the atmosphere reflect the heat radiation back to Earth, preventing it from escaping. The gradual warming of the atmosphere is known as the greenhouse effect. This effect is compounded by the increase of certain gases (greenhouse gases) such as carbon dioxide, methane, and nitrous oxide. Although these gases occur naturally and are needed to keep the surface of Earth at a temperature that is hospitable to life, the recent buildup of these gases is due primarily to human activities. According to the Natural Resources Defense Council, carbon dioxide concentrations have increased by 30% globally over the past century. In Exercise 54 on page 655, we will use a quadratic function to analyze the U.S. level of emissions of methane gas.

Name _____ Section _____ Date _____

Chapter 8 Pretest

1. _____

2. _____

3. _____

4. _____

5. _____

6. _____

7. _____

8. _____

9. _____

10. _____

11. _____

12. see graph _____

13. see graph _____

14. see graph _____

15. see graph _____

16. _____

586

Solve each equation for the variable.

1. $x^2 = 54$

2. $(3y + 2)^2 = 12$

Solve each equation for the variable by completing the square.

3. $x^2 + 4x = 10$

4. $3y^2 + 18y - 1 = 0$

Use the quadratic formula to solve each equation.

5. $2x^2 - x + 4 = 0$

6. $3y^2 + \dfrac{1}{2}y - \dfrac{1}{3} = 0$

Solve.

7. $x^3 = 64$

8. $x^{2/3} - 5x^{1/3} + 4 = 0$

Solve each inequality for x. Write the solution set in interval notation.

9. $(x - 6)(x + 5) \geq 0$ **10.** $x^2 < -x$ **11.** $\dfrac{x + 2}{x + 3} > 1$

Graph each quadratic function. Label the vertex, y-intercept, and x-intercepts (if any).

12. $f(x) = x^2 - 5$

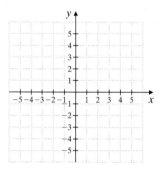

13. $g(x) = (x - 2)^2$

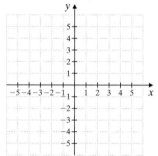

14. $h(x) = (x + 3)^2 - 1$

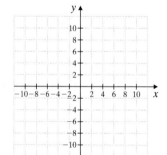

15. $f(x) = -x^2 - 4x + 2$

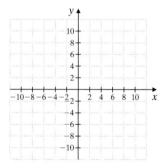

Solve.

△ **16.** The width of a rectangle is $\dfrac{1}{4}$ its length. If its area is 100 square centimeters, find its length and width.

8.1 Solving Quadratic Equations by Completing the Square

OBJECTIVES

Ⓐ Use the square root property to solve quadratic equations.

Ⓑ Write perfect square trinomials.

Ⓒ Solve quadratic equations by completing the square.

Ⓓ Use quadratic equations to solve problems.

SSM   TUTOR CENTER   SG   CD & VIDEO   MATH PRO   WEB

Ⓐ Using the Square Root Property

In Chapter 5, we solved quadratic equations by factoring. Recall that a **quadratic**, or **second-degree**, **equation** is an equation that can be written in the form $ax^2 + bx + c = 0$, where a, b, and c are real numbers and a is not 0. To solve a quadratic equation such as $x^2 = 9$ by factoring, we use the zero-factor property. To use the zero-factor property, the equation must first be written in the standard form $ax^2 + bx + c = 0$.

$$x^2 = 9$$
$$x^2 - 9 = 0 \qquad \text{Subtract 9 from both sides to write in standard form.}$$
$$(x + 3)(x - 3) = 0 \qquad \text{Factor.}$$
$$x + 3 = 0 \quad \text{or} \quad x - 3 = 0 \qquad \text{Set each factor equal to 0.}$$
$$x = -3 \qquad\qquad x = 3 \qquad \text{Solve.}$$

The solution set is $\{-3, 3\}$, the positive and negative square roots of 9.

Not all quadratic equations can be solved by factoring, so we need to explore other methods. Notice that the solutions of the equation $x^2 = 9$ are two numbers whose square is 9:

$$3^2 = 9 \quad \text{and} \quad (-3)^2 = 9$$

Thus, we can solve the equation $x^2 = 9$ by taking the square root of both sides. Be sure to include both $\sqrt{9}$ and $-\sqrt{9}$ as solutions since both $\sqrt{9}$ and $-\sqrt{9}$ are numbers whose square is 9.

$$x^2 = 9$$
$$x = \pm\sqrt{9} \qquad \text{The notation } \pm\sqrt{9} \text{ (read as plus or minus } \sqrt{9}\text{) indicates the pair of}$$
$$x = \pm 3 \qquad\qquad \text{numbers } +\sqrt{9} \text{ and } -\sqrt{9}.$$

This illustrates the square root property.

> **Helpful Hint**
>
> The notation ± 3, for example, is read as "plus or minus 3." It is a shorthand notation for the pair of numbers $+3$ and -3.

Square Root Property

If b is a real number and if $a^2 = b$, then $a = \pm\sqrt{b}$.

EXAMPLE 1   Use the square root property to solve $x^2 = 50$.

Solution:
$$x^2 = 50$$
$$x = \pm\sqrt{50} \qquad \text{Use the square root property.}$$
$$x = \pm 5\sqrt{2} \qquad \text{Simplify the radical.}$$

Check: Let $x = 5\sqrt{2}$.   Let $x = -5\sqrt{2}$.
$$x^2 = 50 \qquad\qquad x^2 = 50.$$
$$(5\sqrt{2})^2 \overset{?}{=} 50 \qquad (-5\sqrt{2})^2 \overset{?}{=} 50$$
$$25 \cdot 2 \overset{?}{=} 50 \qquad\qquad 25 \cdot 2 \overset{?}{=} 50$$
$$50 = 50 \quad \text{True.} \qquad\quad 50 = 50 \quad \text{True.}$$

The solution set is $\{5\sqrt{2}, -5\sqrt{2}\}$.

Practice Problem 1

Use the square root property to solve $x^2 = 20$.

Answer

1. $\{2\sqrt{5}, -2\sqrt{5}\}$

Practice Problem 2

Use the square root property to solve $5x^2 = 55$.

EXAMPLE 2   Use the square root property to solve $2x^2 = 14$.

Solution: First we get the squared variable alone on one side of the equation.

$$2x^2 = 14$$
$$x^2 = 7 \qquad \text{Divide both sides by 2.}$$
$$x = \pm\sqrt{7} \qquad \text{Use the square root property.}$$

Check: Let $x = \sqrt{7}$.   Let $x = -\sqrt{7}$.

$$2x^2 = 14 \qquad\qquad\qquad\qquad 2x^2 = 14$$
$$2(\sqrt{7})^2 \overset{?}{=} 14 \qquad\qquad\qquad 2(-\sqrt{7})^2 \overset{?}{=} 14$$
$$2 \cdot 7 \overset{?}{=} 14 \qquad\qquad\qquad\qquad 2 \cdot 7 \overset{?}{=} 14$$
$$14 = 14 \quad \text{True.} \qquad\qquad\qquad 14 = 14 \quad \text{True.}$$

The solution set is $\{\sqrt{7}, -\sqrt{7}\}$.   ●

Practice Problem 3

Use the square root property to solve $(x + 2)^2 = 18$.

EXAMPLE 3   Use the square root property to solve $(x + 1)^2 = 12$.

Solution: $(x + 1)^2 = 12$
$$x + 1 = \pm\sqrt{12} \qquad \text{Use the square root property.}$$
$$x + 1 = \pm 2\sqrt{3} \qquad \text{Simplify the radical.}$$
$$x = -1 \pm 2\sqrt{3} \qquad \text{Subtract 1 from both sides.}$$

Check: Below is a check for $-1 + 2\sqrt{3}$. The check for $-1 - 2\sqrt{3}$ is almost the same and is left for you to do on your own.

$$(x + 1)^2 = 12$$
$$(-1 + 2\sqrt{3} + 1)^2 \overset{?}{=} 12$$
$$(2\sqrt{3})^2 \overset{?}{=} 12$$
$$4 \cdot 3 \overset{?}{=} 12$$
$$12 = 12 \quad \text{True.}$$

The solution set is $\{-1 + 2\sqrt{3}, -1 - 2\sqrt{3}\}$.   ●

Practice Problem 4

Use the square root property to solve $(3x - 1)^2 = -4$.

EXAMPLE 4   Use the square root property to solve $(2x - 5)^2 = -16$.

Solution: $(2x - 5)^2 = -16$
$$2x - 5 = \pm\sqrt{-16} \qquad \text{Use the square root property.}$$
$$2x - 5 = \pm 4i \qquad \text{Simplify the radical.}$$
$$2x = 5 \pm 4i \qquad \text{Add 5 to both sides.}$$
$$x = \frac{5 \pm 4i}{2} \qquad \text{Divide both sides by 2.}$$

Check each proposed solution in the original equation to see that the solution set is $\left\{\dfrac{5 + 4i}{2}, \dfrac{5 - 4i}{2}\right\}$.   ●

Try the Concept Check in the margin.

Concept Check

How do you know just by looking that $(x - 2)^2 = -4$ has complex solutions?

Answers

2. $\{\sqrt{11}, -\sqrt{11}\}$,   **3.** $\{-2 + 3\sqrt{2}, -2 - 3\sqrt{2}\}$,
4. $\left\{\dfrac{1 - 2i}{3}, \dfrac{1 + 2i}{3}\right\}$

Concept Check: answers may vary

(B)   **Writing Perfect Square Trinomials**

Notice from Examples 3 and 4 that, if we write a quadratic equation so that one side is the square of a binomial, we can solve by using the square root property. To write the square of a binomial, we must have a perfect square trinomial. Recall that a perfect square trinomial is a trinomial that can be factored into two identical binomial factors, that is, as a binomial squared.

Perfect Square Trinomials   **Factored Form**

$$x^2 + 8x + 16$$ $$(x + 4)^2$$
$$x^2 - 6x + 9$$ $$(x - 3)^2$$
$$x^2 + 3x + \frac{9}{4}$$ $$\left(x + \frac{3}{2}\right)^2$$

Notice that for each perfect square trinomial, *the constant term of the trinomial is the square of half the coefficient of the x-term.* For example,

$$x^2 + 8x + 16 \qquad\qquad x^2 - 6x + 9$$

$$\frac{1}{2}(8) = 4 \text{ and } 4^2 = 16 \qquad \frac{1}{2}(-6) = -3 \text{ and } (-3)^2 = 9$$

EXAMPLE 5

Add the proper constant to $x^2 + 6x$ so that the result is a perfect square trinomial. Then factor.

Solution: We add the square of half the coefficient of x.

$$x^2 + 6x + 9 \quad = \quad (x + 3)^2 \quad \text{In factored form}$$

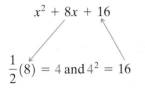

$$\frac{1}{2}(6) = 3 \text{ and } 3^2 = 9$$

EXAMPLE 6

Add the proper constant to $x^2 - 3x$ so that the result is a perfect square trinomial. Then factor.

Solution: We add the square of half the coefficient of x.

$$x^2 - 3x + \frac{9}{4} \quad = \quad \left(x - \frac{3}{2}\right)^2 \quad \text{In factored form}$$

$$\frac{1}{2}(-3) = -\frac{3}{2} \text{ and } \left(-\frac{3}{2}\right)^2 = \frac{9}{4}$$

(c)   Solving by Completing the Square

The process of writing a quadratic equation so that one side is a perfect square trinomial is called **completing the square**. We will use this process in the next examples.

EXAMPLE 7   Solve $p^2 + 2p = 4$ by completing the square.

Solution: First we add the square of half the coefficient of p to both sides so that the resulting trinomial will be a perfect square trinomial. The coefficient of p is 2.

$$\frac{1}{2}(2) = 1 \qquad \text{and} \qquad 1^2 = 1$$

Practice Problem 5

Add the proper constant to $x^2 + 12x$ so that the result is a perfect square trinomial. Then factor.

Practice Problem 6

Add the proper constant to $y^2 - 5y$ so that the result is a perfect square trinomial. Then factor.

Practice Problem 7

Solve $x^2 + 8x = 1$ by completing the square.

Answers

5. $x^2 + 12x + 36 = (x + 6)^2$,

6. $y^2 - 5y + \frac{25}{4} = \left(x - \frac{5}{2}\right)^2$,

7. $\{-4 - \sqrt{17}, -4 + \sqrt{17}\}$

Now we add 1 to both sides of the original equation.

$$p^2 + 2p = 4$$

$$p^2 + 2p + 1 = 4 + 1 \qquad \text{Add 1 to both sides.}$$

$$(p + 1)^2 = 5 \qquad\quad \text{Factor the trinomial; simplify the right side.}$$

We may now use the square root property and solve for p.

$$p + 1 = \pm\sqrt{5} \qquad \text{Use the square root property.}$$

$$p = -1 \pm \sqrt{5} \qquad \text{Subtract 1 from both sides.}$$

Notice that there are two solutions: $-1 + \sqrt{5}$ and $-1 - \sqrt{5}$. The solution set is $\{-1 + \sqrt{5}, -1 - \sqrt{5}\}$. ●

Practice Problem 8

Solve $y^2 - 5y + 2 = 0$ by completing the square.

EXAMPLE 8   Solve $m^2 - 7m - 1 = 0$ by completing the square.

Solution: First we add 1 to both sides of the equation so that the left side has no constant term. We can then add the constant term on both sides that will make the left side a perfect square trinomial.

$$m^2 = 7m - 1 = 0$$

$$m^2 - 7m = 1$$

Now we find the constant term that makes the left side a perfect square trinomial by squaring half the coefficient of m. We add this constant to both sides of the equation.

$$\frac{1}{2}(-7) = -\frac{7}{2} \qquad \text{and} \qquad \left(-\frac{7}{2}\right)^2 = \frac{49}{4}$$

$$m^2 - 7m + \frac{49}{4} = 1 + \frac{49}{4} \qquad \text{Add } \frac{49}{4} \text{ to both sides of the equation.}$$

$$\left(m - \frac{7}{2}\right)^2 = \frac{53}{4} \qquad \text{Factor the perfect square trinomial and simplify the right side.}$$

$$m - \frac{7}{2} = \pm\sqrt{\frac{53}{4}} \qquad \text{Use the square root property.}$$

$$m = \frac{7}{2} \pm \frac{\sqrt{53}}{2} \qquad \text{Add } \frac{7}{2} \text{ to both sides and simplify } \sqrt{\frac{53}{4}}.$$

$$m = \frac{7 \pm \sqrt{53}}{2} \qquad \text{Simplify.}$$

The solution set is $\left\{\dfrac{7 + \sqrt{53}}{2}, \dfrac{7 - \sqrt{53}}{2}\right\}$. ●

Answer

8. $\left\{\dfrac{5 - \sqrt{17}}{2}, \dfrac{5 + \sqrt{17}}{2}\right\}$

The following steps may be used to solve a quadratic equation such as $ax^2 + bx + c = 0$ by completing the square. This method may be used whether or not the polynomial $ax^2 + bx + c$ is factorable.

Solving a Quadratic Equation in x by Completing the Square

Step 1. If the coefficient of x^2 is 1, go to Step 2. Otherwise, divide both sides of the equation by the coefficient of x^2.

Step 2. Get all variable terms alone on one side of the equation.

Step 3. Complete the square for the resulting binomial by adding the square of half of the coefficient of x to both sides of the equation.

Step 4. Factor the resulting perfect square trinomial and write it as the square of a binomial.

Step 5. Use the square root property to solve for x.

EXAMPLE 9   Solve $4x^2 - 24x + 41 = 0$ by completing the square.

Solution: First we divide both sides of the equation by 4 so that the coefficient of x^2 is 1.

$$4x^2 - 24x + 41 = 0$$

Step 1. $x^2 - 6x + \dfrac{41}{4} = 0$ Divide both sides of the equation by 4.

Step 2. $x^2 - 6x = -\dfrac{41}{4}$ Subtract $\dfrac{41}{4}$ from both sides.

Since $\dfrac{1}{2}(-6) = -3$ and $(-3)^2 = 9$, we add 9 to both sides of the equation.

Step 3. $x^2 - 6x + 9 = -\dfrac{41}{4} + 9$ Add 9 to both sides.

Step 4. $(x - 3)^2 = -\dfrac{41}{4} + \dfrac{36}{4}$ Factor the perfect square trinomial.

$(x - 3)^2 = -\dfrac{5}{4}$

Step 5. $x - 3 = \pm\sqrt{-\dfrac{5}{4}}$ Use the square root property.

$x - 3 = \pm\dfrac{i\sqrt{5}}{2}$ Simplify the radical.

$x = 3 \pm \dfrac{i\sqrt{5}}{2}$ Add 3 to both sides.

$= \dfrac{6}{2} \pm \dfrac{i\sqrt{5}}{2}$ Find a common denominator.

$= \dfrac{6 \pm i\sqrt{5}}{2}$ Simplify.

The solution set is $\left\{\dfrac{6 + i\sqrt{5}}{2}, \dfrac{6 - i\sqrt{5}}{2}\right\}$.

Practice Problem 9

Solve $2x^2 - 2x + 7 = 0$ by completing the square.

Answer

9. $\left\{\dfrac{1 + i\sqrt{13}}{2}, \dfrac{1 - i\sqrt{13}}{2}\right\}$

⒟ Solving Problems Modeled by Quadratic Equations

Recall the **simple interest** formula $I = Prt$, where I is the interest earned, P is the principal, r is the rate of interest, and t is time. If \$100 is invested at a simple interest rate of 5% annually, at the end of 3 years the total interest I earned is

$$I = P \cdot r \cdot t$$

or

$$I = 100 \cdot 0.05 \cdot 3 = \$15$$

and the new principal is

$$\$100 + \$15 = \$115$$

Most of the time, the interest computed on money borrowed or money deposited is **compound interest**. Unlike simple interest, compound interest is computed on original principal *and* on interest already earned. To see the difference between simple interest and compound interest, suppose that \$100 is invested at a rate of 5% compounded annually. To find the total amount of money at the end of 3 years, we calculate as follows:

$$I = P \cdot r \cdot t$$

First year:	Interest = \$100 · 0.05 · 1 = \$5.00
	New principal = \$100.00 + \$5.00 = \$105.00
Second year:	Interest = \$105.00 · 0.05 · 1 = \$5.25
	New principal = 105.00 + \$5.25 = \$110.25
Third year:	Interest = \$110.25 · 0.05 · 1 ≈ \$5.51
	New principal = \$110.25 + \$5.51 = \$115.76

At the end of the third year, the total compound interest earned is \$15.76, whereas the total simple interest earned is \$15.

It is tedious to calculate compound interest as we did above, so we use a compound interest formula. The formula for calculating the total amount of money when interest is compounded annually is

$$A = P(1 + r)^t$$

where P is the original investment, r is the interest rate per compounding period, and t is the number of periods. For example, the amount of money A at the end of 3 years if \$100 is invested at 5% compounded annually is

$$A = \$100(1 + 0.05)^3 \approx 100(1.1576) = \$115.76$$

as we previously calculated.

Practice Problem 10

Use the formula from Example 10 to find the interest rate r if \$1600 compounded annually grows to \$1764 in 2 years.

Answer

10. 5%

EXAMPLE 10   Finding an Interest Rate

Find the interest rate r if \$2000 compounded annually grows to \$2420 in 2 years.

Solution:

1. **UNDERSTAND** the problem. For this example, make sure that you understand the formula for compounding interest annually.

2. **TRANSLATE.** We substitute the given values into the formula.

$$A = P(1 + r)^t$$
$$2420 = 2000(1 + r)^2 \qquad \text{Let } A = 2420, P = 2000, \text{ and } t = 2.$$

3. SOLVE. We now solve the equation for r.

$$2420 = 2000(1 + r)^2$$

$$\frac{2420}{2000} = (1 + r)^2 \qquad \text{Divide both sides by 2000.}$$

$$\frac{121}{100} = (1 + r)^2 \qquad \text{Simplify the fraction.}$$

$$\pm\sqrt{\frac{121}{100}} = 1 + r \qquad \text{Use the square root property.}$$

$$\pm\frac{11}{10} = 1 + r \qquad \text{Simplify.}$$

$$-1 \pm \frac{11}{10} = r$$

$$-\frac{10}{10} \pm \frac{11}{10} = r$$

$$\frac{1}{10} = r \quad \text{or} \quad -\frac{21}{10} = r$$

4. INTERPRET. The rate cannot be negative, so we reject $-\dfrac{21}{10}$.

Check: $\dfrac{1}{10} = 0.10 = 10\%$ per year. If we invest \$2000 at 10% compounded annually, in 2 years the amount in the account would be $2000(1 + 0.10)^2 = 2420$ dollars, the desired amount.

State: The interest rate is 10% compounded annually. ●

GRAPHING CALCULATOR EXPLORATIONS

In Section 5.8, we showed how we can use a grapher to approximate real number solutions of a quadratic equation written in standard form. We can also use a grapher to solve a quadratic equation when it is not written in standard form. For example, to solve $(x + 1)^2 = 12$, the quadratic equation in Example 3, we graph the following on the same set of axes. We use Xmin $= -10$, Xmax $= 10$, Ymin $= -13$, and Ymax $= 13$.

$$Y_1 = (x + 1)^2 \quad \text{and} \quad Y_2 = 12$$

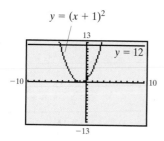

We use the INTERSECT feature or the ZOOM and TRACE features to locate the points of intersection of the graphs. The x-values of these points are the solutions of $(x + 1)^2 = 12$. The solutions, rounded to two decimal points, are 2.46 and -4.46.

Check to see that these numbers are approximations of the exact solutions, $-1 \pm 2\sqrt{3}$.

Use a grapher to solve each quadratic equation. Round all solutions to the nearest hundredth.

1. $x(x - 5) = 8$
2. $x(x + 2) = 5$
3. $x^2 + 0.5x = 0.3x + 1$
4. $x^2 - 2.6x = -2.2x + 3$
5. Use a grapher to solve $(2x - 5)^2 = -16$, (Example 4) using the window

   Xmin $= -20$

   Xmax $= 20$

   Xscl $= 1$

   Ymin $= -20$

   Ymax $= 20$

   Yscl $= 1$

   Explain the results. Compare your results with the solution found in Example 4.

6. What are the advantages and disadvantages of using a grapher to solve quadratic equations?

EXERCISE SET 8.1

(A) *Use the square root property to solve each equation. See Examples 1 through 4.*

1. $x^2 = 16$

2. $x^2 = 49$

3. $x^2 - 7 = 0$

4. $x^2 - 11 = 0$

5. $x^2 = 18$

6. $y^2 = 20$

7. $3z^2 - 30 = 0$

8. $2x^2 = 4$

9. $(x + 5)^2 = 9$

10. $(y - 3)^2 = 4$

11. $(z - 6)^2 = 18$

12. $(y + 4)^2 = 27$

13. $(2x - 3)^2 = 8$

14. $(4x + 9)^2 = 6$

15. $x^2 + 9 = 0$

16. $x^2 + 4 = 0$

17. $x^2 - 6 = 0$

18. $y^2 - 10 = 0$

19. $2z^2 + 16 = 0$

20. $3p^2 + 36 = 0$

21. $(x - 1)^2 = -16$

22. $(y + 2)^2 = -25$

23. $(z + 7)^2 = 5$

24. $(x + 10)^2 = 11$

25. $(x + 3)^2 = -8$

26. $(y - 4)^2 = -18$

(B) *Add the proper constant to each binomial so that the resulting trinomial is a perfect square trinomial. Then factor the trinomial. See Examples 5 and 6.*

27. $x^2 + 16x$

28. $y^2 + 2y$

29. $z^2 - 12z$

30. $x^2 - 8x$

31. $p^2 + 9p$

32. $n^2 + 5n$

33. $r^2 - 3r$

34. $p^2 - 7p$

(C) *Solve each equation by completing the square. See Examples 7 through 9.*

35. $x^2 + 8x = -15$

36. $y^2 + 6y = -8$

37. $x^2 + 6x + 2 = 0$

38. $x^2 - 2x - 2 = 0$

39. $x^2 + x - 1 = 0$

40. $x^2 + 3x - 2 = 0$

41. $x^2 + 2x - 5 = 0$

42. $y^2 + y - 7 = 0$

43. $3p^2 - 12p + 2 = 0$

44. $2x^2 + 14x - 1 = 0$

45. $2x^2 + 7x = 4$

46. $3x^2 - 4x = 4$

47. $x^2 + 8x + 1 = 0$

48. $x^2 - 10x + 2 = 0$

49. $3y^2 + 6y - 4 = 0$

50. $2y^2 + 12y + 3 = 0$

51. $y^2 + 2y + 2 = 0$

52. $x^2 + 4x + 6 = 0$

53. $x^2 - 6x + 3 = 0$

54. $x^2 - 7x - 1 = 0$

55. $2a^2 + 8a = -12$

56. $3x^2 + 12x = -14$

57. $2x^2 - x + 6 = 0$

58. $4x^2 - 2x + 5 = 0$

59. $x^2 + 10x + 28 = 0$

60. $y^2 + 8y + 18 = 0$

61. $z^2 + 3z - 4 = 0$

62. $y^2 + y - 2 = 0$

63. $2x^2 - 4x + 3 = 0$

64. $9x^2 - 36x = -40$

65. $3x^2 + 3x = 5$

66. $5y^2 - 15y = 1$

D *Use the formula $A = P(1 + r)^t$ to solve Exercises 67 through 70. See Example 10.*

67. Find the rate r at which $3000 grows to $4320 in 2 years.

68. Find the rate r at which $800 grows to $882 in 2 years.

69. Find the rate r at which $810 grows to approximately $1000 in 2 years.

70. Find the rate r at which $2000 grows to $2880 in 2 years.

71. In your own words, what is the difference between simple interest and compound interest?

 72. If you are depositing money in an account that pays 4%, would you prefer the interest to be simple or compound? Explain your answer.

73. If you are borrowing money at a rate of 10%, would you prefer the interest to be simple or compound? Explain your answer.

Review and Preview

Simplify each expression. See Section 7.5.

74. $\dfrac{6 + 4\sqrt{5}}{2}$

75. $\dfrac{10 - 20\sqrt{3}}{2}$

76. $\dfrac{3 - 9\sqrt{2}}{6}$

77. $\dfrac{12 - 8\sqrt{7}}{16}$

Evaluate $\sqrt{b^2 - 4ac}$ for each set of values. See Section 7.3.

78. $a = 2, b = 4, c = -1$ **79.** $a = 1, b = 6, c = 2$ **80.** $a = 3, b = -1, c = -2$ **81.** $a = 1, b = -3, c = -1$

Combining Concepts

Find two possible missing terms so that each is a perfect square trinomial.

82. $x^2 + \underline{} + 16$

83. $y^2 + \underline{} + 9$

Neglecting air resistance, the distance $s(t)$ in feet traveled by a freely falling object is given by the function $s(t) = 16t^2$, where t is time in seconds. Use this formula to solve Exercises 84 through 87. Round answers to two decimal places.

84. The Petronas Towers in Kuala Lumpur, built in 1997, are the tallest buildings in Malaysia. Each tower is 1483 feet tall. How long would it take an object to fall to the ground from the top of one of the towers? (*Source*: Council on Tall Buildings and Urban Habitat, Lehigh University)

85. The height of the Chicago Beach Tower Hotel, built in 1998 in Dubai, United Arab Emirates, is 1053 feet. How long would it take an object to fall to the ground from the top of the building? (*Source*: Council on Tall Buildings and Urban Habitat, Lehigh University)

86. The height of the Nurek Dam in Tajikistan (part of the former USSR that borders Afghanistan) is 984 feet. How long would it take an object to fall from the top to the base of the dam? (*Source*: U.S. Committee on Large Dams of the International Commission on Large Dams)

87. The Hoover Dam, located on the Colorado River on the border of Nevada and Arizona near Las Vegas, is 725 feet tall. How long would it take an object to fall from the top to the base of the dam? (*Source*: U.S. Committee on Large Dams of the International Commission on Large Dams)

Solve.

△ **88.** The area of a square room is 225 square feet. Find the dimensions of the room.

△ **89.** The area of a circle is 36π square inches. Find the radius of the circle.

△ **90.** An isosceles right triangle has legs of equal length. If the hypotenuse is 20 centimeters long, find the length of each leg.

△ **91.** A 27-inch TV is advertised in the *Daily Sentry* newspaper. If 27 inches is the measure of the diagonal of the picture tube, find the measure of the side of the picture tube.

Wednesday February 20, 2002   *Daily Sentry*

Aaron's Appliance & Electronics

TVs
VCRs
DVDs
they all must GO!

FEBRUARY SALE

27 in.

📖 *A common equation used in business is a demand equation. It expresses the relationship between the unit price of some commodity and the quantity demanded. For Exercises 92 and 93, p represents the unit price and x represents the quantity demanded in thousands.*

92. A manufacturing company has found that the demand equation for a certain type of scissors is given by the equation $p = -x^2 + 47$. Find the demand for the scissors if the price is $11 per pair.

93. Acme, Inc., sells desk lamps and has found that the demand equation for a certain style of desk lamp is given by the equation $p = -x^2 + 15$. Find the demand for the desk lamp if the price is $7 per lamp.

Internet Excursions

 WWW   Go To: http://www.prenhall.com/martin-gay_interm   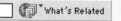 What's Related

This World Wide Web address will direct you to a Web site that contains current interest rates—both composite rates and those offered by individual savings institutions—on a wide variety of financial products such as savings deposits and auto loans.

94. Choose a financial product. Using actual data for a current interest rate on that type of product, write a problem similar to Exercises 67 through 70. When you have finished writing your problem, trade with another student in your class to solve. Then check each other's work.

95. Choose a different financial product from this Web site. Write another interest rate problem using current interest rates. Then exchange problems with another student in your class to solve. Check each other's work.

8.2 Solving Quadratic Equations by Using the Quadratic Formula

OBJECTIVES

Ⓐ Solve quadratic equations by using the quadratic formula.

Ⓑ Determine the number and type of solutions of a quadratic equation by using the discriminant.

Ⓒ Solve geometric problems modeled by quadratic equations.

SSM
TUTOR CENTER   SG   CD & VIDEO   MATH PRO   WEB

Ⓐ Solving Equations by Using the Quadratic Formula

Any quadratic equation can be solved by completing the square. Since the same sequence of steps is repeated each time we complete the square, let's complete the square for a general quadratic equation, $ax^2 + bx + c = 0$. By doing so, we will find a pattern for the solutions of a quadratic equation known as the **quadratic formula**.

Recall that to complete the square for an equation such as $ax^2 + bx + c = 0, a \neq 0$, we first divide both sides by the coefficient of x^2.

$$ax^2 + bx + c = 0$$

$$x^2 + \frac{b}{a}x + \frac{c}{a} = 0 \qquad \text{Divide both sides by } a, \text{ the coefficient of } x^2.$$

$$x^2 + \frac{b}{a}x = -\frac{c}{a} \qquad \text{Subtract the constant } \frac{c}{a} \text{ from both sides.}$$

Next we find the square of half $\frac{b}{a}$, the coefficient of x.

$$\frac{1}{2}\left(\frac{b}{a}\right) = \frac{b}{2a} \qquad \text{and} \qquad \left(\frac{b}{2a}\right)^2 = \frac{b^2}{4a^2}$$

Now we add this result to both sides of the equation.

$$x^2 + \frac{b}{a}x + \frac{b^2}{4a^2} = -\frac{c}{a} + \frac{b^2}{4a^2} \qquad \text{Add } \frac{b^2}{4a^2} \text{ to both sides.}$$

$$x^2 + \frac{b}{a}x + \frac{b^2}{4a^2} = \frac{-c \cdot 4a}{a \cdot 4a} + \frac{b^2}{4a^2} \qquad \text{Find a common denominator on the right side.}$$

$$x^2 + \frac{b}{a}x + \frac{b^2}{4a^2} = \frac{b^2 - 4ac}{4a^2} \qquad \text{Simplify the right side.}$$

$$\left(x + \frac{b}{2a}\right)^2 = \frac{b^2 - 4ac}{4a^2} \qquad \text{Factor the perfect square trinomial on the left side.}$$

$$x + \frac{b}{2a} = \pm\sqrt{\frac{b^2 - 4ac}{4a^2}} \qquad \text{Use the square root property.}$$

$$x + \frac{b}{2a} = \pm\frac{\sqrt{b^2 - 4ac}}{2a} \qquad \text{Simplify the radical.}$$

$$x = -\frac{b}{2a} \pm \frac{\sqrt{b^2 - 4ac}}{2a} \qquad \text{Subtract } \frac{b}{2a} \text{ from both sides.}$$

$$x = \frac{-b \pm \sqrt{b^2 - 4ac}}{2a} \qquad \text{Simplify.}$$

The resulting equation identifies the solutions of the general quadratic equation in standard form and is called the quadratic formula. It can be used to solve any equation written in standard form $ax^2 + bx + c = 0$ as long as a is not 0.

Quadratic Formula

A quadratic equation written in the form $ax^2 + bx + c = 0, a \neq 0$, has the solutions

$$x = \frac{-b \pm \sqrt{b^2 - 4ac}}{2a}$$

Practice Problem 1

Solve: $2x^2 + 9x + 10 = 0$

EXAMPLE 1 Solve: $3x^2 + 16x + 5 = 0$

Solution: This equation is in standard form with $a = 3$, $b = 16$, and $c = 5$. We substitute these values into the quadratic formula.

$$x = \frac{-b \pm \sqrt{b^2 - 4ac}}{2a} \quad \text{Quadratic formula}$$

$$= \frac{-16 \pm \sqrt{16^2 - 4(3)(5)}}{2(3)} \quad \text{Let } a = 3, b = 16, \text{ and } c = 5.$$

$$= \frac{-16 \pm \sqrt{256 - 60}}{6}$$

$$= \frac{-16 \pm \sqrt{196}}{6} = \frac{-16 \pm 14}{6}$$

$$x = \frac{-16 + 14}{6} = -\frac{1}{3} \quad \text{or} \quad x = \frac{-16 - 14}{6} = -\frac{30}{6} = -5$$

The solution set is $\left\{ -\dfrac{1}{3}, -5 \right\}$.

> **Helpful Hint**
>
> To replace a, b, and c correctly in the quadratic formula, write the quadratic equation in standard form $ax^2 + bx + c = 0$.

Practice Problem 2

Solve: $2x^2 - 6x - 1 = 0$

EXAMPLE 2 Solve: $2x^2 - 4x = 3$

Solution: First we write the equation in standard form by subtracting 3 from both sides.

$$2x^2 - 4x - 3 = 0$$

Now $a = 2$, $b = -4$, and $c = -3$. We substitute these values into the quadratic formula.

$$x = \frac{-b \pm \sqrt{b^2 - 4ac}}{2a}$$

$$= \frac{-(-4) \pm \sqrt{(-4)^2 - 4(2)(-3)}}{2(2)}$$

$$= \frac{4 \pm \sqrt{16 + 24}}{4}$$

$$= \frac{4 \pm \sqrt{40}}{4} = \frac{4 \pm 2\sqrt{10}}{4}$$

$$= \frac{2(2 \pm \sqrt{10})}{2 \cdot 2} = \frac{2 \pm \sqrt{10}}{2}$$

The solution set is $\left\{ \dfrac{2 + \sqrt{10}}{2}, \dfrac{2 - \sqrt{10}}{2} \right\}$.

Concept Check

For the quadratic equation $x^2 = 7$, which substitution is correct?

a. $a = 1, b = 0$, and $c = -7$
b. $a = 1, b = 0$, and $c = 7$
c. $a = 0, b = 0$, and $c = 7$
d. $a = 1, b = 1$, and $c = -7$

Answers

1. $\left\{ -\dfrac{5}{2}, -2 \right\}$, **2.** $\left\{ \dfrac{3 + \sqrt{11}}{2}, \dfrac{3 - \sqrt{11}}{2} \right\}$

Concept Check: a

Try the Concept Check in the margin.

Helpful Hint

To simplify the expression $\dfrac{4 \pm 2\sqrt{10}}{4}$ in Example 2, note that we factored 2 out of both terms of the numerator *before* simplifying.

$$\frac{4 \pm 2\sqrt{10}}{4} = \frac{\boxed{2}(2 \pm \sqrt{10})}{\boxed{2}\cdot 2} = \frac{2 \pm \sqrt{10}}{2}$$

EXAMPLE 3   Solve: $\dfrac{1}{4}m^2 - m + \dfrac{1}{2} = 0$

Solution: We could use the quadratic formula with $a = \dfrac{1}{4}$, $b = -1$, and $c = \dfrac{1}{2}$. Instead, let's find a simpler, equivalent, standard-form equation whose coefficients are not fractions.

First we multiply both sides of the equation by 4 to clear the fractions.

$$4\left(\frac{1}{4}m^2 - m + \frac{1}{2}\right) = 4\cdot 0$$

$$m^2 - 4m + 2 = 0 \qquad \text{Simplify.}$$

Now we can substitute $a = 1$, $b = -4$, and $c = 2$ into the quadratic formula and simplify.

$$m = \frac{-(-4) \pm \sqrt{(-4)^2 - 4(1)(2)}}{2(1)}$$

$$= \frac{4 \pm \sqrt{16 - 8}}{2}$$

$$= \frac{4 \pm \sqrt{8}}{2} = \frac{4 \pm 2\sqrt{2}}{2} = \frac{2\,(2 \pm \sqrt{2})}{2} = 2 \pm \sqrt{2}$$

The solution set is $\{2 + \sqrt{2}, 2 - \sqrt{2}\}$. ●

EXAMPLE 4   Solve: $p = -3p^2 - 3$

Solution: The equation in standard form is $3p^2 + p + 3 = 0$. Thus, $a = 3$, $b = 1$, and $c = 3$ in the quadratic formula.

$$p = \frac{-1 \pm \sqrt{1^2 - 4(3)(3)}}{2(3)} = \frac{-1 \pm \sqrt{1 - 36}}{6}$$

$$= \frac{-1 \pm \sqrt{-35}}{6} = \frac{-1 \pm i\sqrt{35}}{6}$$

The solution set is $\left\{\dfrac{-1 + i\sqrt{35}}{6}, \dfrac{-1 - i\sqrt{35}}{6}\right\}$. ●

Try the Concept Check in the margin.

Practice Problem 3

Solve: $\dfrac{1}{6}x^2 - \dfrac{1}{2}x - 1 = 0$

Practice Problem 4

Solve: $x = -4x^2 - 4$

Concept Check

What is the first step in solving $-3x^2 = 5x - 4$ using the quadratic formula?

Answers

3. $\left\{\dfrac{3 + \sqrt{33}}{2}, \dfrac{3 - \sqrt{33}}{2}\right\}$,

4. $\left\{\dfrac{-1 - 3i\sqrt{7}}{8}, \dfrac{-1 + 3i\sqrt{7}}{8}\right\}$

Concept Check: Write the equation in standard form.

B Using the Discriminant

In the quadratic formula $x = \dfrac{-b \pm \sqrt{b^2 - 4ac}}{2a}$, the radicand $b^2 - 4ac$ is called the **discriminant** because when we know its value, we can **discriminate** among the possible number and type of solutions of a quadratic equation. Possible values of the discriminant and their meanings are summarized next.

Discriminant

The following table relates the discriminant $b^2 - 4ac$ of a quadratic equation of the form $ax^2 + bx + c = 0$ with the number and type of solutions of the equation.

$b^2 - 4ac$	Number and Type of Solutions
Positive	Two real solutions
Zero	One real solution
Negative	Two complex but not real solutions

Practice Problem 5

Use the discriminant to determine the number and type of solutions of $x^2 + 4x + 4 = 0$.

EXAMPLE 5

Use the discriminant to determine the number and type of solutions of $x^2 + 2x + 1$.

Solution: In $x^2 + 2x + 1 = 0$, $a = 1$, $b = 2$, and $c = 1$. Thus,

$$b^2 - 4ac = 2^2 - 4(1)(1) = 0$$

Since $b^2 - 4ac = 0$, this quadratic equation has one real solution. ●

Practice Problem 6

Use the discriminant to determine the number and type of solutions of $5x^2 + 7 = 0$.

EXAMPLE 6

Use the discriminant to determine the number and type of solutions of $3x^2 + 2 = 0$.

Solution: In this equation, $a = 3$, $b = 0$, and $c = 2$. Then

$$b^2 - 4ac = 0^2 - 4(3)(2) = -24$$

Since $b^2 - 4ac$ is negative, this quadratic equation has two complex but not real solutions. ●

Practice Problem 7

Use the discriminant to determine the number and type of solutions of $3x^2 - 2x - 2 = 0$.

EXAMPLE 7

Use the discriminant to determine the number and type of solutions of $2x^2 - 7x - 4 = 0$.

Solution: In this equation, $a = 2$, $b = -7$, and $c = -4$. Then

$$b^2 - 4ac = (-7)^2 - 4(2)(-4) = 81$$

Since $b^2 - 4ac$ is positive, this quadratic equation has two real solutions. ●

C Solving Problems Modeled by Quadratic Equations

The quadratic formula is useful in solving problems that are modeled by quadratic equations.

Answers

5. one real solution, **6.** two complex but not real solutions, **7.** two real solutions

EXAMPLE 8   Calculating Distance Saved

At a local university, students often leave the sidewalk and cut across the lawn to save walking distance. Given the diagram below of a favorite place to cut across the lawn, approximate to the nearest foot how many feet of walking distance a student saves by cutting across the lawn instead of walking on the sidewalk.

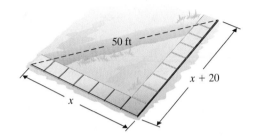

Solution:

1. UNDERSTAND. Read and reread the problem. You may want to review the Pythagorean theorem.
2. TRANSLATE. By the Pythagorean theorem, we have

    In words: $(\text{leg})^2 + (\text{leg})^2 = (\text{hypotenuse})^2$

    Translate: $x^2 + (x + 20)^2 = 50^2$

3. SOLVE. Use the quadratic formula to solve.

    $x^2 + x^2 + 40x + 400 = 2500$ Square $(x + 20)$ and 50.

    $2x^2 + 40x - 2100 = 0$ Set the equation to 0.

    $x^2 + 20x - 1050 = 0$ Divide by 2.

    Here, $a = 1$, $b = 20$, and $c = -1050$. By the quadratic formula,

    $$x = \frac{-20 \pm \sqrt{20^2 - 4(1)(-1050)}}{2 \cdot 1}$$

    $$= \frac{-20 \pm \sqrt{400 + 4200}}{2} = \frac{-20 \pm \sqrt{4600}}{2}$$

    $$= \frac{-20 \pm \sqrt{100 \cdot 46}}{2} = \frac{-20 \pm 10\sqrt{46}}{2}$$

    $$= -10 \pm 5\sqrt{46}$$ Simplify.

Check:

4. INTERPRET. We check our calculations from the quadratic formula. The length of a side of a triangle can't be negative, so we reject $-10 - 5\sqrt{46}$. Since $-10 + 5\sqrt{46} \approx 24$ feet, the walking distance along the sidewalk is

    $x + (x + 20) \approx 24 + (24 + 20) = 68$ feet.

State: A person saves about $68 - 50$ or 18 feet of walking distance by cutting across the lawn.

Practice Problem 8   △

Given the diagram below, approximate to the nearest foot how many feet of walking distance a person saves by cutting across the lawn instead of walking on the sidewalk.

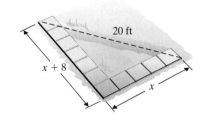

Answer

8. 8 ft

Practice Problem 9

How long after the object in Example 9 is thrown will it be 100 feet from the ground? Round to the nearest tenth of a second.

EXAMPLE 9   Calculating Landing Time

An object is thrown upward from the top of a 200-foot cliff with a velocity of 12 feet per second. The height h in feet of the object after t seconds is

$$h = -16t^2 + 12t + 200$$

How long after the object is thrown will it strike the ground? Round to the nearest tenth of a second.

200 ft

Solution:

1. UNDERSTAND. Read and reread the problem.
2. TRANSLATE. Since we want to know when the object strikes the ground, we want to know when the height $h = 0$, or

$$0 = -16t^2 + 12t + 200$$

3. SOLVE. First we divide both sides of the equation by -4.

$$0 = 4t^2 - 3t - 50 \qquad \text{Divide both sides by } -4.$$

Here, $a = 4$, $b = -3$, and $c = -50$. By the quadratic formula,

$$t = \frac{-(-3) \pm \sqrt{(-3)^2 - 4(4)(-50)}}{2 \cdot 4}$$

$$= \frac{3 \pm \sqrt{9 + 800}}{8}$$

$$= \frac{3 \pm \sqrt{809}}{8}$$

Check:

4. INTERPRET. We check our calculations from the quadratic formula. Since the time won't be negative, we reject the proposed solution

$$\frac{3 - \sqrt{809}}{8}.$$

State: The time it takes for the object to strike the ground is exactly $\frac{3 + \sqrt{809}}{8}$ seconds ≈ 3.9 seconds.

●

Answer

9. 1.7 sec

Name _____ Section _____ Date _____

Mental Math

Identify the values of a, b, and c in each quadratic equation.

1. $x^2 + 3x + 1 = 0$

2. $2x^2 - 5x - 7 = 0$

3. $7x^2 - 4 = 0$

4. $x^2 + 9 = 0$

5. $6x^2 - x = 0$

6. $5x^2 + 3x = 0$

EXERCISE SET 8.2

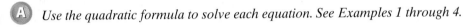 **A** Use the quadratic formula to solve each equation. See Examples 1 through 4.

1. $m^2 + 5m - 6 = 0$

2. $p^2 + 11p - 12 = 0$

3. $2y = 5y^2 - 3$

4. $5x^2 - 3 = 14x$

5. $x^2 - 6x + 9 = 0$

6. $y^2 + 10y + 25 = 0$

7. $x^2 + 7x + 4 = 0$

8. $y^2 + 5y + 3 = 0$

9. $8m^2 - 2m = 7$

10. $11n^2 - 9n = 1$

11. $3m^2 - 7m = 3$

12. $x^2 - 13 = 5x$

13. $\frac{1}{2}x^2 - x - 1 = 0$

14. $\frac{1}{6}x^2 + x + \frac{1}{3} = 0$

15. $\frac{2}{5}y^2 + \frac{1}{5}y = \frac{3}{5}$

16. $\frac{1}{8}x^2 + x = \frac{5}{2}$

17. $\frac{1}{3}y^2 - y - \frac{1}{6} = 0$

18. $\frac{1}{2}y^2 = y + \frac{1}{2}$

19. $10y^2 + 10y + 3 = 0$

20. $3y^2 + 6y + 5 = 0$

21. $x^2 + 5x = -2$

22. $y^2 - 8 = 4y$

23. $(m + 2)(2m - 6) = 5(m - 1) - 12$

605

24. $7p(p - 2) + 2(p + 4) = 3$

25. $\dfrac{x^2}{3} - x = \dfrac{5}{3}$

26. $\dfrac{x^2}{2} - 3 = -\dfrac{9}{2}x$

27. $x(6x + 2) - 3 = 0$

28. $x(7x + 1) = 2$

29. Solve Exercise 1 by factoring. Explain the result.

30. Solve Exercise 2 by factoring. Explain the result.

Use the quadratic formula to solve each equation. See Examples 1 through 4.

31. $6 = -4x^2 + 3x$

32. $9x^2 + x + 2 = 0$

33. $(x + 5)(x - 1) = 2$

34. $x(x + 6) = 2$

35. $x^2 + 6x + 13 = 0$

36. $x^2 + 2x + 2 = 0$

37. $\dfrac{2}{5}y^2 + \dfrac{1}{5}y + \dfrac{3}{5} = 0$

38. $\dfrac{1}{8}x^2 + x + \dfrac{5}{2} = 0$

39. $\dfrac{1}{2}y^2 = y - \dfrac{1}{2}$

40. $\dfrac{2}{3}x^2 - \dfrac{20}{3}x = -\dfrac{100}{6}$

41. $(n - 2)^2 = 2n$

42. $\left(p - \dfrac{1}{2}\right)^2 = \dfrac{p}{2}$

B *Use the discriminant to determine the number and types of solutions of each equation. See Examples 5 through 7.*

43. $9x - 2x^2 + 5 = 0$

44. $5 - 4x + 12x^2 = 0$

45. $4x^2 + 12x = -9$

46. $9x^2 + 1 = 6x$

47. $3x = -2x^2 + 7$

48. $3x^2 = 5 - 7x$

49. $6 = 4x - 5x^2$

50. $8x = 3 - 9x^2$

Solve. See Examples 8 and 9.

51. Nancy, Thelma, and John Varner live on a corner lot. Often, neighborhood children cut across their lot to save walking distance. Given the diagram below, approximate to the nearest foot how many feet of walking distance children save by cutting across their property instead of walking around the lot.

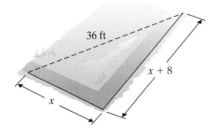

36 ft

x + 8

x

△ **52.** Given the diagram below, approximate to the nearest foot how many feet of walking distance a person saves by cutting across the lawn instead of walking on the sidewalk.

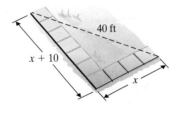

40 ft

x + 10

x

△ **53.** The hypotenuse of an isosceles right triangle is 2 centimeters longer than either of its legs. Find the exact length of each side. (*Hint:* An isosceles right triangle is a right triangle whose legs are the same length.)

△ **54.** The hypotenuse of an isosceles right triangle is one meter longer than either of its legs. Find the length of each side.

△ **55.** Uri Chechov's rectangular dog pen for his Irish setter must have an area of 400 square feet. Also, the length must be 10 feet longer than the width. Find the dimensions of the pen.

△ **56.** An entry in the Peach Festival Poster Contest must be rectangular and have an area of 1200 square inches. Furthermore, its length must be 20 inches longer than its width. Find the dimensions each entry must have.

△ **57.** A holding pen for cattle must be square and have a diagonal length of 100 meters.
   a. Find the length of a side of the pen.
   b. Find the area of the pen.

△ **58.** A rectangle is three times longer than it is wide. It has a diagonal of length 50 centimeters.
   a. Find the dimensions of the rectangle.
   b. Find the perimeter of the rectangle.

50 cm

59. The heaviest reported door in the world is the 708.6 ton radiation shield door in the National Institute for Fusion Science at Toki, Japan. If the height of the door is 1.1 feet longer than its width, and its front area (neglecting depth) is 1439.9 square feet, find its width and height [Interesting note: the door is 6.6 feet thick.] (*Source: Guiness World Records*, 2001)

60. Christi and Robbie Wegmann are constructing a rectangular stained glass window whose length is 7.3 inches longer than its width. If the area of the window is 569.9 square inches, find its width and length.

61. If a point B divides a line segment such that the smaller portion is to the larger portion as the larger is to the whole, the whole is the length of the *golden ratio*.

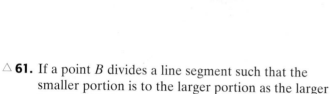

The golden ratio was thought by the Greeks to be the most pleasing to the eye, and many of their buildings contained numerous examples of the golden ratio. The value of the golden ratio is the positive solution of the following equation.

$$\text{(smaller)} \quad \frac{x-1}{1} = \frac{1}{x} \quad \text{(larger)}$$
$$\text{(larger)} \qquad\qquad\qquad \text{(whole)}$$

Find this value.

62. The base of a triangle is four more than twice its height. If the area of the triangle is 42 square centimeters, find its base and height.

The Wollomombi Falls in Australia have a height of 1100 feet. A pebble is thrown upward from the top of the falls with an initial velocity of 20 feet per second. The height of the pebble h in feet after t seconds is given by the equation $h = -16t^2 + 20t + 1100$. Use this equation for Exercises 63 and 64.

63. How long after the pebble is thrown will it hit the ground? Round to the nearest tenth of a second.

64. How long after the pebble is thrown will it be 550 feet from the ground? Round to the nearest tenth of a second.

A ball is thrown downward from the top of a 180-foot building with an initial velocity of 20 feet per second. The height of the ball h in feet after t seconds is given by the equation $h = -16t^2 - 20t + 180$. Use this equation to answer Exercises 65 and 66.

65. How long after the ball is thrown will it strike the ground? Round the result to the nearest tenth of a second.

66. How long after the ball is thrown will it be 50 feet from the ground? Round the result to the nearest tenth of a second.

Review and Preview

Solve each equation. See Sections 6.4 and 7.6.

67. $\sqrt{5x - 2} = 3$

68. $\sqrt{y + 2} + 7 = 12$

69. $\dfrac{1}{x} + \dfrac{2}{5} = \dfrac{7}{x}$

70. $\dfrac{10}{z} = \dfrac{5}{z} - \dfrac{1}{3}$

Factor. See Section 5.6.

71. $x^4 + x^2 - 20$

72. $2y^4 + 11y^2 - 6$

73. $z^4 - 13z^2 + 36$

74. $x^4 - 1$

Combining Concepts

Use the quadratic formula and a calculator to approximate each solution to the nearest tenth.

75. $2x^2 - 6x + 3 = 0$

76. $3.6x^2 + 1.8x - 4.3 = 0$

The graph shows the daily low temperatures for one week in New Orleans, Louisiana. Use this graph to answer Exercises 77 through 80.

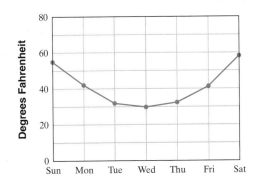

77. Which day of the week shows the greatest decrease in the low temperature?

78. Which day of the week shows the greatest increase in the low temperature?

79. Which day of the week had the lowest temperature?

80. Use the graph to estimate the low temperature on Thursday.

Notice that the shape of the temperature graph is similar to a parabola (see Section 5.9). In fact, this graph can be approximated by the quadratic function $f(x) = 3x^2 - 18x + 57$, where $f(x)$ is the temperature in degrees Fahrenheit and x is the number of days from Sunday. Use this function to answer Exercises 81 and 82.

81. Use the given quadratic function to approximate the low temperature on Thursday. Does your answer agree with the graph above?

82. Use the given function and the quadratic formula to find when the low temperature was 35°F. [*Hint:* Let $f(x) = 35$ and solve for x.] Round your answer to one decimal place and interpret your result. Does your answer agree with the graph above?

83. Use a grapher to solve Exercise 75.

85. Procter & Gamble's net earnings can be modeled by the quadratic function
$f(x) = -199.5x^2 - 21.5x + 3763$, where $f(x)$ is net earnings in millions of dollars and x is the number of years after 1999. (*Source:* Based on data from The Procter & Gamble Co., 1999–2001)
  a. Find Procter & Gamble's net earnings in 2001.
  b. If the trend described by the model continues, predict the year after 1999 in which Procter & Gamble's net earnings will be $485 million.

87. The relationship between body weight and the Recommended Dietary Allowance (RDA) for vitamin A in children up to age 10 is modeled by the quadratic equation $y = 0.149x^2 - 4.475x + 406.478$, where y is the RDA for vitamin A in micrograms for a child whose weight is x pounds. (*Source:* Based on data from the Food and Nutrition Board, National Academy of Sciences—Institute of Medicine, 1989)
  a. Determine the vitamin A requirements of a child who weighs 35 pounds.
  b. What is the weight of a child whose RDA of vitamin A is 600 micrograms? Round your answer to the nearest pound.

84. Use a grapher to solve Exercise 76.

86. The number of inmates in custody in U.S. prisons and jails can be modeled by the quadratic function $p(x) = -x^2 + 93x + 1128$, where $p(x)$ is the number of inmates in thousands and x is the number of years after 1990. (*Source:* Based on data from the Bureau of Justice Statistics, U.S. Department of Justice, 1990–2000)
  a. Find the number of prison and jail inmates in the United States in 1995.
  b. Find the number of prison and jail inmates in the United States in 2000.
  c. If the trend described by the model continues, predict the year in which the number of prisoners will be 3,200,000 (that is, 3200 thousand).

88. The total amount of passenger traffic at Phoenix Sky Harbor International Airport in Phoenix, Arizona, during the period 1960 through 2000 can be modeled by the equation $y = 26x^2 - 136x + 917$, where y is the number of passengers enplaned and deplaned in thousands and x is the number of years after 1960. (*Source:* Based on data from The City of Phoenix Aviation Department, 1960–2000)
  a. Estimate the passenger traffic at Phoenix Sky Harbor International Airport in 2000.
  b. According to this model, in what year will passenger traffic at Phoenix Sky Harbor International Airport reach 60,000,000 passengers?

8.3 Solving Equations by Using Quadratic Methods

(A) Solve various equations that are quadratic in form.

(B) Solve problems that lead to quadratic equations.

SSM
TUTOR CENTER   SG   CD & VIDEO   MATH PRO   WEB

(A) Solving Equations That Are Quadratic in Form

In this section, we discuss various types of equations that can be solved in part by using the methods for solving quadratic equations.

Once each equation is simplified, you may want to use these steps when deciding what method to use to solve the quadratic equation.

Solving a Quadratic Equation

Step 1. If the equation is in the form $(ax + b)^2 = c$, use the square root property and solve. If not, go to Step 2.

Step 2. Write the equation in standard form: $ax^2 + bx + c = 0$.

Step 3. Try to solve the equation by the factoring method. If not possible, go to Step 4.

Step 4. Solve the equation by the quadratic formula.

The first example is a radical equation that becomes a quadratic equation once we square both sides.

EXAMPLE 1   Solve: $x - \sqrt{x} - 6 = 0$

Solution: Recall that to solve a radical equation, we first get the radical alone on one side of the equation. Then we square both sides.

$$x - 6 = \sqrt{x} \qquad \text{Add } \sqrt{x} \text{ to both sides.}$$
$$x^2 - 12x + 36 = x \qquad \text{Square both sides.}$$
$$x^2 - 13x + 36 = 0 \qquad \text{Set the equation equal to 0.}$$
$$(x - 9)(x - 4) = 0$$
$$x - 9 = 0 \quad \text{or} \quad x - 4 = 0 \qquad \text{Set each factor equal to 0.}$$
$$x = 9 \qquad\qquad x = 4 \qquad \text{Solve.}$$

Check: Let $x = 9$.

$$x - \sqrt{x} - 6 = 0$$
$$9 - \sqrt{9} - 6 \stackrel{?}{=} 0$$
$$9 - 3 - 6 \stackrel{?}{=} 0$$
$$0 = 0 \quad \text{True.}$$

Let $x = 4$.

$$x - \sqrt{x} - 6 = 0$$
$$4 - \sqrt{4} - 6 \stackrel{?}{=} 0$$
$$4 - 2 - 6 \stackrel{?}{=} 0$$
$$-4 = 0 \quad \text{False.}$$

The solution set is $\{9\}$.

EXAMPLE 2   Solve: $\dfrac{3x}{x - 2} - \dfrac{x + 1}{x} = \dfrac{6}{x(x - 2)}$

Solution: In this equation, x cannot be either 2 or 0 because these values cause denominators to equal zero. To solve for x, we first multiply both sides of the equation by $x(x - 2)$ to clear the fractions. By the distributive property, this means that we multiply each term by $x(x - 2)$.

$$x(x - 2)\left(\frac{3x}{x - 2}\right) - x(x - 2)\left(\frac{x + 1}{x}\right) = x(x - 2)\left[\frac{6}{x(x - 2)}\right]$$
$$3x^2 - (x - 2)(x + 1) = 6 \qquad \text{Simplify.}$$
$$3x^2 - (x^2 - x - 2) = 6 \qquad \text{Multiply.}$$
$$3x^2 - x^2 + x + 2 = 6$$
$$2x^2 + x - 4 = 0 \qquad \text{Simplify.}$$

Practice Problem 1

Solve: $x - \sqrt{x - 1} - 3 = 0$

Practice Problem 2

Solve: $\dfrac{2x}{x - 1} - \dfrac{x + 2}{x} = \dfrac{5}{x(x - 1)}$

Answers

1. $\{5\}$, **2.** $\left\{\dfrac{1 + \sqrt{13}}{2}, \dfrac{1 - \sqrt{13}}{2}\right\}$

This equation cannot be factored using integers, so we solve by the quadratic formula.

$$x = \frac{-1 \pm \sqrt{1^2 - 4(2)(-4)}}{2 \cdot 2}$$
Let $a = 2$, $b = 1$, and $c = -4$, in the quadratic formula.

$$= \frac{-1 \pm \sqrt{1 + 32}}{4}$$ Simplify.

$$= \frac{-1 \pm \sqrt{33}}{4}$$

Neither proposed solution will make the denominators 0.

The solution set is $\left\{ \dfrac{-1 + \sqrt{33}}{4}, \dfrac{-1 - \sqrt{33}}{4} \right\}$.

EXAMPLE 3 Solve: $p^4 - 3p^2 - 4 = 0$

Solution: First we factor the trinomial.

$$p^4 - 3p^2 - 4 = 0$$
$$(p^2 - 4)(p^2 + 1) = 0$$ Factor.
$$(p - 2)(p + 2)(p^2 + 1) = 0$$ Factor further.

$p - 2 = 0$ or $p + 2 = 0$ or $p^2 + 1 = 0$ Set each factor equal to 0 and solve.
$p = 2$ $p = -2$ $p^2 = -1$
$$p = \pm\sqrt{-1} = \pm i$$

The solution set is $\{2, -2, i, -i\}$.

Try the Concept Check in the margin.

EXAMPLE 4 Solve: $(x - 3)^2 - 3(x - 3) - 4 = 0$

Solution: Notice that the quantity $(x - 3)$ is repeated in this equation. Sometimes it is helpful to substitute a variable (in this case other than x) for the repeated quantity. We will let $y = x - 3$. Then

$$(x - 3)^2 - 3(x - 3) - 4 = 0$$
becomes
$$y^2 - 3y - 4 = 0$$ Let $x - 3 = y$.
$$(y - 4)(y + 1) = 0$$ Factor.

To solve, we use the zero-factor property.

$y - 4 = 0$ or $y + 1 = 0$ Set each factor equal to 0.
$y = 4$ $y = -1$ Solve.

To find values of x, we substitute back. That is, we substitute $x - 3$ for y.

$x - 3 = 4$ or $x - 3 = -1$
$x = 7$ $x = 2$

Both 2 and 7 check. The solution is $\{2, 7\}$.

Practice Problem 3

Solve: $x^4 - 5x^2 - 36 = 0$

Concept Check

a. True or false? The maximum number of solutions that a quadratic equation can have is 2.
b. True or false? The maximum number of solutions that an equation in quadratic form can have is 2.

Practice Problem 4

Solve: $(x + 4)^2 - (x + 4) - 6 = 0$

Helpful Hint

When using substitution, don't forget to substitute back to the original variable.

Answers

3. $\{3, -3, 2i, -2i\}$, **4.** $\{-1, -6\}$

Concept Check: **a.** true, **b.** false

EXAMPLE 5   Solve: $x^{2/3} - 5x^{1/3} + 6 = 0$

Solution: The key to solving this equation is recognizing that $x^{2/3} = (x^{1/3})^2$. We replace $x^{1/3}$ with m so that

$$(x^{1/3})^2 - 5x^{1/3} + 6 = 0$$

becomes

$$m^2 - 5m + 6 = 0$$

Now we solve by factoring.

$$m^2 - 5m + 6 = 0$$
$$(m - 3)(m - 2) = 0 \qquad \text{Factor.}$$
$$m - 3 = 0 \quad \text{or} \quad m - 2 = 0 \qquad \text{Set each factor equal to 0.}$$
$$m = 3 \qquad\qquad m = 2$$

Since $m = x^{1/3}$, we have

$$x^{1/3} = 3 \qquad \text{or} \quad x^{1/3} = 2$$
$$x = 3^3 = 27 \quad \text{or} \quad x = 2^3 = 8$$

Both 8 and 27 check. The solution set is $\{8, 27\}$. ●

Helpful Hint

Example 3 can be solved using substitution also. Think of $p^4 - 3p^2 - 4 = 0$ as

$$(p^2)^2 - 3p^2 - 4 = 0 \qquad \text{Then let } x = p^2 \text{, and solve and substitute back. The solution set will be the same.}$$
$$\downarrow \qquad \swarrow$$
$$x^2 - 3x - 4 = 0$$

B   **Solving Problems That Lead to Quadratic Equations**

The next example is a work problem. This problem is modeled by a rational equation that simplifies to a quadratic equation.

EXAMPLE 6   Finding Work Time

Together, an experienced typist and an apprentice typist can process a document in 6 hours. Alone, the experienced typist can process the document 2 hours faster than the apprentice typist can. Find the time in which each person can process the document alone.

Solution:

1.   UNDERSTAND. Read and reread the problem. The key idea here is the relationship between the *time* (hours) it takes to complete the job and the *part of the job* completed in one unit of time (hour). For example, because they can complete the job together in 6 hours, the *part of the job* they can complete in 1 hour is $\frac{1}{6}$. We let

   $x =$ the time in hours it takes the apprentice typist to complete the job alone

   $x - 2 =$ the time in hours it takes the experienced typist to complete the job alone

Practice Problem 5

Solve: $x^{2/3} - 7x^{1/3} + 10 = 0$

Practice Problem 6

Together, Karen and Doug Lewis can clean a strip of beach in 5 hours. Alone, Karen can clean the strip of beach one hour faster than Doug. Find the time that each person can clean the strip of beach alone. Give an exact answer and a one-decimal-place approximation.

Answers

5. $\{8, 125\}$,   **6.** Doug: $\dfrac{11 + \sqrt{101}}{2} \approx 10.5$ hr;

Karen: $\dfrac{9 + \sqrt{101}}{2} \approx 9.5$ hr

We can summarize in a chart the information discussed.

	Total Hours to Complete Job	Part of Job Completed in 1 Hour
Apprentice Typist	x	$\dfrac{1}{x}$
Experienced Typist	$x - 2$	$\dfrac{1}{x - 2}$
Together	6	$\dfrac{1}{6}$

2. TRANSLATE.

	part of job completed by apprentice typist in 1 hour	added to	part of job completed by experienced typist in 1 hour	is equal to	part of job completed together in 1 hour
In words:	$\downarrow$	$\downarrow$	$\downarrow$	$\downarrow$	$\downarrow$
Translate:	$\dfrac{1}{x}$	$+$	$\dfrac{1}{x - 2}$	$=$	$\dfrac{1}{6}$

3. SOLVE.

$$\frac{1}{x} + \frac{1}{x - 2} = \frac{1}{6}$$

$$6x(x - 2)\left(\frac{1}{x} + \frac{1}{x - 2}\right) = 6x(x - 2) \cdot \frac{1}{6} \qquad \text{Multiply both sides by the LCD } 6x(x - 2).$$

$$6x(x - 2) \cdot \frac{1}{x} + 6x(x - 2) \cdot \frac{1}{x - 2} = 6x(x - 2) \cdot \frac{1}{6} \qquad \text{Use the distributive property.}$$

$$6(x - 2) + 6x = x(x - 2)$$
$$6x - 12 + 6x = x^2 - 2x$$
$$0 = x^2 - 14x + 12$$

Now we can substitute $a = 1$, $b = -14$, and $c = 12$ into the quadratic formula and simplify.

$$x = \frac{-(-14) \pm \sqrt{(-14)^2 - 4(1)(12)}}{2(1)} = \frac{14 \pm \sqrt{148}}{2}$$

Using a calculator or a square root table, we see that $\sqrt{148} \approx 12.2$ rounded to one decimal place. Thus,

$$x \approx \frac{14 \pm 12.2}{2}$$

$$x \approx \frac{14 + 12.2}{2} = 13.1 \quad \text{or} \quad x \approx \frac{14 - 12.2}{2} = 0.9$$

4. INTERPRET.

Check: If the apprentice typist completes the job alone in 0.9 hours, the experienced typist completes the job alone in $x - 2 = 0.9 - 2 = -1.1$ hours. Since this is not possible, we reject the solution 0.9. The approximate solution is thus 13.1 hours.

State: The apprentice typist can complete the job alone in approximately 13.1 hours, and the experienced typist can complete the job alone in approximately $x - 2 = 13.1 - 2 = 11.1$ hours. ●

EXAMPLE 7   **Calculating Driving Speeds**

Beach and Fargo are about 400 miles apart. A salesperson travels from Fargo to Beach one day at a certain speed. She returns to Fargo the next day and drives 10 miles per hour faster. Her total travel time was $14\frac{2}{3}$ hours. Find her speed to Beach and the return speed to Fargo.

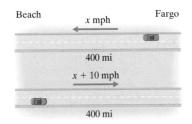

Solution:

1. UNDERSTAND. Read and reread the problem. Let

x = the speed to Beach, so

$x + 10$ = the return speed to Fargo

Then organize the given information in a table.

	Distance	=	Rate	·	Time
To Beach	400		x		$\dfrac{400}{x}$
Return to Fargo	400		$x + 10$		$\dfrac{400}{x + 10}$

2. TRANSLATE.

In words: time to Beach $+$ return time to Fargo $=$ $14\frac{2}{3}$ hours

Translate: $\dfrac{400}{x} + \dfrac{400}{x + 10} = \dfrac{44}{3}$

3. SOLVE.

$$\frac{400}{x} + \frac{400}{x + 10} = \frac{44}{3}$$

$$\frac{100}{x} + \frac{100}{x + 10} = \frac{11}{3} \qquad \text{Divide both sides by 4.}$$

$$3x(x + 10)\left(\frac{100}{x} + \frac{100}{x + 10}\right) = 3x(x + 10) \cdot \frac{11}{3} \qquad \begin{array}{l}\text{Multiply both sides by the LCD, } 3x(x + 10).\end{array}$$

$$3x(x + 10)\frac{100}{x} + 3x(x + 10)\frac{100}{x + 10} = 3x(x + 10) \cdot \frac{11}{3} \qquad \begin{array}{l}\text{Use the distributive property.}\end{array}$$

$$3(x + 10)100 + 3x(100) = x(x + 10)11$$

$$300x + 3000 + 300x = 11x^2 + 110x$$

$$0 = 11x^2 - 490x - 3000 \qquad \begin{array}{l}\text{Set equa-tion equal to 0.}\end{array}$$

$$0 = (11x + 60)(x - 50) \qquad \text{Factor.}$$

$11x + 60 = 0 \qquad$ or $\qquad x - 50 = 0 \qquad$ Set each factor equal to 0.

$x = -\dfrac{60}{11} \qquad$ or $\qquad -5\dfrac{5}{11} \qquad x = 50$

Practice Problem 7

A family drives 500 miles to the beach for a vacation. The return trip was made at a speed that was 10 miles per hour faster. The total traveling time was $18\frac{1}{3}$ hours. Find the speed to the beach and the return speed.

Answer

7. 50 mph to the beach; 60 mph returning

4. INTERPRET.

Check: The speed is not negative, so it's not $-5\dfrac{5}{11}$. The number 50 does check.

State: The speed to Beach was 50 miles per hour and the return speed to Fargo was 60 miles per hour. ●

FOCUS ON **Business and Career**

FLOWCHARTS

We saw in the Focus On Business and Career feature in Chapter 6 that four of the top 10 fastest-growing jobs through 2008 are computer related. A useful skill in computer-related careers is *flowcharting*. A **flowchart** is a diagram showing a sequence of procedures used to complete a task. Flowcharts are commonly used in computer programming to help a programmer plan the steps and commands needed to write a program. Flowcharts are also used in other types of careers, such as manufacturing or finance, to describe the sequence of events needed in a certain process.

A flowchart usually uses the following symbols to represent certain types of actions.

For instance, suppose we want to write a flowchart for the process of computing a household's monthly electric bill. The flowchart might look something like the one here.

CRITICAL THINKING

1. Using the given flowchart as a guide, describe in words this utility company's pricing structure for household electricity usage.

2. Make a flowchart for the process of computing the human-equivalent age for the age of a dog if 1 dog year is equivalent to 7 human years.

3. Make a flowchart for the process of determining the number and type of solutions of a quadratic equation of the form $ax^2 + bx + c = 0$ using the discriminant.

4. (Optional) Write a programmable calculator program for your graphing calculator that determines the number and type of solutions of a quadratic equation of the form $ax^2 + bx + c = 0$.

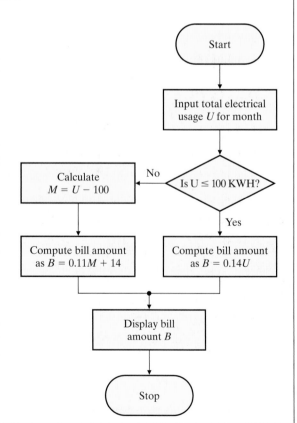

EXERCISE SET 8.3

A *Solve. See Example 1.*

1. $2x = \sqrt{10 + 3x}$

2. $3x = \sqrt{8x + 1}$

3. $x - 2\sqrt{x} = 8$

4. $x - \sqrt{2x} = 4$

5. $\sqrt{9x} = x + 2$

6. $\sqrt{16x} = x + 3$

Solve. See Example 2.

7. $\dfrac{2}{x} + \dfrac{3}{x - 1} = 1$

8. $\dfrac{6}{x^2} = \dfrac{3}{x + 1}$

9. $\dfrac{3}{x} + \dfrac{4}{x + 2} = 2$

10. $\dfrac{5}{x - 2} + \dfrac{4}{x + 2} = 1$

11. $\dfrac{7}{x^2 - 5x + 6} = \dfrac{2x}{x - 3} - \dfrac{x}{x - 2}$

12. $\dfrac{11}{2x^2 + x - 15} = \dfrac{5}{2x - 5} - \dfrac{x}{x + 3}$

Solve. See Example 3.

13. $p^4 - 16 = 0$

14. $x^4 + 2x^2 - 3 = 0$

15. $4x^4 + 11x^2 = 3$

16. $z^4 = 81$

17. $z^4 - 13z^2 + 36 = 0$

18. $9x^4 + 5x^2 - 4 = 0$

Solve. See Examples 4 and 5.

19. $x^{2/3} - 3x^{1/3} - 10 = 0$

20. $x^{2/3} + 2x^{1/3} + 1 = 0$

21. $(5n + 1)^2 + 2(5n + 1) - 3 = 0$

22. $(m - 6)^2 + 5(m - 6) + 4 = 0$

23. $2x^{2/3} - 5x^{1/3} = 3$

24. $3x^{2/3} + 11x^{1/3} = 4$

25. $1 + \dfrac{2}{3t - 2} = \dfrac{8}{(3t - 2)^2}$

26. $2 - \dfrac{7}{x + 6} = \dfrac{15}{(x + 6)^2}$

27. $20x^{2/3} - 6x^{1/3} - 2 = 0$

28. $4x^{2/3} + 16x^{1/3} = -15$

Solve. See Examples 1 through 5.

29. $a^4 - 5a^2 + 6 = 0$

30. $x^4 - 12x^2 + 11 = 0$

31. $\dfrac{2x}{x - 2} + \dfrac{x}{x + 3} = \dfrac{-5}{x + 3}$

32. $\dfrac{5}{x - 3} + \dfrac{x}{x + 3} = \dfrac{19}{x^2 - 9}$

33. $(p + 2)^2 = 9(p + 2) - 20$

34. $2(4m - 3)^2 - 9(4m - 3) = 5$

35. $2x = \sqrt{11x + 3}$

36. $4x = \sqrt{2x + 3}$

37. $x^{2/3} - 8x^{1/3} + 15 = 0$

38. $x^{2/3} - 2x^{1/3} - 8 = 0$

39. $y^3 + 9y - y^2 - 9 = 0$

40. $x^3 + x - 3x^2 - 3 = 0$

41. $2x^{2/3} + 3x^{1/3} - 2 = 0$

42. $6x^{2/3} - 25x^{1/3} - 25 = 0$

43. $x^{-2} - x^{-1} - 6 = 0$

44. $y^{-2} - 8y^{-1} + 7 = 0$

45. $x - \sqrt{x} = 2$

46. $x - \sqrt{3x} = 6$

47. $\dfrac{x}{x - 1} + \dfrac{1}{x + 1} = \dfrac{2}{x^2 - 1}$

48. $\dfrac{x}{x - 5} + \dfrac{5}{x + 5} = \dfrac{-1}{x^2 - 25}$

49. $p^4 - p^2 - 20 = 0$

50. $x^4 - 10x^2 + 9 = 0$

51. $2x^3 = -54$

52. $y^3 - 216 = 0$

618

53. $1 = \dfrac{4}{x-7} + \dfrac{5}{(x-7)^2}$

54. $3 + \dfrac{1}{(2p+4)} = \dfrac{10}{(2p+4)^2}$

55. $27y^4 + 15y^2 = 2$

56. $8z^4 + 14z^2 = -5$

B *Solve. See Examples 6 and 7.*

57. A jogger ran 3 miles, decreased her speed by 1 mile per hour and then ran another 4 miles. If her total time jogging was $1\dfrac{3}{5}$ hours, find her speed for each part of her run.

58. Mark Keaton's workout consists of jogging for 3 miles, and then riding his bike for 5 miles at a speed 4 miles per hour faster than he jogs. If his total workout time is 1 hour, find his jogging speed and his biking speed.

59. A Chinese restaurant in Mandeville, Louisiana, has a large goldfish pond around the restaurant. Suppose that an inlet pipe and a hose together can fill the pond in 8 hours. The inlet pipe alone can complete the job in one hour less time than the hose alone. Find the time that the hose can complete the job alone and the time that the inlet pipe can complete the job alone. Round each to the nearest tenth of an hour.

60. A water tank on a farm in Flatonia, Texas, can be filled with a large inlet pipe and a small inlet pipe in 3 hours. The large inlet pipe alone can fill the tank in 2 hours less time than the small inlet pipe alone. Find the time to the nearest tenth of an hour each pipe can fill the tank alone.

61. Roma Sherry drove 330 miles from her home town to Tucson. During her return trip, she was able to increase her speed by 11 miles per hour. If her return trip took 1 hour less time, find her original speed and her speed returning home.

62. A salesperson drove to Portland, a distance of 300 miles. During the last 80 miles of his trip, heavy rainfall forced him to decrease his speed by 15 miles per hour. If his total driving time was 6 hours, find his original speed and his speed during the rainfall.

63. Bill Shaughnessy and his son Billy can clean the house together in 4 hours. When the son works alone, it takes him an hour longer to clean than it takes his dad alone. Find how long to the nearest tenth of an hour it takes the son to clean alone.

64. Together, Noodles and Freckles eat a 50-pound bag of dog food in 30 days. Noodles by herself eats a 50-pound bag in 2 weeks less time than Freckles does by himself. How many days to the nearest whole day would a 50-pound bag of dog food last Freckles?

65. The product of a number and 4 less than the number is 96. Find the number.

66. A whole number increased by its square is two more than twice itself. Find the number.

△ **67.** Suppose that we want to make an open box from a square sheet of cardboard by cutting out squares from each corner as shown and then folding along the dotted lines. If the box is to have a volume of 300 cubic centimeters, find the original dimensions of the sheet of cardboard.

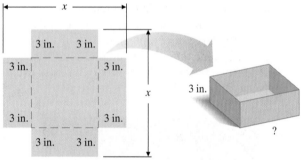

a. The ? in the drawing to the left will be the length (and also the width) of the box as shown in the drawing to the right. Represent this length in terms of x.
b. Use the formula for volume of a box, $V = l \cdot w \cdot h$, to write an equation in x.
c. Solve the equation for x and give the dimensions of the sheet of cardboard. Check your solution.

△ **68.** Suppose that we want to make an open box from a square sheet of cardboard by cutting out squares from each corner as shown and then folding along the dotted lines. If the box is to have a volume of 128 cubic inches, find the original dimensions of the sheet of cardboard. (*Hint:* Use Exercise 67 parts (a), (b), and (c) to help you.)

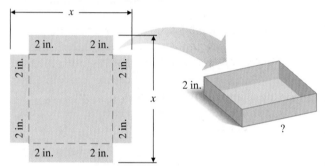

△ **69.** A sprinkler that sprays water in a circular motion is to be used to water a square garden. If the area of the garden is 920 square feet, find the smallest whole number *radius* that the sprinkler can be adjusted to so that the entire garden is watered.

△ **70.** Suppose that a square field has an area of 6270 square feet. See Exercise 69 and find a new sprinkler radius.

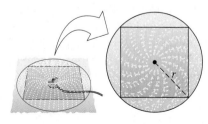

Review and Preview

Solve each inequality. See Section 2.4.

71. $\dfrac{5x}{3} + 2 \le 7$

72. $\dfrac{2x}{3} + \dfrac{1}{6} \ge 2$

73. $\dfrac{y-1}{15} > \dfrac{-2}{5}$

74. $\dfrac{z-2}{12} < \dfrac{1}{4}$

 Combining Concepts

75. Write a polynomial equation that has three solutions: 2, 5, and −7.

76. Write a polynomial equation that has three solutions: 0, 2i, and −2i.

77. During the 2001 American Memorial auto race held at the Eurospeedway in Lausitz, Germany, Tony Kanaan posted the fastest speed but Kenny Brack won the race. The track is 10,682 feet long. Kanaan's fastest speed was 6.326 feet per second faster than Brack's fastest speed. Traveling at these fastest speeds, Brack would have taken 0.73 seconds longer than Kanaan to complete a lap. (*Source:* Based on data from Championship Auto Racing Teams, Inc.)

a. Find Kenny Brack's fastest speed during the race. Round to three decimal places.

b. Find Tony Kanaan's fastest speed during the race. Round to three decimal places.

c. Convert Kanaan's speed to miles per hour. Round to three decimal places.

78. Use a grapher to solve Exercise 29. Compare the solution with the solution from Exercise 29. Explain any differences.

FINANCIAL RATIOS

A financial ratio is a number found with a rational expression that tells something about a company's activities. Such ratios allow a comparison between the financial positions of two companies, even if the values of the companies' financial data are very different. Here are some common financial ratios:

- The **current ratio** gauges a company's ability to pay its short-term debts. It is given by the formula

$$\text{current ratio} = \frac{\text{total current assets}}{\text{total current liabilities}}$$

The higher the value of this ratio, the better able the company is to pay off its short-term debts.

- The **total asset turnover ratio** gauges how effectively a company is using all of its resources to generate sales of its products and services. It is given by the formula

$$\text{total asset turnover ratio} = \frac{\text{sales}}{\text{total assets}}$$

The higher the value of this ratio, the more effective the company is at utilizing its resources for sales generation.

- The **gross profit margin ratio** gauges how effectively the company is making pricing decisions as well as controlling production costs. It is given by the formula

$$\text{gross profit margin ratio} = \frac{\text{sales} - \text{cost of sales}}{\text{sales}}$$

The higher the value of this ratio, the better the company is controlling costs and pricing products.

- The **price-to-earnings (P/E) ratio** gauges the stock market's view of a company with respect to risk. It is given by the formula

$$\text{P/E ratio} = \frac{\text{stock market price per share}}{\text{earnings per share}}$$

A high P/E ratio generally means a company with low risk. A high P/E ratio also translates into better growth potential for the company's earnings.

For all of these ratios, a higher-than-industry-average ratio is generally considered to be a sign of good financial health.

ADDITIONAL DEFINITIONS

- **Assets**—things of value that are owned by a company. *Current assets* include cash and assets that can be converted into cash quickly. *Total assets* are all things of value, including property and equipment, owned by the company. Current assets and total assets may be found on a company's consolidated balance sheet or statement of financial position in an annual report.

- **Liabilities**—what a company owes to creditors. *Current liabilities* include any debts expected to come due within the next year. Current liabilities may be found on a company's consolidated balance sheet or statement of financial position in an annual report.

- **Sales**—the total amount of money collected by a company from the sales of its goods or services. Sales (or sometimes noted as "net sales") may be found on a company's consolidated statement of income/earnings/operations in an annual report.

- **Cost of sales**—a company's cost of inventory actually sold to customers. This is also sometimes referred to as "cost of goods/merchandise sold." Cost of sales may be found on a company's consolidated statement of income/earnings/operations in an annual report.

- **Earnings per share**—the value of a company's earnings available for each share of common stock held by stockholders. Earnings per share (or sometimes noted as net income per share) may be found on a company's consolidated statement of income/earnings/operations in an annual report.

- **Stock market price per share**—the current price of a company's share of stock as given on one of the major stock markets. Current share prices may be found in newspapers or on the World Wide Web.

GROUP ACTIVITY

Locate annual reports for two companies involved in similar industries. Using the information and definitions given above, compute these four financial ratios for each company. Then compare the companies' ratios and discuss what the ratios indicate about the two companies. Which company do you think is in better overall financial health? Why?

Integrated Review–Summary on Solving Quadratic Equations

Use the square root property to solve each equation.

1. $x^2 - 10 = 0$

2. $x^2 - 14 = 0$

3. $(x - 1)^2 = 8$

4. $(x + 5)^2 = 12$

Solve each equation by completing the square.

5. $x^2 + 2x - 12 = 0$

6. $x^2 - 12x + 11 = 0$

7. $3x^2 + 3x = 5$

8. $16y^2 + 16y = 1$

Use the quadratic formula to solve each equation.

9. $2x^2 - 4x + 1 = 0$

10. $\dfrac{1}{2}x^2 + 3x + 2 = 0$

11. $x^2 + 4x = -7$

12. $x^2 + x = -3$

Solve each equation. Use a method of your choice.

13. $x^2 + 3x + 6 = 0$

14. $2x^2 + 18 = 0$

15. $x^2 + 17x = 0$

1. _____

2. _____

3. _____

4. _____

5. _____

6. _____

7. _____

8. _____

9. _____

10. _____

11. _____

12. _____

13. _____

14. _____

15. _____

16. $4x^2 - 2x - 3 = 0$ **17.** $(x - 2)^2 = 27$ **18.** $\frac{1}{2}x^2 - 2x + \frac{1}{2} = 0$

19. $3x^2 + 2x = 8$ **20.** $2x^2 = -5x - 1$ **21.** $x(x - 2) = 5$

22. $x^2 - 31 = 0$ **23.** $5x^2 - 55 = 0$ **24.** $5x^2 + 55 = 0$

25. $x(x + 5) = 66$ **26.** $5x^2 + 6x - 2 = 0$ **27.** $2x^2 + 3x = 1$

△**28.** The diagonal of a square room measures 20 feet. Find the exact length of a side of the room. Then approximate the length to the nearest tenth of a foot.

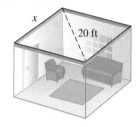

29. Diane Gray and Lucy Hoag together can prepare a crawfish boil for a large party in 4 hours. Lucy alone can complete the job in 2 hours less time than Diane alone. Find the time that each person can prepare the crawfish boil alone. Round each time to the nearest tenth of an hour.

30. Kraig Blackwelder exercises at Total Body Gym. On the treadmill, he runs 5 miles, then increases his speed by 1 mile per hour and runs an additional 2 miles. If his total time on the treadmill is $1\frac{1}{3}$ hours, find his speed during each part of his run.

16. _____

17. _____

18. _____

19. _____

20. _____

21. _____

22. _____

23. _____

24. _____

25. _____

26. _____

27. _____

28. _____

29. _____

30. _____

8.4 Nonlinear Inequalities in One Variable

(A) Solving Polynomial Inequalities

Just as we can solve linear inequalities in one variable, we can also solve quadratic inequalities in one variable. A **quadratic inequality** is an inequality that can be written so that one side is a quadratic expression and the other side is 0. Here are examples of quadratic inequalities in one variable. Each is written in **standard form**.

$$x^2 - 10x + 7 \le 0 \qquad 3x^2 + 2x - 6 > 0$$
$$2x^2 + 9x - 2 < 0 \qquad x^2 - 3x + 11 \ge 0$$

A solution of a quadratic inequality in one variable is a value of the variable that makes the inequality a true statement.

The value of an expression such as $x^2 - 3x - 10$ will sometimes be positive, sometimes negative, and sometimes 0, depending on the value substituted for x. To solve the inequality $x^2 - 3x - 10 < 0$, we look for all values of x that make the expression $x^2 - 3x - 10$ **less than 0**, or **negative**. To understand how we find these values, we'll study the graph of the quadratic function $y = x^2 - 3x - 10$.

OBJECTIVES

(A) Solve polynomial inequalities of degree 2 or greater.

(B) Solve inequalities that contain rational expressions with variables in the denominator.

SSM
TUTOR CENTER   SG   CD & VIDEO   MATH PRO   WEB

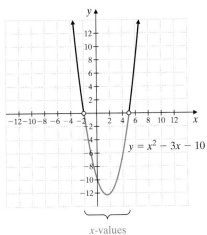

$y = x^2 - 3x - 10$

x-values
corresponding to *negative*
y-values

Notice that the x-values for which y or $x^2 - 3x - 10$ is positive are separated from the x values for which y or $x^2 - 3x - 10$ is negative by the values for which y or $x^2 - 3x - 10$ is 0, the x-intercepts. Thus, the solution set of $x^2 - 3x - 10 < 0$ consists of all real numbers from -2 to 5 or, in interval notation, $(-2, 5)$.

It is not necessary to graph $y = x^2 - 3x - 10$ to solve the related inequality $x^2 - 3x - 10 < 0$. Instead, we can draw a number line representing the x-axis and keep the following in mind: *A region on the number line for which the value of $x^2 - 3x - 10$ is positive is separated from a region on the number line for which the value of $x^2 - 3x - 10$ is negative by a value for which the expression is 0.*

Let's find these values for which the expression is 0 by solving the related equation, $x^2 - 3x - 10 = 0$.

$$x^2 - 3x - 10 = 0$$
$$(x - 5)(x + 2) = 0 \qquad \text{Factor.}$$

$$x - 5 = 0 \quad \text{or} \quad x + 2 = 0 \qquad \text{Set each factor equal to 0.}$$
$$x = 5 \qquad \qquad x = -2 \qquad \text{Solve.}$$

These two numbers -2 and 5 divide the number line into three regions. We will call the regions A, B, and C. These regions are important because if the value of $x^2 - 3x - 10$ is negative when a number from a region is substituted for x, then $x^2 - 3x - 10$ is negative when any number in that region is substituted for x. Similarly, if the value of $x^2 - 3x - 10$ is positive when a number from a region is substituted for x, then $x^2 - 3x - 10$ is positive when any number in that region is substituted for x.

To see whether the inequality $x^2 - 3x - 10 < 0$ is true or false in each region, we choose a test point from each region and substitute its value for x in the inequality $x^2 - 3x - 10 < 0$. If the resulting inequality is true, the region containing the test point is a solution region.

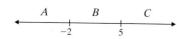

Region	Test Point Value	$(x - 5)(x + 2) < 0$	Result
A	-3	$(-8)(-1) < 0$	False.
B	0	$(-5)(2) < 0$	True.
C	6	$(1)(8) < 0$	False.

The values in region B satisfy the inequality. The numbers -2 and 5 are not included in the solution set since the inequality symbol is $<$. The solution set is $(-2, 5)$, and its graph is shown.

$$
\begin{array}{ccc}
A & B & C \\
F \quad -2 & T \quad 5 & F
\end{array}
$$

Practice Problem 1

Solve: $(x - 2)(x + 4) > 0$

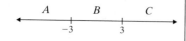

Answer

1. $(-\infty, -4) \cup (2, \infty)$

EXAMPLE 1   Solve: $(x + 3)(x - 3) > 0$

Solution: First we solve the related equation, $(x + 3)(x - 3) = 0$.

$$(x + 3)(x - 3) = 0$$
$$x + 3 = 0 \quad \text{or} \quad x - 3 = 0$$
$$x = -3 \qquad\qquad x = 3$$

The two numbers -3 and 3 separate the number line into three regions, A, B, and C.

Now we substitute the value of a test point from each region. If the test value satisfies the inequality, every value in the region containing the test value is a solution.

Region	Test Point Value	$(x + 3)(x - 3) > 0$	Result
A	-4	$(-1)(-7) > 0$	True.
B	0	$(3)(-3) > 0$	False.
C	4	$(7)(1) > 0$	True.

The points in regions A and C satisfy the inequality. The numbers -3 and 3 are not included in the solution since the inequality symbol is $>$. The solution set is $(-\infty, -3) \cup (3, \infty)$, and its graph is shown.

$$
\begin{array}{ccc}
A & B & C \\
T \quad -3 & F \quad 3 & T
\end{array}
$$

The steps below may be used to solve a polynomial inequality.

Solving a Polynomial Inequality

Step 1. Write the inequality in standard form and then solve the related equation.

Step 2. Separate the number line into regions with the solutions from Step 1.

Step 3. For each region, choose a test point and determine whether its value satisfies the *original inequality*.

Step 4. The solution set includes the regions whose test point value is a solution. If the inequality symbol is ≤ or ≥, the values from Step 1 are solutions; if < or >, they are not.

Try the Concept Check in the margin.

EXAMPLE 2   Solve: $x^2 - 4x \le 0$

Solution: First we solve the related equation, $x^2 - 4x = 0$.

$$x^2 - 4x = 0$$
$$x(x - 4) = 0$$
$$x = 0 \quad \text{or} \quad x = 4$$

The numbers 0 and 4 separate the number line into three regions, A, B and C.

$$
\begin{array}{ccc}
A & B & C \\
\hline
& 0 \qquad 4 &
\end{array}
$$

We check a test value in each region in the original inequality. Values in region B satisfy the inequality. The numbers 0 and 4 are included in the solution since the inequality symbol is ≤. The solution set is $[0, 4]$, and its graph is shown.

$$
\begin{array}{ccccc}
A & & B & & C \\
\hline
F & 0 & T & 4 & F
\end{array}
$$

EXAMPLE 3   Solve: $(x + 2)(x - 1)(x - 5) \le 0$

Solution: First we solve $(x + 2)(x - 1)(x - 5) = 0$. By inspection, we see that the solutions are $-2, 1$, and 5. They separate the number line into four regions, $A, B, C,$ and D. Next we check test points from each region.

Region	Test Point Value	$(x + 2)(x - 1)$ $(x - 5) \le 0$	Result
A	-3	$(-1)(-4)(-8) \le 0$	True.
B	0	$(2)(-1)(-5) \le 0$	False.
C	2	$(4)(1)(-3) \le 0$	True.
D	6	$(8)(5)(1) \le 0$	False.

The solution set is $(-\infty, -2] \cup [1, 5]$, and its graph is shown. We include the numbers $-2, 1$, and 5 because the inequality symbol is ≤.

$$
\begin{array}{cccc}
A & B & C & D \\
\hline
T \;\; -2 & F \;\; 1 & T \;\; 5 & F
\end{array}
$$

Concept Check

When choosing a test point in Step 4, why would the solutions from Step 2 not make good choices for test points?

Practice Problem 2

Solve: $x^2 - 6x \le 0$

Practice Problem 3

Solve: $(x - 2)(x + 1)(x + 5) \le 0$

$$
\begin{array}{cccc}
A & B & C & D \\
\hline
-2 & 1 & 5 &
\end{array}
$$

Answers

2. $[0, 6]$,   **3.** $(-\infty, -5] \cup [-1, 2]$

Concept Check: The solutions found in Step 2 have a value of 0 in the original inequality.

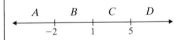

Practice Problem 4

Solve: $\dfrac{x - 3}{x + 5} \le 0$

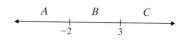

 Solving Rational Inequalities

Inequalities containing rational expressions with variables in the denominator are solved by using a similar procedure.

EXAMPLE 4   Solve: $\dfrac{x + 2}{x - 3} \le 0$

Solution: First we find all values that make the denominator equal to 0. To do this, we solve $x - 3 = 0$, or $x = 3$.

Next, we solve the related equation, $\dfrac{x + 2}{x - 3} = 0$.

$$\dfrac{x + 2}{x - 3} = 0 \qquad \text{Multiply both sides by the LCD, } x - 3.$$

$$x + 2 = 0$$

$$x = -2$$

Now we place these numbers on a number line and proceed as before, checking test point values in the original inequality.

Choose −3 from region A.	**Choose 0 from region B.**	**Choose 4 from region C.**
$\dfrac{x + 2}{x - 3} \le 0$	$\dfrac{x + 2}{x - 3} \le 0$	$\dfrac{x + 2}{x - 3} \le 0$
$\dfrac{-3 + 2}{-3 - 3} \le 0$	$\dfrac{0 + 2}{0 - 3} \le 0$	$\dfrac{4 + 2}{4 - 3} \le 0$
$\dfrac{-1}{-6} \le 0$	$-\dfrac{2}{3} \le 0$ True.	$6 \le 0$ False.
$\dfrac{1}{6} \le 0$ False.		

The solution set is $[-2, 3)$. This interval includes -2 because -2 satisfies the original inequality. This interval does not include 3 because 3 would make the denominator 0.

The steps below may be used to solve a rational inequality with variables in the denominator.

Solving a Rational Inequality

Step 1. Solve for values that make all denominators 0.

Step 2. Solve the related equation.

Step 3. Separate the number line into regions with the solutions from Steps 1 and 2.

Step 4. For each region, choose a test point and determine whether its value satisfies the *original inequality.*

Step 5. The solution set includes the regions whose test point value is a solution. Check whether to include values from Step 2. Be sure *not* to include values that make any denominator 0.

Answer

4. $(-5, 3]$

EXAMPLE 5   Solve: $\dfrac{5}{x + 1} < -2$

Solution: First we find values for x that make the denominator equal to 0.

$$x + 1 = 0$$
$$x = -1$$

Next we solve $\dfrac{5}{x + 1} = -2$.

$$(x + 1) \cdot \frac{5}{x + 1} = (x + 1) \cdot -2 \qquad \text{Multiply both sides by the LCD, } x + 1.$$
$$5 = -2x - 2 \qquad \text{Simplify.}$$
$$7 = -2x$$
$$-\frac{7}{2} = x$$

We use these two solutions to divide a number line into three regions and choose test points. Only a test point value from region B satisfies the *original inequality*. The solution set is $\left(-\dfrac{7}{2}, -1 \right)$, and its graph is shown.

Practice Problem 5

Solve: $\dfrac{3}{x - 2} < 2$

Answer

5. $(-\infty, 2) \cup \left(\dfrac{7}{2}, \infty \right)$

FOCUS ON **History**

THE EVOLUTION OF SOLVING QUADRATIC EQUATIONS

The ancient Babylonians (circa 2000 B.C.) are sometimes credited with being the first to solve quadratic equations. This is only partially true because the Babylonians had no concept of an equation. However, what they did develop was a method for completing the square to apply to problems that today would be solved with a quadratic equation. The Babylonians only recognized positive solutions to such problems and did not acknowledge the existence of negative solutions at all.

Babylonian mathematical knowledge influenced much of the ancient world, most notably Hindu Indians. The Hindus were the first culture to denote debts in everyday business affairs with negative numbers. With this level of comfort with negative numbers, the Indian mathematician Brahmagupta (598–665 A.D.) extended the Babylonian methods and was the first to recognize negative solutions to quadratic equations. Later Hindu mathematicians noted that every positive number has two square roots: a positive square root and a negative square root. Hindus allowed irrational solutions (quite an innovation in the ancient world!) to quadratic equations and were the first to realize that quadratic equations could have 0, 1, or 2 real number solutions. They did not, however, acknowledge complex numbers and, therefore, could not solve equations with solutions requiring the square root of a negative number.

Complex numbers were finally developed by European mathematicians during the 17th and 18th centuries. Up until that time, what we would today consider complex solutions to quadratic equations were routinely ignored by mathematicians.

EXERCISE SET 8.4

(A) *Solve. See Examples 1 through 3.*

1. $(x + 1)(x + 5) > 0$

2. $(x + 1)(x + 5) \le 0$

3. $(x - 3)(x + 4) \le 0$

4. $(x + 4)(x - 1) > 0$

5. $x^2 - 7x + 10 \le 0$

6. $x^2 + 8x + 15 \ge 0$

7. $3x^2 + 16x < -5$

8. $2x^2 - 5x < 7$

9. $(x - 6)(x - 4)(x - 2) > 0$

10. $(x - 6)(x - 4)(x - 2) \le 0$

11. $x(x - 1)(x + 4) \le 0$

12. $x(x - 6)(x + 2) > 0$

13. $(x^2 - 9)(x^2 - 4) > 0$

14. $(x^2 - 16)(x^2 - 1) \le 0$ **15.** $x^2 - x - 56 > 0$

16. $x^2 - 4x - 5 < 0$

17. $6x^2 + 5x \le 4$

18. $12x^2 - 5x \ge 3$

19. $x^2 > x$

20. $x^2 < 25$

21. $(2x - 8)(x + 4)(x - 6) \le 0$

22. $(3x - 12)(x + 5)(2x - 3) \ge 0$

(B) *Solve. See Examples 4 and 5.*

23. $\dfrac{x + 7}{x - 2} < 0$

24. $\dfrac{x - 5}{x - 6} > 0$

25. $\dfrac{5}{x + 1} > 0$

26. $\dfrac{3}{y - 5} < 0$

27. $\dfrac{x + 1}{x - 4} \ge 0$

28. $\dfrac{x + 1}{x - 4} \le 0$

29. $\dfrac{x + 2}{x - 3} < 1$

30. $\dfrac{x - 1}{x + 4} > 2$

31. $\dfrac{x}{x - 10} < 0$

32. $\dfrac{x + 10}{x - 10} > 0$

33. $\dfrac{x - 5}{x + 4} \ge 0$

34. $\dfrac{x - 3}{x + 2} \le 0$

35. $\dfrac{x(x + 6)}{(x - 7)(x + 1)} \ge 0$

36. $\dfrac{(x - 2)(x + 2)}{(x + 1)(x - 4)} \le 0$

37. $\dfrac{-1}{x - 1} > -1$

38. $\dfrac{4}{y + 2} < -2$

39. $\dfrac{x}{x + 4} \le 2$

40. $\dfrac{4x}{x - 3} \ge 5$

Review and Preview

Fill in each table so that each ordered pair is a solution of the given function. See Section 3.5.

41. $f(x) = x^2$

x	y
0	
1	
−1	
2	
−2	

42. $f(x) = 2x^2$

x	y
0	
1	
−1	
2	
−2	

43. $f(x) = -x^2$

x	y
0	
1	
−1	
2	
−2	

44. $f(x) = -3x^2$

x	y
0	
1	
−1	
2	
−2	

Combining Concepts

45. Explain why $\dfrac{x + 2}{x - 3} > 0$ and $(x + 2)(x - 3) > 0$ have the same solutions.

46. Explain why $\dfrac{x + 2}{x - 3} \geq 0$ and $(x + 2)(x - 3) \geq 0$ do not have the same solutions.

Find all numbers that satisfy each statement.

47. A number minus its reciprocal is less than zero. Find the numbers.

48. Twice a number added to its reciprocal is nonnegative. Find the numbers.

49. The total profit $P(x)$ for a company producing x thousand units is given by the function $P(x) = -2x^2 + 26x - 44$. Find the values of x for which the company makes a profit. [*Hint:* The company makes a profit when $P(x) > 0$.]

50. A projectile is fired straight up from the ground with an initial velocity of 80 feet per second. Its height $s(t)$ in feet at any time t in seconds is given by the function $s(t) = -16t^2 + 80t$. Find the interval of time for which the height of the projectile is greater than 96 feet.

Use a graphing calculator to check each exercise.

51. Exercise 15

52. Exercise 16

8.5 Quaratic Functions and Their Graphs

Ⓐ Graphing $f(x) = x^2 + k$

We first graphed the quadratic function $f(x) = x^2$ in Section 5.8. In that section, we discovered that the graph of a quadratic function is a parabola opening upward or downward. Now, as we continue our study, we will discover more details about quadratic functions and their graphs.

First, let's recall the definition of a *quadratic function*.

Quadratic Function

A **quadratic function** is a function that can be written in the form $f(x) = ax^2 + bx + c$, where $a, b,$ and c are real numbers and $a \neq 0$.

Notice that equations of the form $y = ax^2 + bx + c$, where $a \neq 0$, also define quadratic functions since y is a function of x or $y = f(x)$.

Recall that if $a > 0$, the parabola opens upward and if $a < 0$, the parabola opens downward. Also, the vertex of a parabola is the lowest point if the parabola opens upward and the highest point if the parabola opens downward. The axis of symmetry is the vertical line that passes through the vertex.

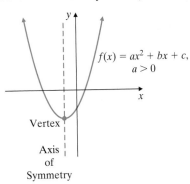

EXAMPLE 1

Graph $f(x) = x^2$ and $g(x) = x^2 + 3$ on the same set of axes.

Solution: First we construct a table of values for f and plot the points. Notice that for each x-value, the corresponding value of $g(x)$ must be 3 more than the corresponding value of $f(x)$ since $f(x) = x^2$ and $g(x) = x^2 + 3$. In other words, the graph of $g(x) = x^2 + 3$ is the same as the graph of $f(x) = x^2$ shifted upward 3 units. The axis of symmetry for both graphs is the y-axis.

x	$f(x) = x^2$	$g(x) = x^2 + 3$
-2	4	7
-1	1	4
0	0	3
1	1	4
2	4	7

Each y-value is increased by 3.

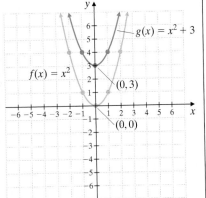

Practice Problem 1

Graph $f(x) = x^2$ and $g(x) = x^2 + 4$ on the same set of axes.

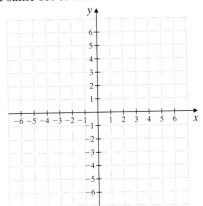

Answer

1.

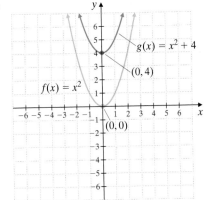

Practice Problems 2–3

Graph each function.

2. $F(x) = x^2 + 1$

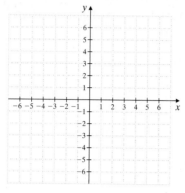

3. $g(x) = x^2 - 2$

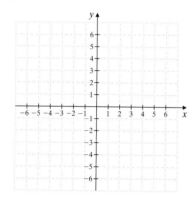

Answers

2.

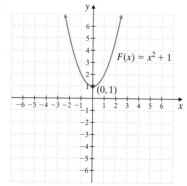

3.

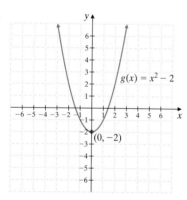

In general, we have the following properties.

> **Graphing the Parabola Defined by $f(x) = x^2 + k$**
>
> If k is positive, the graph of $f(x) = x^2 + k$ is the graph of $y = x^2$ shifted upward k units.
>
> If k is negative, the graph of $f(x) = x^2 + k$ is the graph of $y = x^2$ shifted downward $|k|$ units.
>
> The vertex is $(0, k)$, and the axis of symmetry is the y-axis.

EXAMPLES Graph each function.

2. $F(x) = x^2 + 2$

The graph of $F(x) = x^2 + 2$ is obtained by shifting the graph of $y = x^2$ upward 2 units.

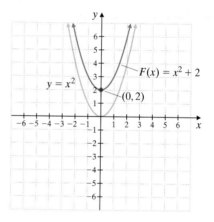

3. $g(x) = x^2 - 3$

The graph of $g(x) = x^2 - 3$ is obtained by shifting the graph of $y = x^2$ downward 3 units.

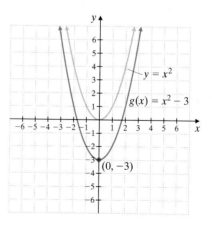

B **Graphing $f(x) = (x - h)^2$**

Now we will graph functions of the form $f(x) = (x - h)^2$.

EXAMPLE 4

Graph $f(x) = x^2$ and $g(x) = (x - 2)^2$ on the same set of axes.

Solution: By plotting points, we see that for each x-value, the corresponding value of $g(x)$ is the same as the value of $f(x)$ when the x-value is increased by 2. Thus, the graph of $g(x) = (x - 2)^2$ is the graph of $f(x) = x^2$ shifted to the right 2 units. The axis of symmetry for the graph of $g(x) = (x - 2)^2$ is also shifted 2 units to the right and is the line $x = 2$.

x	f(x) = x²	x	g(x) = (x − 2)²
−2	4	0	4
−1	1	1	1
0	0	2	0
1	1	3	1
2	4	4	4

Each x-value increased by 2 corresponds to same y-value.

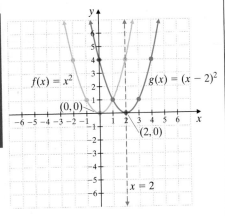

In general, we have the following properties.

Graphing the Parabola Defined by $f(x) = (x - h)^2$

If h is positive, the graph of $f(x) = (x - h)^2$ is the graph of $y = x^2$ shifted to the right h units.

If h is negative, the graph of $f(x) = (x - h)^2$ is the graph of $y = x^2$ shifted to the left $|h|$ units.

The vertex is $(h, 0)$, and the axis of symmetry is the vertical line $x = h$.

EXAMPLES   Graph each function.

5. $G(x) = (x - 3)^2$

The graph of $G(x) = (x - 3)^2$ is obtained by shifting the graph of $y = x^2$ to the right 3 units.

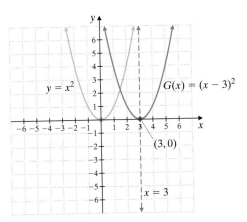

Practice Problem 4

Graph $f(x) = x^2$ and $g(x) = (x - 1)^2$ on the same set of axes.

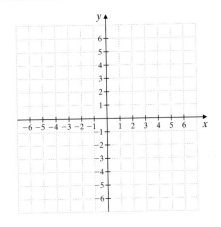

Answer

4.

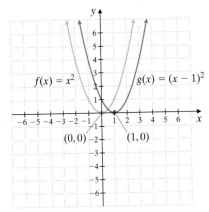

Practice Problems 5–6

Graph each function.

5. $G(x) = (x - 4)^2$

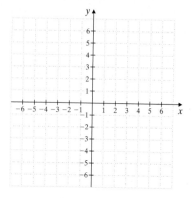

6. $F(x) = (x + 2)^2$

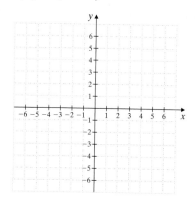

Answers

5.

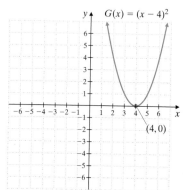

6.

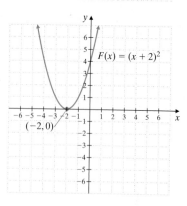

6. $F(x) = (x + 1)^2$

The equation $F(x) = (x + 1)^2$ can be written as $F(x) = [x - (-1)]^2$. The graph of $F(x) = [x - (-1)]^2$ is obtained by shifting the graph of $y = x^2$ to the left 1 unit.

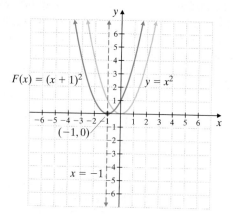

C **Graphing $f(x) = (x - h)^2 + k$**

As we will see in graphing functions of the form $f(x) = (x - h)^2 + k$, it is possible to combine vertical and horizontal shifts.

Graphing the Parabola Defined by $f(x) = (x - h)^2 + k$

The parabola has the same shape as $y = x^2$.
The vertex is (h, k), and the axis of symmetry is the vertical line $x = h$.

EXAMPLE 7 Graph: $F(x) = (x - 3)^2 + 1$

Solution: The graph of $F(x) = (x - 3)^2 + 1$ is the graph of $y = x^2$ shifted 3 units to the right and 1 unit up. The vertex is then $(3, 1)$, and the axis of symmetry is $x = 3$. A few ordered pair solutions are plotted to aid in graphing.

x	$F(x) = (x - 3)^2 + 1$
1	5
2	2
4	2
5	5

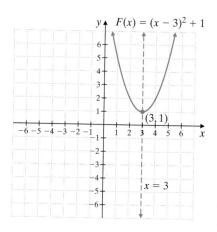

D **Graphing $f(x) = ax^2$**

Next, we discover the change in the shape of the graph when the coefficient of x^2 is not 1.

EXAMPLE 8

Graph $f(x) = x^2$, $g(x) = 3x^2$, and $h(x) = \dfrac{1}{2}x^2$ on the same set of axes.

Solution: Comparing the table of values, we see that for each x-value, the corresponding value of $g(x)$ is triple the corresponding value of $f(x)$. Similarly, the value of $h(x)$ is half the value of $f(x)$.

x	$f(x) = x^2$
-2	4
-1	1
0	0
1	1
2	4

x	$g(x) = 3x^2$
-2	12
-1	3
0	0
1	3
2	12

x	$h(x) = \dfrac{1}{2}x^2$
-2	2
-1	$\dfrac{1}{2}$
0	0
1	$\dfrac{1}{2}$
2	2

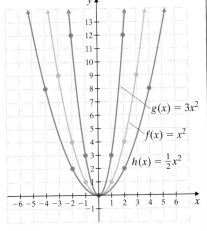

The result is that the graph of $g(x) = 3x^2$ is narrower than the graph of $f(x) = x^2$ and the graph of $h(x) = \dfrac{1}{2}x^2$ is wider. The vertex for each graph is $(0, 0)$, and the axis of symmetry is the y-axis.

> **Graphing the Parabola Defined by $f(x) = ax^2$**
>
> If a is positive, the parabola opens upward, and if a is negative, the parabola opens downward.
> If $|a| > 1$, the graph of the parabola is narrower than the graph of $y = x^2$.
> If $|a| < 1$, the graph of the parabola is wider than the graph of $y = x^2$.

EXAMPLE 9   Graph: $f(x) = -2x^2$

Solution: Because $a = -2$, a negative value, this parabola opens downward. Since $|-2| = 2$ and $2 > 1$, the parabola is narrower than the graph of $y = x^2$. The vertex is $(0, 0)$, and the axis of symmetry is the y-axis. We verify this by plotting a few points.

Practice Problem 7

Graph: $F(x) = (x - 2)^2 + 3$

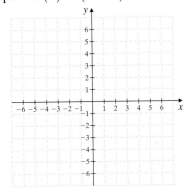

Practice Problem 8

Graph $f(x) = x^2$, $g(x) = 2x^2$, and $h(x) = \dfrac{1}{3}x^2$ on the same set of axes.

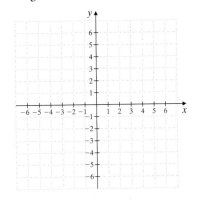

Answers

7.

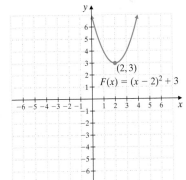

8.

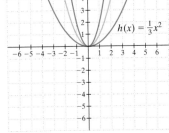

Practice Problem 9

Graph: $f(x) = -3x^2$

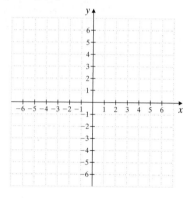

x	$f(x) = -2x^2$
-2	-8
-1	-2
0	0
1	-2
2	-8

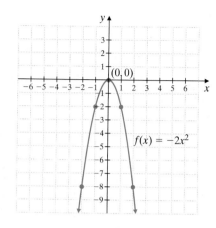

Practice Problem 10

Graph: $f(x) = 2(x + 3)^2 - 4$. Find the vertex and axis of symmetry.

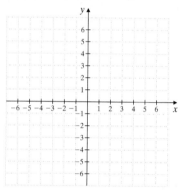

Answers

9.

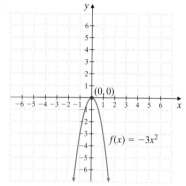

10.

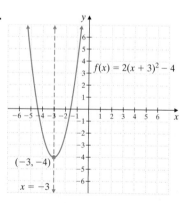

E **Graphing $f(x) = a(x - h)^2 + k$**

Now we will see the shape of the graph of a quadratic function of the form $f(x) = a(x - h)^2 + k$.

EXAMPLE 10

Graph: $g(x) = \dfrac{1}{2}(x + 2)^2 + 5$. Find the vertex and the axis of symmetry.

Solution: The function $g(x) = \dfrac{1}{2}(x + 2)^2 + 5$ may be written as $g(x) = \dfrac{1}{2}[x - (-2)]^2 + 5$. Thus, this graph is the same as the graph of $y = x^2$ shifted 2 units to the left and 5 units upward and widened because a is $\dfrac{1}{2}$. The vertex is $(-2, 5)$, and the axis of symmetry is $x = -2$. We plot a few points to verify.

x	$g(x) = \dfrac{1}{2}(x + 2)^2 + 5$
-4	7
-3	$5\dfrac{1}{2}$
-2	5
-1	$5\dfrac{1}{2}$
0	7

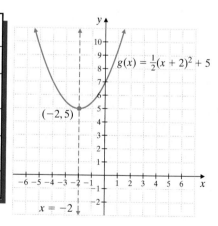

In general, the following holds.

Graphing a Quadratic Function

The graph of a quadratic function written in the form $f(x) = a(x - h)^2 + k$ is a parabola with vertex (h, k).
If $a > 0$, the parabola opens upward.
If $a < 0$, the parabola opens downward. The axis of symmetry is the line whose equation is $x = h$.

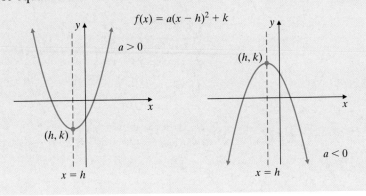

Try the Concept Check in the margin.

GRAPHING CALCULATOR EXPLORATIONS

Use a graphing calculator to graph the first function of each pair. Then use its graph to predict the graph of the second function. Check your prediction by graphing both on the same set of axes.

1. $F(x) = \sqrt{x}; G(x) = \sqrt{x} + 1$

2. $g(x) = x^3; H(x) = x^3 - 2$

3. $H(x) = |x|; f(x) = |x - 5|$

4. $h(x) = x^3 + 2; g(x) = (x - 3)^3 + 2$

5. $f(x) = |x + 4|; F(x) = |x + 4| + 3$

6. $G(x) = \sqrt{x} - 2; g(x) = \sqrt{x - 4} - 2$

Mental Math

State the vertex of the graph of each quadratic function.

1. $f(x) = x^2$

2. $f(x) = -5x^2$

3. $g(x) = (x - 2)^2$

4. $g(x) = (x + 5)^2$

5. $f(x) = 2x^2 + 3$

6. $h(x) = x^2 - 1$

7. $g(x) = (x + 1)^2 + 5$

8. $h(x) = (x - 10)^2 - 7$

EXERCISE SET 8.5

Ⓐ Ⓑ *Graph each quadratic function. Label the vertex and sketch and label the axis of symmetry. See Examples 1 through 6.*

1. $f(x) = x^2 - 1$

2. $g(x) = x^2 + 3$

3. $f(x) = (x - 5)^2$

4. $g(x) = (x + 5)^2$

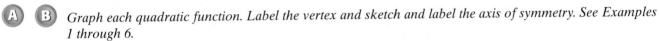

5. $h(x) = x^2 + 5$

6. $h(x) = x^2 - 4$

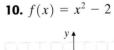 **7.** $h(x) = (x + 2)^2$

8. $H(x) = (x - 1)^2$

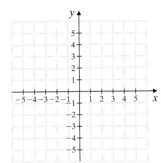
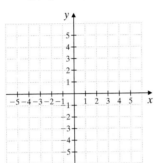
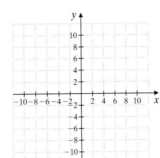
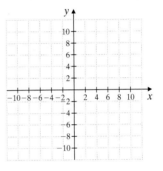

9. $g(x) = x^2 + 7$

10. $f(x) = x^2 - 2$

11. $G(x) = (x + 3)^2$

12. $f(x) = (x - 6)^2$

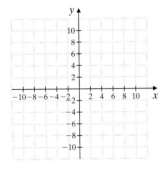
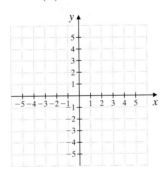
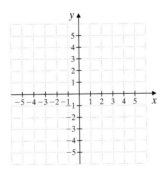
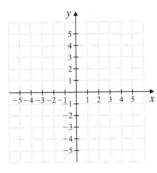

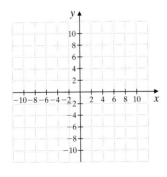
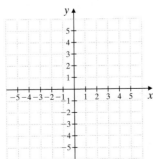
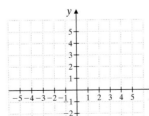
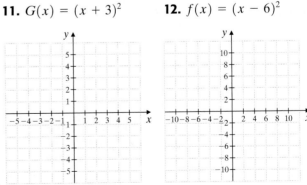

Graph each quadratic function. Label the vertex and sketch and label the axis of symmetry. See Example 7.

13. $f(x) = (x - 2)^2 + 5$

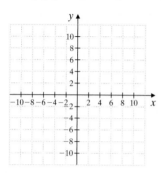

14. $g(x) = (x - 6)^2 + 1$

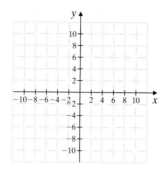

15. $h(x) = (x + 1)^2 + 4$

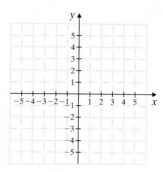

16. $G(x) = (x + 3)^2 + 3$

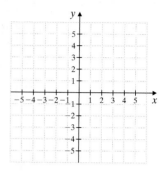

17. $g(x) = (x + 2)^2 - 5$

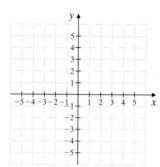

18. $h(x) = (x + 4)^2 - 6$

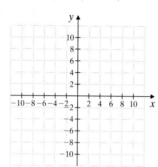

19. $h(x) = (x - 3)^2 + 2$

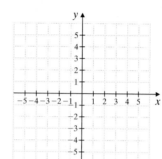

20. $F(x) = (x - 2)^2 - 3$

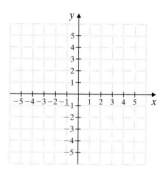

D *Graph each quadratic function. Label the vertex and sketch and label the axis of symmetry. See Examples 8 and 9.*

21. $g(x) = -x^2$

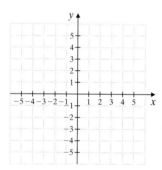

22. $f(x) = 5x^2$

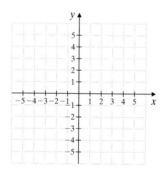

23. $h(x) = \frac{1}{3}x^2$

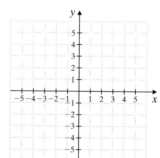

24. $g(x) = -3x^2$

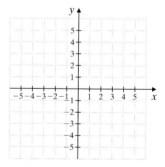

25. $H(x) = 2x^2$

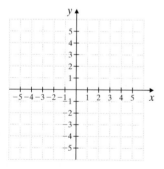

26. $f(x) = -\frac{1}{4}x^2$

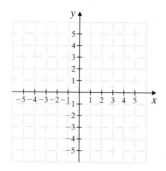

27. $F(x) = -4x^2$

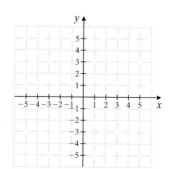

28. $G(x) = \frac{1}{5}x^2$

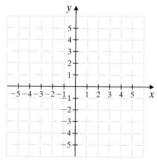

E Graph each quadratic function. Label the vertex and sketch and label the axis of symmetry. See Example 10.

29. $f(x) = 10(x + 4)^2 - 6$

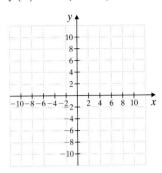

30. $g(x) = 4(x - 4)^2 + 2$

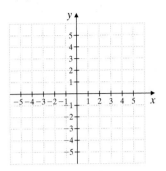

31. $h(x) = -3(x + 3)^2 + 1$

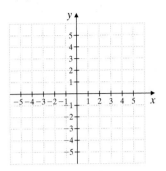

32. $f(x) = -(x - 2)^2 - 6$

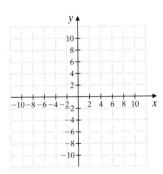

33. $H(x) = \frac{1}{2}(x - 6)^2 - 3$

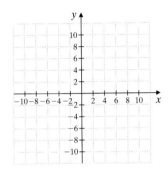

34. $G(x) = \frac{1}{5}(x + 4)^2 + 3$

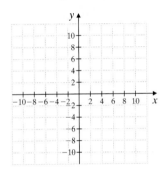

35. $f(x) = -(x - 1)^2$

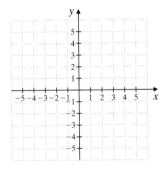

36. $f(x) = 2(x + 3)^2$

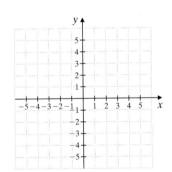

37. $F(x) = \left(x + \frac{1}{2}\right)^2 - 2$

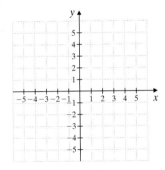

38. $H(x) = \left(x + \frac{1}{4}\right)^2 - 3$

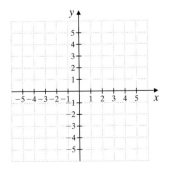

39. $F(x) = -x^2 + 2$

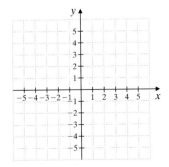

40. $G(x) = 3x^2 + 1$

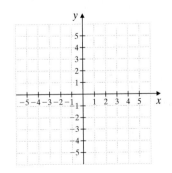

Review and Preview

Add the proper constant to each binomial so that the resulting trinomial is a perfect square trinomial. See Section 8.1.

41. $x^2 + 8x$ **42.** $y^2 + 4y$ **43.** $z^2 - 16z$ **44.** $x^2 - 10x$ **45.** $y^2 + y$ **46.** $z^2 - 3z$

 Combining Concepts

Write the equation of the parabola that has the same shape as $f(x) = 5x^2$ but with each given vertex.

47. $(2, 3)$ **48.** $(1, 6)$ **49.** $(-3, 6)$ **50.** $(4, -1)$

The shifting properties covered in this section apply to the graphs of all functions. Given the accompanying graph of $y = f(x)$, graph each function.

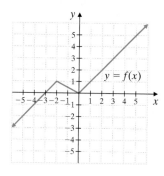

51. $y = f(x) + 1$ **52.** $y = f(x) - 2$ **53.** $y = f(x - 3)$

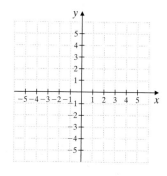   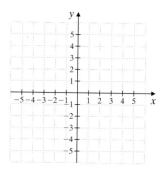   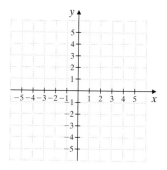

54. $y = f(x + 3)$ **55.** $y = f(x + 2) + 2$ **56.** $y = f(x - 1) + 1$

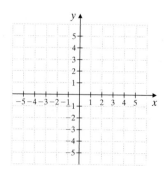   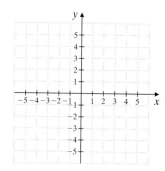   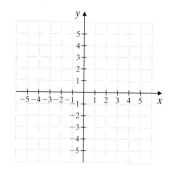

8.6 Further Graphing of Quadratic Functions

(A) Writing Quadratic Functions in the Form $y = a(x - h)^2 + k$

We know that the graph of a quadratic function is a parabola. If a quadratic function is written in the form

$$f(x) = a(x - h)^2 + k$$

we can easily find the vertex (h, k) and graph the parabola. To write a quadratic function in this form, we need to complete the square. (See Section 8.1 for a review of completing the square.)

EXAMPLE 1

Graph: $f(x) = x^2 - 4x - 12$. Find the vertex and any intercepts.

Solution: The graph of this quadratic function is a parabola. To find the vertex of the parabola, we complete the square on the binomial $x^2 - 4x$. To simplify our work, we let $f(x) = y$.

$$y = x^2 - 4x - 12 \quad \text{Let } f(x) = y.$$
$$y + 12 = x^2 - 4x \qquad \text{Add 12 to both sides to get the } x\text{-variable terms alone.}$$

Now we add the square of half of -4 to both sides.

$$\frac{1}{2}(-4) = -2 \quad \text{and} \quad (-2)^2 = 4$$

$$y + 12 + 4 = x^2 - 4x + 4 \qquad \text{Add 4 to both sides.}$$
$$y + 16 = (x - 2)^2 \qquad \text{Factor the trinomial.}$$
$$y = (x - 2)^2 - 16 \qquad \text{Subtract 16 from both sides.}$$
$$f(x) = (x - 2)^2 - 16 \qquad \text{Replace } y \text{ with } f(x).$$

From this equation, we can see that the vertex of the parabola is $(2, -16)$, a point in quadrant IV, and the axis of symmetry is the line $x = 2$.

Notice that $a = 1$. Since $a > 0$, the parabola opens upward. This parabola opening upward with vertex $(2, -16)$ will have two x-intercepts.

To find the x-intercepts, we let $f(x)$ or $y = 0$.

$$0 = x^2 - 4x - 12$$
$$0 = (x - 6)(x + 2)$$
$$0 = x - 6 \quad \text{or} \quad 0 = x + 2$$
$$6 = x \qquad \qquad -2 = x$$

The two x-intercepts are $(6, 0)$ and $(-2, 0)$. To find the y-intercept, we let $x = 0$.

$$f(0) = 0^2 - 4 \cdot 0 - 12 = -12$$

The y-intercept is $(0, -12)$. The sketch of $f(x) = x^2 - 4x - 12$ is shown.

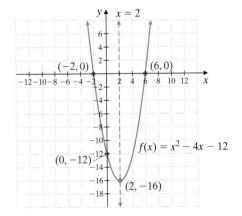

OBJECTIVES

(A) Write quadratic functions in the form $y = a(x - h)^2 + k$.

(B) Derive a formula for finding the vertex of a parabola.

(C) Find the minimum or maximum value of a quadratic function.

SSM
TUTOR CENTER   SG   CD & VIDEO   MATH PRO   WEB

Practice Problem 1

Graph: $f(x) = x^2 - 4x - 5$. Find the vertex and any intercepts.

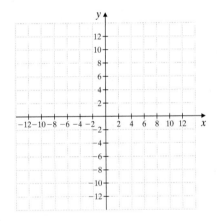

Answer

1. vertex: $(2, -9)$; x-intercepts: $(-1, 0)$, $(5, 0)$; y-intercept: $(0, -5)$

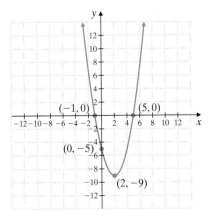

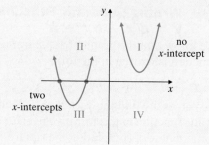

Helpful Hint

Parabola Opens Upward
Vertex in I or II: no x-intercepts
Vertex in III or IV: 2 x-intercepts

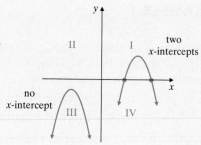

Parabola Opens Downward
Vertex in I or II: 2 x-intercepts
Vertex in III or IV: no x-intercepts

Practice Problem 2

Graph: $f(x) = 2x^2 + 2x + 5$. Find the vertex and any intercepts.

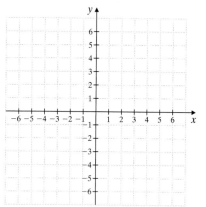

Answer

2. vertex: $\left(-\dfrac{1}{2}, \dfrac{9}{2}\right)$; y-intercept: $(0, 5)$

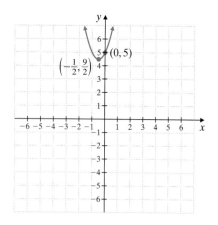

EXAMPLE 2

Graph: $f(x) = 3x^2 + 3x + 1$. Find the vertex and any intercepts.

Solution: We replace $f(x)$ with y and complete the square on x to write the equation in the form $y = a(x - h)^2 + k$.

$$y = 3x^2 + 3x + 1 \qquad \text{Replace } f(x) \text{ with } y.$$

$$y - 1 = 3x^2 + 3x \qquad \text{Get the } x\text{-variable terms alone.}$$

Next we factor 3 from the terms $3x^2 + 3x$ so that the coefficient of x^2 is 1.

$$y - 1 = 3(x^2 + x) \qquad \text{Factor out 3.}$$

The coefficient of x is 1. Then $\dfrac{1}{2}(1) = \dfrac{1}{2}$ and $\left(\dfrac{1}{2}\right)^2 = \dfrac{1}{4}$. Since we are adding $\dfrac{1}{4}$ inside the parentheses, we are really adding $3\left(\dfrac{1}{4}\right)$, so we *must* add $3\left(\dfrac{1}{4}\right)$ to the left side.

$$y - 1 + 3\left(\frac{1}{4}\right) = 3\left(x^2 + x + \frac{1}{4}\right)$$

$$y - \frac{1}{4} = 3\left(x + \frac{1}{2}\right)^2 \qquad \begin{array}{l}\text{Simplify the left side and factor} \\ \text{the right side.}\end{array}$$

$$y = 3\left(x + \frac{1}{2}\right)^2 + \frac{1}{4} \qquad \text{Add } \frac{1}{4} \text{ to both sides.}$$

$$f(x) = 3\left(x + \frac{1}{2}\right)^2 + \frac{1}{4} \qquad \text{Replace } y \text{ with } f(x).$$

Then $a = 3, h = -\dfrac{1}{2}$, and $k = \dfrac{1}{4}$. This means that the parabola opens upward with vertex $\left(-\dfrac{1}{2}, \dfrac{1}{4}\right)$ and that the axis of symmetry is the line $x = -\dfrac{1}{2}$.

To find the y-intercept, we let $x = 0$. Then

$$f(0) = 3(0)^2 + 3(0) + 1 = 1$$

This parabola has no x-intercepts since the vertex is in the second quadrant and it opens upward. We use the vertex, axis of symmetry, and y-intercept to graph the parabola.

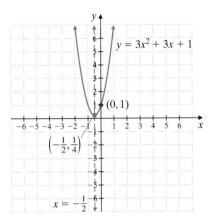

EXAMPLE 3

Graph: $f(x) = -x^2 - 2x + 3$. Find the vertex and any intercepts.

Solution: We write $f(x)$ in the form $a(x - h)^2 + k$ by completing the square. First we replace $f(x)$ with y.

$$f(x) = -x^2 - 2x + 3$$

$$y = -x^2 - 2x + 3$$

$$y - 3 = -x^2 - 2x \qquad \text{Subtract 3 from both sides to get the } x\text{-variable terms alone.}$$

$$y - 3 = -1(x^2 + 2x) \qquad \text{Factor } -1 \text{ from the terms } -x^2 - 2x.$$

The coefficient of x is 2. Then $\frac{1}{2}(2) = 1$ and $1^2 = 1$. We add 1 to the right side inside the parentheses and add $-1(1)$ to the left side.

$$y - 3 - 1(1) = -1(x^2 + 2x + 1)$$

$$y - 4 = -1(x + 1)^2 \qquad \text{Simplify the left side and factor the right side.}$$

$$y = -1(x + 1)^2 + 4 \qquad \text{Add 4 to both sides.}$$

$$f(x) = -1(x + 1)^2 + 4 \qquad \text{Replace } y \text{ with } f(x).$$

Since $a = -1$, the parabola opens downward with vertex $(-1, 4)$ and axis of symmetry $x = -1$.

To find the y-intercept, we let $x = 0$ and solve for y. Then

$$f(0) = -0^2 - 2(0) + 3 = 3$$

Thus, $(0, 3)$ is the y-intercept.

To find the x-intercepts, we let y or $f(x) = 0$ and solve for x.

$$f(x) = -x^2 - 2x + 3$$

$$0 = -x^2 - 2x + 3 \qquad \text{Let } f(x) = 0.$$

Practice Problem 3

Graph: $f(x) = -x^2 - 2x + 8$. Find the vertex and any intercepts.

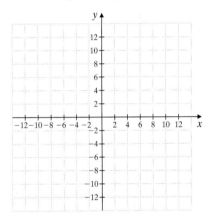

Helpful Hint

This can be written as
$f(x) = -1[x - (-1)]^2 + 4$.
Notice that the vertex is $(-1, 4)$.

Answer

3. vertex: $(-1, 9)$; x-intercepts: $(-4, 0)$, $(2, 0)$; y-intercept: $(0, 8)$

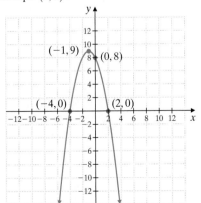

Now we divide both sides by -1 so that the coefficient of x^2 is 1.

$$\frac{0}{-1} = \frac{-x^2}{-1} - \frac{2x}{-1} + \frac{3}{-1}$$ Divide both sides by -1.

$$0 = x^2 + 2x - 3$$ Simplify.

$$0 = (x + 3)(x - 1)$$ Factor.

$$x + 3 = 0 \quad \text{or} \quad x - 1 = 0$$ Set each factor equal to 0.

$$x = -3 \qquad\qquad x = 1$$ Solve.

The x-intercepts are $(-3, 0)$ and $(1, 0)$. We use these points to graph the parabola.

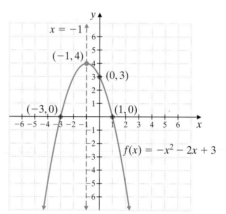

B Deriving a Formula for Finding the Vertex

Recall from Section 5.8 that we introduced a formula for finding the vertex of a parabola. Now that we have practiced completing the square, we will show that the x-coordinate of the vertex of the graph of $f(x)$ or $y = ax^2 + bx + c$ can be found by the formula $x = \dfrac{-b}{2a}$. To do so, we complete the square on x and write the equation in the form $y = (x - h)^2 + k$.

First we get the x-variable terms alone by subtracting c from both sides.

$$y = ax^2 + bx + c$$

$$y - c = ax^2 + bx$$

$$y - c = a\left(x^2 + \frac{b}{a}x\right)$$ Factor a from the terms $ax^2 + bx$.

Now we add the square of half of $\dfrac{b}{a}$, or $\left(\dfrac{b}{2a}\right)^2 = \dfrac{b^2}{4a^2}$, to the right side inside the parentheses. Because of the factor a, what we really added is $a\left(\dfrac{b^2}{4a^2}\right)$ and this must be added to the left side as well.

$$y - c + a\left(\frac{b^2}{4a^2}\right) = a\left(x^2 + \frac{b}{a}x + \frac{b^2}{4a^2}\right)$$

$$y - c + \frac{b^2}{4a} = a\left(x + \frac{b}{2a}\right)^2$$ Simplify the left side and factor the right side. Add c to both sides and subtract $\dfrac{b^2}{4a}$ from both sides.

$$y = a\left(x + \frac{b}{2a}\right)^2 + c - \frac{b^2}{4a}$$

Compare this form with $f(x)$ or $y = a(x - h)^2 + k$ and see that h is $\dfrac{-b}{2a}$, which means that the x-coordinate of the vertex of the graph of $f(x) = ax^2 + bx + c$ is $\dfrac{-b}{2a}$.

Let's use the vertex formula below to find the vertex of the parabola we graphed in Example 1.

Vertex Formula

The graph of $f(x) = ax^2 + bx + c$, when $a \neq 0$, is a parabola with vertex

$$\left(\frac{-b}{2a}, f\left(\frac{-b}{2a} \right) \right)$$

EXAMPLE 4   Find the vertex of the graph of $f(x) = x^2 - 4x - 12$.

Solution: In the quadratic function $f(x) = x^2 - 4x - 12$, notice that $a = 1, b = -4$, and $c = -12$.

$$\frac{-b}{2a} = \frac{-(-4)}{2(1)} = 2$$

The x-value of the vertex is 2. To find the corresponding $f(x)$ or y-value, find $f(2)$. Then

$$f(2) = 2^2 - 4(2) - 12 = 4 - 8 - 12 = -16$$

The vertex is $(2, -16)$. These results agree with our findings in Example 1.●

C Finding Minimum and Maximum Values

The quadratic function whose graph is a parabola that opens upward has a minimum value, and the quadratic function whose graph is a parabola that opens downward has a maximum value. The $f(x)$- or y-value of the vertex is the minimum or maximum value of the function.

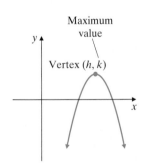
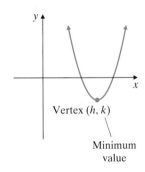

Recall from Section 8.2 that the discriminant, $b^2 - 4ac$, tells us how many solutions the quadratic equation $0 = ax^2 + bx + c$ has. It also tells us how many x-intercepts the graph of a quadratic equation $y = ax^2 + bx + c$ has.

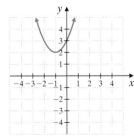

$$y = x^2 + 2x + 3$$
$$b^2 - 4ac < 0$$
No x-intercepts

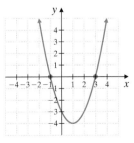

$$y = x^2 - 2x + 1$$
$$b^2 - 4ac = 0$$
One x-intercept

$$y = x^2 - 2x - 3$$
$$b^2 - 4ac > 0$$
Two x-intercepts

Practice Problem 4

Find the vertex of the graph of $f(x) = x^2 - 4x - 5$. Compare your result with the result of Practice Problem 1.

Answer

4. $(2, -9)$

Concept Check

Without making any calculations, tell whether the graph of $f(x) = 7 - x - 0.3x^2$ has a maximum value or a minimum value. Explain your reasoning.

Practice Problem 5

An object is thrown upward from the top of a 100-foot cliff. Its height in feet above ground after t seconds is given by the function $f(t) = -16t^2 + 10t + 100$. Find the maximum height of the object and the number of seconds it took for the object to reach its maximum height.

Try the Concept Check in the margin.

EXAMPLE 5 Finding Maximum Height

A rock is thrown upward from the ground. Its height in feet above ground after t seconds is given by the function $f(t) = -16t^2 + 20t$. Find the maximum height of the rock and the number of seconds it took for the rock to reach its maximum height.

Solution:

1. UNDERSTAND. The maximum height of the rock is the largest value of $f(t)$. Since the function $f(t) = -16t^2 + 20t$ is a quadratic function, its graph is a parabola. It opens downward since $-16 < 0$. Thus, the maximum value of $f(t)$ is the $f(t)$- or y-value of the vertex of its graph.

2. TRANSLATE. To find the vertex (h, k), we notice that for $f(t) = -16t^2 + 20t$, $a = -16$, $b = 20$, and $c = 0$. We will use these values and the vertex formula

$$\left(\frac{-b}{2a}, f\left(\frac{-b}{2a} \right) \right)$$

3. SOLVE. $h = \dfrac{-b}{2a} = \dfrac{-20}{-32} = \dfrac{5}{8}$

$$f\left(\frac{5}{8} \right) = -16\left(\frac{5}{8} \right)^2 + 20\left(\frac{5}{8} \right) = -16\left(\frac{25}{64} \right) + \frac{25}{2} = -\frac{25}{4} + \frac{50}{4} = \frac{25}{4}$$

4. INTERPRET. The graph of $f(t)$ is a parabola opening downward with vertex $\left(\dfrac{5}{8}, \dfrac{25}{4} \right)$. This means that the rock's maximum height is $\dfrac{25}{4}$ feet, or $6\dfrac{1}{4}$ feet, which was reached in $\dfrac{5}{8}$ second. ●

Answers

5. maximum height: $101\dfrac{9}{16}$ ft in $\dfrac{5}{16}$ sec

Concept Check: $f(x)$ has a maximum value since it opens downward.

EXERCISE SET 8.6

  Find the vertex of the graph of each quadratic function by completing the square or using the vertex formula. See Examples 1 through 4.

1. $f(x) = x^2 + 8x + 7$ **2.** $f(x) = x^2 + 6x + 5$ **3.** $f(x) = -x^2 + 10x + 5$ **4.** $f(x) = -x^2 - 8x + 2$

5. $f(x) = 5x^2 - 10x + 3$ **6.** $f(x) = -3x^2 + 6x + 4$ **7.** $f(x) = -x^2 + x + 1$ **8.** $f(x) = x^2 - 9x + 8$

Match each function with its graph. See Examples 1 through 4.

A.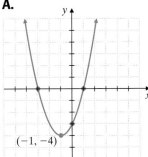

B.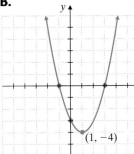

C.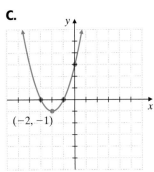

D.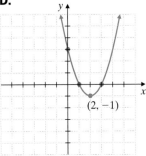

9. $f(x) = x^2 - 4x + 3$ **10.** $f(x) = x^2 + 2x - 3$ **11.** $f(x) = x^2 - 2x - 3$ **12.** $f(x) = x^2 + 4x + 3$

Find the vertex of the graph of each quadratic function. Determine whether the graph opens upward or downward, find any intercepts, and graph the function. See Examples 1 through 4.

13. $f(x) = x^2 + 4x - 5$ **14.** $f(x) = x^2 + 2x - 3$ **15.** $f(x) = -x^2 + 2x - 1$

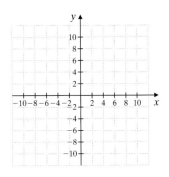

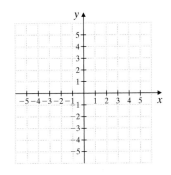

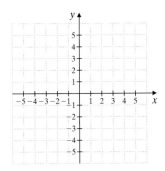

16. $f(x) = -x^2 + 4x - 4$

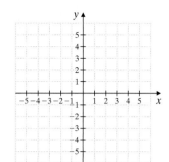

17. $f(x) = x^2 - 4$

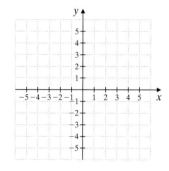

18. $f(x) = x^2 - 1$

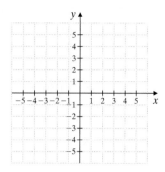

19. $f(x) = 4x^2 + 4x - 3$

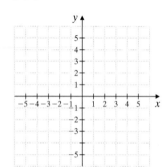

20. $f(x) = 2x^2 - x - 3$

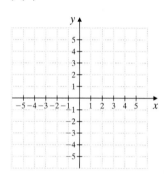

21. $f(x) = \dfrac{1}{2}x^2 + 4x + \dfrac{15}{2}$

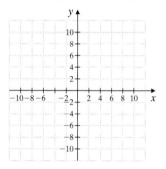

22. $f(x) = \dfrac{1}{5}x^2 + 2x + \dfrac{9}{5}$

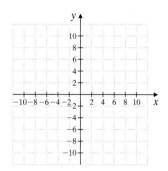

23. $f(x) = x^2 - 4x + 5$

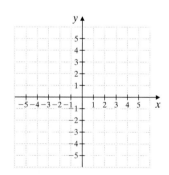

24. $f(x) = x^2 - 6x + 11$

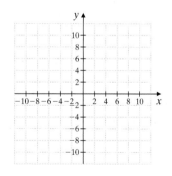

25. $f(x) = 2x^2 + 4x + 5$

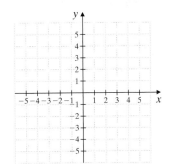

26. $f(x) = 3x^2 + 12x + 16$

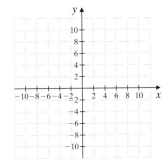

27. $f(x) = -2x^2 + 12x$

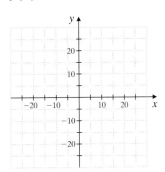

28. $f(x) = -4x^2 + 8x$

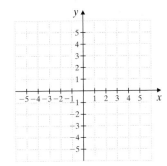

C *Solve. See Example 5.*

29. If a projectile is fired straight upward from the ground with an initial speed of 96 feet per second, then its height h in feet after t seconds is given by the function $h(t) = -16t^2 + 96t$. Find the maximum height of the projectile.

30. The cost C in dollars of manufacturing x bicycles at Holladay's Production Plant is given by the function $C(x) = 2x^2 - 800x + 92,000$.
  a. Find the number of bicycles that must be manufactured to minimize the cost.
  b. Find the minimum cost.

31. If Rheam Gaspar throws a ball upward with an initial speed of 32 feet per second, then its height h in feet after t seconds is given by the function $h(t) = -16t^2 + 32t$. Find the maximum height of the ball.

32. The Utah Ski Club sells calendars to raise money. The profit P, in cents, from selling x calendars is given by the function $P(x) = 360x - x^2$.
  a. Find how many calendars must be sold to maximize profit.
  b. Find the maximum profit.

33. Find two numbers whose sum is 60 and whose product is as large as possible. [*Hint:* Let x and $60 - x$ be the two positive numbers. Their product can be described by the function $f(x) = x(60 - x)$.]

34. Find two numbers whose sum is 11 and whose product is as large as possible. (Use the hint for Exercise 33.)

35. Find two numbers whose difference is 10 and whose product is as small as possible. (Use the hint for Exercise 33.)

36. Find two numbers whose difference is 8 and whose product is as small as possible.

△ **37.** The length and width of a rectangle must have a sum of 40. Find the dimensions of the rectangle that will have the maximum area. (Use the hint for Exercise 33.)

△ **38.** The length and width of a rectangle must have a sum of 50. Find the dimensions of the rectangle that will have maximum area.

Review and Preview

Find the vertex of the graph of each function. See Section 8.5.

39. $f(x) = x^2 + 2$

40. $f(x) = (x - 3)^2$

41. $g(x) = x + 2$

42. $h(x) = x - 3$

43. $f(x) = (x + 5)^2 + 2$

44. $f(x) = 2(x - 3)^2 + 2$

45. $f(x) = 3(x - 4)^2 + 1$

46. $f(x) = (x + 1)^2 + 4$

 Combining Concepts

Find the vertex of the graph of each quadratic function. Determine whether the graph opens upward or downward, find the y-intercept, approximate the x-intercepts to one decimal place, and graph the function.

47. $f(x) = x^2 + 10x + 15$

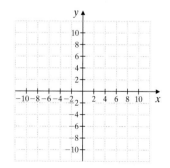

48. $f(x) = 2x^2 + 4x - 1$

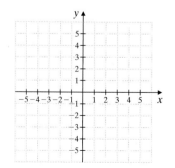

Use a graphing calculator to verify the graph of each exercise.

49. Exercise 21.

50. Exercise 22.

Find the maximum or minimum value of each function. Approximate to two decimal places.

51. $f(x) = 2.3x^2 - 6.1x + 3.2$

52. $f(x) = 7.6x^2 + 9.8x - 2.1$

53. The number of inmates in custody in U.S. prisons and jails can be modeled by the quadratic function

$$p(x) = -x^2 + 93x + 1128$$

where $p(x)$ is the number of inmates in thousands and x is the number of years after 1990. (*Source:* Based on data from the Bureau of Justice Statistics, U.S. Department of Justice, 1990–2000)

  a. Will this function have a maximum or a minimum? How can you tell?

  b. According to this model, in what year will the number of prison and jail inmates in custody in the United States be at its maximum/minimum?

  c. What is the maximum/minimum number of inmates predicted?

54. Methane is a gas produced by landfills, natural gas systems, and coal mining that contributes to the greenhouse effect and global warming. Projected methane emissions in the United States can be modeled by the quadratic function

$$f(x) = -0.072x^2 + 1.93x + 173.9$$

where $f(x)$ is the amount of methane produced in million metric tons and x is the number of years after 2000. (*Source:* Based on data from the U.S. Environmental Protection Agency, 2000–2020)

  a. According to this model, what will U.S. emissions of methane be in 2009?

  b. Will this function have a maximum or a minimum? How can you tell?

  c. In what year will methane emissions in the United States be at their maximum/minimum? Round to the nearest whole year.

  d. What is the level of methane emissions for that year? (Use your rounded answer from part c.)

STUDY SKILLS REMINDER

Are you prepared for your final exam?

To prepare for your final exam, try the following study techniques.

- Review the material that you will be responsible for on your exam. Also check your notebook for any lecture notes that you highlighted.

- Review any formulas that you may need to memorize.

- Check to see if your instructor or math department will be conducting a final exam review.

- Check with your instructor to see whether there are final exams from previous semesters/quarters that are available to students for study.

- Use your previously taken tests as a practice final exam. To do so, rewrite the test questions in mixed order on blank sheets of paper. This will help you prepare for exam conditions.

- If you are unsure of a few topics, see your instructor or visit a learning lab for further assistance. Also, viewing the video segment of a troublesome section will help.

- If you need further exercises to work, try the chapter tests at the end of appropriate chapters.

Good luck! I hope you have enjoyed this textbook and your intermediate algebra course.

This activity may be completed by working in groups or individually.

We have seen in this and previous chapters that data can be modeled by both linear models and quadratic models. However, when we are given a set of data to model, how can we tell if a linear or quadratic model is appropriate? The best answer depends on looking at a scatter diagram of the data. If the plotted data points fall roughly on a line, a linear model is usually the better choice. If the plotted data points seem to fall on a definite curve or if a maximum or minimum point is apparent, a quadratic model is usually the better choice.

One of the sets of data shown in the tables is best modeled by a linear function and one is best modeled by a quadratic function. In each case, the variable x represents the number of years after 1994.

Saturn Vehicle Sales (in thousands)

Year	1996	1997	1998	1999	2000
x	2	3	4	5	6
Number of Saturns sold, y (in thousands)	279	251	232	233	272

(*Source:* General Motors)

Number of Domestic Wal-Mart Stores and Supercenters

Year	1994	1995	1996	1997	1998	1999	2000	2001
x	0	1	2	3	4	5	6	7
Number of stores, y	2022	2132	2234	2304	2362	2433	2522	2624

(*Source:* Wal-Mart Stores, Inc.)

1. Make a scatter diagram for each set of data. Which type of model should be used for each set of data?

2. For the set of data that you have determined to be linear, find a linear function that fits the data points. Explain the method that you used. (*Hint:* See the Focus on Business and Career: Linear Modeling in Section 3.4 or the Focus on the Real World: Linear Modeling in Section 4.5 for more information.)

3. For the set of data that you have determined to be quadratic, identify the point on your scatter diagram that appears to be the vertex of the parabola. Use the coordinates of this vertex in the quadratic model $f(x) = a(x - h)^2 + k$.

4. Solve for the remaining unknown constant in the quadratic model by substituting the coordinates for another data point into the function. Write the final form of the quadratic model for this data set.

5. Use your models to estimate the number of Saturns sold and the number of domestic Wal-Mart stores and supercenters in 2002.

6. (Optional) For each set of data, enter the data from the table into a graphing calculator and use either the linear regression feature or the quadratic regression feature to find an appropriate function that models the data.* Compare these functions with the ones you found by hand. How are they alike or different?

*To find out more about using your graphing calculator to find a regression equation, consult your user's manual.

Chapter 8 VOCABULARY CHECK

Fill in each blank with one of the words or phrases listed below.

quadratic formula	quadratic	discriminant	$\pm\sqrt{b}$
completing the square	quadratic inequality		
(h, k) $(0, k)$ $(h, 0)$	$\dfrac{-b}{2a}$		

1. The _____ helps us know find the number and type of solutions of a quadratic equation.

2. If $a^2 = b$, then $a =$ _____.

3. The graph of $f(x) = ax^2 + bx + c$, where a is not 0, is a parabola whose vertex has an *x*-value of _____.

4. A(n) _____ is an inequality that can be written so that one side is a quadratic expression and the other side is 0.

5. The process of writing a quadratic equation so that one side is a perfect square trinomial is called _____.

6. The graph of $f(x) = x^2 + k$ has vertex _____.

7. The graph of $f(x) = (x - h)^2$ has vertex _____.

8. The graph of $f(x) = (x - h)^2 + k$ has vertex _____.

9. The formula $x = \dfrac{-b \pm \sqrt{b^2 - 4ac}}{2a}$ is called the _____.

10. A _____ equation is one that can be written in the form $ax^2 + bx + c = 0$, where $a, b,$ and c are real numbers and a is not 0.

CHAPTER # Highlights

DEFINITIONS AND CONCEPTS	**EXAMPLES**
Section 8.1 Solving Quadratic Equations by Completing the Square	

SQUARE ROOT PROPERTY If b is a real number and if $a^2 = b$, then $a = \pm\sqrt{b}$.	Solve: $(x + 3)^2 = 14$ $x + 3 = \pm\sqrt{14}$ $x = -3 \pm \sqrt{14}$
SOLVING A QUADRATIC EQUATION IN X BY COMPLETING THE SQUARE **Step 1.** If the coefficient of x^2 is not 1, divide both sides of the equation by the coefficient of x^2. **Step 2.** Get the variable terms alone. **Step 3.** Complete the square by adding the square of half of the coefficient of x to both sides. **Step 4.** Write the resulting trinomial as the square of a binomial. **Step 5.** Use the square root property.	Solve: $3x^2 - 12x - 18 = 0$ **1.** $x^2 - 4x - 6 = 0$ **2.** $\quad x^2 - 4x = 6$ **3.** $\dfrac{1}{2}(-4) = -2$ and $(-2)^2 = 4$ $\quad x^2 - 4x + 4 = 6 + 4$ **4.** $(x - 2)^2 = 10$ **5.** $x - 2 = \pm\sqrt{10}$ $\quad x = 2 \pm \sqrt{10}$

Section 8.2 Solving Quadratic Equations by Using the Quadratic Formula	

QUADRATIC FORMULA A quadratic equation written in the form $ax^2 + bx + c = 0$ has solutions $x = \dfrac{-b \pm \sqrt{b^2 - 4ac}}{2a}$	Solve: $x^2 - x - 3 = 0$ $a = 1, b = -1, c = -3$ $x = \dfrac{-(-1) \pm \sqrt{(-1)^2 - 4(1)(-3)}}{2 \cdot 1}$ $x = \dfrac{1 \pm \sqrt{13}}{2}$

| **DEFINITIONS AND CONCEPTS** | **EXAMPLES** |

Section 8.3 Solving Equations by Using Quadratic Methods

Substitution is often helpful in solving an equation that contains a repeated variable expression.

Solve: $(2x + 1)^2 - 5(2x + 1) + 6 = 0$

Let $m = 2x + 1$. Then

$$m^2 - 5m + 6 = 0 \qquad \text{Let } m = 2x + 1.$$
$$(m - 3)(m - 2) = 0$$
$$m = 3 \quad \text{or} \quad m = 2$$
$$2x + 1 = 3 \qquad 2x + 1 = 2 \qquad \text{Substitute back.}$$
$$x = 1 \qquad x = \frac{1}{2}$$

Section 8.4 Nonlinear Inequalities in One Variable

SOLVING A POLYNOMIAL INEQUALITY

Step 1. Write the inequality in standard form and solve the related equation.

Step 2. Use solutions from Step 1 to separate the number line into regions.

Step 3. Use a test point to determine whether values in each region satisfy the original inequality.

Step 4. Write the solution set as the union of regions whose test point values are solutions.

Solve: $x^2 \geq 6x$

1. $x^2 - 6x \geq 0$
2. $x^2 - 6x = 0$
   $x(x - 6) = 0$
   $x = 0 \quad \text{or} \quad x = 6$
3.

4.

Region	Test Point Value	$x^2 \geq 6x$	Result
A	-2	$(-2)^2 \geq 6(-2)$	True.
B	1	$1^2 \geq 6(1)$	False.
C	7	$7^2 \geq 6(7)$	True.

5.

The solution set is $(-\infty, 0] \cup [6, \infty)$.

SOLVING A RATIONAL INEQUALITY

Step 1. Solve for values that make all denominators 0.

Step 2. Solve the related equation.

Step 3. Use solutions from Steps 1 and 2 to separate the number line into regions.

Step 4. Use a test point to determine whether values in each region satisfy the original inequality.

Step 5. Write the solution set as the union of regions whose test point value is a solution.

Solve: $\dfrac{6}{x - 1} < -2$

1. $x - 1 = 0 \qquad$ Set the denominator equal to 0.
   $x = 1$

2. $\dfrac{6}{x - 1} = -2$
   $6 = -2(x - 1) \qquad$ Multiply by $(x - 1)$.
   $6 = -2x + 2$
   $4 = -2x$
   $-2 = x$

3.

4. Only a test value from region B satisfies the original inequality.

5.

The solution set is $(-2, 1)$.

Section 8.5   Quadratic Functions and Their Graphs

GRAPHING A QUADRATIC FUNCTION

The graph of a quadratic function written in the form $f(x) = a(x - h)^2 + k$ is a parabola with vertex (h, k).
If $a > 0$, the parabola opens upward.
If $a < 0$, the parabola opens downward.
The axis of symmetry is the line whose equation is $x = h$.

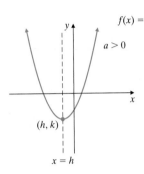 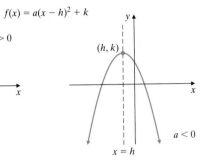

Graph: $g(x) = 3(x - 1)^2 + 4$
The graph is a parabola with vertex $(1, 4)$ and axis of symmetry $x = 1$. Since $a = 3$ is positive, the graph opens upward.

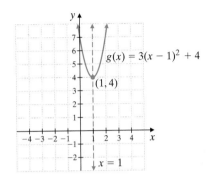

Section 8.6   Further Graphing of Quadratic Functions

The graph of $f(x) = ax^2 + bx + c$, $a \neq 0$, is a parabola with vertex

$$\left(\frac{-b}{2a}, f\left(\frac{-b}{2a}\right)\right).$$

Graph: $f(x) = x^2 - 2x - 8$. Find the vertex and x- and y-intercepts.

$$\frac{-b}{2a} = \frac{-(-2)}{2 \cdot 1} = 1$$

$$f(1) = 1^2 - 2(1) - 8 = -9$$

The vertex is $(1, -9)$.

$$0 = x^2 - 2x - 8$$
$$0 = (x - 4)(x + 2)$$
$$x = 4 \quad \text{or} \quad x = -2$$

The x-intercepts are $(4, 0)$ and $(-2, 0)$.

$$f(0) = 0^2 - 2 \cdot 0 - 8 = -8$$

The y-intercept is $(0, -8)$.

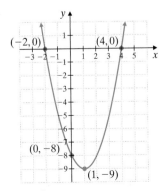

Are you preparing for a test on Chapter 8?

Below I have listed some common trouble areas for students in Chapter 8. After studying for your test—but before taking your test—read these.

■ Don't forget that to solve a quadratic equation such as $x^2 + 6x = 1$, by completing the square, add the square of half of 6 to *both* sides.

$$x^2 + 6x = 1$$

$$x^2 + 6x + 9 = 1 + 9 \qquad \text{Add 9 to both sides.} \left(\frac{1}{2}(6) = 3 \text{ and } 3^2 = 9\right)$$

$$(x + 3)^2 = 10$$

$$x + 3 = \pm\sqrt{10}$$

$$x = -3 \pm \sqrt{10}$$

■ Remember to write a quadratic equation in standard form ($ax^2 + bx + c = 0$) before using the quadratic formula of solve.

$$x(4x - 1) = 1$$

$$4x^2 - x - 1 = 0 \qquad \text{Write in standard form.}$$

$$x = \frac{-(-1) \pm \sqrt{(-1)^2 - 4(4)(-1)}}{2 \cdot 4} \qquad \text{Use the quadratic formula with } a = 4, b = -1, \text{ and } c = -1.$$

$$x = \frac{1 \pm \sqrt{17}}{8} \qquad \text{Simplify.}$$

■ Review the steps for solving a quadratic equation in general on page 611.

■ Don't forget how to graph a quadratic function in the form $f(x) = a(x - h)^2 + k$. The graph of
$f(x) = -2(x - 3)^2 - 1$

opens downward

narrower

shift 3 units right

shift 1 units down

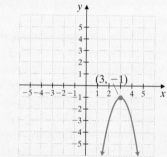

Remember: This is simply a checklist of common trouble areas. For a review of Chapter 8, see the Highlights and Chapter Review at the end of this chapter.

Chapter 8 Review

(8.1) *Solve by factoring.*

1. $x^2 - 15x + 14 = 0$
2. $x^2 - x - 30 = 0$
3. $10x^2 = 3x + 4$
4. $7a^2 = 29a + 30$

Use the square root property to solve each equation.

5. $4m^2 = 196$
6. $9y^2 = 36$
7. $(9n + 1)^2 = 9$
8. $(5x - 2)^2 = 2$

Solve by completing the square.

9. $z^2 + 3z + 1 = 0$
10. $x^2 + x + 7 = 0$
11. $(2x + 1)^2 = x$
12. $(3x - 4)^2 = 10x$

13. If P dollars are invested, the formula $A = P(1 + r)^2$ gives the amount A in an account paying interest rate r compounded annually after 2 years. Find the interest rate r such that \$2500 increases to \$2717 in 2 years. Round the result to the nearest hundredth of a percent.

14. Two ships leave a port at the same time and travel at the same speed. One ship is traveling due north and the other due east. In a few hours, the ships are 150 miles apart. How many miles has each ship traveled? Give an exact answer and a one-decimal-place approximation.

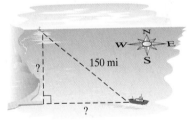

(8.2) *If the discriminant of a quadratic equation has the given value, determine the number and type of solutions of the equation.*

15. -8
16. 48
17. 100
18. 0

Use the quadratic formula to solve each equation.

19. $x^2 - 16x + 64 = 0$
20. $x^2 + 5x = 0$
21. $x^2 + 11 = 0$
22. $2x^2 + 3x = 5$

23. $6x^2 + 7 = 5x$ **24.** $9a^2 + 4 = 2a$ **25.** $(5a - 2)^2 - a = 0$ **26.** $(2x - 3)^2 = x$

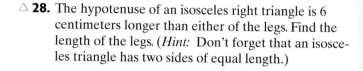

27. Cadets graduating from military school usually toss their hats high into the air at the end of the ceremony. One cadet threw his hat so that its distance $d(t)$ in feet above the ground t seconds after it was thrown was $d(t) = -16t^2 + 30t + 6$.

   a. Find the distance above the ground of the hat 1 second after it was thrown.

   b. Find the time it took the hat to hit the ground. Give an exact time and a one-decimal-place approximation.

△ **28.** The hypotenuse of an isosceles right triangle is 6 centimeters longer than either of the legs. Find the length of the legs. (*Hint:* Don't forget that an isosceles triangle has two sides of equal length.)

(8.3) *Solve each equation.*

29. $x^3 = 27$ **30.** $y^3 = -64$ **31.** $\dfrac{5}{x} + \dfrac{6}{x - 2} = 3$ **32.** $\dfrac{7}{8} = \dfrac{8}{x^2}$

33. $x^4 - 21x^2 - 100 = 0$ **34.** $5(x + 3)^2 - 19(x + 3) = 4$ **35.** $x^{2/3} - 6x^{1/3} + 5 = 0$

36. $x^{2/3} - 6x^{1/3} = -8$ **37.** $a^6 - a^2 = a^4 - 1$ **38.** $y^{-2} + y^{-1} = 20$

39. Two postal workers, Jerome Grant and Tim Bozik, can sort a stack of mail in 5 hours. Working alone, Tim can sort the mail in 1 hour less time than Jerome can. Find the time that each postal worker can sort the mail alone. Round the result to one decimal place.

40. A negative number decreased by its reciprocal is $-\dfrac{24}{5}$. Find the number.

(8.4) *Solve each inequality for x. Write each solution set in interval notation.*

41. $2x^2 - 50 \le 0$ **42.** $\dfrac{1}{4}x^2 < \dfrac{1}{16}$ **43.** $(2x - 3)(4x + 5) \ge 0$

44. $(x^2 - 16)(x^2 - 1) > 0$

45. $\dfrac{x - 5}{x - 6} < 0$

46. $\dfrac{x(x + 5)}{4x - 3} \geq 0$

47. $\dfrac{(4x + 3)(x - 5)}{x(x + 6)} > 0$

48. $(x + 5)(x - 6)(x + 2) \leq 0$

49. $x^3 + 3x^2 - 25x - 75 > 0$

50. $\dfrac{x^2 + 4}{3x} \leq 1$

51. $\dfrac{(5x + 6)(x - 3)}{x(6x - 5)} < 0$

52. $\dfrac{3}{x - 2} > 2$

(8.5) *Graph each function. Label the vertex and the axis of symmetry of each graph.*

53. $f(x) = x^2 - 4$

54. $g(x) = x^2 + 7$

55. $H(x) = 2x^2$

56. $h(x) = -\dfrac{1}{3}x^2$

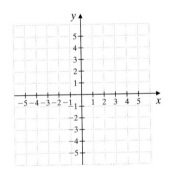
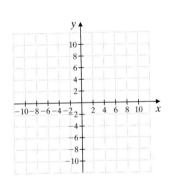
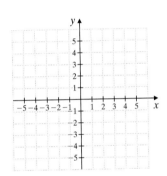
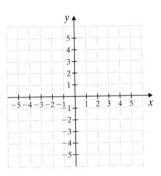

57. $F(x) = (x - 1)^2$

58. $G(x) = (x + 5)^2$

59. $f(x) = (x - 4)^2 - 2$

60. $f(x) = -3(x - 1)^2 + 1$

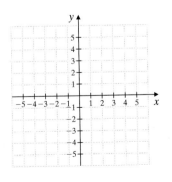
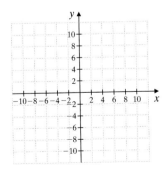
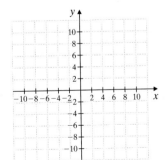
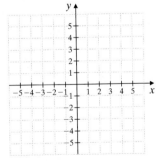

(8.6) *Graph each function. Find the vertex and any intercepts of each graph.*

61. $f(x) = x^2 + 10x + 25$

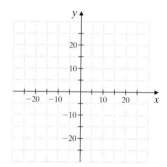

62. $f(x) = -x^2 + 6x - 9$

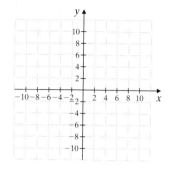

63. $f(x) = 4x^2 - 1$

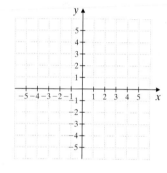

64. $f(x) = -5x^2 + 5$

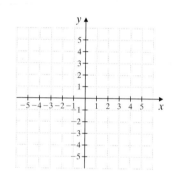

 **65.** Find the vertex of the graph of $f(x) = -3x^2 - 5x + 4$. Determine whether the graph opens upward or downward, find the y-intercept, approximate the x-intercepts to one decimal place, and graph the function.

 **66.** The function $h(t) = -16t^2 + 120t + 300$ gives the height in feet of a projectile fired from the top of a building at t seconds.

  a. When will the object reach a height of 350 feet? Round your answer to one decimal place.

  b. Explain why part (a) has two answers.

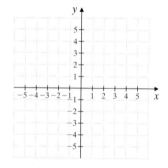

67. Find two numbers whose sum is 420 and whose product is as large as possible.

Name _____

Chapter 8 Test

Solve each equation.

1. $5x^2 - 2x = 7$

2. $(x + 1)^2 = 10$

3. $m^2 - m + 8 = 0$

4. $u^2 - 6u + 2 = 0$

5. $7x^2 + 8x + 1 = 0$

6. $a^2 - 3a = 5$

7. $\dfrac{4}{x + 2} + \dfrac{2x}{x - 2} = \dfrac{6}{x^2 - 4}$

8. $x^4 - 8x^2 - 9 = 0$

9. $x^6 + 1 = x^4 + x^2$

10. $(x + 1)^2 - 15(x + 1) + 56 = 0$

Solve by completing the square.

11. $x^2 - 6x = -2$

12. $2a^2 + 5 = 4a$

Solve each inequality. Write each solution set in interval notation.

13. $2x^2 - 7x > 15$

14. $(x^2 - 16)(x^2 - 25) > 0$

15. $\dfrac{5}{x + 3} < 1$

16. $\dfrac{7x - 14}{x^2 - 9} \le 0$

Graph each function. Label the vertex for each graph.

17. $f(x) = 3x^2$

18. $G(x) = -2(x - 1)^2 + 5$

2. _____

3. _____

4. _____

5. _____

6. _____

7. _____

8. _____

9. _____

10. _____

11. _____

12. _____

13. _____

14. _____

15. _____

16. _____

17. see graph _____

18. see graph _____

Graph each function. Find and label the vertex, y-intercept, and x-intercepts (if any) for each graph.

19. $h(x) = x^2 - 4x + 4$

20. $F(x) = 2x^2 - 8x + 9$

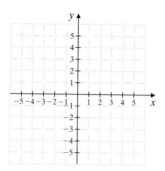

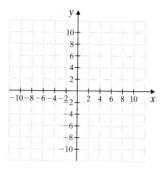

△**21.** A 10-foot ladder is leaning against a house. The distance from the bottom of the ladder to the house is 4 feet less than the distance from the top of the ladder to the ground. Find how far the top of the ladder is from the ground. Give an exact answer and a one-decimal-place approximation.

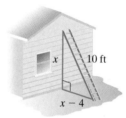

x 10 ft

$x - 4$

22. Dave and Sandy Hartranft can paint a room together in 4 hours. Working alone, Dave can paint the room in 2 hours less time than Sandy can. Find how long it takes Sandy to paint the room alone. Give an exact answer and a two-decimal-place approximation.

23. A stone is thrown upward from a bridge. The stone's height $s(t)$ in feet, above the water t seconds after the stone is thrown is given by the function $s(t) = -16t^2 + 32t + 256$.

a. Find the maximum height of the stone.

b. Find the time it takes the stone to hit the water.

256 ft

Cumulative Review

Answers

1. _____

2. _____

3. _____

4. _____

5. _____

6. _____

7. _____

8. _____

9. _____

10. _____

11. _____

12. _____

13. _____

14. _____

15. _____

16. _____

17. _____

18. _____

1. Write 730,000 in scientific notation.

2. Solve: $|2x - 1| + 5 = 6$

3. Determine whether the relation $x = y^2$ is also a function.

4. Use the elimination method to solve the system: $\begin{cases} 3x + \dfrac{y}{2} = 2 \\ 6x + y = 5 \end{cases}$

5. Subtract vertically:
$(10x^3 - 7x^2) - (4x^3 - 3x^2 + 2)$

Use the FOIL order to multiply.

6. $(2x - 7)(3x - 4)$

7. $(3x + y)(5x - 2y)$

8. Divide $3x^4 + 2x^3 - 8x + 6$ by $x^2 - 1$.

9. Factor: $6x^2 - 3x^3$

10. Factor: $2(a + 3)^2 - 5(a + 3) - 7$

11. Factor: $3a^2x - 12abx + 12b^2x$

12. Solve: $2x^2 = \dfrac{17}{3}x + 1$

Multiply.

13. $\dfrac{3n + 1}{2n} \cdot \dfrac{2n - 4}{3n^2 - 2n - 1}$

14. $\dfrac{x^3 - 1}{-3x + 3} \cdot \dfrac{15x^2}{x^2 + x + 1}$

15. Add: $\dfrac{2}{x^2} + \dfrac{5}{3x^3}$

16. Simplify: $\dfrac{\dfrac{x}{y^2} + \dfrac{1}{y}}{\dfrac{y}{x^2} + \dfrac{1}{x}}$

17. Solve: $\dfrac{2x}{x - 3} + \dfrac{6 - 2x}{x^2 - 9} = \dfrac{x}{x + 3}$

18. Solve $\dfrac{1}{x} + \dfrac{1}{y} = \dfrac{1}{z}$ for x.

19. _____

20. _____

21. _____

22. _____

23. _____

24. _____

25. _____

26. _____

27. _____

28. _____

29. _____

△**19.** The lateral surface area of a cylinder varies jointly as its radius and height. Express surface area S in terms of radius r and height h.

Find each cube root.

20. $\sqrt[3]{-64}$

21. $\sqrt[3]{\dfrac{8}{125}}$

Multiply.

22. $\sqrt[3]{4} \cdot \sqrt[3]{2}$

23. $\sqrt{\dfrac{2}{a}} \cdot \sqrt{\dfrac{b}{3}}$

Add or subtract.

24. $\sqrt[3]{54} - 5\sqrt[3]{16} + \sqrt[3]{2}$

25. $\sqrt[3]{\dfrac{7x}{8}} + 2\sqrt[3]{7x}$

26. Solve $p^2 + 2p = 4$ by completing the square.

27. Use the quadratic formula to solve $2x^2 - 4x = 3$.

28. An object is thrown upward from the top of a 200-foot cliff with a velocity of 12 feet per second. The height h in feet of the object after t seconds is $h = -16t^2 + 12t + 200$. How long after the object is thrown will it strike the ground? Round to the nearest tenth of a second.

29. Beach and Fargo are about 400 miles apart. A salesperson travels from Fargo to Beach one day at a certain speed. She returns to Fargo the next day and drives 10 miles per hour faster. Her total travel time was $14\dfrac{2}{3}$ hours. Find her speed to Beach and the return speed to Fargo.

Conic Sections

In Chapter 8, we analyzed some of the important connections between a parabola and its equation. Parabolas are interesting in their own right but are more interesting still because they are part of a collection of curves known as conic sections. This chapter is devoted to quadratic equations in two variables and their conic section graphs: the parabola, circle, ellipse, and hyperbola.

9.1 The Parabola and the Circle

9.2 The Ellipse and the Hyperbola

Integrated Review—Graphing Conic Sections

9.3 Graphing Nonlinear Functions

9.4 Solving Nonlinear Systems of Equations

9.5 Nonlinear Inequalities and Systems of Inequalities

The shapes of conic sections are used in a variety of applications. They are used in architecture in the design of bridges, arches, and vaults. They are also used in astronomy to model the orbits of planets, comets, and satellites. Conic sections are used also in engineering in the design of certain gears and reflectors. In blueprints or diagrams of any of these situations, the exact shape of the conic section involved must be depicted. In Exercises 68 and 69 on page 685, we will see how architects and engineers use conic sections in their work.

Name _____ Section _____ Date _____

Chapter 9 Pretest

1. Find the distance between the points $(-4, 6)$ and $(1, 9)$.

2. Find the midpoint of the line segment whose endpoints are $(9, -15)$ and $(10, 22)$.

Sketch the graph of each equation.

3. $x = (y - 2)^2 + 1$

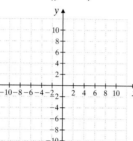

4. $y = x^2 + 4x - 5$

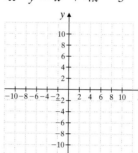

5. $x^2 + y^2 = 4$

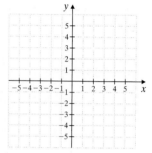

6. $(x + 2)^2 + (y - 3)^2 = 9$

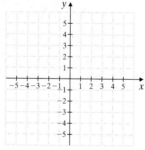

7. $4x^2 + 25y^2 = 100$

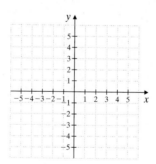

8. $\dfrac{x^2}{25} - \dfrac{y^2}{25} = 1$

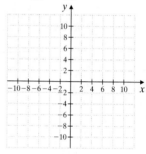

Graph each function.

9. $f(x) = |x| - 5$

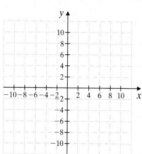

10. $f(x) = \sqrt{x + 1} + 3$

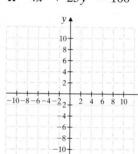

11. Solve the system:
$$\begin{cases} y = x^2 + 4x + 3 \\ x + y = 3 \end{cases}$$

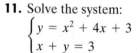

12. Graph the system:
$$\begin{cases} y \geq \dfrac{1}{4}x^2 + 2 \\ x^2 + (y - 3)^2 \leq 4 \end{cases}$$

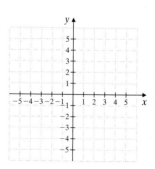

13. Write an equation of the circle with center $(2, 5)$ and radius 8.

14. Find the center and radius of the circle defined by $x^2 + y^2 + 6x - 8y = -16$.

9.1 The Parabola and the Circle

Conic sections are called such because each conic section is the intersection of a right circular cone and a plane. The circle, parabola, ellipse, and hyperbola are the conic sections.

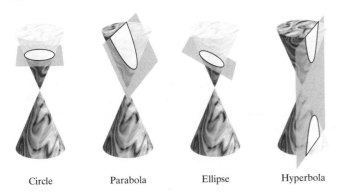

Circle   Parabola   Ellipse   Hyperbola

OBJECTIVES

Ⓐ Graph parabolas of the forms
$y = a(x - h)^2 + k$ and
$x = a(y - k)^2 + h$.

Ⓑ Use the distance formula and the midpoint formula.

Ⓒ Graph circles of the form
$(x - h)^2 + (y - k)^2 = r^2$.

Ⓓ Write the equation of a circle, given its center and radius.

Ⓔ Find the center and the radius of a circle, given its equation.

SSM
TUTOR CENTER   SG   CD & VIDEO   MATH PRO   WEB

Ⓐ Graphing Parabolas

Thus far, we have seen that $f(x)$ or $y = a(x - h)^2 + k$ is the equation of a parabola that opens upward if $a > 0$ or downward if $a < 0$. Parabolas can also open left or right, or even on a slant. Equations of these parabolas are not functions of x, of course, since a parabola opening any way other than upward or downward fails the vertical line test. In this section, we introduce parabolas that open to the left and to the right. Parabolas opening on a slant will not be developed in this book.

Just as $y = a(x - h)^2 + k$ is the equation of a parabola that opens upward or downward, $x = a(y - k)^2 + h$ is the equation of a parabola that opens to the right or to the left. The parabola opens to the right if $a > 0$ and to the left if $a < 0$. The parabola has vertex (h, k), and its axis of symmetry is the line $y = k$.

Parabolas

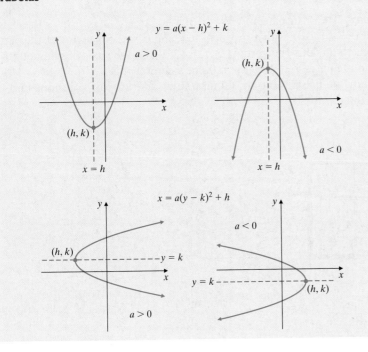

Concept Check

Does the graph of the parabola given by the equation $x = -3y^2$ open to the left, to the right, upward, or downward?

Practice Problem 1

Graph: $x = 3y^2$

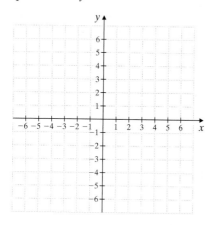

Practice Problem 2

Graph: $x = -2(y - 3)^2 + 1$

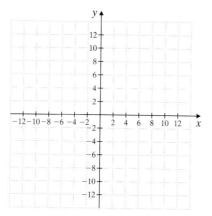

Answers

1.

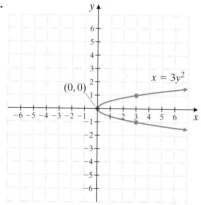

Concept Check: to the left

The forms $y = a(x - h)^2 + k$ and $x = a(y - k)^2 + h$ are called **standard forms**.

Try the Concept Check in the margin.

EXAMPLE 1   Graph: $x = 2y^2$

Solution: Written in standard form, the equation $x = 2y^2$ is $x = 2(y - 0)^2 + 0$ with $a = 2$, $h = 0$, and $k = 0$. Its graph is a parabola with vertex $(0, 0)$, and its axis of symmetry is the line $y = 0$. Since $a > 0$, this parabola opens to the right. We use a table to obtain a few more ordered pair solutions to help us graph $x = 2y^2$.

x	y
8	-2
2	-1
0	0
2	1
8	2

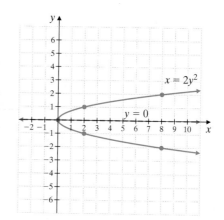

EXAMPLE 2   Graph: $x = -3(y - 1)^2 + 2$

Solution: The equation $x = -3(y - 1)^2 + 2$ is in the form $x = a(y - k)^2 + h$ with $a = -3$, $k = 1$, and $h = 2$. Since $a < 0$, the parabola opens to the left. The vertex (h, k) is $(2, 1)$, and the axis of symmetry is the horizontal line $y = 1$. When $y = 0$, the x-intercept is $x = (-1, 0)$.

Again, we use a table to obtain a few ordered pair solutions and then graph the parabola.

x	y
2	1
-1	0
-1	2

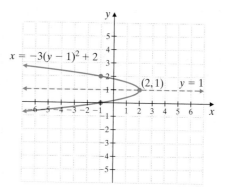

EXAMPLE 3   Graph: $y = -x^2 - 2x + 15$

Solution: There are two methods that we can use to find the vertex. The first method is completing the square.

$$y - 15 = -x^2 - 2x \qquad \text{Subtract 15 from both sides.}$$

$$y - 15 = -1(x^2 + 2x) \qquad \text{Factor } -1 \text{ from the terms } -x^2 - 2x.$$

The coefficient of x is 2, so we find the square of half of 2.

$$\frac{1}{2}(2) = 1 \quad \text{and} \quad 1^2 = 1$$

$$y - 15 - 1(1) = -1(x^2 + 2x + 1) \qquad \text{Add } -1(1) \text{ to both sides.}$$

$$y - 16 = -1(x + 1)^2 \qquad \begin{array}{l}\text{Simplify the left side,} \\ \text{and factor the right side.}\end{array}$$

$$y = -(x + 1)^2 + 16 \qquad \text{Add 16 to both sides.}$$

The vertex is $(-1, 16)$.

The second method for finding the vertex is by using the expression $\dfrac{-b}{2a}$. Since the equation is quadratic in x, the expression gives us the x-value of the vertex.

$$x = \frac{-(-2)}{2(-1)} = \frac{2}{-2} = -1$$

To find the corresponding y-value of the vertex, replace x with -1 in the original equation.

$$y = -(-1)^2 - 2(-1) + 15 = -1 + 2 + 15 = 16$$

Again, we see that the vertex is $(-1, 16)$, and the axis of symmetry is the vertical line $x = -1$. The y-intercept is $(0, 15)$. Now we can use a few more ordered pair solutions to graph the parabola.

x	y
−5	0
−3	12
−2	15
−1	16
0	15
1	12
3	0

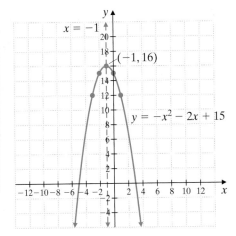

Practice Problem 3

Graph: $y = -x^2 - 4x + 12$

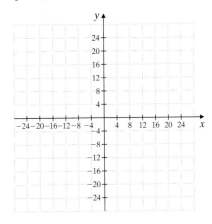

Answers

2.

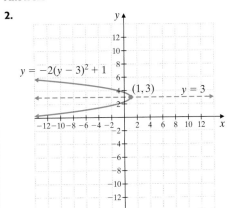

$y = -2(y - 3)^2 + 1$ $(1, 3)$ $y = 3$

3.

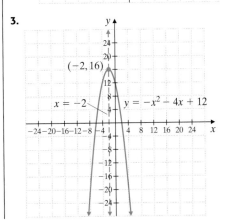

$(-2, 16)$ $x = -2$ $y = -x^2 - 4x + 12$

Practice Problem 4

Graph: $x = 3y^2 + 12y + 13$

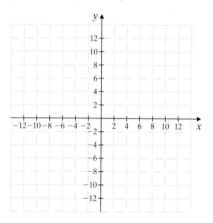

EXAMPLE 4 Graph: $x = 2y^2 + 4y + 5$

Solution: We notice that this equation is quadratic in y, so its graph is a parabola that opens to the left or the right. We can complete the square on y or we can use the expression $\dfrac{-b}{2a}$ to find the vertex.

Since the equation is quadratic in y, the expression gives us the y-value of the vertex.

$$y = \frac{-4}{2 \cdot 2} = \frac{-4}{4} = -1$$

$$x = 2(-1)^2 + 4(-1) + 5 = 2 \cdot 1 - 4 + 5 = 3$$

The vertex is $(3, -1)$, and the axis of symmetry is the line $y = -1$. The parabola opens to the right since $a > 0$. The x-intercept is $(5, 0)$.

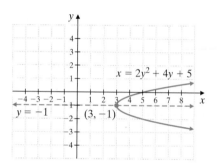

B Using the Distance and Midpoint Formulas

The Cartesian coordinate system helps us visualize a distance between points. To find the distance between two points, we use the distance formula, which is derived from the Pythagorean theorem.

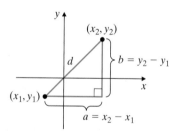

Answer

4.

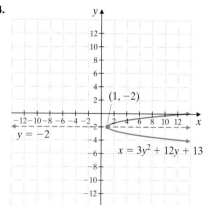

To find the distance d between two points (x_1, y_1) and (x_2, y_2), draw vertical and horizontal lines so that a right triangle is formed, as shown. Notice that the length of leg a is $x_2 - x_1$ and that the length of leg b is $y_2 - y_1$. Thus, the Pythagorean theorem tell us that

$$d^2 = a^2 + b^2$$

or

$$d^2 = (x_2 - x_1)^2 + (y_2 - y_1)^2$$

or

$$d = \sqrt{(x_2 - x_1)^2 + (y_2 - y_1)^2}$$

This formula gives us the distance between any two points on the real plane.

Distance Formula

The distance d between two points (x_1, y_1) and (x_2, y_2) is given by

$$d = \sqrt{(x_2 - x_1)^2 + (y_2 - y_1)^2}$$

EXAMPLE 5

Find the distance between $(2, -5)$ and $(1, -4)$. Give an exact distance and a three-decimal-place approximation.

Solution: To use the distance formula, it makes no difference which point we call (x_1, y_1) and which point we call (x_2, y_2). We will let $(x_1, y_1) = (2, -5)$ and $(x_2, y_2) = (1, -4)$.

$$
\begin{aligned}
d &= \sqrt{(x_2 - x_1)^2 + (y_2 - y_1)^2} \\
&= \sqrt{(1 - 2)^2 + [-4 - (-5)]^2} \\
&= \sqrt{(-1)^2 + (1)^2} \\
&= \sqrt{1 + 1} \\
&= \sqrt{2} \approx 1.414
\end{aligned}
$$

The distance between the two points is exactly $\sqrt{2}$ units, or approximately 1.414 units. ●

The **midpoint** of a line segment is the **point** located exactly halfway between the two endpoints of the line segment. On the following graph, the point M is the midpoint of line segment PQ. Thus, the distance between M and P equals the distance between M and Q.

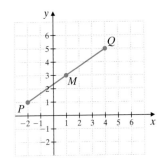

The x-coordinate of M is at half the distance between the x-coordinates of P and Q, and the y-coordinate of M is at half the distance between the y-coordinates of P and Q. That is, the x-coordinate of M is the average of the x-coordinates of P and Q; the y-coordinate of M is the average of the y-coordinates of P and Q.

Midpoint Formula

The midpoint of the line segment whose endpoints are (x_1, y_1) and (x_2, y_2) is the point with coordinates

$$\left(\frac{x_1 + x_2}{2}, \frac{y_1 + y_2}{2} \right)$$

Practice Problem 5

Find the distance between $(-1, 3)$ and $(-2, 6)$. Give an exact distance and a three-decimal-place approximation.

Answer

5. $\sqrt{10} \approx 3.162$

Practice Problem 6

Find the midpoint of the line segment that joins points $P(-2, 5)$ and $Q(4, -6)$.

EXAMPLE 6

Find the midpoint of the line segment that joins points $P(-3, 3)$ and $Q(1, 0)$.

Solution: To use the midpoint formula, it makes no difference which point we call (x_1, y_1) and which point we call (x_2, y_2). We will let $(x_1, y_1) = (-3, 3)$ and $(x_2, y_2) = (1, 0)$.

$$\text{midpoint} = \left(\frac{x_1 + x_2}{2}, \frac{y_1 + y_2}{2} \right)$$

$$= \left(\frac{-3 + 1}{2}, \frac{3 + 0}{2} \right)$$

$$= \left(\frac{-2}{2}, \frac{3}{2} \right)$$

$$= \left(-1, \frac{3}{2} \right)$$

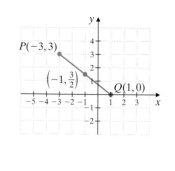

The midpoint of the segment is $\left(-1, \frac{3}{2} \right)$.

Ⓒ Graphing Circles

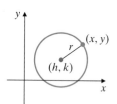

Another conic section is the **circle**. A circle is the set of all points in a plane that are the same distance from a fixed point called the **center**. The distance is called the **radius** of the circle. To find a standard equation for a circle, let (h, k) represent the center of the circle, and let (x, y) represent any point on the circle. The distance between (h, k) and (x, y) is defined to be the radius, r units. We can find this distance r by using the distance formula.

$$r = \sqrt{(x - h)^2 + (y - k)^2}$$

$$r^2 = (x - h)^2 + (y - k)^2 \qquad \text{Square both sides.}$$

Circle

The graph of $(x - h)^2 + (y - k)^2 = r^2$ is a circle with center (h, k) and radius r.

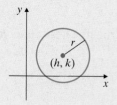

Answer

6. $\left(1, -\frac{1}{2} \right)$

The form $(x - h)^2 + (y - k)^2 = r^2$ is called **standard form**.

If an equation can be written in the standard form

$$(x - h)^2 + (y - k)^2 = r^2$$

then its graph is a circle, which we can draw by graphing the center (h, k) and using the radius r.

> **Helpful Hint**
>
> Notice that the radius is the *distance* from the center of the circle to any point of the circle. Also notice that the *midpoint* of a diameter of a circle is the center of the circle.
>
>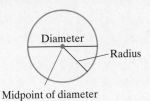

EXAMPLE 7   Graph: $x^2 + y^2 = 4$

Solution: The equation can be written in standard form as

$$(x - 0)^2 + (y - 0)^2 = 2^2$$

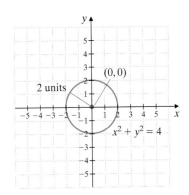

The center of the circle is $(0, 0)$, and the radius is 2. The graph of the circle is shown above.

> **Helpful Hint**
>
> Notice the difference between the equation of a circle and the equation of a parabola. The equation of a circle contains both x^2- and y^2-terms on the same side of the equation with equal coefficients. The equation of a parabola has either an x^2-term or a y^2-term but not both.

Practice Problem 7

Graph: $x^2 + y^2 = 9$

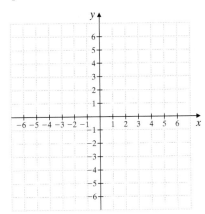

Answer

7.

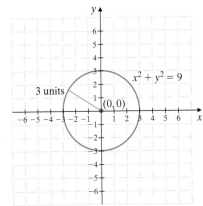

Practice Problem 8

Graph: $x^2 + (y + 2)^2 = 6$

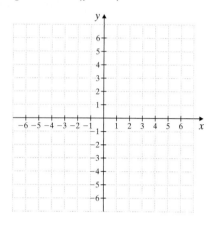

Concept Check

In the graph of the equation $(x - 3)^2 + (y - 2)^2 = 5$, what is the distance between the center of the circle and any point on the circle?

Practice Problem 9

Write an equation of the circle with the center $(2, -5)$ and radius 7.

Answers

8.

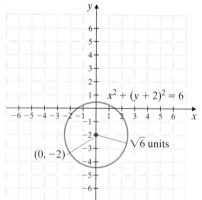

$x^2 + (y + 2)^2 = 6$

$\sqrt{6}$ units

$(0, -2)$

9. $(x - 2)^2 + (y + 5)^2 = 49$

Concept Check: $\sqrt{5}$ units

EXAMPLE 8 Graph: $(x + 1)^2 + y^2 = 8$

Solution: The equation can be written as $(x - (-1))^2 + (y - 0)^2 = 8$ with $h = -1$, $k = 0$, and $r = \sqrt{8}$. The center is $(-1, 0)$, and the radius is $\sqrt{8} = 2\sqrt{2} \approx 2.8$.

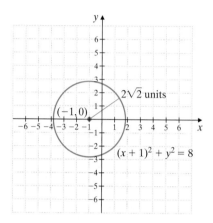

$2\sqrt{2}$ units

$(-1, 0)$

$(x + 1)^2 + y^2 = 8$

Try the Concept Check in the margin.

D Writing Equations of Circles

Since a circle is determined entirely by its center and radius, this information is all we need to write the equation of a circle.

EXAMPLE 9

Write an equation of the circle with center $(-7, 3)$ and radius 10.

Solution: Using the given values $h = -7$, $k = 3$, and $r = 10$, we write the equation

$$(x - h)^2 + (y - k)^2 = r^2$$

or

$$(x - (-7))^2 + (y - 3)^2 = 10^2 \qquad \text{Substitute the given values.}$$

or

$$(x + 7)^2 + (y - 3)^2 = 100$$

E Finding the Center and the Radius of a Circle

To find the center and the radius of a circle from its equation, we write the equation in standard form. To write the equation of a circle in standard form, we complete the square on both x and y.

EXAMPLE 10   Graph: $x^2 + y^2 + 4x - 8y = 16$

Solution: Since this equation contains x^2- and y^2-terms on the same side of the equation with equal coefficients, its graph is a circle. To write the equation in standard form, we group the terms involving x and the terms involving y, and then complete the square on each variable.

$$(x^2 + 4x) + (y^2 - 8y) = 16$$

Thus, $\frac{1}{2}(4) = 2$ and $2^2 = 4$. Also, $\frac{1}{2}(-8) = -4$ and $(-4)^2 = 16$. We add 4 and then 16 to both sides.

$$(x^2 + 4x + 4) + (y^2 - 8y + 16) = 16 + 4 + 16$$
$$(x + 2)^2 + (y - 4)^2 = 36 \qquad \text{Factor.}$$

This circle has the center $(-2, 4)$ and radius 6, as shown.

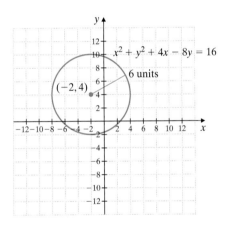

Practice Problem 10

Graph: $x^2 + y^2 - 2x + 6y = 6$

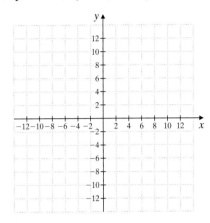

●

Answer

10.

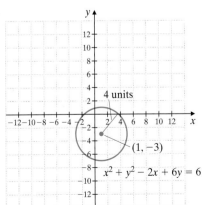

GRAPHING CALCULATOR EXPLORATIONS

To graph an equation such as $x^2 + y^2 = 25$ with a graphing calculator, we first solve the equation for y.

$$x^2 + y^2 = 25$$
$$y^2 = 25 - x^2$$
$$y = \pm\sqrt{25 - x^2}$$

The graph of $y = \sqrt{25 - x^2}$ will be the top half of the circle, and the graph of $y = -\sqrt{25 - x^2}$ will be the bottom half of the circle.

To graph, we press $\boxed{\text{Y=}}$ and enter $Y_1 = \sqrt{25 - x^2}$ and $Y_2 = -\sqrt{25 - x^2}$. We insert parentheses about $25 - x^2$ so that $\sqrt{25 - x^2}$ and not $\sqrt{25} - x^2$ is graphed.

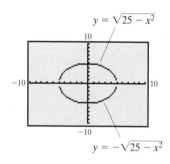

The graph does not appear to be a circle because we are currently using a standard window and the screen is rectangular. This causes the tick marks on the x-axis to be farther apart than the tick marks on the y-axis and thus creates the distorted circle. If we want the graph to appear circular, we define a square window by using a feature of the graphing calculator or redefine the window to show the x-axis from -15 to 15 and the y-axis from -10 to 10. Using a square window, the graph appears as follows:

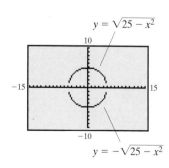

Use a graphing calculator to graph each circle.

1. $x^2 + y^2 = 55$

2. $x^2 + y^2 = 20$

3. $7x^2 + 7y^2 - 89 = 0$

4. $3x^2 + 3y^2 - 35 = 0$

Name _____ Section _____ Date _____

Mental Math

The graph of each equation is a parabola. Determine whether the parabola opens upward, downward, to the left, or to the right.

1. $y = x^2 - 7x + 5$

2. $y = -x^2 + 16$

3. $x = -y^2 - y + 2$

4. $x = 3y^2 + 2y - 5$

5. $y = -x^2 + 2x + 1$

6. $x = -y^2 + 2y - 6$

EXERCISE SET 9.1

Ⓐ *The graph of each equation is a parabola. Find the vertex of the parabola and then graph it. See Examples 1 through 4.*

1. $x = 3y^2$

2. $x = -2y^2$

3. $x = (y - 2)^2 + 3$

4. $x = (y - 4)^2 - 1$

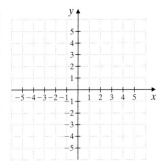
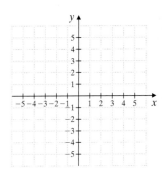
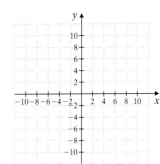
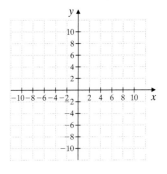

5. $y = 3(x - 1)^2 + 5$

6. $x = -4(y - 2)^2 + 2$

7. $x = y^2 + 6y + 8$

8. $x = y^2 - 6y + 6$

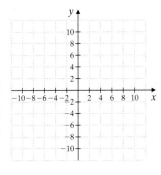
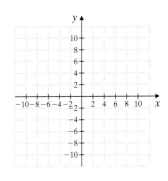
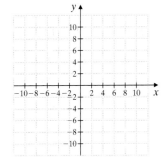
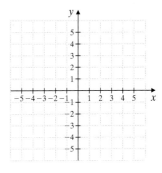

9. $y = x^2 + 10x + 20$

10. $y = x^2 + 4x - 5$

11. $x = -2y^2 + 4y + 6$

12. $x = 3y^2 + 6y + 7$

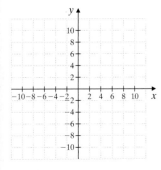
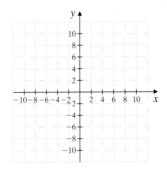
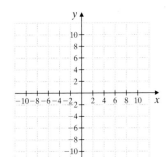
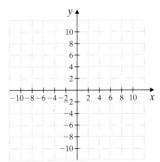

681

B *Find the distance between each pair of points. Give an exact distance and a three-decimal-place approximation. See Example 5.*

13. $(5, 1)$ and $(8, 5)$ **14.** $(2, 3)$ and $(14, 8)$ **15.** $(-3, 2)$ and $(1, -3)$ **16.** $(3, -2)$ and $(-4, 1)$

17. $(-9, 4)$ and $(-8, 1)$ **18.** $(-5, -2)$ and $(-6, -6)$ **19.** $(0, -\sqrt{2})$ and $(\sqrt{3}, 0)$ **20.** $(-\sqrt{5}, 0)$ and $(0, \sqrt{7})$

21. $(1.7, -3.6)$ and $(-8.6, 5.7)$ **22.** $(9.6, 2.5)$ and $(-1.9, -3.7)$

23. $(2\sqrt{3}, \sqrt{6})$ and $(-\sqrt{3}, 4\sqrt{6})$ **24.** $(5\sqrt{2}, -4)$ and $(-3\sqrt{2}, -8)$

Find the midpoint of each line segment whose endpoints are given. See Example 6.

25. $(6, -8); (2, 4)$ **26.** $(3, 9); (7, 11)$ **27.** $(-2, -1); (-8, 6)$ **28.** $(-3, -4); (6, -8)$

29. $(7, 3); (-1, -3)$ **30.** $(-2, 5); (-1, 6)$ **31.** $\left(\frac{1}{2}, \frac{3}{8}\right); \left(-\frac{3}{2}, \frac{5}{8}\right)$ **32.** $\left(-\frac{2}{5}, \frac{7}{15}\right); \left(-\frac{2}{5}, -\frac{4}{15}\right)$

33. $(\sqrt{2}, 3\sqrt{5}); (\sqrt{2}, -2\sqrt{5})$ **34.** $(\sqrt{8}, -\sqrt{12}); (3\sqrt{2}, 7\sqrt{3})$ **35.** $(4.6, -3.5); (7.8, -9.8)$ **36.** $(-4.6, 2.1); (-6.7, 1.9)$

The graph of each equation is a circle. Find the center and the radius, and then graph the circle. See Examples 7, 8, and 10.

37. $x^2 + y^2 = 9$

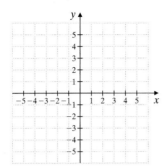

38. $x^2 + y^2 = 25$

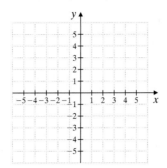

39. $x^2 + (y - 2)^2 = 1$

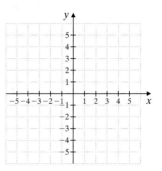

40. $(x - 3)^2 + y^2 = 9$

41. $(x - 5)^2 + (y + 2)^2 = 1$

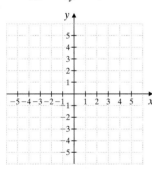

42. $(x + 3)^2 + (y + 3)^2 = 4$

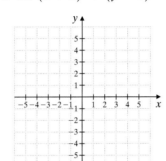

43. $x^2 + y^2 + 6y = 0$

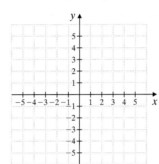

44. $x^2 + 10x + y^2 = 0$

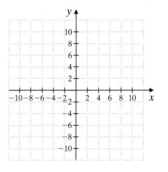

45. $x^2 + y^2 + 2x - 4y = 4$

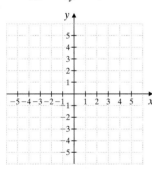

46. $x^2 + y^2 + 6x - 4y = 3$

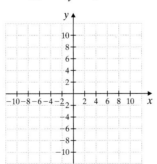

47. $x^2 + y^2 - 4x - 8y - 2 = 0$

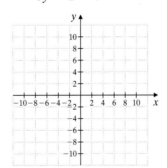

48. $x^2 + y^2 - 2x - 6y - 5 = 0$

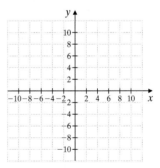

D Write an equation of the circle with the given center and radius. See Example 9.

49. $(2, 3); 6$

50. $(-7, 6); 2$

51. $(0, 0); \sqrt{3}$

52. $(0, -6); \sqrt{2}$ **53.** $(-5, 4); 3\sqrt{5}$ **54.** The origin; $4\sqrt{7}$

Review and Preview

Graph each equation. See Section 3.3.

55. $y = 2x + 5$ **56.** $y = -3x + 3$ **57.** $y = 3$ **58.** $x = -2$

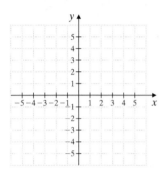
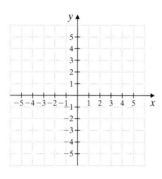
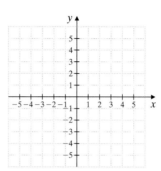
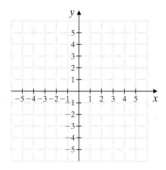

Rationalize each denominator and simplify if possible. See Section 7.5.

59. $\dfrac{1}{\sqrt{3}}$ **60.** $\dfrac{\sqrt{5}}{\sqrt{8}}$ **61.** $\dfrac{4\sqrt{7}}{\sqrt{6}}$ **62.** $\dfrac{10}{\sqrt{5}}$

 Combining Concepts

63. In 1893, Pittsburgh bridge builder George Ferris designed and built a gigantic revolving steel wheel whose height was 264 feet and diameter was 250 feet. This Ferris wheel opened at the 1893 exposition in Chicago. It had 36 wooden cars, each capable of holding 60 passengers. (*Source:* The Handy Science Answer Book)
   a. What was the radius of this Ferris wheel?
   b. How close is the wheel to the ground?
   c. How high is the center of the wheel from the ground?
   d. Using the axes in the drawing, what are the coordinates of the center of the wheel?
   e. Use parts **a** and **d** to write the equation of the wheel.

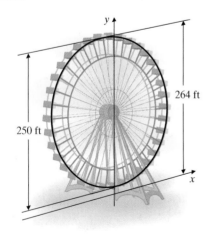

64. The world's largest-diameter Ferris wheel currently operating is the Cosmoclock 21 at Yokohama City, Japan. It has a 60-armed wheel, its diameter is 100 meters and it has a height of 105 meters. (*Source:* The Handy Science Answer Book)
   a. What is the radius of this Ferris wheel?
   b. How close is the wheel to the ground?
   c. How high is the center of the wheel from the ground?
   d. Using the axes in the drawing, what are the coordinates of the center of the wheel?
   e. Use parts **a** and **d** to write the equation of the wheel.

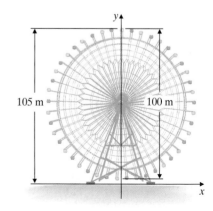

65. If you are given a list of equations of circles and parabolas and none are in standard form, explain how you would determine which is an equation of a circle and which is an equation of a parabola. Explain also how you would distinguish the upward or downward parabolas from the left-opening or right-opening parabolas.

△ **66.** Determine whether the triangle with vertices $(2, 6)$, $(0, -2)$, and $(5, 1)$ is an isosceles triangle.

Solve.

67. Two surveyors need to find the distance across a lake. They place a reference pole at point A in the diagram. Point B is 3 meters east and 1 meter north of the reference point A. Point C is 19 meters east and 13 meters north of point A. Find the distance across the lake, from B to C.

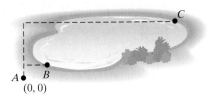

68. Cindy Brown, an architect, is drawing plans on grid paper for a circular pool with a fountain in the middle. The paper is marked off in centimeters, and each centimeter represents 1 foot. On the paper, the diameter of the "pool" is 20 centimeters, and "fountain" is the point $(0, 0)$.

  a. Sketch the architect's drawing. Be sure to label the axes.

  b. Write an equation that describes the circular pool.

  c. Cindy plans to place a circle of lights around the fountain such that each light is 5 feet from the fountain. Write an equation for the circle of lights and sketch the circle on your drawing.

69. A bridge constructed over a bayou has a supporting arch in the shape of a parabola. Find an equation of the parabolic arch if the length of the road over the arch is 100 meters and the maximum height of the arch is 40 meters.

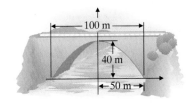

STUDY SKILLS REMINDER

How are you doing?

If you haven't done so yet, take a few moments and think about how you are doing in this course. Are you working toward your goal of successfully completing this course? Is your performance on homework, quizzes, and tests satisfactory? If not, you might want to see your instructor to see if he/she has any suggestions on how you can improve your performance. Let me once again remind you that, in addition to your instructor, there are many places to get help with your mathematics course. A few suggestions are below.

- This text has an accompanying video lesson for every section in this text.
- The back of this book contains answers to odd-numbered exercises and selected solutions.
- MathPro is available with this text. It is a tutorial software program with lessons corresponding to each section in the text.
- There is a student solutions manual available that contains worked-out solutions to odd-numbered exercises as well as solutions to every exercise in the Chapter Pretests, Integrated Reviews, Chapter Reviews, Chapter Tests, and Cumulative Reviews.
- Don't forget to check with your instructor for other local resources available to you, such as a tutor center.

9.2 The Ellipse and the Hyperbola

Ⓐ Graphing Ellipses

An **ellipse** can be thought of as the set of points in a plane such that the sum of the distances of each of those points from two fixed points is constant. Each of the two fixed points is called a **focus**. The plural of focus is **foci**. The point midway between the foci is called the **center**.

An ellipse may be drawn by hand by using two tacks, a piece of string, and a pencil. Secure the two tacks into a piece of cardboard, for example, and tie each end of the string to a tack. Use your pencil to pull the string tight and draw the ellipse. The two tacks are the foci of the drawn ellipse.

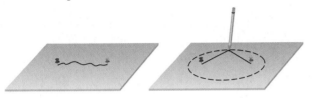

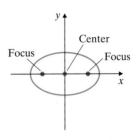

Ellipse with Center $(0, 0)$

The graph of an equation of the form $\dfrac{x^2}{a^2} + \dfrac{y^2}{b^2} = 1$ is an ellipse with center $(0,0)$. The x-intercepts are a and $-a$, and the y-intercepts are b and $-b$.

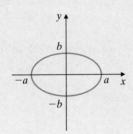

The **standard form** of an ellipse with center $(0, 0)$ is $\dfrac{x^2}{a^2} + \dfrac{y^2}{b^2} = 1$.

EXAMPLE 1   Graph: $\dfrac{x^2}{9} + \dfrac{y^2}{16} = 1$

Solution: The equation is of the form $\dfrac{x^2}{a^2} + \dfrac{y^2}{b^2} = 1$ with $a = 3$ and $b = 4$, so its graph is an ellipse with center $(0, 0)$, x-intercepts $(3, 0)$ and $(-3, 0)$, and y-intercepts $(0, 4)$ and $(0, -4)$.

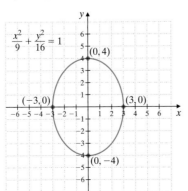

Practice Problem 1

Graph: $\dfrac{x^2}{4} + \dfrac{y^2}{9} = 1$

Answer

1.

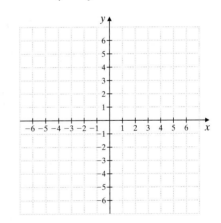

Practice Problem 2

Graph: $4x^2 + 25y^2 = 100$

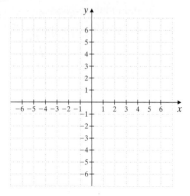

Practice Problem 3

Graph: $\dfrac{(x-1)^2}{9} + \dfrac{(y-3)^2}{16} = 1$

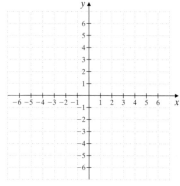

Answers

2.

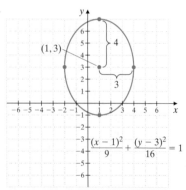

3.

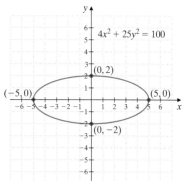

EXAMPLE 2 Graph: $4x^2 + 16y^2 = 64$

Solution: Although this equation contains a sum of squared terms in x and y on the same side of an equation, this is not the equation of a circle since the coefficients of x^2 and y^2 are not the same. When this happens, the graph is an ellipse. Since the standard form of the equation of an ellipse has 1 on one side, we divide both sides of this equation by 64 to get it in standard form.

$$4x^2 + 16y^2 = 64$$
$$\frac{4x^2}{64} + \frac{16y^2}{64} = \frac{64}{64} \qquad \text{Divide both sides by 64.}$$
$$\frac{x^2}{16} + \frac{y^2}{4} = 1 \qquad \text{Simplify.}$$

We now recognize the equation of an ellipse with center $(0, 0)$, x-intercepts $(4, 0)$ and $(-4, 0)$, and y-intercepts $(0, 2)$ and $(0, -2)$.

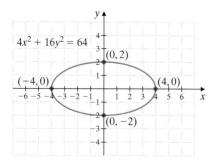

The center of an ellipse is not always $(0, 0)$, as shown in the next example. The standard form of an ellipse with center (h, k) is

$$\frac{(x-h)^2}{a^2} + \frac{(y-k)^2}{b^2} = 1$$

EXAMPLE 3 Graph: $\dfrac{(x+3)^2}{25} + \dfrac{(y-2)^2}{36} = 1$

Solution: This ellipse has center $(-3, 2)$. Notice that $a = 5$ and $b = 6$. To find four points on the graph of the ellipse, we first graph the center, $(-3, 2)$. Since $a = 5$, we count 5 units right and then 5 units left of the point with coordinates $(-3, 2)$. Next, since $b = 6$, we start at $(-3, 2)$ and count 6 units up and then 6 units down to find two more points on the ellipse.

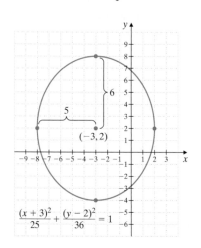

Try the Concept Check in the margin.

B   Graphing Hyperbolas

The final conic section is the **hyperbola**. A hyperbola is the set of points in a plane such that for each point in the set, the absolute value of the difference of the distances from two fixed points is constant. Each of the two fixed points is called a **focus**. The point midway between the foci is called the **center**.

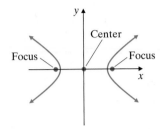

Using the distance formula, we can show that the graph of $\frac{x^2}{a^2} - \frac{y^2}{b^2} = 1$ is a hyperbola with center $(0,0)$ and x-intercepts $(a, 0)$ and $(-a, 0)$. Also, the graph of $\frac{y^2}{b^2} - \frac{x^2}{a^2} = 1$ is a hyperbola with center $(0,0)$ and y-intercepts $(0, b)$ and $(0, -b)$.

Hyperbola with Center $(0, 0)$

The graph of an equation of the form $\frac{x^2}{a^2} - \frac{y^2}{b^2} = 1$ is a hyperbola with center $(0, 0)$ and x-intercepts $(a, 0)$ and $(-a, 0)$.

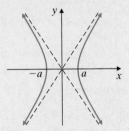

The graph of an equation of the form $\frac{y^2}{b^2} - \frac{x^2}{a^2} = 1$ is a hyperbola with center $(0, 0)$ and y-intercepts $(0, b)$ and $(0, -b)$.

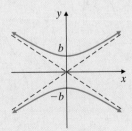

The equations $\frac{x^2}{a^2} - \frac{y^2}{b^2} = 1$ and $\frac{y^2}{b^2} - \frac{x^2}{a^2} = 1$ are the **standard forms** for the equation of a hyperbola.

Concept Check

In the graph of the equation $\frac{x^2}{64} + \frac{y^2}{36} = 1$, which distance is longer: the distance between the x-intercepts or the distance between the y-intercepts? How much longer? Explain.

Answer

Concept Check: x-intercepts, by 4 units

Helpful Hint

Notice the difference between the equation of an ellipse and a hyperbola. The equation of the ellipse contains x^2- and y^2-terms on the same side of the equation with same-sign coefficients. For a hyperbola, the coefficients on the same side of the equation have different signs.

Graphing a hyperbola such as $\dfrac{y^2}{b^2} - \dfrac{x^2}{a^2} = 1$ is made easier by recognizing one of its important characteristics. Examining the figure below, notice how the sides of the branches of the hyperbola extend indefinitely and seem to approach, but not intersect, the dashed lines in the figure. These dashed lines are called the **asymptotes** of the hyperbola.

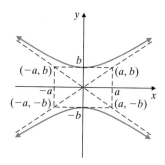

Practice Problem 4

Graph: $\dfrac{x^2}{4} - \dfrac{y^2}{9} = 1$

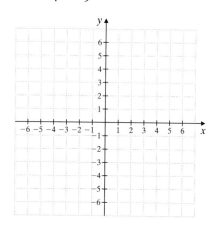

To sketch these lines, or asymptotes, draw a rectangle with vertices $(a, b), (-a, b), (a, -b)$, and $(-a, -b)$. The asymptotes of the hyperbola are the extended diagonals of this rectangle.

EXAMPLE 4 Graph: $\dfrac{x^2}{16} - \dfrac{y^2}{25} = 1$

Solution: This equation has the form $\dfrac{x^2}{a^2} - \dfrac{y^2}{b^2} = 1$, with $a = 4$ and $b = 5$.

Thus, its graph is a hyperbola with center $(0, 0)$ and x-intercepts of $(4, 0)$ and $(-4, 0)$. To aid in graphing the hyperbola, we first sketch its asymptotes. The extended diagonals of the rectangle with coordinates $(4, 5), (4, -5), (-4, 5)$, and $(-4, -5)$ are the asymptotes of the hyperbola. Then we use the asymptotes to aid in graphing the hyperbola.

Answer

4.

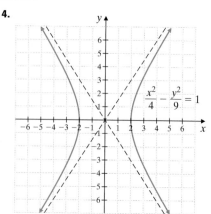

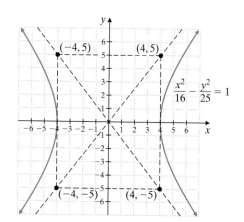

EXAMPLE 5   Graph: $4y^2 - 9x^2 = 36$

Solution: Since this is a difference of squared terms in x and y on the same side of the equation, its graph is a hyperbola, as opposed to an ellipse or a circle. The standard form of the equation of a hyperbola has a 1 on one side, so we divide both sides of the equation by 36 to get it in standard form.

$$4y^2 - 9x^2 = 36$$

$$\frac{4y^2}{36} - \frac{9x^2}{36} = \frac{36}{36}$$ Divide both sides by 36.

$$\frac{y^2}{9} - \frac{x^2}{4} = 1$$ Simplify.

The equation is of the form $\dfrac{y^2}{b^2} - \dfrac{x^2}{a^2} = 1$ with $a = 2$ and $b = 3$, so the hyperbola is centered at $(0, 0)$ with y-intercepts $(0, 3)$ and $(0, -3)$.

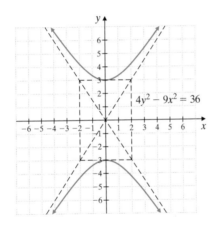

Practice Problem 5

Graph: $9y^2 - 16x^2 = 144$

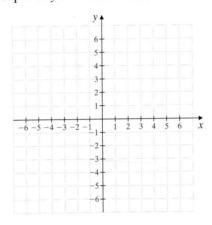

Answer

5.

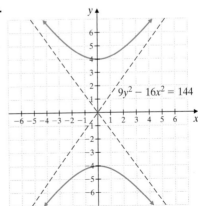

GRAPHING CALCULATOR EXPLORATIONS

To find the graph of an ellipse by using a graphing calculator, use the same procedure as for graphing a circle. For example, to graph $x^2 + 3y^2 = 22$, first solve for y.

$$3y^2 = 22 - x^2$$

$$y^2 = \frac{22 - x^2}{3}$$

$$y = \pm\sqrt{\frac{22 - x^2}{3}}$$

Next press the $\boxed{Y=}$ key and enter $Y_1 = \sqrt{\dfrac{22 - x^2}{3}}$ and

$Y_2 = -\sqrt{\dfrac{22 - x^2}{3}}$. (Insert two sets of parentheses in

the radicand as $\sqrt{((22 - x^2)/3)}$ so that the desired graph is obtained.) The graph appears as follows:

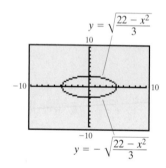

Use a graphing calculator to graph each ellipse.

1. $10x^2 + y^2 = 32$

2. $20x^2 + 5y^2 = 100$

3. $7.3x^2 + 15.5y^2 = 95.2$

4. $18.8x^2 + 36.1y^2 = 205.8$

Mental Math

Identify the graph of each equation as an ellipse or a hyperbola.

1. $\dfrac{x^2}{16} + \dfrac{y^2}{4} = 1$

2. $\dfrac{x^2}{16} - \dfrac{y^2}{4} = 1$

3. $x^2 - 5y^2 = 3$

4. $-x^2 + 5y^2 = 3$

5. $-\dfrac{y^2}{25} + \dfrac{x^2}{36} = 1$

6. $\dfrac{y^2}{25} + \dfrac{x^2}{36} = 1$

EXERCISE SET 9.2

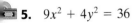 **A** *Graph each equation. See Examples 1 and 2.*

1. $\dfrac{x^2}{4} + \dfrac{y^2}{25} = 1$

2. $\dfrac{x^2}{9} + y^2 = 1$

3. $\dfrac{x^2}{16} + \dfrac{y^2}{9} = 1$

4. $x^2 + \dfrac{y^2}{4} = 1$

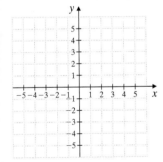
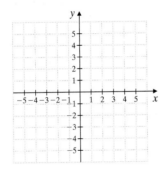
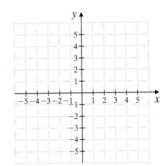
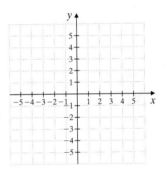

5. $9x^2 + 4y^2 = 36$

6. $x^2 + 4y^2 = 16$

7. $4x^2 + 25y^2 = 100$

8. $36x^2 + y^2 = 36$

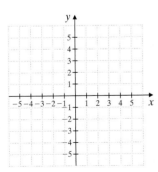
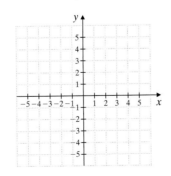
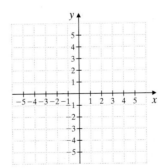
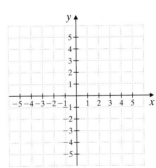

Graph each equation. See Example 3.

9. $\dfrac{(x+1)^2}{36} + \dfrac{(y-2)^2}{49} = 1$ **10.** $\dfrac{(x-3)^2}{9} + \dfrac{(y+3)^2}{16} = 1$ **11.** $\dfrac{(x-1)^2}{4} + \dfrac{(y-1)^2}{25} = 1$ **12.** $\dfrac{(x+3)^2}{16} + \dfrac{(y+2)^2}{4} = 1$

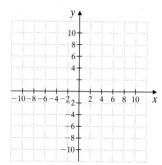
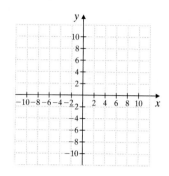
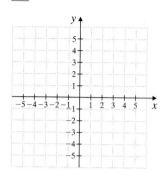
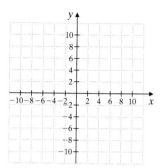

B *Graph each equation. See Examples 4 and 5.*

13. $\dfrac{x^2}{4} - \dfrac{y^2}{9} = 1$ **14.** $\dfrac{x^2}{36} - \dfrac{y^2}{36} = 1$ **15.** $\dfrac{y^2}{25} - \dfrac{x^2}{16} = 1$ **16.** $\dfrac{y^2}{25} - \dfrac{x^2}{49} = 1$

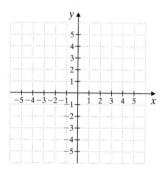
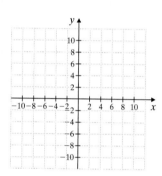
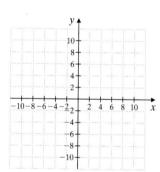
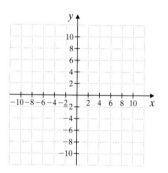

17. $x^2 - 4y^2 = 16$ **18.** $4x^2 - y^2 = 36$ **19.** $16y^2 - x^2 = 16$ **20.** $4y^2 - 25x^2 = 100$

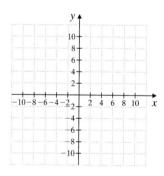
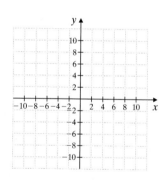
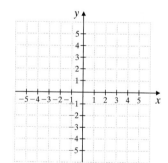
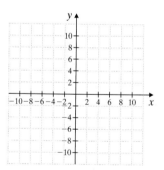

Review and Preview

Perform each indicated operation. See Sections 5.1 and 5.2.

21. $(2x^3)(-4x^2)$ **22.** $2x^3 - 4x^3$ **23.** $-5x^2 + x^2$ **24.** $(-5x^2)(x^2)$

25. We know that $x^2 + y^2 = 25$ is the equation of a circle. Rewrite the equation so that the right side is equal to 1. Which type of conic section does this equation form resemble? In fact, the circle is a special case of this type of conic section. Describe the conditions under which this type of conic section is a circle.

The orbits of stars, planets, comets, asteroids, and satellites all have the shape of one of the conic sections. Astronomers use a measure called eccentricity *to describe the shape and elongation of an orbital path. For the circle and ellipse, eccentricity*

e *is calculated with the formula* $e = \dfrac{c}{d}$, *where* $c^2 = |a^2 - b^2|$ *and* d *is the larger value of* a *or* b. *For a hyperbola, eccentricity*

e *is calculated with the formula* $e = \dfrac{c}{d}$, *where* $c^2 = a^2 + b^2$ *and the value of* d *is equal to* a *if the hyperbola has x-intercepts*

or equal to b *if the hyperbola has y-intercepts. Use equations A–H to answer Exercises 26–35.*

A $\dfrac{x^2}{36} - \dfrac{y^2}{13} = 1$ **B** $\dfrac{x^2}{4} + \dfrac{y^2}{4} = 1$ **C** $\dfrac{x^2}{25} + \dfrac{y^2}{16} = 1$ **D** $\dfrac{y^2}{25} - \dfrac{x^2}{39} = 1$

E $\dfrac{x^2}{17} + \dfrac{y^2}{81} = 1$ **F** $\dfrac{x^2}{36} + \dfrac{y^2}{36} = 1$ **G** $\dfrac{x^2}{16} - \dfrac{y^2}{65} = 1$ **H** $\dfrac{x^2}{144} + \dfrac{y^2}{140} = 1$

26. Identify the type of conic section represented by each of the equations A–H.

27. For each of the equations A–H, identify the values of a^2 and b^2.

28. For each of the equations A–H, calculate the value of c^2 and c.

29. For each of the equations A–H, find the value of d.

30. For each of the equations A–H, calculate the eccentricity e.

31. What do you notice about the values of e for the equations you identified as ellipses?

32. What do you notice about the values of e for the equations you identified as circles?

33. What do you notice about the values of e for the equations you identified as hyperbolas?

34. The eccentricity of a parabola is exactly 1. Use this information and the observations you made in Exercises 31, 32, and 33 to describe a way that could be used to identify the type of conic section based on its eccentricity value.

35. Graph each of the conic sections given in equations A–H. What do you notice about the shape of the ellipses for increasing values of eccentricity? Which is the most elliptical? Which is the least elliptical, that is, the most circular?

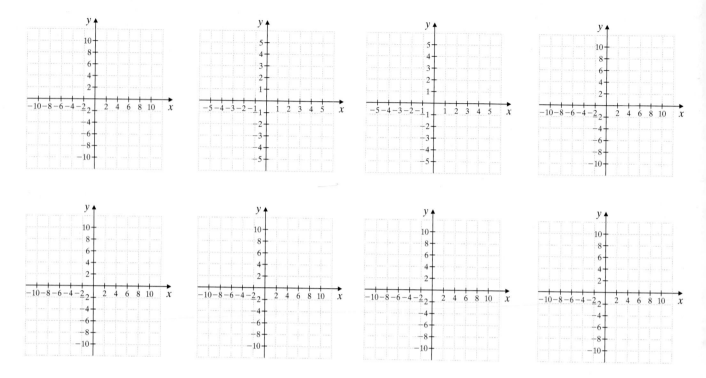

Internet Excursions

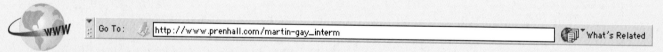

By going to this World Wide Web address, you will be directed to a Web site where you can look up information to help you answer the questions below.

36. Under Planets, select the Mean Orbital Elements option on this homepage. This gives data about the nine planets of our solar system, including the eccentricities of their orbits. Using the data for the eccentricities of the planets' orbits and your conclusions from Exercise 34, decide which type of conic describes the orbital paths of all the planets. Which planet has the most circular path? Which planet has the most elliptical path?

37. Return to the JPL Solar System Dynamics page. Under Comets and Asteroids, select the Orbital Elements option. Then choose the Comets option. This gives data about the orbits of comets known to pass through our solar system, including their eccentricities. Using the data for the eccentricities of the comets' orbits and your conclusions from Exercise 34, decide which type of conic section describes the orbital paths of the majority of the comets. Which comets are the exceptions? What type of orbital paths do these comets have?

Integrated Review–Graphing Conic Sections

Following is a summary of conic sections.

Conic Sections

	Standard Form	**Graph**
Parabola	$y = a(x - h)^2 + k$	
Parabola	$x = a(y - k)^2 + h$	
Circle	$(x - h)^2 + (y - k)^2 = r^2$	
Ellipse	$\dfrac{x^2}{a^2} + \dfrac{y^2}{b^2} = 1$	
Hyperbola	$\dfrac{x^2}{a^2} - \dfrac{y^2}{b^2} = 1$	
Hyperbola	$\dfrac{y^2}{b^2} - \dfrac{x^2}{a^2} = 1$	

Identify whether each equation, when graphed, will be a parabola, circle, ellipse, or hyperbola. Then graph each equation.

1. $(x - 7)^2 + (y - 2)^2 = 4$ **2.** $y = x^2 + 4$ **3.** $y = x^2 + 12x + 36$

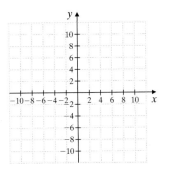 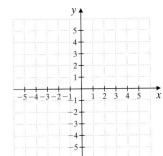 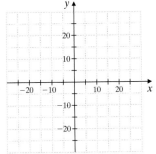

4. _____

5. _____

6. _____

7. _____

8. _____

9. _____

10. _____

11. _____

12. _____

13. _____

14. _____

15. _____

4. $\dfrac{x^2}{4} + \dfrac{y^2}{9} = 1$

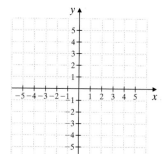

5. $\dfrac{y^2}{9} - \dfrac{x^2}{9} = 1$

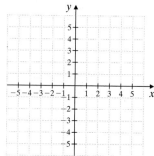

6. $\dfrac{x^2}{16} - \dfrac{y^2}{4} = 1$

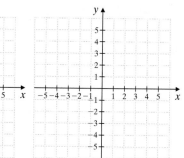

7. $\dfrac{x^2}{16} + \dfrac{y^2}{4} = 1$

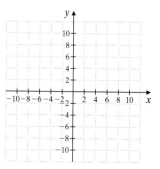

8. $x^2 + y^2 = 16$

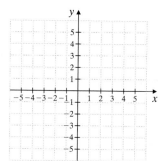

9. $x = y^2 + 4y - 1$

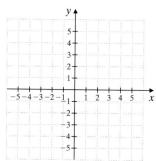

10. $x = -y^2 + 6y$

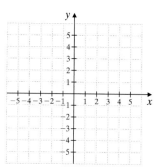

11. $9x^2 - 4y^2 = 36$

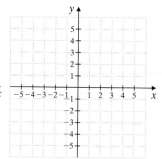

12. $9x^2 + 4y^2 = 36$

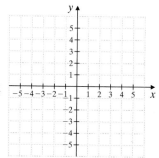

13. $\dfrac{(x-1)^2}{49} + \dfrac{(y+2)^2}{25} = 1$ **14.** $y^2 = x^2 + 16$ **15.** $\left(x + \dfrac{1}{2}\right)^2 + \left(y - \dfrac{1}{2}\right)^2 = 1$

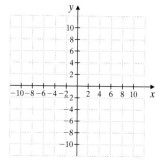

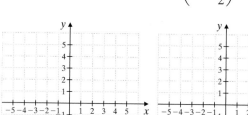

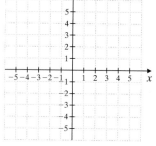

9.3 Graphing Nonlinear Functions

(A) Graphing Nonlinear Functions

Recall that the graph of $f(x) = x^2$ is a parabola with vertex $(0,0)$. How does the graph of $g(x) = (x - 3)^2 + 2$ compare? Its graph is a parabola of the same shape as f, but with vertex $(3,2)$. In other words, the graph of g is the same as the graph of f, except that it has been shifted 3 units to the right and 2 units up. Keep this in mind as we graph other elementary functions. Remember, we are graphing functions, so all graphs should pass the vertical line test.

OBJECTIVE

(A) Graph the nonlinear functions $f(x) = |x|$ and $f(x) = \sqrt{x}$.

SSM
TUTOR CENTER   SG   CD & VIDEO   MATH PRO   WEB

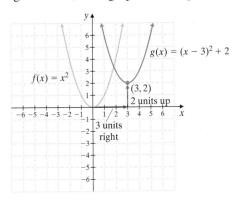

EXAMPLE 1   Graph: $f(x) = |x|$

Solution: This is not a linear function, and its graph is not a line. Because we do not know the shape of this graph, we find many ordered pair solutions. We will choose x-values and substitute to find corresponding y-values. Recall that

If $x = -3$, then $y = |-3|$, or 3.
If $x = -2$, then $y = |-2|$, or 2.
If $x = -1$, then $y = |-1|$, or 1.
If $x = 0$, then $y = |0|$, or 0.
If $x = 1$, then $y = |1|$, or 1.
If $x = 2$, then $y = |2|$, or 2.
If $x = 3$, then $y = |3|$, or 3.

x	y
-3	3
-2	2
-1	1
0	0
1	1
2	2
3	3

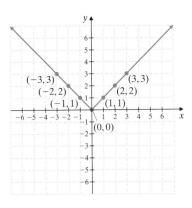

Study the table of values for a moment and notice any patterns. Since the absolute value of a real number is a nonnegative number, notice that the domain of this function is $\{x \mid x$ is a real number,$\}$ but the range is $\{y \mid y \geq 0\}$. From the plotted ordered pairs, we see that the graph of this absolute value function is V-shaped.

Practice Problem 1

Graph: $f(x) = |x| + 1$

Answer

1.

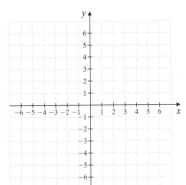

Practice Problem 2

Graph $f(x) = |x| - 1$

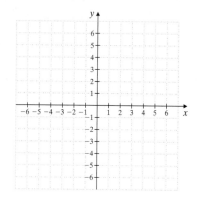

Practice Problem 3

Graph: $f(x) = |x - 1|$

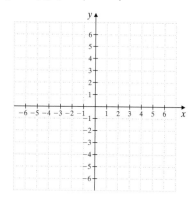

Answers

2.

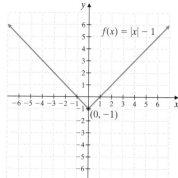

3.

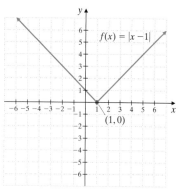

EXAMPLE 2   Graph: $f(x) = |x| - 3$

Solution: To graph $f(x)$ or $y = |x| - 3$, choose x-values and substitute to find corresponding y-values.

x	y
-3	0
-2	-1
-1	-2
0	-3
1	-2
2	-1
3	0

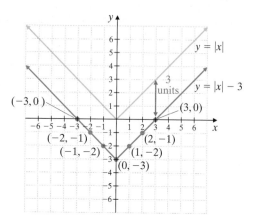

Recall that the graph of $y = x^2 - 3$ is the same as the graph of $y = x^2$ lowered 3 units. Now compare the graph of $y = |x|$ with the graph of $y = |x| - 3$. The graph of $y = |x| - 3$ is the same as the graph of $y = |x|$ lowered 3 units.   ●

EXAMPLE 3   Graph: $f(x) = |x - 2|$

Solution: First let's think about the graph of $y = (x - 2)^2$. The vertex of this graph is $(2, 0)$. In other words, the graph of $y = (x - 2)^2$ is the same as the graph of $y = x^2$ shifted to the right 2 units.

In the same manner, the graph of $y = |x - 2|$ is the same as the graph of $y = |x|$ shifted to the right 2 units. We use this knowledge along with a table of ordered pair solutions to graph the function.

x	y
0	2
1	1
2	0
3	1
4	2

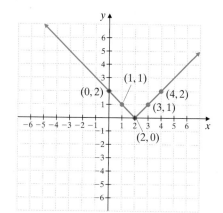

●

EXAMPLE 4 Graph: $f(x) = |x - 1| + 2$

Solution: The graph of $y = (x - 1)^2 + 2$ has vertex $(1, 2)$. In other words, it is the graph of $y = x^2$ shifted 1 unit to the right and 2 units up. Similarly, the graph of $y = |x - 1| + 2$ is the graph of $y = |x|$ shifted 1 unit to the right and 2 units up. We use this knowledge along with a table of ordered pair solutions to graph the function.

x	y
−1	4
0	3
1	2
2	3
3	4

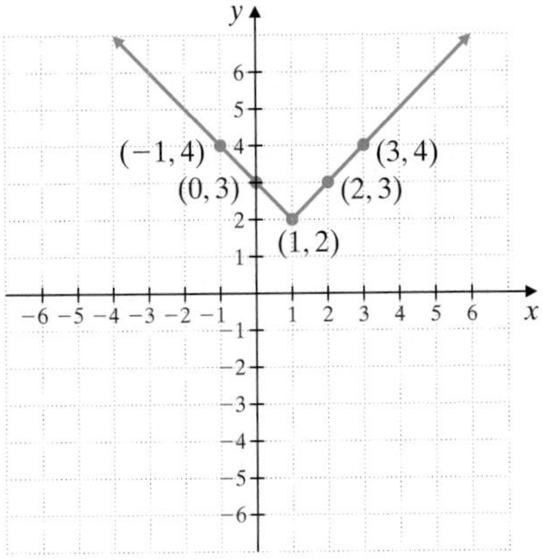

Recall that the domain of a function is basically the set of all possible x-values for that function. The domains of the functions thus far in this section have been the set of all real numbers. This is not the case for our next function, the square root function.

EXAMPLE 5 Graph the square root function $f(x) = \sqrt{x}$.

Solution: Recall that the square root of a negative number is not a real number. This means that the domain of this function is the set of all nonnegative numbers, or $\{x | x \geq 0\}$. To graph this function, evaluate the function for several values of x, plot the resulting points, and connect the points with a smooth curve.

x	y
0	0
1	1
3	$\sqrt{3} \approx 1.7$
4	2
9	3

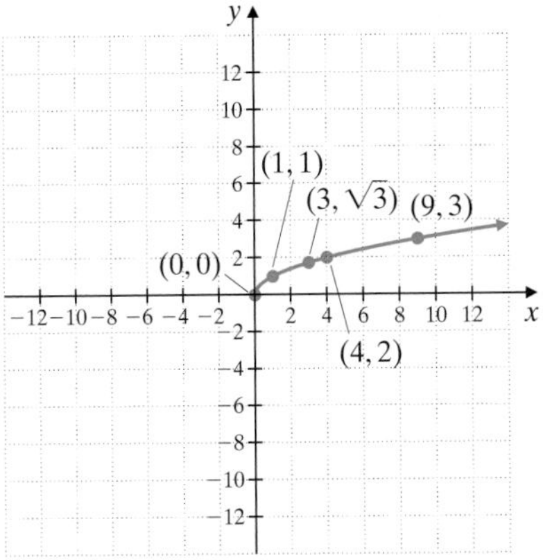

Practice Problem 4

Graph: $f(x) = |x - 3| + 2$

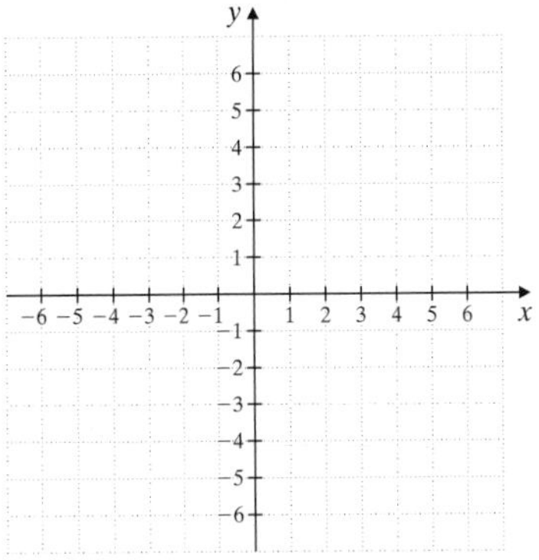

Practice Problem 5

Graph: $f(x) = \sqrt{x} + 2$

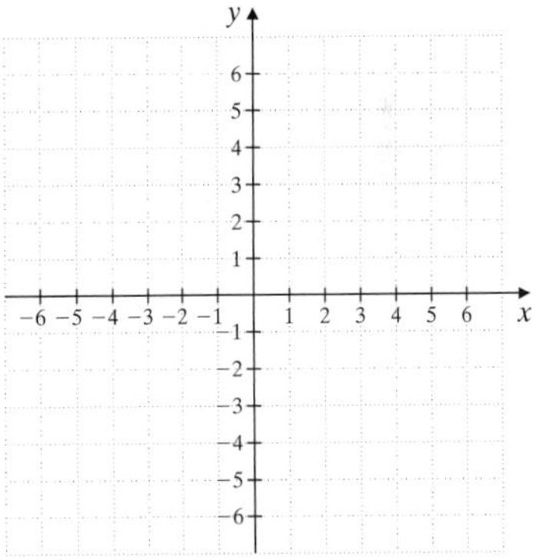

Answers

4.

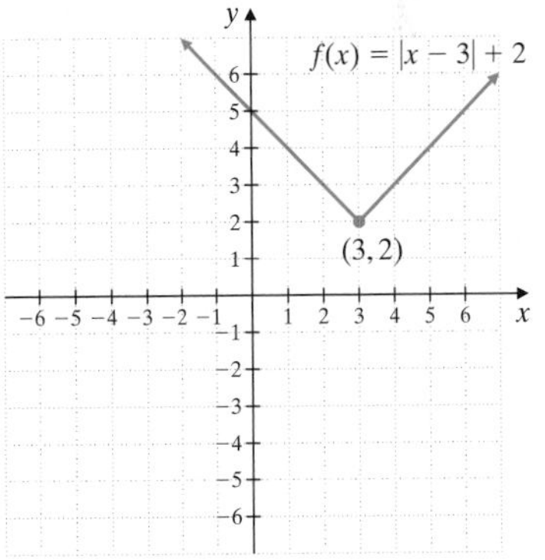

5.

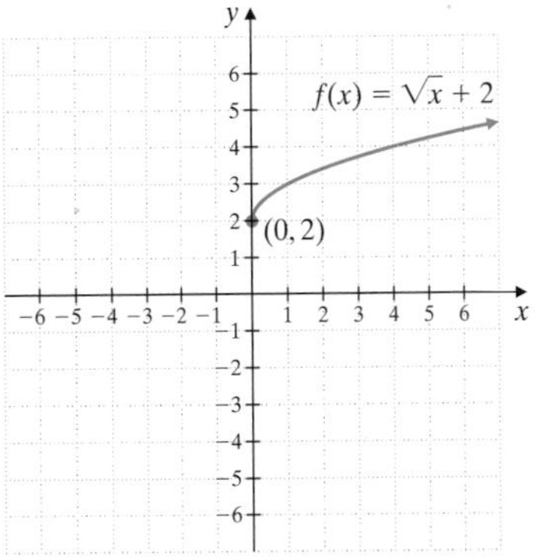

Practice Problem 6

Graph: $f(x) = \sqrt{x + 1} + 2$

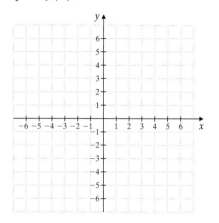

EXAMPLE 6   Graph: $f(x) = \sqrt{x + 2} + 3$

Solution: The graph of $y = (x + 2)^2 + 3$ has vertex $(-2, 3)$ and is the graph of $y = x^2$ shifted 2 units left and 3 units up. Similarly, the graph of $y = \sqrt{x + 2} + 3$ is the graph of $y = \sqrt{x}$ shifted 2 units left and 3 units up. We use this knowledge along with a table of values to graph the function.

x	y
−2	3
−1	4
2	5

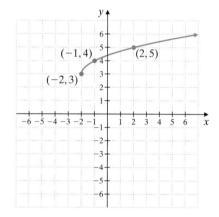

This vertical and horizontal shifting works for any function.

Graphing $F(x) = f(x - h) + k$

The graph of the function

$$F(x) = f(x - h) + k$$

is the same as the graph of the function $y = f(x)$ except that it has been shifted left or right h units and up or down k units. It is shifted to the right if $h > 0$ and left if $h < 0$. It is shifted up if $k > 0$ and down if $k < 0$.

Recall from Chapter 7 that the domain of $f(x) = \sqrt{x + 2} + 3$ includes all real numbers that make the radicand ≥ 0. To see what numbers these are, solve $x + 2 \geq 0$ and find that $x \geq -2$. The domain of f is $\{x | x \geq -2\}$. This can be verified by observing the graph of Example 6.

Answer

6.

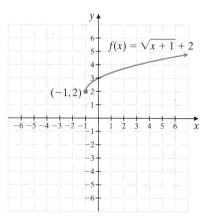

EXERCISE SET 9.3

A *Graph each function. See Examples 1 through 6.*

1. $f(x) = |x| + 3$

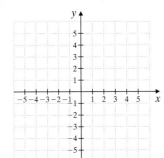

2. $f(x) = |x| - 2$

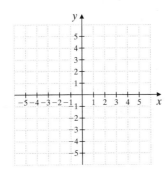

3. $f(x) = \sqrt{x} - 2$

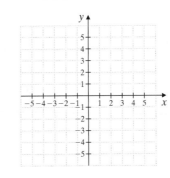

4. $f(x) = \sqrt{x} + 3$

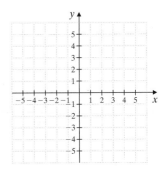

5. $f(x) = |x - 4|$

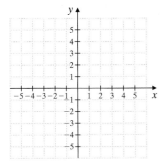

6. $f(x) = |x + 3|$

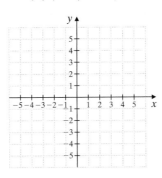

7. $f(x) = \sqrt{x + 2}$

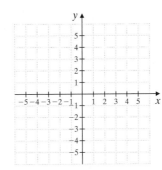

8. $f(x) = \sqrt{x - 2}$

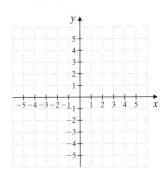

9. $f(x) = \sqrt{x - 2} + 3$

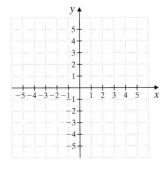

10. $f(x) = \sqrt{x - 1} + 3$

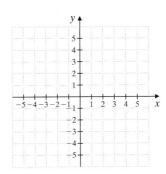

11. $f(x) = |x - 1| + 5$

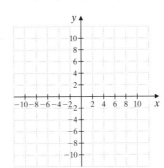

12. $f(x) = |x - 3| + 2$

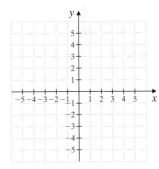

13. $f(x) = \sqrt{x + 1} + 1$

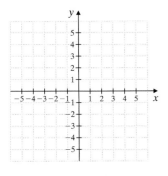

14. $f(x) = \sqrt{x + 3} + 2$

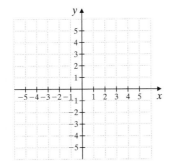

15. $f(x) = |x + 3| - 1$

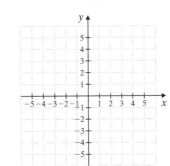

16. $f(x) = |x + 1| - 4$

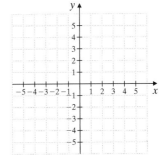

703

Review and Preview

Solve each system of equations. See Section 4.1.

17. $\begin{cases} x + y = 6 \\ x - y = 10 \end{cases}$

18. $\begin{cases} x + y = -2 \\ -x + y = -8 \end{cases}$

19. $\begin{cases} 2x + 3y = 7 \\ -x + 4y = 13 \end{cases}$

20. $\begin{cases} 4x - 3y = -4 \\ 3x - y = 10 \end{cases}$

 Combining Concepts

Graph each function. Recall that the domain of the cube function and the cube root function is the set of all real numbers. For Exercises 22 and 24, predict the location and appearance of the graph and then use a graphing calculator to verify.

21. $f(x) = x^3$

22. $f(x) = (x - 1)^3 + 2$

23. $f(x) = \sqrt[3]{x}$

24. $f(x) = \sqrt[3]{x - 3} + 1$

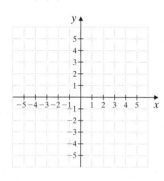 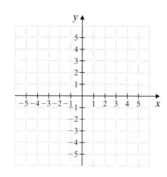 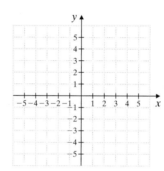 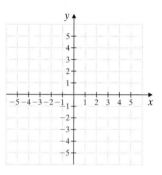

Without graphing, find the domain of each function.

25. $f(x) = 5\sqrt{x - 20} + 1$

26. $g(x) = -3\sqrt{x + 5}$

27. $h(x) = 5|x - 20| + 1$

28. $f(x) = -3|x + 5.7|$

29. $g(x) = 9 - \sqrt{x + 103}$

30. $h(x) = \sqrt{x - 17} - 3$

9.4 Solving Nonlinear Systems of Equations

In Section 4.1, we used graphing, substitution, and elimination methods to find solutions of systems of linear equations in two variables. We now apply these same methods to nonlinear systems of equations in two variables. A **nonlinear system of equations** is a system of equations at least one of which is not linear. Since we will be graphing the equations in each system, we are interested in real number solutions only.

Ⓐ Solving Nonlinear Systems by Substitution

First we solve nonlinear systems by the substitution method.

EXAMPLE 1   Solve the system:

$$\begin{cases} x^2 - 3y = 1 \\ x - y = 1 \end{cases}$$

Solution: We can solve this system by substitution if we solve one equation for one of the variables. Solving the first equation for x is not the best choice since doing so introduces a radical. Also, solving for y in the first equation introduces a fraction. Thus, we solve the second equation for y.

$x - y = 1$ Second equation
$x - 1 = y$ Solve for y.

Now we replace y with $x - 1$ in the first equation, and then solve for x.

$$x^2 - 3y = 1 \qquad \text{First equation}$$

$$x^2 - 3(x - 1) = 1 \qquad \text{Replace } y \text{ with } x - 1.$$
$$x^2 - 3x + 3 = 1$$
$$x^2 - 3x + 2 = 0$$
$$(x - 2)(x - 1) = 0$$
$$x = 2 \quad \text{or} \quad x = 1$$

Now we let $x = 2$ and then $x = 1$ in the equation $y = x - 1$ to find corresponding y-values.

Let $x = 2$.   Let $x = 1$.
$\quad y = x - 1$ $\quad y = x - 1$
$\quad y = 2 - 1 = 1$ $\quad y = 1 - 1 = 0$

When we check $(2, 1)$ and $(1, 0)$ in the equations, we find that both ordered pairs satisfy both equations. Thus, the solution set for the system is $\{(2, 1), (1, 0)\}$. The graph of each equation in the system is shown.

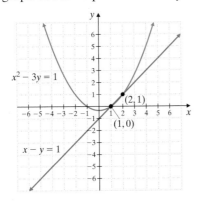

OBJECTIVES

Ⓐ Solve a nonlinear system by substitution.

Ⓑ Solve a nonlinear system by elimination.

SSM
TUTOR CENTER   SG   CD & VIDEO   MATH PRO   WEB

Practice Problem 1

Solve the system: $\begin{cases} x^2 - 2y = 5 \\ x + y = -1 \end{cases}$

Answer
1. $\{(-3, 2), (1, -2)\}$

Practice Problem 2

Solve the system: $\begin{cases} y = \sqrt{x} \\ x^2 + y^2 = 12 \end{cases}$

EXAMPLE 2   Solve the system:

$$\begin{cases} y = \sqrt{x} \\ x^2 + y^2 = 6 \end{cases}$$

Solution: This system is ideal for the substitution method since y is expressed in terms of x in the first equation. Notice that if $y = \sqrt{x}$, then both x and y must be nonnegative if they are real numbers. Let's substitute $\sqrt{x}$ for y in the second equation, and solve for x.

$$x^2 + y^2 = 6$$
$$x^2 + (\sqrt{x})^2 = 6 \qquad \text{Let } y = \sqrt{x}.$$
$$x^2 + x = 6$$
$$x^2 + x - 6 = 0$$
$$(x + 3)(x - 2) = 0$$
$$x = -3 \quad \text{or} \quad x = 2$$

The solution -3 is discarded because we have noted that x must be nonnegative. To see this, we let $x = -3$ and $x = 2$ in the first equation to find the corresponding y-values.

Let $x = -3$.

$$y = \sqrt{x}$$
$$y = \sqrt{-3} \quad \text{Not a real number}$$

Let $x = 2$.

$$y = \sqrt{x}$$
$$y = \sqrt{2}$$

Since we are interested only in real number solutions, the only solution is $(2, \sqrt{2})$. The solution set is $\{(2, \sqrt{2})\}$. Check to see that this solution satisfies both equations. The graph of each equation in this system is shown.

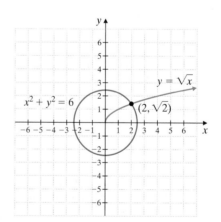

Practice Problem 3

Solve the system: $\begin{cases} x^2 + y^2 = 1 \\ x + y = 4 \end{cases}$

EXAMPLE 3   Solve the system:

$$\begin{cases} x^2 + y^2 = 4 \\ x + y = 3 \end{cases}$$

Solution: We use the substitution method and solve the second equation for x.

$$x + y = 3 \qquad \text{Second equation}$$
$$x = 3 - y$$

Answers

2. $\{(3, \sqrt{3})\}$ **3.** no solution

Now we let $x = 3 - y$ in the first equation.

$$x^2 + y^2 = 4 \qquad \text{First equation}$$

$$(3 - y)^2 + y^2 = 4 \qquad \text{Let } x = 3 - y.$$

$$9 - 6y + y^2 + y^2 = 4$$

$$2y^2 - 6y + 5 = 0$$

By the quadratic formula, where $a = 2$, $b = -6$, and $c = 5$, we have

$$y = \frac{6 \pm \sqrt{(-6)^2 - 4 \cdot 2 \cdot 5}}{2 \cdot 2} = \frac{6 \pm \sqrt{-4}}{4}$$

Since $\sqrt{-4}$ is not a real number, there is no solution. Graphically, the circle and the line do not intersect, as shown.

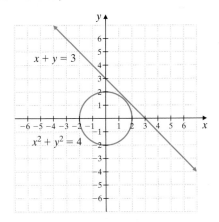

Try the Concept Check in the margin.

B **Solving Nonlinear Systems by Elimination**

Some nonlinear systems may be solved by the elimination method.

EXAMPLE 4   Solve the system:

$$\begin{cases} x^2 + 2y^2 = 10 \\ x^2 - y^2 = 1 \end{cases}$$

Solution: We will use the elimination, or addition, method to solve this system. To eliminate x^2 when we add the two equations, we multiply both sides of the second equation by -1. Then

$$\begin{cases} x^2 + 2y^2 = 10 \\ (-1)(x^2 - y^2) = -1 \cdot 1 \end{cases} \begin{matrix} \text{is} \\ \text{equivalent} \\ \text{to} \end{matrix} \begin{cases} x^2 + 2y^2 = 10 \\ -x^2 + y^2 = -1 \end{cases}$$

$$\begin{aligned} 3y^2 &= 9 \qquad \text{Add.} \\ y^2 &= 3 \\ y &= \pm\sqrt{3} \quad \begin{matrix} \text{Divide both} \\ \text{sides by 3.} \end{matrix} \end{aligned}$$

Concept Check

Without solving, how can you tell that $x^2 + y^2 = 9$ and $x^2 + y^2 = 16$ do not have any points of intersection?

Practice Problem 4

Solve the equation: $\begin{cases} x^2 + 3y^2 = 21 \\ x^2 - y^2 = 1 \end{cases}$

Answers

4. $\{(\sqrt{6}, \sqrt{5}), (\sqrt{6}, -\sqrt{5}), (-\sqrt{6}, \sqrt{5}), (-\sqrt{6}, -\sqrt{5})\}$

Concept Check: $x^2 + y^2 = 9$ is a circle inside the circle $x^2 + y^2 = 16$, therefore they do not have any points of intersection.

To find the corresponding x-values, we let $y = \sqrt{3}$ and $y = -\sqrt{3}$ in either original equation. We choose the second equation.

Let $y = \sqrt{3}$.

$$x^2 - y^2 = 1$$

$$x^2 - (\sqrt{3})^2 = 1$$

$$x^2 - 3 = 1$$

$$x^2 = 4$$

$$x = \pm\sqrt{4} = \pm 2$$

Let $y = -\sqrt{3}$.

$$x^2 - y^2 = 1$$

$$x^2 - (-\sqrt{3})^2 = 1$$

$$x^2 - 3 = 1$$

$$x^2 = 4$$

$$x = \pm\sqrt{4} = \pm 2$$

The solution set is $\{(2, \sqrt{3}), (-2, \sqrt{3}), (2, -\sqrt{3}), (-2, -\sqrt{3})\}$. Check all four ordered pairs in both equations of the system. The graph of each equation in this system is shown.

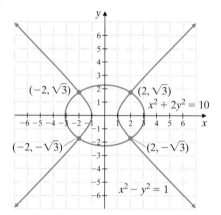

EXERCISE SET 9.4

A **B** *Solve each nonlinear system of equations. See Examples 1 through 4.*

1. $\begin{cases} x^2 + y^2 = 25 \\ 4x + 3y = 0 \end{cases}$

2. $\begin{cases} x^2 + y^2 = 25 \\ 3x + 4y = 0 \end{cases}$

3. $\begin{cases} x^2 + 4y^2 = 10 \\ y = x \end{cases}$

4. $\begin{cases} 4x^2 + y^2 = 10 \\ y = x \end{cases}$

5. $\begin{cases} y^2 = 4 - x \\ x - 2y = 4 \end{cases}$

6. $\begin{cases} x^2 + y^2 = 4 \\ x + y = -2 \end{cases}$

7. $\begin{cases} x^2 + y^2 = 9 \\ 16x^2 - 4y^2 = 64 \end{cases}$

8. $\begin{cases} 4x^2 + 3y^2 = 35 \\ 5x^2 + 2y^2 = 42 \end{cases}$

9. $\begin{cases} x^2 + 2y^2 = 2 \\ x - y = 2 \end{cases}$

10. $\begin{cases} x^2 + 2y^2 = 2 \\ x^2 - 2y^2 = 6 \end{cases}$

11. $\begin{cases} y = x^2 - 3 \\ 4x - y = 6 \end{cases}$

12. $\begin{cases} y = x + 1 \\ x^2 - y^2 = 1 \end{cases}$

13. $\begin{cases} y = x^2 \\ 3x + y = 10 \end{cases}$

14. $\begin{cases} 6x - y = 5 \\ xy = 1 \end{cases}$

15. $\begin{cases} y = 2x^2 + 1 \\ x + y = -1 \end{cases}$

16. $\begin{cases} x^2 + y^2 = 9 \\ x + y = 5 \end{cases}$

17. $\begin{cases} y = x^2 - 4 \\ y = x^2 - 4x \end{cases}$

18. $\begin{cases} x = y^2 - 3 \\ x = y^2 - 3y \end{cases}$

19. $\begin{cases} 2x^2 + 3y^2 = 14 \\ -x^2 + y^2 = 3 \end{cases}$

20. $\begin{cases} 4x^2 - 2y^2 = 2 \\ -x^2 + y^2 = 2 \end{cases}$

21. $\begin{cases} x^2 + y^2 = 1 \\ x^2 + (y + 3)^2 = 4 \end{cases}$

22. $\begin{cases} x^2 + 2y^2 = 4 \\ x^2 - y^2 = 4 \end{cases}$

23. $\begin{cases} y = x^2 + 2 \\ y = -x^2 + 4 \end{cases}$

24. $\begin{cases} x = -y^2 - 3 \\ x = y^2 - 5 \end{cases}$

25. $\begin{cases} 3x^2 + y^2 = 9 \\ 3x^2 - y^2 = 9 \end{cases}$

26. $\begin{cases} x^2 + y^2 = 25 \\ x = y^2 - 5 \end{cases}$

27. $\begin{cases} x^2 + 3y^2 = 6 \\ x^2 - 3y^2 = 10 \end{cases}$

28. $\begin{cases} x^2 + y^2 = 1 \\ y = x^2 - 9 \end{cases}$

29. $\begin{cases} x^2 + y^2 = 36 \\ y = \dfrac{1}{6}x^2 - 6 \end{cases}$

30. $\begin{cases} x^2 + y^2 = 16 \\ y = -\dfrac{1}{4}x^2 + 4 \end{cases}$

31. How many real solutions are possible for a system of equations whose graphs are a circle and a parabola?

32. How many real solutions are possible for a system of equations whose graphs are an ellipse and a line?

Review and Preview

Graph each inequality in two variables. See Section 3.6.

33. $x > -3$ **34.** $y \le 1$ **35.** $y < 2x - 1$ **36.** $3x - y \le 4$

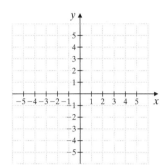 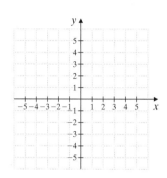 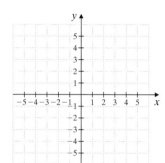 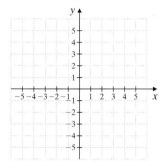

 Combining Concepts

Solve.

37. The sum of the squares of two numbers is 130. The difference of the squares of the two numbers is 32. Find the two numbers.

38. The sum of the squares of two numbers is 20. Their product is 8. Find the two numbers.

△ **39.** During the development stage of a new rectangular keypad for a security system, it was decided that the area of the rectangle should be 285 square centimeters and the perimeter should be 68 centimeters. Find the dimensions of the keypad.

△ **40.** A rectangular holding pen for cattle is to be designed so that its perimeter is 92 feet and its area is 525 feet. Find the dimensions of the holding pen.

*Recall that in business, a demand function expresses the quantity of a commodity demanded as a function of the commodity's unit price. A supply function expresses the quantity of a commodity supplied as a function of the commodity's unit price. When the quantity produced and supplied is equal to the quantity demanded, then we have what is called **market equilibrium**. Use this information for Exercises 41–42.*

41. The demand function for a certain compact disc is given by the function $p(x) = -0.01x^2 - 0.2x + 9$ and the corresponding supply function is given by $p(x) = 0.01x^2 - 0.1x + 3$, where $p(x)$ is in dollars and x is in thousands of units. Find the equilibrium quantity and the corresponding price by solving the system consisting of the two given equations.

42. The demand function for a certain style of picture frame is given by the function $p(x) = -2x^2 + 90$ and the corresponding supply function is given by $p(x) = 9x + 34$, where $p(x)$ is in dollars and x is in thousands of units. Find the equilibrium quantity and the corresponding price by solving the system consisting of the two given equations.

Use a grapher to verify the results of each exercise.

43. Exercise 3

44. Exercise 4

45. Exercise 23

46. Exercise 24

9.5 Nonlinear Inequalities and Systems of Inequalities

OBJECTIVES

Ⓐ Graph a nonlinear inequality.

Ⓑ Graph a system of nonlinear inequalities.

SSM TUTOR CENTER   SG   CD & VIDEO   MATH PRO   WEB

Ⓐ Graphing Nonlinear Inequalities

We can graph a nonlinear inequality in two variables such as $\frac{x^2}{9} + \frac{y^2}{16} \leq 1$ in a way similar to the way we graphed a linear inequality in two variables in Section 3.6. First, we graph the related equation $\frac{x^2}{9} + \frac{y^2}{16} = 1$. The graph of the equation is our boundary. Then, using test points, we determine and shade the region whose points satisfy the inequality.

EXAMPLE 1 Graph: $\frac{x^2}{9} + \frac{y^2}{16} \leq 1$

Solution: First we graph the equation $\frac{x^2}{9} + \frac{y^2}{16} = 1$. We sketch a solid curve because of the inequality symbol $\leq$. It means that the graph of $\frac{x^2}{9} + \frac{y^2}{16} \leq 1$ includes the graph of $\frac{x^2}{9} + \frac{y^2}{16} = 1$. The graph is an ellipse, and it divides the plane into two regions, the "inside" and the "outside" of the ellipse. Recall from Section 3.6 that to determine which region contains the solutions, we select a test point in either region and determine whether the coordinates of the point satisfy the inequality. We choose $(0, 0)$ as the test point.

$$\frac{x^2}{9} + \frac{y^2}{16} \leq 1$$

$$\frac{0^2}{9} + \frac{0^2}{16} \leq 1 \qquad \text{Let } x = 0 \text{ and } y = 0.$$

$$0 \leq 1 \qquad \text{True.}$$

Since this statement is true, the solution set is the region containing $(0, 0)$. The graph of the solution set includes the points on and inside the ellipse, as shaded in the figure.

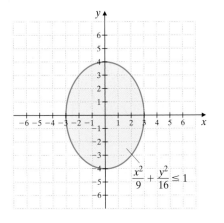

$$\frac{x^2}{9} + \frac{y^2}{16} \leq 1$$

Practice Problem 1

Graph: $\frac{x^2}{25} + \frac{y^2}{4} \leq 1$

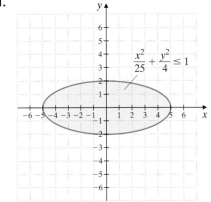

Answer

1.

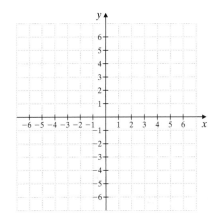

$$\frac{x^2}{25} + \frac{y^2}{4} \leq 1$$

Practice Problem 2

Graph: $9x^2 > 4y^2 + 144$

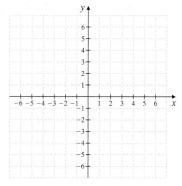

Practice Problem 3

Graph the system:

$$\begin{cases} y \ge x^2 \\ y \le -4x + 2 \end{cases}$$

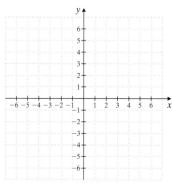

Answers

2.

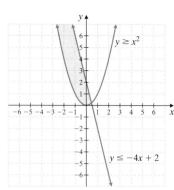

3.

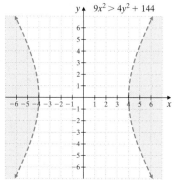

EXAMPLE 2   Graph: $4y^2 > x^2 + 16$

Solution: The related equation is $4y^2 = x^2 + 16$, or $\dfrac{y^2}{4} - \dfrac{x^2}{16} = 1$, which is a hyperbola. We graph the hyperbola as a dashed curve because of the inequality symbol $>$. It means that the graph of $4y^2 > x^2 + 16$ does *not* include the graph of $4y^2 = x^2 + 16$. The hyperbola divides the plane into three regions. We select a test point in each region—not on a boundary line—to determine whether that region contains solutions of the inequality.

Test Region A with $(0, 4)$	Test Region B with $(0, 0)$	Test Region C with $(0, -4)$
$4y^2 > x^2 + 16$	$4y^2 > x^2 + 16$	$4y^2 > x^2 + 16$
$4(4)^2 > 0^2 + 16$	$4(0)^2 > 0^2 + 16$	$4(-4)^2 > 0^2 + 16$
$64 > 16$ True.	$0 > 16$ False.	$64 > 16$ True.

The graph of the solution set includes the shaded regions A and C only, not the boundary.

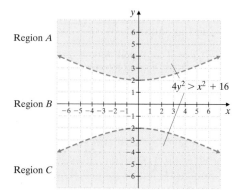

B   **Graphing Systems of Nonlinear Inequalities**

In Section 4.5 we graphed systems of linear inequalities. Recall that the graph of a system of inequalities is the intersection of the graphs of the inequalities.

EXAMPLE 3   Graph the system:

$$\begin{cases} x \le 1 - 2y \\ y \le x^2 \end{cases}$$

Solution: We graph each inequality on the same set of axes. The intersection is the darkest shaded region along with its boundary lines. The coordinates of the points of intersection can be found by solving the related system.

$$\begin{cases} x = 1 - 2y \\ y = x^2 \end{cases}$$

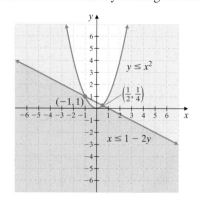

EXAMPLE 4   Graph the system:

$$\begin{cases} x^2 + y^2 < 25 \\ \dfrac{x^2}{9} - \dfrac{y^2}{25} < 1 \\ \qquad y < x + 3 \end{cases}$$

Solution: We graph each inequality. The graph of $x^2 + y^2 < 25$ contains points "inside" the circle that has center $(0,0)$ and radius 5. The graph of $\dfrac{x^2}{9} - \dfrac{y^2}{25} < 1$ is the region between the two branches of the hyperbola with x-intercepts $(-3, 0)$ and $(3, 0)$ and center $(0, 0)$. The graph of $y < x + 3$ is the region "below" the line with the slope 1 and y-intercept $(0, 3)$. The graph of the solution set of the system is the intersection of all the graphs, the darkest shaded region shown. The boundary of this region is not part of the solution.

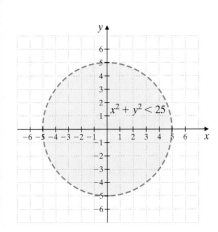

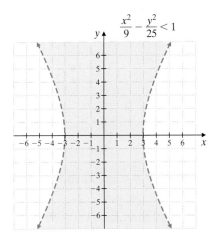

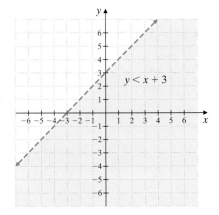

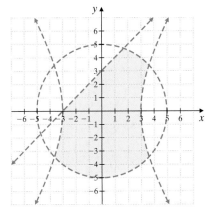

Practice Problem 4

Graph the system:

$$\begin{cases} x^2 + y^2 < 9 \\ \dfrac{x^2}{9} - \dfrac{y^2}{4} < 1 \\ \qquad y > x - 2 \end{cases}$$

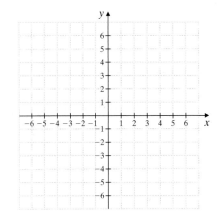

Answer

4.

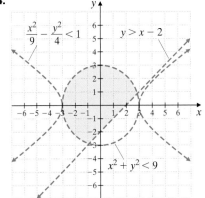

STUDY SKILLS REMINDER

Are you preparing for a test on Chapter 9?

Below I have listed some common trouble areas for students in Chapter 9. After studying for your test—but before taking your test—read these.

■ Don't forget to review all the standard forms for the conic sections.

■ Remember that the midpoint of a segment is a *point*. The x-coordinate is the average of the x-coordinates of the endpoints of the segment and the y-coordinate is the average of the y-coordinates of the endpoints of the segment.

The midpoint of the segment joining $(-1, 5)$ and $(3, 4)$ is $\left(\dfrac{-1 + 3}{2}, \dfrac{5 + 4}{2} \right)$ or $\left(1, \dfrac{9}{2} \right)$.

■ Remember that the distance formula gives the *distance* between two points.

The distance between $(-1, 5)$ and $(3, 4)$ is

$$\sqrt{(3 - (-1))^2 + (4 - 5)^2} = \sqrt{4^2 + (-1)^2} = \sqrt{16 + 1} = \sqrt{17} \text{ units}$$

■ Don't forget that both methods, substitution and elimination, are available for solving nonlinear systems of equations.

$$\begin{cases} x^2 + y^2 = 7 \\ 2x^2 - 3y^2 = 4 \end{cases} \quad \text{is equivalent to} \quad \begin{cases} 3x^2 + 3y^2 = 21 \\ 2x^2 - 3y^2 = 4 \end{cases}$$
$$\begin{aligned} 5x^2 &= 25 \\ x^2 &= 5 \\ x &= \pm\sqrt{5} \end{aligned}$$

Let $x = \pm\sqrt{5}$ in either original equation, and $y = \pm\sqrt{2}$, the solution set is $\{(\sqrt{5}, \sqrt{2}), (-\sqrt{5}, \sqrt{2}), (\sqrt{5}, -\sqrt{2}), (\sqrt{5}, -\sqrt{2})\}$.

Remember: This is simply a checklist of common trouble areas. For a review of Chapter 9, see the Highlights and Chapter Review at the end of this chapter.

Name _____ Section _____ Date _____

EXERCISE SET 9.5

 A *Graph each inequality. See Examples 1 and 2.*

1. $y < x^2$

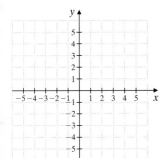

2. $y < -x^2$

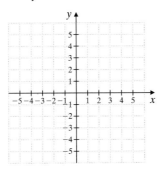

3. $x^2 + y^2 \geq 16$

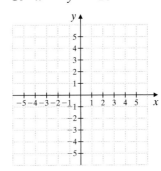

4. $x^2 + y^2 < 36$

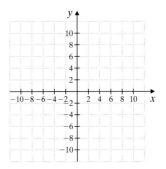

5. $\dfrac{x^2}{4} - y^2 < 1$

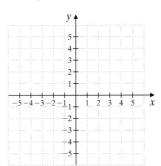

6. $x^2 - \dfrac{y^2}{9} \geq 1$

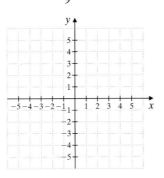

7. $y > (x - 1)^2 - 3$

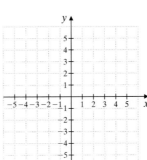

8. $y > (x + 3)^2 + 2$

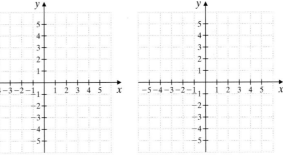

9. $x^2 + y^2 \leq 9$

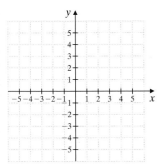

10. $x^2 + y^2 > 4$

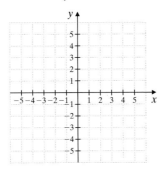

11. $y > -x^2 + 5$

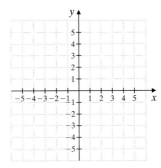

12. $y < -x^2 + 5$

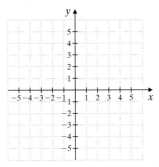

13. $\dfrac{x^2}{4} + \dfrac{y^2}{9} \leq 1$

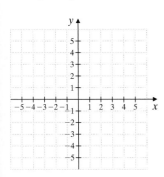

14. $\dfrac{x^2}{25} + \dfrac{y^2}{4} \geq 1$

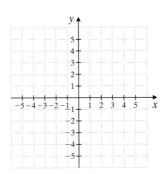

15. $\dfrac{y^2}{4} - x^2 \leq 1$

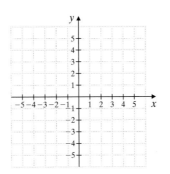

16. $\dfrac{y^2}{16} - \dfrac{x^2}{9} > 1$

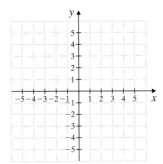

17. $y < (x - 2)^2 + 1$

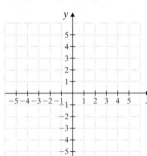

18. $y > (x - 2)^2 + 1$

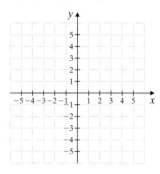

19. $y \le x^2 + x - 2$

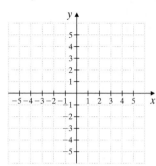

20. $y > x^2 + x - 2$

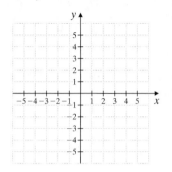

B *Graph each system. See Examples 3 and 4.*

21. $\begin{cases} 4x + 3y \ge 12 \\ x^2 + y^2 < 16 \end{cases}$

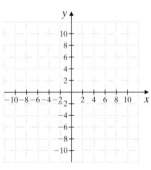

22. $\begin{cases} 3x - 4y \le 12 \\ x^2 + y^2 < 16 \end{cases}$

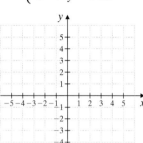

23. $\begin{cases} x^2 + y^2 \le 9 \\ x^2 + y^2 \ge 1 \end{cases}$

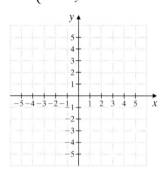

24. $\begin{cases} x^2 + y^2 \ge 9 \\ x^2 + y^2 \ge 16 \end{cases}$

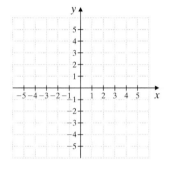

25. $\begin{cases} y > x^2 \\ y \ge 2x + 1 \end{cases}$

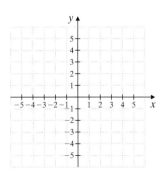

26. $\begin{cases} y \le -x^2 + 3 \\ y \le 2x - 1 \end{cases}$

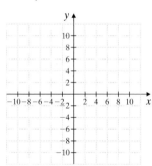

27. $\begin{cases} x^2 + y^2 > 9 \\ y > x^2 \end{cases}$

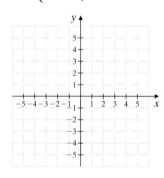

28. $\begin{cases} x^2 + y^2 \le 9 \\ y < x^2 \end{cases}$

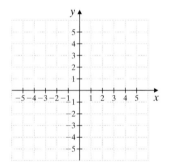

29. $\begin{cases} \dfrac{x^2}{4} + \dfrac{y^2}{9} \ge 1 \\ x^2 + y^2 \ge 4 \end{cases}$

30. $\begin{cases} x^2 + (y - 2)^2 \ge 9 \\ \dfrac{x^2}{4} + \dfrac{y^2}{25} < 1 \end{cases}$

31. $\begin{cases} x^2 - y^2 \ge 1 \\ y \ge 0 \end{cases}$

32. $\begin{cases} x^2 - y^2 \ge 1 \\ x \ge 0 \end{cases}$

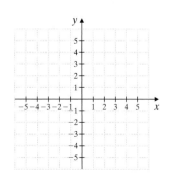

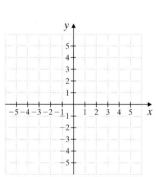

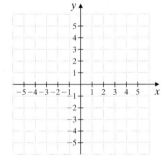

33. $\begin{cases} x + y \geq 1 \\ 2x + 3y < 1 \\ x > -3 \end{cases}$

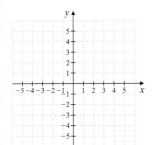

34. $\begin{cases} x - y < -1 \\ 4x - 3y > 0 \\ y > 0 \end{cases}$

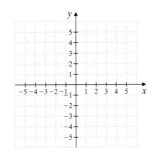

35. $\begin{cases} x^2 - y^2 < 1 \\ \dfrac{x^2}{16} + y^2 \leq 1 \\ x \geq -2 \end{cases}$

36. $\begin{cases} x^2 - y^2 \geq 1 \\ \dfrac{x^2}{16} + \dfrac{y^2}{4} \leq 1 \\ y \geq 1 \end{cases}$

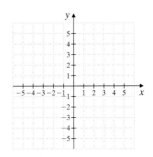

Review and Preview

Determine whether each graph is the graph of a function. See Section 3.5.

37.

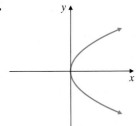

38.

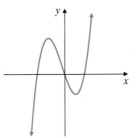

39.

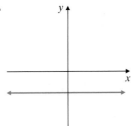

40.

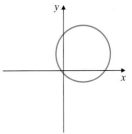

 Combining Concepts

41. Discuss how graphing a linear inequality such as $x + y < 9$ is similar to graphing a nonlinear inequality such as $x^2 + y^2 < 9$.

42. Discuss how graphing a linear inequality such as $x + y < 9$ is different from graphing a nonlinear inequality such as $x^2 + y^2 < 9$.

43. Graph the system:

$\begin{cases} y \leq x^2 \\ y \geq x + 2 \\ x \geq 0 \\ y \geq 0 \end{cases}$

see graph

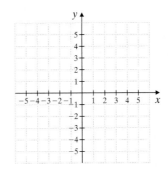

717

Modeling Conic Sections

MATERIALS

- two thumbtacks (or nails)
- graph paper
- cardboard
- tape
- string
- pencil
- ruler

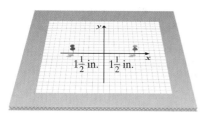

Figure 1

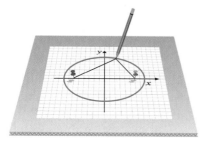

Figure 2

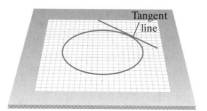

Figure 3

This activity may be completed by working in groups or individually.

1. Draw an *x*-axis and a *y*-axis on the graph paper as shown in Figure 1.

2. Place the graph paper on the cardboard and use tape to attach.

3. Locate two points on the *x*-axis each about $1\frac{1}{2}$ inches from the origin and on opposite sides of the origin (see Figure 1). Insert thumbtacks (or nails) at each of these locations.

4. Fasten a 9-inch piece of string to the thumbtacks as shown in Figure 2. Use your pencil to draw and keep the string taut while you carefully move the pencil in a path all around the thumbtacks.

5. Using the grid of the graph paper as a guide, find an approximate equation of the ellipse you drew.

6. Experiment by moving the tacks closer together or farther apart and drawing new ellipses. What do you observe?

7. Write a paragraph explaining why the figure drawn by the pencil is an ellipse. How might you use the same materials to draw a circle?

8. (Optional) Choose one of the ellipses you drew with the string and pencil. Use a ruler to draw any six tangent lines to the ellipse. (A line is tangent to the ellipse if it intersects, or just touches, the ellipse at only one point. See Figure 3.) Extend the tangent lines to yield six points of intersection among the tangents. Use a straight edge to draw a line connecting each pair of opposite points of intersection. What do you observe? Repeat with a different ellipse. Can you make a conjecture about the relationship among the lines that connect opposite points of intersection?

Chapter 9 VOCABULARY CHECK

Fill in each blank with one of the words or phrases listed below.

circle	midpoint	radius	distance
center	ellipse	hyperbola	nonlinear system of equations

1. The _____ formula is $d = \sqrt{(x_2 - x_1)^2 + (y_2 - y_1)^2}$.
2. A(n) _____ is the set of all points in a plane that are the same distance from a fixed point, called the _____.
3. A _____ is a system of equations at least one of which is not linear.
4. A(n) _____ is the set of points on a plane such that the sum of the distances of those points from two fixed points is a constant.
5. In a circle, the distance from the center to a point of the circle is called its _____.
6. A(n) _____ is the set of points in a plane such that the absolute value of the difference of the distance from two fixed points is constant.
7. The _____ formula is $\left(\dfrac{x_1 + x_2}{2}, \dfrac{y_1 + y_2}{2} \right)$.

CHAPTER 9 Highlights

DEFINITIONS AND CONCEPTS

EXAMPLES

Section 9.1   The Parabola and the Circle

PARABOLAS

$$y = a(x - h)^2 + k$$

$$x = a(y - k)^2 + h$$

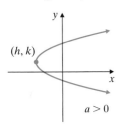

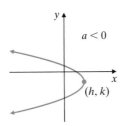

Graph: $x = 3y^2 - 12y + 13$

$$x - 13 = 3(y^2 - 4y)$$
$$x - 13 + 3(4) = 3(y^2 - 4y + 4)$$
$$x = 3(y - 2)^2 + 1$$

Since $a = 3$, this parabola opens to the right with vertex $(1, 2)$. Its axis of symmetry is $y = 2$. The x-intercept is $(13, 0)$.

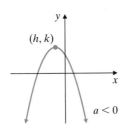

DISTANCE FORMULA

The distance d between two points (x_1, y_1) and (x_2, y_2) is given by

$$d = \sqrt{(x_2 - x_1)^2 + (y_2 - y_1)^2}$$

Find the distance between points $(-1, 6)$ and $(-2, -4)$. Let $(x_1, y_1) = (-1, 6)$ and $(x_2, y_2) = (-2, -4)$.

$$d = \sqrt{(x_2 - x_1)^2 + (y_2 - y_1)^2}$$
$$= \sqrt{(-2 - (-1))^2 + (-4 - 6)^2}$$
$$= \sqrt{1 + 100} = \sqrt{101}$$

| **DEFINITIONS AND CONCEPTS** | **EXAMPLES** |

Section 9.1   The Parabola and the Circle *(continued)*

MIDPOINT FORMULA

The midpoint of the line segment whose endpoints are (x_1, y_1) and (x_2, y_2) is the point with coordinates

$$\left(\frac{x_1 + x_2}{2}, \frac{y_1 + y_2}{2} \right)$$

Find the midpoint of the line segment whose endpoints are $(-1, 6)$ and $(-2, -4)$.

$$\left(\frac{-1 + (-2)}{2}, \frac{6 + (-4)}{2} \right)$$

The midpoint is $\left(-\frac{3}{2}, 1 \right)$.

CIRCLE

The graph $(x - h)^2 + (y - k)^2 = r^2$ is a circle with center (h, k) and radius r.

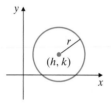

Graph: $x^2 + (y + 3)^2 = 5$

This equation can be written as

$$(x - 0)^2 + (y + 3)^2 = 5$$

with $h = 0$, $k = -3$, and $r = \sqrt{5}$. The center of this circle is $(0, -3)$, and the radius is $\sqrt{5}$.

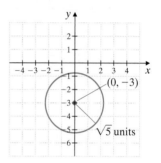

Section 9.2   The Ellipse and the Hyperbola

ELLIPSE WITH CENTER $(0, 0)$

The graph of an equation of the form $\dfrac{x^2}{a^2} + \dfrac{y^2}{b^2} = 1$ is an ellipse with center $(0, 0)$. The x-intercepts are $(a, 0)$ and $(-a, 0)$, and the y-intercepts are $(0, b)$ and $(0, -b)$.

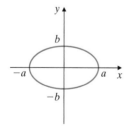

Graph: $4x^2 + 9y^2 = 36$

$$\frac{x^2}{9} + \frac{y^2}{4} = 1 \qquad \text{Divide both sides by 36.}$$

$$\frac{x^2}{3^2} + \frac{y^2}{2^2} = 1$$

The ellipse has center $(0, 0)$, x-intercepts $(3, 0)$ and $(-3, 0)$, and y-intercepts $(0, 2)$ and $(0, -2)$.

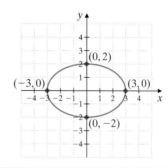

Section 9.2 The Ellipse and the Hyperbola *(continued)*	

HYPERBOLA WITH CENTER $(0, 0)$

The graph of an equation of the form $\dfrac{x^2}{a^2} - \dfrac{y^2}{b^2} = 1$ is a hyperbola with center $(0,0)$ and x-intercepts $(a, 0)$ and $(-a, 0)$.

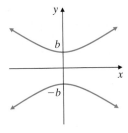

The graph of an equation of the form $\dfrac{y^2}{b^2} - \dfrac{x^2}{a^2} = 1$ is a hyperbola with center $(0,0)$ and y-intercepts $(0, b)$ and $(0, -b)$.

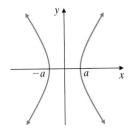

Graph: $\dfrac{x^2}{9} - \dfrac{y^2}{4} = 1$. Here $a = 3$ and $b = 2$.

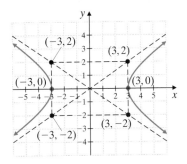

Section 9.3   Graphing Nonlinear Functions

$f(x) = |x|$

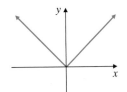

$f(x) = \sqrt{x}$

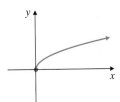

Graph: $f(x) = |x - 1| + 2$

The graph of $y = |x - 1| + 2$ is the same as the graph of $y = |x|$ shifted 1 unit to the right and 2 units up.

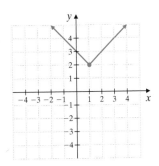

DEFINITIONS AND CONCEPTS	EXAMPLES

Section 9.4 Solving Nonlinear Systems of Equations

A **nonlinear system of equations** is a system of equations at least one of which is not linear. Both the substitution method and the elimination method may be used to solve a nonlinear system of equations.

Solve the nonlinear system: $\begin{cases} y = x + 2 \\ 2x^2 + y^2 = 3 \end{cases}$

Substitute $x + 2$ for y in the second equation:

$$2x^2 + y^2 = 3$$

$$2x^2 + (x + 2)^2 = 3$$

$$2x^2 + x^2 + 4x + 4 = 3$$

$$3x^2 + 4x + 1 = 0$$

$$(3x + 1)(x + 1) = 0$$

$$x = -\frac{1}{3} \quad \text{or} \quad x = -1$$

If $x = -\frac{1}{3}, y = x + 2 = -\frac{1}{3} + 2 = \frac{5}{3}.$

If $x = -1, y = x + 2 = -1 + 2 = 1.$

The solution set is $\left\{ \left(-\frac{1}{3}, \frac{5}{3} \right), (-1, 1) \right\}$

Section 9.5 Nonlinear Inequalities and Systems of Inequalities

The **graph of a system of inequalities** is the intersection of the graphs of the inequalities.

Graph the system: $\begin{cases} x \geq y^2 \\ x + y \leq 4 \end{cases}$

The graph of the system is the darkest shaded region along with its boundary lines.

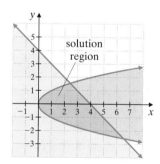

Chapter 9 Review

(9.1) *Find the distance between each pair of points. Give an exact value and a three-decimal-place approximation.*

1. $(-6, 3)$ and $(8, 4)$

2. $(3, 5)$ and $(8, 9)$

3. $(-4, -6)$ and $(-1, 5)$

4. $(-1, 5)$ and $(2, -3)$

5. $(-\sqrt{2}, 0)$ and $(0, -4\sqrt{6})$

6. $(-\sqrt{5}, -\sqrt{11})$ and $(-\sqrt{5}, -3\sqrt{11})$

7. $(7.4, -8.6)$ and $(-1.2, 5.6)$

8. $(2.3, 1.8)$ and $(10.7, -9.2)$

Find the midpoint of each line segment whose endpoints are given.

9. $(2, 6); (-12, 4)$

10. $(-3, 8); (11, 24)$

11. $(-6, -5); (-9, 7)$

12. $(4, -6); (-15, 2)$

13. $\left(0, -\dfrac{3}{8}\right); \left(\dfrac{1}{10}, 0\right)$

14. $\left(\dfrac{3}{4}, -\dfrac{1}{7}\right); \left(-\dfrac{1}{4}, -\dfrac{3}{7}\right)$

15. $(\sqrt{3}, -2\sqrt{6})$ and $(\sqrt{3}, -4\sqrt{6})$

16. $(-5\sqrt{3}, 2\sqrt{7}); (-3\sqrt{3}, 10\sqrt{7})$

Write an equation of each circle with the given center and radius or diameter.

17. Center $(-4, 4)$, radius 3

18. Center $(5, 0)$, diameter 10

19. Center $(-7, -9)$, radius $\sqrt{11}$

20. Center $(0, 0)$, diameter 7

Graph each equation. If the graph is a circle, find its center and radius. If the graph is a parabola, find its vertex.

21. $x^2 + y^2 = 7$

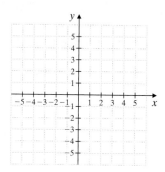

22. $x = 2(y - 5)^2 + 4$

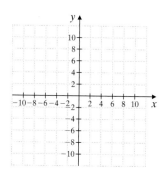

23. $x = -(y + 2)^2 + 3$

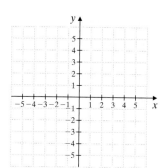

24. $(x - 1)^2 + (y - 2)^2 = 4$

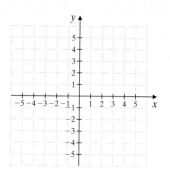

25. $y = -x^2 + 4x + 10$

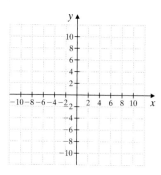

26. $x = -y^2 - 4y + 6$

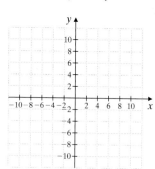

27. $x = \dfrac{1}{2}y^2 + 2y + 1$

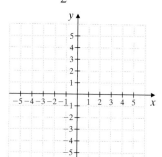

28. $y = -3x^2 + \dfrac{1}{2}x + 4$

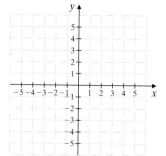

29. $x^2 + y^2 + 2x + y = \dfrac{3}{4}$

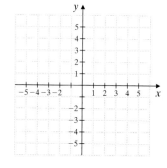

30. $x^2 + y^2 - 3y = \dfrac{7}{4}$

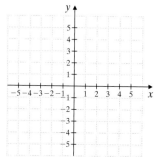

31. $4x^2 + 4y^2 + 16x + 8y = 1$

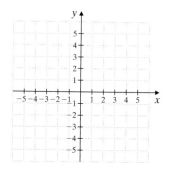

32. $3x^2 + 6x + 3y^2 = 9$

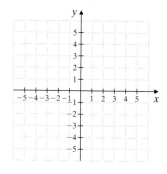

33. $y = x^2 + 6x + 9$

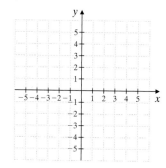

34. $x = y^2 + 6y + 9$

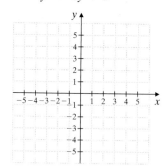

(9.2) *Graph each equation.*

35. $x^2 + \dfrac{y^2}{4} = 1$

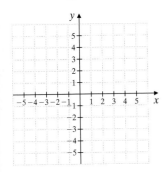

36. $x^2 - \dfrac{y^2}{4} = 1$

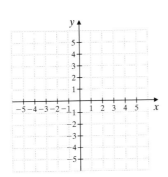

37. $\dfrac{y^2}{4} - \dfrac{x^2}{16} = 1$

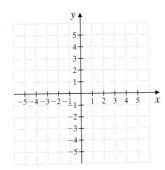

38. $\dfrac{y^2}{4} + \dfrac{x^2}{16} = 1$

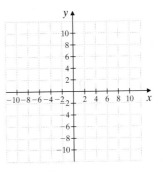

39. $-5x^2 + 25y^2 = 125$

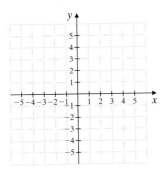

40. $4y^2 + 9x^2 = 36$

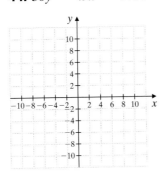

41. $\dfrac{(x-2)^2}{4} + (y-1)^2 = 1$ **42.** $\dfrac{(x+3)^2}{9} + \dfrac{(y-4)^2}{25} = 1$

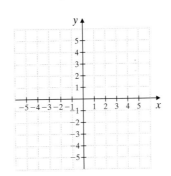

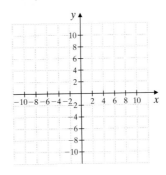

43. $x^2 - y^2 = 1$

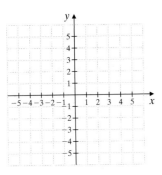

44. $36y^2 - 49x^2 = 1764$

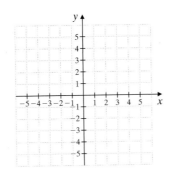

45. $y = x^2 + 9$

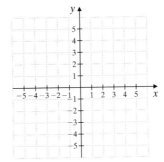

46. $x = 4y^2 - 16$

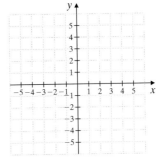

Graph each equation.

47. $y = x^2 + 4x + 6$

48. $y^2 = x^2 + 6$

49. $y^2 + x^2 = 4x + 6$

50. $y^2 + 2x^2 = 4x + 6$

51. $x^2 + y^2 - 8y = 0$

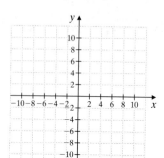

52. $x - 4y = y^2$

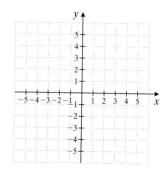

53. $x^2 - 4 = y^2$

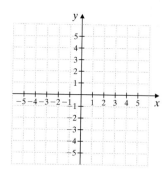

54. $x^2 = 4 - y^2$

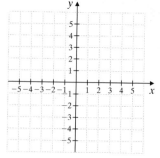

55. $6(x - 2)^2 +$
  $9(y + 5)^2 = 36$

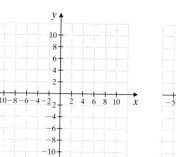

56. $36y^2 = 576 + 16x^2$

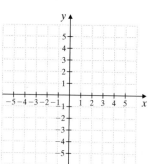

57. $\dfrac{x^2}{16} - \dfrac{y^2}{25} = 1$

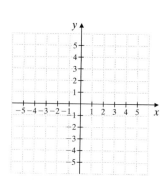

58. $3(x - 7)^2 +$
  $3(y + 4)^2 = 1$

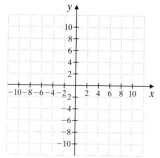

(9.3) *Graph each function.*

59. $f(x) = |x| + 2$

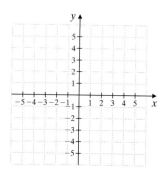

60. $f(x) = \sqrt{x} - 1$

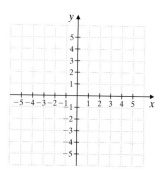

61. $f(x) = \sqrt{x - 4}$

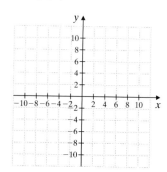

62. $f(x) = |x + 1|$

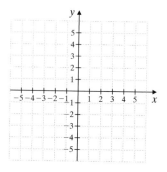

63. $f(x) = \sqrt{x - 3} + 1$

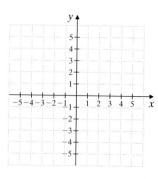

64. $f(x) = |x - 1| + 3$

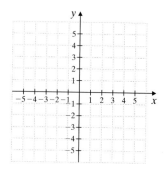

65. $f(x) = |x + 2| + 2$

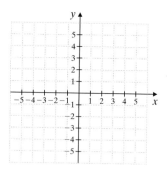

66. $f(x) = \sqrt{x + 2} + 2$

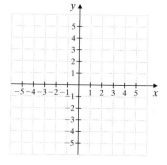

(9.4) *Solve each system of equations.*

67. $\begin{cases} y = 2x - 4 \\ y^2 = 4x \end{cases}$

68. $\begin{cases} x^2 + y^2 = 4 \\ x - y = 4 \end{cases}$

69. $\begin{cases} y = x + 2 \\ y = x^2 \end{cases}$

70. $\begin{cases} y = x^2 - 5x + 1 \\ y = -x + 6 \end{cases}$

71. $\begin{cases} 4x - y^2 = 0 \\ 2x^2 + y^2 = 16 \end{cases}$

72. $\begin{cases} x^2 + 4y^2 = 16 \\ x^2 + y^2 = 4 \end{cases}$

73. $\begin{cases} x^2 + y^2 = 10 \\ 9x^2 + y^2 = 18 \end{cases}$

74. $\begin{cases} x^2 + 2y = 9 \\ 5x - 2y = 5 \end{cases}$

75. $\begin{cases} y = 3x^2 + 5x - 4 \\ y = 3x^2 - x + 2 \end{cases}$

76. $\begin{cases} x^2 - 3y^2 = 1 \\ 4x^2 + 5y^2 = 21 \end{cases}$

△ **77.** Find the length and the width of a room whose area is 150 square feet and whose perimeter is 50 feet.

78. What is the greatest number of real number solutions possible for a system of two equations whose graphs are an ellipse and a hyperbola?

(9.5) *Graph each inequality or system of inequalities.*

79. $y \leq -x^2 + 3$

80. $x^2 + y^2 < 9$

81. $x^2 = y^2 < 1$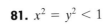

82. $\dfrac{x^2}{4} + \dfrac{y^2}{9} \geq 1$

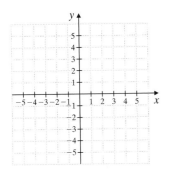
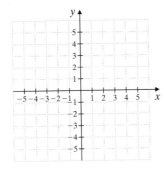
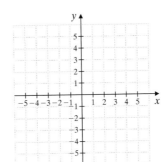
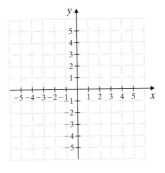

83. $\begin{cases} 2x \leq 4 \\ x + y \geq 1 \end{cases}$

84. $\begin{cases} 3x + 4y \leq 12 \\ x - 2y > 6 \end{cases}$

85. $\begin{cases} y > x^2 \\ x + y \geq 3 \end{cases}$

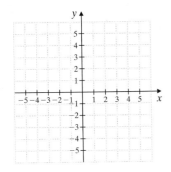
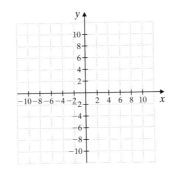
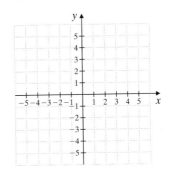

86. $\begin{cases} x^2 + y^2 \le 16 \\ x^2 + y^2 \ge 4 \end{cases}$

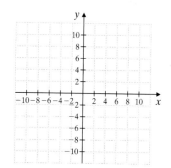

87. $\begin{cases} x^2 + y^2 < 4 \\ x^2 - y^2 \le 1 \end{cases}$

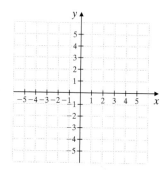

88. $\begin{cases} x^2 + y^2 < 4 \\ y \ge x^2 - 1 \\ x \ge 0 \end{cases}$

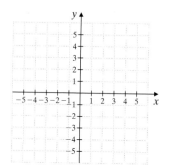

STUDY SKILLS REMINDER

Are you satisfied with your performance on a particular quiz or exam?

If not, don't forget to analyze your quiz or exam and look for common errors.

Were most of your errors a result of

- *Carelessness?* If your errors were careless, did you turn in your work before the allotted time expired? If so, resolve next time to use the entire time allotted. Any extra time can be spent checking your work.

- *Running out of time?* If so, make a point to better manage your time on your next exam. A few suggestions are to work any questions that you are unsure of last and to check your work after all questions have been answered.

- *Not understanding a concept?* If so, review that concept and correct your work. Remember next time to make sure that all concepts on a quiz or exam are understood before the exam.

Chapter 9 Test

1. Find the distance between the points $(-6, 3)$ and $(-8, -7)$.

2. Find the distance between the points $(-2\sqrt{5}, \sqrt{10})$ and $(-\sqrt{5}, 4\sqrt{10})$.

3. Find the midpoint of the line segment whose endpoints are $(-2, -5)$ and $(-6, 12)$.

4. Find the midpoint of the line segment whose endpoints are $\left(-\dfrac{2}{3}, -\dfrac{1}{5}\right)$ and $\left(-\dfrac{1}{3}, \dfrac{4}{5}\right)$.

Graph each equation.

5. $x^2 + y^2 = 36$

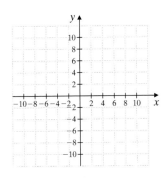

6. $x^2 - y^2 = 36$

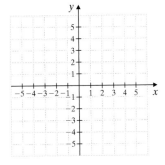

7. $16x^2 + 9y^2 = 144$

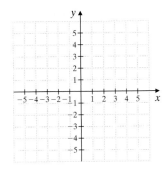

8. $y = x^2 - 8x + 16$

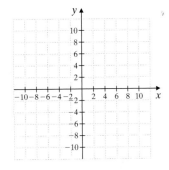

9. $x^2 + y^2 + 6x = 16$

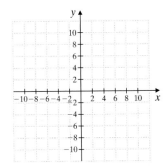

10. $x = y^2 + 8y - 3$

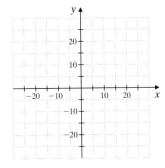

Answers

1. _____

2. _____

3. _____

4. _____

5. see graph

6. see graph

7. see graph

8. see graph

9. see graph

10. see graph

11. $\dfrac{(x-4)^2}{16} + \dfrac{(y-3)^2}{9} = 1$

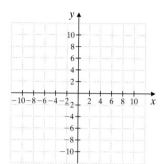

12. $y^2 - x^2 = 1$

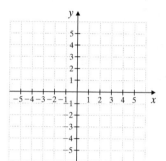

Solve each system.

13. $\begin{cases} x^2 + y^2 = 169 \\ 5x + 12y = 0 \end{cases}$

14. $\begin{cases} x^2 + y^2 = 26 \\ x^2 - y^2 = 24 \end{cases}$

15. $\begin{cases} y = x^2 - 5x + 6 \\ y = 2x \end{cases}$

16. $\begin{cases} x^2 + 4y^2 = 5 \\ y = x \end{cases}$

Graph each system.

17. $\begin{cases} 2x + 5y \ge 10 \\ y \ge x^2 + 1 \end{cases}$

18. $\begin{cases} \dfrac{x^2}{4} + y^2 \le 1 \\ x + y > 1 \end{cases}$

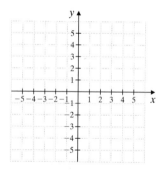

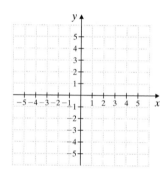

19. $\begin{cases} x^2 + y^2 > 1 \\ \dfrac{x^2}{4} - y^2 \ge 1 \end{cases}$

20. $\begin{cases} x^2 + y^2 \ge 4 \\ x^2 + y^2 < 16 \\ y \ge 0 \end{cases}$

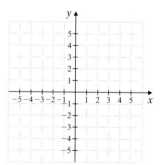

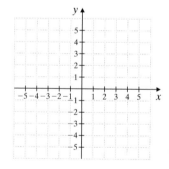

21. Which graph best resembles the graph of $x = a(y - k)^2 + h$ if $a > 0, h < 0$, and $k > 0$?

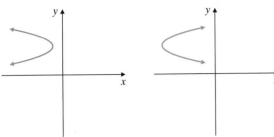

A

B

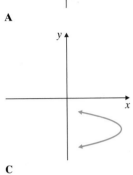

C

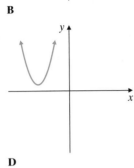

D

21. _____

22. _____

22. A bridge has an arch in the shape of half an ellipse. If the equation of the ellipse, measured in feet, is $100x^2 + 225y^2 = 22{,}500$, find the height of the arch from the road and the width of the arch.

23. see graph _____

Graph each function.

23. $f(x) = \sqrt{x - 2}$

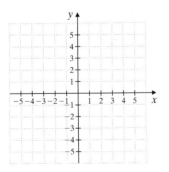

24. $f(x) = |x + 1| - 3$

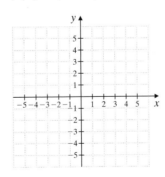

24. see graph _____

CONIC SECTIONS

It is believed that the conic sections were discovered by the Greek mathematician Menaechmus (380 B.C. –320 B.C.). He was the first to realize that the shapes of the parabola, ellipse, and hyperbola are formed by cutting a right circular cone with a plane in various ways. However, these conic sections were not given their names until later. Another Greek mathematician, Apollonius of Perga (262 B.C.–190 B.C.), was responsible for coining the terms *parabola*, *ellipse*, and *hyperbola* in his set of eight texts titled *Treatise on Conic Sections*. In these texts, Apollonius discussed the basic properties of conic sections as well as how they are drawn.

A contemporary of Apollonius, Archimedes of Syracuse (287 B.C.–212 B.C.), is probably the most famous of the Greek mathematicians who made contributions to the base of knowledge on conic sections. His detailed study of circles led to an important contribution: a calculation of the value of π as being between $3\frac{10}{71}$ and $3\frac{1}{7}$. He also studied the areas of conic sections, including parabolas and ellipses, and other shapes and solids that arise from the conic sections. He developed a special method for finding such areas by dividing the area of a figure up into infinitely many narrow rectangles and then summing these individual areas to find the area of the entire figure. This revolutionary method eventually led to the discovery of the branch of advanced mathematics called *calculus* nearly 2000 years later.

As important as was his work with conics, Archimedes is probably best remembered for his work on practical matters for the king of Syracuse. For instance, Archimedes is credited with inventing the catapult, at the king's request, as a defense measure against a Roman invasion. Another time, the king asked Archimedes to help prove that a gold crown that he had commissioned was made partially from silver as well. The story goes that while Archimedes pondered this question, he was taking a bath and noticed that the amount of water that overflowed the tub was proportional to the portion of his body that was under water. He had discovered what is now known as Archimedes' Principle of Buoyancy: that an object immersed in water is buoyed up by a force that is equal to the weight of the water it displaces. Archimedes was so excited by his discovery that he supposedly ran naked straight from his bath through the streets of Syracuse shouting "Eureka, eureka!" ("I have found it!"). He immediately applied this discovery to the crown problem by comparing the amount of water displaced by a crown made from the same weight of pure gold to the amount of water displaced by the suspect crown. Because these amounts of water were not the same, Archimedes proved that the maker of the crown had cheated the king by using silver, a cheaper metal, in place of some of the gold.

Cumulative Review

1. Add $11x^3 - 12x^2 + x - 3$ and $x^3 - 10x + 5$.

2. Multiply: $(x + 3)(2x + 5)$

3. Use synthetic division to divide $x^4 - 2x^3 - 11x^2 + 34$ by $x + 2$.

4. Factor: $2(x - 5) + 3a(x - 5)$

5. Factor: $x^2 - 12x + 35$

6. Factor: $(x + 3)^2 - 36$

7. Solve: $2x^2 + 9x - 5 = 0$

Simplify each rational expression.

8. $\dfrac{2 + x}{x + 2}$

9. $\dfrac{2 - x}{x - 2}$

Subtract.

10. $\dfrac{x^2}{x + 7} - \dfrac{49}{x + 7}$

11. $\dfrac{x}{3y^2} - \dfrac{x + 1}{3y^2}$

12. Simplify: $\dfrac{\dfrac{5x}{x + 2}}{\dfrac{10}{x - 2}}$

13. Solve: $\dfrac{2x}{x - 3} + \dfrac{6 - 2x}{x^2 - 9} = \dfrac{x}{x + 3}$

14. If a certain number is subtracted from the numerator and added to the denominator of $\dfrac{9}{19}$, the new fraction is equivalent to $\dfrac{1}{3}$. Find the number.

Find each square root.

15. $\sqrt{0}$

16. $\sqrt{0.25}$

Answers

1. _____
2. _____
3. _____
4. _____
5. _____
6. _____
7. _____
8. _____
9. _____
10. _____
11. _____
12. _____
13. _____
14. _____
15. _____
16. _____

17.
_____

18.
_____

19.
_____

20.
_____

21.
_____

22.
_____

23.
_____

24.
_____

25.
_____

26.
_____

27. see graph
_____

28.
_____

29.
_____

734

Use rational exponents to simplify. Assume that all variables represent positive real numbers.

17. $\sqrt[8]{x^4}$

18. $\sqrt[6]{r^2 s^4}$

Simplify.

19. $\sqrt[3]{24}$

20. $\sqrt[4]{32}$

21. Rationalize the denominator of $\dfrac{2}{\sqrt{5}}$.

22. Solve: $\sqrt{-10x - 1} + 3x = 0$

23. Multiply: $(2 - 5i)(4 + i)$

24. Solve $p^2 + 2p = 4$ by completing the square.

25. Solve: $2x^2 - 4x = 3$

26. Find the distance between $(2, -5)$ and $(1, -4)$. Give an exact distance and a three-decimal-place approximation.

27. Graph: $\dfrac{x^2}{9} + \dfrac{y^2}{16} = 1$

Solve each system.

28. $\begin{cases} x^2 + y^2 = 4 \\ x + y = 3 \end{cases}$

29. $\begin{cases} x^2 + 2y^2 = 10 \\ x^2 - y^2 = 1 \end{cases}$

Exponential and Logarithmic Functions

CHAPTER 10

In this chapter, we discuss two closely related functions: exponential and logarithmic functions. These functions are vital in applications in economics, finance, engineering, the sciences, education, and other fields. Models of tumor growth and learning curves are two examples of the uses of exponential and logarithmic functions.

10.1 The Algebra of Functions

10.2 Inverse Functions

10.3 Exponential Functions

10.4 Logarithmic Functions

10.5 Properties of Logarithms

Integrated Review—Functions and Properties of Logarithms

10.6 Common Logarithms, Natural Logarithms, and Change of Base

10.7 Exponential and Logarithmic Equations and Problem Solving

An earthquake is a series of vibrations in the crust of Earth. The size, or magnitude, of an earthquake is measured on the Richter scale. The magnitude of an earthquake can range broadly, from barely detectable (2.5 or less on the Richter scale) to massively destructive (7.0 or greater on the Richter scale). According to the United States Geological Survey, earthquakes are an everyday occurrence. In 2000, there were a total of 22,309 earthquakes around the world, or an average of 61 earthquakes per day. However, most of these were minor tremors with magnitudes of 3.9 or less, and many could only be detected by seismographs. Only 0.08% of the earthquakes occurring during 2000 could be classified as major earthquakes (7.0 or greater on the Richter scale). Even so, earthquakes were responsible for 231 deaths that year. In Exercises 75 through 78 on page 791 and Exercises 63 and 64 on page 801, we will investigate the role of logarithms in finding the magnitude of an earthquake.

Name _____ Section _____ Date _____

Chapter **10** Pretest

If $f(x) = x^2 - 2x$ and $g(x) = 5x + 3$, find the following.

1. $(f + g)(x)$ **2.** $(f \circ g)(x)$

Determine whether the functions in Exercises 3 and 4 are one-to-one.

3. $\{(-2, 7), (7, -3), (2, 1), (5, -8)\}$ **5.** Given $f(x) = 7x - 12$, find $f^{-1}(x)$.

4.   **6.** Graph $y = 2^x - 1$.

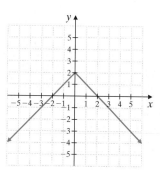 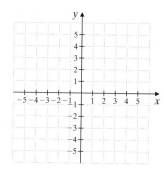

Solve each equation for x.

7. $6^x = 216$ **8.** $27^{4x+1} = 3$ **9.** $\log_5 x = 4$ **10.** $\log_2 \dfrac{1}{64} = x$

11. $\log_x 1000 = 3$ **12.** Simplify: $\log_7 7^5$ **13.** Graph: $y = \log_4 x$

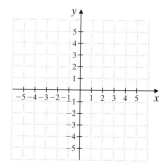

Use the properties of logarithms to write each expression as a single logarithm.

14. $\log_5 3 + \log_5 2$ **15.** $2 \log_6 a - 7 \log_6(a + 1)$

16. Write the expression $\log_7 \dfrac{3y}{5x^2}$ as the sum or difference of multiples of logarithms.

Find the exact value.

17. $\log \dfrac{1}{100}$ **18.** $\ln e^8$

19. Approximate $\log_3 15$ to four decimal places.   **20.** Solve: $\log_2 10 + \log_2(x + 5) = 3$

Answers column:

1. _____

2. _____

3. _____

4. _____

5. _____

6. see graph

7. _____

8. _____

9. _____

10. _____

11. _____

12. _____

13. see graph

14. _____

15. _____

16. _____

17. _____

18. _____

19. _____

20. _____

10.1 The Algebra of Functions

OBJECTIVES

A Add, subtract, multiply, and divide functions.

B Compose functions.

SSM   TUTOR CENTER   SG   CD & VIDEO   MATH PRO   WEB

A Adding, Subtracting, Multiplying, and Dividing Functions

As we have seen in earlier chapters, it is possible to add, subtract, multiply, and divide functions. Although we have not stated them as such, the sums, differences, products, and quotients of functions are themselves functions. For example, if $f(x) = 3x$ and $g(x) = x + 1$, their product, $f(x) \cdot g(x) = 3x(x + 1) = 3x^2 + 3x$, is a new function. We can use the notation $(f \cdot g)(x)$ to denote this new function. Finding the sum, difference, product, and quotient of functions to generate new functions is called the **algebra of functions**.

Algebra of Functions

Let f and g be functions. New functions from f and g are defined as follows:

Sum	$(f + g)(x) = f(x) + g(x)$
Difference	$(f - g)(x) = f(x) - g(x)$
Product	$(f \cdot g)(x) = f(x) \cdot g(x)$
Quotient	$\left(\dfrac{f}{g}\right)(x) = \dfrac{f(x)}{g(x)}$

EXAMPLE 1   If $f(x) = x - 1$ and $g(x) = 2x - 3$, find the following.

a. $(f + g)(x)$

b. $(f - g)(x)$

c. $(f \cdot g)(x)$

d. $\left(\dfrac{f}{g}\right)(x)$

Solution: Use the algebra of functions and replace $f(x)$ by $x - 1$ and $g(x)$ by $2x - 3$. Then simplify.

a. $(f + g)(x) = f(x) + g(x)$
$= (x - 1) + (2x - 3)$
$= 3x - 4$

b. $(f - g)(x) = f(x) - g(x)$
$= (x - 1) - (2x - 3)$
$= x - 1 - 2x + 3$
$= -x + 2$

c. $(f \cdot g)(x) = f(x) \cdot g(x)$
$= (x - 1)(2x - 3)$
$= 2x^2 - 5x + 3$

d. $\left(\dfrac{f}{g}\right)(x) = \dfrac{f(x)}{g(x)} = \dfrac{x - 1}{2x - 3}$ where $x \neq \dfrac{3}{2}$

There is an interesting but not surprising relationship between the graphs of functions and the graphs of their sum, difference, product, and quotient. For example, the graph of $(f + g)$ can be found by adding the graph of f to the graph of g. We add two graphs by adding corresponding y-values.

Practice Problem 1

If $f(x) = x + 3$ and $g(x) = 3x - 1$, find:

a. $(f + g)(x)$

b. $(f - g)(x)$

c. $(f \cdot g)(x)$

d. $\left(\dfrac{f}{g}\right)(x)$

Answers

1. a. $4x + 2$,   **b.** $-2x + 4$,   **c.** $3x^2 + 8x - 3$,

d. $\dfrac{x + 3}{3x - 1}$ where $x \neq \dfrac{1}{3}$

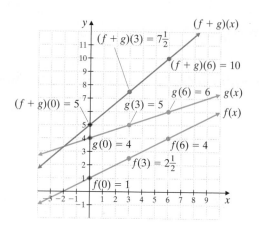

B Composition of Functions

Another way to combine functions is called **function composition**. To understand this new way of combining functions, study the tables below. They show degrees Fahrenheit converted to equivalent degrees Celsius, and then degrees Celsius converted to equivalent degrees Kelvin. (The Kelvin scale is a temperature scale devised by Lord Kelvin in 1848.)

x = Degrees Fahrenheit (**Input**)	-31	-13	32	68	149	212
$C(x)$ = Degrees Celsius (**Output**)	-35	-25	0	20	65	100

C = Degrees Celsius (**Input**)	-35	-25	0	20	65	100
$K(C)$ = Kelvins (**Output**)	238.15	248.15	273.15	293.15	338.15	373.15

Suppose that we want a table that shows a direct conversion from degrees Fahrenheit to kelvins. In other words, suppose that a table is needed that shows kelvins as a function of degrees Fahrenheit. This can easily be done because in the tables, the output of the first table is the same as the input of the second table. The new table is as follows.

x = Degrees Fahrenheit (**Input**)	-31	-13	32	68	149	212
$K(C(x))$ = Kelvins (**Output**)	238.15	248.15	273.15	293.15	338.15	373.15

Since the output of the first table is used as the input of the second table, we write the new function as $K(C(x))$. The new function is formed from the composition of the other two functions. The mathematical symbol for this composition is $(K \circ C)(x)$. Thus, $(K \circ C)(x) = K(C(x))$.

It is possible to find an equation for the composition of the two functions $C(x)$ and $K(x)$. In other words, we can find a function that converts degrees Fahrenheit directly to kelvins. The function $C(x) = \dfrac{5}{9}(x - 32)$ converts degrees Fahrenheit to degrees Celsius, and the function $K(C) = C + 273.15$ converts degrees Celsius to kelvins. Thus,

$$(K \circ C)(x) = K(C(x)) = K\left(\frac{5}{9}(x - 32)\right) = \frac{5}{9}(x - 32) + 273.15$$

In general, the notation $f(g(x))$ means "f composed with g" and can be written as $(f \circ g)(x)$. Also $g(f(x))$, or $(g \circ f)(x)$, means "g composed with f."

Composite Functions

The composition of functions f and g is

$$(f \circ g)(x) = f(g(x))$$

Helpful Hint

$(f \circ g)(x)$ does not mean the same as $(f \cdot g)(x)$.

$$(f \circ g)(x) = f(g(x)) \quad \text{while} \quad (f \cdot g)(x) = f(x) \cdot g(x)$$

EXAMPLE 2   If $f(x) = x^2$ and $g(x) = x + 3$, find each composition.

a. $(f \circ g)(2)$ and $(g \circ f)(2)$
b. $(f \circ g)(x)$ and $(g \circ f)(x)$

Solution:

a. $(f \circ g)(2) = f(g(2))$
$\qquad\qquad\quad = f(5) \qquad\qquad$ Since $g(x) = x + 3$, then $g(2) = 2 + 3 = 5$.
$\qquad\qquad\quad = 5^2 = 25$
$\quad (g \circ f)(2) = g(f(2))$
$\qquad\qquad\quad = g(4) \qquad\qquad$ Since $f(x) = x^2$, then $f(2) = 2^2 = 4$.
$\qquad\qquad\quad = 4 + 3 = 7$

b. $(f \circ g)(x) = f(g(x))$

$\qquad\qquad\quad = f(x + 3) \qquad$ Replace $g(x)$ with $x + 3$.

$\qquad\qquad\quad = (x + 3)^2 \qquad f(x + 3) = (x + 3)^2$

$\qquad\qquad\quad = x^2 + 6x + 9 \qquad$ Square $(x + 3)$.

$\quad (g \circ f)(x) = g(f(x))$

$\qquad\qquad\quad = g(x^2) \qquad\qquad$ Replace $f(x)$ with x^2.

$\qquad\qquad\quad = x^2 + 3 \qquad\qquad g(x^2) = x^2 + 3$

EXAMPLE 3   If $f(x) = |x|$ and $g(x) = x - 2$, find each composition.

a. $(f \circ g)(x)$
b. $(g \circ f)(x)$

Solution:

a. $(f \circ g)(x) = f(g(x)) = f(x - 2) = |x - 2|$
b. $(g \circ f)(x) = g(f(x)) = g(|x|) = |x| - 2$

Helpful Hint

In Examples 2 and 3, notice that $(g \circ f)(x) \neq (f \circ g)(x)$. In general, $(g \circ f)(x)$ *may* or *may not* equal $(f \circ g)(x)$.

EXAMPLE 4

If $f(x) = 5x$, $g(x) = x - 2$, and $h(x) = \sqrt{x}$, write each function as a composition with f, g, or h.

a. $F(x) = \sqrt{x - 2}$
b. $G(x) = 5x - 2$

Practice Problem 2

If $f(x) = x^2$ and $g(x) = 2x + 1$, find each composition.

a. $(f \circ g)(3)$ and $(g \circ f)(3)$
b. $(f \circ g)(x)$ and $(g \circ f)(x)$

Practice Problem 3

If $f(x) = \sqrt{x}$ and $g(x) = x + 1$, find each composition.

a. $(f \circ g)(x)$
b. $(g \circ f)(x)$

Practice Problem 4

If $f(x) = 2x$, $g(x) = x + 5$, and $h(x) = |x|$, write each function as a composition of f, g, or h.

a. $F(x) = |x + 5|$
b. $G(x) = 2x + 5$

Answers

2. a. $49; 19$,   **b.** $4x^2 + 4x + 1; 2x^2 + 1$,
3. a. $\sqrt{x + 1}$,   **b.** $\sqrt{x} + 1$,
4. a. $(h \circ g)(x)$,   **b.** $(g \circ f)(x)$

Solution:

a. Notice the order in which the function F operates on an input value x. First, 2 is subtracted from x, and then the square root of that result is taken. This means that $F = h \circ g$. To check, we find $h \circ g$

$$(h \circ g)(x) = h(g(x)) = h(x - 2) = \sqrt{x - 2}$$

b. Notice the order in which the function G operates on an input value x. First, x is multiplied by 5, and then 2 is subtracted from the result. This means that $G = g \circ f$. To check, we find $g \circ f$.

$$(g \circ f)(x) = g(f(x)) = g(5x) = 5x - 2 \qquad \bullet$$

GRAPHING CALCULATOR EXPLORATIONS

If $f(x) = \dfrac{1}{2}x + 2$ and $g(x) = \dfrac{1}{3}x^2 + 4$, then

$$(f + g)(x) = f(x) + g(x)$$

$$= \left(\frac{1}{2}x + 2\right) + \left(\frac{1}{3}x^2 + 4\right)$$

$$= \frac{1}{3}x^2 + \frac{1}{2}x + 6.$$

To visualize this addition of functions with a grapher, graph

$$Y_1 = \frac{1}{2}x + 2, \quad Y_2 = \frac{1}{3}x^2 + 4, \quad \text{and} \quad Y_3 = \frac{1}{3}x^2 + \frac{1}{2}x + 6$$

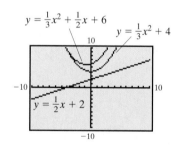

Use a TABLE feature to verify that for a given x value, $Y_1 + Y_2 = Y_3$. For example, verify that when $x = 0$, $Y_1 = 2$, $Y_2 = 4$ and $Y_3 = 2 + 4 = 6$.

EXERCISE SET 10.1

A *For the functions f and g, find* **a.** $(f + g)(x),$ **b.** $(f - g)(x),$ **c.** $(f \cdot g)(x),$ *and* **d.** $\left(\dfrac{f}{g}\right)(x).$ *See Example 1.*

1. $f(x) = x - 7; g(x) = 2x + 1$ **2.** $f(x) = x + 4; g(x) = 5x - 2$ **3.** $f(x) = x^2 + 1; g(x) = 5x$

4. $f(x) = x^2 - 2; g(x) = 3x$ **5.** $f(x) = \sqrt{x}; g(x) = x + 5$ **6.** $f(x) = \sqrt[3]{x}; g(x) = x - 3$

7. $f(x) = -3x; g(x) = 5x^2$ **8.** $f(x) = 4x^3; g(x) = -6x$

B *If* $f(x) = x^2 - 6x + 2,$ $g(x) = -2x,$ *and* $h(x) = \sqrt{x},$ *find each composition. See Example 2.*

9. $(f \circ g)(2)$ **10.** $(h \circ f)(-2)$ **11.** $(g \circ f)(-1)$

12. $(f \circ h)(1)$ **13.** $(g \circ h)(0)$ **14.** $(h \circ g)(0)$

Find $(f \circ g)(x)$ *and* $(g \circ f)(x).$ *See Examples 2 and 3.*

15. $f(x) = x^2 + 1; g(x) = 5x$ **16.** $f(x) = x - 3; g(x) = x^2$

17. $f(x) = 2x - 3; g(x) = x + 7$ **18.** $f(x) = x + 10; g(x) = 3x + 1$

19. $f(x) = x^3 + x - 2; g(x) = -2x$ **20.** $f(x) = -4x; g(x) = x^3 + x^2 - 6$

21. $f(x) = \sqrt{x}; g(x) = -5x + 2$

22. $f(x) = 7x - 1; g(x) = \sqrt[3]{x}$

If $f(x) = 3x, g(x) = \sqrt{x}$, and $h(x) = x^2 + 2$, write each function as a composition with f, g, or h. See Example 4.

 23. $H(x) = \sqrt{x^2 + 2}$

24. $G(x) = \sqrt{3x}$

25. $F(x) = 9x^2 + 2$

26. $H(x) = 3x^2 + 6$

27. $G(x) = 3\sqrt{x}$

28. $F(x) = x + 2$

Find $f(x)$ and $g(x)$ so that the given function $h(x) = (f \circ g)(x)$.

 29. $h(x) = (x + 2)^2$

30. $h(x) = |x - 1|$

31. $h(x) = \sqrt{x + 5} + 2$

32. $h(x) = (3x + 4)^2 + 3$

33. $h(x) = \dfrac{1}{2x - 3}$

34. $h(x) = \dfrac{1}{x + 10}$

Review and Preview

Solve each equation for y. See Section 2.3.

35. $x = y + 2$

36. $x = y - 5$

37. $x = 3y$

38. $x = -6y$

39. $x = -2y - 7$

40. $x = 4y + 7$

Combining Concepts

41. Business people are concerned with cost functions, revenue functions, and profit functions. Recall that the profit $P(x)$ obtained from selling x units of a product is equal to the revenue $R(x)$ from selling the x units minus the cost $C(x)$ of manufacturing the x units. Write an equation expressing this relationship among $C(x)$, $R(x)$, and $P(x)$.

42. Suppose the revenue $R(x)$ for x units of a product can be described by $R(x) = 25x$, and the cost $C(x)$ can be described by $C(x) = 50 + x^2 + 4x$. Find the profit $P(x)$ for x units.

43. If you are given $f(x)$ and $g(x)$, explain in your own words how to find $(f \circ g)(x)$, and then how to find $(g \circ f)(x)$.

44. Given $f(x)$ and $g(x)$, describe in your own words the difference between $(f \circ g)(x)$ and $(f \cdot g)(x)$.

10.2   Inverse Functions

In the next section, we begin a study of two new functions: exponential and logarithmic functions. As we learn more about these functions, we will discover that they share a special relation to each other; they are inverses of each other.

Before we study these functions, we need to learn about inverses. We begin by defining one-to-one functions.

A   Determining Whether a Function Is One-to-One

Study the following table.

Degrees Fahrenheit (Input)	-31	-13	32	68	149	212
Degrees Celsius (Output)	-35	-25	0	20	65	100

Recall that since each degrees Fahrenheit (input) corresponds to exactly one degrees Celsius (output), this table of inputs and outputs does describe a function. Also notice that each output corresponds to a different input. This type of function is given a special name—a *one-to-one function*.

Does the set $f = \{(0,1),(2,2),(-3,5),(7,6)\}$ describe a one-to-one function? It is a function since each x-value corresponds to a unique y-value. For this particular function f, each y-value corresponds to a unique x-value. Thus, this function is also a one-to-one function.

One-to-One Function

For a **one-to-one function**, each x-value (input) corresponds to only one y-value (output) and each y-value (output) corresponds to only one x-value (input).

EXAMPLES

Determine whether each function described is one-to-one.

1. $f\{(6,2),(5,4),(-1,0),(7,3)\}$

The function f is one-to-one since each y-value corresponds to only one x-value.

2. $g = \{(3,9),(-4,2),(-3,9),(0,0)\}$

The function g is not one-to-one because the y-value 9 in $(3,9)$ and $(-3,9)$ corresponds to two different x-values.

3. $h = \{(1,1),(2,2),(10,10),(-5,-5)\}$

The function h is one-to-one since each y-value corresponds to only one x-value.

4.

Mineral (Input)	Talc	Gypsum	Diamond	Topaz	Stibnite
Hardness on the Mohs Scale (Output)	1	2	10	8	2

OBJECTIVES

Ⓐ Determine whether a function is a one-to-one function.

Ⓑ Use the horizontal line test to decide whether a function is a one-to-one function.

Ⓒ Find the inverse of a function.

Ⓓ Find the equation of the inverse of a function.

Ⓔ Graph functions and their inverses.

SSM   TUTOR CENTER   SG   CD & VIDEO   MATH PRO   WEB

Practice Problems 1–5

Determine whether each function described is one-to-one.

1. $f = \{(7,3),(-1,1),(5,0),(4,-2)\}$
2. $g = \{(-3,2),(6,3),(2,14),(-6,2)\}$
3. $h = \{(0,0),(1,2),(3,4),(5,6)\}$
4.

State (Input)	Colorado	Mississippi	Nevada	New Mexico	Utah
Number of Colleges and Universities (Output)	9	44	13	44	21

Source: The Chronicle of Higher Education, Vol. XLV, No. 1, August 28, 1998.

5.

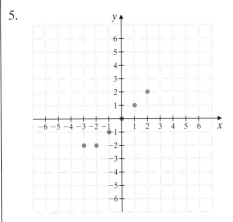

Answers

1. one-to-one,   **2.** not one-to-one,   **3.** one-to-one,
4. not one-to-one,   **5.** not one-to-one

This table does not describe a one-to-one function since the output 2 corresponds to two different inputs, gypsum and stibnite.

5.

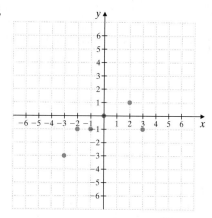

This graph does not describe a one-to-one function since the y-value -1 corresponds to three different x-values, $-2, -1$ and 3, as shown to the right.

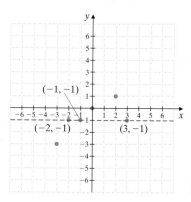

B Using the Horizontal Line Test

Recall that we recognize the graph of a function when it passes the vertical line test. Since every x-value of the function corresponds to exactly one y-value, each vertical line intersects the function's graph at most once. The graph shown next, for instance, is the graph of a function.

Is this function a *one-to-one* function? The answer is no. To see why not, notice that the y-value of the ordered pair $(-3, 3)$, for example, is the same as the y-value of the ordered pair $(3, 3)$. This function is therefore not one-to-one.

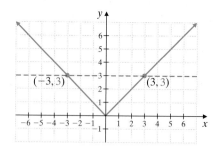

To test whether a graph is the graph of a one-to-one function, we can apply the vertical line test to see whether it is a function, and then apply a similar **horizontal line test** to see whether it is a one-to-one function.

Horizontal Line Test

If every horizontal line intersects the graph of a function at most once, then the function is a one-to-one function.

EXAMPLE 6

Use the vertical and horizontal line tests to determine whether each graph is the graph of a one-to-one function.

a.

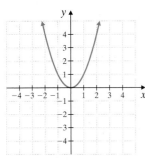

b.

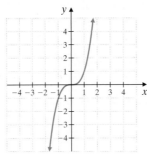

c.

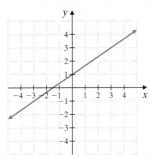

d.

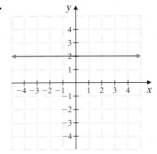

e.

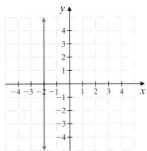

Solution: Graphs **a**, **b**, **c**, and **d** all pass the vertical line test, so only these graphs are graphs of functions. But, of these, only **b** and **c** pass the horizontal line test, so only **b** and **c** are graphs of one-to-one functions.   ●

Helpful Hint

All linear equations are one-to-one functions except those whose graphs are horizontal or vertical lines. A vertical line does not pass the vertical line test and hence is not the graph of a function. A horizontal line is the graph of a function but does not pass the horizontal line test and hence is not the graph of a one-to-one function.

(C) Finding the Inverse of a Function

One-to-one functions are special in that their graphs pass the vertical and horizontal line tests. They are special, too, in another sense: We can find the **inverse function** for any one-to-one function by switching the coordinates of

Practice Problem 6

Use the vertical and horizontal line tests to determine whether each graph is the graph of a one-to-one function.

a.

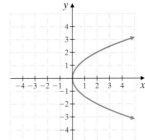

b.

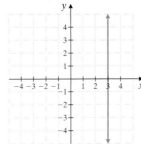

c.

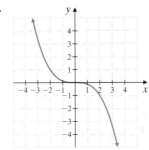

d.

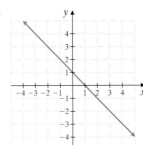

e.

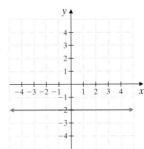

Answers

6. a. not one-to-one,   **b.** not one-to-one,
c. one-to-one,   **d.** one-to-one,   **e.** not one-to-one

the ordered pairs of the function, or the inputs and the outputs. For example, the inverse of the one-to-one function

Degrees Fahrenheit (Input)	−31	−13	32	68	149	212
Degrees Celsius (Output)	−35	−25	0	20	65	100

is the function

Degrees Celsius (Input)	−35	−25	0	20	65	100
Degrees Fahrenheit (Output)	−31	−13	32	68	149	212

Notice that the ordered pair $(-31, -35)$ of the function, for example, becomes the ordered pair $(-35, -31)$ of its inverse.

Also, the inverse of the one-to-one function $f = \{(2, -3), (5, 10), (9, 1)\}$ is $\{(-3, 2), (10, 5), (1, 9)\}$. For a function f, we use the notation f^{-1}, read "f inverse," to denote its inverse function. Notice that since the coordinates of each ordered pair have been switched, the domain (set of inputs) of f is the range (set of outputs) of f^{-1}, and the range of f is the domain of f^{-1}.

Inverse Function

The inverse of a one-to-one function f is the one-to-one function f^{-1} that consists of the set of all ordered pairs (y, x) where (x, y) belongs to f.

Practice Problem 7

Find the inverse of the one-to-one function: $f = \{(2, -4), (-1, 13), (0, 0), (-7, -8)\}$

Concept Check

Suppose that f is a one-to-one function. If the ordered pair $(1, 5)$ belongs to f, name one point that we know must belong to the inverse function f^{-1}.

EXAMPLE 7   Find the inverse of the one-to-one function:

$$f = \{(0, 1), (-2, 7), (3, -6), (4, 4)\}$$

Solution: $f^{-1} = \{(1, 0), (7, -2), (-6, 3), (4, 4)\}$

Switch coordinates of each ordered pair.   ●

Try the Concept Check in the margin.

ⓓ Finding the Equation of the Inverse of a Function

If a one-to-one function f is defined as a set of ordered pairs, we can find f^{-1} by interchanging the x- and y-coordinates of the ordered pairs. If a one-to-one function f is given in the form of an equation, we can find the equation of f^{-1} by using a similar procedure.

Finding an Equation of the Inverse of a One-to-One Function f

Step 1.   Replace $f(x)$ with y.
Step 2.   Interchange x with y.
Step 3.   Solve the equation for y.
Step 4.   Replace y with the notation $f^{-1}(x)$.

Helpful Hint

The symbol f^{-1} is the single symbol used to denote the inverse of the function f. It is read as "f inverse." This symbol *does not mean* $\dfrac{1}{f}$.

Answers

7. $f^{-1} = \{(-4, 2), (13, -1), (0, 0), (-8, -7)\}$

Concept Check: $(5, 1)$

EXAMPLE 8   Find the equation of the inverse of $f(x) = x + 3$.

Solution: $f(x) = x + 3$

Step 1. $y = x + 3$ Replace $f(x)$ with y.

Step 2. $x = y + 3$ Interchange x and y.

Step 3. $x - 3 = y$ Solve for y.

Step 4. $f^{-1}(x) = x - 3$ Replace y with $f^{-1}(x)$.

The inverse of $f(x) = x + 3$ is $f^{-1}(x) = x - 3$. Notice that, for example,

$$f(1) = 1 + 3 = 4 \quad \text{and} \quad f^{-1}(4) = 4 - 3 = 1$$

Ordered pair: $(1, 4)$ Ordered pair: $(4, 1)$

The coordinates are switched, as expected.

EXAMPLE 9

Find the equation of the inverse of $f(x) = 3x - 5$. Graph f and f^{-1} on the same set of axes.

Solution: $f(x) = 3x - 5$

Step 1. $y = 3x - 5$ Replace $f(x)$ with y.

Step 2. $x = 3y - 5$ Interchange x and y.

Step 3. $3y = x + 5$ Solve for y.

$$y = \frac{x + 5}{3}$$

Step 4. $f^{-1}(x) = \dfrac{x + 5}{3}$ Replace y with $f^{-1}(x)$.

Now we graph f and f^{-1} on the same set of axes. Both $f(x) = 3x - 5$ and $f^{-1}(x) = \dfrac{x + 5}{3}$ are linear functions, so each graph is a line.

$f(x) = 3x - 5$

x	$y = f(x)$
1	-2
0	-5
$\dfrac{5}{3}$	0

$f^{-1}(x) = \dfrac{x + 5}{3}$

x	$y = f^{-1}(x)$
-2	1
-5	0
0	$\dfrac{5}{3}$

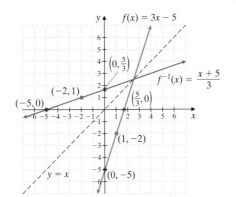

E   Graphing Inverse Functions

Notice that the graphs of f and f^{-1} in Example 9 are mirror images of each other, and the "mirror" is the dashed line $y = x$. This is true for every function and its inverse. For this reason, we say that the *graphs of f and f^{-1} are symmetric about the line $y = x$.*

Practice Problem 8

Find the equation of the inverse of $f(x) = x - 6$.

Practice Problem 9

Find the equation of the inverse of $f(x) = 2x + 3$. Graph f and f^{-1} on the same set of axes.

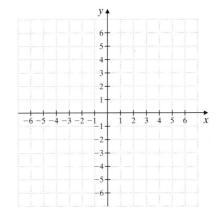

Answers

8. $f^{-1}(x) = x + 6$,

9. $f^{-1}(x) = \dfrac{x - 3}{2}$

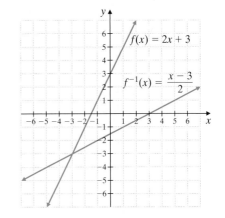

To see why this happens, study the graph of a few ordered pairs and their switched coordinates.

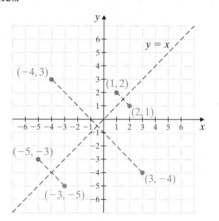

Practice Problem 10

Graph the inverse of each function.

a.

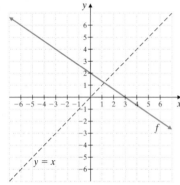

b.

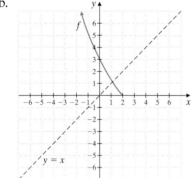

EXAMPLE 10 Graph the inverse of each function.

Solution: The function is graphed in blue and the inverse is graphed in red.

a.

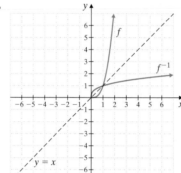

b.

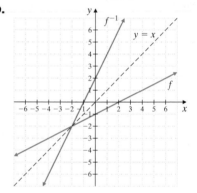

Answers

10. a.

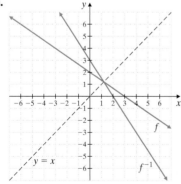

b.

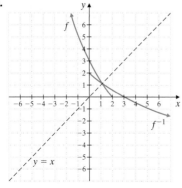

GRAPHING CALCULATOR EXPLORATIONS

A grapher can be used to visualize functions and their inverses. Recall that the graph of a function f and its inverse f^{-1} are mirror images of each other across the line $y = x$. To see this for the function $f(x) = 3x + 2$, use a square window and graph

the given function: $Y_1 = 3x + 2$

its inverse: $Y_2 = \dfrac{x - 2}{3}$

and the line: $Y_3 = x$

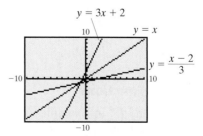

Exercises will follow in Exercise Set 10.2.

EXERCISE SET 10.2

  Determine whether each function is a one-to-one function. If it is one-to-one, list the inverse function by switching coordinates, or inputs and outputs. See Examples 1 through 5, and 7.

1. $f = \{(-1, -1), (1, 1), (0, 2), (2, 0)\}$

2. $g = \{(8, 6), (9, 6), (3, 4), (-4, 4)\}$

3. $h = \{(10, 10)\}$

4. $r = \{(1, 2), (3, 4), (5, 6), (6, 7)\}$

5. $f = \{(11, 12), (4, 3), (3, 4), (6, 6)\}$

6. $g = \{(0, 3), (3, 7), (6, 7), (-2, -2)\}$

7.

Month of 2001 (Input)	January	February	March	April	May	June
Unemployment Rate in Percent (Output)	4.2	4.2	4.3	4.5	4.4	4.5

(*Source:* U.S. Bureau of Labor Statistics)

8.

State (Input)	Wisconsin	Ohio	Georgia	Colorado	California	Arizona
Electoral Votes (Output)	10	20	15	9	55	10

(*Source:* National Archives and Records Administration, based on the 2000 Census)

9.

State (Input)	California	Alaska	Indiana	Louisiana	New Mexico
Rank in Population (Output)	1	48	14	22	36

(*Source:* U.S. Bureau of the Census)

10.

Shape (Input)	Triangle	Pentagon	Quadrilateral	Hexagon	Decagon
Number of Sides (Output)	3	5	4	6	10

Given the one-to-one function $f(x) = x^3 + 2$, find the following. (Hint: You do not need to find the equation for f^{-1}.)

11. a. $f(1)$
  b. $f^{-1}(3)$

12. a. $f(0)$
  b. $f^{-1}(2)$

13. a. $f(-1)$
  b. $f^{-1}(1)$

14. a. $f(-2)$
  b. $f^{-1}(-6)$

B *Determine whether the graph of each function is the graph of a one-to-one function. See Example 6.*

15.

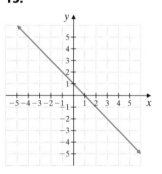

16.

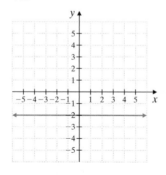

17.

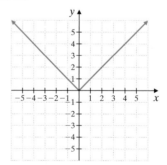

18.

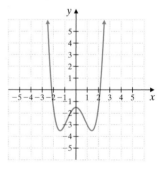

19.

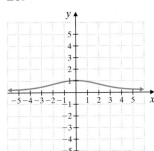

20.

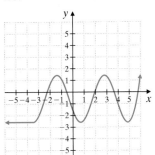

21.

22.

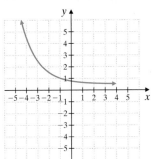

  Each of the following functions is one-to-one. Find the inverse of each function and graph the function and its inverse on the same set of axes. See Examples 8 and 9.

23. $f(x) = x + 4$

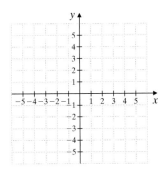

24. $f(x) = x - 5$

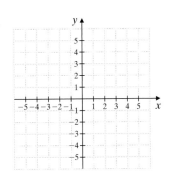

25. $f(x) = 2x - 3$

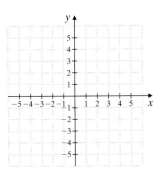

26. $f(x) = 4x + 9$

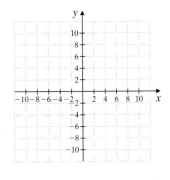

27. $f(x) = \frac{1}{2}x - 1$

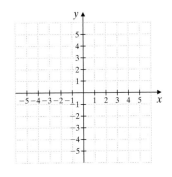

28. $f(x) = -\frac{1}{2}x + 2$

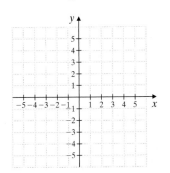

29. $f(x) = x^3$

30. $f(x) = x^3 - 1$

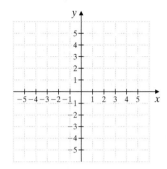

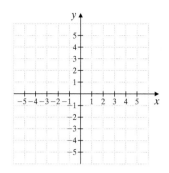

Find the inverse of each one-to-one function. See Examples 8 and 9.

31. $f(x) = \dfrac{x - 2}{5}$

32. $f(x) = \dfrac{4x - 3}{2}$

33. $f(x) = \sqrt[3]{x}$

34. $f(x) = \sqrt[3]{x + 1}$

35. $f(x) = \dfrac{5}{3x + 1}$

36. $f(x) = \dfrac{7}{2x + 4}$

37. $f(x) = (x + 2)^3$

38. $f(x) = (x - 5)^3$

Graph the inverse of each function on the same set of axes. See Example 10.

39.

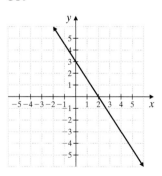

40.

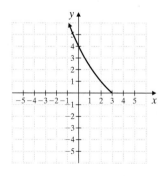

41.

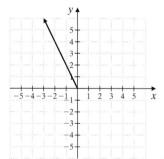

42.

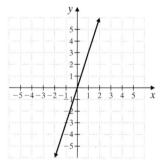

Review and Preview

Evaluate each exponential expression. See Section 7.2.

43. $25^{1/2}$

44. $49^{1/2}$

45. $16^{3/4}$

46. $27^{2/3}$

47. $9^{-3/2}$

48. $81^{-3/4}$

If $f(x) = 3^x$, find each value. In Exercises 51 and 52, give an exact answer and a two-decimal-place approximation. See Section 3.5.

49. $f(2)$

50. $f(0)$

51. $f\left(\dfrac{1}{2}\right)$

52. $f\left(\dfrac{2}{3}\right)$

 Combining Concepts

For Exercises 53 and 54,

   a. Write the ordered pairs for f whose points are highlighted. (Include the points whose coordinates are given.)
   b. Write the corresponding ordered pairs for the inverse of f, f^{-1}.
   c. Graph the ordered pairs for f^{-1} found in part (b).
   d. Graph f^{-1} by drawing a smooth curve through the plotted points.

53. a.

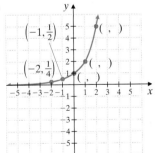

54. a.

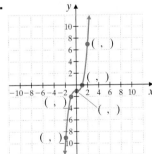

b. c. d.

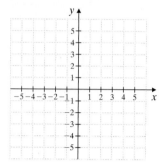

b. c. d.

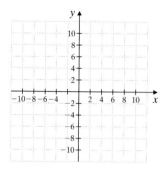

Find the inverse of each one-to-one function. Then graph the function and its inverse in a square window.

55. $f(x) = 3x + 1$

56. $f(x) = -2x - 6$

57. $f(x) = \sqrt[3]{x + 3}$

58. $f(x) = x^3 - 3$

59. If you are given the graph of a function, describe how you can tell from the graph whether a function has an inverse.

60. Describe the appearance of the graphs of a function and its inverse.

10.3 Exponential Functions

In earlier chapters, we gave meaning to exponential expressions such as 2^x, where x is a rational number. Recall the following examples.

$$2^3 = 2 \cdot 2 \cdot 2 \qquad \text{Three factors; each factor is 2}$$

$$2^{3/2} = (2^{1/2})^3 = \sqrt{2} \cdot \sqrt{2} \cdot \sqrt{2} \qquad \text{Three factors; each factor is } \sqrt{2}$$

When x is an irrational number (for example, $\sqrt{3}$), what meaning can we give to $2^{\sqrt{3}}$?

It is beyond the scope of this book to give precise meaning to 2^x if x is irrational. We can confirm your intuition and say that $2^{\sqrt{3}}$ is a real number, and since $1 < \sqrt{3} < 2$, then $2^1 < 2^{\sqrt{3}} < 2^2$. We can also use a calculator and approximate $2^{\sqrt{3}}$: $2^{\sqrt{3}} \approx 3.321997$. In fact, as long as the base b is positive, b^x is a real number for all real numbers x. Finally, the rules of exponents apply whether x is rational or irrational, as long as b is positive.

In this section, we are interested in functions of the form $f(x) = b^x$, where $b > 0$. A function of this form is called an *exponential function*.

Exponential Function

A function of the form

$$f(x) = b^x$$

is called an **exponential function** if $b > 0$, b is not 1, and x is a real number.

OBJECTIVES

Ⓐ Graph exponential functions.

Ⓑ Solve equations of the form $b^x = b^y$.

Ⓒ Solve problems modeled by exponential equations.

SSM
TUTOR CENTER   SG   CD & VIDEO   MATH PRO   WEB

Ⓐ Graphing Exponential Functions

Now let's practice graphing exponential functions.

EXAMPLE 1

Graph the exponential functions $f(x) = 2^x$ and $g(x) = 3^x$ on the same set of axes.

Solution: To graph these functions, we find some ordered pair solutions, plot the points, and connect them with a smooth curve.

$f(x) = 2^x$

x	0	1	2	3	-1	-2
$f(x)$	1	2	4	8	$\dfrac{1}{2}$	$\dfrac{1}{4}$

$g(x) = 3^x$

x	0	1	2	3	-1	-2
$g(x)$	1	3	9	27	$\dfrac{1}{3}$	$\dfrac{1}{9}$

Practice Problem 1

Graph the exponential function $f(x) = 4^x$.

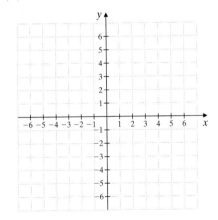

Answer

1.

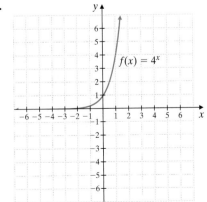

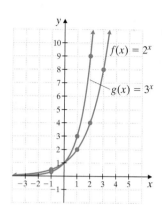

A number of things should be noted about the two graphs of exponential functions in Example 1. First, the graphs show that $f(x) = 2^x$ and $g(x) = 3^x$ are one-to-one functions since each graph passes the vertical and horizontal line tests. The y-intercept of each graph is $(0, 1)$, but neither graph has an x-intercept. From the graph, we can also see that the domain of each function is all real numbers and that the range is $(0, \infty)$. We can also see that as x-values are increasing, y-values are increasing also.

Practice Problem 2

Graph the exponential function $f(x) = \left(\dfrac{1}{5}\right)^x$.

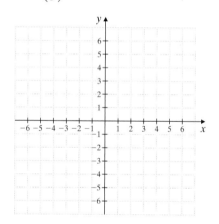

EXAMPLE 2

Graph the exponential functions $y = \left(\dfrac{1}{2}\right)^x$ and $y = \left(\dfrac{1}{3}\right)^x$ on the same set of axes.

Solution: As before, we find some ordered pair solutions, plot the points, and connect them with a smooth curve.

$y = \left(\dfrac{1}{2}\right)^x$

x	0	1	2	3	-1	-2
y	1	$\dfrac{1}{2}$	$\dfrac{1}{4}$	$\dfrac{1}{8}$	2	4

$y = \left(\dfrac{1}{3}\right)^x$

x	0	1	2	3	-1	-2
y	1	$\dfrac{1}{3}$	$\dfrac{1}{9}$	$\dfrac{1}{27}$	3	9

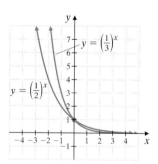

Answer

2.

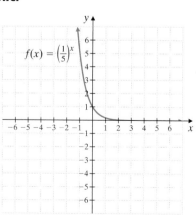

Each function in Example 2 again is a one-to-one function. The y-intercept of both is $(0, 1)$. The domain is the set of all real numbers, and the range is $(0, \infty)$.

Notice the difference between the graphs of Example 1 and the graphs of Example 2. An exponential function is always increasing if the base is greater than 1. When the base is between 0 and 1, the graph is always decreasing. The following figures summarize these characteristics of exponential functions.

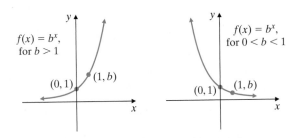

EXAMPLE 3   Graph the exponential function $f(x) = 3^{x+2}$.

Solution: As before, we find and plot a few ordered pair solutions. Then we connect the points with a smooth curve.

$y = 3^{x+2}$

x	0	−1	−2	−3	−4
y	9	3	1	$\frac{1}{3}$	$\frac{1}{9}$

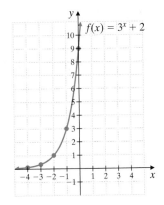

Try the Concept Check in the margin.

B   Solving Equations of the Form $b^x = b^y$

We have seen that an exponential function $y = b^x$ is a one-to-one function. Another way of stating this fact is a property that we can use to solve exponential equations.

Uniqueness of b^x

Let $b > 0$ and $b \neq 1$. Then $b^x = b^y$ is equivalent to $x = y$.

Practice Problem 3

Graph the exponential function $f(x) = 2^{x-1}$.

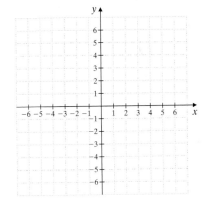

Concept Check

Which functions are exponential functions?

a. $f(x) = x^3$ b. $g(x) = \left(\frac{2}{3}\right)^x$

c. $h(x) = 5^{x-2}$ d. $w(x) = (2x)^2$

Answers

3.

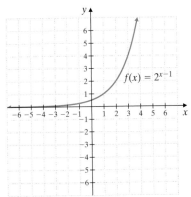

Concept Check: b and c

Practice Problem 4

Solve: $5^x = 125$

EXAMPLE 4 Solve: $2^x = 16$

Solution: We write 16 as a power of 2 and then use the uniqueness of b^x to solve.

$$2^x = 16$$
$$2^x = 2^4$$

Since the bases are the same and are nonnegative, by the uniqueness of b^x we then have that the exponents are equal. Thus,

$$x = 4$$

To check, we replace x with 4 in the original equation.
The solution set is $\{4\}$. ●

Practice Problem 5

Solve: $4^x = 8$

EXAMPLE 5 Solve: $9^x = 27$

Solution: Since both 9 and 27 are powers of 3, we can use the uniqueness of b^x.

$$9^x = 27$$
$$(3^2)^x = 3^3 \qquad \text{Write 9 and 27 as powers of 3.}$$
$$3^{2x} = 3^3$$
$$2x = 3 \qquad \text{Use the uniqueness of } b^x.$$
$$x = \frac{3}{2} \qquad \text{Divide both sides by 2.}$$

To check, we replace x with $\frac{3}{2}$ in the original equation.
The solution set is $\left\{\frac{3}{2}\right\}$. ●

Practice Problem 6

Solve: $9^{x-1} = 27^x$

EXAMPLE 6 Solve: $4^{x+3} = 8^x$

Solution: We write both 4 and 8 as powers of 2, and then use the uniqueness of b^x.

$$4^{x+3} = 8^x$$
$$(2^2)^{x+3} = (2^3)^x$$
$$2^{2x+6} = 2^{3x}$$
$$2x + 6 = 3x \qquad \text{Use the uniqueness of } b^x.$$
$$6 = x \qquad \text{Subtract } 2x \text{ from both sides.}$$

Check to see that the solution set is $\{6\}$. ●

There is one major problem with the preceding technique. Often the two sides of an equation, $4 = 3^x$ for example, cannot easily be written as powers of a common base. We explore how to solve such an equation with the help of *logarithms* later.

Ⓐ **Solving Problems Modeled by Exponential Equations**

The bar graph on the next page shows the increase in the number of cellular phone users. Notice that the graph of the exponential function $y = 25.759(1.277)^x$ approximates the heights of the bars. This is just one ex-

Answers

4. $\{3\}$, **5.** $\left\{\frac{3}{2}\right\}$, **6.** $\{-2\}$

ample of how the world abounds with patterns that can be modeled by exponential functions. To make these applications realistic, we use numbers that warrant a calculator. Another application of an exponential function has to do with interest rates on loans.

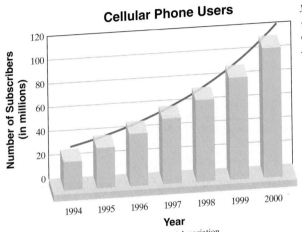

Cellular Phone Users

$y = 25.759(1.277)^x$

where $x = 0$ corresponds to 1994, $x = 1$ corresponds to 1995, and so on

Source: Cellular Telecommunications & Internet Association

The exponential function defined by $A = P\left(1 + \dfrac{r}{n}\right)^{nt}$ models the pattern relating the dollars A accrued (or owed) after P dollars are invested (or loaned) at an annual rate of interest r compounded n times each year for t years. This function is known as the *compound interest formula*.

EXAMPLE 7   Using the Compound Interest Formula

Find the amount owed at the end of 5 years if $1600 is loaned at a rate of 9% compounded monthly.

Solution: Use the formula $A = P\left(1 + \dfrac{r}{n}\right)^{nt}$, with the following values:

$P = \$1600$ (the amount of the loan)

$r = 9\% = 0.09$ (the annual rate of interest)

$n = 12$ (the number of times interest is compounded each year)

$t = 5$ (the duration of the loan, in years)

$$A = P\left(1 + \frac{r}{n}\right)^{nt} \qquad \text{Compound interest formula}$$

$$= 1600\left(1 + \frac{0.09}{12}\right)^{12(5)} \qquad \text{Substitute known values.}$$

$$= 1600(1.0075)^{60}$$

To approximate A, use the $\boxed{y^x}$ or $\boxed{\;\land\;}$ key on your calculator.

$\boxed{2505.0896}$

Thus, the amount A owed is approximately $2505.09.

Practice Problem 7

As a result of the Chernobyl nuclear accident, radioactive debris was carried through the atmosphere. One immediate concern was the impact that debris had on the milk supply. The percent y of radioactive material in raw milk t days after the accident is estimated by $y = 100(2.7)^{-0.1t}$. Estimate the expected percent of radioactive material in the milk after 30 days.

Answer

7. approximately 5.08%

GRAPHING CALCULATOR EXPLORATIONS

We can use a graphing calculator and its TRACE feature to solve Practice Problem 7 graphically.

To estimate the percent of radioactive material in the milk after 30 days, enter $Y_1 = 100(2.7)^{-0.1x}$. The graph does not appear on a standard viewing window, so we need to determine an appropriate viewing window. Because it doesn't make sense to look at radioactivity *before* the Chernobyl nuclear accident, we use Xmin = 0. We are interested in finding the percent of radioactive material in the milk when $x = 30$, so we choose Xmax = 35 to leave enough space to see the graph at $x = 30$. Because the values of y are percents, it seems appropriate that $0 \le y \le 100$. (We also use Xscl = 1 and Yscl = 10.) Now we graph the function.

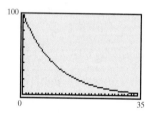

We can use the TRACE feature to obtain an approximation of the expected percent of radioactive material in the milk when $x = 30$. (A TABLE feature may also be used to approximate the percent.) To obtain a better approximation, let's use the ZOOM feature several times to zoom in near $x = 30$.

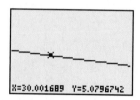

The percent of radioactive material in the milk 30 days after the Chernobyl accident was 5.08%, accurate to two decimal places.

Use a grapher to find each percent. Approximate your solutions so that they are accurate to two decimal places.

1. Estimate the percent of radioactive material in the milk 2 days after the Chernobyl nuclear accident.

2. Estimate the percent of radioactive material in the milk 10 days after the Chernobyl nuclear accident.

3. Estimate the percent of radioactive material in the milk 15 days after the Chernobyl nuclear accident.

4. Estimate the percent of radioactive material in the milk 25 days after the Chernobyl nuclear accident.

EXERCISE SET 10.3

Ⓐ   *Graph each exponential function. See Examples 1 through 3.*

1. $y = 4^x$

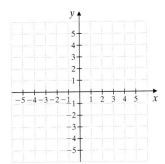

2. $y = 5^x$

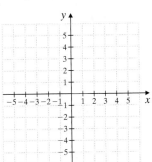

3. $y = 1 + 2^x$

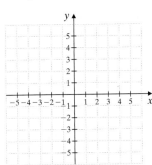

4. $y = 3^x - 1$

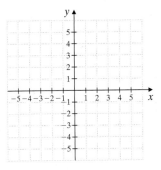

5. $y = \left(\dfrac{1}{4}\right)^x$

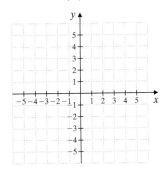

6. $y = \left(\dfrac{1}{5}\right)^x$

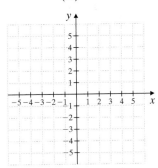

7. $y = \left(\dfrac{1}{2}\right)^x - 2$

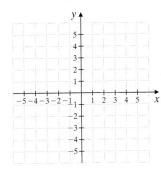

8. $y = \left(\dfrac{1}{3}\right)^x + 2$

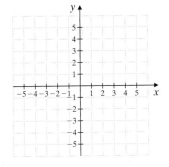

9. $y = -2^x$

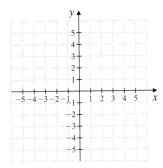

10. $y = -3^x$

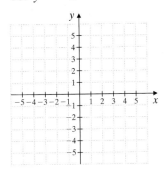

11. $y = 3^x - 2$

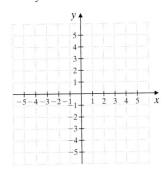

12. $y = 2^x - 3$

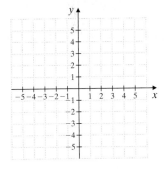

13. $y = -\left(\dfrac{1}{4}\right)^x$

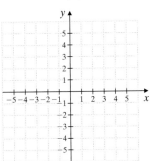

14. $y = -\left(\dfrac{1}{5}\right)^x$

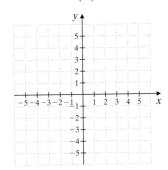

15. $y = \left(\dfrac{1}{3}\right)^x + 1$

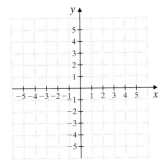

16. $y = \left(\dfrac{1}{2}\right)^x - 2$

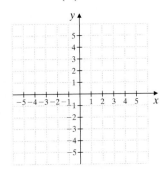

Solve. See Examples 4 through 6.

17. $3^x = 27$ **18.** $6^x = 36$ **19.** $16^x = 8$ **20.** $64^x = 16$ **21.** $32^{2x-3} = 2$

22. $9^{2x+1} = 81$ **23.** $\dfrac{1}{4} = 2^{3x}$ **24.** $\dfrac{1}{27} = 3^{2x}$ **25.** $4^x = 8$ **26.** $32^x = 4$

27. $27^{x+1} = 9$ **28.** $125^{x-2} = 25$ **29.** $81^{x-1} = 27^{2x}$ **30.** $4^{3x-7} = 32^{2x}$

Solve. Unless otherwise indicated, round results to one decimal place. See Example 7.

31. One type of uranium has a daily radioactive decay rate of 0.4%. If 30 pounds of this uranium is available today, how much will still remain after 50 days? Use $y = 30(2.7)^{-0.004t}$, and let t be 50.

32. The nuclear waste from an atomic energy plant decays at a rate of 3% each century. If 150 pounds of nuclear waste is disposed of, how much of it will still remain after 10 centuries? Use $y = 150(2.7)^{-0.03t}$, and let t be 10.

33. National Park Service personnel are trying to increase the size of the bison population of Theodore Roosevelt National Park. If 260 bison currently live in the park, and if the population's rate of growth is 2.5 % annually, how many bison (rounded to the nearest whole) should there be in 10 years? Use $y = 260(2.7)^{0.025t}$.

34. The equation $y = 120.882(1.012)^x$ models the population of the United States from 1930 through 2000. In the equation, y is the population in millions and x represents the number of years after 1930. Round answers to the nearest tenth of a million. (*Source:* Based on data from the U.S. Bureau of the Census)
  a. Estimate the population of the United States in 1970.
  b. Assuming this equation continues to be valid in the future, use the equation to predict the population of the United States in 2020.

35. Retail revenue from shopping on the Internet is currently growing at a rate of 56% per year. In 2000, a total of $42.1 billion in revenue was collected through Internet retail sales. Answer the following questions using $y = 42.1(1.56)^t$ where y is Internet revenue in billions of dollars and t is the number of years after 2000. Round answers to the nearest tenth of a billion dollars. (*Source:* Based on data from eMarketer)
  a. According to the model, what level of retail revenues from Internet shopping was expected in 2001?
  b. Predict the level of Internet shopping revenues in 2009.

36. Carbon dioxide (CO_2) is a greenhouse gas that contributes to global warming. Partially due to the combustion of fossil fuels, the amount of CO_2 in Earth's atmosphere has been increasing by 0.4% annually over the past century. In 2000, the concentration of CO_2 in the atmosphere was 369.4 parts per million by volume. To make the following predictions, use $y = 369.4(1.004)^t$ where y is the concentration of CO_2 in parts per million and t is the number of years after 2000. Round answers to the nearest tenth. (*Sources:* Based on data from the United Nations Environment Programme and the Carbon Dioxide Information Analysis Center)

a. Predict the concentration of CO_2 in the atmosphere in the year 2006.

b. Predict the concentration of CO_2 in the atmosphere in the year 2030.

The equation $y = 25.759(1.277)^x$ gives the number of cellular phone users y (in millions) in the United States for the years 1994 through 2000. In this equation, $x = 0$ corresponds to 1994, $x = 1$ corresponds to 1995, and so on. Use this model to solve Exercises 37 and 38. Round answers to the nearest tenth of a million.

37. Predict the number of cellular phone users in the year 2007.

38. Predict the number of cellular phone users in the year 2015.

Solve. Use $A = P\left(1 + \dfrac{r}{n}\right)^{nt}$. Round answers to two decimal places. See Example 7.

39. Find the amount Erica Entada owes at the end of 3 years if $6000 is loaned to her at a rate of 8% compounded monthly.

40. Find the amount owed at the end of 5 years if $3000 is loaned at a rate of 10% compounded quarterly.

Review and Preview

Solve each equation. See Sections 2.1 and 5.7.

41. $5x - 2 = 18$

42. $3x - 7 = 11$

43. $3x - 4 = 3(x + 1)$

44. $2 - 6x = 6(1 - x)$

Match each exponential function with its graph.

45. $f(x) = \left(\dfrac{1}{2}\right)^x$ ___ **46.** $f(x) = 2^x$ ___ **47.** $f(x) = \left(\dfrac{1}{4}\right)^x$ ___ **48.** $f(x) = 3^x$

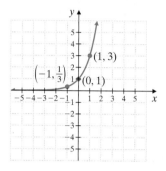
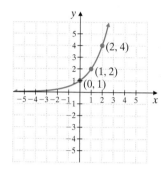
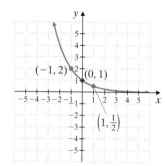
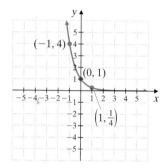

49. Explain why the graph of an exponential function $y = b^x$ contains the point $(1, b)$.

50. Explain why an exponential function $y = b^x$ has a y-intercept of $(0, 1)$.

Use a graphing calculator to solve. Estimate your results to two decimal places.

51. Verify the results of Exercise 31.

52. From Exercise 31, estimate the number of pounds of uranium that will be available after 100 days.

53. From Exercise 31, estimate the number of pounds of uranium that will be available after 120 days.

10.4 Logarithmic Functions

Ⓐ Using Logarithmic Notation

Since the exponential function $f(x) = 2^x$ is a one-to-one function, it has an inverse. We can create a table of values for f^{-1} by switching the coordinates in the accompanying table of values for $f(x) = 2^x$.

x	$y = f(x)$
-3	$\dfrac{1}{8}$
-2	$\dfrac{1}{4}$
-1	$\dfrac{1}{2}$
0	1
1	2
2	4
3	8

x	$y = f^{-1}(x)$
$\dfrac{1}{8}$	-3
$\dfrac{1}{4}$	-2
$\dfrac{1}{2}$	-1
1	0
2	1
4	2
8	3

The graphs of f and its inverse are shown in the margin. Notice that the graphs of f and f^{-1} are symmetric about the line $y = x$, as expected.

Now we would like to be able to write an equation for f^{-1}. To do so, we follow the steps for finding the equation of an inverse.

$$f(x) = 2^x$$

Step 1. Replace $f(x)$ by y. $y = 2^x$

Step 2. Interchange x and y. $x = 2^y$

Step 3. Solve for y.

At this point, we are stuck. To solve this equation for y, a new notation, **logarithmic notation**, is needed.

The symbol $\log_b x$ means "the power to which b is raised to produce a result of x." In other words,

$$\log_b x = y \quad \text{means} \quad b^y = x$$

We say that $\log_b x$ is "the logarithm of x to the base b" or "the log of x to the base b."

> ### Logarithmic Definition
>
> If $b > 0$, and $b \neq 1$, then
> $$y = \log_b x \quad \text{means} \quad x = b^y$$
> for every $x > 0$ and every real number y.

Before returning to the function $x = 2^y$ and solving it for y in terms of x, let's practice using the new notation $\log_b x$.

It is important to be able to write exponential equations with logarithmic notation, and vice versa. The following table shows examples of both forms.

Logarithmic Equation	Corresponding Exponential Equation
$\log_3 9 = 2$	$3^2 = 9$
$\log_6 1 = 0$	$6^0 = 1$
$\log_2 8 = 3$	$2^3 = 8$
$\log_4 \dfrac{1}{16} = -2$	$4^{-2} = \dfrac{1}{16}$
$\log_8 2 = \dfrac{1}{3}$	$8^{1/3} = 2$

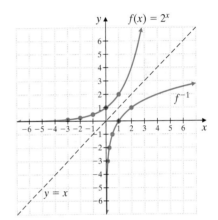

Helpful Hint

Notice that a *logarithm* is an *exponent*. In other words, $\log_3 9$ is the *power* that we raise 3 to in order to get 9.

Practice Problems 1–3

Write as an exponential equation.

1. $\log_7 49 = 2$ 2. $\log_8 \dfrac{1}{8} = -1$

3. $\log_3 \sqrt{3} = \dfrac{1}{2}$

Practice Problems 4–6

Write as a logarithmic equation.

4. $3^4 = 81$ 5. $2^{-3} = \dfrac{1}{8}$

6. $7^{1/3} = \sqrt[3]{7}$

Practice Problem 7

Find the value of each logarithmic expression.

a. $\log_2 8$ b. $\log_3 \dfrac{1}{9}$

c. $\log_{25} 5$

Practice Problem 8

Solve: $\log_2 x = 4$

Practice Problem 9

Solve: $\log_x 9 = 2$

Answers

1. $7^2 = 49$, 2. $8^{-1} = \dfrac{1}{8}$, 3. $3^{1/2} = \sqrt{3}$,

4. $\log_3 81 = 4$, 5. $\log_2 \dfrac{1}{8} = -3$,

6. $\log_7 \sqrt[3]{7} = \dfrac{1}{3}$, 7. a. 3, b. -2, c. $\dfrac{1}{2}$,

8. $\{16\}$, 9. $\{3\}$

EXAMPLES   Write as an exponential equation.

1. $\log_5 25 = 2$ means $5^2 = 25$.

2. $\log_6 \dfrac{1}{6} = -1$ means $6^{-1} = \dfrac{1}{6}$.

3. $\log_2 \sqrt{2} = \dfrac{1}{2}$ means $2^{1/2} = \sqrt{2}$.

EXAMPLES   Write as a logarithmic equation.

4. $9^3 = 729$ means $\log_9 729 = 3$.

5. $6^{-2} = \dfrac{1}{36}$ means $\log_6 \dfrac{1}{36} = -2$.

6. $5^{1/3} = \sqrt[3]{5}$ means $\log_5 \sqrt[3]{5} = \dfrac{1}{3}$.

EXAMPLE 7   Find the value of each logarithmic expression.

a. $\log_4 16$

b. $\log_{10} \dfrac{1}{10}$

c. $\log_9 3$

Solution:

a. $\log_4 16 = 2$ because $4^2 = 16$.

b. $\log_{10} \dfrac{1}{10} = -1$ because $10^{-1} = \dfrac{1}{10}$.

c. $\log_9 3 = \dfrac{1}{2}$ because $9^{1/2} = \sqrt{9} = 3$.

B   **Solving Logarithmic Equations**

The ability to interchange the logarithmic and exponential forms of a statement is often the key to solving logarithmic equations.

EXAMPLE 8   Solve: $\log_5 x = 3$

Solution: $\log_5 x = 3$
$5^3 = x$ Write as an exponential equation.
$125 = x$

The solution set is $\{125\}$.

EXAMPLE 9   Solve: $\log_x 25 = 2$

Solution: $\log_x 25 = 2$
$x^2 = 25$ Write as an exponential equation.
$x = 5$

Even though $(-5)^2 = 25$, the base b of a logarithm must be positive. The solution set is $\{5\}$.

EXAMPLE 10   Solve: $\log_3 1 = x$

Solution: $\log_3 1 = x$

$\qquad\quad 3^x = 1$ Write as an exponential equation.

$\qquad\quad 3^x = 3^0$ Write 1 as 3^0.

$\qquad\quad\; x = 0$ Use the uniqueness of b^x.

The solution set is $\{0\}$.   ●

In Example 10, we illustrated an important property of logarithms. That is, $\log_b 1$ is always 0. This property as well as two important others are given below.

> **Properties of Logarithms**
>
> If b is a real number, $b > 0$ and $b \neq 1$, then
>
> **1.** $\log_b 1 = 0$
> **2.** $\log_b b^x = x$
> **3.** $b^{\log_b x} = x$

To see that $\log_b b^x = x$, we change the logarithmic form to exponential form. Then, $\log_b b^x = x$ means $b^x = b^x$. In exponential form, the statement is true, so in logarithmic form, the statement is also true.

EXAMPLE 11   Simplify.

a. $\log_3 3^2$ **b.** $\log_7 7^{-1}$

c. $5^{\log_5 3}$ **d.** $2^{\log_2 6}$

Solution:

a. From property 2, $\log_3 3^2 = 2$.
b. From property 2, $\log_7 7^{-1} = -1$.
c. From property 3, $5^{\log_5 3} = 3$.
d. From property 3, $2^{\log_2 6} = 6$.   ●

Ⓒ Graphing Logarithmic Functions

Let us now return to the function $f(x) = 2^x$ and write an equation for its inverse, f^{-1}. Recall our earlier work.

$$f(x) = 2^x$$

Step 1. Replace $f(x)$ by y. $y = 2^x$
Step 2. Interchange x and y. $x = 2^y$

Having gained proficiency with the notation $\log_b x$, we can now complete the steps for writing the inverse equation.

Step 3. Solve for y. $y = \log_2 x$
Step 4. Replace y with $f^{-1}(x)$. $f^{-1}(x) = \log_2 x$

Thus, $f^{-1}(x) = \log_2 x$ defines a function that is the inverse function of the function $f(x) = 2^x$. The function $f^{-1}(x)$ or $y = \log_2 x$ is called a *logarithmic function.*

Practice Problem 10

Solve: $\log_2 1 = x$

Practice Problem 11

Simplify.

a. $\log_6 6^3$ b. $\log_{11} 11^{-4}$

c. $7^{\log_7 5}$ d. $3^{\log_3 10}$

Answers

10. $\{0\}$,   **11. a.** 3,   **b.** -4,   **c.** 5,   **d.** 10

Concept Check

Let $f(x) = \log_3 x$ and $g(x) = 3^x$. These two functions are inverses of each other. Since $(2, 9)$ is an ordered pair solution of $g(x)$, what ordered pair do we know to be a solution of $f(x)$? Explain why.

Practice Problem 12

Graph the logarithmic function $y = \log_4 x$.

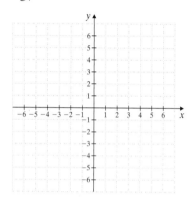

Practice Problem 13

Graph the logarithmic function $f(x) = \log_{1/2} x$.

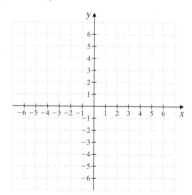

Answers

12.

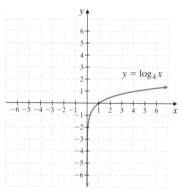

Logarithmic Function

If x is a positive real number, b is a constant positive real number, and b is not 1, then a **logarithmic function** is a function that can be defined by

$$f(x) = \log_b x$$

The domain of f is the set of positive real numbers, and the range of f is the set of real numbers.

Try the Concept Check in the margin.

We can explore logarithmic functions by graphing them.

EXAMPLE 12 Graph the logarithmic function $y = \log_2 x$.

Solution: First we write the equation with exponential notation as $2^y = x$. Then we find some ordered pair solutions that satisfy this equation. Finally, we plot the points and connect them with a smooth curve. The domain of this function is $(0, \infty)$, and the range is all real numbers.

Since $x = 2^y$ is solved for x, we choose y-values and compute corresponding x-values.

If $y = 0$, $x = 2^0 = 1$.
If $y = 1$, $x = 2^1 = 2$.
If $y = 2$, $x = 2^2 = 4$.

If $y = -1$, $x = 2^{-1} = \dfrac{1}{2}$.

$x = 2^y$	y
1	0
2	1
4	2
$\dfrac{1}{2}$	-1

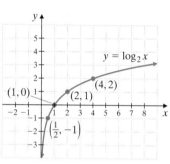

EXAMPLE 13 Graph the logarithmic function $f(x) = \log_{1/3} x$.

Solution: We can replace $f(x)$ with y, and write the result with exponential notation.

$$f(x) = \log_{1/3} x$$

$$y = \log_{1/3} x \qquad \text{Replace } f(x) \text{ with } y.$$

$$\left(\frac{1}{3}\right)^y = x \qquad \text{Write in exponential form.}$$

Now we can find ordered pair solutions that satisfy $\left(\dfrac{1}{3}\right)^y = x$, plot these points, and connect them with a smooth curve.

If $y = 0$, $x = \left(\dfrac{1}{3}\right)^0 = 1$.

If $y = 1$, $x = \left(\dfrac{1}{3}\right)^1 = \dfrac{1}{3}$.

If $y = -1$, $x = \left(\dfrac{1}{3}\right)^{-1} = 3$.

If $y = -2$, $x = \left(\dfrac{1}{3}\right)^{-2} = 9$.

$x = \left(\dfrac{1}{3}\right)^y$	y
1	0
$\dfrac{1}{3}$	1
3	-1
9	-2

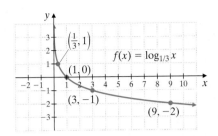

$$f(x) = \log_{1/3} x$$

The domain of this function is $(0, \infty)$, and the range is the set of all real numbers.

The following figures summarize characteristics of logarithmic functions.

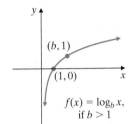

$f(x) = \log_b x,$
if $b > 1$

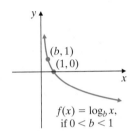

$f(x) = \log_b x,$
if $0 < b < 1$

Answer

13.

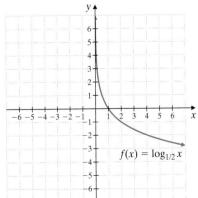

$$f(x) = \log_{1/2} x$$

FOCUS ON **On History**

THE INVENTION OF LOGARITHMS

Logarithms were the invention of John Napier (1550–1617), a Scottish land owner and theologian. Napier was also fascinated by mathematics and made it his hobby. Over a period of 20 years in his spare time, he developed his theory of logarithms, which were explained in his Latin text *Mirifici Logarithmorum Canonis Descriptio* (A Description of an Admirable Table of Logarithms), published in 1614. He hoped that his discovery would help to simplify the many time-consuming calculations required in astronomy. In fact, Napier's logarithms revolutionized astronomy and many other advanced mathematical fields by replacing "the multiplications, divisions, square and cubical extractions of great numbers, which besides the tedious expense of time are for the most part subject to many slippery errors" with related numbers that can be easily added and subtracted instead. His discovery was a great time-saving device. Some historians suggest that the use of logarithms to simplify calculations enabled German astronomer Johannes Kepler to develop his three laws of planetary motion, which in turn helped English physicist Sir Isaac Newton develop his theory of gravitation. Two hundred years after Napier's discovery, the French mathematician Pierre de Laplace wrote that logarithms, "by shortening the labors, doubled the life of the astronomer."

Napier's original logarithm tables had several flaws: They did not actually use a particular logarithmic base per se and log 1 was not defined to be equal to 0. An English mathematician, Henry Briggs, read Napier's Latin text soon after it was published and was very impressed by his ideas. Briggs wrote to Napier, asking to meet in person to discuss his wonderful discovery and to offer several improvements. The two mathematicians met in the summer of 1615. Briggs suggested redefining logarithms to base 10 and defining log 1 = 0. Napier had also thought of using base 10 but hadn't been well enough to start a new set of tables. He asked Briggs to undertake the construction of a new set of base 10 tables. And so it was that the first table of common logarithms was constructed by Briggs over the next two years. Napier died in 1617 before Briggs was able to complete his new tables.

CRITICAL THINKING

Locate a table of common logarithms and describe how to use it. Give several examples. Explain why a table of common logarithms would have been invaluable to many calculations before the invention of the hand-held calculator.

EXERCISE SET 10.4

A *Write each as an exponential equation. See Examples 1 through 3.*

1. $\log_6 36 = 2$

2. $\log_2 32 = 5$

3. $\log_3 \dfrac{1}{27} = -3$

4. $\log_5 \dfrac{1}{25} = -2$

5. $\log_{10} 1000 = 3$

6. $\log_{10} 10 = 1$

7. $\log_e x = 4$

8. $\log_e \dfrac{1}{e} = -1$

9. $\log_e \dfrac{1}{e^2} = -2$

10. $\log_e y = 7$

11. $\log_7 \sqrt{7} = \dfrac{1}{2}$

12. $\log_{11} \sqrt[4]{11} = \dfrac{1}{4}$

Write each as a logarithmic equation. See Examples 4 through 6.

13. $2^4 = 16$

14. $5^3 = 125$

15. $10^2 = 100$

16. $10^4 = 1000$

17. $e^3 = x$

18. $e^5 = y$

19. $10^{-1} = \dfrac{1}{10}$

20. $10^{-2} = \dfrac{1}{100}$

21. $4^{-2} = \dfrac{1}{16}$

22. $3^{-4} = \dfrac{1}{81}$

23. $5^{1/2} = \sqrt{5}$

24. $4^{1/3} = \sqrt[3]{4}$

Find the value of each logarithmic expression. See Example 7.

25. $\log_2 8$

26. $\log_3 9$

27. $\log_3 \dfrac{1}{9}$

28. $\log_2 \dfrac{1}{32}$

29. $\log_{25} 5$

30. $\log_8 \dfrac{1}{2}$

31. $\log_{1/2} 2$

32. $\log_{2/3} \dfrac{4}{9}$

33. $\log_6 1$

34. $\log_9 9$

35. $\log_2 2^4$

36. $\log_6 6^{-2}$

37. $\log_{10} 100$

38. $\log_{10} \dfrac{1}{10}$

39. $3^{\log_3 5}$

40. $5^{\log_5 7}$

41. $\log_3 81$

42. $\log_2 16$

43. $\log_4 \dfrac{1}{64}$

44. $\log_3 \dfrac{1}{9}$

45. Explain why negative numbers are not included as logarithmic bases.

46. Explain why 1 is not included as a logarithmic base.

B *Solve. See Examples 8 through 10.*

47. $\log_3 9 = x$

48. $\log_2 8 = x$

49. $\log_3 x = 4$

50. $\log_2 x = 3$

51. $\log_x 49 = 2$

52. $\log_x 8 = 3$

53. $\log_2 \dfrac{1}{8} = x$

54. $\log_3 \dfrac{1}{81} = x$

55. $\log_3 \dfrac{1}{27} = x$

56. $\log_5 \dfrac{1}{125} = x$

57. $\log_8 x = \dfrac{1}{3}$

58. $\log_9 x = \dfrac{1}{2}$

59. $\log_4 16 = x$

60. $\log_2 16 = x$

61. $\log_{3/4} x = 3$

62. $\log_{2/3} x = 2$

63. $\log_x 100 = 2$

64. $\log_x 27 = 3$

Simplify. See Example 11.

65. $\log_5 5^3$

66. $\log_6 6^2$

67. $2^{\log_2 3}$

68. $7^{\log_7 4}$

69. $\log_9 9$

70. $\log_8 (8)^{-1}$

C *Graph each logarithmic function. See Examples 12 and 13.*

71. $y = \log_3 x$

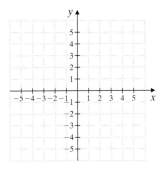

72. $y = \log_2 x$

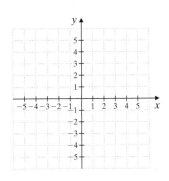

73. $f(x) = \log_{1/4} x$

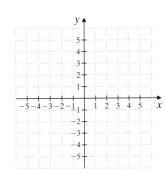

74. $f(x) = \log_{1/2} x$

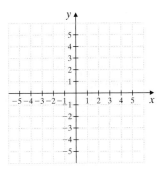

75. $f(x) = \log_5 x$

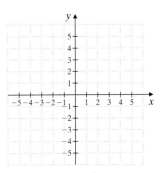

76. $f(x) = \log_6 x$

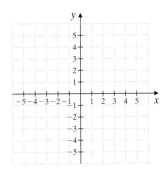

77. $f(x) = \log_{1/6} x$

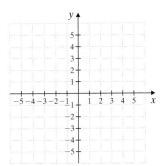

78. $f(x) = \log_{1/5} x$

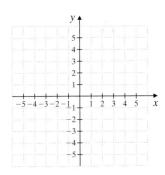

Review and Preview

Simplify each rational expression. See Section 6.1.

79. $\dfrac{x + 3}{3 + x}$

80. $\dfrac{x - 5}{5 - x}$

81. $\dfrac{x^2 - 8x + 16}{2x - 8}$

82. $\dfrac{x^2 - 3x - 10}{2 + x}$

 Combining Concepts

Graph each function and its inverse on the same set of axes.

83. $y = 4^x$; $y = \log_4 x$

84. $y = 3^x$; $y = \log_3 x$

85. $y = \left(\dfrac{1}{3}\right)^x$; $y = \log_{1/3} x$

86. $y = \left(\dfrac{1}{2}\right)^x$; $y = \log_{1/2} x$

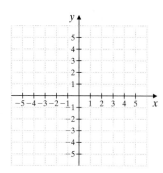 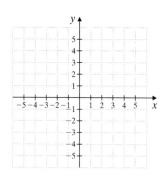 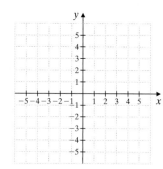 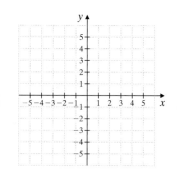

87. The formula $\log_{10}(1 - k) = \dfrac{-0.3}{H}$ models the relationship between the half-life H of a radioactive material and its rate of decay k. Find the rate of decay of the iodine isotope I-131 if its half-life is 8 days. Round to four decimal places.

88. Explain why the graph of the function $y = \log_b x$ contains the point $(1, 0)$ no matter what b is.

89. $\text{Log}_3 10$ is between which two integers? Explain your answer.

10.5 Properties of Logarithms

In the previous section we explored some basic properties of logarithms. We now introduce and study additional properties. Because a logarithm is an exponent, logarithmic properties are just restatements of exponential properties.

OBJECTIVES

Ⓐ Use the product property of logarithms.

Ⓑ Use the quotient property of logarithms.

Ⓒ Use the power property of logarithms.

Ⓓ Use the properties of logarithms together.

SSM
TUTOR CENTER   SG   CD & VIDEO   MATH PRO   WEB

Ⓐ Using the Product Property

The first of these properties is called the **product property of logarithms** because it deals with the logarithm of a product.

Product Property of Logarithms

If x, y, and b are positive real numbers and $b \neq 1$, then

$$\log_b xy = \log_b x + \log_b y$$

To prove this, we let $\log_b x = M$ and $\log_b y = N$. Now we write each logarithm with exponential notation.

$\log_b x = M$ is equivalent to $b^M = x$

$\log_b y = N$ is equivalent to $b^N = y$

When we multiply the left sides and the right sides of the exponential equations, we have that

$$xy = (b^M)(b^N) = b^{M+N}$$

If we write the equation $xy = b^{M+N}$ in equivalent logarithmic form, we have

$$\log_b xy = M + N$$

But since $M = \log_b x$ and $N = \log_b y$, we can write

$$\log_b xy = \log_b x + \log_b y \quad \text{Let } M = \log_b x \text{ and } N = \log_b y.$$

In other words, the logarithm of a product is the sum of the logarithms of the factors. This property is sometimes used to simplify logarithmic expressions.

EXAMPLE 1   Write as a single logarithm: $\log_{11} 10 + \log_{11} 3$

Solution: $\log_{11} 10 + \log_{11} 3 = \log_{11}(10 \cdot 3)$ Use the product property.

$$= \log_{11} 30$$

●

EXAMPLE 2   Write as a single logarithm: $\log_2(x + 2) + \log_2 x$

Solution:

$$\log_2(x + 2) + \log_2 x = \log_2[(x + 2) \cdot x] = \log_2(x^2 + 2x)$$

●

Ⓑ Using the Quotient Property

The second property is the **quotient property of logarithms**.

Practice Problem 1

Write as a single logarithm:
$\log_2 7 + \log_2 5$

Practice Problem 2

Write as a single logarithm:
$\log_3 x + \log_3(x - 9)$

Answers

1. $\log_2 35$,   **2.** $\log_3(x^2 - 9x)$

Concept Check

Which of the following is the correct way to rewrite $\log_5 \frac{7}{2}$?

a. $\log_5 7 - \log_5 2$
b. $\log_5(7 - 2)$
c. $\dfrac{\log_5 7}{\log_5 2}$
d. $\log_5 14$

Practice Problem 3

Write as a single logarithm:
$\log_7 40 - \log_7 8$

Practice Problem 4

Write as a single logarithm:
$\log_3(x^3 + 4) - \log_3(x^2 + 2)$

Practice Problems 5–6

Use the power property to rewrite each expression.

5. $\log_3 x^5$ 6. $\log_7 \sqrt[3]{4}$

Practice Problems 7–8

Write as a single logarithm.

7. $3 \log_4 2 + 2 \log_4 5$
8. $5 \log_2(2x - 1) - \log_2 x$

Answers

3. $\log_7 5$,   **4.** $\log_3 \dfrac{x^3 + 4}{x^2 + 2}$,   **5.** $5 \log_3 x$,

6. $\dfrac{1}{3} \log_7 4$,   **7.** $\log_4 200$,   **8.** $\log_2 \dfrac{(2x - 1)^5}{x}$

Concept Check: a

Quotient Property of Logarithms

If x, y, and b are positive real numbers and $b \neq 1$, then

$$\log_b \frac{x}{y} = \log_b x - \log_b y$$

The proof of the quotient property of logarithms is similar to the proof of the product property. Notice that the quotient property says that the logarithm of a quotient is the difference of the logarithms of the dividend and divisor.

Try the Concept Check in the margin.

EXAMPLE 3   Write as a single logarithm: $\log_{10} 27 - \log_{10} 3$

Solution: $\log_{10} 27 - \log_{10} 3 = \log_{10} \dfrac{27}{3}$ Use the quotient property.

$$= \log_{10} 9 \qquad \bullet$$

EXAMPLE 4   Write as a single logarithm: $\log_3(x^2 + 5) - \log_3(x^2 + 1)$

Solution:

$$\log_3(x^2 + 5) - \log_3(x^2 + 1) = \log_3 \frac{x^2 + 5}{x^2 + 1} \qquad \text{Use the quotient property.} \bullet$$

C   Using the Power Property

The third and final property we introduce is the **power property of logarithms**.

Power Property of Logarithms

If x and b are positive real numbers, $b \neq 1$, and r is a real number, then

$$\log_b x^r = r \log_b x$$

EXAMPLES   Use the power property to rewrite each expression.

5. $\log_5 x^3 = 3 \log_5 x$

6. $\log_4 \sqrt{2} = \log_4 2^{1/2} = \dfrac{1}{2} \log_4 2$ $\bullet$

D   Using More Than One Property

Many times we must use more than one property of logarithms to simplify logarithmic expression.

EXAMPLES   Write as a single logarithm.

7. $2 \log_5 3 + 3 \log_5 2 = \log_5 3^2 + \log_5 2^3$ Use the power property.

$$= \log_5 9 + \log_5 8$$
$$= \log_5(9 \cdot 8) \qquad \text{Use the product property.}$$
$$= \log_5 72$$

8. $3 \log_9 x - \log_9(x + 1) = \log_9 x^3 - \log_9(x + 1)$ Use the power property.

$$= \log_9 \frac{x^3}{x + 1}$$ Use the quotient property.

EXAMPLES   Write each expression as sums or differences of logarithms.

9. $\log_3 \dfrac{5 \cdot 7}{4} = \log_3(5 \cdot 7) - \log_3 4$ Use the quotient property.

$$= \log_3 5 + \log_3 7 - \log_3 4$$ Use the product property.

10. $\log_2 \dfrac{x^5}{y^2} = \log_2(x^5) - \log_2(y^2)$ Use the quotient property.

$$= 5 \log_2 x - 2 \log_2 y$$ Use the power property.

Helpful Hint

Notice that we are not able to simplify further a logarithmic expression such as $\log_5(2x - 1)$. None of the basic properties gives a way to write the logarithm of a difference (or sum) in some equivalent form.

Try the Concept Check in the margin.

EXAMPLES

If $\log_b 2 = 0.43$ and $\log_b 3 = 0.68$, use the properties of logarithms to evaluate each expression.

11. $\log_b 6 = \log_b(2 \cdot 3)$ Write 6 as $2 \cdot 3$.

$\qquad = \log_b 2 + \log_b 3$ Use the product property.

$\qquad = 0.43 + 0.68$ Substitute given values.

$\qquad = 1.11$ Simplify.

12. $\log_b 9 = \log_b 3^2$ Write 9 as 3^2.

$\qquad = 2 \log_b 3$ Use the power property.

$\qquad = 2(0.68)$ Substitute the given value.

$\qquad = 1.36$ Simplify.

13. $\log_b \sqrt{2} = \log_b 2^{1/2}$ Write $\sqrt{2}$ as $2^{1/2}$.

$\qquad = \dfrac{1}{2} \log_b 2$ Use the power property.

$\qquad = \dfrac{1}{2}(0.43)$ Substitute the given value.

$\qquad = 0.215$ Simplify.

Practice Problems 9–10

Write each expression as sums or differences of logarithms.

9. $\log_7 \dfrac{6 \cdot 2}{5}$

10. $\log_3 \dfrac{x^4}{y^3}$

Concept Check

What is wrong with the following?
$$\log_{10}(x^2 + 5) = \log_{10} x^2 + \log_{10} 5$$
$$= 2 \log_{10} x + \log_{10} 5$$
Use a numerical example to demonstrate that the result is incorrect.

Practice Problems 11–13

If $\log_b 4 = 0.86$ and $\log_b 7 = 1.21$, use the properties of logarithms to evaluate each expression.

11. $\log_b 28$
12. $\log_b 49$
13. $\log_b \sqrt[3]{4}$

Answers

9. $\log_7 6 + \log_7 2 - \log_7 5$,
10. $4 \log_3 x - 3 \log_3 y$, **11.** 2.07, **12.** 2.42,
13. $0.28\overline{6}$

Concept Check: The properties do not give any way to simplify the logarithm of a sum; answers may vary.

STUDY SKILLS REMINDER

Are you preparing for a test on Chapter 10?

Below I have listed some common trouble areas for students in Chapter 10. After studying for your test—but before taking your test—read these.

- Don't forget how to find the composition of two functions.

    If $f(x) = x^2 + 5$ and $g(x) = 3x$, then
    $(f \circ g)(x) = f[g(x)] = f(3x) = (3x)^2 + 5 = 9x^2 + 5$
    $(g \circ f)(x) = g[f(x)] = g(x^2 + 5) = 3(x^2 + 5) = 3x^2 + 15$

- Don't forget that f^{-1} is a special notation used to denote the inverse of a function.

Let's find the inverse of the invertible function $f(x) = 3x - 5$.

$$f(x) = 3x - 5$$
$$y = 3x - 5 \qquad \text{Replace } f(x) \text{ with } y.$$
$$x = 3y - 5 \qquad \text{Switch variables.}$$
$$x + 5 = 3y$$
$$\frac{x + 5}{3} = y \qquad \text{Solve for } y.$$
$$f^{-1}(x) = \frac{x + 5}{3} \qquad \text{Replace } y \text{ with } f^{-1}(x).$$

- Don't forget that $y = \log_b x$ means $b^y = x$.

    Thus, $3 = \log_5 125$ means $5^3 = 125$.

- Remember rules for logarithms.

    $\log_b 3x = \log_b 3 + \log_b x$

    $\log_b(3 + x)$ cannot be simplified in the same manner.

Remember: This is simply a checklist of common trouble areas. For a review of Chapter 10, see the Highlights and Chapter Review at the end of this chapter.

EXERCISE SET 10.5

Ⓐ *Write each sum as a single logarithm. Assume that variables represent positive numbers. See Examples 1 and 2.*

1. $\log_5 2 + \log_5 7$
2. $\log_3 8 + \log_3 4$
3. $\log_4 9 + \log_4 x$
4. $\log_2 x + \log_2 y$

5. $\log_{10} 5 + \log_{10} 2 + \log_{10}(x^2 + 2)$
6. $\log_6 3 + \log_6(x + 4) + \log_6 5$

Ⓑ *Write each difference as a single logarithm. Assume that variables represent positive numbers. See Examples 3 and 4.*

7. $\log_5 12 - \log_5 4$
8. $\log_7 20 - \log_7 4$
9. $\log_2 x - \log_2 y$

10. $\log_3 12 - \log_3 z$
11. $\log_3 8 - \log_3 2$
12. $\log_5 12 - \log_5 3$

Ⓒ *Use the power property to rewrite each expression. See Examples 5 and 6.*

13. $\log_3 x^2$
14. $\log_2 x^5$
15. $\log_4 5^{-1}$

16. $\log_6 7^{-2}$
17. $\log_5 \sqrt{y}$
18. $\log_5 \sqrt[3]{x}$

Ⓓ *Write each as a single logarithm. Assume that variables represent positive numbers. See Examples 7 and 8.*

19. $2 \log_2 5$
20. $3 \log_5 2$
21. $3 \log_5 x + 6 \log_5 z$
22. $2 \log_7 y + 6 \log_7 z$

23. $\log_{10} x - \log_{10}(x + 1) + \log_{10}(x^2 - 2)$

24. $\log_9(4x) - \log_9(x - 3) + \log_9(x^3 + 1)$

25. $\log_4 2 + \log_4 10 - \log_4 5$

26. $\log_6 18 + \log_6 2 - \log_6 9$

27. $\log_7 6 + \log_7 3 - \log_7 4$

28. $\log_8 5 + \log_8 15 - \log_8 20$

29. $3 \log_4 2 + \log_4 6$

30. $2 \log_3 5 + \log_3 2$

31. $3 \log_2 x + \dfrac{1}{2} \log_2 x - 2 \log_2(x + 1)$

32. $2 \log_5 x + \dfrac{1}{3} \log_5 x - 3 \log_5(x + 5)$

33. $2 \log_8 x - \dfrac{2}{3} \log_8 x + 4 \log_8 x$

34. $5 \log_6 x - \dfrac{3}{4} \log_6 x + 3 \log_6 x$

Write each expression as a sum or difference of logarithms. Assume that variables represent positive numbers. See Examples 9 and 10.

35. $\log_3 \dfrac{4y}{5}$

36. $\log_4 \dfrac{2}{9z}$

37. $\log_2 \dfrac{x^3}{y}$

38. $\log_5 \dfrac{x}{y^4}$

39. $\log_b \sqrt{7x}$

40. $\log_b \sqrt{\dfrac{3}{y}}$

41. $\log_7 \dfrac{5x}{4}$

42. $\log_9 \dfrac{7}{y}$

43. $\log_5 x^3(x + 1)$

44. $\log_2 y^3 z$

45. $\log_6 \dfrac{x^2}{x + 3}$

46. $\log_3 \dfrac{(x + 5)^2}{x}$

If $\log_b 3 = 0.5$ and $\log_b 5 = 0.7$, evaluate each expression. See Examples 11 through 13.

47. $\log_b \dfrac{5}{3}$ **48.** $\log_b 25$ **49.** $\log_b 15$ **50.** $\log_b \dfrac{3}{5}$ **51.** $\log_b \sqrt{5}$ **52.** $\log_b \sqrt[4]{3}$

If $\log_b 2 = 0.43$ and $\log_b 3 = 0.68$, evaluate each expression. See Examples 11 through 13.

53. $\log_b 8$ **54.** $\log_b 81$ **55.** $\log_b \dfrac{3}{9}$ **56.** $\log_b \dfrac{4}{32}$ **57.** $\log_b \sqrt{\dfrac{2}{3}}$ **58.** $\log_b \sqrt{\dfrac{3}{2}}$

Review and Preview

59. Graph the functions $y = 10^x$ and $y = \log_{10} x$ on the same set of axes. See Section 10.4.

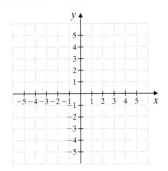

Evaluate each expression. See Section 10.4.

60. $\log_{10} 100$ **61.** $\log_{10} \dfrac{1}{10}$ **62.** $\log_7 7^2$ **63.** $\log_7 \sqrt{7}$

Combining Concepts

Determine whether each statement is true or false.

64. $\log_2 x^3 = 3\log_2 x$ **65.** $\log_3(x + y) = \log_3 x + \log_3 y$ **66.** $\dfrac{\log_7 10}{\log_7 5} = \log_7 2$

67. $\log_7 \dfrac{14}{8} = \log_7 14 - \log_7 8$ **68.** $\dfrac{\log_7 x}{\log_7 y} = (\log_7 x) - (\log_7 y)$ **69.** $(\log_3 6) \cdot (\log_3 4) = \log_3 24$

70. It is true that $\log 8 = \log(8 \cdot 1) = \log 8 + \log 1$.
Explain how $\log 8$ can equal $\log 8 + \log 1$.

Sound Intensity

The decibel (dB) measures sound intensity, or the relative loudness or strength of a sound. One decibel is the smallest difference in sound levels that is detectable by humans. The decibel is a logarithmic unit. This means that for approximately every 3-decibel increase in sound intensity, the relative loudness of the sound is doubled. For example, a 35 dB sound is twice as loud as a 32 dB sound.

In the modern world, noise pollution has increasingly become a concern. Sustained exposure to high sound intensities can lead to hearing loss. Regular exposure to 90 dB sounds can eventually lead to loss of hearing. Sounds of 130 dB and more can cause permanent loss of hearing instantaneously.

The relative loudness of a sound D in decibels is given by the equation

$$D = 10 \log_{10} \frac{I}{10^{-16}}$$

where I is the intensity of a sound given in watts per square centimeter. Some sound intensities of common noises are listed in the table in order of increasing sound intensity.

Group Activity

1. Work together to create a table of the relative loudness (in decibels) of the sounds listed in the table.

Some Sound Intensities of Common Noises	
Noise	**Intensity (watts/cm²)**
Whispering	10^{-15}
Rustling leaves	$10^{-14.2}$
Normal conversation	10^{-13}
Background noise in a quiet residence	$10^{-12.2}$
Typewriter	10^{-11}
Air conditioning	10^{-10}
Freight train at 50 feet	$10^{-8.5}$
Vacuum cleaner	10^{-8}
Nearby thunder	10^{-7}
Air hammer	$10^{-6.5}$
Jet plane at takeoff	10^{-6}
Threshold of pain	10^{-4}

2. Research the loudness of other common noises. Add these sounds and their decibel levels to your table. Be sure to list the sounds in order of increasing sound intensity.

Name _____

Integrated Review—Functions and Properties of Logarithms

If $f(x) = x - 6$ and $g(x) = x^2 + 1$, find each value.

1. $(f + g)(x)$ **2.** $(f - g)(x)$ **3.** $(f \cdot g)(x)$ **4.** $\left(\dfrac{f}{g}\right)(x)$

If $f(x) = \sqrt{x}$ and $g(x) = 3x - 1$, find each value.

5. $(f \circ g)(x)$ **6.** $(g \circ f)(x)$

Determine whether each is a one-to-one function. If it is, find its inverse.

7. $f = \{(-2, 6), (4, 8), (2, -6), (3, 3)\}$ **8.** $g = \{(4, 2), (-1, 3), (5, 3), (7, 1)\}$

Determine whether the graph of each function is one-to-one.

9.
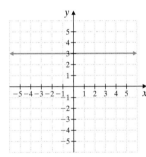

10.
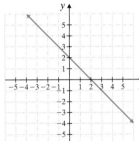

11.
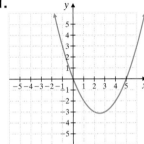

Each function listed is one-to-one. Find the inverse of each function.

12. $f(x) = 3x$ **13.** $f(x) = x + 4$

14. $f(x) = 5x - 1$ **15.** $f(x) = 3x + 2$

Graph each function.

16. $y = \left(\dfrac{1}{2}\right)^x$ **17.** $y = 2^x + 1$

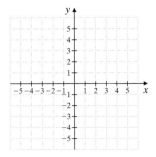

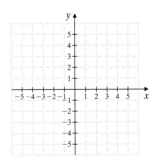

1. _____

2. _____

3. _____

4. _____

5. _____

6. _____

7. _____

8. _____

9. _____

10. _____

11. _____

12. _____

13. _____

14. _____

15. _____

16. see graph

17. see graph

18. see graph

19. see graph

20. _____

21. _____

22. _____

23. _____

24. _____

25. _____

26. _____

27. _____

28. _____

29. _____

30. _____

31. _____

32. _____

33. _____

34. _____

35. _____

36. _____

37. _____

18. $y = \log_3 x$

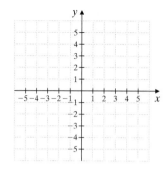

19. $y = \log_{1/3} x$

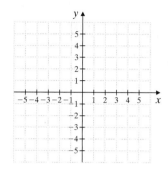

Solve.

20. $2^x = 8$

21. $9 = 3^{x-5}$

22. $4^{x-1} = 8^{x+2}$

23. $25^x = 125^{x-1}$

24. $\log_4 16 = x$

25. $\log_{49} 7 = x$

26. $\log_2 x = 5$

27. $\log_x 64 = 3$

28. $\log_x \dfrac{1}{125} = -3$

29. $\log_3 x = -2$

Write each as a single logarithm.

30. $5 \log_2 x$

31. $x \log_2 5$

32. $3 \log_5 x - 5 \log_5 y$

33. $9 \log_5 x + 3 \log_5 y$

34. $\log_2 x + \log_2(x - 3) - \log_2(x^2 + 4)$

35. $\log_3 y - \log_3(y + 2) + \log_3(y^3 + 11)$

Write each expression as a sum or difference of logarithms.

36. $\log_7 \dfrac{9x^2}{y}$

37. $\log_6 \dfrac{5y}{z^2}$

10.6   Common Logarithms, Natural Logarithms, and Change of Base

In this section we look closely at two particular logarithmic bases. These two logarithmic bases are used so frequently that logarithms to their bases are given special names. **Common logarithms** are logarithms to base 10. **Natural logarithms** are logarithms to base e, which we introduce in this section. The work in this section is based on the use of a calculator that has both the common "log" $\boxed{\text{LOG}}$ and the natural "log" $\boxed{\text{LN}}$ keys.

Ⓐ   Approximating Common Logarithms

Logarithms to base 10—**common logarithms**—are used frequently because our number system is a base 10 decimal system. The notation $\log x$ means the same as $\log_{10} x$.

Common Logarithm
$\log x \quad \text{means} \quad \log_{10} x$

EXAMPLE 1

Use a calculator to approximate $\log 7$ to four decimal places.

Solution: Press the following sequence of keys:

$\boxed{7}\,\boxed{\text{LOG}}$ or $\boxed{\text{LOG}}\,\boxed{7}\,\boxed{\text{ENTER}}$

To four decimal places,

$\log 7 \approx 0.8451$ ●

Ⓑ   Evaluating Common Logarithms of Powers of 10

To evaluate the common log of a power of 10, a calculator is not needed. According to the property of logarithms,

$\log_b b^x = x$

It follows that if b is replaced with 10, we have

$\log 10^x = x$

Helpful Hint

Remember that the base of this logarithm is understood to be 10.

EXAMPLES   Find the exact value of each logarithm.

2. $\log 10 = \log 10^1 = 1$

3. $\log \dfrac{1}{10} = \log 10^{-1} = -1$

4. $\log 1000 = \log 10^3 = 3$

5. $\log \sqrt{10} = \log 10^{1/2} = \dfrac{1}{2}$ ●

As we will soon see, equations containing common logs are useful models of many natural phenomena.

Practice Problem 1

Use a calculator to approximate $\log 21$ to four decimal places.

Practice Problems 2–5

Find the exact value of each logarithm.

2. $\log 100$

3. $\log \dfrac{1}{100}$

4. $\log 10{,}000$

5. $\log \sqrt[3]{10}$

Answers

1. 1.3222,   **2.** 2,   **3.** -2,   **4.** 4,   **5.** $\dfrac{1}{3}$

Practice Problem 6

Solve: $\log x = 2.9$. Give an exact solution and then approximate the solution to four decimal places.

Practice Problem 7

Use a calculator to approximate $\ln 11$ to four decimal places.

Practice Problems 8–9

Find the exact value of each natural logarithm.

8. $\ln e^5$ 9. $\ln \sqrt{e}$

Practice Problem 10

Solve: $\ln 7x = 10$. Give an exact solution and then approximate the solution to four decimal places.

Answers

6. $x = 10^{2.9}; x \approx 794.3282,$ **7.** $2.3979,$ **8.** $5,$

9. $\dfrac{1}{2},$ **10.** $x = \dfrac{e^{10}}{7}; x \approx 3146.6380$

EXAMPLE 6

Solve: $\log x = 1.2$. Give an exact solution and then approximate the solution to four decimal places.

Solution: Remember that the base of a common log is understood to be 10.

Helpful Hint
The understood base is 10.

$$\log x = 1.2$$
$$10^{1.2} = x \quad \text{Write with exponential notation.}$$

The exact solution is $10^{1.2}$. To four decimal places, $x \approx 15.8489$. ●

C Approximating Natural Logarithms

Natural logarithms are also frequently used, especially to describe natural events; hence the label "natural logarithm." **Natural logarithms** are logarithms to the base e, which is a constant approximately equal to 2.7183. The number e is an irrational number, as is π. The notation $\log_e x$ is usually abbreviated to $\ln x$. (The abbreviation ln is read "el en.")

Natural Logarithm

$\ln x$ means $\log_e x$

EXAMPLE 7   Use a calculator to approximate $\ln 8$ to four decimal places.

Solution: Press the following sequence of keys:

$\boxed{8}\ \boxed{\ln}$ or $\boxed{\ln}\ \boxed{8}\ \boxed{\text{ENTER}}$

To four decimal places,
$$\ln 8 \approx 2.0794$$ ●

D Evaluating Natural Logarithms of Powers of e

As a result of the property $\log_b b^x = x$, we know that $\log_e e^x = x$, or $\ln e^x = x$.

EXAMPLES   Find the exact value of each natural logarithm.

8. $\ln e^3 = 3$

9. $\ln \sqrt[5]{e} = \ln e^{1/5} = \dfrac{1}{5}$ ●

EXAMPLE 10

Solve: $\ln 3x = 5$. Give an exact solution and then approximate the solution to four decimal places.

Solution: Remember that the base of a natural logarithm is understood to be e.

Helpful Hint
The understood base is e.

$$\ln 3x = 5$$
$$e^5 = 3x \quad \text{Write with exponential notation.}$$
$$\dfrac{e^5}{3} = x \quad \text{Solve for } x.$$

The exact solution is $\dfrac{e^5}{3}$. To four decimal places, $x \approx 49.4711$. ●

Recall from Section 10.3 the formula $A = P\left(1 + \dfrac{r}{n}\right)^{nt}$ for compound interest, where n represents the number of compoundings per year. When interest is compounded continuously, we use the formula $A = Pe^{rt}$, where r is the annual interest rate and interest is compounded continuously for t years.

EXAMPLE 11   Finding the Amount Owed on a Loan

Find the amount owed at the end of 5 years if $1600 is loaned at a rate of 9% compounded continuously.

Solution: We use the formula $A = Pe^{rt}$ and the following values of the variables.

$P = \$1600$ (the amount of the loan)

$r = 9\% = 0.09$ (the rate of interest)

$t = 5$ (the 5-year duration of the loan)

$A = Pe^{rt}$

$\quad = 1600e^{0.09(5)}$ Substitute known values.

$\quad = 1600e^{0.45}$

Now we can use a calculator to approximate the solution.

$\quad A \approx 2509.30$

The total amount of money owed is approximately $2509.30. ●

E   Using the Change of Base Formula

Calculators are handy tools for approximating natural and common logarithms. Unfortunately, some calculators cannot be used to approximate logarithms to bases other than e or 10—at least not directly. In such cases, we use the **change of base formula.**

Change of Base

If $a, b,$ and c are positive real numbers and neither b nor c is 1, then

$$\log_b a = \frac{\log_c a}{\log_c b}$$

EXAMPLE 12   Approximate $\log_5 3$ to four decimal places.

Solution: We use the change of base property to write $\log_5 3$ as a quotient of logarithms to base 10.

$\log_5 3 = \dfrac{\log 3}{\log 5}$ Use the change of base property.

$\quad\quad \approx \dfrac{0.4771213}{0.69897}$ Approximate the logarithms by calculator.

$\quad\quad \approx 0.6826063$ Simplify by calculator.

To four decimal places, $\log_5 3 \approx 0.6826$. ●

Try the Concept Check in the margin.

Practice Problem 11

Find the amount owed at the end of 3 years if $1200 is loaned at a rate of 8% compounded continuously.

Practice Problem 12

Approximate $\log_7 5$ to four decimal places.

Concept Check

If a graphing calculator cannot directly evaluate logarithms to base 5, describe how you could use the graphing calculator to graph the function $f(x) = \log_5 x$.

Answers

11. $1525.50,   **12.** 0.8271

Concept Check: $f(x) = \dfrac{\log x}{\log 5}$

STUDY SKILLS REMINDER

Are you prepared for your final exam?

To prepare for your final exam, try the following study techniques.

- Review the material that you will be responsible for on your exam. Also check your notebook for any lecture notes that you highlighted.

- Review any formulas that you may need to memorize.

- Check to see if your instructor or math department will be conducting a final exam review.

- Check with your instructor to see whether there are final exams from previous semesters/quarters that are available to students for study.

- Use your previously taken tests as a practice final exam. To do so, rewrite the test questions in mixed order on blank sheets of paper. This will help you prepare for exam conditions.

- If you are unsure of a few topics, see your instructor or visit a learning lab for further assistance. Also, viewing the video segment of a troublesome section will help.

- If you need further exercises to work, try the chapter tests at the end of appropriate chapters.

Good luck! I hope you have enjoyed this textbook and your intermediate algebra course.

EXERCISE SET 10.6

 A **C** *Use a calculator to approximate each logarithm to four decimal places. See Examples 1 and 7.*

1. $\log 8$

2. $\log 6$

3. $\log 2.31$

4. $\log 4.86$

5. $\ln 2$

6. $\ln 3$

7. $\ln 0.0716$

8. $\ln 0.0032$

9. $\log 12.6$

10. $\log 25.9$

11. $\ln 5$

12. $\ln 7$

13. $\log 41.5$

14. $\ln 41.5$

15. Use a calculator to try to approximate $\log 0$. Describe what happens and explain why.

16. Use a calculator to try to approximate $\ln 0$. Describe what happens and explain why.

B **D** *Find the exact value of each logarithm. See Examples 2 through 5, 8, and 9.*

17. $\log 100$

18. $\log 10{,}000$

19. $\log \dfrac{1}{1000}$

20. $\log \dfrac{1}{10}$

21. $\ln e^2$

22. $\ln e^4$

23. $\ln \sqrt[4]{e}$

24. $\ln \sqrt[5]{e}$

25. $\log 10^3$

26. $\ln e^5$

27. $\ln e^2$

28. $\log 10^7$

29. $\log 0.0001$

30. $\log 0.001$

31. $\ln \sqrt{e}$

32. $\log \sqrt{10}$

Solve each equation. Give an exact solution and a four-decimal-place approximation. See Examples 6 and 10.

33. $\log x = 1.3$

34. $\log x = 2.1$

35. $\log 2x = 1.1$

36. $\log 3x = 1.3$

37. $\ln x = 1.4$

38. $\ln x = 2.1$

39. $\ln (3x - 4) = 2.3$

40. $\ln (2x + 5) = 3.4$

41. $\log x = 2.3$

42. $\log x = 3.1$

43. $\ln x = -2.3$

44. $\ln x = -3.7$

45. $\log (2x + 1) = -0.5$

46. $\log (3x - 2) = -0.8$

47. $\ln 4x = 0.18$

48. $\ln 3x = 0.76$

Use the formula $A = Pe^{rt}$ to solve. See Example 11.

49. How much money does Dana Jones have after 12 years if she invests $1400 at 8% interest compounded continuously?

50. Determine the size of an account in which $3500 earns 6% interest compounded continuously for 1 year.

51. How much money does Barbara Mack owe at the end of 4 years if 6% interest is compounded continuously on her $2000 debt?

52. Find the amount of money for which a $2500 certificate of deposit is redeemable if it has been paying 10% interest compounded continuously for 3 years.

E *Approximate each logarithm to four decimal places. See Example 12.*

53. $\log_2 3$

54. $\log_3 2$

55. $\log_{1/2} 5$

56. $\log_{1/3} 2$

57. $\log_4 9$

58. $\log_9 4$

59. $\log_3 \dfrac{1}{6}$

60. $\log_6 \dfrac{2}{3}$

61. $\log_8 6$

62. $\log_6 8$

Review and Preview

Solve for x. See Sections 2.1 and 5.7.

63. $6x - 3(2 - 5x) = 6$

64. $2x + 3 = 5 - 2(3x - 1)$

65. $2x + 3y = 6x$

66. $4x - 8y = 10x$

67. $x^2 + 7x = -6$

68. $x^2 + 4x = 12$

Combining Concepts

Graph each function by finding ordered pair solutions, plotting the solutions, and then drawing a smooth curve through the plotted points.

69. $f(x) = e^x$

70. $f(x) = e^{2x}$

71. $f(x) = \ln x$

72. $f(x) = \log x$

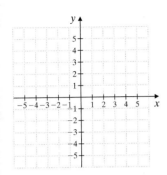

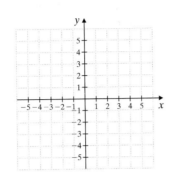

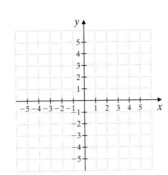

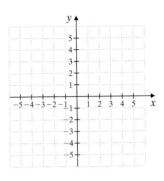

73. Without using a calculator, explain which of $\log 50$ or $\ln 50$ must be larger.

74. Without using a calculator, explain which of $\log 50^{-1}$ or $\ln 50^{-1}$ must be larger.

The Richter scale measures the intensity, or magnitude, of an earthquake. The formula for the magnitude R of an earthquake is $R = \log\left(\dfrac{a}{T}\right) + B$, where a is the amplitude in micrometers of the vertical motion of the ground at the recording station, T is the number of seconds between successive seismic waves, and B is an adjustment factor that takes into account the weakening of the seismic wave as the distance increases from the epicenter of the earthquake.

Use the Richter scale formula to find the magnitude R of the earthquake that fits the description given. Round answers to one decimal place.

75. Amplitude a is 200 micrometers, time T between waves is 1.6 seconds, and B is 2.1.

76. Amplitude a is 150 micrometers, time T between waves is 3.6 seconds, and B is 1.9.

77. Amplitude a is 400 micrometers, time T between waves is 2.6 seconds, and B is 3.1.

78. Amplitude a is 450 micrometers, time T between waves is 4.2 seconds, and B is 2.7.

Do you remember what to do the day of an exam?

On the day of an exam, don't forget to try the following:

- Allow yourself plenty of time to arrive.
- Read the directions on the test carefully.
- Read each problem carefully as you take your test. Make sure that you answer the question asked.
- Watch your time and pace yourself so that you may attempt each problem on your test.
- If you have time, check your work and answers.
- Do not turn your test in early. If you have extra time, spend it double-checking your work.

Good luck!

10.7 Exponential and Logarithmic Equations and Problem Solving

OBJECTIVES

Ⓐ Solve exponential equations.

Ⓑ Solve logarithmic equations.

Ⓒ Solve problems that can be modeled by exponential and logarithmic equations.

SSM   TUTOR CENTER   SG   CD & VIDEO   MATH PRO   WEB

Ⓐ Solving Exponential Equations

In Section 10.3 we solved exponential equations such as $2^x = 16$ by writing 16 as a power of 2 and using the uniqueness of b^x.

$$2^x = 16$$
$$2^x = 2^4 \qquad \text{Write 16 as } 2^4.$$
$$x = 4 \qquad \text{Use the uniqueness of } b^x.$$

To solve an equation such as $3^x = 7$, we use the fact that $f(x) = \log_b x$ is a one-to-one function. Another way of stating this fact is as a property of equality.

Logarithm Property of Equality

Let $a, b,$ and c be real numbers such that $\log_b a$ and $\log_b c$ are real numbers and b is not 1. Then

$$\log_b a = \log_b c \quad \text{is equivalent to} \quad a = c$$

EXAMPLE 1

Solve: $3^x = 7$. Give an exact answer and a four-decimal-place approximation.

Solution: We use the logarithm property of equality and take the logarithm of both sides. For this example, we use the common logarithm.

$$3^x = 7$$
$$\log 3^x = \log 7 \qquad \text{Take the common log of both sides.}$$
$$x \log 3 = \log 7 \qquad \text{Use the power property of logarithms.}$$
$$x = \frac{\log 7}{\log 3} \qquad \text{Divide both sides by log 3.}$$

The exact solution is $\dfrac{\log 7}{\log 3}$. When we approximate to four decimal places, we have

$$\frac{\log 7}{\log 3} \approx \frac{0.845098}{0.4771213} \approx 1.7712$$

The solution set is $\left\{ \dfrac{\log 7}{\log 3} \right\}$, or {approximately 1.7712}. ●

Ⓑ Solving Logarithmic Equations

By applying the appropriate properties of logarithms, we can solve a broad variety of logarithmic equations.

EXAMPLE 2   Solve: $\log_4(x - 2) = 2$

Solution: Notice that $x - 2$ must be positive, so x must be greater than 2. With this in mind, we first write the equation with exponential notation.

Practice Problem 1

Solve: $2^x = 5$. Give an exact answer and a four-decimal-place approximation.

Practice Problem 2

Solve: $\log_3(x + 5) = 2$

Answers

1. $\left\{ \dfrac{\log 5}{\log 2} \right\}$; {2.3219},   **2.** {4}

$$\log_4(x - 2) = 2$$
$$4^2 = x - 2$$
$$16 = x - 2$$
$$18 = x \qquad \text{Add 2 to both sides.}$$

To check, we replace x with 18 in the original equation.

$$\log_4(x - 2) = 2$$
$$\log_4(18 - 2) \overset{?}{=} 2 \qquad \text{Let } x = 18.$$
$$\log_4 16 \overset{?}{=} 2$$
$$4^2 = 16 \qquad \text{True.}$$

The solution set is $\{18\}$.

Practice Problem 3

Solve: $\log_6 x + \log_6 (x + 1) = 1$

EXAMPLE 3   Solve: $\log_2 x + \log_2(x - 1) = 1$

Solution: Notice that $x - 1$ must be positive, so x must be greater than 1. We use the product property on the left side of the equation.

$$\log_2 x + \log_2(x - 1) = 1$$
$$\log_2[x(x - 1)] = 1 \qquad \text{Use the product property.}$$
$$\log_2(x^2 - x) = 1$$

Next we write the equation with exponential notation and solve for x.

$$2^1 = x^2 - x$$
$$0 = x^2 - x - 2 \qquad \text{Subtract 2 from both sides.}$$
$$0 = (x - 2)(x + 1) \qquad \text{Factor.}$$
$$0 = x - 2 \quad \text{or} \quad 0 = x + 1 \qquad \text{Set each factor equal to 0.}$$
$$2 = x \qquad\qquad -1 = x$$

Recall that -1 cannot be a solution because x must be greater than 1. If we forgot this, we would still reject -1 after checking. To see this, we replace x with -1 in the original equation.

$$\log_2 x + \log_2(x - 1) = 1$$
$$\log_2(-1) + \log_2(-1 - 1) \overset{?}{=} 1 \qquad \text{Let } x = -1.$$

Because the logarithm of a negative number is undefined, -1 is rejected. Check to see that the solution set is $\{2\}$.

Practice Problem 4

Solve: $\log (x - 1) - \log x = 1$

EXAMPLE 4   Solve: $\log(x + 2) - \log x = 2$

Solution: We use the quotient property of logarithms on the left side of the equation.

$$\log(x + 2) - \log x = 2$$
$$\log \frac{x + 2}{x} = 2 \qquad \text{Use the quotient property.}$$
$$10^2 = \frac{x + 2}{x} \qquad \text{Write using exponential notation.}$$
$$100 = \frac{x + 2}{x}$$
$$100x = x + 2 \qquad \text{Multiply both sides by } x.$$
$$99x = 2 \qquad \text{Subtract } x \text{ from both sides.}$$
$$x = \frac{2}{99} \qquad \text{Divide both sides by 99.}$$

Answers

3. $\{2\}$,   **4.** $\varnothing$

Check to see that the solution set is $\left\{ \dfrac{2}{99} \right\}$.

(c) Solving Problems Modeled by Exponential and Logarithmic Equations

Logarithmic and exponential functions are used in a variety of scientific, technical, and business settings. A few examples follow.

EXAMPLE 5   Estimating Population Size

The population size y of a community of lemmings varies according to the relationship $y = y_0 e^{0.15t}$. In this formula, t is time in months, and y_0 is the initial population at time 0. Estimate the population after 6 months if there were originally 5000 lemmings.

Solution: We substitute 5000 for y_0 and 6 for t.

$$y = y_0 e^{0.15t}$$
$$= 5000 e^{0.15(6)} \qquad \text{Let } t = 6 \text{ and } y_0 = 5000.$$
$$= 5000 e^{0.9} \qquad \text{Multiply.}$$

Using a calculator, we find that $y \approx 12,298.016$. In 6 months the population will be approximately 12,300 lemmings. ●

EXAMPLE 6   Doubling an Investment

How long does it take an investment of $2000 to double if it is invested at 5% interest compounded quarterly? The necessary formula is $A = P\left(1 + \dfrac{r}{n}\right)^{nt}$, where A is the accrued amount, P is the principal invested, r is the annual rate of interest, n is the number of compounding periods per year, and t is the number of years.

Solution: We are given that $P = \$2000$ and $r = 5\% = 0.05$. Compounding quarterly means 4 times a year, so $n = 4$. The investment is to double, so A must be $4000. We substitute these values and solve for t.

Practice Problem 5

Use the equation in Example 5 to estimate the lemming population in 8 months.

Practice Problem 6

How long does it take an investment of $1000 to double if it is invested at 6% interest compounded quarterly?

Answers

5. approximately 16,601 lemmings,   **6.** $11\dfrac{3}{4}$ yr

$$A = P\left(1 + \frac{r}{n}\right)^{nt}$$

$$4000 = 2000\left(1 + \frac{0.05}{4}\right)^{4t} \qquad \text{Substitute known values.}$$

$$4000 = 2000(1.0125)^{4t} \qquad \text{Simplify } 1 + \frac{0.05}{4}.$$

$$2 = (1.0125)^{4t} \qquad \text{Divide both sides by 2000.}$$

$$\log 2 = \log 1.0125^{4t} \qquad \text{Take the logarithm of both sides.}$$

$$\log 2 = 4t(\log 1.0125) \qquad \text{Use the power property.}$$

$$\frac{\log 2}{4\log 1.0125} = t \qquad \text{Divide both sides by } 4\log 1.0125.$$

$$13.949408 \approx t \qquad \text{Approximate by calculator.}$$

It takes approximately 14 years for the money to double in value. ●

GRAPHING CALCULATOR EXPLORATIONS

Use a grapher to find how long it takes an investment of $1500 to triple if it is invested at 8% interest compounded monthly. First, let $P = \$1500$, $r = 0.08$, and $n = 12$ (for 12 months) in the formula

$$A = P\left(1 + \frac{r}{n}\right)^{nt}$$

Notice that when the investment has tripled, the accrued amount A is $4500. Thus,

$$4500 = 1500\left(1 + \frac{0.08}{12}\right)^{12t}$$

Determine an appropriate viewing window and enter and graph the equations

$$Y_1 = 1500\left(1 + \frac{0.08}{12}\right)^{12x}$$

and

$$Y_2 = 4500$$

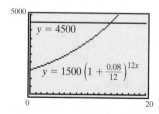

The point of intersection of the two curves is the solution. The x-coordinate tells how long it takes for the investment to triple.

Use a TRACE feature or an INTERSECT feature to approximate the coordinates of the point of intersection of the two curves. It takes approximately 13.78 years, or 13 years and 10 months, for the investment to triple in value to $4500.

Use this graphical solution method to solve each problem. Round each answer to the nearest hundredth.

1. Find how long it takes an investment of $5000 to grow to $6000 if it is invested at 5% interest compounded quarterly.

2. Find how long it takes an investment of $1000 to double if it is invested at 4.5% interest compounded daily. (Use 365 days in a year.)

3. Find how long it takes an investment of $10,000 to quadruple if it is invested at 6% interest compounded monthly.

4. Find how long it takes $500 to grow to $800 if it is invested at 4% interest compounded semiannually.

EXERCISE SET 10.7

A *Solve each equation. Give an exact solution and a four-decimal-place approximation. See Example 1.*

1. $3^x = 6$

2. $4^x = 7$

3. $3^{2x} = 3.8$

4. $5^{3x} = 5.6$

5. $2^{x-3} = 5$

6. $8^{x-2} = 12$

7. $9^x = 5$

8. $3^x = 11$

9. $4^{x+7} = 3$

10. $6^{x+3} = 2$

11. $7^{3x-4} = 11$

12. $5^{2x-6} = 12$

13. $e^{6x} = 5$

14. $e^{2x} = 8$

B *Solve each equation. See Examples 2 through 4.*

15. $\log_2(x + 5) = 4$

16. $\log_6(x^2 - x) = 1$

17. $\log_4 2 + \log_4 x = 0$

18. $\log_3 5 + \log_3 x = 1$

19. $\log_2 6 - \log_2 x = 3$

20. $\log_4 10 - \log_4 x = 2$

21. $\log_4 x + \log_4(x + 6) = 2$

22. $\log_3 x + \log_3(x + 6) = 3$

23. $\log_5(x + 3) - \log_5 x = 2$

24. $\log_6(x + 2) - \log_6 x = 2$

25. $\log_3(x - 2) = 2$

26. $\log_2(x - 5) = 3$

27. $\log_4(x^2 - 3x) = 1$

28. $\log_8(x^2 - 2x) = 1$

29. $\log_2 x + \log_2(3x + 1) = 1$

30. $\log_3 x + \log_3(x - 8) = 2$

Solve. See Example 5.

31. The size of the wolf population at Isle Royale National Park increases at a rate of 4.3% per year. If the size of the current population is 83 wolves, find how many there should be in 5 years. Use $y = y_0 e^{0.043t}$ and round to the nearest whole number.

32. The number of victims of a flu epidemic is increasing at a rate of 7.5% per week. If 20,000 people are currently infected, in how many days can we expect 45,000 people to have the flu? Use $y = y_0 e^{0.075t}$ and round to the nearest whole number.

33. The size of the population of Paraguay is increasing at a rate of 2.7% per year. The population of Paraguay in 2001 was approximately 5,700,000. Use $y = y_0 e^{0.027t}$ to estimate the population of Paraguay in 2008. Round to the nearest whole number. (*Source:* Population Reference Bureau)

34. In 2001, 171.8 million people lived in Brazil. The population of Brazil is growing at a rate of 1.5% per year. Find how long it will take the Brazilian population to reach a size of 200 million people. Use $y = y_0 e^{0.015t}$ and round to the nearest tenth. (*Source:* Population Reference Bureau)

35. In 2001, Hungary had a population of about 10,000,000. At that time, Hungary's population was declining at a rate of 0.4% per year. How long will it take for Hungary's population to decline to 9,000,000? Use $y = y_0 e^{-0.04t}$ and round to the nearest tenth. (*Source:* Population Reference Bureau)

36. The population of Russia has been decreasing at the rate of 0.7% per year. There were about 144,400,000 people living in Russia in 2001. How many inhabitants will there be in 2016? Use $y = y_0 e^{-0.007t}$. Round to the nearest whole number. (*Source:* Population Reference Bureau)

Use the formula $A = P\left(1 + \dfrac{r}{n}\right)^{nt}$ to solve these compound interest problems. Round to the nearest tenth. See Example 6.

37. How long does it take for $600 to double if it is invested at 7% interest compounded monthly?

38. How long does it take for $600 to double if it is invested at 12% interest compounded monthly?

39. How long does it take for a $1200 investment to earn $200 interest if it is invested at 9% interest compounded quarterly?

40. How long does it take for a $1500 investment to earn $200 interest if it is invested at 10% interest compounded semiannually?

41. How long does it take for $1000 to double if it is invested at 8% interest compounded semiannually?

42. How long does it take for $1000 to double if it is invested at 8% interest compounded monthly?

The formula $w = 0.00185h^{2.67}$ is used to estimate the normal weight w in pounds of a boy h inches tall. Use this formula to solve Exercises 43 and 44. Round to the nearest tenth.

43. Find the expected height of a boy who weighs 85 pounds.

44. Find the expected height of a boy who weighs 140 pounds.

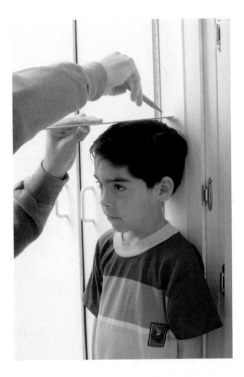

The formula $P = 14.7e^{-0.21x}$ gives the average atmospheric pressure P, in pounds per square inch, at an altitude x, in miles above sea level. Use this formula to solve Exercises 45 through 58. Round to the nearest tenth.

45. Find the average atmospheric pressure of Denver, which is 1 mile above sea level.

46. Find the average atmospheric pressure of Pikes Peak, which is 2.7 miles above sea level.

47. Find the elevation of a Delta jet if the atmospheric pressure outside the jet is 7.5 pounds per square inch.

48. Find the elevation of a remote Himalayan peak if the atmospheric pressure atop the peak is 6.5 pounds per square inch.

Psychologists call the graph of the formula $t = \dfrac{1}{c}\ln\left(\dfrac{A}{A - N}\right)$ the learning curve since the formula relates time t passed, in weeks, to a measure N of learning achieved, to a measure A of maximum learning possible, and to a measure c of an individual's learning style. Use this formula to answer Exercises 49 through 52. Round to the nearest whole number.

49. Norman Weidner is learning to type. If he wants to type at a rate of 50 words per minute (N is 50) and his expected maximum rate is 75 words per minute (A is 75), how many weeks should it take him to achieve his goal? Assume that c is 0.09.

50. An experiment on teaching chimpanzees sign language shows that a typical chimp can master a maximum of 65 signs. How many weeks should it take a chimpanzee to master 30 signs if c is 0.03?

51. Janine Jenkins is working on her dictation skills. She wants to take dictation at a rate of 150 words per minute and believes that the maximum rate she can hope for is 210 words per minute. How many weeks should it take her to achieve the 150-word level if c is 0.07?

52. A psychologist is measuring human capability to memorize nonsense syllables. How many weeks should it take a subject to learn 15 nonsense syllables if the maximum possible to learn is 24 syllables and c is 0.17?

Review and Preview

If $x = -2$, $y = 0$, and $z = 3$, find the value of each expression. See Section 1.5.

53. $\dfrac{x^2 - y + 2z}{3x}$

54. $\dfrac{x^3 - 2y + z}{2z}$

55. $\dfrac{3z - 4x + y}{x + 2z}$

56. $\dfrac{4y - 3x + z}{2x + y}$

 Combining Concepts

The formula $y = y_0 e^{kt}$ gives the population size y of a population that experiences an annual rate of population growth k (given as a decimal). In this formula, t is time in years and y_0 is the initial population at time 0. Use this formula to solve Exercises 57 and 58.

57. In 1990, the population of Arizona was 3,665,228. By 2000, the population had grown to 5,130,632. Find the annual rate of population growth over this period. Round your answer to the nearest tenth of a percent. (*Source:* U.S. Bureau of the Census)

58. In 1990, the population of Nevada was 1,201,833. By 2000, the population had grown to 1,998,257. Find the annual rate of population growth over this period. Round your answer to the nearest tenth of a percent. (*Source:* U.S. Bureau of the Census)

59. When solving a logarithmic equation, explain why you must check possible solutions in the original equation.

60. Solve $5^x = 9$ by taking the common logarithm of both sides of the equation. Next, solve this equation by taking the natural logarithm of both sides. Compare your solutions. Are they the same? Why or why not?

Use a graphing calculator to solve. Round your answers to two decimal places.

61. $e^{0.3x} = 8$

62. $10^{0.5x} = 7$

Internet Excursions

Go To: http://www.prenhall.com/martin-gay_interm   What's Related

In Section 10.6 Combining Concepts, we learned that the Richter scale measures the magnitude of an earthquake. The relationship between a Richter scale reading R and an intensity I of the earthquake's shock wave is given by the equation $R = \log I$. Given the Richter scale magnitudes of two earthquakes, we can compare their intensities. First we use this relationship to find the intensity of each earthquake. Then we can use a ratio of the resulting intensities to conclude that one earthquake was so many times more intense than the other.

By going to the World Wide Web address listed above, you will gain access to a Web site where you can look up current earthquake information to help you answer Exercises 63 and 64.

63. Scan the list of the recent earthquakes to find the earthquake events with the highest and lowest Richter scale magnitudes. Report the date, time, location, and magnitude of each. How many times more intense was the earthquake with the highest Richter scale reading than the one with the lowest reading?

64. Scan the list of recent earthquakes to find the most recent and least recent earthquake events that are listed. Report the date, time, location, and magnitude of each. How many times more intense was the earthquake with the higher Richter scale reading than the other earthquake?

CHAPTER 10 ACTIVITY   Modeling Temperature

METHOD 1 MATERIALS:

- a container of either cold or hot liquid
- thermometer
- stopwatch
- grapher with curve-fitting capabilities (optional)

METHOD 2 MATERIALS:

- a container of either cold or hot liquid
- TI-82, TI-83, or TI-85 graphing calculator with unit-to-unit link cable
- TI-CBL (Calculator-Based Laboratory) unit with temperature probe

This activity may be completed by working in groups or individually.

Newton's law of cooling relates the temperature of an object to the time elapsed since its warming or cooling began. In this activity you will investigate experimental data to find a mathematical model for this relationship. You may collect the temperature data by using either Method 1 (stopwatch and thermometer) or Method 2 (CBL).

Method 1

a. Insert the thermometer into the liquid and allow a thermometer reading to register. Take a temperature reading T as you start the stopwatch (at $t = 0$) and record it in the accompanying data table.

t	T
0	

b. Continue taking temperature readings at uniform intervals anywhere between 5 and 10 minutes long. At each reading, use the stopwatch to measure the length of time that has elapsed *since the temperature readings started.* Record your time t and liquid temperature T in the data table. Gather data for six to twelve readings.

c. Plot the data from the data table. Plot t on the horizontal axis and T on the vertical axis.

Method 2

a. Prepare the CBL unit and TI-82, TI-83, or TI-85 graphing calculator. Insert the temperature probe into the liquid.

b. Start the HEAT program on the TI graphing calculator and follow its instructions to begin collecting data. The program will collect 36 temperature readings in degrees Celsius and plot them in real time with t on the horizontal axis and T on the vertical axis.

1. Which of the following mathematical models best fits the data you collected? Explain your reasoning. (Assume $a > 0$.)
   a. $T = ab^t + c$
   b. $T = ab^{-t} + c$
   c. $T = -ab^{-t} + c$
   d. $T = \ln(-ax + b) + c$
   e. $T = -\ln(-ax + b) + c$

2. What does the constant c represent in the model you chose? What is the value of c in this activity?

3. (Optional) Subtract the value of c from each of your observations of T. Enter the new ordered pairs $(t, T - c)$ into a grapher. Use the exponential or logarithmic curve-fitting feature to find a model for your experimental data. Graph the ordered pairs $(t, T - c)$ with the model you found. How well does the model fit the data? How does the model compare with your selection from Question 1?

Chapter 10 VOCABULARY CHECK

Fill in each blank with one of the words or phrases listed below.

inverse	common	composition	symmetric	exponential
vertical	logarithmic	natural	horizontal	

1. For each one-to-one function, we can find its _____ function by switching the coordinates of the ordered pairs of the function.

2. The _____ of functions f and g is $(f \circ g)(x) = f(g(x))$.

3. A function of the form $f(x) = b^x$ is called an _____ function if $b > 0$, b is not 1, and x is a real number.

4. The graphs of f and f^{-1} are _____ about the line $y = x$.

5. _____ logarithms are logarithms to base e.

6. _____ logarithms are logarithms to base 10.

7. To see whether a graph is the graph of a one-to-one function, apply the _____ line test to see whether it is a function, and then apply the _____ line test to see whether it is a one-to-one function.

8. A _____ function is a function that can be defined by $f(x) = \log_b x$ where x is a positive real number, b is a constant positive real number, and b is not 1.

CHAPTER

Highlights

DEFINITIONS AND CONCEPTS	EXAMPLES

Section 10.1   The Algebra of Functions

ALGEBRA OF FUNCTIONS

Sum $\quad (f + g)(x) = f(x) + g(x)$

Difference $\quad (f - g)(x) = f(x) - g(x)$

Product $\quad (f \cdot g)(x) = f(x) \cdot g(x)$

Quotient $\quad \left(\dfrac{f}{g}\right)(x) = \dfrac{f(x)}{g(x)}$

If $f(x) = 7x$ and $g(x) = x^2 + 1$,

$$(f + g)(x) = f(x) + g(x) = 7x + x^2 + 1$$
$$(f - g)(x) = f(x) - g(x) = 7x - (x^2 + 1)$$
$$= 7x - x^2 - 1$$
$$(f \cdot g)(x) = f(x) \cdot g(x) = 7x(x^2 + 1)$$
$$= 7x^3 + 7x^2$$
$$\left(\frac{f}{g}\right)(x) = \frac{f(x)}{g(x)} = \frac{7x}{x^2 + 1}$$

COMPOSITE FUNCTIONS

The notation $(f \circ g)(x)$ means "f composed with g."

$$(f \circ g)(x) = f(g(x))$$
$$(g \circ f)(x) = g(f(x))$$

If $f(x) = x^2 + 1$ and $g(x) = x - 5$, find $(f \circ g)(x)$.

$$(f \circ g)(x) = f(g(x))$$
$$= f(x - 5)$$
$$= (x - 5)^2 + 1$$
$$= x^2 - 10x + 26$$

DEFINITIONS AND CONCEPTS	**EXAMPLES**

Section 10.2　Inverse Functions

If f is a function, then f is a **one-to-one function** only if each y-value (output) corresponds to only one x-value (input).

HORIZONTAL LINE TEST

If every horizontal line intersects the graph of a function at most once, then the function is a one-to-one function.

Determine whether each graph is a one-to-one function.

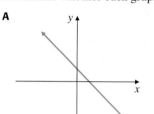

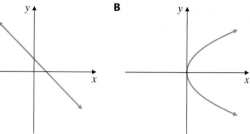

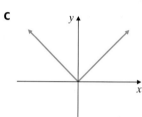

Graphs A and C pass the vertical line test, so only these are graphs of functions. Of graphs A and C, only graph A passes the horizontal line test, so only graph A is the graph of a one-to-one function.

The **inverse** of a one-to-one function f is the one-to-one function f^{-1} that is the set of all ordered pairs (b, a) such that (a, b) belongs to f.

FINDING THE INVERSE OF A ONE-TO-ONE FUNCTION f

Step 1. Replace $f(x)$ with y.

Step 2. Interchange x and y.

Step 3. Solve for y.

Step 4. Replace y with $f^{-1}(x)$.

Find the inverse of $f(x) = 2x + 7$.

$$y = 2x + 7 \qquad \text{Replace } f(x) \text{ with } y.$$

$$x = 2y + 7 \qquad \text{Interchange } x \text{ and } y.$$

$$2y = x - 7 \qquad \text{Solve for } y.$$

$$y = \frac{x - 7}{2}$$

$$f^{-1}(x) = \frac{x - 7}{2} \qquad \text{Replace } y \text{ with } f^{-1}(x).$$

The inverse of $f(x) = 2x + 7$ is $f^{-1}(x) = \dfrac{x - 7}{2}$.

Section 10.3　Exponential Functions

A function of the form $f(x) = b^x$ is an **exponential function**, where $b > 0$, $b \neq 1$, and x is a real number.

Graph the exponential function $y = 4^x$.

x	y
-2	$\dfrac{1}{6}$
-1	$\dfrac{1}{4}$
0	1
1	4
2	16

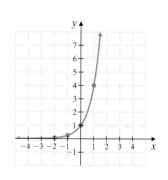

DEFINITIONS AND CONCEPTS	**EXAMPLES**

Section 10.3 *(continued)*

UNIQUENESS OF b^x

If $b > 0$ and $b \neq 1$, then $b^x = b^y$ is equivalent to $x = y$.

Solve: $2^{x+5} = 8$

$2^{x+5} = 2^3$ Write 8 as 2^3.

$x + 5 = 3$ Use the uniqueness of b^x.

$x = -2$ Subtract 5 from both sides.

Section 10.4   Logarithmic Functions

LOGARITHMIC DEFINITION

If $b > 0$ and $b \neq 1$, then

$$y = \log_b x \quad \text{means} \quad x = b^y$$

for any positive number x and real number y.

LOGARITHMIC FORM	CORRESPONDING EXPONENTIAL STATEMENT
$\log_5 25 = 2$	$5^2 = 25$
$\log_9 3 = \dfrac{1}{2}$	$9^{1/2} = 3$

PROPERTIES OF LOGARITHMS

If b is a real number, $b > 0$ and $b \neq 1$, then

$$\log_b 1 = 0, \quad \log_b b^x = x, \quad \text{and} \quad b^{\log_b x} = x$$

$\log_5 1 = 0, \quad \log_7 7^2 = 2, \quad \text{and} \quad 3^{\log_3 6} = 6$

LOGARITHMIC FUNCTION

If $b > 0$ and $b \neq 1$, then a **logarithmic function** is a function that can be defined as

$$f(x) = \log_b x$$

The domain of f is the set of positive real numbers, and the range of f is the set of real numbers.

Graph: $y = \log_3 x$

Write $y = \log_3 x$ as $3^y = x$. Plot the ordered pair solutions listed in the table, and connect them with a smooth curve.

x	y
3	1
1	0
$\dfrac{1}{3}$	-1
$\dfrac{1}{9}$	-2

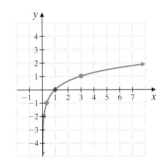

Section 10.5   Properties of Logarithms

Let x, y, and b be positive numbers and $b \neq 1$.

PRODUCT PROPERTY

$$\log_b xy = \log_b x + \log_b y$$

QUOTIENT PROPERTY

$$\log_b \frac{x}{y} = \log_b x - \log_b y$$

POWER PROPERTY

$$\log_b x^r = r \log_b x$$

Write as a single logarithm:

$2 \log_5 6 + \log_5 x - \log_5 (y + 2)$

$= \log_5 6^2 + \log_5 x - \log_5 (y + 2)$ Power property

$= \log_5 36 \cdot x - \log_5 (y + 2)$ Product property

$= \log_5 \dfrac{36x}{y + 2}$ Quotient property

DEFINITIONS AND CONCEPTS	EXAMPLES

Section 10.6 Common Logarithms, Natural Logarithms, and Change of Base

COMMON LOGARITHMS

$\log x$ means $\log_{10} x$

$\log 5 = \log_{10} 5 \approx 0.69897$

$\ln 7 = \log_e 7 \approx 1.94591$

NATURAL LOGARITHMS

$\ln x$ means $\log_e x$

Find the amount in an account at the end of 3 years if $1000 is invested at an interest rate of 4% compounded continuously.

CONTINUOUSLY COMPOUNDED INTEREST FORMULA

$A = Pe^{rt}$

where r is the annual interest rate for P dollars invested for t years.

Here, $t = 3$ years, $P = \$1000$, and $r = 0.04$.

$$A = Pe^{rt}$$
$$= 1000e^{0.04(3)}$$
$$\approx \$1127.50$$

Section 10.7 Exponential and Logarithmic Equations and Problem Solving

LOGARITHM PROPERTY OF EQUALITY

Let $\log_b a$ and $\log_b c$ be real numbers and $b \neq 1$. Then

$\log_b a = \log_b c$ is equivalent to $a = c$

Solve: $2^x = 5$

$\log 2^x = \log 5$	Log property of equality
$x \log 2 = \log 5$	Power property
$x = \dfrac{\log 5}{\log 2}$	Divide both sides by log 2.
$x \approx 2.3219$	Use a calculator.

Name _____ Section _____ Date _____

Chapter 10 Review

(10.1) *If $f(x) = x - 5$ and $g(x) = 2x + 1$, find the following.*

1. $(f + g)(x)$ **2.** $(f - g)(x)$ **3.** $(f \cdot g)(x)$ **4.** $\left(\dfrac{g}{f}\right)(x)$

If $f(x) = x^2 - 2$, $g(x) = x + 1$, and $h(x) = x^3 - x^2$, find each composition.

5. $(f \circ g)(x)$ **6.** $(g \circ f)(x)$ **7.** $(h \circ g)(2)$

8. $(f \circ f)(x)$ **9.** $(f \circ g)(-1)$ **10.** $(h \circ h)(2)$

(10.2) *Determine whether each function is a one-to-one function. If it is one-to-one, list the elements of its inverse.*

11. $h = \{(-9, 14), (6, 8), (-11, 12), (15, 15)\}$ **12.** $f = \{(-5, 5), (0, 4), (13, 5), (11, -6)\}$

13.

U.S. Region (Input)	West	Midwest	South	Northeast
Rank in Automobile Thefts (Output)	2	4	1	3

14.

Shape (Input)	Square	Triangle	Parallelogram	Rectangle
Number of Sides (Output)	4	3	4	4

Determine whether each function is a one-to-one function.

15.

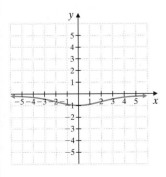

16.

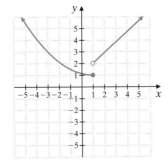

17.

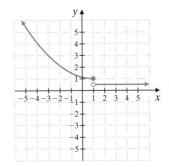

18.

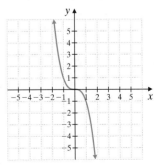

Find an equation defining the inverse function of each one-to-one function.

19. $f(x) = 6x + 11$ **20.** $f(x) = 12x$ **21.** $f(x) = 3x - 5$ **22.** $f(x) = 2x + 1$

Graph each one-to-one function and its inverse on the same set of axes.

23. $f(x) = -2x + 3$ **24.** $f(x) = 5x - 5$

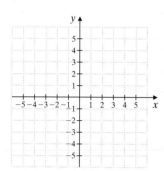
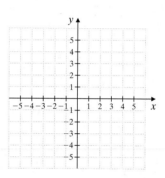

(10.3) *Solve each equation.*

25. $4^x = 64$ **26.** $3^x = \dfrac{1}{9}$ **27.** $2^{3x} = \dfrac{1}{16}$ **28.** $5^{2x} = 125$ **29.** $9^{x+1} = 243$ **30.** $8^{3x-2} = 4$

Graph each exponential function.

31. $y = 3^x$ **32.** $y = \left(\dfrac{1}{3}\right)^x$ **33.** $y = 4 \cdot 2^x$ **34.** $y = 2^x + 4$

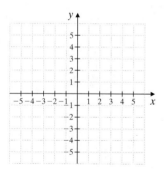
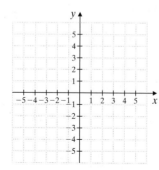
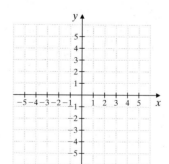
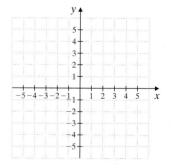

Use the formula $A = P\left(1 + \dfrac{r}{n}\right)^{nt}$ to solve Exercises 35 and 36. In this formula,

A = amount accrued (or owed)
P = principal invested (or loaned)
r = rate of interest
n = number of compounding periods per year
t = time in years

35. Find the amount accrued if $1600 is invested at 9% interest compounded semiannually for 7 years.

36. A total of $800 is invested in a 7% certificate of deposit for which interest is compounded quarterly. Find the value that this certificate will have at the end of 5 years.

(10.4) *Write each exponential equation with logarithmic notation.*

37. $49 = 7^2$

38. $2^{-4} = \dfrac{1}{16}$

Write each logarithmic equation with exponential notation.

39. $\log_{1/2} 16 = -4$

40. $\log_{0.4} 0.064 = 3$

Solve.

41. $\log_4 x = -3$ **42.** $\log_3 x = 2$ **43.** $\log_3 1 = x$ **44.** $\log_4 64 = x$ **45.** $\log_x 64 = 2$

46. $\log_x 81 = 4$ **47.** $\log_4 4^5 = x$ **48.** $\log_7 7^{-2} = x$ **49.** $5^{\log_5 4} = x$ **50.** $2^{\log_2 9} = x$

51. $\log_2(3x - 1) = 4$ **52.** $\log_3(2x + 5) = 2$ **53.** $\log_4(x^2 - 3x) = 1$ **54.** $\log_8(x^2 + 7x) = 1$

Graph each pair of equations on the same set of axes.

55. $y = 2^x$; $y = \log_2 x$

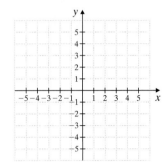

56. $y = \left(\dfrac{1}{2}\right)^x$; $y = \log_{1/2} x$

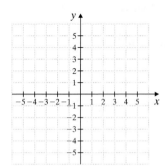

(10.5) *Write each expression as a single logarithm.*

57. $\log_3 8 + \log_3 4$

58. $\log_2 6 + \log_2 3$

59. $\log_7 15 - \log_7 20$

60. $\log 18 - \log 12$

61. $\log_{11} 8 + \log_{11} 3 - \log_{11} 6$

62. $\log_5 14 + \log_5 3 - \log_5 21$

63. $2 \log_5 x - 2 \log_5(x + 1) + \log_5 x$

64. $4 \log_3 x - \log_3 x + \log_3(x + 2)$

Use properties of logarithms to write each expression as a sum or difference of logarithms.

65. $\log_3 \dfrac{x^3}{x + 2}$

66. $\log_4 \dfrac{x + 5}{x^2}$

67. $\log_2 \dfrac{3x^2 y}{z}$

68. $\log_7 \dfrac{yz^3}{x}$

If $\log_b 2 = 0.36$ and $\log_b 5 = 0.83$, evaluate each expression.

69. $\log_b 50$

70. $\log_b \dfrac{4}{5}$

(10.6) *Use a calculator to approximate each logarithm to four decimal places.*

71. $\log 3.6$

72. $\log 0.15$

73. $\ln 1.25$

74. $\ln 4.63$

Find the exact value of each logarithm.

75. $\log 1000$

76. $\log \dfrac{1}{10}$

77. $\ln\left(\dfrac{1}{e}\right)$

78. $\ln(e^4)$

Solve each equation.

79. $\ln(2x) = 2$

80. $\ln(3x) = 1.6$

81. $\ln(2x - 3) = -1$

82. $\ln(3x + 1) = 2$

Approximate each logarithm to four decimal places.

83. $\log_5 1.6$

84. $\log_3 4$

Use the formula $A = Pe^{rt}$ to solve Exercises 85 and 86, in which interest is compounded continuously. In this formula,

A = amount accrued (or owed)
P = principal invested (or loaned)
r = rate of interest
t = time in years

85. Bank of New York offers a 5-year 6% continuously compounded investment option. Find the amount accrued if $1450 is invested.

86. Find the amount to which a $940 investment grows if it is invested at 11% interest compounded continuously for 3 years.

(10.7) *Solve each exponential equation. Given an exact solution and a four-decimal-place approximation.*

87. $3^{2x} = 7$

88. $6^{3x} = 5$

89. $3^{2x+1} = 6$

90. $4^{3x+2} = 9$

91. $5^{3x-5} = 4$

92. $8^{4x-2} = 3$

93. $2 \cdot 5^{x-1} = 1$

94. $3 \cdot 4^{x+5} = 2$

Solve each equation.

95. $\log_5 2 + \log_5 x = 2$

96. $\log_3 x + \log_3 10 = 2$

97. $\log(5x) - \log(x + 1) = 4$

98. $\ln(3x) - \ln(x - 3) = 2$

99. $\log_2 x + \log_2 2x - 3 = 1$

100. $-\log_6(4x + 7) + \log_6 x = 1$

Use the formula $y = y_0 e^{kt}$ to solve Exercises 101 through 105. In this formula,

y = size of population

y_0 = initial count of population

k = rate of growth

t = time

Round each answer to the nearest whole number.

101. The population of mallard ducks in Nova Scotia is expected to grow at a rate of 6% per week during the spring migration. If 155,000 ducks are already in Nova Scotia, how many are expected by the end of 4 weeks?

102. The population of Sierra Leone is growing at a rate of 2.6% per year. In 2001, the population of Sierra Leone was about 5,400,000. Find the expected population by 2009. (*Source:* Population Reference Bureau)

103. France is experiencing an annual growth rate of 0.4%. In 2001, the population of France was 59,200,000. How long will it take for the population to reach 65,000,000? (*Source:* Population Reference Bureau)

104. In 2001, Australia had a population of 19,400,000. How long will it take Australia to double in population if its growth rate is 0.6% annually? (*Source:* Population Reference Bureau)

105. Mexico's population is increasing at a rate of 1.9% per year. How long will it take for its 2001 population of 99,600,000 to double in size? (*Source:* Population Reference Bureau)

Use the compound interest equation $A = P\left(1 + \dfrac{r}{n}\right)^{nt}$ to solve Exercises 106 and 107. (See the directions for Exercises 35 and 36 for an explanation of this formula.) Round answers to the nearest tenth.

106. How long does it take for a $5000 investment to grow to $10,000 if it is invested at 8% interest compounded quarterly?

107. An investment of $6000 has grown to $10,000 while the money was invested at 6% interest compounded monthly. How long was it invested?

Name _____

Chapter 10 Test

If $f(x) = x$ and $g(x) = 2x - 3$, find the following.

1. $(f \cdot g)(x)$ **2.** $(f - g)(x)$

If $f(x) = x$, $g(x) = x - 7$, and $h(x) = x^2 - 6x + 5$, find each composition.

3. $(f \circ h)(0)$ **4.** $(g \circ f)(x)$ **5.** $(g \circ h)(x)$

Graph the one-to-one function and its inverse on the same set of axes.

6. $f(x) = 7x - 14$

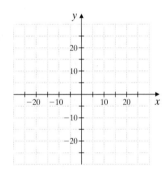

Determine whether each graph is the graph of a one-to-one function.

7.

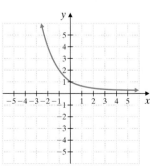

8.

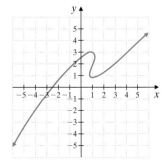

Determine whether each function is one-to-one. If it is one-to-one, find an equation or a set of ordered pairs that defines the inverse function of the given function.

9. $y = 6 - 2x$

10. $f = \{(0, 0), (2, 3), (-1, 5)\}$

11.

Word (Input)	Dog	Cat	House	Desk	Circle
First Letter of Word (Output)	d	c	h	d	c

Use the properties of logarithms to write each expression as a single logarithm.

12. $\log_3 6 + \log_3 4$

13. $\log_5 x + 3\log_5 x - \log_5(x + 1)$

14. Write the expression $\log_6 \dfrac{2x}{y^3}$ as the sum or difference of logarithms.

15. If $\log_b 3 = 0.79$ and $\log_b 5 = 1.16$, find the value of $\log_b\left(\dfrac{3}{25}\right)$.

16. Approximate $\log_7 8$ to four decimal places.

17. Solve $8^{x-1} = \dfrac{1}{64}$ for x. Give an exact solution.

18. Solve $3^{2x+5} = 4$ for x. Give an exact solution and a four-decimal-place approximation.

Solve each logarithmic equation. Give an exact solution.

19. $\log_3 x = -2$

20. $\ln \sqrt{e} = x$

21. $\log_8(3x - 2) = 2$

22. $\log_5 x + \log_5 3 = 2$

23. $\log_4(x + 1) - \log_4(x - 2) = 3$

24. Solve $\ln(3x + 7) = 1.31$ accurate to four decimal places.

25. Graph $y = \left(\dfrac{1}{2}\right)^x + 1$.

26. Graph the functions $y = 3^x$ and $y = \log_3 x$ on the same set of axes.

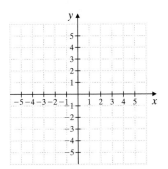

Use the formula $A = P\left(1 + \dfrac{r}{n}\right)^{nt}$ *to solve Exercises 27 and 28.*

27. Find the amount in an account in which $4000 is invested for 3 years at 9% interest compounded monthly.

28. How long will it take $2000 to grow to $3000 if the money is invested at 7% interest compounded semiannually? Round to the nearest whole.

19. _____

20. _____

21. _____

22. _____

23. _____

24. _____

25. see graph _____

26. see graph _____

27. _____

28. _____

29. _____

Use the population growth formula $y = y_0 e^{kt}$ to solve Exercises 29 and 30.

29. The prairie dog population of the Grand Forks area now stands at 57,000 animals. If the population is growing at a rate of 2.6% annually, how many prairie dogs will there be in that area 5 years from now?

30. In an attempt to save an endangered species of wood duck, naturalists would like to increase the wood duck population from 400 to 1000 ducks. If the annual population growth rate is 6.2%, how long will it take the naturalists to reach their goal? Round to the nearest whole year.

30. _____

The reliability of a new model of CD player can be described by the exponential function $R(t) = 2.7^{-(1/3)t}$, where the reliability R is the probability (as a decimal) that the CD player is still working t years after it is manufactured. Round answers to the nearest hundredth. Then write your answers as a percent.

31. What is the probability that the CD player will still work half a year after it is manufactured?

32. When is the probability that the CD player will still work 2 years after it is manufactured?

31. _____

32. _____

33. The world population is currently growing at a rate of 1.3% annually. In 2001, the midyear population of the world was 6137 million people. Predict the midyear world population in 2012. Use $y = 6137(2.7)^{0.013t}$, where y is the world population in millions and t is the number of years after 2001. Round to the nearest million. (*Source:* Based on data from the Population Reference Bureau)

33. _____

Cumulative Review

Find each root.

1. $\sqrt[3]{27}$

2. $\sqrt[4]{16}$

△ **3.** The measure of the largest angle of a triangle is 80° more than the measure of the smallest angle, and the measure of the remaining angle is 10° more than the measure of the smallest angle. Find the measure of each angle.

4. Factor: $7x(x^2 + 5y) - (x^2 + 5y)$

5. Subtract: $\dfrac{5k}{k^2 - 4} - \dfrac{2}{k^2 + k - 2}$

If $f(x) = \sqrt{x - 4}$ and $g(x) = \sqrt[3]{x + 2}$, find each function value.

6. $f(8)$

7. $g(-1)$

Use the properties of exponents to simplify.

8. $x^{1/2}x^{1/3}$

9. $\dfrac{(2x^{2/5})^5}{x^2}$

Use the quotient rule to simplify.

10. $\sqrt{\dfrac{x}{9}}$

11. $\sqrt[4]{\dfrac{3}{16y^4}}$

Multiply.

12. $(\sqrt{2x} + 5)(\sqrt{2x} - 5)$

13. $(\sqrt{3} - 1)^2$

14. Rationalize the numerator of $\dfrac{\sqrt{7}}{\sqrt{45}}$.

15. Solve: $\sqrt{4 - x} = x - 2$

16. Add: $(2 + 3i) + (-3 + 2i)$

17. Solve $4x^2 - 24x + 41 = 0$ by completing the square.

18. Solve: $\dfrac{1}{4}m^2 - m + \dfrac{1}{2} = 0$

19. Solve: $x^{2/3} - 5x^{1/3} + 6 = 0$

Answers

1. _____
2. _____
3. _____
4. _____
5. _____
6. _____
7. _____
8. _____
9. _____
10. _____
11. _____
12. _____
13. _____
14. _____
15. _____
16. _____
17. _____
18. _____
19. _____

20. _____

21. see graph _____

22. _____

23. see graph _____

24. _____

25. a. _____

b. _____

26. _____

27. _____

28. _____

29. _____

30. _____

31. _____

20. Solve: $\dfrac{5}{x+1} < -2$

21. Graph $f(x) = 3x^2 + 3x + 1$. Find the vertex and any intercepts.

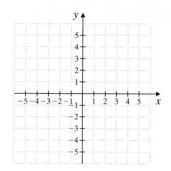

22. Find the midpoint of the line segment that joins points $P(-3, 3)$ and $Q(1, 0)$.

23. Graph: $4y^2 - 9x^2 = 36$

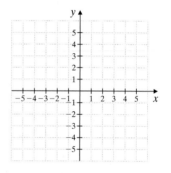

24. Solve the system: $\begin{cases} y = \sqrt{x} \\ x^2 + y^2 = 6 \end{cases}$

25. If $f(x) = |x|$ and $g(x) = x - 2$, find
 a. $(f \circ g)(x)$
 b. $(g \circ f)(x)$

26. Find an equation of the inverse of $f(x) = 3x - 5$

27. Solve: $4^{x+3} = 8^x$

28. Solve: $\log_x 25 = 2$

29. Write as a single logarithm: $2 \log_5 3 + 3 \log_5 2$

30. Find the amount owed at the end of 5 years if $1600 is loaned at a rate of 9% compounded continuously.

31. Solve: $\log(x + 2) - \log x = 2$

APPENDIX A

An Introduction to Using a Graphing Utility

Ⓐ Viewing Window and Interpreting Window Settings

In this appendix, we will use the term **graphing utility** to mean a graphing calculator or a computer software graphing package. All graphing utilities graph equations by plotting points on a screen. While plotting several points can be slow and sometimes tedious for us, a graphing utility can quickly and accurately plot hundreds of points. How does a graphing utility show plotted points? A computer or calculator screen is made up of a grid of small rectangular areas called **pixels**. If a pixel contains a point to be plotted, the pixel is turned "on"; otherwise, the pixel remains "off." The graph of an equation is then a collection of pixels turned "on." The graph of $y = 3x + 1$ from a graphing calculator is shown in Figure A-1. Notice the irregular shape of the line caused by the rectangular pixels.

The portion of the coordinate plane shown on the screen in Figure A-1 is called the **viewing window** or the **viewing rectangle**. Notice the x-axis and the y-axis on the graph. While tick marks are shown on the axes, they are not labeled. This means that from this screen alone, we do not know how many units each tick mark represents. To see what each tick mark represents and the minimum and maximum values on the axes, check the *window setting* of the graphing utility. It defines the viewing window. The window of the graph of $y = 3x + 1$ shown in Figure A-1 has the following setting (Figure A-2):

Xmin $= -10$	The minimum x-value is -10.
Xmax $= 10$	The maximum x-value is 10.
Xscl $= 1$	The x-axis scale is 1 unit per tick mark.
Ymin $= -10$	The minimum y-value is -10.
Ymax $= 10$	The maximum y-value is 10.
Yscl $= 1$	The y-axis scale is 1 unit per tick mark.

By knowing the scale, we can find the minimum and the maximum values on the axes simply by counting tick marks. For example, if both the Xscl (x-axis scale) and the Yscl are 1 unit per tick mark on the graph in Figure A-3, we can count the tick marks and find that the minimum x-value is -10 and the maximum x-value is 10. Also, the minimum y-value is -10 and the maximum y-value is 10. If the Xscl (x-axis scale) changes to 2 units per tick mark (shown in Figure A-4), by counting tick marks, we see that the minimum x-value is now -20 and the maximum x-value is now 20.

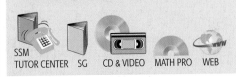
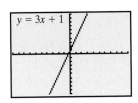

$y = 3x + 1$

Figure A-1

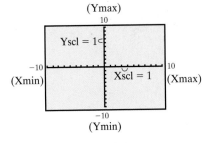

Figure A-2

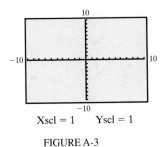

Xscl $= 1$ Yscl $= 1$

FIGURE A-3

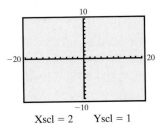

Xscl $= 2$ Yscl $= 1$

FIGURE A-4

819

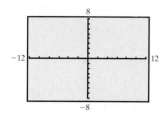

FIGURE A-5

It is also true that if we know the Xmin and the Xmax values, we can calculate the Xscl by the displayed axes. For example, the Xscl of the graph in Figure A-5 must be 3 units per tick mark for the maximum and minimum x-values to be as shown. Also, the Yscl of that graph must be 2 units per tick mark for the maximum and minimum y-values to be as shown.

We will call the viewing window in Figure A-3 a *standard* viewing window or rectangle. Although a standard viewing window is sufficient for much of this text, special care must be taken to ensure that all key features of a graph are shown. Figures A-6, A-7, and A-8 show the graph of $y = x^2 + 11x - 1$ on three different viewing windows. Note that certain viewing windows for this equation are misleading.

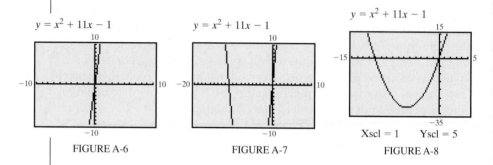

FIGURE A-6   FIGURE A-7   FIGURE A-8

How do we ensure that all distinguishing features of the graph of an equation are shown? It helps to know about the equation that is being graphed. For example, the equation $y = x^2 + 11x - 1$ is not a linear equation, and its graph is not a line. This equation is a quadratic equation, and therefore its graph is a parabola. By knowing this information, we know that the graph shown in Figure A-6, although correct, is misleading. Of the three viewing rectangles shown, the graph in Figure A-8 is best because it shows more of the distinguishing features of the parabola. Properties of equations needed for graphing will be studied in this text.

B   Graphing Equations and Square Viewing Window

In general, the following steps may be used to graph an equation on a standard viewing window.

> **To Graph an Equation in x and y with a Graphing Utility on a Standard Viewing Window**
>
> **Step 1.** Solve the equation for y.
> **Step 2.** Use your graphing utility and enter the equation in the form
> $Y = $ *expression involving x*
> **Step 3.** Activate the graphing utility.

Special care must be taken when entering the *expression involving x* in *Step 2*. You must be sure that the graphing utility you are using interprets the expression as you want it to. For example, let's graph $3y = 4x$. To do so,

Step 1. Solve the equation for y.

$$3y = 4x \qquad \frac{3y}{3} = \frac{4x}{3} \qquad y = \frac{4}{3}x$$

Step 2. Using your graphing utility, enter the expression $\frac{4}{3}x$ after the Y = prompt. In order for your graphing utility to correctly interpret the expression, you may need to enter $(4/3)x$ or $(4 \div 3)x$.

Step 3. Activate the graphing utility. The graph should appear as in Figure A-9.

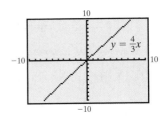

FIGURE A-9

Distinguishing features of the graph of a line include showing all the intercepts of the line. For example, the window of the graph of the line in Figure A-10 does not show both intercepts of the line, but the window of the graph of the same line in Figure A-11 does show both intercepts. Notice the notation below each graph. This is a shorthand notation of the range setting of the graph. This notation means [Xmin, Xmax] by [Ymin, Ymax].

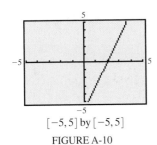

$[-5, 5]$ by $[-5, 5]$

FIGURE A-10

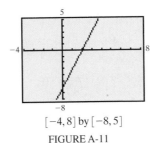

$[-4, 8]$ by $[-8, 5]$

FIGURE A-11

On a standard viewing window, the tick marks on the y-axis are closer than the tick marks on the x-axis. This happens because the viewing window is a rectangle, and so 10 equally spaced tick marks on the positive y-axis will be closer together than 10 equally spaced tick marks on the positive x-axis. This causes the appearance of graphs to be distorted.

For example, notice the different appearances of the same line graphed using different viewing windows. The line in Figure A-12 is distorted because the tick marks along the x-axis are farther apart than the tick marks along the y-axis. The graph of the same line in Figure A-13 is not distorted because the viewing rectangle has been selected so that there is equal spacing between tick marks on both axes.

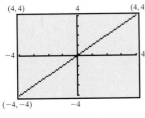

FIGURE A-12

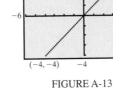

FIGURE A-13

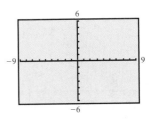

FIGURE A-14

We say that the line in Figure A-13 is graphed on a *square* setting. Some graphing utilities have a built-in program that, if activated, will automatically provide a square setting. A square setting is especially helpful when we are graphing perpendicular lines, circles, or when a true geometric perspective is desired. Some examples of square screens are shown in Figures A-14 and A-15.

Other features of a graphing utility such as Trace, Zoom, Intersect, and Table are discussed in appropriate Graphing Calculator Explorations in this text.

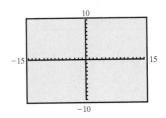

FIGURE A-15

APPENDIX A EXERCISE SET

 In Exercises 1–4, determine whether all ordered pairs listed will lie within a standard viewing rectangle.

1. $(-9, 0), (5, 8), (1, -8)$

2. $(4, 7), (0, 0), (-8, 9)$

3. $(-11, 0), (2, 2), (7, -5)$

4. $(3, 5), (-3, -5), (15, 0)$

In Exercises 5–10, choose an Xmin, Xmax, Ymin, and Ymax so that all ordered pairs listed will lie within the viewing rectangle.

5. $(-90, 0), (55, 80), (0, -80)$

6. $(4, 70), (20, 20), (-18, 90)$

7. $(-11, 0), (2, 2), (7, -5)$

8. $(3, 5), (-3, -5), (15, 0)$

9. $(200, 200), (50, -50), (70, -50)$

10. $(40, 800), (-30, 500), (15, 0)$

Write the window setting for each viewing window shown. Use the following format:

Xmin =   Ymin =

Xmax =   Ymax =

Xscl =   Yscl =

11.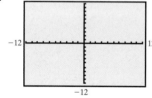

12.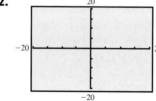

13.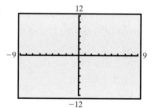

14.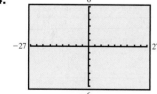

15.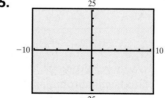

16.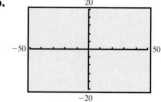

17.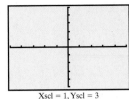

Xscl = 1, Yscl = 3

18.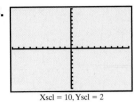

Xscl = 10, Yscl = 2

19.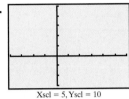

Xscl = 5, Yscl = 10

20.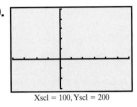

Xscl = 100, Yscl = 200

B *Graph each linear equation in two variables, using the two different range settings given. Determine which setting shows all intercepts of a line.*

1. $y = 2x + 12$

Setting A: $[-10, 10]$ by $[-10, 10]$
Setting B: $[-10, 10]$ by $[-10, 15]$

2. $y = -3x + 25$

Setting A: $[-5, 5]$ by $[-30, 10]$
Setting B: $[-10, 10]$ by $[-10, 30]$

3. $y = -x - 41$

Setting A: $[-50, 10]$ by $[-10, 10]$
Setting B: $[-50, 10]$ by $[-50, 15]$

4. $y = 6x - 18$

Setting A: $[-10, 10]$ by $[-20, 10]$
Setting B: $[-10, 10]$ by $[-10, 10]$

5. $y = \dfrac{1}{2}x - 15$

Setting A: $[-10, 10]$ by $[-20, 10]$
Setting B: $[-10, 35]$ by $[-20, 15]$

6. $y = -\dfrac{2}{3}x - \dfrac{29}{3}$

Setting A: $[-10, 10]$ by $[-10, 10]$
Setting B: $[-15, 5]$ by $[-15, 5]$

The graph of each equation is a line. Use a graphing utility and a standard viewing window to graph each equation.

7. $3x = 5y$

8. $7y = -3x$

9. $9x - 5y = 30$

10. $4x + 6y = 20$

11. $y = -7$ **12.** $y = 2$ **13.** $x + 10y = -5$ **14.** $x - 5y = 9$

Graph the following equations using the square setting given. Some keystrokes that may be helpful are given.

15. $y = \sqrt{x}$ $[-12, 12]$ by $[-8, 8]$
Suggested keystrokes: $\sqrt{\ } x$

16. $y = \sqrt{2x}$ $[-12, 12]$ by $[-8, 8]$
Suggested keystrokes: $\sqrt{\ } (2x)$

17. $y = x^2 + 2x + 1$ $[-15, 15]$ by $[-10, 10]$
Suggested keystrokes: $x \wedge 2 + 2x + 1$

18. $y = x^2 - 5$ $[-15, 15]$ by $[-10, 10]$
Suggested keystrokes: $x \wedge 2 - 5$

19. $y = |x|$ $[-9, 9]$ by $[-6, 6]$
Suggested keystrokes: ABS x

20. $y = |x - 2|$ $[-9, 9]$ by $[-6, 6]$
Suggested keystrokes: ABS $(x - 2)$

Graph the line on a single set of axes. Use a standard viewing window; then, if necessary, change the viewing window so that all intercepts of the line show.

21. $x + 2y = 30$

22. $1.5x - 3.7y = 40.3$

APPENDIX B

Solving Systems of Equations Using Determinants

We have solved systems of two linear equations in two variables in four different ways: graphically, by substitution, by elimination, and by matrices. Now we analyze another method called **Cramer's rule.**

(A) Evaluating 2 × 2 Determinants

Recall that a matrix is a rectangular array of numbers. If a matrix has the same number of rows and columns, it is called a **square matrix.** Examples of square matrices are

$$\begin{bmatrix} 1 & 6 \\ 5 & 2 \end{bmatrix} \qquad \begin{bmatrix} 2 & 4 & 1 \\ 0 & 5 & 2 \\ 3 & 6 & 9 \end{bmatrix}$$

A **determinant** is a real number associated with a square matrix. The determinant of a square matrix is denoted by placing vertical bars about the array of numbers. Thus,

The determinant of the square matrix $\begin{bmatrix} 1 & 6 \\ 5 & 2 \end{bmatrix}$ is $\begin{vmatrix} 1 & 6 \\ 5 & 2 \end{vmatrix}$.

The determinant of the square matrix $\begin{bmatrix} 2 & 4 & 1 \\ 0 & 5 & 2 \\ 3 & 6 & 9 \end{bmatrix}$ is $\begin{vmatrix} 2 & 4 & 1 \\ 0 & 5 & 2 \\ 3 & 6 & 9 \end{vmatrix}$.

We define the determinant of a 2 × 2 matrix first. (Recall that 2 × 2 is read "two by two." It means that the matrix has 2 rows and 2 columns.)

Determinant of a 2 × 2 Matrix

$$\begin{vmatrix} a & b \\ c & d \end{vmatrix} = ad - bc$$

EXAMPLE 1   Evaluate each determinant.

a. $\begin{vmatrix} -1 & 2 \\ 3 & -4 \end{vmatrix}$ **b.** $\begin{vmatrix} 2 & 0 \\ 7 & -5 \end{vmatrix}$

Solution: First we identify the values of a, b, c, and d. Then we perform the evaluation.

a. Here $a = -1, b = 2, c = 3$, and $d = -4$.

$$\begin{vmatrix} -1 & 2 \\ 3 & -4 \end{vmatrix} = ad - bc = (-1)(-4) - (2)(3) = -2$$

b. In this example, $a = 2, b = 0, c = 7$, and $d = -5$.

$$\begin{vmatrix} 2 & 0 \\ 7 & -5 \end{vmatrix} = ad - bc = 2(-5) - (0)(7) = -10$$

Practice Problem 1

Evaluate each determinant.

a. $\begin{vmatrix} -3 & 6 \\ 2 & 1 \end{vmatrix}$

b. $\begin{vmatrix} 4 & 5 \\ 0 & -5 \end{vmatrix}$

Answers

1. a. -15, **b.** -20

825

B Using Cramer's Rule to Solve a System of Two Linear Equations

To develop Cramer's rule, we solve the system $\begin{cases} ax + by = h \\ cx + dy = k \end{cases}$ using elimination. First, we eliminate y by multiplying both sides of the first equation by d and both sides of the second equation by $-b$ so that the coefficients of y are opposites. The result is that

$$\begin{cases} d(ax + by) = d \cdot h \\ -b(cx + dy) = -b \cdot k \end{cases} \quad \text{simplifies to} \quad \begin{cases} adx + bdy = hd \\ -bcx - bdy = -kb \end{cases}$$

We now add the two equations and solve for x.

$$\begin{array}{ll} adx + bdy = hd & \\ \underline{-bcx - bdy = -kb} & \\ adx - bcx = hd - kb & \text{Add the equations.} \\ (ad - bc)x = hd - kb & \\ x = \dfrac{hd - kb}{ad - bc} & \text{Solve for } x. \end{array}$$

When we replace x with $\dfrac{hd - kb}{ad - bc}$ in the equation $ax + by = h$ and solve for y, we find that $y = \dfrac{ak - ch}{ad - bc}$.

Notice that the numerator of the value of x is the determinant of

$$\begin{vmatrix} h & b \\ k & d \end{vmatrix} = hd - kb$$

Also, the numerator of the value of y is the determinant of

$$\begin{vmatrix} a & h \\ c & k \end{vmatrix} = ak - hc$$

Finally, the denominators of the values of x and y are the same and are the determinant of

$$\begin{vmatrix} a & b \\ c & d \end{vmatrix} = ad - bc$$

This means that the values of x and y can be written in determinant notation:

$$x = \dfrac{\begin{vmatrix} h & b \\ k & d \end{vmatrix}}{\begin{vmatrix} a & b \\ c & d \end{vmatrix}} \quad \text{and} \quad y = \dfrac{\begin{vmatrix} a & h \\ c & k \end{vmatrix}}{\begin{vmatrix} a & b \\ c & d \end{vmatrix}}$$

For convenience, we label the determinants D, D_x, and D_y.

$$\underset{\substack{\uparrow \\ x\text{-coefficients}}}{\overset{}{\begin{vmatrix} a & b \\ c & d \end{vmatrix}}} = D \qquad \begin{vmatrix} h & b \\ k & d \end{vmatrix} = D_x \qquad \begin{vmatrix} a & h \\ c & k \end{vmatrix} = D_y$$

x-coefficients

y-coefficients

x-column replaced by constants

y-column replaced by constants

These determinant formulas for the coordinates of the solution of a system are known as **Cramer's rule.**

Cramer's Rule for Two Linear Equations in Two Variables

The solution of the system $\begin{cases} ax + by = h \\ cx + dy = k \end{cases}$ is given by

$$x = \frac{\begin{vmatrix} h & b \\ k & d \end{vmatrix}}{\begin{vmatrix} a & b \\ c & d \end{vmatrix}} = \frac{D_x}{D} \qquad y = \frac{\begin{vmatrix} a & h \\ c & k \end{vmatrix}}{\begin{vmatrix} a & b \\ c & d \end{vmatrix}} = \frac{D_y}{D}$$

as long as $D = ad - bc$ is not 0.

When $D = 0$, the system is either inconsistent or the equations are dependent. When this happens, we need to use another method to see which is the case.

EXAMPLE 2   Use Cramer's rule to solve the system:

$$\begin{cases} 3x + 4y = -7 \\ x - 2y = -9 \end{cases}$$

Solution: First we find D, D_x, and D_y.

$$\begin{array}{ccc} a & b & h \\ \downarrow & \downarrow & \downarrow \end{array}$$

$$\begin{cases} 3x + 4y = -7 \\ x - 2y = -9 \end{cases}$$

$$\begin{array}{ccc} \uparrow & \uparrow & \uparrow \\ c & d & k \end{array}$$

$$D = \begin{vmatrix} a & b \\ c & d \end{vmatrix} = \begin{vmatrix} 3 & 4 \\ 1 & -2 \end{vmatrix} = 3(-2) - 4(1) = -10$$

$$D_x = \begin{vmatrix} h & b \\ k & d \end{vmatrix} = \begin{vmatrix} -7 & 4 \\ -9 & -2 \end{vmatrix} = (-7)(-2) - 4(-9) = 50$$

$$D_y = \begin{vmatrix} a & h \\ c & d \end{vmatrix} = \begin{vmatrix} 3 & -7 \\ 1 & -9 \end{vmatrix} = 3(-9) - (-7)(1) = -20$$

Then $x = \dfrac{D_x}{D} = \dfrac{50}{-10} = -5$ and $y = \dfrac{D_y}{D} = \dfrac{-20}{-10} = 2$.

The ordered pair solution is $(-5, 2)$.
As always, check the solution in both original equations.

EXAMPLE 3   Use Cramer's rule to solve the system:

$$\begin{cases} 5x + y = 5 \\ -7x - 2y = -7 \end{cases}$$

Solution: First we find D, D_x, and D_y.

$$D = \begin{vmatrix} 5 & 1 \\ -7 & -2 \end{vmatrix} = 5(-2) - (-7)(1) = -3$$

$$D_x = \begin{vmatrix} 5 & 1 \\ -7 & -2 \end{vmatrix} = 5(-2) - (-7)(1) = -3$$

$$D_y = \begin{vmatrix} 5 & 5 \\ -7 & -7 \end{vmatrix} = 5(-7) - 5(-7) = 0$$

Practice Problem 2

Use Cramer's rule to solve the system.

$$\begin{cases} x - y = -4 \\ 2x + 3y = 2 \end{cases}$$

Practice Problem 3

Use Cramer's rule to solve the system.

$$\begin{cases} 4x + y = 3 \\ 2x - 3y = -9 \end{cases}$$

Answers

2. $(-2, 2)$,   **3.** $(0, 3)$

Then

$$x = \frac{D_x}{D} = \frac{-3}{-3} = 1 \qquad y = \frac{D_y}{D} = \frac{0}{-3} = 0$$

The ordered pair solution is $(1, 0)$.

C Evaluating 3 × 3 Determinants

A 3 × 3 determinant can be used to solve a system of three equations in three variables. The determinant of a 3 × 3 matrix, however, is considerably more complex than a 2 × 2 one.

Determinant of a 3 × 3 Matrix

$$\begin{vmatrix} a_1 & b_1 & c_1 \\ a_2 & b_2 & c_2 \\ a_3 & b_3 & c_3 \end{vmatrix} = a_1 \cdot \begin{vmatrix} b_2 & c_2 \\ b_3 & c_3 \end{vmatrix} - a_2 \cdot \begin{vmatrix} b_1 & c_1 \\ b_3 & c_3 \end{vmatrix} + a_3 \cdot \begin{vmatrix} b_1 & c_1 \\ b_2 & c_2 \end{vmatrix}$$

Notice that the determinant of a 3 × 3 matrix is related to the determinants of three 2 × 2 matrices. Each determinant of these 2 × 2 matrices is called a **minor**, and every element of a 3 × 3 matrix has a minor associated with it. For example, the minor of c_2 is the determinant of the 2 × 2 matrix found by deleting the row and column containing c_2.

The minor of c_2 is $\begin{vmatrix} a_1 & b_1 \\ a_3 & b_3 \end{vmatrix}$

Also, the minor of element a_1 is the determinant of the 2 × 2 matrix that has no row or column containing a_1.

The minor of a_1 is $\begin{vmatrix} b_2 & c_2 \\ b_3 & c_3 \end{vmatrix}$

So the determinant of a 3 × 3 matrix can be written as:

$$a_1 \cdot (\text{minor of } a_1) - a_2 \cdot (\text{minor of } a_2) + a_3 \cdot (\text{minor of } a_3)$$

Finding the determinant by using minors of elements in the first column is called **expanding** by the minors of the first column. *The value of a determinant can be found by expanding by the minors of any row or column.* The following **array of signs** is helpful in determining whether to add or subtract the product of an element and its minor.

$$\begin{matrix} + & - & + \\ - & + & - \\ + & - & + \end{matrix}$$

If an element is in a position marked $+$, we add. If marked $-$, we subtract.

Try the Concept Check in the margin.

EXAMPLE 4

Evaluate by expanding by the minors of the given row or column.

$$\begin{vmatrix} 0 & 5 & 1 \\ 1 & 3 & -1 \\ -2 & 2 & 4 \end{vmatrix}$$

a. First column   **b.** Second row

Concept Check

Suppose you are interested in finding the determinant of a 4 × 4 matrix. Study the pattern shown in the array of signs for a 3 × 3 matrix. Use the pattern to expand the array of signs for use with a 4 × 4 matrix.

Practice Problem 4

Evaluate by expanding by the minors of the given row or column.

a. First column   b.   Third row

$$\begin{vmatrix} 2 & 0 & 1 \\ -1 & 3 & 2 \\ 5 & 1 & 4 \end{vmatrix}$$

Answers

4. a. 4,   **b.** 4

Concept Check:
$$\begin{matrix} + & - & + & - \\ - & + & - & + \\ + & - & + & - \\ - & + & - & + \end{matrix}$$

Solution:

a. The elements of the first column are $0, 1,$ and -2. The first column of the array of signs is $+, -, +$.

$$\begin{vmatrix} 0 & 5 & 1 \\ 1 & 3 & -1 \\ -2 & 2 & 4 \end{vmatrix} = 0 \cdot \begin{vmatrix} 3 & -1 \\ 2 & 4 \end{vmatrix} - 1 \cdot \begin{vmatrix} 5 & 1 \\ 2 & 4 \end{vmatrix} + (-2) \cdot \begin{vmatrix} 5 & 1 \\ 3 & -1 \end{vmatrix}$$

$$= 0(12 - (-2)) - 1(20 - 2) + (-2)(-5 - 3)$$

$$= 0 - 18 + 16 = -2$$

b. The elements of the second row are $1, 3,$ and -1. This time, the signs begin with $-$ and again alternate.

$$\begin{vmatrix} 0 & 5 & 1 \\ 1 & 3 & -1 \\ -2 & 2 & 4 \end{vmatrix} = -1 \cdot \begin{vmatrix} 5 & 1 \\ 2 & 4 \end{vmatrix} + 3 \cdot \begin{vmatrix} 0 & 1 \\ -2 & 4 \end{vmatrix} - (-1) \cdot \begin{vmatrix} 0 & 5 \\ -2 & 2 \end{vmatrix}$$

$$= -1(20 - 2) + 3(0 - (-2)) - (-1)(0 - (-10))$$

$$= -18 + 6 + 10 = -2$$

Notice that the determinant of the 3×3 matrix is the same regardless of the row or column you select to expand by. ●

Try the Concept Check in the margin.

D **Using Cramer's Rule To Solve a System of Three Linear Equations**

A system of three equations in three variables may be solved with Cramer's rule also. Using the elimination process to solve a system with unknown constants as coefficients leads to the following.

Cramer's Rule for Three Equations in Three Variables

The solution of the system $\begin{cases} a_1 x + b_1 y + c_1 z = k_1 \\ a_2 x + b_2 y + c_2 z = k_2 \\ a_3 x + b_3 y + c_3 z = k_3 \end{cases}$ is given by

$$x = \frac{D_x}{D} \qquad y = \frac{D_y}{D} \qquad \text{and} \qquad z = \frac{D_z}{D}$$

where

$$D = \begin{vmatrix} a_1 & b_1 & c_1 \\ a_2 & b_2 & c_2 \\ a_3 & b_3 & c_3 \end{vmatrix} \quad D_x = \begin{vmatrix} k_1 & b_1 & c_1 \\ k_2 & b_2 & c_2 \\ k_3 & b_3 & c_3 \end{vmatrix}$$

$$D_y = \begin{vmatrix} a_1 & k_1 & c_1 \\ a_2 & k_2 & c_2 \\ a_3 & k_3 & c_3 \end{vmatrix} \quad D_z = \begin{vmatrix} a_1 & b_1 & k_1 \\ a_2 & b_2 & k_2 \\ a_3 & b_3 & k_3 \end{vmatrix}$$

as long as D is not 0.

Concept Check

Why would expanding by minors of the second row be a good choice for the

$$\text{determinant } \begin{vmatrix} 3 & 4 & -2 \\ 5 & 0 & 0 \\ 6 & -3 & 7 \end{vmatrix}?$$

Answer

Concept Check: Two elements of the second row are 0, which, makes calculations easier.

Practice Problem 5

Use Cramer's rule to solve the system:

$$\begin{cases} x + 2y - z = 3 \\ 2x - 3y + z = -9 \\ -x + y - 2z = 0 \end{cases}$$

EXAMPLE 5   Use Cramer's rule to solve the system:

$$\begin{cases} x - 2y + z = 4 \\ 3x + y - 2z = 3 \\ 5x + 5y + 3z = -8 \end{cases}$$

Solution: First we find $D, D_x, D_y,$ and D_z. Beginning with D, we expand by the minors of the first column.

$$D = \begin{vmatrix} 1 & -2 & 1 \\ 3 & 1 & -2 \\ 5 & 5 & 3 \end{vmatrix} = 1 \cdot \begin{vmatrix} 1 & -2 \\ 5 & 3 \end{vmatrix} - 3 \cdot \begin{vmatrix} -2 & 1 \\ 5 & 3 \end{vmatrix} + 5 \cdot \begin{vmatrix} -2 & 1 \\ 1 & -2 \end{vmatrix}$$

$$= 1(3 - (-10)) - 3(-6 - 5) + 5(4 - 1)$$

$$= 13 + 33 + 15 = 61$$

$$D_x = \begin{vmatrix} 4 & -2 & 1 \\ 3 & 1 & -2 \\ -8 & 5 & 3 \end{vmatrix} = 4 \cdot \begin{vmatrix} 1 & -2 \\ 5 & 3 \end{vmatrix} - 3 \cdot \begin{vmatrix} -2 & 1 \\ 5 & 3 \end{vmatrix} + (-8) \cdot \begin{vmatrix} -2 & 1 \\ 1 & -2 \end{vmatrix}$$

$$= 4(3 - (-10)) - 3(-6 - 5) + (-8)(4 - 1)$$

$$= 52 + 33 - 24 = 61$$

$$D_y = \begin{vmatrix} 1 & 4 & 1 \\ 3 & 3 & -2 \\ 5 & -8 & 3 \end{vmatrix} = 1 \cdot \begin{vmatrix} 3 & -2 \\ -8 & 3 \end{vmatrix} - 3 \cdot \begin{vmatrix} 4 & 1 \\ -8 & 3 \end{vmatrix} + 5 \cdot \begin{vmatrix} -4 & 1 \\ 3 & -2 \end{vmatrix}$$

$$= 1(9 - 16) - 3(12 - (-8)) + 5(-8 - 3)$$

$$= -7 - 60 - 55 = -122$$

$$D_z = \begin{vmatrix} 1 & -2 & 4 \\ 3 & 1 & 3 \\ 5 & 5 & -8 \end{vmatrix} = 1 \cdot \begin{vmatrix} 1 & 3 \\ 5 & -8 \end{vmatrix} - 3 \cdot \begin{vmatrix} -2 & 4 \\ 5 & -8 \end{vmatrix} + 5 \cdot \begin{vmatrix} -2 & 4 \\ 1 & 3 \end{vmatrix}$$

$$= 1(-8 - 15) - 3(16 - 20) + 5(-6 - 4)$$

$$= -23 + 12 - 50 = -61$$

From these determinants, we calculate the solution:

$$x = \frac{D_x}{D} = \frac{61}{61} = 1 \quad y = \frac{D_y}{D} = \frac{-122}{61} = -2 \quad z = \frac{D_z}{D} = \frac{-61}{61} = -1$$

The ordered triple solution is $(1, -2, -1)$. Check this solution by verifying that it satisfies each equation of the system.   ●

Answer

5. $(-1, 3, 2)$

Mental Math

Evaluate each determinant mentally.

1. $\begin{vmatrix} 7 & 2 \\ 0 & 8 \end{vmatrix}$

2. $\begin{vmatrix} 6 & 0 \\ 1 & 2 \end{vmatrix}$

3. $\begin{vmatrix} -4 & 2 \\ 0 & 8 \end{vmatrix}$

4. $\begin{vmatrix} 5 & 0 \\ 3 & -5 \end{vmatrix}$

5. $\begin{vmatrix} -2 & 0 \\ 3 & -10 \end{vmatrix}$

6. $\begin{vmatrix} -1 & 4 \\ 0 & -18 \end{vmatrix}$

APPENDIX B EXERCISE SET

(A) *Evaluate each determinant. See Example 1.*

1. $\begin{vmatrix} 3 & 5 \\ -1 & 7 \end{vmatrix}$

2. $\begin{vmatrix} -5 & 1 \\ 1 & -4 \end{vmatrix}$

3. $\begin{vmatrix} 9 & -2 \\ 4 & -3 \end{vmatrix}$

4. $\begin{vmatrix} 4 & -1 \\ 9 & 8 \end{vmatrix}$

5. $\begin{vmatrix} -2 & 9 \\ 4 & -18 \end{vmatrix}$

6. $\begin{vmatrix} -40 & 8 \\ 70 & -14 \end{vmatrix}$

7. $\begin{vmatrix} \frac{3}{4} & \frac{5}{2} \\ -\frac{1}{6} & \frac{7}{3} \end{vmatrix}$

8. $\begin{vmatrix} \frac{5}{7} & \frac{1}{3} \\ \frac{6}{7} & \frac{2}{3} \end{vmatrix}$

(B) *Use Cramer's rule, if possible, to solve each system of linear equations. See Examples 2 and 3.*

9. $\begin{cases} 2y - 4 = 0 \\ x + 2y = 5 \end{cases}$

10. $\begin{cases} 4x - y = 5 \\ 3x - 3 = 0 \end{cases}$

11. $\begin{cases} 3x + y = 1 \\ 2y = 2 - 6x \end{cases}$

12. $\begin{cases} y = 2x - 5 \\ 8x - 4y = 20 \end{cases}$

13. $\begin{cases} 5x - 2y = 27 \\ -3x + 5y = 18 \end{cases}$

14. $\begin{cases} 4x - y = 9 \\ 2x + 3y = -27 \end{cases}$

15. $\begin{cases} 2x - 5y = 4 \\ x + 2y = -7 \end{cases}$

16. $\begin{cases} 3x - y = 2 \\ -5x + 2y = 0 \end{cases}$

17. $\begin{cases} \frac{2}{3}x - \frac{3}{4}y = -1 \\ -\frac{1}{6}x + \frac{3}{4}y = \frac{5}{2} \end{cases}$

18. $\begin{cases} \frac{1}{2}x - \frac{1}{3}y = -3 \\ \frac{1}{8}x + \frac{1}{6}y = 0 \end{cases}$

C *Evaluate. See Example 4.*

19. $\begin{vmatrix} 2 & 1 & 0 \\ 0 & 5 & -3 \\ 4 & 0 & 2 \end{vmatrix}$

20. $\begin{vmatrix} -6 & 4 & 2 \\ 1 & 0 & 5 \\ 0 & 3 & 1 \end{vmatrix}$

21. $\begin{vmatrix} 4 & -6 & 0 \\ -2 & 3 & 0 \\ 4 & -6 & 1 \end{vmatrix}$

22. $\begin{vmatrix} 5 & 2 & 1 \\ 3 & -6 & 0 \\ -2 & 8 & 0 \end{vmatrix}$

23. $\begin{vmatrix} 1 & 0 & 4 \\ 1 & -1 & 2 \\ 3 & 2 & 1 \end{vmatrix}$

24. $\begin{vmatrix} 0 & 1 & 2 \\ 3 & -1 & 2 \\ 3 & 2 & -2 \end{vmatrix}$

25. $\begin{vmatrix} 3 & 6 & -3 \\ -1 & -2 & 3 \\ 4 & -1 & 6 \end{vmatrix}$

26. $\begin{vmatrix} 2 & -2 & 1 \\ 4 & 1 & 3 \\ 3 & 1 & 2 \end{vmatrix}$

D *Use Cramer's rule, if possible, to solve each system of linear equations. See Example 5.*

27. $\begin{cases} 3x + z = -1 \\ -x - 3y + z = 7 \\ 3y + z = 5 \end{cases}$

28. $\begin{cases} 4y - 3z = -2 \\ 8x - 4y = 4 \\ -8x + 4y + z = -2 \end{cases}$

29. $\begin{cases} x + y + z = 8 \\ 2x - y - z = 10 \\ x - 2y + 3z = 22 \end{cases}$

30. $\begin{cases} 5x + y + 3z = 1 \\ x - y - 3z = -7 \\ -x + y = 1 \end{cases}$

31. $\begin{cases} 2x + 2y + z = 1 \\ -x + y + 2z = 3 \\ x + 2y + 4z = 0 \end{cases}$

32. $\begin{cases} 2x - 3y + z = 5 \\ x + y + z = 0 \\ 4x + 2y + 4z = 4 \end{cases}$

33. $\begin{cases} x - 2y + z = -5 \\ 3y + 2z = 4 \\ 3x - y = -2 \end{cases}$

34. $\begin{cases} 4x + 5y = 10 \\ 3y + 2z = -6 \\ x + y + z = 3 \end{cases}$

Combining Concepts

Find the value of x that will make each a true statement.

35. $\begin{vmatrix} 1 & x \\ 2 & 7 \end{vmatrix} = -3$

36. $\begin{vmatrix} 6 & 1 \\ -2 & x \end{vmatrix} = 26$

37. If all the elements in a single row of a determinant are zero, what is the value of the determinant? Explain your answer.

38. If all the elements in a single column of a determinant are 0, what is the value of the determinant? Explain your answer.

APPENDIX C

Fractions

A quotient of two numbers such as $\frac{2}{9}$ is called a **fraction**. The parts of a fraction are:

$$\text{Fraction bar} \rightarrow \frac{2}{9} \begin{array}{l} \leftarrow \text{Numerator} \\ \leftarrow \text{Denominator} \end{array}$$

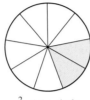

$\frac{2}{9}$ of the circle
is shaded.

A fraction may be used to refer to part of a whole. For example, $\frac{2}{9}$ of the circle in the figure is shaded. The denominator 9 tells us how many equal parts the whole circle is divided into and the numerator 2 tells us how many equal parts are shaded.

In this section, we will use numerators that are **whole numbers** and denominators that are nonzero whole numbers. The whole numbers consist of 0 and the natural numbers.

Whole Numbers: $0, 1, 2, 3, 4, 5,$ and so on

A Writing Equivalent Fractions

More than one fraction can be used to name the same part of a whole. Such fractions are called **equivalent fractions**.

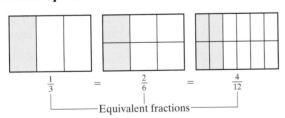

$$\frac{1}{3} \quad = \quad \frac{2}{6} \quad = \quad \frac{4}{12}$$

Equivalent fractions

Equivalent Fractions

Fractions that represent the same portion of a whole are called **equivalent fractions**.

To write equivalent fractions, we use the **fundamental principle of fractions**. This principle guarantees that, if we multiply both the numerator and the denominator by the same nonzero number, the result is an equivalent fraction. For example, if we multiply the numerator and denominator of $\frac{1}{3}$ by the same number, 2, the result is the equivalent fraction $\frac{2}{6}$.

$$\frac{1 \cdot 2}{3 \cdot 2} = \frac{2}{6}$$

Fundamental Principle of Fractions

If a, b, and c are numbers, then

$$\frac{a}{b} = \frac{a \cdot c}{b \cdot c} \quad \text{or} \quad \frac{a \cdot c}{b \cdot c} = \frac{a}{b}$$

as long as b and c are not 0.

Practice Problem 1

Write $\frac{1}{4}$ as an equivalent fraction with a denominator of 20.

EXAMPLE 1   Write $\frac{2}{5}$ as an equivalent fraction with a denominator of 15.

Solution: Since $5 \cdot 3 = 15$, we use the fundamental principle of fractions and multiply the numerator and denominator of $\frac{2}{5}$ by 3.

$$\frac{2}{5} = \frac{2 \cdot 3}{5 \cdot 3} = \frac{6}{15}$$

Thus $\frac{2}{5}$ is equivalent to $\frac{6}{15}$. They both represent the same part of a whole. ●

B   Simplifying Fractions

A fraction is said to be **simplified** or in **lowest terms** when the numerator and the denominator have no factors in common other than 1. For example, the fraction $\frac{5}{11}$ is in lowest terms since 5 and 11 have no common factors other than 1.

One way to simplify fractions is to write both the numerator and the denominator as a product of primes and then apply the fundamental principle of fractions.

Practice Problem 2

Simplify: $\frac{20}{35}$

EXAMPLE 2   Simplify: $\frac{42}{49}$

Solution: We write the numerator and the denominator as products of primes. Then we apply the fundamental principle of fractions to the common factor 7.

$$\frac{42}{49} = \frac{2 \cdot 3 \cdot 7}{7 \cdot 7} = \frac{2 \cdot 3}{7} = \frac{6}{7}$$ ●

Concept Check

Explain the error in the following steps.

a. $\dfrac{15}{55} = \dfrac{1\cancel{5}}{5\cancel{5}} = \dfrac{1}{5}$

b. $\dfrac{6}{7} = \dfrac{5+1}{5+2} = \dfrac{1}{2}$

Try the Concept Check in the margin.

Practice Problems 3–4

Simplify each fraction.

3. $\dfrac{7}{20}$

4. $\dfrac{12}{40}$

EXAMPLES   Simplify each fraction.

3. $\dfrac{11}{27} = \dfrac{11}{3 \cdot 3 \cdot 3}$ There are no common factors other than 1, so $\dfrac{11}{27}$ is already simplified.

4. $\dfrac{88}{20} = \dfrac{2 \cdot 2 \cdot 2 \cdot 11}{2 \cdot 2 \cdot 5} = \dfrac{22}{5}$ ●

A **proper fraction** is a fraction whose numerator is less than its denominator. The fraction $\dfrac{22}{5}$ from Example 4 is called an improper fraction. An **improper fraction** is a fraction whose numerator is greater than or equal to its denominator.

Answers

1. $\dfrac{5}{20}$,   **2.** $\dfrac{4}{7}$,   **3.** $\dfrac{7}{20}$,   **4.** $\dfrac{3}{10}$

Concept Check: answers may vary

The improper fraction $\frac{22}{5}$ may be written as the mixed number $4\frac{2}{5}$.

Notice that a **mixed number** has a whole number part and a fraction part. We review operations on mixed numbers in objective E in this section.

We may simplify some fractions by recalling that the fraction bar means division.

$$\frac{6}{6} = 6 \div 6 = 1 \quad \text{and} \quad \frac{3}{1} = 3 \div 1 = 3$$

EXAMPLES   Simplify by dividing the numerator by the denominator.

5. $\frac{3}{3} = 1$ Since $3 \div 3 = 1$.

6. $\frac{4}{2} = 2$ Since $4 \div 2 = 2$.

7. $\frac{7}{7} = 1$ Since $7 \div 7 = 1$.

8. $\frac{8}{1} = 8$ Since $8 \div 1 = 8$.   ●

In general, if the numerator and the denominator are the same nonzero number, the fraction is equivalent to 1. Also, if the denominator of a fraction is 1, the fraction is equivalent to the numerator.

If a is any number other than 0, then $\frac{a}{a} = 1$.

Also, if a is any number, $\frac{a}{1} = a$.

ⓒ **Multiplying and Dividing Fractions**

To multiply two fractions, we multiply numerator times numerator to obtain the numerator of the product. Then we multiply denominator times denominator to obtain the denominator of the product.

Multiplying Fractions

$$\frac{a}{b} \cdot \frac{c}{d} = \frac{a \cdot c}{b \cdot d} \quad \text{if } b \neq 0 \text{ and } d \neq 0$$

EXAMPLE 9   Multiply: $\frac{2}{15} \cdot \frac{5}{13}$. Simplify the product if possible.

Solution: $\frac{2}{15} \cdot \frac{5}{13} = \frac{2 \cdot 5}{15 \cdot 13}$ Multiply numerators.
Multiply denominators.

To simplify the product, we divide the numerator and the denominator by any common factors.

$$\frac{2}{15} \cdot \frac{5}{13} = \frac{2 \cdot 5}{3 \cdot 5 \cdot 13}$$

$$= \frac{2}{39}$$

Practice Problems 5–8

Simplify by dividing the numerator by the denominator.

5. $\frac{4}{4}$ **6.** $\frac{9}{3}$

7. $\frac{10}{10}$ **8.** $\frac{5}{1}$

Practice Problem 9

Multiply: $\frac{3}{7} \cdot \frac{3}{5}$. Simplify the product if possible.

Answers

5. 1, **6.** 3, **7.** 1, **8.** 5, **9.** $\frac{9}{35}$ ●

Before we divide fractions, we first define **reciprocals**. Two numbers are reciprocals of each other if their product is 1.

The reciprocal of $\frac{2}{3}$ is $\frac{3}{2}$ because $\frac{2}{3} \cdot \frac{3}{2} = \frac{6}{6} = 1$.

The reciprocal of 5 is $\frac{1}{5}$ because $5 \cdot \frac{1}{5} = \frac{5}{1} \cdot \frac{1}{5} = \frac{5}{5} = 1$.

To divide fractions, we multiply the first fraction by the reciprocal of the second fraction. For example,

$$\frac{1}{2} \div \frac{5}{7} = \frac{1}{2} \cdot \frac{7}{5} = \frac{1 \cdot 7}{2 \cdot 5} = \frac{7}{10}$$

Helpful Hint

To divide, multiply by the reciprocal.

Dividing Fractions

$$\frac{a}{b} \div \frac{c}{d} = \frac{a}{b} \cdot \frac{d}{c}, \qquad \text{if } b \neq 0, d \neq 0, \text{ and } c \neq 0$$

EXAMPLES Divide and simplify.

10. $\frac{4}{5} \div \frac{5}{16} = \frac{4}{5} \cdot \frac{16}{5} = \frac{4 \cdot 16}{5 \cdot 5} = \frac{64}{25}$

11. $\frac{7}{10} \div 14 = \frac{7}{10} \div \frac{14}{1} = \frac{7}{10} \cdot \frac{1}{14} = \frac{7 \cdot 1}{2 \cdot 5 \cdot 2 \cdot 7} = \frac{1}{20}$

12. $\frac{3}{8} \div \frac{3}{10} = \frac{3}{8} \cdot \frac{10}{3} = \frac{3 \cdot 2 \cdot 5}{2 \cdot 2 \cdot 2 \cdot 3} = \frac{5}{4}$

D **Adding and Subtracting Fractions**

To add or subtract fractions with the same denominator, we combine numerators and place the sum or difference over the common denominator.

Adding and Subtracting Fractions with the Same Denominator

$$\frac{a}{b} + \frac{c}{b} = \frac{a+c}{b}, \qquad \text{if } b \neq 0$$
$$\frac{a}{b} - \frac{c}{b} = \frac{a-c}{b}, \qquad \text{if } b \neq 0$$

EXAMPLES Add or subtract as indicated. Then simplify if possible.

13. $\frac{2}{7} + \frac{4}{7} = \frac{2+4}{7} = \frac{6}{7}$

14. $\frac{3}{10} + \frac{2}{10} = \frac{3+2}{10} = \frac{5}{10} = \frac{5}{2 \cdot 5} = \frac{1}{2}$

Practice Problems 10–12

Divide and simplify.

10. $\frac{2}{9} \div \frac{3}{4}$

11. $\frac{8}{11} \div 24$

12. $\frac{5}{4} \div \frac{5}{8}$

Practice Problems 13–16

Add or subtract as indicated. Then simplify if possible.

13. $\frac{2}{11} + \frac{5}{11}$

14. $\frac{1}{8} + \frac{3}{8}$

15. $\frac{13}{10} - \frac{3}{10}$

16. $\frac{7}{6} - \frac{2}{6}$

Answers

10. $\frac{8}{27}$, **11.** $\frac{1}{33}$, **12.** 2, **13.** $\frac{7}{11}$, **14.** $\frac{1}{2}$

15. $\frac{9}{7} - \frac{2}{7} = \frac{9-2}{7} = \frac{7}{7} = 1$

16. $\frac{5}{3} - \frac{1}{3} = \frac{5-1}{3} = \frac{4}{3}$ ●

To add or subtract with different denominators, we first write the fractions as **equivalent fractions** with the same denominator. We use the smallest or **least common denominator**, or **LCD**. The LCD is the same as the least common multiple we reviewed in Section R.1.

EXAMPLE 17 Add: $\frac{2}{5} + \frac{1}{4}$

Solution: We first must find the least common denominator before the fractions can be added. The least common multiple for the denominators 5 and 4 is 20. This is the LCD we will use.

We write both fractions as equivalent fractions with denominators of 20. Since

$$\frac{2}{5} = \frac{2 \cdot 4}{5 \cdot 4} = \frac{8}{20} \quad \text{and} \quad \frac{1}{4} = \frac{1 \cdot 5}{4 \cdot 5} = \frac{5}{20}$$

then

$$\frac{2}{5} + \frac{1}{4} = \frac{8}{20} + \frac{5}{20} = \frac{13}{20}$$ ●

EXAMPLE 18 Subtract and simplify: $\frac{19}{6} - \frac{23}{12}$

Solution: The LCD is 12. We write both fractions as equivalent fractions with denominators of 12.

$$\frac{19}{6} - \frac{23}{12} = \frac{19 \cdot 2}{6 \cdot 2} - \frac{23}{12}$$
$$= \frac{38}{12} - \frac{23}{12}$$
$$= \frac{15}{12} = \frac{3 \cdot 5}{2 \cdot 2 \cdot 3} = \frac{5}{4}$$ ●

ⓔ Performing Operations on Mixed Numbers

To perform operations on mixed numbers, first write each mixed number as an improper fraction. To recall how this is done, let's write $3\frac{1}{5}$ as an improper fraction.

$$3\frac{1}{5} = 3 + \frac{1}{5} = \frac{15}{5} + \frac{1}{5} = \frac{16}{5}$$

Because of the steps above, notice we can use a shortcut process for writing a mixed number as an improper fraction.

$$3\frac{1}{5} = \frac{5 \cdot 3 + 1}{5} = \frac{16}{5}$$

Practice Problem 17

Add: $\frac{3}{8} + \frac{1}{20}$

Practice Problem 18

Subtract and simplify: $\frac{8}{15} - \frac{1}{3}$

Practice Problem 19

Multiply: $5\dfrac{1}{6} \cdot 4\dfrac{2}{5}$

Practice Problem 20

Subtract: $7\dfrac{3}{8} - 6\dfrac{1}{4}$

Practice Problem 21

Add: $76\dfrac{1}{9} + 35\dfrac{3}{4}$

EXAMPLE 19   Divide: $2\dfrac{1}{8} \div 1\dfrac{2}{3}$

Solution: First write each mixed number as an improper fraction.

$$2\dfrac{1}{8} = \dfrac{8 \cdot 2 + 1}{8} = \dfrac{17}{8}; \qquad 1\dfrac{2}{3} = \dfrac{3 \cdot 1 + 2}{3} = \dfrac{5}{3}$$

Now divide as usual.

$$2\dfrac{1}{8} \div 1\dfrac{2}{3} = \dfrac{17}{8} \div \dfrac{5}{3} = \dfrac{17}{8} \cdot \dfrac{3}{5} = \dfrac{51}{40} \quad \text{or} \quad 1\dfrac{11}{40}$$

As a general rule, if the original exercise contains mixed numbers, write the result as a mixed number, if possible.

EXAMPLE 20   Add: $2\dfrac{1}{8} + 1\dfrac{2}{3}$

Solution: $2\dfrac{1}{8} + 1\dfrac{2}{3} = \dfrac{17}{8} + \dfrac{5}{3} = \dfrac{51}{24} + \dfrac{40}{24} = \dfrac{91}{24} \quad \text{or} \quad 3\dfrac{19}{24}$

When adding or subtracting larger mixed numbers, you might want to use the following method.

EXAMPLE 21   Subtract: $50\dfrac{1}{6} - 38\dfrac{1}{3}$

Solution:

$$\begin{aligned}
50\dfrac{1}{6} &= 50\dfrac{1}{6} = 49\dfrac{7}{6} \qquad 50\dfrac{1}{6} = 49 + 1 + \dfrac{1}{6} = 49\dfrac{7}{6}\\
-38\dfrac{1}{3} &= -38\dfrac{2}{6} = -38\dfrac{2}{6}\\
&\qquad\qquad\qquad\quad 11\dfrac{5}{6}
\end{aligned}$$

Answers

19. $22\dfrac{11}{15}$,   **20.** $1\dfrac{1}{8}$,   **21.** $111\dfrac{31}{36}$

APPENDIX C EXERCISE SET

Ⓐ *Write each fraction as an equivalent fraction with the given denominator. See Example 1.*

1. $\dfrac{7}{10}$ with a denominator of 30

2. $\dfrac{2}{3}$ with a denominator of 9

3. $\dfrac{2}{9}$ with a denominator of 18

4. $\dfrac{8}{7}$ with a denominator of 56

5. $\dfrac{4}{5}$ with a denominator of 20

6. $\dfrac{4}{5}$ with a denominator of 25

Ⓑ *Simplify each fraction. See Examples 2 through 8.*

7. $\dfrac{2}{4}$

8. $\dfrac{3}{6}$

9. $\dfrac{10}{15}$

10. $\dfrac{15}{20}$

11. $\dfrac{3}{7}$

12. $\dfrac{5}{9}$

13. $\dfrac{20}{20}$

14. $\dfrac{24}{24}$

15. $\dfrac{35}{7}$

16. $\dfrac{42}{6}$

17. $\dfrac{18}{30}$

18. $\dfrac{42}{45}$

19. $\dfrac{16}{20}$

20. $\dfrac{8}{40}$

21. $\dfrac{66}{48}$

22. $\dfrac{64}{24}$

23. $\dfrac{120}{244}$

24. $\dfrac{360}{700}$

25. $\dfrac{192}{264}$

26. $\dfrac{455}{525}$

Ⓒ **Ⓔ** *Multiply or divide as indicated. See Examples 9 through 12 and 19.*

27. $\dfrac{1}{2} \cdot \dfrac{3}{4}$

28. $\dfrac{10}{6} \cdot \dfrac{3}{5}$

29. $\dfrac{2}{3} \cdot \dfrac{3}{4}$

30. $\dfrac{7}{8} \cdot \dfrac{3}{21}$

31. $5\dfrac{1}{9} \cdot 3\dfrac{2}{3}$

32. $2\dfrac{3}{4} \cdot 1\dfrac{7}{8}$

33. $7\dfrac{2}{5} \div \dfrac{1}{5}$

34. $9\dfrac{5}{6} \div \dfrac{1}{6}$

35. $\dfrac{1}{2} \div \dfrac{7}{12}$

36. $\dfrac{7}{12} \div \dfrac{1}{2}$

37. $\dfrac{3}{4} \div \dfrac{1}{20}$

38. $\dfrac{3}{5} \div \dfrac{9}{10}$

39. $\dfrac{7}{10} \cdot \dfrac{5}{21}$

40. $\dfrac{3}{35} \cdot \dfrac{10}{63}$

41. $\dfrac{9}{20} \div 12$

42. $\dfrac{25}{36} \div 10$

43. $4\dfrac{2}{11} \cdot 2\dfrac{1}{2}$

44. $6\dfrac{6}{7} \cdot 3\dfrac{1}{2}$

45. $8\dfrac{3}{5} \div 2\dfrac{9}{10}$

46. $1\dfrac{7}{8} \div 3\dfrac{8}{9}$

D **E** *Add or subtract as indicated. See Examples 13 through 18, 20 and 21.*

47. $\dfrac{4}{5} + \dfrac{1}{5}$

48. $\dfrac{6}{7} + \dfrac{1}{7}$

49. $\dfrac{4}{5} - \dfrac{1}{5}$

50. $\dfrac{6}{7} - \dfrac{1}{7}$

51. $\dfrac{23}{105} + \dfrac{4}{105}$

52. $\dfrac{13}{132} + \dfrac{35}{132}$

53. $\dfrac{17}{21} - \dfrac{10}{21}$

54. $\dfrac{18}{35} - \dfrac{11}{35}$

55. $9\dfrac{7}{8} + 2\dfrac{3}{8}$

56. $8\dfrac{1}{8} - 6\dfrac{3}{8}$

57. $5\dfrac{2}{5} - 3\dfrac{4}{5}$

58. $7\dfrac{3}{4} + 2\dfrac{1}{4}$

59. $\dfrac{2}{3} + \dfrac{3}{7}$

60. $\dfrac{3}{4} + \dfrac{1}{6}$

61. $\dfrac{10}{3} - \dfrac{5}{21}$

62. $\dfrac{11}{7} - \dfrac{3}{35}$

63. $\dfrac{10}{21} + \dfrac{5}{21}$

64. $\dfrac{11}{35} + \dfrac{3}{35}$

65. $\dfrac{5}{22} - \dfrac{5}{33}$

66. $\dfrac{7}{10} - \dfrac{8}{15}$

67. $8\dfrac{11}{12} - 1\dfrac{5}{6}$

68. $4\dfrac{7}{8} - 2\dfrac{3}{16}$

69. $17\dfrac{2}{5} + 30\dfrac{2}{3}$

70. $26\dfrac{11}{20} + 40\dfrac{7}{10}$

71. $\dfrac{12}{5} - 1$

72. $2 - \dfrac{3}{8}$

73. $\dfrac{2}{3} - \dfrac{5}{9} + \dfrac{5}{6}$

74. $\dfrac{8}{11} - \dfrac{1}{4} + \dfrac{1}{2}$

 75. In your own words, describe how to add or subtract fractions.

76. In your own words, describe how to divide fractions.

◥ Combining Concepts

Each circle below represents a whole, or 1. Determine the unknown part of the circle.

77.

78.

79.

80.

81. During the 2000 Summer Olympic Games, Ellina Zvereva of Belarus took the gold medal in the women's discus throw with a distance of $224\dfrac{5}{12}$ feet. However, the Olympic record for the women's discus throw was set in 1988 by Martina Hellmann of East Germany with a distance of $237\dfrac{1}{6}$ feet. How much longer was the Olympic record discus throw than the gold medal throw in 2000? (*Source: World Almanac and Book of Facts, 2001*)

82. Approximately $\dfrac{41}{50}$ of all American adults agree that the U.S. federal government should support basic scientific research. What fraction of American adults do *not* agree that the U.S. federal government should support such research? (*Source:* National Science Foundation)

APPENDIX D

Review of Angles, Lines, and Special Triangles

The word **geometry** is formed from the Greek words, **geo**, meaning earth, and **metron**, meaning measure. Geometry literally means to measure the earth.

This appendix contains a review of some basic geometric ideas. It will be assumed that fundamental ideas of geometry such as point, line, ray, and angle are known. In this appendix, the notation $\angle 1$ is read "angle 1" and the notation $m \angle 1$ is read "the measure of angle 1."

We first review types of angles.

Angles

An angle whose measure is greater than $0°$ but less than $90°$ is called an **acute angle**.

A **right angle** is an angle whose measure is $90°$. A right angle can be indicated by a square drawn at the vertex of the angle, as shown below.

An angle whose measure is greater than $90°$ but less than $180°$ is called an **obtuse angle**.

An angle whose measure is $180°$ is called a **straight angle**.

Two angles are said to be **complementary** if the sum of their measures is $90°$. Each angle is called the **complement** of the other.

Two angles are said to be **supplementary** if the sum of their measures is $180°$. Each angle is called the **supplement** of the other.

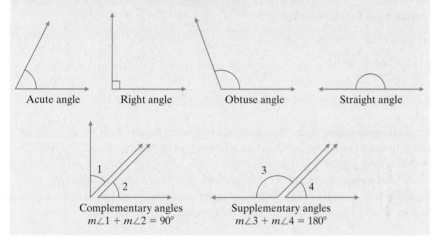

Acute angle   Right angle   Obtuse angle   Straight angle

Complementary angles
$m \angle 1 + m \angle 2 = 90°$

Supplementary angles
$m \angle 3 + m \angle 4 = 180°$

EXAMPLE 1   If an angle measures $28°$, find its complement.

Solution: Two angles are complementary if the sum of their measures is $90°$. The complement of a $28°$ angle is an angle whose measure is $90° - 28° = 62°$. To check, notice that $28° + 62° = 90°$.   ●

Plane is an undefined term that we will describe. A plane can be thought of as a flat surface with infinite length and width, but no thickness. A plane is two dimensional. The arrows in the following diagram indicate that a plane extends indefinitely and has no boundaries.

841

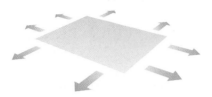

Figures that lie on a plane are called **plane figures**. (See the description of common plane figures in Appendix B.) Lines that lie in the same plane are called **coplanar**.

Lines

Two lines are **parallel** if they lie in the same plane but never meet. **Intersecting lines** meet or cross in one point.

Two lines that form right angles when they intersect are said to be **perpendicular**.

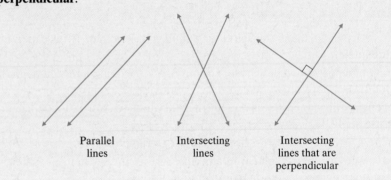

| Parallel lines | Intersecting lines | Intersecting lines that are perpendicular |

Two intersecting lines form **vertical angles**. Angles 1 and 3 are vertical angles. Also angles 2 and 4 are vertical angles. It can be shown that **vertical angles have equal measures**.

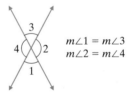

$$m\angle 1 = m\angle 3$$
$$m\angle 2 = m\angle 4$$

Adjacent angles have the same vertex and share a side. Angles 1 and 2 are adjacent angles. Other pairs of adjacent angles are angles 2 and 4, angles 3 and 4, and angles 3 and 1.

A **transversal** is a line that intersects two or more lines in the same plane. Line l is a transversal that intersects lines m and n. The eight angles formed are numbered and certain pairs of these angles are given special names.

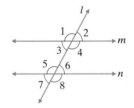

Corresponding angles: $\angle 1$ and $\angle 5$, $\angle 3$ and $\angle 7$, $\angle 2$ and $\angle 6$, and $\angle 4$ and $\angle 8$.

Exterior angles: $\angle 1$, $\angle 2$, $\angle 7$, and $\angle 8$.

Interior angles: $\angle 3$, $\angle 4$, $\angle 5$, and $\angle 6$.

Alternate interior angles: $\angle 3$ and $\angle 6$, $\angle 4$ and $\angle 5$.

These angles and parallel lines are related in the following manner.

> **Parallel Lines Cut by a Transversal**
>
> 1. If two parallel lines are cut by a transversal, then
>   **a. corresponding angles are equal** and
>   **b. alternate interior angles are equal**.
> 2. If corresponding angles formed by two lines and a transversal are equal, then the lines are parallel.
> 3. If alternate interior angles formed by two lines and a transversal are equal, then the lines are parallel.

EXAMPLE 2

Given that lines *m* and *n* are parallel and that the measure of angle 1 is 100°, find the measures of angles 2, 3, and 4.

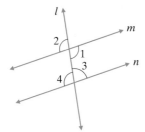

Solution:

$m\angle 2 = 100°$ since angles 1 and 2 are vertical angles

$m\angle 4 = 100°$ since angles 1 and 4 are alternate interior angles

$m\angle 3 = 180° - 100° = 80°$ since angles 4 and 3 are supplementary angles

●

    A **polygon** is the union of three or more coplanar line segments that intersect each other only at each end point, with each end point shared by exactly two segments.

    A **triangle** is a polygon with three sides. The sum of the measures of the three angles of a triangle is 180°. In the following figure, $m\angle 1 + m\angle 2 + m\angle 3 = 180°$.

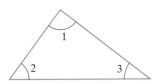

EXAMPLE 3   Find the measure of the third angle of the triangle shown.

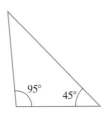

Solution: The sum of the measures of the angles of a triangle is 180°. Since one angle measures 45° and the other angle measures 95°, the third angle measures $180° - 45° - 95° = 40°$. ●

Two triangles are **congruent** if they have the same size and the same shape. In congruent triangles, the measures of corresponding angles are equal and the lengths of corresponding sides are equal. The following triangles are congruent.

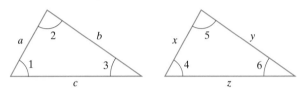

Corresponding angles are equal: $m\angle 1 = m\angle 4$, $m\angle 2 = m\angle 5$, and $m\angle 3 = m\angle 6$. Also, lengths of corresponding sides are equal: $a = x$, $b = y$, and $c = z$.

Any one of the following may be used to determine whether two triangles are congruent.

Congruent Triangles

1. If the measures of two angles of a triangle equal the measures of two angles of another triangle and the lengths of the sides between each pair of angles are equal, the triangles are congruent.

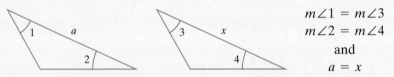

$$m\angle 1 = m\angle 3$$
$$m\angle 2 = m\angle 4$$
and
$$a = x$$

2. If the lengths of the three sides of a triangle equal the lengths of corresponding sides of another triangle, the triangles are congruent.

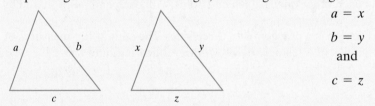

$$a = x$$
$$b = y$$
and
$$c = z$$

3. If the lengths of two sides of a triangle equal the lengths of corresponding sides of another triangle, and the measures of the angles between each pair of sides are equal, the triangles are congruent.

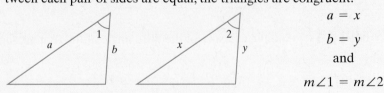

$$a = x$$
$$b = y$$
and
$$m\angle 1 = m\angle 2$$

Two triangles are **similar** if they have the same shape but not necessarily the same size. In similar triangles, the measures of corresponding angles are equal and corresponding sides are in proportion. The following triangles are similar. (All similar triangles drawn in this appendix will be oriented the same.)

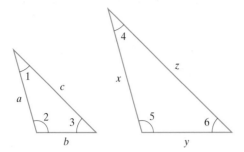

Corresponding angles are equal: $m\angle 1 = m\angle 4, m\angle 2 = m\angle 5,$ and $m\angle 3 = m\angle 6$. Also, corresponding sides are proportional: $\dfrac{a}{x} = \dfrac{b}{y} = \dfrac{c}{z}$.

Any one of the following may be used to determine whether two triangles are similar.

Similar Triangles

1. If the measures of two angles of a triangle equal the measures of two angles of another triangle, the triangles are similar.

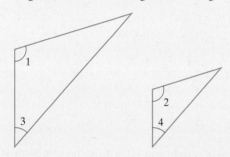

$$m\angle 1 = m\angle 2$$
$$\text{and}$$
$$m\angle 3 = m\angle 4$$

2. If three sides of one triangle are proportional to three sides of another triangle, the triangles are similar.

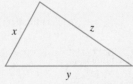

$$\frac{a}{x} = \frac{b}{y} = \frac{c}{z}$$

3. If two sides of a triangle are proportional to two sides of another triangle and the measures of the included angles are equal, the triangles are similar.

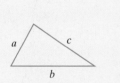

$$m\angle 1 = m\angle 2$$
$$\text{and}$$
$$\frac{a}{x} = \frac{b}{y}$$

EXAMPLE 4

Given that the following triangles are similar, find the missing length x.

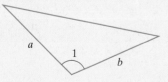

Solution: Since the triangles are similar, corresponding sides are in proportion. Thus, $\frac{2}{3} = \frac{10}{x}$. To solve this equation for x, we cross multiply.

$$\frac{2}{3} = \frac{10}{x}$$
$$2x = 30$$
$$x = 15$$

The missing length is 15 units. ●

A **right triangle** contains a right angle. The side opposite the right angle is called the **hypotenuse**, and the other two sides are called the **legs**. The **Pythagorean theorem** gives a formula that relates the lengths of the three sides of a right triangle.

The Pythagorean Theorem

If a and b are the lengths of the legs of a right triangle, and c is the length of the hypotenuse, then $a^2 + b^2 = c^2$.

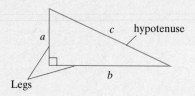

EXAMPLE 5

Find the length of the hypotenuse of a right triangle whose legs have lengths of 3 centimeters and 4 centimeters.

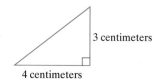

Solution: Because we have a right triangle, we use the Pythagorean theorem. The legs are 3 centimeters and 4 centimeters, so let $a = 3$ and $b = 4$ in the formula.

$$a^2 + b^2 = c^2$$
$$3^2 + 4^2 = c^2$$
$$9 + 16 = c^2$$
$$25 = c^2$$

Since c represents a length, we assume that c is positive. Thus, if c^2 is 25, c must be 5. The hypotenuse has a length of 5 centimeters. ●

Name _____ Section _____ Date _____

Find the complement of each angle. See Example 1.

1. 19° **2.** 65° **3.** 70.8° **4.** $45\frac{2}{3}°$ **5.** $11\frac{1}{4}°$ **6.** 19.6°

Find the supplement of each angle.

7. 150° **8.** 90° **9.** 30.2° **10.** 81.9° **11.** $79\frac{1}{2}°$ **12.** $165\frac{8}{9}°$

13. If lines *m* and *n* are parallel, find the measures of angles 1 through 7. See Example 2.

14. If lines *m* and *n* are parallel, find the measures of angles 1 through 5. See Example 2.

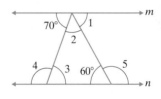

In each of the following, the measures of two angles of a triangle are given. Find the measure of the third angle. See Example 3.

15. 11°, 79° **16.** 8°, 102° **17.** 25°, 65° **18.** 44°, 19° **19.** 30°, 60° **20.** 67°, 23°

In each of the following, the measure of one angle of a right triangle is given. Find the measures of the other two angles.

21. 45° **22.** 60° **23.** 17° **24.** 30° **25.** $39\frac{3}{4}°$ **26.** 72.6°

Given that each of the following pairs of triangles is similar, find the missing length x. See Example 4.

27.

28.

29.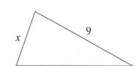

30.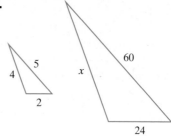

Use the Pythagorean theorem to find the missing lengths in the right triangles. See Example 5.

31.

32.

33.
13
5

34.
12
20

APPENDIX E

Review of Geometric Figures

	Plane Figures Have Length and Width but No Thickness or Depth.	
Name	**Description**	**Figure**
Polygon	Union of three or more coplanar line segments that intersect with each other only at each end point, with each end point shared by two segments.	
Triangle	Polygon with three sides (sum of measures of three angles is 180°).	
Scalene Triangle	Triangle with no sides of equal length.	
Isosceles Triangle	Triangle with two sides of equal length.	
Equilateral Triangle	Triangle with all sides of equal length.	
Right Triangle	Triangle that contains a right angle.	leg, hypotenuse, leg
Quadrilateral	Polygon with four sides (sum of measures of four angles is 360°).	
Trapezoid	Quadrilateral with exactly one pair of opposite sides parallel.	base, leg, parallel sides, leg, base
Isosceles Trapezoid	Trapezoid with legs of equal length.	
Parallelogram	Quadrilateral with both pairs of opposite sides parallel.	
Rhombus	Parallelogram with all sides of equal length.	

(continued)

Name	Description	Figure
Rectangle	Parallelogram with four right angles.	
Square	Rectangle with all sides of equal length.	
Circle	All points in a plane the same distance from a fixed point called the **center**.	

Solid Figures Have Length, Width, and Height or Depth.		
Name	**Description**	**Figure**
Rectangular Solid	A solid with six sides, all of which are rectangles.	
Cube	A rectangular solid whose six sides are squares.	
Sphere	All points the same distance from a fixed point, called the **center**.	
Right Circular Cylinder	A cylinder having two circular bases that are perpendicular to its altitude.	
Right Circular Cone	A cone with a circular base that is perpendicular to its altitude.	

Answers to Selected Exercises

Chapter 1 Answers to Selected Exercises

Chapter 1 Pretest
1. -2; 1.2A **2.** $7 + 2x$; 1.2C **3.** $>$; 1.3B **4.** $<$; 1.3B **5.** -9.25; 1.3C **6.** $\dfrac{7}{8}$; 1.3C **7.** -21; 1.4A **8.** 9; 1.4B **9.** -3; 1.4C

10. -25; 1.4D **11.** $\dfrac{5}{11}$; 1.4E **12.** 27; 1.5A **13.** $10x - 18$; 1.5C **14.** $-20b^{14}$; 1.6A **15.** $\dfrac{5a}{b^5 c}$; 1.6C **16.** z^{15}; 1.6D **17.** $\dfrac{5}{y^{12} z^3}$; 1.7A

18. $\dfrac{81 x^{12} z^{20}}{y^{12}}$; 1.7B **19.** 5; 1.6B **20.** 6.54×10^{10}; 1.6E

Exercise Set 1.2
1. $\{1, 2, 3, 4, 5\}$ **3.** $\{11, 12, 13, 14, 15, 16\}$ **5.** $\{0\}$ **7.** $\{0, 2, 4, 6, 8\}$ **9.** $\{3, 0, \sqrt{36}\}$ **11.** $\{3, \sqrt{36}\}$ **13.** $\{\sqrt{7}\}$ **15.** $\in$ **17.** $\notin$

19. $\notin$ **21.** true **23.** true **25.** false **27.** $2x$ **29.** $x - 10$ **31.** $x + 2$ **33.** $\dfrac{x}{11}$ or $x \div 11$ **35.** $x - 4$ **37.** $x + 20$

39. $10 - x$ **41.** $9x$ **43.** $x + 9$ **45.** $2x + 5$ **47.** $12 - 3x$ **49.** $1 + 2x$ **51.** $5x - 10$ **53.** $2(x + 3)$ **55.** $\dfrac{5}{4 - x}$

57. 5.4; 2.1; 1.7; 2.6; 1.4; 0.7 **59.** answers may vary

Exercise Set 1.3
1. $4c = 7$ **3.** $2x + 5 = -14$ **5.** $\dfrac{n}{5} = 4n$ **7.** $z - 2 = 2z$ **9.** $>$ **11.** $=$ **13.** $<$ **15.** $=$ **17.** $>$ **19.** $<$ **21.** $<$ **23.** $>$

25. true **27.** true **29.** false **31.** true **33.** 6.2 **35.** $-\dfrac{4}{7}$ **37.** $\dfrac{2}{3}$ **39.** 0 **41.** $\dfrac{1}{5}$ **43.** $-\dfrac{1}{8}$ **45.** -4 **47.** undefined

49. $\dfrac{8}{7}$ **51.** $-5; \dfrac{1}{5}$ **53.** $-\dfrac{2}{3}; \dfrac{3}{2}$ **55.** zero; answers may vary **57.** $y + 7x$ **59.** $w \cdot z$ **61.** $\dfrac{x}{5} \cdot \dfrac{1}{3}$ **63.** $(5 \cdot 7)x$ **65.** $x + (1.2 + y)$

67. $14(z \cdot y)$ **69.** $3x + 15$ **71.** $16a + 8b$ **73.** $-12x - 10y - 4z$ **75.** $4z - 24$ **77.** $-6xy + 24x$ **79.** $0.8x + 2y$ **81.** $2x - \dfrac{9}{2}y$

83. $6 + 3x$ **85.** 0 **87.** 7 **89.** $(10 \cdot 2)y$ **91.** $3x + 12$ **93.** $8y + 4$ **95.** no; answers may vary **97.** answers may vary

Exercise Set 1.4
1. 2 **3.** 4 **5.** 0 **7.** -3 **9.** -2 **11.** 5 **13.** -24 **15.** -11 **17.** -4 **19.** $\dfrac{4}{3}$ **21.** -2 **23.** -21 **25.** $-\dfrac{1}{2}$ **27.** -6

29. 3 **31.** -12 **33.** 15 **35.** -12.2 **37.** -1 **39.** -22 **41.** -60 **43.** -18 **45.** 80 **47.** -72 **49.** 0 **51.** 0 **53.** -3

55. -8 **57.** 3 **59.** -3 **61.** 0 **63.** undefined **65.** -8 **67.** 35 **69.** $-\dfrac{3}{7}$ **71.** $\dfrac{1}{21}$ **73.** $-\dfrac{5}{27}$ **75.** -1 **77.** 7.2 **79.** 0

81. 4 **83.** 5 **85.** $-\dfrac{3}{2}$ **87.** -7 **89.** -49 **91.** 36 **93.** -8 **95.** $-\dfrac{1}{27}$ **97.** answers may vary **99.** 7 **101.** 8 **103.** $\dfrac{1}{3}$

105. $\dfrac{1}{4}$ **107.** 4 **109.** 3 **111.** 2 **113.** $\dfrac{13}{35}$ **115.** 4205 m **117.** b **119.** d

121. yes; two players have 6 points each (the third player has 0 points), or two players have 5 points each (the third player has 2 points)
123. 16.5227 **125.** 4.4272 **127.** 13.2% **129.** 10.8% **131.** answers may vary

Integrated Review
1. $\{1, 2, 3\}$ **2.** $\{1, 3, 5\}$ **3.** $\{8, 10, 12, \ldots\}$ **4.** $\{11, 12, 13, 14\}$ **5.** $2(x - 3)$ **6.** $\dfrac{6}{x + 10}$ **7.** $>$ **8.** $=$ **9.** $<$ **10.** $<$

11. $5x = 20$ **12.** $a + 12 = 14$ **13.** $\dfrac{y}{10} = y \cdot 10$ **14.** $x + 1 = x - 1$ **15.** 3 **16.** 9 **17.** -28 **18.** -220 **19.** 5 **20.** -28

21. 5 **22.** -3 **23.** -25 **24.** 25 **25.** 13 **26.** -5 **27.** 0 **28.** undefined **29.** -24 **30.** 30 **31.** $-\dfrac{3}{7}$ **32.** $-\dfrac{1}{10}$

33. $-\dfrac{1}{2}$ **34.** $-\dfrac{11}{12}$ **35.** -8 **36.** -0.3 **37.** 8 **38.** 4.4 **39.** $6; -\dfrac{1}{6}$ **40.** $-4; \dfrac{1}{4}$ **41.** $\dfrac{5}{7}; -\dfrac{5}{7}$ **42.** $-\dfrac{2}{3}; -\dfrac{3}{2}$ **43.** $9m + 45$

44. $77 + 11r$ **45.** $-6y + 9x$ **46.** $-32m + 56n$ **47.** $-0.6a + 1.4$ **48.** $-1.2n + 3$ **49.** $2x - \dfrac{19}{5}y + 4$ **50.** $5x - \dfrac{19}{2}y + 10$

Exercise Set 1.5
1. 48 **3.** -1 **5.** -3 **7.** 14.4 **9.** 17 **11.** -24 **13.** -102 **15.** 40 **17.** -2 **19.** 11 **21.** -56 **23.** -6 **25.** -26

27. 37 **29.** $-\dfrac{3}{4}$ **31.** 7 **33.** 3 **35.** -11 **37.** -2.1 **39.** $-\dfrac{1}{3}$ **41.** $-\dfrac{79}{15}$ **43.** $-\dfrac{4}{5}$ **45.** -81 **47.** -235.5 **49.** 12.25

51. a. 18; 22; 28; 208   **b.** increase; answers may vary   **53. a.** 600; 150; 105   **b.** decrease; answers may vary   **55.** $2381.148   **57.** $8x$

59. $8x$ **61.** $18y$ **63.** $12x$ **65.** $-x - 8$ **67.** $14a + 15$ **69.** $-6x + 9$ **71.** $4a - 13b$ **73.** $2x - 2y$ **75.** $0.8x - 3.6$

77. $\frac{11}{12}b - \frac{7}{6}$ **79.** $6x + 14$ **81.** $-5x + 5$ **83.** $6a - 9b + 12$ **85.** $2k + 10$ **87.** $-3x + 5$ **89.** $4x + 9$ **91.** $4n - 8$ **93.** -24

95. $-2x + \frac{6}{5}y - 1$ **97.** $2x + 10$ **99.** $3a + \frac{3}{35}$ **101.** $1.91x + 4.32$ **103.** $15.4z + 31.11$ **105.** $(2 + 7) \cdot (1 + 3)$ **107.** 20 million

109. 70 million   **111.** increasing   **113.** -0.5876

Calculator Explorations

1. 6×10^{43} **3.** 3.796×10^{28}

Mental Math

1. $\frac{5}{xy^2}$ **3.** $\frac{a^2}{bc^5}$ **5.** $\frac{x^4}{y^2}$

Exercise Set 1.6

1. 4^5 **3.** x^8 **5.** $-140x^{12}$ **7.** $-20x^2y$ **9.** $-16x^6y^3p^2$ **11.** x^{15} **13.** $10x^{10}$ **15.** -1 **17.** 1   **19.** 6   **21.** 9   **23.** -2

25. answers may vary   **27.** a^3 **29.** x **31.** $-13z^4$ **33.** $-6a^4b^4c^6$ **35.** $\frac{1}{z^3}$ **37.** $\frac{1}{16}$ **39.** $\frac{1}{x^8}$ **41.** $\frac{5}{a^4}$ **43.** $\frac{1}{x^7}$ **45.** $4r^8$ **47.** 1

49. $\frac{13}{36}$ **51.** y^4 **53.** $\frac{3}{x}$ **55.** r^8 **57.** $\frac{1}{x^9y^4}$ **59.** $\frac{b^7}{9a^7}$ **61.** $\frac{6x^{16}}{5}$ **63.** x^{7a+5} **65.** x^{2t-1} **67.** x^{4a+7} **69.** z^{6x-7} **71.** x^{6t-1}

73. 3.125×10^7 **75.** 1.6×10^{-2} **77.** 6.7413×10^4 **79.** 1.25×10^{-2} **81.** 5.3×10^{-5} **83.** 7.78×10^8 **85.** 6.137×10^9
87. 1.3×10^9 **89.** 1.0×10^{-3} **91.** 0.0000000036   **93.** 93,000,000   **95.** 1,278,000   **97.** 7,350,000,000,000   **99.** 0.000000403
101. 200,000,000   **103.** 4,900,000,000   **105.** answers may vary

Mental Math

1. x^{20} **3.** x^9 **5.** y^{42} **7.** z^{36} **9.** z^{18}

Exercise Set 1.7

1. $\frac{1}{9}$ **3.** $\frac{1}{x^{36}}$ **5.** $\frac{1}{y^5}$ **7.** $9x^4y^6$ **9.** $16x^{20}y^{12}$ **11.** $\frac{c^{18}}{a^{12}b^6}$ **13.** $\frac{y^{15}}{x^{35}z^{20}}$ **15.** $\frac{1}{125}$ **17.** $\frac{1}{x^{63}}$ **19.** $\frac{343}{512}$ **21.** $16x^4$ **23.** $-\frac{y^3}{64}$

25. $4^8x^2y^6$ **27.** $\frac{x^9}{8y^3}$ **29.** $\frac{1}{a^2}$ **31.** $4a^8b^4$ **33.** $\frac{x^4}{4z^2}$ **35.** $\frac{36}{p^{12}}$ **37.** $-\frac{a^6}{512x^3y^9}$ **39.** $\frac{x^{14}y^{14}}{a^{21}}$ **41.** $\frac{x^4}{16}$ **43.** 64   **45.** $\frac{1}{y^{15}}$

47. $\frac{16\,a^2b^9}{9}$ **49.** $\frac{3}{8x^8y^7}$ **51.** $\frac{1}{x^{30}b^6c^6}$ **53.** $\frac{25}{8x^5y^4}$ **55.** $\frac{2}{x^4y^{10}}$ **57.** x^{9a+18} **59.** x^{12a+2} **61.** b^{10x^2-4x} **63.** y^{15a+3} **65.** $16x^{4t+4}$

67. $5x^{a+2b}y^{a-2b}$ **69.** 1.45×10^9 **71.** 8×10^{15} **73.** 4×10^{-7} **75.** 3×10^{-1} **77.** 2×10^1 **79.** 1×10^1 **81.** 8×10^{-5}

83. 1.1×10^7 **85.** $8.877840909 \times 10^{20}$ **87.** $0.002 = 2 \times 10^{-3}$ sec   **89.** 6.232×10^{-11} cu m   **91.** $\frac{15y^3}{x^8}$ sq ft   **93.** 1.331928×10^{13}

95. no   **97.** 81 people per sq mi   **99.** 4.1 times

Chapter 1 Review

1. $\frac{x}{7}$ **2.** $7x$ **3.** $4(x + 10)$ **4.** $3x - 9$ **5.** $\{-1, 1, 3\}$ **6.** $\{-2, 0, 2, 4, 6\}$ **7.** $\varnothing$ **8.** $\varnothing$ **9.** $\{6, 7, 8, \ldots\}$ **10.** $\{\ldots, -1, 0, 1, 2\}$

11. true   **12.** false   **13.** true   **14.** true   **15.** false   **16.** true   **17.** false   **18.** true   **19.** $\left\{5, \frac{8}{2}, \sqrt{9}\right\}$ **20.** $\left\{5, \frac{8}{2}, \sqrt{9}\right\}$

21. $\left\{5, -\frac{2}{3}, \frac{8}{2}, \sqrt{9}, 0.3, 1\frac{5}{8}, -1\right\}$ **22.** $\{\sqrt{7}, \pi\}$ **23.** $\left\{5, -\frac{2}{3}, \frac{8}{2}, \sqrt{9}, 0.3, \sqrt{7}, 1\frac{5}{8}, -1, \pi\right\}$ **24.** $\left\{5, \frac{8}{2}, \sqrt{9}, -1\right\}$ **25.** $12 = -4x$

26. $n + 2n = -15$ **27.** $4(y + 3) = -1$ **28.** $6(t - 5) = 4$ **29.** $z - 7 = 6$ **30.** $9x - 10 = 5$ **31.** $x - 5 = 12$ **32.** $-4 = 7y$

33. $\frac{2}{3} = 2\left(n + \frac{1}{4}\right)$ **34.** $t + 6 = -12$ **35.** $\frac{3}{4}$ **36.** -0.6 **37.** 0   **38.** -1 **39.** $-\frac{4}{3}$ **40.** $\frac{1}{0.6}$ **41.** undefined   **42.** 1

43. associative property of addition   **44.** distributive property   **45.** additive inverse property   **46.** commutative property of addition
47. associative and commutative properties of multiplication   **48.** multiplicative inverse property   **49.** multiplication property of zero
50. associative property of multiplication   **51.** additive identity property   **52.** multiplicative identity property   **53.** $5x - 15z$

54. $(3 + x) + (7 + y)$ **55.** $2 + (-2)$, for example   **56.** $2 \cdot \frac{1}{2}$, for example   **57.** $(3.4)[(0.7)5]$ **58.** $7 + 0$ **59.** $>$ **60.** $>$ **61.** $<$

62. $=$ **63.** $<$ **64.** $>$ **65.** $-6x + 21y$ **66.** $-90a - 36b$ **67.** $9m - 4n + \frac{1}{2}$ **68.** $4x - 11y + \frac{2}{3}$ **69.** -4 **70.** -35 **71.** -2

72. 0.31 **73.** 8   **74.** 13.3   **75.** -4 **76.** -22 **77.** undefined   **78.** 0   **79.** 4   **80.** -5 **81.** $-\frac{2}{15}$ **82.** 4   **83.** $\frac{5}{12}$

84. 29,852 ft below sea level   **85.** 9   **86.** 13   **87.** 3   **88.** 54   **89.** $-\frac{32}{135}$ **90.** $-\frac{15}{56}$ **91.** $-\frac{5}{4}$ **92.** $-\frac{5}{2}$ **93.** $\frac{5}{8}$ **94.** $-6\frac{1}{2}$

95. -1 **96.** 24 **97.** 1 **98.** 18 **99.** -4 **100.** $\dfrac{7}{3}$ **101.** $\dfrac{5}{7}$ **102.** $-\dfrac{8}{25}$ **103.** $\dfrac{1}{5}$ **104.** 1 **105.** $6.28; 62.8; 628$ **106.** increase

107. $3x - 13$ **108.** $80y - 1$ **109.** $a + 7y - 6$ **110.** $-23b + 13x - 10$ **111.** $n + 5$ **112.** $y - 4$ **113.** 4 **114.** 81 **115.** -4

116. -81 **117.** 1 **118.** -1 **119.** $-\dfrac{1}{16}$ **120.** $\dfrac{1}{16}$ **121.** $-x^2 y^7 z$ **122.** $12x^2 y^3 b$ **123.** $\dfrac{1}{a^9}$ **124.** $\dfrac{1}{a}$ **125.** $\dfrac{1}{x^{11}}$ **126.** $\dfrac{1}{2a^{17}}$

127. $\dfrac{1}{y^5}$ **128.** $9x^{4a+2b} y^{-6b}$ **129.** 3.689×10^7 **130.** -3.62×10^{-4} **131.** 0.000001678 **132.** $410,000$ **133.** 8^{15} **134.** $\dfrac{a^2}{16}$

135. $27x^3$ **136.** $\dfrac{1}{16x^2}$ **137.** $\dfrac{36x^2}{25}$ **138.** $\dfrac{1}{8^{18}}$ **139.** $\dfrac{9}{16}$ **140.** $-\dfrac{1}{8x^9}$ **141.** $\dfrac{1}{4p^4}$ **142.** $-\dfrac{27y^6}{x^6}$ **143.** $x^{25} y^{15} z^{15}$ **144.** $\dfrac{xz}{4}$

145. $\dfrac{x^2}{625 y^4 z^4}$ **146.** $\dfrac{2}{27z^3}$ **147.** $27x^{19a}$ **148.** $2y^{x-7}$

CHAPTER 1 TEST
1. false **2.** false **3.** false **4.** false **5.** true **6.** false **7.** -3 **8.** 43 **9.** -225 **10.** -2 **11.** 1 **12.** 12 **13.** 1
14. a. $5.75; 17.25; 57.50; 115.00$ **b.** increase **15.** $3\left(\dfrac{n}{5}\right) = -n$ **16.** $20 = 2x - 6$ **17.** $-2 = \dfrac{x}{x+5}$ **18.** distributive property
19. associative property of addition **20.** additive inverse property **21.** multiplication property of zero **22.** $6x + 12y + 14$
23. $4x - 9y + \dfrac{7}{3}$ **24.** $\dfrac{1}{81x^2}$ **25.** $\dfrac{3a^7}{2b^5}$ **26.** $-\dfrac{y^{40}}{z^5}$ **27.** $\dfrac{3x^7}{2y^2}$ **28.** x^{4w-8} **29.** 6.3×10^8 **30.** 1.2×10^{-2} **31.** 0.000005
32. 5.76×10^4 **33.** 9×10^{-4} **34.** 5.78×10^8

Chapter 2

CHAPTER 2 PRETEST
1. $\{19\}$; 2.1B **2.** $\{2\}$; 2.1C **3.** $\left\{\dfrac{45}{11}\right\}$; 2.1D **4.** $\{\ \}$; 2.1E **5.** $\left\{1, \dfrac{13}{3}\right\}$; 2.6A **6.** $\left\{-3, \dfrac{5}{3}\right\}$; 2.6A **7.** $y = \dfrac{-5x + 6}{7}$; 2.3A
8. $L = \dfrac{S - 2WH}{2W + 2H}$; 2.3A **9.** $(-\infty, -20]$; 2.4B **10.** $(28, \infty)$; 2.4C **11.** $[-12, \infty)$; 2.4D **12.** $\left(-\infty, -\dfrac{1}{3}\right)$; 2.4D **13.** $[2, \infty)$; 2.5B
14. $[-5, 4]$; 2.5B **15.** $(-\infty, 5)$; 2.5D **16.** $(-15, -3]$; 2.6B **17.** $(-\infty, 2] \cup [4, \infty)$; 2.6B **18.** $18, 32$; 2.2A **19.** 162; 2.2A **20.** 4 ft; 2.3B

MENTAL MATH
1. $\{6\}$ **3.** $\{17\}$ **5.** $\{8\}$ **7.** $\{10\}$

EXERCISE SET 2.1
1. -24 is a solution **3.** -3 is not a solution **5.** -2 is a solution **7.** 5 is not a solution **9.** 5 is not a solution **11.** -8 is a solution
13. $\{6\}$ **15.** $\{-2\}$ **17.** $\{-0.9\}$ **19.** $\{6\}$ **21.** $\{-1.1\}$ **23.** $\{-5\}$ **25.** $\{0\}$ **27.** $\{2\}$ **29.** $\{-9\}$ **31.** $\{-2\}$ **33.** $\left\{-\dfrac{10}{7}\right\}$
35. $\{4\}$ **37. a.** $4x + 5$ **b.** $\{-3\}$ **c.** answers may vary **39.** $\left\{\dfrac{1}{6}\right\}$ **41.** $\{4\}$ **43.** $\{1\}$ **45.** $\{5\}$ **47.** $\left\{\dfrac{40}{3}\right\}$ **49.** $\{17\}$
51. $\{n | n$ is a real number$\}$ **53.** $\varnothing$ **55.** $\{8\}$ **57.** $\{x | x$ is a real number$\}$ **59.** $\left\{\dfrac{1}{8}\right\}$ **61.** $\{2\}$ **63.** $\varnothing$ **65.** $\left\{\dfrac{4}{5}\right\}$ **67.** $\varnothing$
69. $\{-8\}$ **71.** $\{-2\}$ **73.** $\{23\}$ **75.** answers may vary **77.** $\dfrac{8}{x}$ **79.** $8x$ **81.** $3x + 2$ **83.** $\{5.217\}$ **85.** $\{1\}$ **87.** $K = -11$

EXERCISE SET 2.2
1. $4y$ **3.** $3z + 3$ **5.** $(15x + 30)$ cents **7.** $10x + 3$ **9.** -5 **11.** $45,225$ **13.** ~ 658.59 million acres **15.** 1991 earthquakes
17. 860 shoppers **19.** 17% **21.** 6750 users **23.** Mile High Stadium: 76,125 seats; Heinz Field: 64,450 seats **25.** 40.5 ft; 202.5 ft; 240 ft
27. Tokyo: 29.9 million; Mexico City: 27.8 million; New York: 14.6 million **29.** B767-300ER: 216 seats; B737-200: 112 seats **31.** $430.00
33. 28.6 million **35. a.** 214,866 operators **b.** answers may vary **37.** 75, 76, 77 **39.** $64°, 32°, 84°$ **41.** height: 48 in.; width: 108 in.
43. length: 14 cm; width: 6 cm **45.** 32,700,000 **47.** Fallon: 89406; Fernley: 89408; Gardnerville Ranchos: 89410 **49.** $80°, 100°$ **51.** $15°, 75°$
53. $40°, 140°$ **55.** width: 8.4 m; height: 47 m **57.** Incandescent: 1500 bulb hr; Fluorescent: 100,000 bulb hr; Halogen: 4000 bulb hr
59. Thome: 49; Palmeiro: 47; Sexson: 45 **61.** 6 **63.** 208 **65.** -55 **67.** 3195 **69.** 11 million trees **71. a.** during 2033 **b.** 1828.75
c. 5; no: This is the average daily number of cigarettes for all American adults—smokers and non smokers. **73.** no such odd integers exist
75. 500 boards; $30,000 **77.** company makes a profit

MENTAL MATH
1. $y = 5 - 2x$ **3.** $a = 5b + 8$ **5.** $k = h - 5j + 6$

EXERCISE SET 2.3
1. $t = \dfrac{d}{r}$ **3.** $R = \dfrac{I}{PT}$ **5.** $c = P - a - b$ **7.** $y = \dfrac{9x - 16}{4}$ or $y = \dfrac{9}{4}x - 4$ **9.** $l = \dfrac{P - 2w}{2}$ **11.** $r = \dfrac{E}{I} - R$
13. $y = \dfrac{20 - 5x}{4}$ or $y = 5 - \dfrac{5}{4}x$ **15.** $H = \dfrac{S - 2LW}{2L + 2W}$ **17.** $r = \dfrac{C}{2\pi}$ **19.** $F = \dfrac{9}{5}C + 32$ **21.** $4703.71; $4713.99; $4719.22; $4722.74; $4724.45

23. a. $7313.97 **b.** $7321.14 **c.** $7325.98   **25.** $40°C$ **27.** 3 hr, 36 min   **29.** 171 packages   **31.** 9 ft   **33.** 2 gal

35. a. 1174.86 cu m   **b.** 310.34 cu m   **c.** 1485.20 cu m   **37.** 164,921 mi   **39.** 0.42 ft   **41.** 41.125π ft; 129.1325 ft   **43.** $f = \dfrac{C - 4h - 4p}{9}$

45. 178 cal   **47.** 1.5 g   **49.** $\{-3, -2, -1\}$ **51.** $\{-3, -2, -1, 0, 1\}$ **53.** answers may vary   **55.** Mercury: 0.388; Venus: 0.723; Earth: 1.00;

Mars: 1.523; Jupiter: 5.202; Saturn: 9.538; Uranus: 19.193; Neptune: 30.065; Pluto: 39.505   **57.** $6.80 per person   **59.** answers may vary

61. $n_e = \dfrac{N}{R^* \times f_p \times f_l \times f_i \times f_c \times L}$

MENTAL MATH

1. $\{x \mid x < 6\}$ **3.** $\{x \mid x \geq 10\}$ **5.** $\{x \mid x > 4\}$ **7.** $\{x \mid x \leq 2\}$

EXERCISE SET 2.4

1. ; $(-\infty, -3)$ **3.** ; $(0.3, \infty)$ **5.** ; $(5, \infty)$

7. ; $(-2, 5)$ **9.** ; $(-1, 5)$ **11.** answers may vary   **13.** ; $[-2, \infty)$

15. ; $(-\infty, 1)$ **17.** ; $(-\infty, 2]$ **19.** ; $\left[\dfrac{8}{3}, \infty\right)$

21. ; $(-\infty, -4.7)$ **23.** ; $(-\infty, -3]$ **25.** ; $(4, \infty)$

27. $(-\infty, -1]$ **29.** $(-\infty, 11]$ **31.** $(-13, \infty)$ **33.** $(-\infty, 7]$ **35.** $[0, \infty)$ **37.** $(-\infty, -5]$ **39.** $[3, \infty)$ **41.** $(0, \infty)$

43. $\left[-\dfrac{79}{3}, \infty\right)$ **45.** $(-\infty, -1]$ **47.** $\left(-\infty, -\dfrac{35}{6}\right)$ **49.** $[-31, \infty)$ **51.** $(-\infty, -2]$ **53.** 30   **55.** 1040 lb   **57.** 12 oz

59. more than 200 calls   **61.** $F \geq 932°$ **63. a.** 2000   **b.** answers may vary   **65.** decreasing   **67.** 59.35 lb   **69.** 2010

71. answers may vary   **73.** 2013   **75.** $\{0, 1, 2, 3, 4, 5, 6, 7\}$ **77.** $\{\ldots, -9, -8, -7, -6\}$ **79.** ; $(-7, 1]$

81. ; $[-2.5, 5.3]$ **83.** $\varnothing$ **85.** $(-\infty, \infty)$ **87.** answers may vary

INTEGRATED REVIEW

1. $\{-5\}$ **2.** $(-5, \infty)$ **3.** $\left[\dfrac{8}{3}, \infty\right)$ **4.** $[-1, \infty)$ **5.** $\{0\}$ **6.** $\left[-\dfrac{1}{10}, \infty\right)$ **7.** $\left(-\infty, -\dfrac{1}{6}\right]$ **8.** $\{0\}$ **9.** $\varnothing$ **10.** $\left[-\dfrac{3}{5}, \infty\right)$

11. $\{4.2\}$ **12.** $\{6\}$ **13.** $\{-8\}$ **14.** $(-\infty, -16)$ **15.** $\left\{\dfrac{20}{11}\right\}$ **16.** $\{1\}$ **17.** $(38, \infty)$ **18.** $\{-5, -5.5\}$ **19.** $\left\{\dfrac{3}{5}\right\}$ **20.** $(-\infty, \infty)$

21. $\{29\}$ **22.** $\{x \mid x \text{ is a real number}\}$ **23.** $(-\infty, 5)$ **24.** $\left\{\dfrac{9}{13}\right\}$ **25.** $(23, \infty)$ **26.** $(-\infty, 6]$ **27.** $\left(-\infty, \dfrac{3}{5}\right]$ **28.** $\left(-\infty, -\dfrac{19}{32}\right)$

EXERCISE SET 2.5

1. $\{2, 4\}$ **3.** $\varnothing$ **5.** $\{3, 5\}$ **7.**   **9.**   **11.**   **13.** $(-2, 5)$

15. $[6, \infty)$ **17.** $(-\infty, -3]$ **19.** $\varnothing$ **21.** $(11, 17)$ **23.** $[1, 4]$ **25.** $\left[-3, \dfrac{3}{2}\right]$ **27.** $[-21, -9]$ **29.** $\left[\dfrac{3}{2}, 6\right]$ **31.** $\left(0, \dfrac{14}{3}\right]$

33. $\{1, 2, 3, 4, 5, 6, 7, 8\}$ **35.** $\{1, 5, 6\}$ **37.** $\{2, 4, 6, 8\}$ **39.**   **41.**

43.   **45.** $(-\infty, -1) \cup (0, \infty)$ **47.** $[2, \infty)$ **49.** $(-\infty, \infty)$ **51.** $(-\infty, 1] \cup \left(\dfrac{29}{7}, \infty\right)$ **53.** $(-7, \infty)$

55. $(-\infty, \infty)$ **57.** -12 **59.** -4 **61.** $-7, 7$ **63.** 0   **65.** 1993, 1994, 1995   **67.** answers may vary   **69.** $(-3, 5)$ **71.** $(2, \infty)$

73. answers may vary

MENTAL MATH

1. 7   **3.** -5 **5.** -6 **7.** 12

EXERCISE SET 2.6

1. $\{7, -7\}$ **3.** $\varnothing$ **5.** $\{4.2, -4.2\}$ **7.** $\{-4, 4\}$ **9.** $\{-9, 9\}$ **11.** $\{-5, 23\}$ **13.** $\{7, -2\}$ **15.** $\{8, 4\}$ **17.** $\{5, -5\}$ **19.** $\{3, -3\}$

21. $\{-3, 6\}$ **23.** $\{0\}$ **25.** $\varnothing$ **27.** $\left\{-\dfrac{1}{3}, \dfrac{7}{3}\right\}$ **29.** $\left\{-\dfrac{1}{2}, 9\right\}$ **31.** $\left\{-\dfrac{5}{2}\right\}$ **33.** $\{3, 2\}$ **35.** $\{-4, 16\}$ **37.** $\{4\}$ **39.** $\left\{\dfrac{3}{2}\right\}$

41. $\left\{\dfrac{32}{21}, \dfrac{38}{9}\right\}$ **43.** $\left\{-8, \dfrac{2}{3}\right\}$ **45.** ; $[-4, 4]$ **47.** ; $(-\infty, -3) \cup (3, \infty)$

49. ; $(-5, -1)$ **51.** ; $(-\infty, -1] \cup [13, \infty)$ **53.** ; $(-5, 1)$

55. ; $[-5,5]$ **57.** ; $(-\infty,-4)\cup(4,\infty)$ **59.** ; $[-10,3]$

61. ; $(-\infty,-24]\cup[4,\infty)$ **63.** ; $[-2,9]$ **65.** ; $(-\infty,\infty)$

67. ; $\left[-\dfrac{1}{2},1\right]$ **69.** ; $\left(-\infty,\dfrac{2}{3}\right)\cup(2,\infty)$ **71.** ; $\varnothing$

73. ; $(-\infty,-12)\cup(0,\infty)$ **75.** $\{-13,13\}$ **77.** $(-\infty,-13)\cup(13,\infty)$ **79.** $\varnothing$ **81.** $[-10,10]$ **83.** $(-2,5)$

85. $\{5,-2\}$ **87.** $(-\infty,-7]\cup[17,\infty)$ **89.** $\left\{-\dfrac{9}{4}\right\}$ **91.** $(-2,1)$ **93.** $(-\infty,-18)\cup(12,\infty)$ **95.** $\left\{2,\dfrac{4}{3}\right\}$ **97.** $\varnothing$ **99.** $\left\{-\dfrac{17}{2},\dfrac{19}{2}\right\}$

101. $\left(-\infty,-\dfrac{25}{3}\right)\cup\left(\dfrac{35}{3},\infty\right)$ **103.** $\left\{4,-\dfrac{1}{5}\right\}$ **105.** 32%   **107.** $28.8°$ **109.** -1.5 **111.** 0   **113.** $|x|=5$ **115.** $|x|<7$
117. $|x|\le 5$; answers may vary   **119.** $3.45 < x < 3.55$

CHAPTER 2 REVIEW

1. $\{3\}$ **2.** $\left\{-\dfrac{23}{3}\right\}$ **3.** $\left\{-\dfrac{45}{14}\right\}$ **4.** $\left\{-\dfrac{7}{11}\right\}$ **5.** $\{0\}$ **6.** $\varnothing$ **7.** $\{6\}$ **8.** $\{7.8\}$ **9.** $\{x\,|\,x \text{ is a real number}\}$ **10.** $\{0\}$ **11.** $\varnothing$

12. $\{p\,|\,p \text{ is a real number}\}$ **13.** $\{-3\}$ **14.** $\{0\}$ **15.** $\left\{\dfrac{96}{5}\right\}$ **16.** $\{-3\}$ **17.** $\{32\}$ **18.** $\{-8\}$ **19.** $\{8\}$ **20.** $\{1\}$ **21.** $\varnothing$

22. $\{11\}$ **23.** $\{2\}$ **24.** $\left\{\dfrac{37}{8}\right\}$ **25.** -7 **26.** $140, 145$ **27.** 52   **28.** 0.12   **29.** $39,136   **30.** $10, 11, 12, 13$

31. no such integers exist   **32.** width: 40 m; length: 75 m   **33.** 258 mi   **34.** 250 calculators   **35.** China: 137 million; USA: 102 million;
France: 93 million   **36.** $W=\dfrac{V}{LH}$ **37.** $r=\dfrac{C}{2\pi}$ **38.** $y=\dfrac{5x+12}{4}$ **39.** $x=\dfrac{4y-12}{5}$ **40.** $n=\dfrac{y-y_1}{x-x_1}$ **41.** $x=\dfrac{y-y_1+mx_1}{m}$

42. $r=\dfrac{E-IR}{I}$ **43.** $g=\dfrac{S-vt}{t^2}$ **44.** $g=\dfrac{T}{r+vt}$ **45.** $P=\dfrac{I}{1+rt}$ **46.** $B=\dfrac{2A-hb}{h}$ **47.** $h=\dfrac{3V}{\pi r^2}$ **48. a.** $3695.27   **b.** $3700.81

49. $\left(\dfrac{290}{9}\right)°C \approx 32.2°C$ **50.** length: 10 in.; width: 8 in.   **51.** 16 packages   **52.** cylinder holds more ice cream   **53.** 58 mph   **54.** $(3,\infty)$

55. $(-\infty,-4]$ **56.** $(-4,\infty)$ **57.** $(-17,\infty)$ **58.** $(-\infty,7]$ **59.** $(-\infty,4]$ **60.** $\left(\dfrac{1}{2},\infty\right)$ **61.** $(-\infty,1)$ **62.** $[-19,\infty)$

63. $(2,\infty)$ **64.** more economical to use housekeeper for more than 35 lb per week   **65.** $260° \le C \le 538°$ **66.** 9.6   **67.** $1750 to $3750

68. $\left[2,\dfrac{5}{2}\right]$ **69.** $\left[-2,-\dfrac{9}{5}\right)$ **70.** $\left(\dfrac{1}{8},2\right)$ **71.** $\left(-\dfrac{3}{5},0\right)$ **72.** $\left(\dfrac{7}{8},\dfrac{27}{20}\right]$ **73.** $\left[-\dfrac{4}{3},\dfrac{7}{6}\right]$ **74.** $(-5,2]$ **75.** $(-\infty,\infty)$ **76.** $\left(\dfrac{11}{3},\infty\right)$

77. $(5,\infty)$ **78.** $\{16,-2\}$ **79.** $\{5,11\}$ **80.** $\{0,-9\}$ **81.** $\left\{-1,\dfrac{11}{3}\right\}$ **82.** $\left\{2,-\dfrac{2}{3}\right\}$ **83.** $\left\{-\dfrac{1}{6}\right\}$ **84.** $\varnothing$ **85.** $\varnothing$ **86.** $\{3,-3\}$

87. $\{1,5\}$ **88.** $\left\{5,-\dfrac{1}{3}\right\}$ **89.** $\varnothing$ **90.** $\left\{7,-\dfrac{8}{5}\right\}$ **91.** $\left\{-10,-\dfrac{4}{3}\right\}$ **92.** ; $\left(-\dfrac{8}{5},2\right)$

93. ; $(-\infty,-4]\cup[1,\infty)$ **94.** ; $(-\infty,-3)\cup(3,\infty)$ **95.** ; $(-3,3)$

96. ; $\varnothing$ **97.** ; $(-\infty,\infty)$ **98.** ; $\left(-\infty,-\dfrac{22}{15}\right]\cup\left[\dfrac{6}{5},\infty\right)$

99. ; $\left(-\dfrac{1}{2},2\right)$ **100.** ; $(-\infty,-27)\cup(-9,\infty)$ **101.** ; $\varnothing$

CHAPTER 2 TEST

1. $\{10\}$ **2.** $\{1\}$ **3.** $\varnothing$ **4.** $\{n\,|\,n \text{ is a real number}\}$ **5.** $\{12\}$ **6.** $\left\{-\dfrac{80}{29}\right\}$ **7.** $\left\{1,\dfrac{2}{3}\right\}$ **8.** $\varnothing$ **9.** $\left\{-4,-\dfrac{1}{3}\right\}$ **10.** $y=\dfrac{3x-8}{4}$

11. $n=\dfrac{9}{7}m$ **12.** $g=\dfrac{S}{t^2+vt}$ **13.** $C=\dfrac{5}{9}(F-32)$ **14.** $(5,\infty)$ **15.** $[2,\infty)$ **16.** $\left(\dfrac{3}{2},5\right]$ **17.** $(-\infty,-2)\cup\left(\dfrac{4}{3},\infty\right)$ **18.** $[1,11]$

19. $[5,\infty)$ **20.** $[4,\infty)$ **21.** $[-3,-1)$ **22.** $(-\infty,\infty)$ **23.** 9.6   **24.** 6,824,000 vehicles   **25.** approximately 8 hunting dogs
26. more than 850 sunglasses   **27.** $3542.27   **28.** Florida: $17 billion; California: $13 billion; New York: $9 billion

CUMULATIVE REVIEW

1. $\{101,102,103,\ldots\}$ Sec. 1.2, Ex. 2   **2.** $x+5=20$; Sec. 1.3, Ex. 1   **3.** $\dfrac{z}{9}=9+z$; Sec. 1.3, Ex. 3   **4.** -8; Sec. 1.3, Ex. 16

5. $\dfrac{1}{5}$; Sec. 1.3, Ex. 17   **6.** -14; Sec. 1.4, Ex. 6   **7.** 5; Sec. 1.4, Ex. 8   **8.** $-\dfrac{5}{21}$, Sec. 1.4, Ex. 11   **9.** -5; Sec. 1.4, Ex. 25   **10.** 0; Sec. 1.4, Ex. 29

11. 0.125; Sec. 1.4, Ex. 30   **12.** 63; Sec. 1.5, Ex. 1   **13.** $-2x + 4$; Sec. 1.5, Ex. 7   **14.** $4y$; Sec. 1.5, Ex. 8   **15.** x^3; Sec. 1.6, Ex. 10

16. $5x$; Sec. 1.6, Ex. 12   **17.** $\dfrac{z^2}{9x^4y^{20}}$; Sec. 1.7, Ex. 14   **18.** $\dfrac{27a^4x^6}{2}$; Sec. 1.7, Ex. 15   **19.** $\{2\}$; Sec. 2.1, Ex. 3

20. $\{x \mid x \text{ is a real number}\}$; Sec. 2.1, Ex. 11   **21.** \$2350; Sec. 2.2, Ex. 2   **22.** $\dfrac{V}{LW} = H$; Sec. 2.3, Ex. 1

23. $(-\infty, 6]$, ⟵———┤———⟶ ; Sec. 2.4, Ex. 7   **24.** $(-\infty, \infty)$; Sec. 2.5, Ex. 8   **25.** $\{3, -3\}$; Sec. 2.6, Ex. 1
                              6

Chapter 3

Chapter 3 Pretest

1. no, yes; 3.1B   **2. a.** quadrant IV   **b.** y-axis   **c.** quadrant III   **d.** quadrant I; 3.1A

3.
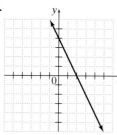
3.1C, 3.3A

4.
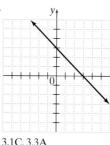
3.1C, 3.3A

5.
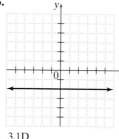
3.1D

6.
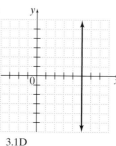
3.1D

7. $\dfrac{2}{11}$; 3.2A   **8.** $m = \dfrac{4}{5}, b = -\dfrac{2}{5}$; 3.2B   **9.** 0; 3.2C   **10.** $y = \dfrac{1}{3}x + 6$; 3.3B   **11.** $y = -7x$; 3.3B   **12.** $2x - y = 1$; 3.4A

13. $x + 3y = 17$; 3.4A   **14.** $y = 10$; 3.4B   **15.** $3x + y = 24$; 3.4C   **16.** domain: $\{-2, 3, 2\}$; range: $\{5, -7\}$; function; 3.5A, B

17. -22; 3.5E   **18.** -5; 3.5E   **19.** 3.6A

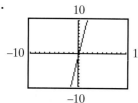

20. 3.6B

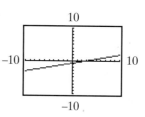

Graphing Calculator Explorations

1.
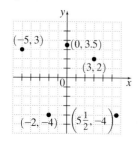

3.
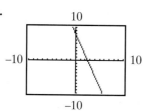

5.
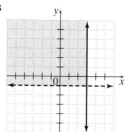

Mental Math

1. $(5, 2)$ **3.** $(3, -1)$ **5.** $(-5, -2)$ **7.** $(-1, 0)$

Exercise Set 3.1

1. quadrant I; quadrant II;
quadrant IV; y-axis;
quadrant III

3. quadrant IV
5. x-axis
7. quadrant III

9.
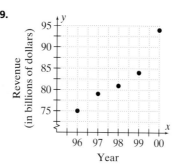

11. no; yes   **13.** yes; yes   **15.** yes; no

17. 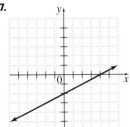   **19.** 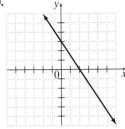   **21.** 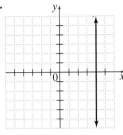   **23.**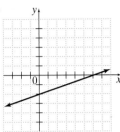

25. 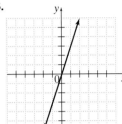   **27.** 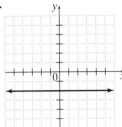   **29.** 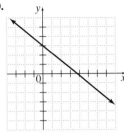   **31.**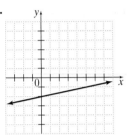

33. 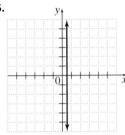   **35.** 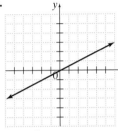   **37.** 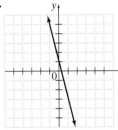   **39.**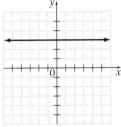

41. C   **43.** A   **45.** $\dfrac{3}{2}$ **47.** 6   **49.** $-\dfrac{0}{5}$ **51.** B   **53.** C   **55.** 1991

57. answers may vary   **a.** $(0, 500)$; 0 tables and 500 chairs are produced   **b.** $(750, 0)$; 750 tables and 0 chairs are produced   **c.** 466 chairs

61. 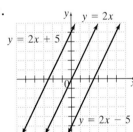   **63.** 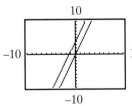   **65.**   **67.** vertical line $x = 0$ has y-intercepts

GRAPHING CALCULATOR EXPLORATIONS

1.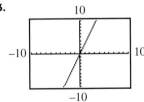
answers may vary

3.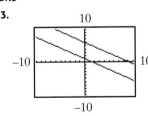
answers may vary

5.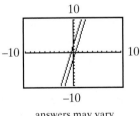
answers may vary

MENTAL MATH

1. upward   **3.** horizontally

EXERCISE SET 3.2

1. $\dfrac{9}{5}$ **3.** $-\dfrac{7}{2}$ **5.** $-\dfrac{5}{6}$ **7.** $\dfrac{1}{3}$ **9.** $-\dfrac{4}{3}$ **11.** 0   **13.** $\dfrac{2}{3}$ **15.** $\dfrac{3}{20}$ **17.** $m = 5; (0, -2)$ **19.** $m = -2; (0, 7)$ **21.** $m = \dfrac{2}{3}; \left(0, -\dfrac{10}{3}\right)$

23. $m = \dfrac{1}{2}; (0, 0)$ **25.** $m = 3; (0, 9)$ **27.** A   **29.** B   **31.** undefined   **33.** -1 **35.** $\dfrac{6}{5}$ **37.** undefined   **39.** 7   **41.** undefined

43. l_2 **45.** l_2 **47.** l_2 **49.** neither   **51.** parallel   **53.** perpendicular   **55.** neither   **57.** parallel   **59.** $-\dfrac{7}{2}$ **61.** $\dfrac{5}{2}$

63. $\{9, -3\}$ **65.** $(-\infty, -4) \cup (-1, \infty)$ **67.** $\left[\dfrac{2}{3}, 2\right]$ **69. a.** $l_1: -2; l_2: -1; l_3: -\dfrac{2}{3}$ **b.** lesser **71.** $(10, 13)$ **73.** $\dfrac{3}{2}$ yd per sec

75. answers may vary **77. a.**

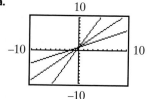

b.

 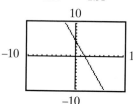

c. true

GRAPHING CALCULATOR EXPLORATIONS

1. $y = \dfrac{1}{3.5}x$

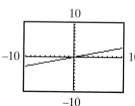

3. $y = -\dfrac{5.78}{2.31}x + \dfrac{10.98}{2.31}$

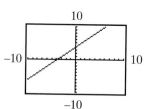

5. $y = x + 3.78$

7. $y = 13.3x + 1.5$

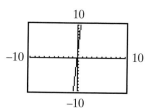

MENTAL MATH

1. $m = -4; (0, 12)$ **3.** $m = 5; (0, 0)$ **5.** $m = \dfrac{1}{2}; (0, 6)$

EXERCISE SET 3.3

1.

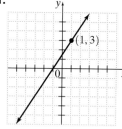

3.

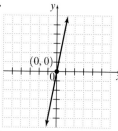

5.

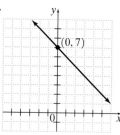

7.

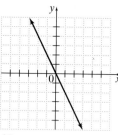

9.

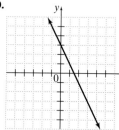

11.

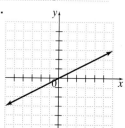

13.

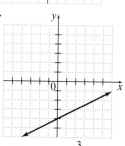

15.

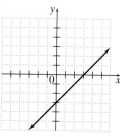

17.

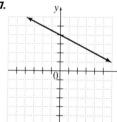

19. C **21.** D **23.** $y = -x + 1$ **25.** $y = 2x + \dfrac{3}{4}$ **27.** $y = \dfrac{2}{7}x$ **29. a.** \$44,640.50 **b.** $m = 1765.1$; annual income increases \$1765.10 every year **c.** $(0, 35,815.0)$ at year $x = 0$, or 1995, annual average income was \$35,815.00 **31. a.** $m = 44, (0, 4, 429)$ **b.** number of people employed as computer support specialists increases 44 thousand for every 1 year **c.** 429 thousand computer support specialists employed in 1998
33. a. \$4734.40 **b.** 2012 **c.** answers may vary **35.** $y = 5x + 32$ **37.** $y = 2x - 1$ **39.** answers may vary
41. $y = -7x + 500$, where y is the height at time x

GRAPHING CALCULATOR EXPLORATIONS

1. 18.4 **3.** -1.5 **5.** 8.7, 7.6

MENTAL MATH

1. $m = -2; (1, 4)$ **3.** $m = \dfrac{1}{4}; (2, 0)$ **5.** $m = 5; (3, -2)$

EXERCISE SET 3.4

1. $y = 3x - 1$ **3.** $y = -2x - 1$ **5.** $y = \dfrac{1}{2}x + 5$ **7.** $y = -\dfrac{9}{10}x - \dfrac{27}{10}$ **9.** $y = 2x + 7$ **11.** $y = -\dfrac{4}{3}x - \dfrac{20}{3}$ **13.** $y = 3x - 6$

15. $y = -2x + 1$ **17.** $y = -\frac{1}{2}x - 5$ **19.** $y = \frac{1}{3}x - 7$ **21.** $y = -\frac{2}{7}x - 6$ **23.** $y = -\frac{3}{8}x + \frac{5}{8}$ **25.** $x = 2$ **27.** $y = 1$ **29.** $x = 0$ **31.** $y = 4x - 4$ **33.** $y = \frac{1}{2}x - 6$ **35.** $y = 4$ **37.** $y = -\frac{3}{2}x - 6$ **39.** $y = -\frac{1}{2}x + \frac{13}{2}$ **41.** $y = -5$ **43.** $y = -\frac{1}{2}x + 1$ **45.** $y = -4x + 1$ **47. a.** $y = 32x$ **b.** 128 ft per sec **49. a.** $y = 12{,}000x + 18{,}000$ **b.** \$102,000 **c.** 9 yrs **51. a.** $y = 6600x + 115{,}800$ **b.** \$175,200 **53. a.** $y = 3{,}704{,}642x + 1{,}089{,}261$ **b.** 27,021,755 DVD players **55.** 31 **57.** −8.4 **59.** 4 **61.** true **63.** $x = 5$
65.

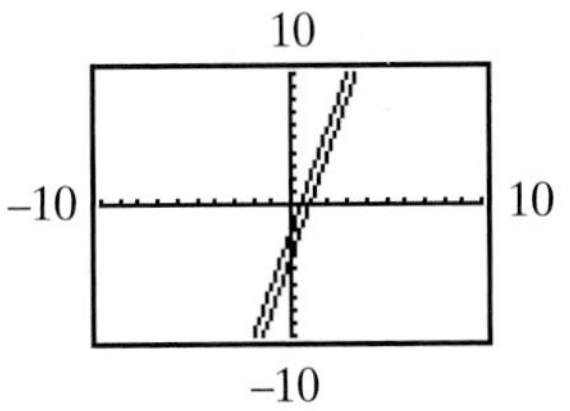

67. answers may vary

Integrated Review

1.

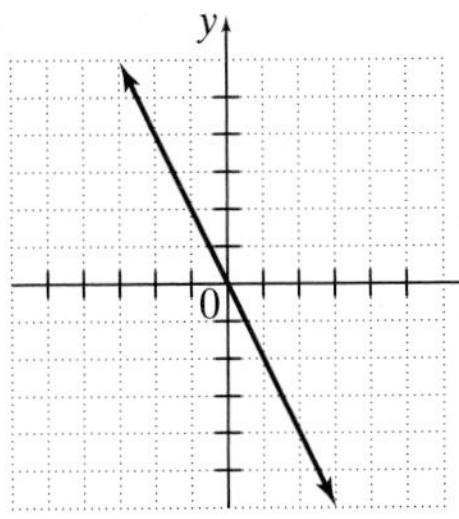

2.

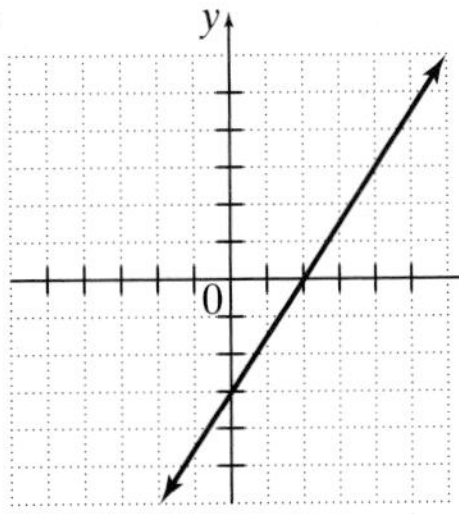

3.

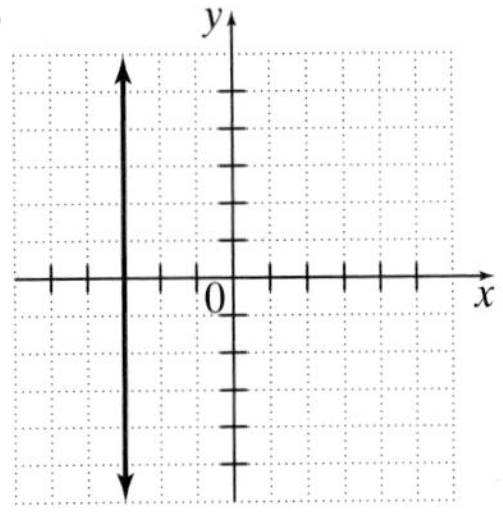

4.

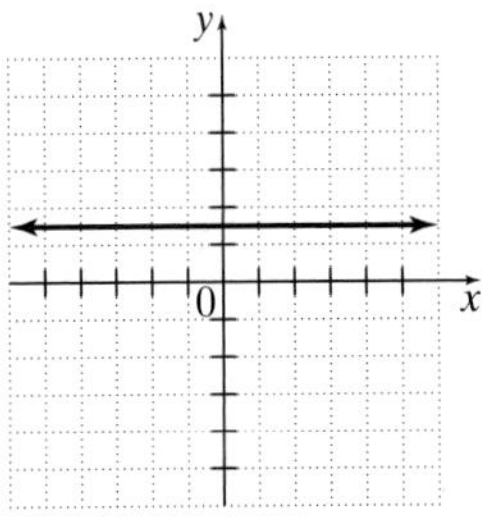

5. 0 **6.** $-\frac{3}{5}$ **7.** $m = 3$; $(0, -5)$ **8.** $m = \frac{5}{2}$; $\left(0, -\frac{7}{2}\right)$ **9.** parallel **10.** perpendicular **11.** $y = -x + 7$ **12.** $x = -2$ **13.** $y = 0$ **14.** $y = -\frac{3}{8}x - \frac{29}{4}$ **15.** $y = -5x - 6$ **16.** $y = -4x + \frac{1}{3}$ **17.** $y = \frac{1}{2}x - 1$ **18.** $y = 3x - \frac{3}{2}$ **19.** $y = 3x - 2$ **20.** $y = -\frac{5}{4}x + 4$ **21.** $y = \frac{1}{4}x - \frac{7}{2}$ **22.** $y = -\frac{5}{2}x - \frac{5}{2}$ **23.** $x = -1$ **24.** $y = 3$

Exercise Set 3.5

1. domain: $\{-1, 0, -2, 5\}$; range: $\{7, 6, 2\}$; function **3.** domain: $\{-2, 6, -7\}$; range: $\{4, -3, -8\}$; not a function **5.** domain: $\{1\}$; range: $\{1, 2, 3, 4\}$; not a function **7.** domain: $\left\{\frac{3}{2}, 0\right\}$; range: $\left\{\frac{1}{2}, -7, \frac{4}{5}\right\}$; not a function **9.** domain: $\{-3, 0, 3\}$; range: $\{-3, 0, 3\}$; function **11.** domain: $\{-1, 1, 2, 3\}$; range: $\{2, 1\}$; function **13.** domain: {Colorado, Alaska, Delaware, Illinois, Connecticut, Texas}; range: $\{6, 1, 20, 30\}$; function **15.** domain: $\{32°, 104°, 212°, 50°\}$; range: $\{0°, 40°, 10°, 100°\}$; function **17.** domain: $\{2, -1, 5, 100\}$; range: $\{0\}$; function **19.** function **21.** yes **23.** no **25.** yes **27.** function **29.** not a function **31.** function **33.** domain: $[0, \infty)$; range: $(-\infty, \infty)$; not a function **35.** domain: $[-1, 1]$; range: $(-\infty, \infty)$; not a function **37.** domain: $(-\infty, \infty)$; range: $(-\infty, -3] \cup [3, \infty)$; not a function **39.** domain: $[1, 7]$; range: $[1, 7]$; not a function **41.** domain: $\{-2\}$; range: $(-\infty, \infty)$; not a function **43.** domain: $(-\infty, \infty)$; range: $(-\infty, 3]$; function **45.** 15 **47.** 38 **49.** 7 **51.** 3 **53. a.** 0 **b.** 1 **c.** −1 **55. a.** −5 **b.** −5 **c.** −5 **57.** 25π sq cm **59.** 2744 cu in. **61.** 166.38 cm **63.** 163.2 mg **65. a.** 95.99; per capita consumption of poultry was 95.99 lb in 2000 **b.** 99.37 lb

67.

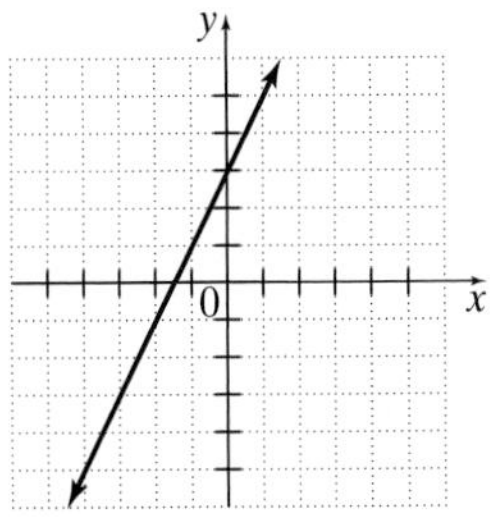

69.

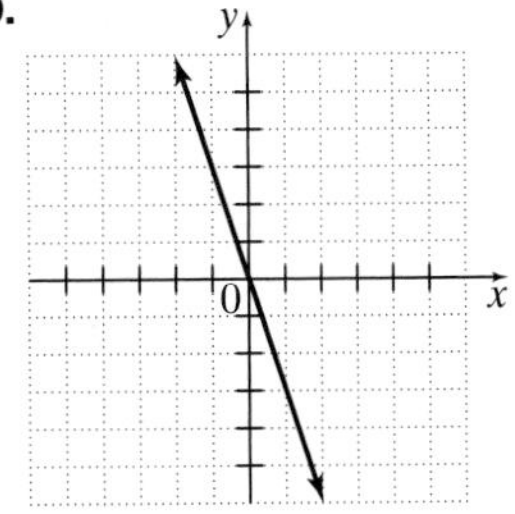

71.

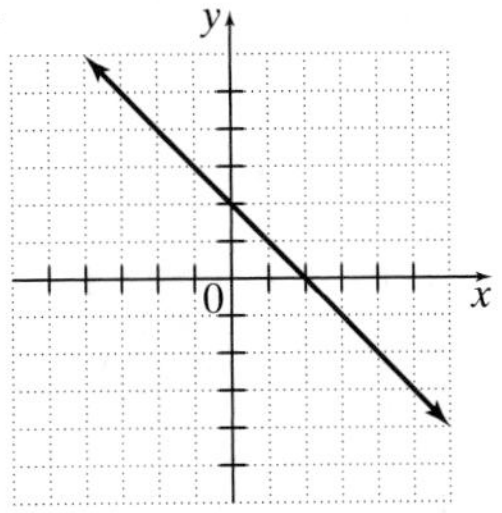

73. $(-\infty, 14]$ **75.** $\left[\frac{7}{2}, \infty\right)$ **77.** $\left(-\infty, -\frac{1}{4}\right)$ **79. a.** 5.1 **b.** 15.5 **c.** 9.533 **81. a.** 132 **b.** $a^2 - 12$ **83.** answers may vary **85.** answers may vary

Exercise Set 3.6

1.

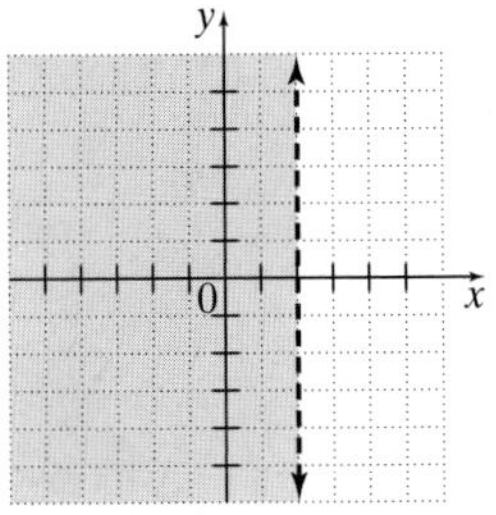

3.

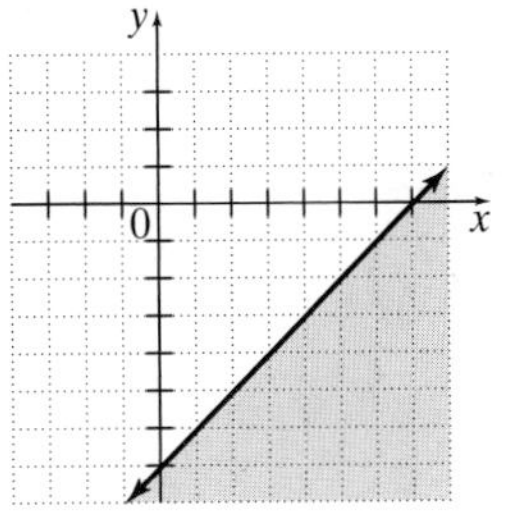

5.

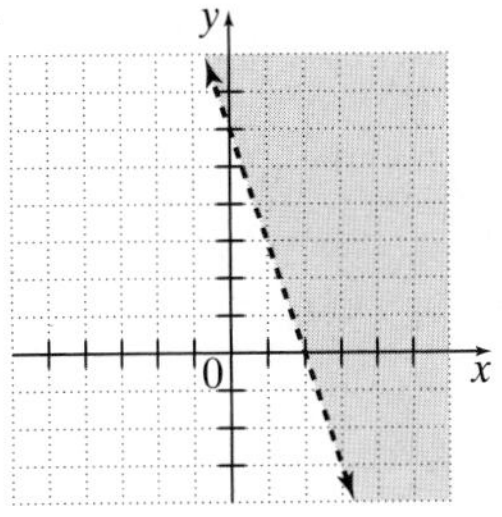

7.

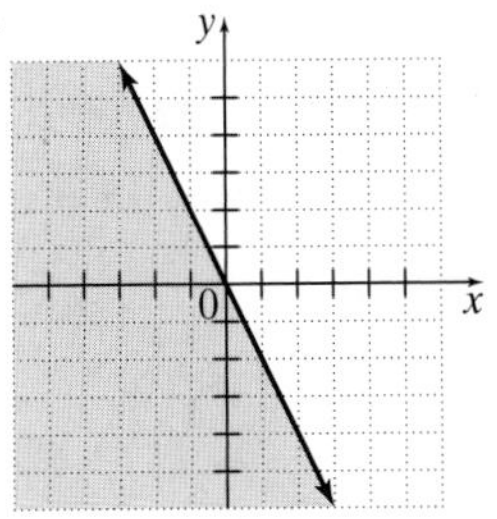

9.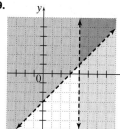

11.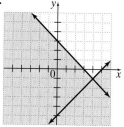

13. with $<$ or $>$ **15.**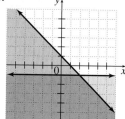

17.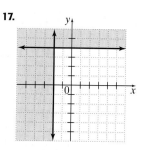

19.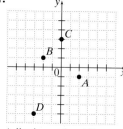

21.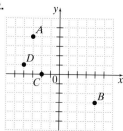

23.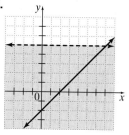

25.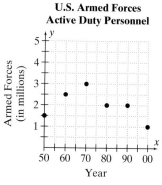

27. D   **29.** A   **31.** yes   **33.** no   **35.**

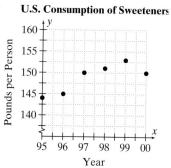

CHAPTER 3 REVIEW

1.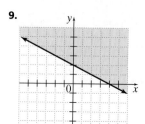

A lies in quadrant II
B lies in quadrant IV
C lies on the y-axis, no quadrant
D lies in quadrant III

2.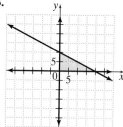

A lies in quadrant IV
B lies in quadrant II
C lies on the y-axis, no quadrant
D lies in quadrant IV

3.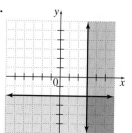

U.S. Consumption of Sweeteners

4.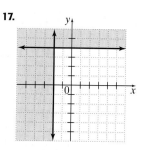

U.S. Armed Forces
Active Duty Personnel

5. no; yes   **6.** yes; no   **7.** yes; yes   **8.** yes; no   **9.**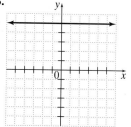

10.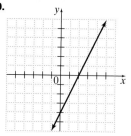

11.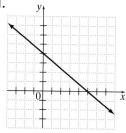

12.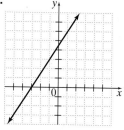

13.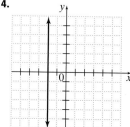

14.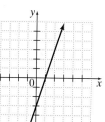

15.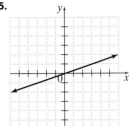

16.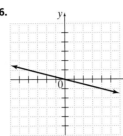

17. -3 **18.** $\dfrac{1}{2}$ **19.** $\dfrac{5}{2}$ **20.** $-\dfrac{3}{4}$ **21.** $\dfrac{3}{2}$ **22.** -3 **23.** $-\dfrac{1}{2}$

24. 1 **25.** l_2 **26.** l_2 **27.** l_2 **28.** l_1 **29.** $m = -3; \left(0, \dfrac{1}{2}\right)$

30. $m = 2; (0,4)$ **31.** $m = \dfrac{2}{5}; \left(0, -\dfrac{4}{3}\right)$ **32.** $m = -\dfrac{2}{7}; \left(0, \dfrac{3}{2}\right)$

33. 0 **34.** undefined   **35.** neither   **36.** parallel   **37.** perpendicular

38. neither

39.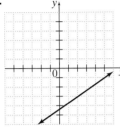

40.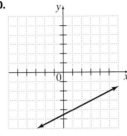

41.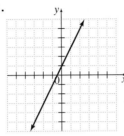

42.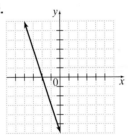

43.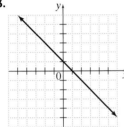

44.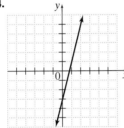

45.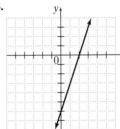

46.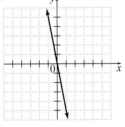

47. a. $C(150) = 87$, $87   **b.** $m = 0.3$; cost increases by $0.30 for each additional mile driven   **c.** $(0, 42)$; cost for 0 miles driven is $42   **48.** $y = -1$

49. $x = -2$ **50.** $x = -4$ **51.** $y = 5$ **52.** $y = 3x + 14$ **53.** $y = 2x - 12$ **54.** $y = -\dfrac{1}{2}x - 4$ **55.** $y = -11x - 52$ **56.** $y = -2x - 2$

57. $y = -\dfrac{3}{2}x - 8$ **58.** $y = \dfrac{3}{4}x + \dfrac{7}{2}$ **59.** $y = -\dfrac{3}{2}x - 1$ **60. a.** $y = -800x + 4200$ **b.** $200   **61. a.** $y = 3000x + 144{,}000$ **b.** $219,000

62. domain: $\left\{-\dfrac{1}{2}, 6, 0, 25\right\}$; range: $\left\{\dfrac{3}{4}, 0.65, -12, 25\right\}$; function   **63.** domain: $\left\{\dfrac{3}{4}, 0.65, -12, 25\right\}$; range: $\left\{-\dfrac{1}{2}, 6, 0, 25\right\}$; function   **64.** domain:

$\{2, 4, 6, 8\}$; range: $\{2, 4, 5, 6\}$; not a function   **65.** domain: {triangle, square, rectangle, parallelogram}; range: $\{3, 4\}$; function   **66.** domain:

$(-\infty, \infty)$; range: $(-\infty, -1] \cup [1, \infty)$; not a function   **67.** domain: $\{-3\}$; range: $(-\infty, \infty)$; not a function   **68.** domain: $(-\infty, \infty)$; range: $\{4\}$;

function   **69.** domain; $[-1, 1]$; range; $[-1, 1]$; not a function   **70.** -3 **71.** 0 **72.** 18 **73.** 9 **74.** -3 **75.** 0 **76.** 381 lb   **77.** 5080 lb

78.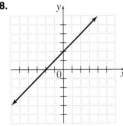

79.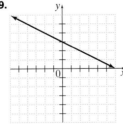

80.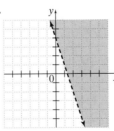

81.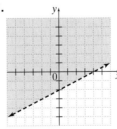

82.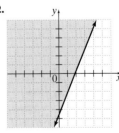

83.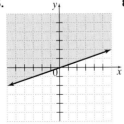

84.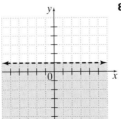

85.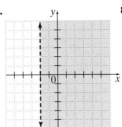

86.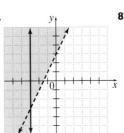

87.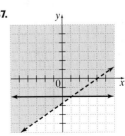

CHAPTER 3 TEST

1.
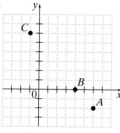

A is in quadrant IV
B is on the *x*-axis, no
  quadrant
C is in quadrant II

2.
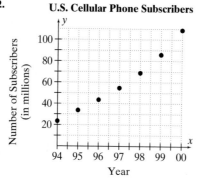

U.S. Cellular Phone Subscribers

3.
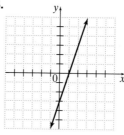

4.
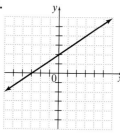

5.
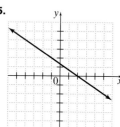

6.
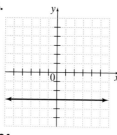

7. $m = -\dfrac{3}{2}$ **8.** $m = -\dfrac{1}{4}; \left(0, \dfrac{2}{3}\right)$ **9.** C   **10.** A   **11.** B   **12.** D

13. $y = -8$ **14.** $x = -4$ **15.** $y = -2$ **16.** $y = -3x + 11$

17. $y = 5x - 2$ **18.** $y = -\dfrac{1}{2}x$ **19.** $y = -\dfrac{1}{3}x + \dfrac{5}{3}$ **20.** $y = -\dfrac{1}{2}x - \dfrac{1}{2}$

21. neither   **22. a.** 82   **b.** 85   **c.** $114.million   **d.** Every million dollars spent on payroll increases winnings by 0.154 game.

23.
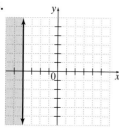

24.
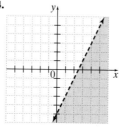

25.
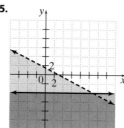

26. domain: $(-\infty, \infty)$; range: $\{5\}$; function

27. domain: $\{-2\}$; range: $(-\infty, \infty)$; not a function   **28.** domain: $(-\infty, \infty)$, range: $[0, \infty)$;

function   **29.** domain: $(-\infty, \infty)$; range: $(-\infty, \infty)$; function

CUMULATIVE REVIEW

1. $\{2, 3, 4, 5\}$; Sec. 1.2, Ex. 4   **2.** $\{101, 102, 103, \dots\}$; Sec. 1.2, Ex. 5   **3.** $-\dfrac{1}{9}$; Sec. 1.3, Ex.21   **4.** $\dfrac{4}{7}$; Sec. 1.3, Ex. 22   **5.** 3; Sec. 1.4, Ex. 1

6. -2; Sec. 1.4, Ex. 4   **7.** 13; Sec. 1.5, Ex. 5   **8.** 2^7; Sec. 1.6, Ex. 1   **9.** y^7; Sec. 1.6, Ex. 3   **10.** $125x^6$; Sec. 1.7, Ex. 5   **11.** $64y^2$; Sec. 1.7, Ex. 8

12. $c = 0.4$; Sec. 2.1, Ex. 4   **13.** 23, 49; Sec. 2.2, Ex. 1   **14.** $11,607.55; Sec. 2.3, Ex. 4   **15.** ; $[2, \infty)$; Sec. 2.4, Ex. 1

16. ; $(0.5, 3]$; Sec. 2.4, Ex. 3 **17.** $(-\infty, 4)$; Sec. 2.5, Ex. 2   **18.** $\{4\}$; Sec. 2.6, Ex. 8

19. $(-\infty, -4) \cup (10, \infty)$; Sec. 2.6, Ex. 11   **20.**
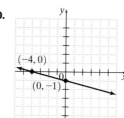

21. $m = 3$; Sec. 3.2, Ex. 3   **22.** $y = \dfrac{1}{4}x - 3$; Sec. 3.3, Ex. 3

23. $y = 3$; Sec. 3.4, Ex. 3   **24.** 5; Sec. 3.5, Ex. 20

25. -2; Sec. 3.5, Ex. 19

Chapter 4

CHAPTER 4 PRETEST

1. $(2, -5)$; 4.1B   **2.** $(-1, -3)$; 4.1B   **3.** $(5, 6)$; 4.1C   **4.** $(0, -4)$; 4.1C   **5.** $(7, 9)$; 4.1D   **6.** $(-5, -8)$; 4.1D   **7.** $(2, -1, -3)$; 4.2A
8. $(1, 0, -2)$; 4.2A   **9.** $(10, -10)$; 4.4A   **10.** $\emptyset$; 4.4A   **11.** $(-3, -1, 2)$; 4.4B   **12.** $(0, 8, 0)$; 4.4B
13. ; 4.5A   **14.** ; 4.5A   **15.** 4 and 12; 4.3A   **16.** $20°, 60°, 100°$; 4.3B

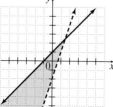

GRAPHING CALCULATOR EXPLORATIONS

1. $(2.11, 0.17)$ **3.** $(-8.21, -6.30)$

MENTAL MATH

1. B   **3.** A

EXERCISE SET 4.1

1. yes   **3.** no   **5.** yes

7. 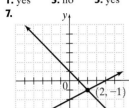   **9.** 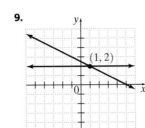   **11.** $\varnothing$ 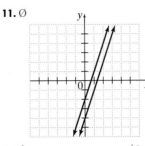   **13.**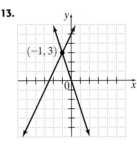

15. $(2, 8)$ **17.** $(0, -9)$ **19.** $(1, -1)$ **21.** $(-5, 3)$ **23.** $\left(-\frac{1}{4}, \frac{1}{2}\right)$ **25.** $(3, 2)$ **27.** $\left(\frac{5}{2}, \frac{5}{4}\right)$ **29.** $(1, -2)$ **31.** $(9, 9)$

33. $(7, 2)$ **35.** $\varnothing$ **37.** $\{(x, y) | 3x + y = 1\}$ **39.** $\varnothing$ **41.** $\varnothing$ **43.** $(8, 2)$ **45.** $(3, 4)$ **47.** $(-2, 1)$ **49.** $\left(\frac{1}{2}, \frac{1}{5}\right)$

51. $\{(x, y) | x = 3y + 2\}$ **53.** true   **55.** false   **57.** $6y - 4z = 25$ **59.** $x + 10y = 2$ **61.** no   **63.** 5000 DVDs; \$21   **65.** supply greater than demand   **67. a.** consumption of red meat decreasing while consumption of poultry increasing   **b.** $(17, 108)$ **c.** In the year 2015, red meat and poultry consumption will each be about 108 lb per person.

EXERCISE SET 4.2

1. $(-2, 5, 1)$ **3.** $(-2, 3, -1)$ **5.** $\{(x, y, z) | x - 2y + z = -5\}$ **7.** $\varnothing$ **9.** $(-10, 6, 2)$ **11.** $(-3, -35, -7)$ **13.** $(6, 22, -20)$ **15.** $\varnothing$
17. $(3, 2, 2)$ **19.** $\{(x, y, z) | x + 2y - 3z = 4\}$ **21.** $(-3, -4, -5)$ **23.** $(12, 6, 4)$ **25.** $(1, 1, -1)$ **27.** 15 and 30   **29.** $\{5\}$
31. $\left\{-\frac{15}{9}\right\}$ **33.** answers may vary   **35.** answers may vary   **37.** $(1, 1, 0, 2)$ **39.** $(1, -1, 2, 3)$

EXERCISE SET 4.3

1. 10 and 8   **3. a.** Enterprise class: 1101 ft; Nimitz class: 1092 ft   **b.** 3.67 football fields   **5.** plane: 520 mph; wind: 40 mph   **7.** 20 qt of 4%;
40 qt of 1%   **9.** United Kingdom: 27,720 students; Spain: 12,292 students   **11.** 9 large frames; 13 small frames   **13.** -10 and -8 **15.** 2005
17. tablets: \$0.80; pens: \$0.20   **19.** speed of plane: 630 mph; speed of wind: 90 mph   **21. a.** answers may vary but notice the slope of each function;   **b.** 2006   **23.** 28 cm; 28 cm; 37 cm   **25.** 600 mi   **27.** $x = 75; y = 105$ **29.** 2 units of mix A; 3 units of mix B; 1 unit of mix C
31. 5 in.; 7 in.; 7 in.; 10 in.   **33.** 18, 13, 9   **35.** 246 free throws; 119 two-point field goals; 85 three-point field goals   **37.** $x = 60; y = 55; z = 65$
39. 625 units   **41.** 3000 units   **43.** 1280 units   **45. a.** $R(x) = 450x$;   **b.** $C(x) = 200x + 6000$;   **c.** 24 desks   **47.** $5x + 5z = 10$

49. $-5y + 2z = 2$ **51.** 1980: 300,000; 2001: 1,400,000   **53.** $a = 3, b = 4, c = -1$ **55.** $a = 8\frac{1}{6}, b = -109\frac{5}{6}, c = 1290$; 1783 students in 2007

INTEGRATED REVIEW

1. C   **2.** D   **3.** A   **4.** B   **5.** $(1, 3)$ **6.** $\left(\frac{4}{3}, \frac{16}{3}\right)$ **7.** $(2, -1)$ **8.** $(5, 2)$ **9.** $\left(\frac{3}{2}, 1\right)$ **10.** $\left(-2, \frac{3}{4}\right)$ **11.** $\varnothing$

12. $\{(x, y) | 2x - 5y = 3\}$ **13.** $(-1, 3, 2)$ **14.** $(1, -3, 0)$ **15.** $\varnothing$ **16.** $\{(x, y, z) | x - y + 3z = 2\}$ **17.** $\left(2, 5, \frac{1}{2}\right)$ **18.** $\left(1, 1, \frac{1}{3}\right)$

19. 19 and 27   **20.** 70°; 70°; 100°; 120°

EXERCISE SET 4.4

1. $(2, -1)$ **3.** $(-4, 2)$ **5.** $\varnothing$ **7.** $\{(x, y) | x - y = 3\}$ **9.** $(4, -3)$ **11.** $(-2, 5, -2)$ **13.** $(1, -2, 3)$ **15.** $(2, 1, -1)$ **17.** $(1, -4, 3)$
19. -13 **21.** -36 **23.** 0   **25. a.** end of 1984   **b.** black-and-white sets; microwave ovens; percent of households owning black-and-white television sets decreasing and percent of households owning microwave ovens increasing; answers may vary   **c.** in 2002

EXERCISE SET 4.5

1.    **3.** 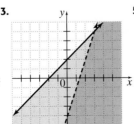   **5.** 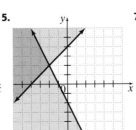   **7.** 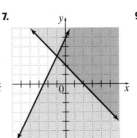   **9.**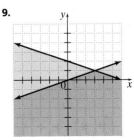

11. 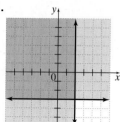   **13.** 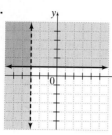   **15.** 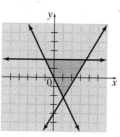   **17.** 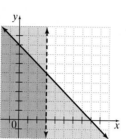   **19.**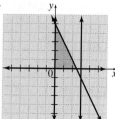

21. C   **23.** D   **25.** 9   **27.** $\frac{4}{9}$ **29.** 5   **31.** 59   **33. a.** $\begin{cases} x + y \leq 8 \\ x < 3 \end{cases}$ **b.**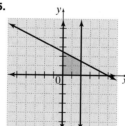

CHAPTER 4 REVIEW

1. $(-3, 1)$ **2.** $\left(0, \frac{2}{3}\right)$ **3.** $\varnothing$ **4.** $\{(x, y) | 3x - 6y = 12\}$ **5.** $\left(3, \frac{8}{3}\right)$ **6.** 1500 backpacks   **7.** $(2, 0, 2)$ **8.** $(2, 0, -3)$ **9.** $\left(-\frac{1}{2}, \frac{3}{4}, 1\right)$
10. $(-1, 2, 0)$ **11.** $\varnothing$ **12.** $(5, 3, 0)$ **13.** $(1, 1, -2)$ **14.** $(3, 1, 1)$ **15.** 10, 40, and 48   **16.** 63 and 21   **17.** 58 mph; 65 mph
18. width: 37 ft; length: 111 ft   **19.** 20 L of 10% solution; 30 L of 60% solution
20. 30 lb of creme-filled; 5 lb of chocolate-covered nuts; 10 lb of chocolate-covered raisins   **21.** 17 pennies; 20 nickels; 16 dimes
22. larger investment: 9.5%; smaller investment: 7.5%   **23.** two sides: 22 cm each; third side: 29 cm   **24.** 120, 115, and 60   **25.** $(-3, 1)$
26. $\{(x, y) | x - 2y = 4\}$ **27.** $\left(-\frac{2}{3}, 3\right)$ **28.** $\left(\frac{1}{3}, \frac{7}{6}\right)$ **29.** $\left(\frac{5}{4}, \frac{5}{8}\right)$ **30.** $(-7, -15)$ **31.** $(1, 3)$ **32.** $(2, 1)$ **33.** $(1, 2, 3)$
34. $(2, 0, -3)$ **35.** $(3, -2, 5)$ **36.** $(-1, 2, 0)$ **37.** $(1, 1, -2)$ **38.** $\varnothing$

39. 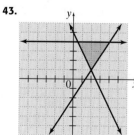   **40.** 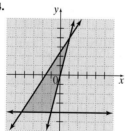   **41.**   **42.**

43.   **44.**   **45.**   **46.**

CHAPTER 4 TEST

1. $(1, 3)$ **2.** $\varnothing$ **3.** $(2, -3)$ **4.** $\{(x, y) | 10x + 4y = 10\}$ **5.** $(-1, -2, 4)$ **6.** $\varnothing$ **7.** $\left(\frac{7}{2}, -10\right)$ **8.** $(2, -1)$ **9.** $(3, 6)$
10. $(3, -1, 2)$ **11.** $(5, 0, -4)$ **12.** $\{(x, y) | x - y = -2\}$ **13.** $(5, -3)$ **14.** $(-1, -1, 0)$ **15.** $\varnothing$ **16.** 53 double rooms; 27 single rooms
17. 5 gal of 10%; 15 gal of 20%   **18.** 275 frames   **19.**   **20.**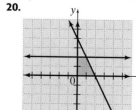

CUMULATIVE REVIEW

1. true; Sec. 1.2, Ex. 7 **2.** false; Sec. 1.2, Ex. 10 **3.** $<$; Sec. 1.3, Ex. 4 **4.** $<$; Sec. 1.3, Ex. 7 **5.** $<$; Sec. 1.3, Ex. 11 **6.** 4; Sec. 1.4; Ex. 3

7. -8; Sec. 1.4, Ex. 5 **8.** -1; Sec. 1.5, Ex. 3 **9.** $\dfrac{1}{3x}$; Sec. 1.6, Ex. 16 **10.** $\dfrac{11}{18}$; Sec. 1.6, Ex. 19 **11.** solution; Sec. 2.1, Ex. 1 **12.** \$2350; Sec.

2.2, Ex. 2 **13.** $b = \dfrac{2A - Bh}{h}$; Sec. 2.3, Ex. 3 **14.** $(-2, \infty)$; Sec. 2.4, Ex. 5 **15.** $\{4, 6\}$, Sec. 2.5, Ex. 1 **16.** $\{0\}$; Sec. 2.6, Ex. 3

17. 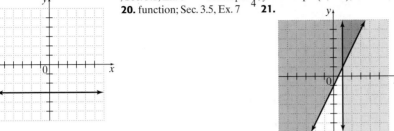 ; Sec. 3.1, Ex. 8 **18.** slope: $\dfrac{3}{4}$; y-intercept: $(0, -1)$; Sec. 3.2, Ex. 4 **19.** $y = -3x - 2$; Sec. 3.4, Ex. 1

20. function; Sec. 3.5, Ex. 7 **21.** ; Sec. 3.6, Ex. 3 **22.** $(0, -5)$; Sec. 4.1, Ex. 10

23. $\varnothing$; Sec. 4.2, Ex. 2

24. 7 and 11; Sec. 4.3, Ex. 1

Chapter 5

CHAPTER 5 PRETEST

1. 6; 5.1A **2.** 3; 5.1E **3.** $3x^2 + 4x + 1$; 5.1C **4.** $-12y^2 + 3y + 6$; 5.1D **5.** $6x^2 + 7x - 5$; 5.2B **6.** $64y^2 + 48y + 9$; 5.2C

7. $4m^2 - 25$; 5.2D **8.** $3t^2 - 2t + \dfrac{5}{2}$; 5.3A **9.** $x^2 - 4x + 1 - \dfrac{7}{2x + 3}$; 5.3B **10.** $2x^2(3x^2 - 6x + 5)$; 5.4A **11.** $(a + 2b)(3c - 4d)$; 5.4B

12. $(x + 9)(x - 7)$; 5.5A **13.** $(6x + 1)(x + 3)$; 5.5B **14.** $(x - 9)(x - 6)$; 5.5C **15.** $(2t - 1)(4t^2 + 2t + 1)$; 5.6C

16. $3x(2x + 3y)(2x - 3y)$; 5.6B **17.** $\left\{ -\dfrac{3}{2}, -5 \right\}$; 5.7A **18.** $\{-5, 5, -1\}$; 5.7A **19.** 8 and 15 or -8 and -15; 5.7B

20. ; 5.8B, C

GRAPHING CALCULATOR EXPLORATIONS

1. $x^3 - 4x^2 + 7x - 8$ **3.** $-2.1x^2 - 3.2x - 1.7$ **5.** $7.69x^2 - 1.26x + 5.3$

MENTAL MATH

1. $10x$ **3.** $5y$ **5.** $-9z$

EXERCISE SET 5.1

1. 0 **3.** 2 **5.** 3 **7.** binomial of degree 1 **9.** trinomial of degree 2 **11.** monomial of degree 3 **13.** degree 3; none of these

15. answers may vary **17.** $6y$ **19.** $11x - 3$ **21.** $xy + 2x - 1$ **23.** $18y^2 - 17$ **25.** $3x^2 - 3xy + 6y^2$ **27.** $x^2 - 4x + 8$

29. $5x^2 + 22x + 16$ **31.** $-3x^2 + 3$ **33.** $2y^4 - 5y^2 + x^2 + 1$ **35.** $4x - 13$ **37.** $-x^3 + 8a - 12$ **39.** $8xy^2 + 2x^3 + 3x^2 - 3$

41. $12x^3y + 8x + 8$ **43.** $4.5x^3 + 0.2x^2 - 3.8x + 9.1$ **45.** $y^2 + 3$ **47.** $-2x^2 + 5x$ **49.** $-2x^2 - 4x + 15$ **51.** $7y^2 - 3$

53. $5x^2 - 9x - 3$ **55.** $x^2 + 12$ **57.** $7x^3 + 4x^2 + 8x - 10$ **59.** $-20y^2 + 3yx$ **61.** $15x^2 + 8x - 6$ **63.** $14ab + 10a^2b - 18a^2 + 12b^2$

65. $\dfrac{1}{3}x^2 - x + 1$ **67.** 57 **69.** 499 **71.** 1 **73.** 202 sq in. **75. a.** 284 ft **b.** 536 ft **c.** 756 ft **d.** 944 ft **e.** answers may vary **f.** 19 sec

77. \$80,000 **79.** \$40,000 **81.** $15x - 10$ **83.** $-2x^2 + 10x - 12$ **85. a.** 22 stations **b.** 3519 stations **c.** 11,391 stations **d.** answers may vary

87. a. 3.1 million SUV's **b.** 5.8 million SUV's **89.** $4x^2 - 3x + 6$ **91.** $2a - 3; -2x - 3; 2x + 2h - 3$ **93.** $12z^{5x} + 13z^{2x} - 2z$

95. $(2x^2 + 7xy + 10y^2)$ units

GRAPHING CALCULATOR EXPLORATIONS

1. $x^2 - 16$ **3.** $9x^2 - 42x + 49$ **5.** $5x^3 - 14x^2 - 13x - 2$

EXERCISE SET 5.2

1. $-12x^5$ **3.** $12x^2 + 21x$ **5.** $-24x^2y - 6xy^2$ **7.** $-4a^3bx - 4a^3by + 12ab$ **9.** $2x^2 - 2x - 12$ **11.** $2x^4 + 3x^3 - 2x^2 + x + 6$

13. $15x^2 - 7x - 2$ **15.** $15m^3 + 16m^2 - m - 2$ **17.** $9x^3 + 30x^2 + 12x - 24$ **19.** $-30a^4b^4 + 36a^3b^2 + 36a^2b^3$

21. $10x^5 + 8x^4 + 2x^3 + 25x^2 + 20x + 5$ **23.** $9x^4 + 12x^3 - 2x^2 - 4x + 1$ **25.** $12x^3 - 2x^2 + 13x + 5$ **27.** answers may vary

29. $x^2 + x - 12$ **31.** $10x^2 + 11xy - 8y^2$ **33.** $3x^2 + 8x - 3$ **35.** $2a^2 - 12a + 16$ **37.** $y^2 - 7y + 12$ **39.** $9x^2 + 18x + 5$

41. $16x^2 - \dfrac{2}{3}x - \dfrac{1}{6}$ **43.** $5x^4 - 17x^2y^2 + 6y^4$ **45.** $x^2 + 8x + 16$ **47.** $36y^2 - 1$ **49.** $9x^2 - 6xy + y^2$ **51.** $49a^2b^2 - 9c^2$

53. $m^2 - 8m + 16$ **55.** $9x^2 + 6x + 1$ **57.** $9b^2 - 36y^2$ **59.** $49x^2 - 9$ **61.** $9x^2 - \dfrac{1}{4}$ **63.** $36x^2 + 12x + 1$ **65.** $x^4 - 4y^2$

67. $16b^2 + 32b + 16$ **69.** $4s^2 - 12s + 8$ **71.** $x^2y^2 - 4xy + 4$ **73.** $2x^3 + 2x^2y + x^2 + xy - x - y$ **75.** $x^4 - 8x^3 + 24x^2 - 32x + 16$

77. $x^4 - 625$ **79.** $2x^2$ **81.** $\dfrac{10a^2b^3}{9}$ **83.** $\dfrac{2m^3}{3}$ **85.** $\pi(25x^2 - 20x + 4)$ sq km   **87.** $(8x^2 - 12x + 4)$ sq in.   **89.** $30x^2y^{2n+1} - 10x^2y^n$

91. $x^{3a} + 5x^{2a} - 3x^a - 15$ **93.** $6x + 12; 9x^2 + 36x + 35$; one operation is addition, the other is multiplication   **95.** $5x^2 + 25x$ **97.** $a^2 - 3a$

Exercise Set 5.3

1. $2a + 4$ **3.** $3ab$ **5.** $2y + \dfrac{3y}{x} - \dfrac{2y}{x^2}$ **7.** $x + 1$ **9.** $2x - 8$ **11.** $x - \dfrac{1}{2}$ **13.** $2x^2 - \dfrac{1}{2}x + 5$ **15.** $2x^2 - 6$

17. $2x^2 + 2x + 8 + \dfrac{28}{x - 4}$ **19.** $5x^2 - 6 - \dfrac{5}{2x - 1}$ **21.** $3x^3 + 5x + 4 - \dfrac{2x}{x^2 - 2}$ **23.** $2x^3 + \dfrac{9}{2}x^2 + 10x + 21 + \dfrac{42}{x - 2}$ **25.** $x + 8$

27. $x - 1$ **29.** $x^2 - 5x - 23 - \dfrac{41}{x - 2}$ **31.** $4x + 8 + \dfrac{7}{x - 2}$ **33.** $2x^3 - 3x^2 + x - 4$ **35.** $3x^2 + 4x - 8 + \dfrac{20}{x + 1}$ **37.** $3x^2 + 3x - 3$

39. $x^2 + x + 1$ **41.** $6x^2 + 23x + 15$ **43.** $(-9, -1)$ **45.** $(-\infty, -8] \cup [1, \infty)$ **47.** $(x^4 + 2x^2 - 6)$ m   **49.** $(3x - 7)$ in.

51. 4; answers may vary   **53.** $x^3 + 2x^2 + 7x + 28$ **55.** answers may vary   **57. a.** answers may vary   **b.** answers may vary

Mental Math

1. 6   **3.** 5   **5.** x **7.** $7x$

Exercise Set 5.4

1. $6(3x - 2)$ **3.** $4y^2(1 - 4xy)$ **5.** $2x^3(3x^2 - 4x + 1)$ **7.** $4ab(2a^2b^2 - ab + 1 + 4b)$ **9.** $(x + 3)(6 + 5a)$ **11.** $(z + 7)(2x + 1)$
13. $(x^2 + 5)(3x - 2)$ **15.** $2\pi r(r + h)$ **17.** $A = P(1 + rt)$ **19.** answers may vary   **21.** $(a + 2)(b + 3)$ **23.** $(a - 2)(c + 4)$
25. $(x - 2)(2y - 3)$ **27.** $(4x - 1)(3y - 2)$ **29.** $(x^2 + 4)(x + 3)$ **31.** $(x^2 - 2)(x - 1)$ **33.** $(2x + 3y)(x + 2)$
35. $(5x - 3)(x + y)$ **37.** $(2x + 3)(3y + 5)$ **39.** $(x + 3)(y - 5)$ **41.** $3b(3ac^2 + 2a^2c - 2a + c)$ **43.** $x^2 - 3x - 10$
45. $x^2 + 5x + 6$ **47.** $y^2 - 4y + 3$ **49.** none   **51.** $f(x) = 2(49x^2 + 257x + 2373)$ **53.** $y^n(3 + 3y^n + 5y^{7n})$ **55.** $3x^{2a}(x^{3a} - 2x^a + 3)$
57. a. $h(t) = -16(t^2 - 14)$ **b.** 160 ft   **c.** answers may vary

Mental Math

1. 5 and 2   **3.** 8 and 3

Exercise Set 5.5

1. $(x + 3)(x + 6)$ **3.** $(x - 8)(x - 4)$ **5.** $(x + 12)(x - 2)$ **7.** $(x - 6)(x + 4)$ **9.** $3(x - 2)(x - 4)$ **11.** $4z(x + 2)(x + 5)$
13. $2(x + 18)(x - 3)$ **15.** $(x - 27)(x + 3)$ **17.** $(x - 18)(x + 3)$ **19.** $3(x - 1)^2$ **21.** $2(x + 3)(x - 2)$ **23.** $(x + 5y)(x + y)$
25. $x(x + 4)(x - 2)$ **27.** $\pm 5, \pm 7$ **29.** $(5x + 1)(x + 3)$ **31.** $(2x - 3)(x - 4)$ **33.** prime polynomial   **35.** $(2x - 3)^2$
37. $2(3x - 5)(2x + 5)$ **39.** $y^2(3y + 5)(y - 2)$ **41.** $2x(3x^2 + 4x + 12)$ **43.** $(x + 7z)(x + z)$ **45.** $(2x + y)(x - 3y)$
47. $(x - 4)(x + 3)$ **49.** $2(7y + 2)(2y + 1)$ **51.** $(2x - 3)(x + 9)$ **53.** $(3x + 1)(x - 2)$ **55.** $(4x - 3)(2x - 5)$
57. $3x^2(2x + 1)(3x + 2)$ **59.** $3(a + 2b)^2$ **61.** $x(3x + 1)(2x - 1)$ **63.** $(4a - 3b)(3a - 5b)$ **65.** $(3x + 5)^2$ **67.** $y(3x - 8)(x - 1)$
69. $(x^2 + 3)(x^2 - 2)$ **71.** $(5x + 8)(5x + 2)$ **73.** $(x^3 - 4)(x^3 - 3)$ **75.** $(a - 3)(a + 8)$ **77.** $(x + 2)(x - 7)$ **79.** $(2x^3 - 3)(x^3 + 3)$
81. $(2x + 13)(x + 3)$ **83.** $(x^2 - 6)(x^2 + 1)$ **85.** $x^2 - 9$ **87.** $4x^2 + 4x + 1$ **89.** $x^3 - 8$ **91. a.** 576 ft; 672 ft; 640 ft; 480 ft
b. answers may vary   **c.** $-16(t + 4)(t - 9)$ **93.** $(x^n + 2)(x^n + 8)$ **95.** $(x^n - 6)(x^n + 3)$ **97.** $(2x^n + 1)(x^n + 5)$ **99.** $(2x^n - 3)^2$
101. $x^2(x + 5)(x + 1)$ **103.** $3x(5x - 1)(2x + 1)$

Exercise Set 5.6

1. $(x + 3)^2$ **3.** $(2x - 3)^2$ **5.** $3(x - 4)^2$ **7.** $x^2(3y + 2)^2$ **9.** $(2a + 3)^2$ **11.** $(x + 5)(x - 5)$ **13.** $(3 + 2z)(3 - 2z)$
15. $(y + 9)(y - 5)$ **17.** $4(4x + 5)(4x - 5)$ **19.** $2y(3x + 1)(3x - 1)$ **21.** $(3x + 7)(3x - 7)$ **23.** $(x^2 + 9)(x + 3)(x - 3)$
25. $(x + 2y + 3)(x + 2y - 3)$ **27.** $(x + 8 + x^2)(x + 8 - x^2)$ **29.** $(x - 5 + y)(x - 5 - y)$ **31.** $(2x + 1 + z)(2x + 1 - z)$
33. $(x + 3)(x^2 - 3x + 9)$ **35.** $(z - 1)(z^2 + z + 1)$ **37.** $(m + n)(m^2 - mn + n^2)$ **39.** $y^2(x - 3)(x^2 + 3x + 9)$
41. $b(a + 2b)(a^2 - 2ab + 4b^2)$ **43.** $(5y - 2x)(25y^2 + 10xy + 4x^2)$ **45.** $(x^2 - y)(x^4 + x^2y + y^2)$ **47.** $(2x + 3y)(4x^2 - 6xy + 9y^2)$

49. $(x - 1)(x^2 + x + 1)$ **51.** $(x + 5)(x^2 - 5x + 25)$ **53.** $3y^2(x^2 + 3)(x^4 - 3x^2 + 9)$ **55.** $\{5\}$ **57.** $\left\{-\dfrac{1}{3}\right\}$ **59.** $\{0\}$

61. $\{5\}$ **63.** $\pi R^2 - \pi r^2 = \pi(R + r)(R - r)$ **65.** $x^3 - y^2x; x(x + y)(x - y)$ **67.** $c = 9$ **69.** $c = 49$
71. a. $(x + 1)(x^2 - x + 1)(x - 1)(x^2 + x + 1)$ **b.** $(x + 1)(x - 1)(x^4 + x^2 + 1)$ **c.** answers may vary   **73.** $(x^n + 6)(x^n - 6)$
75. $(5x^n + 9)(5x^n - 9)$ **77.** $(x^{2n} + 25)(x^n + 5)(x^n - 5)$

Integrated Review

1. $2y^2 + 2y - 11$ **2.** $-2z^4 - 6z^2 + 3z$ **3.** $x^2 - 7x + 7$ **4.** $7x^2 - 4x - 5$ **5.** $25x^2 - 30x + 9$ **6.** $x - 3$ **7.** $2x^3 - 4x^2 + 5x - 5 + \dfrac{8}{x + 2}$

8. $4x^3 - 13x^2 - 5x + 2$ **9.** $(x - 4 + y)(x - 4 - y)$ **10.** $2(3x + 2)(2x - 5)$ **11.** $x(x - 1)(x^2 + x + 1)$ **12.** $2x(2x - 1)$
13. $2xy(7x - 1)$ **14.** $6ab(4b - 1)$ **15.** $4(x + 2)(x - 2)$ **16.** $9(x + 3)(x - 3)$ **17.** $(3x - 11)(x + 1)$ **18.** $(5x + 3)(x - 1)$
19. $4(x + 3)(x - 1)$ **20.** $6(x + 1)(x - 2)$ **21.** $(2x + 9)^2$ **22.** $(5x + 4)^2$ **23.** $(2x + 5y)(4x^2 - 10xy + 25y^2)$
24. $(3x - 4y)(9x^2 + 12xy + 16y^2)$ **25.** $8x^2(2y - 1)(4y^2 + 2y + 1)$ **26.** $27x^2y(xy - 2)(x^2y^2 + 2xy + 4)$
27. $(x + 5 + y)(x^2 + 10x - xy - 5y + y^2 + 25)$ **28.** $(y - 1 + 3x)(y^2 - 2y + 1 - 3xy + 3x + 9x^2)$ **29.** $(5a - 6)^2$ **30.** $(4r + 5)^2$

31. $7x(x - 9)$ **32.** $(4x + 3)(5x + 2)$ **33.** $(a + 7)(b - 6)$ **34.** $20(x - 6)(x - 5)$ **35.** $(x^2 + 1)(x - 1)(x + 1)$ **36.** $5x(3x - 4)$
37. $(5x - 11)(2x + 3)$ **38.** $9m^2n^2(5mn - 3)$ **39.** $5a^3b(b^2 - 10)$ **40.** $x(x + 1)(x^2 - x + 1)$ **41.** prime   **42.** $20(x + y)(x^2 - xy + y^2)$
43. $10x(x - 10)(x - 11)$ **44.** $(3y - 7)^2$ **45.** $a^3b(4b - 3)(16b^2 + 12b + 9)$ **46.** $(y^2 + 4)(y + 2)(y - 2)$ **47.** $2(x - 3)(x^2 + 3x + 9)$
48. $(2s - 1)(r + 5)$ **49.** $(y^4 + 2)(3y - 5)$ **50.** prime   **51.** $100(z + 1)(z^2 - z + 1)$ **52.** $2x(5x - 2)(25x^2 + 10x + 4)$
53. $(2b - 9)^2$ **54.** $(a^4 + 3)(2a - 1)$ **55.** $(y - 4)(y - 5)$ **56.** $(c - 3)(c + 1)$ **57.** $A = 9 - 4x^2 = (3 + 2x)(3 - 2x)$

MENTAL MATH

1. $\{3, -5\}$ **3.** $\{3, -7\}$ **5.** $\{0, 9\}$

EXERCISE SET 5.7

1. $\left\{-3, \dfrac{4}{3}\right\}$ **3.** $\left\{-\dfrac{3}{4}, \dfrac{5}{2}\right\}$ **5.** $\{-3, -8\}$ **7.** $\left\{\dfrac{1}{4}, -\dfrac{2}{3}\right\}$ **9.** $\{1, 9\}$ **11.** $\left\{\dfrac{3}{5}, -1\right\}$ **13.** $\{0\}$ **15.** $\{6, -3\}$ **17.** $\left\{\dfrac{2}{5}, -\dfrac{1}{2}\right\}$
19. $\left\{\dfrac{3}{4}, -\dfrac{1}{2}\right\}$ **21.** $\left\{-2, 7, \dfrac{8}{3}\right\}$ **23.** $\{0, 3, -3\}$ **25.** $\{-1, 1, 2\}$ **27.** answers may vary   **29.** $\left\{-\dfrac{7}{2}, 10\right\}$ **31.** $\{0, 5\}$ **33.** $\{-3, 5\}$
35. $\left\{-\dfrac{1}{2}, \dfrac{1}{3}\right\}$ **37.** $\{-4, 9\}$ **39.** $\left\{\dfrac{4}{5}\right\}$ **41.** $\{-5, 0, 2\}$ **43.** $\left\{-3, 0, \dfrac{4}{5}\right\}$ **45.** $\varnothing$ **47.** $\{1\}$ **49.** -11 and -6, or 6 and 11 **51.** 75 ft
53. 6 ft and 8 ft   **55.** 12 and 14, or -12 and -14 **57.** 10 sec   **59.** width: 7 ft; length: 13 ft   **61.** 2 in.   **63.** 10-in. square   **65.** 9 sec
67. $(-3, 0), (0, 2)$ **69.** $(-4, 0), (4, 0), (0, 2), (0, -2)$ **71.** $\left\{-3, -\dfrac{1}{3}, 2, 5\right\}$ **73.** no; answers may vary   **75.** answers may vary

GRAPHING CALCULATOR EXPLORATIONS

1. $\{-3.562, 0.562\}$ **3.** $\{-0.874, 2.787\}$ **5.** $\{-0.465, 1.910\}$

MENTAL MATH

1. upward   **3.** downward

EXERCISE SET 5.8

1. a. domain: $(-\infty, \infty)$; range: $(-\infty, 5]$ **b.** x-intercepts: $(-2, 0), (6, 0)$; y-intercept: $(0, 5)$ **c.** $(0, 5)$ **d.** no such point   **3. a.** domain: $(-\infty, \infty)$;
range: $[-4, \infty)$ **b.** x-intercepts: $(-3, 0), (1, 0)$; y-intercept: $(0, -3)$ **c.** no such point   **d.** $(-1, -4)$ **5. a.** domain: $(-\infty, \infty)$; range: $(-\infty, \infty)$
b. x-intercepts: $(-2, 0), (0, 0), (2, 0)$; y-intercept: $(0, 0)$ **c.** no such point   **d.** no such point

7.
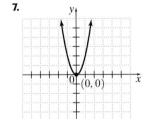

9.
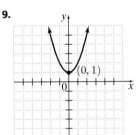

11.
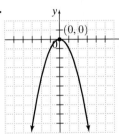

13.
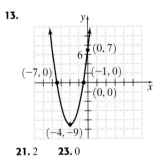

15.
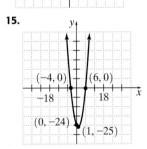

17.
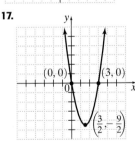

19.
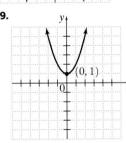

21. 2   **23.** 0

25. no; answers may vary   **27.** $-\dfrac{4}{5}$ **29.** $\dfrac{x^4}{y^5}$ **31.** $\dfrac{m^3}{2n^{10}}$ **33.** C   **35.** B   **37.**
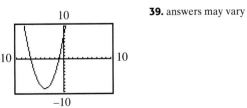
39. answers may vary

CHAPTER 5 REVIEW

1. 5   **2.** 1   **3.** $12x - 6x^2 - 6x^2y$ **4.** $-4xy^3 - 3x^3y$ **5.** $4x^2 + 8y + 6$ **6.** $-4x^2 + 10y^2$ **7.** $8x^2 + 2b - 22$
8. $-4x^3 + 4x^2 + 16xy - 9x + 18$ **9.** $12x^2y - 7xy + 3$ **10.** $x^2 - 6x + 3$ **11.** $x^3 + x - 2xy^2 - y - 7$ **12.** 290   **13.** 58
14. 110   **15.** $x^2 + 4x - 6$ **16.** $-x^2 + 2x + 3$ **17.** $(6x^2y - 12x + 12)$cm   **18.** $-24x^3 + 36x^2 - 6x$ **19.** $-12a^2b^5 - 28a^2b^3 - 4ab^2$
20. $2x^2 + x - 36$ **21.** $9x^2a^2 - 24xab + 16b^2$ **22.** $36x^3 - 11x^2 - 8x - 3$ **23.** $15x^2 + 18xy - 81y^2$ **24.** $x^2 + \dfrac{1}{3}x - \dfrac{2}{9}$

25. $x^4 + 18x^3 + 83x^2 + 18x + 1$ **26.** $2x^3 + 3x^2 - 12x + 5$ **27.** $9x^2 - 6xy + y^2$ **28.** $16x^2 + 72x + 81$ **29.** $x^2 - 9y^2$

30. $-9a^2 + 6ab - b^2 + 16$ **31.** $(9y^2 - 49z^2)$sq units   **32.** $1 + \dfrac{x}{2y} - \dfrac{9}{4xy}$ **33.** $\dfrac{3}{b} + 4b$ **34.** $3x^3 + 9x^2 + 2x + 6 - \dfrac{2}{x-3}$

35. $2x^3 + 6x^2 + 17x + 56 + \dfrac{156}{x-3}$ **36.** $2x^3 + 2x - 2$ **37.** $x^2 + \dfrac{7}{2}x - \dfrac{1}{4} + \dfrac{15}{8\left(x - \frac{1}{2}\right)}$ **38.** $3x^2 + 2x - 1$ **39.** $3x^2 + 6$

40. $3x^2 + 6x + 24 + \dfrac{44}{x-2}$ **41.** $4x^2 - 4x + 2 - \dfrac{5}{x + \frac{3}{2}}$ **42.** $x^4 - x^3 + x^2 - x + 1 - \dfrac{2}{x+1}$ **43.** $x^2 + 3x + 9 - \dfrac{54}{x-3}$

44. $3x^3 + 13x^2 + 51x + 204 + \dfrac{814}{x-4}$ **45.** $3x^3 - 6x^2 + 10x - 20 + \dfrac{50}{x+2}$ **46.** $8x^2(2x - 3)$ **47.** $12y(3 - 2y)$ **48.** $2ab(3b + 4 - 2ab)$

49. $7ab(2ab - 3b + 1)$ **50.** $(a + 3b)(6a - 5)$ **51.** $(x - 2y)(4x - 5)$ **52.** $(x - 6)(y + 3)$ **53.** $(a - 8)(b + 4)$

54. $(p - 5)(q - 3)$ **55.** $(x^2 - 2)(x - 1)$ **56.** $x(2y - x)$ **57.** $(x - 18)(x + 4)$ **58.** $(x - 4)(x + 20)$ **59.** $2(x - 2)(x - 7)$

60. $3(x + 2)(x + 9)$ **61.** $x(2x - 9)(x + 1)$ **62.** $(3x + 8)(x - 2)$ **63.** $(6x + 5)(x + 2)$ **64.** $(15x - 1)(x - 6)$

65. $2(2x - 3)(x + 2)$ **66.** $3(x - 2)(3x + 2)$ **67.** $(x + 6)^2(y - 3)(y + 1)$ **68.** $(x + 7)(x + 9)$ **69.** $(x^2 - 8)(x^2 + 2)$

70. $(x^2 - 2)(x^2 + 10)$ **71.** $(x + 10)(x - 10)$ **72.** $(x - 9)(x + 9)$ **73.** $2(x + 4)(x - 4)$ **74.** $6(x - 3)(x + 3)$

75. $(9 + x^2)(3 + x)(3 - x)$ **76.** $(4 + y^2)(2 - y)(2 + y)$ **77.** $(y + 7)(y - 3)$ **78.** $(x - 7)(x + 1)$ **79.** $(x + 6)(x^2 - 6x + 36)$

80. $(y + 8)(y^2 - 8y + 64)$ **81.** $(2 - 3y)(4 + 6y + 9y^2)$ **82.** $(1 - 4y)(1 + 4y + 16y^2)$ **83.** $6xy(x + 2)(x^2 - 2x + 4)$

84. $2x^2(x + 2y)(x^2 - 2xy + 4y^2)$ **85.** $(x - 1 + y)(x - 1 - y)$ **86.** $(x - 3 - 2y)(x - 3 + 2y)$ **87.** $(2x + 3)^2$ **88.** $(4a - 5b)^2$

89. $\pi h(R + r)(R - r)$ cu units   **90.** $\left\{\dfrac{1}{3}, -7\right\}$ **91.** $\left\{-5, \dfrac{3}{8}\right\}$ **92.** $\left\{0, 4, \dfrac{9}{2}\right\}$ **93.** $\left\{-3, -\dfrac{1}{5}, 4\right\}$ **94.** $\{0, 6\}$ **95.** $\{-3, 0, 3\}$

96. $\left\{-\dfrac{1}{3}, 2\right\}$ **97.** $\{2, 10\}$ **98.** $\{-4, 1\}$ **99.** $\left\{\dfrac{7}{2}, -5\right\}$ **100.** $\{0, 6, -3\}$ **101.** $\{-21, 0, 2\}$ **102.** $\{0, -2, 1\}$ **103.** $\left\{-\dfrac{3}{2}, 0, \dfrac{1}{4}\right\}$

104. $-\dfrac{15}{2}$, or 7 **105.** width: 2 m; length: 8 m   **106.** 5 sec   **107.** domain: $(-\infty, \infty)$; range: $(-\infty, 4]$

108. x-intercepts: $(-4, 0)$, $(2, 0)$; y-intercept: $(0, 3)$ **109.** $(-1, 4)$ **110.** between $x = -4$ and $x = 2$

111.
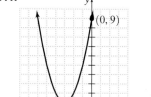

112.
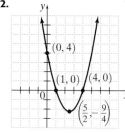

113.
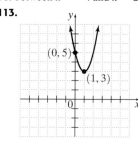

114.
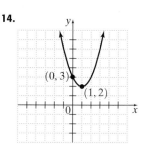

Chapter 5 Test

1. $-5x^3 - 11x - 9$ **2.** $-12x^2y - 3xy^2$ **3.** $12x^2 - 5x - 28$ **4.** $25a^2 - 4b^2$ **5.** $36m^2 + 12mn + n^2$ **6.** $2x^3 - 13x^2 + 14x - 4$

7. $\dfrac{4xy}{3z} + \dfrac{3}{z} + \dfrac{1}{3x}$ **8.** $2x^4 + 2x - 2 + \dfrac{1}{2x - 1}$ **9.** $4x^3 - 15x^2 + 47x - 142 + \dfrac{425}{x+3}$ **10.** $4x^2y(4x - 3y^3)$ **11.** $(x - 15)(x + 2)$

12. $(2y + 5)^2$ **13.** $3(2x + 1)(x - 3)$ **14.** $(2x + 5)(2x - 5)$ **15.** $(x + 4)(x^2 - 4x + 16)$ **16.** $3y(x + 3y)(x - 3y)$ **17.** $6(x^2 + 4)$

18. $(x + 3)(x - 3)(y - 3)$ **19.** $\left\{4, -\dfrac{8}{7}\right\}$ **20.** $\{-3, 8\}$ **21.** $\left\{-\dfrac{5}{2}, -2, 2\right\}$ **22.** $(x + 2y)(x - 2y)$ **23. a.** 960 ft **b.** 953.44 ft

c. $-16(t - 11)(t + 5)$ **d.** 11 sec   **24.**
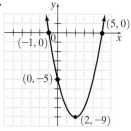

Cumulative Review

1. $-2x - 6$; Sec. 1.5, Ex. 12   **2.** $2x + 23$; Sec. 1.5, Ex. 13   **3.** 6×10^{-5}; Sec. 1.7, Ex. 20   **4.** $\{1\}$; Sec. 2.1, Ex. 6

5. $y = \dfrac{2x + 7}{3}$ or $y = \dfrac{2x}{3} + \dfrac{7}{3}$; Sec. 2.3, Ex. 2   **6.** $[4, \infty)$; Sec. 2.4, Ex. 9   **7.** $\left[-9, -\dfrac{9}{2}\right]$; Sec. 2.5, Ex. 5   **8.** $\left\{\dfrac{3}{4}, 5\right\}$; Sec. 2.6, Ex. 7

9. undefined; Sec. 3.2, Ex. 5   **10.**

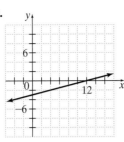

; Sec. 3.3, Ex. 1   **11.** $y = \dfrac{5}{3}x + \dfrac{13}{3}$; Sec. 3.4, Ex. 6

12. domain: $\{2, 0, 3\}$; range: $\{3, 4, -1\}$; Sec. 3.5, Ex. 1

13. -2; Sec. 3.5, Ex. 19   **14.** 5; Sec. 3.5, Ex. 20

15.

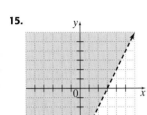

; Sec. 3.6, Ex. 1

16. $\left(-4, \dfrac{1}{2}\right)$; Sec. 4.1, Ex. 6

17. $(-4, 2, -1)$; Sec. 4.2, Ex. 1

18. $(-1, 2)$; Sec. 4.4, Ex. 1

19.

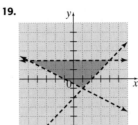

; Sec. 4.5, Ex. 2

20. 4; Sec. 5.1, Ex. 9

21. $10x^2 - 8x$; Sec. 5.2, Ex. 3

22. $-7x^3y^2 - 3x^2y^2 + 11xy$; Sec. 5.2, Ex. 5

23. $2x^2 - x + 4$; Sec. 5.3, Ex. 1

Chapter 6

CHAPTER 6 PRETEST

1. $x = 2$ or $x = 3$; (6.1A)   **2.** $1 - 2x$; (6.1B)   **3.** $\dfrac{x + 2}{x - 4}$; (6.1B)   **4.** $-\dfrac{3}{2}$; (6.1C)   **5.** $\dfrac{2x - 7}{2x - 5}$; (6.1C)   **6.** $\dfrac{2mn^2}{m - 5}$; (6.1D)   **7.** -1; (6.1D)

8. $\dfrac{7 + x}{x - 7}$; (6.2A)   **9.** $\dfrac{20 - 6a}{15a^2}$; (6.2C)   **10.** $\dfrac{3(3y + 5)}{(y - 3)(y + 3)}$; (6.2C)   **11.** $\dfrac{x^2 + 11x + 5}{(x + 6)(x + 1)(2x + 1)}$; (6.2C)   **12.** $\{12\}$; (6.3A)

13. $\dfrac{4y + 5}{16 - 25y}$; (6.3B)   **14.** $\dfrac{y}{xy + x^2}$; (6.3C)   **15.** $\{60\}$; (6.4A)   **16.** $\{-2\}$; (6.4A)   **17.** $\dfrac{2S - na}{n} = L$; (6.5A)   **18.** 2 and 3; (6.5B)

19. $y = \dfrac{15}{2}$; (6.6A)   **20.** $W = 32$; (6.6B)

GRAPHING CALCULATOR EXPLORATIONS

1. $\{x \mid x \text{ is a real number and } x \neq 6\}$ **3.** $\{x \mid x \text{ is a real number and } x \neq -2, x \neq 2\}$ **5.** $\left\{x \mid x \text{ is a real number and } x \neq -4, x \neq \dfrac{1}{2}\right\}$

7. $\{x \mid x \text{ is a real number}\}$

MENTAL MATH

1. $\dfrac{xy}{10}$ **3.** $\dfrac{2y}{3x}$ **5.** $\dfrac{m^2}{36}$

EXERCISE SET 6.1

1. 2   **3.** $-\dfrac{1}{5}$ **5.** 0   **7.** no real number   **9.** $0, -2, 1$ **11.** $2, -2$ **13.** $\dfrac{5x^2}{9}$ **15.** $\dfrac{x^4}{2y^2}$ **17.** $1 - 2x$ **19.** $x + 3$ **21.** $\dfrac{9}{7}$

23. $x - 4$ **25.** -1 **27.** $-(x + 7)$ **29.** $\dfrac{2x + 1}{x - 1}$ **31.** $\dfrac{x^2 + 5x + 25}{2}$ **33.** $\dfrac{x - 2}{2x^2 + 1}$ **35.** $\dfrac{1}{3x + 5}$ **37.** $\dfrac{x}{2}$ **39.** $-\dfrac{4}{5}$ **41.** $-\dfrac{6a}{2a + 1}$

43. $\dfrac{3}{2(x - 1)}$ **45.** $\dfrac{x + 2}{x + 3}$ **47.** $\dfrac{3a}{5(a - b)}$ **49.** $\dfrac{1}{6}$ **51.** $\dfrac{5}{x - 2}$ sq m   **53.** $\dfrac{x}{2}$ **55.** $\dfrac{x}{3}$ **57.** $\dfrac{4a^2}{a - b}$ **59.** $\dfrac{4}{(x + 2)(x + 3)}$ **61.** $\dfrac{1}{2}$

63. -1 **65.** $\dfrac{8(a - 2)}{3(a + 2)}$ **67.** $\dfrac{8}{x^2 y}$ **69.** $\dfrac{(y + 5)(2x - 1)}{(y + 2)(5x + 1)}$ **71.** $\dfrac{10}{3}, -8, -\dfrac{7}{3}$ **73. a.** \$200 million   **b.** \$500 million   **c.** \$300 million

d. $\{x \mid x \text{ is a real number}\}$ **75.** $\dfrac{7}{5}$ **77.** $\dfrac{1}{12}$ **79.** $\dfrac{11}{16}$ **81.** $\dfrac{(x + 2)(x - 1)^2}{x^5}$ ft   **83.** answers may vary   **85. a.** 1   **b.** -1 **c.** neither   **d.** -1

e. -1 **f.** 1   **87.** $(x - 5)(2x + 7)$ **89.** 1   **91.** $x^k - 3$

GRAPHING CALCULATOR EXPLORATIONS

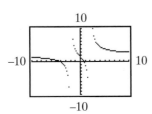

Exercise Set 6.2

1. $-\dfrac{3}{x}$ **3.** $\dfrac{x+2}{x-2}$ **5.** $x-2$ **7.** $\dfrac{-1}{x-2}$ or $\dfrac{1}{2-x}$ **9.** $-\dfrac{5}{x}$ **11.** $\dfrac{4x}{x+5}$ ft   **13.** $35x$ **15.** $x(x+1)$ **17.** $(x+7)(x-7)$

19. $6(x+2)(x-2)$ **21.** $(a+b)(a-b)^2$ **23.** $-4x(x+3)(x-3)$ **25.** answers may vary   **27.** $\dfrac{17}{6x}$ **29.** $\dfrac{35-4y}{14y^2}$

31. $\dfrac{-13x+4}{(x+4)(x-4)}$ **33.** $\dfrac{3}{x+4}$ **35.** $\dfrac{25}{6(x+5)}$ **37.** $\dfrac{-2x-1}{x^2(x-3)}$ or $-\dfrac{2x+1}{x^2(x-3)}$ **39.** 0 **41.** $\dfrac{1-x}{x-2}$ **43.** $\dfrac{y^2+2y+10}{(y+4)(y-4)(y-2)}$

45. $\dfrac{x^2+5x+21}{(x-2)(x+1)(x+3)}$ **47.** $\dfrac{5(x^2+x-4)}{(3x+2)(x+3)(2x-5)}$ **49.** $\dfrac{5a+1}{(a+1)^2(a-1)}$ **51.** $\dfrac{3ab-b^2}{2(a+b)(a-b)}$ **53.** $\dfrac{x+18}{(x+2)^2(x-2)}$

55. $\dfrac{5-2x}{2(x+1)}$ **57.** $\dfrac{2(x^2+x-21)}{(x+3)^2(x-3)}$ **59.** $\dfrac{6x}{(x+3)(x-3)^2}$ **61.** $\dfrac{4}{3}$ **63.** 10 **65.** $4+x^2$ **67.** answers may vary

69. answers may vary   **71.** $\dfrac{2x^2+9x-18}{6x^2}$ **73.** $\dfrac{4a^2}{9(a-1)}$ **75.** 4 **77.** $-\dfrac{4}{x-1}$ **79.** $-\dfrac{32}{x(x+2)(x-2)}$ **81.** $\dfrac{3}{2x}$ **83.** $\dfrac{4-3x}{x^2}$

Exercise Set 6.3

1. 4 **3.** $\dfrac{7}{13}$ **5.** $\dfrac{4}{x}$ **7.** $\dfrac{9(x-2)}{9x^2+4}$ **9.** $2x+y$ **11.** $\dfrac{2(x+1)}{2x-1}$ **13.** $\dfrac{2x+3}{4-9x}$ **15.** $\dfrac{1}{x^2-2x+4}$ **17.** $\dfrac{x}{5x-10}$ **19.** $\dfrac{x-2}{2x-1}$

21. $\dfrac{x}{2-3x}$ **23.** $-\dfrac{y}{x+y}$ **25.** $-\dfrac{2x^3}{y(x-y)}$ **27.** $\dfrac{2x+1}{y}$ **29.** $\dfrac{x-3}{9}$ **31.** $\dfrac{1}{x+2}$ **33.** $\dfrac{xy^2}{x^2+y^2}$ **35.** $\dfrac{2b^2+3a}{b(b-a)}$

37. $\dfrac{x}{(x+1)(x-1)}$ **39.** $\dfrac{1+a}{1-a}$ **41.** $\dfrac{x(x+6y)}{2y}$ **43.** $\dfrac{5a}{2a+4}$ **45.** $5xy^2+2x^2y$ **47.** $\dfrac{xy}{2x+5y}$ **49.** $\left\{-\dfrac{5}{6}\right\}$ **51.** $\{2\}$ **53.** $\{54\}$

55. $\dfrac{770a}{770-s}$ **57.** a, b   **59.** $\dfrac{1+x}{2+x}$ **61.** x^2+x **63.** $\dfrac{x-3y}{x+3y}$

Mental Math

1. equation   **3.** expression   **5.** equation

Exercise Set 6.4

1. $\{72\}$ **3.** $\{3\}$ **5.** $\{-1\}$ **7.** $\{2\}$ **9.** $\{6\}$ **11.** $\{2\}$ **13.** $\left\{-\dfrac{28}{3}\right\}$ **15.** $\{3\}$ **17.** $\{3\}$ **19.** $\varnothing$ **21.** $\{1\}$ **23.** $\left\{\dfrac{1}{3}\right\}$

25. $\{-5,5\}$ **27.** $\{3\}$ **29.** $\varnothing$ **31.** $\left\{\dfrac{4}{3}\right\}$ **33.** $\{-12\}$ **35.** $\left\{1,\dfrac{11}{4}\right\}$ **37.** $\{-5,-1\}$ **39.** $\left\{-\dfrac{7}{5}\right\}$ **41.** 5 **43.** length: 15 in.;

width: 10 in.   **45.** 10% **47.** 25–29 and 30–34   **49.** 5560   **51.** answers may vary   **53.** $\left\{\dfrac{1}{16},\dfrac{1}{3}\right\}$ **55.** $\{-0.17\}$ **57.** 3000 game disks

Integrated Review

1. $\left\{\dfrac{1}{2}\right\}$ **2.** $\{10\}$ **3.** $\dfrac{1+2x}{8}$ **4.** $\dfrac{15+x}{10}$ **5.** $\dfrac{2(x-4)}{(x+2)(x-1)}$ **6.** $-\dfrac{5(x-8)}{(x-2)(x+4)}$ **7.** $\{4\}$ **8.** $\{8\}$ **9.** $\{-5\}$ **10.** $\left\{-\dfrac{2}{3}\right\}$

11. $\dfrac{2x+5}{x(x-3)}$ **12.** $\dfrac{5}{2x}$ **13.** $\{-2\}$ **14.** $-\dfrac{y}{x}$ **15.** $\dfrac{(a+3)(a+1)}{a+2}$ **16.** $\dfrac{-a^2+31a+10}{5(a-6)(a+1)}$ **17.** $\left\{-\dfrac{1}{5}\right\}$ **18.** $\left\{-\dfrac{3}{13}\right\}$ **19.** $\dfrac{4a+1}{(3a+1)(3a-1)}$

20. $-\dfrac{a+8}{4a(a-2)}$ **21.** $\left\{-1,\dfrac{3}{2}\right\}$ **22.** $\dfrac{x^2-3x+10}{2(x+3)(x-3)}$ **23.** $\dfrac{3}{x+1}$ **24.** $\{x\,|\,x$ is a real number and $x\neq 2, x\neq -1\}$ **25.** $\{-1\}$

26. $\dfrac{22z-45}{3z(z-3)}$ **27. a.** $\dfrac{x}{5}-\dfrac{x}{4}+\dfrac{1}{10}$ **b.** Write each rational expression term so that each denominator is the LCD, 20.   **c.** $\dfrac{-x+2}{20}$

28. a. $\dfrac{x}{5}-\dfrac{x}{4}=\dfrac{1}{10}$ **b.** Clear the equation of fractions by multiplying each term by the LCD, 20.   **c.** $\{-2\}$

Exercise Set 6.5

1. $C=\dfrac{5}{9}(F-32)$ **3.** $I=A-QL$ **5.** $R=\dfrac{R_1R_2}{R_1+R_2}$ **7.** $n=\dfrac{2S}{a+L}$ **9.** $b=\dfrac{2A-ah}{h}$ **11.** $T_2=\dfrac{P_2V_2T_1}{P_1V_1}$ **13.** $f_2=\dfrac{f_1f}{f_1-f}$

15. $L=\dfrac{n\lambda}{2}$ **17.** $c=\dfrac{2Lw}{\theta}$ **19.** 1 and 5   **21.** 5   **23.** 4.5 gal   **25.** 3643 women   **27.** 15.6 hr   **29.** 10 min   **31.** 200 mph

33. 15 mph   **35.** -8 and -7 **37.** 36 min   **39.** 45 mph and 60 mph   **41.** 5.9 hr   **43.** 2 hr   **45.** 135 mph   **47.** 12 mi   **49.** $2\dfrac{2}{9}$ hr

51. $\dfrac{7}{8}$ **53.** $1\dfrac{1}{2}$ min   **55.** 63 mph   **57.** 1 hr   **59.** 2 hr   **61.** $\{6\}$ **63.** $\{22\}$ **65.** 60 in. or 5 ft   **67.** 6 ohms

69. $\dfrac{1}{R}=\dfrac{1}{R_1}+\dfrac{1}{R_2}+\dfrac{1}{R_3}$; $R=\dfrac{15}{13}$ ohms   **71.** answers may vary

Exercise Set 6.6

1. $k=\dfrac{1}{5}$; $y=\dfrac{1}{5}x$ **3.** $k=\dfrac{3}{2}$; $y=\dfrac{3}{2}x$ **5.** $k=14$; $y=14x$ **7.** $k=0.25$; $y=0.25x$ **9.** 4.05 lb   **11.** 204,706 tons   **13.** $k=30$; $y=\dfrac{30}{x}$

15. $k=700$; $y=\dfrac{700}{x}$ **17.** $k=2$; $y=\dfrac{2}{x}$ **19.** $k=0.14$; $y=\dfrac{0.14}{x}$ **21.** $R=54$ mph   **23.** 72 amps   **25.** divided by 4   **27.** $x=kyz$

29. $r = kst^3$ **31.** $k = \dfrac{1}{3}$; $y = \dfrac{1}{3}x^3$ **33.** $k = 0.2$; $y = 0.2\sqrt{x}$ **35.** $k = 1.3$; $y = \dfrac{1.3}{x^2}$ **37.** $k = 3$; $y = 3xz^3$ **39.** 22.5 tons   **41.** 15π cu in.

43. 8 ft   **45.** $C = 12\pi$ cm; $A = 36\pi$ sq cm   **47.** $C = 14\pi$ m; $C = 49\pi$ sq m   **49.** 0   **51.** -1 **53.** multiplied by 2   **55.** multiplied by 4

CHAPTER 6 REVIEW

1. none   **2.** none   **3.** 5   **4.** 4   **5.** $0, -8$ **6.** $-4, 4$ **7.** $\dfrac{x^2}{3}$ **8.** 1   **9.** $\dfrac{9m^2p}{5}$ **10.** -1 **11.** $\dfrac{1}{5}$ **12.** 2   **13.** $\dfrac{1}{x-1}$

14. $\dfrac{1}{x-7}$ **15.** $\dfrac{2(x-3)}{x-4}$ **16.** $\dfrac{y-3}{x+2}$ **17. a.** \$119   **b.** \$77   **c.** decrease   **18.** $\dfrac{1}{3x}$ **19.** $\dfrac{2x^3}{z^3}$ **20.** $-\dfrac{3}{2}$ **21.** $\dfrac{2}{5}$ **22.** $\dfrac{a-b}{2a}$ **23.** $\dfrac{1}{6}$

24. $\dfrac{4(x+2)}{5(x-2)}$ **25.** $\dfrac{3x}{16}$ **26.** $\dfrac{12}{5}$ **27.** $\dfrac{3c^2}{14a^2b}$ **28.** $\dfrac{3}{5x}$ **29.** $\dfrac{(x+4)(x+5)}{3}$ **30.** $\dfrac{a-b}{5a}$ **31.** $\dfrac{7(x-4)}{2(x-2)}$ **32.** $\dfrac{3(x+1)}{x-7}$ **33.** $-\dfrac{1}{x}$

34. $\dfrac{5(a-2)}{7}$ **35.** $\dfrac{8}{9a^2}$ **36.** $-\dfrac{x+3}{2(x+2)}$ **37.** $\dfrac{6}{a}$ **38.** 18   **39.** $60x^2y^5$ **40.** $2x(x-2)$ **41.** $5x(x-5)$

42. $10x^3(x-4)(x+7)(x-3)$ **43.** $\dfrac{2}{5}$ **44.** $\dfrac{4+x}{x-4}$ **45.** $\dfrac{2}{x^2}$ **46.** $\dfrac{3}{2(x-2)}$ **47.** $\dfrac{1}{x-2}$ **48.** $\dfrac{13}{3x}$ **49.** $\dfrac{5x^2-3y^2}{15x^4y^3}$ **50.** $\dfrac{x-2}{x-10}$

51. $\dfrac{-x+5}{(x+1)(x-1)}$ **52.** $\dfrac{-7x-6}{5(x-3)(x+3)}$ **53.** $\dfrac{2x^2-5x-4}{x-3}$ **54.** $\dfrac{5a-1}{(a-1)^2(a+1)}$ **55.** $\dfrac{3x^2-7x-4}{(3x-4)(9x^2+12x+16)}$ **56.** $\dfrac{5-2x}{2(x-1)}$

57. $\dfrac{2(7x-20)}{(x+4)^2(x-4)}$ **58.** $\dfrac{11}{x}$ **59.** $\dfrac{2}{3}$ **60.** $\dfrac{4-3x}{8+x}$ **61.** $\dfrac{2}{15-2x}$ **62.** $\dfrac{x^3}{3(x+1)}$ **63.** $\dfrac{y}{2}$ **64.** $\dfrac{(x+2)(x-2)}{15}$ **65.** $\dfrac{5(4x-3)}{2(5x^2-2)}$

66. $\dfrac{y}{x-y}$ **67.** $\dfrac{x(5y+1)}{3y}$ **68.** $\dfrac{x-3}{3}$ **69.** $\dfrac{1+x}{1-x}$ **70.** $\dfrac{2x}{x-2}$ **71.** $\dfrac{x-1}{3x-1}$ **72.** $\dfrac{2(x-1)}{x+6}$ **73.** $-\dfrac{x^2+9}{6x}$ **74.** $\dfrac{5x}{2}$ **75.** $\{6\}$

76. $\{6\}$ **77.** $\{2\}$ **78.** $\{3\}$ **79.** $\left\{\dfrac{3}{2}\right\}$ **80.** $\{-7, 7\}$ **81.** $\left\{\dfrac{5}{3}\right\}$ **82.** $\dfrac{2x+5}{x(x-7)}$ **83.** $\left\{-\dfrac{1}{3}, 2\right\}$ **84.** $\dfrac{-5x-30}{2x(x-3)}$

85. $a = \dfrac{2A}{h} - b$ **86.** $R_2 = \dfrac{RR_1}{R_1 - R}$ **87.** $R = \dfrac{E}{I} - r$ **88.** $r = \dfrac{A-P}{Pt}$ **89.** $A = \dfrac{HL}{k(T_1 - T_2)}$ **90.** $\{1, 2\}$ **91.** 7   **92.** $\dfrac{23}{25}$

93. -10 and -8 **94.** $1\dfrac{23}{37}$ hr   **95.** 12 hr   **96.** 10 hr   **97.** 490 mph   **98.** 6 ohms   **99.** 8 mph   **100.** 45 mph   **101.** 4 mph

102. $63\dfrac{2}{3}$ mph and 45 mph   **103.** 9   **104.** 4   **105.** 3.125 cu ft   **106.** 64π sq in.

CHAPTER 6 TEST

1. 1   **2.** $-1, -3$ **3.** $\dfrac{5x^3}{3}$ **4.** $-\dfrac{7}{8}$ **5.** $\dfrac{x}{x+9}$ **6.** $\dfrac{x+2}{5}$ **7.** $\dfrac{5}{3x}$ **8.** $\dfrac{4a^3b^4}{c^6}$ **9.** $\dfrac{x+2}{2(x+3)}$ **10.** $\dfrac{-4(2x+9)}{5}$ **11.** $\dfrac{3}{x^3}$

12. -1 **13.** $\dfrac{5x-2}{(x-3)(x+2)(x-2)}$ **14.** $\dfrac{-x+30}{6(x-7)}$ **15.** $\dfrac{3}{2}$ **16.** $\dfrac{1}{5}$ **17.** $\dfrac{64}{3}$ **18.** $\dfrac{(x-3)^2}{x-2}$ **19.** $\{7\}$ **20.** $\{2, -2\}$ **21.** $\{8\}$

22. $x = \dfrac{7a^2 + b^2}{4a - b}$ **23.** 5   **24.** $\dfrac{6}{7}$ hr   **25.** 16   **26.** 9   **27.** 256 ft   **28.** 42

CUMULATIVE REVIEW

1. 8; Sec. 1.1, Ex. 3   **2.** -1; Sec. 1.5, Ex. 7   **3.** 1; Sec. 1.5, Ex. 8   **4.** $\varnothing$; Sec. 2.5, Ex. 3   **5.** $\{4, -10\}$; Sec. 2.6, Ex. 5

6. $(1, 9)$, not a solution; $(0, -12)$, solution; $(2, -6)$, solution; Sec. 3.1, Ex. 3   **7.** parallel; Sec. 3.2, Ex. 7   **8.** $y = \dfrac{5}{8}x - \dfrac{5}{2}$; Sec. 3.4, Ex. 2

9. domain: $(-\infty, \infty)$; range: $[0, \infty)$; Sec. 3.5, Ex. 15   **10.** domain: $[-4, 4]$; range: $[-2, 2]$; Sec. 3.5, Ex. 16   **11.** $\left(-\dfrac{21}{10}, \dfrac{3}{10}\right)$; Sec. 4.1, Ex. 7

12. $\left(\dfrac{1}{2}, 0, \dfrac{3}{4}\right)$; Sec. 4.2, Ex. 3   **13.** 52 mph and 47 mph; Sec. 4.3, Ex. 2   **14.** $(1, -1, 3)$; Sec. 4.4, Ex. 3   **15.** $(1, 0)$; Sec. 4.5, Ex. 3

16. 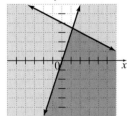   ; Sec. 4.6, Ex. 1   **17.** $-5x^2 - 6x$; Sec. 5.1, Ex. 10   **18.** $8xy - 3x$; Sec. 5.1, Ex. 11

19. $4x^4 + 8x^3 + 39x^2 + 14x + 56$; Sec. 5.2, Ex. 8   **20.** $2x - 5$; Sec. 5.3, Ex. 3   **21.** $xy(-3x^2 + 2x - 5)$ or $-xy(3x^2 - 2x + 5)$; Sec. 5.4, Ex. 5

22. $(x + 2)(x + 8)$; Sec. 5.5, Ex. 1   **23.** $(p^2 + 4)(p + 2)(p - 2)$; Sec. 5.6, Ex. 9   **24.** $\{-2, 6\}$; Sec. 5.7, Ex. 1   **25.** $\dfrac{1}{9xy^2}$; Sec. 6.3, Ex. 1

CHAPTER 7

PRETEST

1. $9x$; (7.1A)   **2.** $-3y^3$; (7.1B)   **3.** x^6; (7.1C)   **4.** 4; (7.1E)   **5.** 12; (7.2A)   **6.** 27; (7.2B)   **7.** $\dfrac{3m^{5/6}}{2}$; (7.2C)   **8.** $\dfrac{27x^6}{y^9}$; (7.2D)

9. $\sqrt{39x}$; (7.3A) **10.** $\dfrac{\sqrt{6}}{7}$; (7.3B) **11.** $-21\sqrt{2}$; (7.4A) **12.** $8x - 10\sqrt{x} - 3$; (7.4B) **13.** $5 - 2\sqrt{5}y + y^2$; (7.4B) **14.** $3b^4\sqrt[3]{5a^2}$; (7.3C)

15. $\dfrac{\sqrt{3x}}{3x}$; (7.5A) **16.** $-4 + 2\sqrt{11}$; (7.5C) **17.** $\{12\}$; (7.6A) **18.** $4 - 8i$; (7.7B) **19.** $-21 - 20i$; (7.7C) **20.** $\dfrac{8}{13} + \dfrac{1}{13}i$; (7.7D)

EXERCISE SET 7.1

1. $2, -2$ **3.** not a real number **5.** $10, -10$ **7.** 10 **9.** $\dfrac{1}{2}$ **11.** 0.01 **13.** -6 **15.** x^5 **17.** $4y^3$ **19.** 2.646 **21.** 6.164

23. 14.142 **25.** 4 **27.** $\dfrac{1}{2}$ **29.** -1 **31.** x^4 **33.** $-3x^3$ **35.** -2 **37.** not a real number **39.** -2 **41.** x^4 **43.** $2x^2$ **45.** $9x^2$

47. $4x^2$ **49.** 8 **51.** -8 **53.** $2|x|$ **55.** x **57.** $|x - 2|$ **59.** $|x + 2|$ **61.** $\sqrt{3}$ **63.** -1 **65.** -3 **67.** $\sqrt{7}$ **69.** $-32x^{15}y^{10}$

71. $-60x^7y^{10}z^5$ **73.** $\dfrac{x^9y^5}{2}$ **75.** answers may vary **77.** 13 **79.** 18 **81.** 1.69 sq m **83.** answers may vary

EXERCISE SET 7.2

1. 7 **3.** 3 **5.** $\dfrac{1}{2}$ **7.** 13 **9.** $2\sqrt[3]{m}$ **11.** $3x^2$ **13.** -3 **15.** -2 **17.** 8 **19.** 16 **21.** not a real number **23.** $\sqrt[5]{(2x)^3}$

25. $\sqrt[3]{(7x + 2)^2}$ **27.** $\dfrac{64}{27}$ **29.** $\dfrac{1}{16}$ **31.** $\dfrac{1}{16}$ **33.** not a real number **35.** $\dfrac{1}{x^{1/4}}$ **37.** $a^{2/3}$ **39.** $\dfrac{5x^{3/4}}{7}$ **41.** answers may vary

43. $a^{7/3}$ **45.** x **47.** $3^{5/8}$ **49.** $y^{1/6}$ **51.** $8u^3$ **53.** $-b$ **55.** $27x^{2/3}$ **57.** $\dfrac{y}{z^{1/6}}$ **59.** $\dfrac{1}{x^{7/4}}$ **61.** $\sqrt{x}$ **63.** $\sqrt[3]{2}$ **65.** $2\sqrt{x}$

67. $\sqrt{x + 3}$ **69.** $\sqrt{xy}$ **71.** $\sqrt[3]{a^2b}$ **73.** $\sqrt[15]{y^{11}}$ **75.** $\sqrt[12]{b^5}$ **77.** $\sqrt[24]{x^{23}}$ **79.** $\sqrt{a}$ **81.** $\sqrt[6]{432}$ **83.** $\sqrt[15]{343y^5}$ **85.** $\sqrt[6]{125r^3s^2}$

87. $25 \cdot 3$ **89.** $16 \cdot 3$ or $4 \cdot 12$ **91.** $8 \cdot 2$ **93.** $27 \cdot 2$ **95.** 1509 calories **97.** 99.5 million **99.** $a^{1/3}$ **101.** $x^{1/5}$ **103.** 1.6818 **105.** $\dfrac{t^{1/2}}{u^{1/2}}$

EXERCISE SET 7.3

1. $\sqrt{14}$ **3.** 2 **5.** $\sqrt[3]{36}$ **7.** $\sqrt{6x}$ **9.** $\sqrt{\dfrac{14}{xy}}$ **11.** $\sqrt[4]{20x^3}$ **13.** $\dfrac{\sqrt{6}}{7}$ **15.** $\dfrac{\sqrt{2}}{7}$ **17.** $\dfrac{\sqrt[4]{x^3}}{2}$ **19.** $\dfrac{\sqrt[3]{4}}{3}$ **21.** $\dfrac{\sqrt[4]{8}}{x^2}$ **23.** $\dfrac{\sqrt[3]{2x}}{3y^4\sqrt[3]{3}}$

25. $\dfrac{x\sqrt{y}}{10}$ **27.** $\dfrac{x\sqrt{5}}{13y}$ **29.** $-\dfrac{z^2\sqrt[3]{z}}{5x}$ **31.** $4\sqrt{2}$ **33.** $4\sqrt[3]{3}$ **35.** $25\sqrt{3}$ **37.** $2\sqrt{6}$ **39.** $10x^2\sqrt{x}$ **41.** $2y^2\sqrt[3]{2y}$ **43.** $a^2b\sqrt[4]{b^3}$

45. $y^2\sqrt{y}$ **47.** $5ab\sqrt{b}$ **49.** $-2x^2\sqrt[5]{y}$ **51.** $x^4\sqrt[3]{50x^2}$ **53.** $-4a^4b^3\sqrt{2b}$ **55.** $3x^3y^4\sqrt{xy}$ **57.** $5r^3s^4$ **59.** $\sqrt{2}$ **61.** 2 **63.** 10

65. x^2y **67.** $24m^2$ **69.** $\dfrac{15x\sqrt{2x}}{2}$ or $\dfrac{15x}{2}\sqrt{2x}$ **71.** $2a^2\sqrt[4]{2}$ **73.** $14x$ **75.** $2x^2 - 7x - 15$ **77.** y^2 **79.** $-3x - 15$

81. $x^2 - 8x + 16$ **83. a.** 20π sq cm **b.** 211.57 sq ft **85. a.** 3.8 times **b.** 2.9 times **c.** answers may vary

MENTAL MATH

1. $6\sqrt{3}$ **3.** $3\sqrt{x}$ **5.** $12\sqrt[3]{x}$ **7.** 3

EXERCISE SET 7.4

1. $-2\sqrt{2}$ **3.** $10x\sqrt{2x}$ **5.** $17\sqrt{2} - 15\sqrt{5}$ **7.** $-\sqrt[3]{2x}$ **9.** $5b\sqrt{b}$ **11.** $\dfrac{31\sqrt{2}}{15}$ **13.** $\dfrac{\sqrt[3]{11}}{3}$ **15.** $\dfrac{5\sqrt{5x}}{9}$ **17.** $14 + \sqrt{3}$ **19.** $7 - 3y$

21. $6\sqrt{3} - 6\sqrt{2}$ **23.** $-23\sqrt[3]{5}$ **25.** $2b\sqrt{b}$ **27.** $20y\sqrt{2y}$ **29.** $2y\sqrt[3]{2x}$ **31.** $6\sqrt[3]{11} - 4\sqrt{11}$ **33.** $4x\sqrt[4]{x^3}$ **35.** $\dfrac{2\sqrt{3}}{3}$ **37.** $\dfrac{5x\sqrt[3]{x}}{7}$

39. $\dfrac{5\sqrt{7}}{2x}$ **41.** $\dfrac{\sqrt[3]{2}}{6}$ **43.** $\dfrac{14x\sqrt[3]{2x}}{9}$ **45.** $15\sqrt{3}$ in. **47.** $\sqrt{35} + \sqrt{21}$ **49.** $7 - 2\sqrt{10}$ **51.** $3\sqrt{x} - x\sqrt{3}$ **53.** $6x - 13\sqrt{x} - 5$

55. $\sqrt[3]{a^2} + \sqrt[3]{a} - 20$ **57.** $6\sqrt{2} - 12$ **59.** $2 + 2x\sqrt{3}$ **61.** $-16 - \sqrt{35}$ **63.** $x - y^2$ **65.** $3 + 2x\sqrt{3} + x^2$

67. $5x - 3\sqrt{10x} - 3\sqrt{15x} + 9\sqrt{6}$ **69.** $2\sqrt[3]{2} - \sqrt[3]{4}$ **71.** $-4\sqrt[6]{x^5} + \sqrt[3]{x^2} + 8\sqrt[3]{x} - 4\sqrt{x} + 7$ **73.** $x + 24 + 10\sqrt{x - 1}$

75. $2x + 6 - 2\sqrt{2x + 5}$ **77.** $x - 7$ **79.** $\dfrac{7}{x + y}$ **81.** $2a - 3$ **83.** $\dfrac{-2 + \sqrt{3}}{3}$ **85.** $22\sqrt{5}$ ft; 150 sq ft

87. a. $2\sqrt{3}$ **b.** 3 **c.** answers may vary

MENTAL MATH

1. $\sqrt{2} - x$ **3.** $5 + \sqrt{a}$ **5.** $7\sqrt{4} - 8\sqrt{x}$

EXERCISE SET 7.5

1. $\dfrac{\sqrt{14}}{7}$ **3.** $\dfrac{\sqrt{5}}{5}$ **5.** $\dfrac{4\sqrt[3]{9}}{3}$ **7.** $\dfrac{3\sqrt{2x}}{4x}$ **9.** $\dfrac{3\sqrt[3]{2x}}{2x}$ **11.** $\dfrac{3\sqrt{3a}}{a}$ **13.** $\dfrac{3\sqrt[3]{4}}{2}$ **15.** $\dfrac{2\sqrt{21}}{7}$ **17.** $\dfrac{\sqrt{10xy}}{5y}$ **19.** $\dfrac{\sqrt[3]{75}}{5}$ **21.** $\dfrac{\sqrt{6x}}{10}$

23. $\dfrac{\sqrt{3z}}{6z}$ **25.** $\dfrac{\sqrt[3]{6xy^2}}{3x}$ **27.** $\dfrac{2\sqrt[3]{9x}}{3x^2}$ **29.** $\dfrac{5a\sqrt[5]{4ab^4}}{2a^2b^3}$ **31.** $-2(2 + \sqrt{7})$ **33.** $\dfrac{7(3 + \sqrt{x})}{9 - x}$ **35.** $-5 + 2\sqrt{6}$ **37.** $\dfrac{2a + 2\sqrt{a} + \sqrt{ab} + \sqrt{b}}{4a - b}$

39. $-\dfrac{8(1 - \sqrt{10})}{9}$ **41.** $\dfrac{x - \sqrt{xy}}{x - y}$ **43.** $\dfrac{5 + 3\sqrt{2}}{7}$ **45.** $\dfrac{5}{\sqrt{15}}$ **47.** $\dfrac{6}{\sqrt{10}}$ **49.** $\dfrac{4x}{7\sqrt{4x}}$ **51.** $\dfrac{5y}{\sqrt[3]{100xy}}$ **53.** $\dfrac{2}{\sqrt{10}}$ **55.** $\dfrac{2x}{11\sqrt{2x}}$

57. $\dfrac{7}{2\sqrt[3]{49}}$ **59.** $\dfrac{3x^2}{10\sqrt[3]{9x}}$ **61.** $\dfrac{6x^2y^3}{\sqrt{6z}}$ **63.** answers may vary **65.** $\dfrac{-7}{12 + 6\sqrt{11}}$ **67.** $\dfrac{3}{10 + 5\sqrt{7}}$ **69.** $\dfrac{x - 9}{x - 3\sqrt{x}}$ **71.** $\dfrac{x - 1}{x - 2\sqrt{x} + 1}$

73. $\{5\}$ **75.** $\left\{-\dfrac{1}{2}, 6\right\}$ **77.** $\{2, 6\}$ **79.** $r = \dfrac{\sqrt{A\pi}}{2\pi}$ **81.** answers may vary

INTEGRATED REVIEW

1. 9   **2.** -2 **3.** $\dfrac{1}{2}$ **4.** x^3 **5.** y^3 **6.** $2y^5$ **7.** $-2y$ **8.** $3b^3$ **9.** 6   **10.** $\sqrt[4]{3y}$ **11.** $\dfrac{1}{16}$ **12.** $\sqrt[5]{(x+1)^3}$ **13.** y **14.** $16x^{1/2}$

15. $x^{5/4}$ **16.** $4^{11/15}$ **17.** $2x^2$ **18.** $\sqrt[4]{a^3b^2}$ **19.** $\sqrt[4]{x^3}$ **20.** $\sqrt[6]{500}$ **21.** $2\sqrt{10}$ **22.** $2xy^2\sqrt[4]{x^3y^2}$ **23.** $3x\sqrt[3]{2x}$ **24.** $-2b^2\sqrt{2}$

25. $\sqrt{5x}$ **26.** $4x$ **27.** $7y^2\sqrt{y}$ **28.** $2a^2\sqrt[4]{3}$ **29.** $2\sqrt{5}-5\sqrt{3}+5\sqrt{7}$ **30.** $y\sqrt[3]{2y}$ **31.** $\sqrt{15}-\sqrt{6}$ **32.** $10+2\sqrt{21}$

33. $4x^2-5$ **34.** $x+2-2\sqrt{x+1}$ **35.** $\dfrac{\sqrt{21}}{3}$ **36.** $\dfrac{5\sqrt[3]{4x}}{2x}$ **37.** $\dfrac{13-3\sqrt{21}}{5}$ **38.** $\dfrac{7}{\sqrt{21}}$ **39.** $\dfrac{3y}{\sqrt[3]{33y^2}}$ **40.** $\dfrac{x-4}{x+2\sqrt{x}}$

CALCULATOR EXPLORATIONS

1. 3.19   **3.** $\varnothing$ **5.** $\{3.23\}$

EXERCISE SET 7.6

1. $\{8\}$ **3.** $\{7\}$ **5.** $\varnothing$ **7.** $\{7\}$ **9.** $\{6\}$ **11.** $\left\{\dfrac{-9}{2}\right\}$ **13.** $\{29\}$ **15.** $\{4\}$ **17.** $\{-4\}$ **19.** $\varnothing$ **21.** $\{7\}$ **23.** $\{9\}$

25. $\{50\}$ **27.** $\varnothing$ **29.** $\left\{\dfrac{15}{4}\right\}$ **31.** $\{7\}$ **33.** $\{5\}$ **35.** $\{-12\}$ **37.** $\{9\}$ **39.** $\{-3\}$ **41.** $\{1\}$ **43.** $\{1\}$ **45.** $\left\{\dfrac{1}{2}\right\}$

47. $\{0,4\}$ **49.** $\left\{\dfrac{37}{4}\right\}$ **51.** $3\sqrt{5}$ ft   **53.** $2\sqrt{10}$ m   **55.** $2\sqrt{131}$ m ≈ 22.9 m   **57.** $\sqrt{100.84}$ mm ≈ 10.0 mm   **59.** 17 ft   **61.** 13 ft

63. 14,657,415 sq mi   **65.** 100 ft   **67.** 100   **69.** $\dfrac{\pi}{2}$ sec ≈ 1.57 sec   **71.** 12.97 ft   **73.** answers may vary   **75.** $15\sqrt{3}$ sq mi ≈ 25.98 sq mi

77. answers may vary   **79.** 0.51 km   **81.** $\dfrac{x}{4x+3}$ **83.** $-\dfrac{4z+2}{3z}$ **85.** $\{1\}$ **87.** 2743 deliveries

MENTAL MATH

1. $9i$ **3.** $i\sqrt{7}$ **5.** -4 **7.** $8i$

EXERCISE SET 7.7

1. $2i\sqrt{6}$ **3.** $-6i$ **5.** $24i\sqrt{7}$ **7.** $-3\sqrt{6}$ **9.** $-\sqrt{14}$ **11.** $-5\sqrt{2}$ **13.** $4i$ **15.** $i\sqrt{3}$ **17.** $2\sqrt{2}$ **19.** $6-4i$ **21.** $2+6i$

23. $-2-4i$ **25.** $2-i$ **27.** $5-10i$ **29.** $8-i$ **31.** -12 **33.** 63   **35.** -40 **37.** $18+12i$ **39.** $27+3i$ **41.** $18+13i$

43. 7   **45.** $12-16i$ **47.** 20   **49.** 2   **51.** $17+144i$ **53.** $-2i$ **55.** $-4i$ **57.** $\dfrac{28}{25}-\dfrac{21}{25}i$ **59.** $-\dfrac{12}{5}+\dfrac{6}{5}i$ **61.** $4+i$

63. $-\dfrac{5}{2}-2i$ **65.** $-5+\dfrac{16}{3}i$ **67.** $\dfrac{3}{5}-\dfrac{1}{5}i$ **69.** $\dfrac{1}{5}-\dfrac{8}{5}i$ **71.** 1   **73.** i **75.** $-i$ **77.** -1 **79.** -64 **81.** $-243i$ **83.** 5 people

85. 14 people   **87.** 16.7%   **89.** $1-i$ **91.** 0   **93.** $2+3i$ **95.** $2+i\sqrt{2}$ **97.** $\dfrac{1}{2}-\dfrac{\sqrt{3}}{2}i$ **99.** answers may vary   **101.** $6-6i$ **103.** yes

CHAPTER 7 REVIEW

1. 9   **2.** 3   **3.** -2 **4.** not a real number   **5.** $-\dfrac{1}{7}$ **6.** x^{32} **7.** -6 **8.** 4   **9.** $-a^2b^3$ **10.** $4a^2b^6$ **11.** $2ab^2$ **12.** $-2x^3y^4$

13. $\dfrac{x^6}{6y}$ **14.** $\dfrac{3y}{z^4}$ **15.** $|x|$ **16.** $|x^2-4|$ **17.** -27 **18.** -5 **19.** $-x$ **20.** $2|2y+z|$ **21.** $5|x-y|$ **22.** y **23.** $|x|$ **24.** $3,6$

25. $2,\sqrt[3]{17}$ **26.** $\dfrac{1}{3}$ **27.** $-\dfrac{1}{3}$ **28.** $-\dfrac{1}{3}$ **29.** $-\dfrac{1}{4}$ **30.** -27 **31.** $\dfrac{1}{4}$ **32.** not a real number   **33.** $\dfrac{343}{125}$ **34.** $\dfrac{9}{4}$ **35.** not a real number

36. $x^{2/3}$ **37.** $5^{1/5}x^{2/5}y^{3/5}$ **38.** $\sqrt[5]{y^4}$ **39.** $5\sqrt[3]{xy^2z^5}$ **40.** $\dfrac{1}{\sqrt{x+2y}}$ **41.** $a^{13/6}$ **42.** $\dfrac{1}{b}$ **43.** $\dfrac{1}{a^{9/2}}$ **44.** $\dfrac{y^2}{x}$ **45.** a^4b^6 **46.** $\dfrac{1}{x^{11/12}}$

47. $\dfrac{b^{5/6}}{49a^{1/4}c^{5/3}}$ **48.** $a-a^2$ **49.** 4.472   **50.** -3.391 **51.** 5.191   **52.** 3.826   **53.** -26.246 **54.** 0.045   **55.** $\sqrt[6]{1372}$ **56.** $\sqrt[12]{81x^3}$

57. $2\sqrt{6}$ **58.** $\sqrt[3]{7x^2yz}$ **59.** $2x$ **60.** ab^3 **61.** $2\sqrt{15}$ **62.** $-5\sqrt{3}$ **63.** $3\sqrt[3]{6}$ **64.** $-2\sqrt[3]{4}$ **65.** $6x^3\sqrt{x}$ **66.** $2ab^2\sqrt[3]{3a^2b}$

67. $\dfrac{p^8\sqrt{p}}{11}$ **68.** $\dfrac{y\sqrt[3]{y^2}}{3x^2}$ **69.** $\dfrac{y\sqrt[4]{xy^2}}{3}$ **70.** $\dfrac{x\sqrt{2x}}{7y^2}$ **71. a.** $\dfrac{5}{\sqrt{\pi}}m$ **b.** 5.75 in.   **72.** $-2\sqrt{5}$ **73.** $2x\sqrt{3x}$ **74.** $9\sqrt[3]{2}$ **75.** $3a\sqrt[4]{2a}$

76. $\dfrac{15+2\sqrt{3}}{6}$ **77.** $\dfrac{3\sqrt{2}}{4x}$ **78.** $17\sqrt{2}-15\sqrt{5}$ **79.** $-4ab\sqrt[4]{2b}$ **80.** 6   **81.** $x-6\sqrt{x}+9$ **82.** $-8\sqrt{5}$ **83.** $4x-9y$ **84.** $a-9$

85. $\sqrt[3]{a^2}+4\sqrt[3]{a}+4$ **86.** $\sqrt[3]{25x^2}-81$ **87.** $a+64$ **88.** $\dfrac{3\sqrt{7}}{7}$ **89.** $\dfrac{\sqrt{3x}}{6}$ **90.** $\dfrac{5\sqrt[3]{2}}{2}$ **91.** $\dfrac{2x^2\sqrt{2x}}{y}$ **92.** $\dfrac{x^2y^2\sqrt[3]{15yz}}{z}$

93. $\dfrac{3\sqrt[4]{2x^2}}{2x^3}$ **94.** $\dfrac{3\sqrt{y}+6}{y-4}$ **95.** $-5+2\sqrt{6}$ **96.** $\dfrac{11}{3\sqrt{11}}$ **97.** $\dfrac{6}{\sqrt{2y}}$ **98.** $\dfrac{3}{7\sqrt[3]{3}}$ **99.** $\dfrac{4x^3}{y\sqrt{2x}}$ **100.** $\dfrac{xy}{\sqrt[3]{10x^2yz}}$ **101.** $\dfrac{x-25}{-3\sqrt{x}+15}$

102. $\{32\}$ **103.** $\varnothing$ **104.** $\{35\}$ **105.** $\varnothing$ **106.** $\{9\}$ **107.** $\{16\}$ **108.** $3\sqrt{2}$ cm   **109.** $\sqrt{241}$ ft   **110.** 51.2 ft   **111.** 4.24 ft

112. $2i\sqrt{2}$ **113.** $-i\sqrt{6}$ **114.** $6i$ **115.** $-\sqrt{10}$ **116.** $15-4i$ **117.** $-13-3i$ **118.** -64 **119.** $10+4i$ **120.** $-12-18i$

121. $1+5i$ **122.** $-5-12i$ **123.** 87   **124.** $\dfrac{3}{2}-i$ **125.** $-\dfrac{1}{3}+\dfrac{1}{3}i$

CHAPTER 7 TEST

1. $6\sqrt{6}$ **2.** $-x^{16}$ **3.** $\dfrac{1}{5}$ **4.** 5   **5.** $\dfrac{4x^2}{9}$ **6.** $-a^6b^3$ **7.** $\dfrac{8a^{1/3}c^{2/3}}{b^{5/12}}$ **8.** $a^{7/12}-a^{7/3}$ **9.** $|4xy|$ or $4|xy|$ **10.** -27 **11.** $\dfrac{3\sqrt{y}}{y}$

12. $\dfrac{8 - 6\sqrt{x} + x}{8 - 2x}$ **13.** $\dfrac{2\sqrt[3]{3x^2}}{3x}$ **14.** $\dfrac{6 - x^2}{8(\sqrt{6} - x)}$ **15.** $-x\sqrt{5x}$ **16.** $4\sqrt{3} - \sqrt{6}$ **17.** $x + 2\sqrt{x} + 1$ **18.** $\sqrt{6} - 4\sqrt{3} + \sqrt{2} - 4$

19. -20 **20.** 23.685 **21.** 0.019 **22.** $\{2, 3\}$ **23.** $\varnothing$ **24.** $\{6\}$ **25.** $i\sqrt{2}$ **26.** $-2i\sqrt{2}$ **27.** $-3i$ **28.** 40 **29.** $7 + 24i$

30. $-\dfrac{3}{2} + \dfrac{5}{2}i$ **31.** $x = \dfrac{5\sqrt{2}}{2}$ in.   **32.** $\sqrt{2}, 5$ **33.** 27 mph   **34.** 360 ft

CUMULATIVE REVIEW

1. $4x^b$; Sec. 1.7, Ex. 16   **2.** y^{5a+6}; Sec. 1.7, Ex. 17   **3.** $\{-7, 5\}$; Sec. 2.6, Ex. 12   **4.** $y = -2x + 12$; Sec. 3.4, Ex. 5   **5.** $(-2, 2)$; Sec. 4.1, Ex. 8

6. $\{(x, y, z)|x - 5y - 2z = 6\}$; Sec. 4.2, Ex. 4   **7.** no solution; Sec. 4.4, Ex. 2   **8.** $13x^3y - xy^3 + 7$; Sec. 5.1, Ex. 14   **9.** -4; Sec. 5.1, Ex. 18

10. 11; Sec. 5.1, Ex. 19   **11.** $x^2 + 10x + 25$; Sec. 5.2, Ex. 12   **12.** $16m^4 - 24m^2n + 9n^2$; Sec. 5.2, Ex. 15   **13.** $2x^2 + 5x + 2 + \dfrac{7}{x - 3}$; Sec. 5.3, Ex. 7

14. $(b - 6)(a + 2)$; Sec. 5.4, Ex. 8   **15.** $2(n^2 - 19n + 40)$; Sec. 5.5, Ex. 4   **16.** $(4x + 3y)^2$; Sec. 5.5, Ex. 8   **17.** $2(5 + 2y)(5 - 2y)$; Sec. 5.6, Ex. 7

18. $\left\{-\dfrac{1}{2}, 4\right\}$; Sec. 5.7, Ex. 3   **19.** 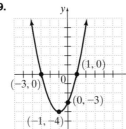 ; Sec. 5.8, Ex. 4   **20.** $\dfrac{3y^4}{x}$; Sec. 6.1, Ex. 4   **21.** $\dfrac{1}{5x - 1}$; Sec. 6.1, Ex. 5

22. $\dfrac{5 + x}{7}$; Sec. 6.2, Ex. 1

23. $\{-2\}$; Sec. 6.4, Ex. 2

24. $\dfrac{1}{8}$; Sec. 7.2, Ex. 12

25. $\dfrac{1}{9}$; Sec. 7.2, Ex. 13

Chapter 8

CHAPTER 8 PRETEST

1. $\{-3\sqrt{6}, 3\sqrt{6}\}$; (8.1A)   **2.** $\left\{\dfrac{-2 - 2\sqrt{3}}{3}, \dfrac{-2 + 2\sqrt{3}}{3}\right\}$; (8.1A)   **3.** $\{-2 - \sqrt{14}, -2 + \sqrt{14}\}$; (8.1C)

4. $\left\{\dfrac{-9 - 2\sqrt{21}}{3}, \dfrac{-9 + 2\sqrt{21}}{3}\right\}$; (8.1C)   **5.** $\left\{\dfrac{1 - i\sqrt{31}}{4}, \dfrac{1 + i\sqrt{31}}{4}\right\}$; (8.2A)   **6.** $\left\{\dfrac{-1 - \sqrt{17}}{12}, \dfrac{-1 + \sqrt{17}}{12}\right\}$; (8.2A)

7. $\{4, -2 - 2i\sqrt{3}, -2 + 2i\sqrt{3}\}$; (8.3A)   **8.** $\{64, 1\}$; (8.3A)   **9.** $(-\infty, -5] \cup [6, \infty)$; (8.4A)   **10.** $(-1, 0)$; (8.4A)   **11.** $(-\infty, -3)$; (8.4B)

12. 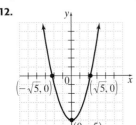   **13.** 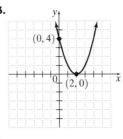   **14.** 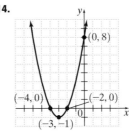   **15.**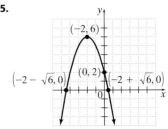

16. length: 20 cm; width: 5 cm; (8.2C)

CALCULATOR EXPLORATIONS

1. $\{-1.27, 6.27\}$ **3.** $\{-1.10, 0.90\}$ **5.** $\varnothing$

EXERCISE SET 8.1

1. $\{-4, 4\}$ **3.** $\{-\sqrt{7}, \sqrt{7}\}$ **5.** $\{-3\sqrt{2}, 3\sqrt{2}\}$ **7.** $\{-\sqrt{10}, \sqrt{10}\}$ **9.** $\{-8, -2\}$ **11.** $\{6 - 3\sqrt{2}, 6 + 3\sqrt{2}\}$

13. $\left\{\dfrac{3 - 2\sqrt{2}}{2}, \dfrac{3 + 2\sqrt{2}}{2}\right\}$ **15.** $\{-3i, 3i\}$ **17.** $\{-\sqrt{6}, \sqrt{6}\}$ **19.** $\{-2i\sqrt{2}, 2i\sqrt{2}\}$ **21.** $\{1 - 4i, 1 + 4i\}$ **23.** $\{-7 - \sqrt{5}, -7 + \sqrt{5}\}$

25. $\{-3 - 2i\sqrt{2}, -3 + 2i\sqrt{2}\}$ **27.** $x^2 + 16x + 64 = (x + 8)^2$ **29.** $z^2 - 12z + 36 = (z - 6)^2$ **31.** $p^2 + 9p + \dfrac{81}{4} = \left(p + \dfrac{9}{2}\right)^2$

33. $r^2 - 3r + \dfrac{9}{4} = \left(r - \dfrac{3}{2}\right)^2$ **35.** $\{-5, -3\}$ **37.** $\{-3 - \sqrt{7}, -3 + \sqrt{7}\}$ **39.** $\left\{\dfrac{-1 - \sqrt{5}}{2}, \dfrac{-1 + \sqrt{5}}{2}\right\}$ **41.** $\{-1 - \sqrt{6}, -1 + \sqrt{6}\}$

43. $\left\{\dfrac{6 - \sqrt{30}}{3}, \dfrac{6 + \sqrt{30}}{3}\right\}$ **45.** $\left\{-4, \dfrac{1}{2}\right\}$ **47.** $\{-4 - \sqrt{15}, -4 + \sqrt{15}\}$ **49.** $\left\{\dfrac{-3 - \sqrt{21}}{3}, \dfrac{-3 + \sqrt{21}}{3}\right\}$ **51.** $\{-1 - i, -1 + i\}$

53. $\{3 - \sqrt{6}, 3 + \sqrt{6}\}$ **55.** $\{-2 - i\sqrt{2}, -2 + i\sqrt{2}\}$ **57.** $\left\{\dfrac{1 - i\sqrt{47}}{4}, \dfrac{1 + i\sqrt{47}}{4}\right\}$ **59.** $\{-5 - i\sqrt{3}, -5 + i\sqrt{3}\}$ **61.** $\{-4, 1\}$

63. $\left\{\dfrac{2 - i\sqrt{2}}{2}, \dfrac{2 + i\sqrt{2}}{2}\right\}$ **65.** $\left\{\dfrac{-3 - \sqrt{69}}{6}, \dfrac{-3 + \sqrt{69}}{6}\right\}$ **67.** 20% **69.** 11% **71.** answers may vary   **73.** simple; answers may vary

75. $5 - 10\sqrt{3}$ **77.** $\dfrac{3 - 2\sqrt{7}}{4}$ **79.** $2\sqrt{7}$ **81.** $\sqrt{13}$ **83.** $-6y, 6y$ **85.** 8.11 sec   **87.** 6.73 sec   **89.** 6 in.   **91.** $\dfrac{27\sqrt{2}}{2}$ in.

93. 2.828 thousand units   **95.** answers may vary

MENTAL MATH

1. $a = 1, b = 3, c = 1$ **3.** $a = 7, b = 0, c = -4$ **5.** $a = 6, b = -1, c = 0$

EXERCISE SET 8.2

1. $\{-6, 1\}$ **3.** $\left\{-\dfrac{3}{5}, 1\right\}$ **5.** $\{3\}$ **7.** $\left\{\dfrac{-7 - \sqrt{33}}{2}, \dfrac{-7 + \sqrt{33}}{2}\right\}$ **9.** $\left\{\dfrac{1 - \sqrt{57}}{8}, \dfrac{1 + \sqrt{57}}{8}\right\}$ **11.** $\left\{\dfrac{7 - \sqrt{85}}{6}, \dfrac{7 + \sqrt{85}}{6}\right\}$

13. $\{1 - \sqrt{3}, 1 + \sqrt{3}\}$ **15.** $\left\{-\dfrac{3}{2}, 1\right\}$ **17.** $\left\{\dfrac{3 - \sqrt{11}}{2}, \dfrac{3 + \sqrt{11}}{2}\right\}$ **19.** $\left\{\dfrac{-5 - i\sqrt{5}}{10}, \dfrac{-5 + i\sqrt{5}}{10}\right\}$ **21.** $\left\{\dfrac{-5 - \sqrt{17}}{2}, \dfrac{-5 + \sqrt{17}}{2}\right\}$

23. $\left\{\dfrac{5}{2}, 1\right\}$ **25.** $\left\{\dfrac{3 - \sqrt{29}}{2}, \dfrac{3 + \sqrt{29}}{2}\right\}$ **27.** $\left\{\dfrac{-1 - \sqrt{19}}{6}, \dfrac{-1 + \sqrt{19}}{6}\right\}$ **29.** answers may vary   **31.** $\left\{\dfrac{3 - i\sqrt{87}}{8}, \dfrac{3 + i\sqrt{87}}{8}\right\}$

33. $\{-2 - \sqrt{11}, -2 + \sqrt{11}\}$ **35.** $\{-3 - 2i, -3 + 2i\}$ **37.** $\left\{\dfrac{-1 - i\sqrt{23}}{4}, \dfrac{-1 + i\sqrt{23}}{4}\right\}$ **39.** $\{1\}$ **41.** $\{3 + \sqrt{5}, 3 - \sqrt{5}\}$

43. two real solutions   **45.** one real solution   **47.** two real solutions   **49.** two complex but not real solutions   **51.** 14 ft
53. $2 + 2\sqrt{2}$ cm, $2 + 2\sqrt{2}$ cm, $4 + 2\sqrt{2}$ cm   **55.** width: $-5 + 5\sqrt{17}$ ft; length: $5 + 5\sqrt{17}$ ft   **57. a.** $50\sqrt{2}$ m **b.** 5000 sq m

59. 37.4 ft by 38.5 ft   **61.** $\dfrac{1 + \sqrt{5}}{2}$ **63.** 8.9 sec   **65.** 2.8 sec   **67.** $\left\{\dfrac{11}{5}\right\}$ **69.** $\{15\}$ **71.** $(x^2 + 5)(x + 2)(x - 2)$

73. $(z + 3)(z - 3)(z + 2)(z - 2)$ **75.** $\{0.6, 2.4\}$ **77.** Sunday to Monday   **79.** Wednesday   **81.** $f(4) = 33$; answers may vary
85. a. \$2922 million **b.** 2003   **87. a.** 432.378 micrograms **b.** 54 lb

EXERCISE SET 8.3

1. $\{2\}$ **3.** $\{16\}$ **5.** $\{1, 4\}$ **7.** $\{3 - \sqrt{7}, 3 + \sqrt{7}\}$ **9.** $\left\{\dfrac{3 - \sqrt{57}}{4}, \dfrac{3 + \sqrt{57}}{4}\right\}$ **11.** $\left\{\dfrac{1 - \sqrt{29}}{2}, \dfrac{1 + \sqrt{29}}{2}\right\}$ **13.** $\{-2, 2, -2i, 2i\}$

15. $\left\{-\dfrac{1}{2}, \dfrac{1}{2}, -i\sqrt{3}, i\sqrt{3}\right\}$ **17.** $\{-3, 3, -2, 2\}$ **19.** $\{125, -8\}$ **21.** $\left\{-\dfrac{4}{5}, 0\right\}$ **23.** $\left\{-\dfrac{1}{8}, 27\right\}$ **25.** $\left\{-\dfrac{2}{3}, \dfrac{4}{3}\right\}$ **27.** $\left\{-\dfrac{1}{125}, \dfrac{1}{8}\right\}$

29. $\{-\sqrt{2}, \sqrt{2}, -\sqrt{3}, \sqrt{3}\}$ **31.** $\left\{\dfrac{-9 - \sqrt{201}}{6}, \dfrac{-9 + \sqrt{201}}{6}\right\}$ **33.** $\{2, 3\}$ **35.** $\{3\}$ **37.** $\{27, 125\}$ **39.** $\{1, -3i, 3i\}$ **41.** $\left\{\dfrac{1}{8}, -8\right\}$

43. $\left\{-\dfrac{1}{2}, \dfrac{1}{3}\right\}$ **45.** $\{4\}$ **47.** $\{-3\}$ **49.** $\{-\sqrt{5}, \sqrt{5}, -2i, 2i\}$ **51.** $\left\{-3, \dfrac{3 - 3i\sqrt{3}}{2}, \dfrac{3 + 3i\sqrt{3}}{2}\right\}$ **53.** $\{6, 12\}$

55. $\left\{-\dfrac{1}{3}, \dfrac{1}{3}, \dfrac{-i\sqrt{6}}{3}, \dfrac{i\sqrt{6}}{3}\right\}$ **57.** 5 mph, then 4 mph   **59.** inlet pipe: 15.5 hr; hose: 16.5 hr   **61.** 55 mph; 66 mph   **63.** 8.5 hr   **65.** 12 or -8
67. a. $x - 6$ **b.** $300 = (x - 6) \cdot (x - 6) \cdot 3$ **c.** 16 cm by 16 cm   **69.** 22 ft   **71.** $(-\infty, 3]$ **73.** $(-5, \infty)$ **75.** answers may vary
77. a. 301.103 ft per sec **b.** 307.429 ft per sec **c.** 209.611 mph

INTEGRATED REVIEW

1. $\{-\sqrt{10}, \sqrt{10}\}$ **2.** $\{-\sqrt{14}, \sqrt{14}\}$ **3.** $\{1 - 2\sqrt{2}, 1 + 2\sqrt{2}\}$ **4.** $\{-5 - 2\sqrt{3}, -5 + 2\sqrt{3}\}$ **5.** $\{-1 - \sqrt{13}, -1 + \sqrt{13}\}$

6. $\{1, 11\}$ **7.** $\left\{\dfrac{-3 - \sqrt{69}}{6}, \dfrac{-3 + \sqrt{69}}{6}\right\}$ **8.** $\left\{\dfrac{-2 - \sqrt{5}}{4}, \dfrac{-2 + \sqrt{5}}{4}\right\}$ **9.** $\left\{\dfrac{2 - \sqrt{2}}{2}, \dfrac{2 + \sqrt{2}}{2}\right\}$ **10.** $\{-3 - \sqrt{5}, -3 + \sqrt{5}\}$

11. $\{-2 + i\sqrt{3}, -2 - i\sqrt{3}\}$ **12.** $\left\{\dfrac{-1 - i\sqrt{11}}{2}, \dfrac{-1 + i\sqrt{11}}{2}\right\}$ **13.** $\left\{\dfrac{-3 + i\sqrt{15}}{2}, \dfrac{-3 - i\sqrt{15}}{2}\right\}$ **14.** $\{3i, -3i\}$ **15.** $\{0, -17\}$

16. $\left\{\dfrac{1 + \sqrt{13}}{4}, \dfrac{1 - \sqrt{13}}{4}\right\}$ **17.** $\{2 + 3\sqrt{3}, 2 - 3\sqrt{3}\}$ **18.** $\{2 + \sqrt{3}, 2 - \sqrt{3}\}$ **19.** $\left\{-2, \dfrac{4}{3}\right\}$ **20.** $\left\{\dfrac{-5 + \sqrt{17}}{4}, \dfrac{-5 - \sqrt{17}}{4}\right\}$

21. $\{1 - \sqrt{6}, 1 + \sqrt{6}\}$ **22.** $\{-\sqrt{31}, \sqrt{31}\}$ **23.** $\{-\sqrt{11}, \sqrt{11}\}$ **24.** $\{-i\sqrt{11}, i\sqrt{11}\}$ **25.** $\{-11, 6\}$ **26.** $\left\{\dfrac{-3 + \sqrt{19}}{5}, \dfrac{-3 - \sqrt{19}}{5}\right\}$

27. $\left\{\dfrac{-3 + \sqrt{17}}{4}, \dfrac{-3 - \sqrt{17}}{4}\right\}$ **28.** $10\sqrt{2}$ ft ≈ 14.1 ft   **29.** Diane: 9.1 hr; Lucy: 7.1 hr   **30.** 5 mph during the first part, then 6 mph

EXERCISE SET 8.4

1. $(-\infty, -5) \cup (-1, \infty)$ **3.** $[-4, 3]$ **5.** $[2, 5]$ **7.** $\left(-5, -\dfrac{1}{3}\right)$ **9.** $(2, 4) \cup (6, \infty)$ **11.** $(-\infty, -4] \cup [0, 1]$

13. $(-\infty, -3) \cup (-2, 2) \cup (3, \infty)$ **15.** $(-\infty, -7) \cup (8, \infty)$ **17.** $\left[-\dfrac{4}{3}, \dfrac{1}{2}\right]$ **19.** $(-\infty, 0) \cup (1, \infty)$ **21.** $(-\infty, -4] \cup [4, 6]$

23. $(-7, 2)$ **25.** $(-1, \infty)$ **27.** $(-\infty, -1] \cup (4, \infty)$ **29.** $(-\infty, 3)$ **31.** $(0, 10)$ **33.** $(-\infty, -4) \cup [5, \infty)$
35. $(-\infty, -6] \cup (-1, 0] \cup (7, \infty)$ **37.** $(-\infty, 1) \cup (2, \infty)$ **39.** $(-\infty, -8] \cup (-4, \infty)$ **41.** $0, 1, 1, 4, 4$ **43.** $0, -1, -1, -4, -4$
45. answers may vary   **47.** $(-\infty, -1) \cup (0, 1)$ **49.** when x is between 2 and 11   **51.**

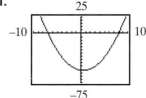

GRAPHING CALCULATOR EXPLORATIONS

1. 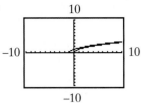   **3.** 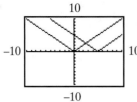   **5.**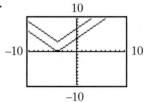

MENTAL MATH

1. $(0, 0)$ **3.** $(2, 0)$ **5.** $(0, 3)$ **7.** $(-1, 5)$

EXERCISE SET 8.5

1. 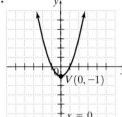   **3.** 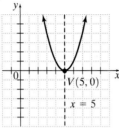   **5.** 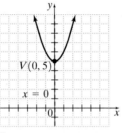   **7.**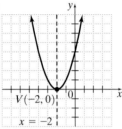

9. 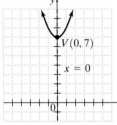   **11.** 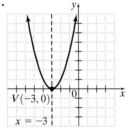   **13.** 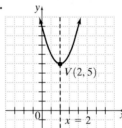   **15.**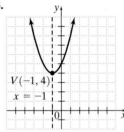

17. 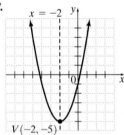   **19.** 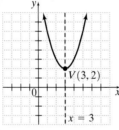   **21.** 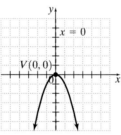   **23.**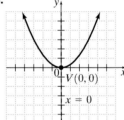

25. 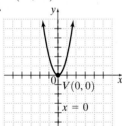   **27.** 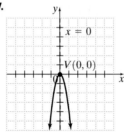   **29.** 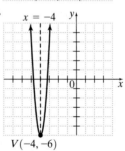   **31.**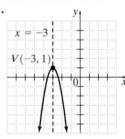

33. 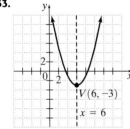   **35.** 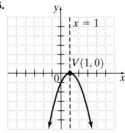   **37.** 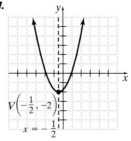   **39.**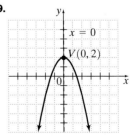

41. $x^2 + 8x + 16$ **43.** $z^2 - 16z + 64$ **45.** $y^2 + y + \dfrac{1}{4}$ **47.** $5(x-2)^2 + 3$ **49.** $5(x+3)^2 + 6$

51.
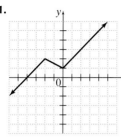

53.
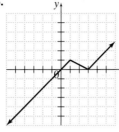

55.
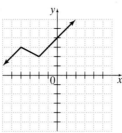

EXERCISE SET 8.6

1. $(-4, -9)$ **3.** $(5, 30)$ **5.** $(1, -2)$ **7.** $\left(\dfrac{1}{2}, \dfrac{5}{4}\right)$ **9.** D   **11.** B

13.   vertex: $(-2, -9)$;
opens upward;
x-intercepts; $(-5, 0)$, $(1, 0)$;
y-intercept: $(0, -5)$

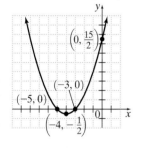

15.   vertex: $(1, 0)$;
opens downward;
x-intercept: $(1, 0)$;
y-intercept: $(0, -1)$

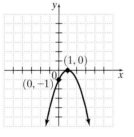

17.   vertex: $(0, -4)$;
opens upward;
x-intercepts: $(-2, 0)$, $(2, 0)$;
y-intercept: $(0, -4)$

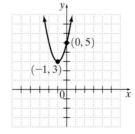

19.   vertex: $\left(-\dfrac{1}{2}, -4\right)$;
opens upward;
x-intercepts: $\left(-\dfrac{3}{2}, 0\right)$, $\left(\dfrac{1}{2}, 0\right)$;
y-intercept: $(0, -3)$

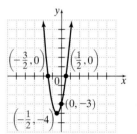

21.   vertex: $\left(-4, -\dfrac{1}{2}\right)$;
opens upward;
x-intercepts: $(-5, 0)$ $(-3, 0)$;
y-intercept: $\left(0, \dfrac{15}{2}\right)$

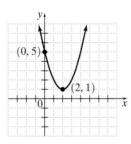

23.   vertex: $(2, 1)$;
opens upward;
y-intercept: $(0, 5)$

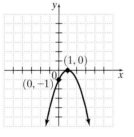

25.   vertex: $(-1, 3)$;
opens upward;
y-intercept: $(0, 5)$

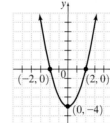

27.   vertex: $(3, 18)$;
opens downward;
x-intercepts: $(0, 0)$, $(6, 0)$;
y-intercept: $(0, 0)$

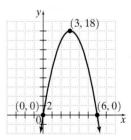

29. 144 ft   **31.** 16 ft   **33.** 30 and 30   **35.** -5 and 5 **37.** length: 20 units; width: 20 units   **39.** $(0, 2)$ **41.** undefined   **43.** $(-5, 2)$

45. $(4, 1)$ **47.**   vertex: $(-5, -10)$;
opens upward;
y-intercept: $(0, 15)$;
x-intercepts:
$(-1.8, 0)$, $(-8.2, 0)$

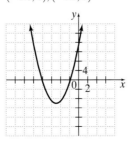

49.
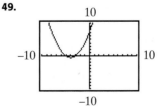

51. -0.84

53. a. maximum; answers may vary
b. 2036
c. 32,902,500
or about 3290.3 thousands

CHAPTER 8 REVIEW

1. $\{14, 1\}$ **2.** $\{-5, 6\}$ **3.** $\left\{\dfrac{4}{5}, -\dfrac{1}{2}\right\}$ **4.** $\left\{-\dfrac{6}{7}, 5\right\}$ **5.** $\{-7, 7\}$ **6.** $\{-2, 2\}$ **7.** $\left\{-\dfrac{4}{9}, \dfrac{2}{9}\right\}$ **8.** $\left\{\dfrac{2 - \sqrt{2}}{5}, \dfrac{2 + \sqrt{2}}{5}\right\}$

9. $\left\{\dfrac{-3 - \sqrt{5}}{2}, \dfrac{-3 + \sqrt{5}}{2}\right\}$ **10.** $\left\{\dfrac{-1 - 3i\sqrt{3}}{2}, \dfrac{-1 + 3i\sqrt{3}}{2}\right\}$ **11.** $\left\{\dfrac{-3 - i\sqrt{7}}{8}, \dfrac{-3 + i\sqrt{7}}{8}\right\}$ **12.** $\left\{\dfrac{17 - \sqrt{145}}{9}, \dfrac{17 + \sqrt{145}}{9}\right\}$

13. 4.25% **14.** $75\sqrt{2}$ mi; 106.1 mi **15.** two complex solutions **16.** two real solutions **17.** two real solutions

18. one real solution **19.** $\{8\}$ **20.** $\{-5, 0\}$ **21.** $\{-i\sqrt{11}, i\sqrt{11}\}$ **22.** $\left\{-\dfrac{5}{2}, 1\right\}$ **23.** $\left\{\dfrac{5 - i\sqrt{143}}{12}, \dfrac{5 + i\sqrt{143}}{12}\right\}$

24. $\left\{\dfrac{1 - i\sqrt{35}}{9}, \dfrac{1 + i\sqrt{35}}{9}\right\}$ **25.** $\left\{\dfrac{21 - \sqrt{41}}{50}, \dfrac{21 + \sqrt{41}}{50}\right\}$ **26.** $\left\{1, \dfrac{9}{4}\right\}$ **27. a.** 20 ft **b.** $\dfrac{15 + \sqrt{321}}{16}$ sec; 2.1 sec **28.** $(6 + 6\sqrt{2})$ cm

29. $\left\{3, \dfrac{-3 + 3i\sqrt{3}}{2}, \dfrac{-3 - 3i\sqrt{3}}{2}\right\}$ **30.** $\{-4, 2 - 2i\sqrt{3}, 2 + 2i\sqrt{3}\}$ **31.** $\left\{\dfrac{2}{3}, 5\right\}$ **32.** $\left\{\dfrac{-8\sqrt{7}}{7}, \dfrac{8\sqrt{7}}{7}\right\}$ **33.** $\{-5, 5, -2i, 2i\}$

34. $\left\{-\dfrac{16}{5}, 1\right\}$ **35.** $\{1, 125\}$ **36.** $\{8, 64\}$ **37.** $\{-1, 1, -i, i\}$ **38.** $\left\{-\dfrac{1}{5}, \dfrac{1}{4}\right\}$ **39.** Jerome: 10.5 hr; Tim: 9.5 hr **40.** -5

41. $[-5, 5]$ **42.** $\left(-\dfrac{1}{2}, \dfrac{1}{2}\right)$ **43.** $\left(-\infty, -\dfrac{5}{4}\right] \cup \left[\dfrac{3}{2}, \infty\right)$ **44.** $(-\infty, -4) \cup (-1, 1) \cup (4, \infty)$ **45.** $(5, 6)$ **46.** $[-5, 0] \cup \left(\dfrac{3}{4}, \infty\right)$

47. $(-\infty, -6) \cup \left(-\dfrac{3}{4}, 0\right) \cup (5, \infty)$ **48.** $(-\infty, -5] \cup [-2, 6]$ **49.** $(-5, -3) \cup (5, \infty)$ **50.** $(-\infty, 0)$ **51.** $\left(-\dfrac{6}{5}, 0\right) \cup \left(\dfrac{5}{6}, 3\right)$ **52.** $\left(2, \dfrac{7}{2}\right)$

53.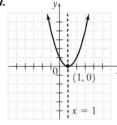

54.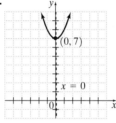

55.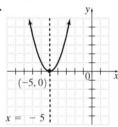

56.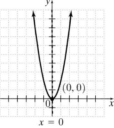

57.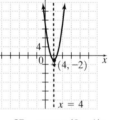

58.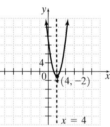

59.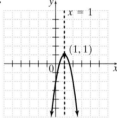

60.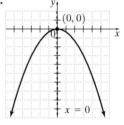

61. vertex: $(-5, 0)$;
x-intercept: $(-5, 0)$;
y-intercept: $(0, 25)$

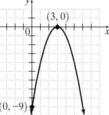

62. vertex: $(3, 0)$;
x-intercept: $(3, 0)$;
y-intercept: $(0, -9)$

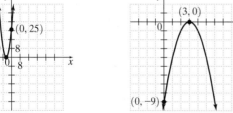

63. vertex: $(0, -1)$;
x-intercepts: $\left(-\dfrac{1}{2}, 0\right), \left(\dfrac{1}{2}, 0\right)$;
y-intercept: $(0, -1)$

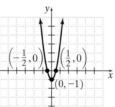
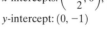

64. vertex: $(0, 5)$;
x-intercepts: $(-1, 0), (1, 0)$;
y-intercept: $(0, 5)$

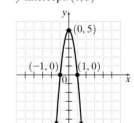

65. vertex: $\left(-\dfrac{5}{6}, \dfrac{73}{12}\right)$;
opens downward;
x-intercepts: $(-2.3, 0), (0.6, 0)$;
y-intercept: $(0, 4)$

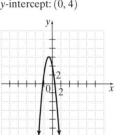

66. a. 0.4 sec and 7.1 sec
b. answers may vary

67. The numbers are both 210.

CHAPTER 8 TEST

1. $\left\{\dfrac{7}{5}, -1\right\}$ **2.** $\{-1 - \sqrt{10}, -1 + \sqrt{10}\}$ **3.** $\left\{\dfrac{1 + i\sqrt{31}}{2}, \dfrac{1 - i\sqrt{31}}{2}\right\}$ **4.** $\{3 - \sqrt{7}, 3 + \sqrt{7}\}$ **5.** $\left\{-\dfrac{1}{7}, -1\right\}$

6. $\left\{\dfrac{3 + \sqrt{29}}{2}, \dfrac{3 - \sqrt{29}}{2}\right\}$ **7.** $\{-2 - \sqrt{11}, -2 + \sqrt{11}\}$ **8.** $\{-3, 3, -i, i\}$ **9.** $\{-1, 1, -i, i\}$ **10.** $\{6, 7\}$ **11.** $\{3 - \sqrt{7}, 3 + \sqrt{7}\}$

12. $\left\{\dfrac{2 - i\sqrt{6}}{2}, \dfrac{2 + i\sqrt{6}}{2}\right\}$ **13.** $\left(-\infty, -\dfrac{3}{2}\right) \cup (5, \infty)$ **14.** $(-\infty, -5) \cup (-4, 4) \cup (5, \infty)$ **15.** $(-\infty, -3) \cup (2, \infty)$ **16.** $(-\infty, -3) \cup [2, 3)$

17.
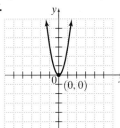

18.
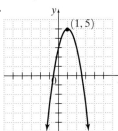

19.
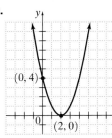

20.
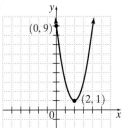

21. $(2 + \sqrt{46})$ ft ≈ 8.8 ft   **22.** $(5 + \sqrt{17})$ hr ≈ 9.12 hr   **23. a.** 272 ft   **b.** 5.12 sec

CUMULATIVE REVIEW

1. 7.3×10^5; (Sec.1.5, Ex. 28)   **2.** $\{0, 1\}$; (Sec. 2.6, Ex. 6)   **3.** not a function; (Sec. 3.5, Ex. 8)   **4.** $\varnothing$; (Sec. 4.1, Ex. 9)

5. $6x^3 - 4x^2 - 2$; Sec. 5.1, Ex. 16   **6.** $6x^2 - 29x + 28$; (Sec. 5.2, Ex. 10)   **7.** $15x^2 - xy - 2y^2$; (Sec. 5.2, Ex. 11)

8. $3x^2 + 2x + 3 + \dfrac{-6x + 9}{x^2 - 1}$; (Sec. 5.3, Ex. 5)   **9.** $3x^2(2 - x)$; (Sec. 5.4, Ex. 2)   **10.** $(2a - 1)(a + 4)$; (Sec. 5.5, Ex. 10)

11. $3x(a - 2b)^2$; (Sec. 5.6, Ex. 4)   **12.** $\left\{-\dfrac{1}{6}, 3\right\}$; (Sec. 5.7, Ex. 5)   **13.** $\dfrac{n - 2}{n(n - 1)}$; (Sec. 6.1, Ex. 11)   **14.** $-5x^2$; (Sec. 6.1, Ex. 12)

15. $\dfrac{6x + 5}{3x^3}$; (Sec. 6.2, Ex. 6)   **16.** $\dfrac{x^2}{y^2}$; (Sec. 6.3, Ex. 5)   **17.** $\{-6, -1\}$; (Sec. 6.4, Ex. 5)   **18.** $x = \dfrac{yz}{y - z}$; (Sec. 6.5, Ex. 1)

19. $S = krh$; (Sec. 6.6, Ex. 5)   **20.** -4; (Sec. 7.1, Ex. 14)   **21.** $\dfrac{2}{5}$; (Sec. 7.1, Ex. 15)   **22.** 2; (Sec. 7.3, Ex. 3)

23. $\sqrt{\dfrac{2b}{3a}}$; (Sec. 7.3, Ex. 5)   **24.** $-6\sqrt[3]{2}$; (Sec. 7.4, Ex. 5)   **25.** $\dfrac{5\sqrt[3]{7x}}{2}$; (Sec. 7.4, Ex. 10)   **26.** $\{-1 + \sqrt{5}, -1 - \sqrt{5}\}$ (Sec. 8.1, Ex. 7)

27. $\left\{\dfrac{2 + \sqrt{10}}{2}, \dfrac{2 - \sqrt{10}}{2}\right\}$ (Sec. 8.2, Ex. 2)   **28.** 3.9 sec (Sec. 8.2, Ex. 9)   **29.** to Beach: 50 mph; to Fargo: 60 mph (Sec. 8.3, Ex. 7)

Chapter 9

CHAPTER 9 PRETEST

1. $\sqrt{34}$ units; 9.1B   **2.** $\left(\dfrac{19}{2}, \dfrac{7}{2}\right)$; 9.1B

3.
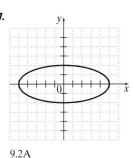
9.1A

4.
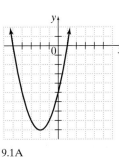
9.1A

5.
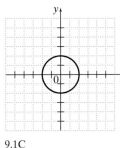
9.1C

6.
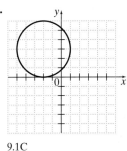
9.1C

7.
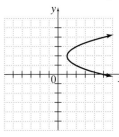
9.2A

8.
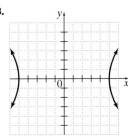
9.2B

9.
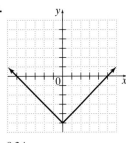
9.3A

10.
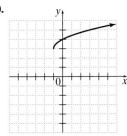
9.3A

11. $\{(0,3),(-5,8)\}$; 9.4A,B   **12.**

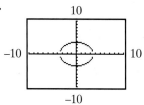

9.5B

13. $(x-2)^2 + (y-5)^2 = 64$; 9.1D   **14.** center: $(-3,4)$; radius: 3

Graphing Calculator Explorations

1.

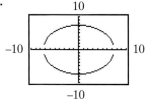

3.

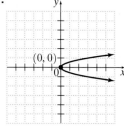

Mental Math

1. upward   **3.** to the left   **5.** downward

Exercise Set 9.1

1.

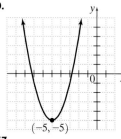

3.

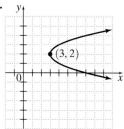

5.

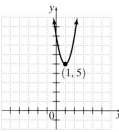

7.

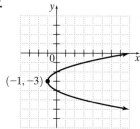

9.

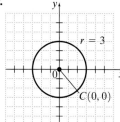

11.

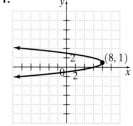

13. 5 units   **15.** $\sqrt{41}$ units ≈ 6.403 **17.** $\sqrt{10}$ units ≈ 3.162

19. $\sqrt{5}$ units ≈ 2.236 **21.** $\sqrt{192.58}$ units ≈ 13.877 **23.** 9 units

25. $(4,-2)$ **27.** $\left(-5,\frac{5}{2}\right)$ **29.** $(3,0)$ **31.** $\left(-\frac{1}{2},\frac{1}{2}\right)$ **33.** $\left(\sqrt{2},\frac{\sqrt{5}}{2}\right)$

35. $(6.2,-6.65)$

37.

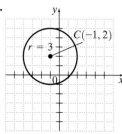

39.

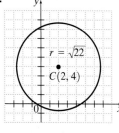

41.

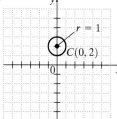

43.

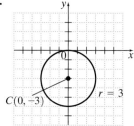

45.

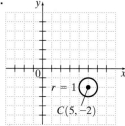

47.

49. $(x-2)^2 + (y-3)^2 = 36$

51. $x^2 + y^2 = 3$

53. $(x+5)^2 + (y-4)^2 = 45$

55.

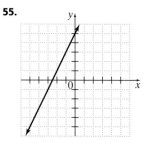

57.

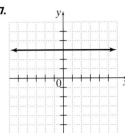

59. $\dfrac{\sqrt{3}}{3}$ **61.** $\dfrac{2\sqrt{42}}{3}$ **63. a.** 125 ft   **b.** 14 ft   **c.** 139 ft   **d.** $(0, 139)$ **e.** $x^2 + (y - 139)^2 = 125^2$

65. answers may vary   **67.** 20 m   **69.** $y = -\dfrac{2}{125}x^2 + 40$

GRAPHING CALCULATOR EXPLORATIONS

1.

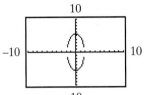

3.

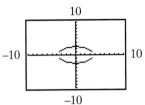

MENTAL MATH

1. ellipse   **3.** hyperbola   **5.** hyperbola

EXERCISE SET 9.2

1.

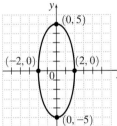

3.

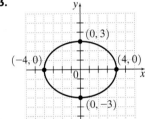

5.

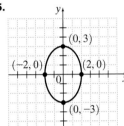

7.

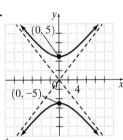

9.

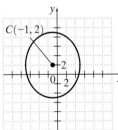

11.

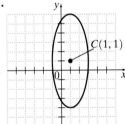

13.

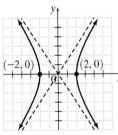

15.

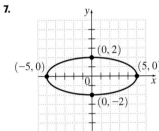

17.

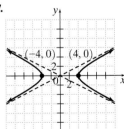

19.

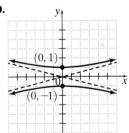

21. $-8x^5$ **23.** $-4x^2$ **25.** $\dfrac{x^2}{25} + \dfrac{y^2}{25} = 1$; ellipse; when $a = b$

27. A: 36, 13; B: 4, 4; C: 25, 16; D: 39, 25; E: 17, 81; F: 36, 36; G: 16, 65; H: 144, 140

29. A: 6; B: 2; C: 5; D: 5; E: 9; F: 6; G: 4; H: 12

31. greater than 0 and less than 1   **33.** greater than 1

35. answers may vary

A:

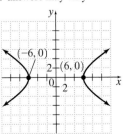

B:

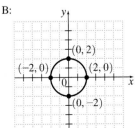

C:

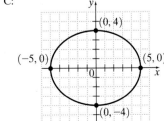

D:

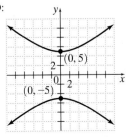

E: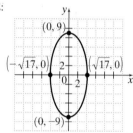

F: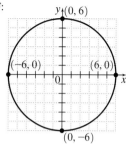

G: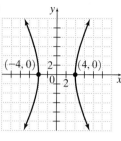

H: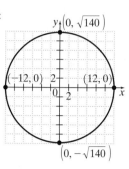

INTEGRATED REVIEW

1. circle;
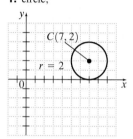

2. parabola;
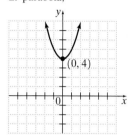

3. parabola;
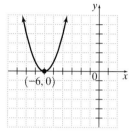

4. ellipse;
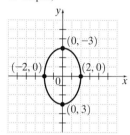

5. hyperbola;
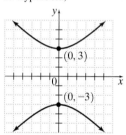

6. hyperbola;
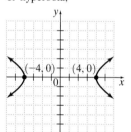

7. ellipse;
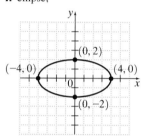

8. circle;
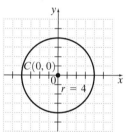

9. parabola;
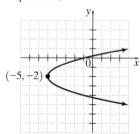

10. parabola;
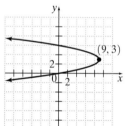

11. hyperbola;
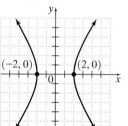

12. ellipse;
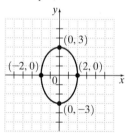

13. ellipse;
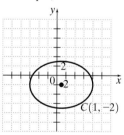

14. hyperbola;
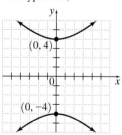

15. circle;
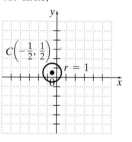

EXERCISE SET 9.3

1.
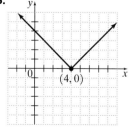

3.
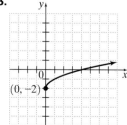

5.
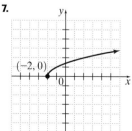

7.

9.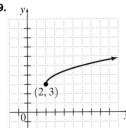

11.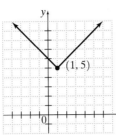

13.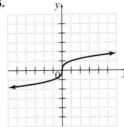

15.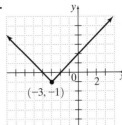

17. $(8, -2)$ **19.** $(-1, 3)$ **21.** 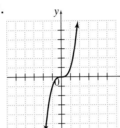   **23.**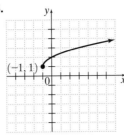

25. $\{x | x \geq 20\}$ **27.** $\{x | x \text{ is a real number}\}$
29. $\{x | x \geq -103\}$

EXERCISE SET 9.4

1. $\{(3, -4), (-3, 4)\}$ **3.** $\{(\sqrt{2}, \sqrt{2}), (-\sqrt{2}, -\sqrt{2})\}$ **5.** $\{(4, 0), (0, -2)\}$ **7.** $\{(-\sqrt{5}, -2), (-\sqrt{5}, 2), (\sqrt{5}, -2), (\sqrt{5}, 2)\}$
9. $\varnothing$ **11.** $\{(1, -2), (3, 6)\}$ **13.** $\{(2, 4), (-5, 25)\}$ **15.** $\varnothing$ **17.** $\{(1, -3)\}$ **19.** $\{(-1, -2), (-1, 2), (1, -2), (1, 2)\}$ **21.** $\{(0, -1)\}$
23. $\{(-1, 3), (1, 3)\}$ **25.** $\{(\sqrt{3}, 0), (-\sqrt{3}, 0)\}$ **27.** $\varnothing$ **29.** $\{(-6, 0), (6, 0), (0, -6)\}$ **31.** $0, 1, 2, 3,$ or 4
37. 9 and 7; 9 and -7; -9 and 7; -9 and -7 **39.** 15 cm by 19 cm
41. 15 thousand compact discs; price: $3.75

33.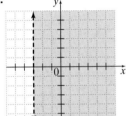

35.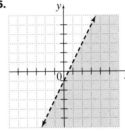

EXERCISE SET 9.5

1.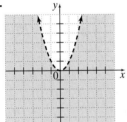

3.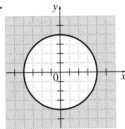

5.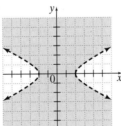

7.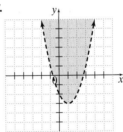

9.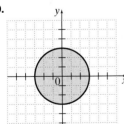

11.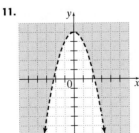

13.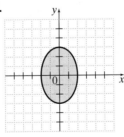

15.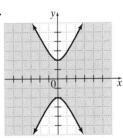

17.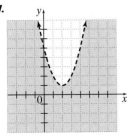

19.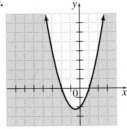

21.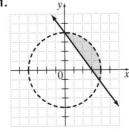

23.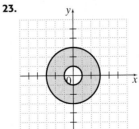

25.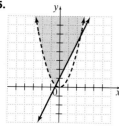

27.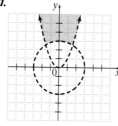

29.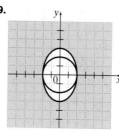

31.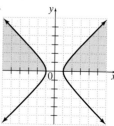

33.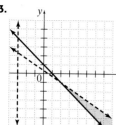

35.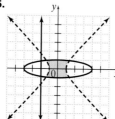

37. not a function **39.** function **41.** answers may vary

43.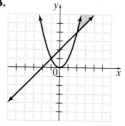

CHAPTER 9 REVIEW

1. $\sqrt{197}$ units ≈ 14.036 **2.** $\sqrt{41}$ units ≈ 6.403 **3.** $\sqrt{130}$ units ≈ 11.402 **4.** $\sqrt{73}$ units ≈ 8.544 **5.** $7\sqrt{2}$ units ≈ 9.899

6. $2\sqrt{11}$ units ≈ 6.633 **7.** $\sqrt{275.6}$ units ≈ 16.601 **8.** $\sqrt{191.56}$ units ≈ 13.841 **9.** $(-5, 5)$ **10.** $(4, 16)$ **11.** $\left(-\dfrac{15}{2}, 1\right)$

12. $\left(-\dfrac{11}{2}, -2\right)$ **13.** $\left(\dfrac{1}{20}, -\dfrac{3}{16}\right)$ **14.** $\left(\dfrac{1}{4}, -\dfrac{2}{7}\right)$ **15.** $(\sqrt{3}, -3\sqrt{6})$ **16.** $(-4\sqrt{3}, 6\sqrt{7})$

17. $(x + 4)^2 + (y - 4)^2 = 9$ **18.** $(x - 5)^2 + y^2 = 25$ **19.** $(x + 7)^2 + (y + 9)^2 = 11$ **20.** $x^2 + y^2 = \dfrac{49}{4}$

21.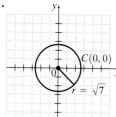

22.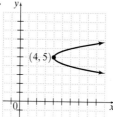

23.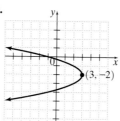

24.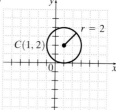

25.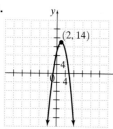

26.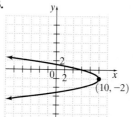

27.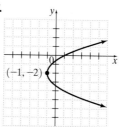

28.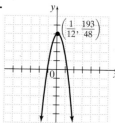

29.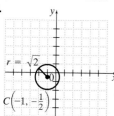

30.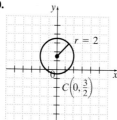

31.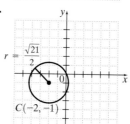

32.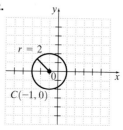

33.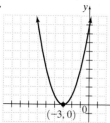

34.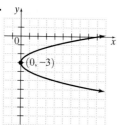

35.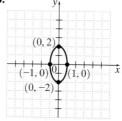

36.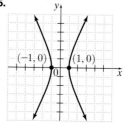

37.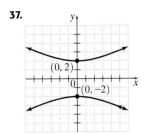

38.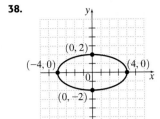

39.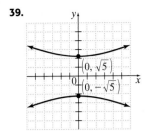

40.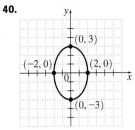

41.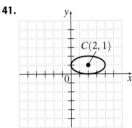

42.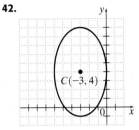

43.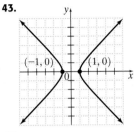

44.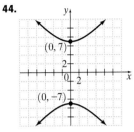

45.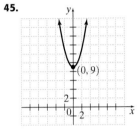

46.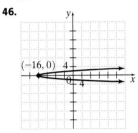

47.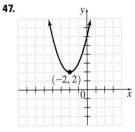

48.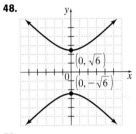

49.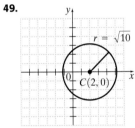

50.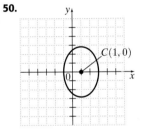

51.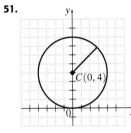

52.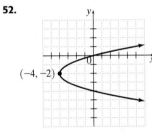

53.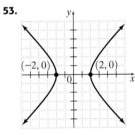

54.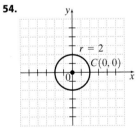

55.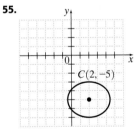

56.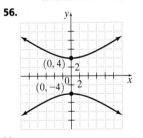

57.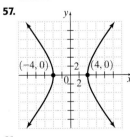

58.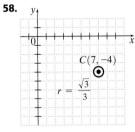

59.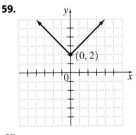

60.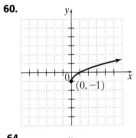

61.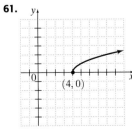

62.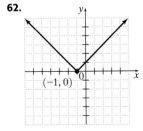

63.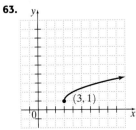

64.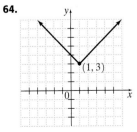

65.
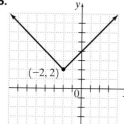

66.
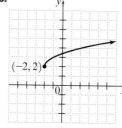

67. $\{(1, -2), (4, 4)\}$ **68.** $\varnothing$ **69.** $\{(-1, 1), (2, 4)\}$

70. $\{(5, 1), (-1, 7)\}$ **71.** $\{(2, 2\sqrt{2}), (2, -2\sqrt{2})\}$ **72.** $\{(0, 2), (0, -2)\}$

73. $\{(-1, 3), (-1, -3), (1, 3), (1, -3)\}$ **74.** $\left\{\left(2, \dfrac{5}{2}\right), (-7, -20)\right\}$

75. $\{(1, 4)\}$ **76.** $\{(-2, -1), (-2, 1), (2, -1), (2, 1)\}$ **77.** length: 15 ft;

width: 10 ft   **78.** 4

79.
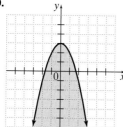

80.
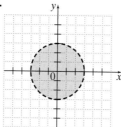

81.
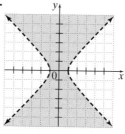

82.
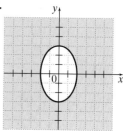

83.
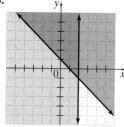

84.
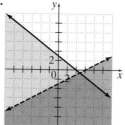

85.
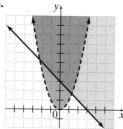

86.
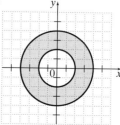

87.
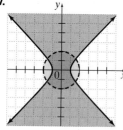

88.
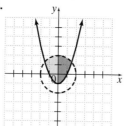

CHAPTER 9 TEST

1. $2\sqrt{26}$ units   **2.** $\sqrt{95}$ units   **3.** $\left(-4, \dfrac{7}{2}\right)$ **4.** $\left(-\dfrac{1}{2}, \dfrac{3}{10}\right)$

5.
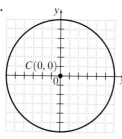

6.
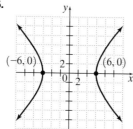

7.
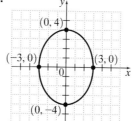

8.
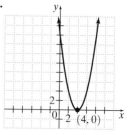

9.
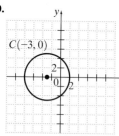

10.
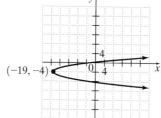

11.
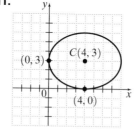

12.
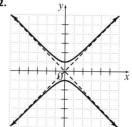

13. $\{(-12, 5), (12, -5)\}$ **14.** $\{(-5, -1), (-5, 1), (5, -1), (5, 1)\}$ **15.** $\{(6, 12), (1, 2)\}$ **16.** $\{(1, 1), (-1, -1)\}$

17.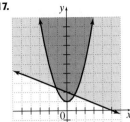

18.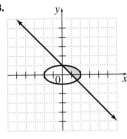

19.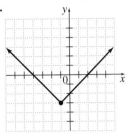

20.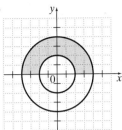

21. B   **22.** height: 10 ft; width: 30 ft

23.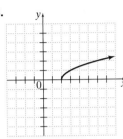

24.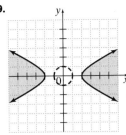

Cumulative Review

1. $12x^3 - 12x^2 - 9x + 2$; Sec. 5.1, Ex. 12   **2.** $2x^2 + 11x + 15$; Sec. 5.2, Ex. 6   **3.** $x^3 - 4x^2 - 3x + 6 + \dfrac{22}{x+2}$; Sec. 5.3, Ex. 8

4. $(x - 5)(2 + 3a)$; Sec. 5.4, Ex. 6   **5.** $(x - 5)(x - 7)$; Sec. 5.5, Ex. 2   **6.** $(x + 9)(x - 3)$; Sec. 5.6, Ex. 10   **7.** $\left\{-5, \dfrac{1}{2}\right\}$; Sec. 5.7, Ex. 2

8. 1; Sec. 6.1, Ex. 6   **9.** -1; Sec. 6.1, Ex. 7   **10.** $x - 7$; Sec. 6.2, Ex. 3   **11.** $-\dfrac{1}{3y^2}$; Sec. 6.2, Ex. 4   **12.** $\dfrac{x(x-2)}{2(x+2)}$; Sec. 6.3, Ex. 2

13. $\{-6, -1\}$; Sec. 6.4, Ex. 5   **14.** 2; Sec. 6.5, Ex. 2   **15.** 0; Sec. 7.1, Ex. 5   **16.** 0.5; Sec. 7.1, Ex. 7   **17.** $\sqrt{x}$; Sec. 7.2, Ex. 17

18. $\sqrt[3]{rs^2}$; Sec. 7.2, Ex. 19   **19.** $2\sqrt[3]{3}$; Sec. 7.3, Ex. 11   **20.** $2\sqrt{2}$; Sec. 7.3, Ex. 13   **21.** $\dfrac{2\sqrt{5}}{5}$; Sec. 7.5, Ex. 1   **22.** $\left\{-\dfrac{1}{9}, -1\right\}$; Sec. 7.6, Ex. 2

23. $13 - 18i$; Sec. 7.7, Ex. 13   **24.** $\{-1 + \sqrt{5}, -1 - \sqrt{5}\}$; Sec. 8.1, Ex. 7   **25.** $\left\{\dfrac{2 + \sqrt{10}}{2}, \dfrac{2 - \sqrt{10}}{2}\right\}$; Sec. 8.2, Ex. 2

26. $\sqrt{2} \approx 1.414$; Sec. 9.1, Ex. 5   **27.** 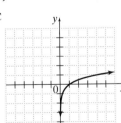   Sec. 9.2, Ex. 1   **28.** No solution; Sec. 9.4, Ex. 3

29. $\{(2, \sqrt{3}), (-2, \sqrt{3}), (2, -\sqrt{3}), (-2, -\sqrt{3})\}$; Sec. 9.4, Ex. 4

Chapter 10

Chapter 10 Pretest

1. $x^2 + 3x + 3$; 10.1A   **2.** $25x^2 + 20x + 3$; 10.1B   **3.** one-to-one; 10.2A   **4.** not one-to-one; 10.2B   **5.** $f^{-1}(x) = \dfrac{x + 12}{7}$; 10.2C

6. 10.3A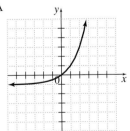

7. $\{3\}$; 10.3 B   **8.** $\left\{-\dfrac{1}{6}\right\}$; 10.3B   **9.** $\{625\}$; 10.4B   **10.** $\{-6\}$; 10.4B   **11.** $\{10\}$; 10.4B

12. 5; 10.4B   **13.** 10.4C

14. $\log_5 6$; 10.5A   **15.** $\log_6 \dfrac{a^2}{(a+1)^7}$; 10.5D

16. $\log_7 3 + \log_7 y - \log_7 5 - 2\log_7 x$; 10.5D

17. -2; 10.6B   **18.** 8; 10.6B   **19.** 2.4650; 10.6E

20. $\left\{-\dfrac{21}{5}\right\}$; 10.7B

Exercise Set 10.1

1. a. $3x - 6$ **b.** $-x - 8$ **c.** $2x^2 - 13x - 7$ **d.** $\dfrac{x - 7}{2x + 1}$ where $x \neq -\dfrac{1}{2}$ **3. a.** $x^2 + 5x + 1$ **b.** $x^2 - 5x + 1$ **c.** $5x^3 + 5x$ **d.** $\dfrac{x^2 + 1}{5x}$ where $x \neq 0$

5. a. $\sqrt{x} + x + 5$ **b.** $\sqrt{x} - x - 5$ **c.** $x\sqrt{x} + 5\sqrt{x}$ **d.** $\dfrac{\sqrt{x}}{x + 5}$ where $x \neq -5$

7. a. $5x^2 - 3x$ **b.** $-5x^2 - 3x$ **c.** $-15x^3$ **d.** $-\dfrac{3}{5x}$ where $x \neq 0$ **9.** 42   **11.** -18 **13.** 0

15. $(f \circ g)(x) = 25x^2 + 1; (g \circ f)(x) = 5x^2 + 5$ **17.** $(f \circ g)(x) = 2x + 11; (g \circ f)(x) = 2x + 4$

19. $(f \circ g)(x) = -8x^3 - 2x - 2; (g \circ f)(x) = -2x^3 - 2x + 4$ **21.** $(f \circ g)(x) = \sqrt{-5x + 2}; (g \circ f)(x) = -5\sqrt{x} + 2$

23. $H(x) = (g \circ h)(x)$ **25.** $F(x) = (h \circ f)(x)$ **27.** $G(x) = (f \circ g)(x)$ **29.** answers may vary   **31.** answers may vary

33. answers may vary   **35.** $y = x - 2$ **37.** $y = \dfrac{x}{3}$ **39.** $y = -\dfrac{x + 7}{2}$ **41.** $P(x) = R(x) - C(x)$ **43.** answers may vary

EXERCISE SET 10.2

1. one-to-one; $f^{-1} = \{(-1, -1), (1, 1), (2, 0), (0, 2)\}$ **3.** one-to-one; $h^{-1} = \{(10, 10)\}$ **5.** one-to-one; $f^{-1} = \{(12, 11), (3, 4), (4, 3), (6, 6)\}$
7. not one-to-one   **9.** one-to-one   **11. a.** 3   **b.** 1   **13. a.** 1   **b.** -1 **15.** one-to-one   **17.** not one-to-one
19. one-to-one   **21.** not one-to-one

23. $f^{-1}(x) = x - 4$ **25.** $f^{-1}(x) = \dfrac{x + 3}{2}$ **27.** $f^{-1}(x) = 2x + 2$ **29.** $f^{-1}(x) = \sqrt[3]{x}$

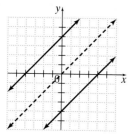   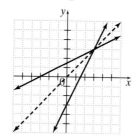   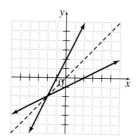   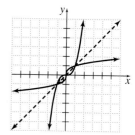

31. $f^{-1}(x) = 5x + 2$ **39.**   **41.**

33. $f^{-1}(x) = x^3$

35. $f^{-1}(x) = \dfrac{5}{3x} - \dfrac{1}{3}$

37. $f^{-1}(x) = \sqrt[3]{x} - 2$

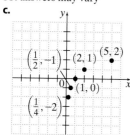   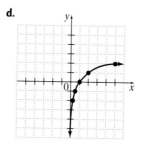

43. 5   **45.** 8   **47.** $\dfrac{1}{27}$ **49.** 9   **51.** $3^{1/2} \approx 1.73$ **53. a.** $\left(-2, \dfrac{1}{4}\right), \left(-1, \dfrac{1}{2}\right), (0, 1), (1, 2), (2, 5)$ **b.** $\left(\dfrac{1}{4}, -2\right), \left(\dfrac{1}{2}, -1\right), (1, 0), (2, 1), (5, 2)$

55. answers may vary
c.   **d.**

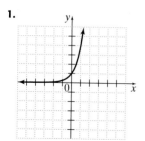   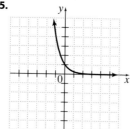

CALCULATOR EXPLORATIONS
1. 81.98%   **2.** 22.54%

EXERCISE SET 10.3

1.   **3.**   **5.**   **7.**

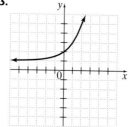   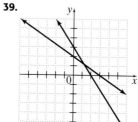   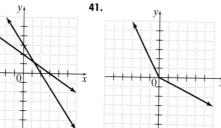   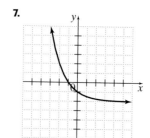

9. 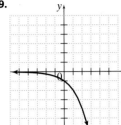 **11.** 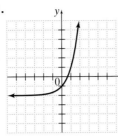 **13.** 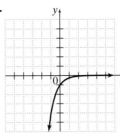 **15.**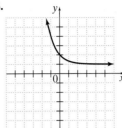

17. $\{3\}$ **19.** $\left\{\dfrac{3}{4}\right\}$ **21.** $\left\{\dfrac{8}{5}\right\}$ **23.** $\left\{-\dfrac{2}{3}\right\}$ **25.** $\left\{\dfrac{3}{2}\right\}$ **27.** $\left\{-\dfrac{1}{3}\right\}$ **29.** $\{-2\}$ **31.** 24.6 lb   **33.** 333 bison   **35. a.** $65.7 billion
b. $2303.6 billion   **37.** approximately 618.6 million cellular phone users   **39.** $7621.42   **41.** $\{4\}$ **43.** $\varnothing$ **45.** C   **47.** D
49. answers may vary   **53.** 18.62 lb

EXERCISE SET 10.4

1. $6^2 = 36$ **3.** $3^{-3} = \dfrac{1}{27}$ **5.** $10^3 = 1000$ **7.** $e^4 = x$ **9.** $e^{-2} = \dfrac{1}{e^2}$ **11.** $7^{1/2} = \sqrt{7}$ **13.** $\log_2 16 = 4$ **15.** $\log_{10} 100 = 2$

17. $\log_e x = 3$ **19.** $\log_{10} \dfrac{1}{10} = -1$ **21.** $\log_4 \dfrac{1}{16} = -2$ **23.** $\log_5 \sqrt{5} = \dfrac{1}{2}$ **25.** 3   **27.** -2 **29.** $\dfrac{1}{2}$ **31.** -1 **33.** 0   **35.** 4   **37.** 2
39. 5   **41.** 4   **43.** -3 **45.** answers may vary   **47.** $\{2\}$ **49.** $\{81\}$ **51.** $\{7\}$ **53.** $\{-3\}$ **55.** $\{-3\}$ **57.** $\{2\}$ **59.** $\{2\}$
61. $\left\{\dfrac{27}{64}\right\}$ **63.** $\{10\}$ **65.** 3   **67.** 3   **69.** 3   **69.** 1

71.   **73.** 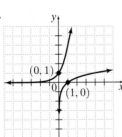   **75.** 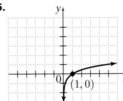   **77.**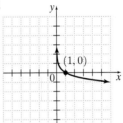

79. 1   **81.** $\dfrac{x-4}{2}$ **83.**   **85.**   **87.** 0.0827   **89.** 2 and 3

EXERCISE SET 10.5

1. $\log_5 14$ **3.** $\log_4 9x$ **5.** $\log_{10}(10x^2 + 20)$ **7.** $\log_5 3$ **9.** $\log_2 \dfrac{x}{y}$ **11.** $\log_3 4$ **13.** $2 \log_3 x$ **15.** $-1 \log_4 5$ **17.** $\dfrac{1}{2} \log_5 y$

19. $\log_2 25$ **21.** $\log_5 x^3 z^6$ **23.** $\log_{10} \dfrac{x^3 - 2x}{x+1}$ **25.** $\log_4 4$, or 1   **27.** $\log_7 \dfrac{9}{2}$ **29.** $\log_4 48$ **31.** $\log_2 \dfrac{x^{7/2}}{(x+1)^2}$ **33.** $\log_8 x^{16/3}$

35. $\log_3 4 + \log_3 y - \log_3 5$ **37.** $3 \log_2 x - \log_2 y$ **39.** $\dfrac{1}{2} \log_b 7 + \dfrac{1}{2} \log_b x$ **41.** $\log_7 5 + \log_7 x - \log_7 4$ **43.** $3 \log_5 x + \log_5(x+1)$
45. $2 \log_6 x - \log_6(x+3)$ **47.** 0.2   **49.** 1.2   **51.** 0.35   **53.** 1.29   **55.** -0.68 **57.** -0.125 **59.**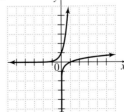
61. $\{-1\}$ **63.** $\left\{\dfrac{1}{2}\right\}$ **65.** false   **67.** true   **69.** false

INTEGRATED REVIEW

1. $x^2 + x - 5$ **2.** $-x^2 + x - 7$ **3.** $x^3 - 6x^2 + x - 6$ **4.** $\dfrac{x - 6}{x^2 + 1}$ **5.** $\sqrt{3x - 1}$ **6.** $3\sqrt{x} - 1$ **7.** one-to-one;
$\{(6, -2), (8, 4), (-6, 2), (3, 3)\}$ **8.** not one-to-one **9.** not one-to-one **10.** one-to-one **11.** not one-to-one **12.** $f^{-1}(x) = \dfrac{x}{3}$
13. $f^{-1}(x) = x - 4$ **14.** $f^{-1}(x) = \dfrac{x + 1}{5}$ **15.** $f^{-1}(x) = \dfrac{x - 2}{3}$

16. 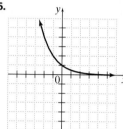 **17.** 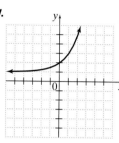 **18.** 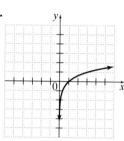 **19.**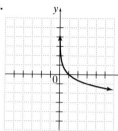

20. $\{3\}$ **21.** $\{7\}$ **22.** $\{-8\}$ **23.** $\{3\}$ **24.** $\{2\}$ **25.** $\left\{\dfrac{1}{2}\right\}$ **26.** $\{32\}$ **27.** $\{4\}$ **28.** $\{5\}$ **29.** $\left\{\dfrac{1}{9}\right\}$ **30.** $\log_2 x^5$
31. $\log_2 5^x$ **32.** $\log_5 \dfrac{x^3}{y^5}$ **33.** $\log_5 x^9 y^3$ **34.** $\log_2 \dfrac{x^2 - 3x}{x^2 + 4}$ **35.** $\log_3 \dfrac{y(y^3 + 11)}{y + 2}$ **36.** $\log_7 9 + 2\log_7 x - \log_7 y$
37. $\log_6 5 + \log_6 y - 2\log_6 z$

EXERCISE SET 10.6

1. 0.9031 **3.** 0.3636 **5.** 0.6931 **7.** -2.6367 **9.** 1.1004 **11.** 1.6094 **13.** 1.6180 **15.** answers may vary **17.** 2 **19.** -3
21. 2 **23.** $\dfrac{1}{4}$ **25.** 3 **27.** 2 **29.** -4 **31.** $\dfrac{1}{2}$ **33.** $\{10^{1.3}\}$; $\{19.9526\}$ **35.** $\left\{\dfrac{10^{1.1}}{2}\right\}$; $\{6.2946\}$ **37.** $\{e^{1.4}\}$; $\{4.0552\}$
39. $\left\{\dfrac{4 + e^{2.3}}{3}\right\}$; $\{4.6581\}$ **41.** $\{10^{2.3}\}$; $\{199.5262\}$ **43.** $\{e^{-2.3}\}$; $\{0.1003\}$ **45.** $\left\{\dfrac{10^{-0.5} - 1}{2}\right\}$; $\{-0.3419\}$ **47.** $\left\{\dfrac{e^{0.18}}{4}\right\}$; $\{0.2993\}$
49. $\$3656.38$ **51.** $\$2542.50$ **53.** 1.5850 **55.** -2.3219 **57.** 1.5850 **59.** -1.6309 **61.** 0.8617 **63.** $\left\{\dfrac{4}{7}\right\}$ **65.** $x = \dfrac{3y}{4}$
67. $\{-6, -1\}$ **69.** 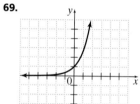 **71.** 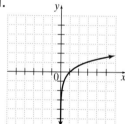 **73.** answers may vary **75.** 4.2 **77.** 5.3

CALCULATOR EXPLORATIONS

1. 3.67 yr, or 3 yr and 8 mo **3.** 23.16 yr, or 23 yr and 2 mo

EXERCISE SET 10.7

1. $\left\{\dfrac{\log 6}{\log 3}\right\}$; $\{1.6309\}$ **3.** $\left\{\dfrac{\log 3.8}{2\log 3}\right\}$; $\{0.6076\}$ **5.** $\left\{3 + \dfrac{\log 5}{\log 2}\right\}$; $\{5.3219\}$ **7.** $\left\{\dfrac{\log 5}{\log 9}\right\}$; $\{0.7325\}$ **9.** $\left\{\dfrac{\log 3}{\log 4} - 7\right\}$; $\{-6.2075\}$
11. $\left\{\dfrac{1}{3}\left(4 + \dfrac{\log 11}{\log 7}\right)\right\}$; $\{1.7441\}$ **13.** $\left\{\dfrac{\ln 5}{6}\right\}$; $\{0.2682\}$ **15.** $\{11\}$ **17.** $\left\{\dfrac{1}{2}\right\}$ **19.** $\left\{\dfrac{3}{4}\right\}$ **21.** $\{2\}$ **23.** $\left\{\dfrac{1}{8}\right\}$ **25.** $\{11\}$
27. $\{4, -1\}$ **29.** $\left\{\dfrac{2}{3}\right\}$ **31.** 103 wolves **33.** $6{,}885{,}833$ people **35.** 26.3 yr **37.** 9.9 yr **39.** 1.7 yr **41.** 8.8 yr **43.** 55.7 in.
45. 11.9 lb per sq in. **47.** 3.2 mi **49.** 12 weeks **51.** 18 weeks **53.** $-\dfrac{5}{3}$ **55.** $\dfrac{17}{4}$ **57.** 3.4% **59.** answers may vary

61. $\{6.93\}$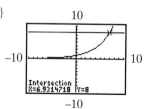

CHAPTER 10 REVIEW

1. $3x - 4$ **2.** $-x - 6$ **3.** $2x^2 - 9x - 5$ **4.** $\dfrac{2x + 1}{x - 5}$ if $x \neq 5$ **5.** $x^2 + 2x - 1$ **6.** $x^2 - 1$ **7.** 18 **8.** $x^4 - 4x^2 + 2$ **9.** -2

10. 48

11. one-to-one; $h^{-1} = \{(14, -9), (8, 6), (12, -11), (15, 15)\}$ **12.** not one-to-one

13. one-to-one;

Rank in Automobile Thefts (Input)	2	4	1	3
U.S. Region (Output)	West	Midwest	South	Northeast

14. not one-to-one   **15.** not one-to-one

16. not one-to-one   **17.** not one-to-one

18. one-to-one   **19.** $f^{-1}(x) = \dfrac{x - 11}{6}$

20. $f^{-1}(x) = \dfrac{x}{12}$ **21.** $f^{-1}(x) = \dfrac{x + 5}{3}$ **22.** $f^{-1}(x) = \dfrac{x - 1}{2}$

33. $f^{-1}(x) = x^3$

35. $f^{-1}(x) = \dfrac{5}{3}x - \dfrac{1}{3}$

37. $f^{-1}(x) = \sqrt[3]{x} - 2$

23.
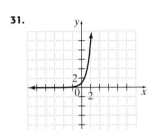

24.
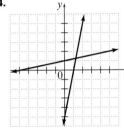

25. $\{3\}$ **26.** $\{-2\}$ **27.** $\left\{-\dfrac{4}{3}\right\}$ **28.** $\left\{\dfrac{3}{2}\right\}$ **29.** $\left\{\dfrac{3}{2}\right\}$ **30.** $\left\{\dfrac{8}{9}\right\}$

31.
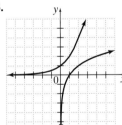

32.
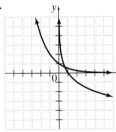

33.
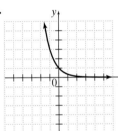

34.
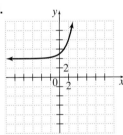

35. $\$2963.11$ **36.** $\$1131.82$ **37.** $\log_7 49 = 2$ **38.** $\log_2\left(\dfrac{1}{16}\right) = -4$ **39.** $\left(\dfrac{1}{2}\right)^{-4} = 16$ **40.** $0.4^3 = 0.064$ **41.** $\left\{\dfrac{1}{64}\right\}$ **42.** $\{9\}$

43. $\{0\}$ **44.** $\{3\}$ **45.** $\{8\}$ **46.** $\{3\}$ **47.** $\{5\}$ **48.** $\{-2\}$ **49.** $\{4\}$ **50.** $\{9\}$ **51.** $\left\{\dfrac{17}{3}\right\}$ **52.** $\{2\}$ **53.** $\{-1, 4\}$ **54.** $\{-8, 1\}$

55.
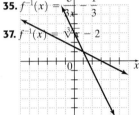

56.
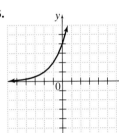

57. $\log_3 32$ **58.** $\log_2 18$ **59.** $\log_7 \dfrac{3}{4}$ **60.** $\log\left(\dfrac{3}{2}\right)$ **61.** $\log_{11} 4$

62. $\log_5 2$ **63.** $\log_5 \dfrac{x^3}{(x + 1)^2}$ **64.** $\log_3 (x^4 + 2x^3)$

65. $3 \log_3 x - \log_3 (x + 2)$ **66.** $\log_4 (x + 5) - 2 \log_4 x$

67. $\log_2 3 + 2 \log_2 x + \log_2 y - \log_2 z$ **68.** $\log_7 y + 3 \log_7 z - \log_7 x$

69. 2.02 **70.** -0.11 **71.** 0.5563 **72.** -0.8239 **73.** 0.2231 **74.** 1.5326 **75.** 3 **76.** -1 **77.** -1 **78.** 4 **79.** $\left\{\dfrac{e^2}{2}\right\}$

80. $\left\{\dfrac{e^{1.6}}{3}\right\}$ **81.** $\left\{\dfrac{e^{-1} + 3}{2}\right\}$ **82.** $\left\{\dfrac{e^2 - 1}{3}\right\}$ **83.** 0.2920 **84.** 1.2619 **85.** $\$1957.30$ **86.** $\$1307.51$ **87.** $\left\{\dfrac{\log 7}{2 \log 3}\right\}$; $\{0.8856\}$

88. $\left\{\dfrac{\log 5}{3 \log 6}\right\}$; $\{0.2994\}$ **89.** $\left\{\dfrac{1}{2}\left(\dfrac{\log 6}{\log 3} - 1\right)\right\}$; $\{0.3155\}$ **90.** $\left\{\dfrac{1}{3}\left(\dfrac{\log 9}{\log 4} - 2\right)\right\}$; $\{-0.1383\}$ **91.** $\left\{\dfrac{1}{3}\left(\dfrac{\log 4}{\log 5} + 5\right)\right\}$; $\{1.9538\}$

92. $\left\{\frac{1}{4}\left(\frac{\log 3}{\log 8} + 2\right)\right\}$; {0.6321}   **93.** $\left\{-\frac{\log 2}{\log 5} + 1\right\}$; {0.5693}   **94.** $\left\{\frac{\log \frac{2}{3}}{\log 4} - 5\right\}$; {−5.2925}   **95.** $\left\{\frac{25}{2}\right\}$ **96.** $\left\{\frac{9}{10}\right\}$ **97.** $\varnothing$

98. $\left\{\frac{3e^2}{e^2 - 3}\right\}$ **99.** $\{2\sqrt{2}\}$ **100.** $\varnothing$ **101.** 197,044 ducks   **102.** 6,648,551 people   **103.** 23 yr   **104.** 116 yr   **105.** 36 yr
106. 8.8 yr   **107.** 8.5 yr

CHAPTER 10 TEST

1. $2x^2 - 3x$ **6.**
2. $3 - x$
3. 5
4. $x - 7$
5. $x^2 - 6x - 2$

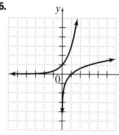

7. one-to-one   **8.** not one-to-one   **9.** one-to-one; $f^{-1}(x) = \frac{-x + 6}{2}$ **10.** one-to-one;

$f^{-1} = \{(0, 0), (3, 2), (5, -1)\}$ **11.** not one-to-one   **12.** $\log_3 24$ **13.** $\log_5\left(\frac{x^4}{x + 1}\right)$

14. $\log_6 2 + \log_6 x - 3\log_6 y$ **15.** −1.53   **16.** 1.0686   **17.** {−1}
18. $\left\{\frac{1}{2}\left(\frac{\log 4}{\log 3} - 5\right)\right\}$; {−1.8691}   **19.** $\left\{\frac{1}{9}\right\}$ **20.** $\left\{\frac{1}{2}\right\}$ **21.** {22}   **22.** $\left\{\frac{25}{3}\right\}$
23. $\left\{\frac{43}{21}\right\}$ **24.** {−1.0979}

25.

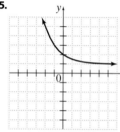

26.

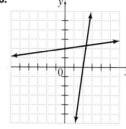

27. $5234.58   **28.** 6 yr   **29.** 64,913 prairie dogs   **30.** 15 yr   **31.** 85%
32. 52%   **33.** 7074 million people

CUMULATIVE REVIEW

1. 3; Sec. 1.4, Ex. 42   **2.** 2; Sec. 1.4, Ex. 44   **3.** 30°, 40°, and 110°; Sec. 4.3, Ex. 4   **4.** $(x^2 + 5y)(7x - 1)$; Sec. 5.4, Ex. 7
5. $\frac{5k^2 - 7k + 4}{(k + 2)(k - 2)(k - 1)}$; Sec. 6.2, Ex. 9   **6.** 2; Sec. 7.1, Ex. 30   **7.** 1; Sec. 7.1, Ex. 32   **8.** $x^{5/6}$; Sec. 7.2, Ex. 14   **9.** 32; Sec. 7.2, Ex. 16
10. $\frac{\sqrt{x}}{3}$; Sec. 7.3, Ex. 7   **11.** $\frac{\sqrt[4]{3}}{2y}$; Sec. 7.3, Ex. 9   **12.** $2x - 25$; Sec. 7.4, Ex. 13   **13.** $4 - 2\sqrt{3}$; Sec. 7.4, Ex. 14   **14.** $\frac{7}{3\sqrt{35}}$; Sec. 7.5, Ex. 6

15. {3}; Sec. 7.6, Ex. 4   **16.** $-1 + 5i$; Sec. 7.7, Ex. 8   **17.** $\left\{\frac{6 + i\sqrt{5}}{2}, \frac{6 - i\sqrt{5}}{2}\right\}$; Sec. 8.1, Ex. 9   **18.** $\{2 + \sqrt{2}, 2 - \sqrt{2}\}$; Sec. 8.2, Ex. 3

19. {8, 27}; Sec. 8.3, Ex. 5   **20.** $\left(-\frac{7}{2}, -1\right)$; Sec. 8.4, Ex. 5   **21.**   **22.** $\left(-1, \frac{3}{2}\right)$; Sec. 9.1, Ex. 6
23.   **24.** $\{(2, \sqrt{2})\}$; Sec. 9.4, Ex. 2

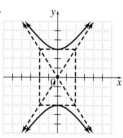

Sec. 9.2, Ex. 5   Sec. 8.6, Ex. 2

25. a. $(f \circ g)(x) = |x - 2|$ **b.** $(g \circ f)(x) = |x| - 2$; (Sec. 10.1, Ex. 3)   **26.** $f^{-1}(x) = \frac{x + 5}{3}$; (Sec. 10.2, Ex. 9)   **27.** {6}; (Sec. 10.3, Ex. 6)

28. {5}; (Sec. 10.4, Ex. 9)   **29.** $\log_5 72$; (Sec. 10.5, Ex. 7)   **30.** $2509.30; (Sec. 10.6, Ex. 11)   **31.** $\left\{\frac{2}{99}\right\}$; Sec. 10.7, Ex. 4

SOLUTIONS TO SELECTED EXERCISES

Chapter 1

Exercise Set 1.2

1. $\{1, 2, 3, 4, 5\}$

5. $\{0\}$

9. $\{3, 0, \sqrt{36}\}$

13. $\{\sqrt{7}\}$

17. $0 \notin \{x \mid x \text{ is a positive integer}\}$

21. true

25. false

29. $x - 10$

33. $\dfrac{x}{11}$ or $x \div 11$

37. $x + 20$

41. $9x$

45. $2x + 5$

49. $1 + 2x$

53. $2(x + 3)$

57. 1950: $71.1 - 65.7 = 5.4$
1960: $73.2 - 71.1 = 2.1$
1970: $74.9 - 73.2 = 1.7$
1980: $77.5 - 74.9 = 2.6$
1990: $78.9 - 77.5 = 1.4$
2000: $79.6 - 78.9 = 0.7$

Exercise Set 1.3

1. $4c = 7$

5. $\dfrac{n}{5} = 4n$

9. $0 > -2$ since 0 is to the right of -2

13. $-7.9 < -7.09$ since -7.9 is to the left of -7.09

17. $8.6 > -3.5$ since 8.6 is to the right of -3.5

21. $\dfrac{1}{2} < \dfrac{5}{8}$ since $\dfrac{1}{2} = 0.5$ and $\dfrac{5}{8} = 0.625$ with 0.5 to the left of 0.625

25. true since -6 is to the left of 0

29. false since -14 is to the left of -1

33. 6.2

37. $\dfrac{2}{3}$ **41.** $\dfrac{1}{5}$

45. $-\dfrac{4}{1} = -4$ **49.** $\dfrac{8}{7}$

53. opposite is $-\dfrac{2}{3}$; reciprocal is $\dfrac{3}{2}$

57. $y + 7x$

61. $\dfrac{x}{5} \cdot \dfrac{1}{3}$ **65.** $x + (1.2 + y)$

69. $3(x + 5) = 3 \cdot x + 3 \cdot 5$
$ = 3x + 15$

73. $-2(6x + 5y + 2z) = -2 \cdot 6x + (-2) \cdot 5y + (-2) \cdot 2z$
$ = -12x - 10y - 4z$

77. $-6x(y - 4) = -6x \cdot y - (-6x) \cdot 4$
$ = -6xy + 24x$

81. $\dfrac{1}{2}(4x - 9y) = \dfrac{1}{2} \cdot 4x - \dfrac{1}{2} \cdot 9y$
$\phantom{\dfrac{1}{2}(4x - 9y)} = 2x - \dfrac{9}{2}y$

85. $\dfrac{2}{3} + \left(-\dfrac{2}{3}\right) = 0$

89. $10(2y) = (10 \cdot 2)y$

93. $4 + 8y = 8y + 4$

97. $24 \div (6 \div 3) = 24 \div 2 = 12$
$(24 \div 6) \div 3 = 4 \div 3 = 4/3$
Division is not associative

Exercise Set 1.4

1. 2 **5.** 0 **9.** -2

13. $-14 + (-10) = -24$

17. $13 - 17 = 13 + (-17)$
$ = -4$

21. $19 - 10 - 11 = 19 + (-10) + (-11)$
$ = 9 + (-11)$
$ = -2$

25. $-\dfrac{4}{5} - \left(-\dfrac{3}{10}\right) = -\dfrac{4}{5} + \dfrac{3}{10}$
$\phantom{-\dfrac{4}{5} - \left(-\dfrac{3}{10}\right)} = -\dfrac{4 \cdot 2}{5 \cdot 2} + \dfrac{3}{10}$
$\phantom{-\dfrac{4}{5} - \left(-\dfrac{3}{10}\right)} = -\dfrac{8}{10} + \dfrac{3}{10}$
$\phantom{-\dfrac{4}{5} - \left(-\dfrac{3}{10}\right)} = -\dfrac{5}{10}$
$\phantom{-\dfrac{4}{5} - \left(-\dfrac{3}{10}\right)} = -\dfrac{1}{2}$

29. $-4 + 7 = 3$

33. $-4 - (-19) = -4 + 19$
$ = 15$

37. $16 - 8 - 9 = 16 + (-8) + (-9)$
$ = 8 + (-9)$
$ = -1$

41. $-5 \cdot 12 = -60$

45. $-8(-10) = 80$

49. $-17 \cdot 0 = 0$

53. $\dfrac{-9}{3} = -3$

57. $\dfrac{-12}{-4} = 3$

61. $\dfrac{0}{-5} = 0$

65. $-4(-2)(-1) = 8(-1)$
$= -8$

69. $\dfrac{-6}{7} \div 2 = \dfrac{-6}{7} \cdot \dfrac{1}{2}$

$= \dfrac{-6}{14}$ or $-\dfrac{6}{14}$ or $-\dfrac{3}{7}$

73. $-\dfrac{1}{6} \div \dfrac{9}{10} = -\dfrac{1}{6} \cdot \dfrac{10}{9} = -\dfrac{10}{54}$ or $-\dfrac{5}{27}$

77. $-2(-3.6) = 7.2$

81. $\dfrac{-5.2}{-1.3} = 4$

85. $\dfrac{3}{5} \div \left(-\dfrac{2}{5}\right) = \dfrac{3}{5} \cdot \left(-\dfrac{5}{2}\right) = -\dfrac{15}{10}$ or $-\dfrac{3}{2}$

89. $-7^2 = -(7 \cdot 7) = -49$

93. $(-2)^3 = (-2)(-2)(-2) = -8$

97. $-3^2 = -(3 \cdot 3) = -9$
$(-3)^2 = (-3)(-3) = 9$

101. $\sqrt{64} = 8$

105. $\sqrt{\dfrac{1}{16}} = \dfrac{1}{4}$

109. $\sqrt[4]{81} = 3$

113. $1 - \dfrac{1}{5} - \dfrac{3}{7} = \dfrac{1 \cdot 35}{1 \cdot 35} - \dfrac{1 \cdot 7}{5 \cdot 7} - \dfrac{3 \cdot 5}{7 \cdot 5}$

$= \dfrac{35}{35} - \dfrac{7}{35} - \dfrac{15}{35}$

$= \dfrac{35}{35} + \left(-\dfrac{7}{35}\right) + \left(-\dfrac{15}{35}\right)$

$= \dfrac{28}{35} + \left(-\dfrac{15}{35}\right)$

$= \dfrac{13}{35}$

117. Spinner b

121. Yes; two players have 6 points each (the third player has 0 points), or two players have 5 points each (the third player has 2 points)

125. $\sqrt{19.6} \approx 4.4272$

129. difference is $12.9\% - 2.1\% = 10.8\%$

Exercise Set 1.5

1. $3(5 - 7)^4 = 3(-2)^4 = 3(16)$
$= 48$

5. $\dfrac{3 - (-12)}{-5} = \dfrac{3 + 12}{-5} = \dfrac{15}{-5} = -3$

9. $(-3)^2 + 2^3 = 9 + 8 = 17$

13. $-9 \cdot 8 + 5(-6) = -72 - 30$
$= -102$

17. $-8\left(-\dfrac{3}{4}\right) - 8 = \dfrac{24}{4} - 8 = 6 - 8 = -2$

21. $5^2 - 3^4 = 25 - 81 = -56$

25. $2 \cdot (7 - 4 \cdot 5) = 2 \cdot (7 - 20)$
$= 2 \cdot (-13)$
$= -26$

29. $\dfrac{(-9 + 6)(-1^2)}{-2 - 2} = \dfrac{(-9 + 6)(-1)}{-2 - 2}$

$= \dfrac{(-3)(-1)}{-4}$

$= \dfrac{3}{-4} = -\dfrac{3}{4}$

33. $12 + \{6 - [5 - 2(-5)]\}$
$= 12 + \{6 - [5 + 10]\}$
$= 12 + (6 - 15)$
$= 12 - 9$
$= 3$

37. $\dfrac{(3 - \sqrt{9}) - (-5 - 1.3)}{-3} = \dfrac{(3 - 3) - (-5 - 1.3)}{-3}$

$= \dfrac{0 - (-6.3)}{-3}$

$= \dfrac{6.3}{-3}$

$= -2.1$

41. $\dfrac{3(-2 + 1)}{5} - \dfrac{-7(2 - 4)}{1 - (-2)} = \dfrac{3(-1)}{5} - \dfrac{-7(-2)}{1 + 2}$

$= \dfrac{-3}{5} - \dfrac{14}{3}$

$= -\dfrac{3}{5} - \dfrac{14}{3}$

$= -\dfrac{9}{15} - \dfrac{70}{15} = -\dfrac{79}{15}$

45. $3\{-2 + 5[1 - 2(-2 + 5)]\}$
$= 3\{-2 + 5[1 - 2(3)]\}$
$= 3\{-2 + 5[1 - 6]\}$
$= 3[-2 + 5(-5)] = 3[-2 - 25] = 3(-27) = -81$

49. $\left(\dfrac{5.6 - 8.4}{1.9 - 2.7}\right)^2 = \left(\dfrac{-2.8}{-0.8}\right)^2 = (3.5)^2 = 12.25$

53. a.

$x = 10$: $\dfrac{100(10) + 5000}{10} = \dfrac{1000 + 5000}{10} = \dfrac{6000}{10} = 600$

$x = 100$: $\dfrac{100(100) + 5000}{100} = \dfrac{10{,}000 + 5000}{100} = \dfrac{15{,}000}{100} = 150$

$x = 1000$: $\dfrac{100(1000) + 5000}{1000} = \dfrac{100{,}000 + 5000}{1000} = \dfrac{105{,}000}{1000} = 105$

b. decrease

57. $6x + 2x = (6 + 2)x = 8x$

61. $19y - y = (19 - 1)y = 18y$

65. $9x - 8 - 10x = 9x - 10x - 8$
$= (9 - 10)x - 8$
$= -1x - 8$
$= -x - 8$

69. $-9 + 4x + 18 - 10x$
$= 4x - 10x - 9 + 18$
$= (4 - 10)x + (-9 + 18)$
$= -6x + 9$

73. $x - y + x - y = x + x - y - y$
$= 1x + 1x - 1y - 1y$
$= (1 + 1)x + (-1 - 1)y$
$= 2x - 2y$

77. $\dfrac{3}{4}b - \dfrac{1}{2} + \dfrac{1}{6}b - \dfrac{2}{3}$

$= \dfrac{3}{4}b + \dfrac{1}{6}b - \dfrac{1}{2} - \dfrac{2}{3}$

$$= \left(\frac{3}{4} + \frac{1}{6}\right)b + \left(-\frac{1}{2} - \frac{2}{3}\right)$$
$$= \left(\frac{9}{12} + \frac{2}{12}\right)b + \left(-\frac{3}{6} - \frac{4}{6}\right)$$
$$= \frac{11}{12}b + \left(-\frac{7}{6}\right)$$
$$= \frac{11}{12}b - \frac{7}{6}$$

81. $-5(x - 1) = -5(x) - (-5)(1)$
$$= -5x + 5$$

85. $5k - (3k - 10) = 5k - 1(3k - 10)$
$$= 5k - 3k + 10$$
$$= 2k + 10$$

89. $3(x - 2) + x + 15 = 3x - 6 + x + 15$
$$= 3x + x - 6 + 15$$
$$= 4x + 9$$

93. $4(6n - 3) - 3(8n + 4) = 24n - 12 - 24n - 12$
$$= 24n - 24n - 12 - 12$$
$$= 0n - 24$$
$$= -24$$

97. $3x - 2(x - 5) + x = 3x - 2x + 10 + x$
$$= 3x - 2x + x + 10$$
$$= 2x + 10$$

101. $-1.2(5.7x - 3.6) + 8.75x$
$$= -6.84x + 4.32 + 8.75x$$
$$= -6.84x + 8.75x + 4.32$$
$$= 1.91x + 4.32$$

105. $(2 + 7) \cdot (1 + 3) = (9) \cdot (4) = 36$

109. 70 million

113. $\dfrac{-1.682 - 17.895}{(-7.102)(-4.691)} = \dfrac{-1.682 - 17.895}{33.315482} = \dfrac{-19.577}{33.315482} \approx -0.5876$

Exercise Set 1.6

1. $4^2 \cdot 4^3 = 4^{2+3} = 4^5$

5. $-7x^3 \cdot 20x^9 = (-7)(20)x^3x^9$
$$= -140x^{3+9}$$
$$= -140x^{12}$$

9. $(-4x^3p^2)(4y^3x^3)$
$$= (-4)(4)x^3x^3y^3p^2$$
$$= -16x^{3+3}y^3p^2$$
$$= -16x^6y^3p^2$$

13. $2x^3 \cdot 5x^7 = 2 \cdot 5 \cdot x^3x^7$
$$= 10x^{3+7}$$
$$= 10x^{10}$$

17. $(4x + 5)^0 = 1$

21. $4x^0 + 5 = 4 \cdot 1 + 5$
$$= 4 + 5$$
$$= 9$$

25. answers may vary

29. $\dfrac{x^9y^6}{x^8y^6} = x^{9-8} \cdot y^{6-6}$
$$= x^1 \cdot y^0$$
$$= x^1 \cdot 1$$
$$= x$$

33. $\dfrac{-36a^5b^7c^{10}}{6ab^3c^4} = \left(-\dfrac{36}{6}\right)a^{5-1} \cdot b^{7-3} \cdot c^{10-4}$
$$= -6a^6b^4c^6$$

37. $4^{-2} = \dfrac{1}{4^2} = \dfrac{1}{16}$

41. $5a^{-4} = 5 \cdot \dfrac{1}{a^4} = \dfrac{5}{a^4}$

45. $\dfrac{8r^4}{2r^{-4}} = \left(\dfrac{8}{2}\right)r^{4-(-4)}$
$$= 4r^{4+4}$$
$$= 4r^8$$

49. $4^{-1} + 3^{-2} = \dfrac{1}{4} + \dfrac{1}{3^2} = \dfrac{1}{4} + \dfrac{1}{9} = \dfrac{9}{36} + \dfrac{4}{36} = \dfrac{13}{36}$

53. $3x^{-1} = 3 \cdot \dfrac{1}{x} = \dfrac{3}{x}$

57. $\dfrac{x^{-7}y^{-2}}{x^2y^2} = x^{-7-2} \cdot y^{-2-2}$
$$= x^{-9} \cdot y^{-4}$$
$$= \dfrac{1}{x^9} \cdot \dfrac{1}{y^4} = \dfrac{1}{x^9y^4}$$

61. $\dfrac{(24x^8)x}{20x^{-7}} = \dfrac{24x^9}{20x^{-7}} = \left(\dfrac{24}{20}\right)x^{9-(-7)}$
$$= \left(\dfrac{6}{5}\right)x^{9+7}$$
$$= \left(\dfrac{6}{5}\right)x^{16} \quad \text{or} \quad \dfrac{6x^{16}}{5}$$

65. $\dfrac{x^{3t-1}}{x^t} = x^{3t-1-t} = x^{2t-1}$

69. $\dfrac{z^{6x}}{z^7} = z^{6x-7}$

73. $31{,}250{,}000 = 3.125 \times 10^7$

77. $67{,}413 = 6.7413 \times 10^4$

81. $0.000053 = 5.3 \times 10^{-5}$

85. $6{,}137{,}000{,}000 = 6.137 \times 10^9$

89. $0.001 = 1 \times 10^{-3}$

93. $9.3 \times 10^7 = 93{,}000{,}000$

97. $7.35 \times 10^{12} = 7{,}350{,}000{,}000{,}000$

101. $2.0 \times 10^8 = 200{,}000{,}000$

105. answers may vary

Exercise Set 1.7

1. $(3^{-1})^2 = 3^{-1 \cdot 2} = 3^{-2} = \dfrac{1}{3^2} = \dfrac{1}{9}$

5. $(y)^{-5} = y^{-5} = \dfrac{1}{y^5}$

9. $\left(\dfrac{2x^5}{y^{-3}}\right)^4 = \dfrac{2^4 \cdot (x^5)^4}{(y^{-3})^4} = \dfrac{16x^{20}}{y^{-12}}$
$$= 16x^{20}y^{12}$$

13. $\left(\dfrac{x^7y^{-3}}{z^{-4}}\right)^{-5} = \dfrac{(x^7)^{-5} \cdot (y^{-3})^{-5}}{(z^{-4})^{-5}}$
$$= \dfrac{x^{-35} \cdot y^{15}}{z^{20}} = \dfrac{y^{15}}{x^{35}z^{20}}$$

17. $(x^7)^{-9} = x^{-63} = \dfrac{1}{x^{63}}$

21. $(4x^2)^2 = 4^2 \cdot (x^2)^2 = 16x^4$

25. $\left(\dfrac{4^{-4}}{y^3x}\right)^{-2} = \dfrac{(4^{-4})^{-2}}{(y^3)^{-2} \cdot x^{-2}} = \dfrac{4^8}{y^{-6} \cdot x^{-2}} = 4^8 x^2 y^6$

29. $\left(\dfrac{a^{-4}}{a^{-5}}\right)^{-2} = a^{-4-(-5)} = (a^1)^{-2} = a^{-2} = \dfrac{1}{a^2}$

33. $\dfrac{4^{-1}x^2yz}{x^{-2}yz^3} = 4^{-1}x^{2-(-2)}y^{1-1}z^{1-3}$

$= 4^{-1}x^4y^0z^{-2}$

$= \dfrac{x^4}{4z^2}$

37. $(-8y^3xa^{-2})^{-3} = (-8)^{-3}(y^3)^{-3}(x)^{-3}(a^{-2})^{-3}$

$= (-8)^{-3}y^{-9}x^{-3}a^6$

$= \dfrac{a^6}{(-8)^3x^3y^9} = -\dfrac{a^6}{512x^3y^9}$

41. $\left(\dfrac{3x^5}{6x^4}\right)^4 = \left(\dfrac{x}{2}\right)^4 = \dfrac{x^4}{2^4} = \dfrac{x^4}{16}$

45. $\dfrac{(y^3)^{-4}}{y^3} = \dfrac{y^{-12}}{y^3} = y^{-12-3} = y^{-15} = \dfrac{1}{y^{15}}$

49. $(4x^6y^5)^{-2}(6x^4y^3)$

$= 4^{-2}(x^6)^{-2}(y^5)^{-2} \cdot 6x^4y^3$

$= 4^{-2}x^{-12}y^{-10} \cdot 6x^4y^3$

$= \dfrac{6}{16}x^{-12+4}y^{-10+3}$

$= \dfrac{3}{8}x^{-8}y^{-7} = \dfrac{3}{8x^8y^7}$

53. $\dfrac{2^{-3}x^2y^{-5}}{5^{-2}x^7y^{-1}} = \left(\dfrac{1}{2^3}\right)(5^2)x^{2-7}y^{-5-(-1)}$

$= \left(\dfrac{1}{8}\right)(25)x^{-5}y^{-4}$

$= \dfrac{25}{8x^5y^4}$

57. $(x^{3a+6})^3 = x^{(3a+6)3} = x^{9a+18}$

61. $(b^{5x-2})^{2x} = b^{(5x-2)2x} = b^{10x^2-4x}$

65. $\left(\dfrac{2x^{3t}}{x^{2t-1}}\right)^4 = [2x^{3t-(2t-1)}]^4$

$= (2x^{t+1})^4 = 2^4(x^{t+1})^4$

$= 16x^{(t+1)4}$

$= 16x^{4t+4}$

69. $(5 \times 10^{11})(2.9 \times 10^{-3})$

$= 5 \times 2.9 \times 10^{11} \times 10^{-3}$

$= 14.5 \times 10^8$

$= 1.45 \times 10^9$

73. $\dfrac{3.6 \times 10^{-4}}{9 \times 10^2} = \left(\dfrac{3.6}{9}\right)\left(\dfrac{10^{-4}}{10^2}\right)$

$= 0.4 \times 10^{-4-2}$

$= 0.4 \times 10^{-6}$

$= 4 \times 10^{-7}$

77. $\dfrac{18{,}200 \times 100}{91{,}000} = \dfrac{(1.82 \times 10^4) \times (1 \times 10^2)}{9.1 \times 10^4}$

$= \dfrac{(1.82)(1)}{9.1} \cdot \dfrac{10^4 \cdot 10^2}{10^4}$

$= 0.2 \times 10^{4+2-4}$

$= 0.2 \times 10^2$

$= 2 \times 10^1$

81. $\dfrac{0.00064 \times 2000}{16{,}000} = \dfrac{(6.4 \times 10^{-4}) \times (2 \times 10^3)}{1.6 \times 10^4}$

$= \dfrac{(6.4)(2)}{1.6} \cdot \dfrac{10^{-4} \cdot 10^3}{10^4}$

$= 8 \times 10^{-4+3-4}$

$= 8 \times 10^{-5}$

85. $\dfrac{1.25 \times 10^{15}}{(2.2 \times 10^{-2})(6.4 \times 10^{-5})} = \dfrac{1.25}{(2.2)(6.4)} \cdot \dfrac{10^{15}}{10^{-2} \cdot 10^{-5}}$

$\approx 0.08877840909 \times 10^{15-(-2)-(-5)}$

$= 0.08877840909 \times 10^{22}$

$= 8.877840909 \times 10^{20}$

89. $(1.64 \times 10^{-5})(3.8 \times 10^{-6})$

$= 1.64 \times 3.8 \times 10^{-5} \times 10^{-6}$

$= 6.232 \times 10^{-11}$

93. $M = DV$

$= (3.12 \times 10^{-2})(4.269 \times 10^{14})$

$= 3.12 \times 4.269 \times 10^{-2} \times 10^{14}$

$= 13.31928 \times 10^{12}$

$= 1.331928 \times 10^{13}$

97. $D = \dfrac{2.854 \times 10^8}{3.536 \times 10^6} = \left(\dfrac{2.854}{3.536}\right)\left(\dfrac{10^8}{10^6}\right)$

$\approx 0.81 \times 10^{8-6} = 0.81 \times 10^2 = 81$ people per sq mi

Chapter 1 Test

1. false　　**5.** true

9. $(4-9)^3 - |-4-6|^2$

$= (-5)^3 - |-10|^2$

$= (-5)^3 - (10)^2$

$= -125 - 100$

$= -225$

13. $\dfrac{5t-3q}{3r-1} = \dfrac{5(1)-3(4)}{3(-2)-1} = \dfrac{5-12}{-6-1} = \dfrac{-7}{-7} = 1$

17. $-2 = \dfrac{x}{x+5}$

21. multiplication property of 0

25. $\dfrac{6^{-1}a^2b^{-3}}{3^{-2}a^{-5}b^2} = \left(\dfrac{1}{6}\right)(3^2)a^{2-(-5)}b^{-3-2}$

$= \dfrac{9}{6}a^{2+5}b^{-3-2}$

$= \dfrac{3}{2}a^7b^{-5}$

$= \dfrac{3a^7}{2b^5}$

29. $630{,}000{,}000 = 6.3 \times 10^8$

33. $\dfrac{(0.0024)(0.00012)}{0.00032}$

$= \dfrac{(2.4 \times 10^{-3})(1.2 \times 10^{-4})}{3.2 \times 10^{-4}}$

$= \dfrac{(2.4)(1.2)}{3.2} \cdot \dfrac{10^{-3} \cdot 10^{-4}}{10^{-4}}$

$= 0.9 \times 10^{-3-4-(-4)}$

$= 0.9 \times 10^{-3}$

$= 9 \times 10^{-4}$

Chapter 2

Exercise Set 2.1

1. $\dfrac{x}{-6} = 4$

$\dfrac{-24}{-6} \overset{?}{=} 4$

$4 = 4$ True

-24 is a solution.

5.
$$5 + 3x = -1$$
$$5 + 3(-2) \overset{?}{=} -1$$
$$5 - 6 \overset{?}{=} -1$$
$$-1 = -1 \quad \text{True}$$
-2 is a solution.

9. $4(x - 3) = 12$
$$4(5 - 3) \overset{?}{=} 12$$
$$4(2) \overset{?}{=} 12$$
$$8 = 12 \quad \text{False}$$
5 is not a solution.

13. $-5x = -30$
$$\frac{-5x}{-5} = \frac{-30}{-5}$$
$$x = 6$$
Check:
$$-5(6) \overset{?}{=} -30$$
$$-30 = -30 \quad \text{True}$$
The solution set is $\{6\}$.

17.
$$x + 2.8 = 1.9$$
$$x + 2.8 - 2.8 = 1.9 - 2.8$$
$$x = -0.9$$
Check:
$$-0.9 + 2.8 \overset{?}{=} 1.9$$
$$1.9 = 1.9 \quad \text{True}$$
The solution set is $\{-0.9\}$.

21.
$$-4.1 - 7z = 3.6$$
$$-4.1 - 7z + 4.1 = 3.6 + 4.1$$
$$-7z = 7.7$$
$$\frac{-7z}{-7} = \frac{7.7}{-7}$$
$$z = -1.1$$
Check:
$$-4.1 - 7(-1.1) \overset{?}{=} 3.6$$
$$-4.1 + 7.7 \overset{?}{=} 3.6$$
$$3.6 = 3.6 \quad \text{True}$$
The solution set is $\{-1.1\}$.

25. $8x - 5x + 3 = x - 7 + 10$
$$3x + 3 = x + 3$$
$$2x = 0$$
$$x = 0$$
Check:
$$8(0) - 5(0) + 3 \overset{?}{=} 0 - 7 + 10$$
$$0 - 0 + 3 \overset{?}{=} 3$$
$$3 = 3 \quad \text{True}$$
The solution set is $\{0\}$.

29. $3(x - 6) = 5x$
$$3x - 18 = 5x$$
$$-18 = 2x$$
$$x = -9$$
Check:
$$3(-9 - 6) \overset{?}{=} 5(-9)$$
$$3(-15) \overset{?}{=} -45$$
$$-45 = -45 \quad \text{True}$$
The solution set is $\{-9\}$.

33. $-2(5y - 1) - y = -4(y - 3)$
$$-10y + 2 - y = -4y + 12$$
$$-11y + 2 = -4y + 12$$
$$-7y = 10$$
$$y = -\frac{10}{7}$$

Check:
$$-2\left[5\left(-\frac{10}{7}\right) - 1\right] - \left(-\frac{10}{7}\right) \overset{?}{=} -4\left(-\frac{10}{7} - 3\right)$$
$$-2\left(-\frac{50}{7} - 1\right) + \frac{10}{7} \overset{?}{=} -4\left(-\frac{31}{7}\right)$$
$$-2\left(-\frac{57}{7}\right) + \frac{10}{7} \overset{?}{=} \frac{124}{7}$$
$$\frac{114}{7} + \frac{10}{7} \overset{?}{=} \frac{124}{7}$$
$$\frac{124}{7} = \frac{124}{7} \quad \text{True}$$
The solution set is $\left\{-\frac{10}{7}\right\}$.

37. a. $4(x + 1) + 1 = 4x + 4 + 1 = 4x + 5$

b. $4(x + 1) + 1 = -7$
$$4x + 5 = -7$$
$$4x = -12$$
$$x = -3$$
The solution set is $\{-3\}$.

c. answers may vary

41.
$$\frac{3t}{4} - \frac{t}{2} = 1$$
$$4\left(\frac{3t}{4} - \frac{t}{2}\right) = 4(1)$$
$$3t - 2t = 4$$
$$t = 4$$
Check:
$$\frac{3(4)}{4} - \frac{4}{2} \overset{?}{=} 1$$
$$3 - 2 \overset{?}{=} 1$$
$$1 = 1 \quad \text{True}$$
The solution set is $\{4\}$.

45. $0.6x - 10 = 1.4x - 14$
$$4 = 0.8x$$
$$x = 5$$
Check:
$$0.6(5) - 10 \overset{?}{=} 1.4(5) - 14$$
$$3 - 10 \overset{?}{=} 7 - 14$$
$$-7 = -7 \quad \text{True}$$
The solution set is $\{5\}$.

49.
$$1.5(4 - x) = 1.3(2 - x)$$
$$10[1.5(4 - x)] = 10[1.3(2 - x)]$$
$$15(4 - x) = 13(2 - x)$$
$$60 - 15x = 26 - 13x$$
$$-2x = -34$$
$$x = 17$$
Check:
$$1.5(4 - 17) \overset{?}{=} 1.3(2 - 17)$$
$$1.5(-13) \overset{?}{=} 1.3(-15)$$
$$-19.5 = -19.5 \quad \text{True}$$
The solution set is $\{17\}$.

53. $3(x + 1) + 5 = 3x + 2$
$$3x + 3 + 5 = 3x + 2$$
$$3x + 8 = 3x + 2$$
$$3x + 8 - 3x = 3x + 2 - 3x$$
$$8 = 2$$
This is false for any x. Therefore, the solution set is $\varnothing$.

57. $2(x - 8) + x = 3(x - 6) + 2$
$$2x - 16 + x = 3x - 18 + 2$$
$$3x - 16 = 3x - 16$$
This is true for all x.
Therefore, all real numbers are solutions.
The solution set is $\{x | x \text{ is a real number}\}$.

61. $x - 10 = -6x + 4$
$$7x = 14$$
$$x = 2$$
Check:
$$2 - 10 \overset{?}{=} -6(2) + 4$$
$$-8 \overset{?}{=} -12 + 4$$
$$-8 = -8 \qquad \text{True}$$
The solution set is $\{2\}$.

65. $\dfrac{1}{4}(a + 2) = \dfrac{1}{6}(5 - a)$
$$12 \cdot \frac{1}{4}(a + 2) = 12 \cdot \frac{1}{6}(5 - a)$$
$$3(a + 2) = 2(5 - a)$$
$$3a + 6 = 10 - 2a$$
$$5a = 4$$
$$a = \frac{4}{5}$$
Check:
$$\frac{1}{4}\left(\frac{4}{5} + 2\right) \overset{?}{=} \frac{1}{6}\left(5 - \frac{4}{5}\right)$$
$$\frac{1}{4}\left(\frac{14}{5}\right) \overset{?}{=} \frac{1}{6}\left(\frac{21}{5}\right)$$
$$\frac{7}{10} = \frac{7}{10} \qquad \text{True}$$
The solution set is $\left\{\dfrac{4}{5}\right\}$.

69. $\dfrac{m - 4}{3} - \dfrac{3m - 1}{5} = 1$
$$15\left(\frac{m - 4}{3} - \frac{3m - 1}{5}\right) = 15(1)$$
$$5(m - 4) - 3(3m - 1) = 15$$
$$5m - 20 - 9m + 3 = 15$$
$$-4m - 17 = 15$$
$$-4m = 32$$
$$m = -8$$
Check:
$$\frac{-8 - 4}{3} - \frac{3(-8) - 1}{5} \overset{?}{=} 1$$
$$\frac{-12}{3} - \frac{-24 - 1}{5} \overset{?}{=} 1$$
$$-4 - \frac{-25}{5} \overset{?}{=} 1$$
$$-4 + 5 \overset{?}{=} 1$$
$$1 = 1 \quad \text{True}$$
The solution set is $\{-8\}$.

73. $\dfrac{1}{3}(y + 4) + 6 = \dfrac{1}{4}(3y - 1) - 2$
$$12\left[\frac{1}{3}(y + 4) + 6\right] = 12\left[\frac{1}{4}(3y - 1) - 2\right]$$
$$4(y + 4) + 72 = 3(3y - 1) - 24$$
$$4y + 16 + 72 = 9y - 3 - 24$$
$$4y + 88 = 9y - 27$$
$$4y + 88 - 9y = 9y - 27 - 9y$$
$$-5y + 88 = -27$$
$$-5y + 88 - 88 = -27 - 88$$
$$-5y = -115$$
$$\frac{-5y}{-5} = \frac{-115}{-5}$$
$$y = 23$$
Check:
$$\frac{1}{3}(23 + 4) + 6 \overset{?}{=} \frac{1}{4}(3 \cdot 23 - 1) - 2$$
$$\frac{1}{3}(27) + 6 \overset{?}{=} \frac{1}{4}(68) - 2$$

$$9 + 6 \overset{?}{=} 17 - 2$$
$$15 = 15 \qquad \text{True}$$
The solution set is $\{23\}$.

77. $\dfrac{8}{x}$ **81.** $3x + 2$

85. $x(x - 6) + 7 = x(x + 1)$
$$x^2 - 6x + 7 = x^2 + x$$
$$x^2 - 6x + 7 - x^2 = x^2 + x - x^2$$
$$-6x + 7 = x$$
$$-7x = -7$$
$$x = 1$$
Check:
$$1(1 - 6) + 7 \overset{?}{=} 1(1 + 1)$$
$$-5 + 7 \overset{?}{=} 2$$
$$2 = 2 \qquad \text{True}$$
The solution set is $\{1\}$.

Exercise Set 2.2

1. $4y$

5. $5x + 10(x + 3)$
$$= 5x + 10x + 30$$
$$= 15x + 30$$
$(15x + 30)$ cents

9. $4(x - 2) = 2 + 6x$
$$4x - 8 = 2 + 6x$$
$$-10 = 2x$$
$$-5 = x$$

13. $29\% \cdot 2271 = 0.29 \cdot 2271 = 658.59$
Approximately 658.59 million acres are federally owned.

17. Let x = number of online shoppers who don't spend more than they intend.
$$100\% - 33\frac{1}{3}\% = 67\frac{2}{3}\%.$$
Also, $67\frac{2}{3}\% = 0.67\frac{2}{3} = \frac{2}{3}$
$$x = \frac{2}{3} \cdot 1290$$
$$x = 860$$
There are 860 shoppers.

21. Let x = the number of users who check their e-mail about once a week.
$$x = 6\% \cdot 112{,}500$$
$$x = 0.06 \cdot 112{,}500$$
$$x = 6750$$
There are 6750 users.

25. Let $x, 5x,$ and $6x - 3$ be the lengths of the three sides.
$$x + 5x + (6x - 3) = 483$$
$$12x - 3 = 483$$
$$12x = 486$$
$$x = \frac{486}{12} = 40.5$$
$$x = 40.5$$
$$5x = 5(40.5) = 202.5$$
$$6x - 3 = 6 \cdot 40.5 - 3$$
$$= 243 - 3$$
$$= 240$$
The sides are 40.5 feet, 202.5 feet, and 240 feet.

29. Let x = seats in B737-200, the $x + 104$ = seats in B767-300ER
$$x + (x + 104) = 328$$
$$2x + 104 = 328$$
$$2x = 224$$

$$x = 112$$
$$x + 104 = 216$$

B737-200 has 112 seats.
B767-300ER has 216 seats.

33. Let x = the population in 2000.
$$x + 2\% \cdot x = 29.2$$
$$x + 0.02x = 29.2$$
$$1.02x = 29.2$$
$$x = \frac{29.2}{1.02} \approx 28.6$$

The population was about 28.6 million people.

37. Let $x, x + 1,$ and $x + 2$ be the consecutive integers.
$$x + (x + 1) + (x + 2) = 228$$
$$3x + 3 = 228$$
$$3x = 225$$
$$x = 75$$
$$x = 75$$

41. Let x = the height. Then $2x + 12$ is the width.
$$2(x) + 2(2x + 12) = 312$$
$$2x + 4x + 24 = 312$$
$$6x + 24 = 312$$
$$6x = 288$$
$$x = 48$$
$$2x + 12 = 2 \cdot 48 + 12 = 96 + 12$$
$$= 108$$

Height is 48 inches.
Width is 108 inches.

45. Let x = the population in 2050.
$$x = 43,600,000 - 25\% \cdot 43,600,000$$
$$x = 43,600,000 - 0.25 \cdot 43,600,000$$
$$x = 43,600,000 - 10,900,000$$
$$x = 32,700,000$$

The expected population in 2050 is 32,700,000 people.

49.
$$x + (x + 20) = 180$$
$$2x + 20 = 180$$
$$2x = 160$$
$$x = 80$$
$$x + 20 = 80 + 20 = 100$$

The angles measure 80° and 100°.

53. Let x = measure of first angle, then $3x + 20$ = measure of second angle.
$$x + (3x + 20) = 180$$
$$4x + 20 = 180$$
$$4x = 160$$
$$x = 40$$
$$3x + 20 = 3 \cdot 40 + 20 = 120 + 20 = 140$$

The angles measure 40° and 140°.

57. Let x = number of hours for a halogen bulb, then $25x$ = number of hours for a fluorescent bulb and $x - 2500$ = number of hours for an incandescent bulb.
$$x + 25x + (x - 2500) = 105,500$$
$$27x + 2500 = 105,500$$
$$27x = 108,000$$
$$x = 4000$$
$$25x = 25 \cdot 4000 = 100,000$$
$$x - 2500 = 4000 - 2500 = 1500$$

A halogen bulb has 4000 bulb hours, a fluorescent bulb has 100,000 bulb hours, and an incandescent bulb has 1500 bulb hours.

61. $2a + b - c = 2(5) + (-1) - (3)$
$$= 10 - 1 - 3$$
$$= 9 - 3$$
$$= 6$$

65. $n^2 - m^2 = (-3)^2 - (-8)^2$
$$= 9 - 64$$
$$= -55$$

69. Let x = number of trees' worth of newsprint recycled each year.
$$\frac{x}{30} = \frac{0.27}{0.73}$$
$$x = \frac{0.27 \cdot 30}{0.73} \approx 11$$

About 11 million trees are needed.

73. Let x = first odd integer, then $x + 2$ = next odd integer.
$$7x - 5(x + 2) = 54$$
$$7x - 5x - 10 = 54$$
$$2x - 10 = 54$$
$$2x = 64$$
$$x = 32 \quad \text{(which is an even integer)}$$

No such odd integers exists.

Exercise Set 2.3

1. $d = rt$
$$\frac{d}{r} = \frac{rt}{r}$$
$$\frac{d}{r} = t$$

5.
$$P = a + b + c$$
$$P - a - b = a + b + c - a - b$$
$$P - a - b = c$$

9.
$$P = 2l + 2w$$
$$P - 2w = 2l + 2w - 2w$$
$$P - 2w = 2l$$
$$\frac{P - 2w}{2} = \frac{2l}{2}$$
$$\frac{P - 2w}{2} = l$$

13. $5x + 4y = 20$
$$4y = 20 - 5x$$
$$\frac{4y}{4} = \frac{20 - 5x}{4}$$
$$y = \frac{20 - 5x}{4} \quad or \quad 5 - \frac{5}{4}x$$

17. $C = 2\pi r$
$$\frac{C}{2\pi} = \frac{2\pi r}{2\pi}$$
$$\frac{C}{2\pi} = r$$

21. $A = P\left(1 + \dfrac{r}{n}\right)^{nt}$

$n = 1: \quad A = 3500\left(1 + \dfrac{0.03}{1}\right)^{1 \cdot 10}$

$A = 3500(1.03)^{10}$
$A \approx \$4703.71$

$n = 2: \quad A = 3500\left(1 + \dfrac{0.03}{2}\right)^{2 \cdot 10}$

$A = 3500(1.015)^{20}$
$A \approx \$4713.99$

$n = 4: \quad A = 3500\left(1 + \dfrac{0.03}{4}\right)^{4 \cdot 10}$

$A = 3500(1.0075)^{40}$
$A \approx \$4719.22$

$n = 12: \quad A = 3500\left(1 + \dfrac{0.03}{12}\right)^{12 \cdot 10}$

$A = 3500(1.0025)^{120}$
$A \approx \$4722.74$

$n = 365: \quad A = 3500\left(1 + \dfrac{0.03}{365}\right)^{365 \cdot 10}$

$A \approx \$4724.45$

25. $C = \dfrac{5}{9}(F - 32)$

$C = \dfrac{5}{9}(104 - 32)$

$C = \dfrac{5}{9}(72)$

$C = 40$
40°C

29. $A = l \cdot w$
$A = 64 \cdot 64$
$A = 4096 \text{ sq ft}$

Packages $= \dfrac{4096}{24} = 170{,}6$

Buy 171 packages of tiles.

33. $A = 2(l \cdot h) + 2(w \cdot h)$
$A = 2(14 \cdot 8) + 2(16 \cdot 8)$
$A = 224 + 256 = 480 \text{ sq ft}$
Two coats $= 2 \cdot 480$
$\qquad\qquad = 960 \text{ sq ft}$

Gallons $= \dfrac{960}{500} = 1.92$

Buy 2 gallons of paint.

37. Radius of satellite orbit $= 22{,}248 + 4000 = 26{,}248 \text{ mi}$
$C = 2\pi r = 2 \cdot \pi \cdot 26{,}248$
$\qquad \approx 164{,}921 \text{ mi}$

41. $C = \pi d = \pi \cdot 41.125 = 41.125\pi \text{ ft}$
$\qquad\qquad\qquad \approx 41.125 \cdot 3.14$
$\qquad\qquad\qquad = 129.1325 \text{ ft}$

45. $C = 4h + 9f + 4p$
$C = 4 \cdot 7 + 9 \cdot 14 + 4 \cdot 6$
$C = 28 + 126 + 24 = 178$
178 cal

49. $\{-3, -2, -1\}$

53. answers may vary

57. $C = \dfrac{1{,}700{,}000{,}000}{250{,}000{,}000} = 6.8$

Cost is \$6.80 per person.

61. $N = R^* \times f_p \times n_e \times f_l \times f_i \times f_c \times L$

$\dfrac{N}{R^* \times f_p \times f_l \times f_i \times f_c \times L} = \dfrac{R^* \times f_p \times n_e \times f_l \times f_i \times f_c \times L}{R^* \times f_p \times f_l \times f_i \times f_c \times L}$

$\dfrac{N}{R^* \times f_p \times f_l \times f_i \times f_c \times L} = n_e$

Exercise Set 2.4

1. $(-\infty, -3)$

5. $(5, \infty)$

9. $(-1, 5)$

13. $x - 7 \geq -9$
$\quad x \geq -2$
$[-2, \infty)$

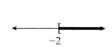

17. $8x - 7 \leq 7x - 5$
$\quad x - 7 \leq -5$
$\qquad x \leq 2$
$(-\infty, 2]$

21. $5x < -23.5$
$\quad x < -4.7$
$(-\infty, -4.7)$

25. $-x < -4$
$\quad x > 4$
$(4, \infty)$

29. $15 + 2x \geq 4x - 7$
$\qquad 15 \geq 2x - 7$
$\qquad 22 \geq 2x$
$\qquad 11 \geq x$
$(-\infty, 11]$

33. $\dfrac{1}{2} + \dfrac{2}{3} \geq \dfrac{x}{6}$

$6\left(\dfrac{1}{2} + \dfrac{2}{3}\right) \geq 6\left(\dfrac{x}{6}\right)$

$\qquad 3 + 4 \geq x$
$\qquad\qquad 7 \geq x$
$(-\infty, 7]$

37. $\dfrac{1}{4}(x - 7) \geq x + 2$

$4\left[\dfrac{1}{4}(x - 7)\right] \geq 4(x + 2)$

$\qquad x - 7 \geq 4x + 8$
$\qquad\quad -15 \geq 3x$
$\qquad\qquad -5 \geq x$
$(-\infty, -5]$

41. $4(2x + 1) > 4$
$\quad 8x + 4 > 4$
$\qquad 8x > 0$
$\qquad\, x > 0$
$(0, \infty)$

45. $4(x - 6) + 2x - 4 \geq 3(x - 7) + 10x$
$\quad 4x - 24 + 2x - 4 \geq 3x - 21 + 10x$
$\qquad\qquad 6x - 28 \geq 13x - 21$
$\qquad\qquad\qquad -7x \geq 7$
$\qquad\qquad\qquad\quad x \leq -1$
$(-\infty, -1]$

49. $14 - (5x - 6) \geq -6(x + 1) - 5$
$14 - 5x + 6 \geq -6x - 6 - 5$
$-5x + 20 \geq -6x - 11$
$x + 20 \geq -11$
$x \geq -31$
$[-31, \infty)$

53. Let $x =$ minimum score.
$\dfrac{2x + 72 + 67 + 82 + 79}{6} \geq 60$

$\dfrac{2x + 300}{6} \geq 60$

$2x + 300 \geq 360$
$2x \geq 60$
$x \geq 30$
The minimum score is 30.

57. Let $x =$ the maximum weight of a package.
$23x + 34 \leq 310$
$23x \leq 276$
$x \leq 12$
The maximum weight of a package is 12 oz.

61. $F \geq \dfrac{9}{5} C + 32$

$F \geq \dfrac{9}{5} \cdot 500 + 32$

$F \geq 900 + 32$
$F \geq 932$
The minimum temperature is 932°F.

65. decreasing; answers may vary

69. Let $t =$ time for consumption to be less than 55 pounds.
$-1.43t + 79.37 \leq 55$
$-1.43t \leq -24.37$
$t \geq 17.042$
This will occur in the 17th year beyond 1993:
$1993 + 17 = 2010$
The year is 2010 for the consumption to be less than 55 pounds per person.

73. $w = s$
$-1.43t + 79.37 = 1.18t + 28.31$
$-1.43t + 51.06 = 1.18t$
$51.06 = 2.61t$
$19.6 \approx t$
or $t \approx 20$ years
$1993 + 20 = 2013$
The year is 2013 when $w = s$.

77. $x < 6$ and $x < -5$ means $x < -5$.
$\{\ldots -9, -8, -7, -6\}$

81. $\{x | -2.5 \leq x < 5.3\}$
$[-2.5, 5.3)$

85. $8(x + 3) \leq 7(x + 5) + x$
$8x + 24 \leq 7x + 35 + x$
$8x + 24 \leq 8x + 35$
$24 \leq 35$
This is true for all x. Thus, the solution is the entire real number line $(-\infty, \infty)$.

Exercise Set 2.5

1. $A \cap C = \{2, 4\}$ **5.** $B \cap C = \{3, 5\}$

9. $x \leq -3$

$x \geq -2$

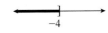

$x \leq -3$ and $x \geq -2$

The solution set is Ø.

13. $x < 5$ and $x > -2$
$-2 < x < 5$
The solution set is $(-2, 5)$.

17. $4x + 2 \leq -10$ and $2x < 0$
$4x \leq -12$ and $x < 0$
$x \leq -3$ and $x < 0$
The solution set is $(-\infty, -3)$.

21. $5 < x - 6 < 11$
$11 < x < 17$
The solution set is $(11, 17)$.

25. $1 \leq \dfrac{2}{3} x + 3 \leq 4$

$-2 \leq \dfrac{2}{3} x \leq 1$

$-3 \leq x \leq \dfrac{3}{2}$

The solution set is $\left[-3, \dfrac{3}{2} \right]$.

29. $0 \leq 2x - 3 \leq 9$
$3 \leq 2x \leq 12$
$\dfrac{3}{2} \leq x \leq 6$

The solution set is $\left[\dfrac{3}{2}, 6 \right]$.

33. $A \cup B = \{1, 2, 3, 4, 5, 6, 7, 8\}$

37. $C \cup D = \{2, 4, 6, 8\}$

41. $x \leq -4$

$x \geq 1$

$x \leq -4$ or $x \geq 1$

45. $x < -1$ or $x > 0$
The solution set is $(-\infty, -1) \cup (0, \infty)$.

49. $3(x - 1) < 12$ or $x + 7 > 10$
$x - 1 < 4$ or $x > 3$
$x < 5$ or $x > 3$
The solution set is all real numbers or $(-\infty, \infty)$.

53. $3x \geq 5$ or $-x - 6 < 1$
$x \geq \dfrac{5}{3}$ or $-x < 7$

$x \geq \dfrac{5}{3}$ or $x > -7$

The solution set is $(-7, \infty)$.

57. $|-7| - |19| = 7 - 19 = -12$

61. $|x| = 7$
   $x = -7, 7$

65. The years that the consumption of pork was greater than 48
   pounds per person were 1992, 1993, 1994, 1995, 1998, and 1999.
   The years that the consumption of chicken was greater than 48
   pounds per person were 1993, 1994, 1995, 1996, 1997, 1998, and 1999.
   The years in common are 1993, 1994, 1995, 1998, and 1999.

69. $x - 6 < 3x < 2x + 5$
   $x - 6 < 3x$ $\qquad$ and $\quad 3x < 2x + 5$
   $\quad -6 < 2x$ $\qquad$ and $\quad x < 5$
   $\quad -3 < x$ $\qquad$ and $\quad x < 5$
   The solution set is $(-3, 5)$.

Exercise Set 2.6

1. $|x| = 7$
   $x = 7$ or $x = -7$
   The solution set is $\{7, -7\}$.

5. $|3x| = 12.6$
   $3x = 12.6$ or $3x = -12.6$
   $x = 4.2$ or $x = -4.2$
   The solution set is $\{4.2, -4.2\}$.

9. $\qquad -6|x| + 44 = -10$
   $\qquad\qquad -6|x| = -54$
   $\qquad\qquad\quad |x| = 9$
   $x = 9$ or $x = -9$
   The solution set is $\{-9, 9\}$.

13. $|2x - 5| = 9$
   $2x - 5 = 9$ or $2x - 5 = -9$
   $\quad 2x = 14$ or $\quad 2x = -4$
   $\quad\; x = 7$ or $\quad\; x = -2$
   The solution set is $\{7, -2\}$.

17. $|z| + 4 = 9$
   $\quad |z| = 5$
   $\quad\; z = 5$ or $z = -5$
   The solution set is $\{5, -5\}$.

21. $\left| \dfrac{4x - 6}{3} \right| = 6$

   $\dfrac{4x - 6}{3} = 6$ or $\dfrac{4x - 6}{3} = -6$

   $4x - 6 = 18$ or $4x - 6 = -18$
   $\quad 4x = 24$ or $\quad 4x = -12$
   $\quad\; x = 6$ or $\quad\; x = -3$
   The solution set is $\{-3, 6\}$.

25. $|4n + 1| + 10 = 4$
   $|4n + 1| = -6$
   Since $|4n + 1| \geq 0$, There are no values satisfying
   $|4n + 1| = -6$
   The solution set is $\{\ \}$ or $\varnothing$.

29. $|5x - 7| = |3x + 11|$
   $5x - 7 = 3x + 11$ or $5x - 7 = -(3x + 11)$
   $\quad 2x - 7 = 11$ or $5x - 7 = -3x - 11$
   $\quad\quad 2x = 18$ or $8x - 7 = -11$
   $\quad\quad\; x = 9$ or $\quad\; 8x = -4$
   $\qquad\qquad\qquad\qquad\qquad x = -\dfrac{4}{8} = -\dfrac{1}{2}$

   The solution set is $\left\{ -\dfrac{1}{2}, 9 \right\}$.

33. $|2y - 3| = |9 - 4y|$
   $2y - 3 = 9 - 4y$ or $2y - 3 = -(9 - 4y)$
   $\quad 6y = 12$ or $2y - 3 = -9 + 4y$

$y = 2$ $\qquad$ or $\qquad 6 = 2y$
$y = 2$ $\qquad$ or $\qquad 3 = y$
The solution set is $\{2, 3\}$.

37. $|2x - 6| = |10 - 2x|$
   $2x - 6 = 10 - 2x$ or $2x - 6 = -(10 - 2x)$
   $4x - 6 = 10$ or $2x - 6 = -10 + 2x$
   $\quad 4x = 16$ or $\quad -6 = -10$ $\qquad$ false
   $\quad\; x = 4$
   The solution set is $\{4\}$.

41. $\left| \dfrac{2x + 1}{5} \right| = \left| \dfrac{3x - 7}{3} \right|$

   $\dfrac{2x + 1}{5} = \dfrac{3x - 7}{3}$ $\quad$ or $\quad \dfrac{2x + 1}{5} = -\left(\dfrac{3x - 7}{3} \right)$

   $15 \cdot \dfrac{2x + 1}{5} = 15 \cdot \dfrac{3x - 7}{3}$ $\quad$ or $\quad \dfrac{2x + 1}{5} = \dfrac{-3x + 7}{3}$

   $3(2x + 1) = 5(3x - 7)$ $\quad$ or $\quad 15 \cdot \dfrac{2x + 1}{5} = 15 \cdot \dfrac{-3x + 7}{3}$

   $6x + 3 = 15x - 35$ $\quad$ or $\quad 3(2x + 1) = 5(-3x + 7)$
   $-9x + 3 = -35$ $\quad$ or $\quad 6x + 3 = -15x + 35$
   $\quad -9x = -38$ $\quad$ or $\quad 21x + 3 = 35$

   $\quad\; x = \dfrac{-38}{-9} = \dfrac{38}{9}$ $\quad$ or $\quad 21x = 32$

   $\qquad\qquad\qquad\qquad\qquad\qquad\qquad x = \dfrac{32}{21}$

   The solution set is $\left\{ \dfrac{32}{21}, \dfrac{38}{9} \right\}$.

45. $|x| \leq 4$
   $-4 \leq x \leq 4$
   The solution set is $[-4, 4]$.

49. $|x + 3| < 2$
   $-2 < x + 3 < 2$
   $-5 < x < -1$
   The solution set is $(-5, -1)$.

53. $\left| \dfrac{x + 2}{3} \right| < 1$

   $-1 < \dfrac{x + 2}{3} < 1$

   $-3 < x + 2 < 3$
   $-5 < x < 1$
   The solution set is $(-5, 1)$.

57. $|x| + 2 > 6$
   $\quad |x| > 4$
   $x < -4$ or $x > 4$
   The solution set is $(-\infty - 4) \cup (4, \infty)$.

61. $|x + 10| \geq 14$
   $x + 10 \leq -14$ or $x + 10 \geq 14$
   $\quad x \leq -24$ or $\qquad x \geq 4$
   The solution set is $(-\infty, -24] \cup [4, \infty)$.

65. $|x| > -4$

The absolute value of a number is always nonnegative. Thus, it will always be greater than -4.

The solution set is all real numbers or $(-\infty, \infty)$.

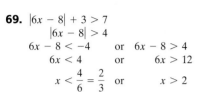

69. $|6x - 8| + 3 > 7$

$$|6x - 8| > 4$$

$6x - 8 < -4 \quad$ or $\quad 6x - 8 > 4$

$6x < 4 \quad$ or $\quad 6x > 12$

$x < \dfrac{4}{6} = \dfrac{2}{3} \quad$ or $\quad x > 2$

The solution set is $\left(-\infty, \dfrac{2}{3}\right) \cup (2, \infty)$.

73. $\left|\dfrac{x + 6}{3}\right| > 2$

$\dfrac{x + 6}{3} < -2 \quad$ or $\quad \dfrac{x + 6}{3} > 2$

$x + 6 < -6 \quad$ or $\quad x + 6 > 6$

$x < -12 \quad$ or $\quad x > 0$

The solution set is $(-\infty, -12) \cup (0, \infty)$.

77. $|x| > 13$

$x < -13 \quad$ or $\quad x > 13$

The solution set is $(-\infty, -13) \cup (13, \infty)$.

81. $2|x| - 9 \le 11$

$2|x| \le 20$

$|x| \le 10$

$-10 \le x \le 10$

The solution set is $[-10, 10]$.

85. $|2x - 3| = 7$

$2x - 3 = 7 \quad$ or $\quad 2x - 3 = -7$

$2x = 10 \quad$ or $\quad 2x = -4$

$x = 5 \quad$ or $\quad x = -2$

The solution set is $\{-2, 5\}$.

89. $|9 + 4x| = 0$

$9 + 4x = 0$

$4x = -9$

$x = \dfrac{-9}{4} \quad$ or $\quad -\dfrac{9}{4}.$

The solution set is $\left\{-\dfrac{9}{4}\right\}$.

93. $\left|\dfrac{1}{3}x + 1\right| > 5$

$\dfrac{1}{3}x + 1 < -5 \quad$ or $\quad \dfrac{1}{3}x + 1 > 5$

$\dfrac{1}{3}x < -6 \quad$ or $\quad \dfrac{1}{3}x > 4$

$x < -18 \quad$ or $\quad x > 12$

The solution set is $(-\infty, -18) \cup (12, \infty)$.

97. $|x + 11| = -1$

The absolute value of a number is always nonnegative. Thus, it cannot be equal to -1.

The solution set is $\{\ \}$ or $\varnothing$.

101. $\left|\dfrac{3x - 5}{6}\right| > 5$

$\dfrac{3x - 5}{6} < -5 \quad$ or $\quad \dfrac{3x - 5}{6} > 5$

$3x - 5 < -30 \quad$ or $\quad 3x - 5 > 30$

$3x < -25 \quad$ or $\quad 3x > 35$

$x < -\dfrac{25}{3} \quad$ or $\quad x > \dfrac{35}{3}$

The solution set is $\left(-\infty, -\dfrac{25}{3}\right) \cup \left(\dfrac{35}{3}, \infty\right)$.

105. $100 - (3 + 8 + 10 + 13 + 34)$

$= 100 - 68$

$= 32$

Mozzarella made up 32% of U.S. cheese production.

109. $3x - 4y = 12$

$3 \cdot 2 - 4y = 12$

$6 - 4y = 12$

$-4y = 6$

$y = \dfrac{6}{-4} = -\dfrac{6}{4} \quad$ or $\quad -1.5$

113. $|x| = 5$

117. $|x| \le 5$; answers may vary

Chapter 2 Test

1. $8x + 14 = 5x + 44$

$3x = 30$

$x = 10$

The solution set is $\{10\}$.

5. $\dfrac{z}{2} + \dfrac{z}{3} = 10$

$6\left(\dfrac{z}{2} + \dfrac{z}{3}\right) = 6(10)$

$3z + 2z = 60$

$5z = 60$

$z = 12$

The solution set is $\{12\}$.

9. $|2x - 3| = |4x + 5|$

$2x - 3 = -(4x + 5) \quad$ or $\quad 2x - 3 = 4x + 5$

$2x - 3 = -4x - 5 \quad$ or $\quad -2x = 8$

$6x = -2 \quad$ or $\quad x = -4$

$x = -\dfrac{1}{3} \quad$ or $\quad x = -4$

The solution set is $\left\{-4, -\dfrac{1}{3}\right\}$.

13. $F = \dfrac{9}{5}C + 32$

$F - 32 = \dfrac{9}{5}C$

$C = \dfrac{5}{9}(F - 32)$

17. $|3x + 1| > 5$

$3x + 1 < -5 \quad$ or $\quad 3x + 1 > 5$

$3x < -6 \quad$ or $\quad 3x > 4$

$x < -2 \quad$ or $\quad x > \dfrac{4}{3}$

The solution set is $(-\infty, -2) \cup \left(\dfrac{4}{3}, \infty\right)$.

21. $-x > 1$ and $3x + 3 \geq x - 3$
   $x < -1$ and $2x \geq -6$
   $x < -1$ and $x \geq -3$
   $-3 \leq x < -1$
   The solution set is $[-3, -1)$.

25. Recall that $C = 2\pi r$. Here $C = 78.5$.
   $78.5 = 2\pi r$
   $r = \dfrac{78.5}{2\pi} = \dfrac{39.25}{\pi}$
   Also, recall that $A = \pi r^2$.
   $A = \pi \left(\dfrac{39.25}{\pi}\right)^2 = \dfrac{39.25^2}{3.14} \approx 490.63$

   Dividing this by 60 yields approximately 8.18. Therefore, about 8 hunting dogs could safely be kept in the pen.

Chapter 3

Exercise Set 3.1

1.

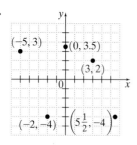

(3, 2) lies in quadrant I.
(−5, 3) lies in quadrant II.
$\left(5\dfrac{1}{2}, -4\right)$ lies quadrant IV.
(0, 3.5) lies on the y-axis.
(−2, −4) lies in quadrant III.

5. $(x, 0)$ lies on the x-axis.

9.

Year, x	Revenue (billions of dollars), y
1996	75
1997	79
1998	81
1999	84
2000	94

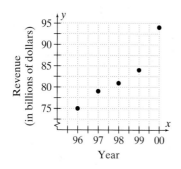

13. Let $x = 1, y = 0$.
   $-6x + 5y = -6$
   $-6(1) + 5(0) \overset{?}{=} -6$

$-6 + 0 \overset{?}{=} -6$
   $-6 \overset{?}{=} -6$
True; yes

Let $x = 2, y = \dfrac{6}{5}$.
   $-6x + 5y = -6$
   $-6(2) + 5\left(\dfrac{6}{5}\right) \overset{?}{=} -6$
   $-12 + 6 \overset{?}{=} -6$
   $-6 \overset{?}{=} -6$
True; yes

17. $x - 2y = 4$
   Find three ordered pair solutions.

x	y
4	0
0	−2
−2	−3

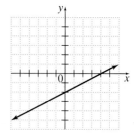

21. $x = 4$
   Any ordered pair with an x-coordinate of 4 is a solution to $x = 4$.

x	y
4	−1
4	0
4	1

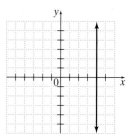

25. $y = 3x$
   Find three ordered pair solutions.

x	y
0	0
1	3
−1	−3

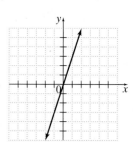

29. $4x + 5y = 15$
   Find three ordered pair solutions.

x	y
0	3
5	−1
−5	7

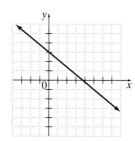

33. $x = \dfrac{1}{2}$

This is a vertical line with an x-intercept at $\left(\dfrac{1}{2}, 0\right)$ and no y-intercept.

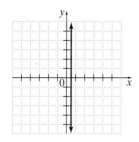

37. $y = -4x + 1$

Find three ordered pair solutions.

x	y
0	1
1	-3
-1	5

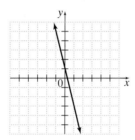

41. $y = 2$ matches graph C.

45. $\dfrac{-6 - 3}{2 - 8} = \dfrac{-9}{-6} = \dfrac{3}{2}$

49. $\dfrac{0 - 6}{5 - 0} = \dfrac{-6}{5} = -\dfrac{6}{5}$

53. The description matches graph C.

57. answers may vary

61.

x	$y = 2x$	$y = 2x - 5$	$y = 2x + 5$
0	0	-5	5
1	2	-3	7
-1	-2	-7	3

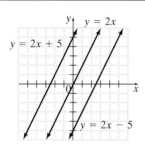

The lines are parallel.

65. $y = -4x + 1$

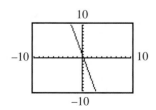

Exercise Set 3.2

1. $m = \dfrac{11 - 2}{8 - 3} = \dfrac{9}{5}$

5. $m = \dfrac{3 - 8}{4 - (-2)} = -\dfrac{5}{6}$

9. $m = \dfrac{11 - (-1)}{-12 - (-3)} = \dfrac{12}{-9} = -\dfrac{4}{3}$

13. $m = \dfrac{8}{12} = \dfrac{2}{3}$

17. $y = 5x - 2$

$m = 5$, y-intercept $= (0, -2)$

21. $2x - 3y = 10$

$-3y = -2x + 10$

$y = \dfrac{2}{3}x - \dfrac{10}{3}$

$m = \dfrac{2}{3}$, y-intercept $= \left(0, -\dfrac{10}{3}\right)$

25. $3x + 9 = y$

$y = 3x + 9$

$m = 3$, y-intercept $= (0, 9)$

29. $y = -2x + 3$

$m = -2$, y-intercept $= (0, 3)$

B

33. $y = -x + 5$

$m = -1$

37. $x = 4$

m is undefined

41. $x + 2 = 0$

$x = -2$

m is undefined

45. l_2 has the greater slope because the slope of l_2 is 0 and the slope of l_1 is negative.

49.

$y = -3x + 6$ $y = 3x + 5$

$m = -3$ $m = 3$

y-int $= (0, 6)$ y-int $= (0, 5)$

Neither, since their slopes are not equal, nor does their product equal -1.

53.

$-2x + 3y = 1$ $3x + 2y = 12$

$3y = 2x + 1$ $2y = -3x + 12$

$y = \dfrac{2}{3}x + \dfrac{1}{3}$ $y = -\dfrac{3}{2}x + 6$

$m = \dfrac{2}{3}$ $m = -\dfrac{3}{2}$

y-int $= \left(0, \dfrac{1}{3}\right)$ y-int $= (0, 6)$

Perpendicular, since the product of their slopes is -1.

57.

$y = 12x + 6$ $y = 12x - 2$

$m = 12$ $m = 12$

y-int $= (0, 6)$ y-int $= (0, -2)$

Parallel, since they have the same slope but different y-intercepts.

61. $5x - 2y = 6$

$-2y = -5x + 6$

$y = \dfrac{5}{2}x - 3$

$m = \dfrac{5}{2}$

The slope of a parallel line is $\dfrac{5}{2}$.

65. $|2x + 5| > 3$

$2x + 5 < -3$ or $2x + 5 > 3$

$2x < -8$ or $2x > -2$

$x < -4$ or $x > -1$

$(-\infty, -4) \cup (-1, \infty)$

69. m for $l_1 = \dfrac{-2 - 4}{2 - (-1)} = \dfrac{-6}{3} = -2$

m for $l_2 = \dfrac{2 - 6}{-4 - (-8)} = \dfrac{-4}{4} = -1$

m for $l_3 = \dfrac{-4 - 0}{0 - (-6)} = \dfrac{-4}{6} = -\dfrac{2}{3}$

73. $F(22, 2), G(26, 8)$

$m = \dfrac{8 - 2}{26 - 22} = \dfrac{6}{4} = \dfrac{3}{2}$,

or $\dfrac{3}{2}$ yards per second.

77. a. $y = \dfrac{1}{2}x + 1$

$y = x + 1$

$y = 2x + 1$

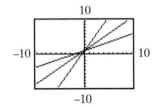

b. $y = -\dfrac{1}{2}x + 1$

$y = -x + 1$

$y = -2x + 1$

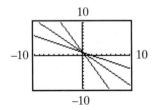

c. True

Exercise Set 3.3

1. Point: $(1, 3)$

Slope: $\dfrac{3}{2}$

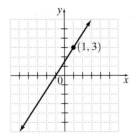

5. Point: $(0, 7)$

Slope: -1

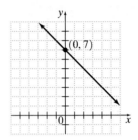

9. $y = -2x + 3$

$m = -2$, y-intercept $= (0, 3)$

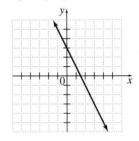

13. $y = \dfrac{1}{2}x - 4$

$m = \dfrac{1}{2}$, y-intercept $= (0, -4)$

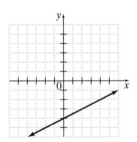

17. $x + 2y = 8$

$2y = -x + 8$

$y = -\dfrac{1}{2}x + 4$

$m = -\dfrac{1}{2}$, y-intercept $= (0, 4)$

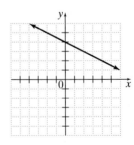

21. $y = 5x + 1$

$m = 5$, y-intercept $= (0, 1)$

graph D

25. $y = mx + b$

$y = 2x + \dfrac{3}{4}$

29. $y = 1765.1x + 35,815.0$

   a. $x = 2000 - 1995 = 5$

       $y = 1765.1(5) + 35,815.0$

       $y = 8825.5 + 35,815.0$

       $y = 44,640.5$

       The average income is $44,640.50

   b. $m = 1765.1$; The annual average income increases $1765.10 each year.

   c. y-intercept $= (0, 35, 815.0)$; At $x = 0$, or 1995, the annual average income was $35,815.00.

33. $y = 136.2x + 2827.6$

   a. $x = 2010 - 1996 = 14$

       $y = 136.2(14) + 2827.6$

       $y = 1906.8 + 2827.6$

       $y = 4734.4$

       The cost is approximately $4734.40.

   b. $y = 5000$

      $5000 = 136.2x + 2827.6$

      $2172.4 = 136.2x$

       $15.95 \approx x$

      $1996 + 15.95 = 2011.95$

      By 2012, the cost is expected to exceed $5000.

   c. answers may vary

37. $y - (-1) = 2(x - 0)$

      $y + 1 = 2x - 0$

   $y + 1 - 1 = 2x - 1$

       $y = 2x - 1$

41. $y = -7x + 500$

    where y is the height at time x.

Exercise Set 3.4

1. $y - y_1 = m(x - x_1)$

   $y - 2 = 3(x - 1)$

   $y - 2 = 3x - 3$

     $y = 3x - 1$

5. $y - y_1 = m(x - x_1)$

   $y - 2 = \dfrac{1}{2}[x - (-6)]$

   $y - 2 = \dfrac{1}{2}(x + 6)$

   $y - 2 = \dfrac{1}{2}x + 3$

     $y = \dfrac{1}{2}x + 5$

9. $y - y_1 = m(x - x_1)$

   $y - 3 = 2[x - (-2)]$

   $y - 3 = 2(x + 2)$

   $y - 3 = 2x + 4$

     $y = 2x + 7$

13. $m = \dfrac{6 - 0}{4 - 2} = \dfrac{6}{2} = 3$

   $y - 0 = 3(x - 2)$

     $y = 3x - 6$

17. $m = \dfrac{-3 - (-4)}{-4 - (-2)} = \dfrac{1}{-2} = -\dfrac{1}{2}$

   $y - (-4) = -\dfrac{1}{2}[x - (-2)]$

   $2(y + 4) = -(x + 2)$

    $2y + 8 = -x - 2$

      $y = -\dfrac{1}{2}x - \dfrac{10}{2}$

      $y = -\dfrac{1}{2}x - 5$

21. $m = \dfrac{-6 - (-4)}{0 - (-7)} = \dfrac{-2}{7} = -\dfrac{2}{7}$

   $y - (-6) = -\dfrac{2}{7}(x - 0)$

    $y + 6 = -\dfrac{2}{7}x$

      $y = -\dfrac{2}{7}x - 6$

25. Every vertical line is in the form $x = c$. Since the line passes through the point $(2, 6)$, its equation is $x = 2$.

29. A line with undefined slope is vertical. Every vertical line is in the form $x = c$. Since the line passes through the point $(0, 5)$, its equation is $x = 0$.

33. $y = -2x - 6, m = -2$

   so perpendicular slope $= \dfrac{1}{2}$

   $y - (-5) = \dfrac{1}{2}(x - 2)$

    $y + 5 = \dfrac{1}{2}x - 1$

      $y = \dfrac{1}{2}x - 6$

37. $3x + 2y = 5$

    $2y = -3x + 5$

     $y = -\dfrac{3}{2}x + \dfrac{5}{2}, m = -\dfrac{3}{2}$

   so parallel slope $= -\dfrac{3}{2}$

   $y - (-3) = -\dfrac{3}{2}[x - (-2)]$

    $y + 3 = -\dfrac{3}{2}(x + 2)$

    $y + 3 = -\dfrac{3}{2}x - 3$

      $y = -\dfrac{3}{2}x - 6$

41. $x = 3$; perpendicular line is $y = -5$

45. $x - 4y = 4$

    $-4y = -x + 4$

     $y = \dfrac{1}{4}x - 1, m = \dfrac{1}{4}$

   so perpendicular slope $= -4$

   $y - 5 = -4(x + 1)$

   $y - 5 = -4x - 4$

     $y = -4x + 1$

49. **a.** $(1, 30,000), (4, 66,000)$

     $m = \dfrac{66,000 - 30,000}{4 - 1} = 12,000$

     $y - 30,000 = 12,000(x - 1)$

           $y = 12,000x + 18,000$

b. $y = 12,000(7) + 18,000 = \$102,000$

c. $126,000 = 12,000x + 18,000$

$$x = \frac{126,000 - 18,000}{12,000}$$

$x = 9\text{ years}$

53. a. $(0, 1,089,261), (2, 8,498,545)$

$$m = \frac{8,498,545 - 1,089,261}{2 - 0}$$

$$m = \frac{7,409,284}{2} = 3,704,642$$

$y - 1,089,261 = 3,704,642(x - 0)$

$\quad\quad\quad\quad y = 3,704,642x + 1,089,261$

b. $x = 2005 - 1998 = 7$

$y = 3,704,642(7) + 1,089,261$

$y = 25,932,494 + 1,089,261$

$y = 27,021,755$

About 27,021,755 DVD players will be sold.

57. $y = 4.2x$
$y = 4.2(-2)$
$y = -8.4$
$(-2, -8.4)$

61. True

65. $y = 4x - 2, y = 4x - 4$

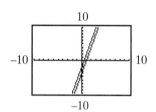

Exercise Set 3.5

1. Domain: $= \{-1, 0, -2, 5\}$
Range $= \{7, 6, 2\}$
The relation is a function.

5. Domain: $= \{1\}$
Range $= \{1, 2, 3, 4\}$
The relation is not a function since 1 is paired with $1, 2, 3$ and 4.

9. Domain: $\{-3, 0, 3\}$
Range $= \{-3, 0, 3\}$
The relation is a function.

13. Domain $= \{$Colorado, Alaska, Delaware, Illinois, Connecticut, Texas$\}$
Range $= \{6, 1, 20, 30\}$
The relation is a function.

17. Domain $= \{2, -1, 5, 100\}$
Range $= \{0\}$
The relation is a funcion.

21. Yes   **25.** Yes

29. Not a function

33. Domain $= [0, \infty)$
Range $=$ All real numbers
The relation is not a function since it fails the vertical line test (try $x = 1$).

37. Domain $= (-\infty, \infty)$
Range $= (-\infty, -3] \cup [3, \infty)$
The relation is not a function since it fails the vertical line test (try $x = 3$).

41. Domain $= \{-2\}$
Range $= (-\infty, \infty)$
The relation is not a function since it fails the vertical line test.

45. $f(x) = 3x + 3$
$f(4) = 3(4) + 3 = 12 + 3 = 15$

49. $g(x) = 4x^2 - 6x + 3$
$g(2) = 4(2)^2 - 6(2) + 3$
$\quad\quad = 4(4) - 12 + 3$
$\quad\quad = 16 - 12 + 3$
$\quad\quad = 7$

53. $f(x) = \dfrac{1}{2}x$

a. $f(0) = \dfrac{1}{2}(0) = 0$

b. $f(2) = \dfrac{1}{2}(2) = 1$

c. $f(-2) = \dfrac{1}{2}(-2) = -1$

57. $A(r) = \pi r^2$
$A(5) = \pi(5)^2 = 25\pi$
25π square centimeters

61. $H(f) = 2.59f + 47.24$
$H(46) = 2.59(46) + 47.24$
$\quad\quad\quad = 119.14 + 47.24$
$\quad\quad\quad = 166.38$
166.38 centimeters

65. $C(x) = 1.69x + 87.54$

a. $C(5) = 1.69(5) + 87.54 = 8.45 + 87.54 = 95.99$

The per capita consumption of poultry was about 95.99 pounds in 2000.

b. $x = 2002 - 1995 = 7$

$C(7) = 1.69(7) + 87.54 = 11.83 + 87.54 = 99.37$

The per capita consumption is about 99.37 pounds in 2002.

69. $f(x) = -3x$
or
$y = -3x$
where $m = -3$ and y-intercetpt $= (0, 0)$

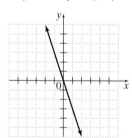

73. $2x - 7 \le 21$
$\quad\quad 2x \le 28$
$\quad\quad\quad x \le 14$
$(-\infty, 14]$

77. $\dfrac{x}{2} + \dfrac{1}{4} < \dfrac{1}{8}$

$\dfrac{x}{2} < \dfrac{1}{8} - \dfrac{1}{4}$

$\dfrac{x}{2} < -\dfrac{1}{8}$

$x < -\dfrac{1}{4}$

$\left(-\infty, -\dfrac{1}{4}\right)$

81. $f(x) = x^2 - 12$

   a. $f(12) = (12)^2 - 12 = 144 - 12 = 132$

   b. $f(a) = a^2 - 12$

85. answers may vary

Exercise Set 3.6

1. $x < 2$

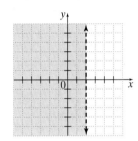

5. $3x + y > 6$
Test $(0,0)$
$3(0) + 0 > 6$
    $0 > 6$ False
Shade the half-plane that does not contain $(0,0)$.

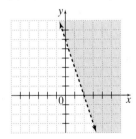

9. $2x + 4y \geq 8$
Test $(0,0)$
$2(0) + 4(0) \geq 8$
    $0 \geq 8$ False
Shade the half-plane that does not contain the point $(0,0)$.

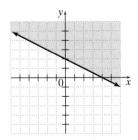

13. A dashed boundary line should be used when the inequality contains a $<$ or $>$.

17. $x \leq -2$ or $y \geq 4$

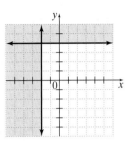

21. $x + y \leq 3$ or $x - y \geq 5$

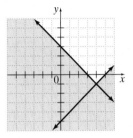

25. $x + y \leq 1$ and $y \leq -1$

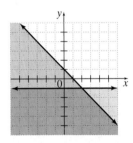

29. $y > 2x + 3$
dashed line
$(0,0)$ results in a false inequality graph A

33.
$$3x + 2y = -12 \qquad\qquad x = 4y$$
$$3(-4) + 2(0) \overset{?}{=} -12 \qquad -4 \overset{?}{=} 4(0)$$
$$-12 = -12 \quad \text{True} \qquad -4 = 0 \qquad \text{False}$$
No, the ordered pair is not a solution of both equations.

Chapter 3 Test

1.

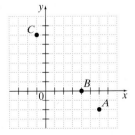

A is in quadrant IV.
B is on the x-axis, no quadrant.
C is in quadrant II.

5. $4x + 6y = 8$
$$6y = -4x + 8$$
$$y = -\frac{2}{3}x + \frac{4}{3}$$
$$m = -\frac{2}{3}, y\text{-intercept} = \left(0, \frac{4}{3}\right)$$

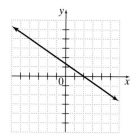

9. $f(x) = 3x + 1$
    $m = 3, y\text{-intercept} = (0, 1)$
    graph C

13. $y = -8$

17. $y = 5x + (-2)$
    $y = 5x - 2$

21. $2x - 5y = 8$
$$5y = 2x - 8$$
$$y = \frac{2}{5}x - \frac{8}{5}, \text{so } m_1 = \frac{2}{5},$$
$$m_2 = \frac{-1 - 4}{-1 - 1} = \frac{-5}{-2} = \frac{5}{2}$$
    Therefore, line L_1 and L_2 are neither parallel nor perpendicular.

25. $2x + 4y < 6$ and $y \leq -4$

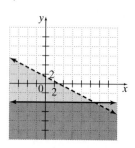

29. Domain: $(-\infty, \infty)$
    Range: $(-\infty, \infty)$
    Function

Chapter 4

Exercise Set 4.1

1. $\begin{cases} x - y = 3 \\ 2x - 4y = 8 \end{cases}$

  $x - y = 3$ $2x - 4y = 8$
$2 - (-1) = 3$ $2(2) - 4(-1) = 8$
$2 + 1 = 3$ $4 + 4 = 8$
         $3 = 3$ True $8 = 8$ True
  Yes, $(2, -1)$ is a solution.

5. $\begin{cases} y = -5x \\ x = -2 \end{cases}$

  $y = -5x$ $x = -2$
$10 = -5(-2)$ $-2 = -2$ True

             $10 = 10$ True
          Yes, $(-2, 10)$ is a solution.

9. $\begin{cases} 2y - 4 = 0 \\ x + 2y = 5 \end{cases}$

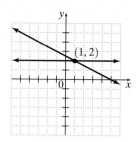

  The solution is $(1, 2)$.

13. $\begin{cases} y = -3x \\ 2x - y = -5 \end{cases}$

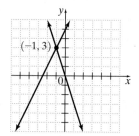

  The solution is $(-1, 3)$.

17. $\begin{cases} 4x - y = 9 \\ 2x + 3y = -27 \end{cases}$
  Solve the first equation for y.
  $4x - y = 9$
        $y = 4x - 9$
  Replace y with $4x - 9$ in the second equation.
  $2x + 3(4x - 9) = -27$
  $2x + 12x - 27 = -27$
            $14x = 0$
              $x = 0$
  Replace x with 0 in the first equation.
  $4(0) - y = 9$
          $y = -9$
  The solution is $(0, -9)$.

21. $\begin{cases} \dfrac{x}{3} + y = \dfrac{4}{3} \\ -x + 2y = 11 \end{cases}$

  Clear fractions by multiplying the first equation by 3.
  $\begin{cases} x + 3y = 4 \\ -x + 2y = 11 \end{cases}$
  Solve the second equation for x.
  $2y - 11 = x$
        $x = 2y - 11$
  Replace x with $2y - 11$ in the first equation.
  $2y - 11 + 3y = 4$
           $5y = 15$
            $y = 3$
  Replace y with 3 in the equation $x = 2y - 11$.
  $x = 2(3) - 11$
  $x = -5$
  The solution is $(-5, 3)$.

25. $\begin{cases} 2x = 6 \\ y = 5 - x \end{cases}$

Solve the first equation for x.

$x = 3$

Replace x with 3 in the second equation.

$y = 5 - 3$

$y = 2$

The solution is $(3, 2)$.

29. $\begin{cases} 5x + 2y = 1 \\ x - 3y = 7 \end{cases}$

Multiply the second equation by -5.

$\begin{cases} 5x + 2y = 1 \\ -5x + 15y = -35 \end{cases}$

Add the equations.

$\begin{array}{r} 5x + 2y = 1 \\ -5x + 15y = -35 \\ \hline 17y = -34 \\ y = -2 \end{array}$

Replace y with -2 in the second equation.

$x - 3(-2) = 7$

$x + 6 = 7$

$x = 1$

The solution is $(1, -2)$.

33. $\begin{cases} 3x - 5y = 11 \\ 2x - 6y = 2 \end{cases}$

Multiply the second equation by $\dfrac{1}{2}$.

$\begin{cases} 3x - 5y = 11 \\ x - 3y = 1 \end{cases}$

Multiply the second equation by -3.

$\begin{cases} 3x - 5y = 11 \\ -3x + 9y = -3 \end{cases}$

Add the equations.

$\begin{array}{r} 3x - 5y = 11 \\ -3x + 9y = -3 \\ \hline 4y = 8 \\ y = 2 \end{array}$

Replace y with 2 in the equation $x - 3y = 1$.

$x - 3(2) = 1$

$x - 6 = 1$

$x = 7$

The solution is $(7, 2)$.

37. $\begin{cases} 3x + y = 1 \\ 2y = 2 - 6x \end{cases}$

$\begin{cases} 3x + y = 1 \\ 6x + 2y = 2 \end{cases}$

Multiply the first equation by -2.

$\begin{cases} -6x - 2y = -2 \\ 6x + 2y = 2 \end{cases}$

Add the equations.

$\begin{array}{r} -6x - 2y = -2 \\ 6x + 2y = 2 \\ \hline 0 = 0 \quad \text{True} \end{array}$

Dependent system

$\{(x, y) | 3x + y = 1\}$

41. $\begin{cases} 4x + 2y = 5 \\ 2x + y = -1 \end{cases}$

Multiply the second equation by -2.

$\begin{cases} 4x + 2y = 5 \\ -4x - 2y = 2 \end{cases}$

Add the equations.

$\begin{array}{r} 4x + 2y = 5 \\ -4x - 2y = 2 \\ \hline 0 = 7 \quad \text{False} \end{array}$

Inconsistent system

The solution is $\varnothing$.

45. $\begin{cases} \dfrac{2}{3}x - \dfrac{3}{4}y = -1 \\ -\dfrac{1}{6}x + \dfrac{3}{8}y = 1 \end{cases}$

Multiply the first equation by 12 and the second equation by 24.

$\begin{cases} 8x - 9y = -12 \\ -4x + 9y = 24 \end{cases}$

Add the equations.

$\begin{array}{r} 8x - 9y = -12 \\ -4x + 9y = 24 \\ \hline 4x = 12 \\ x = 3 \end{array}$

Replace x with 3 in the equation $-4x + 9y = 24$.

$-4(3) + 9y = 24$

$-12 + 9y = 24$

$9y = 36$

$y = 4$

The solution is $(3, 4)$.

49. $\begin{cases} 10y - 2x = 1 \\ 5y = 4 - 6x \end{cases}$

Multiply the second equation by -2.

$\begin{cases} 10y = 2x + 1 \\ -10y = 12x - 8 \end{cases}$

Add the equations.

$\begin{array}{r} 10y = 2x + 1 \\ -10y = 12x - 8 \\ \hline 0 = 14x - 7 \\ 14x = 7 \\ x = \dfrac{1}{2} \end{array}$

Replace x with $\dfrac{1}{2}$ in the first equation.

$10y - 2\left(\dfrac{1}{2}\right) = 1$

$10y - 1 = 1$

$10y = 2$

$y = \dfrac{1}{5}$

The solution is $\left(\dfrac{1}{2}, \dfrac{1}{5}\right)$.

53. $\begin{aligned} 3x - 4y + 2z &= 5 \\ 3(1) - 4(2) + 2(5) &\overset{?}{=} 5 \\ 3 - 8 + 10 &\overset{?}{=} 5 \\ 5 &= 5 \quad \text{True} \end{aligned}$

57. $\begin{array}{r} 3x + 2y - 5z = 10 \\ -3x + 4y + z = 15 \\ \hline 6y - 4z = 25 \end{array}$

61. no

65. Supply is greater than demand because the supply line is above the demand line and the supply line is increasing linearly while the demand line is decreasing linearly.

Exercise Set 4.2

1. $\begin{cases} x + y & = 3 \\ 2y & = 10 \\ 3x + 2y - 3z = 1 \end{cases}$

Solve the second equation for y.

$y = 5$

Replace y with 5 in the first equation.

$x + 5 = 3$

$x = -2$

Replace x with -2 and y with 5 in the third equation.

$3(-2) + 2(5) - 3z = 1$

$-6 + 10 - 3z = 1$

$4 - 3z = 1$

$-3z = -3$

$z = 1$

The solution is $(-2, 5, 1)$.

5. $\begin{cases} x - 2y + z = -5 & (1) \\ -3x + 6y - 3z = 15 & (2) \\ 2x - 4y + 2z = -10 & (3) \end{cases}$

Multiply equation (2) by $-\dfrac{1}{3}$ and equation (3) by $\dfrac{1}{2}$.

$\begin{cases} x - 2y + z = -5 \\ x - 2y + z = -5 \\ x - 2y + z = -5 \end{cases}$

All three equations are identical. There are infinitely many solutions.

The solution is $\{(x, y, z) | x - 2y + z = -5\}$.

9. $\begin{cases} x + 5z = 0 & (1) \\ 5x + y + 10z = -24 & (2) \\ y - 3z = 0 & (3) \end{cases}$

Solve equation (1) for x.

$x = -5z$ (4)

Solve equation (3) for y.

$y = 3z$ (5)

Now substitute $-5z$ for x and $3z$ for y in equation (2).

$5(-5z) + 3z + 10z = -24$

$-25z + 3z + 10z = -24$

$-12z = -24$

$z = 2$

Replace z with 2 in equation (4).

$x = -5(2)$

$x = -10$

Replace z with 2 in equation (5)

$y = 3(2)$

$y = 6$

The solution is $(-10, 6, 2)$.

13. $\begin{cases} x + y + z = 8 & (1) \\ 2x - y - z = 10 & (2) \\ x - 2y - 3z = 22 & (3) \end{cases}$

Add equations (1) and (2).

$3x = 18$ or $x = 6$

Add twice equation 1 to equation 3.

$2x + 2y + 2z = 16$

$\underline{x - 2y - 3z = 22}$

$3x \qquad\quad - z = 38$

Replace x with 6 in this equation.

$3(6) - z = 38$

$18 - z = 38$

$-z = 20$

$z = -20$

Replace x with 6 and z with -20 in equation (1).

$6 + y + (-20) = 8$

$y - 14 = 8$

$y = 22$

The solution is $(6, 22, -20)$.

17. $\begin{cases} 2x - 3y + z = 2 & (1) \\ x - 5y + 5z = 3 & (2) \\ 3x + y - 3z = 5 & (3) \end{cases}$

Add -2 times equation (2) to equation (1).

$2x - 3y + z = 2$

$\underline{-2x + 10y - 10z = -6}$

$7y - 9z = -4$

Add -3 times equation (2) to equation (3).

$-3x + 15y - 15z = -9$

$\underline{3x + y - 3z = 5}$

$16y - 18z = -4$

We now have the system:

$\begin{cases} 7y - 9z = -4 & (4) \\ 16y - 18z = -4 & (5) \end{cases}$

Multiply equation (4) by -2 and add to equation (5).

$-14y + 18z = 8$

$\underline{16y - 18z = -4}$

$2y = 4$

$y = 2$

Replace y with 2 in equation (4).

$7(2) - 9z = -4$

$-9z = -18$

$z = 2$

Replace y with 2 and z with 2 in equation (1).

$2x - 3(2) + 2 = 2$

$2x = 6$

$x = 3$

The solution is $(3, 2, 2)$.

21. $\begin{cases} 2x + 2y - 3z = 1 & (1) \\ y + 2z = -14 & (2) \\ 3x - 2y = -1 & (3) \end{cases}$

Add equations (1) and (3).

$5x - 3z = 0$ (4)

Multiply equation (2) by 2 and add to equation (3).

$2y + 4z = -28$

$\underline{3x - 2y = -1}$

$3x + 4z = -29$ (5)

Multiply equation (4) by 4, multiply equation (5) by 3, and add.

$20x - 12z = 0$

$\underline{9x + 12z = -87}$

$29x = -87$

$x = -3$

Replace x with -3 in equation (4).

$5(-3) - 3z = 0$

$3z = -15$

$z = -5$

Replace z with -5 in equation (2).

$y + 2(-5) = -14$

$y - 10 = -14$

$y = -4$

The solution is $(-3, -4, -5)$.

25. $\begin{cases} x + y + z = 1 & (1) \\ 2x - y + z = 0 & (2) \\ -x + 2y + 2z = -1 & (3) \end{cases}$

Multiply equation (3) by 2 and add to equation (2).

$2x - y + z = 0$

$\underline{-2x + 4y + 4z = -2}$

$3y + 5z = -2$ (4)

Multiply equation (1) by -2 and add to equation (2).

$-2x - 2y - 2z = -2$

$\underline{2x - y + z = 0}$

$-3y - z = -2$ (5)

Add equations (4) and (5).

$$3y + 5z = -2$$
$$\underline{-3y - z = -2}$$
$$4z = -4$$
$$z = -1$$

Replace z with -1 in equation (4).

$$3y + 5(-1) = -2$$
$$3y = 3$$
$$y = 1$$

Replace y with 1 and z with -1 in equation (1).

$$x + 1 + (-1) = 1$$
$$x = 1$$

The solution is $(1, 1, -1)$.

$$\frac{1}{24} = \frac{x}{8} + \frac{y}{4} + \frac{z}{3}$$
$$\frac{1}{24} = \frac{1}{8} + \frac{1}{4} - \frac{1}{3}$$
$$\frac{1}{24} = \frac{3}{24} + \frac{6}{24} - \frac{8}{24}$$
$$\frac{1}{24} = \frac{1}{24} \quad \text{True}$$

29. $2(x - 1) - 3x = x - 12$

$$2x - 2 - 3x = x - 12$$
$$-2 - x = x - 12$$
$$10 = 2x$$
$$5 = x$$

The solution is $\{5\}$.

33. answers may vary

37.
$$\begin{cases} x + y \quad\quad - w = 0 & (1) \\ \quad\quad y + 2z + w = 3 & (2) \\ x \quad\quad - z \quad\quad = 1 & (3) \\ 2x - y \quad\quad - w = -1 & (4) \end{cases}$$

Add equation (2) to equation (4).

$$2x + 2z = 2$$
$$x = 1$$

Replace x with 1 in equation (3).

$$1 - z = 1$$
$$-z = 0$$
$$z = 0$$

Add equation (1) to equation (2).

$$x + 2y + 2z = 3$$

Replace x with 1 and z with 0.

$$1 + 2y + 2(0) = 3$$
$$1 + 2y = 3$$
$$2y = 2$$
$$y = 1$$

Replace x with 1 and y with 1 in equation (1).

$$1 + 1 - w = 0$$
$$2 - w = 0$$
$$2 = w$$

The solution is $(1, 1, 0, 2)$.

Exercise Set 4.3

1. Let m = the first number

n = the second number

$$\begin{cases} m = n + 2 \\ 2m = 3n - 4 \end{cases}$$

Substitute $m = n + 2$ in the second equation.

$$2(n + 2) = 3n - 4$$
$$2n + 4 = 3n - 4$$
$$n = 8$$

Replace n with 8 in the first equation.

$$m = 8 + 2 = 10$$

The numbers are 10 and 8.

5. Let p = speed of the plane in still air,

w = speed to the wind

$$\begin{cases} p + w = 560 \\ p - w = 480 \end{cases}$$

Add the equations.

$$2p = 1040$$
$$p = 520$$

Replace p with 520 in the first equation.

$$520 + w = 560$$
$$w = 40$$

The speed of the plane is 520 mph and the speed of the wind is 40 mph.

9. Let x be the number of students in the United Kingdom,

y = the number of students in Spain

$$x + y = 40{,}012 \quad (1)$$
$$x = y + 15{,}428 \quad (2)$$

Substitute $x = y + 15{,}428$ in the first equation.

$$(y + 15{,}428) + y = 40{,}012$$
$$2y + 15{,}428 = 40{,}012$$
$$2y = 24{,}584$$
$$y = 12{,}292$$

Replace y with 12,292 in equation (2).

$$x = 12{,}292 + 15{,}428$$
$$x = 27{,}720$$

27,720 students were in the United Kingdom and 12,292 students were in Spain.

13. Let m = the first number

n = the second number

$$\begin{cases} m = n - 2 \\ 2m = 3n + 4 \end{cases}$$

Substitute $m = n - 2$ in the second equation.

$$2(n - 2) = 3n + 4$$
$$2n - 4 = 3n + 4$$
$$n = -8$$

Replace n with -8 in the first equation.

$$m = -8 - 2 = -10$$

The numbers are -10 and -8.

19. Let p = the speed of the plane in still air

w = the speed of the wind

First note:

$$\frac{2160 \text{ miles}}{3 \text{ hours}} = 720 \text{ mph and}$$

$$\frac{2160 \text{ miles}}{4 \text{ hours}} = 540 \text{ mph}$$

Now,

$$\begin{cases} p + w = 720 \\ p - w = 540 \end{cases}$$

Add the equations.

$$2p = 1260$$
$$p = 630$$

Replace p with 630 in the first equation.

$$630 + w = 720$$
$$w = 90$$

The speed of the plane in still air is 630 mph and the speed of the wind is 90 mph.

21. a. answers may vary, but, notice the slopes are different.

b. Equate y values.

$$0.02x + 7.2 = -0.52x + 64.5$$
$$0.54x + 7.2 = 64.5$$
$$0.54x = 57.3$$
$$x = \frac{57.3}{54} \approx 106$$

The year is $1900 + 106 = 2006$.

25. Let m be the number of miles.

Hertz $= 25 + 0.10\,m$

Budget $= 20 + 0.25\,m$

Using Budget $= 2 \cdot$ Hertz gives

$$20 + 0.25\,m = 2(25 + 0.10\,m)$$
$$20 + 0.25\,m = 50 + 0.20\,m$$
$$0.25\,m = 30 + 0.20\,m$$
$$0.05\,m = 30$$
$$m = \frac{30}{0.05} = 600$$

The daily mileage must be 600 miles.

29. Let $x =$ number of units of Mix A,

$y =$ number of units of Mix B,

$z =$ number of units of Mix C

$$\begin{cases} 4x + 6y + 4z = 30 \ (1) \\ 6x + y + z = 16 \ (2) \\ 3x + 2y + 12z = 24 \ (3) \end{cases}$$

Multiply equation (2) by -6 and add to equation (1).

$$-32x - 2z = -66$$

or $16x + z = 33$ (4)

Multiply equation (2) by -2 and add to equation (3).

$$-9x + 10z = -8 \ (5)$$

Multiply equation (4) by -10 and add to equation (5).

$$-169x = -338$$
$$x = 2$$

Replace x with 2 in equation (4).

$$16(2) + z = 33$$
$$32 + z = 33$$
$$z = 1$$

Replace x with 2 and z with 1 in equation (2).

$$6(2) + y + 1 = 16$$
$$12 + y + 1 = 16$$
$$y + 13 = 16$$
$$y = 3$$

Combine 2 units of Mix A, 3 units of Mix B, and 1 unit of Mix C.

33. Let $x =$ the first number

$y =$ the second number

$z =$ the third number

$$\begin{cases} x + y + z = 40 \\ x = y + 5 \\ x = 2z \end{cases}$$

$$\begin{cases} x + y + z = 40 \\ y = x - 5 \\ z = \frac{1}{2}x \end{cases}$$

Substitute $y = x - 5$ and $z = \frac{1}{2}x$ in the first equation.

$$x + x - 5 + \frac{1}{2}x = 40$$
$$\frac{5}{2}x - 5 = 40$$
$$\frac{5}{2}x = 45$$
$$x = \frac{2}{5}(45) = 18$$
$$y = 18 - 5 = 13$$
$$z = \frac{1}{2}(18) = 9$$

The three numbers are 18, 13, and 9.

37. $z = 180 - (2x - 5)$

$z = 180 - 2x + 5$

$z = 185 - 2x.$

$y = 180 - (2x + 5)$

$= 180 - 2x - 5$

$= 175 - 2x.$

The sum of the angles is 180.

$$x + (175 - 2x) + (185 - 2x) = 180$$
$$360 - 3x = 180$$
$$-3x = -180$$
$$x = 60$$

$z = 185 - 2(60) = 185 - 120 = 65$

$y = 175 - 2(60) = 175 - 120 = 55$

The values are $x = 60$, $y = 55$, and $z = 65$.

41. $C(x) = 1.2x + 1500$

$R(x) = 1.7x$

$1.7x = 1.2x + 1500$

$0.5x = 1500$

$$x = \frac{1500}{0.5} = 3000$$

3000 units

45. a. $R(x) = 450x$

b. $C(x) = 200x + 6000$

c. $450x = 200x + 6000$

$250x = 6000$

$x = 24$

24 desks

47. $\begin{cases} 3x - y + z = 2 \\ -x + 2y + 3z = 6 \end{cases}$

$6x - 2y + 2z = 4$

$\underline{-x + 2y + 3z = 6}$

$5x \qquad + 5z = 10$

53. $y = ax^2 + bx + c$

For $(1, 6)$, use $x = 1$ and $y = 6$.

$6 = a + b + c$ (1)

For $(-1, -2)$, use $x = -1$ and $y = -2$.

$-2 = a - b + c$ (2)

For $(0, -1)$, use $x = 0$ and $y = -1$.

$-1 = a \cdot 0 + b \cdot 0 + c$

$-1 = c$ (3)

The system is

$6 = a + b + c$ (1)

$-2 = a - b + c$ (2)

$-1 = c$ (3)

From equation (3), we see that $c = -1$. Multiply equation (2) by -1 and add to equation (1).

$8 = 2b$

$4 = b$

Replace b with 4 and c with -1 in equation (1).

$6 = a + 4 - 1$

$6 = a + 3$

$3 = a$

The solution is $a = 3$, $b = 4$, and $c = -1$.

Exercise Set 4.4

1. $\begin{cases} x + y = 1 \\ x - 2y = 4 \end{cases}$

$$\begin{bmatrix} 1 & 1 & | & 1 \\ 1 & -2 & | & 4 \end{bmatrix}$$

Multiply row 1 by -1 and add to row 2.

$$\begin{bmatrix} 1 & 1 & | & 1 \\ 0 & -3 & | & 3 \end{bmatrix}$$

Divide row 2 by -3.

$$\begin{bmatrix} 1 & 1 & | & 1 \\ 0 & 1 & | & -1 \end{bmatrix}$$

This corresponds to $\begin{cases} x + y = 1 \\ y = -1 \end{cases}$.

$x + (-1) = 1$
$x - 1 = 1$
$x = 2$

The solution is $(2, -1)$.

5. $\begin{cases} x - 2y = 4 \\ 2x - 4y = 4 \end{cases}$

$\begin{bmatrix} 1 & -2 & | & 4 \\ 2 & -4 & | & 4 \end{bmatrix}$

Multiply row 1 by -2 and add to row 2.

$\begin{bmatrix} 1 & -2 & | & 4 \\ 0 & 0 & | & -4 \end{bmatrix}$

This is an inconsistent system.

The solution is $\varnothing$.

9. $\begin{cases} x - 4 = 0 \\ x + y = 1 \end{cases}$ or $\begin{cases} x = 4 \\ x + y = 1 \end{cases}$

$\begin{bmatrix} 1 & 0 & | & 4 \\ 1 & 1 & | & 1 \end{bmatrix}$

Multiply row 1 by -1 and add to row 2.

$\begin{bmatrix} 1 & 0 & | & 4 \\ 0 & 1 & | & -3 \end{bmatrix}$

This corresponds to $\begin{cases} x = 4 \\ y = -3 \end{cases}$.

The solution is $(4, -3)$.

13. $\begin{cases} 2y - z = -7 \\ x + 4y + z = -4 \\ 5x - y + 2z = 13 \end{cases}$

$\begin{bmatrix} 0 & 2 & -1 & | & -7 \\ 1 & 4 & 1 & | & -4 \\ 5 & -1 & 2 & | & 13 \end{bmatrix}$

Interchange rows 1 and 2.

$\begin{bmatrix} 1 & 4 & 1 & | & -4 \\ 0 & 2 & -1 & | & -7 \\ 5 & -1 & 2 & | & 13 \end{bmatrix}$

Multiply row 1 by -5 and add to row 3.

$\begin{bmatrix} 1 & 4 & 1 & | & -4 \\ 0 & 2 & -1 & | & -7 \\ 0 & -21 & -3 & | & 33 \end{bmatrix}$

Divide row 2 by 2.

$\begin{bmatrix} 1 & 4 & 1 & | & -4 \\ 0 & 1 & -\frac{1}{2} & | & -\frac{7}{2} \\ 0 & -21 & -3 & | & 33 \end{bmatrix}$

Multiply row 2 by 21 and add to row 3.

$\begin{bmatrix} 1 & 4 & 1 & | & -4 \\ 0 & 1 & -\frac{1}{2} & | & -\frac{7}{2} \\ 0 & 0 & -\frac{27}{2} & | & -\frac{81}{2} \end{bmatrix}$

Multiply row 3 by $-\dfrac{2}{27}$.

$\begin{bmatrix} 1 & 4 & 1 & | & -4 \\ 0 & 1 & -\frac{1}{2} & | & -\frac{7}{2} \\ 0 & 0 & 1 & | & 3 \end{bmatrix}$

This corresponds to $\begin{cases} x + 4y + z = -4 \\ y - \frac{1}{2}z = -\frac{7}{2} \\ z = 3 \end{cases}$.

$y - \dfrac{1}{2}(3) = -\dfrac{7}{2}$

$y - \dfrac{3}{2} = -\dfrac{7}{2}$

$y = -2$

$x + 4(-2) + 3 = -4$

$x - 8 + 3 = -4$

$x = 1$

The solution is $(1, -2, 3)$.

17. $\begin{cases} 4x + y + z = 3 \\ -x + y - 2z = -11 \\ x + 2y + 2z = -1 \end{cases}$

$\begin{bmatrix} 4 & 1 & 1 & | & 3 \\ -1 & 1 & -2 & | & -11 \\ 1 & 2 & 2 & | & -1 \end{bmatrix}$

Interchange rows 1 and 3.

$\begin{bmatrix} 1 & 2 & 2 & | & -1 \\ -1 & 1 & -2 & | & -11 \\ 4 & 1 & 1 & | & 3 \end{bmatrix}$

Multiply row 1 by 1 and add to row 2.

Multiply row 1 by -4 and add to row 3.

$\begin{bmatrix} 1 & 2 & 2 & | & -1 \\ 0 & 3 & 0 & | & -12 \\ 0 & -7 & -7 & | & 7 \end{bmatrix}$

Divide row 2 by 3.

$\begin{bmatrix} 1 & 2 & 2 & | & -1 \\ 0 & 1 & 0 & | & -4 \\ 0 & -7 & -7 & | & 7 \end{bmatrix}$

Multiply row 2 by 7 and add to row 3.

$\begin{bmatrix} 1 & 2 & 2 & | & -1 \\ 0 & 1 & 0 & | & -4 \\ 0 & 0 & -7 & | & -21 \end{bmatrix}$

Divide row 3 by -7.

$\begin{bmatrix} 1 & 2 & 2 & | & -1 \\ 0 & 1 & 0 & | & -4 \\ 0 & 0 & 1 & | & 3 \end{bmatrix}$

This corresponds to $\begin{cases} x + 2y + 2z = -1 \\ y = -4 \\ z = 3 \end{cases}$.

$x + 2(-4) + 2(3) = -1$

$x - 8 + 6 = -1$

$x = 1$

The solution is $(1, -4, 3)$.

21. $(4)(-10) - (2)(-2) = -40 + 4 = -36$

25. a. Solve the system $\begin{cases} 2.3x + y = 52 \\ -5.4x + y = 14 \end{cases}$.

$\begin{bmatrix} 2.3 & 1 & | & 52 \\ -5.4 & 1 & | & 14 \end{bmatrix}$

Since getting a 1 in the first column would lead to repeating decimals, we multiply row 1 by -1 and add to row 2.

$\begin{bmatrix} 2.3 & 1 & | & 52 \\ -7.7 & 0 & | & -38 \end{bmatrix}$

This corresponds to $\begin{cases} 2.3x + y = 52 \\ -7.7x = -38 \end{cases}$.

From the second equation, $x = \dfrac{-38}{-77} \approx 4.935$.

Thus, the percent of U.S. households owning black-and-white television sets was the same as the percent of U.S. households owning a microwave oven in the end of 1984 (about 4.9 years after 1980).

b. Solving the television equation for y, we get

$y = -2.3x + 52$. Thus, for

1980, $y = -2.3x + 52 = 52$, and for

1993, $y = -2.3(13) + 52 = 22.1$.

Solving the microwave oven equation for y, we get

$y = 5.4x + 14$. Thus, for

1980, $y = 5.4(0) + 14 = 14$, and for

1993, $y = 5.4(13) + 14 = 84.2$.

In 1980, a greater percent of U.S. households, hence more households, owned black-and-white television sets. In 1993, more households owned a microwave oven. The percent of households owning black-and-white television sets is decreasing and the percent of households owning microwave ovens is increasing.
answers may vary

c. The percent will reach 0% when $y = 0$ in the equation $2.3x + y = 52$.

$$2.3x + 0 = 52$$

$$x = \dfrac{52}{2.3}$$

$$x \approx 22.6$$

According to this model, the percent of U.S. households owning a black-and-white television set will be 0% about 22.6 years after 1980, or sometime in 2002.

Exercise Set 4.5

1. $\begin{cases} y \geq x + 1 \\ y \geq 3 - x \end{cases}$

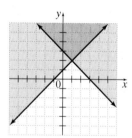

5. $\begin{cases} y \leq -2x - 2 \\ y \geq x + 4 \end{cases}$

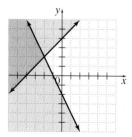

9. $\begin{cases} x \geq 3y \\ x + 3y \leq 6 \end{cases}$

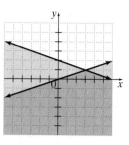

13. $\begin{cases} y \geq 1 \\ x < -3 \end{cases}$

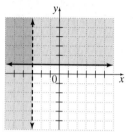

17. $\begin{cases} 3x - 4y \geq -6 \\ 2x + y \leq 7 \\ y \geq -3 \end{cases}$

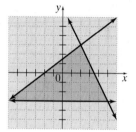

21. $\begin{cases} y < 5 \\ x > 3 \end{cases}$
graph C

25. $(-3)^2 = (-3)(-3) = 9$

29. $(-2)^2 - (-3) + 2(-1) = 4 + 3 - 2 = 5$

33. a. $\begin{cases} x + y \leq 8 \\ x < 3 \end{cases}$

b.

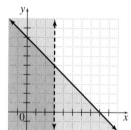

Chapter 4 Test

1. $2x - y = -1$
$5x + 4y = 17$

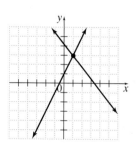

The solution is $(1, 3)$.

5. $\begin{cases} 2x - 3y \quad\;\; = 4 \quad (1) \\ \quad\;\; 3y + 2z = 2 \quad (2) \\ x \qquad\; - z = -5 \; (3) \end{cases}$

Add equation (1) to equation (2).

$$2x + 2z = 6$$

or $\qquad x + z = 3 \;\; (5)$

Add equation (3) to equation (5).

$$2x = -2$$
$$x = -1$$

Replace x with -1 in equation (3).

$$-1 - z = -5$$
$$-z = -4$$
$$z = 4$$

Replace x with -1 in equation (1).

$$2(-1) - 3y = 4$$
$$-2 - 3y = 4$$
$$-3y = 6$$
$$y = -2$$

The solution is $(-1, -2, 4)$.

9. $\begin{cases} 4x - 3y = -6 \quad (1) \\ -2x + \;\; y = 0 \qquad (2) \end{cases}$

Multiply equation (2) by 3 and add to equation (1).

$$-2x = -6$$
$$x = 3$$

Replace x with 3 in equation (2).

$$-2(3) + y = 0$$
$$-6 + y = 0$$
$$y = 6$$

The solution is $(3, 6)$.

13. $\begin{cases} x + 2y = -1 \\ 2x + 5y = -5 \end{cases}$

$$\begin{bmatrix} 1 & 2 & | & -1 \\ 2 & 5 & | & -5 \end{bmatrix}$$

Multiply row 1 by -2 and add to row 2.

$$\begin{bmatrix} 1 & 2 & | & -1 \\ 0 & 1 & | & -3 \end{bmatrix}$$

This corresponds to $\begin{cases} x + 2y = -1 \\ \qquad\; y = -3 \end{cases}$

$$x + 2(-3) = -1$$
$$x - 6 = -1$$
$$x = 5$$

The solution is $(5, -3)$.

17. Let x = number of gallons of 10% solution, y = number of gallons of 20% solution.

$\begin{cases} \qquad x + y = 20 \qquad\qquad (1) \\ 0.10x + 0.20y = 20(0.175) \quad (2) \end{cases}$

Multiply equation (2) by 10.

$$x + 2y = 35 \qquad (3)$$

The system is

$\begin{cases} x + y = 20 \qquad (1) \\ x + 2y = 35 \qquad (3) \end{cases}$

Multiply equation (1) by -1 and add to equation (2).

$$y = 15$$

Replace y with 15 in equation (1).

$$x + 15 = 20$$
$$x = 5$$

Mix 5 gallons of the 10% solution with 15 gallons of the 20% solution.

Chapter 5

Exercise Set 5.1

1. 4 has degree 0. **5.** $-3xy^2$ has degree $1 + 2 = 3$.

9. $3x^2 - 2x + 5$ has degree 2 and is a trinomial.

13. $x^2y - 4xy^2 + 5x + y$ has degree $2 + 1 = 3$ and is none of these.

17. $5y + y = 6y$

21. $4xy + 2x - 3xy - 1 = xy + 2x - 1$

25. $\begin{array}{r} x^2 + xy - \;\; y^2 \\ + 2x^2 - 4xy + 7y^2 \\ \hline 3x^2 - 3xy + 6y^2 \end{array}$

29. $\begin{array}{r} 3x^2 + 15x + \;\; 8 \\ + (2x^2 + \;\; 7x + \;\; 8) \\ \hline 5x^2 + 22x + 16 \end{array}$

33. $(5y^4 - 7y^2 + x^2 - 3) + (-3y^4 + 2y^2 + 4)$
$= 5y^4 - 7y^2 + x^2 - 3 - 3y^4 + 2y^2 + 4$
$= 5y^4 - 3y^4 - 7y^2 + 2y^2 + x^2 - 3 + 4$
$= 2y^4 - 5y^2 + x^2 + 1$

37. $\begin{array}{r} 3x^3 - b + 2a - \;\; 6 \\ - 4x^3 + b + 6a - \;\; 6 \\ \hline -x^3 + 8a - 12 \end{array}$

41. $(7x^3y - 4xy + 8) + (5x^3y + 4xy + 8x)$
$= 7x^3y + 5x^3y - 4xy + 4xy + 8x + 8$
$= 12x^3y + 8x + 8$

45. $(9y^2 - 7y + 5) - (8y^2 - 7y + 2)$
$= 9y^2 - 7y + 5 - 8y^2 + 7y - 2$
$= y^2 + 3$

49. $\begin{array}{r} 3x^2 - 4x + 8 \\ - \qquad\;\; (5x^2 - 7) \end{array}$ or $\begin{array}{r} 3x^2 - 4x + 8 \\ - 5x^2 \qquad\;\; + 7 \\ \hline -2x^2 - 4x + 15 \end{array}$

53. $\begin{array}{r} 4x^2 - 6x + 2 \\ - (-x^2 + 3x + 5) \end{array}$ or $\begin{array}{r} 4x^2 - 6x + 2 \\ x^2 - 3x - 5 \\ \hline 5x^2 - 9x - 3 \end{array}$

57. $(9x^3 - 2x^2 + 4x - 7) - (2x^3 - 6x^2 - 4x + 3)$
$= 9x^3 - 2x^2 + 4x - 7 - 2x^3 + 6x^2 + 4x - 3$
$= 7x^3 + 4x^2 + 8x - 10$

61. $\begin{array}{r} 7x^2 + 4x + 9 \\ + (8x^2 + 7x - 8) \\ \hline 15x^2 + 11x + 1 \end{array}$ or $\begin{array}{r} 15x^2 + 11x + 1 \\ - \qquad\;\; (3x + 7) \end{array}$ $\begin{array}{r} 15x^2 + 11x + 1 \\ - 3x - 7 \\ \hline 15x^2 + \;\; 8x - 6 \end{array}$

65. $\left(\dfrac{2}{3}x^2 - \dfrac{1}{6}x + \dfrac{5}{6}\right) - \left(\dfrac{1}{3}x^2 + \dfrac{5}{6}x - \dfrac{1}{6}\right)$

$= \dfrac{2}{3}x^2 - \dfrac{1}{6}x + \dfrac{5}{6} - \dfrac{1}{3}x^2 - \dfrac{5}{6}x + \dfrac{1}{6}$

$= \dfrac{1}{3}x^2 - x + 1$

69. $Q(x) = 5x^2 - 1$
$Q(-10) = 5(-10)^2 - 1$
$= 5(100) - 1$
$= 500 - 1$
$= 499$

73. Surface area $= 2HL + 2LW + 2HW$
$= 2(5)(4) + 2(4)(9) + 2(5)(9)$
$= 40 + 72 + 90$
$= 202$

The surface area is 202 sq in.

77. $P(x) = 45x - 100,000$
$P(4000) = 45(4000) - 100,000$
$= 180,000 - 100,000$
$= 80,000$
The profit is $80,000.

81. $5(3x - 2) = 5(3x) - 5(2)$
$= 15x - 10$

85. a. $f(x) = 136.7x^2 + 327.6x + 21.6$

$f(0) = 136.7(0)^2 + 327.6(0) + 21.6$

$= 21.6$

≈ 22

In 1996, there were about 22 stations.

b. $f(4) = 136.7(4)^2 + 327.6(4) + 21.6$

$= 136.7(16) + 327.6(4) + 21.6$

$= 2187.2 + 1310.4 + 21.6$

$= 3519.2$

≈ 3519

In 2000, there were about 3519 stations.

c. $f(8) = 136.7(8)^2 + 327.6(8) + 21.6$

$= 136.7(64) + 327.6(8) + 21.6$

$= 8748.8 + 2620.8 + 21.6$

$= 11,391.2$

$\approx 11,391$

In 2004, there will be about 11,391 stations.

d. increasing at a steady rate; answers may vary

89. $P(x) + Q(x) = 3x + 3 + (4x^2 - 6x + 3)$
$= 4x^2 + 3x - 6x + 3 + 3$
$= 4x^2 - 3x + 6$

93. $(14z^{5x} + 3z^{2x} + z) - (2z^{5x} - 10z^{2x} + 3z)$
$= 14z^{5x} + 3z^{2x} + z - 2z^{5x} + 10z^{2x} - 3z$
$= 14z^{5x} - 2z^{5x} + 3z^{2x} + 10z^{2x} + z - 3z$
$= 12z^{5x} + 13z^{2x} - 2z$

Exercise Set 5.2

1. $(-4x^3)(3x^2) = -12x^5$

5. $-6xy(4x + y) = -6xy(4x) - 6xy(y)$
$= -24x^2y - 6xy^2$

9. $(x - 3)(2x + 4) = x(2x + 4) - 3(2x + 4)$
$= 2x^2 + 4x - 6x - 12$
$= 2x^2 - 2x - 12$

13.
$$
\begin{array}{r}
3x - 2 \\
5x + 1 \\
\hline
3x - 2 \\
15x^2 - 10x \\
\hline
15x^2 - 7x - 2
\end{array}
$$

17.
$$
\begin{array}{r}
3x^2 + 4x - 4 \\
3x + 6 \\
\hline
18x^2 + 24x - 24 \\
9x^3 + 12x^2 - 12x \\
\hline
9x^3 + 30x^2 + 12x - 24
\end{array}
$$

21. $(2x^3 + 5)(5x^2 + 4x + 1)$
$= 2x^3(5x^2 + 4x + 1) + 5(5x^2 + 4x + 1)$
$= 10x^5 + 8x^4 + 2x^3 + 25x^2 + 20x + 5$

25.
$$
\begin{array}{r}
4x^2 - 2x + 5 \\
3x + 1 \\
\hline
4x^2 - 2x + 5 \\
12x^3 - 6x^2 + 15x \\
\hline
12x^3 - 2x^2 + 13x + 5
\end{array}
$$

29. $(x - 3)(x + 4) = x^2 + 4x - 3x - 12$
$= x^2 + x - 12$

33. $(3x - 1)(x + 3) = 3x^2 + 9x - x - 3$
$= 3x^2 + 8x - 3$

37. $(y - 4)(y - 3) = y^2 - 3y - 4y + 12$
$= y^2 - 7y + 12$

41. $\left(4x + \dfrac{1}{3}\right)\left(4x - \dfrac{1}{2}\right) = 16x^2 - 2x + \dfrac{4}{3}x - \dfrac{1}{6}$

$= 16x^2 - \dfrac{2}{3}x - \dfrac{1}{6}$

45. $(x + 4)^2 = x^2 + 2(x)(4) + 4^2$
$= x^2 + 8x + 16$

49. $(3x - y)^2 = (3x)^2 - 2(3x)y + y^2$
$= 9x^2 - 6xy + y^2$

53. $(m - 4)^2 = m^2 - 2(m)(4) + 4^2$
$= m^2 - 8m + 16$

57. $(3b - 6y)(3b + 6y) = (3b)^2 - (6y)^2 = 9b^2 - 36y^2$

61. $\left(3x + \dfrac{1}{2}\right)\left(3x - \dfrac{1}{2}\right) = (3x)^2 - \left(\dfrac{1}{2}\right)^2$

$= 9x^2 - \dfrac{1}{4}$

65. $(x^2 + 2y)(x^2 - 2y) = (x^2)^2 - (2y)^2$
$= x^4 - 4y^2$

69. $[(2s - 3) - 1][(2s - 3) + 1]$
$= (2s - 3)^2 - 1^2$
$= (2s)^2 - 2(2s)(3) + 3^2 - 1$
$= 4s^2 - 12s + 9 - 1$
$= 4s^2 - 12s + 8$

73.
$$
\begin{array}{r}
x + y \\
2x - 1 \\
\hline
- x - y \\
2x^2 + 2xy \\
\hline
2x^2 + 2xy - x - y \\
x + 1 \\
\hline
2x^2 + 2xy - x - y \\
2x^3 + 2x^2y - x^2 - xy \\
\hline
2x^3 + 2x^2y + x^2 + xy - x - y
\end{array}
$$

77. $(x - 5)(x + 5)(x^2 + 25) = (x^2 - 25)(x^2 + 25)$
$= (x^2)^2 - (25)^2$
$= x^4 - 625$

81. $\dfrac{20a^3b^5}{18ab^2} = \dfrac{2ab^2(10a^2b^3)}{2ab^2(9)} = \dfrac{10a^2b^3}{9}$

85. $A = \pi r^2$
$A = \pi(5x - 2)^2$
$A = \pi[(5x)^2 - 2(5x)(2) + 2^2]$
$A = \pi(25x^2 - 20x + 4)$ square kilometers

89. $5x^2y^n(6y^{n+1} - 2)$
$= 5x^2y^n(6y^{n+1}) - (5x^2y^n)(2)$
$= 30x^2y^{2n+1} - 10x^2y^n$

93. a. $(3x + 5) + (3x + 7) = 3x + 5 + 3x + 7$
$= 3x + 3x + 5 + 7$
$= 6x + 12$

This is addition.

b. $(3x + 5)(3x + 7) = 3x(3x) + 3x(7) + 5(3x) + 5(7)$
$= 9x^2 + 21x + 15x + 35$
$= 9x^2 + 36x + 35$

This is multiplication.

97. $f(x) = x^2 - 3x$
$f(a) = a^2 - 3a$

Exercise Set 5.3

1. $\dfrac{4a^2 + 8a}{2a} = \dfrac{4a^2}{2a} + \dfrac{8a}{2a} = 2a + 4$

5. $\dfrac{4x^2y^2 + 6xy^2 - 4y^2}{2x^2y} = \dfrac{4x^2y^2}{2x^2y} + \dfrac{6xy^2}{2x^2y} - \dfrac{4y^2}{2x^2y}$

$$= 2y + \dfrac{3y}{x} - \dfrac{2y}{x^2}$$

9.
$$
\begin{array}{r}
2x - 8 \\
x + 1 \overline{)2x^2 - 6x - 8} \\
\underline{2x^2 + 2x} \\
-8x - 8 \\
\underline{-8x - 8} \\
0
\end{array}
$$

Answer: $2x - 8$

13.
$$
\begin{array}{r}
2x^2 - \frac{1}{2}x + 5 \\
2x + 4 \overline{)4x^3 + 7x^2 + 8x + 20} \\
\underline{4x^3 + 8x^2} \\
-x^2 + 8x \\
\underline{-x^2 - 2x} \\
10x + 20 \\
\underline{10x + 20} \\
0
\end{array}
$$

Answer: $2x^2 - \dfrac{1}{2}x + 5$

17.
$$
\begin{array}{r}
2x^2 + 2x + 8 \\
x - 4 \overline{)2x^3 - 6x^2 + 0x - 4} \\
\underline{2x^3 - 8x^2} \\
2x^2 + 0x \\
\underline{2x^2 - 8x} \\
8x - 4 \\
\underline{8x - 32} \\
28
\end{array}
$$

Answer: $2x^2 + 2x + 8 + \dfrac{28}{x - 4}$

21.
$$
\begin{array}{r}
3x^3 \qquad + 5x + 4 \\
x^2 - 2 \overline{)3x^5 + 0x^4 - x^3 + 4x^2 - 12x - 8} \\
\underline{3x^5 \qquad - 6x^3} \\
5x^3 + 4x^2 - 12x \\
\underline{5x^3 \qquad - 10x} \\
4x^2 - 2x - 8 \\
\underline{4x^2 \qquad - 8} \\
-2x
\end{array}
$$

Answer: $3x^3 + 5x + 4 - \dfrac{2x}{x^2 - 2}$

25.
$$
\begin{array}{r|rrr}
5 & 1 & 3 & -40 \\
& & 5 & 40 \\
\hline
& 1 & 8 & 0
\end{array}
$$

$x + 8$

29.
$$
\begin{array}{r|rrrr}
2 & 1 & -7 & -13 & 5 \\
& & 2 & -10 & -46 \\
\hline
& 1 & -5 & -23 & -41
\end{array}
$$

$x^2 - 5x - 23 - \dfrac{41}{x - 2}$

33.
$$
\begin{array}{r|rrrrr}
5 & 2 & -13 & 16 & -9 & 20 \\
& & 10 & -15 & 5 & -20 \\
\hline
& 2 & -3 & 1 & -4 & 0
\end{array}
$$

$2x^3 - 3x^2 + x - 4$

37.
$$
\begin{array}{r|rrrr}
\frac{1}{3} & 3 & 2 & -4 & 1 \\
& & 1 & 1 & -1 \\
\hline
& 3 & 3 & -3 & 0
\end{array}
$$

$3x^2 + 3x - 3$

41. $6x(x + 3) + 5(x + 3) = 6x^2 + 18x + 5x + 15$
$$= 6x^2 + 23x + 15$$

45. $|2x + 7| \geq 9$
$2x + 7 \leq -9$ or $2x + 7 \geq 9$
$2x \leq -16$ or $2x \geq 2$
$x \leq -8$ or $x \geq 1$
$(-\infty, -8] \cup [1, \infty)$

49. Recall that $A = l \cdot w$ so
$$w = \dfrac{A}{l} = \dfrac{15x^2 - 29x - 14}{5x + 2}$$

$$
\begin{array}{r}
3x - 7 \\
5x + 2 \overline{)15x^2 - 29x - 14} \\
\underline{15x^2 + 6x} \\
-35x - 14 \\
\underline{-35x - 14} \\
0
\end{array}
$$

The width is $(3x - 7)$ in.

53. Multiply $(x^2 - x + 10)$ by $(x + 3)$ and add the remainder, -2.
$(x^2 - x + 10)(x + 3)$
$x^3 - x^2 + 10x + 3x^2 - 3x + 30$
$x^3 + 2x^2 + 7x + 30$
$$\dfrac{-2}{x^3 + 2x^2 + 7x + 28}$$

57. a. answers may vary

b. answers may vary

Exercise Set 5.4

1. $18x - 12 = 6(3x - 2)$

5. $6x^5 - 8x^4 + 2x^3 = 2x^3(3x^2 - 4x + 1)$

9. $6(x + 3) + 5a(x + 3) = (6 + 5a)(x + 3)$

13. $3x(x^2 + 5) - 2(x^2 + 5) = (3x - 2)(x^2 + 5)$

17. $A = P + Prt$
$A = P(1 + rt)$

21. $ab + 3a + 2b + 6 = a(b + 3) + 2(b + 3)$
$= (a + 2)(b + 3)$

25. $2xy - 3x - 4y + 6 = x(2y - 3) - 2(2y - 3)$
$= (x - 2)(2y - 3)$

29. $x^3 + 3x^2 + 4x + 12 = x^2(x + 3) + 4(x + 3)$
$= (x^2 + 4)(x + 3)$

33. $2x^2 + 3xy + 4x + 6y = x(2x + 3y) + 2(2x + 3y)$
$= (x + 2)(2x + 3y)$

37. $6xy + 10x + 9y + 15 = 2x(3y + 5) + 3(3y + 5)$
$= (2x + 3)(3y + 5)$

41. $9abc^2 + 6a^2bc - 6ab + 3bc$
$= 3b(3ac^2 + 2a^2c - 2a + c)$

45. $(x + 3)(x + 2) = x^2 + 2x + 3x + 6$
$= x^2 + 5x + 6$

49. a. $(2 - x)(3 - y) = 6 - 2y - 3x + xy$
$= xy - 3x - 2y + 6$

b. $(-2 + x)(-3 + y) = 6 - 2y - 3x + xy$
$= xy - 3x - 2y + 6$

c. $(x - 2)(y - 3) = xy - 3x - 2y + 6$

d. $(-x + 2)(-y + 3) = xy - 3x - 2y + 6$

The answer is none.

53. $3y^n + 3y^{2n} + 5y^{8n} = y^n(3 + 3y^n + 5y^{7n})$

57. a. $h(t) = -16t^2 + 224 = -16(t^2 - 14)$

b. $h(2) = -16(2)^2 + 224$
$= -16(4) + 224$
$= -64 + 224$
$= 160 \text{ ft}$
$h(2) = -16(2^2 - 14)$
$= -16(4 - 14)$
$= -16(-10)$
$= 160 \text{ ft}$

c. answers may varyy

Exercise Set 5.5

1. $x^2 + 9x + 18 = (x + 6)(x + 3)$

5. $x^2 + 10x - 24 = (x + 12)(x - 2)$

9. $3x^2 - 18x + 24 = 3(x^2 - 6x + 8)$
$= 3(x - 2)(x - 4)$

13. $2x^2 + 30x - 108 = 2(x^2 + 15x - 54)$
$= 2(x + 18)(x - 3)$

17. $x^2 - 15x - 54 = (x - 18)(x + 3)$

21. $2x^2 + 2x - 12 = 2(x^2 + x - 6)$
$= 2(x + 3)(x - 2)$

25. $x^3 + 2x^2 - 8x = x(x^2 + 2x - 8)$
$= x(x + 4)(x - 2)$

29. $5x^2 + 16x + 3 = (5x + 1)(x + 3)$

33. $2x^2 + 25x - 20$ is prime.

37. $12x^2 + 10x - 50 = 2(6x^2 + 5x - 25)$
$= 2(3x - 5)(2x + 5)$

41. $6x^3 + 8x^2 + 24x = 2x(3x^2 + 4x + 12)$

45. $2x^2 - 5xy - 3y^2 = (2x + y)(x - 3y)$

49. $28y^2 + 22y + 4 = 2(14y^2 + 11y + 2)$
$= 2(7y + 2)(2y + 1)$

53. $3x^2 - 5x - 2 = (3x + 1)(x - 2)$

57. $18x^4 + 21x^3 + 6x^2 = 3x^2(6x^2 + 7x + 2)$
$= 3x^2(3x + 2)(2x + 1)$

61. $6x^3 - x^2 - x = x(6x^2 - x - 1)$
$= x(3x + 1)(2x - 1)$

65. $9x^2 + 30x + 25 = (3x + 5)^2$

69. Let $y = x^2$. Then $y^2 = x^4$.
$x^4 + x^2 - 6 = y^2 + y - 6$
$= (y + 3)(y - 2)$
$= (x^2 + 3)(x^2 - 2)$

73. Let $y = x^3$. Then $y^2 = x^6$.
$x^6 - 7x^3 + 12 = y^2 - 7y + 12$
$= (y - 4)(y - 3)$
$= (x^3 - 4)(x^3 - 3)$

77. Let $y = x - 4$.
$(x - 4)^2 + 3(x - 4) - 18$
$= y^2 + 3y - 18$
$= (y + 6)(y - 3)$
$= [(x - 4) + 6][(x - 4) - 3]$
$= (x + 2)(x - 7)$

81. Let $y = x + 4$.
$2(x + 4)^2 + 3(x + 4) - 5$
$= 2y^2 + 3y - 5$
$= (2y + 5)(y - 1)$
$= [2(x + 4) + 5][(x + 4) - 1]$
$= [2x + 8 + 5][x + 3]$
$= (2x + 13)(x + 3)$

85. $(x - 3)(x + 3) = x^2 - 3^2$
$= x^2 - 9$

89.
$$x^2 + 2x + 4$$
$$\underline{\qquad x - 2}$$
$$-2x^2 - 4x - 8$$
$$\underline{x^3 + 2x^2 + 4x \qquad}$$
$$x^3 \qquad\qquad - 8$$

93. $x^{2n} + 10x^n + 16 = (x^n + 2)(x^n + 8)$

97. $2x^{2n} + 11x^n + 5 = (2x^n + 1)(x^n + 5)$

101. $x^4 + 6x^3 + 5x^2 = x^2(x^2 + 6x + 5)$
$= x^2(x + 5)(x + 1)$

Exercise Set 5.6

1. $x^2 + 6x + 9 = x^2 + 2(x)(3) + 3^2 = (x + 3)^2$

5. $3x^2 - 24x + 48 = 3(x^2 - 8x + 16)$
$= 3[x^2 - 2(x)(4) + 4^2]$
$= 3(x - 4)^2$

9. $4a^2 + 12a + 9 = (2a)^2 + 2(2a)(3) + 3^2$
$= (2a + 3)^2$

13. $9 - 4z^2 = (3 + 2z)(3 - 2z)$

17. $64x^2 - 100 = 4(16x^2 - 25)$
$= 4(4x + 5)(4x - 5)$

21. $9x^2 - 49 = (3x + 7)(3x - 7)$

25. $(x + 2y)^2 - 9 = (x + 2y - 3)(x + 2y + 3)$

29. $x^2 - 10x + 25 - y^2 = (x^2 - 10x + 25) - y^2$
$= (x - 5)^2 - y^2$
$= (x - 5 - y)(x - 5 + y)$

33. $x^3 + 27 = x^3 + 3^3$
$= (x + 3)(x^2 - 3x + 9)$

37. $m^3 + n^3 = (m + n)(m^2 - mn + n^2)$

41. $a^3b + 8b^4 = b(a^3 + 8b^3)$
$= b(a^3 + (2b)^3)$
$= b(a + 2b)(a^2 - 2ab + 4b^2)$

45. $x^6 - y^3 = (x^2)^3 - y^3$
$= (x^2 - y)(x^4 + x^2y + y^2)$

49. $x^3 - 1 = x^3 - 1^3$
$= (x - 1)(x^2 + x + 1)$

53. $3x^6y^2 + 81y^2 = 3y^2(x^6 + 27)$
$= 3y^2((x^2)^3 + 3^3)$
$= 3y^2(x^2 + 3)(x^4 - 3x^2 + 9)$

57. $3x + 1 = 0$
$3x = -1$
$x = -\dfrac{1}{3}$
$\left\{-\dfrac{1}{3}\right\}$

61. $-5x + 25 = 0$
$-5x = -25$
$x = 5$
$\{5\}$

65. $V = x \cdot x \cdot x - x \cdot y \cdot y$
$= x^3 - xy^2$
$= x(x^2 - y^2)$
$= x(x + y)(x - y)$

69. $c = \left(\dfrac{-14}{2}\right)^2 = (-7)^2 = 49$
$m^2 - 14m + 49 = (m - 7)^2$

73. $x^{2n} - 36 = (x^n)^2 - 6^2$
$= (x^n + 6)(x^n - 6)$

77. $x^{4n} - 625 = (x^{2n})^2 - 25^2$
$= (x^{2n} + 25)(x^{2n} - 25)$
$= (x^{2n} + 25)[(x^n)^2 - 5^2]$
$= (x^{2n} + 25)(x^n + 5)(x^n - 5)$

Exercise Set 5.7

1. $(x + 3)(3x - 4) = 0$
$x + 3 = 0$ or $3x - 4 = 0$
$3x = 4$
$x = -3$ or $x = \dfrac{4}{3}$
$\left\{-3, \dfrac{4}{3}\right\}$

5. $x^2 + 11x + 24 = 0$
$(x + 3)(x + 8) = 0$
$x + 3 = 0$ or $x + 8 = 0$
$x = -3$ or $x = -8$
$\{-3, -8\}$

9. $z^2 + 9 = 10z$
$z^2 - 10z + 9 = 0$
$(z - 1)(z - 9) = 0$
$z - 1 = 0$ or $z - 9 = 0$
$z = 1$ or $z = 9$
$\{1, 9\}$

13. $x^2 - 6x = x(8 + x)$
$x^2 - 6x = 8x + x^2$
$0 = 14x$
$x = 0$
$\{0\}$

17. $\dfrac{x^2}{2} + \dfrac{x}{20} = \dfrac{1}{10}$
$10x^2 + x = 2$
$10x^2 + x - 2 = 0$
$(5x - 2)(2x + 1) = 0$
$5x - 2 = 0$ or $2x + 1 = 0$
$5x = 2$ or $2x = -1$
$x = \dfrac{2}{5}$ or $x = -\dfrac{1}{2}$
$\left\{\dfrac{2}{5}, -\dfrac{1}{2}\right\}$

21. $(x + 2)(x - 7)(3x - 8) = 0$
$x + 2 = 0$ or $x - 7 = 0$ or $3x - 8 = 0$
$x = -2$ or $x = 7$ or $x = \dfrac{8}{3}$
$\left\{-2, 7, \dfrac{8}{3}\right\}$

25. $x^3 - x = 2x^2 - 2$
$x^3 - 2x^2 - x + 2 = 0$
$x^2(x - 2) - (x - 2) = 0$
$(x^2 - 1)(x - 2) = 0$
$(x + 1)(x - 1)(x - 2) = 0$
$x + 1 = 0$ or $x - 1 = 0$ or $x - 2 = 0$
$x = -1$ or $x = 1$ or $x = 2$
$\{-1, 1, 2\}$

29. $(2x + 7)(x - 10) = 0$
$2x + 7 = 0$ or $x - 10 = 0$
$2x = -7$
$x = -\dfrac{7}{2}$ or $x = 10$
$\left\{-\dfrac{7}{2}, 10\right\}$

33. $x^2 - 2x - 15 = 0$
$(x - 5)(x + 3) = 0$
$x - 5 = 0$ or $x + 3 = 0$
$x = 5$ or $x = -3$
$\{5, -3\}$

37. $w^2 - 5w = 36$
$w^2 - 5w - 36 = 0$
$(w - 9)(w + 4) = 0$
$w - 9 = 0$ or $w + 4 = 0$
$w = 9$ or $w = -4$
$\{9, -4\}$

41. $2r^3 + 6r^2 = 20r$
$r^3 + 3r^2 = 10r$
$r^3 + 3r^2 - 10r = 0$
$r(r^2 + 3r - 10) = 0$
$r(r + 5)(r - 2) = 0$
$r = 0$ or $r + 5 = 0$ or $r - 2 = 0$
$r = 0$ or $r = -5$ or $r = 2$
$\{0, -5, 2\}$

45. $2z(z + 6) = 2z^2 + 12z - 8$
$2z^2 + 12z = 2z^2 + 12z - 8$
$0 = -8$ False
No solution exists.
$\varnothing$

49. Let n = one number and $n + 5$ = the other number.
$n(n + 5) = 66$
$n^2 + 5n - 66 = 0$
$(n + 11)(n - 6) = 0$
$n + 11 = 0$ or $n - 6 = 0$
$n = -11$ or $n = 6$
$n + 5 = -6$ $n + 5 = 11$
There are two solutions: -11 and -6 and 6 and 11.

53. Let x = longer leg
$x - 2$ = shorter leg
$x^2 + (x - 2)^2 = (10)^2$
$x^2 + x^2 - 4x + 4 = 100$
$2x^2 - 4x - 96 = 0$
$2(x - 8)(x + 6) = 0$
$x = 8$ or $x = -6$ (disregard)
$x - 2 = 6$
The legs are 6 feet and 8 feet.

57. $h = -16t^2 + 1600$
$0 = -16(t^2 - 100)$
$0 = -16(t + 10)(t - 10)$
$t + 10 = 0$ or $t - 10 = 0$
$t = -10$ (disregard)   or $t = 10$
They will hit the ground in 10 seconds.

61. Let x = width of border
$(2x + 16)(2x + 12) - 12(16) = 128$
$4x^2 + 24x + 32x + 192 - 192 = 128$
$4x^2 + 56x - 128 = 0$
$4(x^2 + 14x - 32) = 0$
$4(x - 2)(x + 16) = 0$
$x - 2 = 0$ or $x + 16 = 0$
$x = 2$ or $x = -16$ (disregard)
The border is 2 inches.

65. $y = -16t^2 + 80t + 576$
Let $y = 0$ to obtain
$0 = -16t^2 + 80t + 576$
$0 = -16(t^2 - 5t - 36)$
$0 = -16(t - 9)(t + 4)$
$t - 9 = 0$ or $t + 4 = 0$
$t = 9$ or $t = -4$ (disregard)
It takes 9 seconds for the object to strike the ground.

71. $(x^2 + x - 6)(3x^2 - 14x - 5) = 0$
$(x + 3)(x - 2)(3x + 1)(x - 5) = 0$
$x + 3 = 0$ or $x - 2 = 0$ or $3x + 1 = 0$ or $x - 5 = 0$
$x = -3$ or $x = 2$ or $x = -\dfrac{1}{3}$ or $x = 5$
$\left\{-3, -\dfrac{1}{3}, 2, 5\right\}$

73. answers may vary

Exercise Set 5.8

1. a. Domain, $(-\infty, \infty)$; Range, $(-\infty, 5]$

b. x-intercepts, $(-2, 0)$, $(6, 0)$; y-intercept, $(0, 5)$

c. $(0, 5)$

d. There is no such point.

5. a. Domain, $(-\infty, \infty)$; Range, $(-\infty, \infty)$

b. x-intercepts $(-2, 0)$, $(0, 0)$, $(2, 0)$; y-intercept, $(0, 0)$

c. There is no such point.

d. There is no such point.

9. $f(x) = x^2 + 1$

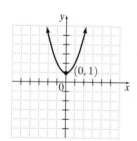

13. $f(x) = x^2 + 8x + 7$
$-\dfrac{b}{2a} = -\dfrac{8}{2(1)} = -4$
$f(-4) = (-4)^2 + 8(-4) + 7 = -9$
The vertex is $(-4, -9)$.
$x^2 + 8x + 7 = 0$
$(x + 7)(x + 1) = 0$
$x + 7 = 0$ or $x + 1 = 0$
$x = -7$ or $x = -1$
The x-intercepts are $(-7, 0)$ and $(-1, 0)$.
If $x = 0$, then $y = f(0) = 7$.
The y-intercept is $(0, 7)$.

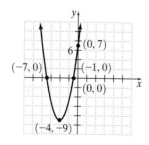

17. $f(x) = 2x^2 - 6x$
$-\dfrac{b}{2a} = -\dfrac{(-6)}{2(2)} = \dfrac{3}{2}$
$f\left(\dfrac{3}{2}\right) = 2\left(\dfrac{3}{2}\right)^2 - 6\left(\dfrac{3}{2}\right) = -\dfrac{9}{2}$
The vertex is $\left(\dfrac{3}{2}, -\dfrac{9}{2}\right)$.
$2x^2 - 6x = 0$
$2x(x - 3) = 0$
$2x = 0$ or $x - 3 = 0$
$x = 0$ or $x = 3$
The x-intercepts are $(0, 0)$ and $(3, 0)$.
If $x = 0$, $y = 2(0)^2 - 6(0) = 0$.
The y-intercept is $(0, 0)$.

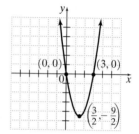

21. Two

25. no; answers may vary

29. $\dfrac{x^7 y^{10}}{x^3 y^{15}} = \dfrac{x^{7-3}}{y^{15-10}} = \dfrac{x^4}{y^5}$

33. $f(x) = (x - 2)(x + 5)$
Graph E because the x-intercepts are at $(2, 0)$ and $(-5, 0)$.

37.

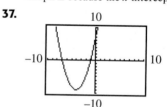

41. answers may vary

Chapter 5 Test

1. $(4x^3 - 3x - 4) - (9x^3 + 8x + 5)$
$= 4x^3 - 3x - 4 - 9x^3 - 8x - 5$
$= 4x^3 - 9x^3 - 3x - 8x - 4 - 5$
$= -5x^3 - 11x - 9$

5. $(6m + n)^2 = 36m^2 + 12mn + n^2$

9.
$\begin{array}{r|rrrrr} -3 & 4 & -3 & 2 & -1 & -1 \\ & & -12 & 45 & -141 & 426 \\ \hline & 4 & -15 & 47 & -142 & 425 \end{array}$
$(4x^4 - 3x^3 + 2x^2 - x - 1) \div (x + 3)$
$= 4x^3 - 15x^2 + 47x - 142 + \dfrac{425}{x + 3}$

13. $6x^2 - 15x - 9 = 3(2x^2 - 5x - 3)$
$= 3(2x + 1)(x - 3)$

17. $6x^2 + 24 = 6(x^2 + 4)$

21. $2x^3 - 8x + 5x^2 - 20 = 0$
$(2x + 5)(x^2 - 4) = 0$
$(2x + 5)(x + 2)(x - 2) = 0$
$2x + 5 = 0$ or $x + 2 = 0$ or $x - 2 = 0$
$x = -\dfrac{5}{2}$ or $x = -2$ or $x = 2$
$\left\{-\dfrac{5}{2}, -2, 2\right\}$

Chapter 6

Exercise Set 6.1

1. $\dfrac{x+3}{x-2}$ is undefined when $x - 2 = 0$ or $x = 2$.

5. $\dfrac{x^2+1}{3x}$ is undefined when $3x = 0$ or $x = 0$.

9. $\dfrac{3x+2}{x^3+x^2-2x}$ is undefined when

$$x^3 + x^2 - 2x = 0$$
$$x(x^2 + x - 2) = 0$$
$$x(x+2)(x-1) = 0$$
$$x = 0 \quad \text{or} \quad x + 2 = 0 \quad \text{or} \quad x - 1 = 0$$
$$x = 0 \quad \text{or} \quad x = -2 \quad \text{or} \quad x = 1$$

13. $\dfrac{10x^3}{18x} = \dfrac{(2x)5x^2}{(2x)9} = \dfrac{5x^2}{9}$

17. $\dfrac{8x - 16x^2}{8x} = \dfrac{8x(1-2x)}{8x} = 1 - 2x$

21. $\dfrac{9y - 18}{7y - 14} = \dfrac{9(y-2)}{7(y-2)} = \dfrac{9}{7}$

25. $\dfrac{x-9}{9-x} = \dfrac{-1(9-x)}{9-x} = -1$

29. $\dfrac{2x^2 - 7x - 4}{x^2 - 5x + 4} = \dfrac{(2x+1)(x-4)}{(x-1)(x-4)}$

$$= \dfrac{2x+1}{x-1}$$

33. $\dfrac{3x^2 - 5x - 2}{6x^3 + 2x^2 + 3x + 1} = \dfrac{(3x+1)(x-2)}{2x^2(3x+1) + 1(3x+1)}$

$$= \dfrac{(3x+1)(x-2)}{(3x+1)(2x^2+1)}$$

$$= \dfrac{x-2}{2x^2+1}$$

37. $\dfrac{4}{x} \cdot \dfrac{x^2}{8} = \dfrac{2 \cdot 2 \cdot x \cdot x}{x \cdot 2 \cdot 2 \cdot 2} = \dfrac{x}{2}$

41. $\dfrac{18a - 12a^2}{4a^2 + 4a + 1} \cdot \dfrac{4a^2 + 8a + 3}{4a^2 - 9}$

$$= \dfrac{6a(3-2a)(2a+3)(2a+1)}{(2a+1)^2(2a+3)(2a-3)}$$

$$= \dfrac{(6a)[-(2a-3)]}{(2a+1)(2a-3)}$$

$$= \dfrac{-6a}{2a+1}$$

45. $\dfrac{2x^3 - 16}{6x^2 + 6x - 36} \cdot \dfrac{9x + 18}{3x^2 + 6x + 12}$

$$= \dfrac{2(x^3-8) \cdot 9(x+2)}{6(x^2+x-6) \cdot 3(x^2+2x+4)}$$

$$= \dfrac{18(x-2)(x^2+2x+4)(x+2)}{18(x+3)(x-2)(x^2+2x+4)}$$

$$= \dfrac{x+2}{x+3}$$

49. $\dfrac{x^2 - 6x - 16}{2x^2 - 128} \cdot \dfrac{x^2 + 16x + 64}{3x^2 + 30x + 48}$

$$= \dfrac{(x-8)(x+2)(x+8)^2}{2(x^2-64)[3(x^2+10x+16)]}$$

$$= \dfrac{(x-8)(x+2)(x+8)^2}{2(x+8)(x-8)3(x+8)(x+2)}$$

$$= \dfrac{1}{6}$$

53. $\dfrac{4}{x} \div \dfrac{8}{x^2} = \dfrac{4}{x} \cdot \dfrac{x^2}{8} = \dfrac{2 \cdot 2 \cdot x \cdot x}{x \cdot 2 \cdot 2 \cdot 2} = \dfrac{x}{2}$

57. $\dfrac{a+b}{ab} \div \dfrac{a^2-b^2}{4a^3b} = \dfrac{a+b}{ab} \cdot \dfrac{4a^3b}{(a+b)(a-b)}$

$$= \dfrac{4a^2}{a-b}$$

61. $\dfrac{x^2 - 6x - 16}{2x^2 - 128} \div \dfrac{x^2 + 10x + 16}{x^2 + 16x + 64}$

$$= \dfrac{(x-8)(x+2)}{2(x^2-64)} \cdot \dfrac{(x+8)^2}{(x+2)(x+8)}$$

$$= \dfrac{(x-8)(x+8)}{2(x+8)(x-8)}$$

$$= \dfrac{1}{2}$$

65. $\dfrac{8b + 24}{3a + 6} \div \dfrac{ab - 2b + 3a - 6}{a^2 - 4a + 4}$

$$= \dfrac{8(b+3)}{3(a+2)} \cdot \dfrac{(a-2)^2}{(a-2)(b+3)}$$

$$= \dfrac{8(a-2)}{3(a+2)}$$

69. $\dfrac{3x^2 - 5x - 2}{y^2 + y - 2} \cdot \dfrac{y^2 + 4y - 5}{12x^2 + 7x + 1} \div \dfrac{5x^2 - 9x - 2}{8x^2 - 2x - 1}$

$$= \dfrac{3x^2 - 5x - 2}{y^2 + y - 2} \cdot \dfrac{y^2 + 4y - 5}{12x^2 + 7x + 1} \cdot \dfrac{8x^2 - 2x - 1}{5x^2 - 9x - 2}$$

$$= \dfrac{(3x+1)(x-2)(y+5)(y-1)(4x+1)(2x-1)}{(y+2)(y-1)(4x+1)(3x+1)(5x+1)(x-2)}$$

$$= \dfrac{(y+5)(2x-1)}{(y+2)(5x+1)}$$

73. $R(x) = \dfrac{1000x^2}{x^2 + 4}$

  a. $R(1) = \dfrac{1000(1)^2}{1^2 + 4} = \$200 \text{ million}$

  b. $R(2) = \dfrac{1000(2)^2}{2^2 + 4} = \$500 \text{ million}$

  c. $\$500 \text{ million} - \$200 \text{ million} = \$300 \text{ million}$

  d. $\{x \mid x \text{ is a real number}\}$

77. $\dfrac{5}{28} - \dfrac{2}{21}$

The LCD is 84.

$$\dfrac{5}{28} \cdot \dfrac{3}{3} - \dfrac{2}{21} \cdot \dfrac{4}{4} = \dfrac{15}{84} - \dfrac{8}{84} = \dfrac{7}{84} = \dfrac{7 \cdot 1}{7 \cdot 12} = \dfrac{1}{12}$$

81. Since $A = b \cdot h$, $b = \dfrac{A}{h}$. Now,

$$b = \dfrac{\dfrac{x^2 + x - 2}{x^3}}{\dfrac{x^2}{x-1}}$$

$$b = \dfrac{(x+2)(x-1)}{x^3} \cdot \dfrac{(x-1)}{x^2}$$

$$b = \dfrac{(x+2)(x-1)^2}{x^5} \text{ feet}$$

85. **a.** $\dfrac{x+5}{5+x} = \dfrac{x+5}{x+5} = 1$

  b. $\dfrac{x-5}{5-x} = \dfrac{x-5}{-(x-5)} = -1$

  c. $\dfrac{x+5}{x-5}$ neither

d. $\dfrac{-x-5}{x+5} = \dfrac{-(x+5)}{x+5} = -1$

e. $\dfrac{x-5}{-x+5} = \dfrac{x-5}{-(x-5)} = -1$

f. $\dfrac{-5+x}{x-5} = \dfrac{x-5}{x-5} = 1$

89. $\dfrac{3+q^n}{q^n+3} = \dfrac{q^n+3}{q^n+3} = 1$

Exercise Set 6.2

1. $\dfrac{2}{x} - \dfrac{5}{x} = \dfrac{2-5}{x} = \dfrac{-3}{x} = -\dfrac{3}{x}$

5. $\dfrac{x^2}{x+2} - \dfrac{4}{x+2} = \dfrac{x^2-4}{x+2} = \dfrac{(x+2)(x-2)}{x+2} = x-2$

9. $\dfrac{x-5}{2x} - \dfrac{x+5}{2x} = \dfrac{x-5-x-5}{2x} = \dfrac{-10}{2x} = -\dfrac{5}{x}$

13. $7 = 7$

$5x = 5 \cdot x$

$\text{LCD} = 7 \cdot 5 \cdot x = 35x$

17. $x+7 = x+7$

$x-7 = x-7$

$\text{LCD} = (x+7)(x-7)$

21. $a^2 - b^2 = (a+b)(a-b)$

$a^2 - 2ab + b^2 = (a-b)(a-b)$

$\text{LCD} = (a+b)(a-b)(a-b) = (a+b)(a-b)^2$

25. answers may vary

29. $\dfrac{5}{2y^2} - \dfrac{2}{7y} = \dfrac{5 \cdot 7}{2y^2 \cdot 7} - \dfrac{2 \cdot 2y}{7y \cdot 2y} = \dfrac{35}{14y^2} - \dfrac{4y}{14y^2} = \dfrac{35-4y}{14y^2}$

33. $\dfrac{1}{x-5} + \dfrac{2x-19}{x^2-x-20} = \dfrac{1}{x-5} + \dfrac{2x-19}{(x-5)(x+4)}$

$= \dfrac{1 \cdot (x+4)}{(x-5) \cdot (x+4)} + \dfrac{2x-19}{(x-5)(x+4)}$

$= \dfrac{(x+4) + 2x - 19}{(x-5)(x+4)} = \dfrac{3x-15}{(x-5)(x+4)}$

$= \dfrac{3(x-5)}{(x-5)(x+4)} = \dfrac{3}{x+4}$

37. $\dfrac{-2}{x^2-3x} - \dfrac{1}{x^3-3x^2} = \dfrac{-2}{x(x-3)} - \dfrac{1}{x^2(x-3)}$

$= \dfrac{-2 \cdot x}{x(x-3) \cdot x} - \dfrac{1}{x^2(x-3)} = \dfrac{-2x-1}{x^2(x-3)}$ or $-\dfrac{2x+1}{x^2(x-3)}$

41. $\dfrac{5}{x-2} + \dfrac{x+4}{2-x} = \dfrac{5}{x-2} + \dfrac{(x+4)(-1)}{(2-x)(-1)}$

$= \dfrac{5}{x-2} + \dfrac{-x-4}{x-2} = \dfrac{5-x-4}{x-2} = \dfrac{-x+1}{x-2}$ or $\dfrac{1-x}{x-2}$

45. $\dfrac{7}{x^2-x-2} + \dfrac{x}{x^2+4x+3} = \dfrac{7}{(x-2)(x+1)} + \dfrac{x}{(x+3)(x+1)}$

$= \dfrac{7 \cdot (x+3)}{(x-2)(x+1) \cdot (x+3)} + \dfrac{x \cdot (x-2)}{(x+3)(x+1) \cdot (x-2)}$

$= \dfrac{7(x+3) + x(x-2)}{(x-2)(x+1)(x+3)} = \dfrac{7x+21+x^2-2x}{(x-2)(x+1)(x+3)}$

$= \dfrac{x^2+5x+21}{(x-2)(x+1)(x+3)}$

49. $\dfrac{2}{a^2+2a+1} + \dfrac{3}{a^2-1} = \dfrac{2}{(a+1)(a+1)} + \dfrac{3}{(a+1)(a-1)}$

$= \dfrac{2 \cdot (a-1)}{(a+1)(a+1) \cdot (a-1)} + \dfrac{3 \cdot (a+1)}{(a+1)(a-1) \cdot (a+1)}$

$= \dfrac{2(a-1) + 3(a+1)}{(a+1)^2(a-1)} = \dfrac{2a-2+3a+3}{(a+1)^2(a-1)} = \dfrac{5a+1}{(a+1)^2(a-1)}$

53. $\dfrac{5}{x^2-4} - \dfrac{4}{x^2+4x+4} = \dfrac{5}{(x+2)(x-2)} - \dfrac{4}{(x+2)(x+2)}$

$= \dfrac{5 \cdot (x+2)}{(x+2)(x-2) \cdot (x+2)} - \dfrac{4 \cdot (x-2)}{(x+2)(x+2)(x-2)}$

$= \dfrac{5(x+2) - 4(x-2)}{(x+2)^2(x-2)} = \dfrac{5x+10-4x+8}{(x+2)^2(x-2)} = \dfrac{x+18}{(x+2)^2(x-2)}$

57. $\dfrac{3}{x+3} + \dfrac{5}{x^2+6x+9} - \dfrac{x}{x^2-9} = \dfrac{3}{x+3} + \dfrac{5}{(x+3)(x+3)} - \dfrac{x}{(x+3)(x-3)}$

$= \dfrac{3 \cdot (x+3)(x-3)}{(x+3) \cdot (x+3)(x-3)} + \dfrac{5 \cdot (x-3)}{(x+3)(x+3)(x-3)} - \dfrac{x \cdot (x+3)}{(x+3)(x-3) \cdot (x+3)}$

$= \dfrac{3(x+3)(x-3) + 5(x-3) - x(x+3)}{(x+3)^2(x-3)} = \dfrac{3x^2-27+5x-15-x^2-3x}{(x+3)^2(x-3)} = \dfrac{2x^2+2x-42}{(x+3)^2(x-3)}$ or $\dfrac{2(x^2+x-21)}{(x+3)^2(x-3)}$

61. $\left(\dfrac{1}{x} + \dfrac{2}{3}\right) - \left(\dfrac{1}{x} - \dfrac{2}{3}\right) = \dfrac{1}{x} + \dfrac{2}{3} - \dfrac{1}{x} + \dfrac{2}{3} = \dfrac{2}{3} + \dfrac{2}{3} = \dfrac{4}{3}$

65. $x^2\left(\dfrac{4}{x^2} + 1\right) = x^2 \cdot \dfrac{4}{x^2} + x^2 \cdot 1 = 4 + x^2$

69. answers may vary

73. $\left(\dfrac{2a}{3}\right)^2 \div \left(\dfrac{a^2}{a + 1} - \dfrac{1}{a + 1}\right) = \dfrac{4a^2}{9} \div \dfrac{a^2 - 1}{a + 1} = \dfrac{4a^2}{9} \div \dfrac{(a + 1)(a - 1)}{a + 1} = \dfrac{4a^2}{9} \div (a - 1)$

$$= \dfrac{4a^2}{9} \cdot \dfrac{1}{a - 1} = \dfrac{4a^2}{9(a - 1)}$$

77. $\left(\dfrac{x}{x + 1} - \dfrac{x}{x - 1}\right) \div \dfrac{x}{2x + 2} = \left(\dfrac{x \cdot (x - 1)}{(x + 1) \cdot (x - 1)} - \dfrac{x \cdot (x + 1)}{(x - 1)(x + 1)}\right) \div \dfrac{x}{2(x + 1)} = \dfrac{x(x - 1) - x(x + 1)}{(x + 1)(x - 1)} \div \dfrac{x}{2(x + 1)}$

$= \dfrac{x^2 - x - x^2 - x}{(x + 1)(x - 1)} \div \dfrac{x}{2(x + 1)} = \dfrac{-2x}{(x + 1)(x - 1)} \cdot \dfrac{2(x + 1)}{x} = -\dfrac{4}{x - 1}$

81. $x^{-1} + (2x)^{-1} = \dfrac{1}{x} + \dfrac{1}{2x} = \dfrac{1 \cdot 2}{x \cdot 2} + \dfrac{1}{2x}$

$$= \dfrac{2 + 1}{2x} = \dfrac{3}{2x}$$

85. answers may vary

Exercise Set 6.3

1. $\dfrac{\dfrac{10}{3x}}{\dfrac{5}{6x}} = \dfrac{10}{3x} \cdot \dfrac{6x}{5} = \dfrac{2}{1} \cdot \dfrac{2}{1} = 4$

5. $\dfrac{\dfrac{4}{x - 1}}{\dfrac{x}{x - 1}} = \dfrac{4}{x - 1} \cdot \dfrac{x - 1}{x} = \dfrac{4}{x}$

9. $\dfrac{\dfrac{4x^2 - y^2}{xy}}{\dfrac{2}{y} - \dfrac{1}{x}} = \dfrac{\left(\dfrac{4x^2 - y^2}{xy}\right) \cdot xy}{\left(\dfrac{2}{y} - \dfrac{1}{x}\right) \cdot xy}$

$$= \dfrac{4x^2 - y^2}{2x - y}$$

$$= \dfrac{(2x - y)(2x + y)}{2x - y}$$

$$= 2x + y$$

13. $\dfrac{\dfrac{2}{x} + \dfrac{3}{x^2}}{\dfrac{4}{x^2} - \dfrac{9}{x}} = \dfrac{\left(\dfrac{2}{x} + \dfrac{3}{x^2}\right)x^2}{\left(\dfrac{4}{x^2} - \dfrac{9}{x}\right)x^2}$

$$= \dfrac{2x + 3}{4 - 9x}$$

17. $\dfrac{\dfrac{4}{5 - x} + \dfrac{5}{x - 5}}{\dfrac{2}{x} + \dfrac{3}{x - 5}} = \dfrac{-\dfrac{4}{x - 5} + \dfrac{5}{x - 5}}{\dfrac{2(x - 5) + 3x}{x(x - 5)}}$

$$= \dfrac{\dfrac{1}{x - 5}}{\dfrac{2x - 10 + 3x}{x(x - 5)}}$$

$$= \dfrac{1}{x - 5} \cdot \dfrac{x(x - 5)}{5x - 10}$$

$$= \dfrac{x}{5x - 10}$$

21. $\dfrac{\dfrac{2}{x} + 3}{\dfrac{4}{x^2} - 9} = \dfrac{\left(\dfrac{2}{x} + 3\right) \cdot x^2}{\left(\dfrac{4}{x^2} - 9\right) \cdot x^2}$

$$= \dfrac{\dfrac{2}{x} \cdot x^2 + 3 \cdot x^2}{\dfrac{4}{x^2} \cdot x^2 - 9 \cdot x^2} = \dfrac{2x + 3x^2}{4 - 9x^2} = \dfrac{x(2 + 3x)}{(2 + 3x)(2 - 3x)}$$

$$= \dfrac{x}{2 - 3x}$$

25. $\dfrac{\dfrac{-2x}{x - y}}{\dfrac{y}{x^2}} = \dfrac{\dfrac{-2x}{(x - y)} \cdot x^2(x - y)}{\dfrac{y}{x^2} \cdot x^2(x - y)} = \dfrac{-2x \cdot x^2}{y(x - y)} = -\dfrac{2x^3}{y(x - y)}$

29. $\dfrac{\dfrac{x}{9} - \dfrac{1}{x}}{1 + \dfrac{3}{x}} = \dfrac{\left(\dfrac{x}{9} - \dfrac{1}{x}\right) \cdot 9x}{\left(1 + \dfrac{3}{x}\right) \cdot 9x} = \dfrac{\dfrac{x}{9} \cdot 9x - \dfrac{1}{x} \cdot 9x}{1 \cdot 9x + \dfrac{3}{x} \cdot 9x}$

$$= \dfrac{x^2 - 9}{9x + 27} = \dfrac{(x + 3)(x - 3)}{9(x + 3)} = \dfrac{x - 3}{9}$$

33. $\dfrac{x^{-1}}{x^{-2} + y^{-2}} = \dfrac{\dfrac{1}{x}}{\dfrac{1}{x^2} + \dfrac{1}{y^2}}$

$\qquad = \dfrac{\dfrac{1}{x}(x^2y^2)}{\left(\dfrac{1}{x^2} + \dfrac{1}{y^2}\right)(x^2y^2)}$

$\qquad = \dfrac{xy^2}{y^2 + x^2}$

37. $\dfrac{1}{x - x^{-1}} = \dfrac{1}{x - \dfrac{1}{x}}$

$\qquad = \dfrac{1 \cdot x}{\left(x - \dfrac{1}{x}\right) \cdot x}$

$\qquad = \dfrac{x}{x^2 - 1}$

$\qquad = \dfrac{x}{(x + 1)(x - 1)}$

41. $\dfrac{3x^{-1} + (2y)^{-1}}{x^{-2}} = \dfrac{\dfrac{3}{x} + \dfrac{1}{2y}}{\dfrac{1}{x^2}}$

$\qquad = \dfrac{\left(\dfrac{3}{x} + \dfrac{1}{2y}\right) \cdot 2x^2y}{\dfrac{1}{x^2} \cdot 2x^2y} = \dfrac{\dfrac{3}{x} \cdot 2x^2y + \dfrac{1}{2y} \cdot 2x^2y}{\dfrac{1}{x^2} \cdot 2x^2y}$

$\qquad = \dfrac{6xy + x^2}{2y} = \dfrac{x(x + 6y)}{2y}$

45. $\dfrac{5x^{-1} + 2y^{-1}}{x^{-2}y^{-2}} = \dfrac{\dfrac{5}{x} + \dfrac{2}{y}}{\dfrac{1}{x^2y^2}} = \dfrac{\left(\dfrac{5}{x} + \dfrac{2}{y}\right) \cdot x^2y^2}{\dfrac{1}{x^2y^2} \cdot x^2y^2}$

$\qquad = \dfrac{\dfrac{5}{x} \cdot x^2y^2 + \dfrac{2}{y} \cdot x^2y^2}{\dfrac{1}{x^2y^2} \cdot x^2y^2} = \dfrac{5xy^2 + 2x^2y}{1} = 5xy^2 + 2x^2y$

49. $7x + 2 = x - 3$

$\qquad 6x + 2 = -3$

$\qquad 6x = -5$

$\qquad x = -\dfrac{5}{6}$

$\qquad \left\{-\dfrac{5}{6}\right\}$

53. $\dfrac{x}{3} - 5 = 13$

$\qquad \dfrac{x}{3} = 18$

$\qquad x = 54$

$\qquad \{54\}$

57. $\dfrac{\dfrac{1}{x}}{\dfrac{3}{y}} = \dfrac{1}{x} \div \dfrac{3}{y} = \dfrac{1}{x} \cdot \dfrac{y}{3}$

a and b

61. $\dfrac{x}{1 - \dfrac{1}{1 + \dfrac{1}{x}}} = \dfrac{x}{1 - \dfrac{1 \cdot x}{\left(1 + \dfrac{1}{x}\right) \cdot x}}$

$\qquad = \dfrac{x}{1 - \dfrac{x}{x + 1}}$

$\qquad = \dfrac{x(x + 1)}{\left(1 - \dfrac{x}{x + 1}\right)(x + 1)}$

$\qquad = \dfrac{x(x + 1)}{(x + 1) - x}$

$\qquad = \dfrac{x(x + 1)}{x + 1 - x}$

$\qquad = \dfrac{x(x + 1)}{1}$

$\qquad = x(x + 1) = x^2 + x$

Exercise Set 6.4

1. $\dfrac{x}{2} - \dfrac{x}{3} = 12$

$\qquad 3x - 2x = 72$

$\qquad x = 72$

$\qquad \{72\}$

5. $1 - \dfrac{4}{a} = 5$

$\qquad a - 4 = 5a$

$\qquad -4a = 4$

$\qquad a = -1$

$\qquad \{-1\}$

9. $\dfrac{2}{x} + \dfrac{1}{2} = \dfrac{5}{x}$

$\qquad \dfrac{1}{2} = \dfrac{3}{x}$

$\qquad x = 6$

$\qquad \{6\}$

13. $\dfrac{5}{x - 2} - \dfrac{2}{x + 4} = -\dfrac{4}{x^2 + 2x - 8}$

$(x - 2)(x + 4)\left(\dfrac{5}{x - 2} - \dfrac{2}{x + 4}\right) = (x - 2)(x + 4)\left[\dfrac{-4}{(x + 4)(x - 2)}\right]$

$\qquad 5(x + 4) - 2(x - 2) = -4$

$\qquad 5x + 20 - 2x + 4 = -4$

$\qquad 3x = -28$

$\qquad x = -\dfrac{28}{3}$

$\qquad \left\{-\dfrac{28}{3}\right\}$

17. $\dfrac{1}{x - 4} - \dfrac{3x}{x^2 - 16} = \dfrac{2}{x + 4}$

$(x + 4)(x - 4)\left[\dfrac{1}{x - 4} - \dfrac{3x}{(x + 4)(x - 4)}\right] = (x + 4)(x - 4)\left(\dfrac{2}{x + 4}\right)$

$\qquad x + 4 - 3x = 2x - 8$

$\qquad -4x = -12$

$\qquad x = 3$

$\qquad \{3\}$

21.
$$\frac{1}{x-2} - \frac{2}{x^2 - 2x} = 1$$
$$x(x-2)\left[\frac{1}{x-2} - \frac{2}{x(x-2)}\right] = x(x-2)\cdot 1$$
$$x - 2 = x^2 - 2x$$
$$0 = x^2 - 3x + 2$$
$$0 = (x-2)(x-1)$$
$$x - 2 = 0 \quad \text{or} \quad x - 1 = 0$$
$$x = 2 \quad \text{or} \quad x = 1$$

We discard 2 as extraneous.

$\{1\}$

25.
$$\frac{1}{x} - \frac{x}{25} = 0$$
$$25x\left(\frac{1}{x} - \frac{x}{25}\right) = 25x \cdot 0$$
$$25 - x^2 = 0$$
$$x^2 - 25 = 0$$
$$(x+5)(x-5) = 0$$
$$x + 5 = 0 \quad \text{or} \quad x - 5 = 0$$
$$x = -5 \quad \text{or} \quad x = 5$$

$\{-5, 5\}$

29.
$$\frac{x+3}{x+2} = \frac{1}{x+2}$$
$$(x+2)\left(\frac{x+3}{x+2}\right) = (x+2)\left(\frac{1}{x+2}\right)$$
$$x + 3 = 1$$
$$x = -2$$

Which we discard as extraneous.

No solution

$\varnothing$

33.
$$\frac{64}{x^2 - 16} + 1 = \frac{2x}{x-4}$$
$$(x+4)(x-4)\left[\frac{64}{(x+4)(x-4)} + 1\right] = (x+4)(x-4)\left(\frac{2x}{x-4}\right)$$
$$64 + (x+4)(x-4) = 2x(x+4)$$
$$64 + x^2 - 16 = 2x^2 + 8x$$
$$x^2 + 8x - 48 = 0$$
$$(x+12)(x-4) = 0$$
$$x + 12 = 0 \quad \text{or} \quad x - 4 = 0$$
$$x = -12 \quad \text{or} \quad x = 4$$

We discard 4 as extraneous.

$\{-12\}$

37.
$$\frac{28}{x^2 - 9} + \frac{2x}{x-3} + \frac{6}{x+3} = 0$$
$$(x^2 - 9)\left(\frac{28}{x^2 - 9} + \frac{2x}{x-3} + \frac{6}{x+3}\right) = (x^2 - 9)\cdot 0$$
$$28 + 2x(x+3) + 6(x-3) = 0$$
$$28 + 2x^2 + 6x + 6x - 18 = 0$$
$$2x^2 + 12x + 10 = 0$$
$$2(x+5)(x+1) = 0$$
$$x + 5 = 0 \quad \text{or} \quad x + 1 = 0$$
$$x = -5 \quad \text{or} \quad x = -1$$

$\{-5, -1\}$

41. Let x = the number.
$$3x + 4 = 19$$
$$3x = 15$$
$$x = 5$$
The number is 5.

45. 10% (reading from the graph)

49. 19% of 35,047
$$0.19(35,047) \approx 6659$$
6659 inmates

53.
$$x^{-2} - 19x^{-1} + 48 = 0$$
$$\frac{1}{x^2} - \frac{19}{x} + 48 = 0$$
$$\left(\frac{1}{x^2} - \frac{19}{x} + 48\right)x^2 = 0 \cdot x^2$$
$$1 - 19x + 48x^2 = 0$$
$$48x^2 - 19x + 1 = 0$$
$$(16x - 1)(3x - 1) = 0$$
$$16x - 1 = 0 \quad \text{or} \quad 3x - 1 = 0$$
$$x = \frac{1}{16} \quad \text{or} \quad x = \frac{1}{3}$$

$\left\{\frac{1}{16}, \frac{1}{3}\right\}$

57.
$$f(x) = 3.3 + \frac{5400}{x}$$
$$5.10 = 3.3 + \frac{5400}{x}$$
$$5.1 - 3.3 = \frac{5400}{x}$$
$$1.8 = \frac{5400}{x}$$
$$1.8x = 5400$$
$$x = \frac{5400}{1.8}$$
$$x = 3000$$
3000 game disks

Exercise Set 6.5

1.
$$F = \frac{9}{5}C + 32$$
$$F - 32 = \frac{9}{5}C$$
$$C = \frac{5}{9}(F - 32)$$

5.
$$\frac{1}{R} = \frac{1}{R_1} + \frac{1}{R_2}$$
$$\frac{1}{R} = \frac{R_2 + R_1}{R_1 R_2}$$
$$R = \frac{R_1 R_2}{R_1 + R_2}$$

9.
$$A = \frac{h(a+b)}{2}$$
$$2A = ah + bh$$
$$2A - ah = bh$$
$$\frac{2A - ah}{h} = b$$

13.
$$f = \frac{f_1 f_2}{f_1 + f_2}$$
$$(f_1 + f_2)f = \left(\frac{f_1 f_2}{f_1 + f_2}\right)(f_1 + f_2)$$
$$f_1 f + f_2 f = f_1 f_2$$
$$f_1 f = f_1 f_2 - f f_2$$
$$f_1 f = f_2(f_1 - f)$$
$$\frac{f_1 f}{f_1 - f} = f_2$$

17.
$$\frac{\theta}{\omega} = \frac{2L}{c}$$
$$c\omega\left(\frac{\theta}{\omega}\right) = \frac{2L}{c}(c\omega)$$
$$c\theta = 2L\omega$$
$$c = \frac{2L\omega}{\theta}$$

21. Let x = the number.
$$\frac{12 + x}{41 + 2x} = \frac{1}{3}$$
$$3(12 + x) = (41 + 2x) \cdot 1$$
$$36 + 3x = 41 + 2x$$
$$x = 5$$
The number is 5.

25. Let x = number of women.
$$\frac{x}{35{,}712} = \frac{10.2}{100}$$
$$100x = (10.2)(35{,}712)$$
$$100x = 364{,}262.4$$
$$x = \frac{364{,}262.4}{100}$$
$$x \approx 3643$$
There are approximately 3643 women.

29. Convert each time to a rate.
$$\frac{1 \text{ task}}{20 \text{ minutes}} = \frac{1}{20}\frac{\text{task}}{\text{minute}}$$
$$\frac{1 \text{ task}}{30 \text{ minutes}} = \frac{1}{30}\frac{\text{task}}{\text{minute}}$$
$$\frac{1 \text{ task}}{60 \text{ minutes}} = \frac{1}{60}\frac{\text{task}}{\text{minute}}$$
Let x = amount of time required for all three computers to complete the task.
So $\dfrac{1 \text{ task}}{x \text{ minute}} = \dfrac{1}{x}\dfrac{\text{task}}{\text{minute}}$. Adding we get,
$$\frac{1}{20} + \frac{1}{30} + \frac{1}{60} = \frac{1}{x}$$
$$\frac{3 + 2 + 1}{60} = \frac{1}{x}$$
$$\frac{6}{60} = \frac{1}{x}$$
$$\frac{1}{10} = \frac{1}{x} \text{ so } x = 10 \text{ minutes}$$
It takes 10 minutes for the three computers to complete the task.

33. Let r = the speed of the boat in still water.

Recall that $d = rt$ or $t = \dfrac{d}{r}$. Using the latter equation we get
$$\frac{20}{r + 5} = \frac{10}{r - 5}$$ where $r + 5$ is the rate of the boat traveling downstream and $r - 5$ is the rate of the boat traveling upstream.

Now,
$$\frac{2}{r + 5} = \frac{1}{r - 5}$$

$$2(r - 5) = r + 5$$
$$2r - 10 = r + 5$$
$$r = 15$$
Thus, the speed of the boat in still water is 15 mph.

37. Convert times to rates.
$$\frac{1 \text{ pond}}{45 \text{ minutes}} = \frac{1}{45}\frac{\text{pond}}{\text{minute}}$$
$$\frac{1 \text{ pond}}{20 \text{ minutes}} = \frac{1}{20}\frac{\text{pond}}{\text{minute}}$$
Let x = the number of minutes required for the second hose to fill the pond. Then,
$$\frac{1 \text{ pond}}{x \text{ minute}} = \frac{1}{x}\frac{\text{pond}}{\text{minute}}. \text{ Now,}$$
$$\frac{1}{20} = \frac{1}{45} + \frac{1}{x}$$
$$\frac{1}{20} = \frac{x + 45}{45x}$$
$$45x = 20(x + 45)$$
$$45x = 20x + 900$$
$$25x = 900$$
$$x = \frac{900}{25}$$
$$= 36$$
Thus, the second hose alone can fill the pond in 36 minutes.

41. Let x = time.
$$\frac{x \text{ hour}}{1 \text{ mile}} = \frac{1 \text{ hour}}{0.17 \text{ mile}}$$
$$\text{or } \frac{x}{1} = \frac{1}{0.17}$$
$$0.17x = 1 \text{ cross multiply}$$
$$x = \frac{1}{0.17} \approx 5.9$$
It takes approximately 5.9 hours to travel 1 mile.

45.

	distance	= rate	· time
with wind	465	$x + 20$	$\frac{465}{x + 20}$
against wind	345	$x - 20$	$\frac{345}{x - 20}$

$$\frac{465}{x + 20} = \frac{345}{x - 20}$$
$$465(x - 20) = 345(x + 20)$$
$$465x - 9300 = 345x + 6900$$
$$120x = 16{,}200$$
$$x = 135 \text{ mph}$$

49.

	Time	In one hour
Experienced	4	$\frac{1}{4}$
Apprentice	5	$\frac{1}{5}$
Together	x	$\frac{1}{x}$

$$\frac{1}{4} + \frac{1}{5} = \frac{1}{x}$$
$$20x\left(\frac{1}{4} + \frac{1}{5}\right) = 20x\left(\frac{1}{x}\right)$$
$$20x\left(\frac{1}{4}\right) + 20x\left(\frac{1}{5}\right) = 20$$
$$5x + 4x = 20$$
$$9x = 20$$
$$x = \frac{20}{9}$$
$$= 2\frac{2}{9} \text{ hours}$$

53.

	Time	In one minute
Belt	2	$\frac{1}{2}$
Smaller	6	$\frac{1}{6}$
Together	x	$\frac{1}{x}$

$$\frac{1}{2} + \frac{1}{6} = \frac{1}{x}$$

$$6x\left(\frac{1}{2} + \frac{1}{6}\right) = 6x\left(\frac{1}{x}\right)$$

$$6x\left(\frac{1}{2}\right) + 6x\left(\frac{1}{6}\right) = 6$$

$$3x + x = 6$$

$$4x = 6$$

$$x = \frac{6}{4}$$

$$= \frac{3}{2}$$

$$= 1\frac{1}{2}\text{ minutes}$$

57.

	distance	= rate · time	
upstream	3	$x - 6$	$\frac{3}{x-6}$
downstream	9	$x + 6$	$\frac{9}{x+6}$

$$\frac{3}{x - 6} = \frac{9}{x + 6}$$

$$3(x + 6) = 9(x - 6)$$

$$3x + 18 = 9x - 54$$

$$3x + 72 = 9x$$

$$72 = 6x$$

$$12 = x$$

Time upstream $= \dfrac{3}{12 - 6} = \dfrac{3}{6} = \dfrac{1}{2}$ hour.

Total time is $2\left(\dfrac{1}{2}\right) = 1$ hour

61. $\dfrac{x}{4} = \dfrac{x + 3}{6}$

$$6x = 4(x + 3)$$

$$6x = 4x + 12$$

$$2x = 12$$

$$x = 6$$

$$\{6\}$$

65. $B = \dfrac{705\omega}{h^2}$

$$47 = \dfrac{705 \cdot 240}{h^2}$$

$$47h^2 = 169{,}200$$

$$h^2 = \dfrac{169{,}200}{47} = 3600$$

$$h = \sqrt{3600} = 60$$

60 inches or 5 feet

answers may vary

69. For three resistances R_1, R_2, and R_3 wired in a parallel circuit, the combined resistance R is given by

$$\frac{1}{R} = \frac{1}{R_1} + \frac{1}{R_2} + \frac{1}{R_3}$$

$$\frac{1}{R} = \frac{1}{5} + \frac{1}{6} + \frac{1}{2}$$

$$\frac{1}{R} = \frac{6 + 5 + 15}{30}$$

$$\frac{1}{R} = \frac{26}{30} = \frac{13}{15}$$

$$R = \frac{15}{13}\text{ ohms}$$

Exercise Set 6.6

1. $y = kx$

$$4 = k(20)$$

$$k = \frac{1}{5}$$

$$y = \frac{1}{5}x$$

5. $y = kx$

$$7 = k\left(\frac{1}{2}\right)$$

$$k = 14$$

$$y = 14x$$

9. $W = kr^3$

$$1.2 = k \cdot 2^3$$

$$k = \frac{1.2}{8}$$

$$= 0.15$$

$$W = 0.15r^3$$

$$= 0.15(3)^3$$

$$W = 0.15(27)$$

$$= 4.05\text{ pounds}$$

13. $y = \dfrac{k}{x}$

$$6 = \frac{k}{5}$$

$$k = 30$$

$$y = \frac{30}{x}$$

17. $y = \dfrac{k}{x}$

$$\frac{1}{8} = \frac{k}{16}$$

$$k = 2$$

$$y = \frac{2}{x}$$

21. $R = \dfrac{k}{T}$

$$45 = \frac{k}{6}$$

$$k = 270$$

$$R = \frac{270}{5}$$

$$R = 54\text{ mph}$$

25. $I_1 = \dfrac{k}{d^2}$

Replace d by $2d$.

$$I_2 = \frac{k}{(2d)^2} = \frac{k}{4d^2} = \frac{1}{4}I_1$$

Thus, the intensity is divided by 4.

29. $r = kst^3$

33. $y = k\sqrt{x}$

$$0.4 = k\sqrt{4}$$

$$0.4 = 2k$$

$$\frac{0.4}{2} = k$$

$$0.2 = k$$

$$y = 0.2\sqrt{x}$$

37. $y = kxz^3$

$$120 = k(5)(2^3)$$

$$120 = k(5)(8)$$

$$120 = 40k$$
$$3 = k$$
$$y = 3xz^3$$

41. $\quad V = kr^2h$
$$32\pi = k(4)^2(6)$$
$$32\pi = k(16)(6)$$
$$32\pi = 96k$$
$$\frac{32\pi}{96} = k$$
$$\frac{\pi}{3} = k$$
$$V = \frac{\pi}{3}r^2h$$
$$V = \frac{\pi}{3}(3)^2(5)$$
$$V = 15\pi \text{ cu in.}$$

45. $\quad C = 2\pi r$
$$C = 2\pi \cdot 6$$
$$C = 12\pi \text{ cm}$$
$$A = \pi r^2$$
$$A = \pi(6)^2$$
$$A = 36\pi \text{ sq cm}$$

49. $(3, 6), (-2, 6)$
$$m = \frac{y_2 - y_1}{x_2 - x_1}$$
$$m = \frac{6 - 6}{-2 - 3}$$
$$= \frac{0}{-5}$$
$$= 0$$

53. $\quad V_1 = khr^2$
$$V_2 = k\left(\frac{1}{2}h\right)(2r)^2$$
$$V_2 = 2khr^2 = 2V_1$$
It is multiplied by 2.

Chapter 6 Test

1. $\dfrac{5x^2}{1 - x}$
Undefined when
$$1 - x = 0$$
$$1 = x$$

5. $\dfrac{x^2 - 4x}{x^2 + 5x - 36} = \dfrac{x(x - 4)}{(x + 9)(x - 4)} = \dfrac{x}{x + 9}$

9. $\dfrac{3x^2 - 12}{x^2 + 2x - 8} \div \dfrac{6x + 18}{x + 4}$
$$= \frac{3(x^2 - 4)}{(x + 4)(x - 2)} \cdot \frac{x + 4}{6(x + 3)}$$
$$= \frac{(x + 2)(x - 2)}{x - 2} \cdot \frac{1}{2(x + 3)}$$
$$= \frac{x + 2}{2(x + 3)}$$

13. $\dfrac{3}{x^2 - x - 6} + \dfrac{2}{x^2 - 5x + 6}$
$$= \frac{3}{(x - 3)(x + 2)} + \frac{2}{(x - 3)(x - 2)}$$
$$= \frac{3(x - 2) + 2(x + 2)}{(x - 3)(x + 2)(x - 2)}$$
$$= \frac{3x - 6 + 2x + 4}{(x - 3)(x + 2)(x - 2)}$$
$$= \frac{5x - 2}{(x - 3)(x + 2)(x - 2)}$$

17. $\dfrac{\dfrac{5}{x} - \dfrac{7}{3x}}{\dfrac{9}{8x} - \dfrac{1}{x}} = \dfrac{\dfrac{15 - 7}{3x}}{\dfrac{9 - 8}{8x}} = \dfrac{8}{3x} \cdot \dfrac{8x}{1} = \dfrac{64}{3}$

21.
$$\frac{x}{x - 4} = 3 - \frac{4}{x - 4}$$
$$\frac{x}{x - 4} + \frac{4}{x - 4} = 3$$
$$\frac{x + 4}{x - 4} = 3$$
$$x + 4 = 3(x - 4)$$
$$x + 4 = 3x - 12$$
$$16 = 2x$$
$$x = 8$$
$\{8\}$

25. $\quad W = \dfrac{k}{V}$
$$20 = \frac{k}{12}$$
$$k = 240$$
$$\text{so } W = \frac{240}{V}$$
$$= \frac{240}{15}$$
$$= 16$$

Chapter 7

Exercise Set 7.1

1. Since $2^2 = 4$ and $(-2)^2 = 4$, the square roots of 4 are 2 and -2.

5. Since $10^2 = 100$ and $(-10)^2 = 100$, the square roots of 100 are 10 and -10.

9. $\sqrt{\dfrac{1}{4}} = \dfrac{1}{2}$ because $\left(\dfrac{1}{2}\right)^2 = \dfrac{1}{4}$.

13. $-\sqrt{36} = -6$ because $(6)^2 = 36$.

17. $\sqrt{16y^6} = \sqrt{16}\sqrt{y^6} = 4y^3$ because $(4y^3)^2 = 16y^6$.

21. $\sqrt{38} \approx 6.164$
Since $36 < 38 < 49$, then $\sqrt{36} < \sqrt{38} < \sqrt{49}$ or $6 < \sqrt{38} < 7$. The approximation is between 6 and 7 and this is reasonable.

25. $\sqrt[3]{64} = 4$ because $(4)^3 = 64$.

29. $\sqrt[3]{-1} = -1$ because $(-1)^3 = -1$.

33. $\sqrt[3]{-27x^9} = -3x^3$ because $(-3x^3)^3 = -27x^9$.

37. $\sqrt[4]{-16}$ is not a real number.

41. $\sqrt[5]{x^{20}} = x^4$ because $(x^4)^5 = x^{20}$.

45. $\sqrt{81x^4} = 9x^2$ because $(9x^2)^2 = 81x^4$.

49. $\sqrt{(-8)^2} = |-8| = 8$

53. $\sqrt{4x^2} = 2|x|$

57. $\sqrt[4]{(x - 2)^4} = |x - 2|$

61. $f(x) = \sqrt{2x + 3}$
$f(0) = \sqrt{2 \cdot 0 + 3} = \sqrt{3}$

65. $g(x) = \sqrt[3]{x - 8}$
$g(-19) = \sqrt[3]{-19 - 8} = \sqrt[3]{-27} = -3$

69. $(-2x^3y^2)^5 = (-2)^5x^{3 \cdot 5}y^{2 \cdot 5} = -32x^{15}y^{10}$

73. $\dfrac{7x^{-1}y}{14(x^5y^2)^{-2}} = \dfrac{7x^{-1}y}{14x^{-10}y^{-4}} = \dfrac{x^9y^5}{2}$

77. b. Since $\sqrt{160}$ is close to $\sqrt{169} = 13$.

81. $B = \sqrt{\dfrac{\text{hw}}{3131}} = \sqrt{\dfrac{66 \cdot 135}{3131}}$

$\quad\quad = \sqrt{\dfrac{8910}{3131}} \approx \sqrt{2.8457}$

$\quad\quad \approx 1.69 \text{ sq m}$

Exercise Set 7.2

1. $49^{1/2} = \sqrt{49} = 7$

5. $\left(\dfrac{1}{16}\right)^{1/4} = \sqrt[4]{\dfrac{1}{16}} = \dfrac{1}{2}$

9. $2m^{1/3} = 2\sqrt[3]{m}$

13. $(-27)^{1/3} = \sqrt[3]{-27} = -3$

17. $16^{3/4} = (\sqrt[4]{16})^3 = 2^3 = 8$

21. $(-16)^{3/4} = (\sqrt[4]{-16})^3$ is not a real number.

25. $(7x + 2)^{2/3} = \sqrt[3]{(7x + 2)^2}$ or $(\sqrt[3]{7x + 2})^2$

29. $8^{-4/3} = \dfrac{1}{8^{4/3}} = \dfrac{1}{(8^{1/3})^4} = \dfrac{1}{2^4} = \dfrac{1}{16}$

33. $(-4)^{-3/2} = \dfrac{1}{(-4)^{3/2}} = \dfrac{1}{[(-4)^{1/2}]^3}$ is not a real number.

37. $\dfrac{1}{a^{-2/3}} = a^{2/3}$ 　　**41.** answers may vary

45. $x^{-2/5} \cdot x^{7/5} = x^{-2/5+7/5} = x^{5/5} = x$

49. $\dfrac{y^{1/3}}{y^{1/6}} = y^{1/3-1/6} = y^{2/6-1/6} = y^{1/6}$

53. $\dfrac{b^{1/2}b^{3/4}}{-b^{1/4}} = -b^{1/2+3/4-1/4} = -b^{1/2+1/2} = -b^1 = -b$

57. $\dfrac{(y^3z)^{1/6}}{y^{-1/2}z^{1/3}} = \dfrac{y^{3/6}z^{1/6}}{y^{-1/2}z^{1/3}}$

$\quad\quad = y^{3/6-(-1/2)}z^{1/6-1/3}$

$\quad\quad = y^{1/2+1/2}z^{1/6-2/6}$

$\quad\quad = y^1 z^{-1/6} = \dfrac{y}{z^{1/6}}$

61. $\sqrt[6]{x^3} = x^{3/6} = x^{1/2} = \sqrt{x}$

65. $\sqrt[4]{16x^2} = 16^{1/4}x^{2/4} = 2x^{1/2} = 2\sqrt{x}$

69. $\sqrt[8]{x^4y^4} = x^{4/8}y^{4/8} = x^{1/2}y^{1/2} = \sqrt{xy}$

73. $\sqrt[3]{y} \cdot \sqrt[5]{y^2} = y^{1/3} \cdot y^{2/5} = y^{5/15} \cdot y^{6/15} = y^{11/15} = \sqrt[15]{y^{11}}$

77. $\sqrt[3]{x} \cdot \sqrt[4]{x} \cdot \sqrt[8]{x^3} = x^{1/3} \cdot x^{1/4} \cdot x^{3/8}$

$\quad\quad = x^{8/24} \cdot x^{6/24} \cdot x^{9/24}$

$\quad\quad = x^{23/24}$

$\quad\quad = \sqrt[24]{x^{23}}$

81. $\sqrt{3} \cdot \sqrt[3]{4} = 3^{1/2} \cdot 4^{1/3}$

$\quad\quad = 3^{3/6} \cdot 4^{2/6}$

$\quad\quad = (3^3 \cdot 4^2)^{1/6}$

$\quad\quad = (432)^{1/6}$

$\quad\quad = \sqrt[6]{432}$

85. $\sqrt{5r} \cdot \sqrt[3]{s} = (5r)^{1/2} \cdot s^{1/3}$

$\quad\quad = (5r)^{3/6} \cdot s^{2/6}$

$\quad\quad = [(5r)^3 \cdot s^2]^{1/6}$

$\quad\quad = (125r^3s^2)^{1/6}$

$\quad\quad = \sqrt[6]{125r^3s^2}$

89. $48 = 16 \cdot 3$ or $4 \cdot 12$ 　　**93.** $54 = 27 \cdot 2$

97. $f(x) = 2.5x^{8/5}$

$\quad x = 2000 - 1990 = 10$

$\quad$ Replace x by 10.

$\quad f(x) = 2.5(10)^{8/5}$

$\quad\quad \approx 99.5$ million subscriptions

101. $x^{1/5}$

$\quad\quad \dfrac{x^{1/5}}{x^{-2/5}} = x^{1/5+2/5} = x^{3/5}$

105. $\dfrac{\sqrt{t}}{\sqrt{u}} = \dfrac{t^{1/2}}{u^{1/2}}$

Exercise Set 7.3

1. $\sqrt{7} \cdot \sqrt{2} = \sqrt{7 \cdot 2} = \sqrt{14}$

5. $\sqrt[3]{4} \cdot \sqrt[3]{9} = \sqrt[3]{4 \cdot 9} = \sqrt[3]{36}$

9. $\sqrt{\dfrac{7}{x}} \cdot \sqrt{\dfrac{2}{y}} = \sqrt{\dfrac{7 \cdot 2}{x \cdot y}} = \sqrt{\dfrac{14}{xy}}$

13. $\sqrt{\dfrac{6}{49}} = \dfrac{\sqrt{6}}{\sqrt{49}} = \dfrac{\sqrt{6}}{7}$

17. $\sqrt[4]{\dfrac{x^3}{16}} = \dfrac{\sqrt[4]{x^3}}{\sqrt[4]{16}} = \dfrac{\sqrt[4]{x^3}}{2}$

21. $\sqrt[4]{\dfrac{8}{x^8}} = \dfrac{\sqrt[4]{8}}{\sqrt[4]{x^8}} = \dfrac{\sqrt[4]{8}}{x^2}$

25. $\sqrt{\dfrac{x^2y}{100}} = \dfrac{\sqrt{x^2} \cdot \sqrt{y}}{\sqrt{100}} = \dfrac{x\sqrt{y}}{10}$

29. $-\sqrt[3]{\dfrac{z^7}{125x^3}} = \dfrac{-\sqrt[3]{z^7}}{\sqrt[3]{125x^3}}$

$\quad\quad = \dfrac{-\sqrt[3]{z^6z}}{\sqrt[3]{125} \cdot \sqrt[3]{x^3}}$

$\quad\quad = \dfrac{-\sqrt[3]{z^6} \cdot \sqrt[3]{z}}{5x}$

$\quad\quad = -\dfrac{z^2\sqrt[3]{z}}{5x}$

33. $\sqrt[3]{192} = \sqrt[3]{64(3)} = \sqrt[3]{64} \cdot \sqrt[3]{3} = 4\sqrt[3]{3}$

37. $\sqrt{24} = \sqrt{4 \cdot 6} = \sqrt{4} \cdot \sqrt{6} = 2\sqrt{6}$

41. $\sqrt[3]{16y^7} = \sqrt[3]{(8y^6)(2y)}$

$\quad\quad = \sqrt[3]{8} \cdot \sqrt[3]{y^6} \cdot \sqrt[3]{2y}$

$\quad\quad = 2y^2\sqrt[3]{2y}$

45. $\sqrt{y^5} = \sqrt{y^4y} = \sqrt{y^4} \cdot \sqrt{y} = y^2\sqrt{y}$

49. $\sqrt[5]{-32x^{10}y} = \sqrt[5]{-32} \cdot \sqrt[5]{x^{10}} \cdot \sqrt[5]{y} = -2x^2\sqrt[5]{y}$

53. $-\sqrt{32a^8b^7} = -\sqrt{16a^8b^6(2b)}$

$\quad\quad = -\sqrt{16} \cdot \sqrt{a^8} \cdot \sqrt{b^6} \cdot \sqrt{2b}$

$\quad\quad = -4a^4b^3\sqrt{2b}$

57. $\sqrt[3]{125r^9s^{12}} = 5r^3s^4$

61. $\dfrac{\sqrt[3]{24}}{\sqrt[3]{3}} = \sqrt[3]{\dfrac{24}{3}} = \sqrt[3]{8} = 2$

65. $\dfrac{\sqrt{x^5y^3}}{\sqrt{xy}} = \sqrt{\dfrac{x^5y^3}{xy}}$

$\quad\quad = \sqrt{x^4y^2}$

$\quad\quad = x^2y$

69. $\dfrac{3\sqrt{100x^2}}{2\sqrt{2x^{-1}}} = \dfrac{3}{2}\sqrt{\dfrac{100x^2}{2x^{-1}}}$

$\quad\quad = \dfrac{3}{2}\sqrt{50x^3}$

$\quad\quad = \dfrac{3}{2}\sqrt{25x^2 \cdot 2x}$

$\quad\quad = \dfrac{3}{2} \cdot 5x\sqrt{2x}$

$\quad\quad = \dfrac{15x}{2}\sqrt{2x}$

73. $6x + 8x = 14x$ **77.** $9y^2 - 8y^2 = y^2$

81. $(x - 4)^2 = x^2 - 2x(4) + 4^2 = x^2 - 8x + 16$

85. $F(x) = 0.6\sqrt{49 - x^2}$

  a. $F(3) = 0.6\sqrt{49 - 3^2}$ **b.** $F(5) = 0.6\sqrt{49 - 5^2}$
  $= 0.6\sqrt{49 - 9}$ $= 0.6\sqrt{49 - 25}$
  $= 0.6\sqrt{40}$ $= 0.6\sqrt{24}$
  ≈ 3.8 ≈ 2.9

  c. answers may vary

Exercise Set 7.4

1. $\sqrt{8} - \sqrt{32} = \sqrt{4(2)} - \sqrt{16(2)}$
  $= \sqrt{4}\sqrt{2} - \sqrt{16}\sqrt{2}$
  $= 2\sqrt{2} - 4\sqrt{2}$
  $= -2\sqrt{2}$

5. $2\sqrt{50} - 3\sqrt{125} + \sqrt{98}$
  $= 2\sqrt{25(2)} - 3\sqrt{25(5)} + \sqrt{49(2)}$
  $= 2\sqrt{25}\sqrt{2} - 3\sqrt{25}\sqrt{5} + \sqrt{49}\sqrt{2}$
  $= 2(5)\sqrt{2} - 3(5)\sqrt{5} + 7\sqrt{2}$
  $= 10\sqrt{2} - 15\sqrt{5} + 7\sqrt{2}$
  $= 17\sqrt{2} - 15\sqrt{5}$

9. $\sqrt{9b^3} - \sqrt{25b^3} + \sqrt{49b^3}$
  $= \sqrt{9b^2(b)} - \sqrt{25b^2(b)} + \sqrt{49b^2(b)}$
  $= \sqrt{9b^2}\sqrt{b} - \sqrt{25b^2}\sqrt{b} + \sqrt{49b^2}\sqrt{b}$
  $= 3b\sqrt{b} - 5b\sqrt{b} + 7b\sqrt{b}$
  $= 5b\sqrt{b}$

13. $\sqrt[3]{\dfrac{11}{8}} - \dfrac{\sqrt[3]{11}}{6} = \dfrac{\sqrt[3]{11}}{\sqrt[3]{8}} - \dfrac{\sqrt[3]{11}}{6}$
  $= \dfrac{\sqrt[3]{11}}{2} - \dfrac{\sqrt[3]{11}}{6}$
  $= \dfrac{3\sqrt[3]{11} - \sqrt[3]{11}}{6}$
  $= \dfrac{2\sqrt[3]{11}}{6}$
  $= \dfrac{\sqrt[3]{11}}{3}$

17. $7\sqrt{9} - 7 + \sqrt{3} = 7(3) - 7 + \sqrt{3}$
  $= 21 - 7 + \sqrt{3}$
  $= 14 + \sqrt{3}$

21. $3\sqrt{108} - 2\sqrt{18} - 3\sqrt{48}$
  $= 3\sqrt{36}\sqrt{3} - 2\sqrt{9}\sqrt{2} - 3\sqrt{16}\sqrt{3}$
  $= 3(6)\sqrt{3} - 2(3)\sqrt{2} - 3(4)\sqrt{3}$
  $= 18\sqrt{3} - 6\sqrt{2} - 12\sqrt{3}$
  $= 6\sqrt{3} - 6\sqrt{2}$

25. $\sqrt{9b^3} - \sqrt{25b^3} + \sqrt{16b^3}$
  $= \sqrt{9b^2}\sqrt{b} - \sqrt{25b^2}\sqrt{b} + \sqrt{16b^2}\sqrt{b}$
  $= 3b\sqrt{b} - 5b\sqrt{b} + 4b\sqrt{b}$
  $= (3 - 5 + 4)b\sqrt{b}$
  $= 2b\sqrt{b}$

29. $\sqrt[3]{54xy^3} - 5\sqrt[3]{2xy^3} + y\sqrt[3]{128x}$
  $= \sqrt[3]{27y^3}\sqrt[3]{2x} - 5\sqrt[3]{y^3}\sqrt[3]{2x} + y\sqrt[3]{64}\sqrt[3]{2x}$
  $= 3y\sqrt[3]{2x} - 5y\sqrt[3]{2x} + y(4)\sqrt[3]{2x}$
  $= -2y\sqrt[3]{2x} + 4y\sqrt[3]{2x}$
  $= 2y\sqrt[3]{2x}$

33. $-2\sqrt[4]{x^7} + 3\sqrt[4]{16x^7} = -2\sqrt[4]{x^4}\sqrt[4]{x^3} + 3\sqrt[4]{16x^4}\sqrt[4]{x^3}$
  $= -2x\sqrt[4]{x^3} + 3(2x)\sqrt[4]{x^3}$
  $= -2x\sqrt[4]{x^3} + 6x\sqrt[4]{x^3}$
  $= 4x\sqrt[4]{x^3}$

37. $\dfrac{\sqrt[3]{8x^4}}{7} + \dfrac{3x\sqrt[3]{x}}{7} = \dfrac{\sqrt[3]{8x^3}\sqrt[3]{x} + 3x\sqrt[3]{x}}{7}$
  $= \dfrac{2x\sqrt[3]{x} + 3x\sqrt[3]{x}}{7}$
  $= \dfrac{5x\sqrt[3]{x}}{7}$

41. $\sqrt[3]{\dfrac{16}{27}} - \dfrac{\sqrt[3]{54}}{6} = \dfrac{\sqrt[3]{16}}{\sqrt[3]{27}} - \dfrac{\sqrt[3]{27}\sqrt[3]{2}}{6}$
  $= \dfrac{\sqrt[3]{8}\sqrt[3]{2}}{3} - \dfrac{3\sqrt[3]{2}}{6}$
  $= \dfrac{2(2)\sqrt[3]{2}}{6} - \dfrac{3\sqrt[3]{2}}{6}$
  $= \dfrac{4\sqrt[3]{2} - 3\sqrt[3]{2}}{6}$
  $= \dfrac{\sqrt[3]{2}}{6}$

45. $P = 2\sqrt{12} + \sqrt{12} + 2\sqrt{27} + 3\sqrt{3}$
  $= 2\sqrt{4}\sqrt{3} + \sqrt{4}\sqrt{3} + 2\sqrt{9}\sqrt{3} + 3\sqrt{3}$
  $= 2\cdot2\sqrt{3} + 2\sqrt{3} + 2\cdot3\sqrt{3} + 3\sqrt{3}$
  $= (4 + 2 + 6 + 3)\sqrt{3}$
  $= 15\sqrt{3}$
  The perimeter of the trapezoid is $15\sqrt{3}$ in.

49. $(\sqrt{5} - \sqrt{2})^2 = \sqrt{5}^2 - 2\sqrt{5}\sqrt{2} + \sqrt{2}^2$
  $= 5 - 2\sqrt{10} + 2$
  $= 7 - 2\sqrt{10}$

53. $(2\sqrt{x} - 5)(3\sqrt{x} + 1)$
  $= (2\sqrt{x})(3\sqrt{x}) + (2\sqrt{x})1 - 5(3\sqrt{x}) - 5\cdot1$
  $= 6x + 2\sqrt{x} - 15\sqrt{x} - 5$
  $= 6x - 13\sqrt{x} - 5$

57. $6(\sqrt{2} - 2) = 6\sqrt{2} - 6\cdot2 = 6\sqrt{2} - 12$

61. $(2\sqrt{7} + 3\sqrt{5})(\sqrt{7} - 2\sqrt{5})$
  $= 2\sqrt{7}^2 - (2\sqrt{7})(2\sqrt{5}) + (3\sqrt{5})\sqrt{7} - 3\cdot2\sqrt{5}^2$
  $= 2\cdot7 - 4\sqrt{35} + 3\sqrt{35} - 6\cdot5$
  $= 14 - \sqrt{35} - 30$
  $= -16 - \sqrt{35}$

65. $(\sqrt{3} + x)^2 = \sqrt{3}^2 + 2\sqrt{3}x + x^2$
  $= 3 + 2x\sqrt{3} + x^2$

69. $(\sqrt[3]{4} + 2)(\sqrt[3]{2} - 1)$
  $= \sqrt[3]{4}\sqrt[3]{2} - \sqrt[3]{4}\cdot1 + 2\sqrt[3]{2} - 2\cdot1$
  $= \sqrt[3]{8} - \sqrt[3]{4} + 2\sqrt[3]{2} - 2$
  $= 2 - \sqrt[3]{4} + 2\sqrt[3]{2} - 2$
  $= -\sqrt[3]{4} + 2\sqrt[3]{2}$

73. $(\sqrt{x - 1} + 5)^2 = (\sqrt{x - 1})^2 + 2\sqrt{x - 1}(5) + 5^2$
  $= x - 1 + 10\sqrt{x - 1} + 25$
  $= x + 24 + 10\sqrt{x - 1}$

77. $\dfrac{2x - 14}{2} = \dfrac{2(x - 7)}{2} = x - 7$

81. $\dfrac{6a^2b - 9ab}{3ab} = \dfrac{3ab(2a - 3)}{3ab} = 2a - 3$

85. $P = 2(3\sqrt{20}) + 2\sqrt{125}$
  $= 6\sqrt{4}\sqrt{5} + 2\sqrt{25}\sqrt{5}$
  $= 6(2)\sqrt{5} + 2(5)\sqrt{5}$
  $= 12\sqrt{5} + 10\sqrt{5}$
  $= 22\sqrt{5}$ feet
  $A = 3\sqrt{20}\cdot\sqrt{125}$
  $= 3\cdot2\sqrt{5}\cdot5\sqrt{5}$
  $= 30(\sqrt{5})^2$
  $= 30\cdot5$
  $= 150$ square feet

Exercise Set 7.5

1. $\dfrac{\sqrt{2}}{\sqrt{7}} = \dfrac{\sqrt{2}\cdot\sqrt{7}}{\sqrt{7}\cdot\sqrt{7}} = \dfrac{\sqrt{14}}{7}$

5. $\dfrac{4}{\sqrt[3]{3}}\cdot\dfrac{\sqrt[3]{9}}{\sqrt[3]{9}} = \dfrac{4\sqrt[3]{9}}{\sqrt[3]{27}} = \dfrac{4\sqrt[3]{9}}{3}$

9. $\dfrac{3}{\sqrt[3]{4x^2}} = \dfrac{3}{\sqrt[3]{4x^2}}\cdot\dfrac{\sqrt[3]{2x}}{\sqrt[3]{2x}} = \dfrac{3\sqrt[3]{2x}}{\sqrt[3]{8x^3}} = \dfrac{3\sqrt[3]{2x}}{2x}$

13. $\dfrac{3}{\sqrt[3]{2}} = \dfrac{3}{\sqrt[3]{2}}\cdot\dfrac{\sqrt[3]{4}}{\sqrt[3]{4}} = \dfrac{3\sqrt[3]{4}}{\sqrt[3]{8}} = \dfrac{3\sqrt[3]{4}}{2}$

17. $\sqrt{\dfrac{2x}{5y}} = \dfrac{\sqrt{2x}}{\sqrt{5y}} = \dfrac{\sqrt{2x}\cdot\sqrt{5y}}{\sqrt{5y}\cdot\sqrt{5y}} = \dfrac{\sqrt{10xy}}{5y}$

21. $\sqrt{\dfrac{3x}{50}} = \dfrac{\sqrt{3x}}{\sqrt{50}} = \dfrac{\sqrt{3x}}{5\sqrt{2}} = \dfrac{\sqrt{3x}\cdot\sqrt{2}}{5\sqrt{2}\cdot\sqrt{2}} = \dfrac{\sqrt{6x}}{5\cdot2} = \dfrac{\sqrt{6x}}{10}$

25. $\dfrac{\sqrt[3]{2y^2}}{\sqrt[3]{9x^2}} = \dfrac{\sqrt[3]{2y^2}\cdot\sqrt[3]{3x}}{\sqrt[3]{9x^2}\cdot\sqrt[3]{3x}} = \dfrac{\sqrt[3]{6xy^2}}{3x}$

29. $\dfrac{5a}{\sqrt[5]{8a^9b^{11}}} = \dfrac{5a}{ab^2\sqrt[5]{8a^4b}}$

$= \dfrac{5a\sqrt[5]{4ab^4}}{ab^2\sqrt[5]{8a^4b}\cdot\sqrt[5]{4ab^4}}$

$= \dfrac{5a\sqrt[5]{4ab^4}}{2a^2b^3}$

33. $\dfrac{-7}{\sqrt{x}-3} = \dfrac{(-7)(\sqrt{x}+3)}{(\sqrt{x}-3)(\sqrt{x}+3)}$

$= \dfrac{-7(\sqrt{x}+3)}{\sqrt{x^2}-3^2}$

$= \dfrac{-7(\sqrt{x}+3)}{x-9} = \dfrac{\sqrt{x}+3}{9-x}$

37. $\dfrac{\sqrt{a}+1}{2\sqrt{a}-\sqrt{b}}$

$= \dfrac{(\sqrt{a}+1)(2\sqrt{a}+\sqrt{b})}{(2\sqrt{a}-\sqrt{b})(2\sqrt{a}+\sqrt{b})}$

$= \dfrac{\sqrt{a}(2\sqrt{a})+\sqrt{a}\sqrt{b}+1(2\sqrt{a})+1\sqrt{b}}{(2\sqrt{a})^2-\sqrt{b}^2}$

$= \dfrac{2a+\sqrt{ab}+2\sqrt{a}+\sqrt{b}}{4a-b}$

41. $\dfrac{\sqrt{x}}{\sqrt{x}+\sqrt{y}} = \dfrac{(\sqrt{x})(\sqrt{x}-\sqrt{y})}{(\sqrt{x}+\sqrt{y})(\sqrt{x}-\sqrt{y})}$

$= \dfrac{\sqrt{x}\sqrt{x}-\sqrt{x}\sqrt{y}}{\sqrt{x^2}-\sqrt{y^2}}$

$= \dfrac{x-\sqrt{xy}}{x-y}$

45. $\sqrt{\dfrac{5}{3}} = \dfrac{\sqrt{5}}{\sqrt{3}} = \dfrac{\sqrt{5}\cdot\sqrt{5}}{\sqrt{3}\cdot\sqrt{5}} = \dfrac{5}{\sqrt{15}}$

49. $\dfrac{\sqrt{4x}}{7} = \dfrac{2\sqrt{x}}{7} = \dfrac{2\sqrt{x}\cdot\sqrt{x}}{7\cdot\sqrt{x}} = \dfrac{2x}{7\sqrt{x}}$

53. $\sqrt{\dfrac{2}{5}} = \dfrac{\sqrt{2}}{\sqrt{5}} = \dfrac{\sqrt{2}\cdot\sqrt{2}}{\sqrt{5}\cdot\sqrt{2}} = \dfrac{2}{\sqrt{10}}$

57. $\sqrt[3]{\dfrac{7}{8}} = \dfrac{\sqrt[3]{7}}{\sqrt[3]{8}} = \dfrac{\sqrt[3]{7}}{2} = \dfrac{\sqrt[3]{7}\cdot\sqrt[3]{7^2}}{2\cdot\sqrt[3]{7^2}} = \dfrac{\sqrt[3]{7^3}}{2\sqrt[3]{7^2}} = \dfrac{7}{2\sqrt[3]{49}}$

61. $\sqrt{\dfrac{18x^4y^6}{3z}} = \dfrac{\sqrt{18x^4y^6}}{\sqrt{3z}} = \dfrac{3x^2y^3\sqrt{2}}{\sqrt{3z}} = \dfrac{3x^2y^3\sqrt{2}\cdot\sqrt{2}}{\sqrt{3z}\cdot\sqrt{2}} = \dfrac{6x^2y^3}{\sqrt{6z}}$

65. $\dfrac{(2-\sqrt{11})}{6}\cdot\dfrac{(2+\sqrt{11})}{(2+\sqrt{11})} = \dfrac{(2-\sqrt{11})(2+\sqrt{11})}{6(2+\sqrt{11})}$

$= \dfrac{4-\sqrt{121}}{12+6\sqrt{11}}$

$= \dfrac{4-11}{12+6\sqrt{11}}$

$= \dfrac{-7}{12+6\sqrt{11}}$

69. $\dfrac{(\sqrt{x}+3)}{\sqrt{x}}\cdot\dfrac{(\sqrt{x}-3)}{(\sqrt{x}-3)} = \dfrac{(\sqrt{x}+3)(\sqrt{x}-3)}{\sqrt{x}(\sqrt{x}-3)}$

$= \dfrac{\sqrt{x^2}-9}{\sqrt{x^2}-3\sqrt{x}}$

$= \dfrac{x-9}{x-3\sqrt{x}}$

73. $\begin{aligned}2x-7 &= 3(x-4)\\ 2x-7 &= 3x-12\\ 2x-7+12 &= 3x-12+12\\ 2x+5 &= 3x\\ 2x-2x+5 &= 3x-2x\\ 5 &= x\end{aligned}$

$\{5\}$

77. $\begin{aligned}x^2-8x &= -12\\ x^2-8x+12 &= 0\\ (x-2)(x-6) &= 0\\ x-2=0 \quad &\text{or} \quad x-6=0\\ x=2 \qquad\quad & \qquad x=6\end{aligned}$

$\{2,6\}$

81. answers may vary

Exercise Set 7.6

1. $\begin{aligned}\sqrt{2x} &= 4\\ 2x &= 4^2\\ 2x &= 16\\ x &= 8\end{aligned}$

The solution set is $\{8\}$.

5. $\sqrt{2x} = -4$

No solution since a principle square root does not yield a negative number. The solution set is $\varnothing$.

9. $\begin{aligned}\sqrt{2x-3}-2 &= 1\\ \sqrt{2x-3} &= 3\\ 2x-3 &= 3^2\\ 2x-3 &= 9\\ 2x &= 12\\ x &= 6\end{aligned}$

The solution set is $\{6\}$.

17. $\begin{aligned}x-\sqrt{4-3x} &= -8\\ x+8 &= \sqrt{4-3x}\\ (x+8)^2 &= 4-3x\\ x^2+16x+64 &= 4-3x\\ x^2+16x+64 &= 4-3x\\ x^2+19x+60 &= 0\\ (x+4)(x+15) &= 0\\ x+4=0 \quad &\text{or} \quad x+15=0\\ x=-4 \qquad & \qquad x=-15\end{aligned}$

We discard the -15 as extraneous, leaving $x=-4$ as the only solution. The solution set is $\{-4\}$.

21. $\begin{aligned}\sqrt{x-3}+\sqrt{x+2} &= 5\\ \sqrt{x-3} &= 5-\sqrt{x+2}\\ x-3 &= 25-10\sqrt{x+2}+x+2\\ -3 &= 27-10\sqrt{x+2}\\ -30 &= -10\sqrt{x+2}\\ 3 &= \sqrt{x+2}\\ 9 &= x+2\\ 7 &= x\end{aligned}$

The solution set is $\{7\}$.

25. $-\sqrt{2x} + 4 = -6$
$$10 = \sqrt{2x}$$
$$10^2 = 2x$$
$$100 = 2x$$
$$x = 50$$

The solution set is $\{50\}$.

29. $\sqrt[4]{4x + 1} - 2 = 0$
$$\sqrt[4]{4x + 1} = 2$$
$$4x + 1 = 2^4$$
$$4x + 1 = 16$$
$$4x = 15$$
$$x = \frac{15}{4}$$

The solution set is $\left\{\frac{15}{4}\right\}$.

33. $\sqrt[3]{6x - 3} - 3 = 0$
$$\sqrt[3]{6x - 3} = 3$$
$$6x - 3 = 3^3$$
$$6x - 3 = 27$$
$$6x = 30$$
$$x = 5$$

The solution set is $\{5\}$.

37. $\sqrt{x + 4} = \sqrt{2x - 5}$
$$x + 4 = 2x - 5$$
$$9 = x$$
$$x = 9$$

The solution set is $\{9\}$.

41. $\sqrt[3]{-6x - 1} = \sqrt[3]{-2x - 5}$
$$-6x - 1 = -2x - 5$$
$$4 = 4x$$
$$x = 1$$

The solution set is $\{1\}$.

45. $\sqrt{2x - 1} = \sqrt{1 - 2x}$
$$\sqrt{2x - 1} = \sqrt{-(2x - 1)}$$

It follows that $2x - 1 = 0$ (Otherwise one of the radicands would be negative).

So $2x = 1$

$$x = \frac{1}{2}$$

The solution set is $\left\{\frac{1}{2}\right\}$.

49. $\sqrt{y + 3} - \sqrt{y - 3} = 1$
$$\sqrt{y + 3} = 1 + \sqrt{y - 3}$$
$$(\sqrt{y + 3})^2 = (1 + \sqrt{y - 3})^2$$
$$y + 3 = 1 + 2\sqrt{y - 3} + y - 3$$
$$5 = 2\sqrt{y - 3}$$
$$25 = 4(y - 3)$$
$$\frac{25}{4} = y - 3$$
$$\frac{25}{4} + \frac{12}{4} = y$$
$$\frac{37}{4} = y$$

The solution set is $\left\{\frac{37}{4}\right\}$.

53. Let b = the length of the unknown leg of the right triangle. By the Pythagorean theorem,
$$7^2 = 3^2 + b^2$$
$$49 = 9 + b^2$$
$$b^2 = 40$$
$$b = \sqrt{40}$$
$$= \sqrt{4}\sqrt{10}$$
$$= 2\sqrt{10} \text{ meters}$$

57. Let c = the length of the hypotenuse of the right triangle. By the Pythagorean theorem,
$$c^2 = 7^2 + (7.2)^2 = 100.84$$
so $c = \sqrt{100.84} \approx 10.0$ millimeters

61. $x^2 = (5)^2 + (12)^2$
$$x^2 = 25 + 144$$
$$x^2 = 169$$
$$x = \sqrt{169}$$
$$x = 13$$

The answer is 13 feet.

65. $v = \sqrt{2gh}$
$$80 = \sqrt{2(32)h}$$
$$(80)^2 = \left(\sqrt{2(32)h}\right)^2$$
$$6400 = 2(32) \cdot h$$
$$100 = h$$

The object fell 100 feet.

69. $P = 2\pi\sqrt{\dfrac{l}{32}}$
$$= 2\pi\sqrt{\frac{2}{32}} = 2\pi\sqrt{\frac{1}{16}}$$
$$= 2\pi\left(\frac{1}{\sqrt{16}}\right) = 2\pi\left(\frac{1}{4}\right)$$
$$= \frac{2\pi}{4} = \frac{\pi}{2} \text{ sec} \approx 1.57 \text{ sec}$$

73. answers may vary

77. answers may vary

81. $\dfrac{\dfrac{x}{6}}{\dfrac{2x}{3} + \dfrac{1}{2}} = \dfrac{\left(\dfrac{x}{6}\right)6}{\left(\dfrac{2x}{3} + \dfrac{1}{2}\right)6} = \dfrac{x}{\left(\dfrac{2x}{3}\right)6 + \left(\dfrac{1}{2}\right)6} = \dfrac{x}{4x + 3}$

85. $\sqrt{\sqrt{x + 3} + \sqrt{x}} = \sqrt{3}$
$$\left(\sqrt{\sqrt{x + 3} + \sqrt{x}}\right)^2 = (\sqrt{3})^2$$
$$\sqrt{x + 3} + \sqrt{x} = 3$$
$$\sqrt{x + 3} = 3 - \sqrt{x}$$
$$(\sqrt{x + 3})^2 = (3 - \sqrt{x})^2$$
$$x + 3 = 9 - 6\sqrt{x} + x$$
$$-6 = -6\sqrt{x}$$
$$1 = \sqrt{x}$$
$$1^2 = (\sqrt{x})^2$$
$$1 = x \text{ and it checks}$$

$\{1\}$

Exercise Set 7.7

1. $\sqrt{-24} = \sqrt{4}\sqrt{6}\sqrt{-1} = 2i\sqrt{6}$

5. $8\sqrt{-63} = 8\sqrt{9}\sqrt{7}\sqrt{-1} = 8 \cdot 3\sqrt{7}i = 24i\sqrt{7}$

9. $\sqrt{-2}\sqrt{-7} = (\sqrt{2}i)(\sqrt{7}i)$
$$= \sqrt{14}i^2$$
$$= \sqrt{14}(-1)$$
$$= -\sqrt{14}$$

13. $\sqrt{16}\sqrt{-1} = 4i$

17. $\dfrac{\sqrt{-80}}{\sqrt{-10}} = \dfrac{\sqrt{80}i}{\sqrt{10}i} = \sqrt{\dfrac{80}{10}} = \sqrt{8} = \sqrt{4}\sqrt{2} = 2\sqrt{2}$

21. $(6 + 5i) - (8 - i) = (6 - 8) + [5 - (-1)]i$
$$= -2 + 6i$$

25. $(6 - 3i) - (4 - 2i) = 6 - 3i - 4 + 2i = 2 - i$

29. $(2 + 4i) + (6 - 5i) = (2 + 6) + (4 - 5)i$
$$= 8 - i$$

33. $(-9i)(7i) = -63i^2 = -63(-1) = 63$

37. $6i(2 - 3i) = 6i(2) - 6i(3i)$
$$= 12i - 18i^2$$
$$= 12i - 18(-1)$$
$$= 18 + 12i$$

41. $(4 + i)(5 + 2i) = 20 + 8i + 5i + 2i^2$
$= 20 + 13i - 2$
$= 18 + 13i$

45. $(4 - 2i)^2 = 16 - 16i + 4i^2$
$= 16 - 16i + 4(-1)$
$= 16 - 4 - 16i$
$= 12 - 16i$

49. $(1 - i)(1 + i) = 1^2 - i^2 = 1^2 + 1^2 = 1 + 1 = 2$

53. $(1 - i)^2 = 1^2 - 2(1)(i) + i^2 = 1 - 2i - 1 = -2i$

57. $\dfrac{7}{4 + 3i} = \dfrac{7}{4 + 3i} \cdot \dfrac{4 - 3i}{4 - 3i}$
$= \dfrac{28 - 21i}{4^2 + 3^2}$
$= \dfrac{28 - 21i}{16 + 9}$
$= \dfrac{28 - 21i}{25}$
$= \dfrac{28}{25} - \dfrac{21}{25}i$

61. $\dfrac{3 + 5i}{1 + i} = \dfrac{3 + 5i}{1 + i} \cdot \dfrac{1 - i}{1 - i}$
$= \dfrac{3 + 5i - 3i - 5i^2}{1^2 + 1^2}$
$= \dfrac{3 + 2i - 5(-1)}{1 + 1}$
$= \dfrac{3 + 5 + 2i}{2}$
$= \dfrac{8 + 2i}{2}$
$= 4 + i$

65. $\dfrac{16 + 15i}{-3i} = \dfrac{(16 + 15i)i}{-3i^2}$
$= \dfrac{16i + 15i^2}{-3(-1)}$
$= \dfrac{16i + 15(-1)}{3}$
$= \dfrac{-15}{3} + \dfrac{16}{3}i$
$= -5 + \dfrac{16}{3}i$

69. $\dfrac{2 - 3i}{2 + i} = \dfrac{(2 - 3i)(2 - i)}{(2 + i)(2 - i)}$
$= \dfrac{4 - 6i - 2i + 3i^2}{2^2 + 1^2}$
$= \dfrac{4 - 8i + 3(-1)}{4 + 1}$
$= \dfrac{4 - 3 - 8i}{5}$
$= \dfrac{1}{5} - \dfrac{8}{5}i$

73. $i^{21} = i^{20}i = (i^4)^5 i = 1^5 i = 1i = i$

77. $i^{-6} = (i^2)^{-3} = (-1)^{-3} = \dfrac{1}{(-1)^3} = \dfrac{1}{-1} = -1$

81. $(-3i)^5 = (-3)^5 i^5$
$= -243 \cdot i^4 \cdot i$
$= -243(1)(i)$
$= -243i$

85. $5 + 9 = 14$

89. $i^3 + i^4 = -i + 1 = 1 - i$

93. $2 + \sqrt{-9} = 2 + 3i$

97. $\dfrac{5 - \sqrt{-75}}{10} = \dfrac{5 - 5i\sqrt{3}}{10} = \dfrac{5}{10} - \dfrac{5i\sqrt{3}}{10} = \dfrac{1}{2} - \dfrac{\sqrt{3}}{2}i$

101. $(8 - \sqrt{-4}) - (2 + \sqrt{-16})$
$= (8 - 2i) - (2 + 4i)$
$= (8 - 2) + (-2i - 4i)$
$= 6 - 6i$

Chapter 7 Test

1. $\sqrt{216} = \sqrt{36 \cdot 6} = 6\sqrt{6}$

5. $\left[\dfrac{8x^3}{27}\right]^{2/3} = \dfrac{8^{2/3}(x^3)^{2/3}}{27^{2/3}}$
$= \dfrac{(8^{1/3})^2 x^2}{(27^{1/3})^2}$
$= \dfrac{2^2 x^2}{3^2}$
$= \dfrac{4x^2}{9}$

9. $\sqrt[4]{(4xy)^4} = |4xy|$ or $4|xy|$

13. $\sqrt[3]{\dfrac{8}{9x}} = \dfrac{\sqrt[3]{8}}{\sqrt[3]{9x}}$
$= \dfrac{2 \cdot \sqrt[3]{3x^2}}{\sqrt[3]{9x} \cdot \sqrt[3]{3x^2}}$
$= \dfrac{2\sqrt[3]{3x^2}}{3x}$

17. $(\sqrt{x} + 1)^2 = \sqrt{x}^2 + 2\sqrt{x} + 1$
$= x + 2\sqrt{x} + 1$

21. $386^{-2/3} \approx 0.019$

25. $\sqrt{-2} = i\sqrt{2}$

29. $(4 + 3i)^2 = 16 + 24i + 9i^2$
$= 16 + 24i + 9(-1)$
$= (16 - 9) + 24i$
$= 7 + 24i$

33. $V = \sqrt{2.5(300)} \approx 27\text{ mph}$

Chapter 8

Exercise Set 8.1

1. $x^2 = 16$
$x = \pm\sqrt{16}$
$x = \pm 4$
The solution set is $\{-4, 4\}$.

5. $x^2 = 18$
$x = \pm\sqrt{18}$
$x = \pm\sqrt{9}\sqrt{2}$
$x = \pm 3\sqrt{2}$
The solution set is $\{-3\sqrt{2}, 3\sqrt{2}\}$.

9. $(x + 5)^2 = 9$
$x + 5 = \pm\sqrt{9}$
$x + 5 = \pm 3$
$x = -5 \pm 3$
$x = -8$ or $x = -2$
The solution set is $\{-8, -2\}$.

13. $(2x - 3)^2 = 8$

$2x - 3 = \pm\sqrt{8}$

$2x - 3 = \pm\sqrt{4}\sqrt{2}$

$2x - 3 = \pm 2\sqrt{2}$

$2x = 3 \pm 2\sqrt{2}$

$x = \dfrac{3 \pm 2\sqrt{2}}{2}$

The solution set is $\left\{ \dfrac{3 - 2\sqrt{2}}{2}, \dfrac{3 + 2\sqrt{2}}{2} \right\}$

17. $x^2 - 6 = 0$

$x^2 = 6$

$x = \pm\sqrt{6}$

The solution set is $\{-\sqrt{6}, \sqrt{6}\}$.

21. $(x - 1)^2 = -16$

$x - 1 = \pm\sqrt{-16}$

$x - 1 = \pm 4i$

$x = 1 \pm 4i$

The solution set is $\{1 - 4i, 1 + 4i\}$.

25. $(x + 3)^2 = -8$

$x + 3 = \pm\sqrt{-8}$

$x + 3 = \pm\sqrt{4}\sqrt{2}\sqrt{-1}$

$x + 3 = \pm 2i\sqrt{2}$

$x = -3 \pm 2i\sqrt{2}$

The solution set is $\{-3 - 2i\sqrt{2}, -3 + 2i\sqrt{2}\}$.

29. $z^2 - 12z + \left(-\dfrac{12}{2}\right)^2 = z^2 - 12z + 36 = (z - 6)^2$

33. $r^2 - 3r + \left(-\dfrac{3}{2}\right)^2 = r^2 - 3r + \dfrac{9}{4}$

$= \left(r - \dfrac{3}{2}\right)^2$

37. $x^2 + 6x + 2 = 0$

$x^2 + 6x + \left(\dfrac{6}{2}\right)^2 = -2 + 9$

$(x + 3)^2 = 7$

$x + 3 = \pm\sqrt{7}$

$x = -3 \pm \sqrt{7}$

The solution set is $\{-3 - \sqrt{7}, -3 + \sqrt{7}\}$.

41. $x^2 + 2x - 5 = 0$

$x^2 + 2x + \left(\dfrac{2}{2}\right)^2 = 5 + 1$

$x^2 + 2x + 1 = 6$

$(x + 1)^2 = 6$

$x + 1 = \pm\sqrt{6}$

$x = -1 \pm \sqrt{6}$

The solution set is $\{-1 - \sqrt{6}, -1 + \sqrt{6}\}$.

45. $2x^2 + 7x = 4$

$x^2 + \dfrac{7}{2}x = 2$

$x^2 + \dfrac{7}{2}x + \left(\dfrac{\frac{7}{2}}{2}\right)^2 = 2 + \dfrac{49}{16}$

$\left(x + \dfrac{7}{4}\right)^2 = \dfrac{81}{16}$

$x + \dfrac{7}{4} = \pm\sqrt{\dfrac{81}{16}}$

$x = -\dfrac{7}{4} \pm \dfrac{9}{4}$

$x = -4 \qquad \text{or } x = \dfrac{1}{2}$

The solution set is $\left\{-4, \dfrac{1}{2}\right\}$.

49. $3y^2 + 6y - 4 = 0$

$3y^2 + 6y = 4$

$y^2 + 2y = \dfrac{4}{3}$

$y^2 + 2y + \left(\dfrac{2}{2}\right)^2 = \dfrac{4}{3} + 1$

$(y + 1)^2 = \dfrac{7}{3}$

$y + 1 = \pm\sqrt{\dfrac{7}{3}}$

$y + 1 = \pm\dfrac{\sqrt{7}}{\sqrt{3}} \cdot \dfrac{\sqrt{3}}{\sqrt{3}}$

$y + 1 = \pm\dfrac{\sqrt{21}}{3}$

$y = -1 \pm \dfrac{\sqrt{21}}{3}$

$= \dfrac{-3 \pm \sqrt{21}}{3}$

The solution set is $\left\{ \dfrac{-3 - \sqrt{21}}{3}, \dfrac{-3 + \sqrt{21}}{3} \right\}$.

53. $x^2 - 6x + 3 = 0$

$x^2 - 6x + \left(-\dfrac{6}{2}\right)^2 = -3 + 9$

$(x - 3)^2 = 6$

$x - 3 = \pm\sqrt{6}$

$x = 3 \pm \sqrt{6}$

The solution set is $\{3 - \sqrt{6}, 3 + \sqrt{6}\}$.

57. $2x^2 - x + 6 = 0$

$2x^2 - x = -6$

$x^2 - \dfrac{1}{2}x = -3$

$x^2 - \dfrac{1}{2}x + \left(\dfrac{-\frac{1}{2}}{2}\right)^2 = -3 + \dfrac{1}{16}$

$\left(x - \dfrac{1}{4}\right)^2 = -\dfrac{47}{16}$

$x - \dfrac{1}{4} = \pm\sqrt{-\dfrac{47}{16}}$

$x - \dfrac{1}{4} = \pm\dfrac{\sqrt{47}\sqrt{-1}}{\sqrt{16}}$

$x - \dfrac{1}{4} = \pm\dfrac{i\sqrt{47}}{4}$

$x = \dfrac{1 \pm i\sqrt{47}}{4}$

The solution set is $\left\{ \dfrac{1 + i\sqrt{47}}{4}, \dfrac{1 - i\sqrt{47}}{4} \right\}$.

61. $z^2 + 3z - 4 = 0$

$z^2 + 3z = 4$

$z^2 + 3z + \left(\dfrac{3}{2}\right)^2 = 4 + \dfrac{9}{4}$

$\left(z + \dfrac{3}{2}\right)^2 = \dfrac{25}{4}$

$z + \dfrac{3}{2} = \pm\sqrt{\dfrac{25}{4}}$

$z + \dfrac{3}{2} = \pm\dfrac{5}{2}$

$z = -\dfrac{3}{2} \pm \dfrac{5}{2}$

$z = -4 \qquad \text{or} \quad z = 1$

The solution set is $\{-4, 1\}$.

65.
$$3x^2 + 3x = 5$$
$$x^2 + x = \frac{5}{3}$$
$$x^2 + x + \left(\frac{1}{2}\right)^2 = \frac{5}{3} + \frac{1}{4}$$
$$\left(x + \frac{1}{2}\right)^2 = \frac{23}{12}$$
$$x + \frac{1}{2} = \pm\sqrt{\frac{23}{12}}$$
$$x + \frac{1}{2} = \pm\frac{\sqrt{23}}{\sqrt{4}\sqrt{3}}$$
$$x + \frac{1}{2} = \pm\frac{\sqrt{23}}{2\sqrt{3}}$$
$$x + \frac{1}{2} = \pm\frac{\sqrt{23}\sqrt{3}}{2\sqrt{3^2}}$$
$$x + \frac{1}{2} = \pm\frac{\sqrt{69}}{2\cdot 3}$$
$$x + \frac{1}{2} = \pm\frac{\sqrt{69}}{6}$$
$$x = -\frac{1}{2} \pm \frac{\sqrt{69}}{6}$$
$$x = \frac{-3 \pm \sqrt{69}}{6}$$
The solution set is $\left\{\dfrac{-3 - \sqrt{69}}{6}, \dfrac{-3 + \sqrt{69}}{6}\right\}$.

69.
$$A = P(1 + r)^t$$
$$1000 = 810(1 + r)^2$$
$$\frac{1000}{810} = (1 + r)^2$$
$$\frac{100}{81} = (1 + r)^2$$
$$\pm\sqrt{\frac{100}{81}} = 1 + r$$
$$\pm\frac{10}{9} = 1 + r$$
$$-1 \pm \frac{10}{9} = r$$
$$r = -1 + \frac{10}{9} \quad \text{or} \quad r = -1 - \frac{10}{9}$$
$$r = \frac{1}{9} \quad\quad \text{or} \quad r = -\frac{19}{9}$$
Rate cannot be negative, so $r = \dfrac{1}{9}$, or about 11%.

73. Simple; answers may vary

77.
$$\frac{12 - 8\sqrt{7}}{16} = \frac{12}{16} - \frac{8\sqrt{7}}{16}$$
$$= \frac{3}{4} - \frac{\sqrt{7}}{2}$$
$$= \frac{3}{4} - \frac{2\sqrt{7}}{4}$$
$$= \frac{3 - 2\sqrt{7}}{4}$$

81. $\sqrt{b^2 - 4ac}$
$a = 1, b = -3, c = -1$
$\sqrt{(-3)^2 - 4(1)(-1)} = \sqrt{9 + 4} = \sqrt{13}$

85. $s(t) = 16t^2$
$$1053 = 16t^2$$
$$t = \pm\sqrt{\frac{1053}{16}}$$
$t \approx 8.11$ or -8.11 (disregard)
It would take about 8.11 seconds.

89.
$$A = \pi r^2$$
$$36\pi = \pi r^2$$
$$36 = r^2$$
$$\pm\sqrt{36} = r$$
$$\pm 6 = r$$
Since the radius cannot be negative,
$r = 6$ inches.

93.
$$p = -x^2 + 15$$
$$7 = -x^2 + 15$$
$$-8 = -x^2$$
$$8 = x^2$$
$$\pm\sqrt{8} = x$$
$$\pm 2.828 \approx x$$
Since x cannot be negative, the demand for each lamp is about 2.828 thousand units.

Exercise Set 8.2

1. $m^2 + 5m - 6 = 0$
$a = 1, b = 5, c = -6$
$$m = \frac{-5 \pm \sqrt{5^2 - 4(1)(-6)}}{2(1)}$$
$$m = \frac{-5 \pm \sqrt{25 + 24}}{2} = \frac{-5 \pm \sqrt{49}}{2}$$
$$m = \frac{-5 \pm 7}{2}$$
$m = -6$ or $m = 1$
The solution set is $\{-6, 1\}$.

5. $x^2 - 6x + 9 = 0$
$a = 1, b = -6, c = 9$
$$x = \frac{6 \pm \sqrt{(-6)^2 - 4(1)(9)}}{2(1)}$$
$$x = \frac{6 \pm \sqrt{36 - 36}}{2} = \frac{6 \pm \sqrt{0}}{2} = \frac{6}{2} = 3$$
The solution set is $\{3\}$.

9. $8m^2 - 2m = 7$
$8m^2 - 2m - 7 = 0$
$a = 8, b = -2, c = -7$
$$m = \frac{2 \pm \sqrt{(-2)^2 - 4(8)(-7)}}{2(8)}$$
$$m = \frac{2 \pm \sqrt{4 + 224}}{16} = \frac{2 \pm \sqrt{228}}{16}$$
$$m = \frac{2 \pm \sqrt{4}\sqrt{57}}{16} = \frac{2 \pm 2\sqrt{57}}{16}$$
$$m = \frac{1 \pm \sqrt{57}}{8}$$
The solution set is $\left\{\dfrac{1 + \sqrt{57}}{8}, \dfrac{1 - \sqrt{57}}{8}\right\}$.

13. $\dfrac{1}{2}x^2 - x - 1 = 0$
$x^2 - 2x - 2 = 0$
$a = 1, b = -2, c = -2$
$$x = \frac{2 \pm \sqrt{(-2)^2 - 4(1)(-2)}}{2(1)}$$
$$x = \frac{2 \pm \sqrt{4 + 8}}{2} = \frac{2 \pm \sqrt{12}}{2}$$
$$x = \frac{2 \pm \sqrt{4}\sqrt{3}}{2} = \frac{2 \pm 2\sqrt{3}}{2}$$
$$x = 1 \pm \sqrt{3}$$
The solution set is $\{1 - \sqrt{3}, 1 + \sqrt{3}\}$.

17. $\frac{1}{3}y^2 - y - \frac{1}{6} = 0$

$2y^2 - 6y - 1 = 0$

$a = 2, b = -6, c = -1$

$y = \dfrac{6 \pm \sqrt{(-6)^2 - 4(2)(-1)}}{2(2)}$

$y = \dfrac{6 \pm \sqrt{36 + 8}}{4} = \dfrac{6 \pm \sqrt{44}}{4}$

$y = \dfrac{6 \pm \sqrt{4}\sqrt{11}}{4} = \dfrac{6 \pm 2\sqrt{11}}{4}$

$y = \dfrac{3 \pm \sqrt{11}}{2}$

The solution set is $\left\{ \dfrac{3 - \sqrt{11}}{2}, \dfrac{3 + \sqrt{11}}{2} \right\}$.

21. $x^2 + 5x = -2$

$x^2 + 5x + 2 = 0$

$a = 1, b = 5, c = 2$

$x = \dfrac{-5 \pm \sqrt{5^2 - 4(1)(2)}}{2(1)}$

$x = \dfrac{-5 \pm \sqrt{25 - 8}}{2} = \dfrac{-5 \pm \sqrt{17}}{2}$

The solution set is $\left\{ \dfrac{-5 - \sqrt{17}}{2}, \dfrac{-5 + \sqrt{17}}{2} \right\}$.

25. $\frac{x^2}{3} - x = \frac{5}{3}$

$x^2 - 3x = 5$

$x^2 - 3x - 5 = 0$

$a = 1, b = -3, c = -5$

$x = \dfrac{3 \pm \sqrt{(-3)^2 - 4(1)(-5)}}{2(1)}$

$x = \dfrac{3 \pm \sqrt{9 + 20}}{2} = \dfrac{3 \pm \sqrt{29}}{2}$

The solution set is $\left\{ \dfrac{3 - \sqrt{29}}{2}, \dfrac{3 + \sqrt{29}}{2} \right\}$.

29. answers may vary

33. $(x + 5)(x - 1) = 2$

$x^2 + 4x - 5 = 2$

$x^2 + 4x - 7 = 0$

$a = 1, b = 4, c = -7$

$x = \dfrac{-4 \pm \sqrt{4^2 - 4(1)(-7)}}{2(1)}$

$x = \dfrac{-4 \pm \sqrt{16 + 28}}{2} = \dfrac{-4 + \sqrt{44}}{2}$

$x = \dfrac{-4 \pm \sqrt{4}\sqrt{11}}{2} = \dfrac{-4 \pm 2\sqrt{11}}{2}$

$x = -2 \pm \sqrt{11}$

The solution set is $\{-2 - \sqrt{11}, -2 + \sqrt{11}\}$.

37. $\frac{2}{5}y^2 + \frac{1}{5}y + \frac{3}{5} = 0$

$2y^2 + y + 3 = 0$

$a = 2, b = 1, c = 3$

$y = \dfrac{-1 \pm \sqrt{1^2 - 4(2)(3)}}{2(2)}$

$y = \dfrac{-1 \pm \sqrt{1 - 24}}{4} = \dfrac{-1 \pm \sqrt{-23}}{4}$

$y = \dfrac{-1 \pm i\sqrt{23}}{4}$

The solution set is $\left\{ \dfrac{-1 - i\sqrt{23}}{4}, \dfrac{-1 + i\sqrt{23}}{4} \right\}$.

41. $(n - 2)^2 = 2n$

$n^2 - 4n + 4 = 2n$

$n^2 - 6n + 4 = 0$

$a = 1, b = -6, c = 4$

$n = \dfrac{6 \pm \sqrt{(-6)^2 - 4(1)(4)}}{2(1)}$

$n = \dfrac{6 \pm \sqrt{36 - 16}}{2} = \dfrac{6 \pm \sqrt{20}}{2}$

$n = \dfrac{6 \pm \sqrt{4}\sqrt{5}}{2} = \dfrac{6 \pm 2\sqrt{5}}{2}$

$n = 3 \pm \sqrt{5}$

The solution set is $\{3 - \sqrt{5}, 3 + \sqrt{5}\}$.

45. $4x^2 + 12x = -9$

$4x^2 + 12x + 9 = 0$

$a = 4, b = 12, c = 9$

$b^2 - 4ac = 12^2 - 4(4)(9)$

$b^2 - 4ac = 144 - 144 = 0$

Therefore, there is 1 real solution.

49. $6 = 4x - 5x^2$

$5x^2 - 4x + 6 = 0$

$a = 5, b = -4, c = 6$

$b^2 - 4ac = (-4)^2 - 4(5)(6)$

$b^2 - 4ac = 16 - 120$

$b^2 - 4ac = -104 < 0$

Therefore, there are 2 complex but not real solutions.

53. Let $x =$ length of leg

$x + 2 =$ length of hypotenuse

$x^2 + x^2 = (x + 2)^2$

$2x^2 = x^2 + 4x + 4$

$x^2 - 4x - 4 = 0$

$a = 1, b = -4, c = -4$

$x = \dfrac{4 \pm \sqrt{(-4)^2 - 4(1)(-4)}}{2(1)}$

$x = \dfrac{4 \pm \sqrt{32}}{2}$

$x = \dfrac{4 \pm 4\sqrt{2}}{2}$

$x = 2 \pm 2\sqrt{2}$ (disregard a negative length)

The sides measure $2 + 2\sqrt{2}$ cm, $2 + 2\sqrt{2}$ cm, and $4 + 2\sqrt{2}$ cm.

57. a. Let $x =$ length

$x^2 + x^2 = 100^2$

$2x^2 - 10,000 = 0$

$a = 2, b = 0, c = -10,000$

$x = \dfrac{0 \pm \sqrt{0^2 - 4(2)(-10,000)}}{2(2)}$

$x = \dfrac{\pm\sqrt{80,000}}{4}$

$x = \dfrac{\pm 200\sqrt{2}}{4}$

$x = \pm 50\sqrt{2}$

Disregard a negative length. The side measures $50\sqrt{2}$ meters.

b. Area $= s^2$

$= (50\sqrt{2})^2$

$= 50^2(\sqrt{2})^2$

$= 2500 \cdot 2$

$= 5000$

The area is 5000 square meters.

61. $\dfrac{x-1}{1}=\dfrac{1}{x}$

$x(x-1)=1\cdot 1$

$x^2-x-1=0$

$a=1, b=-1, c=-1$

$x=\dfrac{-(-1)\pm\sqrt{(-1)^2-4(1)(-1)}}{2}$

$x=\dfrac{1\pm\sqrt{5}}{2}$

Disregard the negative value $\dfrac{1-\sqrt{5}}{2}$.

The value is $\dfrac{1+\sqrt{5}}{2}$.

65. $h=-16t^2-20t+180$

Let $h=0$:

$0=-16t^2-20t+180$

$0=4t^2+5t-45 \quad \div\text{ by }-4$

$a=4, b=5, c=-45$

$t=\dfrac{-5\pm\sqrt{(5)^2-4(4)(-45)}}{2(4)}$

$t=\dfrac{-5\pm\sqrt{745}}{8}$

$t\approx 2.8$ or $t\approx -4.0$ (disregard)

It will take about 2.8 seconds.

69. $\dfrac{1}{x}+\dfrac{2}{5}=\dfrac{7}{x}$

$5x\left(\dfrac{1}{x}+\dfrac{2}{5}\right)=5x\left(\dfrac{7}{x}\right)$ LCD is $5x$

$5+2x=35$

$2x=30$

$x=15$

$\{15\}$

73. $z^4-13z^2+36=(z^2-9)(z^2-4)=(z+3)(z-3)(z+2)(z-2)$

77. Sunday to Monday

81. Thursday corresponds to $x=4$.

$f(x)=3x^2-18x+57$

$f(4)=3(4)^2-18(4)+57$

$=3(16)-18(4)+57$

$=48-72+57$

$=33$

Yes; answers may vary

85. $f(x)=-199.5x^2-21.5x+3763$

   a. $f(2)=-199.5(2)^2-21.5(2)+3763$

   $=-199.5(4)-21.5(2)+3763$

   $=-798-43+3763$

   $=2922$

   The earnings should be \$2922 million.

   b. $485=-199.5x^2-21.5x+3763$

   $199.5x^2+21.5x-3278=0$

   $a=199.5, b=21.5, c=-3278$

   $x=\dfrac{-21.5\pm\sqrt{(21.5)^2-4(199.5)(-3278)}}{2(199.5)}$

   $x=\dfrac{-21.5\pm\sqrt{2,616,306.25}}{399}$

   $x=\dfrac{-21.5\pm 1617.5}{399}$

   $x=4$ or $x\approx -4.1$ (Disregard)

   The year is $1999+4=2003$.

Exercise Set 8.3

1. $\qquad 2x=\sqrt{10+3x}$

$4x^2=10+3x$

$4x^2-3x-10=0$

$(4x+5)(x-2)=0$

$4x+5=0$ or $x-2=0$

$x=-\dfrac{5}{4}$ or $x=2$

Discard $x=-\dfrac{5}{4}$.

The solution set is $\{2\}$.

5. $\sqrt{9x}=x+2$

$9x=x^2+4x+4$

$0=x^2-5x+4$

$0=(x-4)(x-1)$

$x-4=0$ or $x-1=0$

$x=4$ or $x=1$

The solution set is $\{1,4\}$.

9. $\dfrac{3}{x}+\dfrac{4}{x+2}=2$

$\dfrac{3(x+2)+4x}{x(x+2)}=2$

$\dfrac{3x+6+4x}{x^2+2x}=2$

$7x+6=2(x^2+2x)$

$7x+6=2x^2+4x$

$2x^2-3x-6=0$

$x=\dfrac{3\pm\sqrt{(-3)^2-4(2)(-6)}}{2(2)}$

$x=\dfrac{3\pm\sqrt{57}}{4}$

The solution set is $\left\{\dfrac{3+\sqrt{57}}{4},\dfrac{3-\sqrt{57}}{4}\right\}$.

13. $\qquad p^4-16=0$

$(p^2+4)(p^2-4)=0$

$(p+2i)(p-2i)(p+2)(p-2)=0$

$p+2i=0$ or $p-2i=0$

or $p+2=0$ or $p-2=0$

$p=-2i$ or $p=2i$

$p=-2$ or $p=2$

The solution set is $\{-2i,2i,-2,2\}$.

17. $\qquad z^4-13z^2+36=0$

$(z^2-9)(z^2-4)=0$

$(z+3)(z-3)(z+2)(z-2)=0$

$z+3=0$ or $z-3=0$ or $z+2=0$ or $z-2=0$

$z=-3$ or $z=3$ or $z=-2$ or $z=2$

The solution set is $\{-3,3,-2,2\}$.

21. $(5n+1)^2+2(5n+1)-3=0$

Let $y=5n+1$.

$y^2+2y-3=0$

$(y+3)(y-1)=0$

$y+3=0$ or $y-1=0$

$y=-3$ or $y=1$

$5n+1=-3$ or $5n+1=1$

$5n=-4$ or $5n=0$

$n=-\dfrac{4}{5}$ or $n=0$

The solution set is $\left\{-\dfrac{4}{5},0\right\}$.

25.
$$1 + \frac{2}{3t - 2} = \frac{8}{(3t - 2)^2}$$
$(3t - 2)^2 + 2(3t - 2) - 8 = 0$
Let $y = 3t - 2$.
$y^2 + 2y - 8 = 0$
$(y + 4)(y - 2) = 0$
$y + 4 = 0$ or $y - 2 = 0$
$y = -4$ or $y = 2$
$3t - 2 = -4$ or $3t - 2 = 2$
$3t = -2$ or $3t = 4$
$t = -\dfrac{2}{3}$ or $t = \dfrac{4}{3}$
The solution set is $\left\{-\dfrac{2}{3}, \dfrac{4}{3}\right\}$.

29. $a^4 - 5a^2 + 6 = 0$
$(a^2 - 3)(a^2 - 2) = 0$
$a^2 - 3 = 0$ or $a^2 - 2 = 0$
$a^2 = 3$ or $a^2 = 2$
$a = \pm\sqrt{3}$ or $a = \pm\sqrt{2}$
The solution set is $\{\sqrt{3}, -\sqrt{3}, \sqrt{2}, -\sqrt{2}\}$.

33.
$$(p + 2)^2 = 9(p + 2) - 20$$
$(p + 2)^2 - 9(p + 2) + 20 = 0$
Let $x = p + 2$.
$x^2 - 9x + 20 = 0$
$(x - 5)(x - 4) = 0$
$x - 5 = 0$ or $x - 4 = 0$
$x = 5$ or $x = 4$
$p + 2 = 5$ or $p + 2 = 4$
$p = 3$ or $p = 2$
The solution set is $\{2, 3\}$.

37. $x^{2/3} - 8x^{1/3} + 15 = 0$
Let $y = x^{1/3}$.
$y^2 - 8y + 15 = 0$
$(y - 5)(y - 3) = 0$
$y - 5 = 0$ or $y - 3 = 0$
$y = 5$ or $y = 3$
$x^{1/3} = 5$ or $x^{1/3} = 3$
$x = 5^3$ or $x = 3^3$
$x = 125$ or $x = 27$
The solution set is $\{125, 27\}$.

41. $2x^{2/3} + 3x^{1/3} - 2 = 0$
Let $m = x^{1/3}$.
$2m^2 + 3m - 2 = 0$
$(2m - 1)(m + 2) = 0$
$2m - 1 = 0$ or $m + 2 = 0$
$2m = 1$ or $m = -2$
$m = \dfrac{1}{2}$ or $m = -2$
$x^{1/3} = \dfrac{1}{2}$ or $x^{1/3} = -2$
$x = \left(\dfrac{1}{2}\right)^3 = \dfrac{1}{8}$ or $x = (-2)^3 = -8$
The solution set is $\left\{\dfrac{1}{8}, -8\right\}$.

45.
$$x - \sqrt{x} = 2$$
$x - 2 = \sqrt{x}$
$x^2 - 4x + 4 = x$
$x^2 - 5x + 4 = 0$
$(x - 4)(x - 1) = 0$
$x = 4$ or $x = 1$ (discard)
The solution set is $\{4\}$.

49. $p^4 - p^2 - 20 = 0$
$(p^2 - 5)(p^2 + 4) = 0$
$p^2 - 5 = 0$ or $p^2 + 4 = 0$

$p^2 = 5$ or $(p + 2i)(p - 2i) = 0$
$p = \pm\sqrt{5}$ or $p + 2i = 0$ or $p - 2i = 0$
$p = \pm\sqrt{5}$ or $p = -2i$ or $p = 2i$
The solution set is $\{\sqrt{5}, -\sqrt{5}, -2i, 2i\}$.

53. $1 = \dfrac{4}{x - 7} + \dfrac{5}{(x - 7)^2}$
$(x - 7)^2 = 4(x - 7) + 5$
Let $y = x - 7$.
$y^2 - 4y - 5 = 0$
$(y - 5)(y + 1) = 0$
$y - 5 = 0$ or $y + 1 = 0$
$y = 5$ or $y = -1$
$x - 7 = 5$ or $x - 7 = -1$
$x = 12$ or $x = 6$
The solution set is $\{6, 12\}$.

57. Let x = speed on first part
$x - 1$ = speed on second part
$D = r \cdot t$ or $t = \dfrac{D}{r}, 1\dfrac{3}{5} = \dfrac{8}{5}$
$$\frac{3}{x} + \frac{4}{x - 1} = \frac{8}{5}$$
$3 \cdot 5(x - 1) + 4 \cdot 5x = 8 \cdot x(x - 1)$
$15x - 15 + 20x = 8x^2 - 8x$
$0 = 8x^2 - 43x + 15$
$0 = (8x - 3)(x - 5)$
$8x - 3 = 0$ or $x - 5 = 0$
$x = \dfrac{3}{8}$ or $x = 5$
$x - 1 = 4$
Her speeds were 5 mph and then 4 mph.

61. Let x = original speed
$x + 11$ = return speed
$D = r \cdot t$ or $t = \dfrac{D}{r}$
$$\frac{330}{x} - \frac{330}{x + 11} = 1$$
$330(x + 11) - 330x = x(x + 11)$
$330x + 3630 - 330x = x^2 + 11x$
$0 = x^2 + 11x - 3630$
$0 = (x - 55)(x + 66)$
$x = 55$ or $x = -66$ (discard)
$x + 11 = 66$
The speeds are 55 mph and 66 mph.

65. Let x = number
$x(x - 4) = 96$
$x^2 - 4x = 96$
$x^2 - 4x - 96 = 0$
$(x - 12)(x + 8) = 0$
$x - 12 = 0$ or $x + 8 = 0$
$x = 12$ or $x = -8$
The number is 12 or -8.

69. Let r = radius
d = diameter
s = side of square
$A = s^2$
$920 = s^2$
$s = \sqrt{920}$
d is the diagonal of the square
$d = \sqrt{(\sqrt{920})^2 + (\sqrt{920})^2}$
$d = \sqrt{920 + 920}$
$d = \sqrt{1840}$
$d \approx 42.9$

Now, $r = \dfrac{d}{2} = \dfrac{42.9}{2} \approx 22$.
The radius is 22 feet.

73.
$$\frac{y-1}{15} > \frac{-2}{5}$$
$$15\left(\frac{y-1}{15}\right) > 15\left(\frac{-2}{5}\right)$$
$$y - 1 > -6$$
$$y > -5 \text{ or } (-5, \infty)$$

77. Let x = Brack's speed and
$x + 6.326$ = Kanaan's speed.

Brack's time = $\dfrac{10{,}682}{x}$

Kanaan's time = $\dfrac{10{,}682}{x + 6.326}$

Equation is

Kanaan's time = Brack's time + 0.73

$$\frac{10{,}682}{x} = \frac{10{,}682}{x + 6.326} + 0.73$$

Multiply both sides by $x(x + 6.326)$.

$$10{,}682(x + 6.326) = 10{,}682(x) + 0.73(x)(x + 6.326)$$
$$10{,}682(6.326) = 0.73(x)(x + 6.326)$$
$$67{,}574.332 = 0.73x^2 + 4.61798x$$
$$0 = 0.73x^2 + 4.61798x - 67{,}574.332$$
$$a = 0.73, b = 4.61798, c = -67{,}574.332$$

$$x = \frac{-4.61798 \pm \sqrt{(4.61798)^2 - 4(0.73)(-67{,}574.332)}}{2(0.73)}$$

$$= \frac{-4.61798 \pm \sqrt{197{,}338.3759}}{1.46}$$

$$= \frac{-4.61798 \pm 444.2278415}{1.46}$$

$$= \frac{-4.61798 + 444.2278415}{1.46} \text{ (discard negative answer)}$$

$$= 301.1026449$$

$$\approx 301.103$$

a. Brack's speed is 301.103 ft/sec

b. Kanaan's speed is 301.103 + 6.326 = 307.429 ft/sec

c. 307.429 ft/sec = $\dfrac{307.429 \text{ ft}}{1 \text{ sec}} \times \dfrac{1 \text{ mi}}{5280 \text{ ft}} \times \dfrac{3600 \text{ sec}}{1 \text{ hr}} \approx$

209.611 mph

Exercise Set 8.4

1. $(x + 1)(x + 5) > 0$
$(x + 1)(x + 5) = 0$
$x + 1 = 0 \quad \text{or} \quad x + 5 = 0$
$\quad x = -1 \quad \text{or} \quad \quad x = -5$

Region	Interval	Test Point
A	$(-\infty, -5)$	-6
B	$(-5, -1)$	-2
C	$(-1, \infty)$	0

$x = -6$	$(-6 + 1)(-6 + 5) > 0$	True
$x = -2$	$(-2 + 1)(-2 + 5) > 0$	False
$x = 0$	$(0 + 1)(0 + 5) > 0$	True

The solution set is $(-\infty, -5) \cup (-1, \infty)$.

5. $x^2 - 7x + 10 \le 0$
$(x - 5)(x - 2) \le 0$
$(x - 5)(x - 2) = 0$
$x - 5 = 0 \quad \text{or} \quad x - 2 = 0$
$\quad x = 5 \quad \text{or} \quad \quad x = 2$

Region	Interval	Test Point
A	$(-\infty, 2)$	0
B	$(2, 5)$	3
C	$(5, \infty)$	6

$x = 0$	$(0 - 5)(0 - 2) \le 0$	False
$x = 3$	$(3 - 5)(3 - 2) \le 0$	True
$x = 6$	$(6 - 5)(6 - 2) \le 0$	False

The solution set is $[2, 5]$.

9. $(x - 6)(x - 4)(x - 2) > 0$
$(x - 6)(x - 4)(x - 2) = 0$
$x - 6 = 0 \text{ or } x - 4 = 0 \text{ or } x - 2 = 0$
$x = 6 \text{ or } x = 4 \text{ or } x = 2$

Region	Interval	Test Point
A	$(-\infty, 2)$	1
B	$(2, 4)$	3
C	$(4, 6)$	5
D	$(6, \infty)$	7

$x = 1$	$(1 - 6)(1 - 4)(1 - 2) > 0$	False
$x = 3$	$(3 - 6)(3 - 4)(3 - 2) > 0$	True
$x = 5$	$(5 - 6)(5 - 4)(5 - 2) > 0$	False
$x = 7$	$(7 - 6)(7 - 4)(7 - 2) > 0$	True

The solution set is $(2, 4) \cup (6, \infty)$.

13. $(x^2 - 9)(x^2 - 4) > 0$
$(x + 3)(x - 3)(x + 2)(x - 2) > 0$
$(x + 3)(x - 3)(x + 2)(x - 2) = 0$
$x + 3 = 0 \text{ or } x - 3 = 0 \text{ or } x + 2 = 0 \text{ or } x - 2 = 0$
$x = -3 \text{ or } x = 3 \text{ or } x = -2 \text{ or } x = 2$

Region	Interval	Test Point
A	$(-\infty, -3)$	-4
B	$(-3, -2)$	$-\dfrac{5}{2}$
C	$(-2, 2)$	0
D	$(2, 3)$	$\dfrac{5}{2}$
E	$(3, \infty)$	4

$x = -4$	$[(-4)^2 - 9][(-4)^2 - 4] > 0$	True
$x = -\dfrac{5}{2}$	$\left[\left(-\dfrac{5}{2}\right)^2 - 9\right]\left[\left(-\dfrac{5}{2}\right)^2 - 4\right]$ > 0	False
$x = 0$	$(0^2 - 9)(0^2 - 4) > 0$	True
$x = \dfrac{5}{2}$	$\left[\left(\dfrac{5}{2}\right)^2 - 9\right]\left[\left(\dfrac{5}{2}\right)^2 - 4\right] > 0$	False
$x = 4$	$(4^2 - 9)(4^2 - 4) > 0$	True

The solution set is $(-\infty, -3) \cup (-2, 2) \cup (3, \infty)$.

17.
$$6x^2 + 5x \le 4$$
$$6x^2 + 5x - 4 \le 0$$
$$(2x - 1)(3x + 4) = 0$$
$$2x - 1 = 0 \quad \text{or} \quad 3x + 4 = 0$$
$$2x = 1 \quad \text{or} \quad \quad 3x = -4$$
$$x = \frac{1}{2} \quad \text{or} \quad \quad x = -\frac{4}{3}$$

Region	Interval	Test Point
A	$\left(-\infty, -\dfrac{4}{3}\right)$	-2
B	$\left(-\dfrac{4}{3}, \dfrac{1}{2}\right)$	0
C	$\left(\dfrac{1}{2}, \infty\right)$	1

$x = -2$	$6(-2)^2 + 5(-2) \leq 4$	False
$x = 0$	$6(0)^2 + 5(0) \leq 4$	True
$x = 1$	$6(1)^2 + 5(1) \leq 4$	False

The solution set is $\left[-\dfrac{4}{3}, \dfrac{1}{2}\right]$.

21. $(2x - 8)(x + 4)(x - 6) \leq 0$

$(2x - 8)(x + 4)(x - 6) = 0$

$2x - 8 = 0$ or $x + 4 = 0$ or $x - 6 = 0$

$2x = 8$ or $x = -4$ or $x = 6$

$x = 4$ or $x = -4$ or $x = 6$

Region	Interval	Test Point
A	$(-\infty, -4)$	-5
B	$(-4, 4)$	0
C	$(4, 6)$	5
D	$(6, \infty)$	7

$x = -5$	$[2(-5) - 8](-5 + 4)$ $(-5 - 6) \leq 0$	True
$x = 0$	$[2(0) - 8](0 + 4)(0 - 6)$ ≤ 0	False
$x = 5$	$[2(5) - 8](5 + 4)(5 - 6)$ ≤ 0	True
$x = 7$	$[2(7) - 8](7 + 4)(7 - 6)$ ≤ 0	False

The solution set is $(-\infty, -4] \cup [4, 6]$.

25. $\dfrac{5}{x + 1} > 0$

$x + 1 = 0$

$x = -1$

Region	Interval	Test Point
A	$(-\infty, -1)$	-2
B	$(-1, \infty)$	0

$x = -2$	$\dfrac{5}{-2 + 1} > 0$	False
$x = 0$	$\dfrac{5}{0 + 1} > 0$	True

The solution set is $(-1, \infty)$.

29. $\dfrac{x + 2}{x - 3} < 1$

$\dfrac{x + 2}{x - 3} - \dfrac{x - 3}{x - 3} < 0$

$\dfrac{x + 2 - x + 3}{x - 3} < 0$

$\dfrac{5}{x - 3} < 0$

$x = 3$

Region	Interval	Test Point
A	$(-\infty, 3)$	0
B	$(3, \infty)$	4

$x = 0$	$\dfrac{0 + 2}{0 - 3} < 1$	True
$x = 4$	$\dfrac{4 + 2}{4 - 3} < 1$	False

The solution set is $(-\infty, 3)$.

33. $\dfrac{x - 5}{x + 4} \geq 0$

$x - 5 = 0$ or $x + 4 = 0$

$x = 5$ or $x = -4$

Region	Interval	Test Point
A	$(-\infty, -4)$	-5
B	$(-4, 5)$	0
C	$(5, \infty)$	6

$x = -5$	$\dfrac{-5 - 5}{-5 + 4} \geq 0$	True
$x = 0$	$\dfrac{0 - 5}{0 + 4} \geq 0$	False
$x = 6$	$\dfrac{6 - 5}{6 + 4} \geq 0$	True

The solution set is $(-\infty, -4) \cup [5, \infty)$.

37. $\dfrac{-1}{x - 1} > -1$

$\dfrac{1}{x - 1} < 1$

$\dfrac{1}{x - 1} - 1 < 0$

$\dfrac{1 - (x - 1)}{x - 1} < 0$

$\dfrac{1 - x + 1}{x - 1} < 0$

$\dfrac{2 - x}{x - 1} < 0$

$2 - x = 0$ or $x - 1 = 0$

$2 = x$ or $x = 1$

Region	Interval	Test Point
A	$(-\infty, 1)$	0
B	$(1, 2)$	$\dfrac{3}{2}$
C	$(2, \infty)$	3

$x = 0$	$\dfrac{-1}{0 - 1} > -1$	True
$x = \dfrac{3}{2}$	$\dfrac{-1}{\dfrac{3}{2} - 1} > -1$	False
$x = 3$	$\dfrac{-1}{3 - 1} > -1$	True

The solution set is $(-\infty, 1) \cup (2, \infty)$.

41.

x	x^2	y
0	$0^2 = 0$	0
1	$1^2 = 1$	1
-1	$(-1)^2 = 1$	1
2	$(2)^2 = 4$	4
-2	$(-2)^2 = 4$	4

45. answers may vary

49. $P(x) = -2x^2 + 26x - 44$
$$-2x^2 + 26x - 44 > 0$$
$$-2(x^2 - 13x + 22) > 0$$
$$-2(x - 11)(x - 2) > 0$$
$$x - 11 = 0 \text{ or } x - 2 = 0$$
$$x = 11 \text{ or } x = 2$$

Region	Interval	Test Point
A	$(-\infty, 2)$	0
B	$(2, 11)$	3
C	$(11, \infty)$	12

$x = 0$	$-2(0)^2 + 26(0) - 44 > 0$	False
$x = 3$	$-2(3)^2 + 26(3) - 44 > 0$	True
$x = 12$	$-2(12)^2 + 26(12) - 44 > 0$	False

The solution set is $(2, 11)$.
The company makes a profit when x is between 2 and 11.

Exercise Set 8.5

1. $f(x) = x^2 - 1$

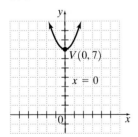

5. $h(x) = x^2 + 5$

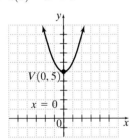

9. $g(x) = x^2 + 7$

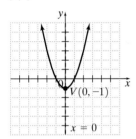

13. $f(x) = (x - 2)^2 + 5$

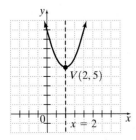

17. $g(x) = (x + 2)^2 - 5$

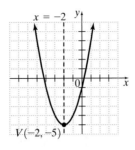

21. $g(x) = -x^2$

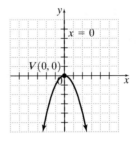

25. $H(x) = 2x^2$

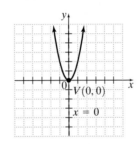

29. $f(x) = 10(x + 4)^2 - 6$

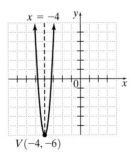

33. $H(x) = \dfrac{1}{2}(x - 6)^2 - 3$

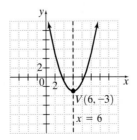

37. $F(x) = \left(x + \dfrac{1}{2}\right)^2 - 2$

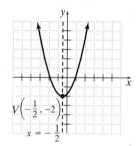

$V\left(-\dfrac{1}{2}, -2\right)$

$x = -\dfrac{1}{2}$

41. $x^2 + 8x$

$\left[\dfrac{1}{2}(8)\right]^2 = [4]^2 = 16$

$x^2 + 8x + 16$

45. $y^2 + y$

$\left[\dfrac{1}{2}(1)\right]^2 = \left[\dfrac{1}{2}\right]^2 = \dfrac{1}{4}$

$y^2 + y + \dfrac{1}{4}$

49. $f(x) = 5[(x - (-3))]^2 + 6$
$f(x) = 5(x + 3)^2 + 6$

53. $y = f(x - 3)$

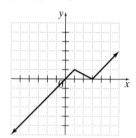

Exercise Set 8.6

1. $f(x) = x^2 + 8x + 7$
$\dfrac{-b}{2a} = \dfrac{-8}{2(1)} = -4$ and $f(-4) = (-4)^2 + 8(-4) + 7$
$f(-4) = 16 - 32 + 7 = -9$
Thus, $V(-4, -9)$.

5. $f(x) = 5x^2 - 10x + 3$
$\dfrac{-b}{2a} = \dfrac{-(-10)}{2(5)} = 1$ and $f(1) = 5(1)^2 - 10(1) + 3$
$f(1) = 5 - 10 + 3 = -2$
Thus, $V(1, -2)$.

9. $f(x) = x^2 - 4x + 3$
$\dfrac{-b}{2a} = \dfrac{-(-4)}{2(1)} = 2$
$f(2) = 2^2 - 4(2) + 3 = -1$
$V(2, -1)$
Graph D

13. $f(x) = x^2 + 4x - 5$
$\dfrac{-b}{2a} = \dfrac{-4}{2(1)} = -2$ and $f(-2) = (-2)^2 + 4(-2) - 5$
$f(-2) = 4 - 8 - 5 = -9$
Thus, $V(-2, -9)$.
The graph opens upward since $a > 0$.
$x^2 + 4x - 5 = 0$
$(x + 5)(x - 1) = 0$

$x + 5 = 0 \quad$ or $\quad x - 1 = 0$
$x = -5 \quad$ or $\qquad x = 1$
The x-intercepts are $(-5, 0)$ and $(1, 0)$.
$f(0) = (0, -5)$ is the y-intercept.

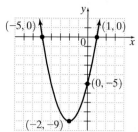

17. $f(x) = x^2 - 4$
$\dfrac{-b}{2a} = \dfrac{-0}{2(1)} = 0$ and $f(0) = -4$
Thus, $V(0, -4)$.
The graph opens upward since $a > 0$.
$x^2 - 4 = 0$
$(x + 2)(x - 2) = 0$
$x + 2 = 0 \quad$ or $\quad x - 2 = 0$
$x = -2 \quad$ or $\qquad x = 2$
The x-intercepts are $(-2, 0)$ and $(2, 0)$.
$f(0) = -4$ so $(0, -4)$ is the y-intercept.

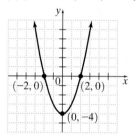

21. $f(x) = \dfrac{1}{2}x^2 + 4x + \dfrac{15}{2}$
$\dfrac{-b}{2a} = \dfrac{-4}{2\left(\dfrac{1}{2}\right)} = -4$

$f(-4) = \dfrac{1}{2}(-4)^2 + 4(-4) + \dfrac{15}{2} = -\dfrac{1}{2}$
$V\left(-4, -\dfrac{1}{2}\right)$
The graph opens upward since $a > 0$.
$\dfrac{1}{2}x^2 + 4x + \dfrac{15}{2} = 0$
$x^2 + 8x + 15 = 0$
$(x + 5)(x + 3) = 0$
$x + 5 = 0 \quad$ or $\quad x + 3 = 0$
$x = -5 \quad$ or $\qquad x = -3$
$x = -5, x = -3$
The x-intercepts are $(-5, 0)$ and $(-3, 0)$.
$f(0) = \dfrac{15}{2}$ so $\left(0, \dfrac{15}{2}\right)$ is the y-intercept.

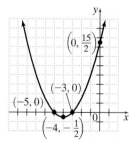

25. $f(x) = 2x^2 + 4x + 5$

$f(x) = 2(x^2 + 2x) + 5$

$f(x) = 2\left[x^2 + 2x + \left(\dfrac{2}{2}\right)^2\right] + 5 - 2$

$f(x) = 2(x + 1)^2 + 3$

Thus, $V(-1, 3)$.

The graph opens upward since $a > 0$.

$2(x + 1)^2 + 3 = 0$

$2(x + 1)^2 = -3$

Hence, there are no x-intercepts.

$f(0) = 2(0 + 1)^2 + 3 = 5$ so $(0, 5)$ is the y-intercept.

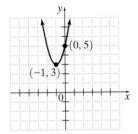

29. $h(t) = -16t^2 + 96t$

$\dfrac{-b}{2a} = \dfrac{-96}{2(-16)} = 3$ and $h(3) = -16(3)^2 + 96(3)$

$h(3) = -144 + 288 = 144$

The maximum height of the projectile is 144 ft.

33. Let $x =$ one number.

$60 - x =$ other number

$f(x) = x(60 - x)$

$f(x) = 60x - x^2$

$f(x) = -x^2 + 60x$

$f(x) = -1(x^2 - 60x)$

$f(x) = -1(x^2 - 60x + 900) + 900$

$f(x) = -(x - 30)^2 + 900$

The maximum will occur at the vertex which is $(30, 900)$. The numbers are 30 and 30.

37. Let $x =$ the width

$40 - x =$ the length

$f(x) = x(40 - x)$

$f(x) = 40x - x^2$

$f(x) = -x^2 + 40x$

$f(x) = -1(x^2 - 40x)$

$f(x) = -1(x^2 - 40x + 400) + 400$

$f(x) = -(x - 20)^2 + 400$

The maximum will occur at the vertex which is $(20, 400)$. The width is 20 units and the length is 20 units.

41. $g(x) = x + 2$

The vertex is undefined.

45. $f(x) = 3(x - 4)^2 + 1$

The vertex is $(4, 1)$

49. $f(x) = \dfrac{1}{2}x^2 + 4x + \dfrac{15}{2}$

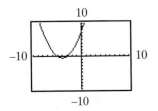

53. a. Because $a < 0$, the parabola will open downward and the function will have a maximum.

b. $x = \dfrac{-b}{2a} = \dfrac{-93}{2(-1)} = \dfrac{93}{2} = 46.5$

The year is $1990 + 46.5 = 2036.5$ which is in the year 2036.

c. $p(46.5) = -(46.5)^2 + 93(46.5) + 1128 = 3290.25$ thousands

Chapter 8 Test

1. $5x^2 - 2x = 7$

$5x^2 - 2x - 7 = 0$

$(5x - 7)(x + 1) = 0$

$5x - 7 = 0$ or $x + 1 = 0$

$5x = 7$ or $x = -1$

$x = \dfrac{7}{5}$ or $x = -1$

The solution set is $\left\{\dfrac{7}{5}, -1\right\}$.

5. $7x^2 + 8x + 1 = 0$

$(7x + 1)(x + 1) = 0$

$7x + 1 = 0$ or $x + 1 = 0$

$7x = -1$ or $x = -1$

$x = -\dfrac{1}{7}$ or $x = -1$

The solution set is $\left\{-\dfrac{1}{7}, -1\right\}$.

9. $x^6 + 1 = x^4 + x^2$

$x^6 - x^4 - x^2 + 1 = 0$

$x^4(x^2 - 1) - (x^2 - 1) = 0$

$(x^4 - 1)(x^2 - 1) = 0$

$(x^2 + 1)(x^2 - 1)(x^2 - 1) = 0$

$(x^2 + 1)(x^2 - 1)^2 = 0$

$(x^2 + 1)[(x + 1)(x - 1)]^2 = 0$

$(x^2 + 1)(x + 1)^2(x - 1)^2 = 0$

$x^2 + 1 = 0$ or $(x + 1)^2 = 0$ or $(x - 1)^2 = 0$

$x^2 = -1$ or $x + 1 = 0$ or $x - 1 = 0$

$x = \pm\sqrt{-1} = \pm i$ or $x = -1$ or $x = 1$

The solution set is $\{1, -1, i, -i\}$.

13. $2x^2 - 7x > 15$

$2x^2 - 7x - 15 > 0$

$(2x + 3)(x - 5) > 0$

$2x + 3 = 0$ or $x - 5 = 0$

$x = -\dfrac{3}{2}$ or $x = 5$

Region	Interval	Test Point
A	$\left(-\infty, -\dfrac{3}{2}\right)$	-2
B	$\left(-\dfrac{3}{2}, 5\right)$	0
C	$(5, \infty)$	6

$x = -2$	$2(-2)^2 - 7(-2) > 15$	True
$x = 0$	$2(0)^2 - 7(0) > 15$	False
$x = 6$	$2(6)^2 - 7(6) > 15$	True

17. $f(x) = 3x^2$
vertex: $(0, 0)$

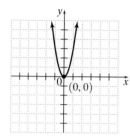

21. $c^2 = a^2 + b^2$
$(10)^2 = x^2 + (x - 4)^2$
$100 = x^2 + x^2 - 8x + 16$
$0 = 2x^2 - 8x - 84$
$0 = x^2 - 4x - 42$
$a = 1, b = -4, c = -42$
$x = \dfrac{-(-4) \pm \sqrt{(-4)^2 - 4(1)(-42)}}{2(1)}$
$x = \dfrac{4 \pm \sqrt{16 + 168}}{2}$
$x = \dfrac{4 \pm \sqrt{184}}{2}$
$x = \dfrac{4 \pm 2\sqrt{46}}{2}$
$x = 2 \pm \sqrt{46}$
Disregard the negative result.
The top of the ladder is $2 + \sqrt{46} \approx 8.8$ feet from the ground.

Chapter 9

Exercise Set 9.1

1. $x = 3y^2$
$x = 3(y - 0)^2$
$V(0, 0)$

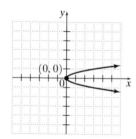

5. $y = 3(x - 1)^2 + 5$
$V(1, 5)$

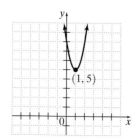

9. $y = x^2 + 10x + 20$
$y = x^2 + 10x + 25 + 20 - 25$
$y = (x + 5)^2 - 5$
$V(-5, -5)$

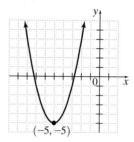

13. $(5, 1), (8, 5)$
$d = \sqrt{(8 - 5)^2 + (5 - 1)^2}$
$d = \sqrt{9 + 16}$
$d = \sqrt{25}$
$d = 5$ units

17. $(-9, 4), (-8, 1)$
$d = \sqrt{[-8 - (-9)]^2 + (1 - 4)^2}$
$d = \sqrt{(-8 + 9)^2 + (-3)^2}$
$d = \sqrt{1 + 9}$
$d = \sqrt{10} \approx 3.162$ units

21. $(1.7, -3.6), (-8.6, 5.7)$
$d = \sqrt{(-8.6 - 1.7)^2 + [5.7 - (-3.6)]^2}$
$d = \sqrt{(-10.3)^2 + (9.3)^2}$
$d = \sqrt{192.58} \approx 13.877$ units

25. $(6, -8), (2, 4)$
$\left(\dfrac{6 + 2}{2}, \dfrac{-8 + 4}{2}\right)$
$(4, -2)$

29. $(7, 3), (-1, -3)$
$\left(\dfrac{7 + (-1)}{2}, \dfrac{3 + (-3)}{2}\right)$
$(3, 0)$

33. $(\sqrt{2}, 3\sqrt{5}), (\sqrt{2}, -2\sqrt{5})$
$\left(\dfrac{\sqrt{2} + \sqrt{2}}{2}, \dfrac{3\sqrt{5} - 2\sqrt{5}}{2}\right)$
$\left(\sqrt{2}, \dfrac{\sqrt{5}}{2}\right)$

37. $x^2 + y^2 = 9$
$(x - 0)^2 + (y - 0)^2 = 3^2$
$C(0, 0)$ and $r = 3$

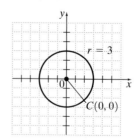

41. $(x - 5)^2 + (y + 2)^2 = 1$
$(x - 5)^2 + (y + 2)^2 = 1^2$
$C(5, -2)$ and $r = 1$

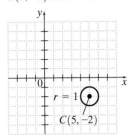

45. $x^2 + y^2 + 2x - 4y = 4$
$x^2 + 2x + 1 + y^2 - 4y + 4 = 4 + 1 + 4$
$(x + 1)^2 + (y - 2)^2 = 9$
$(x + 1)^2 + (y - 2)^2 = 3^2$
$C(-1, 2)$ and $r = 3$

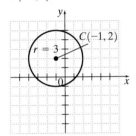

49. $C(2, 3);\ \ r = 6$
$(x - 2)^2 + (y - 3)^2 = 6^2$
$(x - 2)^2 + (y - 3)^2 = 36$

53. $C(-5, 4);\ \ r = 3\sqrt{5}$
$[x - (-5)]^2 + (y - 4)^2 = (3\sqrt{5})^2$
$(x + 5)^2 + (y - 4)^2 = 45$

57. $y = 3$

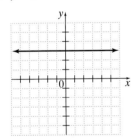

61. $\dfrac{4\sqrt{7}}{\sqrt{6}} = \dfrac{4\sqrt{7}}{\sqrt{6}} \cdot \dfrac{\sqrt{6}}{\sqrt{6}}$

$= \dfrac{4\sqrt{42}}{6}$

$= \dfrac{2\sqrt{42}}{3}$

65. Answers may vary

69. The equation of the arch is $y = ax^2 + k$ passing through the points $(0, 40)$ and $(50, 0)$.
Substitute 0 for x and 40 for y:

$40 = a(0)^2 + k$

$40 = k$
The equation is $y = ax^2 + 40$.
Substitute 50 for x and 0 for y:

$0 = a(50)^2 + 40$

$0 = 2500a + 40$

$-40 = 2500a$

$-\dfrac{40}{2500} = a$ or $a = -\dfrac{2}{125}$

Thus, $y = -\dfrac{2}{125}x^2 + 40$.

Exercise Set 9.2

1. $\dfrac{x^2}{4} + \dfrac{y^2}{25} = 1$

$\dfrac{x^2}{2^2} + \dfrac{y^2}{5^2} = 1$

$C(0, 0)$
x-intercepts: $(-2, 0), (2, 0)$
y-intercepts: $(0, -5), (0, 5)$

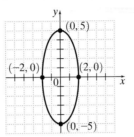

5. $9x^2 + 4y^2 = 36$

$\dfrac{x^2}{4} + \dfrac{y^2}{9} = 1$

$\dfrac{x^2}{2^2} + \dfrac{y^2}{3^2} = 1$

$C(0, 0)$
x-intercepts: $(-2, 0), (2, 0)$
y-intercepts: $(0, -3), (0, 3)$

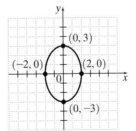

9. $\dfrac{(x + 1)^2}{36} + \dfrac{(y - 2)^2}{49} = 1$

$\dfrac{(x + 1)^2}{6^2} + \dfrac{(y - 2)^2}{7^2} = 1$

$C(-1, 2)$
other points:
$(-1 - 6, 2)$ or $(-7, 2)$
$(-1 + 6, 2)$ or $(5, 2)$
$(-1, 2 - 7)$ or $(-1, -5)$
$(-1, 2 + 7)$ or $(-1, 9)$

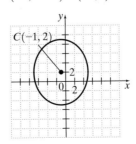

13. $\dfrac{x^2}{4} - \dfrac{y^2}{9} = 1$

$\dfrac{x^2}{2^2} - \dfrac{y^2}{3^2} = 1$

$a = 2, b = 3$

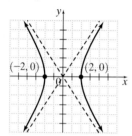

17. $x^2 - 4y^2 = 16$

$\dfrac{x^2}{16} - \dfrac{y^2}{4} = 1$

$\dfrac{x^2}{4^2} - \dfrac{y^2}{2^2} = 1$

$a = 4, b = 2$

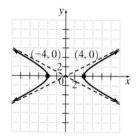

21. $(2x^3)(-4x^2) = 2(-4)x^{3+2}$
$\qquad\qquad\qquad = -8x^5$

25. $x^2 + y^2 = 25$

$\dfrac{x^2}{25} + \dfrac{y^2}{25} = \dfrac{25}{25}$

$\dfrac{x^2}{25} + \dfrac{y^2}{25} = 1$

It resembles an ellipse.
An ellipse is a circle when $a = b$.

29. $A: d = 6$
$B: d = 2$
$C: d = 5$
$D: d = 5$
$E: d = 9$
$F: d = 6$
$G: d = 4$
$H: d = 12$

33. They are greater than 1.

Exercise Set 9.3

1. $f(x) = |x| + 3$
Shift $y = |x|$ up 3 units.

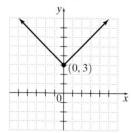

5. $f(x) = |x - 4|$
Shift $y = |x|$ to the right 4 units.

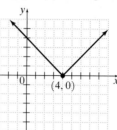

9. $f(x) = \sqrt{x - 2} + 3$
Shift $y = \sqrt{x}$ to the right 2 units and up 3 units.

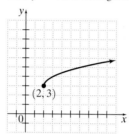

13. $f(x) = \sqrt{x + 1} + 1$
Shift $y = \sqrt{x}$ to the left one unit and up one unit.

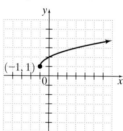

17. $\begin{array}{r} x + y = 6 \\ x - y = 10 \\ \hline 2x = 16 \end{array}$

$x = 8$

Replace x with 8 in the first equation.
$8 + y = 6$
$\quad y = -2$
The solution is $(8, -2)$.

21. $f(x) = x^3$

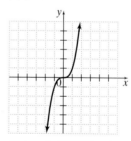

25. $f(x) = 5\sqrt{x - 20} + 1$
For domain, use
$x - 20 \geq 0$
$\qquad x \geq 20$
$\{x | x \geq 20\}$

29. $g(x) = 9 - \sqrt{x + 103}$
For domain, use
$x + 103 \geq 0$
$\qquad x \geq -103$
$\{x | x \geq -103\}$

Exercise Set 9.4

1. $\begin{cases} x^2 + y^2 = 25 \\ 4x + 3y = 0 \end{cases}$

Solve equation 2 for y.

$3y = -4x$

$y = \dfrac{-4x}{3}$

Substitute.

$x^2 + \left(-\dfrac{4x}{3}\right)^2 = 25$

$x^2 + \dfrac{16x^2}{9} = 25$

$\dfrac{25}{9}x^2 = 25$

$\dfrac{x^2}{9} = 1$

$x^2 = 9$

$x = \pm\sqrt{9} = \pm 3$

$x = 3:\quad y = -\dfrac{4}{3}(3) = -4$

$x = -3:\quad y = -\dfrac{4}{3}(-3) = 4$

The solution set is $\{(3, -4), (-3, 4)\}$.

5. $\begin{cases} y^2 = 4 - x \\ x - 2y = 4 \end{cases}$

$-2y = 4 - x$

Substitute.

$y^2 = -2y$

$y^2 + 2y = 0$

$y(y + 2) = 0$

$y = 0$ or $y + 2 = 0$

$\qquad\qquad y = -2$

$y = 0:\quad x - 2(0) = 4$

$\qquad\qquad x = 4$

$y = -2:\quad x - 2(-2) = 4$

$\qquad\qquad x + 4 = 4$

$\qquad\qquad x = 0$

The solution set is $\{(4, 0), (0, -2)\}$.

9. $\begin{cases} x^2 + 2y^2 = 2 \\ x - y = 2 \end{cases}$

$x = y + 2$

Substitute.

$(y + 2)^2 + 2y^2 = 2$

$y^2 + 4y + 4 + 2y^2 = 2$

$3y^2 + 4y + 4 = 2$

$3y^2 + 4y + 2 = 0$

$b^2 - 4ac = 4^2 - 4(3)(2)$

$\qquad\qquad = 16 - 24 = -8 < 0$

Therefore, no real solutions exits.

The solution set is $\varnothing$.

13. $\begin{cases} y = x^2 \\ 3x + y = 10 \end{cases}$

Substitute.

$3x + x^2 = 10$

$x^2 + 3x - 10 = 0$

$(x + 5)(x - 2) = 0$

$x + 5 = 0 \quad$ or $\quad x - 2 = 0$

$x = -5 \qquad$ or $\qquad x = 2$

$x = -5:\quad y = (-5)^2 = 25$

$x = 2:\qquad y = 2^2 = 4$

The solution set is $\{(-5, 25), (2, 4)\}$.

17. $\begin{cases} y = x^2 - 4 \\ y = x^2 - 4x \end{cases}$

Substitute.

$x^2 - 4 = x^2 - 4x$

$\qquad -4 = -4x$

$\qquad\quad x = 1$

$\qquad\quad y = 1^2 - 4 = -3$

The solution set is $\{(1, -3)\}$.

21. $\begin{cases} x^2 + y^2 = 1 \\ x^2 + (y + 3)^2 = 4 \end{cases}$

Subtract equation 1 from equation 2.

$(y + 3)^2 - y^2 = 3$

$y^2 + 6y + 9 - y^2 = 3$

$\qquad\quad 6y + 9 = 3$

$\qquad\qquad 6y = -6$

$\qquad\qquad y = -1$

Substitute back.

$x^2 + (-1)^2 = 1$

$x^2 + 1 = 1$

$\quad x^2 = 0$

$\quad x = 0$

The solution set is $\{(0, -1)\}$.

25. $\begin{cases} 3x^2 + y^2 = 9 \\ 3x^2 - y^2 = 9 \end{cases}$

Subtract.

$2y^2 = 0$

$y^2 = 0$

$y = 0$

Substitute back.

$3x^2 + 0 = 9$

$3x^2 = 9$

$x^2 = 3$

$x = \pm\sqrt{3}$

The solution set is $\{(\sqrt{3}, 0), (-\sqrt{3}, 0)\}$.

29. $\begin{cases} x^2 + y^2 = 36 \\ y = \dfrac{1}{6}x^2 - 6 \end{cases}$

$y + 6 = \dfrac{1}{6}x^2$

$x^2 = 6(y + 6)$

Substitute.

$6(y + 6) + y^2 = 36$

$6y + 36 + y^2 = 36$

$6y + y^2 = 0$

$y(6 + y) = 0$

$y = 0$ or $6 + y = 0$

$\qquad\qquad y = -6$

$y = 0: x^2 + 0^2 = 36$

$\qquad\quad x^2 = 36$

$\qquad\quad x = \pm 6$

$y = -6: x^2 + (-6)^2 = 36$

$\qquad\quad x^2 + 36 = 36$

$\qquad\quad x^2 = 0$

$\qquad\quad x = 0$

The solution set is $\{(6, 0), (-6, 0), (0, -6)\}$.

33. $x > -3$

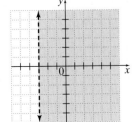

37. $\begin{cases} x^2 + y^2 = 130 \\ x^2 - y^2 = 32 \end{cases}$

Add.

$2x^2 = 162$

$x^2 = 81$

$x = \pm 9$

Substitute back.

$9^2 + y^2 = 130 \qquad (-9)^2 + y^2 = 130$

$\qquad y^2 = 49 \qquad\qquad\quad y^2 = 49$

$\qquad\quad y = \pm 7 \qquad\qquad\quad y = \pm 7$

The numbers are 9 and 7, 9 and -7, -9 and 7, or -9 and -7.

41. $p = -0.01x^2 - 0.2x + 9$

$p = 0.01x^2 - 0.1x + 3$

$-0.01x^2 - 0.2x + 9 = 0.01x^2 - 0.1x + 3$

$0 = 0.02x^2 + 0.1x - 6$

$0 = x^2 + 5x - 300$

$0 = (x + 20)(x - 15)$

$x + 20 = 0 \qquad$ or $\quad x - 15 = 0$

$\qquad x = -20 \quad$ or $\qquad x = 15$

Disregard the negative

$p = -0.01(15)^2 - 0.2(15) + 9$

$p = 3.75$

15 thousand compact discs; price, \$3.75

45. $\begin{cases} y = x^2 + 2 \\ y = -x^2 + 4 \end{cases}$

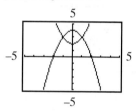

Exercise Set 9.5

1. $y < x^2$

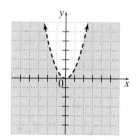

5. $\dfrac{x^2}{4} - y^2 < 1$

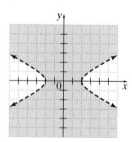

9. $x^2 + y^2 \le 9$

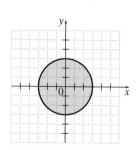

13. $\dfrac{x^2}{4} + \dfrac{y^2}{9} \le 1$

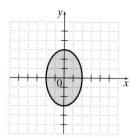

17. $y < (x - 2)^2 + 1$

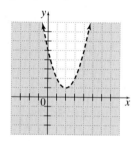

21. $\begin{cases} 4x + 3y \ge 12 \\ x^2 + y^2 < 16 \end{cases}$

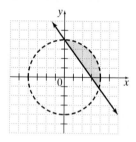

25. $\begin{cases} y > x^2 \\ y \ge 2x + 1 \end{cases}$

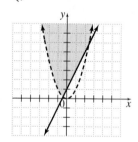

29. $\begin{cases} \dfrac{x^2}{4} + \dfrac{y^2}{9} \ge 1 \\ x^2 + y^2 \ge 4 \end{cases}$

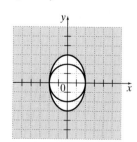

33. $\begin{cases} x + y \geq 1 \\ 2x + 3y < 1 \\ x > -3 \end{cases}$

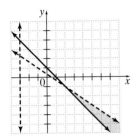

37. This is not a function because a vertical line can cross the graph in two places.

41. answers may vary

Chapter 9 Test

1. $(-6, 3)$ and $(-8, -7)$

$d = \sqrt{(-8 + 6)^2 + (-7 - 3)^2}$

$d = \sqrt{(-2)^2 + (-10)^2}$

$d = \sqrt{4 + 100}$

$d = \sqrt{104}$

$d = 2\sqrt{26}$ units

5. $x^2 + y^2 = 36$ or $(x - 0)^2 + (y - 0)^2 = 6^2$

Circle: $C(0, 0), r = 6$

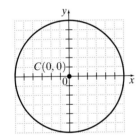

9. $\quad x^2 + y^2 + 6x = 16$

$\quad\quad x^2 + 6x + y^2 = 16$

$(x^2 + 6x + 9) + y^2 = 16 + 9$

$\quad\quad (x + 3)^2 + y^2 = 5^2$

Circle: $C(-3, 0), r = 5$

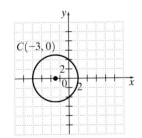

13. $\begin{cases} x^2 + y^2 = 169 \\ 5x + 12y = 0 \end{cases}$

$12y = -5x$

$y = -\dfrac{5x}{12}$

Substitute.

$x^2 + \left(-\dfrac{5x}{12}\right)^2 = 169$

$x^2 + \dfrac{25x^2}{144} = 169$

$\dfrac{169x^2}{144} = 169 = \dfrac{x^2}{144} = 1$

$x^2 = 144$ so $x = \pm 12$

Substitute back.

$x = 12: \quad y = -\dfrac{5}{12}(12) = -5$

$x = -12: \quad y = -\dfrac{5}{12}(-12) = 5$

$\{(12, -5), (-12, 5)\}$

17. $\begin{cases} 2x + 5y \geq 10 \\ y \geq x^2 + 1 \end{cases}$ First graph.

$\begin{cases} 2x + 5y = 10 \\ y = x^2 + 1 \end{cases}$ or

$\begin{cases} y = -\dfrac{2}{5}x + 2 \\ y = 1 \cdot (x - 0)^2 + 1 \end{cases}$

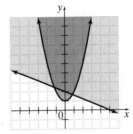

21. Graph B; vertex in third quadrant, opens to the right.

Chapter 10

Exercise Set 10.1

1. a. $(f + g)(x) = x - 7 + 2x + 1$

$\quad\quad\quad\quad\quad = 3x - 6$

b. $(f - g)(x) = x - 7 - (2x + 1)$

$\quad\quad\quad\quad\quad = x - 7 - 2x - 1$

$\quad\quad\quad\quad\quad = -x - 8$

c. $(f \cdot g)(x) = (x - 7)(2x + 1)$

$\quad\quad\quad\quad\quad = 2x^2 - 13x - 7$

d. $\left(\dfrac{f}{g}\right)(x) = \dfrac{x - 7}{2x + 1}$, where $x \neq -\dfrac{1}{2}$

5. a. $(f + g)(x) = \sqrt{x} + x + 5$

b. $(f \circ g)(x) = \sqrt{x} - x - 5$

c. $(f \circ g)(x) = \sqrt{x}(x + 5)$

$\quad\quad\quad\quad\quad = x\sqrt{x} + 5\sqrt{x}$

d. $\left(\dfrac{f}{g}\right)(x) = \dfrac{\sqrt{x}}{x + 5}$, where $x \neq -5$

9. $(f \circ g)(2) = f(g(2))$

$\quad\quad\quad\quad = f(-4)$

$\quad\quad\quad\quad = (-4)^2 - 6(-4) + 2$

$\quad\quad\quad\quad = 16 + 24 + 2$

$\quad\quad\quad\quad = 42$

13. $(g \circ h)(0) = g(h(0))$
$= g(0)$
$= -2(0)$
$= 0$

17. $(f \circ g)(x) = f(g(x))$
$= f(x + 7)$
$= 2(x + 7) - 3$
$= 2x + 14 - 3$
$= 2x + 11$
$(g \circ f)(x) = g(f(x))$
$= g(2x - 3)$
$= (2x - 3) + 7$
$= 2x + 4$

21. $(f \circ g)(x) = f(g(x))$
$= f(-5x + 2)$
$= \sqrt{-5x + 2}$
$(g \circ f)(x) = g(f(x))$
$= g(\sqrt{x})$
$= -5\sqrt{x} + 2$

25. $F(x) = (h \circ f)(x)$
$= h(f(x))$
$= h(3x)$
$= (3x)^2 + 2$
$= 9x^2 + 2$

29. answers may vary

33. answers may vary

37. $x = 3y$
$\dfrac{x}{3} = \dfrac{3y}{3}$
$\dfrac{x}{3} = y$
$y = \dfrac{x}{3}$

41. $P(x) = R(x) - C(x)$

Exercise Set 10.2

1. $f = \{(-1, -1), (1, 1), (0, 2), (2, 0)\}$ is a one-to-one function.
$f^{-1} = \{(-1, -1), (1, 1), (2, 0), (0, 2)\}$

5. $f = \{(11, 12), (4, 3), (3, 4), (6, 6)\}$ is a one-to-one function.
$f^{-1} = \{(12, 11), (3, 4), (4, 3), (6, 6)\}$

9. This function is one-to-one.

Rank (input)	1	48	14	22	36
State (output)	CA	AK	IN	LA	NM

13. $f(x) = x^3 + 2$

a. $f(-1) = (-1)^3 + 2 = 1$

b. $f^{-1}(1) = -1$

17. The graph does not represent a one-to-one function because it does not pass the horizontal line test.

21. The graph does not represent a one-to-one function because it does not pass the horizontal line test.

25. $f(x) = 2x - 3$
$y = 2x - 3$
$x = 2y - 3$
$2y = x + 3$

$y = \dfrac{x + 3}{2}$
$f^{-1}(x) = \dfrac{x + 3}{2}$

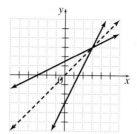

29. $f(x) = x^3$
$y = x^3$
$x = y^3$
$y = \sqrt[3]{x}$
$f^{-1}(x) = \sqrt[3]{x}$

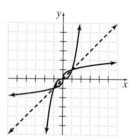

33. $f(x) = \sqrt[3]{x}$
$y = \sqrt[3]{x}$
$x = \sqrt[3]{y}$
$x^3 = y$
$f^{-1}(x) = x^3$

37. $f(x) = (x + 2)^3$
$y = (x + 2)^3$
$x = (y + 2)^3$
$\sqrt[3]{x} = y + 2$
$\sqrt[3]{x} - 2 = y$
$f^{-1}(x) = \sqrt[3]{x} - 2$

41.

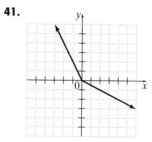

45. $16^{3/4} = (16^3)^{1/4} = (4096)^{1/4} = 8$

49. $f(x) = 3^x$
$f(2) = 3^2$
$= 9$

53. a. $\left(-2, \dfrac{1}{4}\right), \left(-1, \dfrac{1}{2}\right), (0, 1), (1, 2), (2, 5)$

b. $\left(\dfrac{1}{4}, -2\right), \left(\dfrac{1}{2}, -1\right), (1, 0), (2, 1), (5, 2)$

c.

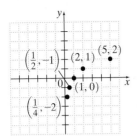

d.

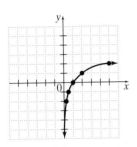

57. $f(x) = \sqrt[3]{x+3}$
$y = \sqrt[3]{x+3}$
$x = \sqrt[3]{y+3}$
$x^3 = y + 3$
$x^3 - 3 = y$
$f^{-1}(x) = x^3 - 3$

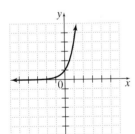

Exercise Set 10.3

1. $y = 4^x$

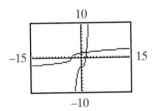

5. $y = \left(\dfrac{1}{4}\right)^x$

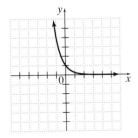

9. $y = -2^x$

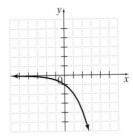

13. $y = -\left(\dfrac{1}{4}\right)^x$

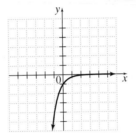

17. $3^x = 27$
$3^x = 3^3$
$x = 3$
The solution set is $\{3\}$.

21. $32^{2x-3} = 2$
$(2^5)^{2x-3} = 2^1$
$10x - 15 = 1$
$10x = 16$
$x = \dfrac{8}{5}$
The solution set is $\left\{\dfrac{8}{5}\right\}$.

25. $4^x = 8$
$(2^2)^x = 2^3$
$2^{2x} = 2^3$
$2x = 3$
$x = \dfrac{3}{2}$
The solution set is $\left\{\dfrac{3}{2}\right\}$.

29. $81^{x-1} = 27^{2x}$
$(3^4)^{x-1} = (3^3)^{2x}$
$3^{4x-4} = 3^{6x}$
$4x - 4 = 6x$
$-4 = 2x$
$x = -2$
The solution set is $\{-2\}$.

33. $y = 260(2.7)^{0.025t}, t = 10$
$y = 260(2.7)^{0.025(10)}$
$y \approx 333$
Approximately 333 bison will remain after 10 years.

37. $y = 25.759(1.277)^x$
$t = 2007 - 1994 = 13$
$y = 25.759(1.277)^{13} \approx 618.6$
There will be approximately 618.6 million cellular phone users in 2007.

41. $5x - 2 = 18$
$5x = 20$
$x = 4$
The solution set is {4}.

45. $f(x) = \left(\dfrac{1}{2}\right)^x$

$b = \dfrac{1}{2}, 0 < b < 1$
graph C

49. answers may vary

53. At $t = 120$, $y \approx 18.62$.

Approximately 18.62 lb of uranium will be available after 120 days.

Exercise Set 10.4

1. $\log_6 36 = 2$
$6^2 = 36$

5. $\log_{10} 1000 = 3$
$10^3 = 1000$

9. $\log_e \dfrac{1}{e^2} = -2$
$e^{-2} = \dfrac{1}{e^2}$

13. $2^4 = 16$
$\log_2 16 = 4$

17. $e^3 = x$
$\log_e x = 3$

21. $4^{-2} = \dfrac{1}{6}$
$\log_4 \dfrac{1}{16} = -2$

25. $\log_2 8 = 3$ since $2^3 = 8$

29. $\log_{25} 5 = \dfrac{1}{2}$ since $25^{1/2} = 5$

33. $\log_6 1 = 0$ since $6^0 = 1$

37. $\log_{10} 100 = 2$ since $10^2 = 100$

41. $\log_3 81 = 4$ since $3^4 = 81$

45. answers may vary

49. $\log_3 x = 4$
$x = 3^4$
$= 81$
The solution set is {81}.

53. $\log_2 \dfrac{1}{8} = x$

$2^x = \dfrac{1}{8}$
$2^x = 2^{-3}$
$x = -3$
The solution set is {−3}.

57. $\log_8 x = \dfrac{1}{3}$
$x = 8^{1/3}$
$= 2$
The solution set is {2}.

61. $\log_{3/4} x = 3$
$\left(\dfrac{3}{4}\right)^3 = x$
$x = \dfrac{3^3}{4^3}$
$= \dfrac{27}{64}$
The solution set is $\left\{\dfrac{27}{64}\right\}$.

65. $\log_5 5^3 = 3$

69. $\log_9 9 = 1$

73. $f(x) = \log_{1/4} x$ or $y = \log_{1/4} x$
$y = 0$:
$0 = \log_{1/4} x$
$x = \left(\dfrac{1}{4}\right)^0 = 1$ so $(1, 0)$ is the x-intercept.
$x = 0$:
$y = \log_{1/4} 0$ which is not defined. No y-intercept exists.

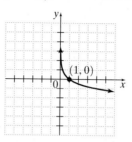

77. $f(x) = \log_{1/6} x$ or $y = \log_{1/6} x$
$x = 0$:
$y = \log_{1/6} 0$ is not defined so there is no y-intercept.
$y = 0$:
$0 = \log_{1/6} x$
$x = \left(\dfrac{1}{6}\right)^0 = 1$ so $(1, 0)$ is the x-intercept.

81. $\dfrac{x^2 - 8x + 16}{2x - 8} = \dfrac{(x - 4)(x - 4)}{2(x - 4)} = \dfrac{x - 4}{2}$

85. $y = \left(\dfrac{1}{3}\right)^x$; $y = \log_{1/3} x$

$x = 0$: $y = \left(\dfrac{1}{3}\right)^0 = 1$ so $(0, 1)$ is the y-intercept of $y = \left(\dfrac{1}{3}\right)^x$,
and hence the x-intercept of $y = \log_{1/3} x$.

$y = 0$: $0 = \left(\dfrac{1}{3}\right)^x$ has no solution so $y = \left(\dfrac{1}{3}\right)^x$ has no
x-intercept: hence $y = \log_{1/3} x$ has no y-intercept.

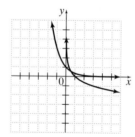

89. $\log_3 10$ is between 2 and 3 because $3^2 = 9$ and $3^3 = 27$

Exercise Set 10.5

1. $\log_5 2 + \log_5 7 = \log_5 (2 \cdot 7) = \log_5 14$

5. $\log_{10} 5 + \log_{10} 2 + \log_{10}(x^2 + 2)$
$= \log_{10}[5 \cdot 2(x^2 + 2)]$
$= \log_{10} (10x^2 + 20)$

9. $\log_2 x - \log_2 y = \log_2\left(\dfrac{x}{y}\right)$

13. $\log_3 x^2 = 2 \log_3 x$

17. $\log_5 \sqrt{y} = \log_5 y^{1/2}$
$= \dfrac{1}{2} \log_5 y$

21. $3\log_5 x + 6\log_5 z = \log_5 x^3 + \log_5 z^6$
$$= \log_5(x^3 z^6)$$

25. $\log_4 2 + \log_4 10 - \log_4 5 = \log_4 2 \cdot 10 - \log_4 5$
$$= \log_4\left(\frac{20}{5}\right)$$
$$= \log_4 4$$
$$= 1$$

29. $3\log_4 2 + \log_4 6 = \log_4 2^3 + \log_4 6$
$$= \log_4 8 + \log_4 6$$
$$= \log_4(8 \cdot 6)$$
$$= \log_4 48$$

33. $2\log_8 x - \dfrac{2}{3}\log_8 x + 4\log_8 x = \left(2 - \dfrac{2}{3} + 4\right)\log_8 x$
$$= \frac{16}{3}\log_x x$$
$$= \log_8 x^{16/3}$$

37. $\log_2\left(\dfrac{x^3}{y}\right) = \log_2 x^3 - \log_2 y$
$$= 3\log_2 x - \log_2 y$$

41. $\log_7\left(\dfrac{5x}{4}\right) = \log_7 5x - \log_7 4$
$$= \log_7 5 + \log_7 x - \log_7 4$$

45. $\log_6\dfrac{x^2}{x+3} = \log_6 x^2 - \log_6(x+3)$
$$= 2\log_6 x - \log_6(x+3)$$

49. $\log_b 15 = \log_b(5 \cdot 3)$
$$= \log_b 5 + \log_b 3 = 0.7 + 0.5 = 1.2$$

53. $\log_b 8 = \log_b 2^3 = 3\log_b 2 = 3(0.43) = 1.29$

57. $\log_b\sqrt{\dfrac{2}{3}} = \log_b\left(\dfrac{2}{3}\right)^{1/2}$
$$= \frac{1}{2}\log_b\frac{2}{3}$$
$$= \frac{1}{2}(\log_b 2 - \log_b 3)$$
$$= \frac{1}{2}(0.43 - 0.68)$$
$$= \frac{1}{2}(-0.25)$$
$$= -0.125$$

61. $\log_{10}\dfrac{1}{10} = x$
$$10^x = \frac{1}{10}$$
$$10^{-1} = \frac{1}{10}$$
$$\log_{10}\frac{1}{10} = -1$$

65. $\log_3(x+y) = \log_3 x + \log_3 y$
false

69. $(\log_3 6)(\log_3 4) = \log_3 24$
false

Exercise Set 10.6

1. $\log 8 \approx 0.9031$

5. $\ln 2 \approx 0.6931$

9. $\log 12.6 \approx 1.1004$

13. $\log 41.5 \approx 1.6180$

17. $\log 100 = \log 10^2 = 2$

21. $\ln e^2 = 2$

25. $\log 10^3 = 3$

29. $\log 0.0001 = \log 10^{-4}$
$$= -4$$

33. $\log x = 1.3$
$$x = 10^{1.3}$$
$$\approx 19.9526$$

37. $\ln x = 1.4$
$$x = e^{1.4}$$
$$\approx 4.0552$$

41. $\log x = 2.3$
$$x = 10^{2.3}$$
$$\approx 199.5262$$

45. $\log(2x+1) = -0.5$
$$2x + 1 = 10^{-0.5}$$
$$2x = 10^{-0.5} - 1$$
$$x = \frac{10^{-0.5} - 1}{2}$$
$$\approx -0.3419$$

49. $A = Pe^{rt}, t = 12, P = 1400,$ and $r = 0.08$
$A = 1400e^{(0.08)12} = 1400e^{0.96} \approx 3656.38$
Dana has \$3656.38 after 12 years.

53. $\log_2 3 = \dfrac{\log 3}{\log 2} \approx 1.5850$

57. $\log_4 9 = \dfrac{\log 9}{\log 4} \approx 1.5850$

61. $\log_8 6 = \dfrac{\log 6}{\log 8} \approx 0.8617$

65. $2x + 3y = 6x$
$$3y = 4x$$
$$x = \frac{3y}{4}$$

69. $f(x) = e^x$

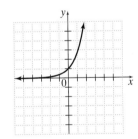

73. answers may vary

77. $R = \log\left(\dfrac{a}{T}\right) + B$
$$= \log\left(\frac{400}{2.6}\right) + 3.1$$
$$\approx 5.3$$

Exercise Set 10.7

1. $3^x = 6$
$$x = \log_3 6$$
$$= \frac{\log 6}{\log 3} \approx 1.6309$$
The solution set is $\left\{\dfrac{\log 6}{\log 3}\right\}$ or approximately $\{1.6309\}$.

5. $2^{x-3} = 5$
$$x - 3 = \log_2 5$$
$$x = 3 + \log_2 5$$

$$= 3 + \frac{\log 5}{\log 2}$$
$$= 5.3219$$

The solution set is $\left\{3 + \frac{\log 5}{\log 2}\right\}$ or approximately $\{5.3219\}$.

9. $\quad 4^{x+7} = 3$
$$x + 7 = \log_4 3$$
$$x = \log_4 3 - 7$$
$$= \frac{\log 3}{\log 4} - 7$$
$$\approx -6.2075$$

The solution set is $\left\{\frac{\log 3}{\log 4} - 7\right\}$ or approximately $\{-6.2075\}$.

13. $e^{6x} = 5$
$$6x = \ln 5$$
$$x = \frac{1}{6}\ln 5$$
$$\approx 0.2682$$

The solution set is $\left\{\frac{1}{6}\ln 5\right\}$ or approximately $\{0.2682\}$.

17. $\log_4 2 + \log_4 x = 0$
$$\log_4(2x) = 0$$
$$2x = 4^0$$
$$2x = 1$$
$$x = \frac{1}{2}$$

The solution set is $\left\{\frac{1}{2}\right\}$.

21. $\log_4 x + \log_4(x + 6) = 2$
$$\log_4 x(x + 6) = 2$$
$$x(x + 6) = 4^2$$
$$x^2 + 6x = 16$$
$$x^2 + 6x - 16 = 0$$
$$(x + 8)(x - 2) = 0$$
$$x + 8 = 0 \quad \text{or} \quad x - 2 = 0$$
$$x = -8 \quad \text{or} \quad x = 2$$

We discard -8 as extraneous. The solution set is $\{2\}$.

25. $\log_3(x - 2) = 2$
$$x - 2 = 3^2$$
$$x - 2 = 9$$
$$x = 11$$

The solution set is $\{11\}$.

29. $\log_2 x + \log_2(3x + 1) = 1$
$$\log_2 x(3x + 1) = 1$$
$$x(3x + 1) = 2^1$$
$$3x^2 + x = 2$$
$$3x^2 + x - 2 = 0$$
$$(3x - 2)(x + 1) = 0$$
$$3x - 2 = 0 \quad \text{or} \quad x + 1 = 0$$
$$3x = 2 \quad \text{or} \quad x = -1$$
$$x = \frac{2}{3}$$

Discard -1 as extraneous.

The solution set is $\left\{\frac{2}{3}\right\}$.

33. $y = y_0 e^{0.027t}$
$y_0 = 5,700,000$ and $t = 7$
$y = 5,700,000e^{0.027(7)}$
$$\approx 6,885,833$$

In 2008, there will be approximately 6,885,833 people in Paraguay.

37. $A = P\left(1 + \frac{r}{n}\right)^{nt}, P = 600.$
$A = 2(600) = 1200, r = 0.07,$ and $n = 12$
$$1200 = 600\left(1 + \frac{0.07}{12}\right)^{12t}$$
$$2 \approx (1.00583)^{12t}$$
$$12t = \log_{1.00583}(2)$$
$$t = \frac{1}{12}\log_{1.00583}(2)$$
$$= \left(\frac{1}{12}\right)\frac{\log 2}{\log (1.00583)} \approx 9.9$$

It would take approximately 9.9 years for the $600 to double.

41. $A = P\left(1 + \frac{r}{n}\right)^{nt}, P = 1000.$
$A = 2(1000) = 2000, r = 0.08,$ and $n = 2.$
$$2000 = 1000\left(1 + \frac{0.08}{2}\right)^{2t}$$
$$2 = (1.04)^{2t}$$
$$2t = \log_{1.04} 2 = \frac{\log 2}{\log 1.04}$$
$$t = \left(\frac{1}{2}\right)\frac{\log 2}{\log 1.04} \approx 8.8$$

It would take approximately 8.8 years for the $1000 to double.

45. $P = 14.7e^{-0.21x}, x = 1$
$$= 14.7e^{-0.21(1)} = 14.7e^{-0.21} \approx 11.9$$

The average atmospheric pressure of Denver is approximately $11.9\,\text{lb/in}^2$.

49. $t = \frac{1}{c}\ln\left(\frac{A}{A - N}\right)$
$$t = \frac{1}{0.09}\ln\left(\frac{75}{75 - 50}\right)$$
$$t = \frac{1}{0.09}\ln(3)$$
$$t \approx 12.21$$

It will take about 12 weeks.

53. $\dfrac{x^2 - y + 2z}{3x} = \dfrac{(-2)^2 - (0) + 2(3)}{3(-2)}$
$$= \frac{4 + 6}{-6}$$
$$= \frac{10}{-6}$$
$$= -\frac{5}{3}$$

57. $y = y_0 e^{kt}$
$$5,130,632 = 3,665,228e^{k \cdot 10}$$
$$k = \frac{1}{10}\ln\frac{5,130,632}{3,665,228}$$
$$k \approx 0.0336$$

The growth rate is about 3.4%.

61. $\{6.93\}$

Chapter 10 Test

1. $(f \cdot g)(x) = f(x) \cdot g(x)$
$$= x(2x - 3)$$
$$= 2x^2 - 3x$$

5. $(g \circ h)(x) = g(h(x))$
$$= g(x^2 - 6x + 5)$$
$$= x^2 - 6x + 5 - 7$$
$$= x^2 - 6x - 2$$

9. $y = 6 - 2x$
$f(x) = -2x + 6$ so the function is one-to-one
$f^{-1}(x) = \dfrac{x - 6}{-2}$ or $f^{-1}(x) = \dfrac{-x + 6}{2}$

13. $\log_5 x + 3 \log_5 x - \log_5(x + 1)$
$= 4 \log_5 x - \log_5(x + 1)$
$= \log_5 x^4 - \log_5(x + 1)$
$= \log_5 \dfrac{x^4}{x + 1}$

17. $8^{x-1} = 8^{-2}$
$x - 1 = -2$
$x = -1$
The solution set is $\{-1\}$.

21. $\log_8(3x - 2) = 2$
$3x - 2 = 8^2$
$3x - 2 = 64$
$3x = 66$
$x = \dfrac{66}{3}$
$= 22$
The solution set is $\{22\}$.

25. $y = \left(\dfrac{1}{2}\right)^x + 1$

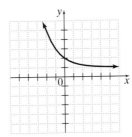

29. $y = y_0 e^{kt}, y_0 = 57{,}000, k = 0.026,$ and $t = 5$
$y = 57{,}000 e^{0.026(5)}$
$= 57{,}000 e^{0.13} \approx 64{,}913$
There will be approximately 64,913 prairie dogs 5 years from now.

SUBJECT INDEX

A

Abacus, 358
Abscissa, 169, 204
Absolute value, 29, 68, 139
Absolute value bars, 140, 506
Absolute value equations, 154–55
    solving, 139–41, 142, 143
Absolute value inequalities, 154–55
    solving, 141–44
Absolute value property, 140
Activities. *See also* Problem solving
    calculating length and period of
      pendulum, 570
    estimating population sizes, 485–86
    finding largest area, 406
    lightning strike locations, 310
    linear and quadratic models, 656
    measuring slope, 238
    modeling conic sections, 718
    modeling temperature, 802
    room redecorating, 150
    searching for patterns, 66
Addition, 9
    associative property of, 23, 68
    commutative property of, 22, 68
    of complex numbers, 562–63, 573
    of functions, 737, 803
    of polynomials, 329, 407
    of radical expressions, 529–30, 572
    of rational expressions, 488
    of rational expressions with different
      denominators, 437–39
    of rational expressions with same
      denominator, 435
    of real numbers, 29–31, 68
Addition method (or elimination method), 485
    solving system of equations using, 264–66
Addition property of equality, 82, 83, 84,
    151, 264
Addition property of inequality, 116–18, 153
Additive identity, 20
Additive inverse (or opposite), 21, 330
Algebra, 1, 262
    history behind, 358
Algebraic expressions, 1, 9, 67, 69
    and combining like terms, 43–44
    evaluating, 42–43
    phrases written as, 13
    writing, 91–92
Algebraic techniques
    elimination (or addition) method, 264
    substitution method, 262
Algebra of functions, 737, 803
Algorithms, source of English language word
    for, 358
Al-jabr wa'l-muqabala (Al-Khwarizmi), 358
Al-Khwarizmi, Muhammad ibn Musa, 358
Americans with Disabilities Act, 238
Amperage (A), 122
Angle measures, finding, 286–87
Annual consumption of cigarettes, finding, 121
Annual reports, 622
AOL, 92
Apollonius of Perga, 732
Applications. *See also* Problem solving
    point-slope form in, 208–9
    with rational functions, 428–29

Approximations, decimal, 504
Archimedes of Syracuse, 732
Archimedes' Principle of Buoyancy, 732
Architecture, conic sections
    used in, 669
Area
    finding largest, 406
    of rectangle, 14, 105
    of trapezoid, 106
Arithmetic, 1
Assets, 622
Associative property, 22
    of addition, 23, 68
    of multiplication, 23, 68
    and multiplication of monomials, 339
Astronomy
    conic sections used in, 669
    logarithms used in, 770
Asymptotes, of hyperbola, 690
Axis of symmetry, 398, 633
    in parabolas, 671, 672, 673, 674, 719

B

Babylonians (ancient), contributions
    by, in solving quadratic
      equations, 630
Back substitution, 612
Base, 33
Base e, logarithms to, 786–87
Bell, Alexander Graham, 79
Bell Telephone Company, 79
Best fit equation, 244
Binomials, 327, 328
    multiplying, 340–41
    squaring, 341
    writing square of, 588
Biology, and problem solving, 79
BLS. *See* U.S. Bureau of Labor Statistics
Body-mass index (BMI), 473
Boston, Massachusetts, America's first
    subway in, 1
Boundary line, 231, 232, 233, 243
Boyle's law, and inverse variation, 478
Braces, 42, 116
Brackets, 42, 116
Brahmagupta, 630
Break-even point, finding, 287–88
Briggs, Henry, 770
Business. *See* Focus on business and career

C

CALCULATE VALUE feature, in
    grapher, 430
Calculator explorations
    graphing equations on same set of
      axes, 192
    scientific notation, 54
Calculus, 732
Capture-recapture method, 485
Career. *See* Focus on business and career
Cartesian coordinate system, 169, 204,
    239, 674
Celsius degrees
    converting Fahrenheit degrees to, 42–43
    Fahrenheit degrees converted to, 738
Centennial Exposition, Philadelphia
    (1876), 79

Center
    of circle, 676, 678–79, 720
    of ellipse, 687
    of hyperbola, 689
Center of gravity, 210
Center of mass, 210
Change of base formula, using, 787
Charles's law, 481
Checking math work, 34
Chernobyl nuclear accident, 759, 760
Circles, 669, 671, 720
    circumference of, 105
    finding center and radius of, 678–79
    graphing, 676–78
    writing equations of, 678
Circular cylinder
    surface area of, 14
    volume of, 14
Circumference, of circle, 105
Class attendance, 34
CNN Financial Network Web site, 38
Coefficients, 327
Column, in matrix, 297
Combined variation, solving problems
    involving, 479–80
Combining like terms, 43–44, 83–84,
    328–29
Common denominator, adding/subtracting
    rational expression with, 435
Common logarithms, 785, 806
Common logarithms of powers of 10,
    evaluating, 785–86
Common "log" key, on calculator, 785
Commutative property
    of addition, 22, 68
    of multiplication, 22, 68
    and multiplication of monomials, 339
Completing the square, 599
    contributions by ancient Babylonians
      about, 630
    solving quadratic equation by, 589–91
    solving quadratic equation in x by, 657
Complex conjugates, 563, 564, 573
Complex fractions
    simplifying: method 1, 445–46, 488
    simplifying: method 2, 446–47, 488
Complex numbers, 501, 573
    adding or subtracting, 562–63, 573
    dividing, 563–64, 573
    finding powers of i, 564–65
    history behind, 630
    multiplying, 563, 573
    subsets of, 562
    writing numbers in form bi, 561–62
Complex rational expressions, 445
Composite functions, 803
Compound inequalities, 131, 154
    solving those containing "and," 131–33
    solving those containing "or," 133–34
Compound interest, 592, 795
Compound interest formula, 107, 759, 787
Conic sections
    circles, 669, 672, 676–79, 697, 720
    ellipses, 669, 671, 687–88, 697, 720
    history behind, 732
    hyperbolas, 669, 671, 689–91, 697, 721
    modeling, 718
    parabolas, 669, 671, 697, 719

Conjugates, 538, 539, 540, 572
  complex, 563, 564, 573
Connected mode, in grapher, 430
Consecutive integers, 91
  even, 91
  finding, 96
  odd, 91
Consistent systems, 260, 261, 262, 273, 312
Constant of proportionality, 475, 477, 479
Constant of variation, 475, 477, 479
Constant term (constant), 327
Continuously compounded interest
    formula, 806
Contradictions, 86
Coordinate, 204
Corner point, finding, 304
Correspondence, 217, 219, 242
Cosmoclock Ferris wheel (Japan), 684
Cost and revenue problems, solving, 287–88
Cost of sales, 622
Crater Al-Khwarizmi, 358
Critical thinking, 24, 452
Cross products, 465
Cube, volume of, 59
Cube roots, 34, 537, 571
  finding, 504–5
  symbol for, 532
Cubic function, degree of, 397
Cubic units, 59
Current ratio, 622

D

Daily Value figure, in nutrition labels, 114
Data
  modeling, 244
  pairs, 171
Decibels, 782
Decimal approximations, finding, 504
Decimals
  rational numbers written as, 12
  solving linear equations containing, 85–86
Decimal system, history behind, 358
Degree
  of polynomial, 328
  of term, 327
Denominators, 41
  rationalizing, 537–38, 572
  rationalizing those having two terms,
    538–40
Dependent equations, 261, 262, 266, 273,
    277, 312
Dependent variables, 222
Descartes, René, 169, 204, 532
Descending order, polynomials written
    in, 327
Diagrams
  flowcharts, 616
  scatter, 171, 172, 244
Die Cross (Rudolff), 532
Difference of cubes, factoring, 379–380
Difference of two cubes, 409
Difference of two squares, 409
  factoring, 378–79
Differences
  of complex numbers, 562
  of functions, 737, 803
  of logarithms, 777
Diffusion, 528
Direct variation, 490
  and Hooke's law, 476
  solving problems involving, 475
Discriminant, using, 602
Distance, rate, and time problems, solving,
    467–68

Distance formula, 105, 674, 719
  and graphing hyperbolas, 689
  and radius of circle, 676
Distance saved, calculating, 603
Distributive property, 23, 44, 68
  and addition of like radicals, 529, 572
  combining like terms by using, 328
  for removing parentheses, 83, 86
Dividend, 349
Division, 9
  of complex numbers, 563–64, 573
  of functions, 737, 803
  long, 349, 408
  by polynomial, 349–52, 408
  of polynomial by monomial, 349
  of polynomial by polynomial other than
    monomial, 408
  of rational expressions, 427–28, 487
  of real numbers, 32–34, 68
  synthetic, 352–54, 408
Divisor, 349
Domain, 218, 242, 397
  defined, 217
  finding from graph, 221–22
  of function, 701, 756
  of function in x, 507
  of logarithmic function, 769
  of rational function, 428
Doppler, Christian, 421
Doppler effect, 421
Dot mode, in grapher, 430
Double negative property, 21
Driving speeds, calculating, 615–16

E

E, evaluating logarithms of, 786–87
Earnings per share, 622
Earthlink, 92
Earthquakes, magnitude of, 735, 801
Economics, and problem solving, 79
Eiffel, Alexandre Gustave, 325
Eiffel Tower (Paris, France), 325
Electric bills, 122
Elementary row operations, 297, 298, 300, 315
Elements
  in matrix, 297
  of set, 10, 67
Elimination method (or addition method)
  nonlinear systems solved by, 707–8
  solving system of equations using, 264–66
  solving system of linear equations by, 312
  solving system of linear inequalities
    with, 304
  solving system of three linear equations
    by, 274, 313
Ellipses, 669, 671, 720, 732
  graphing, 687–88
Ellipsis (...), 10, 67
Empire State Building (New York City), 325
Empty set, 10, 275
Energy audit, 122
Engineering
  conic sections used in, 669
  and problem solving, 79
Equal signs, 19, 453
Equations, 79, 151
  best fit, 244
  of circles, 678
  definition of, 81
  dependent, 261, 262, 266, 273, 277, 312
  equal signs in, 453
  equivalent, 81, 82, 151, 277
  finding slope from, 187–88
  independent, 261, 312

  of inverse of function, 746–47
  with no solution, 86
  point-slope form used for writing, 205–6
  quadratic, 387
  rational, 453
  regression, 244
  sentences written as, 19
  slope-intercept form used for writing, 198
  solving form $b^x = b^y$, 757–58
  solving for specified variable, 106
  solving those containing radical
    expressions, 549–52
  solving those containing rational
    expressions, 453, 488
  solving those quadratic in form, 611–13
  in standard form, 241
  writing for parallel and perpendicular
    lines, 206–8
  writing for vertical and horizontal
    lines, 206
Equivalent equations, 81, 82, 151, 277
Equivalent inequalities, 116, 118
Equivalent matrices, 297, 299, 315
Evaluating algebraic expressions, 42–43, 69
Evaluating an expression, 9
Evaluating polynomial functions, 330–31
Even integers, consecutive, 91
Exams
  day of, 316
  performance on, 728
  preparing for and taking, 5–6
  study tips for, 234
Expenses, 362
Exponential equations, 775
  solving, 757, 793
  solving problems modeled by, 758–59, 795
  writing with logarithmic notation, 765
Exponential expressions, operations on, 49
Exponential functions, 735, 743, 804
  defined, 755
  graphing, 755–57
Exponential notation, 775
Exponent rules, for rational exponents, 571
Exponents, 69
  power rules for, 59
  product rule for, 49
  quotient rule for, 50–51
  simplifying expressions containing, 33–34
  summary of rules for, 60
Expressions, simplifying those containing
    exponents, 33–34
Extraneous solutions, 454, 549, 551

F

Factoring, 359
  difference of two squares, 378–79
  perfect square trinomials, 377–78
  polynomial equations solved by, 387–89,
    409
  polynomials, 385
  polynomials by grouping, 409
  quadratic equations solved by, 587
  by special products, 377–80, 409
  by substitution, 371–72
  sum or difference of two cubes, 379–80
  trinomials, 367–71, 409
Factoring by grouping, 360–61
Factoring method, 387
Factors, 33
Fahrenheit degrees
  Celsius degrees converted to, 738
  converting to Celsius degrees, 42–43
FDA. *See* Food and Drug Administration
Fear of failure, 162

Ferris, George, 684
Final exam preparation, 655, 788
Financial ratios, 622
Finite differences, 344
First differences, and linear functions, 344
Flowcharts, 616
Focus (foci)
    of ellipse, 687
    of hyperbola, 689
Focus on business and career
    business terms, 362
    fastest-growing occupations, 440
    financial ratios, 622
    flowcharts, 616
    linear modeling, 244
Focus on history
    Cartesian coordinate system, 204
    conic sections, 732
    development of radical symbol, 532
    evolution of solving quadratic
        equations, 630
    Heron of Alexandria, 508
    invention of logarithms, 770
    Muhammad Al-Khwarizmi, 358
Focus on mathematical connections
    finite differences, 344
    geometry investigations, 14
    perpendicular bisectors, 224
    solving nonlinear systems, 278
Focus on study skills. See also Study skills
        reminder
    critical thinking, 24, 452
    problem solving, 452
Focus on the real world
    center of mass, 210
    diffusion, 528
    energy audit, 122
    linear modeling, 306
    mathematical models, 268
    nutrition labels, 114
    sound intensity, 782
FOIL order, 340–41, 408
Food and Drug Administration, 114
Formulas. See also Symbols
    atmospheric pressure, 799
    change of base, 787
    compound interest, 592, 759, 787
    continuously compounded interest,
        787, 806
    distance, 674
    for factoring perfect trinomials, 377
    midpoint, 674, 720
    pendulum's period, 570
    population, 800
    for problem solving, 107–8, 153
    quadratic, 599
    simple interest, 592
    slope, 186, 187
    solving for specified variables, 105–7, 153
    vertex, 400, 648–49
    weight, 799
Fourth-degree polynomial, 353
Fourth root, symbol for, 532
Fraction bars, 41
Fractions, 423
    clearing equations of, 263, 264
    complex, 445
    solving linear equations containing, 84
French Revolution, 325
Function composition, 738–40
Function notation, 222–23, 242, 330
Functions, 167, 242
    adding, subtracting, multiplying, and
        dividing, 737–38, 803

algebra of, 737
composition of, 738–40
cubic, 397
identifying, 218–19
inverse, 745–46
linear, 344, 397
profit, 362
quadratic, 397, 409
quartic, 397
rational, 428–29, 487
vertical line test and graphs of, 220–21
Function values, finding, 507
Fundamental principle of fractions, for
    rationalizing the denominator, 537
Fundamental principle of rational
    expressions, 424

G

GCF. See Greatest common factor
Geographic centers, 210
Geometria a Renato Des Cartes (Geometry by
    René Descartes) (van Schooten), 204
Geometry investigations, 14
Gradiometer scale, 238
Graham's law, 528
Graphing
    ellipses, 687–88
    exponential functions, 755–57
    functions, 220–21
    hyperbolas, 689–91
    intersections and unions, 233–34
    inverse functions, 747–48
    linear equations, 173–76
    linear equations in two variables, 240
    linear functions, 223, 242
    linear inequalities, 231–33, 243
    of line using slope and y-intercept, 197–98
    logarithmic functions, 767–69
    nonlinear functions, 699–702, 721
    nonlinear inequalities, 711–12
    numbers on number line, 10
    parabolas, 671–74
    polynomial functions, 397
    quadratic functions, 398–99, 633–39, 659
    solutions of system of linear
        inequalities, 303
    solving system of equations by, 260–62
    systems of linear inequalities, 303–5, 315
    systems of nonlinear inequalities, 712–13
    vertical and horizontal lines, 176–77
Graphing calculator explorations
    approximating solutions of systems of
        equations, 267
    checking addition/subtraction of
        polynomials, 332
    compound interest, 796
    evaluating expressions, 108
    features in graphing calculators, 178
    function composition, 740
    graphing ellipses, 692
    graphing equations, 680
    graphing equations by solving equation
        for y, 200
    graphing functions, 640
    inverse functions, 748
    multiplication of polynomials in one
        variable, 343
    operations on rational expressions, 439
    problem solving, 760
    quadratic equations in standard form, 402
    rational functions, 430
    solving quadratic equations, 594
    solving radical equations, 554
    TRACE feature, 209

Graphs
    domain and range found from, 221–22
    examples of, 169
Gravity, center of, 210
Greatest common factor, 409
    factoring out, 359–60, 368, 385
Greenhouse effect, 585
Gross profit margin ratio, 622
Grouping symbols, 41, 42, 69

H

Half-planes, 231, 232, 233, 243
Height of object, finding, 331
Help
    getting, 5
    with mathematics, 354
Heron of Alexandria, 508
Hindus, negative solutions to quadratic
    equations recognized by, 630
History. See Focus on history
Homework, 34
Hooke's law, and direct variation, 476
Horizontal axis (or x-axis), 169, 172
Horizontal change, as ratio to vertical
    change, 185
Horizontal lines
    graphing, 177
    slope of, 189–90, 215, 241
    writing equations of, 206
Horizontal line test, 804
    and one-to-one functions, 744–45
Hyperbolas, 669, 671, 721, 732
    asymptotes of, 690
    graphing, 689–91
Hypotenuse, of right triangle, 391

I

Identities, 86
Identity property, 68
    of one, 20
    of zero, 20
IIHS. See Insurance Institute for Highway
    Safety
Imaginary numbers, 562
Imaginary unit, 561
Income with commission, calculating,
    120–21
Inconsistent systems, 261, 262, 265, 273,
    275, 299, 312
Independent equations, 261, 312
Independent variables, 222
Index, 505
Inequalities, 79
Inequality symbols, 115
    and graphing linear inequalities, 231, 232
    reversing direction of, 118, 153
    using, 19–20
Infinity symbol, 116
i notation, 561
Inputs, 217, 242
Insurance Institute for Highway Safety, 306
Integers, 10, 12, 67
Interest formula, 105
Interest rates, 93, 592–93, 598
International Astronomical Union, 358
Internet excursions
    body-mass index, 474
    graphing polynomial or quadratic
        functions, 405
    Insurance Institute for Highway
        Safety, 306
    interest rates, 598
    New York Stock Exchange, 38
    planetary data, 512, 696

Internet excursions (*cont.*)
    Richter scale, 801
    target heart rate zone, 138
    U.S. Bureau of Labor Statistics, 214
INTERSECT feature, on grapher, 267, 554, 594, 796
Intersection of two sets, finding, 131
Intersections
    graphing, 233
    of solution sets of two inequalities, 154
Interval notation
    solutions written in, 120
    using, 115–16
Inverse functions, 745–46, 765, 804
    graphing, 747–48
Inverse of function, finding equation of, 746–47
Inverse property, 68
Inverse variation, 490
    and Boyle's law, 478
    solving problems involving, 477–78
Irrational numbers, 67, 504, 786
    identifying, 11
    written as decimals, 12
"Is an element of" symbol, 10
"Is equal to" symbol, 19, 68
"Is greater than or equal to" symbol, 20, 68
"Is greater than" symbol, 19, 68
"Is less than or equal to" symbol, 20, 68
"Is less than" symbol, 19, 68
"Is not an element of" symbol, 10
"Is not equal to" symbol, 19, 68

J

Jobs, fastest-growing, 440. *See also* Focus on business and career
Joint variation, 490
    solving problems involving, 478–79

K

Kelvin, Lord, 738
Kelvin degrees, Celsius degrees converted to, 738
Kelvin scale, 738
Kepler, Johannes, 770
Kilowatt-hour (KWH), 122

L

Labor statistics, 214
La Géométrie (Descartes), 204, 532
Landing time, calculating, 604
Laplace, Pierre de, 770
LCD. *See* Least common denominator
Learning curve, 799
Least common denominator, 84
    finding for rational expressions, 435–37
    simplifying complex fraction by multiplying numerator/denominator by, 446–47
    and solving equation for specified variable, 463
Least squares regression, 198, 244
Legs, of right triangle, 391
Leibniz, Gottfried Wilhelm, 204
Length, of pendulum, 570
Liabilities, 622
Lightning strikes, locating, 257, 310
Like radicals, 529, 572
Like terms, combining, 44, 328–29
Linear equations
    forms of, 215
    graphing, 173–76
    slope-intercept form for graphing, 197
    slope-intercept form of, 240

solving by combining like terms, 83–84
    solving those containing fractions or decimals, 84–86
Linear equations in one variable, 81–86, 151, 152
Linear equations in three variables, 273
Linear equations in two variables, 173
    graphs of, 240
    ordered pair solutions of, 173
Linear functions, 242, 344
    degree of, 397
    graphing, 223
Linear inequalities, 115
    graphing, 231–33, 243
    and problem solving, 120–21
    solving, 116
Linear inequalities in one variable, 115, 119, 153
Linear inequalities in two variables, 231–33, 243
Linear models/modeling, 244, 306, 656
Lines
    graphing, using slope and y-intercept, 197–98
    slope of, 185–87
Line segments, midpoints of, 224, 675
Logarithmic definition, 765, 805
Logarithmic equations
    solving, 766–67, 793–94
    solving problems modeled by, 795
Logarithmic form, and corresponding exponential statement, 805
Logarithmic functions, 735, 743, 805
    graphing, 767–69
    and logarithmic notation, 765–66
Logarithmic notation, exponential equations written with, 765
Logarithm property of equality, 793, 806
Logarithms, 758
    invention of, 770
    properties of, 767, 775–76, 805
London, England, world's first subway system built in, 1
Long division, 349, 408
Lowest terms, simplifying/writing rational expressions in, 424

M

Market equilibrium, 710
Mass, center of, 210
Mathematical Institute (Zurich, Switzerland), 66
Mathematical models, 268
Mathematics
    careers requiring knowledge of, 440
    help with, 354
    and problem solving, 79
    tips for success in, 3–6
Mathematics textbook, reading, 4
Matrix (matrices)
    systems of equations solved using, 314–15
    using to solve a system of three equations, 299–300
    using to solve system of two equations, 297–99
Maximum values, finding, 649–50
Medicine, and problem solving, 79
Menaechmus, 732
Methane gas emissions, and greenhouse effect, 585
Metrica (Heron of Alexandria), 508
Midpoint
    of diameter of circle, 677
    of line segment, 224, 675

Midpoint formula, 674, 720
Minimum values, finding, 649
Mirifici Logarithmorum Canonis Descriptio (Napier), 770
Missing powers, 351, 354
Mistakes in math class, 34
Modeling data, 244
Monomials, 327, 328
    dividing polynomials by, 349
    dividing polynomials by polynomials other than, 408
    multiplying, 339
Mount Vesuvius, 501
MSN, 92
Multiplication, 9
    associative property of, 23, 68
    of binomials, 340–41
    commutative property of, 22, 68
    of complex numbers, 563, 573
    of functions, 737, 803
    of polynomials, 339–40, 408
    of radical expressions, 530–31, 572
    of rational expressions, 426–27, 487
    of real numbers, 31–32, 68
    of three or more polynomials, 342
Multiplication property of equality, 82, 83, 151
Multiplication property of inequality, 118–19, 153
Multiplicative identity, 20
Multiplicative inverses (or reciprocals), 21, 22

N

Napier, John, 770
NASA, 512
National Space Science Data Center, 512
Natural logarithms, 785, 806
    approximating, 786
    of powers of e, 786–87
Natural "log" key, on calculator, 785
Natural numbers, 10, 12, 67
Natural Resources Defense Council, 585
Negative constant, in trinomial, 370
Negative exponent rule, 60
Negative exponents, 51, 69, 448, 515
Negative net earnings, 362
Negative numbers
    history behind, 630
    multiplying both sides of inequality by, 118, 153
    on number line, 9
Negative powers, simplifying expressions raised to, 51–52
Negative radical sign, 503
Negative reciprocal slopes, 207, 208
Negative slope, 190
Negative square root, 571
Net income/loss, 362
Net sales, 622
Newton, Sir Isaac
    gravitation theory of, 770
    law of cooling by, 802
New York City, subway in, 1
New York City Transit, 1
New York Stock Exchange, 38
Noise pollution, 782
Nonlinear functions, graphing, 699–702, 721
Nonlinear inequalities, graphing, 711–12
Nonlinear systems of equations
    solving, 278, 705–8, 722
    solving by elimination, 707–8
    solving by substitution, 705–7
NSSDC. *See* National Space Science Data Center
nth roots, finding, 505–6

Null set, 10
Number line, 9
    numbers graphed on, 10
    opposites on, 21
Number problems, solving those modeled by
    rational equations, 463–64
Numbers
    absolute value of, 29, 139
    percents of, 94
    roots of, 34
    as solutions of equations, 81
    writing in scientific notation, 53
Numerators, 41
    rationalizing, 540–41, 572
Numerical coefficient, 327
Nutrition labels, 93, 114
NYSE. *See* New York Stock Exchange

O

Occupations, fastest-growing, 440
Odd integers, consecutive, 91
One, identity property of, 20, 21
One-to-one functions, 743–44, 793, 804
Operations. *See also* Addition; Division;
    Multiplication; Subtraction
    on exponential expressions, 49
    on numbers written in scientific notation,
      61–62
    on polynomials, 385
    on real numbers, 29–34, 68–69
Opposites (or additive inverses), 21, 330
Ordered pairs, 170, 311
    plotting on rectangular coordinate
      system, 169–72, 239
    as solutions, 172–73, 259
Ordered pair solutions
    finding/plotting with graphers, 178
    of linear equations in two variables, 173
Ordered triple, 273, 275, 276, 298, 300, 313
Order of operations, 41–42, 69
Ordinate, 169, 204
Organization, of class and study materials,
    4, 144
Origin, 9, 169, 239
Outputs, 217, 242

P

Parabolas, 398, 409, 669, 671, 719, 732
    downward opening, 399, 633, 637, 646,
      647, 650, 659
    graphing, 671–74
    leftward opening, 671, 674
    and minimum/maximum values, 649
    rightward opening, 671, 674, 719
    upward opening, 399, 400, 401, 633, 637,
      645, 646, 647, 659
    vertex of, 399–401
Parallel lines
    slope of, 190–91, 215
    writing equations for, 206–7
Parentheses, 41, 69
    distributive property and removal of,
      83, 86
    in interval notation, 116
Partial products, 340
Patterns, searching for, 66
Pendulum, length and period of, 570
Percents, 93–94
Perfect cubes, 523, 537
Perfect 4th powers, 523
Perfect square factors, and simplifying
    radicals, 522, 523
Perfect squares, 504, 523
Perfect square trinomials, 371, 409

factoring, 377–78
    writing, 588–89
Perimeter, of rectangle, 14
Period, of pendulum, 570
Perpendicular bisectors, 224
Perpendicular lines
    slope of, 190–91, 215
    writing equations for, 207–8
Phoenix Sky Harbor International Airport
    (Arizona), 244
Phrases, writing as algebraic expressions, 13
Pi (π), 11, 732, 786
Planetary data, 512, 696
Plotting ordered pairs, on rectangular
    coordinate system, 169–72, 239
Points, 170, 675
Point-slope form, 215
    in applications, 208–9
    of equation of line, 241
    and equations of parallel and
      perpendicular lines, 206–7
    using to write equations, 205–6
Polynomial equations
    solving by factoring, 387–89, 409
    solving problems modeled by, 390–92
Polynomial functions, 325, 344, 407
    analyzing graphs of, 397
    evaluating, 330–31
Polynomial inequalities, solving, 625–27, 658
Polynomials, 325, 407
    adding, 329, 407
    defined, 327
    degree of, 328
    dividing, 349–52, 408
    dividing by monomial, 349
    factoring, 385
    factoring by grouping, 409
    in one variable, 327
    multiplying, 339–40, 408
    multiplying any two, 339
    multiplying three or more, 342
    operations on, 385
    prime, 368
    subtracting, 329–30, 407
    in x, 327
Pompeii, 501
Population sizes, estimating, 485–86
Positive attitude, 3, 162
Positive constant, in trinomial, 370
Positive exponents, 51, 61
Positive net earnings, 362
Positive number(s)
    multiplying or dividing both sides of
      inequality by, 118, 153
    on number line, 9
Positive (or principal) square root, 34, 69,
    506, 571
Positive slope, 188, 190
Power of a product rule, 59, 60
Power of a quotient rule, 59, 60
Power property of logarithms, 776, 805
Power rules, 59–60, 70
    for exponents, 515
    for products and quotients, 515
    for solving radical equations, 549
Powers of i, finding, 564–65
Price-to-earnings (P/E) ratio, 622
Prime polynomials, 368
Principal, 107
Principal (or positive) square root, 34,
    69, 503
Problem solving, 452. *See also* Activities;
    Formulas
    with combined variation, 479–80

and critical thinking, 24
and direct variation, 475–76
with exponential equations, 758–59
formulas for, 107–8, 153
general strategy for, 92
with inverse variation, 477–78
with joint variation, 478–79
leading to quadratic equations, 613–16
linear inequalities and, 120–21
with polynomial equations, 390–92
with proportions, 464–66
with quadratic equations, 592–93, 602–4
and rational equations, 489
strategy for, 152
and systems of linear equations, 314
and systems of three equations, 285–87
and systems of two equations, 281–85
and variation, 490
Product property of logarithms, 775, 805
Product rule, 69
    for exponents, 49, 60, 339, 515
    for radical expressions, 521, 572
Products
    of functions, 737, 803
    of sum and difference of two terms, 342
Profit function, 362
Properties of equality, using, 81–83
Proportions, solving problems modeled by,
    464–66
Pythagorean theorem, 391, 552–53, 674

Q

Quadrantel point, 171
Quadrants, 170
Quadratic equalities in one variable, solution
    of, 625
Quadratic equations, 387, 585
    evolution of solving, 630
    graphers used for finding solutions of, in
      standard form, 402
    solving by completing the square, 589–91
    solving by using quadratic formula,
      599–601
    solving in x by completing the square, 657
    solving problems leading to, 613–16
    solving problems modeled by, 592–93,
      602–4
Quadratic formula, 612
    discriminant in, 602
    quadratic equations solved by using,
      599–601, 657
Quadratic functions, 409, 585, 625
    degree of, 397
    graphing, 398–99, 645–48, 659
    graphing $f(x) = ax^2$, 636–38
    graphing $f(x) = a(x - h)^2 + k$, 638–39
    graphing $f(x) = (x - h)^2$, 634–36
    graphing $f(x) = (x - h)^2 + k$, 636
    graphing $f(x) = x^2 + k$, 633–34
    writing in form $y = a(x - h)^2 + k$,
      645–48
Quadratic inequalities, 625
Quadratic methods, solving equations by
    using, 611–13
Quadratic models, recognizing, 656
Quartic function, degree of, 397
Quizzes, performance on, 728
Quotient property of logarithms, 775–76, 805
Quotient rule, 60, 69
    for exponents, 50–51, 515
    for radical expressions, 521–22, 572
Quotients, 349
    of functions, 737, 803

R

Radical equations, solving, 573, 611
Radical expressions, 503, 571
  adding or subtracting, 529–30, 572
  multiplying, 530–31, 572
  product rule for, 521
  quotient rule for, 521–22
  rational exponents used for simplifying, 516
  simplifying, 522–24, 572
  solving equations containing, 549–52
Radical functions, 507, 571
Radical notation, 501, 503, 516
Radical sign, 503
Radical symbol, development of, 532
Radicand, 503, 523
Radius, of circle, 676, 677, 678–79, 720
Raising to powers, 9
Range, 218, 242, 397
  defined, 217
  finding from graph, 221–22
  of logarithmic function, 769
Rate, 465
Rate of change, and slope, 186
Rate of speed, finding, 283–84
Ratio, 464
Rational equations, 453
  and problem solving, 489
  solving for specified variable, 463
  solving number problems modeled by, 463–64
Rational exponents, 501, 513–15, 571
  definition of $a^{-m/n}$, 514–15
  definition of $a^{m/n}$, 513–14
  definition of $a^{1/n}$, 513
  using rules for, 515
  using to simplify radical expressions, 516
Rational expressions, 421, 423, 487
  adding or subtracting, 488
  adding or subtracting those with different denominators, 437–39
  adding or subtracting those with same denominator, 435
  complex, 445
  dividing, 427–28, 487
  finding least common denominator of, 435–37
  multiplying, 426–27, 487
  simplifying, 424–26, 487
  simplifying with negative exponents, 448
  solving equations containing, 453–56, 488
  undefined, 423
Rational functions, 421, 487
  applications with, 428–29
Rational inequalities, solving, 628–29, 658
Rationalizing the denominator, 537–40, 572
Rationalizing the numerator, 540–41, 572
Rational numbers, 67, 423
  identifying, 11
  writing as decimals, 12
Real numbers, 67
  addition of, 29–31, 68
  division of, 32–34
  identifying, 11
  multiplication of, 31–32
  operations on, 29–34, 68–69
  properties of, 19–23, 68
  as subset of complex numbers, 562
  subsets of, 12
  subtraction of, 29–31, 68
Reasoning skills, 24
Reciprocals (or multiplicative inverses), 21
Rectangle
  area of, 14, 105
  perimeter of, 14

Rectangular coordinate system, 204, 239
  plotting ordered pairs on, 169–72
Rectangular solid, volume of, 105
Regression equation, 244
Relations, 167, 217, 218, 219, 242
Revenue, 362
Richter scale, 735, 791, 801
Right triangle, 391, 552
Rise, 186
Rolle, Michel, 532
Room redecorating, 150
ROOT feature, on grapher, 402
Roots, 9
  of numbers, 34
Roster form, set written in, 10, 67
Row, in matrix, 297
Row operations, elementary, 297, 298, 300, 315
Rudolff, Christoff, 532
Run, 186

S

Sales, 622
Savings accounts, amount in, 107–8
Scatter diagrams, 244
  and paired data, 171, 172
Schedule grid, 8
Scheduling math course, 3
Scientific notation, 69–70
  converting between standard notation and, 52–53
  using to compute, 61–62
Second-degree equations, 587. *See also* Quadratic equations
Second differences, and second-degree polynomial, 344
Sentences, writing as equations, 19
Sequence, 344
Set builder notation, 10, 11, 67
Set notation, 116
Sets of numbers, 67
  elements of, 10
  identifying common, 9–12
Sign patterns, for multiplying real numbers, 31
Sign rules, for division, 32
Simple interest, 592
Simplifying algebraic expressions, by combining like terms, 43–44
Simplifying complex fractions, 488
Simplifying expressions
  with order of operations, 41
  raised to negative powers, 51–52
  raised to zero power, 50
Simplifying radical expressions, 522–24, 572
Simplifying rational expressions, 424–26, 487
Slope, 185
  finding from equation, 187–88
  finding from two points, 185–87
  of horizontal and vertical lines, 189–90
  of line, 240–41
  measuring, 238
  of parallel and perpendicular lines, 190–91
Slope formula, 186, 187
Slope-intercept form, 188, 197–99, 215
  interpreting, 198–99
  of linear equation, 240
  using to write equation, 198
  and writing equation of line given slope and y-intercept, 241
Solid-state diffusion, 528
Solution of an equation, 81, 151, 172–73
Solution of equation in three variables, 313
Solution of equation in two variables, 172

Solution of inequality, 115
Solution of system of linear inequalities, 303
Solution of system of two equations in two variables, 259, 311
Solution region, 231
Solution set of an equation, 81
Solution set of inequality, 115
Solving an equation, 81
Sound intensity, 782
Special products, 408
  factoring by, 377–80, 409
Square of binomial, 341
Square root function, 701
Square root property, 587–88, 657
Square roots, 11
  approximating, 504
  finding, 503–4
  Heron's method of approximating, 508
  positive or principal, 34, 69
  symbol for, 532
Standard form
  of ellipse with center, 687, 688
  for equation of hyperbola, 689
  equations written in, 241, 297
  and graphs of circle, 677
  of linear equation, 215
  linear equation written in, 173
  polynomial equation in, 387
  quadratic equation written in, 587
  quadratic inequalities in one variable written in, 625
Standard notation, converting between scientific notation and, 52–53
Standard window, in graphers, 178
Stock market price per share, 622
Study skills reminder, 6
  course performance, 429
  errors, 366
  exam day, 316, 792
  final exam preparation, 655, 788
  help with mathematics, 354
  homework assignments, 254, 542
  notebook organization, 460
  organization of materials, 144
  performance evaluation, 714
  performance on quiz or exam, 728
  successful completion of course, 162
  test preparation, 18, 156, 243, 410, 456, 566, 660, 686, 778
  textbook features, 316
  tips for studying for exam, 234
Subsets, 12, 67
Substitution
  factoring by, 371–72
  nonlinear systems solved by, 705–7
  for solving equations with repeated variable expressions, 612, 658
Substitution method
  solving system of equations with, 262–64
  solving system of linear equations with, 312
  solving system of linear inequalities with, 304
  solving system of three equations in three variables with, 274, 277
  solving system of three linear equations in three variables with, 277
Subtraction, 9
  of complex numbers, 562–63, 573
  of functions, 737, 803
  of polynomials, 329, 330, 407
  of radical expressions, 529–30, 572
  of rational expressions, 488

of rational expressions with different
denominators, 437–39
of rational expressions with same
denominator, 435
of real numbers, 29–31, 68
Subways, history behind, 1
Success in math class, tips for, 34
Sum of cubes, factoring, 379–80
Sum of two cubes, 409
Sum of two squares, 378
Sums. *See also* Addition
of complex numbers, 562
of functions, 737, 803
of logarithms, 777
Surface area, of circular cylinder, 14
Symbols. *See also* Formulas
cube roots, 504–5
empty set, 10, 275
equal, 19
in flowcharts, 616
function composition, 738
function notation, 222, 242
greater than, 19, 68
greater than or equal to, 20, 68
grouping, 41, 42
inequality, 115, 626
infinity, 116
inverse of function, 746
"is an element of," 10
is equal to, 68
"is not an element of," 10
is not equal to, 68
less than, 19, 68
less than or equal to, 20, 68
logarithmic notation, 765
nth roots, 505
radical, development of, 532
radical signs, 503
square roots, 34
Synthetic division, 352–54, 408
System of equations, 257
elimination method used for solving,
264–66
and matrices, 297
solving by graphing, 260–62
substitution method used for solving,
262–64
System of inequalities, 257
graph of, 722
System of linear equations, 311–12
and problem solving, 314
solving by elimination method, 312
solving by substitution method, 312
System of linear equations in three variables,
solving, 273–77, 313
System of linear inequalities, graphing,
303–5, 315
System of nonlinear inequalities, graphing,
712–13
System of three equations
matrices used for solving, 299–300
and problem solving, 285–87
System of three linear equations, solving by
elimination method, 274, 313
System of three linear equations in three
unknowns, solving, 268
System of three linear equations in three
variables, solving, 274–77
System of two equations
matrices used for solving, 297–99
solving problems modeled by, 281–85
System of two linear equations, possible
solutions to, 262

T

TABLE feature, on grapher, 430, 439,
740, 759
Target heart rate zone, 138
Tax rates, 93
Telecommunications industry, 79
Telephone, history behind, 79
Telephone operators, 79
Temperature
interconversions, 42–43, 738
modeling, 802
Terms, 327
of expression, 43
multiplying sum and difference of
two, 342
Test point, 232, 243
Test point values, 626, 627, 628, 629
Test preparation, 18
Textbook, using, 4–5
Third-degree polynomial, 353
Time management, 6
Total asset turnover ratio, 622
TRACE feature, on grapher, 209, 402, 439,
554, 594, 759, 796
Trachtenberg, Jakow, 66
Trachtenberg's rule for multiplication by
eleven, 66
Traité d'Algébre (Rolle), 532
Transportation, and problem solving, 79
Trapezoid, area of, 106
Treatise on Conic Sections (Apollonius of
Perga), 732
Triangle's sides, finding lengths of, 95–96
Trinomials, 327
factoring, 409
factoring form $ax^2 + bx + c$, 368–70
factoring form $ax^2 + bx + c$ by
grouping, 371
factoring form $x^2 + bx + c$, 367–68
positive/negative constants in, 370
Tuition cost projections, 167
Tutoring services, 4

U

Undefined number, 32
Undefined rational expressions, 423, 454
Undefined slope, 189, 190, 241
Union
graphing, 234
of solution sets of two inequalities, 154
Union of two sets, finding, 133
Uniqueness of b^x, 757–58, 805
Unit cost, finding, 429
Unit distance, 9
United States Geological Survey, 735
Unknown numbers, finding, 93–94,
282–83, 464
Unlike terms, 328–29
U.S. Bureau of Labor Statistics, 214, 440
U.S. Department of Agriculture, 114

V

Value of the expression, 9, 69
Van Schooten, Frans, 204
Variables, 1, 9, 67
dependent, 222
independent, 222
solving formulas for specified, 105–7
solving rational equations for specified,
463
Variation
combined, 479–80

direct, 475–76, 490
inverse, 477–78, 490
joint, 478–79, 490
Vertex
deriving formula for finding, 648–49
of parabola, 398, 399–401, 409, 633, 671,
672, 673, 674, 719
Vertex formula, 400, 401, 650
Vertical change, as ratio to horizontal
change, 185
Vertical lines
graphing, 176, 177
slope of, 189–90, 215, 241
writing equations of, 206
Vertical line test, 223, 242, 398, 699, 804
and graphs of polynomial functions, 397
and one-to-one functions, 744, 745
using, 220–21
Vertical (or y-axis), 169, 172
Vinculum, 532
Voltage (V), 122
Volume
of circular cylinder, 14
of cube, 59
of rectangular solid, 105

W

Wattage (W), 122
Watt hours, 122
Wheelchair ramps, measuring grades
of, 238
Whole numbers, 10, 12, 67
Wildlife population sizes, estimating,
485–86
Window, in graphers, 178
Work problems, solving, 466–67, 489
Work time, finding, 613–14
World War II, 66

X

x-axis, 169, 170, 171, 204, 239
x-coordinate, 170, 171
x-intercepts, finding, 174, 175, 240, 400
x-values, 176
and ordered pair solutions, 174
relations between y-values and, 217

Y

y-axis, 169, 170, 171, 204, 239, 398
y-coordinate, 170, 171
y-intercept
finding, 174, 175, 240, 400
graphing line using slope and, 197–98
and slope-intercept form, 188
writing equation of line given slope
and, 241
y-values, 176
and ordered pair solutions, 174
relations between x-values and, 217

Z

Zero, 20, 31, 358
division by, 32
identity property of, 20
Zero exponent, 50, 69, 515
Zero exponent rule, 60
Zero-factor property, 387, 388, 587, 612
Zero-slope, 189, 190, 241
ZOOM feature, on grapher, 402, 554, 594,
759
ZOOM IN feature, on grapher, 267

PHOTO CREDITS